CONTENTS

P A R T 5   Analyzing Data: Means, Variances, and Proportions   195

Chapter 5   INFERENCES ABOUT μ **197**

Chapter 6   INFERENCES ABOUT $\mu_1 - \mu_2$ **260**

Chapter 15 ANALYSIS OF VARIANCE IN SOME STANDARD EXPERIMENTAL DESIGNS   **842**

Chapter 16 ANALYSIS OF VARIANCE FOR SOME UNBALANCED DESIGNS   **929**

PREFACE

An Introduction to Statistical Methods and Data Analysis, Fourth Edition, was written to give advanced undergraduate statistics students and graduate students from many disciplines a first exposure to statistical methods and data analysis. The first three editions of this textbook have been widely used in both one-term and two-term courses. Students are assumed to have a background in high school algebra, but no prior knowledge of statistics is required. The first few chapters provide a good initial exposure to much of the material covered in an introductory statistics course.

FOCUS OF THIS EDITION

The focus of this edition has changed slightly from previous editions. Certainly the primary objectives of a course in *Statistical Methods and Data Analysis* include developing the student's appreciation and understanding of the role of statistics in his or her field and an ability to apply appropriate statistical methods to summarize and analyze data for some of the more routine experimental settings. While fulfilling these objectives, we also want to focus the student on where these statistical methods fit into the context of making sense of data. To this end we have approached the fourth edition by considering the four steps in making sense of data: (1) gathering data, (2) summarizing data, (3) analyzing data, and (4) communicating the results of data analyses. The text is divided into parts (or sections) which include chapters on the four steps to making sense of data as well as separate chapters which contain the necessary background or connective material. With this organization and emphasis, we want the student to understand that the summarization and analysis of data are steps in the larger problem of making sense of data. Thus, this edition aims at being ever more practical than previous editions by relating the methods and data analysis techniques of the text to the context in which they are used to solve real life, practical problems.

The availability and widespread use of statistical software packages for personal computers, mini-computers and mainframes has allowed us to de-emphasize the calculations and focus on making sense of gathering data, summarizing data, analyzing data and communicating the results of data analyses. We are still interested in teaching the statistical methods appropriate for various experimental conditions, but we don't want to spend needless hours figuring calculations once we understand the methods. If we can identify the appropriate methods for summarizing and analyzing data from an experiment, available statistical software systems can be used to do the actual calculations required. Computer output for many examples and exercises is provided in the text so that the student can interpret the results of analyses after the calculations have been done using one of several widely used statistical software systems.

IMPORTANT NEW FEATURES

The main new features of the fourth edition reflect the emphasis on making sense of data. These include the following:

- ▼ The textbook has been divided into parts containing one or more chapters. Each part addresses one of the four steps in making sense of data or the relevant background and connective material.
- ▼ Chapters have been rearranged to reflect the four, ordered steps in making sense of data—gathering data, summarizing data, analyzing data, and communicating the results of data analyses.
- ▼ A new chapter on gathering data has been added (Chapter 2).
- ▼ The first chapter has been rewritten to appropriately introduce the revised focus of this edition.
- ▼ Quality and process improvement have been introduced in Chapter 1; further emphasis follows in succeeding chapters.
- ▼ All regression chapters have been pulled together as a group (Chapters 9–12).
- ▼ Similarly, all chapters on analysis of variance have been pulled together (Chapters 18–19).

Additional features include:

▼ This edition, as in the previous editions, includes many practical applications of statistical methods and data analysis techniques from a wide range of application areas including: agriculture, business, economics, education, engineering, environmental studies, medicine, law, political science, psychology and sociology.

▼ Exercises are grouped as Basic Techniques and Applications.

▼ Additional drill and thought-provoking exercises have been added to the already extensive list from the previous editions.

▼ Computer output has been updated for Minitab and SAS; output from another software package available on a PC, Execustat, has been included in the textbook.

▼ The more traditional dot notation for analysis of variance formulas has been used in Chapters 13–19.

▼ A clinical trials database, taken from studies conducted for a new prescription drug product, has been added in the Appendix. Exercises are drawn from this database.

▼ As with the previous edition, attention is paid to the underlying assumptions. Data plots, residual plots, probability plots and other tools are used to diagnose potential problems with the assumptions.

▼ Some computer simulations are provided to illustrate the effects of violating certain assumptions.

In addition to these major features and changes, numerous substantive changes have been made in response to constructive comments from reviewers and users.

▼ ACKNOWLEDGMENTS

I wish to express my appreciation to friends and colleagues who have made many constructive suggestions and criticisms at various stages of development of the original manuscript and during the preparation of the second, third and fourth editions. Special thanks to my colleague and fellow author, David Hildebrand, University of Pennsylvania, who provided a long, thoughtful review of the second edition. Similar thanks go to the reviewers who provided extensive, careful reviews of the third edition. I am also deeply indebted to my colleague, Chris Franklin, University of Georgia, for providing careful review of the manuscript for the fourth edition as well as updating output, preparing the solutions manual and the data disk. For this edition, I am deeply appreciative of the constructive comments of

J. Richard Alldredge
Washington State University

Donald Sisson
Utah State University

Steven Bajgier
Drexel University

Charles Sommer
SUNY—Brockport

M.A. Evans
Washington State University

William Swallow
North Carolina State University

Jeffrey Jarrett
University of Rhode Island, Kingston

Krishnanand Verma
University of Wisconsin—Whitewater

Paul Nelson
Kansas State University

Mary Sue Younger
University of Tennessee, Knoxville

Also, thanks are due to A. Hald, E.S. Pearson, H.O. Hartley, R.E. Kirk, the Biometrika Trustees, the Chemical Rubber Company, Lederle Laboratories, the Editors of the Annals of Mathematical Statistics, D.B. Duncan, R.A. Waller, the Editors of Biometrics, the Editors of JASA, and the American Society of Testing and Materials for permission to reprint tables. A special note of appreciation is extended to my typists for their careful translations of my "drafts" into printed form; to Ruth Campbell, Jeanette Beach, and Ellen Evans for the first edition, to Vicki Mason for the second and to Linda Rabe for the third edition, and to Phyllis Switzer for careful attention to detail in the preparation of a word processing document for this edition. I am also indebted to my editor, Michael Payne, and to Kirby Lozyniak and Susan Krikorian, my production editors, and to the many others at Duxbury who worked long and hard behind the scenes to transform my manuscript into a finished product. Finally, I acknowledge the support and encouragement of my wife, Sally, and two children, Curtis and Kathryn, during the long evenings and weekends of work required to write the original manuscript and to prepare subsequent new editions.

R. Lyman Ott

WHAT IS STATISTICS?

1.1 INTRODUCTION

What is statistics? Is it the addition of numbers? Is it graphs, batting averages, percentages of passes completed, percentages of unemployment, and, in general, numerical descriptions of society and nature?

Statistics, as a subject, is the study of *making sense of data*. Almost everyone—including corporate presidents, marketing representatives, social scientists, chemists, and consumers—deals with data. These data could be in the form of quarterly sales figures, expenditures for goods and services, pulse rates for patients undergoing therapy, contamination levels in samples of surface water, or census figures. In this text, we approach the study of statistics by considering the four steps in making sense of data: (1) gathering data, (2) summarizing data, (3) analyzing data, and (4) reporting the results of data analyses.

The text is divided into eight parts, some of which include chapters on the four steps in making sense of data. Other parts contain chapters that provide necessary background or connective material. The relationship between the eight parts of the textbook, the four steps in making sense of data, and the chapters is shown in Table 1.1.

As you can see from this table, a great deal of time is spent discussing how to analyze data using the basic methods (for means, variances, and proportions), regression methods, and analysis of variance methods. However, you must remember that for each data set requiring analysis, someone has developed a plan for gathering the data (Step 1) and prepared the data for analysis. And, following summarization (Step 2) and analysis of the data (Step 3), someone has to communicate the results of the analysis (Step 4) in written or verbal form to the intended audience. All four steps are important in making sense of data; the analysis step,

TABLE 1.1
Organization of the Text

Parts of the Textbook	Steps in Making Sense of Data	Chapters of the Textbook
1 Introduction		1 What is Statistics?
2 Gathering the Data	1	2 Using Surveys and Scientific Studies to Gather Data
3 Summarizing Data	2	3 Data Description
4 Tools and Concepts		4 Probability and Probability Distribution
5 Analyzing Data: Means, Variances, and Proportions	3	5 Inferences About μ
		6 Inferences About $\mu_1 - \mu_2$
		7 Inferences About Population Variances
		8 Categorical Data
6 Analyzing Data: Regression Methods	3	9 Linear Regression and Correlation
		10 Inferences Related to Linear Regression and Correlation
		11 Multiple Regression and the General Linear Model
		12 More on Multiple Regression
7 Analyzing Data: Analysis of Variance Methods	3	13 Introduction to Analysis of Variance
		14 Multiple Comparisons
		15 Analysis of Variance for Some Standard Experimental Designs
		16 Analysis of Variance for Some Unbalanced Designs
		17 Analysis of Variance for Some Fixed-, Random-, and Mixed-Effects Models
		18 Experiments with Repeated Measures
		19 The Analysis of Covariance
8 Communicating and Documenting the Results of Analyses	4	20 Communication and Documentation of Results

while time consuming, is only one of the four steps. Throughout the text, we will try to keep you focused on the bigger picture of making sense of data. Also, periodically, it would be good for you to refer to this table as a reminder of where each chapter fits into the overall scheme of things.

Before we jump into the study of statistics, let's consider three instances in which the application of statistics could help to solve a practical problem.

1. Suppose that a manufacturer of light bulbs produces roughly a half million bulbs per day. Because of some customer reactions to its product, the firm wishes to determine the fraction of bulbs produced on a given day that are defective. It can solve the problem in two ways. The half million bulbs could be inserted into sockets and tested, but the cost of this solution would be substantial and could greatly increase the price per bulb. A second method for determining the fraction of defective bulbs would be to select 1,000 bulbs from the half million produced and test each of the 1,000. The fraction of bulbs defective in the 1,000 tested could be used to estimate the fraction defective in the entire day's production. We will show in later chapters that the fraction defective in the bulbs tested will probably be quite close to the fraction defective for the entire half million bulbs. Also, we will be able to tell by how much this estimate might differ from the fraction of defective bulbs produced on any given day.

   2. A similar application of statistics is brought to mind by the frequent use of the Gallup poll, the Harris poll, and other public opinion polls. How can these pollsters presume to know the opinions of more than 100 million Americans? They certainly cannot reach their conclusions by contacting every voter in the United States. Rather, as we have suggested in the light bulb example, they sample the opinions of a small number of voters, perhaps as few as 1,500, to estimate the re-action of every voter in the country. The amazing result of this process is that the fraction of those people contacted who hold a particular opinion will match very closely the fraction of voters holding that opinion in the total population at that point in time. Most students find this assertion difficult to believe, but we will supply convincing supportive evidence in subsequent chapters.

   3. Another example of a statistical problem is taken from the field of medicine. Suppose a research physician wishes to investigate the effect of a new drug on the stimulation of a patient's heart. The physician is really interested in the effect of the drug on all future heart patients who might be treated with the drug. Fifty heart patients are selected and treated with the drug. The increase in the pulse rate is recorded for each over a period of time. After observing the effect of the drug on the fifty patients, the physician may infer that the drug will have a similar effect on all heart patients in the future.

sample

population

   These problems illustrate the four steps in making sense of data. First, each problem involved a data-gathering stage using sampling. A group (**sample**) of light bulbs was selected from the day's production, a sample of people was obtained from the entire voting **population** in the United States, and a sample of 50 heart patients was obtained. Then a measurement was obtained for each element (bulb, voter, or patient) in the sample. These data are then used to solve the problem.

   Next, in order to make sense of the data collected, someone would have to summarize and analyse it. In the light bulb example, the fraction of defective bulbs could be computed for those bulbs tested. Based on this value and the number of bulbs tested (1,000), one could accurately predict the fraction of defective bulbs in the entire production of a half million bulbs. Similarly, for the voter opinion poll, one could compute the fraction of the sample voters who favored each of the candidates. Based on the results for the 1,500 sample voters, one could accurately predict the voting pattern for the entire voting public in the United States, at that point in time. In the study of 50 heart patients, measurements on variables such as exercise capacity, oxygen consumption, and quality of life could be used to predict the effect (efficacy) of the compound on other patients, who would be candidates for similar treatment.

Finally, having collected, summarized, and analyzed the data, it would be important to report the results in unambiguous terms to interested persons. For the light bulb example, management and technical staff would need to know the quality of their production batches. Based on this information they could determine whether adjustments in the process are necessary. The results of the statistical analyses cannot be presented in ambiguous terms; decisions must be made from a well-defined knowledge base. The results of the voter opinion poll example would be of vital interest to political candidates, campaign managers, and potential campaign contributors and might lead to major shifts in campaign and funding strategies. Finally, the results of the heart patient study would interest the physician treating the patients, the company developing the compound, the Food and Drug Administration, and the medical community in general. The results must be presented clearly so that informed decisions can be made for the future development of the compound in the treatment of heart patients.

DEFINITION 1.1
Population

▼

A **population** is the set of all measurements of interest to the sample collector. See Figure 1.1.

FIGURE 1.1
Population and Sample

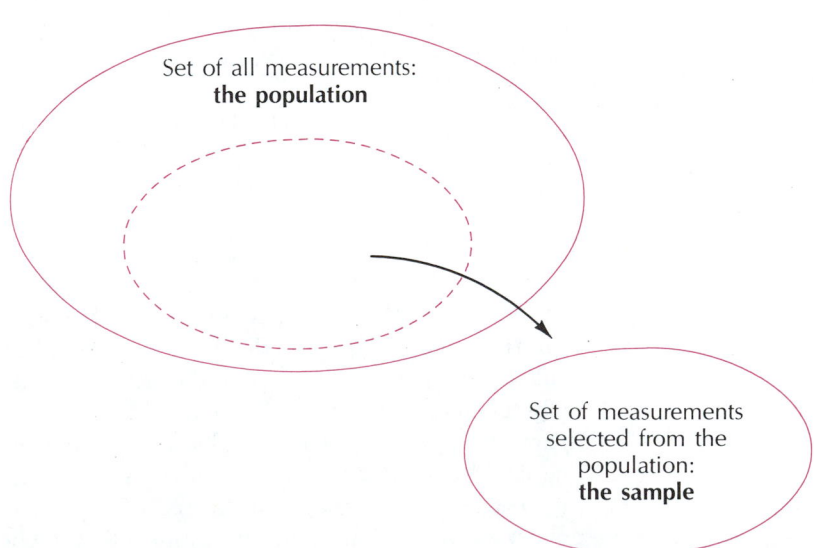

Set of all measurements:
the population

Set of measurements selected from the population:
the sample

DEFINITION 1.2
Sample

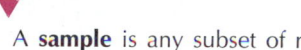

A **sample** is any subset of measurements selected from the population. See Figure 1.1.

1.2 ▼ WHY STUDY STATISTICS?

We can think of two good reasons for taking an introductory course in statistics. One reason is that you need to know how to evaluate published numerical facts. Every person is exposed to manufacturers' claims for products; to the results of sociological, consumer, and political polls; and to the published results of scientific research. Many of these results are inferences based on sampling. Some of the inferences are valid; others are invalid. Some are based on samples of adequate size; others are not. Yet all these published results bear the ring of truth. Some people say that statistics can be made to support almost anything (particularly statisticians). Others say it is easy to lie with statistics. Both statements are true. It is easy, purposely or unwittingly, to distort the truth by using statistics when presenting the results of sampling to the uninformed.

A second reason for studying statistics is that your profession or employment may require you to interpret the results of sampling (surveys or experimentation) or to employ statistical methods of analysis to make inferences in your work. For example, practicing physicians receive large amounts of advertising describing the benefits of new drugs. These advertisements frequently display the numerical results of experiments that compare a new drug with an older one. Do such data really imply that the new drug is more effective, or is the observed difference in results due simply to random variation in the experimental measurements?

Recent trends in the conduct of court trials indicate an increasing use of probability and statistical inference in evaluating the quality of evidence. The use of statistics in the social, biological, and physical sciences is essential because all these sciences make use of observations of natural phenomena, through sample surveys or experimentation, to develop and test new theories. Statistical methods are employed in business when sample data are used to forecast sales and profit. In addition, they are used in engineering and manufacturing to monitor product quality. The sampling of accounts is a useful tool to assist accountants in conducting audits. Thus, statistics plays an important role in almost all areas of science, business, and industry; persons employed in these areas need to know the basic concepts, strengths, and limitations of statistics.

1.3 ▼ SOME CURRENT APPLICATIONS OF STATISTICS

Acid Rain: A Threat to Our Environment

The accepted causes of acid rain are sulfuric and nitric acids; the sources of these acidic components of rain are hydrocarbon fuels, which spew sulfur and nitric oxide into the atmosphere when burned. The effects of acid rain are many. Some of these effects are listed here:

▼ Acid rain, when present in spring snow melts, invades breeding areas for many fish, which prevents successful reproduction. Forms of life that depend on ponds and lakes contaminated by acid rain begin to disappear.

▼ In forests, acid rain is blamed for weakening some varieties of trees, making them more susceptible to insect damage and disease.

▼ In areas surrounded by affected bodies of water, vital nutrients are leached from the soil.

▼ Man-made structures are also affected by acid rain. Experts from the United States estimate that acid rain has caused nearly $15 billion of damage to buildings and other structures thus far.

Solutions to the problems associated with acid rain will not be easy. The National Science Foundation (NSF) has recommended that we strive for a 50% reduction in sulfur-oxide emissions. Perhaps that is easier said than done. High-sulfur coal is a major source of these emissions, but in states dependent on coal for energy, a shift to lower sulfur coal is not always possible. Rather, better scrubbers must be developed to remove these contaminating oxides from the burning process before they are released into the atmosphere. Fuels for internal combustion engines are also major sources of the nitric and sulfur oxides of acid rain. Clearly, better emission control is needed for automobiles and trucks.

Reducing the oxide emissions from coal-burning furnaces and motor vehicles will require greater use of existing scrubbers and emission control devices as well as the development of new technology to allow us to use available energy sources. Developing alternative, cleaner energy sources is also important if we are to meet NSF's goal. Statistics and statisticians will play a key role in monitoring atmosphere conditions, testing the effectiveness of proposed emission control devices, and developing new control technology and alternative energy sources.

Determining the Effectiveness of a New Drug Product

The development and testing of the Salk vaccine for protection against poliomyelitis (polio) provide an excellent example of how statistics can be used in solving practical problems. Most parents and children growing up before 1954 can recall the panic brought on by the outbreak of polio cases during the summer months.

Although relatively few children fell victim to the disease each year, the pattern of outbreak of polio was unpredictable and caused great concern because of the possibility of paralysis or death. The fact that very few of today's youth have even heard of polio demonstrates the great success of the vaccine and the testing program that preceded its release on the market.

It is standard practice in establishing the effectiveness of a particular drug product to conduct an experiment (often called a *clinical trial*) with human subjects. For some clinical trials, assignments of subjects are made at random, with half receiving the drug product and the other half receiving a solution or tablet (called a *placebo*) that does not contain the medication. One statistical problem concerns the determination of the total number of subjects to be included in the clinical trial. This problem was particularly important in the testing of the Salk vaccine because data from previous years suggested that the incidence rate might be less than 50 cases for every 100,000 children. Hence, a large number of subjects had to be included in the clinical trial in order to detect a difference in the incidence rates for those treated with the vaccine and those receiving the placebo.

With the assistance of statisticians, it was decided that a total of 400,000 children should be included in the Salk clinical trial begun in 1954, with half of them randomly assigned the vaccine and the remaining children assigned the placebo. No other clinical trial had ever been attempted on such a large group of subjects. Through a public school inoculation program, the 400,000 subjects were treated and then observed over the summer to determine the number of children contracting polio. Although less than 200 cases of polio were reported for the 400,000 subjects in the clinical trial, more than three times as many cases appeared in the group receiving the placebo. These results together with some statistical calculations were sufficient to indicate the effectiveness of the Salk polio vaccine. However, these conclusions would not have been possible if the statisticians and scientists had not planned for and conducted such a large clinical trial.

The development of the Salk vaccine is not an isolated example of the use of statistics in the testing and developing of drug products. In recent years, the Food and Drug Administration (FDA) has placed stringent requirements on pharmaceutical firms to establish the effectiveness of proposed new drug products. Thus, statistics has played an important role in the development and testing of birth control pills, rubella vaccines, chemotherapeutic agents in the treatment of cancer, and many other preparations.

Applications of Statistics in Our Courts

Libel suits related to consumer products have touched each one of us; you may have been involved as a plaintiff or defendant in a suit or you may know of someone who was involved in such litigation. Certainly we all help to fund the costs of this litigation indirectly through increased insurance premiums and increased costs of goods. The testimony in libel suits concerning a particular product (automobile, drug product, and so on) frequently leans heavily on the interpretation of data from one or more scientific studies involving the product. This is how and why statistics and statisticians have been pulled into the courtroom.

For example, epidemiologists have used statistical concepts applied to data to determine whether there is a statistical "association" between a specific characteristic, such as the use of a brand-name tampon, and a disease condition, such as toxic shock syndrome. An epidemiologist who finds an association should try to determine whether the observed statistical association from the study is due to random variation or whether it reflects an actual association between the characteristic and the disease. Arguments in courtrooms about the interpretations of these types of associations involve data analyses using statistical concepts as well as a clinical interpretation of the data.

The Energy Crisis: A Search for New Sources and a Search for Oil

The OPEC oil crisis of 1973–1974 and the more recent war in the Middle East have brought to our attention a problem that is with us today and will continue to plague us for decades: a shortage of energy. The United States is confronted by staggering annual demands for energy with supplies that may not meet current and future demands, especially if a major supplier "interrupts" service. Such an interruption by OPEC in 1974 led to an energy rush and subsequent supply problems.

Possible sources of energy needed to supply the present and future requirements of the United States include vast coal and oil shale reserves, nuclear reactors, new oil and natural gas reserves, solar energy, and alternative new fuels. For example, methanol (wood alcohol) and ethanol (grain alcohol) may be major contributors to octane boost as leaded fuels are phased out. These alcohols are also likely candidates to reduce our dependence on foreign crude oil.

In which of these resources should we, the American public, invest the capital necessary for development? Which source will yield a given amount of energy at minimum cost? What unfavorable impact will each have on the environment or the quality of life? Which might yield dangerous side effects? These questions and others must be answered by experimentation. Statisticians will assist in designing experiments and in interpreting experimental data.

Opinion and Preference Polls

Public opinion, consumer preference, and election polls are commonly used to assess the opinions or preferences of a segment of the public for issues, products, or candidates of interest. And we, the American public, are exposed to the results of these polls on a daily basis in newspapers, in magazines, on the radio, and on television. For example, the results of polls related to the following subjects were printed in local newspapers over a 2-day period:

- ▼ consumer confidence related to future expectations about the economy
- ▼ preferences for candidates in upcoming elections and caucuses
- ▼ attitudes toward cheating on federal income tax returns
- ▼ preference polls related to specific products (for example, foreign vs. American cars, Coke vs. Pepsi, McDonald's vs. Wendy's)
- ▼ reactions of North Carolina residents toward arguments about the morality of tobacco
- ▼ opinions of voters toward proposed tax increases and proposed changes in the Defense Department budget.

A number of questions can be raised about these polls. How many people were polled? What questions were asked? Was each person asked the same question? How were people chosen or selected for the poll? Can we believe the results of these polls? Do these results "represent" how the general public feels?

Opinion and preference polls are an important, visible application of statistics for the consumer. We will discuss this topic in more detail in Chapter 8. We hope, after studying this material, you will have a better understanding of how to interpret the results of these polls.

1.4 ▼ WHAT DO STATISTICIANS DO?

What do statisticians do? In the context of making sense of data, statisticians are involved with all aspects of gathering, summarizing, and analyzing data, and reporting the results of their analyses. There are both good and bad ways to gather data. Statisticians apply their knowledge of existing survey techniques and scientific study designs or they develop new techniques to provide a guide to good methods of data collection. We will explore these ideas further in Chapter 2.

Once the data are gathered, they must be summarized before any meaningful interpretation can be made. Statisticians can recommend and apply useful methods for summarizing data in graphical, tabular, and numerical forms. Intelligent graphs and tables are useful first steps in making sense of the data. Also, measures of the average (or typical) value and some measure of the range or spread of the data help in interpretation. These topics will be discussed in detail in Chapter 3.

The objective of statistics is to make an inference about a population of interest based on information obtained from a sample of measurements from that population. The analysis stage of making sense of data deals with making inferences. For example, a market research study reaches only a few of the potential buyers of a new product, but probable reaction of the set of potential buyers (population) must be inferred from the reactions of the buyers included in the study (sample). If the market research study has been carefully planned and executed, the reactions of those included in the sample should agree reasonably well (but not necessarily exactly) with the population. The reason we can say this is because the basic concepts of probability allow us to make an inference about the population of interest that includes our best guess plus a statement of the probable error in our best guess.

We will illustrate how inferences are made by way of an example. Suppose an auditor samples 2,000 financial accounts from a set of more than 25,000 accounts and finds that 84 (4.2%) are in error. What can be said about the set of 25,000 accounts? That is, what inference can we make about the percentage of accounts in error for the population of 25,000 accounts based on information obtained from the sample of 2,000 accounts? We will show (in Chapter 8) that our best guess (inference) about the percentage of accounts in error for the population is 4.2%, and this best guess should be within ±.9% of the actual unknown percentage of accounts in error for the population. The plus-or-minus factor is called the *probable error* of our inference. Anyone can make a guess about the percentage of accounts in error; concepts of probability allow us to calculate the (probable) error of our guess.

In dealing with the analyses of data, statisticians can apply existing methods for making inferences; some theoretical statisticians engage in the development of new methods with more advanced mathematics and probability theory. Our study of the methods for analysing sample data will begin in Chapter 5, after we discuss the basic concepts of probability and sampling distributions in Chapter 4.

Finally, statisticians are involved with communicating the results of their analyses as the final stage in making sense of data. The form of the communication varies from an informal conversation to a formal report. The advantage of a more formal verbal presentation with visual aids or study report is that the communication can make use of the graphical, tabular, and numerical displays as well as the analyses done on the data to help convey the "sense" found in the data. Too often this is lost in an information conversation. The report or communication should convey to the intended audience what can be gleaned from the sample data, and it should be conveyed in as nontechnical terms as possible so that there can be no confusion as to what is being inferred. More information about the communication of results is presented in Chapter 20.

1.5 ▼ QUALITY AND PROCESS IMPROVEMENT

One might wonder, at this stage, why we would bring up the subject of quality and process improvement in a statistics textbook. We do so to make you aware of some of the broader issues involved with making sense of data in the business and scientific communities.

The post-World War II years saw U.S. business and the U.S. economy dominate world business, and this lasted for about 30 years. During this time, there was very little attempt to change the ways things were done; the major focus was doing things on a much grander scale, perfecting mass production. However, from the mid–1970s through today, many industries have had to face fierce competition from their counterparts in Japan and, more recently, from other countries in the Far East, such as China and Korea.

Quality, rather than *quantity*, has become the principal buying gauge used by consumers, and American industries have had a difficult time adjusting to this new emphasis. Unless there are drastic changes in the way many American industries approach their businesses, there will be many more casualties to the "quality" revolution.

The Japanese were the first to learn the lessons of quality. They readily used the statistical quality-control and process-control suggestions espoused by Deming (1981) and others and installed total quality programs. Through the organization— from top management down—they had a commitment to improving the quality of their products and the way they did things. They were never satisfied with the way things were and continually looked for new and better ways of doing things.

A number of American companies have now begun the journey toward excellence through the institution of a quality-improvement process. Harrington (1987) articulated the following ten basic requirements that would make a quality-improvement process successful:

Harrington's Fundamental Requirements for a Successful Quality-Improvement Process

1. focus on the customer as the most important part of the process
2. long-term commitment by management to make the quality-improvement process part of the management system
3. belief that there is room to improve
4. belief that preventing problems is better than reacting to problems
5. management focus, leadership, and participation
6. a performance standard (goal) of zero errors
7. participation by all employees, both as groups and as individuals
8. improvement focus on the process, not the people
9. belief that suppliers will work with you if they understand your needs
10. recognition for success.

re-engineering

Embedded in a companywide quality-improvement process or running concurrent with such a process is the idea of improving the work processes or, as it has become known, of **re-engineering** the work processes. For years, companies, in trying to boost and improve performance, have tried to speed up their processes, usually with additional people or technology but without addressing possible deficiencies in the work processes. As Michael Hammer (1990) states, "It is time to stop paving the cow paths." In many cases, we need to rethink completely the processes that have stayed with us in spite of the ever-changing business and technology environment. Fundamental to re-engineering is to identify and break from outdated habits, rules, and assumptions. Dramatic, radical improvement in quality, efficiency, and effectiveness are possible with these re-engineered efforts. Within a work place, a re-engineering effort can begin by developing a case for action with management that focuses on the following five questions:

▼ What is important to our mission?
▼ How are we using people, processes, and technology to perform our mission?
▼ Are there opportunities to use technology or process change to improve significantly our work processes?
▼ What opportunities provide the greatest potential, and what are the costs and payoffs of implementing these opportunities?
▼ How do we get started?

At the heart of any quality-improvement process or re-engineering effort lie the data that reflect the state of health of a process and the reports based on these data. Here's where statistics and statistical tools can be of help. A number of these tools and techniques are listed next:

Statistical Tools, Techniques, and Methods Used in Quality Improvement and Re-Engineering

▼ histograms
▼ numerical descriptive measures (means, standard deviations, proportions, etc.)
▼ scatterplots
▼ line graphs (scatterplot with dots connected)
▼ control charts: $\bar{y}$ (sample mean), r (sample range), and s (sample standard deviation)
▼ sampling schemes
▼ experimental designs.

The statistical tools and concepts listed here and discussed in this textbook comprise only a small component of a total quality-improvement process.

Keep in mind where you think these tools and concepts may have application in a quality-improvement process or re-engineering project as you encounter them in various parts of this text. Quality improvement and process re-engineering are clearly the focus of American industry for the 1990s in world markets characterized by increased competition, more consolidation, and increased specialization. These shifts will have impacts on us all, either as consumers or business participants, and it will be useful to know some of the statistical tools that are part of this revolution.

1.6 ▼ A NOTE TO THE STUDENT

We think with words and concepts. A study of the discipline of statistics requires the memorization of new terms and concepts (as does the study of a foreign language). Commit these definitions, theorems, and concepts to memory.

Also, focus on the broader concept of making sense of data. Do not let details obscure these broader characteristics of the subject. The teaching objective of this text is to identify and amplify these broader concepts of statistics.

1.7 ▼ SUMMARY

The discipline of statistics and those who apply the tools of that discipline deal with making sense of data. As such, statisticians are involved with methods of data collection, data summarization, and data analyses, as well as communicating the results of its analyses.

▼ SUPPLEMENTARY EXERCISES

Basic Techniques

 1.1 Selecting the proper diet for shrimp or other sea animals is an important aspect of sea farming. A researcher wishes to estimate the mean weight of shrimp maintained on a specific diet for a period of six months. One hundred shrimp are randomly selected from an artificial pond and each is weighed.
  a. Identify the population of measurements that is of interest to the researcher.
  b. Identify the sample.
  c. What characteristics of the population would be of interest to the researcher?
  d. If the sample measurements are used to make inferences about certain characteristics of the population, why would a measure of the reliability of the inferences be important?

 1.2 Radioactive waste disposal as well as the production of radioactive material in some mining operations are creating a serious pollution problem in some areas of the United States. State health officials have decided to investigate the radioactivity levels in one suspect area. Two hundred points are randomly selected in the area and the level of radioactivity is measured at each point. Answer questions a, b, c, and d in Exercise 1.1 for this sampling situation.

 1.3 A social researcher in a particular city wishes to obtain information on the number of children in households that receive welfare support. A random sample of 400 households is selected from the welfare rolls of the city. A check on welfare recipient data provides the number of children in each household. Answer questions a, b, c, and d of Exercise 1.1 for this sample survey.

1.4 Search issues of your local newspaper to locate the results of a recent Harris or Gallup survey.

    a. Identify the items that will be observed in order to obtain the sample measurements.

    b. Identify the measurement made on each item.

    c. Clearly identify the population associated with the survey.

    d. What characteristic(s) of the population is (are) of interest to the pollster?

    e. Does the article explain how the sample was selected?

    f. Does the article include the number of measurements in the sample?

    g. What type of inference is made concerning the population characteristics?

    h. Does the article tell you how much faith you can place in the inference about the population characteristic?

GATHERING THE DATA

USING SURVEYS AND SCIENTIFIC STUDIES TO GATHER DATA

2.1 INTRODUCTION

As we've mentioned previously, the first step in making sense of data is to gather data on one or more variables of interest. But *intelligent data gathering* doesn't just happen; it takes a conscious, concerted effort focused on the following steps:

▼ specifying the objective of the data-gathering exercise
▼ identifying the variable(s) of interest
▼ choosing an appropriate design for the survey or scientific study
▼ collecting the data.

To specify the objective of the data-gathering exercise you must understand the problem under study. For example, if the management of a large manufacturing company is considering whether to institute a new incentive pay plan for its production workers, it might want to determine the attitudes of the production supervisors toward the proposed plan. This, then, could be the objective of a data-gathering exercise.

To identify the variable(s) of interest, you must examine the objective of the data-gathering exercise. For the production incentive plan, the variable of interest could be the attitude of the production supervisors. Measurements would consist of preferences (favor, oppose), accompanied by comments as to why a production supervisor might favor or oppose the new incentive plan. Once the objective is determined and the variable(s) of interest specified, you must choose how to collect the data. In statistics, data can be gathered by way of a survey, a study, or a combination of the two. Survey theory and the theory of experimental designs for scientific studies provide good methods for collecting data. Usually, surveys are passive where the aim is to gather (survey) data on existing conditions, attitudes, or behaviors. Thus, the management of the manufacturing company would use a survey to sample the opinions of the production supervisors on the merits of a new incentive plan. Scientific studies, on the other hand, tend to be more active: the person conducting the study would tend to deliberately vary certain conditions in order to reach a conclusion. For example, if a plant manager is interested in the effect of noise level in his manufacturing plant, he could vary the noise level and certain controlled conditions in order to see, directly, the gains or losses in productivity.

In this chapter, we will consider some of the survey methods and designs for scientific studies. We will also make a distinction between a scientific study and an observational study.

2.2 ▼ SURVEYS

Information from surveys affects almost every facet of our daily lives. These surveys determine such government policies as the control of the economy and the promotion of social programs. Opinion polls are the basis of much of the news reported by the various news media. Ratings of television shows determine which shows are to be available for viewing in the future.

One usually thinks of the U.S. Census Bureau as contacting every household in the country. Actually, in the 1980 census only 14 questions were asked of all households. Information on an additional 42 questions was obtained from only a sample of households. The resulting information is used by many agencies and individuals for manifold purposes. For example, the federal government uses it to determine allocations of funds to states and cities; it is used by businesses to forecast sales, to manage personnel, and to establish future site locations; it is used by urban and regional planners to plan land use, transportation networks, and energy consumption. It is used by social scientists to study economic conditions, racial balance, and other aspects of the quality of life.

The U.S. Bureau of Labor Statistics (BLS) routinely conducts more than twenty surveys. Some of the best known and most widely used are the surveys that establish the consumer price index (CPI). The CPI is a measure of price change for a fixed market basket of goods and services over time. It is used as a measure of inflation and serves as an economic indicator for government policies. Businesses have wage rates and pension plans tied to the CPI. Federal health and welfare programs, as well as many state and local programs, tie their bases of eligibility to the CPI. Escalator clauses in rents and mortgages are based on the CPI. So we can see that this one index, determined on the basis of sample surveys, plays a fundamental role in our society.

Many other surveys from the BLS are crucial to society. The monthly Current Population Survey establishes basic information on the labor force, employment, and unemployment. The consumer expenditure surveys collect data on family expenditures for goods and services used in day-to-day living. The Establishment Survey collects information on employment hours and earnings for nonagricultural business establishments. The survey on occupational outlook provides information on future employment opportunities for a variety of occupations, projecting to approximately ten years ahead. Other activities of the BLS are addressed in the *BLS Handbook of Methods* (1982).

Opinion polls are constantly in the news, and the names of Gallup and Harris have become well known to everyone. These polls, or sample surveys, reflect the attitudes and opinions of citizens on everything from politics and religion to sports and entertainment. The Nielsen ratings determine the success or failure of TV shows.

Businesses conduct sample surveys for their internal operations, in addition to using government surveys for crucial management decisions. Auditors estimate account balances and check on compliance with operating rules by sampling accounts. Quality control of manufacturing processes relies heavily on sampling techniques.

One particular area of business activity that depends on detailed sampling activities is marketing. Decisions on which products to market, where to market them, and how to advertise them are often made on the basis of sample survey data. The data may come from surveys conducted by the firm that manufactures the product or may be purchased from survey firms that specialize in marketing data.

Sampling Techniques

simple random sampling

The basic design (**simple random sampling**) consists of selecting a group of n units in such a way that each sample of size n has the same chance of being selected. Thus, we can obtain a random sample of eligible voters in the bond-issue poll by drawing names from the list of registered voters in such a way that each sample of size n has the same probability of selection. The details of simple random sampling are discussed in Section 4.11. At this point, we merely state that a simple random sample will contain as much information on the community preference as any other sample survey design, provided all voters in the community have similar socioeconomic backgrounds.

Suppose, however, that the community consists of people in two distinct income brackets, high and low. Voters in the high bracket may have opinions on the bond issue that are quite different from the opinions of voters in the low bracket. Therefore, to obtain accurate information about the population, we want to sample voters from each bracket. We can divide the population elements into two groups, or strata, according to income and select a simple random sample from **stratified random sample** each group. The resulting sample is called a **stratified random sample.** (See Chapter 5 of Scheaffer et al., 1990.)

ratio estimation

Note that stratification is accomplished by using knowledge of an auxiliary variable, namely, personal income. By stratifying on high and low values of income, we increase the accuracy of our estimator. **Ratio estimation** is a second method for using the information contained in an auxiliary variable. Ratio estimators not only use measurements on the response of interest but they incorporate measurements on an auxiliary variable. Ratio estimation can also be used with stratified random sampling.

cluster sampling

Although individual preferences are desired in the survey, a more economical procedure, especially in urban areas, may be to sample specific families, apartment buildings, or city blocks rather than individual voters. Individual preferences can then be obtained from each eligible voter within the unit sampled. This technique is called **cluster sampling**. Although we divide the population into groups for both cluster sampling and stratified random sampling, the techniques differ. In stratified random sampling, we take a simple random sample within each group, whereas, in cluster sampling, we take a simple random sample of groups and then sample all items within the selected groups (clusters). (See Chapters 8 and 9 of Scheaffer et al., 1990, for details.)

systematic sample

Sometimes, the names of persons in the population of interest are available in a list, such as a registration list, or on file cards stored in a drawer. For this situation, an economical technique is to draw the sample by selecting one name near the beginning of the list and then selecting every tenth or fifteenth name thereafter. If the sampling is conducted in this manner we obtain a **systematic sample**. As you might expect, systematic sampling offers a convenient means of obtaining sample information; unfortunately, we do not necessarily obtain the most information for a specified amount of money. (Details are given in Chapter 7 of Scheaffer et al., 1990.)

The important point to understand is that there are different kinds of surveys that can be used to collect sample data. For the surveys discussed in this text, we will be dealing with simple random sampling and methods for summarizing and analyzing data collected in such a manner. More complicated surveys lead to even more complicated problems at the summarization and analyses stages of statistics.

Data Collection Techniques

Having chosen a particular sample survey, how does one actually collect the data? The most commonly used methods of data collection in sample surveys are personal interviews and telephone interviews. These methods, with appropriately trained interviewers and carefully planned callbacks, commonly achieve response rates of 60% to 75% and sometimes even higher. A mailed questionnaire sent to a specific group of interested persons can achieve good results, but generally the response rates for this type of data collection are so low that all reported results are suspect. Frequently, objective information can be found from direct observation rather than from an interview or mailed questionnaire.

personal interviews

Data are frequently obtained by **personal interviews**. For example, we can use personal interviews with eligible voters to obtain a sample of the public sentiments toward a community bond issue. The procedure usually requires the interviewer to ask prepared questions and to record the respondent's answers. The primary advantage of these interviews is that people will usually respond when confronted in person. In addition, the interviewer can note specific reactions and eliminate misunderstandings about the questions asked. The major limitations of the personal interview (aside from the cost involved) concern the interviewers. If they are not thoroughly trained, they may deviate from the required protocol, thus introducing a bias into the sample data. Any movement, facial expression, or statement by the interviewer can affect the response obtained. For example, a leading question such as "Are you also in favor of the bond issue?" may tend to elicit a positive response. Finally, errors in recording the responses can also lead to erroneous results.

telephone interviews

Information can also be obtained from persons in the sample through **telephone interviews**. With the advent of wide-area telephone service lines (WATS lines), an interviewer can place any number of calls to specified areas of the country for a fixed monthly rate. Surveys conducted through telephone interviews are frequently less expensive than personal interviews, owing to the elimination of travel expenses. The investigator can also monitor the interviews to be certain that the specified interview procedure is being followed.

A major problem with telephone surveys is that it is difficult to find a list or directory that closely corresponds to the population. Telephone directories have many numbers that do not belong to households, and many households have unlisted numbers. A few households have no phone service, although lack of phone service is now only a minor problem for most surveys in the United States. A technique that avoids the problem of unlisted numbers is random-digit dialing. In this method, a telephone exchange number (the first three digits of a seven-digit number) is selected, and then the last four digits are dialed randomly until a fixed number of households of a specified type are reached. This technique produces samples from the target population and avoids many of the problems inherent in sampling a telephone directory.

Telephone interviews generally must be kept shorter than personal interviews because responders tend to get impatient more easily when talking over the telephone. With appropriately designed questionnaires and trained interviewers, telephone interviews can be as successful as personal interviews.

self-administered questionnaire

Another useful method of data collection is the **self-administered questionnaire**, to be completed by the respondent. These questionnaires usually are mailed to the individuals included in the sample, although other distribution methods can be used. The questionnaire must be carefully constructed if it is to encourage participation by the respondents.

The self-administered questionnaire does not require interviewers, and thus its use results in a savings in the survey cost. This savings in cost is usually bought at the expense of a lower response rate. Nonresponse can be a problem in any form of data collection, but since we have the least contact with respondents in a mailed questionnaire, we frequently have the lowest rate of response. The low response rate can introduce a bias into the sample because the people who answer questionnaires may not be representative of the population of interest. To eliminate some of the bias, investigators frequently contact the nonrespondents through follow-up letters, telephone interviews, or personal interviews.

direct observation

The fourth method for collecting data is **direct observation**. For example, if we were interested in estimating the number of trucks that use a particular road during the 4–6 P.M. rush hours, we could assign a person to count the number of trucks passing a specified point during this period. Possibly, electronic counting equipment could also be used. The disadvantage in using an observer is the possibility of error in observation.

Direct observation is used in many surveys that do not involve measurements on people. The U.S. Department of Agriculture, for instance, measures certain variables on crops in sections of fields in order to produce estimates of crop yields. Wildlife biologists may count animals, animal tracks, eggs, or nests in order to estimate the size of animal populations.

A closely related notion to direct observation is that of getting data from objective sources that are not affected by the respondents themselves. For example, health information can sometimes be obtained from hospital records, and income information from employer's records (especially for state and federal government workers). This approach may take more time but can yield large rewards in important surveys.

▼ EXERCISES

Basic Techniques

2.1 An experimenter wants to estimate the average water consumption per family in a city. Discuss the relative merits of choosing individual families, dwelling units (single-family houses, apartment buildings, etc.), and city blocks as sampling units.

2.2 A forester wants to estimate the total number of trees on a tree farm that possess diameters exceeding 12 inches. A map of the farm is available. Discuss the problem of choosing what to sample and how to select the sample.

2.3 A safety expert is interested in estimating the proportion of automobile tires with unsafe treads. Should he use individual cars or collections of cars, such as those in parking lots, in his sample?

2.4 An industry is composed of many small plants located throughout the United States. An executive wants to survey the opinions of the employees on the vacation policy of the industry. What would you suggest she sample?

2.5 A state department of agriculture desires to estimate the number of acres planted in corn within the state. How might one conduct such a survey?

2.6 A political scientist wants to estimate the proportion of adult residents of a state who favor a unicameral legislature. What could be sampled? Also, discuss the relative merits of personal interviews, telephone interviews, and mailed questionnaires as methods of data collection.

2.7 Discuss the relative merits of using personal interviews, telephone interviews, and mailed questionnaires as methods of data collection for each of the following situations:

a. A television executive wants to estimate the proportion of viewers in the country who are watching the network at a certain hour.

b. A newspaper editor wants to survey the attitudes of the public toward the type of news coverage offered by the paper.

c. A city commissioner is interested in determining how homeowners feel about a proposed zoning change.

d. A county health department wants to estimate the proportion of dogs that have had rabies shots within the last year.

 2.8 A Yankelovich, Skelly, and White poll taken in the fall of 1984 showed that one-fifth of the 2,207 people surveyed admitted to having cheated on their federal income taxes. Do you think that this fraction is close to the actual proportion who cheated? Why? (Discuss the difficulties of obtaining accurate information on a question of this type.)

2.3 ▼ SCIENTIFIC STUDIES

The subject of experimental designs for scientific studies cannot be given much justice in the beginning of a statistical methods course, since entire courses at the undergraduate and graduate levels are needed to get a comprehensive understanding of the methods and concepts of experimental design. Even so, we will attempt to give you a brief overview of the subject because much data requiring summarization and analyses arise from scientific studies involving one of a number of experimental designs. We will work by way of examples.

A multinational oil company has been developing an unleaded gasoline that can be competitively priced while delivering higher gasoline mileage. A number of blends of gasolines have been proposed and undergone initial testing. One blend in particular appears to yield good gasoline mileage and can be produced economically. An additional test is planned to obtain an accurate estimate of the gasoline mileage (miles per gallon) for the blend under normal road conditions. To do this, a standard car model is chosen and each of ten cars of this model type is driven over a predetermined course and the miles per gallon recorded. (Later, in Chapter 5, we will show how these data could then be summarized to provide an estimate of the miles per gallon for the blend under these fixed road conditions.)

To change the problem slightly, suppose the oil company had three different gasoline blends for further testing in the gasoline mileage test conducted under normal road conditions. For this study, the company could take nine standard model cars and randomly assign three cars to each gasoline blend and test the cars under the road conditions dictated by the experiment. There would be a recorded gasoline mileage for each car, and three cars per gasoline blend. The methods presented in Chapters 5 and 15 could be used to summarize and analyze the sample mileage data in order to make comparisons (inferences) among the three gasoline blends. One possible inference of interest could be the selection of the best gasoline blend from a gasoline mileage standpoint. Which blend performed better? Can the best performing blend in the sample data be expected to provide better gasoline mileage if the same study were repeated?

Experimental Designs

completely randomized design

The experimental design for this scientific study is called a **completely randomized design**. Table 2.1 displays a completely randomized design for the gasoline blend study.

T A B L E   2.1
Completely Randomized Design of Gasoline Blends

Blend 1	Blend 2	Blend 3
3 cars	3 cars	3 cars

In general a completely randomized design is used when one is interested in comparing t "treatments" (in our $t = 3$ case, the treatments were gasoline blends). For each of the treatments we obtain a sample of observations, and the samples are not necessarily of the same size for the different treatments. The sample of observations from a treatment is assumed to be the result of a simple random sample of observations from the hypothetical population of possible values that could have been obtained for that treatment. In our example, the sample of three gasoline mileages obtained from blend 1 was considered to be the outcome for a simple random sample of three observations selected from the hypothetical population of possible mileages for standard model cars using gasoline blend 1. The same reasoning applies for the samples from blends 2 and 3.

This experimental design could be changed to accommodate the study of the same three blends using each of, say, three different drivers. Because of individual driving habits, not all drivers get the same gasoline mileage for the same blend of gasoline; so it would be desirable to have each of the drivers test each of the blends over the specified road course. Here we avoid having the comparison of blends distorted by differences among drivers. The experimental design is called a **randomized block design** because we have "blocked" out any differences among drivers in order to get a precise comparison of the three blends. See Table 2.2.

randomized block design

T A B L E   2.2
Randomized Block Design of Gasoline Blends

Driver 1	Driver 2	Driver 3
Blend 1	Blend 2	Blend 1
Blend 3	Blend 3	Blend 2
Blend 2	Blend 1	Blend 3

Note that each driver tests each blend, and the order of testing the blends is randomized for each driver (i.e., driver 2 will test blend 2 first, then blend 3, and finally blend 1).

What happens if the order of testing influences a driver's performance and the first blend tested generally receives a higher mileage rating than the ones tested second or third? Then blend 1 could (possibly) look better than blends 2 and 3 simply because it was tested first by two of the three drivers (see Table 2.2).

Latin square design

A variation on the previous randomized block design, called a **Latin square design**, eliminates the order of testing as a factor affecting the comparison of treatments (blends). A Latin square design for our example is shown in Table 2.3.

T A B L E   2.3
Latin Square Design of Gasoline Blends

Order of Testing	Driver 1	Driver 2	Driver 3
1st	Blend 1	Blend 3	Blend 2
2nd	Blend 2	Blend 1	Blend 3
3rd	Blend 3	Blend 2	Blend 1

Note that with this design, each blend is tested first once, second once, and third once, *and* each driver tests all blends.

The randomized block and Latin square designs are both extensions of the completely randomized design where the objective is to compare *t* treatments. The analysis of data collected according to a completely randomized design and the inferences made from these analyses are discussed further in Chapters 15 and 16. A special case of the randomized block design is presented in Chapter 6, where the number of treatments is $t = 2$ and the analysis of data and the inferences from these analyses are discussed.

Factorial Experiments

factors

Suppose that we want to examine the effects of two (or more) variables (**factors**) on a response. For example, suppose that an experimenter is interested in examining the effects of two variables, nitrogen and phosphorus, on the yield of a crop. For simplicity we will assume that two levels have been selected for the study of each factor: 40 and 60 pounds per plot for nitrogen, 10 and 20 pounds per plot for phosphorus. For this study the experimental units are small, relatively homogeneous plots that have been partitioned from the acreage of a farm.

one-at-a-time approach

One approach for examining the effects of two or more factors on a response is called the **one-at-a-time approach**. To examine the effect of a single variable, an experimenter varies the levels of this variable while holding the levels of the other independent variables fixed. This process is continued until the effect of each variable on the response has been examined. See Table 2.4 for an example.

TABLE 2.4
Factor–Level Combination for a One-at-a-Time Approach

Combination	Nitrogen	Phosphorus
1	60	10
2	40	10
3	40	20

Hypothetical yields corresponding to the three factor–level combinations of our experiment are given in Table 2.5. Suppose the experimenter is interested in using the sample information to determine the factor–level combination that will give the maximum yield. From the table, we see that crop yield increases when the nitrogen application is increased from 40 to 60 (holding phosphorus at 10). Yield also increases when the phosphorus setting is changed from 10 to 20 (at a fixed nitrogen setting of 40). Thus, it might seem logical to predict that increasing both the nitrogen and phosphorus applications to the soil will result in a larger crop yield. The fallacy in this argument is that our prediction is based on the assumption that the effect of one factor is the same for both levels of the other factor.

TABLE 2.5
Yields for the Three Factor–Level Combinations

Observation (yield)	Nitrogen	Phosphorus
145	60	10
125	40	10
160	40	20
?	60	20

We know from our investigation what happens to yield when the nitrogen application is increased from 40 to 60 for a phosphorus setting of 10. But will the yield also increase by approximately 20 units when the nitrogen application is changed from 40 to 60 at a setting of 20 for phosphorus?

To answer this question we could apply the factor–level combination of 60 nitrogen–20 phosphorus to another experimental plot and observe the crop yield. If the yield is 180, then the information obtained from the three factor–level combinations would be correct and useful in predicting the factor–level combination that produces the greatest yield. However, suppose the yield obtained from the high settings of nitrogen and phosphorus turns out to be 110. If this happens, the two factors nitrogen and phosphorus are said to **interact**. That is, the effect of one factor on the response does not remain the same for different levels of the second factor, and the information obtained from the one-at-a-time approach would lead to a faulty prediction.

The two outcomes just discussed for the crop yield at 60–20 setting are displayed in Figure 2.1 along with the yields at the initial factor–level settings. Figure 2.1 (top) illustrates a situation with no interaction between the two factors. The effect of nitrogen on yield is the same for both levels of phosphorus. In

interaction

FIGURE 2.1

Yields at the Initial Three Factor–Level Settings and at a Fourth Setting

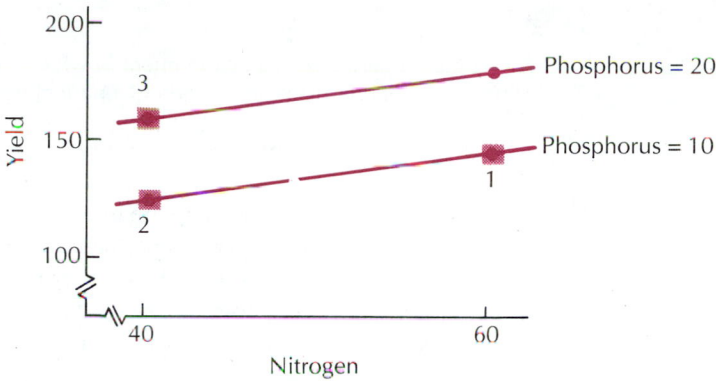

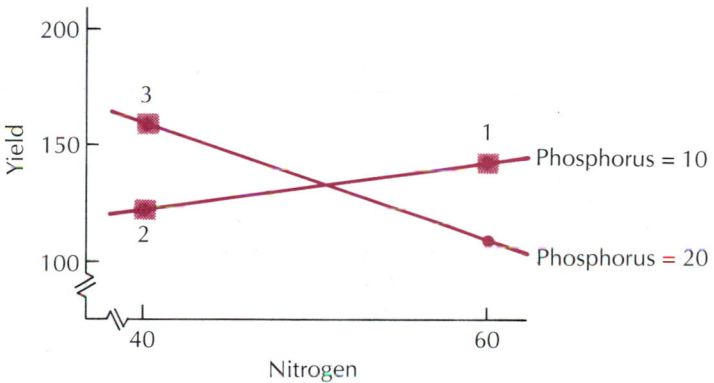

contrast, Figure 2.1 (bottom) illustrates a case in which the two factors nitrogen and phosphorus do interact.

We have seen that the one–at–a–time approach to investigating the effect of two factors on a response is suitable only for situations in which the two factors do not interact. Although this was illustrated for the simple case in which two factors were to be investigated at each of two levels, the inadequacies of a one–at–a–time approach are even more salient when trying to investigate the effects of more than two factors on a response.

factorial experiment

Factorial experiments are useful for examining the effects of two or more factors on a response, whether or not interaction exists. As before, the choice of the number of levels of each variable and the actual settings of these variables is important. When the factor–level combinations are assigned to experimental units at random, we have a completely randomized design with treatments being the factor–level combinations.

▼

A **factorial experiment** is an experiment in which the response is observed at all factor–level combinations of the independent variables.

Using our previous example, if we are interested in examining the effect of two levels of nitrogen at 40 and 60 pounds per plot and two levels of phosphorus at 10 and 20 pounds per plot on the yield of a crop, we could use a completely randomized design where the four factor–level combinations (treatments) of Table 2.6 are assigned at random to the experimental units.

TABLE 2.6
2 × 2 Factorial Experiment for Crop Yield

Factor–Level Combinations	
Nitrogen	Phosphorus
40	10
40	20
60	10
60	20

Similarly, if we wished to examine nitrogen at the two levels 40 and 60 and phosphorus at the three levels 10, 15, and 20, we could use the six factor–level combinations of Table 2.7 as treatments in a completely randomized design.

TABLE 2.7
2 × 3 Factorial Experiment for Crop Yield

Factor–Level Combinations	
Nitrogen	Phosphorus
40	10
40	15
40	20
60	10
60	15
60	20

The examples of factorial experiments presented in this section deal with the effects of two variables (factors) on a response. However, the procedure applies to any number of factors and levels per factor. Thus, if we have four different factors at two, three, three, and four levels, respectively, we could formulate a $2 \times 3 \times 3 \times 4$ factorial experiment by considering all $2 \cdot 3 \cdot 3 \cdot 4 = 72$ factor–level combinations. Analyses and inferences of data obtained from factorial experiments in various designs are discussed in Chapters 15, 16, and 17.

More Complicated Designs

Sometimes the objectives of a study are such that we wish to investigate the effects of certain factors on a response while blocking out certain other extraneous sources of variability. Such situations require a block design with treatments from a factorial experiment. Such a situation can be illustrated with the following example.

An investigator wants to examine the effects of two factors (factors A and B each measured at three levels) on a response y. It is determined that two observations are desired at each factor–level combination, but only nine observations can be done each day. Since nine observations can be obtained each day, it is possible to run a complete replication of the 3×3 factorial experiment on two different days to get the desired number of observations. The design is shown in Table 2.8.

T A B L E 2.8
A Block Design Combined with a Factorial Experiment

	Day 1				Day 2		
		Factor B				Factor B	
Factor A	1	2	3	Factor A	1	2	3
1				1			
2				2			
3				3			

Note that this design is really a randomized block design where the blocks are days and the treatments are the nine factor–level combinations of the 3×3 factorial experiment. Other more complicated combinations of block designs and factorial experiments are possible. As with sample surveys, though, we will deal only with the simplest experimental designs in this text. The point we want to make is that there are many different experimental designs that can be used in scientific studies for designating the collection of sample data. Each has certain advantages and disadvantages. We expand our discussion of experimental designs in Chapters 15–19, where we concentrate on the analysis of data generated from these designs.

2.4 ▼ OBSERVATIONAL STUDIES

observational study

Before leaving the subject of sample data collection, we will draw a distinction between an **observational study** and a scientific study. In experimental designs for scientific studies, the observation conditions are fixed or controlled. For example, with a factorial experiment laid off in a completely randomized design, an observation is made at each factor–level combination. Similarly, with a randomized block design, an observation is obtained on each treatment in every block. These "controlled" studies are very different from observational studies, which are sometimes used because it is not feasible to do a proper scientific study. This can be illustrated by way of an example.

Much research and public interest centers on the effect of cigarette smoking on lung cancer and cardiovascular disease. One possible experimental design would be to randomize a fixed number of individuals (say 1,000) to each of two groups—one group would be required to smoke cigarettes for the duration of the study (say 10 years), while those in the second group would not be allowed to smoke throughout the study. At the end of the study, the two groups would be compared for lung cancer and cardiovascular disease. Even if we ignore ethical questions, this type of study would be impossible to do. Because of the long duration, it would be difficult to follow all participants and make certain that they follow the study plan. And it would be difficult to find nonsmoking individuals willing to take the chance of being assigned to the smoking group.

Another possible study would be to sample a fixed number of smokers and a fixed number of nonsmokers to compare the groups for lung cancer and for cardiovascular disease. Assuming one could obtain willing groups of participants, this study could be done in a *much shorter* period of time.

What has been sacrificed? Well, the fundamental difference between an observational study and a scientific study lies in the inferences(s) that can be drawn. For a scientific study comparing smokers to nonsmokers, assuming the two groups of individuals followed the study plan, the observed differences between the smoking and nonsmoking groups could be attributed to the effects of cigarette smoking because individuals were randomized to the two groups; hence, the groups were assumed to be comparable at the outset.

This type of reasoning does not apply to the observational study of cigarette smoking. Differences between the two groups in the observation could not necessarily be attributed to the effects of cigarette smoking because, for example, there may be hereditary factors that predispose people to smoking and cancer of the lungs and/or cardiovascular disease. Thus, differences between the groups might be due to hereditary factors, smoking, or a combination of the two. Typically, the results of an observational study are reported by way of a statement of association. For our example, if the observational study showed a higher frequency of lung cancer and cardiovascular disease for smokers relative to nonsmokers, it would be stated that this study showed that cigarette smoking was associated with an increased frequency of lung cancer and cardiovascular disease. It is a careful rewording in order not to infer that cigarette smoking *causes* lung cancer and cardiovascular disease.

Many times, however, an observational study is the only type of study that can be run. Our job is to make certain that we understand the type of study run and, hence, understand how the data were collected. Then we can critique inferences drawn from an analysis of the study data.

2.5 ▼ DATA MANAGEMENT: PREPARING DATA FOR SUMMARIZATION AND ANALYSIS

In this section, we concentrate on some important data management procedures that are followed between the time the data are gathered and the time the data are available in computer-readable form for analysis. This is not a complete manual with all tools required; rather, it is an overview—what a manager should know about these steps. As an example, this section reflects standard procedures in the pharmaceutical industry, which is highly regulated. Procedures may differ somewhat in other industries and settings.

We begin with a discussion of the procedures involved in processing data from a study. In practice, these procedures may consume 75% of the total effort from the receipt of the raw data to the presentation of results from the analysis. What are these procedures, why are they so important, and why are they so time consuming?

To answer these questions, let's list the major data-processing procedures in the cycle, which begins with receipt of the data and ends when the statistical analysis begins. Then we'll discuss each procedure separately.

Procedures in Processing Data for Summarization and Analysis

> 1. Receive the raw data source.
> 2. Create the data base from the raw data source.
> 3. Edit the data base.
> 4. Correct and clarify the raw data source.
> 5. Finalize the data base.
> 6. Create data files from the data base.

1. *Receiving the raw data source.* For each study that is to be summarized and analyzed, the data arrive in some form, which we'll refer to as the **raw data source**

raw data source

For a clinical trial, the raw data source is usually case report forms, sheets of $8\frac{1}{2} \times 11$-inch paper that have been used to record study data for each patient entered into the study. For other types of studies, the raw data source may be sheets of paper from a laboratory notebook, a magnetic tape (or any other form of machine-readable data), hand-tabulations, and so on.

data trail

It is important to retain the raw data source, since it is the beginning of the **data trail**, which leads from the raw data to the conclusions drawn from a study. Many consulting operations involved with the analysis and summarization of many different studies keep a log that contains vital information related to the study and raw data source. In a regulated environment such as the pharmaceutical industry, one may have to redo or reproduce data and data analyses based on previous work. Other situations outside the pharmaceutical industry may also require a retrospective review of what was done in the analysis of a study. In these situations, the study log can be an invaluable source of study information. General information contained in a study log is shown next.

Log for Study Data

▼

1. Date received, and from whom
2. Study investigator
3. Statistician (and others) assigned
4. Brief description of study
5. Treatments (compounds, preparations, etc.) studied
6. Raw data source
7. Response(s) measured and how measured
8. Reference number for study
9. Estimated (actual) completion date
10. Other pertinent information

Later, when the study has been analyzed and results have been communicated, additional information can be added to the log on how the study results were communicated, where these results are recorded, what data files have been saved, and where these files are stored.

2. *Creating the data base from the raw data source.* For most studies that are scheduled for a statistical analysis, a machine-readable data base is created. The steps taken to create the data base and the eventual form of the data base vary from one operation to another, depending on the software systems to be used in the statistical analysis. However, we can give a few guidelines based on the form of the entry system.

When the data are to be typed at a terminal, the raw data are first checked for legibility. Any illegible numbers or letters or other problems should be brought to the attention of the study coordinator. Then a coding guide that assigns column numbers and variable names to the data is filled out. Certain codes for missing values (for example, not available) are also defined here. Also, it is helpful to give a brief description of each variable. The data file keyed in at the terminal is referred to as the **machine-readable data base**. A listing (printout) of the data base should be obtained and checked carefully against the raw data source. Any errors should be corrected at the terminal and verified against an updated listing.

Sometimes data are received in machine-readable form. For these situations, the magnetic tape or disk file is considered to be the data base. You must, however, have a coding guide to "read" the data base. Using the coding guide, obtain a listing of the data base and check it *carefully* to see that all numbers and characters look reasonable and that proper formats were used to create the file. Any problems that arise must be resolved before proceeding further.

machine-readable data base

Some data sets are so small that it is not necessary to create a machine-readable data file from the raw data source. Instead, calculations are performed by hand or the data are entered into an electronic calculator. For these situations check any calculations to see that they make sense. Don't believe everything you see; redoing the calculations is not a bad idea.

3. *Editing the data base.* The types of edits done and the completeness of the editing process really depend on the type of study and how concerned you are about the accuracy and completeness of the data prior to the analysis. For example, in using a statistical software package (such as SAS or Minitab), it is wise to examine the minimum, maximum, and frequency distribution for each variable to make certain nothing looks unreasonable.

logic checks

Certain other checks should be made. Plot the data and look for problems. Also, certain **logic checks** should be done depending on the structure of the data. If, for example, data are recorded for patients at several different visits, then the data recorded for visit 2 can't be earlier than the data for visit 1; similarly, if a patient is lost to follow-up after visit 2, we can't have any data for that patient at later visits.

For small data sets, we can do these data edits by hand, but, for large data sets, the job may be too time-consuming and tedious. If machine editing is required, look for a software system that allows the user to specify certain data edits. Even so, for more complicated edits and logic checks it may be necessary to have a customized edit program written in order to machine edit the data. This programming chore can be a time-consuming step; plan for this well in advance of the receipt of the data.

4. *Correcting and clarifying the raw data source.* Questions frequently arise concerning the legibility or accuracy of the raw data during any one of the steps from the receipt of the raw data to the communication of the results from the statistical analysis. We have found it helpful to keep a list of these problems or discrepancies in order to define the data trail for a study. If a correction (or clarification) is required to the raw data source, this should be indicated on the form and the appropriate change made to the raw data source. If no correction is required, this should be indicated on the form as well. Keep in mind that the machine-readable data base should be changed to reflect any changes made to the raw data source.

5. *Finalizing the data base.* You may have been led to believe that all data for a study arrive at one time. This, of course, is not always the case. For example, with a marketing survey, different geographic locations may be surveyed at different times and, hence, those responsible for data processing do not receive all the data at one time. All these subsets of data, however, must be processed through the cycles required to create, edit, and correct the data base. Eventually the study is declared complete and the data is processed into the data base. At this time, the data base should be reviewed again and final corrections made before beginning the analysis. The reason for this is that, for large data sets, the analysis and summarization chores take considerable human labor and computer time. It's better to agree on a final data base analysis than to have to repeat all analyses on a changed data base at a later date.

6. *Creating data files from the data base.* Generally there are one or two sets of data files created from the machine-readable data base. The first set, referred to as **original files** reflects the basic structure of the data base. A listing of the files is checked against the data base listing to verify that the variables have been read with correct formats and missing value codes have been retained. For some studies, the original files are actually used for editing the data base.

A second set of data files, called **work files**, may be created from the original files. Work files are designed to facilitate the analysis. They may require restructuring of the original files, a selection of important variables, or the creation or addition of new variables by insertion, computation, or transformation. A listing of the work files is checked against that of the original files to ensure proper restructuring and variable selection. Computed and transformed variables are checked by hand calculations to verify the program code.

If original and work files are SAS data sets, you should utilize the documentation features provided by SAS. At the time an SAS data set is created, a descriptive label for the data set, of up to 40 characters, should be assigned. The label can be stored with the data set imprinted wherever the contents procedure is used to print the data set's contents. All variables can be given descriptive names, up to 8 characters in length, which are meaningful to those involved in the project. In addition, variable labels up to 40 characters in length can be used to provide additional information. Title statements can be included in the SAS code to identify the project and describe each job. For each file, a listing (proc print) and a dictionary (proc contents) can be retained.

For files created from the data base using other software packages, you should utilize the labeling and documentation features available in the computer program.

Even if appropriate statistical methods are applied to data, the conclusions drawn from the study are only as good as the data on which they are based. So you be the judge. The amount of time you should spend on these data-processing chores before analysis really depends on the nature of the study, the quality of the raw data source, and how confident you want to be about the completeness and accuracy of the data.

original files

work files

2.6 ▼ SUMMARY

The first step in making sense of data involves intelligent data gathering. This involves specifying the objectives of the data-gathering exercise, identifying the variables of interest, and choosing an appropriate design for the survey or scientific study. In this chapter, we discussed various survey designs and experimental designs for scientific studies. Armed with a basic understanding of some design considerations for conducting surveys or scientific studies, one can address how data are to be collected on the variables of interest in order to address the stated objectives of the data-gathering exercise.

We also drew a distinction between observational and scientific studies in terms of the inferences (conclusions) that can be drawn from the sample data. Differences found between treatment groups from an observational study are said to be *associated with* the use of the treatments; on the other hand, differences found between treatments in a scientific study are said to be *due to* the treatments. In the next chapter, we will examine the methods for summarizing the data we collect.

SUMMARIZING DATA

DATA DESCRIPTION

3.1 ▼ INTRODUCTION

In the previous chapter, we discussed how to do intelligent data gathering for an experiment or survey, Step 1 in making sense of data. We turn now to Step 2, summarizing the data.

The field of statistics can be divided into two major branches: descriptive statistics and inferential statistics. In both branches, we work with a set of measurements. For situations in which data description is our major objective, the set of measurements available to us is frequently the entire population. For example, suppose that we wish to describe the distribution of annual incomes for all families registered in the 1990 census. Since all these data are recorded and are available on computer tapes, we do not need to obtain a random sample from the population; the complete set of measurements is at our disposal. Our major problem is in organizing, summarizing, and describing these data—that is, making sense of the data. Similarly, vast amounts of monthly, quarterly, and yearly data are available on sales for the steel industry, broken down by domestic consumption, exports, consumer inventory change, and imports. However, in order to present such data in formats appropriate for audiences of consumers, financial analysts, or managers, it is necessary to organize, summarize, and describe the data. Good descriptive statistics enable us to make sense of the data by reducing a large set of measurements to a few summary measures that provide a good, rough picture of the original measurements.

In situations in which we are concerned with statistical inference, a sample is usually the only set of measurements available to us. We use information in the sample to draw conclusions about the population from which the sample was

41

drawn. Of course, in the process of making inferences, we also need to organize, summarize, and describe the sample data.

For example, the tragedy surrounding isolated incidents of product tampering has brought about federal legislation requiring tamper-resistant packaging for certain drug products sold over the counter. These same incidents also brought about increased awareness by industry of the rigid standards of product and packaging quality that must be maintained by companies involved with delivering these products to the store shelves. In particular, one company was interested in determining the proportion of packages out of total production that are improperly sealed or have been damaged in transit. Obviously, it would be impossible to inspect all packages at all the stores where the product is sold, but a random sample of the production could be obtained and the proportion defective in the sample could be used to estimate the actual proportion of improperly sealed or damaged packages.

Similarly, in developing an economic forecast of new housing starts for the next year, it is necessary to use sample data from various economic indicators in order to make such a prediction (inference).

In both of these examples involving an inference, description of the sample data is an important step leading toward the inference that we make. Thus, no matter what our objective, statistical inference or data description, we must first describe the set of measurements at our disposal.

There are two ways to describe a set of measurements. We can use either graphical techniques or numerical descriptive techniques. Section 3.2 is concerned with graphical methods for describing data on a single variable. In Sections 3.3, 3.4, and 3.6, we discuss numerical techniques for describing data from a single variable. Section 3.5 is optional and deals with coding techniques. The final topics on data description are presented in Section 3.7, where we consider a few techniques for describing (summarizing) data on more than one variable.

3.2 ▼ DESCRIBING DATA ON A SINGLE VARIABLE: GRAPHICAL METHODS

When many measurements are obtained on a variable, the data must first be organized, prior to presentation, by using one of several graphical techniques. As a general rule, the data should be arranged in such a way that *each observation can fall into one and only one category of the variable.* This procedure eliminates any ambiguity that might arise in placing observations into categories and aids in the interpretation of the data. For example, administrative officials of many universities require the parents of students applying for financial aid to file a financial report.

Suppose a particular university required parents to pick one of the following gross-income categories:

less than $15,000
$15,000—$34,999
$35,000—$54,999
$55,000—$74,999
$75,000 or more

Clearly, the adjusted gross income for a family filing a joint return or the combined adjusted gross income for separate returns will fall into one and only one income category. However, if the income categories had been defined as

less than $15,000
$15,000—$35,000
$35,000—$55,000
$55,000—$75,000
$75,000 or more

pie chart

there could be some confusion as to which category should be checked for families with adjusted gross incomes falling on the boundary points.

Having organized the data according to the guideline suggested, there are several ways to display the data graphically. The first and simplest graphical procedure for data organized in this manner is the **pie chart**. It is used to display the percentage of the total number of measurements falling into each of the categories of the variable by partitioning a circle (much as one might slice a pie). The data of Table 3.1 represents a summary of a study to determine paths to authority for individuals occupying top positions of responsibility in key public-interest organizations.

Using biographical information, each of 1,345 individuals was classified according to how she or he was recruited for the current elite position.

TABLE 3.1
Recruitment to Top Public-Interest Positions*

Recruitment From	Number	Percentage
Corporate	501	37.2
Public-interest	683	50.8
Government	94	7.0
Other	67	5.0

* Includes trustees of private colleges and universities, directors of large private foundations, senior partners of top law firms, and directors of certain large cultural and civic organizations.

Source: Thomas R. Dye and L. Harmon Zeigler, *The Irony of Democracy,* 5th ed. (Monterey, Calif.: Duxbury Press, 1981), p. 130.

FIGURE 3.1
**Pie Chart for the Data of
Table 3.1**

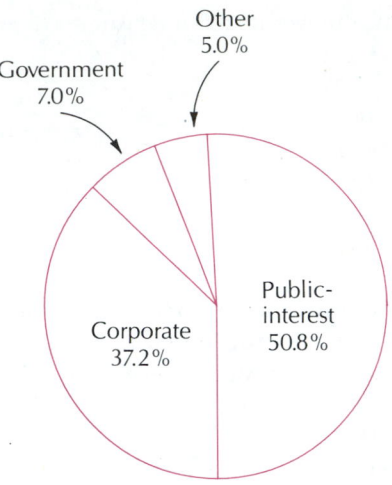

Although you can scan the data in Table 3.1, the results are more easily interpreted by using a pie chart. From Figure 3.1 we could make certain inferences about channels to positions of authority. For example, more people were recruited for elite positions from public–interest organizations (approximately 51%) than from elite positions in other organizations.

Other variations of the pie chart are shown in Figures 3.2 and 3.3. Clearly, from Figure 3.2, cola soft drinks have gained in popularity from 1980 to 1990 at the expense of some of the other types of soft drinks. Also, it's evident from Figure 3.3 that the loss of a major food chain account affected fountain sales for PepsiCo, Inc. In summary, the pie chart can be used to display percentages associated with

FIGURE 3.2
**Approximate Market Share
of Soft Drinks by Type,
1980 and 1990**

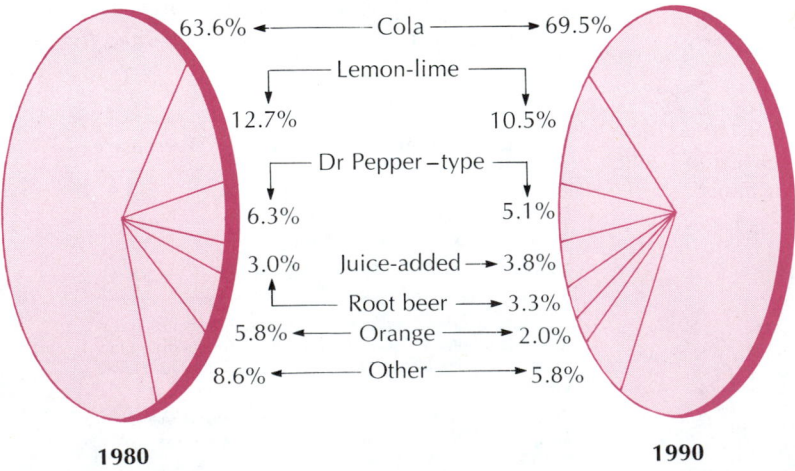

each category of the variable. The following guidelines should help you to obtain clarity of presentation in pie charts.

Guidelines for Constructing Pie Charts

1. Choose a small number of categories for the variable (five or six) because too many make the pie chart difficult to interpret.
2. Whenever possible, construct the pie chart so that percentages are in either ascending or descending order.

bar chart

A second graphical technique for data organized according to the recommended guideline is the **bar chart**, or bar graph. Figure 3.4 displays the number of workers in the Cincinnati, Ohio, area for the largest five foreign investors. The estimated total work force is 680,000. There are many variations of the bar chart. Sometimes the bars are displayed horizontally, as in Figure 3.5(a) and (b). They can also be used to display data across time, as in Figure 3.6. Bar charts are relatively easy to construct if you use the guidelines given.

Guidelines for Constructing Bar Charts

1. Label frequencies on one axis; categories of the variable on the other axis.
2. Construct a rectangle at each category of the variable with a height equal to the frequency (number of observations) in the category.
3. Leave a space between each category to connote distinct, separate categories and to clarify the presentation.

FIGURE 3.3
Estimated U.S. Market Share Before and After Switch in Accounts*

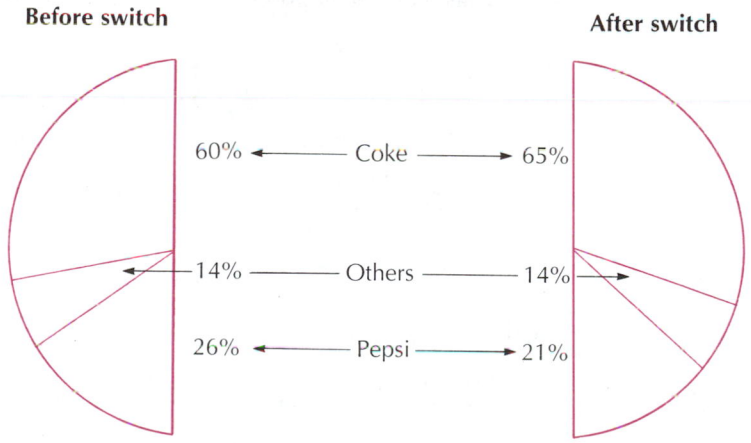

Before switch

After switch

60% ◄──── Coke ────► 65%

14% ──── Others ──── 14%

26% ◄──── Pepsi ───► 21%

A major fast food chain switched its account from Pepsi to Coca-Cola for fountain sales.

FIGURE 3.4
**Number of Workers by
Major Foreign Investors**

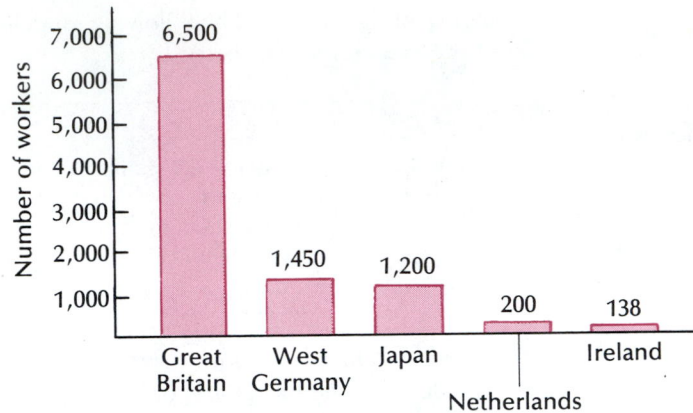

FIGURE 3.4
**Number of Workers by
Major Foreign Investors**

FIGURE 3.5
**Greatest per Capita
Consumption, by Country**

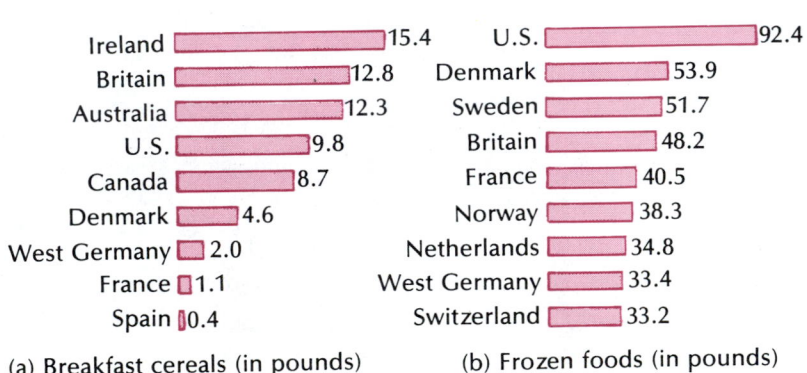

(a) Breakfast cereals (in pounds)   (b) Frozen foods (in pounds)

FIGURE 3.6
**Estimated Direct and
Indirect Costs for
Developing a New Drug,
by Selected Years**

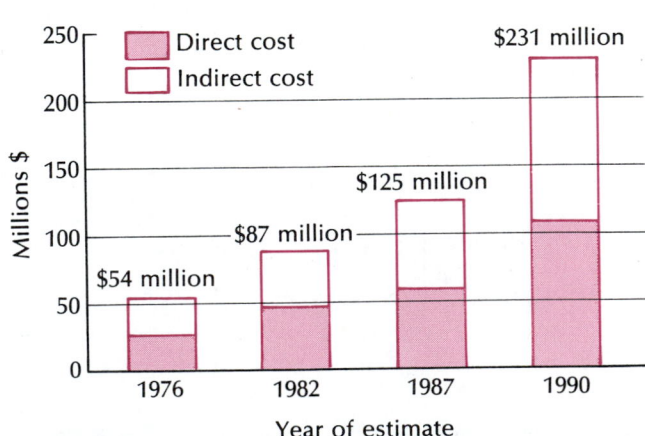

frequency histogram, relative frequency histogram

The next two graphical techniques that we will discuss in this text are the **frequency histogram** and the **relative frequency histogram**. Both of these graphical techniques are applicable only to quantitative (measured) data. As with the previous graphical techniques, we must organize the data before constructing a graph.

Consider the following kind of measurement: weight gain for each of 100 baby chicks fed on a new diet and observed over an 8-week period. These data are recorded in Table 3.2.

In trying to describe the set of measurements recorded in Table 3.2 we note that the largest weight gain is 4.9 and the smallest is 3.6. But although we might examine the table very closely, it is difficult to describe how the measurements are situated along the interval from 3.6 to 4.9. Are most of the measurements near 3.6, near 4.9, or are they evenly distributed along the interval? To answer the questions, we summarize the data in a **frequency table**.

frequency table

class intervals

To construct a frequency table, we begin by dividing the range from 3.6 to 4.9 into an arbitrary number of subintervals called **class intervals**. The number of subintervals chosen depends on the number of measurements in the set, but we generally recommend using from 5 to 20 class intervals. The more data we have, the larger the number of classes we tend to use.

The guidelines given here can be used for constructing the appropriate class intervals.

Guidelines for Constructing Class Intervals

1. Divide the **range** of the measurements (the difference between the largest and the smallest measurements) by the approximate number of class intervals desired. Generally, we will wish to have from 5 to 20 class intervals.
2. After dividing the range by the desired number of subintervals, round the resulting number to a convenient (easy to work with) unit. This unit represents a common width for the class intervals.
3. Choose the first class interval so that it contains the smallest measurement. It is also advisable to choose a starting point for the first interval so that no measurement falls on a point of division between two subintervals. This eliminates any ambiguity in placing measurements into the class intervals. (One way to do this is to choose boundaries to one more decimal place than the data.)

For the data in Table 3.2,

$$\text{range} = 4.9 - 3.6 = 1.3.$$

TABLE 3.2
Weight Gains for Chicks (grams)

3.7	4.2	4.4	4.4	4.3	4.2	4.4	4.8	4.9	4.4
4.2	3.8	4.2	4.4	4.6	3.9	4.3	4.5	4.8	3.9
4.7	4.2	4.2	4.8	4.5	3.6	4.1	4.3	3.9	4.2
4.0	4.2	4.0	4.5	4.4	4.1	4.0	4.0	3.8	4.6
4.9	3.8	4.3	4.3	3.9	3.8	4.7	3.9	4.0	4.2
4.3	4.7	4.1	4.0	4.6	4.4	4.6	4.4	4.9	4.4
4.0	3.9	4.5	4.3	3.8	4.1	4.3	4.2	4.5	4.4
4.2	4.7	3.8	4.5	4.0	4.2	4.1	4.0	4.7	4.1
4.7	4.1	4.8	4.1	4.3	4.7	4.2	4.1	4.4	4.8
4.1	4.9	4.3	4.4	4.4	4.3	4.6	4.5	4.6	4.0

Assume that we want to have approximately 10 subintervals. Dividing the range by 10 and rounding to a convenient unit, we have $1.3/10=.13 \approx .1$. Thus, the class interval width is .1.

It is convenient to choose the first interval to be 3.55—3.65, the second to be 3.65—3.75, and so on. Note that the smallest measurement, 3.6, falls in the first interval and that no measurement falls on the endpoint of a class interval (see Table 3.3).

TABLE 3.3
Frequency Table for the Chick Data

Class	Class Interval	Frequency f_i	Relative frequency f_i/n
1	3.55–3.65	1	.01
2	3.65–3.75	1	.01
3	3.75–3.85	6	.06
4	3.85–3.95	6	.06
5	3.95–4.05	10	.10
6	4.05–4.15	10	.10
7	4.15–4.25	13	.13
8	4.25–4.35	11	.11
9	4.35–4.45	13	.13
10	4.45–4.55	7	.07
11	4.55–4.65	6	.06
12	4.65–4.75	7	.07
13	4.75–4.85	5	.05
14	4.85–4.95	4	.04
Totals		$n = 100$	1.00

Having determined the class interval, we then construct a frequency table for the data. The first column labels the classes by number and the second column indicates the class intervals. We then examine the 100 measurements of Table 3.2, keeping a tally of the number of measurements falling in each interval. The number of measurements falling in a given class interval is called the **class frequency**. These data are recorded in the third column of the frequency table (see Table 3.3).

The **relative frequency** of a class is defined to be the frequency of the class divided by the total number of measurements in the set (total frequency). Thus, if

class frequency

relative frequency

we let f_i denote the frequency for class i and n denote the total number of measurements, the relative frequency for class i is f_i/n. The relative frequencies for all the classes are listed in the fourth column of Table 3.3.

histogram

The data of Table 3.2 have been organized into a frequency table, which can now be used to construct a *frequency histogram* or a *relative frequency* **histogram**. To construct a frequency histogram, draw two axes: a horizontal axis labeled with the class intervals and a vertical axis labeled with the frequencies. Then construct a rectangle over each class interval with a height equal to the number of measurements falling in a given subinterval. The frequency histogram for the data of Table 3.3 is shown in Figure 3.7.

FIGURE 3.7
Frequency Histogram for the Chick Data of Table 3.3

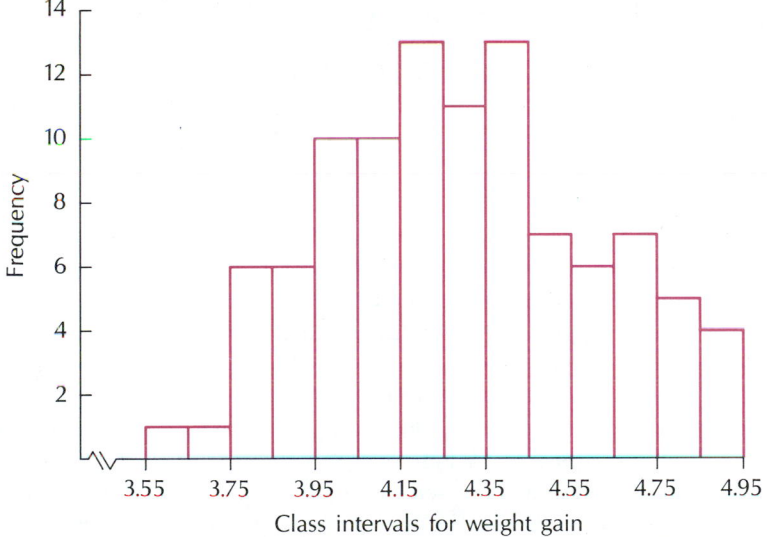

The relative frequency histogram is constructed in much the same way as a frequency histogram. In the relative frequency histogram, however, the vertical axis is labeled as relative frequency, and a rectangle is constructed over each class interval with a height equal to the class relative frequency (the fourth column of Table 3.3). The relative frequency histogram for the data of Table 3.3 is shown in Figure 3.8. Clearly, the two histograms of Figures 3.7 and 3.8 are of the same shape and would be identical if the vertical axes were equivalent. We will frequently refer to either one as simply a histogram.

There are several comments that should be made concerning histograms. First, the distinction between bar charts and histograms is based on the distinction between *qualitative* and *quantitative* variables. Values of qualitative variables vary in kind but not degree and hence are not measurements. For example, the variable political party affiliation can be categorized as Republican, Democrat, or other,

FIGURE 3.8
**Relative Frequency
Histogram for the Chick
Data of Table 3.3**

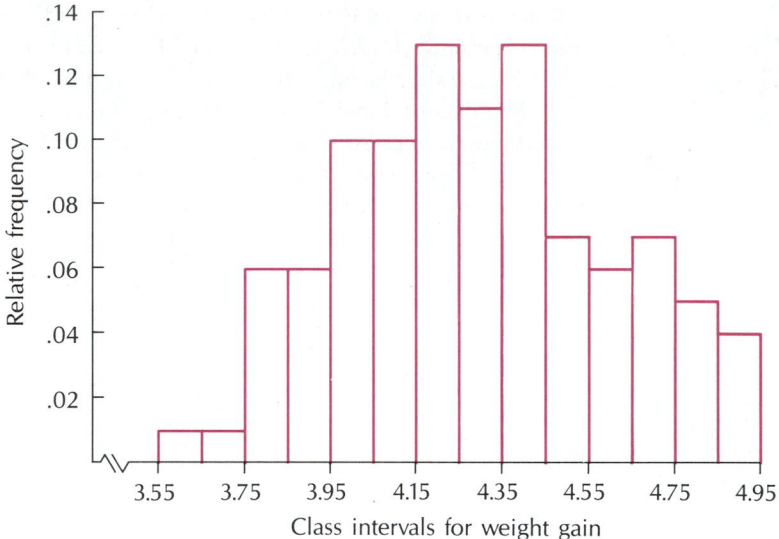

and, although we could label the categories as one, two, or three, these values are only codes and have no quantitative interpretation. In contrast, quantitative variables have actual units of measure. For example, the variable yield (in bushels) per acre of corn can assume specific values. *Bar charts are used to display frequency data from qualitative variables; histograms are appropriate for displaying frequency data for quantitative variables.*

Second, the histogram is the most important graphical technique we will present because of the role it plays in statistical inference, a subject we will discuss in later chapters. Third, if we had an extremely large set of measurements, and if we constructed a histogram using many class intervals, each with a very narrow width, the histogram for the set of measurements would be, for all practical purposes, a smooth curve. Fourth, the fraction of the total number of measurements in an interval is equal to the fraction of the total area under the histogram over the interval. For example, if we consider the interval 3.75 to 4.35 for the chick data of Table 3.3, we see that there are exactly 56 of the 100 measurements in that interval. Thus, .56, the fraction of the total number of sample measurements falling in that interval, is equal to the fraction of the total area under the histogram over that interval, as indicated in Figure 3.9.

Fifth, if a single measurement is selected at random from the set of sample measurements, the chance, or **probability**, that it lies in a particular interval is equal to the fraction of the total number of sample measurements falling in that interval. This same fraction will be used to estimate the probability that a measurement randomly selected from the population lies in the interval of interest. For example, from the sample data of Table 3.2, the chance or probability of selecting a baby chicken with a weight gain in the interval 3.75 to 4.35 is .56. This

probability

F I G U R E 3.9
**Fraction of Measurements
in the Interval 3.75 to 4.35
for the Chick Data**

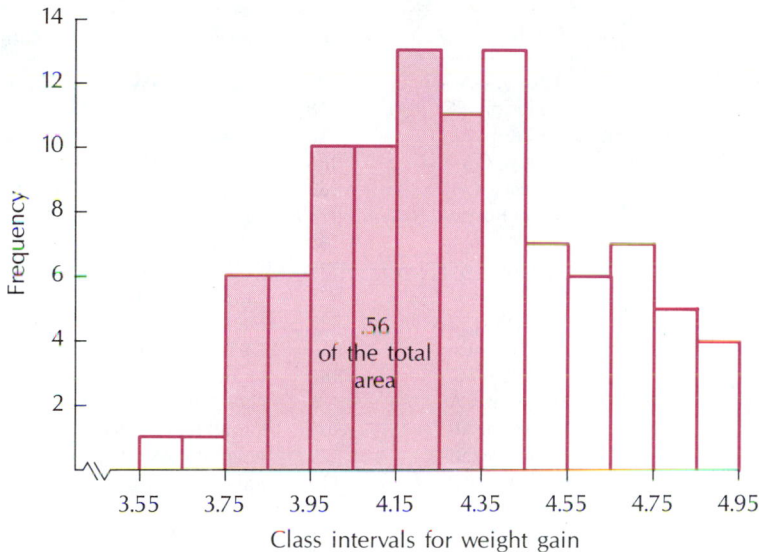

value, .56, is an approximation of the probability of selecting a measurement in the interval 3.75 to 4.35 for the population of all weight gains for baby chickens.

Finally, since we use proportions rather than frequencies in a relative frequency histogram, we can compare two different samples (or populations) by examining their relative frequency histograms even if the samples (populations) are of different sizes.

exploratory data analysis

The next graphical technique to be presented in this section is a display technique taken from an area of statistics called **exploratory data analysis (EDA)**. Professor John Tukey (1977) has been the leading proponent of this practical philosophy of data analysis aimed at exploring and understanding data.

stem-and-leaf plot

The **stem-and-leaf plot** is a clever, simple device for constructing a histogramlike picture of a frequency distribution. It allows us to use the information contained in a frequency distribution to show the range of scores, where the scores are concentrated, the shape of the distribution, whether there are any specific values or scores not represented, and whether there are any stray or extreme scores. The stem-and-leaf plot does not follow the organization principles stated previously for histograms. We will use the data shown in Table 3.4 to illustrate how to construct a stem-and-leaf plot.

The original scores in Table 3.4 are either three- or four-digit numbers. We will use the first, or leading, digit of each score as the stem (see Figure 3.10) and the trailing digits as the leaf. For example, the violent crime rate in Albany is 876. The leading digit is 8 and the trailing digits are 76. In the case of Fresno, the leading digits are 10 and the trailing digits are 20. If our data consisted of 6-digit numbers such as 104,328, we might use the first two digits as stem numbers, use the second two digits as leaf numbers, and ignore the last two digits.

TABLE 3.4
Violent Crime Rates for 90 Standard Metropolitan Statistical Areas Selected from the North, South, and West

South	Rate	North	Rate	West	Rate
Albany, GA	876	Allentown, PA	189	Abilene, TX	570
Anderson, SC	578	Battle Creek, MI	661	Albuquerque, NM	928
Anniston, AL	718	Benton Harbor, MI	877	Anchorage, AK	516
Athens, GA	388	Bridgeport, CT	563	Bakersfield, CA	885
Augusta, GA	562	Buffalo, NY	647	Brownsville, TX	751
Baton Rouge, LA	971	Canton, OH	447	Denver, CO	561
Charleston, SC	698	Cincinnati, OH	336	Fresno, CA	1,020
Charlottesville, VA	298	Cleveland, OH	526	Galveston, TX	592
Chattanooga, TN	673	Columbus, OH	624	Houston, TX	814
Columbus, GA	537	Dayton, OH	605	Kansas City, MO	843
Dothan, AL	642	Des Moines, IA	496	Lawton, OK	466
Florence, SC	856	Dubuque, IA	296	Lubbock, TX	498
Fort Smith, AR	376	Gary, IN	628	Merced, CA	562
Gadsden, AL	508	Grand Rapids, MI	481	Modesto, CA	739
Greensboro, NC	529	Janesville, WI	224	Oklahoma City, OK	562
Hickery, NC	393	Kalamazoo, MI	868	Reno, NV	817
Knoxville, TN	354	Lima, OH	804	Sacramento, CA	690
Lake Charles, LA	735	Madison, WI	210	St. Louis, MO	720
Little Rock, AR	811	Milwaukee, WI	421	Salinas, CA	758
Macon, GA	504	Minneapolis, MN	435	San Diego, CA	731
Monroe, LA	807	Nassau, NY	291	Santa Ana, CA	480
Nashville, TN	719	New Britain, CT	393	Seattle, WA	559
Norfolk, VA	464	Philadelphia, PA	605	Sioux City, IA	505
Raleigh, NC	410	Pittsburgh, PA	341	Stockton, CA	703
Richmond, VA	491	Portland, ME	352	Tacoma, WA	809
Savannah, GA	557	Racine, WI	374	Tucson, AZ	706
Shreveport, LA	771	Reading, PA	267	Victoria, TX	631
Washington, DC	685	Saginaw, MI	684	Waco, TX	626
Wilmington, DE	448	Syracuse, NY	685	Wichita Falls, TX	639
Wilmington, NC	571	Worcester, MA	460	Yakima, WA	585

Note: Rates represent the number of violent crimes (murder, forcible rape, robbery, and aggravated assault) per 100,000 inhabitants, rounded to the nearest whole number.
Source: Department of Justice, Uniform Crime Reports for the United States, 1990.

For the data on violent crime, the smallest rate is 189, the largest is 1,020, and the leading digits are 1, 2, 3, . . . , 10. In the same way that a class interval determines where a measurement is placed in a frequency table, the leading digit (stem of a score) determines the row in which a score is placed in a stem–and–leaf plot. The trailing digits for the score are then written in the appropriate row. In this way, each score is recorded in the stem–and–leaf plot. This has been done in Figure 3.10 for the violent crime data.

We can see that each stem defines a class interval and the limits of each interval are the largest and smallest possible scores for the class. The values represented by each leaf must be between the lower and upper limits of the interval.

Note that a stem–and–leaf plot is a graph that looks much like a histogram turned sideways, as in Figure 3.10. The plot can be made a bit more useful by

FIGURE 3.10
Stem-and-Leaf Plot for Violent Crime Rates of Table 3.4

```
 1    89
 2    98 96 24 10 91 67
 3    88 76 93 54 36 93 41 52 74
 4    64 10 91 48 47 96 81 21 35 60 66 98 80
 5    78 62 37 08 29 04 57 71 63 26 70 16 61 92 62 62 59 05 85
 6    98 73 42 85 61 47 24 05 28 05 84 85 90 31 26 39
 7    18 35 19 71 51 39 20 58 31 03 06
 8    76 56 11 07 77 68 04 85 14 43 17 09
 9    71 28
10    20
```

FIGURE 3.11
Stem-and-Leaf Plot with Ordered Leaves

```
 1    89
 2    10 24 67 91 96 98
 3    36 41 52 54 74 76 88 93 93
 4    10 21 35 47 48 60 64 66 80 81 91 96 98
 5    04 05 08 16 26 29 37 57 59 61 62 62 62 63 70 71 78 85 92
 6    05 05 24 26 28 31 39 42 47 61 73 84 85 85 90 98
 7    03 06 18 19 20 31 35 39 51 58 71
 8    04 07 09 11 14 17 43 56 68 76 77 85
 9    28 71
10    20
```

ordering the data (leaves) within a row (stem) from lowest to highest (Figure 3.11). The advantage of such a graph over the histogram is that it reflects not only frequencies, concentration(s) of scores, and shapes of the distribution but also the actual scores.

Guidelines for Constructing Stem-and-Leaf Plots

▼

1. Split each score or value into two sets of digits. The first or leading set of digits is the stem and the second or trailing set of digits is the leaf.
2. List all possible stem digits from lowest to highest.
3. For each score in the mass of data, write down the leaf values on the line labeled by the appropriate stem number.
4. If the display looks too cramped and narrow, stretch the display by using two lines per stem so that, for example, leaf digits 0, 1, 2, 3, and 4 and placed on the first line of the stem and leaf digits 5, 6, 7, 8, and 9 are placed on the second line.
5. If too many digits are present, such as in a 6- or 7-digit score, drop the right-most trailing digit(s) to maximize the clarity of the display.
6. The rules for developing a stem-and-leaf plot are somewhat different from the rules governing the establishment of class intervals for the traditional frequency distribution and for a variety of other procedures that we will consider in later sections of the text. Class intervals for stem-and-leaf plots are, then, in a sense slightly atypical.

The last graphical technique to be presented in this chapter deals with how certain variables change over time. For both macroeconomic data such as disposable income and microeconomic data such as weekly sales data of one particular product at one particular store, plots of data over time are fundamental to business management. Similarly, social researchers are often interested in showing how variables change over time. They might be interested in changes with time in attitudes toward various racial and ethnic groups or changes in the rate of savings in the United States, or changes in crime rates for various cities. A pictorial method of presenting changes in a variable over time is called a **time series**. Figure 3.12 is a time series showing the percentage of white women aged 30 to 34 who have not had any children. This trend is presented from 1970 to 1986.

time series

F I G U R E 3.12
Percentage of Childless Women Aged 30 to 34, 1970–1986

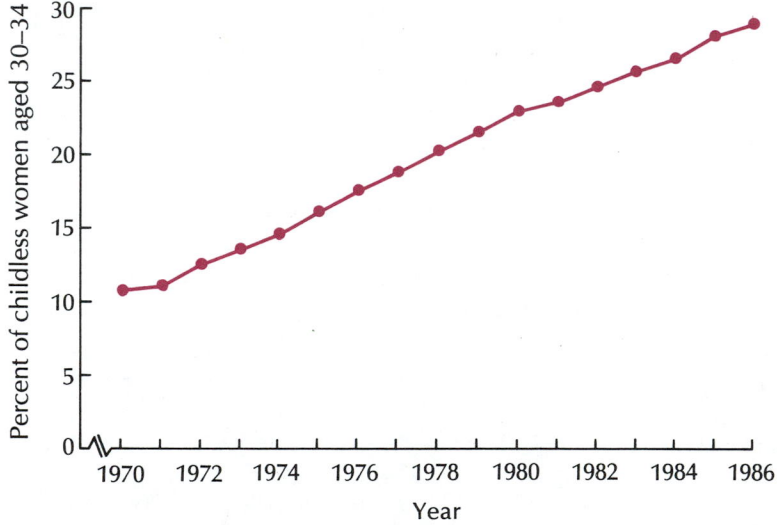

Usually, time points are labeled chronologically across the horizontal axis (abscissa), and the numerical values (frequencies, percentages, rates, etc.) of the variable of interest are labeled along the vertical axis (ordinate). Time can be measured in days, months, years, or whichever unit is most appropriate. As a rule of thumb, a time series should consist of no fewer than four or five time points; typically, these time points are equally spaced. Many more time points than this are desirable, though, in order to show a more complete picture of changes in a variable over time.

How one displays the time axis in a time series frequently depends on the time intervals at which data are available. For example, the U.S. Census Bureau reports average family income in the United States only on a yearly basis. When information about a variable of interest is available in different units of time, one must decide which unit or units are most appropriate for the research. In an election year, a political scientist would most likely examine weekly or monthly changes in candidate preferences among registered voters. On the other hand, a manufacturer of machine-tool equipment might keep track of sales (in dollars and number of units) on a monthly, quarterly, and yearly basis. Figure 3.13 shows the quarterly sales (in thousands of units) of a machine-tool product over the past three years.

FIGURE 3.13

Quarterly Sales (in thousands)

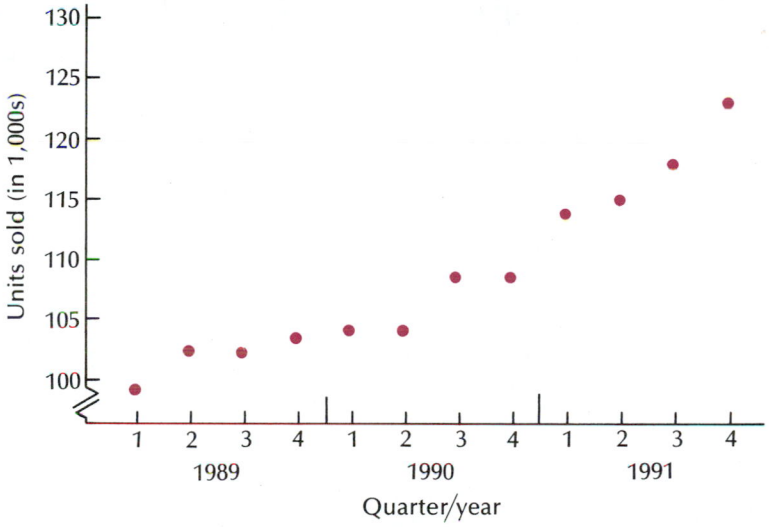

Note that from this time series, it is clear that the company has experienced a gradual but steady growth in the number of units over the past 3 years.

Time series plots are useful for examining general trends and seasonal or cyclic patterns. For example, the "Money and Investing" section of the *Wall Street Journal* gives the daily, workday values for the Dow Jones Industrials, Transportation, and Utilities Averages for a 6-month period. These are displayed in Figure 3.14 for a typical period. A quick glance at these plots shows general trends downward in the Industrials and Transportation indices over the July through September period of 1990. Seasonal or cyclic patterns could possibly be detected if we had weekly (or perhaps monthly) data for several years.

FIGURE 3.14
**Time Series Plots for the
Dow Jones Industrials,
Transportation, and
Utilities Averages**

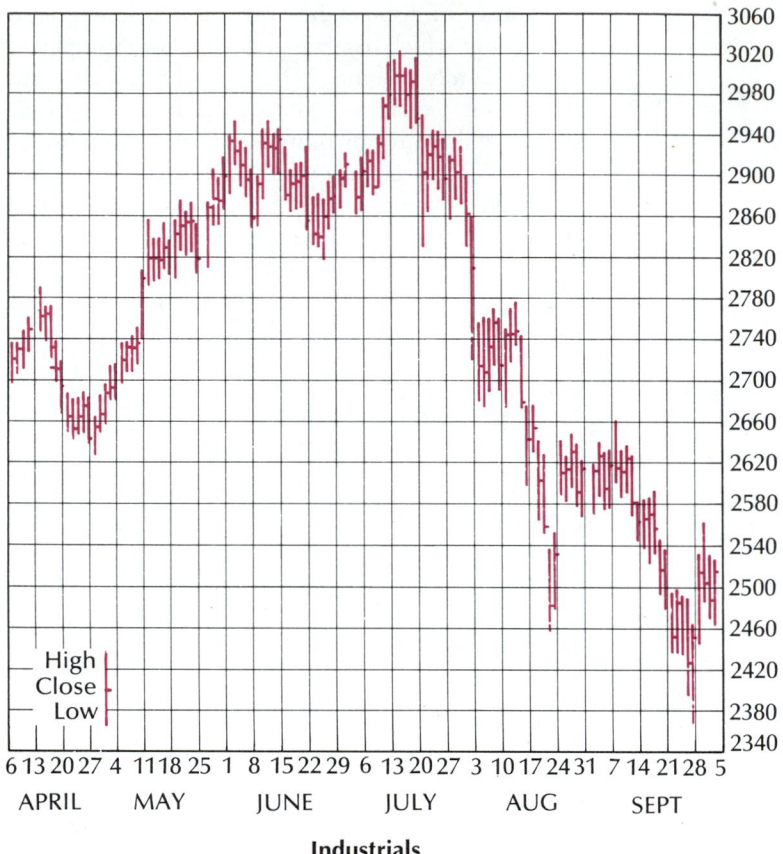

Industrials

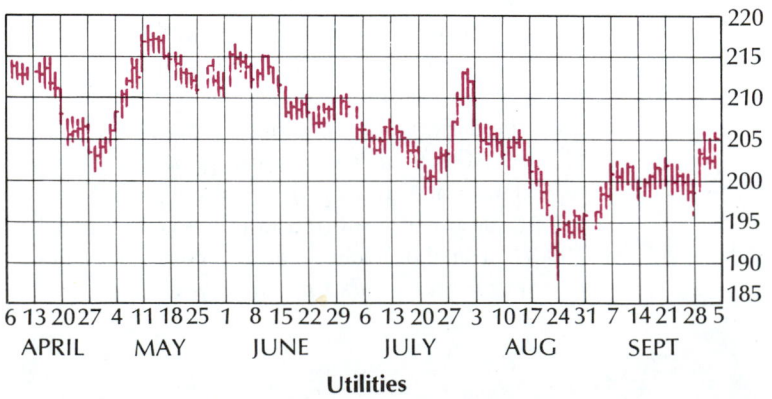

Utilities

FIGURE 3.14
**Time Series Plots for the
Dow Jones Industrials,
Transportation, and
Utilities Averages
(continued)**

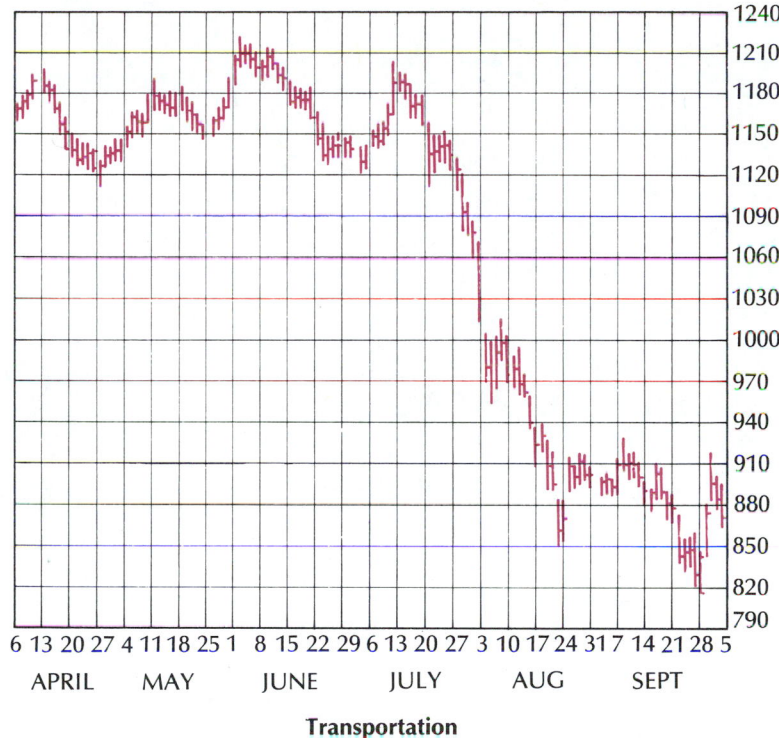

Transportation

Source: *Wall Street Journal*, 5 October 1990

EXAMPLE 3.1 ▼

Monthly data on the number of paid admissions (in thousands) to an indoor sports and entertainments arena for 72 months are given below and plotted in Figure 3.15. Identify any apparent trend, seasonal, and cyclic effects.

J	F	M	A	M	J	J	A	S	O	N	D
89	101	116	111	94	59	44	44	73	78	99	93
96	110	118	116	107	69	52	54	85	92	113	109
120	136	155	155	129	98	69	78	116	143	154	166
183	199	227	219	198	148	94	108	160	201	215	246
242	264	359	308	265	193	150	146	243	260	332	293
323	377	470	422	345	239	176	182	288	342	380	379

FIGURE   3.15
**Admissions to Arena;
Example 3.1**

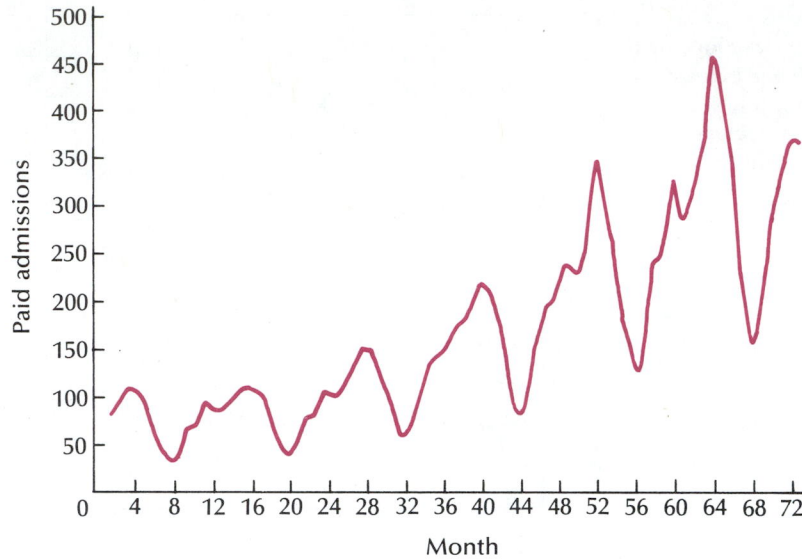

Solution   The upward trend in admissions is obvious in the figure. The trend may not be linear, however. Rather, the trend involves an initial slow growth, a period of more rapid growth, and then a tapering-off. (One consideration suggests that the trend cannot increase forever. Paid admissions to the arena are limited by the available number of dates and seats.) There is a strong seasonal effect; admissions drop dramatically in the summer months. If there's any cyclic component, it's not apparent in the figure.   ▲

Sometimes it is important to compare trends over time in a variable for two or more groups. Figure 3.16 reports the values of two ratios from 1976 to 1988: the ratio of the median family income of African-Americans to the median income of Anglo-Americans and the ratio of the median income of Hispanics to the median income of Anglo-Americans.

Median family income represents the income amount that divides family incomes into two groups—the top half and the bottom half. In 1987, the median family income for African-Americans was $18,098, meaning that 50% of all African-American families had incomes above $18,098, and 50% had incomes below $18,098. The median, one of several measures of central tendency, is discussed more fully later in this chapter.

FIGURE 3.16

Ratio of African-American and Hispanic Median Family Income to Anglo-American Median Family Income, 1976–1988

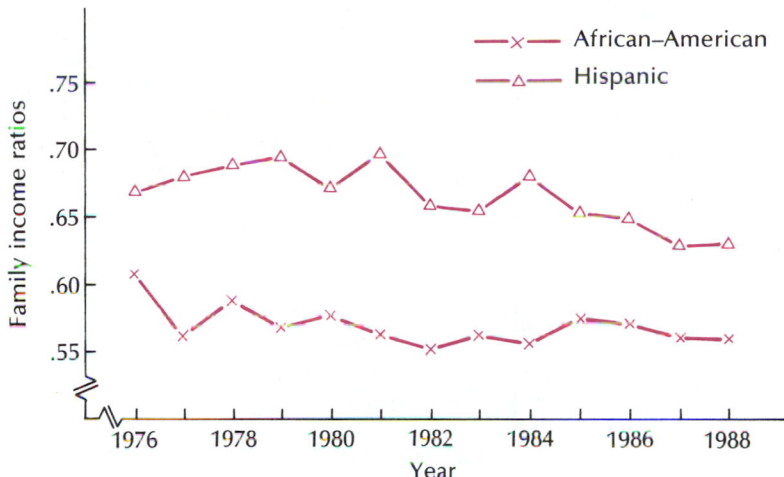

Figure 3.16 shows that the ratio of African-American or Hispanic to Anglo-American family income fluctuated between 1976 and 1988, but the overall trend in both ratios indicates that they declined over this time. A social researcher would interpret these trends to mean that the income of African-American and Hispanic families generally declined relative to the income of Anglo-American families.

Sometimes information is not available in equal time intervals. For example, polling organizations such as Gallup or the National Opinion Research Center do not necessarily ask the American public the same questions about their attitudes or behavior on a yearly basis. Sometimes there is a time gap of more than 2 years before a question is asked again.

When information is not available in equal time intervals, it is important for the interval width between time points (the horizontal axis) to reflect this fact. If, for example, a social researcher is plotting values of a variable for 1985, 1986, 1987, and 1990, the interval width between 1987 and 1990 on the horizontal axis should be three times the width of that between the other years. If these interval widths were spaced evenly, the resulting trend line could be seriously misleading. Other examples of graphic distortion are discussed in Chapter 20.

Figure 3.17 presents the trend in church attendance among American Catholics and Protestants from 1958 to 1988. As can be seen, the width of the intervals between time points reflects the fact that Catholics were not asked about their church attendance every year.

Before leaving graphical methods for describing data, there are several general guidelines that can be helpful in developing graphs with an impact. These guidelines pay attention to the design and presentation techniques and should help you make better, more informative graphs.

FIGURE 3.17
Church Attendance of American Protestants and Catholics in a Typical Week, 1954–1988

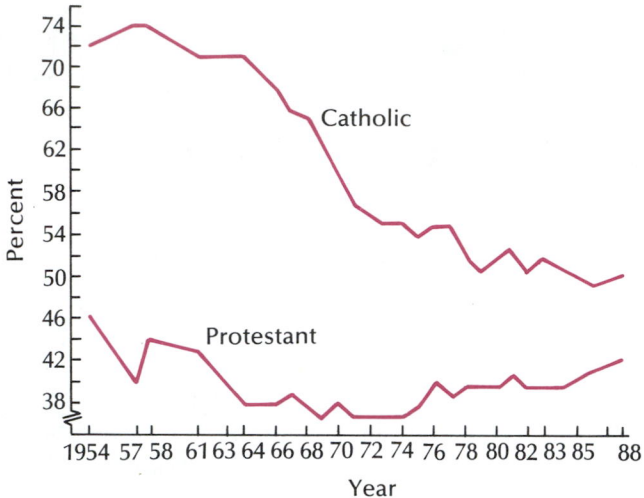

Source: Gallup.

General Guidelines for Successful Graphics

1. Before constructing a graph, set your priorities. What messages should the viewer get?
2. Choose the type of graph (pie chart, bar graph, histogram, and so on).
3. Pay attention to the title. One of the most important aspects of a graph is its title. The title should immediately inform the viewer of the point of the graph and draw the eye toward the most important elements of the graph.
4. Fight the urge to use many type sizes, styles, and color changes. The indiscriminate and possibly excessive use of different type sizes and styles and of numerous colors will confuse the viewer. Generally, we recommend only two typefaces; color changes and italics should be used in only one or two places.
5. Convey the tone of your graph by using colors and patterns. The more intense, warm colors (yellows, oranges, reds) are more dramatic than the blues and purples and, hence, help to stimulate enthusiasm by the viewer. On the other hand, pastels (particularly grays) convey a conservative, business-like tone. Similarly, the simple patterns convey a conservative tone, whereas the busier patterns help to stimulate more excitement.
6. Don't underestimate the effectiveness of a simple, straightforward graph.
7. Practice drawing graphs frequently. As with almost anything, practice improves skill.

▼ EXERCISES

Basic Techniques

 3.1 University officials periodically review the distribution of undergraduate majors within the colleges of the university to help determine a fair allocation of resources to departments within the colleges. At one such review, the following data were obtained:

College	Number of Majors
Agriculture	1,500
Arts and Sciences	11,000
Business Administration	7,000
Education	2,000
Engineering	5,000

a. Construct a pie chart for these data.
b. Use the same data to construct a bar graph.

 3.2 Because of the difficult times many basic industries have endured in recent years, financial analysts have monitored the influx of foreign materials. The data here show steel industry imports (in thousands of tons) for the years 1980 to 1990.

Year	1980	1981	1982	1983	1984	1985	1986	1987	1988	1989	1990
Import	15,491	19,898	16,663	17,061	26,171	23,650	19,650	18,982	17,675	17,050	16,493

a. Would a pie chart be an appropriate graphical method for describing these data? Explain.
b. Construct a bar graph for the data.

 3.3 Graph the data shown here in the allocation of our food dollars to the categories of the table shown here. Try a pie chart and a bar graph. Which seems better?

Where Our Food Dollars Go	Percent
Dairy products	13.4
Cereal and baked goods	12.6
Nonalcoholic beverages	8.9
Poultry and seafood	7.5
Fruit and vegetables	15.6
Meat	24.5
Other foods	17.5

3.4   A large study of employment trends, based on a survey of 45,000 businesses, was conducted by Ohio State University. Assuming an unemployment rate of 5% or less, it was predicted that 2.1 million job openings would be created between 1980 and 1990. This employment growth is shown by major industry groups.

Industry Group	Employment Growth, 1980–1990
Service	33.2%
Manufacturing	25.0%
Retail trade	17.9%
Finance, insurance, real estate	6.6%
Wholesale trade	4.8%
Construction	4.6%
Transportation	3.9%
Government	2.7%
Other	1.3%

Construct a pie chart to display these data.

 3.5   From the same study described in Exercise 3.4, data were obtained on the job openings between 1980 and 1990. Use the data to construct a bar chart.

Occupational Groups	Percentage of Job Openings, 1980–1990
Clerical workers	20.9%
Sales	7.3%
Managers	9.5%
Professional and technical	16.3%
Laborers	3.7%
Service workers	18.1%
Operatives	13.1%
Craft and kindred workers	11.1%

 3.6   The regulations of the board of health in a particular state specify that the fluoride level must not exceed 1.5 parts per million (ppm). The 25 measurements given here represent the fluoride levels for a sample of 25 days. Although fluoride levels are measured more than once per day, these data represent the early morning readings for the 25 days sampled.

.75	.86	.84	.85	.97
.94	.89	.84	.83	.89
.88	.78	.77	.76	.82
.72	.92	1.05	.94	.83
.81	.85	.97	.93	.79

a. Determine the range of the measurements.

b. Dividing the range by 7, the number of subintervals selected, and rounding, we have a class interval width of .05. Using .705 as the lower limit of the first interval, construct a frequency histogram.

c. Compute relative frequencies for each class interval and construct a relative frequency histogram. Note that the frequency and relative frequency histograms for these data have the same shape.

d. If one of these 25 days were selected at random, what would be the chance (probability) that the fluoride reading would be greater than .90 ppm? Guess (predict) what proportion of days in the coming year will have a fluoride reading greater than .90 ppm.

3.7 The National Highway Traffic Safety Administration has studied the use of rear-seat automobile lap and shoulder seat belts. The number of lives potentially saved with the use of lap and shoulder seat belts is shown for various percentages of use.

Percentage of Use	Lives Saved Wearing	
	Lap Belt Only	Lap and Shoulder Belt
100	529	678
80	423	543
60	318	407
40	212	271
20	106	136
10	85	108

Suggest several different ways to graph these data. Which one seems more appropriate and why?

3.8 Construct a frequency histogram for the data of Table 3.4. Compare the histogram to the stem-and-leaf plot of Figure 3.11. Which is more informative?

3.9 Construct a relative frequency histogram for the data in the accompanying table.

Per-Capita Public Welfare Expenses, by Number of States

Dollars	Number of States
50–74	3
75–99	6
100–124	14
125–149	11
150–174	2
175–199	5
200–224	2
225–249	5
250–274	1
275–299	1
Total	50

3.10   Construct a frequency table with suitable class intervals for the data presented here.

32.3	22.8	30.5
31.3	31.3	30.0
29.4	31.1	29.4
31.6	27.6	29.7
30.4	28.8	31.6
31.2	30.7	32.5
29.8	30.3	29.2

3.11   Construct a relative frequency histogram for the data of Exercise 3.10.

3.12   Construct a frequency histogram for the following data.

2.9	3.0	4.4
0.8	2.7	1.6
3.5	3.6	1.2
1.9	3.8	2.2
2.6	3.9	1.5
2.8	4.4	0.9
2.5	4.1	2.3
4.5	3.5	2.5

3.13   Construct a stem-and-leaf diagram for the data in Exercise 3.12. Which plot seems to be more informative for these data?

 3.14   Survival times (in months) are shown for patients with severe chronic left-ventricular heart failure. Construct a stem-and-leaf diagram for these data.

4	15	24	10
1	27	31	14
2	16	32	7
13	36	29	6
12	18	14	15
18	6	13	21
20	8	3	24

3.15   Use the data from Exercise 3.14 to construct a frequency histogram. Which plot (the stem-and-leaf diagram or the frequency histogram) describes these data best? Why?

 3.16   Data from SAT exams are given for selected years. Plot these time series data and give some interpretations to the data.

	Year				
Gender, Type	**1967**	**1970**	**1975**	**1980**	**1990**
Male, math	514	509	495	491	493
Female, math	467	465	449	443	445
Male, verbal	463	459	437	428	430
Female, verbal	468	461	431	420	420

Source: College Entrance Examination Board.

3.17 Using the data shown in the table here, construct a time series plot, taking into account the unequal time points.

Year	Percentage of all Families Headed by a Single Woman
1960	5.3
1965	6.6
1970	9.1
1975	13.4
1980	16.7
1983	18.2
1985	20.0
1987	21.7
1988	22.9
1989	25.4

3.18 Using data from the table in Exercise 3.17, construct a time series plot in which the time points along the horizontal axis are evenly spaced.

3.19 Compare the time series you obtained in Exercises 3.17 and 3.18. Discuss how a time series can be misleading if the interval width between time points does not correspond to the actual length of time between data observation points.

 3.20 Construct a frequency histogram plot for the telephone data in the accompanying table (telephones per 1,000).

State	Telephones	State	Telephones
Alabama	500	Montana	540
Alaska	350	Nebraska	590
Arizona	550	Nevada	720
Arkansas	480	New Hampshire	590
California	610	New Jersey	650
Colorado	570	New Mexico	470
Connecticut	620	New York	530
Delaware	630	N. Carolina	530
Florida	620	N. Dakota	560
Georgia	570	Ohio	550
Hawaii	480	Oklahoma	580
Idaho	550	Oregon	560
Illinois	650	Pennsylvania	610
Indiana	580	Rhode Island	560
Iowa	570	S. Carolina	510
Kansas	600	S. Dakota	540
Kentucky	480	Tennessee	540
Louisiana	520	Texas	570
Maine	540	Utah	560
Maryland	610	Vermont	520
Massachusetts	570	Virginia	530
Michigan	580	Washington	570
Minnesota	560	W. Virginia	450
Mississippi	470	Wisconsin	540
Missouri	570	Wyoming	580

3.21   Construct a stem-and-leaf plot for the data of Exercise 3.20. Interpret the data display.

3.22   Computer output is shown for the data of Exercise 3.20. Compare the stem-and-leaf plot in the output to the one you constructed in Exercise 3.21.

```
MTB > name c1 'phones'
MTB > set c1
MTB > end

MTB > STEM-AND-LEAF 'PHONES'

Stem-and-leaf of phones     N  = 50
Leaf Unit = 10


     1     3 5
     1     4
     7     4 577888
    19     5 012233344444
   (21)    5 5556666677777777888899
    10     6 0111223
     3     6 55
     1     7 2

MTB > STOP
```

$ **3.23**   A supplier of high-quality audio equipment for automobiles accumulates monthly sales data on speakers and receiver–amplifier units for five years. The data (in thousands of units per month) are shown in a table. Plot the sales data. Do you see any overall trend in the data? Do there seem to be any cyclic or seasonal effects?

Year	J	F	M	A	M	J	J	A	S	O	N	D
1	101.9	93.0	93.5	93.9	104.9	94.6	105.9	116.7	128.4	118.2	107.3	108.6
2	109.0	98.4	99.1	110.7	100.2	112.1	123.8	135.8	124.8	114.1	114.9	112.9
3	115.5	104.5	105.1	105.4	117.5	106.4	118.6	130.9	143.7	132.2	120.8	121.3
4	122.0	110.4	110.8	111.2	124.4	112.4	124.9	138.0	151.5	139.5	127.7	128.0
5	128.1	115.8	116.0	117.2	130.7	117.5	131.8	145.5	159.3	146.5	134.0	134.2

⚙   **3.24**   A machine-tool firm that produces a variety of products for manufacturers has quarterly records of total activity for the previous 8 years. The data reflect activity rather than price, so inflation is irrelevant. The data are shown in a table.

		Quarter		
Year	1	2	3	4
1	97.2	100.2	102.8	102.6
2	106.1	107.8	110.5	110.6
3	116.5	117.3	119.9	119.3
4	126.1	125.7	128.3	132.1
5	133.2	133.8	141.1	142.1
6	144.2	146.1	151.6	154.0
7	155.8	158.6	165.8	167.0
8	171.1	172.6	176.5	179.7

a. Plot the data against time (quarters 1–32).
b. Does there appear to be a clear trend? If So, what form of trend equation would you suggest?
c. Can you detect cyclic or seasonal features?

3.3 ▼ DESCRIBING DATA ON A SINGLE VARIABLE: MEASURES OF CENTRAL TENDENCY

Numerical descriptive measures are commonly used to convey a mental image of pictures, objects, and other phenomena. There are two main reasons for this: first, graphical descriptive measures are inappropriate for statistical inference, since it is difficult to describe the similarity of a sample frequency histogram and the corresponding population frequency histogram. The second reason for using numerical descriptive measures is one of expediency—we never seem to carry the appropriate graphs or histograms with us, and so must resort to our powers of verbal communication to convey the appropriate picture. We seek several numbers, called *numerical descriptive measures*, that will create a mental picture of the frequency distribution for a set of measurements.

central tendency
variability

The two most common numerical descriptive measures are measures of **central tendency** and measures of **variability**. That is, we seek to describe the center of the distribution of measurements and also how the measurements vary about the center of the distribution. We will draw a distinction between numerical descriptive measures for a population, called **parameters**, and numerical descriptive measures for a sample, called **statistics**. In problems requiring statistical inference, we will not be able to calculate values for various parameters but we will be able to compute corresponding statistics from the sample and use these quantities to estimate the corresponding population parameters.

parameter
statistic

In this section, we will consider various measures of central tendency, followed in Section 3.4 by a discussion of measures of variability.

mode

The first measure of central tendency we consider is the **mode**.

DEFINITION 3.1
Mode

▼

The **mode** of a set of measurements is defined to be the measurement that occurs most often (with the highest frequency).

We illustrate the use and determination of the mode in an example.

EXAMPLE 3.2

▼

Slaughter weights (in pounds) for a sample of 15 Herefords each with a frame size of 3 (on a 1–7 scale) are shown here.

962	1,005	1,033
980	965	1,030
975	989	955
1,015	1,000	970
1,042	1,005	995

Determine the modal slaughter weight.

Solution   For these data, the weight 1,005 occurs twice and all the others once. Hence, the mode is 1,005.   ▲

Identification of the mode for Example 3.2 was quite easy because we were able to count the number of times each measurement occurred. When dealing with grouped data—data presented in the form of a frequency table—we can define the modal interval to be the class interval with the highest frequency. However, since we would not know the actual measurements but only how many measurements fall into each interval, the mode is taken as the midpoint of the modal interval; it is an approximation to the mode of the actual sample measurements.

The mode is also commonly used as a measure of popularity that reflects central tendency or opinion. For example, we might talk about the most preferred stock, a most preferred model of washing machine, or the most popular candidate. In each case, we would be referring to the mode of the distribution.

It should be noted that some distributions have more than one measurement that occurs with the highest frequency. Thus, we might encounter bimodal, trimodal, and so on, distributions.

median

The second measure of central tendency we consider is the **median**.

DEFINITION 3.2
Median

▼

The **median** of a set of measurements is defined to be the middle value when the measurements are arranged from lowest to highest.

The median is most often used to measure the midpoint of a large set of measurements. For example, we may read about the median wage increase won by union members, the median age of persons receiving social security benefits, and the median weight of cattle prior to slaughter during a given month. Each of these situations involves a large set of measurements, and the median would reflect the central value of the data.

However, we may use the definition of median for small sets of measurements by using the following convention. The median for an even number of measurements is the average of the two middle values when the measurements are arranged from lowest to highest. When there is an odd number of measurements, the median is still the middle value. Thus, whether there is an even or odd number of measurements, there is an equal number of measurements above and below the median.

EXAMPLE 3.3

▼

Each of 10 children in the second grade was given a reading aptitude test. The scores were as follows:

95   86   78   90   62   73   89   92   84   76

Determine the median test score.

Solution   We must first arrange the scores in order of magnitude.

62   73   76   78   84   86   89   90   92   95

Since there is an even number of measurements, the median is the average of the two midpoint scores.

$$\text{median} = \frac{84 + 86}{2} = 85.$$

▲

EXAMPLE 3.4

▼

An experiment was conducted to measure the effectiveness of a new procedure for pruning grapes. Each of 13 workers was assigned the task of pruning an acre of grapes. The productivity, measured in worker-hours/acre, is recorded for each person.

4.4   4.9   4.2   4.4   4.8   4.9   4.8   4.5   4.3   4.8   4.7   4.4   4.2

Determine the mode and median productivity for the group.

Solution   First arrange the measurements in order of magnitude:

4.2   4.2   4.3   4.4   4.4   4.4   4.5   4.7   4.8   4.8   4.8   4.9   4.9

For these data, we have two measurements appearing three times each. Hence, the data are bimodal, with modes of 4.4 and 4.8. The median for the odd number of measurements is the middle score, 4.5.

▲

grouped data median

The **median for grouped data** is slightly more difficult to compute. Since the actual values of the measurements are unknown, we know that the median occurs in a particular class interval, but we do not know where to locate the median within the interval. If we assume that the measurements are spread evenly throughout the interval, we get the following result. Let

L = lower class limit of the interval that contains the median
n = total frequency
cf_b = the sum of frequencies (cumulative frequency) for all classes before the median class
f_m = frequency of the class interval containing the median
w = interval width.

Then, for grouped data,

$$\text{median} = L + \frac{w}{f_m}(.5n - cf_b).$$

The next example illustrates how to find the median for grouped data.

EXAMPLE 3.5

Table 3.5 is the frequency table for the chick data of Table 3.3. Compute the median weight gain for these data.

TABLE 3.5
Frequency Table for the Chick Data, Table 3.3

Class Interval	f_i	Cumulative f_i	f_i/n	Cumulative f_i/n
3.55–3.65	1	1	.01	.01
3.65–3.75	1	2	.01	.02
3.75–3.85	6	8	.06	.08
3.85–3.95	6	14	.06	.14
3.95–4.05	10	24	.10	.24
4.05–4.15	10	34	.10	.34
4.15–4.25	13	47	.13	.47
4.25–4.35	11	58	.11	.58
4.35–4.45	13	71	.13	.71
4.45–4.55	7	78	.07	.78
4.55–4.65	6	84	.06	.84
4.65–4.75	7	91	.07	.91
4.75–4.85	5	96	.05	.96
4.85–4.95	4	100	.04	1.00
Totals	$n = 100$		1.00	

Solution   Let the cumulative relative frequency for class j equal the sum of the relative frequencies for class 1 through class j. To determine the interval that contains the median, we must find the first interval for which the cumulative relative frequency exceeds .50. This interval will be the one containing the median. For these data, the interval from 4.25 to 4.35 is the first interval for which the cumulative relative frequency exceeds .50, as shown in Table 3.5, column 5. So this interval contains the median. Then

$$L = 4.25 \qquad f_m = 11$$
$$n = 100 \qquad w = .1$$
$$cf_b = 47$$

and

$$\text{median} = L + \frac{w}{f_m}(.5n - cf_b) = 4.25 + \frac{.1}{11}(50 - 47) = 4.28.$$

▲

The third, and last, measure of central tendency we will discuss in this text is the arithmetic mean, known simply as the **mean**.

mean

DEFINITION 3.3
Arithmetic Mean
Mean

▼

The **arithmetic mean**, or **mean**, of a set of measurements is defined to be the sum of the measurements divided by the total number of measurements.

When people talk about an "average," they quite often are referring to the mean. Because of the important role that the mean will play in statistical inference in later chapters, we give special symbols to the population mean and the sample mean. The *population mean* is denoted by the Greek letter μ (read "mu"), and the *sample mean* is denoted by the symbol $\bar{y}$ (read "y-bar"). As indicated in Chapter 1, a population of measurements is the complete set of measurements of interest to us; a sample of measurements is a subset of measurements selected from the population of interest. If we let $y_1, y_2, \ldots, y_n$ denote the measurements observed in a sample of size n, then the sample mean $\bar{y}$ can be written as

μ
$\bar{y}$

$$\bar{y} = \frac{\sum_i y_i}{n},$$

where the symbol appearing in the numerator, $\sum_i y_i$, is the notation used to designate a sum of n measurements, y_i:

$$\sum_i y_i = y_1 + y_2 + \cdots + y_n.$$

The corresponding population mean is μ.

In most situations, we will not know the population mean; the sample will be used to make inferences about the corresponding unknown population mean. Details about how this is done and the inferences we can make will be discussed in Chapter 5.

E X A M P L E 3.6

▼

A sample of $n = 15$ overdue accounts in a large department store yields the following amounts due:

$55.20	$ 4.88	$271.95
18.06	180.29	365.29
28.16	399.11	807.80
44.14	97.47	9.98
61.61	56.89	82.73

a. Determine the mean amount due for the 15 accounts sampled.
b. If there are a total of 150 overdue accounts, use the sample mean to predict the total amount overdue for all 150 accounts.

Solution
a. The sample mean is computed as follows:

$$\bar{y} = \frac{\sum_i y_i}{15} = \frac{55.20 + 18.06 + \cdots + 82.73}{15} = \frac{2483.56}{15} = \$165.57.$$

b. From part a we found that the 15 accounts sampled averaged $165.57 overdue. Using this information, we would predict, or estimate, the total amount overdue for the 150 accounts to be $150(165.57) = \$24,835.50$.

▲

The sample mean formula for grouped data is only slightly more complicated than the formula just presented for ungrouped data. Usually, we do not know the individual sample measurements, only the interval to which a measurement is assigned, so this formula will be an approximation to the actual sample mean. Hence, when the sample measurements are known, the formula for ungrouped data should be used. We will use the same symbol $\bar{y}$ to designate the sample mean for grouped data. If there are k class intervals and

y_i = midpoint of the ith class interval
f_i = frequency associated with the ith class interval
n = the total number of measurements

then

$$\bar{y} = \frac{\sum_i f_i y_i}{n}.$$

EXAMPLE 3.7

▼

The data of Example 3.5 are reproduced in Table 3.6, along with two additional columns, y_i and f_iy_i, that will be helpful in computing the mean. Compute the sample mean for this set of grouped data.

TABLE 3.6
Chick Data, Example 3.7

Class Interval	f_i	y_i	f_iy_i
3.55–3.65	1	3.6	3.6
3.65–3.75	1	3.7	3.7
3.75–3.85	6	3.8	22.8
3.85–3.95	6	3.9	23.4
3.95–4.05	10	4.0	40.0
4.05–4.15	10	4.1	41.0
4.15–4.25	13	4.2	54.6
4.25–4.35	11	4.3	47.3
4.35–4.45	13	4.4	57.2
4.45–4.55	7	4.5	31.5
4.55–4.65	6	4.6	27.6
4.65–4.75	7	4.7	32.9
4.75–4.85	5	4.8	24.0
4.85–4.95	4	4.9	19.6
Totals	100		429.2

Solution Adding the entries in the f_iy_i column and substituting into the formula, we find the sample mean to be

$$\bar{y} = \frac{\sum_i f_iy_i}{100} = \frac{429.2}{100} = 4.29.$$

▲

The mean is a useful measure of the central value of a set of measurements, but it is subject to distortion due to the presence of one or more extreme values in the set. In these situations, the extreme values (called **outliers**) pull the mean in the direction of the outliers, thus distorting the mean as a measure of the central value. A variation of the mean, called a **trimmed mean**, drops the highest and lowest extreme values and averages the rest. For example, a 20% trimmed mean drops the highest 20% and the lowest 20% of the measurements and averages the rest. Similarly, a 10% trimmed mean drops the highest and the lowest 10% of the measurements and averages the rest. By trimming the data, we are able to reduce the impact of very large (or small) values on the mean, and thus get a more reliable measure of the central value of the set. This will be particularly important when the sample mean is used to predict the corresponding population central value.

outliers

trimmed mean

In this section, we discussed the mode, median, mean, and trimmed mean. How are these measures of central tendency related for a given set of measurements? The answer depends on the **skewness** of the data. If the distribution is

skewness

mound–shaped and symmetrical about a single peak, the mode (M_o), median (M_d), mean (μ), and trimmed mean (*TM*) will all be the same. This is shown using a smooth curve and population quantities in Figure 3.18(a). If the distribution is skewed, having a long tail in one direction and a single peak, the mean is pulled in the direction of the tail; the median falls between the mode and the mean; and depending on the degree of trimming, the trimmed mean usually falls between the median and the mean. Figure 3.18(b) and (c) illustrates this for distributions skewed to the left and to the right.

FIGURE 3.18

Relation Among the Mean μ, the Trimmed Mean *TM*, the Median M_d, and the Mode M_o

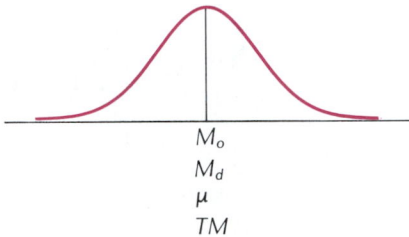

(a) A mound-shaped distribution

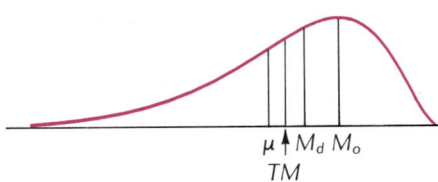

(b) A distribution skewed to the left

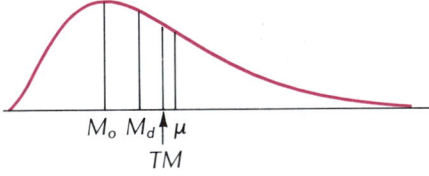

(c) A distribution skewed to the right

    The important thing to remember is that we are not restricted to using only one measure of central tendency. For some data sets, it will be necessary to use more than one of these measures to provide an accurate descriptive summary of central tendency for the data.

Major Characteristics of Each Measure of Central Tendency

▼

Mode

1. It is the most frequent or probable measurement in the data set.
2. There can be more than one mode for a data set.
3. It is not influenced by extreme measurements.
4. Modes of subsets cannot be combined to determine the mode of the complete data set.
5. For grouped data, its value can change depending on the categories used.
6. It is applicable for both qualitative and quantitative data.

Median

1. It is the central value; 50% of the measurements lie above it and 50% fall below it.
2. There is only one median for a data set.
3. It is not influenced by extreme measurements.
4. Medians of subsets cannot be combined to determine the median of the complete data set.
5. For grouped data, its value is rather stable even when the data are organized into different categories.
6. It is applicable to quantitative data only.

Mean

1. It is the arithmetic average of the measurements in a data set.
2. There is only one mean for a data set.
3. Its value is influenced by extreme measurements; trimming can help to reduce the degree of influence.
4. Means of subsets can be combined to determine the mean of the complete data set.
5. It is applicable to quantitative data only.

Measures of central tendency do not provide a complete mental picture of the frequency distribution for a set of measurements. In addition to determining the center of the distribution, we must have some measure of the spread of the data. In the next section, we discuss measures of variability, or dispersion.

▼ EXERCISES

Basic Techniques

3.25   Compute the mean, median, and mode for the following data:

11	17	18	10	22	23	15	17
14	13	10	12	18	18	11	14

3.26 Refer to the data in Exercise 3.25 with the measurements 22 and 23 replaced by 42 and 43. Recompute the mean, median, and mode. Discuss the impact of these extreme measurements on the three measures of central tendency.

3.27 Refer to Exercises 3.25 and 3.26. Compute a 10% trimmed mean for both data sets. Do the extreme values affect the 10% trimmed mean? Would a 5% trimmed mean be affected?

3.28 Determine the mode, median, and mean for the following measurements:

10	2	1	5
1	5	7	10
3	4	8	12
5	6	8	9

3.29 Determine the mean, median, and mode for the data presented in the following frequency table:

Class Interval	Frequency
0–2	1
3–5	3
6–8	5
9–11	4
12–14	2

Applications

3.30 Salaries for 40 recent M.B.A. graduates from a major university are summarized here (in thousands of dollars). Determine the mode, median, and mean for the data shown in this frequency table. What does the relation among the three measures indicate about the shape of the histogram for these data?

Interval	Frequency
24.9–29.9	6
29.9–34.9	10
34.9–39.9	15
39.9–44.9	7
44.9–49.9	2

3.31 Exercise capacity (in seconds) was determined for each of 11 patients being treated for chronic heart failure. Determine the median and mean.

906	1,320
711	1,170
684	1,200
837	1,056
897	882
1,008	

3.32 Daily crude oil output (in millions of barrels) is shown for the years 1971 to 1990. Compute the mean and median daily output for these years.

Year	Output
1971	9.45
1972	9.40
1973	9.25
1974	8.75
1975	8.30
1976	8.10
1977	8.25
1978	8.70
1979	8.55
1980	8.60
1981	8.55
1982	8.65
1983	8.70
1984	8.70
1985	8.91
1986	8.60
1987	8.20
1988	7.70
1989	7.20
1990	6.75

 3.33 Based on the frequency distribution contained in the data in the accompanying table, what is the class interval width?

Normal Daily Mean Temperatures, Annual Average

Temperature	Frequency, f
39–41	3
42–44	2
45–47	8
48–50	10
51–53	9
54–56	10
57–59	8
60–62	7
63–65	3
66–68	3
69–71	2
72–74	0
75–77	2
Total	67

Determine the mode, median, and mean for these data. Would a trimmed mean better describe the center of the distribution than the mean does? Explain.

3.34 Nitrogen is a limiting factor in the yield of many different plants. In particular, the yield of apple trees is directly related to the nitrogen content of apple tree leaves and must be carefully monitored to protect the trees in an orchard. Research has shown that the nitrogen content should be approximately 2.5% for best yield results. (It should be noted that some researchers report their results in parts per million (ppm); hence, 1% would be equivalent to 10,000 ppm.)

To determine the nitrogen content of trees in an orchard, the growing tips of 150 leaves are clipped from trees throughout the orchard. These leaves are ground to form one composite sample, which the researcher assays for percentage of nitrogen. Composite samples obtained from a random sample of 36 orchards throughout the state gave the following nitrogen contents:

2.0968	2.8220	2.1739	1.9928	2.2194	3.0926
2.4685	2.5198	2.7983	2.0961	2.9216	2.1997
1.7486	2.7741	2.8241	2.6691	3.0521	2.9263
2.9367	1.9762	2.3821	2.6456	2.7678	1.8488
1.6850	2.7043	2.6814	2.0596	2.3597	2.2783
2.7507	2.4259	2.3936	2.5464	1.8049	1.9629

a. Round each of these measurements to the nearest hundredth. (Use the convention that 5 is rounded up.)
b. Determine the sample mode for the rounded data.
c. Determine the sample median for the rounded data.
d. Determine the sample mean for the rounded data.

3.35 Refer to the data of Exercise 3.34 rounded to the nearest hundredth. Replace the fourth measurement (2.94) by the value 29.40. Compute the sample mean, median, and mode for these data. Compare these results to those you found in Exercise 3.34.

3.36 Refer to the data of Example 3.6. Since the sample mean is greater than 10 of the 15 observations, suggest a more appropriate measure of central tendency. Compute its value. How does the distribution of amounts for overdue accounts appear to be skewed?

3.37 Effective tax rates (per $100) on residential property for three groups of large cities, ranked by residential property tax rate, are shown here.

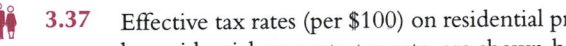

Group 1	Rate	Group 2	Rate	Group 3	Rate
Detroit, MI	4.10	Burlington, VT	1.76	Little Rock, AR	1.02
Milwaukee, WI	3.69	Manchester, NH	1.71	Albuquerque, NM	1.01
Newark, NJ	3.20	Fargo, ND	1.62	Denver, CO	.94
Portland, OR	3.10	Portland, ME	1.57	Las Vegas, NV	.88
Des Moines, IA	2.97	Indianapolis, IN	1.57	Oklahoma City, OK	.81
Baltimore, MD	2.64	Wilmington, DE	1.56	Casper, WY	.70
Sioux Falls, IA	2.47	Bridgeport, CT	1.55	Birmingham, AL	.70
Providence, RI	2.39	Chicago, IL	1.55	Phoenix, AZ	.68
Philadelphia, PA	2.38	Houston, TX	1.53	Los Angeles, CA	.64
Omaha, NE	2.29	Atlanta, GA	1.50	Honolulu, HI	.59

Source: Government of the District of Columbia, Department of Finance and Revenue, Tax Rates and Tax Burdens in the District of Columbia: a nationwide comparison, annual.

a. Compute the mean, median, and mode separately for the three groups.
b. Compute the mean, median, and mode for the complete set of 30 measurements.
c. What measure or measures best summarize the center of these distributions? Explain.

3.38   Refer to Exercise 3.37. Average the three group means, the three group medians, and the three group modes, and compare your results to those of part b. Comment on your findings.

3.4 ▼ DESCRIBING DATA ON A SINGLE VARIABLE: MEASURES OF VARIABILITY

variability

range

The need for some measure of variability is illustrated in the relative frequency histograms of Figure 3.19. All the histograms have the same mean but each has a different spread, or **variability**, about the mean. For purposes of illustration, we have shown the histograms as smooth curves.

The simplest but least useful measure of data variation is the **range**. Recall that we alluded to the range in Section 3.2. We now present its definition.

DEFINITION 3.4
Range

▼

The **range** of a set of measurements is defined to be the difference between the largest and the smallest measurements of the set.

FIGURE 3.19
Relative Frequency Histograms with Different Variabilities but the Same Mean

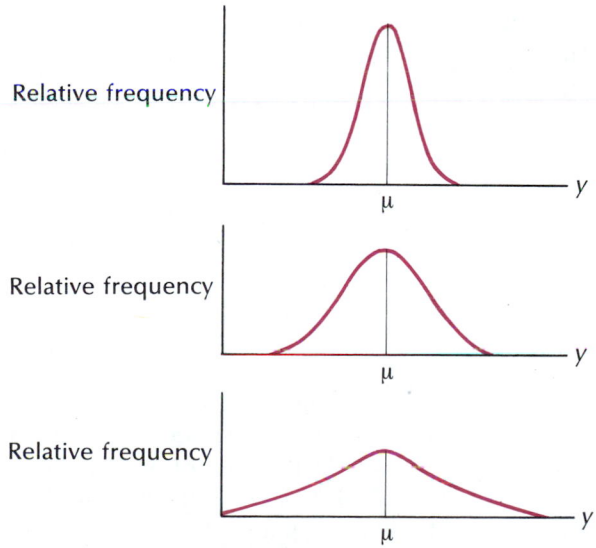

EXAMPLE 3.8

Determine the range of the 15 overdue accounts of Example 3.6.

Solution   The smallest measurement is $4.88 and the largest $807.80. Hence, the range is

807.80 − 4.88 = $802.92.   ▲

grouped data
range

For **grouped data**, since we do not know the individual measurements, the **range** is taken to be the difference between the upper limit of the last interval and the lower limit of the first interval.

Although the range is easy to compute, it is sensitive to outliers since it depends on the most extreme values. It does not give much information about the pattern of variability. In Figure 3.19, the middle and the bottom distributions have the same mean and the same range, yet they differ substantially in their variability about the mean. What we seek is a measure of variability that is more sensitive to the piling up of data about the mean.

percentile

A second measure of variability involves the use of **percentiles**.

DEFINITION 3.5
_p_th Percentile

The **_p_th percentile** of a set of n measurements arranged in order of magnitude is that value that has at most p% of the measurements below it and at most $(100 - p)$% above it.

For example, Figure 3.20 illustrates the 60th percentile of a set of measurements.

FIGURE 3.20
The 60th Percentile of a
Set of Measurements

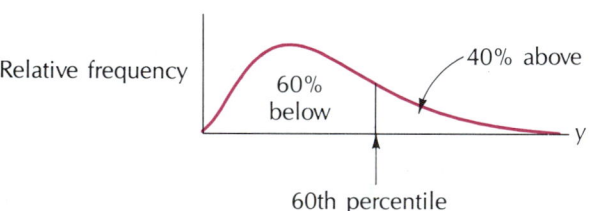

Percentiles are frequently used to describe the results of achievement test scores and the ranking of a person in comparison to the rest of the people taking an examination. Specific percentiles of interest are the 25th, 50th, and 75th percentiles, often called the _lower quartile_, the _middle quartile_ (median), and the _upper quartile_, respectively (see Figure 3.21).

FIGURE 3.21
Quartiles of a Distribution

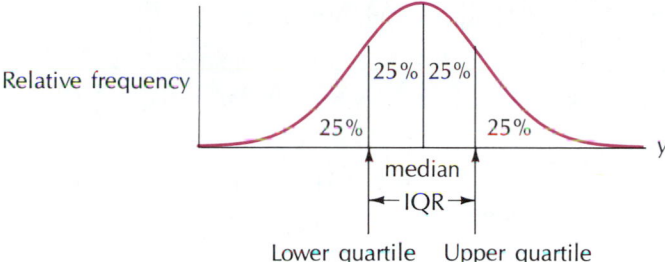

grouped data percentile

Now that we have learned how to compute the median (50th percentile) for grouped data, other percentiles follow immediately. Let

P = percentile of interest
L = lower limit of the class interval that includes percentile of interest
n = total frequency
cf_b = cumulative frequency for all class intervals before the percentile class
f_p = frequency of the class interval that includes the percentile of interest
w = interval width.

Then, for example, the 65th percentile for a set of grouped data would be computed using the formula

$$P = L + \frac{w}{f_p}(.65n - cf_b).$$

To determine L, f_p, and cf_b, begin with the lowest interval and find the first interval for which the cumulative relative frequency exceeds .65. This interval would contain the 65th percentile.

EXAMPLE 3.9 ▼

Refer to the chick data of Table 3.5. Compute the 90th percentile.

Solution Since the twelfth interval is the first interval for which the cumulative relative frequency exceeds .90, we have

$L = 4.65$
$n = 100$
$cf_b = 84$
$f_{90} = 7$
$w = .1$

Thus, the 90th percentile is

$$P_{90} = 4.65 + \frac{.1}{7}[.9(100) - 84] = 4.74.$$

This means that 90% of the measurements lie below this value and 10% lie above it. ▲

interquartile range

The second measure of variability, the **interquartile range** is now defined. A slightly different definition of the interquartile range is given along with the box plot (Section 3.6).

DEFINITION 3.6
**Interquartile
Range (IQR)**

▼

> The **interquartile range (IQR)** of a set of measurements is defined to be the difference between the upper and lower quartiles (see Figure 3.21). That is, IQR = 75th percentile − 25th percentile.

The interquartile range, although more sensitive to data pileup about the midpoint than the range, is still not sufficient for our purposes. In particular, the IQR can be used for comparing the variability of two sets of measurements, but not much useful information is gained from the IQR in interpreting the variability of a single set of measurements.

deviation

We seek now a sensitive measure of variability, not only for comparing the variabilities of two sets of measurements, but also for interpreting the variability of a single set of measurements. To do this, we work with the **deviation** $y - \bar{y}$ of a measurement y from the mean $\bar{y}$ of the set of measurements.

To illustrate, suppose we have five sample measurements $y_1 = 68$, $y_2 = 67$, $y_3 = 66$, $y_4 = 63$, and $y_5 = 61$, which represents the percentages of registered voters in five cities who exercised their right to vote at least once during the past year. These measurements are shown in the dot diagram of Figure 3.22. Each measure-

FIGURE 3.22
**Dot Diagram of the
Percentages of Registered
Voters in Five Cities**

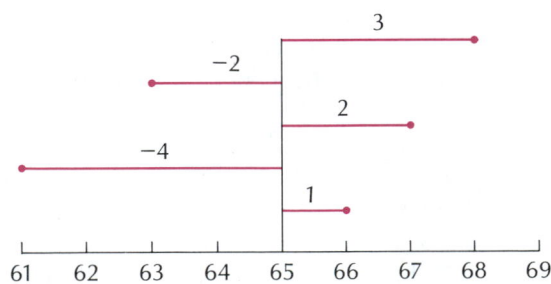

ment is located by a dot above the horizontal axis of the diagram. We use the sample mean

$$\bar{y} = \frac{\sum_i y_i}{n} = \frac{325}{5} = 65$$

to locate the center of the set and we construct horizontal lines in Figure 3.22 to represent the deviations of the sample measurements from their mean. The deviations of the measurements are computed by using the formula $y - \bar{y}$. The five measurements and their deviations are shown in Figure 3.22.

A data set with very little variability will have most of the measurements located near the center of the distribution. Deviations from the mean for a more variable set of measurements would be relatively large.

Many different measures of variability can be constructed by using the deviations $y - \bar{y}$. A first thought would be to use the mean deviation, but this will always equal zero, as it does for our example. A second possibility would be to ignore the minus signs and compute the average of the absolute values. However, a more easily interpreted function of the deviations involves the sum of the squared deviations of the measurements from their mean. This measure is called the **variance**.

variance

DEFINITION 3.7
Variance

The **variance** of a set of n measurements $y_1, y_2, \ldots, y_n$ with mean $\bar{y}$ is the sum of the squared deviations divided by $n - 1$:

$$\frac{\sum_i (y - \bar{y})^2}{n - 1}$$

s^2
σ^2

As with the sample and population means, we have special symbols to denote the sample and population variances. The symbol s^2 represents the sample variance, and the corresponding population variance is denoted by the symbol σ^2.

The definition for the variance of a set of measurements depends on whether the data are regarded as a sample or population of measurements. The definition we have given here assumes we are working with the sample, since the population measurements usually are not available. Many statisticians define the sample variance to be the average of the squared deviations, $\sum (y - \bar{y})^2/n$. However, the use of $(n - 1)$ as the denominator of s^2 is not arbitrary. This definition of the sample variance makes it an *unbiased estimator* of the population variance σ^2. This means roughly that if we were to draw a very large number of samples, each of size n, from the population of interest and if we computed s^2 for each sample, the average sample variance would equal the population variance σ^2. Had we divided by n in the definition of the sample variance s^2, the average sample variance computed from a large number of samples would be less than the population variance; hence, s^2 would tend to underestimate σ^2.

Another useful measure of variability, the **standard deviation**, involves the square root of the variance.

DEFINITION 3.8
Standard Deviation

▼

The **standard deviation** of a set of measurements is defined to be the positive square root of the variance.

s
σ

We then have **s** denoting the sample standard deviation and **σ** denoting the corresponding population standard deviation.

EXAMPLE 3.10   ▼

The time between an electric light stimulus and a bar press to avoid a shock was noted for each of five conditioned rats. Use the data below to compute the sample variance and standard deviation.

shock avoidance times (seconds):   5, 4, 3, 1, 3

Solution   The deviations and the squared deviations are shown here. The sample mean $\bar{y}$ is 3.2.

	y_i	$y_i - \bar{y}$	$(y_i - \bar{y})^2$
	5	1.8	3.24
	4	.8	.64
	3	−.2	.04
	1	−2.2	4.84
	3	−.2	.04
Totals	16	0	8.80

Using the total of the squared deviations column, we find the sample variance to be

$$s^2 = \frac{\sum_i (y_i - \bar{y})^2}{4} = \frac{8.80}{4} = 2.2.$$

▲

Computation of the quantities s^2 and s is sometimes simplified using the following algebraic identity:

$$\sum_i (\bar{y}_i - \bar{y})^2 = \sum_i y_i^2 - \frac{(\sum_i y_i)^2}{n}.$$

Hence, we have the shortcut formula for s^2 (and s) given here.

Shortcut Formula for s^2 and s

$$s^2 = \frac{1}{n-1}\left[\sum_i y_i^2 - \frac{(\sum_i y_i)^2}{n}\right] \quad \text{and} \quad s = \sqrt{s^2}$$

E X A M P L E   3.11   ▼

Use the data of Example 3.10 to compute the sample variance using the shortcut formula.

Solution   It is convenient to construct the following table to perform the calculations:

y_i	y_i^2
5	25
4	16
3	9
1	1
3	9
Totals 16	60

Using the totals from the table, we have

$$s^2 = \frac{1}{4}\left[60 - \frac{(16)^2}{5}\right] = \frac{1}{4}[60 - 51.2] = 2.2$$

which is exactly the result we obtained in Example 3.10.   ▲

We can make a simple modification of our shortcut formula to approximate the sample variance if only grouped data are available. Recall that in approximating the sample mean for grouped data, we let y_i and f_i denote the midpoint and frequency, respectively, for the ith class interval. With this notation, the sample variance for grouped data is

$$s^2 = \frac{1}{n-1}\left[\sum_i f_i y_i^2 - \frac{(\sum_i f_i y_i)^2}{n}\right] \quad \text{or} \quad \frac{\sum f_i(y_i - \bar{y})^2}{n-1}.$$

The sample standard deviation is $\sqrt{s^2}$.

E X A M P L E   3.12   ▼

Refer to the chick data from Table 3.6 of Example 3.7. Calculate the sample variance and standard deviation for these data.

Solution   In addition to the calculations in Table 3.6, we also need the calculations for $f_i y_i^2$. These calculations, formed by multiplying corresponding elements in the y_i and $f_i y_i$ columns, are shown in the following listing.

$f_i y_i^2$	12.96	13.69	86.64	91.26	160.00	168.10	229.32
	203.39	251.68	141.75	126.96	154.63	115.20	96.04

The sum of the $f_i y_i^2$ calculations is 1,851.62. Using this total and the total of $f_i y_i$ in Table 3.6, we can determine s^2 and s.

$$s^2 = \frac{1}{n-1}\left[\sum_i f_i y_i^2 - \frac{(\sum_i f_i y_i)^2}{n}\right]$$

$$= \frac{1}{99}\left[1851.62 - \frac{(429.2)^2}{100}\right] = \frac{9.49}{99} = .10$$

$$s = \sqrt{.10} = .32.$$ ▲

We have now discussed several measures of variability, each of which can be used to compare the variabilities of two or more sets of measurements. The standard deviation is particularly appealing for two reasons: (1) we can compare the variabilities of *two or more* sets of data using the standard deviation, and (2) we can also use the results of the rule that follows to interpret the standard deviation of a single set of measurements. This rule applies to data sets with roughly a "mound-shaped" histogram; that is, a histogram that has a single peak, is symmetrical, and tapers off gradually in the tails. Since so many data sets can be classified as mound-shaped, the rule has wide applicability. For this reason, it is called the *Empirical Rule*.

Empirical Rule   ▼

Give a set of n measurements possessing a mound-shaped histogram, then

the interval $\bar{y} \pm s$ contains approximately 68% of the measurements,
the interval $\bar{y} \pm 2s$ contains approximately 95% of the measurements,
the interval $\bar{y} \pm 3s$ contains approximately all the measurements.

EXAMPLE 3.13   ▼

A sample of 20 days throughout the previous year indicated that the average wholesale price per pound for steers at a particular stockyard was $.61, with a standard deviation of $.07. If the histogram for the measurements is mound-shaped, describe the variability of the data using the Empirical Rule.

Solution   Applying the Empirical Rule, the interval

$$.61 \pm .07 \qquad \text{or} \qquad \$.54 \text{ to } \$.68$$

contains approximately 68% of the measurements. The interval

$$.61 \pm .14 \qquad \text{or} \qquad \$.47 \text{ to } \$.75$$

contains approximately 95% of the measurements. The interval

$$.61 \pm .21 \qquad \text{or} \qquad \$.40 \text{ to } \$.82$$

contains approximately all the measurements.   ▲

In English, approximately 2/3 of the steers sold for between $.54 and $.68 per pound; and 95% sold for between $.47 and $.75 per pound, with minimum and maximum prices being approximately $.40 and $.82.

To increase our confidence in the Empirical Rule, let us see how well it describes the five frequency distributions of Figure 3.23. We calculated the mean and standard deviation for each of the five data sets (not given), and these are shown next to each frequency distribution. Figure 3.23(a) shows the frequency distribution for measurements made on a variable that can take values $y = 0, 1, 2, \ldots, 10$. The mean and standard deviation $\bar{y} = 5.50$ and $s = 1.49$ for this symmetric mound-shaped distribution were used to calculate the interval $\bar{y} \pm 2s$, which is marked below the horizontal axis of the graph. We found 94% of the measurements falling in this interval—that is, lying within two standard deviations of the mean. Note that this percentage is very close to the 95% specified in the Empirical Rule. We also calculated the percentage of measurements lying within one standard deviation of the mean. We found this percentage to be 60%, a figure that is not too far from the 68% specified by the Empirical Rule. Consequently, we think the Empirical Rule provides an adequate description for Figure 3.23(a).

Figure 3.23(b) shows another mound-shaped frequency distribution but one that is less peaked than the distribution of Figure 3.23(a). The mean and standard deviation for this distribution, shown to the right of the figure, are 5.50 and 2.07, respectively. The percentages of measurements lying within one and two standard deviations of the mean are 64% and 96%, respectively. Once again, these percentages agree very well with the Empirical Rule.

FIGURE 3.23

A Demonstration of the Utility of the Empirical Rule

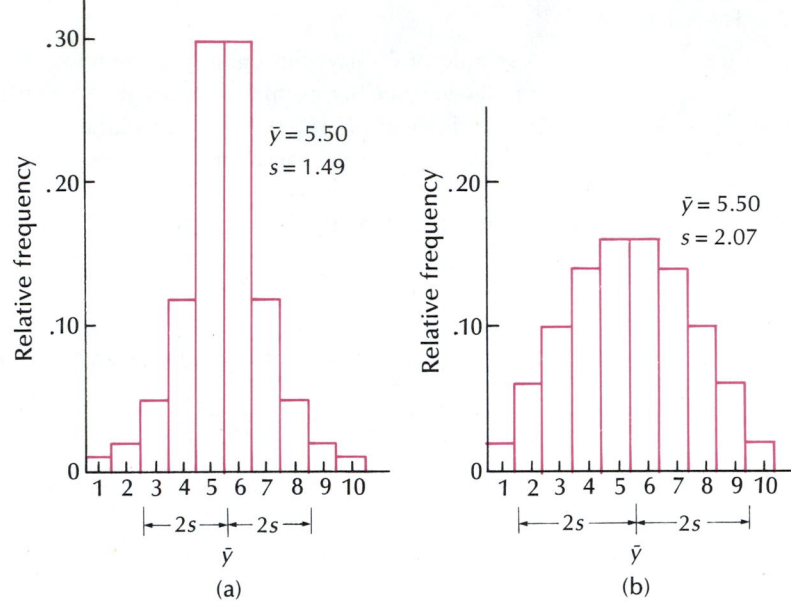

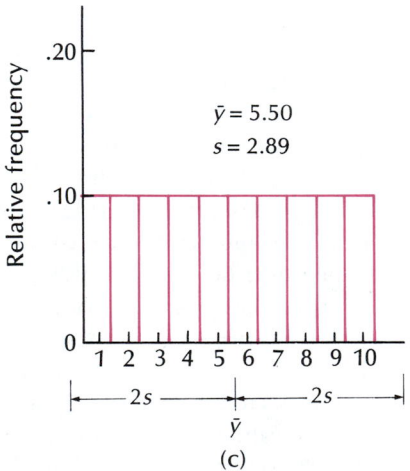

FIGURE 3.23
A Demonstration of the
Utility of the Empirical
Rule (continued)

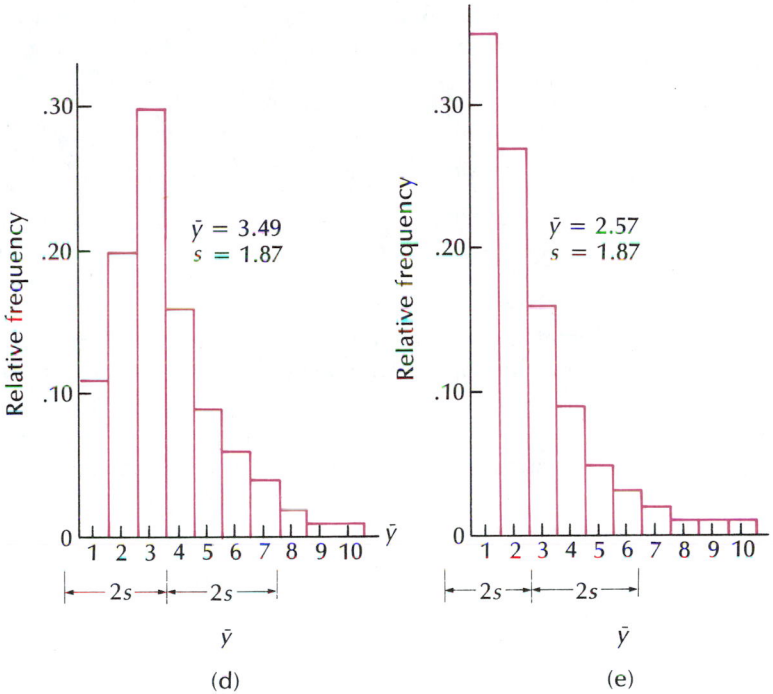

(d)

(e)

Now let us look at three other distributions. The distribution in Figure 3.23(c) is perfectly flat, while the distributions of Figure 3.23(d) and (e) are nonsymmetric and skewed to the right. The percentages of measurements lying within two standard deviations of the mean are 100%, 96%, and 95%, respectively, for these three distributions. All these percentages are reasonably close to the 95% specified by the Empirical Rule. The percentages lying within one standard deviation of the mean (60%, 75%, and 87%, respectively) show some disagreement with the 68% of the Empirical Rule.

To summarize, you can see that the Empirical Rule accurately forecasts the percentage of measurements falling within two standard deviations of the mean for all five distributions of Figure 3.23, even for the distributions that are flat, as in Figure 3.23(c), or highly skewed to the right, as in Figure 3.23(e). The Empirical Rule is less accurate in forecasting the percentages within one standard deviation of the mean, but the forecast, 68%, compares reasonably well for the three distributions that might be called mound-shaped, Figure 3.23(a), (b), and (d).

The results of the Empirical Rule enable us to obtain a quick approximation to the sample standard deviation s. The Empirical Rule states that approximately 95% of the measurements lie in the interval $\bar{y} \pm 2s$. The length of this interval is, therefore, 4*s*. Since the range of the measurements is approximately 4*s*, we obtain an **approximate value for *s*** by dividing the range by 4.

$$\text{approximate value of } s = \frac{\text{range}}{4}$$

Some people might wonder why we did not equate the range to 6*s*, since the interval $\bar{y} \pm 3s$ should contain almost all the measurements. This procedure would yield an approximate value for *s* that is smaller than the one obtained by the procedure above. If we are going to make an error (as we are bound to do with any approximation), it is better to overestimate the sample standard deviation so that we are not led to believe there is less variability than may be the case.

E X A M P L E   3.14

▼

The following data represent the percentages of family income allocated to groceries for a sample of 30 shoppers:

26	28	30	37	33	30
29	39	49	31	38	36
33	24	34	40	29	41
40	29	35	44	32	45
35	26	42	36	37	35

For these data $\sum y_i = 1{,}043$ and $\sum y_i^2 = 37{,}331$.

Compute the mean, variance, and standard deviation of the percentage of income spent on food. Check your calculation of *s*.

Solution   The sample mean is

$$\bar{y} = \frac{\sum_i y_i}{30} = \frac{1043}{30} = 34.77.$$

The corresponding sample variance and standard deviation are

$$s^2 = \frac{1}{n-1}\left[\sum_i y_i^2 - \frac{(\sum_i y_i)^2}{n}\right]$$

$$= \frac{1}{29}[37{,}331 - 36{,}261.63] = \frac{1069.27}{29} = 36.87$$

$$s = \sqrt{36.87} = 6.07.$$

We can check our calculation of s by using the range approximation. The largest measurement is 49 and the smallest is 24. Hence, an approximate value of s is

$$s \approx \frac{\text{range}}{4} = \frac{49 - 24}{4} = 6.25.$$

Note how close the approximation is to our computed value.   ▲

Although there will not always be the close agreement found in Example 3.14, the range approximation provides a useful and quick check on the calculation of s.

▼ EXERCISES

Basic Techniques

3.39 To give you some practice, consider a small set of five measurements, say 5, 4, 1, 2, 3.
  a. So that you can see the variation in the measurements, construct a dot diagram.
  b. Use $\sum y$ and $\sum y^2$ to calculate $\sum (y - \bar{y})^2$.
  c. Calculate s^2 and s.
  d. Because the number of measurements in the sample is so small, the frequency distribution for the sample measurements is not mound-shaped. Nevertheless, note that the interval $(\bar{y} \pm 2s)$ contains all the measurements. (Construct this interval on the dot diagram for the data so that you can see the location of the points within the interval.)

3.40 Repeat the instructions of Exercise 3.39 for the six measurements 1, 0, 3, 1, 2, 2.

3.41 Repeat the instructions of Exercise 3.39 for the ten measurements 4, 1, 3, 5, 2, 3, 1, 4, 0, 2.

Applications

3.42 The treatment times for patients at a health clinic are as follows:

21	20	31	24	15	21	24	18	33	8
26	17	27	29	24	14	29	41	15	11
13	28	22	16	12	15	11	16	18	17
29	16	24	21	19	7	16	12	45	24
21	12	10	13	20	35	32	22	12	10

Use the shortcut formula to calculate s^2 and s. You can verify that $\sum y = 1,016$ and $\sum y^2 = 24,080$ for the 50 treatment times.

3.43   Refer to Exercise 3.42. To increase your confidence in the applicability of the Empirical Rule, construct the intervals $(\bar{y} \pm s)$, $(\bar{y} \pm 2s)$, and $(\bar{y} \pm 3s)$, and count the number of treatment times falling in each of the three intervals. From these frequencies, calculate the corresponding percentage of measurements falling in the three intervals. Does the Empirical Rule give a reasonable approximation to the relative frequencies you have observed?

 3.44   To assist in estimating the amount of lumber in a tract of timber, an owner decided to count the number of trees with diameters exceeding 12 inches in randomly selected 50 × 50-foot squares. Seventy 50 × 50 squares were randomly selected from the tract and the number of trees (with diameters in excess of 12 inches) were counted for each. The data were as follows:

7	8	6	4	9	11	9	9	9	10
9	8	11	5	8	5	8	8	7	8
3	5	8	7	10	7	8	9	8	11
10	8	9	8	9	9	7	8	13	8
9	6	7	9	9	7	9	5	6	5
6	9	8	8	4	4	7	7	8	9
10	2	7	10	8	10	6	7	7	8

a. Construct a relative frequency histogram to describe these data.

b. Calculate the sample mean $\bar{y}$ as an estimate of μ, the mean number of timber trees with diameter exceeding 12 inches for all 50 × 50 squares in the tract.

c. Calculate s for the data. Construct the intervals $(\bar{y} \pm s)$, $(\bar{y} \pm 2s)$, and $(\bar{y} \pm 3s)$. Count the percentages of squares falling in each of the three intervals, and compare these percentages with the corresponding percentages given by the Empirical Rule.

3.5 ▼ CODING TO SIMPLIFY CALCULATIONS (optional)

Data are frequently coded to simplify the calculations of $\bar{y}$ and s^2 (and s). For example, it is much easier to calculate the mean of the five measurements $-.1$, $.2$, $.1$, 0, and $.2$ than of the five measurements 99.9, 100.2, 100.1, 100.0, and 100.2. The first set of measurements was obtained by subtracting 100 from each measurement in the second set. Similarly, we might wish to simplify a set of measurements by multiplying by a constant. It is easier to work with the set 3, 1, 4, 6, 4, 2 than with the set .003, .001, .004, .006, .004, .002. The first set was obtained by multiplying each element of the second set by 1000.

coding

Data are generally **coded** by performing one or both of the following operations: subtracting (or adding) a constant m from each measurement, or multiplying (or dividing) each measurement by a constant k.

How are the mean $(\bar{y}_c)$ and standard deviation (s_c) of the coded measurements related to the mean and standard deviation of the sample measurements $y_1, y_2, \ldots, y_n$? The theorem given next answers this question.

THEOREM 3.1

▼

Let $y_1, y_2, \ldots, y_n$ be n measurements with sample mean $\bar{y}$ and sample standard deviation s. Then we have the following:

1. If we subtract a constant m from each measurement, the mean and standard deviation for the original measurements will be

$$\bar{y} = \bar{y}_c + m \qquad \text{and} \qquad s = s_c.$$

2. If we multiply each measurement by a positive constant k, the mean and standard deviation for the original data will be

$$\bar{y} = \frac{\bar{y}_c}{k} \qquad \text{and} \qquad s = \frac{s_c}{k}.$$

EXAMPLE 3.15

▼

Use the results of Theorem 3.1 to compute the mean and standard deviation of the measurements .003, .001, .004, .006, .004, .002 by multiplying each by 1,000.

Solution   Multiplying each measurement by 1,000, we have the coded data 3, 1, 4, 6, 4, 2. For the coded data, then,

$$\bar{y}_c = \frac{\sum y_c}{6} = \frac{20}{6} = 3.33.$$

Similarly, with $\sum y_c = 20$ and $\sum y_c^2 = 82$, we have

$$s_c^2 = \frac{1}{5}\left[82 - \frac{(20)^2}{6}\right] = 3.07 \qquad \text{and} \qquad s_c = 1.75.$$

Then applying the results of Theorem 3.1, with $k = 1,000$,

$$\bar{y} = \frac{3.33}{1000} = .00333 \qquad \text{and} \qquad s = \frac{1.75}{1000} = .00175.$$

▲

coding grouped data   We can also **code grouped data** to simplify the calculations of $\bar{y}$ and s.

Rules for Coding Grouped Data

1. Based on inspection, select the class interval that you think is likely to contain the mean. Let m denote the midpoint of this interval. (Note: the selection of m is not critical.)

2. Code the interval midpoints as

$$y_c = \frac{y - m}{w},$$

where w is the interval width and y is the midpoint (uncoded units).

3. Compute $\bar{y}_c$ and s_c in the usual way for grouped data.

4. $\bar{y} = w\bar{y}_c + m$ and $s = ws_c$.

EXAMPLE 3.16

Use the frequency distribution of crime rates (violent crimes per 100,000 inhabitants) for the data if Table 3.4 shown here to approximate $\bar{y}$ and s.

Class Limits	Midpoint y	f	$y_c = \dfrac{y-m}{w}$	fy_c	fy_c^2
68.5–137.5	103	1	−7	−7	49
137.5–206.5	172	3	−6	−18	108
206.5–275.5	241	5	−5	−25	125
275.5–344.5	310	6	−4	−24	96
344.5–413.5	379	7	−3	−21	63
413.5–482.5	448	11	−2	−22	44
482.5–551.5	517	11	−1	−11	11
551.5–620.5	586	12	0	0	0
620.5–689.5	655	7	1	7	7
689.5–758.5	724	7	2	14	28
758.5–827.5	793	6	3	18	54
827.5–896.5	862	6	4	24	96
896.5–965.5	931	3	5	15	75
965.5–1034.5	1,000	3	6	18	108
1034.5–1103.5	1,069	1	7	7	49
1103.5–1172.5	1,138	1	8	8	64
Totals		90		−17	977

Solution　For the frequency data shown, we selected the eighth interval from the top as the class interval that we thought would contain the mean; then

$m = 586$. Step 2 of the coding procedure is to code the midpoints of the intervals using the formula $y_c = (y - m)/w$ where the interval width is $w = 69$. The coded midpoints are shown in column 4 of the table. The remaining two columns provide useful computations for $\bar{y}$ and s.   ▲

Using the grouped data formula for $\bar{y}_c$ and the total of column 5 in the table, we have

$$\bar{y}_c = \frac{\sum f y_c}{n} = \frac{-17}{90} = -.189.$$

The corresponding mean for the uncoded measurements can be approximated as

$$\bar{y} = w\bar{y}_c + m = 69(-.189) + 586 = 572.96.$$

The computations for s_c^2 and s_c follow in a similar fashion:

$$s_c^2 = \frac{1}{n-1}\left[\sum f y_c^2 - \frac{(\sum f y_c)^2}{n}\right] = \frac{1}{89}\left[977 - \frac{(-17)^2}{90}\right]$$

$$= \frac{1}{89}(977 - 3.21) = 10.94$$

and $s_c = \sqrt{10.94} = 3.31$. Then multiplying s_c by $w = 69$, we have

$$s = ws_c = 69(3.31) = 228.39.$$

You may think that the additional steps required in coding do not really simplify things, especially when most of us have access to a calculator or a computer. However, the mere fact that the numbers are smaller and simpler to manipulate makes the coding worthwhile. More important, by working with smaller numbers, we are probably less prone to make arithmetic errors and are less likely to face serious rounding errors.

▼ EXERCISES

Basic Techniques

3.45   Three data sets are shown here. Compute $\bar{y}$ and s for each one. Use the shortcut formula for s^2.

Data set 1: 1, 2, 3
Data set 2: .01, .02, .03
Data set 3: 1001, 1002, 1003

3.46   Refer to Exercise 3.45. Compute $\bar{y}$ and s for each data set using the formula

$$s^2 = \frac{\sum (y_i - \bar{y})^2}{n-1}.$$

Compare your results here to what you obtained in Exercise 3.45. Which formula appears to be more accurate? Why? Do you have any words of caution?

3.47   Based on $\bar{y}$ and s from Exercise 3.46, indicate how the coding steps could be used to compute the sample mean and standard deviation for the three measurements 10,000, 10,001, and 10,002.

3.6 ▼ THE BOX PLOT

box plot

As mentioned earlier in this chapter, a stem-and-leaf plot provides a graphical representation of a set of scores that can be used to examine the shape of the distribution, the range of scores, and where the scores are concentrated. The **box plot**, which builds on the information displayed in a stem-and-leaf plot, is more concerned with the symmetry of the distribution and incorporates numerical measures of central tendency and location in order to study the variability of the scores and the concentration of scores in the tails of the distribution.

Before we show how to construct and interpret a box plot, it is necessary to introduce several new terms that are peculiar to the language of exploratory data analysis (EDA). We are familiar with the definitions for the first, second (median), and third quartiles of a distribution presented earlier in this chapter. The box plot

hinges

uses the median and **hinges** of a distribution. Hinges are very similar to quartiles of a distribution, but owing to the method by which they are computed for sample data, the lower and upper hinges of a distribution may differ very slightly from the first and third quartiles of a set of scores.

Having said this, and recognizing the slight distinction, we will compute hinges in this text but refer to them as the lower and upper quartiles of the sample data.

We can now illustrate a *skeletal box plot* by way of an example.

EXAMPLE 3.17   ▼

Use the stem-and-leaf plot in Figure 3.24 for the 90 violent crime rates of Table 3.4 to construct a skeletal box plot.

Solution   When the scores are ordered from lowest to highest, the median score and quartile scores are located as follows:

$$\text{median location} = \frac{n+1}{2}$$

$$\text{quartile location} = \frac{truncated\ median\ location + 1}{2},$$

FIGURE 3.24

Stem-and-Leaf Plot

1	89
2	10 24 67 91 96 98
3	36 41 52 54 74 76 88 93 93
4	10 21 35 47 48 60 64 66 80 81 91 96 98
5	04 05 08 16 26 29 37 57 59 61 62 62 62 63 70 71 78 85 92
6	05 05 24 26 28 31 39 42 47 61 73 84 85 85 90 98
7	03 06 18 19 20 31 35 39 51 58 71
8	04 07 09 11 14 17 43 56 68 76 77 85
9	28 71
10	20

where the truncated median location is simply the median location with the decimal .5 omitted where present. For the distribution of $n = 90$ violent crime rates, we have

$$\text{median location} = \frac{90 + 1}{2} = 45.5$$

$$\text{truncated median location} = 45$$

and

$$\text{quartile location} = \frac{45 + 1}{2} = 23.$$

Since the median location is score 45.5 in the distribution, we average the 45th and 46th scores to compute the median. For these data, the 45th score (counting from the lowest to the highest in Figure 3.24) is 571 and the 46th is 578, hence, the median is

$$M = \frac{571 + 578}{2} = 574.5.$$

Then to find the lower and upper quartiles for this distribution of scores, we determine the 23rd score counting in from the low side of the distribution and from the high side of the distribution, respectively. The 23rd–lowest score and 23rd–highest scores are 464 and 719.

$$\text{lower quartile, } Q_1 = 464$$
$$\text{upper quartile, } Q_3 = 719$$

skeletal box plot

These three descriptive measures and the smallest and largest values in a data set are used to construct a skeletal box plot (see Figure 3.25). The **skeletal box plot** is constructed by drawing a box between the lower and upper quartiles with a solid line drawn across the box to locate the median. A straight

FIGURE 3.25
**Skeletal Box Plot for the
Data of Figure 3.24**

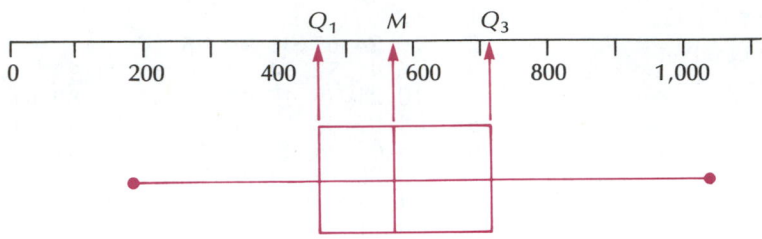

box-and-whiskers plot

line is then drawn connecting the box to the largest value; a second line is drawn from the box to the smallest value. These straight lines are sometimes called whiskers and the entire graph, a **box-and-whiskers plot**.   ▲

With a quick glance at a skeletal box plot, is is easy to obtain an impression about the following aspects of the data:

1. the lower and upper quartiles, Q_1 and Q_3
2. the interquartile range (IQR), the distance between the lower and upper quartiles
3. the most extreme (lowest and highest) values
4. the symmetry or asymmetry of the distribution of scores.

If we had been presented with Figure 3.25 without having seen the original data, we would have observed that

$$Q_1 \approx 475$$
$$Q_3 \approx 725$$
$$IQR \approx 725 - 475 = 250$$
$$M \approx 575$$

most extreme values: 175 and 1,025.

Also, because the median is closer to the lower quartile than the upper quartile and because the upper whisker is a little longer than the lower whisker, the distribution is slightly nonsymmetrical. To see that this conclusion is true, construct a frequency histogram for these data (or refer to your results in Exercise 3.8).

The skeletal box plot can be expanded to include more information about extreme values in the tails of the distribution. To do so, we need the following additional quantities:

lower inner fence: $Q_1 - 1.5(IQR)$
upper inner fence: $Q_3 + 1.5(IQR)$
lower outer fence: $Q_1 - 3(IQR)$
upper outer fence: $Q_3 + 3(IQR)$

Any score beyond an inner fence on either side is called a *mild outlier*, a score beyond an outer fence on either side is called an *extreme outlier*.

E X A M P L E 3.18 ▼

Compute the inner and outer fences for the data of Example 3.17. Identify any mild and extreme outliers.

Solution For these data, we found the lower and upper quartiles to be 464 and 719, respectively; IQR = 719 − 464 = 255. Then

lower inner fence = 464 − 1.5(255) = 81.5
upper inner fence = 719 + 1.5(255) = 1,101.5
lower outer fence = 464 − 3(255) = −301
upper outer fence = 719 + 3(255) = 1,484

Also, from the stem-and-leaf plot, we see that the lower and upper adjacent values are 189 and 1,020. Since the upper and lower fences are 1,101.5 and 81.5, respectively, there are no observations that are beyond the inner fences. Hence, there are no mild or extreme outliers. ▲

We now have all the quantities necessary for constructing a box plot.

Steps in Constructing a Box Plot

▼

1. As with a skeletal box plot, mark off a box from the lower quartile to the upper quartile.
2. Draw a solid line across the box to locate the median.
3. Mark the location of the upper and lower adjacent values with an x.
4. Draw a dashed line between each quartile and its adjacent value.
5. Mark each extreme outlier with the symbol o.

E X A M P L E 3.19 ▼

Construct a box plot for the data of Example 3.17.

Solution The box plot is shown in Figure 3.26.

FIGURE 3.26
The Box Plot for the Data of Example 3.17

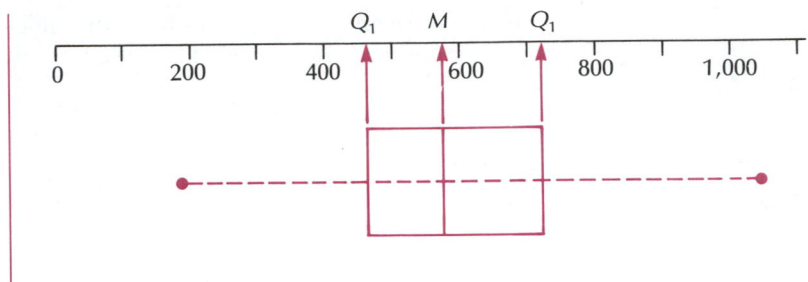

What information can be drawn from a box plot? First, the center of the distribution of scores is indicated by the median line in the box plot. Second, a measure of the variability of the scores is given by the interquartile range, the length of the box. Recall that the box is constructed between the lower and upper quartiles so it contains the middle 50% of the scores in the distribution, with 25% on either side of the median line inside the box. Third, by examining the relative position of the median line, we can gauge the symmetry of the middle 50% of the scores. For example, if the median line is closer to the lower quartile than the upper, there is a greater concentration of scores on the lower side of the median within the box than on the upper side; a symmetric distribution of scores would have the median line located in the center of the box. Fourth, additional information about skewness is obtained from the lengths of the whiskers; the longer one whisker is relative to the other one, the more skewness there is in the tail with the longer whisker. Fifth, a general assessment can be made about the presence of outliers by examining the number of scores classified as mild outliers and the number classified as extreme outliers.

▼ EXERCISES

Basic Techniques

3.48   Find the median and the lower and upper quartiles for the following measurements: 5, 9, 3, 6, 5, 7, 9, 2, 8, 10, 4, 3, 9.

3.49   Repeat Exercise 3.48 for the following measurements: 29, 22, 26, 20, 19, 16, 24, 11, 22, 23, 20, 29, 17, 18, 15.

Applications

3.50   The number of persons who volunteered to give a pint of blood at a central donor center was recorded for each of 20 successive Fridays. The data are shown here:

320	370	386	334	325	315	334	301	270	310
274	308	315	368	332	260	295	356	333	250

a. Construct a stem-and-leaf plot.

b. Construct a skeletal box plot and interpret the results.

3.51 Construct a skeletal box plot for the data of Table 3.2. You may find the following stem-and-leaf plot helpful.

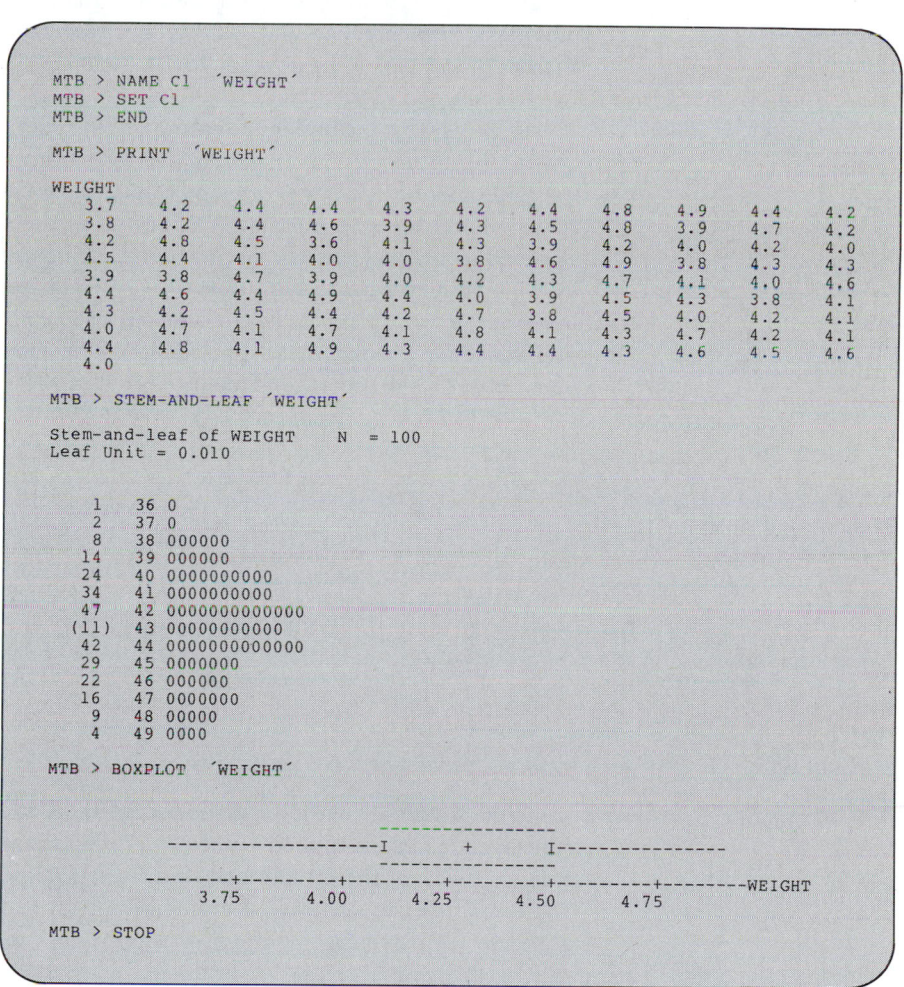

```
MTB > NAME C1 'WEIGHT'
MTB > SET C1
MTB > END

MTB > PRINT 'WEIGHT'

WEIGHT
   3.7    4.2    4.4    4.4    4.3    4.2    4.4    4.8    4.9    4.4    4.2
   3.8    4.2    4.4    4.6    3.9    4.3    4.5    4.8    3.9    4.7    4.2
   4.2    4.8    4.5    3.6    4.1    4.3    3.9    4.2    4.0    4.2    4.0
   4.5    4.4    4.1    4.0    4.0    3.8    4.6    4.9    3.8    4.3    4.3
   3.9    3.8    4.7    3.9    4.0    4.2    4.3    4.7    4.1    4.0    4.6
   4.4    4.6    4.4    4.9    4.4    4.0    3.9    4.5    4.3    3.8    4.1
   4.3    4.2    4.5    4.4    4.2    4.7    3.8    4.5    4.0    4.2    4.1
   4.0    4.7    4.1    4.7    4.1    4.8    4.1    4.3    4.7    4.2    4.1
   4.4    4.8    4.1    4.9    4.3    4.4    4.4    4.3    4.6    4.5    4.6
   4.0

MTB > STEM-AND-LEAF 'WEIGHT'

Stem-and-leaf of WEIGHT    N = 100
Leaf Unit = 0.010

    1    36 0
    2    37 0
    8    38 000000
   14    39 000000
   24    40 0000000000
   34    41 0000000000
   47    42 00000000000000
  (11)   43 00000000000
   42    44 00000000000000
   29    45 0000000
   22    46 000000
   16    47 0000000
    9    48 00000
    4    49 0000

MTB > BOXPLOT 'WEIGHT'


                                          ---------------
     --------------------I      +       I----------------
                                          ---------------
     --------+---------+---------+---------+---------+--------WEIGHT
           3.75      4.00      4.25      4.50      4.75

MTB > STOP
```

3.7 ▼ SUMMARIZING DATA FROM MORE THAN ONE VARIABLE

In the previous sections, we've discussed graphical methods and numerical descriptive methods for summarizing data from a single variable. Frequently, more than one variable is being studied at the same time and, although we might be interested in summarizing the data on each variable separately, we might also be interested in studying relations among the variables. For example, we might be interested in the prime interest rate and in the consumer price index, as well as in the relation between the two. In this section, we'll discuss a few techniques for summarizing data from two (or more) variables. Material in this section will provide a brief preview and introduction to chi-square methods (Chapter 8), analysis of variance (Chapters 13, 15, 16, and 17), and regression (Chapters 9, 10, 11, and 12).

contingency table

Consider first the problem of summarizing data from two qualitative variables. Cross-tabulations can be constructed to form a **contingency table**. The rows of the table identify the categories of one variable, and the columns identify the categories of the other variable. The entries in the table are the number of times each value of one variable occurs with each possible value of the other. For example, a television viewing survey was conducted on 1,500 individuals. Each individual surveyed was asked to state his or her place of residence and network preference for national news. The results of the survey are shown in Table 3.7. As you can see, 144 urban residents preferred ABC, 135 urban residents preferred CBS, and so on.

TABLE 3.7
Data from a Survey of Television Viewing

	Residence			
Nework Preference	Urban	Suburban	Rural	Total
ABC	144	180	90	414
CBS	135	240	96	471
NBC	108	225	54	387
Other	63	105	60	228
Total	450	750	300	1500

The simplest method for looking at relations between variables in a contingency table is to do a percentage comparison based on the row totals, the column totals, or the overall total. If we calculate percentages within each row of Table 3.7, we can compare the distribution of residences within each network preference. A percentage comparison such as this, based on the row totals, is shown in Table 3.8.

TABLE 3.8
Comparing the Distribution of Residences for Each Network

	Residence			
Network Preference	Urban	Suburban	Rural	Total
ABC	34.8	43.5	21.7	100 ($n = 414$)
CBS	28.7	50.9	20.4	100 ($n = 471$)
NBC	27.9	58.1	14.0	100 ($n = 387$)
Other	27.6	46.1	26.3	100 ($n = 228$)

Except for ABC, which has the highest urban percent among the networks, the differences among the residence distributions are in the suburban and rural categories. The percent of suburban preferences rises from 43.5% for ABC to 58.1% for NBC. Corresponding shifts downward occur in the rural category. In Chapter 8, we will use chi-square methods to explore further relations between two (or more) qualitative variables.

An extension of the bar graph provides a convenient method for summarizing joint data from a single qualitative and a single quantitative variable. We will discuss this method by way of an example. Suppose that a company wants to investigate the relative effects of three different employee-incentive systems on productivity. A total of 15 work teams are selected randomly. Of these teams, 7 participate in a released-time plan, by which teams achieving certain goals are allowed to take extra time off, with pay; 5 participate in a bonus-pay plan; and 3 participate in a profit-sharing plan. The company has a standard productivity measure and calculates the increased productivity of each work team over a 3 month period. Suppose that the results are the following:

released time, R :	16.2	15.6	19.4	18.8	16.9	15.9	17.6
bonus pay, B :	12.4	15.8	14.0	9.8	10.0		
profit sharing, P :	4.6	8.0	6.0				

Do the data indicate a strong relationship between plan and productivity gain?

The data summarized in Figure 3.27 clearly indicate that productivity gains are generally largest for the released-time plan, R, gains for the bonus-pay plan, B, are in the middle, and gains for the profit-sharing plan, P, are lowest. For now, we will use a plot such as that in Figure 3.27. Later, we will use analysis of variance methods (Chapters 13, 15, 16, and 17) to examine the relationships between a quantitative variable and one or more qualitative variables.

Finally, we can construct data plots for summarizing the relation between two quantitative variables. Consider the following example. A manager of a small

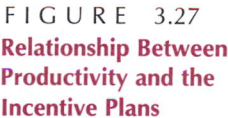

FIGURE 3.27
Relationship Between Productivity and the Incentive Plans

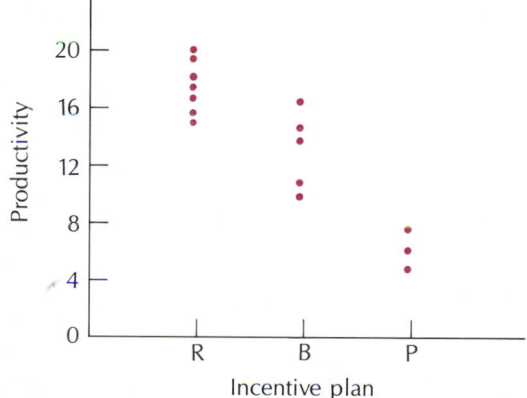

machine shop examined the starting hourly wage y offered to machinists with x years of previous experience. The data are shown here:

y (dollars):	8.90	8.70	9.10	9.00	9.79	9.45	10.00	10.65	11.10	11.05
x (years):	1.25	1.50	2.00	2.00	2.75	4.00	5.00	6.00	8.00	12.00

Is there a relationship between x and y?

scatterplot

One way to summarize these data is to use a **scatterplot**, as shown in Figure 3.28. Each point on the plot represents a machinist with a particular starting wage and years of experience. The point circled corresponds to $y = 9.45$, $x = 4.00$.

**FIGURE 3.28
Scatterplot of Starting
Hourly Wage and Years
Experience**

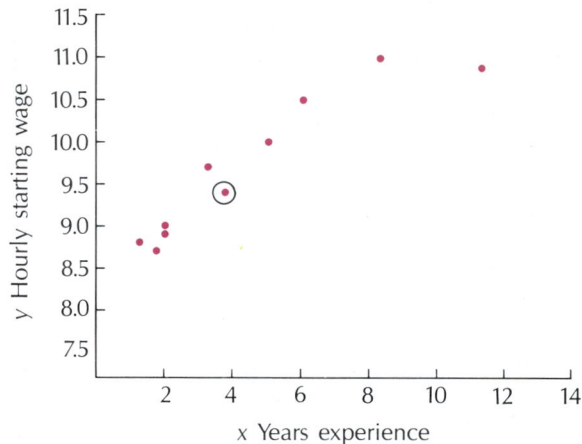

In general, the data displayed in Figure 3.28 indicate that, as the years of previous experience x increases, the hourly starting wage y for machinists increases. This basic idea of relating two quantitative variables is discussed and expanded in the chapters on regression (9, 10, 11, and 12).

▼ EXERCISES

3.52 Refer to the television survey data of Table 3.7. Do a percentage comparison based on the column totals. Interpret the data.

3.53 Data on the age at the time of a job turnover and on the reason for the job turnover are displayed here for 250 job changes in a large corporation.

Reason for Turnover	Age (Years)				
	≤29	30–39	40–49	≥50	Total
Resigned	30	6	4	20	60
Transferred	12	45	4	5	66
Retired/fired	8	9	52	55	124
Total	50	60	60	80	250

Do a percentage comparison based on the row totals and use this to describe the data.

3.54 Refer to Exercise 3.53. What different summary would one get with a percentage comparison based on the column totals? Do this summary and describe your results.

3.55 The lengths of hospital stays were recorded for patients undergoing a particular surgical procedure at each of four hospitals. These data are shown here.

Hospital	Length of Stay (Days)							
A	18	20	22	22	24	26		
B	14	15	17	17	18	19	21	21
C	21	25	27	31				
D	27	33						

a. Compute the mean stay for each hospital.
b. Plot the sample data.
c. Use parts (a) and (b) to describe the data. Which hospital appears to have shorter stays?

3.56 The federal government keeps a close watch on money growth versus targets that have been set for that growth. Below we list two measures of the money supply in the United States, M2 (private checking deposits, cash, and some savings) and M3 (M2 plus some investments), which are given here for 20 consecutive months.

Month	Money Supply (in Trillions of Dollars)	
	M2	M3
1	2.25	2.81
2	2.27	2.84
3	2.28	2.86
4	2.29	2.88
5	2.31	2.90
6	2.32	2.92
7	2.35	2.96
8	2.37	2.99
9	2.40	3.02
10	2.42	3.04
11	2.43	3.05
12	2.42	3.05
13	2.44	3.08
14	2.47	3.10
15	2.49	3.10
16	2.51	3.13
17	2.53	3.17
18	2.53	3.18
19	2.54	3.19
20	2.55	3.20

a. Would a scatterplot describe the relation between M2 and M3?
b. Construct a scatterplot. Is there an obvious relation?

3.57   Refer to Exercise 3.56. What other data plot might be used to describe and summarize these data? Make the plot and interpret your results.

3.8 ▼ CALCULATORS, COMPUTERS, AND SOFTWARE SYSTEMS

Electronic calculators can be great aids in performing some of the calculations mentioned in this chapter, especially for small data sets. For example, many calculators have keys for obtaining the sample mean and standard deviation directly after the data are entered. Others can be used with the shortcut formulas of the previous sections to obtain y, s^2, and s. For larger data sets, even hand calculators are of little use because of the time required for entering data and the inability to update an erroneous entry without reentering the entire data set. In these situations, a computer can be of help. Specific programs or more general software systems can be used to perform statistical analyses almost instantaneously even for very large data sets after the data are entered into the computer from a terminal, a magnetic tape, or disk storage. It is not necessary to have knowledge of computer programming to make use of specific programs or software systems for planned analyses. Most have user's manuals that give detailed directions for their use. Others, developed for use at a terminal, provide program prompts that lead the user through the analysis of choice.

There are many statistical software packages available for use on computers. Three of the more commonly used systems are Minitab, SAS, and SPSS. Each is available in a mainframe version, as well as in a personal computer version. Since a software system is a group of programs that work together, it is possible to obtain plots, data descriptions, and complex statistical analyses in a single job. Most people find that they can use any particular system easily, although they are frustrated by minor errors committed on the first few tries. The ability of such packages to perform complicated analyses on large numbers of data more than repays the initial investment of time and irritation.

In general, to use a system you need not learn everything about it. You need to learn about only the programs in which you are interested. Typical steps in a job involve describing your data to the software system, manipulating your data if they are not in the proper format or if you want a subset of your original data set, and then calling the appropriate set of programs or procedures using the key words particular to the software system you are using. The results obtained from calling a program are then displayed at your terminal or sent to your printer.

If you have access to a computer and are interested in using it, find out how to obtain an account, what programs and software systems are available for doing statistical analyses, and where to obtain instruction on data entry for these programs and software systems.

Because computer configurations, operating systems, and text editors vary from site to site, it is best to talk to someone knowledgeable about gaining access to a software system. Once you have mastered the commands to begin executing programs in a software system, you will find that running a job within a given software system is similar from site to site.

Since this isn't a text on computer usage, we won't spend additional time and space on the mechanics, which are best learned by doing. Our main interest is in interpreting the output from these programs. The designers of these programs tend to include in the output everything that a user could conceivably want to know; as a result, in any particular situation, some of the output is irrelevant. When reading computer output look for the values you want; if you don't need or don't understand an output statistic, don't worry. Of course, as you learn more about statistics, more of the output will be meaningful. In the meantime, look for what you need and disregard the rest.

There are dangers in using such packages carelessly. A computer is a mindless beast, and will do anything asked of it, no matter how absurd the result might be. For instance, suppose that the data include age, gender (1 = female, 2 = male), religion (1 = Catholic, 2 = Jewish, 3 = Protestant, 4 = other or none), and monthly income of a group of people. If we asked the computer to calculate means we would get means for the variables gender and religion, as well as for age and monthly income, even though these averages are meaningless. Furthermore, it is unlikely that a standard program would warn a user that extreme skewness was distorting a mean value or that the data contained a gross outlier. Used intelligently, these packages are convenient, powerful, and useful—but be sure to examine the output from any computer run to make certain the results make sense. Did anything go wrong? Was something overlooked? In other words, be *skeptical*. One of the important acronyms of computer technology still holds; namely, GIGO: garbage in, garbage out.

Throughout the textbook we will use computer software systems to do some of the more tedious calculations of statistics *after* we have explained how the calculations can be done. Used in this way, computers (and associated graphical and statistical analysis packages) will enable us to spend additional time on interpreting the results of the analyses, rather than on doing the analyses.

3.9 ▼ SUMMARY

This chapter was concerned with graphical and numerical description of data. The pie chart and bar graph are particularly appropriate for graphically displaying data obtained from a qualitative variable. The frequency and relative frequency histograms and stem-and-leaf plots are graphical techniques applicable only to quantitative data.

Numerical descriptive measures of data are used to convey a mental image of the distribution of measurements. Measures of central tendency include the mode, the median, and the arithmetic mean. Measures of variability include the range, the interquartile range, the variance, and the standard deviation of a set of measurements. While some disciplines emphasize different measures, we will use the mean and the standard deviation of a set of measurements as the primary numerical measures of central tendency and variability, respectively. One explanation for our choice is that we can not only compare variabilities of the *two* sets of measurements using the standard deviation of each, but we can also interpret the variability of a *single* set of measurements using the mean, the standard deviation, and the Empirical Rule.

We extended the concept of data description to summarizing the relations between two qualitative variables. Here cross-tabulations were used to develop percentage comparisons. We examined plots for summarizing the relations between quantitative and qualitative variables and between two quantitative variables. Material presented here (namely, summarizing relations among variables) will be discussed and expanded in later chapters on chi-square methods, on the analysis of variance, and on regression.

▼ KEY FORMULAS

1. Median, grouped data

$$\text{median} = L + \frac{w}{f_m}(.5n - cf_b)$$

2. Sample mean

$$\bar{y} = \frac{\sum_i y_i}{n}$$

3. Sample mean, grouped data

$$\bar{y} = \frac{\sum_i f_i y_i}{n}$$

4. Sample variance

$$s^2 = \frac{1}{n-1}\left[\sum_i y_i^2 - \frac{(\sum_i y_i)^2}{n}\right]$$

5. Sample variance, grouped data

$$s^2 = \frac{1}{n-1}\left[\sum_i f_i y_i^2 - \frac{(\sum_i f_i y_i)^2}{n}\right]$$

6. Sample standard deviation

$$s = \sqrt{s^2}$$

▼ SUPPLEMENTARY EXERCISES

3.58 The rounded nitrogen contents for the 36 composite apple leaf samples of Exercise 3.35 are presented below.

2.10	2.82	2.17	1.99	2.22	3.09
2.47	2.52	2.80	2.10	2.92	2.20
1.75	2.77	2.82	2.67	3.05	2.93
2.94	1.98	2.38	2.65	2.77	1.85
1.69	2.70	2.68	2.06	2.36	2.28
2.75	2.43	2.39	2.55	1.80	1.96

a. Use the shortcut formula to compute s^2 and s. You can verify that

$$\sum_i y_i = 87.61 \quad \text{and} \quad \sum_i y_i^2 = 218.7297$$

b. Use the range approximation to check your calculation of s.

c. To increase your confidence in the Empirical Rule, construct the intervals $\bar{y} \pm s$, $\bar{y} \pm 2s$, and $\bar{y} \pm 3s$. Count the number of rounded nitrogen content readings falling in each of the three intervals. Convert these numbers to percentages and compare your results to the Empirical Rule.

3.59 The College of Dentistry at the University of Florida has made a commitment to develop its entire curriculum around the use of self-paced instructional materials such as videotapes, slide tapes, syllabi, and so on. It is hoped that each student will proceed at a pace commensurate with his or her ability and that the instructional staff will have more free time for personal consultation in student-faculty interaction. One such instructional module was developed and tested on the first 50 students proceeding through the curriculum. The measurements below represent the number of hours it took these students to complete the required modular material.

16	8	33	21	34	17	12	14	27	6
33	25	16	7	15	18	25	29	19	27
5	12	29	22	14	25	21	17	9	4
12	15	13	11	6	9	26	5	16	5
9	11	5	4	5	23	21	10	17	15

a. Calculate the mode, the median, and the mean for these recorded completion times.
b. Guess the value of s.
c. Compute s by using the shortcut formula and compare your answers to that of part b.
d. Would you expect the Empirical Rule to describe adequately the variability of these data? Explain.

3.60   Refer to the data of Examples 3.7 and 3.12. We previously computed the sample mean and standard deviation to be 4.29 and .32, respectively. Use the coding procedures of Section 3.5 to compute $\bar{y}$ and s. Proceed assuming that you think the eighth class interval contains the mean.

3.61   Repeat Exercise 3.60 by using $m = 4.4$, the midpoint of the ninth interval.

3.62   Repeat Exercise 3.60 by using $m = 4.5$, the midpoint of the 10th interval. Are your answers to Exercises 3.60, 3.61, and 3.62 identical? If not, check your calculations.

3.63   A study was conducted to determine urine flow of sheep (in milliliters/minute) when infused intravenously with the antidiuretic hormone ADH. The urine flows of 10 sheep are recorded here.

0.7   0.5   0.5   0.6   0.5   0.4   0.3   0.9   1.2   0.9

a. Determine the mean, the median, and the mode for these sample data.
b. Suppose that the largest measurement is 6.8 rather than 1.2. How does this affect the mean, the median, and the mode?

3.64   Refer to Exercise 3.63.
a. Compute the range and the sample standard deviation.
b. Check your calculation of s using the range approximation.
c. How are the range and standard deviation affected if the largest measurement is 6.8 rather than 1.2? What about 68?

3.65   Refer to Exercise 3.63. Code the data by multiplying each measurement by 10. Compute the sample mean and standard deviation for the original set using the coded values.

3.66   A stem-and-leaf plot is shown for the telephone data from Exercise 3.20. Compute the mean, median, mode, and standard deviation for the data.

```
MTB > PRINT 'PHONES'

phones
    500    350    550    480    610    570    620    630    620    570    480
    550    650    580    570    600    480    520    540    610    570    580
    560    470    570    540    590    720    590    650    470    530    530
    560    550    580    560    610    560    510    540    540    570    560
    520    530    570    450    540    580

MTB > STEM AND LEAF 'PHONES'

Stem-and-leaf of phones    N  = 50
Leaf Unit = 10


     1    3 5
     1    4
     7    4 577888
    19    5 012233344444
   (21)   5 555666667777777888899
    10    6 0111223
     3    6 55
     1    7 2

MTB > STOP
```

3.67 A box plot was constructed for the data of Exercise 3.66 using Minitab. Describe the data using information conveyed by the box plot.

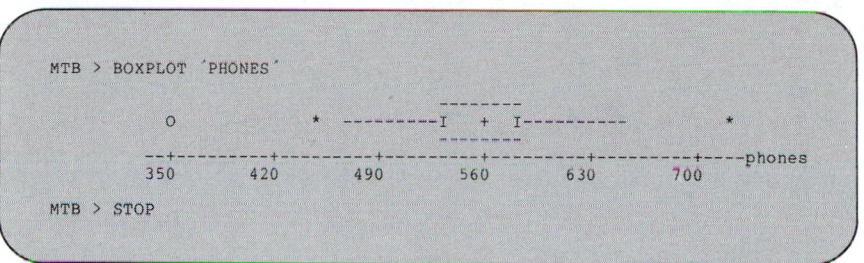

3.68 A random sample of 90 standard metropolitan statistical areas (SMSA) was studied to obtain information on murder rates. The murder rate (number of murders per 100,000 people) was recorded, and these data are summarized in the frequency table displayed below.

Class Interval	f_i	Class Interval	f_i
−.5–1.5	2	13.5–15.5	9
1.5–3.5	18	15.5–17.5	4
3.5–5.5	15	17.5–19.5	2
5.5–7.5	13	19.5–21.5	1
7.5–9.5	9	21.5–23.5	1
9.5–11.5	8	23.5–25.5	1
11.5–13.5	7		

Construct a relative frequency histogram for these data.

3.69 Refer to the data of Exercise 3.68.
   a. Compute the sample median and the mode.
   b. Compute the sample mean.
   c. Which measure of central tendency would you use to describe the center of the distribution of murder rates?

3.70 Refer to the data of Exercise 3.68.
   a. Compute the interquartile range.
   b. Compute the sample standard deviation.

3.71 Refer to the data of Exercise 3.68. If you did not employ coding to compute $\bar{y}$ and s in Exercises 3.69 and 3.70, use the midpoint of the fifth interval as m. Code the sample data to compute y and s. Compare your answers to those of Exercises 3.69 and 3.70.

3.72 Every 20 minutes a sample of 10 transistors is drawn from the outgoing product on a production line and tested. The data are summarized below for the first 500 samples of 10 measurements.

y_i	0	1	2	3	4	5	6	7	8	9	10
f_i	170	185	75	25	15	10	8	5	4	2	1

Construct a relative frequency distribution depicting the interquartile range. (Note: y_i in the table is the number of defectives in a sample of 10.)

3.73 Refer to Exercise 3.72.
  a. Determine the sample median and the mode.
  b. Calculate the sample mean.
  c. Based on the mean, the median, and the mode, how is the distribution skewed?

3.74 Refer to Exercise 3.72.
  a. Code the sample data to compute the sample standard deviation.
  b. Can the Empirical Rule be used to describe this set of measurements?

 3.75 Per capita expenditure (dollars) for health and hospital services by state are shown here.

Dollars	f
45–59	1
60–74	4
75–89	9
90–104	9
105–119	12
120–134	6
135–149	4
150–164	1
165–179	3
180–194	0
195–209	1
Total	50

  a. Construct a relative frequency histogram.
  b. Compute approximate values for $\bar{y}$ and s from the grouped expenditure data.

3.76 Refer to the data of Table 3.4. Eliminate Philadelphia from the north and San Jose and Seattle from the west.
  a. Compute $\bar{y}_i$ for the revised subgroups.
  b. Combine the subgroup means $(\bar{y}_i)$ to obtain the overall sample mean using the formula

$$\bar{y} = \frac{\sum_i n_i \bar{y}_i}{n},$$

  where n_i is the number of observations in subgroup i.
  c. Show that the sample mean computed in part b is identical to that obtained for the 87 measurements in part a.

 3.77 The Insurance Institute for Highway Safety published data on the total damage suffered by compact automobiles in a series of controlled, low-speed collisions. The data, in dollars, with brand names removed are as follows:

| 361 | 393 | 430 | 543 | 566 | 610 | 763 | 851 | 886 | 887 | 976 | 1,039 |
| 1,124 | 1,267 | 1,328 | 1,415 | 1,425 | 1,444 | 1,476 | 1,542 | 1,544 | 2,048 | 2,197 | |

a. Draw a histogram of the data using six or seven categories.
b. On the basis of the histogram, what would you guess the mean to be?
c. Calculate the median and mean.
d. What does the relation between the mean and median indicate about the shape of the data?

3.78 Production records for an automobile manufacturer show the following figures for production per shift (maximum production is 720 cars per shift):

| 688 | 711 | 625 | 701 | 688 | 667 | 694 | 630 | 547 | 703 | 688 | 697 | 703 |
| 656 | 677 | 700 | 702 | 688 | 691 | 664 | 688 | 679 | 708 | 699 | 667 | 703 |

a. Would the mode be a useful summary statistic for these data?
b. Find the median.
c. Find the mean.
d. What does the relation between the mean and median indicate about the shape of the data?

3.79 Draw a stem-and-leaf plot of the data in Exercise 3.78. The stems should include (from highest to lowest) 71, 70, 69, ... Does the shape of the stem-and-leaf display confirm your judgment in part d of Exercise 3.78?

3.80 Refer to Exercise 3.79.
a. Find the median and IQR.
b. Find the inner and outer fences. Are there any outliers?
c. Draw a box plot of the data.

 3.81 Data are collected on the weekly expenditures of a sample of urban households on food (including restaurant expenditures). The data, obtained from diaries kept by each household, are grouped by number of members of the household. The expenditures were as follows:

1 member:	67	62	168	128	131	118	80	53	99	68		
	76	55	84	77	70	140	84	65	67	183		
2 members:	129	116	122	70	141	102	120	75	114	81	106	95
	94	98	85	81	67	69	119	105	94	94	92	
3 members:	79	99	171	145	86	100	116	125				
	82	142	82	94	85	191	100	116				
4 members:	139	251	93	155	158	114	108					
	111	106	99	132	62	129	91					
5+ members:	121	128	129	140	206	111	104	109	135	136		

a. Calculate the mean expenditure separately for each number of members.
b. Calculate the median expenditure separately for each number of members.

3.82 Answer the following for the data in Exercise 3.81:
a. Calculate the mean of the combined data, using the raw data.
b. Can the combined mean be calculated from the means for each number of members?
c. Calculate the median of the combined data using the raw data.
d. Can the combined median be calculated from the medians for each number of members?

 3.83 A company revised a long-standing policy to eliminate the time clocks and cards for nonexempt employees. Along with this change, all employees (exempt and nonexempt) were expected to account for their own time on the job as well as absences due to sickness, vacation, holidays, and so on. The previous policy of allocating a certain number of sick days was eliminated; if an employee was sick, he or she was given time off with pay; otherwise, he or she was expected to be working.

In order to see how well the new program was working, the records of a random sample of 15 employees were examined to determine the number of sick days this year (under the new plan) and the corresponding number for the preceding year. These data are shown here:

Employee	This Year (new policy)	Preceding Year (old policy)
1	0	2
2	0	2
3	0	3
4	0	4
5	2	5
6	1	2
7	1	6
8	3	8
9	1	5
10	0	4
11	5	5
12	6	12
13	1	3
14	2	4
15	12	4

a. Obtain the mean and standard deviation for each column.
b. Based on the sample data, what might you conclude (infer) about the new policies? Explain your reason(s).

3.84 Refer to Exercise 3.83. What happens to $\bar{y}$ and s for each column if we eliminate the two 12s and substitute values of 7? Are the ranges for the old and new policies affected by these substitutions?

3.85 Federal authorities have destroyed considerable amounts of wild and cultivated marijuana plants. The following table shows the number of plants destroyed and the number of arrests for a 12-month period for 15 states.

State	Plants	Arrests
1	110,010	280
2	256,000	460
3	665	6
4	367,000	66
5	4,700,000	15
6	4,500	8
7	247,000	36
8	300,200	300
9	3,100	9
10	1,250	4
11	3,900,200	14
12	68,100	185
13	450	5
14	2,600	4
15	205,844	33

a. Discuss the appropriateness of using the sample mean to describe these two variables.

b. Compute the sample mean, 10% trimmed mean, and 20% trimmed mean. Which trimmed mean seems more appropriate for each variable? Why?

3.86 Refer to Exercise 3.85. Does there appear to be a relation between the number of plants destroyed and the number of arrests? How might you examine this question? What other variable(s) might be related to the number of plants destroyed?

$ 3.87 Monthly readings for the FDC Index, a popular barometer of the health of the pharmaceutical industry, are shown here. As can be seen, the Index has several components—one for pharmaceutical companies, one for diversified companies, one for chain drugstores, and another for drug and medical supply wholesalers.

	Pharmaceuticals	Diversified	Chain	Wholesaler
January	123.1	154.6	393.3	475.5
February	122.4	146.0	407.6	504.1
March	125.2	169.2	405.0	476.6
April	136.1	156.7	415.1	513.3
May	149.3	177.0	418.9	543.5
June	145.7	158.1	443.2	552.6
July	162.4	156.6	419.1	526.2
August	168.0	178.6	404.0	516.3
September	155.6	170.4	391.8	482.1
October	177.0	162.9	410.9	484.0
November	196.6	182.4	459.8	522.6
December	195.2	195.4	431.9	536.8

a. Plot these data on a single graph.

b. Discuss trends within each component and any apparent relations among the separate components of the FDC Index.

3.88   Refer to Exercise 3.87. Compute the percent change for each month of each component of the Index. (Assume that the percent changes in January were 12.3, −.7, 12.1, and 16.1, respectively, for the four components.) Plot these data. Are they more revealing than the original measurements were?

$ **3.89**   Closing New York Stock Exchange (NYSE) prices for the components (as of March 1987) of the Dow Jones Industrial Average (DJIA) are shown here:

	Components of Dow Jones Industrial Average	
	Percent of DJIA*	**Closing NYSE Stock Price 1/24**
Allied-Signal	2.81%	$46.875
Alcoa	2.39	39.875
American Can	3.93	65.500
American Express	3.25	54.125
AT&T	1.35	22.500
Bethlehem Steel	1.06	17.750
Chevron	2.17	36.250
duPont	3.70	61.750
Eastman Kodak	2.82	47.000
Exxon	3.06	51.000
General Electric	4.12	68.750
General Motors	4.22	70.375
Goodyear	1.90	31.625
Inco	0.85	14.250
IBM	8.99	150.000
Intl. Harvester	0.53	8.875
Intl. Paper	2.96	49.375
McDonald's	4.48	74.750
Merck	8.10	135.125
Minnesota Mining	5.18	86.375
Owens-Illinois	3.35	55.875
Phillip Morris	5.49	91.625
Procter & Gamble	3.94	65.750
Sears, Roebuck	2.22	37.000
Texaco	1.72	28.625
Union Carbide	4.95	82.625
United Technologies	2.73	45.500
U.S. Steel	1.40	23.375
Westinghouse	2.00	44.875
Woolworth	3.61	60.250

a. Compute the actual range of the stock prices.

b. The DJIA is actually a weighted average, so only a certain percent of the actual NYSE price is part of the DJIA for each stock. The weighted average can be written as

$$\bar{y}_w = \frac{\sum_i w_i y_i}{n},$$

where y_i is the closing price for stock i, and w_i is the weight attached to stock i. Using the weights (percent of DJIA) listed in the above table, compute the DJIA for this particular day.

c. Refer to part b. Why might the DJIA be a weighted average, rather than a simple average?

3.90 The number of telephones (per 1000 people) is shown by state in the accompanying table. These data were plotted in Exercises 3.20 and 3.21.

State	Telephones	State	Telephones	State	Telephones
Alabama	500	Louisiana	520	Ohio	550
Alaska	350	Maine	540	Oklahoma	580
Arizona	550	Maryland	610	Oregon	560
Arkansas	480	Massachusetts	570	Pennsylvania	610
California	610	Michigan	580	Rhode Island	560
Colorado	570	Minnesota	560	S. Carolina	510
Connecticut	620	Mississippi	470	S. Dakota	540
Delaware	630	Missouri	570	Tennessee	540
Florida	620	Montana	540	Texas	570
Georgia	570	Nebraska	590	Utah	560
Hawaii	480	Nevada	720	Vermont	520
Idaho	550	New Hampshire	590	Virginia	530
Illinois	650	New Jersey	650	Washington	570
Indiana	580	New Mexico	470	W. Virginia	450
Iowa	570	New York	530	Wisconsin	540
Kansas	600	N. Carolina	530	Wyoming	580
Kentucky	480	N. Dakota	560		

a. Might the Empirical Rule be used to describe the data?
b. Compute $\bar{y}$ and s and count the number (percent) of measurements falling in the intervals $\bar{y} \pm s$, $\bar{y} \pm 2s$, $\bar{y} \pm 3s$.

3.91 Refer to Exercise 3.90. Are there many extreme values affecting $\bar{y}$? Should this have been anticipated based on the data plot in Exercise 3.21? Compute the 10% trimmed mean for these data.

3.92 As one part of a review of middle-manager selection procedures, a study was made of the relation between hiring source (promoted from within, hired from related business, hired from unrelated business) and the 3-year job history (additional promotion, same position, resigned, dismissed). The data for 120 middle managers follows.

	Source			
Job History	Within Firm	Related Business	Unrelated Business	Total
Promoted	13	4	10	27
Same position	32	8	18	58
Resigned	9	6	10	25
Dismissed	3	3	4	10
Total	57	21	42	120

a. Calculate job-history percentages within each source.

b. Would you say that there is a strong dependence between source and job history?

 3.93 A survey was taken of 150 residents of major coal-producing states, 200 residents of major oil- and natural-gas–producing states, and 450 residents of other states. Each resident chose a most preferred national energy policy. The results are shown in the following SPSS printout.

COUNT ROW PCT COL PCT TOT PCT	COAL	STATE OIL AND GAS	OTHER	ROW TOTAL
OPINION COAL ENCOURAGED	62 32.8 41.3 7.8	25 13.2 12.5 3.1	102 54.0 22.7 12.8	189 23.6
FUSION DEVELOP	3 7.3 2.0 0.4	12 29.3 6.0 1.5	26 63.4 5.8 3.3	41 5.1
NUCLEAR DEVELOP	8 22.2 5.3 1.0	6 16.7 3.0 0.8	22 61.1 4.9 2.8	36 4.5
OIL DEREGULATION	19 12.6 12.7 2.4	79 52.3 39.5 9.9	53 35.1 11.8 6.6	151 18.9
SOLAR DEVELOP	58 15.1 38.7 7.3	78 20.4 39.0 9.8	247 64.5 54.9 30.9	383 47.9
COLUMN TOTAL	150 18.8	200 25.0	450 56.3	800 100.0

CHI SQUARE = 106.19406 WITH 8 DEGREES OF FREEDOM SIGNIFICANCE = 0.0000
CRAMER'S V = 0.25763
CONTINGENCY COEFFICIENT = 0.34233
LAMBDA = 0.01199 WITH OPINION DEPENDENT, = 0.07429 WITH STATE DEPENDENT.

a. Interpret the values 62, 32.8, 41.3, and 7.8 in the upper left cell of the cross tabulation. Note the labels COUNT, ROW PCT, COL PCT, and TOT PCT at the upper left corner.

b. Which of the percentage calculations seems most meaningful to you?

c. According to the percentage calculations you prefer, does there appear to be a strong dependence between state and opinion?

 3.94 A municipal workers' union that represents sanitation workers in many small midwestern cities studied the contracts that were signed in the previous years. The contracts were subdivided into those settled by negotiation without a strike, those settled by arbitration without a strike, and all those settled after a strike. For each contract, the first-year percentage wage increase was determined. Summary figures follow.

Contract Type	Negotiation	Arbitration	Poststrike
Mean percentage wage increase	8.20	9.42	8.40
Variance	0.87	1.04	1.47
Standard deviation	0.93	1.02	1.21
Sample size	38	16	6

Does there appear to be a relation between contract type and mean percent wage increase? If you were management rather than union affiliated, which posture would you take in future contract negotiations?

TOOLS AND CONCEPTS

PROBABILITY AND PROBABILITY DISTRIBUTIONS

4.1 HOW PROBABILITY CAN BE USED IN MAKING INFERENCES

We stated in Chapter 1 that a scientist uses inferential statistics to make statements about a population based on information contained in a sample. Because populations are sets of measurements, we need a way to state an inference about them. Graphical and numerical descriptive techniques were presented in Chapter 3. Most management decisions must be made in the presence of uncertainty. Price and models for new automobiles must be selected on the basis of shaky forecasts of consumer preference, national economic trends, and competitive actions. The size and allocation of a hospital staff must be decided with limited information on patient load. The inventory of a product must be set in the face of uncertainty about demand. Probability is the language of uncertainty. Now let us examine probability, the mechanism for making inferences. This idea is probably best illustrated by means of an example.

Martha Jones, a candidate for Congress, publicly announces that her forthcoming election is a guaranteed success, and she forecasts victory by a substantial margin in all precincts of her district. Somewhat dubious about her claims, a local television station randomly selects 20 names from the voter registration list, calls each voter, and asks the voters for whom they will vote in the upcoming election. Not one of the 20 voters states that he or she will vote for Jones; all favor her opponent. What do you conclude about Jones's claim to victory in the sampled area?

If Jones had been correct in her claim of victory, at least half the voters in the district would have favored her, and somewhat near this same proportion should have been observed in the sample. As it turned out, none of the voters in the sample favored Jones, a result highly contradictory to her claim. Hence, we infer that the proportion of voters in the population (the district) favoring Jones is less than 1/2 and that she will lose the district. We conclude that Jones will lose because the sample yielded results highly contradictory to her claim. By "contradictory" we do not mean that it is impossible to select at random 0 voters who favor Jones out of the 20 sampled assuming Jones's claim of victory is correct. We mean, rather, that such a draw is highly *improbable*. Thus, we measure the degree of contradiction to Jones's claim of victory in terms of the probability of the observed sample.

To get a better view of the role that probability plays in making this inference, suppose that the sample produced 9 in favor of Jones and 11 in favor of her opponent. Would we consider this result highly improbable and reject Jones's claim? How about 7 in favor and 13 against, or 5 in favor and 15 against? Where do we draw the line? At what point do we decide that the result of the observed sample is so improbable, assuming Jones's claim is correct, that we disagree with her claim? To answer this question we must know how to find the probability of obtaining a particular sample outcome. Knowing this probability we can determine whether we agree or disagree with Jones's claim. Probability is the tool that enables us to make an inference.

Since probability is the tool for making inferences, we might ask: What is probability? In the preceding discussion, we used the term *probability* in its everyday sense. Let us examine this idea more closely.

Observations of phenomena can result in many different outcomes, some of which are more likely than others. Numerous attempts have been made to give a precise definition for the probability of an outcome. We will cite a few of these.

classical interpretation

The first interpretation of probability, called the **classical interpretation of probability** arose from games of chance. Typical probability statements of this type are, for example, "the probability that a flip of a balanced coin will show 'heads' is 1/2" and "the probability of drawing an ace when a single card is drawn from a standard deck of 52 cards is 4/52." The numerical values for these probabilities arise from the nature of the games. A coin flip has two possible outcomes (a head or a tail); the probability of a head should then be 1/2 (1 out of 2). Similarly, there are 4 aces in a standard deck of 52 cards, so the probability of drawing an ace in a single draw is 4/52 or 4 out of 52.

outcome
event

In the classical interpretation of probability, each possible distinct result is called an **outcome**; an **event** is identified as a collection of outcomes. The probability of an event E under the classical interpretation of probability is computed by taking the ratio of the number of outcomes N_e favorable to event

E to the total number N of possible outcomes:

$$P(\text{event } E) = \frac{N_e}{N}.$$

The applicability of this interpretation depends on the assumption that all outcomes are equally likely. If this assumption does not hold, the probabilities indicated by the classical interpretation of probability will be in error.

relative frequency interpretation

A second interpretation of probability is called the **relative frequency concept of probability**; this is an empirical approach to probability. If an experiment is repeated a large number of times and event E occurs 30% of the time, then .30 should be a very good approximation to the probability of event E. Symbolically, if an experiment is conducted n different times and if event E occurs on n_e of these trials, then the probability of event E is approximately

$$P(\text{event } E) \approx \frac{n_e}{n}.$$

We say "approximate" because we think of the actual probability $P(\text{event } E)$ as the relative frequency of the occurrence of event E over a very large number of observations or repetitions of the phenomenon. The fact that we can check probabilities that have a relative frequency interpretation (by simulating many repetitions of the experiment) makes this interpretation very appealing and practical.

The third interpretation of probability can be used for problems in which it is difficult to imagine a repetition of an experiment. These are "one-shot" situations. For example, the director of a state welfare agency who estimates the probability that a proposed revision in eligibility rules will be passed by the state legislature would not be thinking in terms of a long series of trials. Rather, the

subjective interpretation

director would use a **personal** or **subjective probability** to make a one-shot statement of belief regarding the likelihood of passage of the proposed legislative revision. The problem with subjective probabilities is that they can vary from person to person and they cannot be checked.

Of the three interpretations presented, the relative frequency concept seems to be the most reasonable one since it provides a practical interpretation of the probability for most events of interest. Even though we will never run the necessary repetitions of the experiment to determine the exact probability of an event, the fact that we could check the probability of an event gives meaning to the relative frequency concept. Throughout the remainder of this text we will lean heavily on this interpretation of probability.

▼ EXERCISES

Applications

4.1 Indicate which interpretation of the probability statement seems most appropriate.

a. The National Angus Association has stated that there is a 60/40 chance that wholesale beef prices will rise by the summer, that is, a .60 probability of an increase and a .40 probability of a decrease.

b. The quality control section of a large chemical manufacturing company has undertaken an intensive process-validation study. From this study, the QC section claims that the probability that the shelf life of a newly released batch of chemical will exceed the minimal time specified is .998.

c. A new blend of coffee is being contemplated for release by the marketing division of a large corporation. Preliminary marketing survey results indicate that 550 of a random sample of 1,000 potential users rated this new blend better than a brand name competitor. The probability of this happening is approximately .001 assuming that there is actually no difference in consumer preference for the two brands.

d. The probability of receiving a busy signal when attempting to access the company WATS line during the 3:00–5:00 P.M. time period is .58.

e. The probability that it will rain tomorrow is .30.

f. Within a city the probability of selecting a household at random in which the head of the household is unemployed is .12.

4.2 Give your own personal probability for each of the following situations. It would be instructive to tabulate these probabilities for the entire class. In which cases did you have large disagreements?

a. The federal budget will be balanced in the next fiscal year.

b. You will receive a "B" or higher in this course.

c. Two or more individuals in the classroom have the same birthday.

d. The New York Giants will win the Super Bowl next year.

e. The total production of Florida oranges next year will exceed this year's production.

4.2 ▼ FINDING THE PROBABILITY OF AN EVENT

In the preceding section, we discussed three different interpretations of probability. In this section, we will use the classical interpretation and the relative frequency concept to illustrate the computation of the probability of an outcome or event. Consider an experiment that consists of tossing two coins, a penny and a dime, and observing the upturned faces. There are four possible outcomes:

TT: tails for both coins

TH: a tail for the penny,
     a head for the dime

HT: a head for the penny,
     a tail for the dime

HH: heads for both coins.

What is the probability of observing the event exactly one head from the two coins?

This probability can be obtained easily we we can assume that all four outcomes are equally likely. In this case, that seems quite reasonable. There are $N = 4$ possible outcomes and $N_e = 2$ of these are favorable for the event of interest, observing exactly one head. Hence, by the classical interpretation of probability,

$$P(\text{exactly 1 head}) = \frac{2}{4} = \frac{1}{2}.$$

Since the event of interest has a relative frequency interpretation, we could also obtain this same result empirically, using the relative frequency concept. Suppose that a penny and a dime were tossed 2,000 times, with the results shown in Table 4.1. Note that this approach yields approximate probabilities that are in agreement with our intuition. That is, intuitively we might expect these outcomes to be equally likely and each to occur with a probability equal to 1/4, or .25. This assumption was made for the classical interpretation.

T A B L E   4.1
Results of 2,000 Tossings of a Penny and a Dime

Outcome	Frequency	Relative Frequency
TT	474	474/2,000 = .237
TH	502	502/2,000 = .251
HT	496	496/2,000 = .248
HH	528	528/2,000 = .264

If we wish to find the probability of tossing two coins and observing exactly one head, we have, from Table 4.1,

$$P(\text{exactly 1 head}) \approx \frac{502 + 496}{2000} = .499.$$

This is very close to the theoretical probability, which we have shown to be .5.

The probability of an event, say event A, will always satisfy the property

$$0 \le P(A) \le 1.$$

That is, the probability of an event lies anywhere in the interval from 0 (the occurrence of the event is impossible) to 1 (the occurrence of an event is a "sure thing").

4.3 ▼ BASIC EVENT RELATIONS AND PROBABILITY LAWS

either A or B occurs

Suppose that A and B represent two experimental events and that you are interested in a new event, the event that **either A or B occurs.** For example, suppose that we toss a pair of dice and define the following events:

A: A total of 7 shows
B: A total of 11 shows.

Then the event "either A or B occurs" is the event that you toss a total of either 7 or 11 with the pair of dice.

mutually exclusive

Note that, for this example, the events A and B are **mutually exclusive**. That is, if you observe event A (a total of 7), you could not at the same time observe event B (a total of 11). This, if A occurs, B cannot occur (and vice versa).

DEFINITION 4.1
Mutually Exclusive

▼

Two events A and B are said to be **mutually exclusive** if (when the experiment is performed a single time) the occurrence of one of the events excludes the possibility of the occurrence of the other event.

The concept of mutually exclusive events is used to specify a second property that the probabilities of events must satisfy. When two events are mutually exclusive, then the probability that either one of the events will occur is the sum of the event probabilities.

DEFINITION 4.2
Probability
(either A or B)

▼

If two events, A and B, are mutually exclusive, the **probability** that either event occurs is $P(\text{either } A \text{ or } B) = P(A) + P(B)$.

The definition of additivity of probabilities for mutually exclusive events can be extended beyond two events. For example, when we toss a pair of dice, the sum S of the numbers appearing on the dice can assume any one of the values $S = 2, 3, 4, \ldots, 11, 12$. On a single toss of the dice, we can observe only one of these values. Therefore, the values $2, 3, \ldots, 12$ represent mutually exclusive events. If we want to find the probability of tossing a sum less than or equal to 4, this probability is

$$P(S \leq 4) = P(2) + P(3) + P(4).$$

For this particular experiment, the dice can fall in 36 different equally likely ways. We can observe a 1 on die 1 and a 1 on die 2, denoted by the symbol (1, 1). We can observe a 1 on die 1 and a 2 on die 2, denoted by (1, 2). In other words, for this experiment, the possible outcomes are

(1, 1)	(2, 1)	(3, 1)	(4, 1)	(5, 1)	(6, 1)
(1, 2)	(2, 2)	(3, 2)	(4, 2)	(5, 2)	(6, 2)
(1, 3)	(2, 3)	(3, 3)	(4, 3)	(5, 3)	(6, 3)
(1, 4)	(2, 4)	(3, 4)	(4, 4)	(5, 4)	(6, 4)
(1, 5)	(2, 5)	(3, 5)	(4, 5)	(5, 5)	(6, 5)
(1, 6)	(2, 6)	(3, 6)	(4, 6)	(5, 6)	(6, 6)

As you can see, only one of these events, (1, 1), will result in a sum equal to 2. Therefore, we would expect a 2 to occur with a relative frequency of 1/36 in a long series of repetitions of the experiment, and we let $P(2) = 1/36$. The sum $S = 3$ will occur if we observe either of the outcomes (1, 2) or (2, 1). Therefore, $P(3) = 2/36 = 1/18$. Similarly, we find $P(4) = 3/36 = 1/12$. It follows that

$$P(S \leq 4) = P(2) + P(3) + P(4) = \frac{1}{36} + \frac{1}{18} + \frac{1}{12} = \frac{1}{6}.$$

In most practical data-collecting situations, we will make observations on a variable. For example, in recording the diastolic blood pressures of hypertensive patients, the variable of interest is diastolic blood pressure and each patient's diastolic blood pressure represents an observation (measurement) on that variable. The fact that these measurements vary in a seemingly random and unpredictable manner leads us to call the variable a **random variable**.

random variable

Consider a situation in which a single measurement is obtained on a random variable. Since we will observe only one of many possible values of the random variable (for example, only one diastolic blood pressure), it follows that the values of a random variable represent mutually exclusive events (if one value occurs, the others cannot have occurred). We will make particular use of the additive property of the probabilities of mutually exclusive events when we wish to find the probability that a random variable assumes one of two or more values when it is observed in an experiment.

complement

A third property of event probabilities concerns an event and its **complement**.

DEFINITION 4.3
Complement

> ▼
>
> The **complement** of an event A is the event that A *does not* occur. The complement of A is denoted by the symbol $\overline{A}$.

Thus, if we define the complement of an event A as a new event, namely, "A does not occur," it follows that

$$P(A) + P(\bar{A}) = 1.$$

For example, refer again to the two-coin-toss experiment. If, in many repetitions of the experiment, the proportion of times you observe event A, "two heads show," is 1/4, then it follows that the proportion of times you observe the event $\bar{A}$, "two heads do not show," is 3/4. Thus, $P(A)$ and $P(\bar{A})$ will always sum to 1.

The three properties that the probabilities of events must satisfy can be summarized as follows:

Properties of Probabilities

▼

If A and B are any two mutually exclusive events associated with an experiment, then $P(A)$ and $P(B)$ must satisfy the following properties:

1. $0 \leq P(A) \leq 1$ and $0 \leq P(B) \leq 1$
2. $P(\text{either } A \text{ or } B) = P(A) + P(B)$
3. $P(A) + P(\bar{A}) = 1$ and $P(B) + P(\bar{B}) = 1$.

union
intersection

We can now define two additional event relations: the **union** and the **intersection** of two events.

DEFINITION 4.4
Union

▼

The **union** of two events A and B is the set of all outcomes that are included in either A or B (or both). The union is denoted as $A \cup B$.

DEFINITION 4.5
Intersection

▼

The **intersection** of two events A and B is the set of all outcomes that are included in both A and B. The intersection is denoted as $A \cap B$.

These definitions along with the definition of the complement of an event formalize some simple concepts. The event $\bar{A}$ occurs when A *does not*; $A \cup B$ occurs when either A or B occurs; $A \cap B$ occurs when A *and* B occur.

The additivity of probabilities for mutually exclusive events, called the *addition law for mutually exclusive events,* can be extended to give the general addition law.

DEFINITION 4.6
Probability of the Union

▼

Consider two events A and B; the **probability of the union** of A and B is

$$P(A \cup B) = P(A) + P(B) - P(A \cap B).$$

EXAMPLE 4.1

▼

Events and event probabilities are shown in the following Venn diagram.

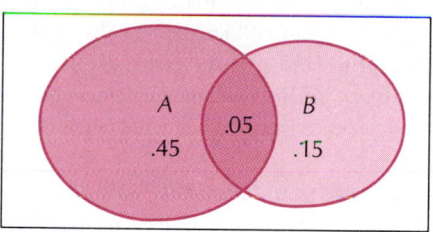

Use this diagram to determine the probabilities listed.
a. $P(A)$, $P(\bar{A})$
b. $P(B)$, $P(\bar{B})$
c. $P(A \cap B)$
d. $P(A \cup B)$

Solution   From the Venn diagram, we are able to determine the following probabilities:
a. $P(A) = .5$, therefore $P(\bar{A}) = 1 - .5 = .5$
b. $P(B) = .2$, therefore $P(\bar{B}) = 1 - .2 = .8$
c. $P(A \cap B) = .05$
d. $P(A \cup B) = P(A) + P(B) - P(A \cap B) = .5 + .2 - .05 = .65.$ ▲

4.4 ▼ CONDITIONAL PROBABILITY AND INDEPENDENCE

Consider the following situation: the examination of a large number of insurance claims, categorized according to type of insurance and whether the claim was fraudulent, produced the results shown in Table 4.2. Suppose you are responsible for checking insurance claims—in particular, for detecting fraudulent claims—and

TABLE 4.2
**Categorization of
Insurance Claims**

	Type of Policy			
Category	Fire	Auto	Other	**Total**
Fraudulent	6%	1%	3%	10%
Nonfraudulent	14%	29%	47%	90%
Total	20%	30%	50%	100%

you examine the next claim that is processed. What is the probability of the event F, "the claim is fraudulent"? To answer the question, you examine Table 4.2 and note that 10% of all claims are fraudulent. Thus, assuming that the percentages given in the table are reasonable approximations to the true probabilities of receiving specific types of claims, it follows that $P(F) = .10$. Would you say that the risk that you face a fraudulent claim has probability .10? We think not, because you have additional information that may affect the assessment of $P(F)$. For example, you would know the type of policy you were examining (fire, auto, or other).

Suppose that you have the additional information that the claim was associated with a fire policy. Checking Table 4.2, we see that 20% (or .20) of all claims are associated with a fire policy and that 6% (or .06) of all claims are fraudulent fire policy claims. Therefore, it follows that the probability that the claim is fraudulent, given that you know the policy is a fire policy, is

$$P(F|\text{fire policy}) = \frac{\text{proportion of claims that are fraudulent fire policy claims}}{\text{proportion of claims that are against fire policies}}$$
$$= \frac{.06}{.20} = .30.$$

conditional probability

This probability, $P(F|\text{fire policy})$, is called a **conditional probability** of the event F, that is, the probability of event F given the fact that the event "fire policy" has already occurred. This tells you that 30% of all fire policy claims are fraudulent. The vertical bar in the expression $P(F|\text{fire policy})$ represents the phrase "given that," or simply "given." Thus, the expression is read, "the probability of the event F given the event fire policy."

unconditional probability

The probability $P(F) = .10$, called the **unconditional**, or **marginal, probability** of the event F, gives the proportion of times a claim is fraudulent; that is, the proportion of times event F occurs in a very large (infinitely large) number of repetitions of the experiment (receiving an insurance claim and determining whether the claim is fraudulent). In contrast, the conditional probability of F, given that the claim is for a fire policy, $P(F|\text{fire policy})$, gives the proportion of fire policy claims that are fraudulent. Clearly, the conditional probabilities of F, given the types of policies, will be of much greater assistance in measuring the risk of fraud than the unconditional probability of F.

DEFINITION 4.7
Conditional Probability

Consider two events A and B with nonzero probabilities, $P(A)$ and $P(B)$. The **conditional probability** of event A given event B is

$$P(A|B) = \frac{P(A \cap B)}{P(B)}.$$

The conditional probability of event B given event A is

$$P(B|A) = \frac{P(A \cap B)}{P(A)}.$$

This definition for conditional probabilities gives rise to what is referred to as the *multiplication law*.

DEFINITION 4.8
Probability of the Intersection

The **probability of the intersection** of two events A and B is

$$P(A \cap B) = P(A)P(B|A)$$
$$= P(B)P(A|B).$$

The only difference between Definitions 4.7 and 4.8, both of which involve conditional probabilities, relates to what probabilities are known and what needs to be calculated. When the intersection probability $P(A \cap B)$ and the individual probability $P(A)$ are known, we can compute $P(B|A)$. When we know $P(A)$ and $P(B|A)$, we can compute $P(A \cap B)$.

EXAMPLE 4.2

Two supervisors are to be selected as safety representatives within the company. Given that there are six supervisors in research and five in development, and each group of two supervisors has the same chance of being selected, find the probability of choosing both supervisors from research.

Solution   Let A be the event that the first supervisor selected is from research and let B be the event that the second supervisor is also from research. Clearly, we want $P(A \cap B) = P(A)P(B|A)$.

For this example,

$$P(A) = \frac{6}{11} \quad \text{and} \quad P(B|A) = \frac{5}{10}.$$

Then

$$P(A \cap B) = \left(\frac{6}{11}\right)\left(\frac{5}{10}\right) = \frac{30}{110} = .27.$$

▲

Suppose that the probability of event A is the same regardless of whether event B has or has not occurred. That is, suppose

$$P(A|B) = P(A).$$

independent events

Then we say that the occurrence of event A is not dependent on the occurrence of event B or, simply, that A and B are **independent events**.

DEFINITION 4.9
Independent Events

▼

Two events A and B are **independent events** if

$$P(A|B) = P(A) \quad \text{or if} \quad P(B|A) = P(B).$$

(Note: You can show that if $P(A|B) = P(A)$, then $P(B|A) = P(B)$, and vice versa.)

The concept of independence is of particular importance in sampling. Subsequently, we will draw samples from two (or more) populations in order to compare the population means, variances, or some other population parameters. For most of these applications, we will select samples in such a way that the observed values in one sample are independent of the values that appear in another **independent samples** sample. We call these **independent samples**.

▼ EXERCISES

Basic Techniques

4.3   A coin is to be flipped three times. List the possible outcomes in the form (result on toss 1, result on toss 2, result on toss 3).

4.4 In Exercise 4.3, assume that each one of the outcomes has probability 1/8 of occurring. Find the probability of
   a. A: observing exactly 1 head
   b. B: observing 1 or more heads
   c. C: observing no heads.

4.5 For Exercise 4.4:
   a. Compute the probability of the complement of event A, event B, and event C.
   b. Determine whether events A and B are mutually exclusive.

4.6 Determine the following conditional probabilities for the events of Exercise 4.4.
   a. $P(A|B)$ **b.** $P(A|C)$ **c.** $P(B|C)$

4.7 Refer to Exercise 4.6. Are events A and B independent? Why or why not? What about A and C? What about B and C?

4.8 A die is to be rolled and we are to observe the number that falls face up. Find the probabilities for these events:
   a. A: observe a 6
   b. B: observe an even number
   c. C: observe a number greater than 2
   d. D: observe an even number and a number greater than 2.

4.9 Refer to Exercise 4.8. Which of the events (A, B, and C) are independent? Which are mutually exclusive?

4.10 Consider the following outcomes for an experiment:

outcome	1	2	3	4	5
probability	.20	.25	.15	.10	.30

Let event A consist of outcomes 1, 3, and 5 and event B consist of outcomes 4 and 5.
   a. Find $P(A)$ and $P(B)$.
   b. Find P(both A and B occur).
   c. Find P(either A or B occurs).

4.11 Refer to Exercise 4.10. Does P(either A or B occurs) $= P(A) + P(B)$? Why or why not?

Applications

4.12 A student has to have an accounting course and an economics course the next term. Assuming there are no schedule conflicts, describe the possible outcomes for selecting one section of the accounting course and one of the economics course if there are four possible accounting sections and three possible economics sections.

4.13 The emergency room of a hospital has two backup generators, either of which can supply enough electricity for basic hospital operations. We define events A and B as follows:

   event A: generator 1 works properly
   event B: generator 2 works properly

Describe the following events in words:
   a. complement of A **b.** $B|A$ **c.** either A or B.

4.14   A survey of a number of large corporations gave the following probability table for events related to the offering of a promotion involving a transfer.

Promotion/ Transfer	Married		Unmarried	Total
	Two-Career Marriage	One-Career Marriage		
Rejected	.184	.0555	.0170	.2565
Accepted	.276	.3145	.1530	.7435
Total	.46	.37	.17	

Use the probabilities to answer the following questions:
a. What is the probability that a professional (selected at random) would accept the promotion? Reject it?
b. What is the probability that a professional (selected at random) is part of a two-career marriage? A one-career marriage?

 4.15   An institutional investor is considering a large investment in two of five companies. Suppose that, unknown to the investor, two of the five firms are on shaky grounds with regard to the development of new products.
a. List the possible outcomes for this situation.
b. Determine the probability of choosing two of the three firms that are on better grounds.
c. What is the probability of choosing one of two firms on shaky grounds?
d. What is the probability of choosing the two shakiest firms?

4.16   A survey of workers in two manufacturing sites of a firm included the following question: How effective is management in responding to legitimate grievances of workers? The results are shown here.

	Number Surveyed	Number Responding "Poor"
Site 1	192	48
Site 2	248	80

Let A be the event the worker comes from Site 1 and B be the event the response is "poor." Compute $P(A)$, $P(B)$, and $P(A \cap B)$.

4.17   Refer to Exercise 4.16.
a. Are events A and B independent?
b. Find $P(B|A)$ and $P(B|\bar{A})$. Are they equal?

4.18   A large corporation has spent considerable time developing employee performance rating scales to evaluate an employee's job performance on a regular basis, so major adjustments can be made when needed and employees who should be considered for a "fast track" can be isolated. Keys to this latter determination are ratings on the ability of an employee to perform to his or her capabilities and on his or her formal training for the job.

Workload Capacity	Formal Training			
	None	Little	Some	Extensive
Low	.01	.02	.02	.04
Medium	.05	.06	.07	.10
High	.10	.15	.16	.22

The probabilities for being placed on a fast track are as indicated for the 12 categories of workload capacity and formal training. The following three events (A, B, and C) are defined:

A: an employee works at the high-capacity level
B: an employee falls into the highest (extensive) formal training category
C: an employee has little or no formal training and works below high capacity.

a. Find $P(A)$, $P(B)$, and $P(C)$.
b. Find $P(A/B)$, $P(A/\bar{B})$, and $P(\bar{B}/C)$.
c. Find $P(A \cup B)$, $P(A \cap C)$, and $P(B \cap C)$.

\$ 4.19 The utility company in a large metropolitan area finds that 70% of its customers pay a given monthly bill in full.
a. Suppose two customers are chosen at random from the list of all customers. What is the probability that both customers will pay their monthly bill in full?
b. What is the probability that at least one of them will pay in full?

4.20 Refer to Exercise 4.19. A more detailed examination of the company records indicates that 95% of the customers who pay one monthly bill in full will pay the next monthly bill in full also; only 10% of those who pay less than the full amount one month will pay in full the next month.
a. Find the probability that a customer selected at random will pay two consecutive months in full.
b. Find the probability that a customer selected at random will pay neither of two consecutive months in full.
c. Find the probability that a customer chosen at random will pay exactly one month in full.

4.5 ▼ BAYES'S FORMULA

In this section, we will show how Bayes's Formula can be used to update conditional probabilities by using sample data when available. These "up-dated" conditional probabilities are useful in decision making. We begin by way of an example. Suppose that .001 (i.e., .1%) of a certain segment of the population has tuberculosis (TB) and that in a reliable screening test, 95% of those with TB will show a positive result and only 2% of those who don't have TB will show a positive result. From this background information, we are given the probabilities shown here for the disease state and test result.

Disease	Test Result
$P(TB) = .001$	$P(\text{positive} \mid TB) = .95$
	$P(\text{negative} \mid TB) = .05$
$P(\text{no } TB) = .999$	$P(\text{positive} \mid \text{no } TB) = .02$
	$P(\text{negative} \mid \text{no } TB) = .98$

Bayes's Formula allows us to answer the following practical question: What is the probability of a person having TB, given a positive test result? A formula and the calculations for this conditional probability are shown here:

$$P(TB \mid \text{positive}) = \frac{P(TB \text{ and positive})}{P(\text{positive})}$$

$$= \frac{P(\text{positive} \mid TB)P(TB)}{P(\text{positive} \mid TB)P(TB) + P(\text{positive} \mid \text{no } TB)P(\text{no } TB)}$$

$$= \frac{(.95)(.001)}{(.95)(.001) + (.02)(.999)} = .045.$$

EXAMPLE 4.3

▼

A book club classifies members as heavy, medium, or light purchasers, and separate mailings are prepared for each of these groups. Overall, 20% of the members are heavy purchasers, 30% medium, and 50% light. A member is not classified into a group until 18 months after joining the club, but a test is made of the feasibility of using the first 3 months' purchases to classify members. The following percentages are obtained from existing records of individuals classified as heavy, medium, or light purchasers.

First 3 Months' Purchases	Group		
	Heavy	Medium	Light
0	5%	15%	60%
1	10%	30%	20%
2	30%	40%	15%
3+	55%	15%	5%

If a member purchases no books in the first 3 months, what is the probability that the member is a light purchaser? (Note: This table contains "conditional" percentages for each column.)

Solution   Using the conditional probabilities in the table, the underlying purchase probabilities, and Bayes's Formula, we can compute this conditional probability.

$P(\text{light}|0)$

$$= \frac{P(0|\text{light})P(\text{light})}{P(0|\text{light})P(\text{light}) + P(0|\text{medium})P(\text{medium}) + P(0|\text{heavy})P(\text{heavy})}$$

$$= \frac{(.60)(.50)}{(.60)(.50) + (.15)(.30) + (.05)(.20)}$$

$$= .845.$$

▲

These examples indicate the basic idea of Bayes's Formula. There is some number k of possible, mutually exclusive, underlying events $A_1, \ldots, A_k$, which are sometimes called the **states of nature.** Unconditional probabilities $P(A_1), \ldots, P(A_k)$, often called **prior probabilities**, are specified. There are m possible, mutually exclusive, **observable events** $B_1, \ldots, B_m$. The conditional probabilities of each observable event given each state of nature, $P(B_i|A_i)$, are also specified and these probabilities are called **likelihoods**. The problem is to find the **posterior probabilities** $P(A_i|B_i)$. Prior and posterior refer to probabilities before and after observing an event B_i.

states of nature

prior probabilities

observable events

likelihoods

posterior probabilities

Bayes's Formula
If $A_1, \ldots, A_k$ are mutually exclusive states of nature and if $B_1, \ldots, B_m$ are m possible mutually exclusive observable events, then

$$P(A_i|B_j) = \frac{P(B_j|A_i)P(A_i)}{P(B_j|A_1)P(A_1) + P(B_j|A_2)P(A_2) + \cdots + P(B_j|A_k)P(A_k)}$$

$$= \frac{P(B_j|A_i)P(A_i)}{\sum_i P(B_j|A_i)P(A_i)}.$$

E X A M P L E   4.4   ▼

a. Specify the states of nature and observable events for Example 4.3.
b. Specify the prior probabilities and likelihoods.
c. Which posterior probability was calculated in Example 4.3?

Solution   a. The states of nature are the three possible groups: $A_1 =$ heavy, $A_2 =$ medium, and $A_3 =$ light. The observable events are the possible 3-month purchases: $B_1 = 0$, $B_2 = 1$, $B_3 = 2$, and $B_4 = 3$ or more.

b. The prior probabilities are $P(A_1) = .20$, $P(A_2) = .30$, and $P(A_3) = .50$. The likelihoods $P(B_i|A_i)$ are shown in the accompanying table.

Purchase B_j	Group A_i		
	Heavy, A_1	Medium, A_2	Light, A_3
0, B_1	.05	.15	.60
1, B_2	.10	.30	.20
2, B_3	.30	.40	.15
3+, B_4	.55	.15	.05

c. The calculated posterior probability was $P(A_3|B_1)$, or equivalently, $P(\text{light}|0)$. The formula given in the solution is exactly Bayes's Formula. ▲

E X A M P L E   4.5   ▼

Of all electronic systems produced in a plant, 80% are nondefective, 15% have defect D_1, and 5% have defect D_2. None of the systems has both defects. Defects can be detected with certainty only by destructive testing, but a fairly reliable nondestructive test has been devised. The test has four possible outcomes. The respective likelihoods are

Test Outcome B_j	Defect A_i		
	None, A_1	D_1, A_2	D_2, A_3
1, B_1	.90	.06	.02
2, B_2	.05	.40	.06
3, B_3	.03	.45	.52
4, B_4	.02	.09	.40

If a particular system yields test result 3, what are the posterior probabilities of no defect, defect D_1, and defect D_2?

Solution   These three posterior probabilities can be computed using Bayes's Formula. For example, the probability of no defect given test result 3 is

$$P(\text{none}|3) = P(A_1|B_3)$$

$$= \frac{P(B_3|A_1)P(A_1)}{P(B_3|A_1)P(A_1) + P(B_3|A_2)P(A_2) + P(B_3|A_3)P(A_3)}$$

$$= \frac{.03(.80)}{.03(.80) + .45(.15) + .52(.05)} = .204.$$

Similarly,

$$P(D_1|3) = P(A_2|B_3) = \frac{P(B_3|A_2)P(A_2)}{P(B_3|A_1)P(A_1) + \cdots + P(B_3|A_3)P(A_3)}$$

$$= \frac{.45(.15)}{.03(.80) + .45(.15) + .52(.05)} = .574$$

and

$$P(D_2|3) = P(A_3|B_3) = \frac{P(B_3|A_3)P(A_3)}{P(B_3|A_1)P(A_1) + \cdots + P(B_3|A_3)P(A_3)}$$

$$= \frac{.52(.05)}{.03(.80) + .45(.15) + .52(.05)} = .221.$$ ▲

▼ EXERCISES

Applications

4.21 One percent of a finance company's loans are defaulted (not completely repaid). The company routinely runs credit checks on all loan applicants. It finds that 30% of defaulted loans went to poor risks, 40% to fair risks, and 30% to good risks. Of the nondefaulted loans, 10% went to poor risks, 40% to fair risks, and 50% to good risks. Use Bayes's Formula to calculate the probability that a poor-risk loan is defaulted.

4.22 Refer to Exercise 4.21. Show that the posterior probability of default, given a fair risk, equals the prior probability of default. Explain why this is a reasonable result.

4.23 A manufacturing firm has three machine operators who produce a certain component. Operator A has a 5% defective rate, B has a 3% defective rate, and C has a 2% defective rate. The three operators produce equal numbers of components. Suppose a randomly selected component is found to be defective. Calculate the posterior probability that the part was produced by A. Compare the result to the prior probability of 1/3.

4.24 Refer to Exercise 4.23. Suppose a sample of 20 components is taken from a lot produced by one operator. If no defective components are found in the sample, find the posterior probability that the lot was produced by operator A. Assume the probability of no defectives in a sample of 20 is .35, .40, and .50 for A, B, and C, respectively.

4.25 An underwriter of home insurance policies studies the problem of home fires resulting from wood-burning furnaces. Of all homes having such furnaces, 30% own a type 1 furnace, 25% a type 2 furnace, 15% a type 3, and 30% other types. Five percent of type 1 furnaces, 3% of type 2, 2% of type 3, and 4% of other types have resulted in fires over three years of operation. If a fire occurs in a particular home, what is the probability that a type 1 furnace is in the home?

4.26 A reviewer of textbooks has a curious "track record." An editor estimates the rating percentages for highly successful, moderately successful, and unsuccessful books as shown in the table below. About 10% of all books are highly successful, 50% are moderately successful, and 40% are unsuccessful. If this reviewer rates a book as good, calculate the posterior probability that the book is unsuccessful. Compare the result to the prior probability, .40.

| | Reviewer's Rating |
Book	Good	Fair	Poor
Highly successful	5%	20%	75%
Moderately successful	15%	40%	45%
Unsuccessful	50%	30%	20%

 4.27 Conditional probabilities can be useful in diagnosing disease. Suppose that three different, closely related diseases (A_1, A_2, and A_3) occur in 25%, 15%, and 12% of the population. In addition, suppose that any one of three mutually exclusive symptom states B_1, B_2, and B_3 may be associated with each of these diseases. Experience shows that the likelihood $P(B_j|A_i)$ of having a given symptom state when the disease is present is as shown in the table below. Find the probability of disease A_2 given symptom B_1, B_2, B_3, and B_4, respectively.

| Symptom State B_j | Disease State A_i | | |
	A_1	A_2	A_3
B_1	.08	.17	.10
B_2	.18	.12	.14
B_3	.06	.07	.08
B_4 (no symptoms)	.68	.64	.68

4.6 ▼ VARIABLES: DISCRETE AND CONTINUOUS

The basic language of probability developed in this chapter deals with many different kinds of events. We are interested in calculating the probabilities associated with both quantitative and qualitative events. For example, we developed techniques that could be used to calculate the probability that a person selected at random for a Nielsen survey of television viewing habits would favor the ABC nightly news program (as opposed to that of CBS or NBC). These same techniques are also applicable to finding the probability that a person selected for the Nielsen survey watches television more than 30 hours per week.

These qualitative and quantitative events can be classified as events (or outcomes) associated with qualitative and quantitative variables. For example, in the Nielsen survey, responses to the question "Which evening television news program do you prefer: ABC, CBS, or NBC?" are observations on a qualitative variable, since the possible responses vary in kind but not in any numerical degree. Because we cannot predict with certainty what a particular person's response will be, the variable is classified as a **qualitative random variable.** Other examples of qualitative random variables that are commonly measured are political party affiliation, socioeconomic status, geographic location, and gender/race classification.

qualitative random variable

There are a finite (and typically quite small) number of possible outcomes associated with any qualitative variable. Using the methods of this chapter, it is possible to calculate the probabilities associated with these events.

quantitative random variable

Many times the events of interest in an experiment are quantitative outcomes associated with a **quantitative random variable**, since the possible responses vary in numerical magnitude. For example, in a Nielsen survey, responses to the question "How many hours a week do you watch television?" are observations on a quantitative random variable. Events of interest, such as viewing television more than 30 hours per week, are measured by this quantitative random variable. Other examples of quantitative random variables are the change in earnings per share of a stock over the next year, the increase in total sales over the next year, and the number of persons voting for the incumbent in an upcoming election. Again, the methods of this chapter can be applied to calculate the probability associated with any particular event.

There are major advantages to dealing with quantitative random variables. The numerical yardstick underlying a quantitative variable makes the mean and standard deviation (for instance) sensible. With qualitative random variables, there isn't much more to be said than has already been said. The methods of this chapter can be used to calculate the probabilities of various events, and that's about all. With quantitative random variables we can do much more: we can average the resulting quantities, find standard deviations, and assess probable errors, among other things. Hereafter, we use the term **random variable** to mean quantitative random variable.

random variable

Most events of interest result in numerical observations or measurements. If a quantitative variable measured (or observed) in an experiment is denoted by the symbol y, we are interested in the values that y can assume. These values are called numerical outcomes. The number of students in a class of 50 who earn an "A" in their biology course is a numerical outcome. The percentage of registered voters who cast ballots in a given election is also a numerical outcome. The quantitative variable y is called a random variable because the value that y assumes in a given experiment is a chance or random outcome.

DEFINITION 4.10
Discrete Random Variable

▼

When observations on a quantitative random variable can assume only a countable number of values, the variable is called a **discrete random variable**.

Examples of discrete variables are these:

1. number of bushels of apples per acre for a given orchard this year
2. number of accidents per year at an intersection
3. number of voters in a sample favoring Jones.

Note that it is possible to count the number of values that each of these random variables can assume.

DEFINITION 4.11
Continuous Random Variable

When observations on a quantitative random variable can assume any one of the uncountable number of values in a line interval, the variable is called a **continuous random variable**.

For example, the daily maximum temperature in Rochester, New York, can assume any of the infinitely many values on a line interval. It could be 89.6, 89.799, or 89.7611114. Typical continuous random variables are temperature, pressure, height, weight, and distance.

discrete random variable
continuous random variable

The distinction between **discrete** and **continuous random** variables is pertinent when we are seeking the probabilities associated with specific values of a random variable. The need for the distinction will be apparent when probability distributions are discussed in later sections of this chapter.

4.7 ▼ PROBABILITY DISTRIBUTIONS FOR DISCRETE RANDOM VARIABLES

As previously stated, we need to know the probability of observing a particular sample outcome in order to make an inference about the population from which the sample was drawn. To do this, we need to know the probability associated with each value of the variable y. Viewed as relative frequencies, these probabilities generate a distribution of theoretical relative frequencies called the **probability distribution** of y. Probability distributions differ for discrete and continuous variables but the interpretation is essentially the same for both.

probability distribution

The *probability distribution for a discrete random variable* displays the probability $P(y)$ associated with each value of y. This display can be presented as a table, a graph, or a formula. To illustrate, consider the tossing of two coins in Section 4.2 and let y be the number of heads observed. Then y can take the values 0, 1, or 2. From the data of Table 4.1, we can determine the approximate probability for each value of y, as given in Table 4.3. We point out that the relative frequencies

T A B L E 4.3

Empirical Sampling Results for *y*: the Number of Heads in 2,000 Tosses of Two Coins

y	Frequency	Relative Frequency
0	474	.237
1	998	.499
2	528	.264

in the table are very close to the theoretical relative frequencies (probabilities), which can be shown to be .25, .50, and .25 using the classical interpretation of probability. If we had employed 2,000,000 tosses of the coins instead of 2,000, the relative frequencies for $y = 0$, 1, and 2 would be indistinguishable from the theoretical probabilities.

The probability distribution for *y*, the number of heads in the toss of two coins, is shown in Table 4.4. It is presented graphically as a *probability histogram* in Figure 4.1.

T A B L E 4.4

Probability Distribution for the Number of Heads When Two Coins Are Tossed

y	P(y)
0	.25
1	.50
2	.25

F I G U R E 4.1

Probability Distribution for the Number of Heads When Two Coins Are Tossed

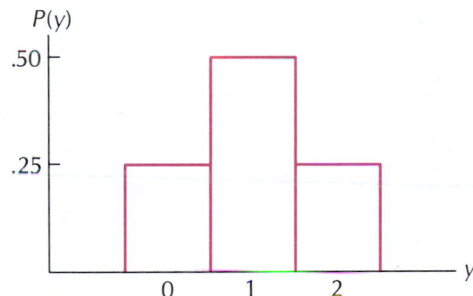

The probability distribution for this simple discrete random variable illustrates three important properties of discrete random variables.

Properties of Discrete Random Variables

1. The probability associated with every value of *y* lies between 0 and 1.
2. The sum of the probabilities for all values of *y* is equal to 1.
3. The probabilities for a discrete random variable are additive. Hence, the probability that $y = 1$ or 2 is equal to $P(1) + P(2)$.

The relevance of the probability distribution to statistical inference will be emphasized when we discuss the probability distribution for the binomial random variable.

4.8 ▼ A USEFUL DISCRETE RANDOM VARIABLE: THE BINOMIAL

Many populations of interest to business persons and scientists can be viewed as large sets of 0s and 1s. For example, consider the set of responses of all adults in the United States to the question, "Do you favor the development of nuclear energy?" If we disallow "no opinion," the responses will constitute a set of "yes" responses and "no" responses. If we assign a 1 to each yes and a 0 to each no, the population will consist of a set of 0s and 1s, and the sum of the 1s will equal the total number of persons favoring the development. The sum of the 1s divided by the number of adults in the United States will equal the proportion of people who favor the development.

Gallup and Harris polls are examples of the sampling of 0, 1 populations. People are surveyed, and their opinions are recorded. Based on the sample responses, Gallup and Harris estimate the proportions of people in the population who favor some particular issue or possess some particular characteristic.

Similar surveys are conducted in the biological sciences, engineering, and business, but they may be called experiments rather than polls. For example, experiments are conducted to determine the effect of new drugs on small animals, such as rats or mice, before progressing to larger animals and, eventually, to human subjects. Many of these experiments bear a marked resemblance to a poll in that the experimenter records only whether the drug was effective. Thus, if 300 rats are injected with a drug and 230 show a favorable response, the experimenter has conducted a "poll"—a poll of rat reaction to the drug, 230 "in favor" and 70 "opposed."

Similar "polls" are conducted by most manufacturers to determine the fraction of a product that is of good quality. Samples of industrial products are collected before shipment and each item in the sample is judged "defective" or "acceptable" according to criteria established by the company's quality control department. Based on the number of defectives in the sample, the company can decide whether the product is suitable for shipment. Note that this example, as well as those preceding, has the practical objective of making an inference about a population based on information contained in a sample.

The public opinion poll, the consumer preference poll, the drug-testing experiment, and the industrial sampling for defectives are all examples of a common, frequently conducted sampling situation known as a *binomial experiment*. The binomial experiment is conducted in all areas of science and business and only differs from one situation to another in the nature of objects being sampled (people, rats, electric light bulbs, oranges). Thus, it is useful to define its characteristics. We can then apply our knowledge of this one kind of experiment to a variety of sampling experiments.

For all practical purposes the binomial experiment is identical to the coin-tossing example of previous sections. Here, n different coins are tossed (or a single coin is tossed n times), and we are interested in the number of heads observed. We assume that the probability of tossing a head on a single trial is π (π may equal .50, as it would for a balanced coin, but in many practical situations π will take some other value between 0 and 1). We also assume that the outcome for any one toss is unaffected by the results of any preceding tosses. These characteristics can be summarized as shown here.

DEFINITION 4.12
Binomial Experiment

A **binomial experiment** is one that has the following properties:

1. The experiment consists of n identical trials.
2. Each trial results in one of two outcomes. We will label one outcome a success and the other a failure.
3. The probability of success on a single trial is equal to π and π remains the same from trial to trial.*
4. The trials are independent; that is, the outcome of one trial does not influence the outcome of any other trial.
5. The random variable y is the number of successes observed during the n trials.

EXAMPLE 4.6

A survey of 500 farmers is conducted to determine the proportion in favor of additional price supports for dairy products. Does this survey satisfy the properties of a binomial experiment?

Solution   To answer this question we check each of the five characteristics of the binomial experiment to determine if they are satisfied.

1. Are there n identical trials? Yes. There are $n = 500$ interviews, all the same.

---

* Some textbooks and computer programs use the letter p rather than π. We have chosen π to avoid confusion with p-values, discussed in Chapter 5.

2. Does each trial result in one of two outcomes? Yes. Each farmer interviewed either favors or does not favor the additional price supports.

3. Is the probability of success the same from trial to trial? Yes. if we let "success" denote a farmer favoring the additional supports, then assuming the list of farmers from which the sample was drawn is large, the probability of success will (for all practical purposes) remain constant from trial to trial.

4. Are the trials independent? Yes. The outcome of one interview is unaffected by the results of the other interviews.

5. Is the random variable of interest to the experimenter the number of successes y in the sample? Yes. We are interested in the number of farmers in the sample of 500 favoring additional price supports for dairy products.

Since all five characteristics are satisfied, the survey represents a binomial experiment.   ▲

EXAMPLE 4.7   ▼

An economist interviews 75 students in a class of 100 to estimate the proportion of students who expect to obtain a "C" or better in the course. Is this a binomial experiment?

Solution   Check this experiment against the five characteristics of a binomial.

1. Are there identical trials? Yes. Each of 75 students is interviewed.
2. Does each trial result in one of two outcomes? Yes. Each student either does or does not expect to obtain a grade of "C" or higher.
3. Is the probability of success the same from trial to trial? No. If we let success denote a student expecting to obtain a "C" or higher, then the probability of success can change considerably from trial to trial. For example, unknown to the professor, suppose that 75 of the 100 students expect to obtain a grade of "C" or higher. Then π, the probability of success for the first student interviewed is $75/100 = .75$. If the student is a failure (does not expect a "C" or higher), the probability of success for the next student is $75/99 = .76$. Suppose that after 70 students have been interviewed, 60 were successes and 10 were failures. Then the probability of success for the next (71st) student is $15/30 = .50$.

This example shows how the probability of success can change substantially from trial to trial in situations where the sample size is a relatively large portion of the total population size. This experiment does not satisfy the properties of a binomial experiment. ▲

It should be noted that very few real–life situations satisfy perfectly the requirements stated in Definition 4.12, but for many the lack of agreement is so small that the binomial experiment still provides a very good model for reality.

Having defined the binomial experiment and suggested several practical applications, we now examine the probability distribution for the binomial random variable y, the number of successes observed in n trials. Although it would be possible to approximate $P(y)$, the probability associated with a value of y in a binomial experiment, by using a relative frequency approach, it is easier to make use of a general formula for binomial probabilities.

Formula for Computing
***P(y)* in a Binomial**
Experiment

The probability of observing y successes in n trials of a binomial experiment is

$$P(y) = \frac{n!}{y!(n-y)!}\, \pi^y (1 - \pi)^{n-y},$$

where

n = number of trials
π = probability of success on a single trial
$1 - \pi$ = probability of failure on a single trial
y = number of successes in n trials
$n! = n(n-1)(n-2)\cdots(3)(2)(1)$.

As indicated above, the notation $n!$ (referred to as n factorial) is used for the product

$$n! = n(n-1)(n-2)\cdots(3)(2)(1).$$

For $n = 3$,

$$n! = 3! = (3)(3-1)(3-2) = (3)(2)(1) = 6.$$

Similarly, for $n = 4$,

$$4! = (4)(3)(2)(1) = 24.$$

We also note that $0!$ is defined to be equal to 1.

To see how the formula for binomial probabilities can be used to calculate the probability for a specific value of y, consider the following examples.

EXAMPLE 4.8 ▼

An experiment consists of tossing a coin two times. If the probability of a head is .5, compute the probability distribution for y, the number of heads, using the binomial formula $P(y)$. Compare your results to those given in Table 4.4.

Solution   Using the formula

$$P(y) = \frac{n!}{y!(n-y)!} \pi^y (1-\pi)^{n-y}$$

and substituting for $n = 2$, $\pi = .5$, $y = 0, 1, 2$, we obtain

$$P(y = 0) = \frac{2!}{0!2!} (.5)^0 (.5)^2 = .25$$

$$P(y = 1) = \frac{2!}{1!1!} (.5)(.5) = .50$$

$$P(y = 2) = \frac{2!}{2!0!} (.5)^2 (.5)^0 = .25.$$

Note that these results are identical to those presented in Table 4.4.   ▲

EXAMPLE 4.9 ▼

A survey is conducted to determine the proportion of adults in a certain locale who favor raising the legal drinking age from 19 to 21. A random sample of 300 adults is selected from the list of registered voters. They are interviewed and the number of those favoring the change is recorded. Is this a binomial experiment?

Solution   To answer this question, we will check each of the five characteristics of a binomial experiment to determine if they are satisfied.

   1.   Are there n identical trials? Yes ($n = 300$ interviews are conducted in an identical manner).

2. Does each trial result in one of two outcomes? Yes, each adult interviewed either favors or does not favor the change.
3. Is the probability of success the same from trial to trial? Yes, if we let success denote a person favoring the change, then, assuming the list of registered voters is large, the probability of selecting an adult favoring the change will remain (for all practical purposes) constant from trial to trial.
4. Are the trials independent? Yes, the outcome of one interview is unaffected by the results of the other interviews.
5. For this experiment, the random variable of interest to the experimenter is y, the number of successes in the sample.

Since all five characteristics are present, the survey represents a binomial experiment.   ▲

E X A M P L E   4.10   ▼

Suppose that a sample of households is randomly selected from all the households in the city in order to estimate the percentage in which the head of the household is unemployed. To illustrate the computation of a binomial probability, suppose that the unknown percentage is actually 10% and that a sample of $n = 5$ (we are selecting a small sample to make the calculation manageable) is selected from the population. What is the probability that all five heads of the households are employed?

Solution   We must carefully define which outcome we wish to call a success. For this example, we will define a success as being employed. Then the probability of success when one person is selected from the population is $\pi = .9$ (because the proportion unemployed is .1). We wish to find the probability that $y = 5$ (all five are employed) in five trials.

$$P(y = 5) = \frac{5!}{5!(5-5)!}(.9)^5(.1)^0$$

$$= \frac{5!}{5!0!}(.9)^5(.1)^0$$

$$= (.9)^5 = .590$$

The binomial probability distribution for $n = 5$, $\pi = .9$ is shown in Figure 4.2. The probability of observing five employed in a sample of five is shaded in the figure.

FIGURE 4.2
The Binomial Probability Distribution for $n = 5$, $\pi = .9$

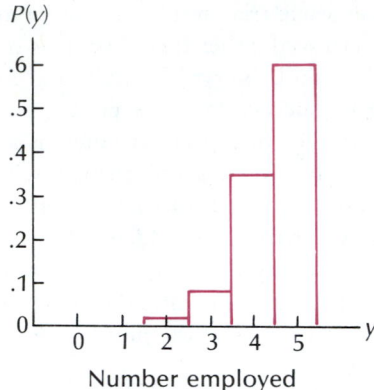

Number employed

▲

EXAMPLE 4.11 ▼

Refer to Example 4.10 and calculate the probability that exactly one person in the sample of five households is unemployed. What is the probability of one or less being unemployed?

Solution Since y is the number of employed in the sample of five, one unemployed person would correspond to four employed ($y = 4$). Then

$$P(4) = \frac{5!}{4!(5-4)!}(.9)^4(.1)^1$$

$$= \frac{(5)(4)(3)(2)(1)}{(4)(3)(2)(1)(1)}(.9)^4(.1)$$

$$= 5(.9)^4(.1)$$

$$= .328.$$

Thus, the probability of selecting four employed heads of households in a sample of five is .328, or, roughly, one chance in three.

The outcome "one or fewer unemployed" is the same as the outcome "4 or 5 employed." Since y represents the number employed, we seek the probability that $y = 4$ or 5. Because the values associated with a random variable represent mutually exclusive events, the probabilities for discrete random variables are additive. Thus, we have

$$P(y = 4 \text{ or } 5) = P(4) + P(5)$$

$$= .328 + .590$$

$$= .918.$$

That is, the probability that a random sample of five households will yield either four or five employed heads of households is .918. This high probability is consistent with our intuition: we could expect the number of employed in the sample to be large if 90% of all heads of households in the city are employed. ▲

Like any relative frequency histogram, a binomial probability distribution possesses a mean μ and a standard deviation σ. Although we omit the derivations, we give the formulas for these parameters.

Mean and Standard Deviation of the Binomial Probability Distribution

▼

Mean and Standard Deviation of the Binomial Probability Distribution

$$\mu = n\pi \quad \text{and} \quad \sigma = \sqrt{n\pi(1 - \pi)},$$

where π is the probability of success in a given trial, and n is the number of trials in the binomial experiment.

Knowing π and the sample size, n, we can calculate μ and σ to locate the center and describe the variability for a particular binomial probability distribution. Thus, we can quickly determine those values of y that are probable and those that are improbable.

EXAMPLE 4.12

▼

Calculate the mean and standard deviation for a binomial probability distribution with $\pi = .5$ and $n = 20$. The probability distribution for the number of successes is shown in Figure 4.3.

Solution Substituting into the formulas, we obtain

$$\mu = n\pi = 20(.5) = 10$$

$$\sigma = \sqrt{n\pi(1 - \pi)} = \sqrt{(20)(.5)(.5)} = \sqrt{5} = 2.24.$$

Note that $y = 0$ is more than 4σ away from the mean $\mu = 10$. If we apply the Empirical Rule to this mound-shaped distribution, we see it is highly improbable that in 20 trials we would observe such a small value of y if π really is equal to .5.

FIGURE 4.3
**Binomial Probability
Distribution for y When
n = 20 and π = .5**

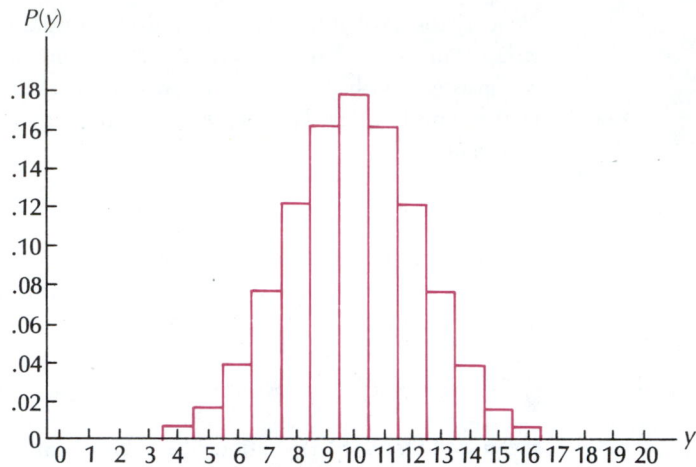

FIGURE 4.3
**Binomial Probability
Distribution for y When
$n = 20$ and $\pi = .5$**

EXAMPLE 4.13

▼

A poll shows that 516 of 1,218 voters favor the reelection of a particular political candidate. Do you think that the candidate will win?

Solution   To win the election, the candidate will need at least 50% of the votes. Let us see whether $y = 516$ is too small a value of y to imply a value of π (the proportion of voters favoring the candidate) equal to .5 or larger. If $\pi = .5$,

$$\mu = n\pi = (1218)(.5) = 609$$

$$\sigma = \sqrt{n\pi(1 - \pi)} = \sqrt{(1218)(.5)(.5)}$$

$$= \sqrt{304.5} = 17.45$$

and $3\sigma = 52.35$.

You can see from Figure 4.4 that $y = 516$ is more than 3σ, or 52.35, away from $\mu = 609$. In fact, if you wish to check, you will see that $y = 516$ is more than 5σ away from $\mu = 609$, the value of μ if π were really equal

FIGURE 4.4
**Location of the Observed
Value of y ($y = 516$)
Relative to μ**

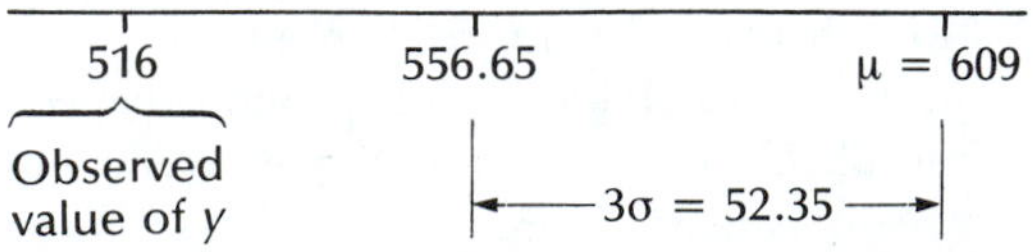

516   556.65   μ = 609

⎨‾‾‾⎬
Observed
value of y |←——— 3σ = 52.35 ———→|

to .5. Thus, it appears that the number of voters in the sample who favor the candidate is much too small if the candidate does, in fact, possess a majority favoring reelection. Consequently, we conclude that he or she will lose. (Note that this conclusion is based on the assumption that the set of voters from which the sample was drawn is the same as the set who will vote. We also must assume that the opinions of the voters will not change between the time of sampling and the date of the election.) ▲

The purpose of this section is to present the binomial probability distribution so that you can see how binomial probabilities are calculated and so that you can calculate them for small values of *n*, if you so desire. In practice, *n* is usually large (in national surveys, sample sizes as large as 1,500 are common), and the computation of the binomial probabilities is very tedious. Later in this chapter, we will present a simple procedure for obtaining approximate values to the probabilities we need in making inferences. We can also use some very rough procedures for evaluating probabilities by using the mean and standard deviation of the binomial random variable *y* along with the Empirical Rule.

The only other discrete random variable that we will discuss in this text is the Poisson (Chapter 8). The interested reader is referred to Hildebrand and Ott (1991) and Hogg and Craig (1978) for more information about discrete random variables. In the next section, we discuss probability distributions with emphasis on the normal distribution.

▼ EXERCISES

Basic Techniques

4.28 Consider the following class experiment: Toss three coins and observe the number of heads *y*. Let each student repeat the experiment ten times, combine the class results, and construct a relative frequency table for *y*. Note that these frequencies give approximations to the actual probabilities that $y = 0$, 1, 2, or 3. (Note: Calculate the actual probabilities by using the binomial formula $P(y)$ to compare the approximate results with the actual probabilities.)

4.29 Let *y* be a binomial random variable; compute $P(y)$ for each of the following situations:
   a. $n = 10$, $\pi = .2$, $y = 3$
   b. $n = 4$, $\pi = .4$, $y = 2$
   c. $n = 16$, $\pi = .7$, $y = 12$

4.30 Let *y* be a binomial random variable with $n = 8$ and $\pi = .4$. Find the following values:
   a. $P(y \leq 4)$
   b. $P(y > 4)$
   c. $P(y \leq 7)$
   d. $P(y > 6)$.

Applications

$ **4.31** An appliance store has the following probabilities for y, the number of major appliances sold on a given day:

y	$P(y)$
0	.100
1	.150
2	.250
3	.140
4	.090
5	.080
6	.060
7	.050
8	.040
9	.025
10	.015

a. Construct a graph of $P(y)$.
b. Find $P(y \leq 2)$.
c. Find $P(y \geq 7)$.
d. Find $P(1 \leq y \leq 5)$.

$ **4.32** The weekly demand for copies of a popular word-processing program at a computer store has the probability distribution shown here.

y	$P(y)$
0	.06
1	.14
2	.16
3	.14
4	.12
5	.10
6	.08
7	.07
8	.06
9	.04
10	.03

a. What is the probability that three or more copies will be demanded in a particular week?
b. What is the probability that the demand will be for at least two but no more than six copies?
c. If the store has eight copies of the program available at the beginning of each week, what is the probability the demand will exceed the supply in a given week?

4.33 A biologist randomly selects ten portions of water, each equal to .1 cm³ in volume, from the local reservoir and counts the number of bacteria present in each portion. The biologist then totals the number of bacteria for the ten portions to obtain an estimate of the number of bacteria per cubic centimeter present in the reservoir water. Is this a binomial experiment?

 4.34 Examine the accompanying newspaper clipping. Does this sampling appear to satisfy the characteristics of a binomial experiment?

Poll Finds Opposition to Phone Taps

New York—People surveyed in a recent poll indicated they are 81% to 13% against having their phones tapped without a court order.

The people in the survey, by 68% to 27%, were opposed to letting the government use a wiretap on citizens suspected of crimes, except with a court order.

The survey was conducted for 1,495 households and also found the following results:

—The people surveyed are 80% to 12% against the use of any kind of electronic spying device without a court order.

—Citizens are 77% to 14% against allowing the government to open their mail without court orders.

—They oppose, by 80% to 12%, letting the telephone company disclose records of long-distance phone calls, except by court order.

For each of the questions, a few of those in the survey had no responses.

 4.35 A survey is conducted to estimate the percentage of pine trees in a forest that are infected by the pine shoot moth. A grid is placed over a map of the forest, dividing the area into 25-foot by 25-foot square sections. One hundred of the squares are randomly selected and the number of infected trees is recorded for each square. Is this a binomial experiment?

 4.36 A survey was conducted to investigate the attitudes of nurses working in Veterans Administration hospitals. A sample of 1,000 nurses was contacted using a mailed questionnaire and the number favoring or opposing a particular issue was recorded. If we confine our attention to the nurses' responses to a single question, would this sampling represent a binomial experiment? As with most mail surveys, some of the nurses will not respond. What effect might nonresponses in the sample have on the estimate of the percentage of all Veterans Administration nurses who favor the particular proposition?

 4.37 A random sample of ten members was obtained to ascertain opinions concerning a new wage package proposal to a local union by union leaders. If we assume that $\pi = .6$ of all the members have disagreements with the wage package, compute the following probabilities.
a. All disagree.
b. Exactly six disagree.
c. Six or more disagree.
d. All agree.
(Note: With Minitab, the binomial variable y is denoted by K and the binomial probability π is denoted by P.)

```
       BINOMIAL PROBABILITIES FOR n=10 AND P(SUCCESS) = 0.6


       ROW      K        PDF        CDF
         1      0     0.000105    0.00010
         2      1     0.001573    0.00168
         3      2     0.010617    0.01229
         4      3     0.042467    0.05476
         5      4     0.111477    0.16624
         6      5     0.200658    0.36690
         7      6     0.250823    0.61772
         8      7     0.214991    0.83271
         9      8     0.120932    0.95364
        10      9     0.040311    0.99395
        11     10     0.006047    1.00000
```

4.38   Refer to Exercise 4.37.
a. Compute the probabilities for parts a through d if $\pi = .3$.
b. Indicate how you would compute $P(y \leq 100)$ for $n = 1,000$ and $\pi = .3$.

4.39   An experiment is conducted to test the effect of an anticoagulant drug on rats. A random sample of four rats is employed in the experiment. If the drug manufacturer claims that 80% of the rats will be favorably affected by the drug, what is the probability that none of the four experimental rats will be favorably affected? One of the four? One or less?

4.40   A criminologist claims that the probability of "reform" for a first-offender embezzler is .9. Suppose that we define "reform" as meaning the person commits no criminal offenses within a 5-year period. Three paroled embezzlers were randomly selected from the prison records, and their behavioral histories were examined for the 5-year period following prison release. If the criminologist's claim is correct, what is the probability that all three were reformed? At least two?

4.41   Consider the following experiment: Toss three coins and observe the number of heads y. Repeat the experiment 100 times and construct a relative frequency table for y. Note that these frequencies give approximations to the exact probabilities that $y = 0$, 1, 2, and 3. (Note: These probabilities can be shown to be 1/8, 3/8, 3/8, and 1/8, respectively.)

4.42   Refer to Exercise 4.41. Use the formula for the binomial probability distribution to show that $P(0) = 1/8$, $P(1) = 3/8$, $P(2) = 3/8$, and $P(3) = 1/8$.

4.43   Suppose you match coins with another person a total of 1,000 times. What is the mean number of matches? The standard deviation? Calculate the interval $(\mu \pm 3\sigma)$. (Hint: The probability of a match in the toss of a single pair of coins is $\pi = .5$.)

4.44   Refer to Exercise 4.37. Indicate how you could compute $P(y \leq 100)$ if $n = 1,000$ for $\pi = .6$.

4.45   Over a long period of time in a large multinational corporation, 10% of all sales trainees are rated as outstanding, 75% are rated as excellent/good, 10% are rated as satisfactory, and 5% are considered unsatisfactory. Find the following probabilities for a sample of ten trainees selected at random:
a. Two are rated as outstanding.
b. Two or more are rated as outstanding.

c. Eight of the ten are rated either outstanding or excellent/good.

d. None of the trainees is rated as unsatisfactory.

4.46 A new technique, balloon angioplasty, is being widely used to open clogged heart valves and vessels. The balloon is inserted via a catheter and is inflated, opening the vessel; thus, no surgery is required. Left untreated, 50% of the people with heart-valve disease die within about 2 years. If experience with this technique suggests that approximately 70% live for more than 2 years, would the next five patients of the patients treated with balloon angioplasty at a hospital constitute a binomial experiment with $n = 5$, $\pi = .70$? Why or why not?

4.47 A prescription drug firm claims that only 12% of all new drugs shown to be effective in animal tests ever make it through a clinical testing program and onto the market. If a firm has 15 new compounds that have shown effectiveness in animal tests, find the following probabilities:

a. None reach the market.

b. One or more reach the market.

c. Two or more reach the market.

4.48 Does Exercise 4.47 satisfy the properties of a binomial experiment? Why or why not?

4.49 A random sample of 50 price changes is selected from the many listed for a large supermarket during a reporting period. If the probability that a price change is posted correctly is .93,

a. Write an expression for the probability that three or fewer changes are posted incorrectly.

b. What assumptions were made for part a?

4.9 ▼ PROBABILITY DISTRIBUTION FOR CONTINUOUS RANDOM VARIABLES

Discrete random variables (such as the binomial) have possible values that are distinct and separate, such as 0 or 1 or 2 or 3. Other random variables are most usefully considered to be *continuous*: their possible values form a whole interval (or range, or continuum). For instance, the 1-year return per dollar invested in a common stock could range from 0 to some quite large value. In practice, virtually all random variables assume a discrete set of values; the return per dollar of a million-dollar common-stock investment could be $1.06219423 or $1.06219424 or $1.06219425 or . . . . But, when there are many, many possible values for a random variable, it is sometimes mathematically useful to treat the random variable as continuous.

Theoretically, then, a continuous random variable is one that can assume values associated with infinitely many points in a line interval. We state, without elaboration, that it is impossible to assign a small amount of probability to each value of y (as was done for a discrete random variable) and retain the property that the probabilities sum to 1.

To overcome this difficulty, we revert to the concept of the relative frequency histogram of Chapter 3, where we talked about the probability of y falling in a given interval. Recall that the relative frequency histogram for a population containing a large number of measurements will almost be a smooth curve because the number of class intervals can be made large and the width of the intervals can be decreased. Thus, we envision a smooth curve that provides a model for the population relative frequency distribution generated by repeated observation of a continuous random variable. This will be similar to the curve shown in Figure 4.5.

FIGURE 4.5

Probability Distribution for a Continuous Random Variable

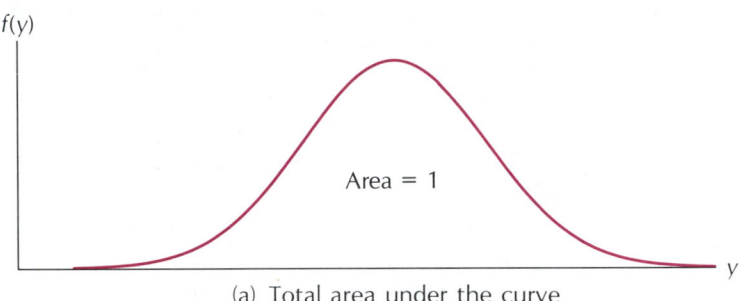

(a) Total area under the curve

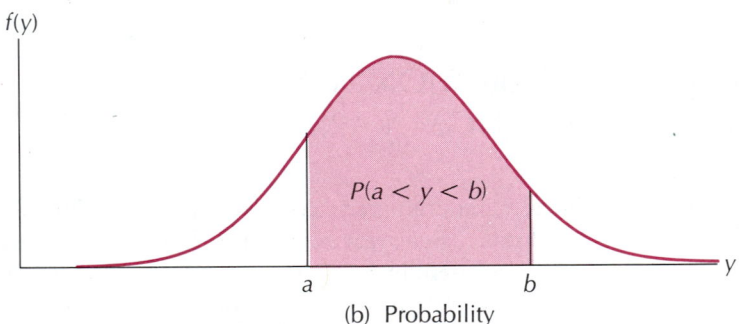

(b) Probability

Recall that the histogram relative frequencies are proportional to areas over the class intervals and that these areas possess a probabilistic interpretation. That is, if a measurement is randomly selected from the set, the probability that it will fall in an interval is proportional to the histogram area above the interval. Since a population is the whole (100%, or 1), we want the total area under the probability curve to equal 1. If we let the total area under the curve equal 1, then areas over intervals are exactly equal to the corresponding probabilities.

The graph for the probability distribution for a continuous random variable is shown in Figure 4.5. The ordinate (height of the curve) for a given value of

y is denoted by the symbol $f(y)$. Many people are tempted to say that $f(y)$, like $P(y)$ for the binomial random variable, designates the probability associated with the continuous random variable y. But, as we mentioned before, it is impossible to assign a probability to each of the infinitely many possible values of a continuous random variable. Thus, all we can say is that $f(y)$ represents the height of the probability distribution for a given value of y.

The probability that a continuous random variable falls in an interval, say between two points a and b, follows directly from the probabilistic interpretation given to the area over an interval for the relative frequency histogram (Section 3.2) and is equal to the area under the curve over the interval a to b, as shown in Figure 4.5. This probability is written $P(a < y < b)$.

There are curves of many shapes that can be used to represent the population relative frequency distribution for measurements associated with a continuous random variable. Fortunately, the areas under these curves have been tabulated and are ready for use. Thus, if we know that student examination scores possess a particular probability distribution, as in Figure 4.6, and if areas under the curve have been tabulated, we could find the probability that a particular student will score more than 80% by looking up the tabulated area, which is shaded in Figure 4.6.

FIGURE 4.6
Hypothetical Probability Distribution for Student Examination Scores

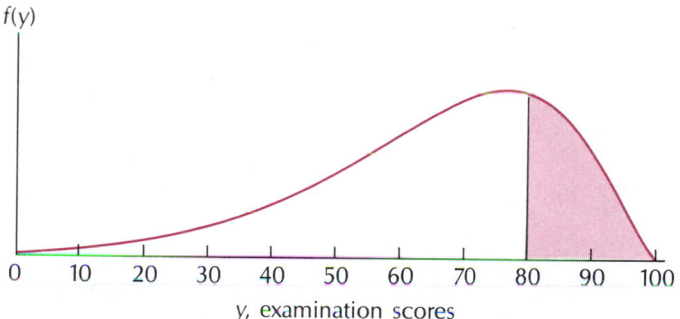

You probably wonder how we know which shape to use for the probability distribution in a given situation. Fortunately, the specific shape chosen for the population frequency distribution (or, equivalently, the probability distribution for the observed variable) will often have little effect on the probability statements associated with the statistical inferences discussed in later chapters. Thus, as far as statistical inferences are concerned, we can relax in the knowledge that the selection of the *exact* shape for the probability distribution for a continuous variable is not crucial.

We will find that data collected on continuous variables often possess mound-shaped frequency distributions and that many of these are nearly bell-shaped. A continuous variable (the normal) and its probability distribution (the bell-shaped, normal curve) provide a good model for these types of data. The normally distributed variable also plays a very important role in statistical inference. We will study its bell-shaped probability distribution in detail in the next section.

4.10 ▼ A USEFUL CONTINUOUS RANDOM VARIABLE: THE NORMAL DISTRIBUTION

Many variables of interest, including several statistics to be discussed in later sections and chapters, have mound-shaped frequency distributions that can be approximated by using a **normal curve**. For example, the distribution of total scores on the Brief Psychiatric Rating Scale for outpatients having a current history of repeated aggressive acts would be mound-shaped. Other practical examples of mound-shaped distributions are social perceptiveness scores of preschool children selected from a particular socioeconomic background, psycho-motor retardation scores for patients with circular-type manic-depressive illness, milk yields for cattle of a particular breed, and perceived anxiety scores for residents of a community. Each of these mound-shaped distributions could be approximated with a normal curve.

Since the normal distribution has been well tabulated, areas under a normal curve—which correspond to probabilities—can be used to approximate probabilities associated with the variables of interest in our experimentation. Thus the normal random variable and its associated distribution will play an important role in statistical inference.

The relative frequency histogram for the normal random variable, called the *normal curve* or *normal* probability distribution, is a smooth bell-shaped curve. Figure 4.7 shows a normal curve. If we let y represent the normal random variable, then the height of the probability distribution for a specific value of y is represented by $f(y)$.* The probabilities associated with a normal curve form the basis for the Empirical Rule.

As we see from Figure 4.7, the normal probability distribution is bell-shaped and symmetrical about the mean μ. Although the normal random variable y may theoretically assume vales from $-\infty$ to $+\infty$, we know from the Empirical Rule that approximately all the measurements are within 3 standard deviations (3σ) of μ. From the Empirical Rule, we also know that if we select a measurement at random from a population of measurements that possesses a mound-shaped distribution, the probability is approximately .68 that the measurement will lie within one standard deviation of its mean (see Figure 4.8).

---

*For the normal distribution, $f(y) = \dfrac{1}{\sqrt{2\pi}\,\sigma}\,e^{-(y-\mu)^2/2\sigma^2}$, where μ and σ are the mean and standard deviation respectively, of the population of y-values.

FIGURE 4.7

A Normal Curve with Mean μ and Standard Deviation σ

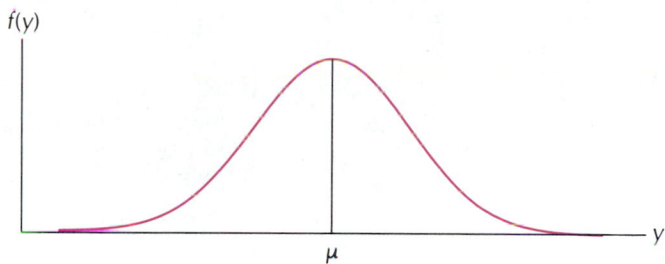

FIGURE 4.8

Area Under a Normal Curve Within One Standard Deviation of the Mean

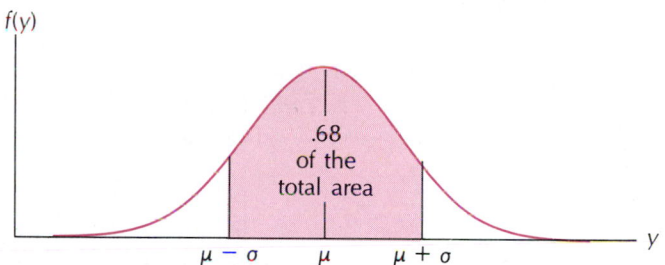

Similarly, we know that the probability is approximately .95 that a value will lie in the interval $\mu \pm 2\sigma$. What we do not know, however, is the probability that the measurement will be within 1.65 standard deviations of its mean, or within 2.58 standard deviations of its mean. The procedure we are going to discuss in this section will enable us to calculate the probability that a measurement falls within any distance of the mean μ for a normal curve.

Since there are many different normal curves (depending on the parameters μ and σ), it might seem to be an impossible task to tabulate areas (probabilities) for all normal curves, especially if each curve requires a separate table. Fortunately, this is not the case. By specifying the probability that a variable y lies within a certain number of standard deviations of its mean (just as we did in using the Empirical Rule), we need only one table of probabilities.

area under a normal curve

z standard deviations

Table 1 in the Appendix gives the **area under a normal curve** from the mean μ to a point z standard deviations $(z\sigma)$ to the right of μ (see Figure 4.9). Because of the symmetry of the normal probability distribution, this would be the same as the area between the mean and a point **z standard deviations** to the left of μ.

The area shown by the shading in Figure 4.9 is the probability listed in Table 1 in the Appendix. Values of z to the nearest tenth are listed along the left-hand column of Table 1, with z to the nearest hundredth along the top of the table. In order to find the probability that a normal random variable will lie in the interval from μ to a point 1.65 standard deviations above the mean, we look up the table entry corresponding to $z = 1.65$. This probability is .4505 (see Figure 4.10).

FIGURE 4.9
**Area Under a Normal
Curve, As Given in Table
1 in the Appendix**

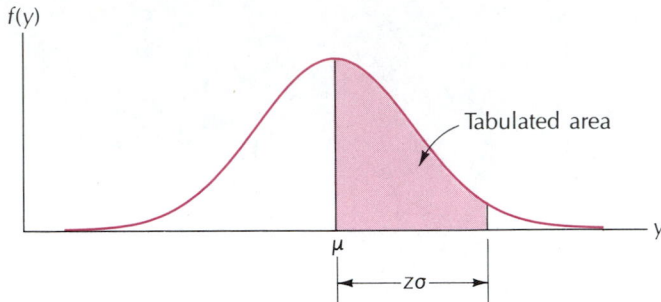

FIGURE 4.10
**Area Under a Normal
Curve from μ to a Point
1.65 Standard Deviations
Above the Mean**

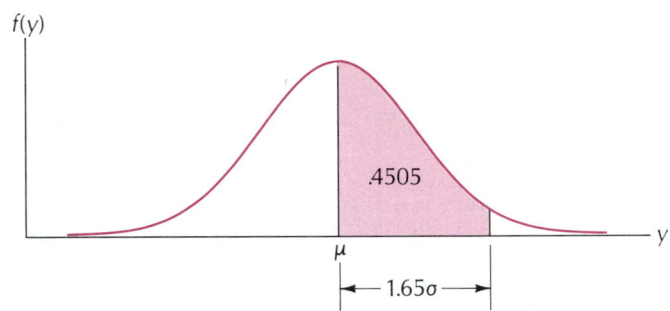

To determine the probability that a measurement will fall in the interval from μ to some value y to the right of μ, we first calculate the number of standard deviations that y lies away from the mean by using the formula

$$z = \frac{y - \mu}{\sigma}.$$

z score

The value of z computed using this formula is sometimes referred to as the **z score** associated with the y-value. Using the computed value of z, we determine the appropriate probability by using Table 1 in the Appendix. Note that we are merely coding the value y by subtracting μ and dividing by σ. (In other words, $y = z\sigma + \mu$.) Figure 4.11 illustrates the values of z

FIGURE 4.11
**Relationship Between
Specific Values of y and
$z = (y - \mu)/\sigma$**

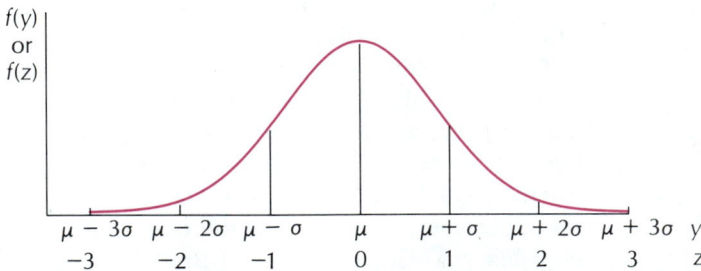

corresponding to specific values of y. Thus, a value of y two standard deviations below (to the left of) μ corresponds to $z = -2$.

EXAMPLE 4.14 ▼

Consider a normal distribution with $\mu = 20$ and $\sigma = 2$. Determine the probability that a measurement will be in the interval from 20 to 23.

Solution When first working problems of this type, it might be a good idea to draw a picture so that you can see the area in question, as we have in Figure 4.12.

FIGURE 4.12
Area Between $\mu = 20$ and $y = 23$, Example 4.14

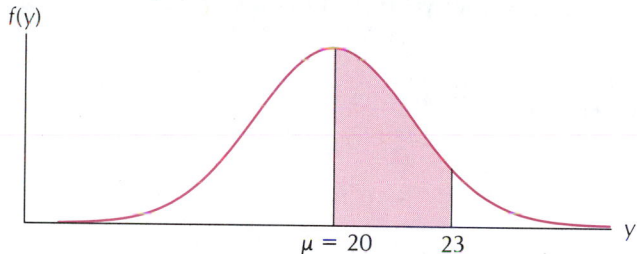

To determine the area under the curve from $\mu = 20$ to the value $y = 23$, we first calculate the number of standard deviations $y = 23$ lies away from the mean.

$$z = \frac{y - \mu}{\sigma} = \frac{23 - 20}{2} = 1.5$$

Thus, $y = 23$ lies 1.5 standard deviations above $\mu = 20$. Referring to Table 1 in the Appendix, we find the area corresponding to $z = 1.5$ to be .4332. This is the probability that a measurement falls in the interval from 20 to 23.   ▲

Similarly, when finding the probability that a measurement lies in the interval from μ to some value of y to the left of the mean, we again compute z, using

$$z = \frac{y - \mu}{\sigma}.$$

negative z

The computed value of z will be negative, but we ignore the negative sign and again refer to Table 1 in the Appendix for the appropriate probability.

E X A M P L E   4.15

For the normal distribution of Example 4.14 with $\mu = 20$ and $\sigma = 2$, find the probability that y will lie in the interval from 16 to 20.

Solution   In determining the area between 16 and 20, we use

$$z = \frac{y - \mu}{\sigma} = \frac{16 - 20}{2} = -2.$$

Ignoring the negative sign, we find the appropriate area from Table 1 to be .4772. thus, 4772 is the probability that a measurement falls in the interval from 16 to 20. The area is shown in Figure 4.13.

F I G U R E   4.13
Area Between $y = 16$ and $\mu = 20$, Example 4.15

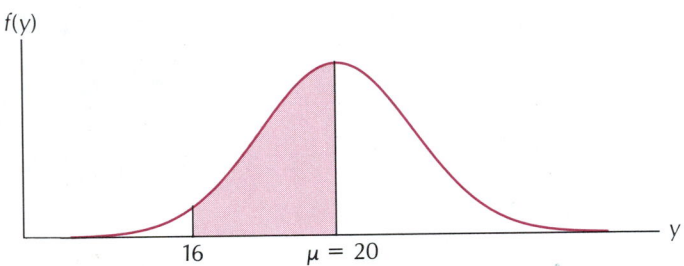

E X A M P L E   4.16

The mean daily milk production of a herd of Guernsey cows is assumed to be normally distributed with $\mu = 70$ pounds and $\sigma = 13$ pounds.
a. What is the probability that the milk production for a cow chosen at random will lie in the interval from 60 pounds to 90 pounds?
b. What is the probability that the milk production for the randomly selected cow will exceed 90 pounds in a given year?

Solution   We begin by drawing a figure to picture the area we are looking for (Figure 4.14). To answer part a, we must compute two areas, the area between 60 and 70 and the area between 70 and 90. The value $y = 60$ corresponds to a z score of

$$z = \frac{y - \mu}{\sigma} = \frac{60 - 70}{13} = -.77.$$

From Table 1, the area between $y = 60$ and $\mu = 70$ is .2794.

FIGURE 4.14

Areas Between 60 and 70 and Between 70 and 90, Example 4.16

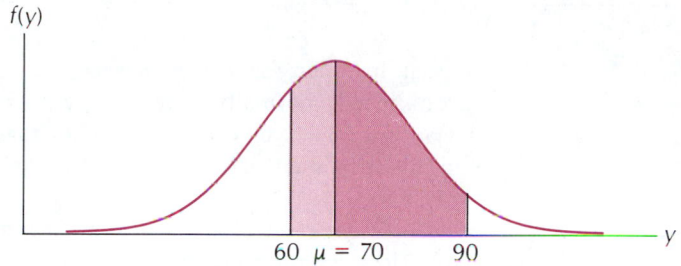

The value $y = 90$ corresponds to a z score of

$$z = \frac{y - \mu}{\sigma} = \frac{90 - 70}{13} = 1.54$$

so the tabulated area between 70 and 90 is .4382. Adding these two areas, we find that the probability that a cow's milk production will lie in the interval from 60 pounds to 90 pounds is .7176.

The probability that the cow's milk production will exceed 90 pounds, part b, can be computed using the second area computed above. Because of the symmetry of a normal curve, we know that the total area under the curve to the right of μ is .5 (similarly, the total area to the left of μ is .5). We computed the area for the interval from 70 to 90 to be .4382. Subtracting this value from .5, we know that the probability of exceeding 90 is $.5 - .4382 = .0618$. (See Figure 4.15.)

FIGURE 4.15

Area Above $y = 90$ for $\mu = 70$ and $\sigma = 13$

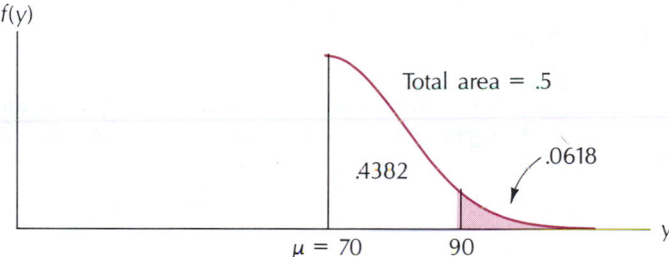

Alternatively, we could use Table 2 in the Appendix (which gives upper-tail probabilities for the normal distribution) to find the answer to part b directly. From Table 2, the probability of exceeding a z-value of 1.54 is .0618. This agrees with our previous answer. ▲

In the future, when we need tail areas of a normal curve, it will be easier to use Table 2. But either Table 1 or Table 2 will do; just draw a picture showing the desired area(s).

E X A M P L E   4.17

Annual incomes for career service employees at a large university are approximately normally distributed with a mean of $17,200 and a standard deviation of $900. Find the probability that an employee chosen at random will have an annual income less than $16,000; an income greater than $18,000.

Solution   First we draw a graph showing the area in question (Figure 4.16). Now we must determine the area between 16,000 and 17,200:

$$z = \frac{y - \mu}{\sigma} = \frac{16,000 - 17,200}{900} = \frac{-1,200}{900} = -1.33$$

In a normal distribution, the area to the right of a value 1.33 standard deviations above the mean is, from Table 2 in the Appendix, .0918. Hence, by symmetry, the probability of observing an annual income less than $16,000 is .0918.

F I G U R E   4.16

Areas Greater Than 18,000 and Less Than 16,000 for μ = 17,200 and σ = 900, Example 4.17

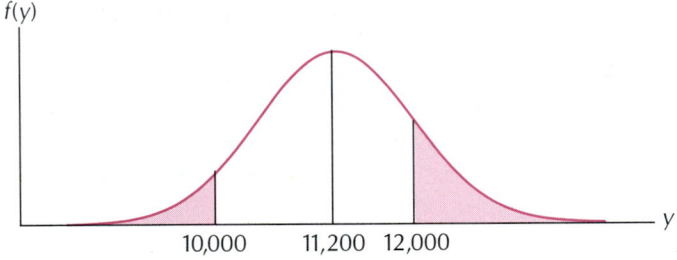

$f(y)$

10,000   11,200 12,000 y

Similarly, to compute the probability of observing a salary above $18,000, we determine the area between 17,200 and 18,000:

$$z = \frac{y - \mu}{\sigma} = \frac{18,000 - 17,200}{900} = .89$$

The area in Table 2 corresponding to $z = .89$ is .1867; this is the desired probability.   ▲

E X A M P L E   4.18

From income tax returns in the previous year, it has been found that for a given income classification, the amount of money owed to the government over and above the amount paid in the estimated tax vouchers for the first three payments is approximately normally distributed with a mean of $530 and a standard deviation of $205. Find the 75th percentile for this distribution of measurements.

Solution   The 75th percentile is, by definition, the value of y such that 75% of the measurements are below it and 25% above it, as shown in Figure 4.17. By referring to Table 2 in the Appendix, we find the value of z corresponding to an area of .25 to the right of $\mu = 530$ is about .67. We find this value of z by first looking for an area in Table 2 that is close to .25. Once we find that, we look across to the z-value and to the top of the table for the hundredths in the z-value. For an area of .25, z is about .67.

FIGURE 4.17

Area Under the Normal Curve for the 75th Percentile, Example 4.18

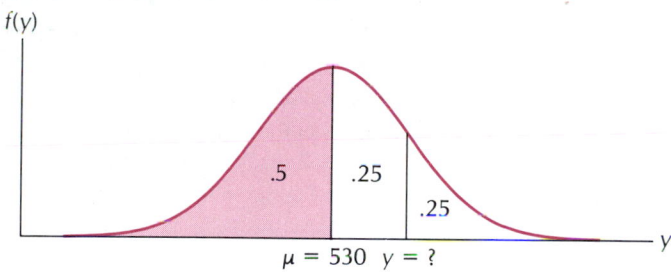

We now substitute in the formula $z = (y - \mu)/\sigma$ to determine y. For this example,

$$.67 = \frac{y - 530}{205}$$

and, hence, $y = .67(205) + 530 = 667.35$. This value is the 75th percentile of the distribution of owed money for this tax classification.   ▲

▼ EXERCISES

Basic Techniques

4.50   Use Table 1 of the Appendix to find the area under the normal curve between these values:
a. $z = 0$ and $z = 1.3$
b. $z = 0$ and $z = -1.9$

4.51   Repeat Exercise 4.50 for these values:
a. $z = 0$ and $z = .7$
b. $z = 0$ and $z = -1.2$

4.52   Repeat Exercise 4.50 for these values:
   a. $z = 0$ and $z = 1.29$
   b. $z = 0$ and $z = -.77$

4.53   Repeat Exercise 4.50 for these values:
   a. $z = -.21$ and $z = 1.35$
   b. $z = .37$ and $z = 1.20$

4.54   Repeat Exercise 4.50 for these values:
   a. $z = 1.43$ and $z = 2.01$
   b. $z = -1.74$ and $z = -.75$

4.55   Find the probability that z is greater than 1.75.

4.56   Find the probability that z is less than 1.14.

4.57   Find a value for z, say z_0, such that $P(z > z_0) = .5$.

4.58   Find a value for z, say z_0, such that $P(z > z_0) = .025$.

4.59   Find a value for z, say z_0, such that $P(z > z_0) = .0089$.

4.60   Find a value for z, say z_0, such that $P(z > z_0) = .05$.

4.61   Find a value for z, say z_0, such that $P(-z_0 < z < z_0) = .95$.

4.62   Let y be a normal random variable with mean equal to 100 and standard deviation equal to 8. Find the following probabilities:
   a. $P(y > 100)$
   b. $P(y > 110)$
   c. $P(y < 115)$
   d. $P(88 < y < 12)$
   e. $P(100 < y < 108)$

4.63   Let y be a normal random variable with $\mu = 500$ and $\sigma = 100$. Find the following probabilities:
   a. $P(500 < y < 696)$
   b. $P(y > 696)$
   c. $P(304 < y < 696)$
   d. k such that $P(500 - k < y < 500 + k) = .60$

4.64   Suppose that y is a normal random variable with $\mu = 100$ and $\sigma = 15$.
   a. Show that $y < 130$ is equivalent to $z < 2$.
   b. Convert $y > 82.5$ to the z-score equivalent.
   c. Find $P(y < 130)$ and $P(y > 82.5)$.
   d. Find $P(y > 106)$, $P(y < 94)$, and $P(94 < y < 106)$.
   e. Find $P(y < 70)$, $P(y > 130)$, and $P(70 < y < 130)$.

4.65   Use Table 1 in the Appendix to calculate the area under the curve between these values.
   a. $z = 0$ and $z = 1.5$
   b. $z = 0$ and $z = 1.8$

4.66   Repeat Exercise 4.65 for these values.
   a. $z = -1.96$ and $z = 1.96$
   b. $z = -2.33$ and $z = 2.33$

4.67   What is the value of z with an area of .05 to its right? To its left? (Hint: Use Table 2 in the Appendix.)

4.68   Find the value of z for these areas.
   a. an area .01 to the right of z
   b. an area .10 to the left of z

4.69 Find the probability of observing a value of z greater than these values.
  a. 1.96
  b. 2.21
  c. 2.86
  d. 0.73

4.70 Find the probability of observing a value of z less than these values.
  a. −1.20
  b. −2.62
  c. 1.84
  d. 2.17

Applications

4.71 Records maintained by the office of budget in a particular state indicate that the amount of time elapsed between the submission of travel vouchers and the final reimbursement of funds has approximately a normal distribution with a mean of 39 days and a standard deviation of 6 days.
  a. What is the probability that the elapsed time between submission and reimbursement will exceed 50 days?
  b. If you had a travel voucher submitted more than 55 days ago, what might you conclude?

4.72 The College Boards, which are administered each year to many thousands of high school students, are scored in such a way as to yield a mean of 500 and a standard deviation of 100. These scores are close to being normally distributed. What percentage of the scores can be expected to satisfy each condition?
  a. greater than 600
  b. greater than 700
  c. less than 450
  d. between 450 and 600

4.73 Sales figures (on a monthly basis) for a particular food industry tend to be normally distributed with mean of 150 (thousand dollars) and a standard deviation of 35 (thousand dollars). Compute the following probabilities:
  a. $P(y > 200)$
  b. $P(y > 220)$
  c. $P(y < 120)$
  d. $P(100 < y < 200)$.

4.74 Refer to Exercise 4.72. An exclusive club wishes to invite those scoring in the top 10% on the College Boards to join.
  a. What score is required to be invited to join the club?
  b. What score separates the top 60% of the population from the bottom 40%? What do we call this value?

4.75 The mean for a normal distribution is 50 and the standard deviation is 10.
  a. What percentile is the value of 38? Choose the appropriate answer.

      88.49   38.49   49.99   0.01   11.51

  b. Which of the following is the z score corresponding to the 67th percentile?

      1.00   0.95   0.44   2.25   none of these

4.76 The distribution of weights of a large group of high-school boys is normal with $\mu = 120$ pounds and $\sigma = 10$ pounds. Which of the following is true?
  a. About 16% of the boys will be over 130 pounds.
  b. Probably fewer than 2.5% of the boys will be below 100 pounds.
  c. Half of the boys can be expected to weigh less than 120 pounds.
  d. All the above are true.

4.11 ▼ RANDOM SAMPLING

So far in the text, we have discussed random samples and introduced various sampling schemes in Chapter 2. What is the importance of random sampling? We must know how the sample was selected so that we can determine probabilities associated with various sample outcomes. The probabilities of samples selected *in a random manner* can be determined and we can use these probabilities to make inferences about the population from which the sample was drawn.

Sample data selected in a nonrandom fashion are frequently distorted by a *selection bias*. A selection bias exists whenever there is a systematic tendency to overrepresent or underrepresent some part of the population. For example, a survey of households conducted during the week entirely between the hours of 9 A.M. and 5 P.M. would be severely biased toward households with at least one member at home. Hence, any inferences made from the sample data would be biased toward the attributes or opinions of those families with at least one member at home and may not be truly representative of the population of households in the region.

random sample

Now we turn to a definition of a **random sample** of n measurements selected from a population containing N measurements $(N > n)$. (Note: This is a simple random sample as discussed in Chapter 2. Since most of the random samples discussed in this text will be simple random samples, we'll drop the adjective unless needed for clarification.)

DEFINITION 4.13
Random Sample

> A sample of n measurements selected from a population is said to be a **random sample** if every different sample of size n from the population has an equal probability of being selected.

EXAMPLE 4.19 ▼

Suppose that a population consists of the six measurements 1, 2, 3, 4, 5, 7. List all possible different samples of two measurements that could be selected from the population. Give the probability associated with each sample in a random sample of $n = 2$ measurements selected from the population.

Solution   All possible samples are listed next.

Sample	Measurements
1	1,2
2	1,3
3	1,4
4	1,5
5	1,7
6	2,3
7	2,4
8	2,5
9	2,7
10	3,4
11	3,5
12	3,7
13	4,5
14	4,7
15	5,7

Now let us suppose that we draw a single sample of $n = 2$ measurements from the 15 possible samples of 2 measurements. The sample selected is called a random sample if every sample had an equal probability ($\frac{1}{15}$) of being selected.

▲

It is rather unlikely that we would ever achieve a truly random sample, because the probabilities of selection will not always be exactly equal. But we do the best we can. One of the simplest and most reliable ways to select a random sample of n measurements from a population is to use a table of random numbers (see Table 8 in the Appendix). **Random number tables** are constructed in such a way that, no matter where you start in the table and no matter what direction you move, the digits occur randomly and with equal probability. Thus, if we wished to choose a random sample of $n = 10$ measurements from a population containing 100 measurements, we could label the measurements in the population from 0 to 99 (or 1 to 100). Then by referring to Table 8 in the Appendix and choosing a random starting point, the next ten two-digit numbers going across the page would indicate the labels of the particular measurements to be included in the random sample. Similarly, by moving up or down the page, we would also obtain a random sample.

random number table

E X A M P L E   4.20

▼

A small community consists of 850 families. We wish to obtain a random sample of 20 families to ascertain public acceptance of a wage and price freeze. Refer to Table 8 in the Appendix to determine which families should be sampled.

Solution   Assuming that a list of all families in the community is available (such as a telephone directory), we could label the families from 0 to 849 (or, equivalently, from 1 to 850). Then referring to Table 8 in the Appendix, we choose a starting point. Suppose we have decided to start at line 1, column 3. Going down the page we will choose the first 20 three-digit numbers between 000 and 849. From Table 8, we have

015	110	482	333
255	564	526	463
225	054	710	337
062	636	518	224
818	533	524	055

These 20 numbers identify the 20 families that are to be included in our sample.   ▲

A telephone directory is not always the best source for names, especially in surveys related to economics or politics. In the 1936 presidential campaign, Franklin Roosevelt was running as the Democratic candidate against the Republican candidate, Governor Alfred Landon of Kansas. This was a difficult time for the nation; the country had not yet recovered from the Great Depression of the 1930s, and there were still 9 million people unemployed.

The *Literary Digest* set out to sample the voting public and predict the winner of the election. Using names and addresses taken from telephone books and club memberships, the *Literary Digest* sent out 10 million questionnaires and got 2.4 million back. Based on the responses to the questionnaire, the *Digest* predicted a Landon victory by 57% to 43%.

At this time, George Gallup was starting his survey business. He conducted two surveys. The first one, based on 3,000 people, predicted what the results of the *Digest* survey would be long before the *Digest* results were published; the second survey, based on 50,000, was used to forecast *correctly* the Roosevelt victory.

Where did the *Literary Digest* go wrong? The first problem was a severe selection bias. By taking the names and addresses from telephone directories and club memberships, its survey systematically excluded the poor. And, unfortunately for the *Digest*, the vote was split along economic lines; the poor gave Roosevelt a large majority, whereas the rich tended to vote for Landon. A second reason for the error could be due to a *nonresponse bias*. Because only 20% of the 10 million people returned their surveys, and approximately half of those responding favored Landon, one might suspect that maybe the nonrespondents had different preferences than did the respondents. This was, in fact, true.

How, then does one achieve a random sample? Careful planning and a certain amount of ingenuity are required to have even a decent chance to approximate random sampling. This is especially true when the universe of interest involves people. People can be difficult to work with; they have a tendency to discard mail questionnaires and refuse to participate in personal interviews. Unless we are very careful, the data we obtain may be full of biases having unknown effects on the inferences we are attempting to make.

We do not have sufficient time to explore the topic of random sampling further in this text; entire courses at the undergraduate and graduate levels can be devoted to sample survey research methodology. The important point to remember is that data from a random sample will provide the foundation for making statistical inferences in later chapters. Random samples are not easy to obtain, but with care we can avoid many potential biases that could affect the inferences we make.

▼ EXERCISES

Basic Techniques

4.77 Define what is meant by a random sample. Is it possible to draw a truly random sample? Comment.

4.78 Suppose that we want to select a random sample of $n = 10$ persons from a population of 800. Use Table 8 in the Appendix to identify the persons to appear in the sample.

4.79 Refer to Exercise 4.78. Identify the elements of a population of $N = 1,000$ to be included in a random sample of $n = 15$.

Applications

 4.80 City officials want to sample the opinions of the homeowners in a community regarding the desirability of increasing local taxes to improve the quality of the public schools. If a random number table is used to identify the homes to be sampled and a home is discarded if the homeowner is not home when visited by the interviewer, is it likely this process will approximate random sampling? Explain.

4.81 A local TV network want to run an informal survey of individuals who exit from a local voting station to ascertain early results on a proposal to raise funds to move the city-owned historical museum to a new location. How might the network sample voters to approximate random sampling?

4.82 A psychologist was interested in studying women who are in the process of obtaining a divorce to determine whether there are significant attitudinal changes after the divorce has been finalized. Existing records from the geographic area in question show that 798 couples have recently filed for divorce. Assume a sample of 25 women is needed for the study, and use Table 8 in the Appendix to determine which women should be asked to participate in the study. (Hint: Begin in column 2, row 1 and proceed down.)

4.83   Refer to Exercise 4.82. As is the case in most surveys, not all persons chosen for a study will agree to participate. Suppose that 5 of the 25 women selected refuse to participate. Determine 5 more women to be included in the study.

 4.84   Suppose you have been asked to run a public opinion poll related to an upcoming election. There are 1,000 registered voters in a specific precinct and you wish to obtain a random sample of 50 persons. Use a computer program to indicate which individuals are to be included in the sample. A Minitab program is shown here for purposes of illustration. (Note: We assume that there is a list of the 1,000 voters, with the numbers 1 to 1,000 corresponding to people on the list.)

```
MTB > RANDOM 50 C1;
SUBC> INTEGER 1 TO 1000.
MTB > PRINT C1

C1
    449       1     113       9     120     883     347     303     865      99     264
    916     475     327     793     120     877     542     721      12     441     100
    393      32     926     638     680     963     362     224     979     293     601
     33      25      54     729      74     199     869     557     618     237     516
    401      52     437     523     364     472

MTB > STOP
```

4.12 ▼ THE SAMPLING DISTRIBUTION FOR $\bar{y}$

We discussed several different measures of central tendency and variability in Chapter 3 and distinguished between numerical descriptive measures of a population (parameters) and numerical descriptive measures of a sample (statistics). Thus, μ and σ are parameters, whereas $\bar{y}$ and s are statistics.

The numerical value that a sample statistic will have cannot be predicted exactly in advance. Even is we knew that a population mean μ was \$216.37 and that the population standard deviation σ was \$32.90—even if we knew the complete population distribution—we could not say that the sample mean $\bar{y}$ would be exactly equal to \$216.37. A sample statistic is a random variable; it is subject to random variation because it is based on a random sample of measurements selected from the population of interest. And, like any other random variable, a sample statistic has a probability distribution. We call the probability distribution of a sample statistic the *sampling distribution* of that statistic. Stated differently, the sampling distribution of a statistic is the population of all possible values for that statistic.

The actual mathematical derivation of sampling distributions is one of the basic problems of mathematical statistics. We will illustrate how the sampling distribution for $\bar{y}$ can be obtained for a simplified population. Later in the chapter, several general results will be presented.

EXAMPLE   4.21

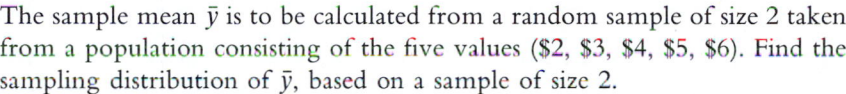

The sample mean $\bar{y}$ is to be calculated from a random sample of size 2 taken from a population consisting of the five values ($2, $3, $4, $5, $6). Find the sampling distribution of $\bar{y}$, based on a sample of size 2.

Solution   One way to find the sampling distribution is by counting. There are ten possible samples of two items from the five items. These are shown here:

Possible Samples of Size 2	Value of $\bar{y}$
2,3	2.5
2,4	3
2,5	3.5
2,6	4
3,4	3.5
3,5	4
3,6	4.5
4,5	4.5
4,6	5
5,6	5.5

Assuming each sample of size 2 is equally likely, it follows that the sampling distribution for $\bar{y}$ based on $n = 2$ observations selected from this population is as indicated here.

$\bar{y}$	$P(\bar{y})$
2.5	1/10
3	1/10
3.5	2/10
4	2/10
4.5	2/10
5	1/10
5.5	1/10

The sampling distribution is shown as a graph in Figure 4.18. Note from Figure 4.18 that the distribution is symmetric with a mean of 4.0 and a standard deviation of approximately 1.0 (the range divided by 4).

FIGURE 4.18
Sampling Distribution for
$\bar{y}$, Example 4.21

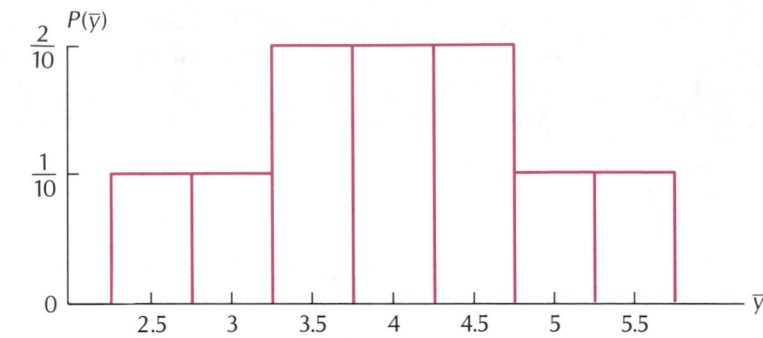

Quite a few of the more common sample statistics have sampling distributions that are normal. A very plausible explanation for this phenomenon is offered by a series of theorems in mathematical statistics called Central Limit Theorems. We will discuss one such theorem (without proof).

THEOREM 4.1
Central Limit Theorem

▼

If random samples of n measurements are repeatedly drawn from a population with a finite mean μ and a standard deviation σ, then, when n is large, the relative frequency histogram for the sample means (calculated from the repeated samples) will be approximately normal (bell-shaped) with mean μ and standard deviation $\sigma/\sqrt{n}$. (Note: The approximation becomes more precise as n increases.)

Figure 4.19 illustrates Theorem 4.1. Figure 4.19(a) shows the distribution of the original measurement y from which the samples are to be drawn. No specific shape need be assigned to the distribution of y. All we know is that its mean is μ and its variance is σ^2. Figure 4.19(b) illustrates the relative frequency histogram,

sampling distribution   called the **sampling distribution**, for the sample mean $\bar{y}$. Repeated samples of size n are to be drawn from the population illustrated in Figure 4.19(a). For each sample drawn, we compute the sample mean $\bar{y}$. If we were to continue this process over and over again, finally plotting the relative frequency histogram for the sample means, it would appear as in Figure 4.19(b). The mean for the sampling distribution of $\bar{y}$ is μ, the same as that for the original y measurements, and the standard deviation of the sampling distribution is equal to the standard deviation of the y measurements (σ) divided by $\sqrt{n}$. This quantity is designated

standard error of $\bar{y}$ by $\sigma_{\bar{y}}$ and called the **standard error of $\bar{y}$**.

We will be able to use the information we have concerning the distribution of $\bar{y}$ to make inferences about the parameter μ. For example, we know from knowledge of the normal curve that 95% of the $\bar{y}$s will be within 1.96 standard

FIGURE 4.19
**The Probability
Distribution of y and the
Sampling Distribution of $\bar{y}$**

(a) Probability distribution of y, with mean μ and standard deviation σ

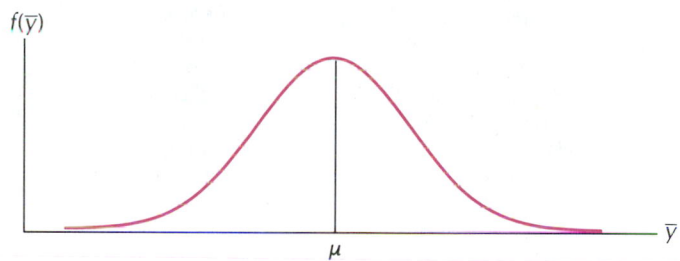

(b) Sampling distribution of $\bar{y}$, with mean μ and standard error $\sigma/\sqrt{n}$

errors $(1.96\sigma_{\bar{y}})$ of their mean. In a given sample, where we calculate a single sample mean $\bar{y}$, we would expect our calculated $\bar{y}$ to be within $1.96\sigma_{\bar{y}}$ of μ (see Figure 4.20). So we not only can use $\bar{y}$ to estimate μ but we also can say how close to μ we expect our estimate to be; that is, we can provide a **measure of goodness** for our estimate.

measure of goodness

We can use the table of random numbers (Table 8 in the Appendix) to illustrate empirically how the Central Limit Theorem applies to sample means. Table 4.5 lists the murder rate (number of murders per 100,000 people) associated with 90 metropolitan areas throughout the United States in the north, the south, and the west in 1980. Although it may be unrealistic to assume that these 90 murder rates represent a population of measurements, we will, for illustration purposes, take the 90 measurements of Table 4.5 as the population of interest.

FIGURE 4.20
**Characterization of the
Sampling Distribution
for $\bar{y}$**

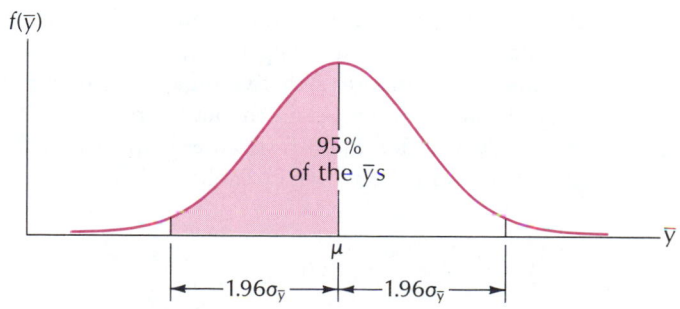

TABLE 4.5

Murder Rates (per 100,000 Inhabitants) for 90 Cities Selected from the South, North, and West

South	Rate	North	Rate	West	Rate
Atlanta, GA	14	Albany, NY	2	Bakersfield, CA	21
Augusta, GA	13	Allentown, PA	2	Boise, ID	3
Baton Rouge, LA	10	Atlantic City, NJ	10	Colorado Springs, CO	6
Beaumont, TX	12	Canton, IL	6	Denver, CO	9
Birmingham, AL	16	Chicago, IL	14	Eugene, OR	2
Charlotte, NC	14	Cincinnati, OH	6	Fresno, CA	20
Chattanooga, TN	10	Cleveland, OH	16	Honolulu, HI	8
Columbia, SC	10	Detroit, MI	16	Kansas City, MO	15
Corpus Christi, TX	16	Evansville, IN	8	Lawton, OK	6
Dallas, TX	18	Grand Rapids, MI	6	Los Angeles, CA	23
El Paso, TX	12	Johnstown, PA	8	Modesto, CA	10
Fort Lauderdale, FL	17	Kankakee, IL	6	Oklahoma City, OK	12
Greensboro, NC	8	Kenosha, WI	4	Oxnard, CA	7
Jackson, MS	17	Lancaster, PA	2	Pueblo, CO	4
Knoxville, TN	7	Lansing, MI	3	Sacramento, CA	9
Lexington, KY	6	Lima, OH	6	St. Louis, MO	15
Lynchburg, VA	12	Madison, WI	2	Salinas, CA	8
Macon, GA	8	Mansfield, OH	3	Salt Lake City, UT	5
Memphis, TN	20	Milwaukee, WI	6	San Diego, CA	10
Monroe, LA	13	Newark, NJ	11	San Francisco, CA	12
Nashville, TN	14	Paterson, NJ	9	San Jose, CA	8
Newport News, VA	11	Philadelphia, PA	12	Seattle, WA	7
Orlando, FL	10	Pittsfield, MA	1	Sioux City, ID	1
Richmond, VA	12	Racine, WI	5	Spokane, WA	4
Roanoke, VA	10	Rockford, IL	5	Stockton, CA	18
San Antonio, TX	18	South Bend, IN	11	Tacoma, WA	5
Shreveport, LA	20	Springfield, IL	6	Topeka, KS	11
Washington, DC	11	Syracuse, NY	3	Tucson, AZ	9
Wichita Falls, KS	13	Vineland, NJ	9	Vallejo, CA	6
Wilmington, DE	8	Youngstown, OH	7	Waco, TX	15

Source: Department of Justice, *Uniform Crime Reports for the United States: 1980* (Washington, DC: U.S. Government Printing Office, 1980), pp. 60–86.

To examine the sampling distribution of the sample mean $\bar{y}$—that is, the population of all $\bar{y}$ values—we randomly draw 50 samples of five scores. For example, numbering the cities from 1 to 90 and using the table of random numbers (Table 8 in the Appendix), we could proceed down column 1 and use the first two digits of each five-digit random number. For our first sample of random numbers we obtain the numbers 10, 22, 24, 42, and 37, and we would select the murder rates from the cities assigned these numbers. The result of this first sample is shown in Table 4.6. The sample mean from the first sample is

$$\bar{y} = \frac{\sum y}{n} = \frac{63}{5} = 12.6.$$

TABLE 4.6

Results of Sample 1

City Number	City	Murder Rate
10	Dallas, TX	18
22	Newport News, VA	11
24	Richmond, VA	12
42	Kankakee, IL	6
37	Cleveland, OH	16
Total		63

TABLE 4.7

Means of 50 Samples of Size 5 Selected from the 90 Murder Rates

12.6	9.5	12.1	6.7	9.6
9.5	13.1	15.6	9.4	9.3
14.9	8.2	12.0	9.1	8.0
11.9	7.1	8.5	10.3	11.0
11.0	9.2	7.9	5.1	9.6
10.7	8.3	6.8	11.3	9.2
9.7	8.7	12.2	12.0	10.3
10.5	9.6	8.3	9.1	7.4
8.2	10.9	14.0	8.3	11.2
7.8	9.4	9.4	5.9	10.3

We repeat this procedure 49 more times to acquire 49 new samples. The 50 sample means are listed in Table 4.7.

To see how the sampling distribution of the sample mean $\bar{y}$ relates to the original distribution for the 90 murder rates, we can construct histograms for both sets of measurements, as shown in Figure 4.21(a) and (b).

As can be seen from Figure 4.21(a), the original population is certainly not mound-shaped or symmetrical. In fact, the distribution is skewed to the right (tails off to the right). But even with a small number of measurements per sample (5), and a small number of samples (50), the sampling distribution of $\bar{y}$ is beginning to look bell-shaped; most of the sample means are grouped closely about the mean of the population (which, in this case, can be computed to be $\mu = 9.71$). See Figure 4.21(b).

The illustration we have presented could have been made more convincing by assuming a much larger population and by taking more samples. The important point to remember is that, in repeated sampling, $\bar{y}$ will be approximately normally distributed with mean μ and standard error $\sigma/\sqrt{n}$. The approximation will be more precise as n, the sample size for each sample, increases. Thus, the frequency histogram for $\bar{y}$ in our example would have been even more bell-shaped if n had been 10 rather than 5, or 15 rather than 10, and so on.

An obvious question is: how large should the sample size be in order for the Central Limit Theorem to hold? Numerous simulation studies have been conducted over the years and the results of these studies suggest that, in general, the Central Limit Theorem holds for $n > 30$. However, one should not apply this

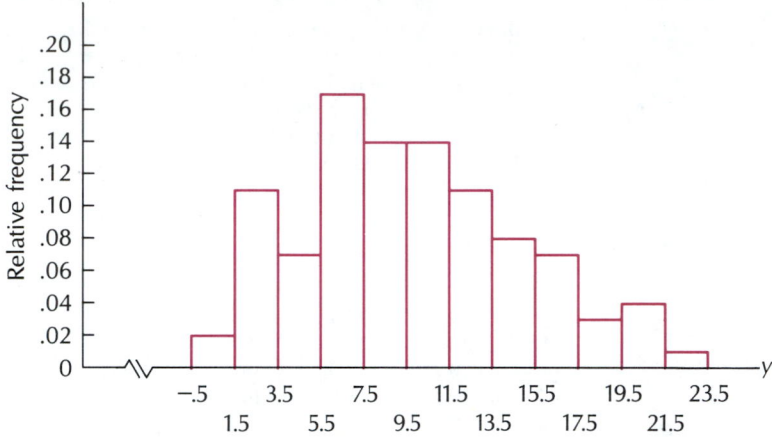

(a) Original population

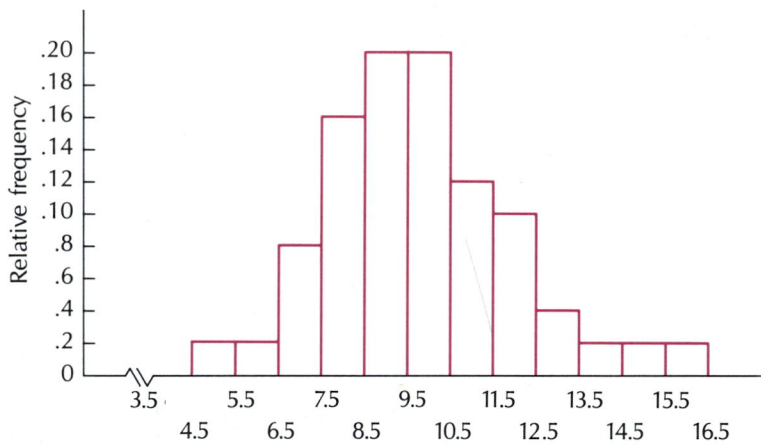

(b) Sampling distribution of ȳ

rule blindly. If the population is heavily skewed, the sampling distribution for ȳ will still be skewed even for $n > 30$. On the other hand, if the population is symmetric, the Central Limit Theorem holds for $n < 30$.

So, take a look at the data. If the sample histogram is clearly skewed, then the population will also probably be skewed. Consequently, a value of n much higher than 30 may be required to have the sampling distribution of ȳ be approximately normal. Any inference based on the normality of ȳ for $n = 30$ under this condition should be examined carefully.

We can use the results of coding, presented in Chapter 3, to extend the Central Limit Theorem to the sample sum $\sum y$. If repeated samples of size n

are drawn from a population, and if we compute $\sum y = n\bar{y}$ for each sample drawn, we have, in essence, coded the sample means for each sample by multiplying by n. Applying our coding results from Chapter 3 to the results of the Central Limit Theorem, the mean and the standard error for $\sum y$ will be, respectively, $n\mu$ and $n\sigma/\sqrt{n} = \sqrt{n}\sigma$. This result is illustrated in Figure 4.22.

Many of the statistics that we will encounter in later chapters will be either sums or averages of variables. Hence, we will employ the Central Limit Theorem discussed in this chapter to specify their sampling distributions.

Usually, a sample statistic is used as an estimate of a population parameter. For example, a sample mean can be used to estimate the corresponding mean μ of the population from which the sample was drawn. The sampling distribution of a sample statistic is then used to determine how accurate the estimate is likely to be. In Example 4.21, the population mean μ is known to be \$4. Obviously, we don't ever know μ in practice. Still, we can use the sampling distribution of $\bar{y}$ to determine the probability that, for example, the computed value of the sample mean will be more than \$.50 away from μ. For Example 4.21, this probability is

$$P(2.5) + P(3) + P(5) + P(5.5) = \frac{4}{10}.$$

In general, a sample statistic is used to make inferences about a population parameter. The sampling distribution of the statistic is crucial in determining how good the inference is likely to be.

FIGURE 4.22

Probability Distribution of y and of the Sample Sum $\sum y$

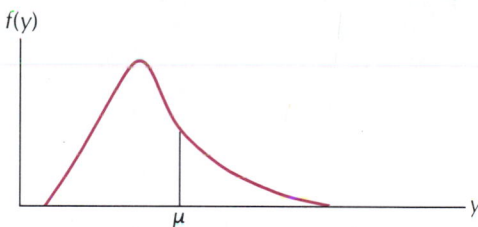

(a) Probability distribution of y, with mean μ and standard deviation σ

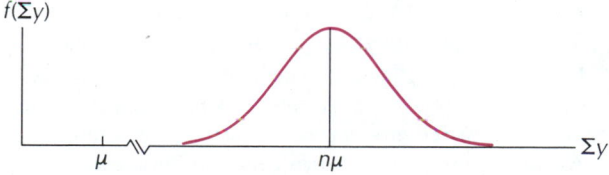

(b) Sampling distribution Σy, with mean $n\mu$ and standard error $\sqrt{n}\sigma$

**interpretations of a
sampling distribution**

Sampling distributions can be **interpreted** in at least two ways. One way uses the long-run relative frequency approach. Imagine taking repeated samples of a fixed size from a given population and calculating the value of the sample statistic for each sample. In the long run, the relative frequencies for the possible values of the sample statistic will approach the corresponding sampling distribution probabilities. For example, if one took a large number of samples from the population distribution corresponding to the probabilities of Example 4.21 and, for each sample, computed the sample mean, approximately 20% would have $\bar{y} = 3.5$.

The other way to interpret a sampling distribution makes use of the classical interpretation of probability. Imagine listing all possible samples that could be drawn from a given population. The probability that a sample statistic will have a particular value (say, that $\bar{y} = 3.5$) is then the proportion of all possible samples that yield that value. In Example 4.21, $P(3.5) = 2/10$ corresponds to the fact that two of the ten samples have a sample mean equal to 3.5. Both the repeated-sampling and the classical approach to finding probabilities for a sample statistic are legitimate.

In practice, though, a sample is taken only once, and only one value of the sample statistic is calculated. A sampling distribution is not something you can see in practice; it is not an empirically observed distribution. Rather, it is a theoretical concept, a set of probabilities derived from assumptions about the population and about the sampling method.

There's an unfortunate similarity between the phrase "sampling distribution," meaning the theoretically derived probability distribution of a statistic, and the phrase "sample distribution," which refers to the histogram of individual values actually observed in a particular sample. The two phrases mean very different things. To avoid confusion, we will refer to the distribution of sample values as the **sample histogram** rather than as the sample distribution.

sample histogram

▼ EXERCISES

Basic Techniques

4.85 A random sample of 16 measurements is drawn from a population with a mean of 60 and a standard deviation of 5. Describe the sampling distribution of $\bar{y}$, the sample mean. Within what interval would you expect $\bar{y}$ to lie approximately 95% of the time?

4.86 Refer to Exercise 4.85. Describe the sampling distribution for the sample sum $\sum y_i$. Is it unlikely (improbable) that $\sum y_i$ would be more than 70 units away from 960? Explain.

4.87 In Exercise 4.85, a random sample of 16 observations was to be selected from a population with $\mu = 60$ and $\sigma = 5$. Assume that the original population of measurements is normal. Use a computer program to simulate the distribution of $\bar{y}$ based on 40 sample means. A Minitab program is shown below in illustration.

```
MTB > RANDOM 16 C1;
SUBC> NORMAL 60 5.
MTB > PRINT C1

C1
   65.8312    64.6174    66.2072    61.9350    50.9882    56.1795    55.8897
   62.1257    69.6186    62.7709    65.8804    60.4337    66.4280    56.2109
   55.6832    55.8120

MTB > MEAN C1
     MEAN    =       61.038
MTB > STOP
```

This set of two instructions would be repeated 39 more times. The complete set of 80 instructions would be entered simultaneously followed by a STOP instruction card.

Applications

4.88 Psychomotor retardation scores for a large group of manic–depressive patients were found to be approximately normal with a mean of 930 and a standard deviation of 130.
  a. What fraction of the patients scored between 800 and 1,100?
  b. Less than 800?
  c. Greater than 1200?

4.89 Refer to Exercise 4.88.
  a. Find the 90th percentile for the distribution of manic–depressive scores. (Hint: Solve for y in the expression $z = (y - \mu)/\sigma$, where z is the number of standard deviations the 90th percentile lies above the mean μ.)
  b. Find the interquartile range.

4.90 Federal resources have been tentatively approved for funding the construction of an outpatient clinic. But in order for the designers to present plans for a facility that will handle patient load requirements while still staying within a limited budget, a study of patient demand was made. From studying a similar facility in the area, it was found that the distribution of the number of patients requiring hospitalization during a week could be approximated by a normal distribution with a mean of 125 and a standard deviation of 32.
  a. Use the Empirical Rule to describe the distribution of y, the number of patients requesting service in a week.
  b. If the facility was built with a 160-patient capacity, what fraction of the weeks might the clinic be unable to handle the demand?

4.91 Refer to Exercise 4.90. What size facility should be built so that the probability of the patient load exceeding the clinic capacity is .05? .01?

4.92 The distribution of the milkfat percentages for Holstein cattle in a particular state during the 1960s was approximately normal with a mean of 3.7 and a standard deviation of .3.
  a. What percentage of the Holsteins had a milkfat percentage less than 3?
  b. Greater than 4.5?

4.93   Refer to Exercise 4.92.
   a. Find the limits within which 90% of the milkfat percentages fell.
   b. Compute the 95th percentile for the distribution of milkfat percentages.

4.94   Refer to Exercise 4.92. Suppose a random sample of $n = 25$ Holsteins is selected from the population of Holstein cattle in the state.
   a. Describe the distribution of $\bar{y}$, the mean milkfat percentage for the sample of 25 cattle.
   b. Compare the distribution of $\bar{y}$ in part a to that for a distribution of $\bar{y}$ from a sample of 100 Holsteins.
   c. What is the probability that the sample mean milkfat percentage would exceed 4 in part a?

4.95   Random samples of size 20 are repeatedly drawn from a normal distribution with a mean of 65 and a standard deviation of 8.
   a. Describe the sampling distribution for $\bar{y}$.
   b. What fraction of the sample means should be in the interval from 60 to 72?

4.96   Refer to Exercise 4.95.
   a. Describe the sampling distribution of the sample sum $\sum_i y_i$.
   b. Locate the 25th and 75th percentiles for the sampling distribution of $\sum_i y_i$.

4.13 ▼ NORMAL APPROXIMATION TO THE BINOMIAL

The Central Limit Theorem discussed in the previous section will enable us to calculate probabilities for a binomial random variable by approximating the binomial distribution with a normal curve and using normal curve areas as approximations to the desired probabilities. We said in Section 4.8 that probabilities associated with values of y can be computed for a binomial experiment for any values of n or π, but the task becomes more difficult when n gets large. For example, suppose a sample of 1,000 voters is polled to determine sentiment toward the consolidation of city and county government. What would be the probability of observing 460 or fewer favoring consolidation if we assume that 50% of the entire population favor the change? Here we have a binomial experiment with $n = 1,000$ and π, the probability of selecting a person favoring consolidation, equal to .5. To determine the probability of observing 460 or fewer favoring consolidation in the random sample of 1,000 voters, we could compute $P(y)$ using the binomial formula for $y = 460, 459, \ldots, 0$. The desired probability would then be

$$P(y = 460) + P(y = 459) + \cdots + P(y = 0).$$

There would be 461 probabilities to calculate with each one being somewhat difficult due to the factorials. For example, the probability of observing 460 favoring consolidation is

$$P(y = 460) = \frac{1000!}{460!540!}(.5)^{460}(.5)^{540}$$

A similar calculation would be needed for all other values of y.

The normal distribution can be used in many situations to approximate the binomial probability distribution, and areas under the normal curve can be used to *approximate* the actual binomial probabilities. The normal distribution that provides the best approximation to the binomial probability distribution has a mean and a standard deviation given by the following formula:

$$\mu = n\pi \qquad \sigma = \sqrt{n\pi(1 - \pi)}.$$

E X A M P L E   4.22

▼

Use the normal approximation to the binomial to compute the probability of observing 460 or fewer in a sample of 1,000 favoring consolidation if we assume that 50% of the entire population favor the change.

Solution   The normal distribution used to approximate the binomial distribution will have

$$\mu = n\pi = 1,000(.5) = 500$$

$$\sigma = \sqrt{n\pi(1 - \pi)} = \sqrt{1,000(.5)(.5)} = 15.8.$$

The desired probability is represented by the shaded area shown in Figure 4.23.

F I G U R E   4.23

Approximating Normal Distribution for the Binomial Distribution of Example 4.22, $\mu = 500$ and $\sigma = 15.8$

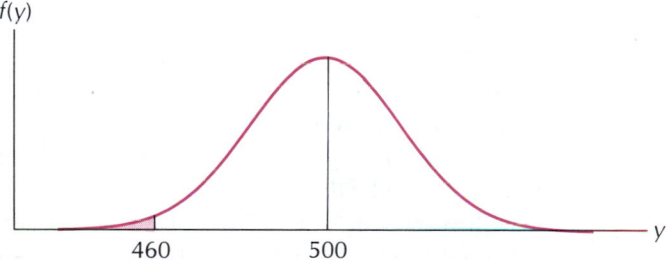

We calculate the desired area by first computing

$$z = \frac{y - \mu}{\sigma} = \frac{460 - 500}{15.8} = -2.53.$$

Referring to Table 1 in the Appendix, we find that the area under the normal curve between 460 and 500 (for $z = 2.53$) is .4943. Thus, the probability of observing 460 or fewer favoring consolidation is approximately .5 − .4943 = .0057. This probability is shown in Table 2 of the Appendix for $z = 2.53$.

▲

The normal approximation to the binomial distribution can be unsatisfactory if $n\pi < 5$ or $n(1 − \pi) < 5$. If π, the probability of success, is small, and n, the sample size, is modest, the actual binomial distribution is seriously skewed to the right. In such a case, the symmetric normal curve will give a bad approximation. If π is near 1, so $n(1 − \pi) < 5$, the actual binomial will be skewed to the left, and again the normal approximation will not be very good. The normal approximation, as described, is quite good when $n\pi$ and $n(1 − \pi)$ exceed about 20. In the middle zone, $n\pi$ or $n(1 − \pi)$ between 5 and 20, a modification called **continuity correction**   a **continuity correction** makes a substantial contribution to the quality of the approximation.

The point of the continuity correction is that we are using the continuous normal curve to approximate a discrete binomial distribution. A picture of the situation is shown in Figure 4.24.

F I G U R E   4.24
Normal Approximation to Binomial

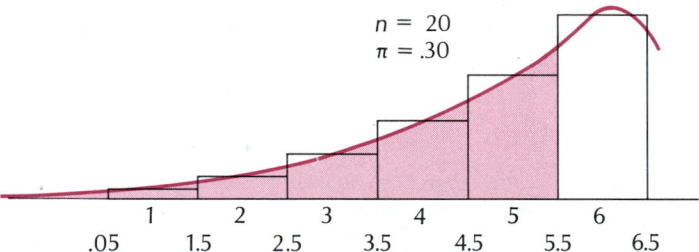

The binomial probability that $y \leq 5$ is the sum of the areas of the rectangle above 5, 4, 3, 2, 1, and 0. This probability (area) is approximated by the area under the superimposed normal curvee to the left of 5. Thus, the normal approximation ignores half of the rectangle above 5. The continuity correction simply includes the area between $y = 5$ and $y = 5.5$. For the binomial distribution with $n = 20$ and $\pi = .30$ (pictured in Figure 4.24), the correction is to take $P(y \leq 5)$ as $P(y \leq 5.5)$. Instead of $P(y \leq 5) = P[z \leq (5 − 20(.3))/\sqrt{20(.3)(.7)}] = P(z \leq −.49) = .3121$,   use $P(y \leq 5.5) = P[z \leq (5.5 − 20(.3))/\sqrt{20(.3)(.7)}] = P(z \leq −.24) = .4052$. The actual binomial probability can be shown to be .4164. The general idea of the continuity correction is to add or subtract .5 from a binomial value before using normal probabilities. The best way to determine whether to add or subtract is to draw a picture like Figure 4.24.

Normal Approximation to the Binomial Probability Distribution

For large n and π not too near 0 or 1, the distribution of a binomial random variable y may be approximated by a normal distribution with $\mu = n\pi$ and $\sigma = \sqrt{n\pi(1 - \pi)}$. This approximation should be used only if $n\pi \geq 5$ and $n(1 - \pi) \geq 5$. A continuity correction will improve the quality of the approximation in cases where n is not overwhelmingly large.

E X A M P L E   4.23

A large drug company has 100 potential new prescription drugs under clinical test. About 20 percent of all drugs that reach this stage are eventually licensed for sale. What is the probability that at least 15 of the 100 drugs are eventually licensed? Assume that the binomial assumptions are satisfied, and use a normal approximation with continuity correction.

Solution   The mean of y is $\mu = 100(.2) = 20$; the standard deviation is $\sigma = \sqrt{100(.2)(.8)} = 4.0$. The desired probability is that 15 or more drugs are approved. Because $y = 15$ is included, the continuity correction is to take the event as y greater than or equal to 14.5.

$$P(y \geq 14.5) = P\left(z \geq \frac{14.5 - 20}{4.0}\right) = P(z \geq -1.375),$$

which is about .92.   ▲

4.14 ▼ SUMMARY

In this chapter, we presented an introduction to probability, probability distributions, and sampling distributions. Knowledge of the probabilities of sample outcomes is vital to a statistical inference. Three different interpretations of the probability of an outome were given: the classical, relative frequency, and subjective interpretations. Although each has a place in statistics, the relative frequency approach has the most intuitive appeal since it can be checked.

Quantitative random variables are classified as either discrete or continuous random variables. The probability distribution for a discrete random variable y is a display of the probability $P(y)$ associated with each value of y. This display may be presented in the form of a histogram, table, or formula.

The binomial is a very important and useful discrete random variable. Many experiments that scientists conduct are similar to a coin-tossing experiment where dichotomous (yes–no) type data are accumulated. The binomial experiment frequently provides an excellent model for computing probabilities of various sample outcomes.

Probabilities associated with a continuous random variable correspond to areas under the probability distribution. Computations of such probabilities were illustrated for areas under the normal curve. The importance of this exercise is borne out by the Central Limit Theorem: any random variable that is expressed as a sum or average will have a normal distribution for sufficiently large sample size. Direct application of the Central Limit Theorem gives the sampling distribution for the sample mean. Since many sample statistics are either sums or averages of random variables, application of the Central Limit Theorem will provide us with information about probabilities of sample outcomes. These probabilities will be vital for the statistical inferences we wish to make.

▼ KEY FORMULAS

1. Binomial probability distribution

$$P(y) = \frac{n!}{y!(n-y)!} \pi^y (1-\pi)^{n-y}$$

2. Sampling distribution for $\bar{y}$

Mean: μ

Standard error: $\sigma_{\bar{y}} = \sigma/\sqrt{n}$

3. Normal approximation to the binomial

$$\mu = n\pi$$

$$\sigma = \sqrt{n\pi(1-\pi)}$$

provided

$n\pi$ and $n(1-\pi)$ greater than or equal to 5,

or, equivalently, if

$$n \geq \frac{5}{\min(\pi, 1-\pi)}$$

▼ SUPPLEMENTARY EXERCISES

$ **4.97** One way to audit expense accounts for a large consulting firm would be to sample all reports dated the last day of each month. Comment on whether such a sample would constitute a random sample.

4.98 Critical key-entry errors in the data-processing operation of a large district bank occur approximately .1% of the time. If a random sample of 10,000 entries is examined, determine the following:
  a. the expected number of errors
  b. the probability of observing less than five errors
  c. the probability of observing less than two errors.

4.99 Use the binomial distribution with $n = 20$, $\pi = .5$ to compare accuracy of the normal approximation to the binomial.
  a. Compute the exact probabilities and corresponding normal approximations for $y < 5$.
  b. The normal approximation can be improved slightly by taking $P(y \leq 4.5)$. Why should this help? Compare your results.
  c. Compute the exact probabilities and corresponding normal approximations with the continuity correction for $P(8 < y < 14)$.

4.100 Let y be a binomial random variable with $n = 10$ and $\pi = .5$.
  a. Calculate $P(4 \leq y \leq 6)$.
  b. Use a normal approximation without the continuity correction to calculate the same probability. Compare your results. How well did the normal approximation work?

4.101 Refer to Exercise 4.100. Use the continuity correction to compute the probability $P(4 \leq y \leq 6)$. Does the continuity correction help?

4.102 A marketing research firm believes that approximately 25% of all persons mailed a "sweepstakes" offer will respond if a preliminary mailing of 5,000 is conducted in a fixed region,
  a. What is the probability that 1,000 or fewer will respond?
  b. What is the probability that 3,000 or more will respond?

4.103 The breaking strengths for 1-foot-square samples of a particular synthetic fabric are approximately normally distributed with a mean of 2,250 pounds per square inch (psi) and a standard deviation of 10.2 psi.
  a. Find the probability of selecting a 1-foot-square sample of material at random that on testing would have a breaking strength in excess of 2,265 psi.
  b. Describe the sampling distribution for $\bar{y}$ based on random samples of 15 one-foot sections.

4.104 Refer to Exercise 4.103. Suppose that a new synthetic fabric has been developed that may have a different mean breaking strength. A random sample of 15 one-foot sections is obtained and each section is tested for breaking strength. If we assume that the population standard deviation for the new fabric is identical to that for the old fabric, give the standard deviation for the sampling distribution of $\bar{y}$ using the new fabric.

4.105 Refer to Exercise 4.104. Suppose that the mean breaking strength for the sample of 15 one-foot sections of the new synthetic fabric is 2268. What is the probability of observing a value of $\bar{y}$ equal to or greater than 2268, assuming that the mean breaking strength for the new fabric is 2250, the same as that for the old?

4.106 Refer to Exercise 4.105. Based on your answer in Exercise 4.105, do you believe the new fabric has the same mean breaking strength as the old? (Assume $\sigma = 10.2$.)

4.107 In Figure 4.21 we visually inspected the relative frequency histogram for 50 sample means, each based on $n = 5$ measurements, and noted its bell shape. Another way to determine whether a set of measurements is bell-shaped (normal) is to construct a plot of the sample data on *probability paper*. This plot is called a *probability plot*. If the probability plot is approximately a straight line, we say the measurements were selected from a normal population. Use a computer program to construct a probability plot for the 50 sample means. A Minitab program is shown below for illustration.

```
MTB > SET C1
MTB > END
MTB > PRINTC1

C1
   12.0     8.8    11.4     5.0     8.8     7.0    12.4    15.0     8.6     8.6    14.2
    7.4    10.2     8.4     7.2    11.2     5.4     7.8     9.6    10.2     9.2     8.8
    6.2     4.4     9.0    10.0     6.4     6.0    10.4     8.4     9.0     7.0    11.0
   11.4     8.6     9.8     8.2     8.2     7.4     5.4     7.4    10.0    13.2     7.4
    8.4     7.0     7.8     7.6     4.2     9.6

MTB > HISTOGRAM C1

Histogram of C1    N = 50

Midpoint    Count
    4         2    **
    5         3    ***
    6         3    ***
    7         8    ********
    8         8    ********
    9         9    *********
   10         8    ********
   11         4    ****
   12         2    **
   13         1    *
   14         1    *
   15         1    *

MTB > NSCORE C1 PUT IN C2
MTB > PLOT C1 VS C2


        -
        -                                                                          *
   14.0+                                                                    *
C1      -                                                               *
        -                                                             *
        -                                                           *
        -                                                     ** 2
   10.5+                                                  2*
        -                                              2*2
        -                                          3 2*
        -                                  23 3
        -                             4 *2
    7.0+                        3  *
        -              ***
        -           2
        -        *
        - *     *
         --------+---------+---------+---------+---------+--------C2
              -1.60     -0.80      0.00      0.80      1.60

MTB > STOP
```

 4.108 A labor union's examining board for the selection of apprentices has a record of admitting 70% of all applicants who satisfy a set of basic requirements. Five members of a minority group recently came before the board, and four out of five were rejected. Find the probability that one or less would be accepted if the record is really .7. Did the board apply a lower probability of acceptance when reviewing the five members of a minority group?

4.109 Suppose that you are a regional director of the IRS office and that you are charged with sampling 1% of the returns with gross income levels above $15,000. How might you go about this? Would you use random sampling? How?

4.110 Experts consider high serum cholesterol levels to be associated with an increased incidence of coronary heart disease. Suppose that the logarithm of cholesterol levels for males in a given age bracket are normally distributed with a mean of 2.35 and a standard deviation of 1.2.

   a. What percent of the males in this age bracket could be expected to have a serum cholesterol level greater than 250 mg/ml, the upper limit of the clinical normal range?

   b. What percent of the males could be expected to have serum cholesterol levels within the clinical normal range of 150–250 mg/ml?

   c. If levels above 300 mg/ml are considered very risky, what percent of the adult males in this age bracket could be expected to exceed 300?

4.111 One of the major soft-drink companies changed the "secret" formula for its leading beverage in order to attract new customers. Recently, a marketing research firm interviewed 1,000 potential new customers and, after giving them a taste of the newly reformulated beverage, determined the number of these individuals planning to buy the reformulated beverage in the near future.

   a. Identify the random variable for the population of $y =$ values of interest.

   b. Can you compute the mean and variance? Why or why not?

   c. How would you calculate $P(y \leq 250)$?

4.112 Many firms are using or exploring the possibility of using telemarketing techniques—that is, marketing their products via the telephone to supplement the more traditional marketing strategies. Assume a firm finds that approximately one in every 100 calls yields a sale.

   a. Find the probability the first sale will occur somewhere in the first 5 calls.

   b. Find the probability the first sale will occur sometime after 10 calls.

4.113 Marketing analysts have determined that a particular advertising campaign should make at least 20% of the adult population aware of the advertised product. After a recent campaign, 25 of 400 adults sampled indicated that they had seen the ad and were aware of the new product.

   a. Find the approximate probability of observing $y \leq 25$ given that 20% of the population is aware of the product through the campaign.

   b. Based on your answer to part a, does it appear the ad was successful? Explain.

4.114 One or more specific, minor birth defects occurs with probability .0001 (that is, 1 in 10,000 births). If 20,000 babies are born in a given geographic area in a given year, can we calculate the probability of observing at least one of the minor defects using the binomial or normal approximation to the binomial? Explain.

4.115 The sample mean is to be calculated from a random sample of size $n = 4$ from a population consisting of six values (0, 1, 2, 4, 6, and 8). Find the sampling distribution of $\bar{y}$ based on a sample of size $n = 4$.

4.116 Plot the sampling distribution of Exercise 4.115. Find the mean and median of the sampling distribution of $\bar{y}$.

4.117 Refer to Exercise 4.115. Use the same population to find the sampling distribution for the sample median based on samples of size $n = 4$.

4.118 Plot the sampling distribution of Exercise 4.117. Find the mean and median of the sampling distribution.

4.119 Random samples of size 20, 40, and 80 are drawn from a population with mean $\mu = 100$ and standard deviation $\sigma = 15$. Give the mean and standard error of the distribution of $\bar{y}$ based on samples of size 20, 40, and 80.

4.120 Refer to Exercise 4.119. For each sampling distribution, find the following probabilities:
a. $P(\bar{y} > 105)$
b. $P(\bar{y} < 96)$
c. $P(96 < \bar{y} < 105)$
d. $1 - P(\bar{y} < 94)$.

4.121 A random sample of $n = 36$ measurements is selected from a population with mean equal to 40 and a standard deviation equal to 12.
a. Describe the sampling distribution of $\bar{y}$.
b. Find $P(\bar{y} > 36)$.
c. Find $P(\bar{y} < 30)$.
d. Find the value of $\bar{y}$ (say k) such that $P(\bar{y} > k) = .05$.

4.122 Refer to Exercise 4.121.
a. Describe the sampling distribution for the sample sum $\sum y_i$.
b. Find $P(\sum y_i > 1,440)$.
c. Find $P(\sum y_i > 1,540)$.
d. Find the value of $\sum y_i$ (say k) such that $P[k < \sum y_i < k_2] = .95$.

4.123 For each of the following situations, find the expected value and standard error of $\bar{y}$ based on a random sample of size n drawn from a population with mean μ and standard deviation σ.
a. $n = 25$, $\mu = 10$, $\sigma = 10$
b. $n = 100$, $\mu = 10$, $\sigma = 10$
c. $n = 25$, $\mu = 10$, $\sigma = 20$
d. $n = 100$, $\mu = 10$, $\sigma = 20$

4.124 Based on the results of Exercise 4.123, speculate on the effect of increasing the sample size and on the effect of an increase in σ on the standard error of $\bar{y}$.

ANALYZING DATA: MEANS, VARIANCES, AND PROPORTIONS

INFERENCES ABOUT μ

5.1 ▾ INTRODUCTION

Inference, specifically decision making and prediction, is centuries old and plays a very important role in our lives. Each of us is faced daily with personal decisions and situations that require predictions concerning the future. The government is concerned with predicting the flow of gold to Europe. A stockbroker wants to know how the stock market will behave. A metallurgist would like to use the results of an experiment to determine whether a new type of steel is more resistant to temperature changes than another. A veterinarian investigates the effectiveness of a new product for treating worms in cattle. The inferences that these individuals make should be based on relevant facts, which we call observations, or data.

In many practical situations the relevant facts are abundant, seemingly inconsistent, and, in many respects, overwhelming. As a result, a careful decision or prediction is often little better than an outright guess. You need only refer to the "Market Views" section of the *Wall Street Journal* to observe the diversity of expert opinion concerning future stock market behavior. Similarly, a visual analysis of data by scientists and engineers often yields conflicting opinions regarding conclusions to be drawn from an experiment.

Many individuals tend to feel that their own built-in inference-making equipment is quite good. But experience suggests that most people are incapable of utilizing large amounts of data, mentally weighing each bit of relevant information, and arriving at a good inference. (You may test your own inference-making ability by using the exercises in Chapters 5 through 8. Scan the data and make an inference before you use the appropriate statistical procedure. Then compare the results.) The statistician, rather than relying upon his or her own intuition, uses statistical results to aid in making inferences. Although we have touched upon some of the notions involved in statistical inference in preceding chapters, we will now collect our ideas in a presentation of some of the basic ideas involved in statistical inference.

The objective of statistics is to make inferences about a population based on information contained in a sample. Populations are characterized by numerical descriptive measures called *parameters*. Typical population parameters are the mean μ, the standard deviation σ, the area under the probability distribution to the right (or left) of some value of the random variable, or the area between two values of the variable. Most practical inferential problems you will encounter can be phrased to imply an inference about one or more parameters of a population. For example, in an experiment in which we wish to predict the average amount of money paid to welfare recipients in a given year, the population of interest is the set of all yearly welfare payments, and we are interested in estimating the value of the population mean μ.

Methods for making inferences about parameters fall into one of two categories. Either we will **estimate** (predict) the value of the population parameter of interest or we will **test a hypothesis** about the value of the parameter. These two methods of statistical inference—estimation and hypothesis testing—involve different procedures, and, more important, they answer two different questions about the parameter. In estimating a population parameter, we are answering the question, "What is the value of the population parameter?" In testing a hypothesis we are answering the question, "Is the parameter value equal to this specific value?"

estimation
hypothesis testing

Consider a study in which an investigator is interested in examining the effectiveness of a drug product in reducing anxiety levels of anxious patients. A screening procedure is employed to identify a group of anxious patients. After the patients are admitted into the study, each one's anxiety level is measured on a rating scale immediately before he or she receives the first dose of the drug and then at the end of one week of drug therapy. These sample data can be used to make inferences about the population from which the sample was drawn either by estimation or by a statistical test:

Estimation: Information from the sample can be used to estimate (or predict) the mean decrease in anxiety ratings for the set of all anxious patients who may conceivably be treated with the drug.

Statistical test: Information from the sample can be used to determine whether the population mean decrease in anxiety ratings is greater than zero.

Notice that the inference related to estimation is aimed at answering the question, "What is the mean decrease in anxiety ratings for the population?" In contrast, the statistical test attempts to answer the question, "Is the mean drop in anxiety ratings greater than zero?"

We will consider estimation of a population mean μ and a statistical test about μ.

▼ EXERCISES

Basic Techniques

 5.1 A researcher is interested in estimating the percentage of registered voters in her state who have voted in at least one election over the past 2 years.
   a. Identify the population of interest to the researcher.
   b. How might you select a sample of voters to gather this information?

5.2 Refer to Exercise 5.1. Is the researcher faced with a problem related to estimation or testing a hypothesis? What is the parameter of interest?

 5.3 A manufacturer claims that the average lifetime of a particular fuse is 1,500 hours. Information from a sample of 35 fuses shows that the average lifetime is 1,380 hours. What can be said about the manufacturer's claim?
   a. Identify the population of interest to us.
   b. Would an answer to the question posed involve estimation or testing a hypothesis?

5.4 Refer to Exercise 5.3. How might you select a sample of fuses from the manufacturer to test the claim?

5.2 ▼ ESTIMATION OF μ

The simplest statistical inference problem is point estimation, where we compute a single value (statistic) from the sample data to estimate a population parameter. Suppose that we are interested in estimating a population mean and that we are willing to assume the underlying population is normal. Then one natural statistic that could be used to estimate the population mean is the sample mean; but we also could use the median and the trimmed mean. Which sample statistic should we use?

A whole branch of mathematical statistics deals with problems related to developing point estimators (the formulas for calculating specific point estimates from sample data) of parameters from various underlying populations and determining whether a particular point estimator has certain desirable properties. Fortunately, we will not have to derive these point estimators— they'll be given to us for each parameter. Then, knowing which point estimator (formula) to use for a given parameter, we can develop confidence intervals (interval estimates) for these same parameters.

In this section, we deal with point and interval estimation of a population mean μ. Tests of hypotheses about μ are covered in Section 5.5.

For most problems in this text, the sample mean $\bar{y}$ will be used as a point estimate of μ; it is also used to form an interval estimate for the population mean μ. From the Central Limit Theorem for the sample mean given in Chapter 4, we know that for large n (crudely $n \geq 30$), $\bar{y}$ will be approximately normally distributed, with a mean μ and a standard error $\sigma_{\bar{y}}$. Then from our knowledge of the Empirical Rule and areas under a normal curve, we know that the interval $\mu + 2\sigma_{\bar{y}}$, or more precisely, the interval $\mu \pm 1.96\sigma_{\bar{y}}$, includes 95% of the $\bar{y}$'s in repeated sampling, as shown in Figure 5.1.

FIGURE 5.1
Sampling Distribution for $\bar{y}$

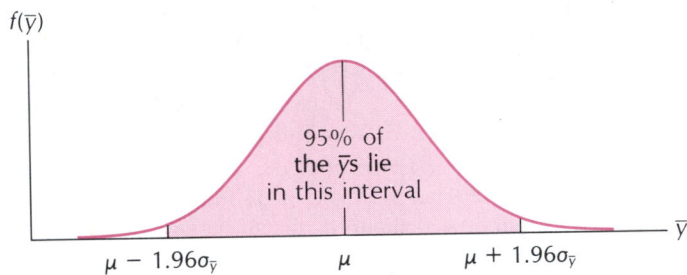

Consider the interval $\bar{y} \pm 1.96\sigma_{\bar{y}}$. Any time $\bar{y}$ lies in the interval $\mu \pm 1.96\sigma_{\bar{y}}$, the interval $\bar{y} \pm 1.96\sigma_{\bar{y}}$ will contain the parameter μ (see Figure 5.2) and this will occur with probability .95. The interval $\bar{y} \pm 1.96\sigma_{\bar{y}}$ represents an interval estimate of μ.

We evaluate the goodness of an interval estimation procedure by examining the fraction of times in repeated sampling that interval estimates would encompass the parameter to be estimated. This fraction, called the **confidence coefficient**, is .95 when using the formula $\bar{y} \pm 1.96\sigma_{\bar{y}}$. That is, 95% of the time in repeated sampling, intervals calculated using the formula $\bar{y} \pm 1.96\sigma_{\bar{y}}$ will contain the mean μ.

confidence coefficient

FIGURE 5.2
When the Observed Value of $\bar{y}$ Lies in the Interval $\mu \pm 1.96\sigma_{\bar{y}}$, the Interval $\bar{y} \pm 1.96\sigma_{\bar{y}}$ Contains the Parameter μ

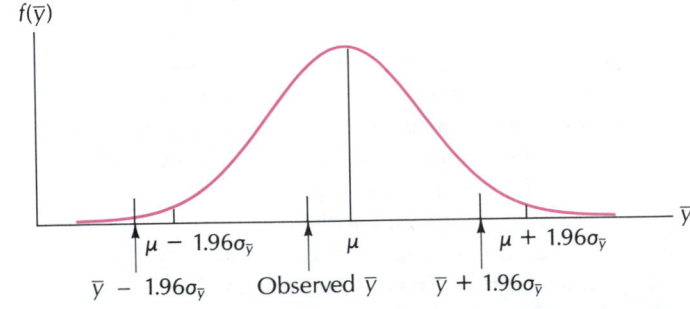

FIGURE 5.3
**Twenty Interval Estimates
Computed by Using
$\bar{y} \pm 1.96\sigma_y$**

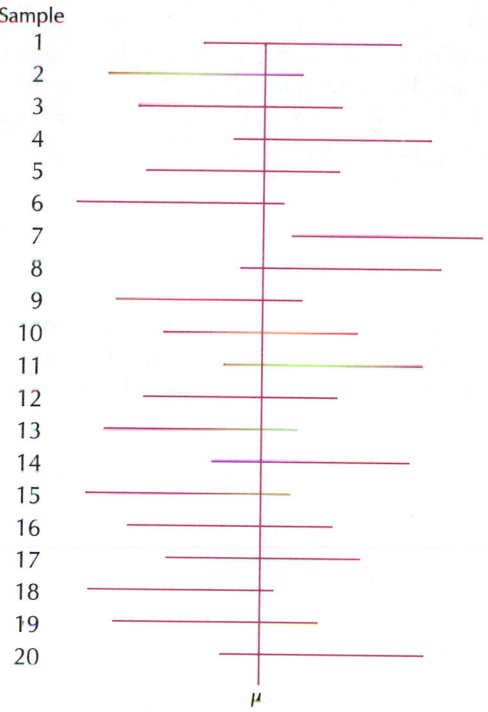

Sample

μ

This idea is illustrated in Figure 5.3. Twenty different samples are drawn from a population with mean μ and variance σ^2. For each sample, an interval estimate is computed using the formula $\bar{y} \pm 1.96\sigma_{\bar{y}}$. Note that although the intervals bob about, most of them capture the parameter μ. In fact, if we repeated the process of drawing samples and computing confidence intervals, 95% of the intervals so formed would contain μ.

In a given experimental situation, we calculate only one such interval. This interval, called a **95% confidence interval**, represents an interval estimate of μ.

95% confidence interval

EXAMPLE 5.1 ▼

In a random sample of $n = 36$ parochial schools throughout the south, the average number of pupils per school is 379.2, with a standard deviation of 124. Use the sample to construct a 95% confidence interval for μ, the mean number of pupils per school for all parochial schools in the south.

Solution The sample data indicate that $\bar{y} = 379.2$ and $s = 124$. The appropriate 95% confidence interval is then computed by using the formula

$$\bar{y} \pm 1.96\sigma_{\bar{y}},$$

where $\sigma_{\bar{y}} = \sigma/\sqrt{n}$. In Section 5.8 we present a procedure for obtaining a confidence interval for μ when σ is unknown. However, for all practical purposes, if the sample size is 30 or more, we can estimate the population standard deviation σ with s in the confidence interval formula. With s replacing σ, our interval is

$$379.2 \pm 1.96 \frac{124}{\sqrt{36}} \quad \text{or} \quad 379.2 \pm 40.51.$$

The interval from 338.69 to 419.71 forms a 95% confidence interval for μ. In other words, we are 95% sure that the average number of pupils per school for parochial schools throughout the south lies between 338.69 and 419.71.   ▲

There are many different confidence intervals for μ, depending on the confidence coefficient we choose. For example, the interval $\mu \pm 2.58\sigma_{\bar{y}}$ includes 99% of the values of $\bar{y}$ in repeated sampling (see Figure 5.4), and the interval $\bar{y} \pm 2.58\sigma_{\bar{y}}$

99% confidence interval   forms a **99% confidence interval** for μ.

FIGURE 5.4
Sampling Distribution of $\bar{y}$

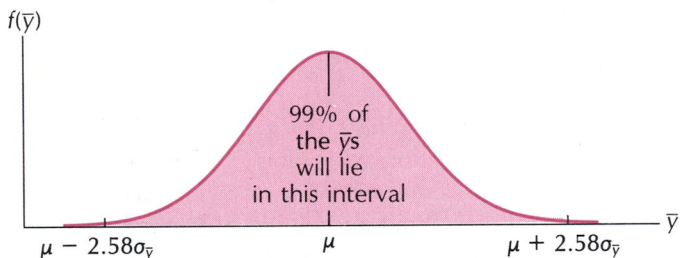

We can state a general formula for a confidence interval for μ with a

(1 − α) = confidence coefficient   **confidence coefficient of (1 − α)**, where α (Greek letter alpha) is between 0 and 1. For a specified value of $(1 − \alpha)$, a $100(1 − \alpha)\%$ confidence interval for μ is given by the following formula. Here we assume that σ is known or that the sample size is large enough to replace σ with s.

Confidence interval for μ; σ Known

$$\bar{y} \pm z_{\alpha/2}\sigma_{\bar{y}}, \text{ where } \sigma_{\bar{y}} = \sigma/\sqrt{n}$$

$z_{\alpha/2}$ The quantity $\mathbf{z}_{\alpha/2}$ is a value of z having a tail area of $\alpha/2$ to its right. In other words, at a distance of $z_{\alpha/2}$ standard deviations to the right of μ, there

FIGURE 5.5
Interpretation of $z_{\alpha/2}$ in the Confidence Interval Formula

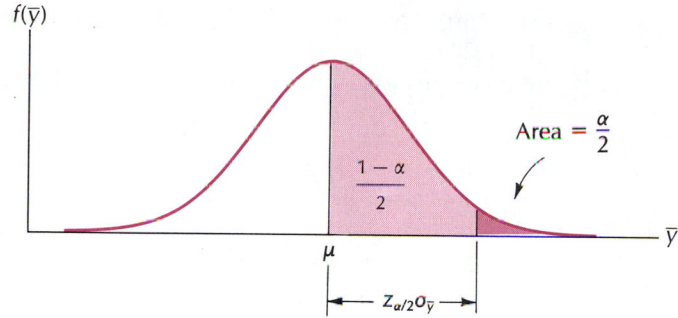

is an area of $\alpha/2$ under the normal curve. Values of $z_{\alpha/2}$ can be obtained from Table 1 in the Appendix by looking up the z-value corresponding to an area of $(1 - \alpha)/2$ (see Figure 5.5). Common values of the confidence coefficient $(1 - \alpha)$ and $z_{\alpha/2}$ are given in Table 5.1.

TABLE 5.1
Common Values of the Confidence Coefficient $(1 - \alpha)$ and the Corresponding z-Value, $z_{\alpha/2}$

Confidence Coefficient $(1 - \alpha)$	Area in Table 1 $(1 - \alpha)/2$	Value of $\alpha/2$	Corresponding z-Value, $z_{\alpha/2}$
.90	.45	.05	1.645
.95	.475	.025	1.96
.98	.49	.01	2.33
.99	.495	.005	2.58

EXAMPLE 5.2

A forester is interested in estimating the average number of "count trees" per acre (trees larger than a specified size) on a 2,000-acre plantation. She can then use this information to determine the total timber volume for trees in the plantation. A random sample of $n = 50$ 1-acre plots is selected and examined. The average (mean) number of count trees per acre is found to be 27.3, with a standard deviation of 12.1. Use this information to construct a 99% confidence interval for μ, the mean number of count trees per acre for the entire plantation.

Solution   We use the general confidence interval with confidence coefficient equal to .99 and a $z_{\alpha/2}$-value equal to 2.58 (see Table 5.1). Substituting into the formula $\bar{y} \pm 2.58\sigma_{\bar{y}}$ and replacing σ with s in $\sigma_{\bar{y}} = \sigma/\sqrt{n}$, we have

$$27.3 \pm 2.58 \frac{12.1}{\sqrt{50}}.$$

> This corresponds to the confidence interval 27.3 ± 4.42, that is, the interval from 22.88 to 31.72. Thus, we are 99% sure that the average number of count trees per acre is between 22.88 and 31.72.   ▲

The discussion in this section has included one rather unrealistic assumption; namely, that the population standard deviation is known. In practice, it's difficulat to find situations in which the population mean is unknown, but the standard deviation is known. Usually both the mean and the standard deviation must be estimated from the sample. Since σ is estimated by the sample standard deviation s, the actual standard error of the mean, $\sigma/\sqrt{n}$, is

substituting s for σ

naturally estimated by $s/\sqrt{n}$. This estimation introduces another source of random error (s will vary randomly, from sample to sample, around σ) and, strictly speaking, invalidates our confidence interval formula. Fortunately, the formula is still a very good approximation for large sample sizes. As a very rough rule, we can use this formula when n is larger than 30; a better way to handle this issue is described in Section 5.3.

Statistical inference-making procedures differ from ordinary procedures in that we not only make an inference, but also provide a measure of how good that inference is. For interval estimation, the width of the confidence interval and the confidence coefficient measure the goodness of the inference. Obviously, for a given confidence coefficient, the smaller the width of the interval, the better the inference. The confidence coefficient, on the other hand, is set by the experimenter to express how much assurance he or she places in whether the interval estimate encompasses the parameter of interest.

▼ EXERCISES

Basic Techniques

5.5 The sample mean and standard deviation based on a sample of 50 measurements are $\bar{y} = 105$ and $s = 11$.
   a. Calculate a 95% confidence interval for μ.
   b. Calculate a 99% confidence interval for μ.

5.6 Give a careful verbal interpretation of the confidence interval in part a of Exercise 5.5.

5.7 Refer to Exercise 5.5.
   a. Discusss the impact of doubling the sample size from $n = 50$ to $n = 100$ on the 95% confidence interval. Assume for discussion purposes that $\bar{y}$ and s are still 105 and 11, respectively.
   b. What impact would quadrupling the sample size have? (Note: Answer this question without doing the calculations.)

Applications

5.8 The caffeine content (in milligrams, mg) was examined for a random sample of 50 cups of black coffee dispensed by a new machine. The mean and standard deviation were 110 mg and 7.1 mg, respectively. Use these data to construct a 98% confidence interval for μ, the mean caffeine content for cups dispensed by the machine.

5.9 A random sample of the year-end statements of 22 small businesses (under $500,000 in sales) in a city shows the mean gross profit margin to be 5.2% (of sales) with a standard deviation of 3.3%. Use these data to place a 90% confidence interval for μ. Assume $\sigma \approx 3.3$.

5.10 Recent data from a national survey of 1,350 women indicated that the average woman goes to a hair salon once every 5 weeks and spends on the average $26.40. With a standard deviation of $12.00, use these data to construct a 99% confidence interval for μ.

5.11 A social worker is interested in estimating the average length of time spent outside of prison for first offenders who later commit a second crime and are sent to prison again. A random sample of $n = 150$ prison records in the county courthouse indicates that the average length of prison-free life between first and second offenses is 3.2 years, with a standard deviation of 1.1 years. Use the sample information to estimate μ, the mean prison-free life between first and second offenses for all prisoners on record in the county courthouse. Construct a 95% confidence interval for μ. Assume that σ can be replaced by s.

5.12 Refer to Exercise 5.9. What impact would a doubling of the sample size have on the confidence interval?

5.13 The rust mite, a major pest of citrus in Florida, punctures the cells of the leaves and fruit. Damage by rust mites is readily recognizable because the injured fruit will display a brownish (rust) color and be somewhat reduced in size depending on the severity of the attack. If the rust mites are not controlled, the affected groves will have a substantial reduction in both the fruit yield and the fruit quality. In either case, the citrus grower suffers financially since the produce will be of a lower grade and sell for less on the fresh fruit market. This year, more and more citrus growers have gone to a preventive program of maintenance spraying for rust mites. In evaluating the effectiveness of the program, a random sample of sixty 10-acre plots, one plot from each of 60 groves, is selected. These show an average yield of 850 boxes, with a standard deviation of 100 boxes. Give a 95% confidence interval for μ, the average (10-acre) yield for all groves utilizing such a maintenance spraying program. Assume that σ can be replaced by s.

5.14 An experiment is conducted to examine the susceptibility of root stocks of a variety of lemon trees to a specific larva. Forty of the plants are subjected to the larvae and examined after a fixed period of time. The response of interest is the logarithm of the number of larvae per gram that is counted on each root stock. For these 40 plants the sample mean is 9.02 and the standard deviation is 1.12. Use these data to construct a 90% confidence interval for μ, the mean susceptibility for the population of lemon tree root stocks from which the sample was drawn. Assume that σ can be replaced by s.

5.15 A mobility study is conducted among a random sample of 900 high-school graduates of a particular state over the past 10 years. For each of the persons sampled, the distance between the high school attended and the present permanent address is recorded. For these data, $\bar{y} = 430$ miles and $s = 262$ miles. Using a 95% confidence interval, estimate the average number of miles between a person's high school and present permanent address for high school graduates of the state over the past 10 years. Assume that σ can be replaced by s.

5.16 A problem of interest to the United States, other governments, and world councils concerned with the critical shortage of food throughout the world is finding a method to estimate the total amount of grain crops that will be produced throughout the world in a particular year.

One method of predicting total crop yields is based on satellite photographs of the earth's surface. Because a scanning device will read the total acreage of a particular type of grain with error, it will be necessary to have the device read many equal-sized plots of a particular planting in order to calibrate the reading on the scanner with the actual acreage. Satellite photographs of 100 50-acre plots of wheat were read by the scanner and gave a sample average and standard deviation

$$\bar{y} = 3.27 \qquad s = .23$$

Find a 95% confidence interval for the mean scanner reading for the population of all 50-acre plots of wheat. Explain the meaning of this interval.

5.17 Another agricultural problem concerns the production of protein, an important component of human and animal diets. Although it is common knowledge that grains and legumes contain high amounts of protein, it is not as well known that certain grasses provide a good source of protein. For example, Bermuda grass contains approximately 20% protein by weight. In a study to verify these results, 100 1-pound samples were analyzed for protein content. The mean and standard deviation of the sample were

$$\bar{y} = .18 \text{ pound} \qquad s = .08 \text{ pound}$$

Estimate the mean protein content per pound for the Bermuda grass from which this sample was selected. Use a 95% confidence interval. Explain the meaning of this interval.

5.3 ▼ CHOOSING THE SAMPLE SIZE FOR ESTIMATING μ

How can we deterine the number of observations to be included in the sample? The implications of such a question are clear. Data collection costs money. If the sample is too large, time and talent are wasted. Conversely, it is wasteful if the sample is too small, because inadequate information has been purchased for the time and effort expended. Also, it may be impossible to increase the sample size at a later time. Hence, the number of observations to be included in the sample will depend on the amount of information the experimenter wants to buy.

Suppose we want to estimate the average amount for accident claims filed against an insurance company. To decide how many claims must be examined, we would have to determine how accurate the company wants to be. For example, the company might indicate that the tolerable error is to be 10 units (± 5 units) or less. Then we would want the confidence interval to be of the form $\bar{y} \pm 5$.

There are two considerations in choosing the appropriate sample size for estimating μ using a confidence interval. The tolerable error establishes the desired width of the confidence interval; the second consideration is the confidence level that should be selected. A wide confidence interval would not be very informative, but the cost of obtaining a narrow confidence interval could be quite large. Similarly, too low a confidence level (say 50%) would mean that the stated confidence interval is likely to be in error, but obtaining a higher level of confidence also would be more expensive.

What constitutes reasonable certainty? In most situations, the confidence level is set at 95% or 90%, partly because of tradition and partly because these levels represent (to some people) a reasonable level of certainty. The 95% (or 90%) level translates into a long-run chance of 1 in 20 (or 1 in 10) of not covering the population parameter. This seems reasonable and is comprehensible, whereas 1 chance in 1,000 or 1 in 10,000 is just too small.

The tolerable error depends heavily on the context of the problem, and only someone who is familiar with the situation can make a reasonable judgment about its magnitude.

When considering a confidence interval for a population mean μ, the plus-or-minus term of the confidence interval is $z_{\alpha/2}\sigma_{\bar{y}}$, where $\sigma_{\bar{y}} = \sigma/\sqrt{n}$. Three quantities determine the value of the plus-or-minus term: the desired confidence level (which determines the z-value used), the standard deviation (σ), and the sample size (which together with σ determines the standard error $\sigma_{\bar{y}}$). Usually, a guess must be made about the size of the population standard deviation. (Sometimes an initial sample is taken to estimate the standard deviation; this estimate provides a basis for determining the additional sample size that will be needed.) For a given tolerable error, once the confidence level is specified and an estimate of σ supplied, the required sample size can be calculated using the formula shown here.

If a 95% confidence interval is to be of the form $\bar{y} \pm E$, then solve the expression

$$1.96\sigma_{\bar{y}} = E$$

for n. The width of the interval is $2E$.

In general, if we want to estimate μ using a $100(1 - \alpha)\%$ confidence interval of the form $\bar{y} \pm E$, where E is specified, then we solve the equation

$$z_{\alpha/2}\sigma_{\bar{y}} = E$$

for n. This is shown here.

Sample Size Required for a 100(1 − α)% Confidence Interval for μ of the Form $\bar{y} \pm E$

▼

$$n = \frac{(z_{\alpha/2})^2 \sigma^2}{E^2}$$

Note that determining a sample size to estimate μ requires knowledge of the population variance σ^2 (or standard deviation σ). We can obtain an approximate sample size by estimating σ^2, using one of these two methods:

1. Employ information from a prior experiment to calculate a sample variance s^2. This value is used to approximate σ^2.
2. Use information on the range of the observations to obtain an estimate of σ.

We would then substitute the estimated value of σ^2 in the sample-size equation to determine an approximate sample size n.

We illustrate the procedure for choosing a sample size with two examples.

EXAMPLE 5.3

▼

Union officials are concerned about reports of inferior wages paid to a company's employees under their jurisdiction. It is decided to take a random sample of n wage sheets from the company to estimate the average hourly wage. If it is known that wages in the company have a range of $10 per hour, determine the sample size required to estimate the average hourly wage μ using a 95% confidence interval with width equal to $1.20.

Solution   Since we want a 95% confidence interval with width $1.20, $E = \$.60$. The value that we use to substitute for σ is range/4 = 2.50. Substituting into the formula for n we have

$$n = \frac{(1.96)^2(2.5)^2}{(.60)^2} = 66.69.$$

To be on the safe side, we will round this number up to the next integer. A sample size of 67 should give a 95% confidence interval with the desired width of $1.20.   ▲

EXAMPLE 5.4 ▼

A federal agency has decided to investigate the advertised weight printed on cartons of a certain brand of cereal. The company in question periodically samples cartons of cereal coming off the production line to check their weight. A summary of 1,500 of the weights made available to the agency indicates a mean weight of 11.80 ounces per carton and a standard deviation of .75 ounce. Use this information to determine the number of cereal cartons the federal agency must examine to estimate the average weight of cartons being produced now, using a 99% confidence interval of width .50.

Solution   The federal agency has specified that the width of the confidence interval is to be .50, so $E = .25$. Assuming that the weights made available to the agency by the company are accurate, we can take $\sigma = .75$. The required sample size with $z_{\alpha/2} = 2.58$ is

$$n = \frac{(2.58)^2(.75)^2}{(.25)^2} = 59.91.$$

That is, the federal agency must obtain a random sample of 60 cereal cartons to estimate the mean weight to within $\pm.25$.   ▲

▼ EXERCISES

Basic Techniques

5.18   Refer to Example 5.3.
   a. How large a sample is needed to obtain a 90% confidence interval with width $.60? $.30? $.15?
   b. In general, for a given confidence level, how much would you increase the sample size to cut the width in half?

Applications

5.19   The giant size of a new "tough cleaning" laundry detergent has a listed net weight of 42 ounces. If the variability in weight has a standard deviation of 2 ounces, how many boxes must be sampled to estimate the average fill weight to within $\pm.25$ ounce using a 95% confidence interval?

5.20   Refer to Exercise 5.19. Determine the effect of a 90% and a 99% confidence level on the required sample size.

5.21   A biologist would like to estimate the effect of an antibiotic on the growth of a particular bacterium by examining the mean amount of bacteria present per plate of culture when a fixed amount of the antibiotic is applied. Previous experimentation with the antibiotic on this type of bacterium indicates that the standard deviation of the amount of bacteria present is approximately 13 cm². Use this information to determine the number of observations (cultures that must be developed and then tested) to estimate the mean amount of bacteria present, using a 99% confidence interval with a half-width of 3 cm².

5.22 Investigators would like to estimate the average annual taxable income of apartment dwellers in a city to within $500 using a 95% confidence interval. If we assume the annual incomes range from $0 to $40,000, determine the number of observations that should be included in the sample.

5.23 Refer to Exercise 5.22. Determine the required sample size if the desired error in a 95% confidence interval is $E = 250$. Do the same for $E = 1,000$. Compare your results to those of Exercise 5.22.

5.24 As part of a much larger study of trends in long-distance telephone usage, a study is to be conducted this month of residential homes with married couples between 25 and 40 years of age. How large a sample should be taken if the mean number of long-distance calls for the month is to be estimated to within one call using a 90% confidence interval? Assume $\sigma \approx 4.0$.

5.4 ▼ QUALITY CONTROL: $\bar{y}$-CHARTS

control chart

We can extend the notion of a confidence interval for μ to obtain a **control chart**. As consumers, we are vitally interested in product quality. We expect product quality for a particular item to be uniform from one time period to another, and we expect it to live up to the product description advertised by the manufacturer. For example, in buying paint from a paint store, we expect different gallons of the same color to be uniform in color and we expect the color to be identical to that advertised in the paint-sample brochure. Similarly, the Food and Drug Administration (FDA) not only expects but also demands that drug products have uniform potency and meet the standards advertised by the pharmaceutical firm.

Consumers are not the only people interested in product quality. Reputable manufacturers are also concerned that their products meet the standards they have claimed. If the quality of a product falls below the standards advertised by the company, then there is a risk that consumers will reject the product and buy from a competitor. Similarly, if product quality drifts above the standards established by the company, then it would be in the company's interest to upgrade their advertising to reflect the increase in quality.

quality control

Quality-control techniques have been developed to monitor the ongoing quality of a manufacturing process in order to maintain uniform quality or at least to detect when product quality has shifted. We can monitor product quality of a production process by using a graph called a *control chart*. Thus, we could graph the sample mean or sample range for samples collected over a period of time to monitor product quality. We discuss the $\bar{y}$-chart in this section; it is used to examine whether the mean output of a process is "in control." *R*-charts and *s*-charts for process variability are discussed in Chapter 7.

Typically a control chart consists of three lines: a center line, an upper control line, and a lower control line. In a control chart for the mean, successive sample means would be plotted much as they appear in Figure 5.6. The sample means are shown by the dots in Figure 5.6. If one of the sample means falls outside either the upper or lower control lines, the process is judged to be out of control; that is, it appears that product quality has shifted. At this point, company officials and production personnel would try to establish the cause of the shift and would initiate corrective changes in the production process.

F I G U R E 5.6
ȳ-Chart for Sample Means

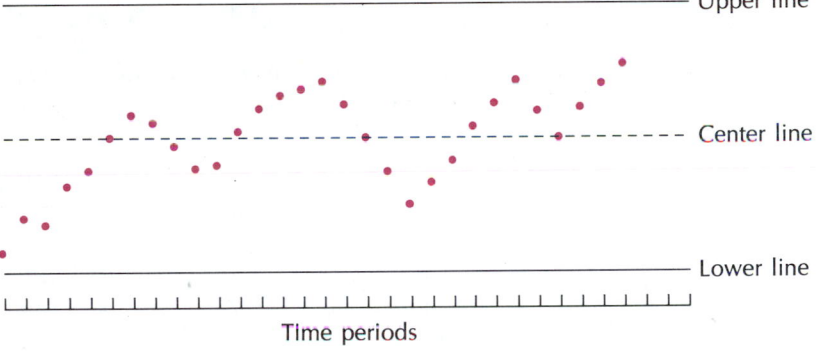

center line

Establishing the three control lines is quite simple. The **center line** (denoted by $\bar{y}_c$) represents the average of k sample means, each based on n observations. We generally recommend taking $k \geq 25$ and $n > 3$. These samples should be taken at some time when the process is judged to be under control (stable and predictable). Then, if we let y_{ij} denote the jth observation in sample i and $\bar{y}_i = \sum_j y_{ij}/n$ denote the mean for sample i, the average of the k sample means is

$$\bar{y}_c = \sum_i \frac{\bar{y}_i}{k} = \frac{\sum_{ij} y_{ij}}{nk}.$$

UCL and LCL for mean quality

The **upper control limit (UCL)** and the **lower control limit (LCL)** are computed as follows:

$$\text{UCL} = \bar{y}_c + 3\frac{\sigma}{\sqrt{n}} \quad \text{and} \quad \text{LCL} = \bar{y}_c - 3\frac{\sigma}{\sqrt{n}}.$$

From knowledge of the Empirical Rule, the interval $\bar{y}_c \pm (3\sigma/\sqrt{n})$ should contain nearly all the sample means $\bar{y}_i$ in repeated sampling. If a sample mean falls outside this interval, we have either observed an extremely unlikely event or the process quality has changed and $\bar{y}_c$ is no longer an accurate measure of the actual mean product quality. This latter conclusion is more realistic and is used to signal a manufacturing process out of control.

The standard deviation σ in the formulas for the upper and lower control lines can be estimated either by using a pooled sample variance from the k samples or, more quickly, by using the k sample ranges. We will employ the latter procedure. Letting r_i denote the range for the n sample measurements in sample i and $\bar{r}$ the average of the k sample ranges, we can estimate σ by

estimate of σ for control chart

$$\hat{\sigma} = \frac{\bar{r}}{d_n},$$

where d_n is obtained from Table 16 of the Appendix. For example, suppose we have $k = 20$ different samples of $n = 7$ observations per sample and $\bar{r} = 5$. Then $d_7 = 2.704$ and $\hat{\sigma} = 5/2.704 = 1.849$.

EXAMPLE 5.5   ▼

A company that dyes rugs is interested in monitoring the color uniformity of its product over time. Although maintaining uniform color is somewhat important for patterned or multicolored rugs, it is much more important for solid–colored rugs, where minor changes in solid colors are readily recognizable. Rug-color quality can be monitored by taking readings on a colorimeter. Twenty-five samples of five measurements each from a rug being dyed red yielded the data listed in Table 5.2. These data were obtained while the manager believed that the process was in control.

Use the data of Table 5.2 to construct a control chart for the mean colorimeter reading.

Solution   From Table 5.2 we have

$$\bar{y}_c = \frac{\sum_{ij} y_{ij}}{nk} = \frac{258.2}{5(25)} = 2.07$$

$$\bar{r} = \frac{\sum_i r_i}{k} = \frac{56.8}{25} = 2.27.$$

From Table 16 in the Appendix, we have $d_5 = 2.326$ and, hence,

$$\hat{\sigma} = \frac{\bar{r}}{d_n} = \frac{2.27}{2.326} = .98.$$

TABLE 5.2
Colorimeter Readings for the 25 Samples of Example 5.5

Sample	Observation	Sample Sum	Sample Range
1	2.4, 1.8, 0.7, 1.0, 2.5	8.4	1.8
2	2.3, 3.0, 2.5, 1.2, 3.1	12.1	1.9
3	1.3, 1.2, 0.9, 1.2, 3.0	7.6	2.1
4	0.5, 2.2, 2.4, 1.5, 3.0	9.6	2.5
5	2.8, 1.9, 2.6, 1.3, 2.9	11.5	1.6
6	2.4, 3.1, 1.7, 3.3, 2.6	13.1	1.6
7	2.5, 2.9, 1.4, 4.0, 2.1	12.9	2.6
8	1.1, 2.9, 3.0, 1.4, 2.8	11.2	1.9
9	3.3, 2.2, 2.7, 2.8, 2.1	13.1	1.2
10	0.8, 4.2, 2.3, 1.4, 2.1	10.8	3.4
11	0.2, 2.6, 2.3, 0.7, 4.2	10.0	4.0
12	1.8, 1.6, 2.3, 2.1, 1.7	9.5	.7
13	0.1, 3.9, 2.3, 1.4, 1.0	8.7	3.8
14	1.1, 3.1, 1.8, 0.9, 1.8	8.7	2.2
15	0.5, 0.9, 4.0, 2.2, 2.8	10.4	3.5
16	2.9, 3.3, 1.9, 3.1, 2.3	13.5	1.4
17	3.5, 2.0, 2.5, 2.0, 0.3	10.3	3.2
18	2.5, 2.1, 2.7, 1.7, 1.5	10.5	1.2
19	1.1, 3.9, 2.7, 1.2, 1.3	10.2	2.8
20	1.4, 2.0, 2.5, 4.2, 2.4	12.5	2.8
21	2.2, 1.9, 0.7, 1.3, 1.4	7.5	1.5
22	1.5, 1.5, 1.1, 2.3, 2.4	8.8	1.3
23	2.2, 1.3, 2.5, 1.9, 0.7	8.6	1.8
24	1.7, 0.1, 1.8, 0.7, 2.1	6.4	2.0
25	3.7, 1.5, 1.9, 0.6, 4.6	12.3	4.0
Totals		258.2	56.8

The center line is then 2.07, with upper and lower control lines given by

$$\text{UCL} = \bar{y}_c + 3\frac{\hat{\sigma}}{\sqrt{n}} = 2.07 + \frac{3(.98)}{\sqrt{5}} = 3.38$$

$$\text{LCL} = \bar{y}_c - 3\frac{\hat{\sigma}}{\sqrt{n}} = 2.07 - \frac{3(.98)}{\sqrt{5}} = .76.$$

Note that this is a conceptually different situation from the classical confidence interval for μ, where there is a single value of μ that we are trying to "discover" (estimate). Here a process can change over time and we are trying to "track" the value of μ. ▲

As stated previously, an observation falling outside one of the control lines is a signal that something has changed. If σ is known and the control lines are computed by using the known value of σ, a value outside a control line would suggest to us that the process mean has shifted. Unfortunately, when σ is unknown and must be estimated, a value outside one of the control lines could suggest a shift in the mean quality, an increase in σ, or both.

To protect ourselves, we should also keep a control chart on product quality variability. Additional information about these control charts for process variability (r-charts and s-charts) is presented in Chapter 7.

▼ EXERCISES

Basic Techniques

5.25 Refer to Example 5.5. Graph the upper and lower control limits and the center line. Plot the sequence of sample means listed here to determine if and when the process is out of control. (Note: Each mean is based on five measurements.)

2.0   1.9   1.6   1.5   1.7   1.8   2.2   2.1   2.0   2.3   2.4   2.7   2.8   2.9

5.26 Fifty random samples of size $n = 8$ are selected from a process. The means of the sample means and ranges $\bar{y}_c$ and $\bar{r}$ are found to be .640 and .015, respectively. Construct a $\bar{y}$-chart for μ for this process.

5.27 Refer to Exercise 5.26. A sample of size $n = 8$ is selected from the process after the control chart was constructed. The sample mean is found to be .647. Does the process seem to be in control? Why or why not?

Applications

5.28 The labeled amount of ingredient A in a marketed cough drop is 1 in 1,500 parts of the total labeled weight (2.2 g). The assay for the ingredient is based on ten cough drops that have been dissolved. A sequence of 48 sample means is shown here, expressed as a percent of the total labeled weight. If $\bar{r} = 4.1(\%)$, determine the center line and the upper and lower control limits for μ.

108	110	110
108	107	112
109	107	111
107	108	111
109	108	110
108	109	110
109	108	112
109	110	110
107	109	110
107	108	112
109	107	110
111	110	111
109	111	113
107	111	111
106	110	110
107	111	110

5.5 ▼ A STATISTICAL TEST FOR μ

The second type of inference-making procedure is statistical testing (or hypothesis testing). As with estimation procedures, we will make an inference about a population parameter, but here the inference will be of a different sort. In this section, we will present a statistical test that will lead to an answer to the question, "Is the population mean equal to a specified value μ_0?" For example, in studying the antipsychotic properties of an experimental compound, we might ask whether the average shock-avoidance response for rats treated with a specific dose is 60, the same value that has been observed after extensive testing using a suitable standard drug.

statistical test

A **statistical test** is based on the concept of proof by contradiction and is composed of the five parts listed here.

Five Parts of a Statistical Test

▼
1. Null hypothesis, denoted by H_0
2. Research hypothesis (also called the alternative hypothesis), denoted by H_a
3. Test statistic, denoted by T.S.
4. Rejection region, denoted by R.R.
5. Conclusion

research hypothesis, H_a

For example, in setting up a statistical test concerning the mean yield per acre (in bushels) for a particular variety of soybeans, we may be interested in the **research hypothesis** that the mean yield per acre μ is greater than 520 bushels, the average observed for farms throughout a particular state in the past several years. To verify the research hypothesis, we try to contradict another hypothesis,

null hypothesis, H_0

called the **null hypothesis**, that $\mu = 520$ (i.e., the highest average yield that is still not part of the research hypothesis).

Having stated the null and research hypotheses, we then obtain a random sample of 1-acre yields from farms throughout the state and compute $\bar{y}$ and s, the sample mean and standard deviation, respectively. The decision to accept the

test statistic, T.S.

null hypothesis or reject it in favor of the research hypothesis is based on a **test statistic** or decision maker computed from the sample data. If the population can be assumed to be more or less mound-shaped, a logical choice as a decision maker for μ would be $\bar{y}$ or some function of the sample mean.

If we choose $\bar{y}$ as the test statistic, we know that the sampling distribution of $\bar{y}$, assuming the null hypothesis is true, is approximately normal with mean $\mu = 520$. Values of $\bar{y}$ that are contradictory to the null hypothesis and are in favor of the research hypothesis will be those that lie in the upper tail of the

rejection region, R.R.

distribution of $\bar{y}$. See Figure 5.7. These contradictory values form a **rejection**

FIGURE 5.7

Assuming That H_0 is True, Contradictory Values of $\bar{y}$ Are in the Upper Tail

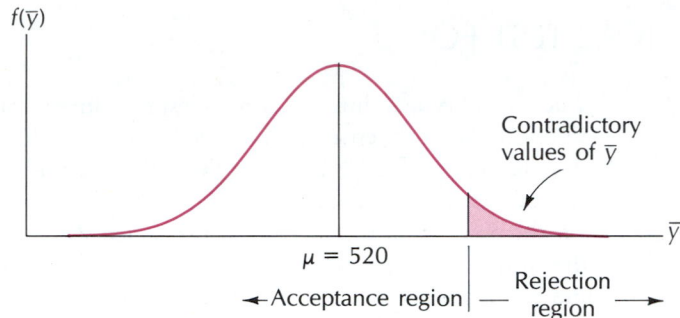

region for our statistical test. If the observed value of $\bar{y}$ falls in the rejection region of Figure 5.7, we would reject the null hypothesis that the mean yield per acre is $\mu = 520$ in favor of the research hypothesis that $\mu > 520$. Note that we are supporting the research hypothesis by contradicting the null hypothesis. If the observed value of $\bar{y}$ falls in the acceptance region rather than in the rejection region, we do not reject the null hypothesis. However, this does not mean that we automatically *accept* the null hypothesis that $H_0: \mu = 520$ (exactly). More will be said on the notion of acceptance of the H_0 after we have discussed the two types of errors that can be made.

As with any two-way decision process, we can make an error by falsely rejecting the null hypothesis or by falsely accepting the null hypothesis. We give these errors the special names **Type I error** and **Type II error**.

Type I error
Type II error

DEFINITION 5.1
Type I Error

▼

A **Type I error** is committed if we reject the null hypothesis when it is true. The probability of a Type I error is denoted by the symbol α.

DEFINITION 5.2
Type II Error

▼

A **Type II error** is committed if we accept the null hypothesis when it is false and the research hypothesis is true. The probability of a Type II error is denoted by the symbol β (Greek letter beta).

The two-way decision process is shown in Table 5.3 with corresponding probabilities associated with each situation.

Although it would be desirable to determine the acceptance and rejection regions to simultaneously minimize both α and β, this is not possible. The probabilities associated with Type I and Type II errors are inversely related.

TABLE 5.3
Two-Way Decision Process

	Null Hypothesis	
Decision	True	False
Reject H_0	Type I error α	Correct $1 - \beta$
Accept H_0	Correct $1 - \alpha$	Type II error β

For a fixed sample size n, as we change the rejection region to increase α, then β decreases, and vice versa.

To alleviate what appears to be an impossible bind, the experimenter specifies a tolerable probability for a Type I error of the statistical test. Thus, the experimenter may choose α to be .01, .05, .10, and so on. Specification of a value for α then locates the rejection region. Determination of the associated probability of a Type II error is more complicated and will be delayed until later in the chapter.

Let us now see how the choice of α locates the rejection region. Returning to our soybean example, we will reject the null hypothesis for large values of the sample mean $\bar{y}$. Suppose we have decided to take a sample of $n = 36$ 1-acre plots, and from these data we compute $\bar{y} = 573$ and $s = 124$. Can we conclude that the mean yield for all farms is above 520?

specifying α

Before answering this question we must **specify α**. If we are willing to take the risk that 1 time in 40 we would incorrectly reject the null hypothesis, then $\alpha = 1/40 = .025$. An appropriate rejection region can be specified for this value of α by referring to the sampling distribution of $\bar{y}$. Assuming that the null hypothesis is true and that σ can be replaced by s, then $\bar{y}$ is normally distributed, with $\mu = 520$ and $\sigma_{\bar{y}} = 124/\sqrt{36} = 20.67$. Since the shaded area of Figure 5.8 corresponds to α, locating a rejection region with an area of .025 in the right tail of the distribution of $\bar{y}$ would be equivalent to determining the value of z that has an area .025 to its right. Referring

FIGURE 5.8
Rejection Region for the Soybean Example When $\alpha = .025$

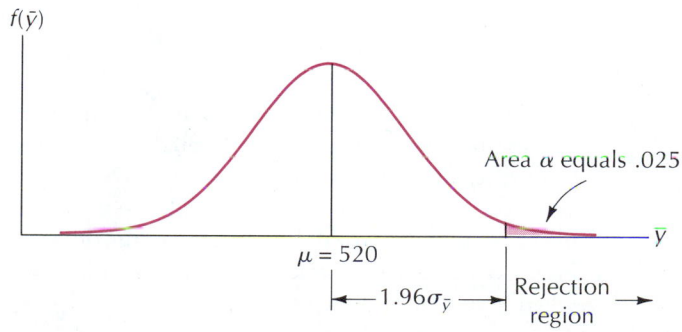

to Table 2 in the Appendix, this value of z is 1.96. Thus, the rejection region for our example would be located 1.96 standard errors $(1.96\sigma_{\bar{y}})$ above the mean $\mu = 520$. If the observed value of $\bar{y}$ is greater than 1.96 standard errors above $\mu = 520$, we reject the null hypothesis, as shown in Figure 5.8.

E X A M P L E   5.6

▼

Set up all the parts of a statistical test for the soybean example and use the sample data to reach a decision on whether to accept or reject the null hypothesis. Set $\alpha = .025$. Assume that σ can be estimated by s.

Solution   The five parts of the test are as follows:

H_0: $\mu = 520$
H_a: $\mu > 520$
T.S.: $\bar{y}$
R.R.:   For $\alpha = .025$, reject the null hypothesis if $\bar{y}$ lies more than 1.96 standard errors above $\mu = 520$.

The computed value of $\bar{y}$ was 573. To determine the number of standard errors that $\bar{y}$ lies above $\mu = 520$, we compute a z score for $\bar{y}$ using the formula

$$z = \frac{\bar{y} - \mu_0}{\sigma_{\bar{y}}},$$

where $\sigma_{\bar{y}} = \sigma/\sqrt{n}$. Substituting into the formula,

$$z = \frac{\bar{y} - \mu_0}{\sigma_{\bar{y}}} = \frac{573 - 520}{124/\sqrt{36}} = 2.56.$$

Conclusion: Since the observed value of $\bar{y}$ lies more than 1.96 standard errors above the hypothesized mean $\mu = 520$, we reject the null hypothesis in favor of the research hypothesis and conclude that the average soybean yield per acre is greater than 520.   ▲

one-tailed test

The statistical test conducted in Example 5.6 is called a **one-tailed test**, because the rejection region is located in only one tail of the distribution of $\bar{y}$. If our research hypothesis is $H_a: \mu < 520$, small values of $\bar{y}$ would indicate rejection of the null hypothesis. This test would also be one-tailed, but the rejection region would be located in the lower tail of the distribution of $\bar{y}$. Figure 5.9 displays the rejection region for the alternative hypothesis $H_a: \mu < 520$ when $\alpha = .025$.

FIGURE 5.9
Rejection Region for H_a: $\mu < 520$ When $\alpha = .025$ for the Soybean Example

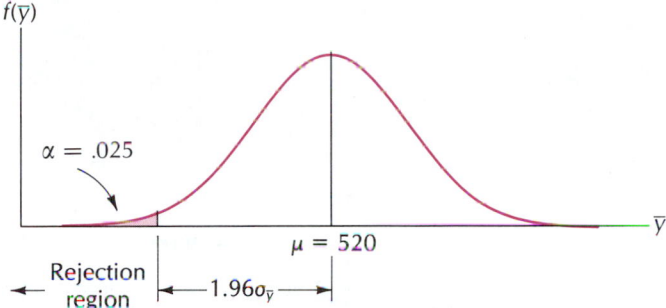

two-tailed test

We can formulate a **two-tailed test** for the research hypothesis $H_a: \mu \neq 520$, where we are interested in detecting whether the mean yield per acre of soybeans is greater or less than 520. Clearly both large and small values of $\bar{y}$ will contradict the null hypothesis, and we would locate the rejection region in both tails of the distribution of $\bar{y}$. A two-tailed rejection region for $H_a: \mu \neq 520$ and $\alpha = .05$ is shown in Figure 5.10.

FIGURE 5.10
Two-Tailed Rejection Region for $H_a: \mu \neq 520$ When $\alpha = .05$ for the Soybean Example

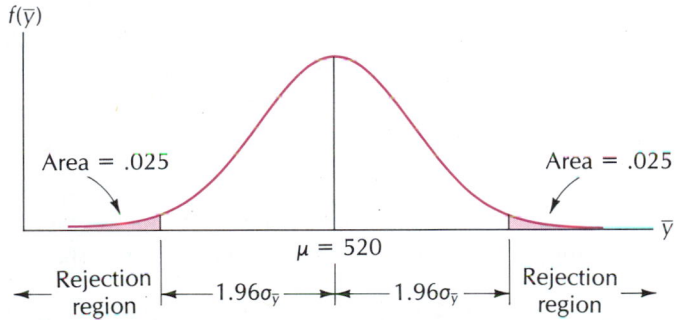

EXAMPLE 5.7

A corporation maintains a large fleet of company cars for its salespeople. To check the average number of miles driven per month per car, a random sample of $n = 40$ cars is examined. The mean and standard deviation for the sample are 2,752 miles and 350 miles, respectively. Records for previous years indicate that the average number of miles driven per car per month was 2,600. Use the sample data to test the research hypothesis that the current mean μ differs from 2,600. Set $\alpha = .05$ and assume that σ can be estimated by s.

Solution   The research hypothesis for this statistical test is $H_a: \mu \neq 2,600$ and the null hypothesis is $H_0: \mu = 2,600$. Using $\alpha = .05$, the two-tailed rejection region for this test is located as shown in Figure 5.11.

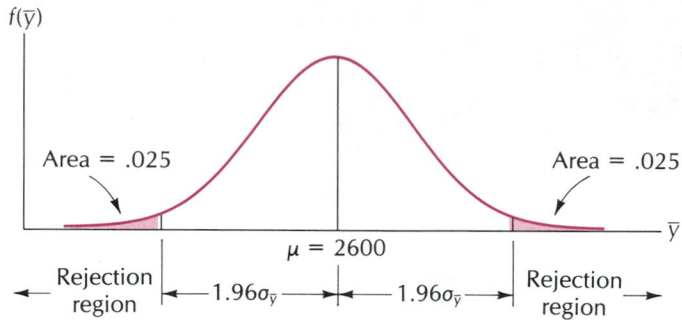

To determine how many standard errors our test statistic $\bar{y}$ lies away from $\mu = 2,600$, we compute

$$z = \frac{\bar{y} - \mu_0}{\sigma/\sqrt{n}} = \frac{2,752 - 2,600}{350/\sqrt{40}} = 2.75.$$

The observed value for $\bar{y}$ lies more than 1.96 standard errors above the mean, so we reject the null hypothesis in favor of the alternative $H_a: \mu \neq 2,600$. Since the computed value of $\bar{y}$ is greater than the hypothesized mean $\mu = 2,600$, we conclude that the mean number of miles driven is greater than 2,600.   ▲

The mechanics of the statistical test for a population mean can be greatly simplified if we use z rather than $\bar{y}$ as a test statistic. Using

H_0: $\mu = \mu_0$ (where μ_0 is some specified value)
H_a: $\mu > \mu_0$

and the test statistic

$$z = \frac{\bar{y} - \mu_0}{\sigma/\sqrt{n}}$$

then for $\alpha = .025$ we would reject the null hypothesis if $z > 1.96$—that is, if $\bar{y}$ lies more than 1.96 standard errors above the mean. Similarly, for the same null hypothesis, $\alpha = .05$, and $H_a: \mu \neq \mu_0$, we would reject the null hypothesis if the computed value of z is greater than 1.96 or less than -1.96, or, equivalently, if $|z| > 1.96$.

test for population mean   The statistical **test for a population mean** is summarized next. For H_0: $\mu = \mu_0$, three different alternatives are given with their corresponding rejection regions. In a given situation, you will choose only one of the three alternatives with its associated rejection region.

Summary of a Statistical Test for μ with σ Known

▼

H_0: $\mu = \mu_0$ (μ_0 is specified)

H_a: 1. $\mu > \mu_0$
2. $\mu < \mu_0$
3. $\mu \neq \mu_0$

T.S.: $z = \dfrac{\bar{y} - \mu_0}{\sigma/\sqrt{n}}$

R.R.: For a probability α of a Type I error,

1. reject H_0 if $z > z_\alpha$
2. reject H_0 if $z < -z_\alpha$
3. reject H_0 if $|z| > z_{\alpha/2}$

Note: For the time being, if σ is unknown but $n \geq 30$, you may replace σ by s in the standard error $\sigma_{\bar{y}} = \sigma/\sqrt{n}$ and proceed with the test. A more detailed discussion of inferences about μ when σ is unknown is presented later in this chapter.

EXAMPLE 5.8 ▼

The average (mean) live weights of a farmer's steers prior to slaughter was 380 pounds in past years. This year his 50 steers were fed on a new diet. Suppose we consider these 50 steers fed on the new diet as a random sample taken from a population of all possible steers that may be fed the diet now or in the future. Use the sample data given here and $\alpha = .01$ to test the research hypothesis that the mean live weight for steers on the new diet is greater than 380. The sample data are $n = 50$; $\bar{y} = 390$; $s = 35.2$.

Solution Using the sample data with $\alpha = .01$, the five parts of a statistical test are as follows:

H_0: $\mu = 380$
H_a: $\mu > 380$

T.S.: $z = \dfrac{\bar{y} - \mu_0}{\sigma/\sqrt{n}} = \dfrac{390 - 380}{35.2/\sqrt{50}} = \dfrac{10}{35.2/7.07} = 2.01$

R.R.: For $\alpha = .01$ and a one-tailed test, we reject H_0 if $z > z_{.01}$, where $z_{.01} = 2.33$.

Conclusion: Since the observed value of z, 2.01, does not exceed 2.33, we might be tempted to accept the null hypothesis that $\mu = 380$. The only problem with this conclusion is that we do not know β, the probability of incorrectly accepting the null hypothesis. To hedge somewhat in situations where z does not fall in the rejection region and β has not been calculated, we recommend stating that there is insufficient evidence to reject the null hypothesis. To reach a conclusion about whether to accept H_0, the experimenter would have to compute β. If β is small for reasonable alternative values of μ, then H_0 is accepted. Otherwise, the experimenter should conclude that there is insufficient evidence to reject the null hypothesis. ▲

computing β

We can illustrate the **computation of β**, the probability of a Type II error or equivalently the *power* $(1 - \beta)$, using the data in Example 5.8. If the null hypothesis is $H_0: \mu = 380$, the probability of incorrectly accepting H_0 will depend on how close the actual mean is to 380. For example, if the actual mean live weight is 400 pounds for steers on the new diet, we would expect β to be much smaller than if the actual mean live weight is 387. The whole process of determining β or the power $(1 - \beta)$ of a test is a "what-if" type of process. We look at β (or $1 - \beta$) for several possible alternative values of μ.

Let us suppose that the actual mean live weight is 395. What is β? With the null and research hypotheses as before,

$$H_0: \quad \mu = 380$$
$$H_a: \quad \mu > 380$$

and with $\alpha = .01$, we use Figure 5.12(a) to display β. The shaded portion of Figure 5.12(a) represents β, since this would be the probability of $\bar{y}$ falling in the acceptance region when the null hypothesis is false and μ is actually 395. Similarly, the power of the test for detecting $H_a: \mu = 395$ is $1 - \beta$, the area in the rejection region.

Let us consider two other possible values for μ, namely, 387 and 400. The corresponding values of β are shown as the shaded portions of Figures 12(b) and (c), respectively; power is the unshaded portion in the rejection region of Figure 5.12(b) and (c). The three situations illustrated in Figure 5.12 confirm what we alluded to earlier, that is, that the probability of a Type II error β decreases (and hence power increases) the further μ lies away from the hypothesized means under H_0.

μ_0, μ_a

We can readily calculate β for a test concerning μ if we adopt the following notation. Let μ_0 denote the hypothesized mean under H_0 and let μ_a denote the actual mean. The procedure for calculating β is then as summarized shortly. Although we never really know the actual mean, we can calculate β for any specified value of μ. The decision whether or not to accept H_0 depends on the magnitude of β for one or more reasonable alternative values. For a one-tailed

FIGURE 5.12

The Probability β of a Type II Error

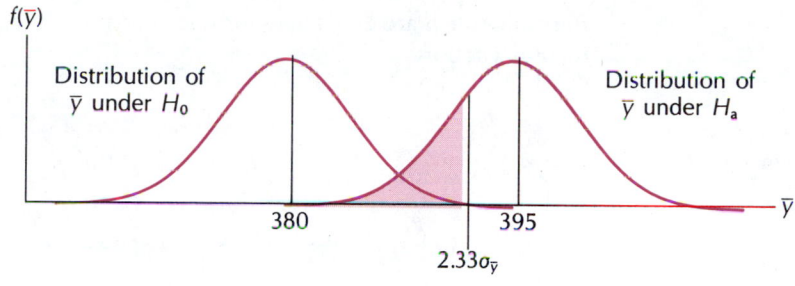

(a) β when H_a is $\mu = 395$

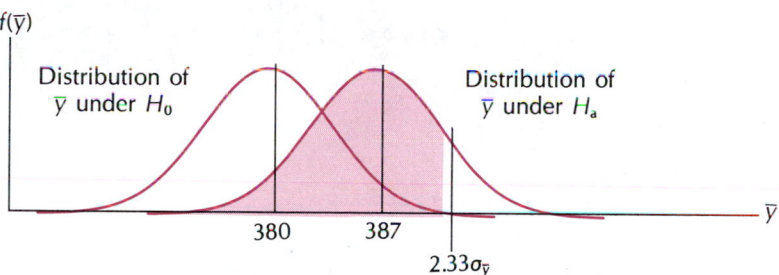

(b) β when H_a is $\mu = 387$

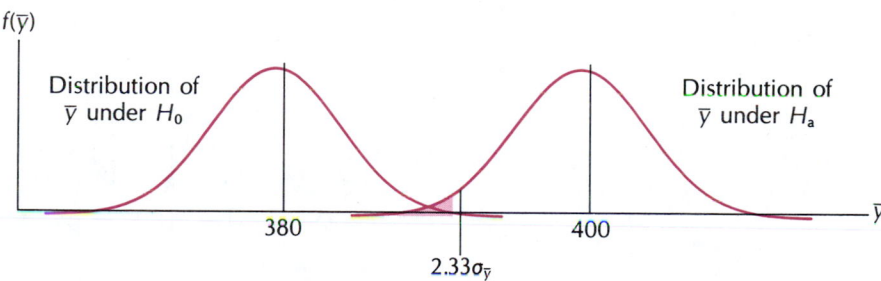

(c) β when H_a is $\mu = 400$

test of $H_0: \mu = \mu_0$, β is the probability that z is less than

$$z_\alpha - \frac{|\mu_0 - \mu_a|}{\sigma_{\bar{y}}}.$$

This probability is written as

$$P\left[z < z_\alpha - \frac{|\mu_0 - \mu_a|}{\sigma_{\bar{y}}}\right].$$

Formulas for β are given here for one- and two-tailed tests. Examples using these formulas follow.

Calculation of β for H_0: $\mu = \mu_0$ When μ_a is the Actual Mean

1. One-tailed test:

$$\beta = P\left(z < z_\alpha - \frac{|\mu_0 - \mu_a|}{\sigma_{\bar{y}}}\right); \qquad \text{power} = 1 - \beta.$$

2. Two-tailed test:

$$\beta \approx P\left(z < z_{\alpha/2} - \frac{|\mu_0 - \mu_a|}{\sigma_{\bar{y}}}\right); \qquad \text{power} = 1 - \beta.$$

E X A M P L E   5.9

Compute β and the power for the test in Example 5.8 if the actual mean live weight of steers is 395.

Solution   The research hypothesis for Example 5.8 was H_a: $\mu > 380$. Using $\alpha = .01$ and the computing formula for β with $\mu_0 = 380$ and $\mu_a = 395$, we have

$$\beta = P\left[z < z_{.01} - \frac{|\mu_0 - \mu_a|}{\sigma_{\bar{y}}}\right] = P\left[z < 2.33 - \frac{|380 - 395|}{35.2/\sqrt{50}}\right]$$

$$= P[z < 2.33 - 3.01] = P[z < -.68].$$

Referring to Table 2 in the Appendix, the area corresponding to $z = .68$ is .2483. Hence, $\beta = .2483$ and power $= 1 - .2483 = .7517$.   ▲

Previously, when $\bar{y}$ did not fall in the rejection region, we concluded that there was insufficient evidence to reject H_0 because β was unknown. Now when $\bar{y}$ falls in the acceptance region, we can compute β corresponding to one (or more) alternative values for μ that appear reasonable in light of the experimental setting. Then provided we are willing to tolerate a probability of falsely accepting the null hypothesis equal to the computed value of β for the alternative value(s) of μ considered, our decision is to accept the null hypothesis. Thus, in Example 5.9, if we are willing to risk a β error of about .25 of falsely accepting the null hypothesis, we would accept the null hypothesis $\mu = 380$.

E X A M P L E   5.10 ▼

Prospective salespeople for an encyclopedia company are now being offered a sales training program. Previous data indicate that the average number of sales per month for those who do not participate in the program is 33. To determine whether the training program is effective, a random sample of 35 new employees is given the sales training and then sent out into the field. One month later, the mean and standard deviation for the number of sets of encyclopedias sold are 35 and 8.4, respectively. Do the data present sufficient evidence to indicate that the training program enhances sales? Use $\alpha = .05$.

Solution   The five parts to our statistical test are as follows:

$$H_0: \quad \mu = 33$$
$$H_a: \quad \mu > 33$$

$$\text{T.S.:} \quad z = \frac{\bar{y} - \mu_0}{\sigma_{\bar{y}}} \approx \frac{35 - 33}{8.4/\sqrt{35}} = 1.41$$

R.R.:   For $\alpha = .05$ we will reject the null hypothesis if $z > z_{.05} = 1.645$.

Conclusion: Since the observed value of z does not fall into the rejection region, we reserve judgment on accepting H_0 until we calculate β. That is, we conclude that there is insufficient evidence to reject the null hypothesis that persons on the sales program have the same mean number of sales per month as those not under the program.   ▲

E X A M P L E   5.11 ▼

Refer to Example 5.10. Suppose that the encyclopedia company thinks that the cost of financing the sales program will be offset by increased sales if those on the program average 38 sales per month. Compute β for $\mu_a = 38$ and, based on the value of β, indicate whether you would accept the null hypothesis.

Solution   Using the computational formula for β with $\mu_0 = 33$, $\mu_a = 38$, and $\alpha = .05$, we have

$$\beta = P\left[z < z_{.05} - \frac{|\mu_0 - \mu_a|}{\sigma_{\bar{y}}}\right] = P\left[z < 1.645 - \frac{|33 - 38|}{8.4/\sqrt{35}}\right]$$

$$= P[z < -1.88].$$

The area corresponding to $z = 1.88$ in Table 2 of the Appendix is .0301. Hence,

$$\beta = .0301; \text{ power} = 1 - .0301 = .9699.$$

Because β is relatively small, we accept the null hypothesis and conclude that the training program has not increased the average sales per month above the point where increased sales would offset the cost of the training program.

▲

In Section 5.2, we discussed how we measure the goodness of interval estimates. The goodness of a statistical test can be measured by the magnitudes of the Type I and Type II errors, α and β. When α is preset at a tolerable level by the experimenter, β is a function of the sample size for a fixed value of μ_a. The larger the sample size, the more information we have concerning μ and, hence, the smaller the value of β. We will consider now the problem of designing an experiment for testing $H_0: \mu = \mu_0$ when α is specified and β is preset for a fixed actual value μ_a. This problem reduces to determining the sample size needed for the fixed values of α and β.

5.6 ▼ CHOOSING THE SAMPLE SIZE FOR TESTING μ

The quantity of information available for a statistical test about μ is measured by the magnitudes of the Type I and II error probabilities, α and β. Suppose that we are interested in testing

$$H_0: \quad \mu = \mu_0$$

against a one-sided alternative

$$H_a: \quad \mu > \mu_0.$$

In addition, suppose that we want the probability of a Type I error to be α and the probability of a Type II error to be β or less when the actual value of μ lies a distance of Δ (delta) or more above μ_0. The sample size necessary to meet these requirements is shown here.

Sample Size for a One-Sided Test of μ

$$n = \sigma^2 \frac{(z_\alpha + z_\beta)^2}{\Delta^2}$$

Note: If σ^2 is unknown, substitute an estimated value to get an approximate sample size.

The same formula applies to the one-sided alternative $H_a: \mu < \mu_0$, with the exception that we want the probability of a Type II error to be of magnitude β or less when the actual value of μ lies a distance of Δ or more below μ_0.

E X A M P L E   5.12

▼

A cereal packager is concerned that one of its machines has a mean fill per package or more than 16 ounces, the labeled net weight. While this is not bad from a public relations standpoint, it could cost the packager a great deal of money. Previous experience suggests that the standard deviation of the package fill weights is approximately .225. For

$$H_0: \quad \mu = 16$$
$$H_a: \quad \mu > 16$$

with $\alpha = .05$, determine the sample size required to make $\beta = .01$ or less if the actual mean is 16.1 ounces or more. By putting this restriction on β, the packager is saying that it wants a very small probability of falsely accepting $H_0: \mu = 16$, when in fact the actual mean is 16.1 ounces or more.

Solution   From previous data the fill weights have a standard deviation approximately equal to .225. The appropriate z-values, $z_{.05}$ and $z_{.01}$, for $\alpha = .05$ and $\beta = .01$ are 1.645 and 2.33, respectively. Using $\Delta = 16.1 - 16 = .1$, the required sample size is

$$n = \frac{(.225)^2(1.645 + 2.33)^2}{(.1)^2} = 79.99 \approx 80.$$

That is, the packager must obtain a random sample of $n = 80$ cartons to conduct this test under the specified conditions.

Suppose that after obtaining the sample, the computed value of

$$z = \frac{\bar{y} - 16}{\sigma_{\bar{y}}}$$

does not fall in the rejection region. What is our conclusion? In similar situations in previous sections, our conclusion would have been that there was insufficient evidence to reject H_0. Now, however, knowing that $\beta \le .01$ when $\mu \ge 16.1$, we would feel safe in our conclusion to accept $H_0: \mu = 16$. No further testing would be required.   ▲

With a slight modification of the sample size formula for the one-tailed tests, we can test

$$H_0: \quad \mu = \mu_0$$
$$H_a: \quad \mu \neq \mu_0$$

for a specified α and β with $\Delta = |\mu - \mu_0|$. A formula for an approximate sample size when testing μ is presented here.

Approximate Sample Size for a Two-Sided Test of H_0: $\mu = \mu_0$

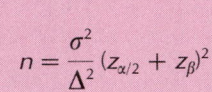

$$n = \frac{\sigma^2}{\Delta^2} (z_{\alpha/2} + z_\beta)^2$$

Note: If σ^2 is unknown, substitute an estimated value to get an approximate sample size.

▼ EXERCISES

Basic Techniques

5.29   Consider the data of Example 5.12 and compute the sample size required for testing $H_0: \mu = 16$ against $H_a: \mu \neq 16$ for $\alpha = .05$ and $\beta \leq .01$ when the actual value of μ lies more than .1 unit away from $\mu_0 = 16$.

5.30   A random sample of 50 measurements from a population yielded $\bar{y} = 40.1$ and $s = 5.6$. Use these data to test the null hypothesis $H_0: \mu = 38$ against the alternative hypothesis $H_a: \mu > 38$. Use $\alpha = .05$ and draw a conclusion. Could you have made a Type II error in this situation? Explain.

5.31   For the data of Exercise 5.30, determine the power of rejecting $H_0: \mu = 38$ given that the alternative hypothesis is true and $\mu_a = 40$. Do the same for $\mu_a = 42$ and 44 in order to sketch the power of this test for the various alternatives.

5.32   The mean and the standard deviation of a random sample of $n = 50$ measurements are $\bar{y} = 63.7$ and $s = 14.2$. Conduct a statistical test of $H_0: \mu = 68$ against the alternative $H_a: \mu < 68$, using $\alpha = .05$.

5.33   Refer to Exercise 5.32. Will your conclusion be different if you select $\alpha = .01$? Explain.

Applications

 5.34   The administrator of a nursing home would like to do a time-and-motion study of staff time spent per day performing nonemergency-type chores. In particular, she would like to test the null hypothesis $H_0: \mu = 16$ (person-hours per day) against $H_a: \mu < 16$. The value of 16 arose from a previous study prior to the introduction of some efficiency measures. How many days must be sampled to test the proposed hypothesis if $\alpha = .05$

and $\beta \leq .10$ when the actual value of μ is 12 hours (a 25% decrease from previous results) or less? Assume $\sigma^2 = 7.64$.

5.35 Refer to Exercise 5.34. Determine the sample size for testing $H_0: \mu = 16$ and $\alpha = .05$ if the power of detecting a mean of 13 or less is .80 or more.

5.36 The increase in exercise capacity (in minutes) was recorded for each of 90 adult male patients following treatment for congestive heart failure. Given that the sample results yield $\bar{y} = 2.2$ and $s = 1.05$, use these data to test the null hypothesis $H_0: \mu = 2.0$ versus $H_a: \mu > 2.0$. Use $\alpha = .05$ to draw a conclusion.

5.37 Refer to Exercise 5.36. Sketch a power curve (power versus μ_a) for this test based on $\mu_a = 2.1, 2.2, 2.3,$ and 2.5.

5.38 To evaluate the success of a 1-year experimental program designed to increase the mathematical achievement of underprivileged high school seniors, the mathematics scores for a sample of $n = 100$ underprivileged seniors were obtained for comparison with the previous year's statewide average of 525 for underprivileged seniors. You wish to examine whether there has been an increase in the mean achievement level over last year's statewide average. Discuss whether you would use a one-tailed or a two-tailed test. Set up all parts of the statistical test for μ, using $\alpha = .05$.

5.39 Refer to Exercise 5.38. Suppose you wish to examine whether the mean achievement has changed (up or down) over the past year. Would you use a two-tailed test? Explain. Set up all parts of the statistical test for μ, using $\alpha = .01$.

5.40 To study the effectiveness of a weight-reducing agent, a clinical trial was conducted in which 35 overweight males were placed on a fixed diet. After a 2-week period, each male was weighed and then given a supply of the weight-reducing agent. The diet was to be maintained; in addition, a single dose of the weight-reducing agent was to be taken each day. At the end of the next 2-week period, weights were again obtained. Set up all parts of the statistical test for the alternative hypothesis that μ, the average weight loss, is greater than 0. Why is a one-tailed test appropriate? Use $\alpha = .05$.

5.41 Refer to Exercise 5.40.

    a. The average weight loss for the second 2-week period was $\bar{y} = 10.3$ pounds, and the standard deviation was $s = 4.6$. Perform a statistical test and draw conclusions. Use $\alpha = .05$.

    b. Based on the results for part a, can you conclude that the weight-reducing agent is effective? Explain.

5.42 Transportation, getting people to their destination and home again, is a national problem. One aspect of this problem currently being studied by the Federal Highway Administration is how to successfully merge automobiles entering at high speed with congested interstate traffic. To study this proble, an automobile merging system was installed on the entrance to I-75 at Tampa, Florida. Through the use of a series of display lights, a driver is told whether he or she is traveling at an appropriate speed to merge successfully into the existing traffic on the highway. Prior to the installation of the system, investigators measured the stress levels of many drivers merging onto the highway during the 4 to 6 P.M. rush hour period. Similar testing on a random sample of 50 drivers was conducted after the merging system was installed.

For the purposes of illustration, suppose that the average stress level prior to the installation of the system was 8.2 (measured on a 10-point scale). Set up appropriate null and alternative hypotheses to test the research hypothesis that the average stress level for drivers under the merging system is less than that observed prior to the installation of the system. Is this a one- or two-tailed test?

5.43 Refer to Exercise 5.42. Suppose the sample mean and standard deviation for the 50 drivers tested using the merging system were, respectively, 7.6 and 1.8. Use these data to test the alternative hypothesis of Exercise 5.42. Use $\alpha = .05$.

▼ **5.44** Tooth decay generally develops first on those teeth that have irregular shapes (typically molars). The most susceptible surfaces on these teeth are the chewing surfaces. Usually the enamel on these surfaces contains tiny pockets that tend to hold food particles. Bacteria begins to eat the food particles to create an environment in which the tooth surface will decay.

Of particular importance in the decay rate of teeth, in addition to the natural hardness of the teeth, is the form of the food eaten by the individual. Some forms of carbohydrates are particularly detrimental to dental health. Many studies have been conducted to verify these findings, and we can imagine how the study might have been run. A random sample of 60 adults was obtained from a given locale. Each person was examined and then maintained a diet supplemented with a sugar solution at all meals. At the end of a one-year period, the average number of newly decayed teeth for the group was .70, and the standard deviation was .4. Do these data present sufficient evidence to indicate that the mean number of newly decayed teeth for people whose diet includes a sugar solution is greater than .30, a rate that had been shown to apply to a person whose diet did not contain the sugar solution supplement? Why would a two-tailed test be inappropriate? Use $\alpha = .05$.

5.7 ▼ THE LEVEL OF SIGNIFICANCE OF A STATISTICAL TEST

In Section 5.6, we introduced hypothesis testing along rather traditional lines: we defined the parts of a statistical test along with the two types of errors and their associated probabilities α and β. In recent years, many statisticians and other users of statistics have objected to this decision-based approach to hypothesis testing. Rather than running a statistical test with a preset value of α, they argue, we should specify the null and alternative hypotheses, collect the sample data, and determine the weight of the evidence for rejecting the null hypothesis. This
level of significance weight, given in terms of a probability, is called the **level of significance** (or p-value) of the statistical test.

We illustrate the calculation of a level of significance with an example.

E X A M P L E   5.13 ▼

Refer to Example 5.8.
a. Rather than specifying a preset value for α, determine the level of significance for the statistical test.
b. How would the level of significance change if $\bar{y}$ had been 397 rather than 390?

Solution   **a.** The null and alternative hypotheses are

$$H_0: \quad \mu = 380$$
$$H_a: \quad \mu > 380.$$

From the sample data, the computed value of the test statistic is

$$z = \frac{\bar{y} - 380}{s/\sqrt{n}} = \frac{390 - 380}{35.2/\sqrt{50}} = 2.01.$$

The level of significance for this test (i.e., the weight of evidence for rejecting H_0) is the probability of observing a value of $\bar{y}$ greater than 390 assuming the null hypothesis is true. This value can be computed by using the z-value of the test statistic, 2.01, and referring to Table 2 in the Appendix to determine the probability of observing a z-value greater than 2.01. This probability, which is sometimes designated by the letter **p**, is seen to be .0222. This value is shown by the shaded area in Figure 5.13. Thus, we would say that the level of significance for this test is .0222.

p

F I G U R E 5.13

Level of Significance for Example 5.13

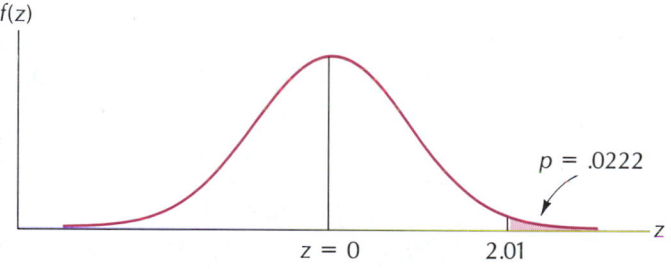

b. For $\bar{y} = 397$ the corresponding value of the z statistic is 3.42. Since the largest value of z in Table 2 of the Appendix is 3.09, the p-value is less than .001. We show this as $p < .001$.   ▲

As we can see from Example 5.13, the level of significance represents the probability of observing a sample outcome more contradictory to H_0 than the observed sample result. *The smaller the value of this probability, the heavier the weight of the sample evidence against H_0.* For example, a statistical test with a level of significance of $p = .01$ shows more evidence for the rejection of H_0 than does another statistical test with $p = .20$.

Suppose the null and alternative hypotheses in Example 5.13 had been

$$H_0: \quad \mu = 380$$
$$H_a: \quad \mu < 380$$

FIGURE 5.14
Level of Significance for
$H_0: \mu = 380,\ H_a: \mu < 380$
and $z = -2.01$

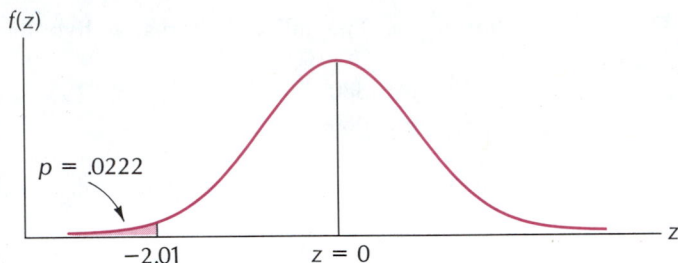

$f(z)$

$p = .0222$

-2.01 $z = 0$ z

and the computed value of z had been $z = -2.01$. The level of significance would still be $p = .0222$ (see Figure 5.14).

***p*-value for one-tailed test**

To summarize, the **level of significance for a one-tailed test** can be computed as follows:

For $H_a: \mu > \mu_0$,

$$p = P[z > \text{computed } z].$$

For $H_a: \mu < \mu_0$,

$$p = P[z < \text{computed } z].$$

***p*-value for two-tailed test**

For two-tailed tests (as determined by the form of H_a), we still compute the probability of obtaining a sample outcome more contradictory to H_0 than the observed result, but the level of significance is commonly taken to be twice this probability. For a two-tailed test, the level of significance can be written as

$$p = 2P[z > |\text{computed } z|].$$

E X A M P L E   5.14   ▼

Determine the level of significance for the data of Example 5.8 if the null hypothesis and alternative hypothesis are

$$H_0: \quad \mu = 380$$
$$H_a: \quad \mu \neq 380.$$

Solution   The computed value of z is 2.01. Since the probability of observing a value of z greater than 2.01 is .0222, the level of significance for the two-tailed statistical test is $p = 2(.0222) = .0444$.   ▲

There is much to be said in favor of this approach to hypothesis testing. Rather than reaching a decision directly, the statistician (or person performing the statistical test) presents the experimenter with the weight of evidence for rejecting the null hypothesis. The experimenter can then draw his or her own conclusion. Some experimenters will reject a null hypothesis if $p = .10$, whereas others will require $p < .05$ or $p < .01$ for rejecting the null hypothesis. The experimenter is left to make the decision based on what he or she believes is enough evidence to indicate rejection of the null hypothesis.

Many professional journals have followed this approach by reporting the results of a statistical test in terms of its level of significance. Thus, we might read that a particular test was significant at the $p = .05$ level or perhaps the $p < .01$ level. By reporting results this way, the reader is left to draw his or her own conclusion.

One word of warning is needed here. The p-value of .05 has become a magic level, and many seem to feel that a particular null hypothesis should not be rejected unless the test achieves the .05 level or lower. This has resulted in part from the decision-based approach with α preset at .05. Try not to fall into this trap when reading journal articles or reporting the results of your statistical tests. After all, statistical significance at a particular level does not dictate importance or practical significance. Rather, it means that a null hypothesis can be rejected with a specified low risk of error. For example, suppose that a company is interested in determining whether the average number of miles driven per car per month for the sales force has risen above 2,600. Sample data from 400 cars show that $\bar{y} = 2,640$ and $s = 35$. For these data the z statistic for H_0: $\mu = 2,600$ is $z = 22.86$ based on $\sigma = 35$; the level of significance is $p < .0000000001$. Thus, even though there has only been a 1.5% increase in the average monthly miles driven for each car, the result is (highly) statistically significant. Is this increase of any practical significance? Probably not. What we have done is proved *conclusively* that the mean μ has increased slightly.

Throughout the text we will conduct statistical tests from both the decision-based approach and from the level-of-significance approach to familiarize you with both avenues of thought. For either approach, remember to consider the practical significance of your findings after drawing conclusions based on the statistical test.

▼ EXERCISES

Basic Techniques

5.45 Sample data for a statistical test of H_0: $\mu = 40$ yielded a z score of 1.86.
  a. Determine the level of significance for a test of H_a: $\mu > 40$.
  b. Determine the level of significance for a test of H_a: $\mu \neq 40$.

Applications

5.46 A random sample of 36 cigarettes of a certain brand was tested for nicotine content. The sample mean and standard deviation (in milligrams) are, respectively, 15.1 and 3.8. Give the level of significance of the statistical test of $H_0: \mu = 14$ (the claimed nicotine content) against the alternative hypothesis $H_a: \mu > 14$.

5.47 A psychological experiment was conducted to investigate the length of time (time delay) between the administration of a stimulus and the observation of a specified reaction. A random sample of 36 persons was subjected to the stimulus and observed for the time delay. The sample mean and standard deviation were 2.2 and .57 seconds, respectively. Test the null hypothesis that the mean time delay for the hypothetical set of all persons who may be subjected to the stimulus is $\mu = 1.6$ against the alternative hypothesis that the mean time delay differs from 1.6. Use $\alpha = .05$.

5.8 ▼ INFERENCES ABOUT μ, σ UNKNOWN

The estimation and test procedures about μ presented earlier in this chapter were based on the assumption that the population variance was known or that we had enough observations to allow s to be a reasonable estimate of σ. In this section, we will present a test that can be applied when σ is unknown, no matter what the sample size. For example, in determining the average concentration of a drug in the bloodstream one hour after patients suffering from a rare disease are treated with the drug, it might be impossible to obtain a random sample of 30 or more observations at a given time. What test procedure could be used in order to make inferences about μ?

W. S. Gosset faced a similar problem around the turn of the century. As a chemist for Guinness Breweries, he was asked to make judgments on the mean quality of various brews, but was not supplied with large sample sizes to reach his conclusions.

Gosset thought that when he used the test statistic

$$z = \frac{\bar{y} - \mu_0}{\sigma/\sqrt{n}}$$

with σ replaced by s for small sample sizes, he was falsely rejecting the null hypothesis $H_0: \mu = \mu_0$ at a slightly higher rate than that specified by α. This problem intrigued him, and he set out to derive the distribution and percentage points of the test statistic

$$\frac{\bar{y} - \mu_0}{s/\sqrt{n}}$$

for $n < 30$.

For example, suppose an experimenter sets α at a nominal level, say .05. Then he or she expects falsely to reject the null hypothesis approximately 1 time in 20. However, Gosset proved that the actual probability of a Type I error for this test was somewhat higher than the nominal level designated by α. He published the results of his study under the pen name Student, because it was against company policy for him to publish his results in his own name at that time. The quantity

$$\frac{\bar{y} - \mu_0}{s/\sqrt{n}}$$

Student's t

is called the t statistic and its distribution is called the *Student's t distribution* or, simply, **Student's** t (see Figure 5.15).

F I G U R E 5.15

A t Distribution with a Normal Distribution Superimposed

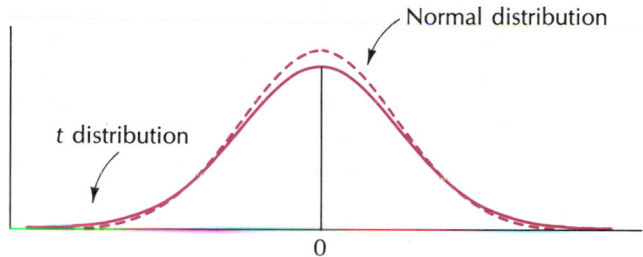

Although the quantity

$$\frac{\bar{y} - \mu_0}{s/\sqrt{n}}$$

will possess a t distribution only when the sample is selected from a normal population, the t distribution provides a reasonable approximation to the distribution of

$$\frac{\bar{y} - \mu_0}{s/\sqrt{n}}$$

when the sample is selected from a population with a mound-shaped distribution. We summarize the properties of t here.

**Properties of Student's
t Distribution**

▼

1. The *t* distribution, like that of *z*, is symmetrical about 0.
2. The *t* distribution is more variable than the *z* distribution (see Figure 5.15).
3. There are many different *t* distributions. We specify a particular one by a parameter called the *degrees of freedom* (df). Thus, we specify

$$t = \frac{\bar{y} - \mu_0}{s/\sqrt{n}} \qquad df = n - 1.$$

4. As *n* (or equivalently df) increases, the distribution of *t* approaches the distribution of *z*.

The phrase "degrees of freedom" sounds awfully mysterious, but the idea will eventually become second nature to you. The technical definition requires advanced mathematics, which we will avoid; on a less technical level the basic idea is that degrees of freedom are pieces of information for estimating σ using *s*. The standard deviation *s* for a sample of *n* measurements is based on the deviations $y_i - \bar{y}$. Because $\sum (y_i - \bar{y}) = 0$ always, if $n - 1$ of the deviations are known, the last (*n*th) is fixed mathematically to make the sum equal 0. It is therefore noninformative. So, in a sample of *n* measurements there are $n - 1$ pieces of information (degrees of freedom) about σ.

Because of the symmetry of *t*, only upper-tail percentage points (probabilities or areas) of the distribution of *t* have been tabulated; these appear in Table 4 in the Appendix. The degrees of freedom (df) are listed along the left column of the page. An entry in the table specifies a value of *t*, say $\boldsymbol{t_a}$, such that an area *a* lies to its right. See Figure 5.16. Various values of *a* appear across the top of Table 4 in the Appendix. Thus, for example, with df = 7, the value of *t* with an area .05 to its right is 1.895 (found in the $a = .05$ column and df = 7 row).

We can use the *t* distribution to make inferences about a population mean μ. The sample test concerning μ is summarized next. The only difference between

t_a

FIGURE 5.16

**Illustration of Area
Tabulated in Table 4 in
the Appendix for the *t*
Distribution**

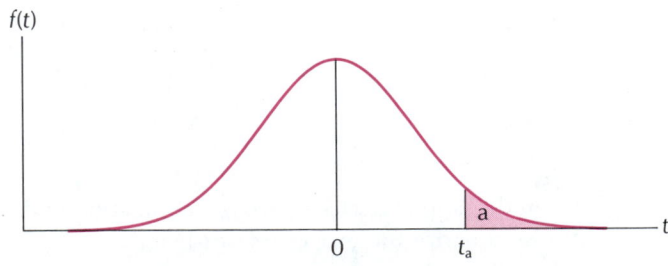

the z test discussed earlier in this chapter and the test given here is that t replaces z. The t test (rather than the z test) should be used any time σ is unknown and the distribution of y-values is mound-shaped.

Statistical Test About μ, σ Unknown

H_0: $\mu = \mu_0$

H_a: 1. $\mu > \mu_0$
        2. $\mu < \mu_0$
        3. $\mu \neq \mu_0$

T.S.: $t = \dfrac{\bar{y} - \mu_0}{s/\sqrt{n}}$

R.R.:   For a probability α of a Type I error and df $= n - 1$,

1. reject H_0 if $t > t_\alpha$
2. reject H_0 if $t < -t_\alpha$
3. reject H_0 if $|t| > t_{\alpha/2}$

Recall that a denotes the area in the tail of the t distribution. For a one-tailed test with the probability of a Type I error equal to α, we locate the rejection region using the value from Table 4 in the Appendix, for $a = \alpha$ and df $= n - 1$. But for a two-tailed test we would use the t-value from Table 4 corresponding to $a = \alpha/2$ and df $= n - 1$.

Thus, for a one-tailed test we reject the null hypothesis if the computed value of t is greater than the t-value from Table 4 in the Appendix, and $a = \alpha$ and df $= n - 1$. Similarly, for a two-tailed test we reject the null hypothesis if $|t|$ is greater than the t-value from Table 4 for $a = \alpha/2$ and df $= n - 1$.

E X A M P L E   5.15

A tire company guarantees that a particular brand of tire has a mean useful lifetime of 42,000 miles or more. A consumer test agency, wishing to verify this claim, observed $n = 10$ tires on a test wheel that simulated normal road conditions. The lifetimes (in thousands of miles) were as follows:

42   36   46   43   41   35   43   45   40   39

Use these data to determine whether there is sufficient evidence to contradict the manufacturer's claim. Set $\alpha = .05$.

Solution   The null and research hypotheses for this example are

$$H_0: \quad \mu = 42$$

and

$$H_a: \quad \mu < 42.$$

Note that we are giving the manufacturer the benefit of the doubt by setting $\mu = 42$ for H_0.

Before setting up the test statistic, and rejection region, we must first compute the sample mean and standard deviation. You can verify that

$$\sum_i y_i = 410 \quad \text{and} \quad \sum_i y_i^2 = 16{,}926.$$

Then

$$\bar{y} = \frac{\sum_i y_i}{10} = \frac{410}{10} = 41.$$

Similarly, substituting into the shortcut formula for s^2, we find

$$s^2 = \frac{1}{9}\left[\sum_i y_i^2 - \frac{\left(\sum_i y_i\right)^2}{10} \right] = 12.89$$

$$s = \sqrt{12.89} = 3.59.$$

The test statistic, then, is

$$t = \frac{\bar{y} - \mu_0}{s/\sqrt{n}} = \frac{41 - 42}{3.59/\sqrt{10}} = -.88$$

and the rejection region is

R.R.:   Reject H_0 if $t < -t_{.05}$.

From Table 4 in the Appendix, the critical t-value with $a = .05$ and df $= 9$ is 1.833, so $-t_{.05}$ is -1.833. Since the observed value of t is not less than -1.833, we have insufficient evidence to indicate that the mean lifetime of this brand of tires is less than 42,000 miles.

At this point, someone might suggest calculating β, the probability of a Type II error, to see whether we can accept the manufacturer's claim. Unfortunately, this is a much more difficult task for a small-sample test than it is for the large-sample test, and it is beyond the scope of this text. (If you are interested in pursuing the topic, consult *Biometrika Tables for Statisticians*, Volume I.) Our conclusion will be that there is insufficient evidence to reject the company's claim and we should continue sampling.   ▲

EXAMPLE 5.16   ▼

Refer to Example 5.15. Rather than performing the statistical test with a preset α level, give the level of significance for the test.

Solution   For the one-tailed lower-tail test, the computed t-value is $t = -.88$. If we had an entire table of t areas for each df, this would be no problem. Because of space limitations we show only a few areas (a) for each df. The best we can do for $t = -.88$ and df $= 9$ is to say that $p > .10$. Based on this probability, the experimenter would probably conclude that there was insufficient evidence to reject the null hypothesis. If you think that the level of significance should be given more precisely, you can refer to more detailed tables of the t distribution in the *Biometrika Tables for Statisticians* or any of several statistical software packages that compute p-values for various test procedures.   ▲

In addition to being able to run a statistical test for μ when σ is unknown, we can construct a confidence interval using t. The confidence interval for μ with σ unknown is identical to the corresponding confidence interval for μ when σ is known, with z replaced by t and σ replaced by s.

100(1 − α)% Confidence Interval for μ, σ Unknown

$$\bar{y} \pm t_{\alpha/2} \frac{s}{\sqrt{n}}$$

Note: df $= n - 1$ and the confidence coefficient is $(1 - \alpha)$.

EXAMPLE 5.17   ▼

In a psychological depth-perception test, a random sample of $n = 14$ airline pilots were asked to judge the distance between two markers at the other end of a laboratory. The sample data (recorded in feet) are listed below.

2.7	2.4	1.9	2.6	2.4	1.9	2.3
2.2	2.5	2.3	1.8	2.5	2.0	2.2

Use the sample data to place a 95% confidence interval on μ, the average recorded distance for this psychological test.

Solution   Before setting up a 95% confidence interval on μ, we must compute $\bar{y}$ and s. You can verify that

$$\sum_i y_i = 31.70 \qquad \text{and} \qquad \sum_i y_i^2 = 72.79.$$

The sample mean, variance, and standard deviation are then

$$\bar{y} = \frac{\sum_i y_i}{14} = \frac{31.70}{14} = 2.26$$

$$s^2 = \frac{1}{13}\left[72.79 - \frac{(31.7)^2}{14}\right] = .078$$

$$s = \sqrt{.078} = .28.$$

Referring to Table 4 in the Appendix, the t-value corresponding to $a = .025$ and df = 13 is 2.160. Hence, the 95% confidence interval is

$$\bar{y} \pm t_{\alpha/2}\frac{s}{\sqrt{n}} \qquad \text{or} \qquad 2.26 \pm \frac{2.160(.28)}{\sqrt{14}},$$

which is the interval $2.26 \pm .16$, or 2.10 to 2.42. Thus, we are 95% confident that the interval from 2.10 to 2.42 will encompass the mean μ.

▲

In this section, we have made the formal mathematical assumption that the population is normally distributed. *In practice, no population has exactly a normal distribution.* How does nonnormality of the population distribution affect inferences based on the t distribution?

There are two issues to consider when populations are assumed to be nonnormal. First, what kind of nonnormality is assumed? Second, what possible effects do these specific forms of nonnormality have on the t-distribution procedures? The most important deviations from normality are **skewed distributions** and **heavy-tailed distributions**. (Heavy-tailed distributions are roughly symmetric but have outliers.)

To evaluate the effect of nonnormality as exhibited by skewness or heavy tails, we will consider whether the t-distribution procedures are still approximately correct for these forms of nonnormality and whether there are other more efficient procedures. For example, even if a confidence interval for μ based on t gave nearly correct results for, say, a heavy-tailed population distribution, it might be possible to get a smaller confidence interval width based on a trimmed mean.

skewness and heavy tails

sensitivity of one-tailed procedures

The question of approximate correctness of t procedures has been studied extensively. In general, probabilities specified by the t procedures, particularly the confidence level (for confidence intervals) and the Type I error (for statistical tests), have been found to be fairly accurate, even when the population distribution is heavy-tailed (or light-tailed). In contrast, skewness, particularly with small sample sizes, can have a nasty effect on these probabilities, particularly in one-tailed procedures. A t distribution is symmetric, of course. When the population distribution is skewed, the actual sampling distribution of a t statistic is skewed. Although this skewness decreases as the sample size increases, there is no magic sample size that completely deskews the actual sampling distribution.

As a consequence, a nominal 95% confidence interval may actually have 80% or lower confidence if the sample size is in the teens and the population distribution looks like that of Figure 5.17.

FIGURE 5.17
Skewed Population Distribution

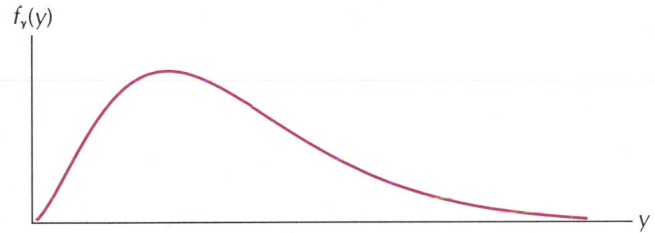

EXAMPLE 5.18

▼

A simulation study takes 1,000 samples of size 30 from a symmetric, moderately outlier-prone population. The following results are obtained:

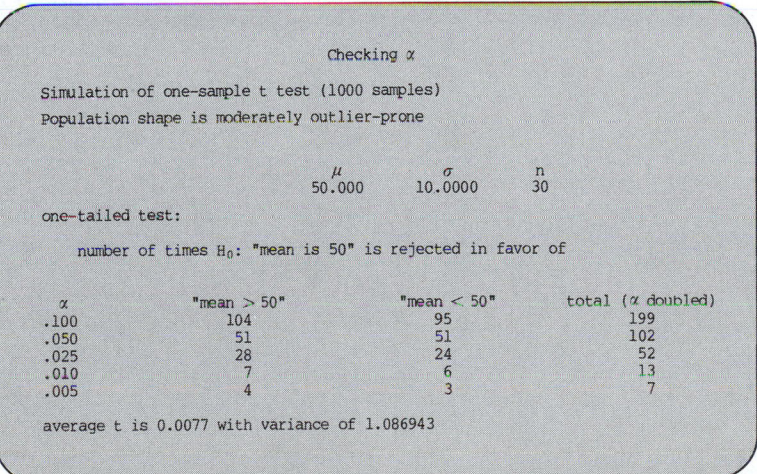

```
                              Checking α

Simulation of one-sample t test (1000 samples)
Population shape is moderately outlier-prone

                           μ            σ           n
                         50.000      10.0000        30
one-tailed test:

    number of times H₀: "mean is 50" is rejected in favor of

     α             "mean > 50"          "mean < 50"       total (α doubled)
    .100              104                   95                 199
    .050               51                   51                 102
    .025               28                   24                  52
    .010                7                    6                  13
    .005                4                    3                   7

average t is 0.0077 with variance of 1.086943
```

Which hypothesis is true in the simulation? Does the outlier-proneness of the population have a serious effect?

Solution   The output indicates that H_0 is $\mu = 50$, and indeed the population mean is 50. Therefore, fractions such as 104/1,000 are approximating α, the probability of a Type I error; the fractions are approximations because they are based on 1,000 samples, not on an infinite number. Notice that all the fractions are very close to the nominal α values. For example, with a one-tailed α of .025, the observed fractions are .028 and .024.   ▲

EXAMPLE   5.19   ▼

A simulation study chooses samples of various sizes from a population that is strongly right-skewed. The following results are obtained for a t test:

	μ	σ	n
	50.000	10.000	10

one-tailed test:

number of times H_0: "mean is 50" is rejected in favor of

α	"mean > 50"	"mean < 50"	total (α doubled)
.100	27	308	335
.050	6	249	255
.025	2	210	212
.010	0	176	176
.005	0	155	155

	μ	σ	n
	50.000	10.000	30

one-tailed test:

number of times H_0: "mean is 50" is rejected in favor of

α	"mean > 50"	"mean < 50"	total (α doubled)
.100	40	237	277
.050	6	189	195
.025	0	156	156
.010	0	120	120
.005	0	99	99

	μ	σ	n
	50.000	10.000	60

one-tailed test:

number of times H_0: "mean is 50" is rejected in favor of

α	"mean > 50"	"mean < 50"	total (α doubled)
.100	53	190	243
.050	9	145	154
.025	2	114	116
.010	1	70	71
.005	0	59	59

What effect does the skewness of this population have? What happens as the sample size increases?

Solution   The skewness causes the nominal α probabilities to be seriously wrong, especially for one-tailed tests. The effect is most severe for the smallest n, 10, but it is still severe when n is 60.   ▲

The second question, that of the *efficiency* of t procedures, has only recently been studied seriously. The near–unanimous conclusion from these studies, when the population distribution is symmetric but heavy-tailed, is that **robust methods** such as the Wilcoxon rank sum test discussed in Chapter 6 are more efficient than the corresponding t test for μ. Therefore, the robust procedures tend to have more accurate estimates with smaller standard errors. Unfortunately, less work has been done on the effectiveness of these robust procedures in cases where the population distribution is skewed.

So what is a nonexpert to do? First, look at the data. A simple histogram of the data, or some other plotting device, will reveal any gross skewness or extreme outliers. If there's no blatant nonnormality, the nominal t-distribution probabilities should be reasonably correct and the t procedures should be reasonably efficient. If the data values are obviously skewed or heavy-tailed, the t-distribution probabilities and the efficiency of the t procedure are highly suspect. In these situations, you may wish to consult other textbooks (such as Hildebrand and Ott, 1991) in which alternative robust procedures are presented. You should at least develop a healthy skepticism toward stated probabilities or confidence levels based on t methods.

Other robust procedures will be mentioned in this text, but we cannot do complete justice to them. We expect that these procedures will be integrated into some of the statistical software systems in the next few years. If such programs are not available, at least a manager can cultivate an alert skepticism about the accuracy of the stated probabilities.

robust methods (margin note)

plotting data (margin note)

▼ EXERCISES

Basic Techniques

5.48   Why is the z-test of Section 5.5 inappropriate for testing H_0: $\mu = \mu_0$ when $n < 30$?

5.49   Set up the rejection region based on t for H_0: $\mu = \mu_0$ when $\alpha = .05$ and for the following conditions:
a. H_a: $\mu < \mu_0$, $n = 15$
b. H_a: $\mu \neq \mu_0$, $n = 23$
c. H_a: $\mu > \mu_0$, $n = 6$

5.50   Repeat Exercise 5.49 with $\alpha = .01$.

5.51   The sample data for a t-test of H_0: $\mu = 15$ and H_a: $\mu > 15$ are $\bar{y} = 16.2$, $s = 3.1$, and $n = 18$. Use $\alpha = .05$ to draw your conclusions.

Applications

5.52 A random sample of ten students in a fourth-grade reading class was thoroughly tested to determine reading speed and reading comprehension. Based on a fixed-length standardized test reading passage, the following speeds (in minutes) and comprehension scores (based on a 100-point scale) were obtained.

Student	1	2	3	4	5	6	7	8	9	10
Reading speed	5	7	15	12	8	7	10	11	13	9
Reading comprehension	60	76	96	100	81	75	85	88	98	83

a. Use the reading speed data to place a 95% confidence interval on μ, the average speed for all fourth-grade students in the large school from which the sample was drawn.
b. Interpret the interval estimate in part a.
c. How would your inference change by using a 98% confidence interval?

5.53 Refer to Exercise 5.52. Using the reading comprehension data, test the research hypothesis that the mean for all fourth graders on the standardized examination is greater than 80, the statewide average for comparable students the previous year. Give the level of significance for your test. Interpret your findings.

5.54 Refer to Exercise 5.53.

a. Set up all parts for a statistical test of the research hypothesis that the mean score for all fourth graders is different from 80, the statewide average the previous year.
b. Give the level of significance for this test.

5.55 Refer to Exercise 5.52. Use a computer program to construct a 90% confidence interval for the mean total reading score (speed plus comprehension). A Minitab program is shown for illustrative purposes.

```
MTB > READ C1 C2
     10 ROWS READ
MTB > END
MTB > LET C3=C1+C2

MTB > PRINT C1-C3

 ROW    C1     C2     C3

   1     5     60     65
   2     7     76     83
   3    15     96    111
   4    12    100    112
   5     8     81     89
   6     7     75     82
   7    10     85     95
   8    11     88     99
   9    13     98    111
  10     9     83     92

MTB > TINTERVAL 90% CONFIDENCE C3

              N      MEAN    STDEV   SE MEAN     90.0 PERCENT C.I.
C3           10     93.90    15.14     4.79   (   85.12,   102.68)

MTB > STOP
```

5.56 The amount of sewage and industrial pollutants dumped into a body of water affects the health of the water by reducing the amount of dissolved oxygen available for aquatic life. Suppose that weekly readings are taken from the same location in a river over a 2-month period. Use the summary data from the computer printout given here to conduct a statistical test of the research hypothesis that the mean dissolved oxygen content is less than 5.0 parts per million, a level some scientists think is marginal for supplying enough dissolved oxygen for fish.

```
5.100000000
4.900000000
5.600000000
4.200000000
4.800000000
4.500000000
5.300000000
5.200000000

8.000000000   sample size
4.950000000   ȳ
 .2028571428   s²
 .4503966505   s
```

 5.57 A dealer in recycled paper places empty trailers at various sites; these are gradually filled by individuals who bring in old newspapers and the like. The trailers are picked up (and replaced by empties) on several schedules. One such schedule involves pickup every second week. This schedule is desirable if the average amount of recycled paper is more than 1,600 cubic feet per 2-week period. The dealer's records for 18 2-week periods show the following volumes (in cubic feet) at a particular site:

1,660	1,820	1,590	1,440	1,730	1,680	1,750	1,720	1,900
1,570	1,700	1,900	1,800	1,770	2,010	1,580	1,620	1,690

($\bar{y} = 1,718.3$, $s = 137.8$)

Assume that these figures represent the results of a random sample. Do they support the research hypothesis that $\mu > 1,600$, using $\alpha = .10$? Write out all parts of the hypothesis-testing procedure.

5.58 Place an upper bound on the p-value of Exercise 5.57. Would you say that $\mu > 1,600$ is strongly supported?

 5.59 A federal regulatory agency is investigating an advertised claim that a certain device can increase the gasoline mileage of cars. Seven such devices are purchased and installed in seven cars belonging to the agency. Gasoline mileage for each of the cars under standard conditions is recorded both before and after installation.

	Car						
	1	2	3	4	5	6	7
Mi/gal before	19.1	19.9	17.6	20.2	23.5	26.8	21.7
Mi/gal after	20.0	23.7	18.7	22.3	23.8	19.2	24.6
Change	.9	3.8	1.1	2.1	.3	−7.6	2.9

The mean change is .50 miles per gallon and the standard deviation is 3.77.

a. Formulate appropriate null and research hypotheses.

b. Is the advertised claim supported at $\alpha = .05$? Carry out the steps of a hypothesis test.

5.60 Use the data of Exercise 5.59 to construct a 90% confidence interval for the mean change. On the basis of this interval, can one reject the hypothesis of no mean change? (Note that the two-sided 90% confidence interval corresponds to a one-tailed $\alpha = .05$ test.)

5.61 Would you say that the agency of Exercises 5.59 and 5.60 has conclusively established that the device has no effect on the average mileage of cars? What does the width of the interval in Exercise 5.60 have to do with your answer?

5.9 ▼ SUMMARY

A population mean can be estimated using point or interval estimation. The goodness of an interval estimate is given by the width of the interval and the confidence coefficient. The general formula for a $100(1 - \alpha)\%$ confidence interval for μ was given along with the sample size formula for planning a study to give a confidence interval of fixed width for μ.

Following the traditional approach to hypothesis testing, a statistical test about μ is composed of five parts: null hypothesis, research hypothesis, test statistic, rejection region, and conclusion. It employs the technique of proof by contradiction. We try to verify the research hypothesis by gathering information to contradict the null hypothesis $H_0: \mu = \mu_0$. As with any two-decision problem, there are two types of errors that can be committed, the rejection of H_0 when H_0 is true—a Type I error—and the acceptance of H_0 when H_0 is false and some alternative is true—a Type II error. The probabilities for these errors, designated by α and β, measure the goodness of the test procedure.

In this chapter, we indicated that for a given sample size, α and β are inversely related; as α is increased, β is decreased, and vice versa. If we specify n and α for a given test procedure, we can compute β for alternative values of μ. Sometimes we may wish to specify both α and β *prior* to conducting the investigation. To do this, we determine the sample size required for the specific values of α and β.

We considered an alternative to the traditional decision-based approach for a statistical test of a hypothesis. Rather than relying on a preset level of α, we compute the weight of evidence for rejecting the null hypothesis. This weight, expressed in terms of a probability, is called the level of significance for the test. Most professional journals summarize the results of a statistical test using the level of significance.

The final topic discussed in this chapter concerned inferences about μ when σ is unknown (which is almost always). Through the use of the t distribution we can construct both confidence intervals and a statistical test for

μ. Since the t-values of the t distribution approach the z-values of a normal distribution and since σ is almost never known, it is convenient to use t results for all inferences about μ (large or small sample).

▼ KEY FORMULAS

Estimation and tests for μ

1. $100(1 - \alpha)\%$ confidence interval for μ (σ known)

$$\bar{y} \pm z_{\alpha/2}\sigma_{\bar{y}}, \quad \text{where} \quad \sigma_{\bar{y}} = \sigma/\sqrt{n}$$

2. $100(1 - \alpha)\%$ confidence interval for μ (σ unknown)

$$\bar{y} \pm t_{\alpha/2}s/\sqrt{n}, \quad df = n - 1$$

3. Sample size for estimating μ with a $100(1 - \alpha)\%$ confidence interval, $\bar{y} \pm E$

$$n = \frac{(z_{\alpha/2})^2\sigma^2}{E^2}$$

4. Statistical test for μ (σ known)

$$H_0: \quad \mu = \mu_0$$

$$\text{T.S.:} \quad z = \frac{\bar{y} - \mu_0}{\sigma/\sqrt{n}}$$

5. Statistical test for μ (σ unknown)

$$H_0: \quad \mu = \mu_0$$

$$\text{T.S.:} \quad t = \frac{\bar{y} - \mu_0}{s/\sqrt{n}}, \quad df = n - 1$$

6. Calculation of β (and equivalently power) for a test on μ
   a. One-tailed test

   $$\beta = P\left(z < z_\alpha - \frac{|\mu_0 - \mu_a|}{\sigma_{\bar{y}}}\right), \quad \text{where} \quad \sigma_{\bar{y}} = \sigma/\sqrt{n}$$

   b. Two-tailed test

   $$\beta \approx P\left(z < z_{\alpha/2} - \frac{|\mu_0 - \mu_a|}{\sigma_{\bar{y}}}\right)$$

7. Sample size for a statistical test on μ
   a. One-tailed test

$$n = \frac{\sigma^2}{\Delta^2}(z_{\alpha/2} + z_\beta)^2$$

   b. Two-tailed test

$$n = \sigma^2 \frac{(z_{\alpha/2} + z_\beta)^2}{\Delta^2}$$

▼ SUPPLEMENTARY EXERCISES

5.62 To test the effectiveness of a new spray for controlling rust mites, we would like to compare the average yield for treated groves with the average yield for untreated groves displayed in previous years. A random sample of thirty 1-acre groves is chosen and sprayed according to a recommended schedule. The average yield for the 30-grove sample was 830 boxes, with a standard deviation of 91. Yields from groves in the same area without rust mite maintenance spraying have averaged 760 boxes over previous years. Do these data present sufficient evidence to indicate that the mean yield for groves sprayed with the new preparation is higher than 760 boxes, the average over previous years without spraying? Is this a one-tailed or two-tailed test? Use $\alpha = .05$.

5.63 A wine manufacturer sells a cabernet wine whose label asserts an alcohol content of 11%. Fifteen bottles of this wine are selected at random and analyzed for alcohol content, with a resulting mean of 10.2% and a standard deviation of 1.2%
   a. Find a 95% confidence interval for the mean alcohol content.
   b. Based on your answer to part a, do you think the label is correct? Explain.

5.64 A paint manufacturer wishes to validate its claim that a gallon of its paint covers 400 square feet, and it sets up a test based on a random sample of 50 1-gallon cans of paint. The hypothesis to be tested is $H_0: \mu = 400$ versus $H_a: \mu > 400$; the significance level is $\alpha = .05$.
   a. In words, what is the parameter of interest (μ)?
   b. Give the rejection region (including critical value) for the test.
   c. If the sample of 50 cans shows an average coverage of 412 square feet and a standard deviation of 38 square feet, what is the conclusion of the test?
   d. Find the p-value of the test.

5.65 A study of the operation of a parking garage showed that, in the past, average parking time was 220 minutes. Recently, the garage was remodeled and charges increased. The management wants to know if the changes have had any effect on mean parking time. Thus, we wish to test $H_0: \mu = 220$ versus $H_a: \mu \neq 220$, and we will use $\alpha = .05$. A random sample of 50 cars had an average parking time of 208 minutes and a standard deviation of 40 minutes.
   a. Give the rejection region, including test-statistic formula and critical value.
   b. Give the observed value of the test statistic.
   c. Give your conclusion about changes in parking time.
   d. Give the significance level (p-value) of the test.

5.66 A random sample of 35 city buses showed the mean number of passengers (per day, per bus) to be 225 with a standard deviation of 60 passengers.
  a. Find a 95% confidence interval for the average number of passengers.
  b. In words, describe the parameter of interest in this problem and give its value (if it is known).
  c. In words, describe a sample statistic in this problem and give its value (if it is known).

 5.67 An office manager wishes to estimate the mean time required to handle a customer complaint. A sample of 38 complaints shows a mean of 28.7 minutes and a standard deviation of 12 minutes.
  a. Give a point estimate for the true mean time required to handle customer complaints.
  b. Construct a 90% confidence interval estimate for the true mean time required to handle customer complaints.

5.68 The concentration of mercury in a lake has been measured many times. This population of measurements has an average of 1.20 mg/m^3 (milligrams per cubic meter) with a standard deviation of 0.30 mg/m^3. Following an accident at a smelter on the shore of the lake, nine more measurements were taken. These have an average mercury concentration of 1.45 mg/m^3. Report the level of significance of the evidence from this sample that the mean mercury concentration in the lake has increased.

5.69 Answer "true" or "false" for each question.
  a. Given any particular random sample, if we form the 95% confidence interval for the sample mean there is a 95% chance that the population mean lies in this confidence interval.
  b. If a large number of random samples are selected and we form the 95% confidence interval for each sample mean, the population mean will lie in about 95% of these confidence intervals.
  c. If a sample size is larger than 30, there is a 95% chance that the sample mean equals the population mean.
  d. If a very large number of random samples are selected, there is a 95% chance that one of the sample means is equal to the population mean.
  e. The 95% confidence interval around a given sample mean is wider than 90% confidence interval around that mean.
  f. In order to prove that $\mu = \mu_0$ with Type I error .05, we must select a sample and fail to reject the null hypothesis $H_0: \mu = \mu_0$ using $\alpha = .05$.
  g. To find the critical value for a *two-tailed* test with Type I error .04, we can look in Table 1 of the Appendix for the z-score corresponding to the area .4800.
  h. To find the critical value for a *one-tailed* test with Type I error .02, we can look in Table 1 of the Appendix for the z-score corresponding to the area .4800.
  i. If we rejected the null hypothesis at the $\alpha = .05$ level, then we would also have rejected it at the $\alpha = .01$ level.

5.70 Answer "true" or "false" for each question. If your answer is "false," change the statement to make it true. Change only the *underlined* words.
  a. A <u>type I error</u> is committed when we fail to reject the null hypothesis H_0 when H_0 is actually false.
  b. If we make a <u>type II error</u>, we have missed detecting an event or effect when there actually was one.
  c. The probability of making a <u>type I error</u> is equal to β.
  d. If we increase the probability of making a Type II error, we will <u>increase</u> the probability of making a Type I error.

5.71 Over the years, projected due dates for expectant mothers have been notoriously bad. In a recent survey of 100 mothers, if was found that the average number of days to birth beyond the projected due date was 9.2, with a standard deviation of 12.4. Use these data to find a 95% confidence interval for the mean number of days to birth beyond the due date.

5.72 Refer to Exercise 5.71. Use these data to find a 90% confidence interval for the mean number of days to birth beyond the due date.

5.73 A corporation maintains a large fleet of company cars for its salespeople. In order to determine the average number of miles driven per month by all salespeople, a random sample of 70 records was obtained. The mean and the standard deviation for the number of miles were 3,250 and 420, respectively. Estimate μ, the average number of miles driven per month for all the salespeople within the corporation, using a 99% confidence interval.

5.74 The length of time to assemble an electronic fuse was measured for 50 assemblers. The mean and the standard deviation were 3.2 minutes and .3 minutes, respectively. Give a 90% confidence interval for the mean length of time to assemble a fuse.

5.75 The diameter of extruded plastic pipe varies about a mean value that is controlled by a machine setting. A random sample of the diameter of 50 pieces of plastic pipe gave a mean and a standard deviation of 4.05 inches and .12 inches, respectively. Do the data present sufficient evidence to indicate that the mean diameter is different from 4 inches? Use $\alpha = .05$.

5.76 The manufacturer of an automatic control device claims that the device will maintain a mean room humidity of 80%. The humidity in a controlled room was recorded for a period of 30 days, and the mean and the standard deviation were found to be 78.3% and 2.9%, respectively. Do the data present sufficient evidence to contradict the manufacturer's claim? Use $\alpha = .05$.

5.77 A buyer wishes to determine whether the mean sugar content per orange shipped from a particular grove is less than .027 pounds. A random sample of 50 oranges produced a mean sugar content of .025 pounds and a standard deviation of .003 pounds. Do the data present sufficient evidence to indicate that the mean sugar content is less than .027 pounds? Use $\alpha = .05$.

5.78 One method for solving the electrical power shortage makes use of floating nuclear power plants located a few miles offshore in the ocean. Because there is great concern about the possibility of a ship colliding with the floating (but anchored) power plant, navigation experts have stated that it would be desirable if the average number of ships per day passing within 10 miles of the proposed power site location were less than 7. To verify this hypothesis for the proposed site, a random sample of 60 days was used throughout the peak shipping months. For each day, the number of ships passing within the 10-mile limit was recorded. The sample mean and standard deviation were 6.3 ships and 2 ships, respectively. Use these data to test the navigation experts' alternative hypothesis. Use $\alpha = .05$.

5.79 Administrative officials for a university are concerned that the freshman students taking advantage of off-campus housing facilities have significantly lower grade point averages (GPA) than all freshmen at the school. After the fall quarter, the all-freshman average GPA was 2.1 (on a 4-point system). Since it was not possible to isolate grades for all students living in off-campus housing by university records, a random sample of 81 off-campus freshmen was obtained by tracing students through their permanent home addresses. The sample mean and standard deviation were found to be 1.92 and .2, respectively. Do these data present sufficient evidence to indicate that the average GPA for all off-campus freshmen is lower than the all-freshman average? Use $\alpha = .05$.

5.80 In a standard dissolution test for tablets of a particular drug product, the manufacturer must obtain the dissolution rate for a batch of tablets prior to release of the batch. Suppose that the dissolution test consists of assays for 36 individual 25-mg tablets. For each test, the tablet is suspended in an acid bath and then assayed after 30 minutes. The sample mean and standard deviation after 30 minutes are 19.8 and .42 mg, respectively. Use these data to test $H_0: \mu = 20$ (80% of the labeled amount in the tablets) against the alternative hypothesis $H_a: \mu < 20$. Use $\alpha = .05$.

5.81 Refer to Exercise 5.80. Give the level of significance for the test when the alternative hypothesis is $H_a: \mu \neq 20$.

5.82 Statistics has become a valuable tool for auditors, especially where large inventories are involved. It would be costly and time consuming for an auditor to inventory each item in a large operation. Thus, the auditor frequently resorts to obtaining a random sample of items and using the sample results to check the validity of a company's financial statement. For example, a hospital financial statement claims an inventory that averages $300 per item. An auditor's random sample of 20 items yielded a mean and standard deviation of $160 and $90, respectively. Do the data contradict the hospital's claimed mean value per inventoried item and indicate that the average is less than $300? Use $\alpha = .05$.

5.83 Over the past 5 years, the mean time for a warehouse to fill a buyer's order has been 25 minutes. Officials of the company believe that the length of time has increased recently, either due to a change in the work force or due to a change in customer purchasing policies. The processing time (in minutes) was recorded for a random sample of 15 orders processed over the past month.

28	25	27	31	10
26	30	15	55	12
24	32	28	42	38

Do the data present sufficient evidence to indicate that the mean time to fill an order has increased? Use the accompanying output to reach a conclusion based on $\alpha = .01$.

```
MTB > NAME C1 'TIME'
MTB > SET C1
MTB > END
MTB > PRINT 'TIME'

TIME
    28    25    27    31    10    26    30    15    55    12    24    32    28
    42    38

MTB > TTEST 25 'TIME';
SUBC> ALTERNATIVE 1.

TEST OF MU = 25.000 VS MU G.T. 25.000

              N      MEAN    STDEV    SE MEAN        T    P VALUE
TIME         15    28.200   11.441      2.954     1.08       0.15

MTB > STOP
```

5.84 Give the level of significance for the statistical test in Exercise 5.83.

5.85 If a new process for mining copper is to be put into full-time operation, it must produce an average of more than 50 tons of ore per day. A 5-day trial period gave the results shown in the accompanying table. Do these figures warrant putting the new process into full-time operation? Test by using $\alpha = .05$.

Day	1	2	3	4	5
Yield in tons	50	47	53	51	52

5.86 A test was conducted to determine the length of time required for a student to read a specified amount of material. All students were instructed to read at the maximum speed at which they could still comprehend the material. Sixteen students took the test, with the following results (in minutes):

25	18	27	29	20	19	25	24
32	21	24	19	23	28	31	22

Estimate the mean length of time required for all students to read the material, using a 95% confidence interval.

5.87 A random sample of eight students participated in a psychological test of depth perception. Two markers, one labeled A and the other B, were arranged a fixed distance apart at the far end of the laboratory. One by one, the students were ushered into the room and asked to judge the distance between the two markers at the other end of the room. The sample data (in feet) were as follows:

2.1	2.2	2.6	2.3
1.8	2.3	2.4	2.5

Construct a 90% confidence interval for μ, the mean judged distance for all students for which the sample is representative.

5.88 The lifetimes (in years) of ten automobile batteries of a certain brand are

2.4   1.9   2.0   2.1   1.8   2.3   2.1   2.3   1.7   2.0

Estimate the mean lifetime, using a 95% confidence interval.

5.89 A drug antibiotic manufacturer randomly sampled 12 different locations in the fermentation vat to determine average potency for the batch of antibiotic being prepared. Readings were as follows:

8.9	9.0	9.1	8.9
9.1	9.0	9.0	9.0
8.9	8.8	9.1	8.8

Use the output shown here to estimate the mean potency for the batch, based on a 95% confidence interval. Interpret the interval.

```
MTB > SET C1
MTB > END
MTB > PRINT C1

C1
   8.9   9.0   9.1   8.9   9.1   9.0   9.0   9.0   8.9   8.8   9.1
   8.8

MTB > TINTERVAL 95% CONFIDENCE C1
                N      MEAN    STDEV   SE MEAN    95.0 PERCENT C.I.
C1             12    8.9667   0.1073   0.0310   ( 8.8985,  9.0349)

MTB > STOP
```

5.90 In a statistical test about μ, the null hypothesis was rejected. Based on this conclusion, which of the following statements are true?
  a. A Type I error was committed.
  b. A Type II error was committed.
  c. A Type I error could have been committed.
  d. A Type II error could have been committed.
  e. It is impossible to have committed both Type I and Type II errors.
  f. It is impossible that neither a Type I nor a Type II error was committed.
  g. Whether any error was committed is not known, but if an error was made, it was Type I.
  h. Whether any error was committed is not known, but if an error was made, it was Type II.

5.91 Answer "true" or "false" for each statement.
  a. In a test of a hypothesis, a test statistic is computed from the sample data.
  b. A statistical test of a hypothesis employs the technique of proof by contradiction. That is, we try to show that the alternative hypothesis is true by showing that the null hypothesis is false.
  c. The sample size n plays an important role in testing hypotheses because it measures the amount of data (and hence information) upon which we base a decision. If the data are quite variable and n is too small, it is unlikely that we will reject the null hypothesis even when the null hypothesis is false.

5.92 Complete the following statements (more than one word may be needed).
  a. If we take all possible samples (of a given sample size) from a population, then the distribution of sample means tends to be _____ and the mean of these sample means is equal _____.
  b. The larger the sample size, other things remaining equal, the _____ the confidence interval.
  c. The larger the confidence coefficient, other things remaining equal, the _____ the confidence interval.
  d. The statement "If random samples of a fixed size are drawn from any population (regardless of the form of the population distribution), as n becomes larger, the distribution of sample means approaches normality," is known as the _____.
  e. By failing to reject a null hypothesis that is false, one makes a _____ error.

5.93 Suppose that the tar content of cigarettes is normally distributed with a mean of 10 and a standard deviation of 2.4 mg. A new manufacturing process is developed for decreasing the tar content. A sample of 16 cigarettes produced by the new process yielded a mean of 8.8 mg. Use $\alpha = .05$.

a. Do a test of hypothesis to determine if the new process has significantly *decreased* the tar content. Use the following outline.

> Null hypothesis
> Alternative hypothesis
> Assumptions
> Rejection region(s)
> Test statistic and computations
> Conclusion
>> In statistical terms
>> In plain English

b. Based on your conclusion, could you have made

> A Type I error?
> A Type II error?
> Neither error?
> Both Type I and Type II errors?

 5.94 The board of health of a particular state was called to investigate claims that raw pollutants were being released into the river flowing past a small residential community. By applying financial pressure, the state was able to get the violating company to make major concessions toward the installation of a new water purification system. In the interim, different production systems were to be initiated to help reduce the pollution level of water entering the stream. To monitor the effect of the interim system, a random sample of 50 water specimens was taken throughout the month at a location downstream from the plant. If $\bar{y} = 5.0$ and $s = .70$, use the sample data to determine whether the mean dissolved oxygen count of the water (in ppm) is less than 5.2, the average reading at this location over the past year.

a. List the five parts of the statistical test, using $\alpha = .05$.

b. Conduct the statistical test and state your conclusion.

5.95 Refer to Figure 5.12. Compute β for (b) and (c) using $H_0: \mu = 380$, $H_a: \mu > 380$. Recall that $n = 50$ and $s = 35.2$.

 5.96 As described in Exercise 5.42, an automatic merge system has been installed at the entrance ramp to a major highway. Prior to the installation of the system, investigators found the average stress level of drivers to be 8.2 on a 10-point scale. After installation, a sample of 50 drivers showed $\bar{y} = 7.6$ and $s = 1.8$. Conduct a statistical test of the research hypothesis that the average stress at peak hours for drivers under the new system is less than 8.2, the average stress level prior to the installation of the automatic merge system. Determine the level of significance of the statistical test. Interpret your findings.

 5.97 The search for alternatives to oil as a major source of fuel and energy will inevitably bring about many environmental challenges. These challenges will require solutions to problems in such areas as strip mining and many others. Let us focus on one. If coal is considered as a major source of fuel and energy, we will have to consider ways to keep large amounts of sulfur dioxide (SO_2) and particulates from getting into the air. This is especially important at large government and industrial operations. Here are several possibilities.

1. Build the smokestack extremely high.
2. Remove the SO_2 and particulates from the coal prior to combustion.
3. Remove the SO_2 from the gases after the coal is burned but before the gases are released into the atmosphere. This is accomplished by using a scrubber.

Several scrubbers have been developed in recent years. Suppose that a new one has been constructed and is set for testing at a given power plant. Fifty samples are obtained at various times from gases emitted from the stack. The mean SO_2 emission is .13 pound per million BTU, with a standard deviation of .05 pound. Use the sample data to construct a statistical test of the null hypothesis $H_0: \mu = .145$, the average emission level for one of the more efficient scrubbers that has been developed. Choose an appropriate alternative hypothesis, with $\alpha = .05$.

5.98 Refer to Exercise 5.15. Construct a 99% confidence interval for μ, the average number of miles between a person's high school and present permanent address for the state sample.

5.99 Refer to Exercise 5.97. Rather than being interested in testing the research hypothesis that $\mu < .145$, the average emission level for one of the more efficient scrubbers, we may wish to estimate the mean emission level for the new scrubber. Use the sample data to construct a 99% confidence interval for μ. Interpret your results.

5.100 As part of an overall evaluation of training methods, an experiment was conducted to determine the average exercise capacity of healthy male army inductees. To do this each male in a random sample of 35 healthy army inductees exercised on a bicycle ergometer (a device for measuring work done by the muscles) under a fixed work load until he tired. Blood pressure, pulse rates, and other indicators were carefully monitored to ensure that no one's health was in danger. The exercise capacities (mean time, in minutes) for the 35 inductees are listed below.

23	19	36	12	41	43	19
28	14	44	15	46	36	25
35	25	29	17	51	33	47
42	45	23	29	18	14	48
21	49	27	39	44	18	13

a. Use these data to construct a 95% confidence interval for μ, the average exercise capacity for healthy male inductees. Interpret your findings.
b. How would your interval change using a 99% confidence interval?

5.101 Using the data of Exercise 5.100, determine the number of sample observations that would be required to estimate μ to within 1 minute, using a 95% confidence interval. (*Hint:* Substitute $s = 12.36$ for σ in your calculations.)

5.102 A study was conducted to examine the effect of a preparation of mosaic virus on tobacco leaves. In a random sample of $n = 32$ leaves, the mean number of lesions was 22, with a standard deviation of 3. Use these data and a 95% confidence interval to estimate the average number of lesions for leaves affected by a preparation of mosaic virus.

5.103 Refer to Exercise 5.102. Use the sample data to form a 99% confidence interval on μ, the average number of lesions for tobacco leaves affected by a preparation of mosaic virus.

5.104 We all remember being told, "Your fever has subsided and your temperature has returned to normal." What do we mean by the word *normal*? Most people use the benchmark 98.6°F, but this does not apply to all people, only the "average" person. Without putting words into someone's mouth, we might define a person's normal temperature to be his or her average temperature when healthy. But even this definition is cloudy because there is variation in a person's temperature throughout the day. To determine a subject's normal temperature, we recorded it for a random sample of 30 days. On each day selected for inclusion in the sample, the temperature reading was made at 7 A.M. The sample mean and standard deviation for these 30 readings were, respectively, 98.4 and .15. Assuming the subject was healthy on the days examined, use these data to estimate the person's 7 A.M. "normal" temperature using a 90% confidence interval.

5.105 Refer to the data of Exercise 5.100. Suppose that the random sample of 35 inductees was selected from a large group of new army personnel being subjected to a new (and hopefully improved) physical fitness program. Assume previous testing with several thousand personnel over the past several years has shown an average exercise capacity of 29 minutes. Run a statistical test for the research hypothesis that the average exercise capacity is improved for the new fitness program. Give the level of significance for the test. Interpret your findings.

5.106 Refer to Exercise 5.105.
   a. How would the research hypothesis change is we were interested in determining whether the new program is better or worse than the physical fitness program for inductees?
   b. What is the level of significance for your test?

5.107 In a random sample of 40 hospitals from a list of hospitals with over 100 semiprivate beds, a researcher collected information on the proportion of persons whose bills are covered by a group policy under a major medical insurance carrier. The sample proportions are given in the following chart.

.67	.74	.68	.63	.91	.81	.79	.73
.82	.93	.92	.59	.90	.75	.76	.88
.85	.90	.77	.51	.67	.67	.92	.72
.69	.73	.71	.76	.84	.74	.54	.79
.71	.75	.70	.82	.93	.83	.58	.84

Use the sample data to construct a 90% confidence interval on μ, the average proportion of patients per hospital with group medical insurance coverage.

5.108 Refer to Exercise 5.107. Use the same data to construct a 99% confidence interval.

5.109 Faculty members in a state university system who resign within 10 years of initial employment are entitled to receive the money paid into a retirement system, plus 4% per annum. Unfortunately, experience has shown that the state is extremely slow in returning this money. Concerned about such a practice, a local teachers' organization decides to investigate. From a random sample of 50 employees who resigned from the state university system over the past five years, the average time between the termination date and reimbursement was 75 days, with a standard deviation of 15 days. Use the data to estimate the mean time to reimbursement, using a 95% confidence interval.

5.110 Refer to Exercise 5.109. After a confrontation with the teachers' union, the state promised to make reimbursements within 60 days. Monitoring of the next 40 resignations yields an average of 58 days, with a standard deviation of 10 days. If we assume that these 40 resignations represent a random sample of the state's future performance, estimate the mean reimbursement time, using a 99% confidence interval.

5.111 Refer to Example 5.11. Compute β for $\mu_a = 40$. What would be your conclusion based on the magnitude of β?

5.112 Refer to Exercise 5.111. Using the values of β computed for $\mu_a = 38$ and $\mu_a = 40$, calculate the probability of a Type II error for several other values of μ_a in order to construct a graph of β against μ_a.

5.113 Refer to the data of Exercise 5.100. It can be shown that the sample standard deviation is 12.36. Use a computer program to construct a 90% confidence interval for μ. A Minitab program is shown here for illustration purposes.

```
MTB > SET C6
MTB > END
MTB > PRINT C6

C6
    23    19    36    12    41    43    19    28    14    44    15    46    36
    25    35    25    29    17    51    33    47    42    45    23    29    18
    14    48    21    49    27    39    44    18    13

MTB > ZINTERVAL 90% CONFIDENCE SIGMA=12.36 C6

THE ASSUMED SIGMA =12.4

                N      MEAN     STDEV   SE MEAN      90.0 PERCENT C.I.
C6             35     30.51     12.36      2.09   (   27.07,    33.95)

MTB > STOP
```

5.114 Use the data of Exercise 5.107 and the Minitab output shown here to respond to the statements that follow concerning the confidence interval for μ.

```
MTB > SET C4
MTB > END
MTB > PRINT C4

C4
   0.67   0.73   0.59   0.93   0.92   0.82   0.75   0.51   0.81   0.54   0.85
   0.68   0.76   0.75   0.58   0.69   0.92   0.82   0.67   0.73   0.71   0.77
   0.91   0.74   0.88   0.74   0.71   0.90   0.83   0.72   0.93   0.70   0.67
   0.79   0.79   0.90   0.63   0.84   0.76   0.84

MTB > STANDARD DEVIATION C4
    ST.DEV. =      0.10861
MTB > ZINTERVAL 98% CONFIDENCE , SIGMA = 0.10861 C4

THE ASSUMED SIGMA =0.109

                N      MEAN     STDEV   SE MEAN      98.0 PERCENT C.I.
C4             40    0.7620    0.1086    0.0172   (  0.7220,   0.8020)

MTB > STOP
```

a. Identify the sample mean for the data shown in the computer output.
b. What is the confidence coefficient for the confidence interval shown?
c. Give the confidence limits and interpret the interval estimate.

5.115 A random sample of birth rates from 40 inner-city areas shows an average of 35 per thousand, with a standard deviation of 6.3. Estimate the mean inner-city birth rate. Use a 95% confidence interval.

5.116 A random sample of 30 standard metropolitan statistical areas (SMSAs) was selected and the ratio (per 1,000) of registered voters to the total number of persons 18 years and over was recorded in each area. Use the data below to test the research hypothesis that μ, the average ratio (per 1,000), is different from 675, last year's average ratio. Give the level of significance for your test.

802	497	653	600	729	812
751	730	635	605	760	681
807	747	728	561	696	710
641	848	672	740	818	725
694	854	674	683	695	803

5.117 Improperly filled orders are a costly problem for mail-order houses. To estimate the mean loss per incorrectly filled order, a large firm plans to sample n incorrectly filled orders and to determine the added cost associated with each one. It is estimated that the added cost is between $40 and $400. How many incorrectly filled orders must be sampled to estimate the mean additional cost using a 95% confidence interval of width $20?

5.118 Records from a particular hospital were examined to determine the average length of stay for patients being treated for lung cancer. Data from a sample of 100 records showed $\bar{y} = 2.1$ months and $s = 2.6$ months.
a. Would a confidence interval for μ based on t be appropriate? Why or why not?
b. Indicate an alternative procedure for estimating the center of the distribution.

5.119 Refer to Exercise 5.28. Graph the center line and control limits for μ and plot the sequence of sample means shown here to determine if the process is out of control:

109	113	108	108	110	108	107	108	103	107
109	109								

5.120 Investigators would like to estimate the average annual taxable income of apartment dwellers in a city to within $500 using a 95% confidence interval. If we assume the annual incomes for apartment dwellers have a range of $40,000, determine the number of observations that should be included in the sample.

5.121 As indicated earlier, the stated weight on the new giant-sized laundry detergent package is 42 ounces. Also displayed on the box is the following statement: "Individual boxes of this product may weigh slightly more or less than the marked weight due to normal variations incurred with high-speed packaging machines, but each day's production of detergent average slightly above the marked weight." Discuss how you might attempt to test this claim. Or would it be simpler to modify this claim slightly for testing purposes? State all parts of your test. Would there be any way to determine in advance the sample size required to pick up a specified alternative with power equal to .90, using $\alpha = .05$?

5.122   Congestive heart failure is known to be fatal in a high percentage of cases. A total of 182 patients with chronic left-ventricular failure who were symptomatic in spite of therapy were followed. The length of survival for these patients ranged from 1 to 41 months with a mean of 12 months and a standard deviation of 10. Would a confidence interval for the mean survival of these patients be appropriate? Why or why not?

5.123   After a decade of steadily increasing popularity, the sales of automatic teller machines (ATMs) have been on the decline. In a recent month, a spot check of a random sample of 40 suppliers indicated that shipments averaged 20% lower than those for the corresponding period 1 year ago. Assume the standard deviation is 6.2% and the percentage data appear mound-shaped. Use these data to construct a 99% confidence interval on the mean percentage decrease in shipments of ATMs.

5.124   Suppose the percentage change in shipments of ATMs from the 40 suppliers of Exercise 5.123 ranged from -40% (a 40% decrease) to $+16\%$ (a 16% increase) with a sample mean of -20%, a median decrease of -10%, and a 10% trimmed mean of -12%. Discuss the appropriateness of the t methods for examining the percentage change in shipments of ATMs.

5.125   Doctors have recommended that we try to keep our caffeine intake at 200 mg or less per day. With the following chart, a sample of 35 office workers were asked to record their caffeine intake for a 7-day period.

coffee (6 oz)	100–150 mg
tea (6 oz)	40–110 mg
cola (12 oz)	30 mg
chocolate cake	20–30 mg
cocoa (6 oz)	5–20 mg
milk chocolate (1 oz)	5–10 mg

After the 7-day period, the average daily intake was obtained for each worker. The sample mean and standard deviation of the daily averages were 560 mg and 160 mg, respectively. Use these data to estimate μ, the average daily intake, using a 90% confidence interval.

5.126   Refer to Exercise 5.125. How many additional observations would be needed to estimate μ to within ± 10 mg with 90% confidence?

5.127   Investigators from the Ohio Department of Agriculture recently selected a junior high school in the area and took samples of the half-pint (8-ounce) milk cartons used for student lunches. Based on 25 containers, the investigators found that the cartons were .067 ounces short of a full half pint on the average, with a standard deviation of .02.
   a. Use these data to test the hypothesis that the average shortfall is zero against a one-sided alternative. Give the p-value for your test.
   b. Although .067 ounces is only a few drops, predict the annual savings (in pints) for the dairy if it sells 3 million 8-ounce cartons of milk each year with this shortweight.

5.128   Refer to the clinical trials data base in the Appendix to construct a 95% confidence interval for the mean HAM-D total score of treatment group C. How would this interval change for a 99% confidence interval?

5.129   Using the clinical trials data base, give a 90% confidence interval for the Hopkins Obrist cluster score of treatment A.

INFERENCES ABOUT $\mu_1 - \mu_2$

6.1 INTRODUCTION

The inferences we have made so far have concerned a parameter from a single population. Quite often we are faced with an inference involving a comparison of parameters from different populations. For example, we might wish to compare the mean corn crop yield for two different varieties of corn, the mean annual income for two ethnic groups, the mean nitrogen content of two different lakes, or the mean length of time between administration and eventual relief for two different antivertigo drugs.

In many sampling situations, we will select independent random samples from two populations in order to compare the population means or proportions. The statistics used to make these inferences will, in many cases, be the difference between the corresponding sample statistics. For example, suppose we select independent random samples of n_1 observations from one population and n_2 observations from a second population. We will use the difference between the sample means, $(\bar{y}_1 - \bar{y}_2)$, to make an inference about the difference between the population means, $(\mu_1 - \mu_2)$.

The following theorem will help in finding the sampling distribution for the difference between sample statistics computed from independent random samples.

THEOREM 6.1

▼

If two independent random variables y_1 and y_2 are normally distributed with means and variances (μ_1, σ_1^2) and (μ_2, σ_2^2), respectively, the difference between the random variables will be normally distributed with mean equal to $(\mu_1 - \mu_2)$ and variance equal to $(\sigma_1^2 + \sigma_2^2)$.

Note: The sum $(y_1 + y_2)$ of the random variables will also be normally distributed with mean $(\mu_1 + \mu_2)$ and variance $(\sigma_1^2 + \sigma_2^2)$.

Theorem 6.1 can be applied directly to find the sampling distribution of the difference between two independent sample means or two independent sample proportions. The Central Limit Theorem (discussed in Chaper 4) implies that if independent samples of sizes n_1 and n_2 are selected from two populations 1 and 2, then, where n_1 and n_2 are large, the sampling distributions of $\bar{y}_1$ and $\bar{y}_2$ will be approximately normal, with means and variances $(\mu_1, \sigma_1^2/n_1)$ and $(\mu_2, \sigma_2^2/n_2)$, respectively. Consequently, since $\bar{y}_1$ and $\bar{y}_2$ are independent, normally distributed random variables, it follows from Theorem 6.1 that the sampling distribution for the difference in the sample means, $(\bar{y}_1 - \bar{y}_2)$, will be approximately normal, with a mean

$$\mu_{\bar{y}_1 - \bar{y}_2} = \mu_1 - \mu_2$$

and a variance

$$\sigma_{\bar{y}_1 - \bar{y}_2}^2 = \sigma_{\bar{y}_1}^2 + \sigma_{\bar{y}_2}^2 = \frac{\sigma_1^2}{n_1} + \frac{\sigma_2^2}{n_2}$$

FIGURE 6.1
Sampling Distribution for the Difference Between Two Sample Means

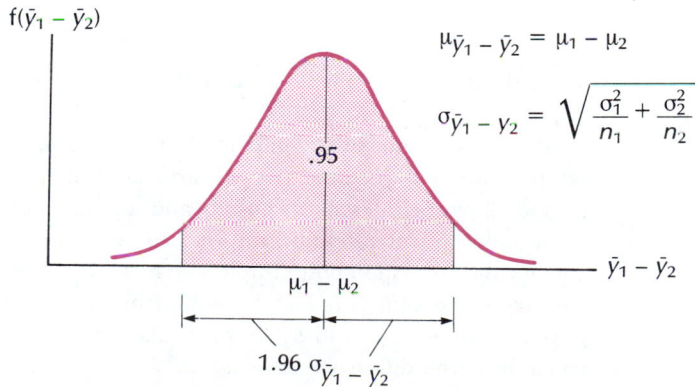

$f(\bar{y}_1 - \bar{y}_2)$

$\mu_{\bar{y}_1 - \bar{y}_2} = \mu_1 - \mu_2$

$\sigma_{\bar{y}_1 - \bar{y}_2} = \sqrt{\dfrac{\sigma_1^2}{n_1} + \dfrac{\sigma_2^2}{n_2}}$

.95

$\mu_1 - \mu_2$

$\bar{y}_1 - \bar{y}_2$

$1.96\ \sigma_{\bar{y}_1 - \bar{y}_2}$

and a standard error

$$\sigma_{\bar{y}_1 - \bar{y}_2} = \sqrt{\frac{\sigma_1^2}{n_1} + \frac{\sigma_2^2}{n_2}}.$$

The sampling distribution of the difference between two independent, normally distributed sample means is shown in Figure 6.1.

Properties of the Sampling Distribution for the Difference Between Two Sample Means, $(\bar{y}_1 - \bar{y}_2)$

1. The sampling distribution of $(\bar{y}_1 - \bar{y}_2)$ is approximately normal for large samples.
2. The mean of the sampling distribution, $\mu_{\bar{y}_1 - \bar{y}_2}$, is equal to the difference between the population means, $(\mu_1 - \mu_2)$.
3. The standard error of the sampling distribution is

$$\sigma_{\bar{y}_1 - \bar{y}_2} = \sqrt{\frac{\sigma_1^2}{n_1} + \frac{\sigma_2^2}{n_2}}.$$

The sampling distribution for the difference between two sample means, $(\bar{y}_1 - \bar{y}_2)$, can be used to answer the same types of questions as were asked about the sampling distribution for $\bar{y}$ in Chapter 4. Since sample statistics are used to make inferences about corresponding population parameters, we can use the sampling distribution of a statistic to calculate the probability that the statistic will be within a specified distance of the population parameter. For example, we could use the sampling distribution of the difference in sample means to calculate the probability that $(\bar{y}_1 - \bar{y}_2)$ will be within a specified distance of the unknown difference in population means $(\mu_1 - \mu_2)$. Inferences (estimations or tests) about $(\mu_1 - \mu_2)$ will be discussed in succeeding sections of this chapter.

6.2 ▼ INFERENCES ABOUT $\mu_1 - \mu_2$: INDEPENDENT SAMPLES

In situations where we are making inferences about $\mu_1 - \mu_2$ based on independent samples, we will assume that we are sampling from two normal populations (1 and 2) with different means μ_1 and μ_2 but identical variances σ^2. We then draw independent random samples of size n_1 and n_2. The sample means are $\bar{y}_1$ and $\bar{y}_2$; the corresponding sample variances are s_1^2 and s_2^2, respectively. Using the data from the two samples, we would like to make a comparison between the population means μ_1 and μ_2. In particular, we will estimate and test a hypothesis concerning the difference $\mu_1 - \mu_2$.

A logical point estimate for the difference in population means is the sample difference $\bar{y}_1 - \bar{y}_2$. The standard error for the difference in sample means is more complicated than for a single sample mean, but the confidence interval has the same form: point estimate $\pm t$ (standard error). A general confidence interval for $\mu_1 - \mu_2$ with confidence coefficient of $(1 - \alpha)$ is given here.

Confidence Interval for $\mu_1 - \mu_2$, Independent Samples

$$(\bar{y}_1 - \bar{y}_2) \pm t_{\alpha/2} s_p \sqrt{\frac{1}{n_1} + \frac{1}{n_2}}$$

where

$$s_p = \sqrt{\frac{(n_1 - 1)s_1^2 + (n_2 - 1)s_2^2}{n_1 + n_2 - 2}}$$

and

$$df = n_1 + n_2 - 2.$$

s_p^2, a weighted average

The quantity s_p in the confidence interval is an estimate of the standard deviation σ for the two populations and is formed by combining (pooling) information from the two samples. In fact, s_p^2 is a **weighted average** of the sample variances s_1^2 and s_2^2. For the special case where the sample sizes are the same ($n_1 = n_2$), the formula for s_p^2 reduces to $s_p^2 = (s_1^2 + s_2^2)/2$, the mean of the two sample variances. The degrees of freedom for the confidence interval are a combination of the degrees of freedom for the two samples; that is, $df = (n_1 - 1) + (n_2 - 1) = n_1 + n_2 - 2$.

Recall that we are assuming that the two populations from which we draw the samples have normal distributions with a common variance σ^2. If the confidence interval presented were valid only when these assumptions were met exactly, the estimation procedure would be of limited use. Fortunately, the confidence coefficient remains relatively stable if both distributions are mound-shaped and the sample sizes are approximately equal. More discussion about these assumptions is presented at the end of this section.

E X A M P L E   6.1   ▼

Company officials were concerned about the length of time a particular drug product retained its potency. A random sample, sample 1, of $n_1 = 10$ bottles of the product was drawn from the production line and analyzed for potency.

A second sample, sample 2, of $n_2 = 10$ bottles was obtained and stored in a regulated environment for a period of one year.

The readings obtained from each sample are given in Table 6.1.

TABLE 6.1
Potency Reading for Two Samples

Sample 1		Sample 2	
10.2	10.6	9.8	9.7
10.5	10.7	9.6	9.5
10.3	10.2	10.1	9.6
10.8	10.0	10.2	9.8
9.8	10.6	10.1	9.9

Suppose we let μ_1 denote the mean potency for all bottles that might be sampled coming off the production line and μ_2 denote the mean potency for all bottles that may be retained for a period of one year. Estimate $\mu_1 - \mu_2$ by using a 95% confidence interval.

Solution   The necessary calculations from the data of Table 6.1 are presented next.

Sample 1	Sample 2
$\sum_j y_{1j} = 103.7$	$\sum_j y_{2j} = 98.3$
$\sum_j y_{1j}^2 = 1076.31$	$\sum_j y_{2j}^2 = 966.81$

Then

$$\bar{y}_1 = \frac{103.7}{10} = 10.37 \qquad\qquad \bar{y}_2 = \frac{98.3}{10} = 9.83$$

$$s_1^2 = \frac{1}{9}\left[1076.31 - \frac{(103.7)^2}{10}\right] = .105 \qquad s_2^2 = \frac{1}{9}\left[966.81 - \frac{(98.3)^2}{10}\right] = .058.$$

The estimate of the common standard deviation σ is

$$s_p = \sqrt{\frac{(n_1 - 1)s_1^2 + (n_2 - 1)s_2^2}{n_1 + n_2 - 2}} = \sqrt{\frac{9(.105) + 9(.058)}{18}}$$

which, for $n_1 = n_2 = 9$, reduces to

$$s_p = \sqrt{\frac{.105 + .058}{2}} = .285.$$

The t-value based on df $= n_1 + n_2 - 2 = 18$ and $a = .025$ is 2.101. A 95% confidence interval for the difference in mean potencies is

$$(10.37 - 9.83) \pm 2.101(.285)\sqrt{1/10 + 1/10}, \quad \text{or} \quad .54 \pm .268.$$

We estimate that the difference in mean potencies for the bottles from the production line and those stored for 1 year, $\mu_1 - \mu_2$, lies in the interval .272 to .808. ▲

EXAMPLE 6.2 ▼

A study was conducted to determine whether persons in suburban district 1 have a different mean income from those in district 2. A random sample of 20 homeowners was taken in district 1. Although 20 homeowners were to be interviewed in district 2 also, 1 person refused to provide the information requested, even though the researcher promised to keep the interview confidential. So only 19 observations were obtained from district 2. The data, recorded in thousands of dollars, produced sample means and variances as shown in Table 6.2. Use these data to construct a 95% confidence interval for $(\mu_1 - \mu_2)$.

TABLE 6.2
Income Data for
Example 6.2

	District 1	District 2
Sample size	20	19
Sample mean	18.27	16.78
Sample variance	8.74	6.58

Solution Histograms plotted for the two samples suggest that the two populations are mound-shaped (near normal). Also, the sample variances are very similar. The difference in the sample means is

$$\bar{y}_1 - \bar{y}_2 = 18.27 - 16.78 = 1.49.$$

The estimate of the common standard deviation σ is

$$s_p = \sqrt{\frac{(n_1 - 1)s_1^2 + (n_2 - 1)s_2^2}{n_1 + n_2 - 2}}$$

$$= \sqrt{\frac{19(8.74) + 18(6.58)}{20 + 19 - 2}} = 2.77.$$

The t-value for $a = \alpha/2 = .025$ and df $= 20 + 19 - 2 = 37$ is not listed in Table 4 of the Appendix, but taking the labeled value for the nearest df (df $= 40$), we have $t = 2.021$. A 95% confidence interval for the difference in mean incomes for the two districts is of the form

$$\bar{y}_1 - \bar{y}_2 \pm t_{\alpha/2} s_p \sqrt{\frac{1}{n_1} + \frac{1}{n_2}}$$

Substituting into the formula we obtain

$$1.49 \pm 2.021(2.77) \sqrt{\frac{1}{20} + \frac{1}{19}}$$

or

$$1.49 \pm 1.79.$$

Thus, we estimate the difference in mean incomes to lie somewhere in the interval from $-.30$ to 3.28. If we multiply these limits by \$1,000, the confidence interval for the difference in mean incomes is $-\$300$ to \$3,280. Since this interval includes both positive and negative values for $\mu_1 - \mu_2$, we are unable to determine whether the mean income for district 1 is larger or smaller than the mean income for district 2. ▲

We can also test a hypothesis about the difference between two population means. As with any test procedure, we begin by specifying a research hypothesis for the difference in population means. Thus, we might, for example, specify that the difference $\mu_1 - \mu_2$ is greater than some value D_0. (Note: D_0 will often be 0.) The entire test procedure is summarized here.

A Statistical Test for $\mu_1 - \mu_2$, Independent Samples

H_0: $\mu_1 - \mu_2 = D_0$ (D_0 is specified)

H_a: **1.** $\mu_1 - \mu_2 > D_0$
       2. $\mu_1 - \mu_2 < D_0$
       3. $\mu_1 - \mu_2 \neq D_0$

T.S.: $t = \dfrac{\bar{y}_1 - \bar{y}_2 - D_0}{s_p \sqrt{1/n_1 + 1/n_2}}$

R.R.:   For a Type I error α and df $= n_1 + n_2 - 2$

       1. reject H_0 if $t > t_\alpha$
       2. reject H_0 if $t < -t_\alpha$
       3. reject H_0 if $|t| > t_{\alpha/2}$.

EXAMPLE 6.3

▼

An experiment was conducted to compare the mean number of tapeworms in the stomachs of sheep that had been treated for worms against the mean number in those that were untreated. A sample of 14 worm–infected lambs was randomly divided into 2 groups. Seven were injected with the drug and the remainder were left untreated. After a 6-month period, the lambs were slaughtered and the following worm counts were recorded:

Drug-treated sheep	18	43	28	50	16	32	13
Untreated sheep	40	54	26	63	21	37	39

a. Test a hypothesis that there is no difference in the mean number of worms between treated and untreated lambs. Assume that the drug cannot increase the number of worms and, hence, use the alternative hypothesis that the mean for treated lambs is less than the mean for untreated lambs. Use $\alpha = .05$.

b. Indicate the level of significance for this test.

Solution

a. The calculations for the samples of treated and untreated sheep are summarized next.

Drug-Treated Sheep	Untreated Sheep
$\sum_i y_{1j} = 200$	$\sum_j y_{2j} = 280$
$\sum_i y_{1j}^2 = 6906$	$\sum_j y_{2j}^2 = 12{,}492$
$\bar{y}_1 = \dfrac{200}{7} = 28.57$	$\bar{y}_2 = \dfrac{280}{7} = 40.0$
$s_1^2 = \dfrac{1}{6}\left[6906 - \dfrac{(200)^2}{7}\right]$	$s_2^2 = \dfrac{1}{6}\left[12{,}492 - \dfrac{(280)^2}{7}\right]$
$= \dfrac{1}{6}[6906 - 5714.29]$	$= \dfrac{1}{6}[12{,}492 - 11{,}200]$
$= 198.62$	$= 215.33$

Under the assumption of equal population variances, the sample variances are combined to form an estimate of the common population standard deviation σ. This assumption appears reasonable based on the sample variances.

$$s_p = \sqrt{\frac{(n_1 - 1)s_1^2 + (n_2 - 1)s_2^2}{n_1 + n_2 - 2}} = \sqrt{\frac{6(198.62) + 6(215.33)}{12}} = 14.39.$$

The test procedure for the research hypothesis that the treated sheep will have a mean infestation level (μ_1) less than the mean level (μ_2) for untreated sheep is as follows:

H_0: $\mu_1 - \mu_2 = 0$ (that is, no difference in the mean infestation levels)

H_a: $\mu_1 - \mu_2 < 0$

T.S.: $t = \dfrac{\bar{y}_1 - \bar{y}_2}{s_p\sqrt{1/n_1 + 1/n_2}} = \dfrac{28.57 - 40}{14.39\sqrt{1/7 + 1/7}} = -1.49$

R.R.: For $\alpha = .05$, the critical t-value for a one-tailed test with df $= n_1 + n_2 - 2 = 12$ can be obtained from Table 4 in the Appendix, using $a = .05$. We will reject H_0 if $t < -1.782$.

Conclusion: Since the observed value of t, -1.49, does not fall in the rejection region, we have insufficient evidence to reject the hypothesis that there is no difference in the mean number of worms in treated and untreated lambs.

b. Using Table 4 in the Appendix with $t = -1.49$ and df $= 12$, we see the level of significance for this test is in the range $.05 < p < .10$.   ▲

The test procedures for comparing two population means presented in this section are based on several assumptions. The first and most critical one is that the two samples are independent. Practically, we mean that the two samples are drawn from two different populations and that the elements of one sample are unrelated to those of the second sample. If this assumption is not valid, then the t methods of this section will likely be in error and other methods (such as those presented in Section 6.3) may be appropriate.

The second assumption that we make is that the samples are drawn from normal populations. Fortunately, this assumption is less critical. The reason for this is that for modest-sized samples the Central Limit Theorem of the previous chapter applies and the sampling distributions for $\bar{y}_1$ and $\bar{y}_2$ are approximately normal. With independent samples and the combined sample size $n_1 + n_2 \geq 30$, the t methods of this section should be reasonably accurate even for modest skewness in the two populations. A nonparametric alternative to the t test for independent samples is presented in the next section; this alternative does not require normality.

The third and final assumption is that the two population variances σ_1^2 and σ_2^2 are equal. For now, just examine the sample variances to see that they are approximately equal; later (in Chapter 7), we'll give a test for this assumption. Many efforts have been made to investigate the effect of deviations from the equal variance assumption on the t methods for independent samples. The general conclusion is that for equal sample sizes, the population variances can differ by as much as a factor of 3 (for example, $\sigma_1^2 = 3\sigma_2^2$) and the t methods will still apply.

This is remarkable and provides a convincing argument to use equal sample sizes. When the sample sizes are different, the more serious case is when the smaller sample size is associated with the larger variance. In this situation and in others where the sample variances (s_1^2 and s_2^2) suggest that $\sigma_1^2 \neq \sigma_2^2$, there is an approximate t test using the test statistic

$$t' = \frac{\bar{y}_1 - \bar{y}_2}{\sqrt{\dfrac{s_1^2}{n_1} + \dfrac{s_2^2}{n_2}}}.$$

Welch (1938) showed that percentage points of a t distribution with modified degrees of freedom can be used to set the rejection region for H_0: $\mu_1 - \mu_2 = D_0$. This t test is summarized here.

Approximate t test for Independent Samples, Unequal Variance

▼

H_0: $\mu_1 - \mu_2 = D_0$

H_a: 1. $\mu_1 - \mu_2 > D_0$
2. $\mu_1 - \mu_2 < D_0$
3. $\mu_1 - \mu_2 \neq D_0$

T.S.: $t' = \dfrac{\bar{y}_1 - \bar{y}_2 - D_0}{\sqrt{\dfrac{s_1^2}{n_1} + \dfrac{s_2^2}{n_2}}}$

R.R.: For a specified value of α,

1. reject H_0 if $t' > t_\alpha$
2. reject H_0 if $t' < -t_\alpha$
3. reject H_0 if $|t'| > t_{\alpha/2}$,

where

$$\text{df} = \frac{(n_1 - 1)(n_2 - 1)}{(n_2 - 1)c^2 + (1 - c)^2(n_1 - 1)}, \qquad \text{where } c = \frac{s_1^2/n_1}{\dfrac{s_1^2}{n_1} + \dfrac{s_2^2}{n_2}}.$$

Note: If the computed value of df is not an integer, *round down* to the nearest integer.

The test based on the t' statistic is sometimes referred to as the *separate-variance t test* because we use the separate sample variances s_1^2 and s_2^2, rather than a pooled sample variance.

E X A M P L E   6.4

▼

Refer to the situation explained in Example 6.3. Suppose that only 13 animals were available for analysis at the end of the treatment period. These data are shown here.

Drug-treated sheep	5	13	18	6	4	2	15
Untreated sheep	40	54	26	63	21	37	

Test the research hypothesis $H_a: \mu_1 - \mu_2 < 0$ under the assumption that the 2 population variances are different. Use $\alpha = .05$.

Solution It is easy to verify that

$$\bar{y}_1 = 9.00 \qquad \bar{y}_2 = 40.17$$
$$s_1^2 = 38.67 \qquad s_2^2 = 258.17.$$

Then the statistical test is set up as follows:

$$H_0: \quad \mu_1 - \mu_2 = 0$$
$$H_a: \quad \mu_1 - \mu_2 < 0$$
$$\text{T.S.:} \quad t' = \frac{\bar{y}_1 - \bar{y}_2}{\sqrt{\dfrac{s_1^2}{n_1} + \dfrac{s_2^2}{n_2}}} = \frac{9 - 40.17}{\sqrt{\dfrac{38.67}{7} + \dfrac{258.17}{6}}} = -4.47.$$

In order to compute the rejection region, we need

$$c = \frac{s_1^2/n_1}{s_1^2/n_1 + s_2^2/n_2} = \frac{38.67/7}{38.67/7 + 258.17/6} = .114$$

$$c^2 = .013$$

and

$$df = \frac{(n_1 - 1)(n_2 - 1)}{(n_2 - 1)c^2 + (1 - c)^2(n_1 - 1)} = 6.283, \text{ which is rounded to 6.}$$

R.R.:   For $\alpha = .05$ and $df = 6$, reject H_0 if $t' < -1.943$.

Since $t' = -4.47$ is less than -1.943, we reject H_0 and conclude that μ_1, the mean worm count for treated sheep, is less than that for untreated sheep.

▲

Computer simulations sometimes can help us to understand some of the assumptions underlying our test procedures. One such study was done to compare the pooled t test and separate-variance t test. We'll illustrate this with an example.

EXAMPLE 6.5

▼

For the simulation study, it was assumed that we were sampling from the independent normal populations shown here:

Population			
1	$\mu_1 = 100$	$\sigma_1 = 15$	$n_1 = 10$
2	$\mu_2 = 100$	$\sigma_2 = 10$	$n_2 = 20$

The study proceeded as follows: A computer program was used to generate a random sample of $n_1 = 10$ observations from population 1 and a random sample of $n_2 = 20$ from population 2. The sample statistics ($\bar{y}_1$, $\bar{y}_2$, s_1^2, and s_2^2) were computed, as were the test statistics t and t'. This process was repeated 999 more times, and for these 1,000 samples, the program kept track of the number of times t and t' rejected at the upper and lower .05 levels. These results are summarized here:

For H_0: $\mu_1 - \mu_2 = 0$ and $\alpha = .05$,

	Pooled t Test	Separate-Variance t Test
H_0 rejected for H_a: $\mu_1 - \mu_2 > 0$	75 (7.5%)	46 (4.6%)
H_0 rejected for H_a: $\mu_1 - \mu_2 < 0$	77 (7.7%)	44 (4.4%)

a. Without running the computer simulations study, which test would you have recommended even without knowing the underlying populations? Explain.

b. What do the computer simulation results tell us about the choice of t or t'? Do they agree with our recommendation in part a?

Solution

a. With samples of sizes $n_1 = 10$ and $n_2 = 20$, we don't have much protection against the possibility of unequal variances. Hence, the separate-variance t test should be somewhat more reliable.

b. Because we know the underlying populations, one of the assumptions underlying the pooled t test—namely, equal population variances—is violated. Because the population means were equal, H_0 is true and we would have expected to reject H_0 approximately 5% of the time in both the upper and lower tails, due to chance. As can be seen, the pooled t test rejected H_0 more frequently (7.5% above and 7.7% below) than would have been expected. The separate-variance t test, on the other hand, rejects about as often as we would expect. These results agree with our conclusion in part a.   ▲

E X A M P L E   6.6   ▼

Another simulation study is done with samples from independent normal populations, with the following results.

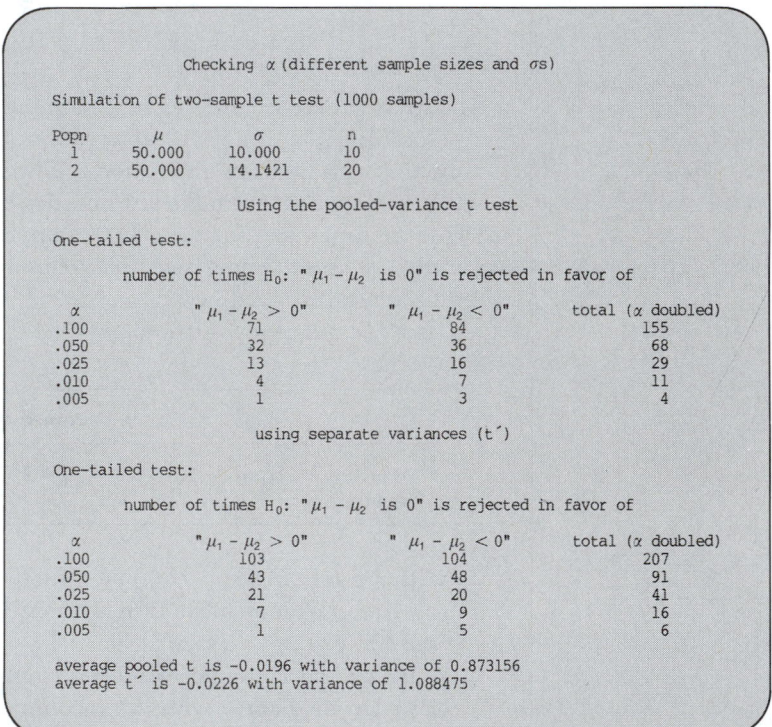

Checking α (different sample sizes and σs)

Simulation of two-sample t test (1000 samples)

Popn	μ	σ	n
1	50.000	10.000	10
2	50.000	14.1421	20

Using the pooled-variance t test

One-tailed test:

number of times H_0: "$\mu_1 - \mu_2$ is 0" is rejected in favor of

α	"$\mu_1 - \mu_2 > 0$"	"$\mu_1 - \mu_2 < 0$"	total (α doubled)
.100	71	84	155
.050	32	36	68
.025	13	16	29
.010	4	7	11
.005	1	3	4

using separate variances (t´)

One-tailed test:

number of times H_0: "$\mu_1 - \mu_2$ is 0" is rejected in favor of

α	"$\mu_1 - \mu_2 > 0$"	"$\mu_1 - \mu_2 < 0$"	total (α doubled)
.100	103	104	207
.050	43	48	91
.025	21	20	41
.010	7	9	16
.005	1	5	6

average pooled t is -0.0196 with variance of 0.873156
average t´ is -0.0226 with variance of 1.088475

What do these results indicate about t versus t'?

Solution   Again, the assumption of equal variances is violated. This time, however, the smaller variance is associated with the smaller sample size; in Example 6.5 the smaller variance was associated with the larger sample size.

> In this example, we find that the number of false rejections of the null hypothesis is consistently *smaller* than would be indicated by the nominal α. Again, the t' test rejects the null hypothesis just about as often as α indicates. ▲

In this section, we developed pooled-variance t methods based on an assumption of equal population variances. In addition, we introduced the t' statistic for an approximate test when the variances are not equal. Confidence intervals and hypothesis tests based on these different procedures (t or t') need not give identical results. Standard computer packages often report the results of both the pooled-variance and separate-variance t tests. Which should you believe?

If the sample sizes are equal it doesn't matter. The alternative t tests give algebraically identical results when $n_1 = n_2$, and, since the t probabilities are robust to nonnormality and unequal population variances, the test results are quite reliable when $n_1 = n_2$. When n_1 and n_2 are nearly equal, the two results are nearly equal. Only when the sample sizes vary greatly (say 1.5 to 1 or worse) will there be a large difference in results. The evidence in such cases indicates that the separate variance methods contained in computer packages are somewhat more reliable and more conservative.

▼ EXERCISES

Basic Techniques

6.1 Set up the rejection regions for testing H_0: $\mu_1 - \mu_2 = 0$ for the following conditions:
a. H_a: $\mu_1 - \mu_2 \neq 0$, $n_1 = 12$, $n_2 = 14$, and $\alpha = .05$
b. H_a: $\mu_1 - \mu_2 > 0$, $n_1 = n_2 = 8$, and $\alpha = .01$
c. H_a: $\mu_1 - \mu_2 < 0$, $n_1 = 6$, $n_2 = 4$, and $\alpha = .05$
What assumptions must be made prior to applying a two-sample t test?

6.2 Conduct a test of H_0: $\mu_1 - \mu_2 = 0$ against the alternative hypothesis H_a: $\mu_1 - \mu_2 < 0$ for the sample data shown here. Use $\alpha = .05$.

	Population	
	1	2
Sample size	16	13
Sample mean	71.5	79.8
Sample variance	68.35	70.26

6.3 Refer to the data of Exercise 6.2. Give the level of significance for your test.

Applications

6.4 In an effort to link cold environments with hypertension in humans, a preliminary experiment was conducted to investigate the effect of cold on hypertension in rats. Two random samples of 6 rats each were exposed to different environments. One sample of rats was held in a normal environment at 26°C. The other sample was held in a cold 5°C environment. Blood pressures and heart rates were measured for rats for both groups. The blood pressures for the 12 rats are shown in the accompanying table. Do the data provide sufficient evidence to indicate that rats exposed to a 5°C environment have a higher mean blood pressure than rats exposed to a 26°C environment? Test by using $\alpha = .05$.

26°		**5°**	
Rat	Blood Pressure	Rat	Blood Pressure
1	152	7	384
2	157	8	369
3	179	9	354
4	182	10	375
5	176	11	366
6	149	12	423

 6.5 A pollution–control inspector suspected that a riverside community was releasing semi-treated sewage into a river and this, as a consequence, was changing the level of dissolved oxygen of the river. To check this, he drew 5 randomly selected specimens of river water at a location above the town and another 5 specimens below. The dissolved oxygen readings, in parts per million, are given in the accompanying table. Do the data provide sufficient evidence to indicate a difference in mean oxygen content between locations above and below the town? Use $\alpha = .05$.

Above town	4.8	5.2	5.0	4.9	5.1
Below town	5.0	4.7	4.9	4.8	4.9

 6.6 A petroleum corporation was interested in running some preliminary tests to compare the performance of a new gasoline mixture to one currently on the market. Ten identical new automobiles were randomly assigned, 5 to gasoline 1 and 5 to gasoline 2. Gasoline 2 contained a mileage additive, and gasoline 1 was regular gasoline. Each automobile was filled with 10 gallons of gasoline and driven over a test course until it stopped. The mileage was recorded for each in the table that follows:

Gasoline 1	**Gasoline 2**
282	284
279	285
280	286
278	277
275	283
$\bar{y}_1 = 278.80$	$\bar{y}_2 = 283.00$

Use these data to construct a 95% confidence interval for the difference in mean mileage for the 2 gasolines.

 6.7 A sociologist gave a current-events test to 4 blue-collar workers and 4 white-collar workers. The blue-collar workers made scores of 23, 18, 22, and 21; the white-collar workers made scores of 17, 22, 19, and 18. Estimate the difference in mean scores for blue-collar and white-collar workers using a 99% confidence interval.

 6.8 Two different emission-control devices were being tested to determine the average amount of nitric oxide being emitted by an automobile over a 1-hour period of time. Twenty cars of the same model and year were selected for the study. Ten cars were randomly selected and equipped with a Type 1 emission-control device, and the remaining cars were equipped with Type II devices. Each of the 20 cars was then monitored for a 1-hour period to determine the amount of nitric oxide emitted.

Use the following data to test the research hypothesis that the mean level of emission for Type 1 devices (μ_1) is greater than the mean emission level for Type II devices (μ_2). Use $\alpha = .01$.

Type I Device		Type II Device	
1.35	1.28	1.01	0.96
1.16	1.21	0.98	0.99
1.23	1.25	0.95	0.98
1.20	1.17	1.02	1.01
1.32	1.19	1.05	1.02

 6.9 It has been estimated that lead poisoning resulting from an unnatural craving (pica) for substances such as paint may affect as many as a quarter of a million children each year, causing them to suffer from severe, irreversible retardation. Explanations for why children voluntarily consume lead range from "improper parental supervision" to "a child's need to mouth objects." Some researchers, however, have been investigating whether the habit of eating such substances has some nutritional explanation. One such study involved a comparison of a regular diet and a calcium-deficient diet on the ingestion of a lead-acetate solution in rats. Each rat in a group of 20 rats was randomly assigned to either an experimental or control group. Those in the control group received a normal diet, while the experimental group received a calcium-deficient diet. Each of the rats occupied a separate cage and was monitored to observe the quantity of a .15% lead-acetate solution consumed during the study period. The sample results are summarized here.

Control group	5.4	6.2	3.1	3.8	6.5	5.8	6.4	4.5	4.9	4.0
Experimental group	8.8	9.5	10.6	9.6	7.5	6.9	7.4	6.5	10.5	8.3

a. Plot the data for the 2 samples separately. Is there reason to think the assumptions for a t test have been violated?

b. Run a test of the research hypothesis that the mean quantity of lead acetate consumed in the experimental group is greater than that consumed in the control group. Use $\alpha = .05$.

6.10 The results of a 3-year study to examine the effect of a variety of ready-to-eat breakfast cereals on dental caries (tooth decay) in adolescent children were reported by Rowe, Anderson, and Wanninger (1974). A sample of 375 adolescent children of both genders from the Ann Arbor, Michigan, public schools was enrolled (after parental consent) in the study. Each child was provided with toothpaste and boxes of different varieties of ready-to-eat cereals. Although these were brand-name cereals, each type of cereal was packaged in plain white 7-ounce boxes and labeled as wheat flakes, corn cereal, oat cereal, fruit-flavored corn puffs, corn puffs, cocoa-flavored cereal, and sugared oat cereal. Note that the last four varieties of cereal had been presweetened and the others had not.

Each child received a dental examination at the beginning of the study, twice during the study, and once at the end. The response of interest was the incremental DMF surfaces, that is, the difference between the final (poststudy) and initial (prestudy) number of decayed, missing, and filled (DMF) tooth surfaces. Careful records for each participant were maintained throughout the 3 years, and at the end of the study, a person was classified as "noneater" if he or she had eaten less than 28 boxes of cereal throughout the study. All others were classified as "eaters." The incremental DMF surface readings for each group are summarized below. Use these data to test the research hypothesis that the mean incremental DMF surface for noneaters is larger than the corresponding mean for eaters. Give the level of significance for your test. Interpret your findings.

	Sample Size	Sample Mean	Sample Standard Deviation
Noneaters	73	6.41	5.62
Eaters	302	5.20	4.67

6.11 Refer to Exercise 6.10. Although complete details of the original study have not been disclosed, critique the procedure that has been discussed.

 6.12 The study of concentrations of atmospheric trace metals in isolated areas of the world has received considerable attention because of the concern that humans might somehow alter the climate of the earth by changing the amount and distribution of trace metals in the atmosphere. Consider a study at the South Pole, where at 10 different sampling periods throughout a 2-month period, 10,000 standard cubic meters (scm) of air were obtained and analyzed for metal concentrations. The results associated with magnesium and europium are listed below. (Note: Magnesium results are in units of 10^{-9} g/scm; europium results are in units of 10^{-15} g/scm.) Note that $s > \bar{y}$ for the magnesium data. Would you expect the data to be normally distributed? Explain.

	Sample Size	Sample Mean	Sample Standard Deviation
Magnesium	10	1.0	2.21
Europium	10	17.0	12.65

6.13 Refer to Exercise 6.12. Could we run a t test comparing the mean metal concentrations for magnesium and europium? Why or why not?

6.14 Refer to Example 6.1.

    a. Does it appear from the sample means and variances that any of the underlying assumptions for pooled t methods have been violated?

    b. Compute t and t' for these data and draw a conclusion based on each statistic. Do these results agree with your preliminary assessment of the underlying assumptions in part a?

6.15 A firm has a generous but rather complicated policy concerning end-of-year bonuses for its lower-level managerial personnel. The policy's key factor is a subjective judgment of "contribution to corporate goals." A personnel officer took samples of 24 female and 36 male managers to see if there was any difference in bonuses, expressed as a percentage of yearly salary. The data are listed here:

Gender	Bonus Percentage								
F	9.2	7.7	11.9	6.2	9.0	8.4	6.9	7.6	7.4
	8.0	9.9	6.7	8.4	9.3	9.1	8.7	9.2	9.1
	8.4	9.6	7.7	9.0	9.0	8.4			
M	10.4	8.9	11.7	12.0	8.7	9.4	9.8	9.0	9.2
	9.7	9.1	8.8	7.9	9.9	10.0	10.1	9.0	11.4
	8.7	9.6	9.2	9.7	8.9	9.2	9.4	9.7	8.9
	9.3	10.4	11.9	9.0	12.0	9.6	9.2	9.9	9.0

A computer program yielded the output shown here.

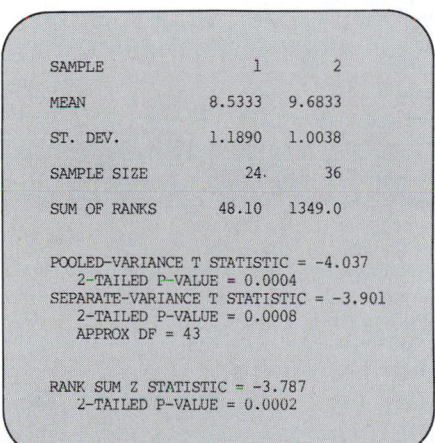

```
SAMPLE              1        2

MEAN           8.5333   9.6833

ST. DEV.       1.1890   1.0038

SAMPLE SIZE       24.      36

SUM OF RANKS    48.10   1349.0


POOLED-VARIANCE T STATISTIC = -4.037
   2-TAILED P-VALUE = 0.0004
SEPARATE-VARIANCE T STATISTIC = -3.901
   2-TAILED P-VALUE = 0.0008
   APPROX DF = 43


RANK SUM Z STATISTIC = -3.787
   2-TAILED P-VALUE = 0.0002
```

    a. Identify the value of the pooled-variance t statistic (the usual t test based on the equal variance assumption).

    b. Identify the value of the t' statistic.

    c. Use both statistics to test the research hypothesis of unequal means at $\alpha = .05$ and at $\alpha = .01$. Does the conclusion depend on which statistic is used?

6.16 The costs of major surgery vary substantially from one state to another due to differences in hospital fees, doctors' fees, malpractice insurance cost, and rent. A study of hysterectomy costs was done in California and Montana. Based on a random sample of 20 patient records from each state, the sample statistics below were obtained. Construct a 95% confidence interval for $\mu_1 - \mu_2$ (the California minus Montana difference).

	Sample Mean	Sample Standard Deviation
Montana	$ 6,458	$520
California	$12,690	$305

 6.17 A national educational organization monitors reading proficiency for American students on a regular basis using a scale that ranges from 0 to 500. Sample results based on 500 students per category are shown here. Use these data to make the inferences listed below. Assume the pooled standard deviation for any comparison is 100.

Age	Gender	Sample Mean*
9	Male	210
	Female	216
13	Male	253
	Female	262
17	Male	283
	Female	293

*What the scale means: 150—Rudimentary reading skills; can follow basic directions. 200—Basic skills; can identify facts from simple paragraphs. 250—Intermediate skills; can organize information in lengthy passages. 300—Adapt skills; can understand and explain complicated information. 350—Advanced skills; can understand and explain specialized materials.

a. Construct a meaningful graph that shows age, gender, and mean proficiency scores.
b. Use the sample data to place a 95% confidence interval on the difference in mean proficiencies for females and males age 17 years.
c. Compare the mean scores for females, age 13 and 17 years, using a 90% confidence interval. Does the interval include 0? Why might these means be different?

6.18 The organization alluded to in Exercise 6.17 also examined the effect of television viewing on reading proficiency scores for students of the same age categories. The sample means and sample sizes are shown here.

Hours of TV Viewing per Day		Age (Years)		
		9	13	17
0–2	$\bar{y}$	220	267	295
	n	300	310	305
3–5	$\bar{y}$	220	262	284
	n	280	260	250
6$^+$	$\bar{y}$	202	246	270
	n	210	220	230

a. Plot the sample means on a graph with age and hours of TV viewing per day. Can you make some general statements about the sample data? What's the effect of TV viewing on reading proficiency within a given age group? What's the influence of age within a given TV viewing category?

b. Construct 99% confidence intervals for the difference between the 0–2 and the 6$^+$ category in each age category. What's your conclusion?

6.19 Although we haven't discussed the design of experiments, think about how you might conduct a survey to examine the effects on reading proficiency of age and hours per day spent watching TV. What factor or factors might affect the results of your survey?

6.3 ▼ A NONPARAMETRIC ALTERNATIVE: THE WILCOXON RANK SUM TEST

The two-sample t test of the previous section was based on several assumptions: independent samples, normality, and equal variances. When the assumptions of normality and equal variances are not valid but the sample sizes are large, the results using a t (or t') test are approximately correct. There is, however, an alternative test procedure that requires less stringent assumptions. This procedure, called the **Wilcoxon rank sum test**, is discussed here.

Wilcoxon rank sum test

The assumptions for this test are that we have independent random samples taken from two populations. The Wilcoxon rank sum test provides a procedure for testing that two populations are identical but not necessarily normal. Since the two populations are assumed to be identical under the null hypothesis, independent random samples from the respective populations should be similar. One way to measure the similarity between the samples is to jointly rank (from lowest to highest) the measurements from the combined samples and examine the sum of the ranks for measurements in sample 1 (or, equivalently, sample 2). Under the null hypothesis of identical populations, the sum of the ranks for a sample will be proportional to the sample size. We let T denote the sum of the ranks for sample 1. Intuitively, if T is extremely small (or large), we would have evidence to reject the null hypothesis that the two populations are identical.

Under the null hypothesis, the statistic T will have a sampling distribution with mean and variance given by

$$\mu_T = \frac{n_1(n_1 + n_2 + 1)}{2}$$

$$\sigma_T^2 = \frac{n_1 n_2}{12}(n_1 + n_2 + 1).$$

If, in addition, both sample sizes are more than 10, the sampling distribution of T is approximately normal; this allows us to use a z statistic in the Wilcoxon rank sum test.

The theory behind the Wilcoxon rank sum test assumes that the population distributions are continuous, so that there is zero probability that any two observations are identical. In practice, there will often be ties—two or more observations with the same value. For these situations, each observation in a set of tied values receives a rank score equal to the average of the ranks for the set. For example, if two observations are tied for the ranks 3 and 4, each is given a rank of 3.5; the next higher value receives a rank of 5, and so on. When there are ties, there is a correction for the variance formula. Then σ_T^2 is as shown here.

$$\sigma_T^2 = \frac{n_1 n_2}{12}\left[(n_1 + n_2 + 1) - \frac{\sum_j t_j(t_j^2 - 1)}{(n_1 + n_2)(n_1 + n_2 - 1)}\right],$$

where t_j denotes the number of tied ranks in the jth group. Note that when there are no tied ranks,

$$\sigma_T^2 = \frac{n_1 n_2(n_1 + n_2 + 1)}{12}.$$

From a practical standpoint, however, unless there are many ties, the correction will have very little effect on the value of σ_T^2. The Wilcoxon rank sum test is summarized here.

Wilcoxon Rank Sum Test*

▼

H_0: The two populations are identical.

H_a:
1. Population 1 is shifted to the right of population 2.
2. Population 1 is shifted to the left of population 2.
3. Populations 1 and 2 have different location parameters.

$(n_1 \leq 10, n_2 \leq 10)$

T.S.: T, the sum of the ranks in sample 1

R.R.: For $\alpha = .05$, use Table 3 in the Appendix to find critical values for T_U and T_L;

1. reject H_0 if $T > T_U$
2. reject H_0 if $T < T_L$
3. reject H_0 if $T > T_U$ or $T < T_L$.

$(n_1, n_2 > 10)$

T.S.: $z = \dfrac{T - \mu_T}{\sigma_T}$,

where T denotes the sum of the ranks in sample 1.

R.R.: For a specified value of α,

1. reject H_0 if $z > z_\alpha$
2. reject H_0 if $z < -z_\alpha$
3. reject H_0 if $|z| > z_{\alpha/2}$.

E X A M P L E   6.7

▼

Environmental engineers were interested in determining whether a cleanup project on a nearby lake was effective. Prior to initiation of the project, 12 water samples had been obtained at random from the lake and analyzed for the amount of dissolved oxygen (in ppm). Due to diurnal fluctuations in the dissolved oxygen, all measurements were obtained at the 2 P.M. peak period. The before and after data are presented in Table 6.3.

---

* This test is equivalent to the Mann–Whitney U test, Conover (1980).

TABLE 6.3
Dissolved Oxygen Measurements (in ppm), Example 6.7

Before Cleanup		After Cleanup	
11.0	11.6	10.2	10.8
11.2	11.7	10.3	10.8
11.2	11.8	10.4	10.9
11.2	11.9	10.6	11.1
11.4	11.9	10.6	11.1
11.5	12.1	10.7	11.3

a. Use $\alpha = .05$ to test the following hypotheses:

H_0: The distributions of measurements for before cleanup and 6 months after the cleanup project began are identical.

H_a: The distribution of dissolved oxygen measurements before the cleanup project is shifted to the right of the corresponding distribution of measurements for 6 months after initiating the cleanup project. (It should be noted that a cleanup project has been effective in one sense if the dissolved oxygen drops over a period of time.)

For convenience, the data have been arranged in ascending order in Table 6.3.

b. Has the correction for ties made much of a difference?

Solution

a. First we must jointly rank the combined sample of 24 observations by assigning the rank of 1 to the smallest observation, the rank of 2 to the next smallest, and so on. When two or more measurements are the same, we assign all of them a rank equal to the average of the ranks they occupy. The sample measurements and associated ranks (shown in parentheses) are listed in Table 6.4.

TABLE 6.4
Dissolved Oxygen Measurements and Ranks for Example 6.7

Before Cleanup		After Cleanup	
11.0	(10)	10.2	(1)
11.2	(14)	10.3	(2)
11.2	(14)	10.4	(3)
11.2	(14)	10.6	(4.5)
11.4	(17)	10.6	(4.5)
11.5	(18)	10.7	(6)
11.6	(19)	10.8	(7.5)
11.7	(20)	10.8	(7.5)
11.8	(21)	10.9	(9)
11.9	(22.5)	11.1	(11.5)
11.9	(22.5)	11.1	(11.5)
12.1	(24)	11.3	(16)
	$T = 216$		

Since n_1 and n_2 are both greater than 10, we will use the test statistic z. If we are trying to detect a shift to the left in the distribution after the cleanup, we would expect the sum of the ranks for the observations in sample 1 to be large. Thus, we will reject H_0 for large values of $z = (T - \mu_T)/\sigma_T$.

Grouping the measurements with tied ranks, we have 18 groups. These groups are listed below with the corresponding values of t_j, the number of tied ranks in the group.

Rank(s)	Group	t_j
1	1	1
2	2	1
3	3	1
4.5, 4.5	4	2
6	5	1
7.5, 7.5	6	2
9	7	1
10	8	1
11.5, 11.5	9	2
14, 14, 14	10	3
16	11	1
17	12	1
18	13	1
19	14	1
20	15	1
21	16	1
22.5, 22.5	17	2
24	18	1

For all groups with $t_j = 1$, there is no contribution for

$$\frac{\sum_j t_j(t_j^2 - 1)}{(n_1 + n_2)(n_1 + n_2 - 1)}$$

in σ_T^2 since $t_j^2 - 1 = 0$. Thus, we will need only $t_j = 2, 3$. Substituting our data in the formulas, we obtain

$$\mu_T = \frac{n_1(n_1 + n_2 + 1)}{2} = \frac{12(12 + 12 + 1)}{2} = 150$$

$$\sigma_T^2 = \frac{n_1 n_2}{12}\left[(n_1 + n_2 + 1) - \frac{\sum t_j(t_j^2 - 1)}{(n_1 + n_2)(n_1 + n_2 - 1)}\right]$$

$$= \frac{12(12)}{12}\left[25 - \frac{6 + 6 + 6 + 24 + 6}{24(23)}\right] = 12(25 - .0870) = 298.956$$

$$\sigma_T = 17.29.$$

The computed value of z is

$$z = \frac{T - \mu_T}{\sigma_T} = \frac{216 - 150}{17.29} = 3.82.$$

Since this value exceeds 1.645, we reject H_0 and conclude that the distribution of before-cleanup measurements is shifted to the right of the corresponding distribution of after-cleanup measurements; that is, the after-cleanup measurements on dissolved oxygen tend to be smaller than the corresponding before-cleanup measurements.

b. The value of σ_T^2 without correcting for ties is

$$\sigma_T^2 = \frac{12(12)(25)}{12} = 300$$

and

$$\sigma_T = 17.32.$$

For this value of $\sigma_T, z = 3.81$ rather than 3.82 found by applying the correction. This should help you understand how little effect the correction has on the final result unless there are very many ties.   ▲

The Wilcoxon rank sum test is an alternative to the two-sample t test that requires fewer assumptions. In particular, Wilcoxon's test does not require normality for the two populations, only that they be identical under H_0. When the assumptions underlying a t test hold, the t test will be more likely to declare an existing difference. This only seems logical since the t test uses the magnitudes of observations rather than just their relative magnitudes (ranks). But when the assumptions for a t test are violated, the Wilcoxon rank sum test is the more informative test and is more likely to declare a difference when it exists. This is particularly true when nonnormality of the populations is present in the form of severe skewness or extreme outliers.

EXAMPLE 6.8   ▼

To investigate the effect of skewness on the pooled-variance t test as well as the Wilcoxon rank sum test, 1,000 samples were drawn from a squared-exponential population; this population was extremely right-skewed. The following results were obtained. What do the results indicate about the effect of skewness on the two tests?

```
                    Checking α (different sample sizes; same σs)

Simulation of two-sample t test (1000 samples)

Popn       μ          σ         n
  1     50.000    10.0000       5
  2     50.000    10.0000      25

                    using the pooled-variance t test
One-tailed test:

           number of times H₀: "μ₁ - μ₂ is 0" is rejected in favor of

     α          "μ₁ - μ₂ > 0"            "μ₁ - μ₂ < 0"           total (α doubled)
   .100             146                        35                      181
   .050              95                         3                       98
   .025              51                         0                       51
   .010              25                         0                       25
   .005              15                         0                       15

Results of Wilcoxon rank sum test using z as test statistic

           number of times H₀: "two populations are identical" rejected in favor of
     α        Popn1 rt of Popn2         Popn1 left of Popn2        total (α doubled)
   .100            102                          93                      195
   .050             37                          51                       88
   .025             15                          30                       45
   .010              5                          14                       19
   .005              4                           3                        7
```

Solution   The null hypothesis is true in this simulation; both means are 50. The actual number of rejections of the null hypothesis by the *t* test is far from what is indicated by the nominal α value for one-tailed probabilities. The Wilcoxon rank sum test, which doesn't assume normal populations, appears to be rejecting the null hypothesis the correct number of times.   ▲

▼ EXERCISES

Applications

 6.20   A plumbing contractor was interested in making her operation more efficient by cutting down on the average distance between service calls while still maintaining at least the same level of business activity. One plumber (plumber 1) was assigned a dispatcher who monitored all his incoming requests for service and outlined a service strategy for that day. Plumber 2 was to continue as she had in the past, by providing service in roughly sequential order for stacks of service calls received. The total daily mileages for these two plumbers are recorded here for a total of 18 days (3 workweeks).

Plumber 1	88.2	94.7	101.8	102.6	89.3	95.7
	78.2	80.1	83.9	86.1	89.4	71.4
	92.4	85.3	87.5	94.6	92.7	84.6
Plumber 2	105.8	117.6	119.5	126.8	108.2	114.7
	90.2	95.6	110.1	115.3	109.6	112.4
	104.6	107.2	109.7	102.9	99.1	111.5

a. Plot the sample data for each plumber and compute $\bar{y}$ and s.

b. Based on your findings in part a, which procedure appears more appropriate for comparing the distributions?

6.21 Computer output is shown below for the data of Exercise 6.20 for a t test of H_0: $\mu_1 - \mu_2 = 0$ and a Wilcoxon rank sum test (which is equivalent to the Mann–Whitney test shown here).

```
MTB > NAME C1 'PLUMBER1'
MTB > NAME C2 'PLUMBER2'
MTB > SET C1
MTB > END
MTB > PRINT C1

PLUMBER1
    88.2    94.7   101.8   102.6    89.3    95.7    78.2    80.1    83.9
    86.1    89.4    71.4    92.4    85.3    87.5    94.6    92.7    84.6

MTB > HISTOGRAM C1

Histogram of PLUMBER1   N = 18

Midpoint    Count
      70       1  *
      75       0
      80       2  **
      85       4  ****
      90       5  *****
      95       4  ****
     100       1  *
     105       1  *


MTB > SET C2
MTB > END
MTB > PRINT C2


PLUMBER2
   105.8   117.6   119.5   126.8   108.2   114.7    90.2    95.6   110.1
   115.3   109.6   112.4   104.6   107.2   109.7   102.9    99.1   111.5


MTB > HISTOGRAM C2

Histogram of PLUMBER2  N = 18
Midpoint    Count
      90       1  *
      95       1  *
     100       1  *
     105       4  ****
     110       6  ******
     115       2  **
     120       2  **
     125       1  *


MTB > TWOSAMPLE 95% CONFIDENCE FOR 'PLUMBER1' AND 'PLUMBER2'

TWOSAMPLE T FOR PLUMBER1 VS PLUMBER2
             N     MEAN    STDEV   SE MEAN
PLUMBER1    18    88.81     7.89      1.9
PLUMBER2    18   108.93     8.73      2.1

95 PCT CI FOR MU PLUMBER1 - MU PLUMBER2: (-25.8, -14.5)

TTEST MU PLUMBER1 = MU PLUMBER2 (VS NE): T= -7.26  P=0.0000  DF= 33

MTB > MANN-WHITNEY 95% CONFIDENCE FOR 'PLUMBER1' AND 'PLUMBER2'
```

(*continues*)

```
Mann-Whitney Confidence Interval and Test

  PLUMBER1   N =  18      MEDIAN =        88.75
  PLUMBER2   N =  18      MEDIAN =       109.65
  POINT ESTIMATE FOR ETA1-ETA2 IS       -20.30
  95.2  PCT C.I. FOR ETA1-ETA2 IS (    -25.40,    -14.90)
  W =    183.0
  TEST OF ETA1 = ETA2   VS.   ETA1 N.E. ETA2 IS SIGNIFICANT AT   0.0000

MTB > STOP
```

a. Compare the results for these two tests and draw a conclusion about the effectiveness of the dispatcher program.

b. Comment on the appropriateness or inappropriateness of the t test based on your findings in Exercise 6.20a and the output shown here.

c. Does it matter which test was used here? Might it be reasonable to run both tests in certain situations? Why?

6.22 An experiment was conducted to compare the weights of the combs of roosters fed two different vitamin-supplemented diets. Twenty-eight healthy roosters were randomly divided into two groups, with one group receiving diet I and the other receiving diet II. After the study period the comb weight (in milligrams) was recorded for each rooster. These data are given here.

Diet I	73	130	115	144	127	126	112	76	68	101	126	49	110	123
Diet II	80	72	73	60	55	74	67	89	75	66	93	75	68	76

a. Use the Wilcoxon rank sum test to determine whether there is a difference in the distributions of comb weights for the two groups. Use $\alpha = .05$.

b. Can you suggest other statistical procedures that might be appropriate for analyzing the same data? Which would you suggest?

6.23 Refer to Exercise 6.22. Suppose the experimenter was interested in determining whether the comb weights for diet I were selected from a distribution shifted above (to the right of) that for comb weights for diet II. Run an appropriate Wilcoxon rank sum test and give the p-value. Draw a conclusion.

6.24 A computer simulation was done to compare the t test to the Wilcoxon rank sum test when sampling from two identical (nonnormal) populations. Chosen for the study were two identical, right-skewed populations with means and standard deviations as shown.

Population 1	$\mu_1 = 100$	$\sigma_1 = 20$	$n_1 = 10$
Population 2	$\mu_2 = 100$	$\sigma_2 = 20$	$n_2 = 20$

The simulation study consisted of 1,000 runs, where random samples of size $n_1 = 10$ and $n_2 = 20$ were drawn from the two populations; after each run, the pooled t test was run to test $H_0: \mu_1 - \mu_2 = 0$ and the Wilcoxon rank sum test was run to test H_0: The populations were identical. The number (%) of times out of 1,000 that H_0 was rejected at the upper and lower .05 levels is recorded for each test.

	Pooled t Test H_0: $\mu_1 - \mu_2 = 0$
H_0 rejected, H_a: $\mu_1 - \mu_2 > 0$	87
H_0 rejected, H_a: $\mu_1 - \mu_2 < 0$	15

	Wilcoxon Rank Sum Test H_0: Populations Are Identical
H_0 rejected, H_a: population 1 is to right of population 2	43
H_0 rejected, H_a: population 1 is to left of population 2	54

What do the simulation results indicate in regard to the effect of skewness in the performance of the pooled t test and the Wilcoxon rank sum test?

6.25   A simulation study evaluates the effect of severe skewness on the relative usefulness of the pooled-variance t test, the t' test, and the rank sum test. Independent samples are taken from the severely skewed distribution. The results are shown below. For any of the tests, are the nominal α values grossly wrong?

```
                    Checking  α (same sample sizes and σs)

      Popn      μ         σ         n
       1     50.000    10.0000     10
       2     50.000    10.0000     10

  using the pooled-variance t test

  One-tailed test:

              number of times H₀: " μ₁ - μ₂ is 0" is rejected in favor of

        α             " μ₁ - μ₂ > 0"      " μ₁ - μ₂ < 0"    total (α doubled)
       .100                99                 108                207
       .050                32                  37                 69
       .025                11                  12                 23
       .010                 3                   3                  6
       .005                 0                   0                  0

  using separate variances (t´)

  One-tailed test:

              number of times H₀: " μ₁ - μ₂ is 0" is rejected in favor of

        α             " μ₁ - μ₂ > 0"      " μ₁ - μ₂ < 0"    total (α doubled)
       .100                94                  98                192
       .050                27                  27                 54
       .025                 6                  10                 16
       .010                 1                   1                  2
       .005                 0                   0                  0

  Results of Wilcoxon rank sum test using z as test statistic

              number of times H₀: "two populations are identical" rejected in favor of
        α          Popn1 rt of Popn2    Popn1 left of Popn2    total (α doubled)
       .100              105                 110                215
       .050               44                  50                 94
       .025               24                  29                 53
       .010                6                  12                 18
       .005                1                   3                  4
```

6.4 ▼ A QUICK, PORTABLE STATISTIC: THE TUKEY–DUCKWORTH TEST (optional)

In addition to providing alternative analyses when underlying assumptions are violated, some nonparametric statistical techniques are so easy to remember and use that they can be quickly applied without a desk calculator, a computer, or a reference table for critical values. In short, they can be carried anywhere and applied in many situations to provide a quick preliminary conclusion.

Throughout the text, we will insert these portable statistics where appropriate to make you aware of how they may help you in situations where it is not possible to do a formal analysis either because time will not permit it or because appropriate reference material (formulas, critical values) or equipment (computer hardware and software) is not available.

Tukey–Duckworth two-sample test

The first technique we will consider is the **Tukey–Duckworth two-sample test** (Tukey, 1959) to determine if two independent samples were drawn from identical populations. This test can be used for sample sizes satisfying the following inequalities:

$$4 \leq n_1 \leq n_2 \leq 30 \qquad n_2 \leq \frac{4n_1}{3} + 3.$$

It should be noted that the designation of population 1 and population 2 is completely arbitrary. Modifications of this procedure have been suggested by other authors (see, for example, Neave (1966 and 1975)).

Tukey–Duckworth Two-Sample Test

H_0: The populations are identical.
H_a: The populations are different (a two-tailed test).

Test procedure

1. Determine the largest and smallest measurement in each sample.
2. For the sample that contains the largest value in the combined samples, count all measurements that are larger than the largest measurement in the other sample.
3. For the other sample, count all measurements that are smaller than the smallest measurement of the first sample.
4. Let C denote the sum of the two counts. For $\alpha = .05, .01,$ or $.001$, reject H_0 if $C \geq 7, 10,$ or 13, respectively.

EXAMPLE 6.9

▼

Thirty different 1–acre plots were randomly divided into two groups, with 15 plots per group. The plots in the first group were fertilized with brand A fertilizer and those in the second with brand B. Each of the thirty 1–acre plots was then planted in corn. Yields (in bushels) are presented in Table 6.5 for each plot. Use these data to determine whether there is a difference in yields for the two brands of fertilizer. Use $\alpha = .05$.

TABLE 6.5
Yields of Corn (in bushels) for Two Different Brands of Fertilizer, Example 6.9

Group 1 (brand A)		Group 2 (brand B)	
96	89	98	92
92	94	94	89
98	80	92	95
82	97	84	92
86	84	99	96
87	85	96	101
93	83	98	103
81		96	

Solution   We can proceed immediately with the Tukey–Duckworth two-sample test, since our sample sizes satisfy the criteria $4 \leq n_1 \leq n_2 \leq 30$ and $n_2 \leq (4n_1/3) + 3$. First we must determine the largest and smallest measurements for each sample, as shown here.

	Group 1	Group 2
Largest	98	103
Smallest	80	84

Group 2 contains the largest measurement (103) for the combined samples. The number of measurements in group 2 larger than 98, which is the largest measurement in group 1, is 3. Similarly, the number of measurements in group 1 less than 84, which is the smallest measurement in group 2, is 4. Since $C = 3 + 4 = 7$, we reject the null hypothesis that the populations of corn yields corresponding to the two fertilizers are identical, at the $\alpha = .05$ level.   ▲

▼ EXERCISES

6.26   a.   Refer to the data of Exercise 6.20 and use the Tukey–Duckworth test to reach a conclusion concerning the utility of a dispatcher.
       b.   Compare your conclusion to the one drawn in Exercise 6.21 after viewing the output for a t test and a Wilcoxon rank sum test.

6.27 Refer to the data of Example 6.7.
  a. What conclusion would you draw regarding the cleanup program using a Tukey–Duckworth test?
  b. Compare your conclusion in part a to the results of Example 6.7.

6.28 Comment on the interchangeability of a two-sample t test, the Wilcoxon rank sum test, and the Tukey–Duckworth test. When might you use one over the others? When might you use two or more for the same data set?

6.5 ▼ INFERENCES ABOUT $\mu_1 - \mu_2$: PAIRED DATA

The methods we presented in the preceding four sections were appropriate for situations in which independent random samples are obtained from two populations. These methods are not appropriate for studies or experiments in which each measurement in one sample is *matched* or *paired* with a particular measurement in the other sample. In this section, we will deal with methods for analyzing "paired" data. We begin with an example.

E X A M P L E   6.10   ▼

Insurance adjusters are concerned about the high estimates they are receiving from garage I for auto repairs compared to garage II. To verify their suspicions each of 15 cars recently involved in an accident was taken to both garages for separate estimates of repair costs. Use a two-sample t test to analyze these data.

Solution   Computer output for these data is shown here.

```
MTB > NAME C1 'GARAGE1'
MTB > NAME C2 'GARAGE2'
MTB > NAME C3 'DIFFER'
MTB > SET C1
MTB > END
MTB > SET C2
MTB > END
MTB > LET C3=C1-C2
MTB > PRINT C1-C3

ROW   GARAGE1   GARAGE2   DIFFER

 1      7.6       7.3      0.3
 2     10.2       9.1      1.1
 3      9.5       8.4      1.1
 4      1.3       1.5     -0.2
 5      3.0       2.7      0.3
 6      6.3       5.8      0.5
 7      5.3       4.9      0.4
 8      6.2       5.3      0.9
 9      2.2       2.0      0.2
10      4.8       4.2      0.6
```

(continues)

```
11      11.3      11.0      0.3
12      12.1      11.0      1.1
13       6.9       6.1      0.8
14       7.6       6.7      0.9
15       8.4       7.5      0.9

MTB > TWOSAMPLE 95% CONFIDENCE FOR 'GARAGE1' AND 'GARAGE2'

TWOSAMPLE T FOR GARAGE1 VS GARAGE2
              N      MEAN     STDEV   SE MEAN
GARAGE1  15          6.85      3.20      0.83
GARAGE2  15          6.23      2.94      0.76

95 PCT CI FOR MU GARAGE1 - MU GARAGE2: (-1.69, 2.92)

TTEST MU GARAGE1 = MU GARAGE2 (VS NE): T= 0.55  P=0.59  DF= 27

MTB > STOP
```

From the output we see there is a consistent difference in the sample means ($\bar{y}_1 - \bar{y}_2 = .62$). But this difference is rather small considering the variability of the measurements ($s_1 = 3.20$, $s_2 = 2.94$). In fact, the computed t-value (.55) has a p-value of .59, indicating very little evidence of a difference in the average claim estimates for the two garages.   ▲

A closer glance at the data in Table 6.6 indicates there is something about the conclusion in Example 6.10 that is inconsistent with our intuition. For all but one of the 15 cars, the estimate from garage I was higher than that from garage II. From our knowledge of the binomial distribution, the probability of observing garage I estimates higher in $y = 14$ or more of the $n = 15$ trials assuming no

T A B L E 6.6

Repair Estimates (in hundreds of dollars), Example 6.10

Car	Garage I	Garage II
1	7.6	7.3
2	10.2	9.1
3	9.5	8.4
4	1.3	1.5
5	3.0	2.7
6	6.3	5.8
7	5.3	4.9
8	6.2	5.3
9	2.2	2.0
10	4.8	4.2
11	11.3	11.0
12	12.1	11.0
13	6.9	6.1
14	7.6	6.7
15	8.4	7.5
Totals	$\bar{y}_1 = 6.85$	$\bar{y}_2 = 6.23$

difference ($\pi = .5$) for garages I and II is

$$P(y = 14 \text{ or } 15) = P(y = 14) + P(y = 15)$$

$$= \binom{15}{14}(.5)^{14}(.5) + \binom{15}{15}(.5)^{15}.$$

This probability is .000 to three decimal places. Using this binomial probability, we would argue that the observed sample results are highly contradictory to the null hypothesis of equality of estimates for the two garages. Where did we go wrong? Why are there such conflicting results?

The explanation of the difference in the conclusion for a t test and the conclusion based on the binomial distribution is that one of the basic assumptions, independent samples, has been violated by the way the experiment was conducted. The adjusters obtained a measurement from both garages for each car rather than having a random sample of 15 cars examined by garage I and a second sample of cars examined by garage II.

As you can see from the data in Table 6.6, the repair estimates for a given car are about the same but vary considerably from car to car. These differences caused large variability among estimates for a given garage and tend to cancel any differences between the two garages. This fact was recognized when the study was planned. By having both garages give an estimate on each car, we can calculate the difference between the two garages for each car and hence cancel out the car-to-car variability.

A proper analysis of the paired data in Example 6.10 makes use of the 15 difference measurements to test the null hypothesis that the mean difference, μ_d, is D_0. This hypothesis is equivalent to $H_0: \mu_1 - \mu_2 = D_0$. A summary of the test procedure is given here.

Paired t Test

$H_0:$ $\mu_d = D_0$

$H_a:$ 1. $\mu_d > D_0$
2. $\mu_d < D_0$
3. $\mu_d \neq D_0$

T.S.: $t = \dfrac{\bar{d} - D_0}{s_d/\sqrt{n}}$, where $\bar{d}$ and s_d are the sample mean and standard deviation of the n differences.

R.R.: For a specified value of α and df $= n - 1$

1. reject H_0 if $t > t_\alpha$
2. reject H_0 if $t < -t_\alpha$
3. reject H_0 if $|t| > t_{\alpha/2}$.

EXAMPLE 6.11

▼

Refer to the data of Example 6.10 and perform a paired t test. Draw a conclusion based on $\alpha = .05$.

Solution   For these data, the parts of the statistical test are

$$H_0: \quad \mu_d = \mu_1 - \mu_2 = 0$$
$$H_a: \quad \mu_d > 0$$

T.S.: $t = \dfrac{\bar{d}}{s_d / \sqrt{n}}$

R.R.:   For df $= n - 1 = 14$, reject H_0 if $t > t_{.05}$.

Before computing t, we must first calculate s_d, the sample standard deviation of the differences. We can calculate s_d by using our shortcut formula for a sample variance or by using a calculator:

$$s_d^2 = \frac{1}{n-1} \left[\sum_i d_i^2 - \frac{(\sum_i d_i)^2}{n} \right].$$

For the data of Table 6.6,

$$\sum_i d_i = .3 + 1.1 + 1.1 + \cdots + .9 = 9.2, \quad \bar{d} = \frac{9.2}{15} = .61 \qquad \text{and}$$

$$\sum_i d_i^2 = (.3)^2 + (1.1)^2 + (1.1)^2 + \cdots + (.9)^2 = 7.82.$$

Hence, for $n = 15$ differences,

$$s_d^2 = \frac{1}{14} \left[7.82 - \frac{(9.2)^2}{15} \right] = .156$$

$$s_d = \sqrt{.156} = .394.$$

Substituting into the test statistic t, we have

$$t = \frac{\bar{d} - 0}{s_d / \sqrt{n}} = \frac{.61}{.394 / \sqrt{15}} = 6.00.$$

Indeed $t = 6.00$ is far beyond all tabulated t values for df $= 14$, so the p-value is less than .005; presumably p is much less than .005. We conclude that the mean repair estimate for garage I is greater than that for garage II. This conclusion agrees with our intuitive finding based on the binomial distribution.

The point of all this discussion is not to suggest that we typically have two or more analyses that may give *very* conflicting results for a given situation. Rather, the point is that the analysis must fit the experimental situation; and for this experiment, the samples are dependent, demanding we use an analysis appropriate for dependent (paired) data. ▲

The corresponding general $100(1 - \alpha)\%$ confidence interval for μ_d based on paired data is shown here.

100(1 − α)% Confidence Interval for μ_d Based on Paired Data

$$\bar{d} \pm \frac{t_{\alpha/2} s_d}{\sqrt{n}},$$

where n is the number of pairs of observations (and hence the number of differences) and df $= n - 1$.

The use of these t procedures depends on the assumption that the population of *differences* is normally distributed. For small samples, plot the sample differences; if severe skewness or outliers are present, the binomial test or the signed-rank test of Section 6.6 should be used.

▼ EXERCISES

Basic Techniques

6.29 Consider the paired data shown here.

Pair	y_1	y_2
1	21	29
2	28	30
3	17	21
4	24	25
5	27	33

a. Run a paired t test and give the p-value for the test.

b. What would your conclusion be using an argument related to the binomial distribution? Does it agree with part a? When might these two approaches not agree?

Applications

6.30 An agricultural experiment station was interested in comparing the yields for two new varieties of corn. Because the investigators thought that there might be a great deal of variability in yield from one farm to another, each variety was randomly assigned to a different 1-acre plot on each of seven farms. The 1-acre plots were planted; the corn was harvested at maturity. The results of the experiment (in bushels of corn) are listed here.

Farm	1	2	3	4	5	6	7
Variety A	48.2	44.6	49.7	40.5	54.6	47.1	51.4
Variety B	41.5	40.1	44.0	41.2	49.8	41.7	46.8

Use these data to test the null hypothesis that there is no difference in mean yields for the two varieties of corn. Use $\alpha = .05$.

6.31 Thirty sets of identical twins were asked to participate in a 1-year study designed to measure certain social attitudes. One twin from each set was randomly assigned to live in the home of a minority family, while the other twin stayed at home. After 1 year, each person was asked to respond to a long questionnaire designed to detect and measure well-defined attitudes. Let sample 1 denote the combined questionnaire scores for those persons who lived at home and sample 2 denote the set of scores for those who lived with a family from a minority class.

Set of Twins	Home Environment, y_1	Minority Environment, y_2	Difference	Set of Twins	Home Environment, y_1	Minority Environment, y_2	Difference
1	78	71	7	16	90	88	2
2	75	70	5	17	89	80	9
3	68	66	2	18	73	65	8
4	92	85	7	19	61	60	1
5	55	60	−5	20	76	74	2
6	74	72	2	21	81	76	5
7	65	57	8	22	89	78	11
8	80	75	5	23	82	78	4
9	98	92	6	24	70	62	8
10	52	56	−4	25	68	73	−5
11	67	63	4	26	74	73	1
12	55	52	3	27	85	75	10
13	49	48	1	28	97	88	9
14	66	67	−1	29	95	94	1
15	75	70	5	30	78	75	3
					$\bar{y}_1 = 75.23$	$\bar{y}_2 = 71.43$	$\bar{d} = \bar{y}_1 - \bar{y}_2 = 3.8$

a. Plot the sample differences. Is there any reason to believe that a *t* test is inappropriate?
b. Test the null hypothesis

$$H_0: \quad \mu_1 - \mu_2 = 0 \text{ (the population mean scores for those not exposed and those exposed to a minority environment are identical)}$$

against the alternative

$$H_a: \quad \mu_1 - \mu_2 \neq 0 \text{ (the population mean scores are different for the two environments).}$$

Use $\alpha = .05$.

6.32 Suppose we wish to estimate the difference between the mean monthly salaries of male and female sales representatives. Since there is a great deal of salary variability from company to company, it was decided to filter out the variability due to companies by making male–female comparisons within each company. One male and one female with the required background and work experience will be selected from each company. If the range of differences in salaries (between males and females) within a company is approximately $300 per month, determine the number of companies that must be examined to estimate the difference in mean monthly salary for males and females. Use a 95% confidence interval with a half width of $5. (Hint: Refer to Section 5.3.)

6.33 Refer to Exercise 6.32. If $n = 35$, $\bar{d} = 120$, and $s_d = 250$, construct a 90% confidence interval for μ_d, the mean difference in salaries for male and female sales representatives.

6.6 ▼ A NONPARAMETRIC ALTERNATIVE: WILCOXON SIGNED-RANK TEST

The Wilcoxon signed-rank test, which makes use of the sign and the magnitude of the rank of the differences between pairs of measurements, provides an alternative to the paired *t* test. The formal null hypothesis for Wilcoxon's signed-rank test is that the population distribution of differences is symmetrical about D_0; the test is sensitive to the distribution of differences being shifted to the right or left of D_0. In most cases, D_0 is 0; otherwise we subtract D_0 from every measurement and proceed as if $D_0 = 0$. The test uses the nonzero differences ranked in absolute value from lowest to highest. If two or more measurements have the same nonzero difference (ignoring sign), we assign each difference a rank equal to the average of the occupied ranks. The appropriate sign is then attached to the rank of each difference.

Before summarizing the Wilcoxon signed-rank test, we define the following notation:

n = the number of pairs of observations with a nonzero difference
T_+ = the sum of the positive ranks; if there are no positive ranks, $T_+ = 0$
T_- = the sum of the negative ranks; if there are no negative ranks, $T_- = 0$
T = the smaller of T_+ and T_-, ignoring their signs

μ_T

$$\mu_T = \frac{n(n+1)}{4}$$

σ_T

$$\sigma_T = \sqrt{\frac{n(n+1)(2n+1)}{24}}.$$

g groups

If we group all differences assigned the same rank together, and there are g such groups, the variance of T is

$$\sigma_T^2 = \frac{1}{24}\left[n(n+1)(2n+1) - \frac{1}{2}\sum_j t_j(t_j - 1)(t_j + 1)\right],$$

t_j

where t_j is the number of tied ranks in the jth group. Note that if there are no tied ranks, $g = n$, and $t_j = 1$ for all groups. The formula then reduces to

$$\sigma_T^2 = \frac{n(n+1)(2n+1)}{24}.$$

The Wilcoxon signed–rank test is presented here.

Wilcoxon Signed-Rank Test

▼

H_0: The distribution of differences is symmetrical around D_0. (D_0 is specified; usually D_0 is 0.)

H_a: **1.** The differences tend to be larger than D_0.
        2. The differences tend to be smaller than D_0.
        3. Either 1 or 2 is true (two-sided H_a).

$(n \leq 50)$

T.S.: **1.** $T = |T_-|$
       2. $T = T_+$
       3. T = smaller of $|T_-|$, T_+

R.R.: For a specified value of α (one-tailed .05, .025, .01, or .005; two-tailed .10, .05, .02, .01) and fixed number of nonzero differences n, reject H_0 if the value of T is less than or equal to the appropriate entry in Table 9 in the Appendix.

(continues)

$(n > 50)$

T.S.:   Compute the test statistic

$$z = \frac{T - \dfrac{n(n + 1)}{4}}{\sqrt{\dfrac{n(n + 1)(2n + 1)}{24}}}.$$

R.R.:   For cases 1 and 2, reject H_0 if $z < -z_\alpha$; for case 3 reject H_0 if $z < -z_{\alpha/2}$.

EXAMPLE 6.12

▼

Two different brands of fertilizer (A and B) were compared on each of 10 different 2-acre plots. Each plot was subdivided into 1-acre subplots, with brand A randomly assigned to one subplot and brand B to the other. Sixty pounds per acre of fertilizers were then applied to subplots. The data for barley yields in bushels per acre are listed in Table 6.7 by fertilizer and plot.

Use the Wilcoxon signed-rank test to test the hypothesis that the distributions of barley yields for the two brands of fertilizer are identical against the alternative that they are different. Use $\alpha = .05$.

TABLE 6.7

Barley Yields (in bushels) by Plot and by Fertilizer, Example 6.12

	Barley Yield		
Plot	Fertilizer A y_1	Fertilizer B y_2	Difference $y_1 - y_2$
1	312	346	−34
2	333	372	−39
3	356	392	−36
4	316	351	−35
5	310	330	−20
6	352	364	−12
7	389	375	14
8	313	315	−2
9	316	327	−11
10	346	378	−32

Solution   First we must rank (from lowest to highest) the absolute values of the $n = 10$ differences. These ranks appear in column 2 of Table 6.8. The appropriate sign is then attached to each rank (see column 3 in Table 6.8). The sum of the positive and negative ranks, are, respectively,

$$T_+ = 4$$
$$T_- = -7 + (-10) + \cdots + (-6) = -51.$$

Thus, T, the smaller of T_+ and T_-, ignoring the sign, is 4.

TABLE 6.8
Rankings for the Data of Table 6.7

Plot	Rank of Difference $\lvert y_1 - y_2 \rvert$	Rank with Appropriate Sign
1	7	−7
2	10	−10
3	9	−9
4	8	−8
5	5	−5
6	3	−3
7	4	4
8	1	−1
9	2	−2
10	6	−6

For a two-tailed test with $n = 10$ and $\alpha = .05$, we see from Table 9 in the Appendix that we will reject H_0 if T is less than or equal to 8. Thus, we reject H_0 and conclude that the distributions of barley yields for the two brands of fertilizers are different. Barley yields for fertilizer A tend to be smaller than (to the left of) corresponding yields for fertilizer B.   ▲

The choice of an appropriate paired-sample test follows the guidelines mentioned for unpaired data in Section 6.2. If the assumptions of the t test are satisfied—in particular, if the distribution of differences is roughly normal—the t test is more powerful. If the distribution of differences is grossly skewed, the nominal t probabilities may be misleading. If the distribution is roughly symmetric but has heavy tails (as indicated by the presence of outliers), the signed-rank test may be more powerful. Often, the tests will yield essentially the same conclusion.

Even with this discussion, you might still be confused as to which statistical test (or confidence interval) to apply in a given situation where there is a choice of two or more methods. When in doubt, do several different tests; computing costs are usually minimal, especially with the availability of many different statistical software packages such as Minitab, SAS, and SPSS. If the results from the different analyses yield different results, you should identify the peculiarities of the data set to understand why the results differ. If the results agree, and there are no blatant violations of assumptions, you should be very confident in your conclusions.

This particular "hedging" strategy is appropriate not only for paired data, but for many of the situations we have discussed. Since computer software makes it easy to run alternative analyses on the same data, the potential concern about assumptions often can be put to rest when the alternative analyses yield essentially the same results.

▼ EXERCISES

Basic Techniques

6.34 Refer to Exercise 6.31.

a. Using the data in the table, run a Wilcoxon signed-rank test. Give the p-value and draw a conclusion.

b. Compare your conclusions here to those in Exercise 6.31. Does it make a difference which test (t or signed rank) is used?

Applications

 6.35 Two judges were asked to rate separately each of 22 inmates on his or her rehabilitative potential. These data appear next.

Inmate	Judge 1	Judge 2	Inmate	Judge 1	Judge 2
1	6	5	12	9	8
2	12	11	13	10	8
3	3	4	14	6	7
4	9	10	15	12	9
5	5	2	16	4	3
6	8	6	17	5	5
7	1	2	18	6	4
8	12	9	19	11	8
9	6	5	20	5	3
10	7	4	21	10	9
11	6	6	22	10	11

Use the computer output shown here to reach a conclusion about the following.

a. H_0: The distribution of differences is symmetrical about 0 versus H_a: The difference tends to be larger than 0. What is your conclusion? What is the p-value?

b. How would the results of part a compare to those from a paired t test? Use the following output to draw conclusions.

```
    OBS    JUDGE-1    JUDGE-2    DIFF
     1        6          5         1
     2       12         11         1
     3        3          4        -1
     4        9         10        -1
     5        5          2         3
     6        8          6         2
     7        1          2        -1
     8       12          9         3
     9        6          5         1
    10        7          4         3
    11        6          6         0
    12        9          8         1
    13       10          8         2
    14        6          7        -1
    15       12          9         3
    16        4          3         1
    17        5          5         0
    18        6          4         2
    19       11          8         3
    20        5          3         2
    21       10          9         1
    22       10         11        -1
   N= 22


                    PAIRED T TEST

                                     STANDARD STD ERROR
  VARIABLE      LABEL            N    MEAN    DEVIATION  OF MEAN     T    PR>|T|
   DIFF    DIFFERENCE IN RATINGS 22 1.09090909 1.47709789 0.31491833 3.46 0.0023


                  WILCOXON SIGNED RANK TEST
          STATISTICAL ANALYSIS SYSTEM—NPAR 360 INTERFACE

  WILCOXON MATCHED-PAIRS SIGNED-RANKS TEST
    WITH ONE-TAIL PROBABILITIES OF THIS OR GREATER T

  HIGHER GROUP  LOWER GROUP  N(TOTAL)  N(SIGNED)  WILCOXON T  PROBABILITY
    JUDGE_1       JUDGE_2       22        20        30.00       0.0026
```

▼ **6.36** The effect of Benzedrine on the heart rate of dogs (in beats per minute) was examined in an experiment on 14 dogs chosen for the study. Each dog was to serve as its own control, with half of the dogs assigned to receive Benzedrine during the first study period and the other half assigned to receive a placebo (saline solution). All dogs were examined to determine the heart rates after 2 hours on the medication. After 2 weeks in which no medication was given, the regimens for the dogs were switched for the second study period. The dogs previously on Benzedrine were given the placebo and the others received Benzedrine. Again heart rates were measured after 2 hours.

The following sample data are not arranged in the order in which they were taken but have been summarized by regimen. Use these data to test the research hypothesis that the distribution of heart rates for the dogs when receiving Benzedrine is shifted to the right of that for the same animals when on the placebo. Use a one-tailed Wilcoxon signed-rank test with $\alpha = .05$.

Dog	Placebo	Benzedrine
1	250	258
2	271	285
3	243	245
4	252	250
5	266	268
6	272	278
7	293	280
8	296	305
9	301	319
10	298	308
11	310	320
12	286	293
13	306	305
14	309	313

6.7 ▼ CHOOSING SAMPLE SIZES FOR INFERENCES ABOUT $\mu_1 - \mu_2$

Sections 5.3 and 5.5 were devoted to sample-size calculations to obtain a confidence interval about μ with a fixed width and specified degree of confidence or to conduct a statistical test concerning μ with predefined levels for α and β. Similar calculations can be made for inferences about $\mu_1 - \mu_2$ with either independent samples or with paired data. Determining the sample size for a $100(1 - \alpha)\%$ confidence interval about $\mu_1 - \mu_2$ of width $2E$ based on independent samples is possible by solving the following expression for n. We will assume that both samples are of the same size.

$$z_{\alpha/2}\sigma \sqrt{\frac{1}{n} + \frac{1}{n}} = E$$

Note that in this formula σ is the common population standard deviation and that we have assumed equal sample sizes.

Sample Sizes for a $100(1 - \alpha)\%$ Confidence Interval for $\mu_1 - \mu_2$ of the Form $\bar{y}_1 - \bar{y}_2 \pm E$, Independent Samples

$$n = \frac{2z_{\alpha/2}^2 \sigma^2}{E^2}$$

Note: If σ is unknown, substitute an estimated value to get an approximate sample size.

The sample sizes obtained using this formula are usually approximate because we have to substitute an estimated value of σ, the common population standard deviation. This estimate will probably be based on an educated guess from information on a previous study or on the range of population values.

Corresponding sample sizes for one- and two-sided tests of $H_0: \mu_1 - \mu_2 = D_0$ based on specific values of α and β are shown here.

Sample Sizes for Testing $H_0: \mu_1 - \mu_2 = D_0$, Independent Samples

▼

One-sided test: $n = 2\sigma^2 \dfrac{(z_\alpha + z_\beta)^2}{\Delta^2}$

Two-sided test: $n = 2\sigma^2 \dfrac{(z_{\alpha/2} + z_\beta)^2}{\Delta^2}$,

where $n_1 = n_2 = n$ and the probability of a Type II error is to be $\leq \beta$ when the true difference $|\mu_1 - \mu_2| \geq \Delta$.

Note: If σ is unknown, substitute an estimated value to obtain an approximate sample size.

E X A M P L E   6.13   ▼

An experiment was done to determine the effect on dairy cattle of a diet supplemented with liquid whey. While no differences were noted in milk production measurements among cattle given a standard diet (7.5 kg of grain plus hay by choice) with water and those on the standard diet and liquid whey only, a considerable difference between the groups was noted in the amount of hay ingested. Suppose that one tests the null hypothesis of no difference in mean hay consumption for the two diet groups of dairy cattle. For a two-tailed test with $\alpha = .05$, determine the approximate number of dairy cattle that should be included in each group if we want $\beta \leq .10$ for $|\mu_1 - \mu_2| \geq .5$. Previous experimentation has shown σ to be approximately .8.

Solution   From the description of the problem, we have $\alpha = .05$, $\beta \leq .10$ for $\Delta = |\mu_1 - \mu_2| \geq .5$ and $\sigma = .8$. Table 2 in the Appendix gives us $z_{.025} = 1.96$ and $z_{0.10} = 1.28$. Substituting into the formula we have

$$n \approx \frac{2(.8)^2(1.96 + 1.28)^2}{(.5)^2} = 53.75, \text{ or } 54.$$

That is, we need 54 cattle per group to run the desired test.   ▲

Sample-size calculation can also be done using the formulas shown when $n_1 \neq n_2$. In this situation, we let n_2 be some multiple m (e.g., $m = .5$) of n_1; then we substitute $(m + 1)/m$ for 2 in the sample size formulas. After solving for n_1, $n_2 = mn_1$.

Sample sizes for estimating μ_d and conducting a statistical test for μ_d based on paired data (differences) are found using the formulas of Chapter 5 for μ. The only change is that we're working with a single sample of differences rather than a single sample of y values. For convenience, the appropriate formulas are shown here.

Sample Size Required for a 100(1 − α)% Confidence Interval for μ_d of the Form $\bar{d} \pm E$

$$n = \frac{z_{\alpha/2}^2 \sigma_d^2}{E^2}$$

Note: If σ_d is unknown, substitute an estimated value to obtain approximate sample size.

Sample Sizes for One- and Two-Sided Tests of H_0: $\mu_d = D_0$

One-sided test: $n = \dfrac{\sigma_d^2 (z_\alpha + z_\beta)^2}{\Delta^2}$

Two-sided test: $n = \dfrac{\sigma_d^2 (z_{\alpha/2} + z_\beta)^2}{\Delta^2}$,

where the probability of a Type II error is β or less if the true difference $\mu_d \geq \Delta$.

Note: If σ_d is unknown, substitute an estimated value to obtain an approximate sample size.

6.8 ▼ SUMMARY

In this chapter, we have considered inferences about $\mu_1 - \mu_2$. The first set of methods was based on independent random samples being selected from the populations of interest. We learned how to sample data to run a statistical test or to construct a confidence interval for $\mu_1 - \mu_2$ using t methods. Wilcoxon's rank sum test, which does not require normality of the underlying populations, was presented as an alternative to the t test. Finally, the Tukey–Duckworth test was introduced as a quick, portable test that can be used as a preliminary test when time or circumstance dictates that a formal test cannot be run immediately.

The second major set of procedures can be used to make comparisons between two populations when the sample measurements are paired. In this situation, we no longer have independent random samples and hence the procedures of Sections 6.2–6.4 (t methods, Wilcoxon's rank sum, and the Tukey–Duckworth test) are inappropriate. The test and estimation methods for paired data are based on the sample differences for the paired measurements or the ranks of the differences. The paired t test and corresponding confidence interval based on the difference measurements were introduced and found to be identical to the single sample t methods of Chapter 5. The nonparametric alternative to the paired t test is Wilcoxon's signed-rank test.

The material presented in Chapters 5 and 6 lays the foundations of statistical inference (estimation and testing) for the remainder of the text. It would be good to review the material in this chapter periodically as new topics are introduced so that you retain the basic elements of statistical inference.

▼ KEY FORMULAS

Inferences about $\mu_1 - \mu_2$

1. $100(1 - \alpha)\%$ confidence interval for $\mu_1 - \mu_2$, independent samples; y_1 and y_2 approximately normal; $\sigma_1^2 = \sigma_2^2$

$$\bar{y}_1 - \bar{y}_2 \pm t_{\alpha/2} s_p \sqrt{\frac{1}{n_1} + \frac{1}{n_2}},$$

where

$$s_p = \sqrt{\frac{(n_1 - 1)s_1^2 + (n_2 - 1)s_2^2}{n_1 + n_2 - 2}} \quad \text{and} \quad df = n_1 + n_2 - 2.$$

2. t test for $\mu_1 - \mu_2$, independent samples; y_1 and y_2 approximately normal; $\sigma_1^2 = \sigma_2^2$

$$H_0: \quad \mu_1 - \mu_2 = D_0$$

$$\text{T.S.:} \quad t = \frac{\bar{y}_1 - \bar{y}_2 - D_0}{s_p \sqrt{1/n_1 + 1/n_2}} \quad df = n_1 + n_2 - 2.$$

3. t' test for $\mu_1 - \mu_2$, unequal variance, independent samples; y_1 and y_2 approximately normal;

$$H_0: \quad \mu_1 - \mu_2 = D_0$$

$$\text{T.S.:} \quad t' = \frac{\bar{y}_1 - \bar{y}_2 - D_0}{\sqrt{\dfrac{s_1^2}{n_1} + \dfrac{s_2^2}{n_2}}} \qquad df = \frac{(n_1 - 1)(n_2 - 1)}{(n_2 - 1)c^2 + (1 - c)^2(n_1 - 1)},$$

where

$$c = \frac{s_1^2/n_1}{\dfrac{s_1^2}{n_1} + \dfrac{s_2^2}{n_2}}.$$

4. Wilcoxon's rank sum test, independent samples

$$H_0: \quad \text{The two populations are identical.}$$

$(n_1 \leq 10, n_2 \leq 10)$

T.S.: T, the sum of the ranks in sample 1

$(n_1, n_2 > 10)$

$$\text{T.S.:} \quad z = \frac{T - \mu_T}{\sigma_T},$$

where T denotes the sum of the ranks in sample 1,

$$\mu_T = \frac{n_1(n_1 + n_2 + 1)}{2} \qquad \text{and} \qquad \sigma_T = \sqrt{\frac{n_1 n_2}{12}(n_1 + n_2 + 1)}$$

provided there are no tied ranks.

5. Paired t test; difference approximately normal

$$H_0: \quad \mu_d = D_0$$

$$\text{T.S.:} \quad t = \frac{\bar{d} - D_0}{s_d/\sqrt{n}} \qquad df = n - 1,$$

where n is the number of differences.

6. $100(1 - \alpha)\%$ confidence interval for μ_d, paired data; differences approximately normal

$$\bar{d} \pm t_{\alpha/2} s_d/\sqrt{n}.$$

7. Wilcoxon's signed-rank test, paired data

H_0:   The distribution of differences is symmetrical about D_0.
T.S.: $n > 50$

$$z = \frac{T - \mu_T}{\sigma_T},$$

where $\mu_T = \dfrac{n(n + 1)}{4}$ and $\sigma_T = \sqrt{\dfrac{n(n + 1)(2n + 1)}{24}}$

provided there are no tied ranks.

8. Independent samples: sample sizes for estimating $\mu_1 - \mu_2$ with a $100(1 - \alpha)\%$ confidence interval, $\bar{y}_1 - \bar{y}_2 \pm E$

$$n = \frac{2z_{\alpha/2}^2 \sigma^2}{E^2}.$$

9. Independent samples: sample sizes for a test of H_0: $\mu_1 - \mu_2 = D_0$
    a. One-sided test:

$$n = \frac{2\sigma^2(z_\alpha + z_\beta)^2}{\Delta^2}$$

    b. Two-sided test:

$$n = \frac{2\sigma^2(z_{\alpha/2} + z_\beta)^2}{\Delta^2}.$$

▼ SUPPLEMENTARY EXERCISES

 6.37   Two alloys, A and B, are used in the manufacture of steel bars. We wish to estimate the difference in load capacity of bars made of each alloy. A sample of 9 bars of alloy A had a mean load capacity of 28.5 tons and a standard deviation of 2.5 tons, whereas a sample of 13 bars of alloy B had an average load capacity of 23.2 tons with a standard deviation of 1.8 tons. Find a 90% confidence interval for the difference.

6.38   It is thought that exposure to ozone increases lung capacity. In order to investigate this possibility, a researcher exposed eight rats to ozone in the amount of 2 parts per million for a period of 30 days. The average lung capacity for these rats at the end of the 30 days was 9.4 mL, and the standard deviation was 0.8 mL. A control group of six rats did not have exposure to ozone and their lung capacity averaged 8.3 mL with a standard deviation of 0.7 mL.
    a. Is there sufficient evidence at a 5% significance level to support the original conjecture? Justify your answer with specific numerical values.
    b. Give the p-value for the hypothesis test.

6.39 In a study of the possible factors that influence the frequency of birds being hit by aircraft (which, ironically, is viewed as a hazard to the aircraft), the noise level of various jets was measured just seconds after their wheels left the ground. The jets were either wide-bodied or narrow-bodied. Twenty-two wide-bodied jets had noise levels averaging 106.4 decibels (dB) and a standard deviation of 3.3 dB, whereas ten narrow-bodied jets had noise levels averaging 114.0 dB with a standard deviation of 2.0 dB. Test whether the average noise levels in the two populations of jets are the same. Report the level of significance of the sample as evidence that the two types of jets have different noise levels.

6.40 A farmer was interested in determining which of two soil fumigants, A or B, is more effective in controlling the number of parasites in a particular agricultural crop. To compare the fumigants, four small fields were divided into equal areas: fumigant A was applied to one part and fumigant B, to the other. Crop samples of equal size were taken from each of the eight plots and the numbers of parasites per square foot were counted. The data are in the following table. Do the data provide sufficient evidence to indicate a difference in the mean level of parasites for the two fumigants?

Field	A	B
1	15	9
2	5	3
3	8	6
4	8	4

6.41 A psychologist was interested in comparing the average length of time it takes individuals to complete two different psychological checklists. From a relatively homogeneous group of 20 individuals, 10 were randomly assigned to list 1 and the other 10 to list 2. The appropriate checklists were then administered, and the amount of time required to complete the task was recorded for each individual. These data are summarized here. Find a 95% confidence interval for $(\mu_1 - \mu_2)$, the difference in the mean completion times. What assumptions must you make?

List 1	List 2
$\bar{y}_1 = 54.3$ minutes	$\bar{y}_2 = 48.1$ minutes
$n_1 = 10$	$n_2 = 10$
$s_1^2 = 16.0$	$s_2^2 = 12.2$

6.42 Refer to the data of Exercise 6.41. Construct a 99% confidence interval for $(\mu_1 - \mu_2)$.

6.43 Use the data of Exercise 6.4 to construct a 90% confidence interval for $(\mu_1 - \mu_2)$, the difference in mean blood pressure in rats subjected to the two environments.

6.44 An experiment was conducted to compare the mean lengths of time required for the bodily absorption of two drugs, A and B. Ten people were randomly selected and assigned to each drug treatment. Each of the ten persons in the sample received an oral dosage of the assigned drug, and the length of time (in minutes) for the drug to reach a specified level in the blood was recorded. The means and variances for the two samples are given in the accompanying table. Find a 95% confidence interval for the difference in mean times for absorption.

	Drug A	Drug B
Sample mean	27.2	33.5
Sample variance	16.36	18.92

 6.45 The accompanying computer output gives the drop in blood pressure for three groups of six rats from a strain of hypertensive rats. The six rats in the first group were treated with a low dose of an antihypertensive product, the second group with a higher dose of the same antihypertensive product, and the third group with an inert control. Note that the variability in blood pressure decreases, even for rats in the control group. Also note that negative values represent increases in blood pressure.

a. Draw conclusions for a comparison of the mean drop for the high-dose group and the control group.

b. Is there evidence to indicate a difference between the low- and high-dose groups? Explain.

```
DESCRIPTIVE STATISTICS

        FILE: Low-Dose Group
                    -51.00000
                     15.00000
                     48.00000
                     65.00000
                    -20.00000
                     75.00000

Low-Dose Group

  NUMBER:           6
    MEAN:      22.00000
  STD DEV:     49.95198
```

```
DESCRIPTIVE STATISTICS

        FILE: High-Dose Group
                     69.00000
                     24.00000
                     63.00000
                     87.50000
                     77.50000
                     40.00000

High-Dose Group

  NUMBER:           6
    MEAN:      60.16667
  STD DEV:     23.86769
```

```
DESCRIPTIVE STATISTICS

    FILE: Control Group
                      9.00000
                     12.00000
                     63.00000
                     77.50000
                     -7.50000
                     32.50000

Control Group

  NUMBER:        6
    MEAN:     26.58333
STD DEV:     29.66381
```

```
UNPAIRED T TEST
              Low-Dose Group   High-Dose Group
                 -51.00000        69.00000
                  15.00000        24.00000
                  48.00000        63.00000
                  65.00000        87.50000
                 -20.00000        77.50000
                  75.00000        40.00000

NO. OF OBSERVATIONS                 6.              6.
MEAN                             22.00000        60.16667
STANDARD DEVIATION               49.95198        23.86769
STANDARD ERROR                   20.39281         9.74394

RATIO OF MEANS (2ND/1ST)                          2.73485
DIFFERENCE OF MEANS (2ND-1ST)                    38.16667
STANDARD ERROR OF DIFFERENCE                     22.60113
95% CONFIDENCE INTERVAL          -12.18865,      88.521991
  FOR DIFFERENCE OF MEANS
RATIO OF VARIANCES (2ND/1ST)                      0.22831

T STATISTIC (EQUAL VARIANCES)                     1.68871
DEGREES OF FREEDOM                               10
PROBABILITY                                       0.12216
```

```
UNPAIRED T TEST
FILES
          High-Dose Group                    Control Group
             69.00000                           9.00000
             24.00000                          12.00000
             63.00000                          36.00000
             87.50000                          77.50000
             77.50000                          -7.50000
             40.00000                          32.50000

NO OF OBSERVATIONS                 6.              6.
MEAN                            60.16667        26.58333
STANDARD DEVIATION              23.86769        29.66381
STANDARD ERROR                   9.74394        12.11020

RATIO OF MEANS (2ND/1ST)                 0.44183
DIFFERENCE OF MEANS (2ND-1ST)          -33.58333
```

(continues)

```
STANDARD ERROR OF DIFFERENCE                    15.54353
95% CONFIDENCE INTERVAL               -68.21432,    1.047661
   FOR DIFFERENCE OF MEANS
RATIO OF VARIANCES (2ND/1ST)                     1.54466

T STATISTIC (EQUAL VARIANCES)                   -2.16060
DEGREES OF FREEDOM                              10
PROBABILITY                                      0.05605
```

6.46 Use the data of Exercise 6.45 to construct a 95% confidence interval for $(\mu_1 - \mu_3)$, the difference in population means for the low-dose group and the control group.

6.47 The elasticity of plastic can vary depending on the process by which the plastic is prepared. To compute the elasticity of plastic produced by two different processes, six samples from each process were analyzed for elasticity. These data are shown in the accompanying table. Do the data present sufficient evidence to indicate a difference in the mean elasticities for the two processes? Use $\alpha = .05$.

Process 1	Process 2
6.1	9.1
9.2	8.2
8.7	8.6
8.9	6.9
7.6	7.5
7.1	7.9
$\bar{y}_1 = 7.93$	$\bar{y}_2 = 8.03$
$s_1^2 = 1.46$	$s_2^2 = .61$

6.48 The purity of ore can vary greatly from one location to another. One determining factor, then, in choosing a site for mining would be the metal content of the ore. Two prospective locations were to be compared. Three ore samples were obtained from each location and analyzed to determine the metal content of the ore; see the accompanying table. Do the data provide sufficient evidence to indicate a difference in mean metal content for the two locations? Use $\alpha = .01$.

Location 1	50.1	49.6	51.2
Location 2	47.0	46.0	46.4

6.49 Refer to Exercise 6.48. Give the approximate level of significance for your test.

6.50 The amount of work accomplished on a construction job is frequently approximated by a visual estimate of the amount of material used per day. Six experienced men were employed to approximate the number of bricks used on two different jobs. Three were randomly assigned to job 1 and three to job 2. Each man, independent of the others, approximated the number of bricks used. The approximations (in thousands of bricks) are shown in the accompanying table. Assume that the men have been randomly selected from a very large set of experienced people. Thus, μ_1 is the mean of the large set of approximations produced by people who visually estimate the number of bricks in job 1. Similarly, μ_2 is a corresponding mean of a large set of approximations that could be acquired for job 2. Do these data provide evidence to indicate that the mean number of bricks approximated for job 1 differs from the mean approximation for job 2? Use $\alpha = .05$.

Job 1	Job 2
107.2	103.2
108.1	105.9
105.7	104.1
$\bar{y}_1 = 107.00$	$\bar{y}_2 = 104.40$
$s_1^2 = 1.47$	$s_2^2 = 1.89$

6.51 Refer to Exercise 6.47. Estimate the difference in the mean elasticities of the two processes, using a 95% confidence interval.

6.52 Refer to Exercise 6.48. Estimate the difference in the mean metal content of the two locations, using a 90% confidence interval.

6.53 Refer to Exercise 6.50. Construct a 95% confidence interval for the difference in the mean estimates for the two jobs.

6.54 An experiment was conducted to investigate the effect of the drug Propranolol in reducing hypertension in rats. Two groups of rats were studied. One group received the drug, and the other group served as the control group. Hypertension was induced in the rats by exposure to a cold environment. The extent of the induced hypertension in a given rat was measured by monitoring its blood pressure. After 6 weeks of cold exposure the sample blood pressure data were summarized for the two groups; see the accompanying table. Use these data to determine whether there is evidence to indicate that rats treated with Propranolol have less hypertension, on the average, than those that are untreated. Use $\alpha = .05$.

	Group 1 (received Propranolol)	Group 2 (control)
Sample size	7	5
Sample mean	129.43	167.60
Sample variance	583.95	249.30

6.55 Refer to the data of Exercise 6.6. Use the Minitab output here to conduct a two-sample t test for $H_0: \mu_1 - \mu_2 = 0$ versus $H_a: \mu_1 - \mu_2 \neq 0$. Give the p-value for your test and draw conclusions.

```
MTB > READ INTO C1 C2
DATA> 282     284
DATA> 279     285
DATA> 280     286
DATA> 278     277
DATA> 275     283
DATA> END

       5 ROWS READ
MTB > PRINT C1 C2

ROW       C1     C2

  1       282    284
  2       279    285
  3       280    286
  4       278    277
  5       275    283

MTB > TWOSAMPLE T FOR C1 VS C2;
SUBC> POOLED.

TWOSAMPLE T FOR C1 VS C2
      N      MEAN      STDEV    SE MEAN
C1    5     278.80     2.59      1.16
C2    5     283.00     3.54      1.58

95 PCT CI FOR MU C1 - MU C2: (-8.720, 0.3200)

TTEST MU C1 = MU C2 (VS NE): T= -2.14   P=0.064   DF= 8

POOLED STDEV =       3.10

MTB > STOP
```

6.56 A processor of recycled aluminum cans is concerned about the levels of impurities (principally other metals) contained in lots from two sources. Laboratory analysis of sample lots yields the following data (in kilograms of impurities per hundred kilograms of product):

Source I: 3.8   3.5   4.1   2.5   3.6   4.3   2.1   2.9   3.2   3.7   2.8   2.7
            (mean = 3.267, standard deviation = .676)
Source II: 1.8   2.2   1.3   5.1   4.0   4.7   3.3   4.3   4.2   2.5   5.4   4.6
            (mean = 3.617, standard deviation = 1.365)

a. Calculate the pooled variance and standard deviation.
b. Calculate a 95% confidence interval for the difference in mean impurity levels.
c. Can the processor conclude, using $\alpha = .05$, that there is a nonzero difference in means?

6.57 To compare the performance of microcomputer spreadsheet programs, teams of three students each choose whatever spreadsheet program they wish. Each team is given the same set of standard accounting and finance problems to solve. The time (in minutes) required for each team to solve the set of problems is recorded. The data shown here were obtained for the two most widely used programs; also displayed are the sample means, sample standard deviations, and sample sizes.

Program	Time										$\bar{y}$	s	n
A	39	57	42	53	41	44	71	56	49	63	51.50	10.46	10
B	43	38	35	45	40	28	50	54	37	29			
	36	27	52	33	31	30					38.00	8.67	16

a. Calculate the pooled variance.

b. Use this variance to find a 99% confidence interval for the difference of population means.

c. According to this interval, can the null hypothesis of equal means be rejected at $\alpha = .01$?

6.58 Redo parts b and c of Exercise 6.57 using a separate-variance (t') method. Which method is more appropriate in this case? How critical is it to use one or the other?

 6.59 Educators compared scores of nursing degree students with scores of students from diploma and associate degree programs on a state licensing board examination. By random sampling procedures, the educators drew a sample of five from those completing the degree program, resulting in a mean score of 400 with a standard deviation of 15. A random sample of five drawn from the associate degree program had a mean of 370 with a standard deviation of 30. Can the licensing board conclude that the mean score of nursing students completing the degree program is higher than the mean score of those who complete the associate program? Base your answer on the results of a statistical test. Give the approximate p-value for your test.

 6.60 An educator wants to compare the effects of two different teaching methods. Two classes of students are selected at random; class 1 receives method 1 and class 2 method 2. A comprehensive standard examination is administered to the two classes to determine the effectiveness of the two methods at the end of the test period. The relevant data are shown here.

	Class 1	Class 2
Sample size	$n_1 = 64$	$n_2 = 64$
Average test score	$\bar{y}_1 = 88$	$\bar{y}_2 = 80$
Sample variance	$s_1^2 = 56$	$s_2^2 = 56$

Determine a 95% confidence interval for the difference between the two population means on the basis of the difference between the two sample means. Would a 90% confidence interval be wider?

 6.61 A study was conducted to see if food prices charged in a ghetto area are higher than those charged in a more affluent suburban area. Food prices were obtained from nine stores in each area and a food price index computed. The summary results for each area were as follows:

Ghetto Area	Suburban Area
$n_1 = 9$	$n_2 = 9$
$\bar{y}_1 = 11.1$	$\bar{y}_2 = 10.5$
$s_2^2 = 2.5$	$s_2^2 = 1.5$

Conduct a statistical test of $H_0: \mu_1 - \mu_2 = 0$ versus $H_a: \mu_1 - \mu_2 > 0$. Show all steps in the test of hypothesis and state your conclusion in *non-statistical terms*. Use $\alpha = .05$. What type of error (Type I or II) could you have made?

6.62 We are given the following data summarizing information on two independent samples taken from populations whose variances are known to be equal:

	n	$\bar{y}$	s^2
Sample 1	6	30	60
Sample 2	4	20	60

From these data, the following was computed:

$$t = \frac{30 - 20}{7.75\sqrt{\dfrac{1}{6} + \dfrac{1}{4}}} = 2.$$

a. Show, by computing it, how the 7.75 in the preceding computation was obtained.
b. Suppose that the null hypothesis had been tested against a two-tailed alternative with $\alpha = .05$.
 i.   What would be the rejection region?
 ii. Would the null hypothesis be rejected?
c. Suppose the null hypothesis had been tested against the one-tailed alternative that the mean of population 1 was larger than the mean of population 2, with $\alpha = .05$.
 i.   What would be the rejection region for this test?
 ii. Would the null hypothesis be rejected?

 6.63 To test the research hypothesis that teacher expectation can improve student performance, two groups of 100 students were compared. Teachers of the experimental group were told that their students would show large IQ gains during the test semester, while teachers of the control group were told nothing. At the end of the semester, IQ change scores were calculated with the following results:

	Mean	Standard Deviation	Sample Size
Experimental	16.5	14.2	100
Control	7.0	13.1	100

a. Test the null hypothesis of no effect on mean IQ change scores.
b. State your conclusion in two ways:
 i.   in statistical terms
 ii. in nontechnical terms as you might explain it to an intelligent person who was not familiar with statistical terminology.

6.64 Those running for public office must now report the amount of money spent in each campaign. It has been reported that women candidates usually find it difficult to raise money and therefore spend less in their campaigns than men candidates. Suppose the accompanying data represent the campaign expenditures of a randomly selected group of men and women candidates who have just completed their campaigns for public office. Do the data support the claim that women candidates generally spend less in their campaigns for public office than men candidates?

a. Would you use a one-tailed test or two-tailed test of hypothesis in this case? Why?

Cost of Campaign (in thousands of dollars)	
Women	Men
138	134
127	137
134	135
125	140
	130
	134
Sum 524	810
Mean 131	135

b. State the null and alternative hypotheses in
  i. statistical terms or symbols
  ii. plain English.

6.65 Refer to Exercise 6.64. Summary data for the two samples are shown in the accompanying Minitab output along with the results of a t test for $\mu_1 - \mu_2$. What assumptions must we make in order to run a t test? Which one (if any) could cause a problem for these data?

```
MTB > SET INTO C1
DATA> 138   127   134   125
DATA> END
MTB > SET INTO C2
DATA> 134   137   135   140   130   134
DATA> END
MTB > PRINT C1 C2

ROW      C1      C2

  1     138     134
  2     127     137
  3     134     135
  4     125     140
  5             130
  6             134

MTB > TWOSAMPLE T FOR C1 VS C2;
SUBC> POOLED;
SUBC> ALTERNATIVE -1

TWOSAMPLE T FOR C1 VS C2
        N      MEAN      STDEV      SE MEAN
C1   4      131.00      6.06         3.03
C2   6      135.00      3.35         1.37

95 PCT CI FOR MU C1 - MU C2: (-10.78, 2.782)

TTEST MU C1 = MU C2 (VS LT): T= -1.36  P=0.11  DF= 8

POOLED STDEV =          4.56

MTB >STOP
```

6.66 Suppose you are the personnel manager for a company and you suspect a difference in the mean length of work time lost due to sickness for two types of employees: those who work at night versus those who work during the day. Particularly, you suspect that the mean time lost for the night shift exceeds the mean for the day shift. To check your theory, you randomly sample the records for ten employees for each shift category and record the number of days lost due to sickness within the past year. The data are shown next.

Night Shift	Day Shift
15	8
10	9
10	2
7	0
7	10
4	9
9	9
6	7
10	3
12	3

a. Would you use a one-tailed test or two-tailed test in your test hypothesis? Why?
b. What is the pooled estimate of σ?
c. Conduct the statistical test and show all parts of the test.
  i. null hypothesis
  ii. alternative hypothesis
  iii. test statistic and computations
  iv. rejection region
  v. conclusion
    a. in statistical terms
    b. in plain English that nonstatisticians can understand

6.67 Refer to Exercise 6.66. Based on your decision could you have made
a. A Type I error?
b. A Type II error?
c. Both Type I and Type II errors?
d. Neither error?
(Answer each with yes or no.)

6.68 Refer to the data of Exercise 6.4. Use the output shown here to determine a 95% confidence interval for $\mu_1 - \mu_2$.

```
MTB > READ INTO C3 C4
DATA> 152    384
DATA> 157    369
DATA> 179    354
DATA> 182    375
DATA> 176    366
DATA> 149    423
DATA> END
                              (continues)
```

```
          6 ROWS READ
MTB > PRINT C3 C4

ROW       C3       C4

  1       152      384
  2       157      369
  3       179      354
  4       182      375
  5       176      366
  6       149      423

MTB > TWOSAMPLE T FOR C3 VS C4;
SUBC> POOLED;
SUBC> ALTERNATIVE -1.

TWOSAMPLE T FOR C3 VS C4
        N       MEAN     STDEV   SE MEAN
C3   6          165.8     14.8     6.03
C4   6          378.5     24.0     9.78

95 PCT CI FOR MU C3 - MU C4: (-238.3, -187.1)

TTEST MU C3 = MU C4 (VS LT): T= -18.51   P=0.0000   DF= 10

POOLED STDEV =          19.9

MTB > STOP
```

6.69 A study was carried out to determine whether nonworking wives from middle-class families have more voluntary association memberships than nonworking wives from working-class families. A random sample of housewives was obtained, and each was asked for information about her husband's occupation and her own memberships in voluntary associations. On the basis of their husbands' occupations, the women were divided into middle-class and working-class groups, and the mean number of voluntary association memberships was computed for each group.

For the 15 middle-class women, the mean number of memberships per woman was $\bar{y}_1 = 3.4$ with $s_1 = 2.5$. For the 15 working-class wives, $\bar{y}_2 = 2.2$ with $s_2 = 2.8$. Use these data to construct a 95% confidence interval for $\mu_1 - \mu_2$.

6.70 A regional IRS auditor ran a test on a sample of returns filed by March 15 to determine whether the average refund for taxpayers is larger this year than last year. Sample data are shown here for a random sample of 100 returns for each year.

	Last Year	This Year
Mean	320	410
Variance	300	350
Sample size	100	100

a. In a test of hypothesis, would you use a one-tailed or two-tailed test? Why?

b. What assumptions are required to conduct a t test of H_0: $\mu_1 - \mu_2 = 0$? Do you think the assumptions hold, and why (or why not)?

6.71 Miss American Pageant officials maintain that their pageant is not a beauty contest and that talent is more important than beauty when it comes to success in the pageant. In an effort to evaluate the assertion, a random sample of 55 preliminary talent-competition winners and a random sample of 53 preliminary swimsuit-competition winners were taken to see if there was a significant difference in the mean amount won for the two groups. For the 55 preliminary talent-competition winners the mean amount was $8,645 with standard deviation of $5,829; for the 53 preliminary swimsuit winners the mean amount won was $9,198 with standard deviation of $8,185. Compute a 95% confidence interval for the difference in the mean amount won by the two groups. Does your confidence interval confirm what the pageant officials contend?

6.72 A visitor to the United States from France insisted that recordings made in Europe are likely to have selections with longer playing times than recordings made in the United States. In order to verify or contradict the contention, a random sample of selections was taken from a group of records produced in France and Germany, and another random sample of selections was taken from American-produced records. The results of the samples were as given.

	Foreign Produced	American Produced
Number in sample	14	14
Mean playing time in seconds	207.45	182.54
Standard deviation	41.43	37.32

Do the foreign-produced selections have longer mean playing times? Use $\alpha = .05$.

6.73 A major federal agency located in Washington, D.C., regularly conducts classes in PL/1, a computer programming language used in the programs written within the agency. One week, the course was taught by an individual associated with an outside consulting firm. The following week, a similar course was taught by a member of the computer staff of the agency. The following results were achieved by the classes:

Taught by outsider	38	42	53	37	36	48	47	47	44
Taught by staff member	46	33	38	60	58	52	44	45	51

The values represent scores aggregated over the 1-week course out of a potential maximum of 64. Do the data present sufficient evidence to indicate a difference in teaching effectiveness, assuming that the scores reflect teaching effectiveness? Use $\alpha = .05$.

6.74 Company officials are concerned about the length of time a particular drug retains its potency. A random sample (sample 1) of 10 bottles of the product is drawn from current production and analyzed for potency. A second sample (sample 2) is obtained, stored for one year, and then analyzed. The readings obtained are as follows:

Sample 1	10.2	0.5	10.3	10.8	9.8	10.6	10.7	10.2	10.0	10.6
Sample 2	9.8	9.6	10.1	10.2	10.1	9.7	9.5	9.6	9.8	9.9

The data are analyzed by a standard program package (SAS). The relevant output is shown here:

```
                        TTEST PROCEDURE

   VARIABLE: POTENCY

   SAMPLE        N        MEAN       STD DEV       STD ERROR

    1           10     10.37000000   0.32335052    0.10125421
    2           10      9.83000000   0.24060110    0.07408475

   MINIMUM       MAXIMUM       VARIANCES       T      DF      PROB > |T|

   9.80000000   10.80000000    UNEQUAL       4.2368   16.6    0.0406
   9.50000000   10.20000000    EQUAL         4.2368   18.0    0.0505

   FOR HO:  VARIANCES ARE EQUAL.    F = 1.81 WITH 9 AND 9 DF
   PROB > F =0.3917
```

a. Identify the sample means and standard deviations
b. Locate the value of the t statistic. Is the pooled-variance t statistic identified as "equal variance" or "unequal variance"?
c. Locate the value of the t' statistic.
d. Why are these two statistics equal in this case?

 6.75 Two possible methods for retrofitting jet engines to reduce noise are being considered. Identical planes are fitted with two systems. Noise-recording devices are installed directly under the flight path of a major airport. Each time one of the planes lands at the airport, a noise level is recorded. The data are analyzed by a computer software package (SAS). The relevant output is:

```
VARIABLE DBREAD

SYSTEM  M      MEAN   STD DEV  STD ERROR    MINIMUM     MAXIMUMVARIANCES        T     DF  PROB > |T|

 H     42 100.90476190 2.99438111 0.46204304 95.00000000 110.00000000UNEQUAL  4.4491  21.5  0.0002
 R     20  92.50000000 8.19178022 1.63173774 79.00000000 111.00000000EQUAL    5.9126  60.0  0.0001
```

a. Locate the t statistic.
b. Locate the t' statistic.
c. Can the research hypothesis of unequal means be supported using $\alpha = .01$? Does it matter which statistic is used?

6.76 A study was conducted on 16 dairy cattle. Eight cows were randomly assigned to a liquid regimen of water only (group 1); the others received liquid whey only (group 2). In addition, each animal was given 7.5 kg of grain per day and allowed to graze on hay at will. Although no significant differences were observed between the groups in the dairy-milk-production gauges, such as milk production and fat content of the milk, the following data on daily hay consumption (in kilograms/cow) were of interest:

Group 1	15.1	14.9	14.8	14.2	13.1	12.8	15.5	15.9
Group 2	6.8	7.5	8.6	8.4	8.9	8.1	9.2	9.5

Use these sample data to test the research hypothesis that there is a difference in mean hay consumption for the two diets. Use $\alpha = .05$.

6.77 Refer to Example 6.13. Suppose we wish to detect only $\mu_1 - \mu_2 > 0$. Determine the sample size required when $\alpha = .05$ and $\beta \leq .10$ when $\mu_1 - \mu_2 \geq .5$.

6.78 An industrial concern has experimented with several different mixtures of the four components magnesium, sodium nitrate, strontium nitrate, and a binder, that comprise a rocket propellant. The company has found that two mixtures in particular give higher flare-illumination values than the others. Mixture 1 consists of a blend composed of the proportions .40, .10, .42, and .08, respectively, for the four components of the mixture; mixture 2 consists of a blend using the proportions .60, .27, .10, and .05. Twenty different blends (10 of each mixture) are prepared and tested to obtain the flare-illumination values. These data appear below (in units of 1,000 candles).

Mixture 1	185	192	201	215	170	190	175	172	198	202
Mixture 2	221	210	215	202	204	196	225	230	214	217

a. Plot the sample data. Which test(s) could be used to compare the mean illumination values for the two mixtures?

b. Give the level of significance of the test and interpret your findings.

6.79 Refer to Exercise 6.78. Instead of conducting a statistical test, use the sample data to answer the question, "What is the difference in mean flare illumination for the two mixtures?"

6.80 Refer to Example 6.3. Suppose the seventh untreated animal died before the study was completed. Analyze the remaining observations to compare the two population means. Assume that $\sigma_1^2 \neq \sigma_2^2$. Give the level of significance for your test.

6.81 Refer to the computer printout for the statistical test of the data in Exercise 6.9.

a. Were there any problems in running a t test?

b. Compare these computer results to your calculations for Exercise 6.9.

```
                                    TWO SAMPLE T-TEST

                                     TTEST PROCEDURE


VARIABLE: LEADPCT

GROUP      N       MEAN      STD DEV     STD ERROR      MINIMUM     MAXIMUM   VARIANCES        T    DF  PROB > | T |

CONTROL   10   5.06000000   1.18902388   0.37600236   3.100000000   6.50000000  UNEQUAL   -5.8507  17.2   0.0001
EXPERMT   10   8.56000000   1.47135614   0.46528366   6.500000000  10.60000000  EQUAL     -5.8507  18.0   0.0001

FOR HO: VARIANCES ARE EQUAL,   F´= 1.53 WITH 9 AND 9 DF      PROB > F´ = 0.5356
```

c. Give the value of the test statistic and the level of significance for a t test of the research hypothesis that the experimental mean is greater than the control mean.

Control group	5.4	6.2	3.1	3.8	6.5	5.8	6.4	4.5	4.9	4.0
Experimental group	8.8	9.5	10.6	9.6	7.5	6.9	7.4	6.5	10.5	8.3

6.82 A study of anxiety was conducted among residents of a southeastern metropolitan area. Each person selected for the study was asked to check a "yes" or a "no" for the presence of each of 12 anxiety symptoms. Anxiety scores ranged from 0 to 12, with higher scores related to higher perceived presence of any anxiety symptoms. The results for a random sample of 50 residents, categorized by gender, are summarized below. Use these data to test the research hypothesis that the mean perceived anxiety score is different for males and females. Give the level of significance for your test.

	Sample Size	Mean	Standard Deviation
Female	26	5.26	3.2
Male	24	7.02	3.9

6.83 A clinical trial was conducted to determine the effectiveness of drug A in the treatment of symptoms associated with alcohol withdrawal. A total of 30 patients were treated (under blinded conditions) with drug and another 30 with an identical-appearing placebo. The average symptom score for the two groups after 1 week of therapy was 1.5 and 6.3, respectively. (Note: Higher symptom scores indicate more withdrawal "problems.") The corresponding standard deviations were 3.1 and 4.2.

a. Compare the mean total symptom scores for the two groups. Give the p-value for a two-sample t test of $H_0: \mu_1 - \mu_2 = 0$ versus $H_a: \mu_1 - \mu_2 < 0$. Draw conclusions.

b. Suppose the average total symptoms scores were 6.8 and 12.2 prior to therapy. How would this affect your conclusions? How could you guard against possible baseline (pretreatment) differences?

6.84 Two analysts, supposedly of identical abilities, each measure the parts per million of a certain type of chemical impurity in drinking water. It is claimed that analyst 1 tends to give higher readings than analyst 2. To test this theory, each of six water samples is divided and then analyzed by both analysts separately. The data are shown in the accompanying table (readings in ppm).

Water Sample	Analyst 1	Analyst 2
1	31.4	28.1
2	37.0	37.1
3	44.0	40.6
4	28.8	27.3
5	59.9	58.4
6	37.6	38.9

a. Is there evidence to indicate that analyst 1 reads higher on the average than analyst 2? Give the level of significance for your test.
b. What would be the conclusion using a Wilcoxon test? Compare your results to part a.

 6.85 A single leaf was taken from each of 11 different tobacco plants. Each was divided in half; one half was chosen at random and treated with preparation I and the other half received preparation II. The object of the experiment was to compare the effects of the two preparations of mosaic virus on the number of lesions on the half leaves after a fixed period of time. These data are recorded below. For $\alpha = .05$, use Wilcoxon's signed-rank test to examine the research hypothesis that the distributions of lesions are different for the two populations.

	Number of Lesions on the Half Leaf	
Tobacco Plant	Preparation I	Preparation II
1	18	14
2	20	15
3	9	6
4	14	12
5	38	32
6	26	30
7	15	9
8	10	2
9	25	18
10	7	3
11	13	6

 6.86 An investigator plans to compare the mean number of particles of effluent in water collected at two different locations in a water treatment plant. If the standard deviation for particle counts is expected to be approximately 6 for the counts in samples taken at each of the locations, determine the sample sizes required to estimate the mean difference in particles of effluent using a 99% confidence interval of width 1 (particle).

6.87 The weight gains for $n_1 = n_2 = 8$ rats tested on diets 1 and 2 are summarized here. Set up a statistical test for $\mu_1 - \mu_2$, the difference in average weight gained for the two diets. Use $\alpha = .05$ and draw conclusions.

Diet 1		Diet 2
25	Σy	26.2
.005	s	.045
8	n	8

6.88 Refer to the computer simulation in Example 6.5. Another computer simulation study of 1,000 samples was run using the same sample sizes but different population standard deviations. The independent normal populations were as shown here:

Population
1        $\mu_1 = 100$ $\sigma_1 = 15$ $n_1 = 10$
2        $\mu_2 = 100$ $\sigma_2 = 10$ $n_2 = 10$

For each run of the simulation study, the pooled t test and separate-variance t test were run. The number (%) of times t and t' were rejected at the upper and lower .05 levels is recorded here.

	Pooled t Test	Separate-Variance t Test
H_0 rejected, $H_a: \mu_1 - \mu_2 < 0$	43 (4.3%)	44 (4.4%)
H_0 rejected, $H_a: \mu_1 - \mu_2 > 0$	48 (4.8%)	46 (4.6%)

a. Before reviewing the results of the simulation study, how well do you think the pooled t test will perform even though one of the assumptions is obviously not satisfied? Explain.

b. Which test performed better? Which would you recommend and why?

6.89 The following memorandum opinion on statistical significance was issued by the judge in a trial involving many scientific issues. The opinion has been stripped of some legal jargon and has been taken out of context. Still, it can give us an understanding of how others deal with the problem of ascertaining the meaning of statistical significance. Read this memorandum and comment on the issues raised regarding statistical significance.

Memorandum Opinion

This matter is before the Court upon two evidentiary issues that were raised in anticipation of trial. First, it is essential to determine the appropriate level of statistical significance for the admission of scientific evidence.

With respect to statistical significance, no statistical evidence will be admitted during the course of the trial unless it meets a confidence level of 95%.

Every relevant study before the court has employed a confidence level of at least 95%. In addition, plaintiffs concede that social scientists routinely utilize a 95% confidence level. Finally, all legal authorities agree that statistical evidence is inadmissible unless it meets the 95% confidence level required by statisticians. Therefore, because plaintiffs advance no reasonable basis to alter the accepted approach of mathematicians to the test of statistical significance, no statistical evidence will be admitted at trial unless it satisfies the 95% confidence level.

6.90 Certain baseline determinations were made on 182 patients entered in a study of survival in males suffering from congestive heart failure. At the time these data were summarized, 88 deaths had been observed. This table summarizes the baseline data for the survivors and nonsurvivors. The variables listed below "heart rate" are measures of the severity of the heart failure. The arrows to the left of each variable indicates the direction of improvement.

a. Discuss these baseline findings.

b. What assumptions have the authors made when doing these t tests?

Baseline Characteristics of Patients with Severe Chronic Left-Ventricular Failure Due to Cardiomyopathy

Variable	Nonsurvivors ($n = 88$)	Survivors ($n = 94$)	t Test p-Value
Age (y)	57 ± 10	56 ± 8	NS
Duration of symptoms (mo)	45 ± 43	39 ± 27	NS
Heart rate (beats/min)	87 ± 15	83 ± 16	NS
↓Mean arterial pressure (mm Hg)	87 ± 13	94 ± 13	<0.001
↓Left-ventricular filling pressure (mm Hg)	29 ± 7	24 ± 9	<0.001
↑Cardiac index (L/min/m^2)	2.0 ± 0.7	2.5 ± 0.8	<0.001
↑Stroke volume (mL/beat)	45 ± 16	59 ± 5	<0.001
↓Systemic vascular resistance (units)	25 ± 10	21 ± 8	<0.01
↑Stroke work (g-m)	35 ± 19	56 ± 33	<0.001

Values are listed as mean $\pm$ standard deviation.

6.91 Hospital administrators studied the patterns of length of hospital stays with particular attention paid to those patients having health-maintenance organization (HMO) payment sources versus those with non–HMO payment sources. The graph shown here summarizes the sample data.

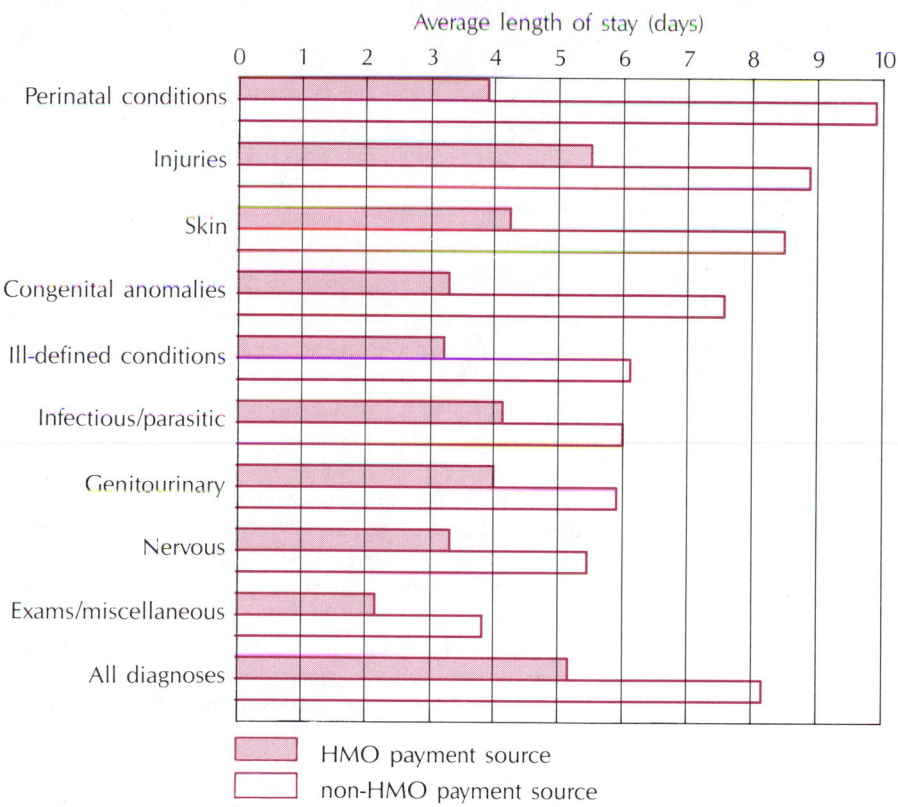

Sources: American Hospital Assn.; Twin Cities Metropolitan Health Board.

a. What general observations would you draw from the graph? What additional information would you need to make more definitive statements regarding these results?

b. Suppose that across all diagnoses, the sample statistics were as shown below. Use these data to test $H_0: \mu_1 - \mu_2 = 0$ versus $H_a: \mu_1 - \mu_2 \neq 0$. Give the p-value for your test.

	Sample Mean	Sample Size	Sample Standard Deviation
HMO	5.0 (days)	120	1.3
Non-HMO	8.1	130	1.9

6.92 Refer to Exercise 6.91. Also run the t' test and compare your results. Which (if any) test is better for these data?

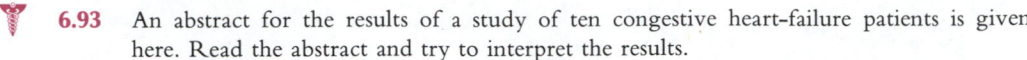

6.93 An abstract for the results of a study of ten congestive heart-failure patients is given here. Read the abstract and try to interpret the results.

Abstract

An experimental compound was studied in ten patients suffering from congestive heart failure. Certain variables were measured at baseline and then four hours after intravenous treatment with the compound. The compound was shown to increase cardiac index from 11.1 to 34.3% from a baseline average of 2.41 ± 0.49 L/min/m^2 ($p < .01$), heart rate by 6–10% from 72 ± 12 beats/min ($p < .02$) and decreased pulmonary capillary wedge pressure by 15.3–24.2% from 18.7 ($p < .001$).

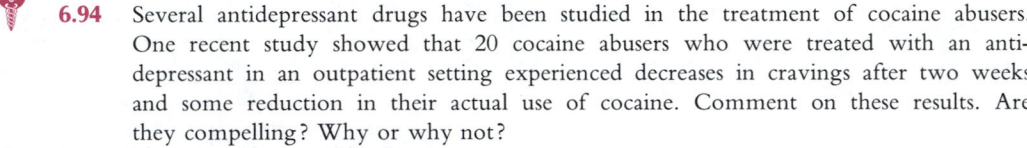

6.94 Several antidepressant drugs have been studied in the treatment of cocaine abusers. One recent study showed that 20 cocaine abusers who were treated with an antidepressant in an outpatient setting experienced decreases in cravings after two weeks and some reduction in their actual use of cocaine. Comment on these results. Are they compelling? Why or why not?

6.95 In April 1986, the *Australian Journal of Statistics* (Vol. 30, No. 1, pp. 23–44) published the results of a study of S. R. Butler and H. W. Marsh on reading and arithmetic achievement for students from non–English-speaking families. All kindergarten students from seven public schools in Sydney, Australia were included in the original sample of 392 children. Reading and arithmetic achievement tests were administered at the start of the study during kindergarten and then at years 1, 2, 3, and 6 of the primary school.

   The table shown here gives the characteristics of the 286 of the original 392 students who were available for testing at year 6 ($n = 226$ students from English-speaking families, and $n = 60$ from non–English-speaking families.)

	Group	
Characteristics	English-Speaking Family ($n = 226$) $\bar{y}$	Non–English-Speaking Family ($n = 60$) $\bar{y}$
Age (in months)	67.17	67.15
Gender (1 = male, 2 = female)	1.50	1.55
Number of children in family	2.54	2.62
Ordinal position in family (1 = oldest child, etc.)	1.89	1.82
Father's occupation (1 = most skilled, 17 = least skilled)	8.26*	11.50
Peabody Picture Vocabulary IQ	99.26*	74.45

*Statistically significant, $p < .01$

    a. Can you suggest better ways to summarize these baseline characteristics?

    b. What test(s) may have been used to compare these characteristics?

    c. What other characteristics could or should have been examined to make a direct comparison of reading and arithmetic achievement?

    d. What effect (if any) might the attrition rate have on the study results? Recall 106 (27%) of the original 392 students were not available for testing at year 6.

6.96   Refer to the clinical trials data base in the Appendix. Use the HAM-D total score data to conduct a statistical test of H_0: $\mu_D - \mu_A = 0$ vs H_a: $\mu_D - \mu_A > 0$; that is, we want to know whether the placebo group (D) has a higher (worse) mean total depression score at the end of the study than the group receiving treatment A. Use $\alpha = .05$. What are your conclusions?

6.97   Refer to Exercise 6.96 and repeat this same comparison with the placebo group for treatment B, and then for treatment C. Give the p-value for each of these tests. Which of the three treatment groups (A, B, or C) appears to have the lowest mean HAM-D total score?

6.98   Use the clinical trials data base to construct a 95% confidence interval for $\mu_D - \mu_A$ based on the HAM-D anxiety score data. What can you conclude about $\mu_D - \mu_A$ based on this interval?

6.99   Refer to the clinical trials data base. Compare the mean ages for treatment groups B and D using a two-sided statistical test. Set up all parts of the test using $\alpha = .05$; draw a conclusion. Why might it be important to have patients with similar ages in the different treatment groups when studying the effects of several drug products on the treatment of depression?

6.100   Refer to Exercise 6.99. What other variables should be comparable among the treatment groups in order to draw conclusions about the effectiveness of the drug products for treating depression?

INFERENCES ABOUT POPULATION VARIANCES

7.1 ▼ INTRODUCTION

When people think of statistical inference, they usually think of inferences concerning population means. However, the population parameter that answers an experimenter's practical questions will vary from one situation to another, and sometimes the variability of a population is more important than its mean. It should also be noted that product quality is often defined in terms of low variability. For example, the producer of a drug product is certainly concerned with controlling the mean potency of tablets, but he or she must also worry about the variation in potency from one tablet to another. Excessive potency or an underdose could be very harmful to a patient. Hence, the manufacturer would like to produce tablets with the desired mean potency and with as little variation in potency (as measured by σ or σ^2) as possible.

Inferential problems about a population variance are similar to those for a population mean. We can estimate or test hypotheses about a single population variance or compare two variances.

7.2 ▼ ESTIMATION AND TESTS FOR A POPULATION VARIANCE

The sample variance

$$s^2 = \frac{\sum (y - \bar{y})^2}{n - 1}$$

FIGURE 7.1

Upper-Tail and Lower-Tail Values of Chi-Square

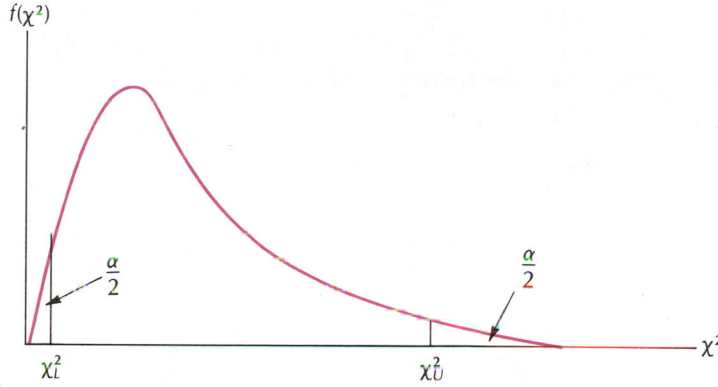

can be used for inferences concerning a population variance σ^2. For a random sample of n measurements drawn from a normal population with mean μ and **point estimate for σ^2** variance σ^2, s^2 provides a **point estimate for σ^2**. In addition, the quantity $(n-1)s^2/\sigma^2$ follows a chi-square distribution with df $= n - 1$. We will not give the mathematical formula for the chi-square (χ^2, where χ is the Greek letter chi) probability distribution but instead will list its properties.

1. The chi-square distribution is a nonsymmetrical distribution (see Figure 7.1).
2. There are many chi-square distributions. We obtain a particular one by specifying the degrees of freedom (df). Figure 7.2 shows a chi-square distribution with df $= 4$.

FIGURE 7.2

Chi-Square Distribution for df $= 4$

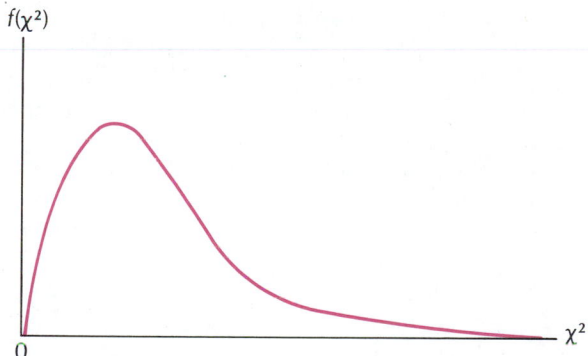

Upper-tail values of the chi-square distribution can be found in Table 5 in the Appendix. Entries in the table are values of χ^2 that have an area a to the right under the curve. The degrees of freedom are specified in the left column of the table, and values of a are listed across the top of the table. Thus, for df $= 14$, the value of chi-square with an area $a = .10$ to its right under the curve is 21.06 (see Figure 7.3).

FIGURE 7.3
Critical Value of the Chi-Square Distribution for $a = .10$ and df $= 14$

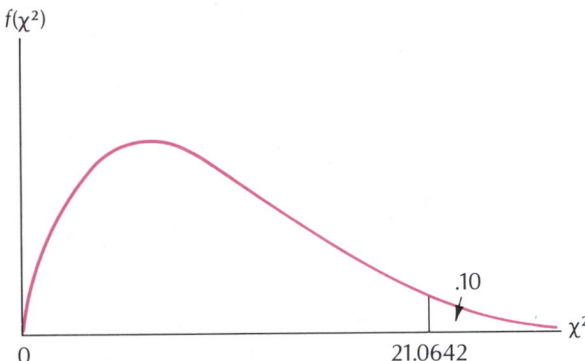

We can use this information to form a confidence interval for σ^2. Because the chi-square distribution is not symmetrical, the confidence intervals based on this distribution don't have the usual form, estimate $\pm$ error, as we saw for μ and $\mu_1 - \mu_2$.

General Confidence Interval for σ^2 (or σ) with Confidence Coefficient $(1 - \alpha)$

$$\frac{(n-1)s^2}{\chi_U^2} < \sigma^2 < \frac{(n-1)s^2}{\chi_L^2},$$

where χ_U^2 is the upper-tail value of chi-square for df $= n - 1$ with area $\alpha/2$ to its right, and χ_L^2 is the lower-tail value with area $\alpha/2$ to its left (see Figure 7.1). We can determine χ_U^2 and χ_L^2 for a specific value of df by obtaining the critical value in Table 5 of the Appendix corresponding to $a = \alpha/2$ and $a = 1 - \alpha/2$, respectively.

Note: The confidence interval for σ is found by taking square roots throughout.

EXAMPLE 7.1 ▼

The variability in milk production for a 305-day lactation period was observed for a random sample of 15 Holstein cows. Use the milk-yield data in Table 7.1 to estimate σ^2, the population variance of milk yields, using a 95% confidence interval.

T A B L E 7.1

Milk Production Data (in 1,000 pounds), Example 7.1

12.928	13.812	11.036
12.120	14.358	9.248
14.972	8.998	9.980
14.044	10.620	11.990
14.788	14.744	14.786

Solution For these data, we find

$$\sum y = 188.424 \qquad \sum y^2 = 2{,}431.470.$$

Substituting into the shortcut formula for s^2 (Chapter 3), we have

$$s^2 = \frac{1}{n-1}\left[\sum y^2 - \frac{(\sum y)^2}{n}\right] = \frac{1}{14}\left[2{,}431.470 - \frac{(188.424)^2}{15}\right] = 4.612.$$

The confidence coefficient for our example is $1 - \alpha = .95$. The upper-tail chi-square value can be obtained from Table 5 in the Appendix, for df $= n - 1 = 14$ and a $= \alpha/2 = .025$. Similarly, the lower-tail chi-square value is obtained from Table 5 with a $= 1 - \alpha/2 = .975$. Thus,

$$\chi_U^2 = 26.12 \qquad \chi_L^2 = 5.629.$$

The 95% confidence interval is then

$$\frac{14(4.612)}{26.12} < \sigma^2 < \frac{14(4.612)}{5.629}$$

or

$$2.472 < \sigma^2 < 11.471$$

thousand pounds. Thus, we estimate the population variance for milk yields to lie between 2,472 and 11,471 pounds. ▲

In addition to estimating a population variance, we can construct a statistical test of the null hypothesis that σ^2 equals a specified value, σ_0^2. This test procedure is summarized here.

Statistical Test for σ^2 (or σ)

H_0: $\sigma^2 = \sigma_0^2$ (σ_0^2 is specified)

H_a: **1.** $\sigma^2 > \sigma_0^2$
     2. $\sigma^2 < \sigma_0^2$
     3. $\sigma^2 \neq \sigma_0^2$

T.S.: $\chi^2 = \dfrac{(n-1)s^2}{\sigma_0^2}$

R.R.:   For a specified value of α,

1. reject H_0 if χ^2 is greater than χ_U^2, the upper-tail value for $a = \alpha$ and df $= n - 1$
2. reject H_0 if χ^2 is less than χ_L^2, the lower-tail value for $a = 1 - \alpha$ and df $= n - 1$
3. reject H_0 if χ^2 is greater than χ_U^2, based on $a = \alpha/2$ and df $= n - 1$, or less than χ_L^2, based on $a = 1 - \alpha/2$ and df $= n - 1$.

EXAMPLE 7.2

A manufacturer of a specific pesticide useful in the control of household bugs claims that its product retains most of its potency for a period of at least 6 months. More specifically, it claims that the drop in potency from 0 to 6 months will vary in the interval from 0% to 8%. To test the manufacturer's claim, a consumer group obtained a random sample of 20 containers of pesticide from the manufacturer. Each can was tested for potency and then stored for a period of 6 months at room temperature. After the storage period, each can was again tested for potency. The drop in potency was recorded for each can and the sample variance for the drops in potencies was computed to be $s^2 = 6.2$. Use these data to determine whether there is sufficient evidence to indicate that the population of potency drops has more variability than that claimed by the manufacturer. Use $\alpha = .05$.

Solution   The manufacturer has claimed that the population of potency reductions has a range of 8%. Dividing the range by 4, we obtain an approximate population standard deviation of $\sigma = 2\%$ (or $\sigma^2 = 4$).

The appropriate null and alternative hypotheses are

H_0: $\sigma^2 = 4$ (i.e., we assume the manufacturer's claim is correct)
H_a: $\sigma^2 > 4$ (i.e., there is more variability than claimed by the manufacturer).

Using the computed sample variance based on 20 observations, the test statistic and rejection region are as follows:

T.S.: $\chi^2 = \dfrac{(n-1)s^2}{\sigma_0^2} = \dfrac{19(6.2)}{4} = 29.45$

R.R.: For $\alpha = .05$, we will reject H_0 if the computed value of chi-square is greater than 30.14, obtained from Table 5 in the Appendix for $a = .05$ and df $= 19$.

Conclusion: Since the computed value of chi-square, 29.45, is less than the critical value, 30.14, there is insufficient evidence to reject the manufacturer's claim, based on $\alpha = .05$. However, the consumer group is not prepared to accept H_0: $\sigma^2 = 4$. Rather, since $s^2 = 6.2$ and the p-value of the test is $.05 < p < .10$, it would be wise to do additional testing with a larger sample size before reaching a definite conclusion. ▲

The χ^2 methods for inferences about σ^2 (and σ) are based on the assumption that the population distribution is normal. This assumption is much more important for inferences about variances than it is for inferences about means. The Central Limit Theorem helps greatly in normalizing the sampling distribution of a mean, but there is no comparable theorem for variances. Population nonnormality, in the form of skewness or heavy tails, can have serious effects on the nominal significance and confidence probabilities for a variance (or standard deviation). If a plot of the sample data shows substantial skewness or outliers, the nominal probabilities given by the χ^2 distribution are suspect. There are some computationally elaborate inference procedures about a variance (such as the so-called jackknife method) that are less sensitive to the normality assumption. These may well replace the χ^2-based methods as computation costs decrease and computer programs become more widely available.

E X A M P L E   7.3   ▼

A simulation study involves 1,000 samples of size 51 each from a moderately outlier-prone population. The population variance is 64.8. A χ^2 test of the variance is performed for each sample. The results are as shown below. What do the results indicate about the test?

```
one-tail:

     number of times H₀: "variance = 64.8 is rejected in favor of
 alpha      "variance < 64.8"        "variance > 64.8"        total (alpha doubled)
 0.100           205                      289                       494
 0.050           162                      221                       383
 0.025           127                      171                       298
 0.010           106                      111                       217
 0.005            87                       86                       173
```

> **Solution**   The nominal α probabilities are much, much smaller than the actual fractions. This is evidence that the claimed α value of the χ^2 test of a variance is quite sensitive to nonnormality.   ▲

▼ EXERCISES

Basic Techniques

7.1   Suppose that Y has a χ^2 distribution with 27 df.
  a. Find $P(Y > 46.96)$.
  b. Find $P(Y > 18.11)$.
  c. Find $P(Y < 12.88)$.
  d. What is $P(12.8786 < Y < 46.9630)$?

7.2   For a χ^2 distribution with 11 df,
  a. Find $\chi^2_{.025}$.
  b. Find $\chi^2_{.975}$.

7.3   Suppose that Y has a χ^2 distribution with 277 df. Find approximate values for $\chi^2_{.025}$ and $\chi^2_{.975}$.

7.4   A sample of 25 observations is drawn from a normal population with unknown mean μ and variance σ^2. Define

$$\chi^2 = \frac{(n-1)s^2}{\sigma^2}.$$

Find the following probabilities:
  a. $P(\chi^2 > 12.4)$
  b. $P(\chi^2 < 36.4)$
  c. $P(9.89 < \chi^2 < 45.56)$

Applications

  7.5   A packaging line fills nominal 32-ounce tomato juice jars with an actual mean of 32.30 ounces. The process should have a standard deviation smaller than .15 ounce per jar (a larger standard deviation leads to too many underweight and overfilled jars). Samples of 61 jars are regularly taken to test the process. One such sample yields a sample mean of 32.28 ounces and a standard deviation of .132 ounce. Does this indicate (using $\alpha = .05$) that $\sigma < .15$? Carry out a formal hypothesis test.

7.6   Suppose that the research hypothesis in Exercise 7.5 is formulated as $\sigma > .15$. Does this reformulation tend to be more or less generous in terms of what sample results cause the packaging line to be shut down for adjustment?

7.7   A certain part for a small assembly should have a diameter of 4.000 mm, and a maximum standard deviation of .011 mm is allowed by specifications. A random sample of 26 parts shows the following diameters:

3.952	3.978	3.979	3.984	3.987	3.991	3.995	3.997	3.999	3.999	3.999
4.000	4.000	4.000	4.001	4.001	4.002	4.002	4.003	4.004	4.006	4.009
4.010	4.012	4.023	4.041							

a. Calculate the sample mean and standard deviation.

b. Can the research hypothesis that $\sigma > .011$ be supported (at $\alpha = .05$) by these data? State all parts of a statistical hypothesis test.

7.8 Calculate 90% confidence intervals for the true variance and for the true standard deviation for the data of Exercise 7.7.

7.9 Plot the data of Exercise 7.7. Does the plot suggest any violation of the assumptions underlying your answers to Exercises 7.7 and 7.8? Would such a violation have a serious effect on the validity of your answers?

7.10 Baseballs vary somewhat in their rebounding coefficient. A "dead ball" has a relatively low rebound, but "rabbit ball" has a high rebound. A standard test has been developed. A purchaser of large quantities of baseballs requires that the mean value be 85 and the standard deviation be less than 2 units. A sample of 81 baseballs is tested. The mean value is 84.91 and the standard deviation is 1.80. Can the research hypothesis that $\sigma < 2$ be supported using $\alpha = .05$? Carry out the steps of a formal hypothesis test.

7.11 Place bounds on the p-value in Exercise 7.10.

7.12 As part of a detailed driver-training program, school officials are requiring teenagers to take a depth-perception test. In one phase of this test, the student is asked to judge the distance between a parked vehicle and a pedestrian stationed a given distance from the student. The recorded distances in feet are listed below for 15 driver-education students.

$$5 \quad 8 \quad 7 \quad 7 \quad 10 \quad 6 \quad 4 \quad 11$$
$$6 \quad 8 \quad 4 \quad 9 \quad 9 \quad 6 \quad 5$$

Use these data to construct a 99% confidence interval for σ^2, the variance of the depth-perception distances.

7.3 ▼ QUALITY CONTROL: r-CHARTS AND s-CHARTS

You will recall that we discussed how to construct and use $\bar{y}$-charts to examine and control the mean output from a process. We can plot them from the samples that are used to construct the $\bar{y}$-chart sample ranges on an r-chart. Of course, the sample range is the difference between the largest and smallest measurements in a sample; hence, it gives a measure of the variability of a process. The smaller the sample ranges are, the smaller the variation in the process output. In contrast, the larger the sample ranges, the more variability in the output of the process.

An r-chart is constructed by plotting the individual sample ranges r_i for successive samples of size n. The sample size is usually kept small, typically less than 10, because the sample range tends to be more volatile and unreliable for larger sample sizes.

The center line of the r-chart is denoted by $\bar{r}$ and is calculated as the average of the sample ranges. The upper and lower control limits for the

r-chart are given by

$$UCL = D'_n \bar{r} \quad \text{and} \quad LCL = D_n \bar{r}$$

where D'_n and D_n are obtained from Table 17 of the Appendix.

E X A M P L E   7.4

▼

Refer to the colorimeter readings for the 25 samples of Example 5.5. The sample ranges for those data are presented in Table 7.2. Construct an r-chart for these data. Plot the sample data on the r-chart as well.

T A B L E   7.2

25 Sample Ranges for Example 5.5

Sample	Sample Range
1	1.8
2	1.9
3	2.1
4	2.5
5	1.6
6	1.6
7	2.6
8	1.9
9	1.2
10	3.4
11	4.0
12	.7
13	3.8
14	2.2
15	3.5
16	1.4
17	3.2
18	1.2
19	2.8
20	2.8
21	1.5
22	1.3
23	1.8
24	2.0
25	4.0
Total	56.8

Solution   The center line for the r-chart is given by

$$\bar{r} = \frac{\sum r_i}{k} = \frac{56.8}{25} = 2.27.$$

For $n = 5$ observations per sample, we find $D'_5 = 2.115$ and $D_5 = 0$ from Table 17 of the Appendix. It follows, then, that for $\bar{r} = 2.27$, the upper and lower control limits for the r-chart are, respectively,

$$\text{UCL} = D'_n \bar{r} = 2.115(2.27) = 4.80$$

and

$$\text{LCL} = D_n \bar{r} = 0(2.27) = 0.$$

The r-chart and the accompanying sample data are shown in Figure 7.4.

FIGURE 7.4
r-Chart for Example 7.4

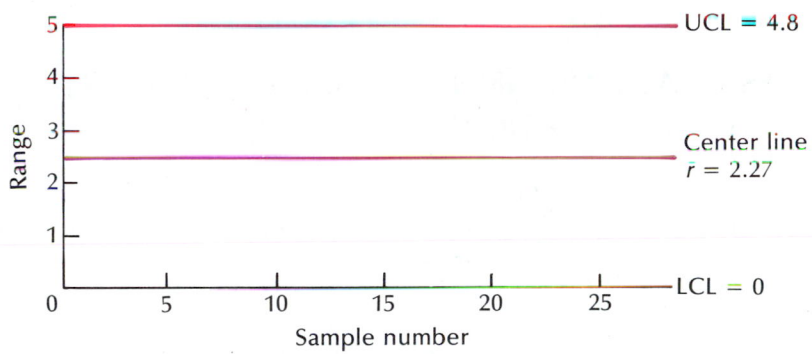

Process variability can also be monitored using an s-chart for the sample standard deviations. Although the s-chart may be preferable to the r-chart because the sample standard deviation is a better measure of variability than the sample range, the r-chart seems to be more popular among production and manufacturing personnel because the sample range is easily understood and, for small sample sizes, tends to be very comparable to the sample standard deviation. Also, r is much easier to compute than s, unless everything is computer-generated. It is for these reasons that we will not present the computations for an s-chart; however, the interpretations for an s-chart follow immediately from our previous discussions about $\bar{y}$ and r-charts.

One final comment should be made about the relationship between $\bar{y}$ and r-charts. Since the center line of an r-chart, $\bar{r}$, is used in constructing the limits of a $\bar{y}$-chart, the $\bar{y}$-chart should never be used without first constructing the corresponding r-chart. If the variability of the process is out of control, as determined by the r-chart, then the average of the sample ranges ($\bar{r}$) will not be reliable, and hence the control limits for the $\bar{y}$-chart will be unreliable.

▼ EXERCISES

Basic Techniques

7.13 Explain the difference between a $\bar{y}$-chart and an r-chart. Give an example of where each might be useful.

7.14   A manufacturing process is monitored for 60 days; on each day a random sample of $n = 6$ items is obtained. If $\bar{r} = 13.0$, determine the center line and upper and lower control limits for an r-chart.

Application

7.15   Refer to Exercise 5.28. Construct an r-chart for these data. Is the process variability in control to permit construction of a $\bar{y}$-chart?

7.4 ▼ ESTIMATION AND TESTS FOR COMPARING TWO POPULATION VARIANCES

checking equal variance assumption

One of the major applications of a test for the equality of two population variances is for **checking** the validity of the **equal variance assumption** (that is, $\sigma_1^2 = \sigma_2^2$) for a two-sample t test. First we hypothesize two populations of measurements that are normally distributed. We label these populations as 1 and 2, respectively. We are interested in comparing the variance of population 1, σ_1^2, to the variance of population 2, σ_2^2.

When independent random samples have been drawn from the respective populations, the ratio

$$\frac{s_1^2/s_2^2}{\sigma_1^2/\sigma_2^2}$$

F distribution

possesses a probability distribution in repeated sampling referred to as an **F distribution**. The formula for the probability distribution is omitted here, but we will specify its properties.

Properties of the F Distribution

▼
1. Unlike t or z but like χ^2, F can assume only positive values.
2. The F distribution, unlike the normal distribution or the t distribution but like the χ^2 distribution, is nonsymmetrical. (See Figure 7.5.)
3. There are many F distributions, and each one has a different shape. We specify a particular one by designating the degrees of freedom associated with s_1^2 and s_2^2. We denote these quantities by df_1 and df_2, respectively.
4. Tail values for the F distribution are tabulated and appear in Table 6 in the Appendix.

FIGURE 7.5
Distribution of s_1^2/s_2^2, the F Distribution

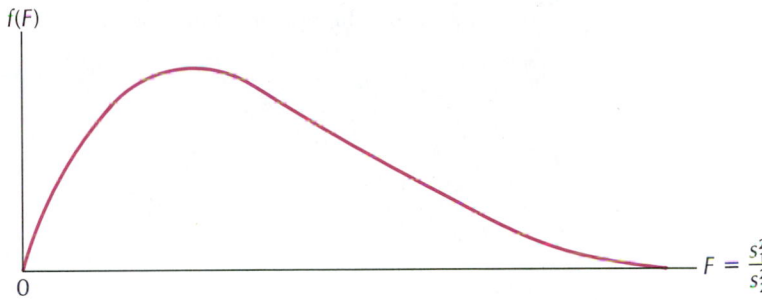

Table 6 in the Appendix records upper-tail values of F corresponding to areas $a = .25, .10, .05, .025, .01, .005,$ and $.001$. The degrees of freedom for s_1^2, designated by df_1 are indicated across the top of the table; df_2, the degrees of freedom for s_2^2, appear in the first column to the left. Values of a are given in the next column. Thus, for $df_1 = 6$ and $df_2 = 4$, the critical values of F corresponding to $a = .25, .10, .05, .025, .01, .005,$ and $.001$ are, respectively, 2.08, 4.01, 6.16, 9.20, 15.21, 21.97, and 50.53. It follows that only 5% of the measurements from an F distribution with $df_1 = 6$ and $df_2 = 4$ would exceed 6.16 in repeated sampling. (See Figure 7.6.) Similarly, for $df_1 = 24$ and $df_2 = 10$, the critical values of F corresponding to tail areas of $a = .01$ and $.001$ are, respectively, 4.33 and 7.64.

A statistical test of the null hypothesis $\sigma_1^2 = \sigma_2^2$ utilizes the test statistic s_1^2/s_2^2. When H_0 is true, s_1^2/s_2^2 follows an F distribution with $df_1 = n_1 - 1$ and $df_2 = n_2 - 1$. If upper-tail and lower-tail values of F were given in Table 6 in the Appendix, we would have no difficulty in performing the test. Unfortunately, only upper-tail values of F are given. To alleviate this situation, we are at liberty to identify either of the two populations as population 1. For a one-tailed alternative hypothesis, the populations are designated 1 and 2 so that H_a is of the form $\sigma_1^2 > \sigma_2^2$. Then the rejection region is located in the upper-tail of the F distribution. For a two-tailed alternative, we designate the population with the larger sample variance as population 1. By this convention, we again are concerned with only upper-tail rejection regions. The upper-tail F-value for a two-tailed test can then be obtained from Table 6 in the Appendix.

FIGURE 7.6
Critical Value for the F Distribution; $df_1 = 6$ and $df_2 = 4$

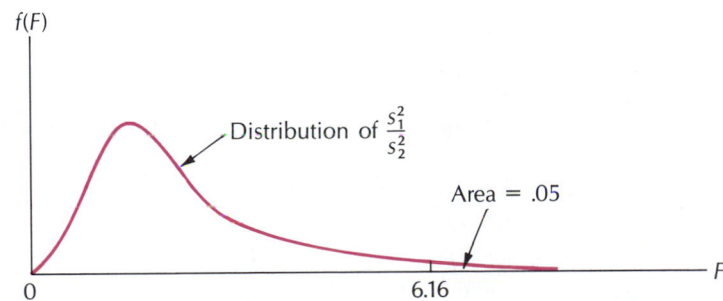

We summarize the test procedure next.

A Statistical Test Comparing σ_1^2 and σ_2^2

H_0: $\sigma_1^2 = \sigma_2^2$

H_a: **1.** $\sigma_1^2 > \sigma_2^2$
  2. $\sigma_1^2 \neq \sigma_2^2$

T.S.: $F = \dfrac{s_1^2}{s_2^2}$

R.R.:   For a specified value of α,

1.   reject H_0 if F exceeds the tabulated value of F for $a = \alpha$, $df_1 = n_1 - 1$, and $df_2 = n_2 - 1$

2.   reject H_0 if F exceeds the tabulated value of F for $a = \alpha/2$, $df_1 = n_1 - 1$, and $df_2 = n_2 - 1$.

EXAMPLE 7.5

Previously, we discussed an experiment in which company officials were concerned about the length of time a particular drug product retained its potency. A random sample of ten bottles was obtained from the production line and each bottle was analyzed to determine its potency. A second sample of ten bottles was obtained and stored in a regulated environment for one year. Potency readings were obtained on these bottles at the end of the year. The sample data were then used to place a confidence interval on $\mu_1 - \mu_2$, the difference in mean potencies for the two time periods.

Although we did not stress this at the time, in order to use t in the confidence interval or in a statistical test, we do require that the samples be drawn from normal populations with possibly different means *but* with a common variance. Use the sample data summarized below to test the equality of the population variances. Use $\alpha = .05$. Sample 1 data are the readings taken immediately after production and sample 2 data are the readings taken one year after production. Draw conclusions.

Sample 1: $\bar{y}_1 = 10.37$, $s_1^2 = 0.105$
Sample 2: $\bar{y}_2 = 9.83$, $s_2^2 = 0.058$

Solution   The four parts of the statistical test of $H_0: \sigma_1^2 = \sigma_2^2$ are shown here.

H_0: $\sigma_1^2 = \sigma_2^2$
H_a: $\sigma_1^2 \neq \sigma_2^2$

T.S.: $F = \dfrac{s_1^2}{s_2^2} = \dfrac{0.105}{0.058} = 1.81$

R.R.: For a two-tailed test with $\alpha = .05$, we will reject H_0 if $F > F_{.025,9,9} = 4.03$. Since 1.81 does not fall in the rejection region, we cannot reject H_0: $\sigma_1^2 = \sigma_2^2$.

Also, since 1.81 is not greater than $F_{.10,9,9} = 2.44$, it appears that the assumption of equality of variances holds for the t-methods used with these data. ▲

We can now formulate a confidence interval for the ratio σ_1^2/σ_2^2.

General Confidence Interval for σ_1^2/σ_2^2 with Confidence Coefficient $(1 - \alpha)$

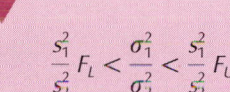

$$\frac{s_1^2}{s_2^2} F_L < \frac{\sigma_1^2}{\sigma_2^2} < \frac{s_1^2}{s_2^2} F_U$$

If F_{df_1, df_2} represents the $\alpha/2$ upper-tail value of an F distribution with df_1 and df_2 degrees of freedom and F_{df_2, df_1} represents the $\alpha/2$ upper-tail value of an F distribution with the degrees of freedom reversed, then

$$F_L = \frac{1}{F_{df_1, df_2}} \quad \text{and} \quad F_U = F_{df_2, df_1},$$

where $df_1 = n_1 - 1$ and $df_2 = n_2 - 1$.

Note: The confidence interval for σ_1/σ_2 is found by taking square roots throughout.

It should be noted that although our estimation procedure for σ_1^2/σ_2^2 is appropriate for any confidence coefficient $(1 - \alpha)$, Table 6 allows us to construct confidence intervals for σ_1^2/σ_2^2 with the more commonly used confidence coefficients, such as .90, .95, .98, .99, and so on. For more detailed tables of the F distribution, see Pearson and Hartley (1966).

EXAMPLE 7.6 ▼

The life length of an electrical component was studied under two operating voltages, V_1 and V_2. Ten different components were randomly assigned to each of the two operating voltages. Use the data below to find a 90% confidence interval for σ_1^2/σ_2^2, the ratio of the variances in life lengths for the two populations, populations 1 and 2, corresponding to the components studied under V_1 and V_2, respectively.

Voltage V_1: $n_1 = 10$, $s_1^2 = .51$
Voltage V_2: $n_2 = 10$, $s_2^2 = .20$

Solution   Before constructing our confidence interval, we must obtain F_{df_1, df_2} and F_{df_2, df_1}. For $n_1 = n_2 = 10$, $df_1 = df_2 = 9$, and hence F_{df_1, df_2} and F_{df_2, df_1} are the same. For a 90% confidence interval (i.e., $1 - \alpha = .90$), we must look up the .05 F-value based on $df_1 = 9$ and $df_2 = 9$. This value is 3.18. The quantities F_L and F_U are

$$F_L = \frac{1}{3.18} \quad \text{and} \quad F_U = 3.18.$$

Substituting into the confidence interval formula, we have

$$\frac{.51}{.20} \left(\frac{1}{3.18} \right) < \frac{\sigma_1^2}{\sigma_2^2} < \frac{.51}{.20} (3.18)$$

$$.80 < \frac{\sigma_1^2}{\sigma_2^2} < 8.11.$$

We are 90% confident that the ratio of population variances corresponding to voltages V_1 and V_2 lies in the interval of .80 to 8.11.   ▲

E X A M P L E   7.7   ▼

Refer to Example 7.6. Suppose one of the components on V_1 was damaged by the experimenter midway through the test period and had to be removed from the study. Then with $n_1 = 9$ and $n_2 = 10$, $df_1 = 8$ and $df_2 = 9$. Assuming s_1^2 and s_2^2 are as given in Example 7.6, set up a 90% confidence interval for σ_1^2 / σ_2^2.

Solution   The appropriate .05 F-values can be obtained from Table 6 in the Appendix.

$$F_{8,9} = 3.23 \quad \text{and} \quad F_L = \frac{1}{3.23}$$

$$F_{9,8} = 3.39 \quad \text{and} \quad F_U = 3.39.$$

We then have the confidence interval

$$\frac{.51}{.20} \left(\frac{1}{3.23} \right) < \frac{\sigma_1^2}{\sigma_2^2} < \frac{.51}{.20} (3.39)$$

$$.79 < \sigma_1^2 / \sigma_2^2 < 8.64.$$ ▲

sensitivity of assumptions

    The inferences about σ_1^2/σ_2^2 based on the F distribution are *very* sensitive to departures from normality of the underlying distributions. The first precaution that you should take is to plot the data for each sample separately. If there is a hint that one or both of the populations may not be normal, be very careful about the inferences you make on σ_1^2/σ_2^2 using the F distribution; the p-value or confidence coefficient may be substantially different from what you found using the test or confidence interval based on F.

    Several alternative procedures are available and are discussed in detail in other textbooks. For example, the Ansari–Bradley test can be used to compare the variances of two populations having the same median (i.e., same location). In contrast, the Moses test and the Miller jackknife procedure are used for comparing the variances of two populations when the populations' medians are unknown and are assumed unequal. The interested reader is referred to Hollander and Wolfe (1973) for details concerning these alternatives to the F-methods of this section.

▼ EXERCISES

Basic Techniques

7.16   Find the value of F that locates an area a in the upper tail of the F-distribution for these conditions:
- **a.** $a = .05$, $df_1 = 7$, $df_2 = 12$
- **b.** $a = .05$, $df_1 = 3$, $df_2 = 10$
- **c.** $a = .05$, $df_1 = 10$, $df_2 = 20$
- **d.** $a = .01$, $df_1 = 8$, $df_2 = 15$
- **e.** $a = .01$, $df_1 = 13$, $df_2 = 25$

7.17   Find approximate values for F_a for these conditions:
- **a.** $a = .05$, $df_1 = 11$, $df_2 = 24$
- **b.** $a = .05$, $df_1 = 14$, $df_2 = 14$
- **c.** $a = .05$, $df_1 = 35$, $df_2 = 22$
- **d.** $a = .01$, $df_1 = 22$, $df_2 = 24$
- **e.** $a = .01$, $df_1 = 17$, $df_2 = 25$

    (Note: Your answers may not agree with those in the back of the book. As long as your answer is close to the recorded answer, it is satisfactory.)

7.18   Random samples of $n_1 = 8$ and $n_2 = 10$ observations were selected from populations 1 and 2, respectively. The corresponding sample variances were $s_1^2 = 7.4$ and $s_2^2 = 12.7$. Do the data provide sufficient evidence to indicate a difference between σ_1^2 and σ_2^2? Test by using $\alpha = .10$. What assumptions have you made?

7.19   An experiment was conducted to determine whether there was sufficient evidence to indicate that data variation within one population, say population A, exceeded the variation within a second population, population B. Random samples of $n_A = n_B = 8$ measurements were selected from the two populations and the sample variances were calculated to be

$$s_A^2 = 2.87 \qquad s_B^2 = .91$$

Do the data provide sufficient evidence to indicate that σ_A^2 is larger than σ_B^2? Test by using $\alpha = .05$.

Applications

 7.20 A soft-drink firm is debating whether it should invest in a new type of canning machine or continue operating with the machines presently in use. The company has already determined that it will be able to fill more cans per day for the same cost if the new machines are installed. However, an important factor as yet unsolved is the variability of fills. (The company would, of course, prefer the model with the smaller variance in fills.) Let σ_1^2 and σ_2^2 denote the variances for fills from the old model and the new model, respectively. Obtaining samples of fills from the two models and utilizing the test statistics s_1^2/s_2^2, we can set up either a one-tailed or a two-tailed rejection region, using the F-distribution.

    a. What type of rejection region would be most favored by the manager of the soft-drink company? Why?

    b. What type of rejection region would be most favored by the salesperson for the company manufacturing the model presently in use? Why?

7.21 Refer to Exercise 7.20. Suppose random samples of $n_1 = n_2 = 11$ cans from the two machines are examined to determine the amount of fill (in ounces). The means and variances are

$$\bar{y}_1 = 11.70 \qquad \bar{y}_2 = 11.60$$
$$s_1^2 = .06 \qquad s_2^2 = .022.$$

Do these data present sufficient evidence to indicate less variability of fills for the new model? Use $\alpha = .10$.

 7.22 In a gasoline economy study, ten 1-gallon samples of a particular brand of gasoline were used for each of two cars (A and B). Both cars averaged approximately 17 miles per gallon, but the sample standard deviations were .95 and 1.56 for cars A and B, respectively. Use these data to test the hypothesis that the variances in miles per gallon for the two cars are identical. Use $\alpha = .05$.

7.5 ▼ SUMMARY

In this chapter, we discussed procedures for making inferences concerning a population variance and the ratio of two population variances. Estimation and statistical tests concerning σ^2 make use of the chi-square probability distribution with df $= n - 1$. Inferences concerning the ratio of two population variances utilize an F distribution with df$_1 = n_1 - 1$ and df$_2 = n_2 - 1$.

    The need for inferences concerning one or more population variances can be traced to our discussion of numerical descriptive measures of a population in Chapter 3. To describe or make inferences about a population of measurements, we cannot always rely on the mean, a measure of central tendency. Many times in evaluating or comparing the performance of individuals on a psychological test, the consistency of manufactured products emerging from a production line, or the yields of a particular variety of corn, we gain important information by studying the population variance.

In the next chapter, we consider additional applications using the chi-square distribution.

▼ KEY FORMULAS

1. $100(1 - \alpha)\%$ confidence interval for σ^2 (or σ)

$$\frac{(n-1)s^2}{\chi_U^2} < \sigma^2 < \frac{(n-1)s^2}{\chi_L^2}$$

or

$$\sqrt{\frac{(n-1)s^2}{\chi_U^2}} < \sigma < \sqrt{\frac{(n-1)s^2}{\chi_L^2}}$$

2. Statistical test for σ^2

$$H_0: \quad \sigma^2 = \sigma_0^2 \ (\sigma_0^2 \text{ is specified})$$

$$\text{T.S.:} \quad \chi^2 = \frac{(n-1)s^2}{\sigma_0^2}$$

3. Statistical test for σ_1^2/σ_2^2

$$H_0: \quad \sigma_1^2 = \sigma_2^2$$

$$\text{T.S.:} \quad F = \frac{s_1^2}{s_2^2}$$

4. $100(1 - \alpha)\%$ confidence interval for σ_1^2/σ_2^2 (or σ_1/σ_2)

$$\frac{s_1^2}{s_2^2} F_L < \frac{\sigma_1^2}{\sigma_2^2} < \frac{s_1^2}{s_2^2} F_U,$$

where

$$F_L = \frac{1}{F_{df_1, df_2}} \quad \text{and} \quad F_U = F_{df_2, df_1}$$

or

$$\sqrt{\frac{s_1^2}{s_2^2}} F_L < \frac{\sigma_1}{\sigma_2} < \sqrt{\frac{s_1^2}{s_2^2}} F_U$$

▼ SUPPLEMENTARY EXERCISES

7.23 Two consumer research groups are vying for a large govenment contract. Since subjective evaluations of consumer products will be made by judges during the study, government officials prefer to award the contract to a company that utilizes judges with consistent ratings (of course, other qualifications are also evaluated before awarding the contract). One measure of consistency is the variability of judges' scores on the same item.

Before issuing the contract, a test is conducted in which 25 judges from each company are asked to rate a single item. The sample variances are given here:

$$\text{company A:} \quad s_1^2 = .50 \qquad \text{company B:} \quad s_2^2 = .15$$

Use these data to test the hypothesis that the variances of the judges' ratings are the same for the two populations. The alternative hypothesis is that the variances are different. Use $\alpha = .10$.

7.24 Refer to Exercise 6.22, in which we were interested in comparing the weights of the combs of roosters fed one of two vitamin-supplemented diets. The Wilcoxon rank sum test was suggested as a test of the hypothesis that the two populations were identical. Would it have been appropriate to run a t test comparing the two population means? Explain.

7.25 A consumer-protection magazine was interested in comparing tires purchased from two different companies, each claiming their tires would last the same number of miles. A sample of five tires of each brand was obtained and tested under simulated road conditions. The number of miles before significant deterioration in tread was recorded for all tires. The data are given next (in 1,000 miles).

Brand I	40.6	35.9	48.5	36.4	38.3
Brand II	40.9	40.2	42.5	39.1	42.6

a. Construct a 98% confidence interval for the ratio of the two population variances.
b. How does the confidence interval change if we're interested in σ_1/σ_2 rather than in σ_1^2/σ_2^2?

7.26 A random sample of 20 patients, each of whom has suffered from depression, was selected from a mental hospital, and each patient was administered the Brief Psychiatric Rating Scale. The scale consists of a series of adjectives that the patient scores according to his or her mood. Extensive testing in the past has shown that ratings in certain mood adjectives tend to be similar and hence are grouped together as jointly measuring one or more components of a person's mood. For example, a group consisting of certain adjectives seems to be measuring depression. Let us suppose that the mean and standard deviation of the 20 patients in the group are 13.2 and 4.6, respectively.
a. Place a 99% confidence interval on σ^2, the variance of the population of patients' scores from which this sample was drawn.
b. What's the critical assumption underlying the inference? Do you know whether this assumption is valid for these data?

7.27 Refer to Exercise 7.26. Suppose that extensive testing in a large number of depressed patients throughout the century has indicated that the population standard deviation of scores for the depression adjectives is 5.9. Use the sample data of Exercise 7.26 to test the research hypothesis that the standard deviation for all patients who might be treated for depression in this hospital is less than 5.9. Give a p-value for these data and draw conclusions.

7.28 A pharmaceutical company manufactures a particular brand of antihistamine tablets. In the quality control division, certain tests are routinely performed to determine whether the product being manufactured meets specific performance criteria prior to release of the product onto the market. In particular, the company requires that the potencies of the tablets lie in the range of 90% to 110% of the labeled drug amount.

    a. If the company is manufacturing 25-mg tablets, within what limits must tablet potencies lie?

    b. A random sample of 30 tablets is obtained from a recent batch of antihistamine tablets. The data for the potencies of the tablets are given below. Is the assumption of normality warranted for inferences about the population variances?

    c. Translate the company's 90% to 110% specifications on the range of the product potency into a statistical test concerning the population variance for potencies. Draw conclusions based on $\alpha = .05$.

24.1	27.2	26.7	23.6	26.4	25.2
25.8	27.3	23.2	26.9	27.1	26.7
22.7	26.9	24.8	24.0	23.4	25.0
24.5	26.1	25.9	25.4	22.9	24.9
26.4	25.4	23.3	23.0	24.3	23.8

7.29 A study was conducted to compare the variabilities in strengths of 1-inch-square sections of a synthetic fiber produced under two different procedures. A random sample of nine squares from each process was obtained and tested.

    a. Plot the data for each sample separately.

    b. Is the assumption of normality warranted?

    c. If permissible from part b, use the following data (psi) to test the research hypothesis that the population variances corresponding to the two procedures are different. Use $\alpha = .10$.

Procedure 1	74	90	103	86	75	102	97	85	69
Procedure 2	59	66	73	68	70	71	82	69	74

7.30 Refer to Example 7.2. Construct a 95% confidence interval for σ^2, and use this interval to help interpret the findings of the consumer group. Does it appear that the test of Example 7.2 had much power to detect an increase in σ^2 of 25% over the claimed value? Explain.

7.31 The risk of an investment is measured in terms of the variance in the return that could be observed. Random samples of 10 yearly returns were obtained from two different portfolios.

| | Portfolio | |
	1	2
Sample mean return (000)	132	146
Sample variance	10.9	25.6
Sample size	10	10

Does Portfolio 2 have a higher risk? Give a *p*-value for your test.

7.32 Refer to Exercise 7.31. Are there any differences in the average returns for the two portfolios? Indicate the method you used in arriving at a conclusion, and explain why you used it.

7.33 Two different modeling techniques for assessing the resale value of houses were considered. A random sample of 12 existing listings was taken and each house was valued using the two techniques. These data are shown here.

| | Assessed Value Listing (000) Technique | |
Listing	1	2
1	155	138
2	137	128
3	248	230
4	136	146
5	102	95
6	87	82
7	63	67
8	129	134
9	144	149
10	270	292
11	157	150
12	51	48

a. Plot the data. Does it appear that the two modeling techniques give similar results?
b. Give an estimate of the mean and standard error of the difference between estimates for the two methods.

7.34 Refer to Exercise 7.33. Place a 90% confidence interval on the variance of the difference in estimates. Give the corresponding interval for σ.

7.35 Refer to Exercises 7.33 and 7.34. What is the critical assumption concerning the sample data? How would you check this assumption? Do the data suggest the assumption holds? Do you have any cautions about the inferences in Exercise 7.34?

7.36 An important consideration in examining the potency of a pharmaceutical product is the amount of drop in potency for a specific shelf life (time on a pharmacist's shelf). In particular, the variability of these drops in potency is very important. Researchers studied the drops in potency for two different drug products over a 6-month period. These data are summarized in the accompanying table. Suppose that drug 1 is an experimental drug product and drug 2 a marketed product. Use a one-tailed test with $\alpha = .01$ to determine whether the data suggest that drug 1 has more variability in potency drop than drug 2.

	Drug 1	Drug 2
Sample size	10	10
Sample mean	58	56
Sample variance	82	23

7.37 Refer to Exercise 7.36. Would your result have changed if you had used a two-tailed test with $\alpha = .10$? Why might a two-tailed test be important?

 7.38 Blood cholesterol levels for randomly selected patients with similar histories were compared for two diets, one a low-fat-content diet and the other a normal diet. The summary data appear in the accompanying table.

	Low-Fat Content	Normal
Sample size	19	24
Sample mean	170	196
Sample variance	198	435

a. Do these data present sufficient evidence to indicate a difference in cholesterol level variabilities for the two diets? Use $\alpha = .10$.

b. What other test might be of interest in comparing the two diets?

 7.39 Sales from weight-reducing agents marketed in the United States represent sizable chunks of income for many of the companies that manufacture these products. Psychological as well as physical effects often contribute to how well a person responds to the recommended therapy. Consider a comparison of two weight-reducing agents, A and B. In particular, consider the variabilities in the lengths of times people remain on the therapy. A total of 26 overweight males, matched as closely as possible physically, were randomly divided into two groups. Those in group 1 received preparation A while those assigned to group 2 received preparation B. Use the summary data to compare the variabilities associated with the lengths of time on therapy. Use a two-tailed test with $\alpha = .10$.

	Preparation A	Preparation B
Sample size	13	13
Sample mean	25 days	35 days
Sample variance	50	16

7.40 Refer to Exercise 7.39. What might the null and alternative hypotheses have been if preparation A had been a placebo (no active medication) and preparation B a marketed product known to be an effective weight-reducing agent?

7.41 A chemist at an iron ore mine suspects that the variance in the amount (weight, in ounces) of iron oxide per pound of ore tends to increase as the mean amount of iron oxide per pound increases. To test this theory, ten 1-pound specimens of iron ore are selected at each of two locations, one, location 1, containing a much higher mean content of iron oxide than the other, location 2. The amounts of iron oxide contained in the ore specimens are shown in the accompanying table.

Location 1	8.1	7.4	9.3	7.5	7.1	8.7	9.1	7.9	8.4	8.8
Location 2	3.9	4.4	4.7	3.6	4.1	3.9	4.6	3.5	4.0	4.2

Do the data provide sufficient evidence to indicate that the amount of iron oxide per pound of ore is more variable at location 1 than at location 2? Use $\alpha = .05$.

7.42 One index of service quality for telephone reservation systems is the waiting time from the first ring until an agent answers, ready to make reservations. The waiting times for a sample of 30 reservation calls during a period of 1 week showed a mean and standard deviation of 28.4 and 17.4 seconds, respectively. Use these data to construct a 95% confidence interval for the standard deviation of waiting time.

7.43 A personnel officer was planning to use a t test to compare the mean number of monthly unexcused absences for two divisions of a multinational company but then noticed a possible difficulty. The variation in the number of unexcused absences per month seemed to differ for the two groups. As a check, a random sample of 5 months was selected at each division, and for each month, the number of unexcused absences was obtained.

Category A	20	14	19	22	25
Category B	37	29	51	40	26

a. What assumption seemed to bother the personnel officer?

b. Do the data provide sufficient evidence to indicate that the variances differ for the populations of absences for the two employee categories? Use $\alpha = .05$.

7.44 A researcher was interested in weather patterns in Phoenix and Seattle. As part of the investigation, the researcher took a random sample of 20 days in July and observed the daily average temperatures. The data were collected over several years to assure independence of daily temperatures. The data collected produced the following information:

	Phoenix Daily Average Temperature	Seattle Daily Average Temperature
Sample size	20	20
Sample mean	95.3	63.3
Sample standard deviation	5.1	7.6

Do the data suggest that there is a difference in the variability of average daily temperatures during July for the two cities? Is there a difference in mean temperatures for the two cities during July? Use $\alpha = .05$ for both tests.

7.45 Refer to the clinical trial data base in the Appendix to calculate the sample variances for the anxiety scores within each treatment group. Use these data to run separate tests comparing each of the treatments A, B, and C to the placebo group D. Use two-sided tests with $\alpha = .05$.

7.46 Do any of these tests in Exercise 7.45 negate the possibility of comparing treatment means for groups A, B, and C to the treatment mean for the placebo group using t tests? Explain.

7.47 Use the sleep disturbance scores from the clinical trial data base to give a 98% confidence interval for σ_B^2/σ_C^2. Do the same for σ_B^2/σ_A^2.

CHAPTER 8

CATEGORICAL DATA

8.1 INTRODUCTION

Up to this point, we have been concerned primarily with sample data measured on a quantitative scale. However, we sometimes encounter situations where levels of the variable of interest are identified by name or rank only and we are interested in the number of observations occurring at each level of the variable. Data obtained from these types of variables are called **categorical** or **count data**. For example, an item coming off an assembly line may be classified into one of three quality classes: acceptable, second, or reject. Similarly, a traffic study might require a count and classification of the type of transportation used by commuters along a major access road into a city. A pollution study might be concerned with the number of different alga species identified in samples from a lake and the number of times each species is identified. A consumer protection group might be interested in the results of a prescription fee survey to compare prices on a checklist of medications in different sections of a large city.

In this chapter, we will examine specific inferences that can be made from experiments involving categorical data.

8.2 THE MULTINOMIAL EXPERIMENT AND CHI-SQUARE GOODNESS-OF-FIT TEST

multinomial experiment

The examples in Section 8.1 all exhibit, to a reasonable degree of approximation, the characteristics of a **multinomial experiment**.

The Multinomial Experiment

1. The experiment consists of n identical trials.
2. Each trial results in one of k outcomes.
3. The probability that a single trial will result in outcome i is π_i, $i = 1, 2, \ldots, k$, and remains constant from trial to trial. (Note: $\sum_i \pi_i = 1$.)
4. The trials are independent.
5. We are interested in n_i, the number of trials resulting in outcome i. (Note: $\sum_i n_i = n$.)

multinomial distribution

The probability distribution for the number of observations resulting in each of the k outcomes, called the **multinomial distribution**, is given by the formula

$$P(n_1, n_2, \ldots, n_k) = \frac{n!}{n_1! n_2! \ldots n_k!} \pi_1^{n_1} \pi_2^{n_2} \cdots \pi_k^{n_k}.$$

Recall from Chapter 4, where we discussed the binomial probability distribution, that

$$n! = n(n-1) \cdots 1$$

and

$$0! = 1.$$

We can use the formula for the multinomial distribution to compute the probability of particular events.

EXAMPLE 8.1 ▼

Previous experience with the breeding of a particular herd of cattle suggests that the probability of obtaining one healthy calf from a mating is .83. Similarly, the probabilities of obtaining zero or two healthy calves are, respectively, .15 and .02. If a farmer breeds three dams from the herd, find the probability of obtaining exactly three healthy calves.

Solution   Assuming the three dams are chosen at random, this experiment can be viewed as a multinomial experiment with $n = 3$ trials and $k = 3$ outcomes. These outcomes are listed below with the corresponding probabilities.

Outcome	Number of Progeny	Probability, π_i
1	0	.15
2	1	.83
3	2	.02

Note that outcomes 1, 2, and 3 refer to the events that a dam produces zero, one, or two healthy calves, respectively. Similarly, n_1, n_2, and n_3 refer to the number of dams producing zero, one, or two healthy progeny, respectively. To obtain exactly three healthy progeny, we must observe one of the following possible events.

$$
A: \begin{cases} 1 \text{ dam gives birth to no healthy progeny: } n_1 = 1 \\ 1 \text{ dam gives birth to 1 healthy progeny: } \quad n_2 = 1 \\ 1 \text{ dam gives birth to 2 healthy progeny: } \quad n_3 = 1 \end{cases}
$$

$$
B: \quad 3 \text{ dams give birth to 1 healthy progeny: } \begin{cases} n_1 = 0 \\ n_2 = 3 \\ n_3 = 0 \end{cases}
$$

For event A with $n = 3$ and $k = 3$,

$$
P(n_1 = 1,\, n_2 = 1,\, n_3 = 1) = \frac{3!}{1!1!1!}\,(.15)^1(.83)^1(.02)^1 \approx .015.
$$

Similarly, for event B,

$$
P(n_1 = 0,\, n_2 = 3,\, n_3 = 0) = \frac{3!}{0!3!0!}\,(.15)^0(.83)^3(.02)^0 = (.83)^3 \approx .572.
$$

Thus, the probability of obtaining exactly three healthy progeny from three dams is the sum of the probabilities for events A and B; namely, $.015 + .572 \approx .59$.   ▲

Our primary interest in the multinomial distribution is as a probability model underlying statistical tests about the probabilities $\pi_1, \pi_2, \ldots, \pi_k$. We will hypothesize specific values for the πs and then determine whether the sample data agree with the hypothesized values. One way to test such a hypothesis is to examine the observed number of trials resulting in each outcome and to compare this to the number we would *expect* to result in each outcome. For instance, in our previous example, we gave the probabilities associated with zero, one, and two progeny as .15, .83, and .02. If we were to examine a sample of 100 mated

expected number of outcomes

dams, we would **expect to observe** 15 dams that produce no healthy progeny. Similarly, we would expect to observe 83 dams that produce one healthy calf and two dams that produce two healthy calves.

DEFINITION 8.1
Expected Number of Outcomes

> ▼
>
> In a multinomial experiment where each trial can result in one of k outcomes, the **expected number of outcomes** of type i in n trials is $n\pi_i$, where π_i is the probability that a single trial results in outcome i.

In 1900, Karl Pearson proposed the following test statistic to test the specified probabilities:

χ^2

$$\chi^2 = \sum_i \left[\frac{(n_i - E_i)^2}{E_i} \right],$$

where n_i represents the number of trials resulting in outcome i and E_i represents the number of trials we would expect to result in outcome i when the hypothesized probabilities represent the actual probabilities assigned to each outcome. Frequently, we will refer to the probabilities $\pi_1, \pi_2, \ldots, \pi_k$ as **cell**

cell probabilities

probabilities, one cell corresponding to each of the k outcomes. The observed

observed cell counts

numbers $n_1, n_2, \ldots, n_k$ corresponding to the k outcomes will be called **observed cell counts**, and the expected numbers $E_1, E_2, \ldots, E_k$ will be referred to as

expected cell counts

expected cell counts.

Suppose that we hypothesize values for the cell probabilities $\pi_1, \pi_2, \ldots, \pi_k$. We can then calculate the expected cell counts by using Definition 8.1 to examine how well the observed data fit, or agree, with what we would expect to observe. Certainly, if the hypothesized π-values are correct, the observed cell counts n_i should not deviate greatly from the expected cell counts E_i, and the computed value of χ^2 should be small. Similarly, when one or more of the hypothesized cell probabilities are incorrect, the observed and expected cell counts will differ substantially, making χ^2 large.

chi-square distribution

The distribution of the quantity χ^2 can be approximated by a **chi-square distribution** provided that the expected cell counts E_i are fairly large.

The chi-square goodness-of-fit test based on k specified cell probabilities will have $k - 1$ degrees of freedom. Upper-tail values of the test statistic

$$\chi^2 = \sum_i \left[\frac{(n_i - E_i)^2}{E_i} \right]$$

can be found in Table 5 in the Appendix. See Figure 8.1 for a chi-square distribution with df = 4.

FIGURE 8.1
**Chi-Square Probability
Distribution for df = 4**

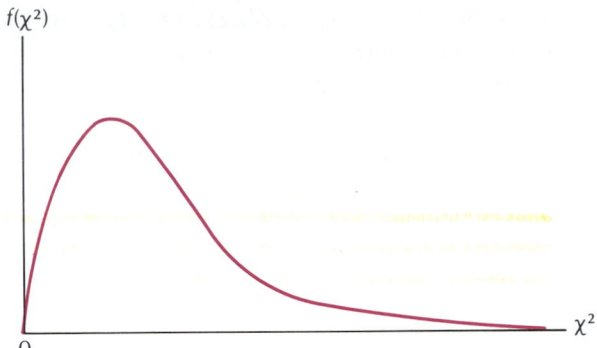

We can now summarize the chi-square goodness-of-fit test concerning k specified cell probabilities.

**Chi-Square
Goodness-of-Fit Test**

> Null hypothesis: $\pi_i = \pi_{i0}$ for categories $i = 1, \ldots, k$, π_{i0} are specified probabilities or proportions.
> Alternative hypothesis: At least one of the cell probabilities differs from the hypothesized value.
> Test statistic: $\chi^2 = \sum_i \left[\dfrac{(n_i - E_i)^2}{E_i} \right]$, where n_i is the observed number in category i and $E_i = n\pi_{i0}$ is the expected number under H_0.
> Rejection region: Reject H_0 if χ^2 exceeds the tabulated critical value for $a = \alpha$ and df $= k - 1$.

Some researchers (see, for example, Siegel (1956) and Dixon and Massey (1969)) recommend that all the E_is should be 5 or more before performing this test. This requirement is perhaps too stringent. Cochran (1954) indicates that the approximation should be quite good if no E_i is less than 1 and no more than 20% of the E_is are less than 5. We recommend applying Cochran's guidelines for determining whether χ^2 can be approximated with a chi-square distribution. We can combine categories if some of the E_is are too small, but care should be taken so that the combination of categories does not change the nature of the hypothesis to be tested.

EXAMPLE 8.2

A test drug is to be compared against a standard drug preparation useful in the maintenance of patients suffering from high blood pressure. Over many clinical trials at many different locations, patients suffering from comparable hypertension (as measured by the New York Heart Association (NYHA) Classification) have been administered the standard therapy. Responses to therapy for this large patient group were classified into one of four response categories. Table 8.1 lists the categories and percentages of patients treated on the standard preparation who have been classified in each category.

TABLE 8.1
Results of Clinical Trials Using the Standard Preparation, Example 8.2

Category	Percentage
Marked decrease in blood pressure	50%
Moderate decrease in blood pressure	25%
Slight decrease in blood pressure	10%
Stationary or slight increase in blood pressure	15%

A clinical trial is conducted with a random sample of 200 patients suffering from high blood pressure. All patients are required to be listed according to the same hypertensive categories of the NYHA Classification as those studied under the standard preparation. Use the sample data in Table 8.2 to test the hypothesis that the cell probabilities associated with the test preparation are identical to those for the standard. Use $\alpha = .05$.

TABLE 8.2
Sample Data for Example 8.2

Category	Observed Cell Counts
1	120
2	60
3	10
4	10

Solution This experiment possesses the characteristics of a multinomial experiment, with $n = 200$ and $k = 4$ outcomes.

Outcome 1: A person's blood pressure will decrease markedly after treatment on the test drug.

Outcome 2: A person's blood pressure will decrease moderately after treatment on the test drug.

Outcome 3: A person's blood pressure will decrease slightly after treatment on the test drug.

Outcome 4: A person's blood pressure will remain stationary or increase slightly after treatment on the test drug.

The null and alternative hypotheses are then

$$H_0: \quad \pi_1 = .50, \pi_2 = .25, \pi_3 = .10, \pi_4 = .15$$

and

H_a: At least one of the cell probabilities is different from the hypothesized value.

Before computing the test statistic, we must determine the expected cell numbers. These data are given in Table 8.3.

T A B L E 8.3
Observed and Expected Cell Numbers for Example 8.2

Category	Observed Cell Number, n_i	Expected Cell Number, E_i
1	120	$200(.50) = 100$
2	60	$200(.25) = 50$
3	10	$200(.10) = 20$
4	10	$200(.15) = 30$

Since all the expected cell numbers are large, we may calculate the chi-square statistic and compare it to a tabulated value of the chi-square distribution.

$$\chi^2 = \sum_i \left[\frac{(n_i - E_i)^2}{E_i} \right]$$

$$= \frac{(120 - 100)^2}{100} + \frac{(60 - 50)^2}{50} + \frac{(10 - 20)^2}{20} + \frac{(10 - 30)^2}{30}$$

$$= 4 + 2 + 5 + 13.33 = 24.33.$$

For the probability of a Type I error set at $\alpha = .05$, we look up the value of the chi-square statistic for a $= .05$ and df $= k - 1 = 3$. The critical value from Table 5 in the Appendix is 7.815.

R.R.:   Reject H_0 if $\chi^2 > 7.815$.

Conclusion: Since the computed value of χ^2 is greater than 7.815, we reject the null hypothesis and conclude that at least one of the cell probabilities differs from that specified under H_0. Practically, it appears that a much higher proportion of patients treated with the test preparation fall into the moderate and marked improvement categories.   ▲

The assumptions needed for running a chi-square goodness-of-fit test are those associated with a multinomial experiment, of which the key ones are independence of the trials and constant cell probabilities. Independence of the trials would be violated if, for example, several patients from the same family were included in the sample since hypertension has a strong hereditary component. The assumption of constant cell probabilities would be violated if the study were conducted over a period of time during which the standards of medical practice shifted, allowing for other "standard" therapies.

The test statistic for the chi–square goodness–of–fit test is the sum of k terms, which is the reason the degrees of freedom depend on k, the number of categories, rather than on n, the total sample size. However, there are only $k - 1$ degrees of freedom, rather than k of them, because the sum of the $n_i - E_i$ terms must be equal to $n - n = 0$; $k - 1$ of the observed minus expected differences are free to vary, but the last one (kth) is determined by the condition that the sum of the $n_i - E_i$ equals zero.

This goodness–of–fit test has been used extensively over the years to test various scientific theories. Unlike previous statistical tests, however, the hypothesis of interest is the null hypothesis, not the research (or alternative) hypothesis. Unfortunately, the logic behind running a statistical test does not hold. In the standard situation where the research (alternative) hypothesis is the one of interest to the scientist, we formulate a suitable null hypothesis and gather data to reject H_0 in favor of H_a. Thus, we "prove" H_a by contradicting H_0.

This is not so with the chi–square goodness–of–fit test. If a scientist has a set theory and wants to show that sample data conforms to or "fit" that theory, she wants to accept H_0. From our previous work, there is the potential for commiting a Type II error in accepting H_0. Here, as with other tests, the calculation of β probabilities is difficult. In general, for a goodness–of–fit test, the potential for committing a Type II error is high if n is small or if k, the number of categories, is large. Even if the expected cell counts E_i conform to our recommendations, the probability of a Type II error could be large. So, the results of a chi–square goodness–of–fit test should be viewed suspiciously. Don't automatically accept the null hypothesis as fact given that H_0 was not rejected.

▼ EXERCISES

Basic Techniques

8.1 List the characteristics of a multinomial experiment.

8.2 How does a binomial experiment relate to a multinomial experiment?

8.3 Determine the rejection region for a chi-square goodness-of-fit test for the following values of k, n, and α.
a. $n = 100$, $k = 4$, $\alpha = .05$
b. $n = 500$, $k = 10$, $\alpha = .01$
c. $n = 200$, $k = 8$, $\alpha = .001$

8.4 Under what conditions is it appropriate to use the chi-square goodness-of-fit test for a multinomial experiment? What qualification(s) might one have to make if the sample data do not yield rejection of the null hypothesis?

Applications

8.5   Hypothetical data are presented here. Use these data to run a chi-square goodness-of-fit test with $H_0: \pi_1 = .2, \pi_2 = .15, \pi_3 = .40, \pi_4 = .15,$ and $\pi_5 = .10$. Use $\alpha = .05$. Do the data fit the hypothesized probabilities?

Category	Observed Cell Number, n_i
1	60
2	50
3	130
4	40
5	20
Total	300

8.6   Use the data of Exercise 8.5 to run a chi-square goodness-of-fit test with this new null hypothesis—$H_0: \pi_1 = .15, \pi_2 = .20, \pi_3 = .45, \pi_4 = .15,$ and $\pi_5 = .05$. Again use $\alpha = .05$. Compare your results to those of Exercise 8.5. How sensitive does this test appear to be for the cell probabilities specified under H_0? What conclusion can be drawn if we do *not* reject H_0?

8.7   Over the past 5 years, an insurance company has had a mix of 40% whole life policies, 20% universal life policies, 25% annual renewable-term (ART) policies, and 15% other types of policies. A change in this mix over the long haul could require a change in the commission structure, reserves, and possibly investments. A sample of 1,000 policies issued over the last few months gave the results below. Use these data to assess whether there has been a shift from the historical percentages. Give the *p*-value for your test. Which policies (if any) seem to be more popular?

Category	Observed Cell Number, n_i
Whole Life	320
Universal Life	280
ART	240
Other	160
Total	1,000

8.8   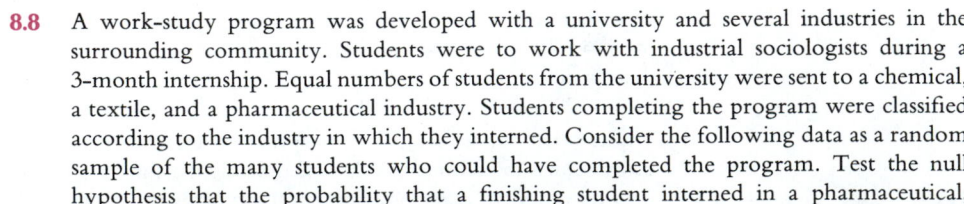 A work-study program was developed with a university and several industries in the surrounding community. Students were to work with industrial sociologists during a 3-month internship. Equal numbers of students from the university were sent to a chemical, a textile, and a pharmaceutical industry. Students completing the program were classified according to the industry in which they interned. Consider the following data as a random sample of the many students who could have completed the program. Test the null hypothesis that the probability that a finishing student interned in a pharmaceutical, chemical, or textile industry is 1/3. Use $\alpha = .01$ with n_i the number of students in group i finishing the program.

Group	n_i
Pharmaceutical	20
Chemical	13
Textile	30

 8.9 An experiment was conducted to determine whether the proportion of mentally ill patients of each social class housed in a county facility agrees with the social class distribution of the county. The observed cell numbers for the 400 patients classified are given here.

Lower: 215   Upper-middle: 60
Lower-middle: 100   Upper: 25

Use these data to test the null hypothesis

$\pi_1 = .25$ $\pi_3 = .20$
$\pi_2 = .48$ $\pi_4 = .07$,

where the πs are the hypothesized proportions of persons in the respective social-class categories in the county. Use $\alpha = .05$ and draw conclusions.

 8.10 In previous presidential elections in a given locality, 50% of the registered voters were Republicans, 40% were Democrats, and 10% were registered as independents. Prior to the upcoming election, a random sample of 200 registered voters showed that 90 were registered as Republicans, 80 as Democrats, and 30 as independents. Test the research hypothesis that the distribution of registered voters is different from that in previous election years. Give the p-value for your test. Draw conclusions.

8.11 A local doctor suspects that there is a seasonal trend in the occurrence of the common cold. He estimates that 40% of the cases each year occur in the winter, 40% in the spring, 10% in the summer, and 10% in the fall. The information below was collected from a random sample of 1,000 cases of patients with the common cold over the past year. Would you agree with the doctor's estimates, based on the sample information? Perform a statistical test using $\alpha = .05$. Draw conclusions.

Season	Frequency
Winter	374
Spring	292
Summer	169
Fall	165

8.12 Refer to Exercise 8.11. What would the null hypothesis be if the doctor claimed that there are no differences in the percentages of cases over the seasons? Test the hypothesis that there is no seasonal trend in the occurrence of the common cold. Give the level of significance of your test. Are there any reservations about your conclusion?

8.13 Previous experimentation with a drug product developed for the relief of depression was conducted with normal adults with no signs of depression. We will assume a large data bank is available from studies conducted with normals and, for all practical purposes, the data bank can represent the population of responses for normals. Each of the adults participating in one of these studies was asked to rate the drug as ineffective, mildly effective, or effective. The percentages of respondents in these categories were 60%, 30%, and 10%, respectively. In a new study of depressed adults, a random sample of 85 adults responded as follows:

Ineffective: 30
Mildly effective: 35
Effective: 20

Is there evidence to indicate a different percentage distribution of responses for depressed adults than for normals? Give the level of significance for your test and draw conclusions.

 8.14 In random sampling, 40 newspaper editors were interviewed to determine their opinions on the degree of future suppression of freedom of the press brought about by recent court decisions. The editors' opinions are summarized below. Use these data to test the null hypothesis that each category is equally preferred. Use $\alpha = .05$. Draw conclusions from these data. What reservation(s), if any, do you have concerning your conclusions?

Degree of Suppression	Frequency
None	8
Very little	8
Moderate	10
Severe	14

8.15 A sample of 125 securities analysts was obtained, and each analyst was asked to select four stocks on the New York Stock Exchange that were expected to outperform the Standard and Poor's Index over a 3-month period. One theory suggests that the securities analysts would be expected to do no better than chance and that the number of correct guesses from the four selected had a multinomial distribution as shown here.

Number correct	0	1	2	3	4
Multinomial probabilities (π_i)	.0625	.2500	.3750	.2500	.0625

If the number of correct guesses from the sample of 125 analysts had a frequency distribution as shown here, use these data to conduct a chi-square goodness-of-fit test. Use $\alpha = .05$. Draw conclusions.

Number correct	0	1	2	3	4
Frequency	3	23	51	39	9

8.16 Refer to Exercise 8.15. Suppose the assumed multinomial probabilities were all .20 (i.e., all $\pi_i = .20$).
a. How would your conclusions change?
b. Can you suggest a problem with the chi-square goodness-of-fit test based on the results of part a?

8.3 ▼ INFERENCES ABOUT THE BINOMIAL PARAMETER π

The binomial experiment discussed in Chapter 4 is a special case of the multinomial experiment, where each trial results in one of two outcomes, which we labeled as either a success or a failure. Recall that we designated π as the probability of a success and $(1 - \pi)$ as the probability of a failure. Then the probability distribution for y, the number of successes in n identical trials, is

$$P(y) = \frac{n!}{y!(n - y)!}\, \pi^y(1 - \pi)^{n-y}.$$

The point estimate of the binomial parameter π is one that we would choose intuitively. In a random sample of n from a population in which the proportion of elements classified as successes is π, the best estimate of the parameter π is the sample proportion of successes. Letting y denote the number of successes in the n sample trials, the sample proportion is

$$\hat{\pi} = \frac{y}{n}.$$

We observed in Section 4.12 that y possesses a mound-shaped probability distribution that can be approximated by using a normal curve when

$$n \geq \frac{5}{\min(\pi,\, 1 - \pi)} \qquad \text{(or equivalently, } n\pi \geq 5 \text{ and } n(1 - \pi) \geq 5\text{).}$$

In a similar way, the distribution of $\hat{\pi} = y/n$ can be approximated by a normal distribution with a mean and a standard error as given below.

Mean and Standard Error of $\hat{\pi}$

$$\mu_{\hat{\pi}} = \pi$$

$$\sigma_{\hat{\pi}} = \sqrt{\frac{\pi(1 - \pi)}{n}}$$

The normal approximation to the distribution of $\hat{\pi}$ can be applied under the same condition as that for approximating y by using a normal distribution. In fact, the approximation for both y and $\hat{\pi}$ becomes more precise for large n. Henceforth, in this text, we will assume that $\hat{\pi}$ can be adequately approximated by using a normal distribution, and we will base all our inferences on results from our previous study of the normal distribution.

A confidence interval can be obtained for π using the methods of Chapter 5 for μ, by replacing $\bar{y}$ with $\hat{\pi}$ and $\sigma_{\bar{y}}$ with $\sigma_{\hat{\pi}}$. A general $100(1 - \alpha)\%$ confidence interval for the binomial parameter is given here.

Confidence Interval for π, with Confidence Coefficient of $(1 - \alpha)$

$n\pi \geq 5$

$n(1-\pi) \geq 5$

$$\hat{\pi} \pm z_{\alpha/2}\sigma_{\hat{\pi}},$$

where

$$\hat{\pi} = \frac{y}{n} \quad \text{and} \quad \sigma_{\hat{\pi}} = \sqrt{\frac{\pi(1 - \pi)}{n}}.$$

Note: Since π is unknown, replace π by $\hat{\pi}$ in $\sigma_{\hat{\pi}}$.

EXAMPLE 8.3 ▼

Response to an advertising display was measured by counting the number of people who purchased the product out of the total number exposed to the display. If 330 purchased the product out of a total of 870 exposed, estimate the proportion of all persons exposed who will buy the product. Use a 90% confidence interval.

Solution For these data,

$$\hat{\pi} = \frac{330}{870} = .38$$

$$\sigma_{\hat{\pi}} = \sqrt{\frac{(.38)(.62)}{870}} = .016.$$

The confidence coefficient for our example is .90. Recall from Chapter 5 that we can obtain $z_{\alpha/2}$ by looking up the z-value in Table 2 in the Appendix corresponding to an area of $(\alpha/2)$. For a confidence coefficient of .90, the z-value corresponding to an area of .05 is 1.645. Hence, the 90% confidence interval on the proportion of persons who will purchase the product after exposure to this display is

$$.38 \pm 1.645(.016) \quad \text{or} \quad .38 \pm .026. \qquad \blacktriangle$$

The confidence interval for π is based on a normal approximation to a binomial, which is appropriate provided n is sufficiently large. The rule we've specified is that both $n\pi$ and $n(1 - \pi)$ should be at least 5, but since π is the unknown parameter, we'll require that $n\hat{\pi}$ and $n(1 - \hat{\pi})$ be at least 5. When the sample size is too small and violates this rule, the confidence interval usually will be too wide to be of any use. For example, with $n = 20$ and $\hat{\pi} = .2$, the rule is not satisfied, since $n\hat{\pi} = 4$. The 95% confidence interval based on these data would be $.025 < \pi < .375$, which is practically useless. Very few product managers would be willing to launch a new product if the expected increase in market share was between .025 and .375.

Keep in mind, however, that a sample size that is sufficiently large to satisfy the rule *does not* guarantee that the interval will be informative. It only judges the adequacy of the normal approximation to the binomial—the basis for the confidence level.

Sample size calculations for estimating π follow very closely the procedures we developed for inferences about μ. In Chapter 5, the required sample size for a $100(1 - \alpha)\%$ confidence interval for π of the form $\hat{\pi} \pm E$ (where E is specified) is found by solving the expression

$$z_{\alpha/2}\sigma_{\hat{\pi}} = E$$

for n. This result is shown here.

Sample Size Required for a $100(1 - \alpha)\%$ Confidence Interval for π of the Form $\hat{\pi} \pm E$

$$n = \frac{z_{\alpha/2}^2 \pi(1 - \pi)}{E^2}$$

Note: Since π is not known, either substitute an educated guess or use $\pi = .5$. Use of $\pi = .5$ will generate the largest possible sample size for the specified confidence interval width, $2E$, and will thus give a conservative answer to the required sample size.

EXAMPLE 8.4 ▼

A large public opinion polling agency plans to conduct a national survey to determine the proportion of employed adults who fear losing their job within the next year. How many workers must be polled to estimate to within .02 using a 95% confidence interval?

Solution   By design, the agency wants the interval of the form $\hat{\pi} \pm .02$. The sample size necessary to achieve this accuracy is given by

$$n = \frac{z_{\alpha/2}^2 \pi(1 - \pi)}{E^2},$$

where $z_{\alpha/2} = 1.96$ and $E = .02$. If a previous survey has been run recently, we could use the sample proportion from that survey to substitute for π; otherwise we could use $\pi = .5$. Using $\pi = .5$ the required sample size is

$$n = \frac{(1.96)^2(.5)(.5)}{(.02)^2} = 2,401.$$

That is, 2,401 workers would have to be surveyed to estimate π to within .02.   ▲

A statistical test about a binomial parameter π is very similar to the large-sample test concerning a population mean presented in Chapter 5. These results are summarized next, with three different alternative hypotheses along with their corresponding rejection regions. Recall that only one alternative is chosen for a particular problem.

Summary of a Statistical Test for π

H_0: $\pi = \pi_0$ (π_0 is specified)

H_a: **1.** $\pi > \pi_0$
         2. $\pi < \pi_0$
         3. $\pi \neq \pi_0$

T.S.: $z = \dfrac{\hat{\pi} - \pi_0}{\sigma_{\hat{\pi}}}$

R.R.:   For a probability α of a Type I error

         1.   reject H_0 if $z > z_\alpha$
         2.   reject H_0 if $z < -z_\alpha$
         3.   reject H_0 if $|z| > z_{\alpha/2}$.

         Note: Under H_0,

$$\sigma_{\hat{\pi}} = \sqrt{\frac{\pi_0(1 - \pi_0)}{n}}$$

EXAMPLE 8.5 ▼

Sports car owners in a town complain that their cars are judged differently from family-style cars at the state vehicle inspection station. Previous records indicate that 30% of all passenger cars fail the inspection on the first time through. In a random sample of 150 sports cars, 60 failed the inspection on the first time through. Is there sufficient evidence to indicate that the percentage of first failures for sports cars is higher than the percentage for all passenger cars? Use $\alpha = .05$.

Solution The appropriate statistical test is as follows.

$$H_0: \quad \pi = .30$$
$$H_a: \quad \pi > .30$$
$$\text{T.S.:} \quad z = \frac{\hat{\pi} - \pi_0}{\sigma_{\hat{\pi}}}$$

R.R.: For $\alpha = .05$, we will reject H_0 if $z > 1.645$.

Using the sample data,

$$\hat{\pi} = \frac{60}{150} = .4 \quad \text{and} \quad \sigma_{\hat{\pi}} = \sqrt{\frac{(.3)(.7)}{150}} = .037.$$

Also,

$$n\pi_0 = 150(.3) = 45$$

and

$$n(1 - \pi_0) = 150(.7) = 105.$$

The test statistic is then

$$z = \frac{.4 - .3}{.037} = 2.7.$$

Since the observed value of z exceeds 1.645, we conclude that sports cars at the vehicle inspection station have a first-failure rate greater than .3. However, we must be careful not to attribute this difference to a difference in standards for sports cars and family-style cars. Parallel testing of sports cars versus other cars would have to be conducted to eliminate other sources of variability that would perhaps account for the higher first-failure rate for sports cars. ▲

The z test for π, like the confidence interval for π based on z, depends on the adequacy of the normal approximation to the binomial. When can you use the z test for π? Generally speaking, you should view the results of a z test for π skeptically if either $n\pi_0$ or $n(1 - \pi_0)$ is 2 or less. If both $n\pi_0$ and $n(1 - \pi_0)$ are at least 5, the z test should be accurate. But for the same sample size n, z tests based on more extreme values of π_0 are less accurate than are those based on values of π_0 closer to .5. For example, for $n = 5,000$, a test of $H_0: \pi = .001$ (for which $n\pi_0 = 5$) would be much more suspect than would a test of $H_0: \pi = .01$ (for which $n\pi_0 = 50$).

▼ EXERCISES

Basic Techniques

8.17 Hypothetical sample results from a binomial experiment with $n = 150$ yielded $\hat{\pi} = .2$.
  a. Does this experiment satisfy the sample-test requirement for a confidence interval for π based on z? What sample sizes would be suspect, given the same sample proportion?
  b. Construct a 90% confidence interval for π.

8.18 Under what conditions can the formula $\hat{\pi} \pm z_{\alpha/2}\sigma_{\hat{\pi}}$ be used to express a confidence interval for π?

8.19 A random sample of 1,500 is drawn from a binomial population. If there are $y = 1,200$ successes,
  a. Construct a 95% confidence interval for π.
  b. Construct a 90% confidence interval for π.

8.20 Refer to the previous exercise. Explain the difference in the interpretation of the two confidence intervals.

Applications

8.21 Experts have predicted that approximately 1 in 12 tractor-trailer units will be involved in an accident this year. One of the reasons for this is that 1 in 3 tractor-trailer units has an imminently hazardous mechanical condition, probably related to the braking systems on the vehicle. A survey of 50 tractor-trailer units passing through a weighing station confirmed that 19 had a potentially serious braking system problem.
  a. Do the binomial assumptions hold?
  b. Can a normal approximation to the binomial be applied here to get a confidence interval for π?
  c. Give a 95% confidence interval for π using these data. Is the interval informative? What could be done to decrease the width of the interval, assuming $\hat{\pi}$ remained the same?

8.22 In a study of self-medication practices, a random sample of 1,230 adults completed a survey. Some of the medical conditions that were self-treated are shown below. Summarize the results of this part of the survey using a 95% confidence interval for each medical condition.

Medical Condition	Home Remedy	% Responding
Sore throat — not related to a cold	Salt water or baking soda mouth wash	30
Burns — other than sunburn	Cold water/butter	28
Overindulgence in alcohol	Homebrew	25
Overweight	Diet	22
Pain associated with injury	Hot or cold compress	21

8.23 In the survey discussed in Exercise 8.22, 441 of the adults reported they had a cough or cold recently and 260 of the respondents said they had treated the condition with an over-the-counter (OTC) remedy. These data are summarized here.

Survey respondents reporting problem	441
Number of patients using any OTC remedy	260
Patients using specific classes of OTC remedies:	
Adult pain relievers	110
Adult cold caps/tabs	57
Cough remedies	44
Allergy/hay fever remedies	9
Liquid cold remedies	35
Sprays/inhalers	4
Children's pain reliever	22
Cough drops	13
Sore-throat lozenges/gum	9
Children's cold caps/tabs	13
Nose drops	9
Chest rubs/ointments	9
Anesthetic throat lozenges	4
Room vaporizers	4
Other product	4

a. How might these data be organized and summarized? Would percentages help? Do the percentages add to 100%? Why or why not?

b. Based on these data, which classes of OTC remedies could be summarized using a 95% confidence interval for π?

8.24 Many individuals over the age of 40 develop an intolerance for milk and milk-based products. A dairy has developed a line of lactose-free products that are more tolerable to such individuals. To assess the potential market for these products, the dairy commissioned a market research study of individuals over 40 in its sales area. A random sample of 250 individuals showed that 86 of them suffer from milk intolerance. Calculate a 90%

confidence interval for the population proportion that suffers milk intolerance based on the sample results.

 8.25 Shortly before April 15 of the previous year, a team of sociologists conducted a survey to study their theory that tax cheaters tend to allay their guilt by holding certain beliefs. A total of 500 adults were interviewed and asked under what situations they think cheating on an income tax return is justified. The responses include:

56% agree that "other people don't report all their income."
50% agree that "the government is often careless with tax dollars."
46% agree that "cheating can be overlooked if one is generally law abiding."

Assuming the data are a simple random sample of the population of taxpayers (or tax-nonpayers), calculate 95% confidence intervals for the population proportion that agrees with each statement.

 8.26 A national columnist recently reported the results of a survey on marriage and the family. Part of the column has been paraphrased here.

The Ingredients of Marriage

The Gallup people offered respondents a list of well-known ingredients. Here in the United States, such elements as faithfulness, mutual respect, and understanding ranked at the top. These were followed by enough money, same background, good housing and agreement in politics. Seventy-five percent of the respondents voted for "a good sex life," 59% for children, 52% for common interests, 48% for "living away from in-laws," and 43% for "sharing household chores." (In West Germany, by contrast, only 52% voted for a good sex life and only 19% for sharing household chores.)

a. How could you display the results of this survey in a graph or table?
b. Would you use a confidence interval to convey more information about the "true" percentages expressing an opinion on the various ingredients of a good marriage? Why or why not?
c. What qualms might you have about the way this survey has been reported?

 8.27 A substantial part of the U.S. population is "technologically illiterate," according to experts at a National Technological Literacy Conference organized by the National Science Foundation and Pennsylvania State University. At this conference, the results of a national survey of 2,000 adults showed that:

▼ 70% do not understand radiation
▼ 40% think space-rocket launchings change the weather and that some unidentified flying objects are actually visitors from other planets
▼ more than 80% do not understand how telephones work
▼ 75% do not have a clear understanding of what computer software is
▼ 72% do not understand the gross national product.

a. How might you display these data in a graph or table? Construct the display.
b. The problem with many newspaper articles reporting the results of a survey is that conclusions are given without sufficient details about the study for the reader to assess the data and reach a separate conclusion. What details are missing here for you to reach your own conclusion?

8.28 More and more people are dining out—so say the results of national surveys. Compared to 1978, here are some figures

Meal Eaten Away from Home	1978	Now
Breakfast	3%	5%
Lunch	18%	20%
Dinner	16%	16%

a. If these data were based on random samples of 1,500 adults in 1978 and at present, what conclusions can be drawn for each meal? Is a normal approximation to the binomial valid here?

b. Can we conclude from the data shown here that more people are eating out? Why or why not?

8.29 The benign mucosal cyst is the most common lesion of a pair of sinuses in the upper jawbone. In a random sample of 800 males, 35 persons were observed to have a benign mucosal cyst.

a. Would it be appropriate to use a normal approximation in conducting a statistical test of the null hypothesis $H_0: \pi = .096$ (the highest incidence in previous studies among males)? Explain.

b. Conduct a statistical test of the research hypothesis $H_a: \pi < .096$. Use $\alpha = .05$.

8.30 National public opinion polls are based on interviews of as few as 1,500 persons in a random sampling of public sentiment toward one or more issues. These interviews are commonly done in person, because mail returns are poor and telephone interviews tend to reach older people, thus biasing the results. Suppose that a random sample of 1,500 persons is surveyed to determine the proportion of the adult public in agreement with recent energy conservation proposals.

a. If 560 indicate they favor the policies set forth by the current administration, estimate π, the proportion of adults holding a "favor" opinion. Use a 95% confidence interval. What is the half width of the confidence interval?

b. How many persons must be surveyed to have a 95% confidence interval with a half width of .01?

8.31 A sample of 20 crayfish of all sizes was obtained from a large lake to estimate the proportion of crayfish that exhibit more than 9 (ppb) units of mercury. Of those sampled, eight exceeded 9 units. Use these data to estimate π, the proportion of all crayfish in the lake with a mercury level greater than 9, using a 95% confidence interval.

8.32 Simulate the binomial distribution for $n = 20$ and $\pi = .4$, using a computer program. Do this by obtaining y, the number of successes in 20 trials when sampling from a binomial distribution with $\pi = .4$. Repeat this experiment 39 more times, for a total of 40 repetitions of the experiment.

a. Plot the sample data (y-values) in a relative frequency histogram.

b. Compute the sample mean and standard deviation. Compare your answers to the *actual* mean and standard deviation of y. (Hint: A Minitab program is given here for purposes of illustration.)

```
40 BINOMIAL EXPERIMENTS WITH n = 20 AND P(SUCCESS)= 0.4.


MTB > NAME C1 ´VALUE´
MTB > RANDOM 40 C1;
SUBC> BINOMIAL 20 0.4.
MTB > TALLY C1

     VALUE   COUNT
        4      3
        5      3
        6      5
        7      6
        8      6
        9      9
       10      4
       12      1
       13      2
       16      1
     N=       40

MTB > HISTOGRAM C1

Histogram of VALUE    N = 40

Midpoint   Count
       4      3    ***
       5      3    ***
       6      5    *****
       7      6    ******
       8      6    ******
       9      9    *********
      10      4    ****
      11      0
      12      1    *
      13      2    **
      14      0
      15      0
      16      1    *

MTB > MEAN C1
     MEAN    =      8.0500
MTB > STANDARD DEVIATION C1
     ST.DEV. =      2.5615
MTB > STOP
```

8.33   Refer to Example 8.4. Suppose a recently done survey resulted in $\hat{\pi} = .15$. Use this guessed value in the computation of an appropriate sample size. Comment on the differences in your answer here and that in Example 8.4.

▼
8.4 OPERATING CHARACTERISTIC CURVES AND CONTROL CHARTS FOR π (optional)

Two techniques are particularly appropriate for monitoring product quality as measured by π, the fraction of items that are defective. Control charts were first discussed in Chapter 5, where we presented the $\bar{y}$-chart; in Chapter 7, we discussed the s-chart and r-chart as well. A similar type of chart can be used when we want to monitor π as a measure of the ongoing quality of a manufacturing process. We are especially interested in detecting a shift in product

lot acceptance sampling

quality. The second technique, called **lot acceptance sampling**, provides a means to screen (sample) incoming raw materials or outgoing production from a plant where the product is shipped in large quantities (often called "lots"). We begin by discussing lot acceptance sampling.

Manufacturers are interested in minimizing not only the amount or proportion of defective raw material to be used in the production process, but also the proportion of defective finished products shipped from the plant. Thus, they would like to sample, or screen, shipments of raw materials entering the plant and reject those shipments (lots) that contain too high a proportion of defectives. Similarly, they must screen the final product to make certain that a shipment does not contain too high a proportion of defectives.

The most obvious type of screen (sampling plan) to employ would be a careful inspection of each item from the lot. Unfortunately, this screen would be both costly and time-consuming. In addition, it would still be subject to errors in reporting brought about by human fatigue.

statistical sampling plan

Another type of screen is called a **statistical sampling plan**. Here we obtain a random sample of n items from the lot. Each item of the sample is inspected, and if y, the number of defectives observed in the sample, is less than or equal to some predetermined number a, we accept the lot. Thus, a statistical sampling plan is designated by n, the sample size, and a, the acceptance number. If the lot is accepted ($y \leq a$), we conclude that the proportion of defectives π in the lot is small and acceptable. However, if $y > a$, we reject the lot and conclude that π is too large (above an acceptable level of defectives).

operating characteristic (OC) curve

We can characterize the goodness of a particular sampling plan (n, a) by constructing an **operating characteristic (OC) curve**. The OC curve for a sampling plan is a graph displaying the probability of accepting a lot for various values of π, the proportion of defective items in the lot (see Figure 8.2). As you can see, the probability of accepting a lot decreases as the proportion of defectives within the lot increases.

FIGURE 8.2
OC Curve for a Sampling Plan

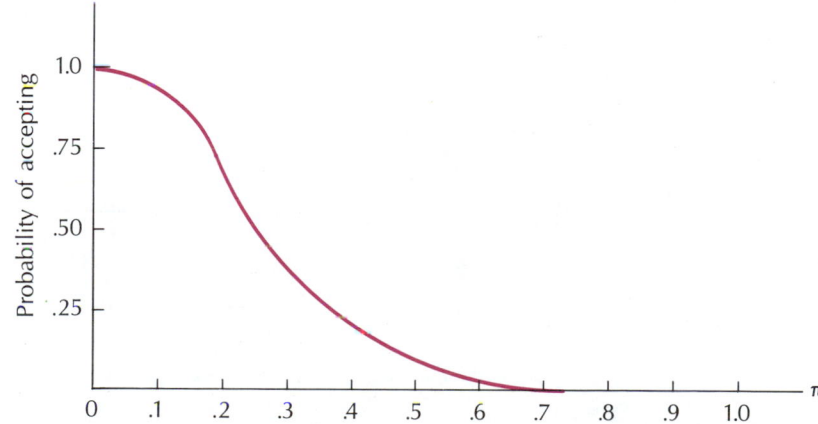

We can construct an OC curve by computing the probability of accepting a lot—namely, $P(y \leq a)$—for several values of π. Consider the sampling plan $n = 4$ and $a = 0$. Here we sample 4 items and accept the lot if y, the number of defectives, is zero. Hence, we must compute $P(y = 0)$ for $n = 4$ and for different values of π to obtain the OC curve. We use the binomial probability distribution

$$P(y) = \frac{n!}{y!(n-y)!} \, \pi^{y}(1 - \pi)^{n-y}.$$

Table 8.4 shows the results of the calculations for $\pi = .1, .2,$ and $.4$. Plotting these three points and connecting them, we have the OC curve as shown in Figure 8.3.

T A B L E 8.4

Calculation of $P(y = 0)$ for $n = 4$ and $\pi = .1, .2,$ and $.4$

Fraction of Defectives, π	Probability of Accepting, $P(y = 0)$
.1	$\dfrac{4!}{0!4!} \, (.1)^{0}(.9)^{4} = .656$
.2	$\dfrac{4!}{0!4!} \, (.2)^{0}(.8)^{4} = .410$
.4	$\dfrac{4!}{0!4!} \, (.4)^{0}(.6)^{4} = .130$

F I G U R E 8.3

OC Curve for the Sampling Plan ($n = 4$, $a = 0$)

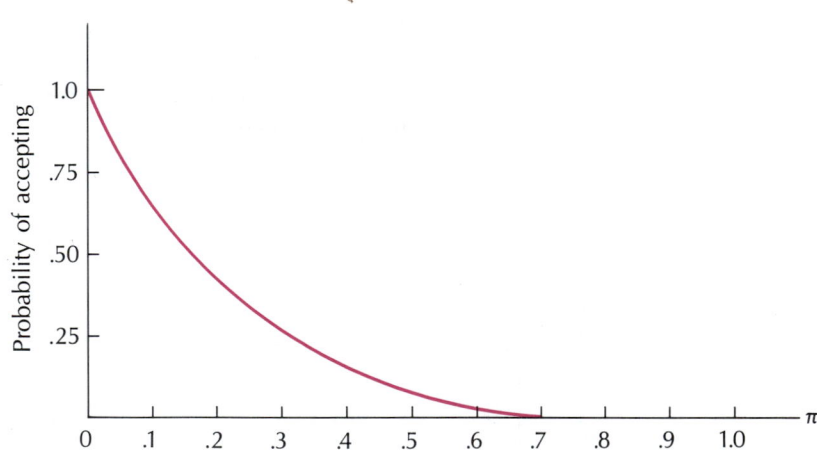

E X A M P L E 8.6

Construct an operating characteristic curve for the sampling plan ($n = 10$, $a = 1$).

Solution The probability of accepting the lot is given by $P(y \leq 1)$. Using the binomial probability distribution, we must calculate $P(y = 0) + P(y = 1)$ for $n = 10$ and various values of π. For $\pi = .1$ and $n = 10$,

$$P(y = 0) = \frac{10!}{0!10!} (.1)^0 (.9)^{10} = .349$$

$$P(y = 1) = \frac{10!}{1!9!} (.1)^1 (.9)^9 = .387.$$

Hence, the probability of accepting the lot is $.349 + .387 = .736$. Similarly, for $\pi = .2$ and $.4$, the probabilities of accepting the lot are found to be $.376$ and $.046$, respectively. Graphing our results, we have the OC curve presented in Figure 8.4.

FIGURE 8.4

OC Curve for the Sampling Plan ($n = 10$, $a = 1$), Example 8.6

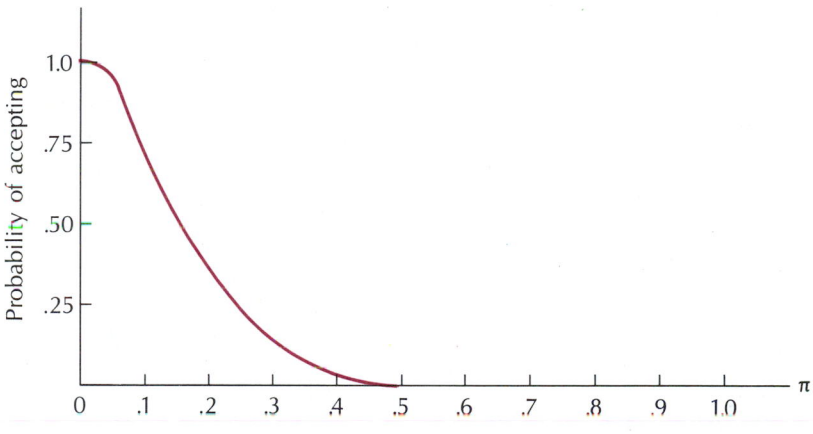

Several comments should be made concerning statistical sampling plans. First, each plan (n, a) is unique, so the inspector must choose a sampling plan that possesses characteristics suitable for his or her particular problem. In general, for a fixed value of a, an increase in n makes the graph of the probability of acceptance drop sharply as π increases; increasing a for a fixed n increases the probability of acceptance for values of π. Second, the OC curve for a particular sampling plan can be thought of as a plot of the probability of a Type II error for the null hypothesis $H_0: \pi = 0$ for various actual values of π, when the rejection region is $y > a$.

The other method of monitoring product quality makes use of a control chart for π. Recall from our previous work that a control chart typically consists

FIGURE 8.5
**Control Chart for the
Sample Proportion $\hat{\pi}$**

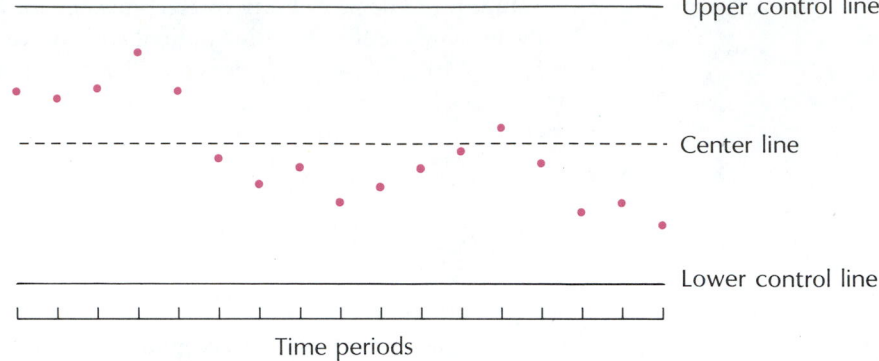

Time periods

of three lines. In a control chart for π, successive sample proportions $\hat{\pi}$ would be plotted and might appear as shown in Figure 8.5. If one of the sample proportions falls outside either the upper or lower control line, the process is judged to be out of control; that is, the proportion of defectives π in the production has shifted.

We can compute the three control lines in the following manner. The center line is designated by $\hat{\pi}_c$. If we obtain k different random samples of n observations each, $\hat{\pi}_c$ is the sample proportion of defectives for the entire set of kn measurements, or, equivalently, $\hat{\pi}_c$ is the average of the sample proportions computed for the k samples of n measurements.

The upper control line (UCL) and lower control line (LCL) are then

UCL

$$\text{UCL} = \hat{\pi}_c + 3 \sqrt{\frac{\hat{\pi}_c(1 - \hat{\pi}_c)}{n}}$$

and

LCL

$$\text{LCL} = \hat{\pi}_c - 3 \sqrt{\frac{\hat{\pi}_c(1 - \hat{\pi}_c)}{n}}.$$

EXAMPLE 8.7

A pharmaceutical firm has been investigating the possibility of having hospital personnel supplied with small disposable vials that can be used to perform many of the standard laboratory analyses. For a particular analysis, such as blood sugar, the technician would insert a measured amount of fluid (perhaps blood) in an appropriate vial and observe its color when thoroughly mixed with the fluid already stored in the vial. By comparing the optical density of the combined fluid to a color-coded chart, the technician would have a reading on the blood sugar level of the patient. Quite obviously, the system must be tightly controlled to ensure that the vials are correctly sealed with the proper amount of fluid prior to shipment to the hospital laboratories. The data in Table 8.5 give the proportion of defectives in 30 different samples (taken from 30 different production hours) of 50 vials each. Use these data to construct the three control lines.

TABLE 8.5
Thirty Sample Proportions Each Based on 50 Observations, Example 8.7

Sample	Sample Proportion Defective	Sample	Sample Proportion Defective
1	.18	16	.18
2	.10	17	.14
3	.18	18	.16
4	.16	19	.14
5	.12	20	.12
6	.12	21	.18
7	.16	22	.16
8	.18	23	.18
9	.12	24	.18
10	.12	25	.14
11	.16	26	.16
12	.18	27	.18
13	.12	28	.16
14	.18	29	.18
15	.14	30	.10

Solution   The sum of the 30 sample proportions is 4.58, so

$$\hat{\pi}_c = \frac{4.58}{30} = .15 \quad \text{and} \quad 1 - \hat{\pi}_c = .85.$$

Substituting into the formulas for the control limits, we have

$$\text{UCL} = .15 + 3\sqrt{\frac{(.15)(.85)}{50}} = .15 + 3(.05) = .30$$

and

$$\text{LCL} = .15 - 3\sqrt{\frac{(.15)(.85)}{50}} = .15 - 3(.05) = 0.$$

▲

We have discussed the binomial probability distribution and various count data problems that utilize the binomial distribution. In the next section, we will consider inferences concerning two binomial parameters.

▼ EXERCISES

Basic Techniques

8.34   Sketch the operating characteristic curve for the sampling plan ($n = 5$, $a = 0$).

8.35   Refer to Exercise 8.34. Superimpose the OC curve for the sampling plan with $n = 10$ and $a = 0$ on that for $n = 5$, $a = 0$. What is the effect of increasing the sample size n while holding the acceptance number a constant?

Application

8.36   Refer to the control limits for π obtained in Example 8.7. Suppose that you are now in charge of quality control for the vial-production line. In the next 15 samples of 50 vials, you observe the following numbers of defectives:

Number	1	2	3	4	5	6	7	8	9	10	11	12	13	14	15
Defective	9	8	6	5	2	4	6	8	9	9	12	13	14	10	12

Determine whether the process has remained in control.

8.5 ▼ COMPARING TWO BINOMIAL PROPORTIONS

Many practical problems involve the comparison of two binomial parameters. For example, social scientists may wish to compare the proportions of women who take advantage of prenatal health services for two communities representing different socioeconomic backgrounds. Or, the director of marketing may wish to compare the public awareness of a new product recently launched and that of a competitor's product.

For comparisons of this type, we assume that independent random samples are drawn from two binomial populations with unknown parameters designated by π_1 and π_2. If y_1 successes are observed for the random sample of size n_1 from population 1 and y_2 successes are observed for the random sample of size n_2 from population 2, then the point estimates of π_1 and π_2 are the observed sample proportions $\hat{\pi}_1$ and $\hat{\pi}_2$, respectively:

$$\hat{\pi}_1 = \frac{y_1}{n_1} \quad \text{and} \quad \hat{\pi}_2 = \frac{y_2}{n_2}.$$

This notation is summarized here.

Notation For Comparing Two Binomial Proportions

▼

	Population	
	1	2
Population proportion	π_1	π_2
Sample size	n_1	n_2
Number of successes	y_1	y_2
Sample proportion	$\hat{\pi}_1 = \dfrac{y_1}{n_1}$	$\hat{\pi}_2 = \dfrac{y_2}{n_2}$

Inferences about two binomal proportions are usually phrased in terms of their difference $\pi_1 - \pi_2$, and we use the difference in sample proportions $\hat{\pi}_1 - \hat{\pi}_2$ as part of a confidence interval or statistical test. The sampling distribution for $\hat{\pi}_1 - \hat{\pi}_2$ can be approximated by a normal distribution with mean and standard error given by

$$\mu_{\hat{\pi}_1 - \hat{\pi}_2} = \pi_1 - \pi_2$$

and

$$\sigma_{\hat{\pi}_1 - \hat{\pi}_2} = \sqrt{\frac{\pi_1(1 - \pi_1)}{n_1} + \frac{\pi_2(1 - \pi_2)}{n_2}}.$$

This approximation is appropriate if we apply the same requirements to both binomal populations that we did in recommending a normal approximation to a binomial (see Chapter 4). Thus, the normal approximation to the distribution of $\hat{\pi}_1 - \hat{\pi}_2$ is appropriate if both $n\pi$ and $n(1 - \pi)$ are 5 or more for *each* population.

Since π_1 and π_2 are never known, make your judgment on the validity of the approximation using $n\hat{\pi}$ and $n(1 - \hat{\pi})$ for each sample.

Confidence intervals and statistical tests about $\pi_1 - \pi_2$ are straightforward and follow the format we used for comparisons using $\mu_1 - \mu_2$. Interval estimation is summarized here; it takes the usual form, point estimate $\pm z$ (standard error).

100(1 − α)% Confidence Interval for $\pi_1 - \pi_2$

$$\hat{\pi}_1 - \hat{\pi}_2 \pm z_{\alpha/2}\sigma_{\hat{\pi}_1-\hat{\pi}_2},$$

where

$$\sigma_{\hat{\pi}_1-\hat{\pi}_2} = \sqrt{\frac{\pi_1(1 - \pi_1)}{n_1} + \frac{\pi_2(1 - \pi_2)}{n_2}}.$$

Note: Substitute $\hat{\pi}_1$ and $\hat{\pi}_2$ for π_1 and π_2 in the formula for $\sigma_{\hat{\pi}_1-\hat{\pi}_2}$. When the normal approximation is valid for $\hat{\pi}_1 - \hat{\pi}_2$, very little error will result from this substitution.

EXAMPLE 8.8

In a survey to analyze the funeral expenditure for various social classes, a random sample of 162 families from the working (blue-collar) class was taken to determine the funeral expenses for a recent death in the family. Of the 162 families contacted, 61 spent over $800 on the funeral. A similar survey was conducted within the middle/upper classes. Of 189 families contacted, 106 spent more than $800. Estimate $\pi_1 - \pi_2$, the difference in the proportions of families who have spent more than $800 for a recent family death. Use a 95% confidence interval to interpret your findings.

Solution The point estimate of $\pi_1 - \pi_2$ is the difference in sample proportions, $\hat{\pi}_1 - \hat{\pi}_2$:

$$\hat{\pi}_1 - \hat{\pi}_2 = \frac{61}{162} - \frac{106}{189} = .376 - .561 = -.185.$$

Note also that $n\hat{\pi}$ and $n(1 - \hat{\pi})$ are 5 or more for both samples, implying that the normal approximation to the binomial is appropriate.

The standard error for $\hat{\pi}_1 - \hat{\pi}_2$ is estimated by

$$\sqrt{\frac{\hat{\pi}_1(1 - \hat{\pi}_1)}{n_1} + \frac{\hat{\pi}_2(1 - \hat{\pi}_2)}{n_2}} = \sqrt{\frac{.376(.624)}{162} + \frac{.561(.439)}{189}} = .052.$$

A 95% confidence interval for $\pi_1 - \pi_2$ has $z_{\alpha/2} = 1.96$ and is of the form

Point estimate $\pm z_{\alpha/2}$ (standard error).

Substituting into this formula we have

$$-.185 + 1.96(.052) \qquad \text{or} \qquad -.185 \pm .102.$$

This interval indicates that π_2 is larger than π_1; we are 95% confident that the difference in the proportions of families paying more than $800 per funeral for the working class (π_1) and the middle/upper class (π_2) lies in the interval $-.287$ to $-.083$. ▲

We can readily formulate a statistical test for the equality of two binomial parameters. The test statistic for testing $H_0: \pi_1 - \pi_2 = 0$ is a z statistic having the familiar form

$$z = \frac{\text{point estimate}}{\text{standard error}} = \frac{\hat{\pi}_1 - \hat{\pi}_2}{\sigma_{\hat{\pi}_1 - \hat{\pi}_2}}.$$

The standard error is slightly different from what we used for a confidence interval. When H_0 is true, $\pi_1 = \pi_2$; we'll call the common value π. Then

$$\sigma_{\hat{\pi}_1 - \hat{\pi}_2} = \sqrt{\frac{\pi_1(1 - \pi_1)}{n_1} + \frac{\pi_2(1 - \pi_2)}{n_2}} = \sqrt{\pi(1 - \pi)\left(\frac{1}{n_1} + \frac{1}{n_2}\right)}.$$

The best estimate of π, the proportion of successes common to both populations, is

$$\hat{\pi} = \frac{\text{total number of successes}}{\text{total number of trials}} = \frac{y_1 + y_2}{n_1 + n_2}.$$

We have summarized the test procedure here.

Statistical Test for Comparing Two Binomial Proportions

H_0: $\pi_1 - \pi_2 = 0$

H_a: 1. $\pi_1 - \pi_2 > 0$
2. $\pi_1 - \pi_2 < 0$
3. $\pi_1 - \pi_2 \neq 0$

T.S.: $z = \dfrac{\hat{\pi}_1 - \hat{\pi}_2}{\sigma_{\hat{\pi}_1 - \hat{\pi}_2}}$, where $\sigma_{\hat{\pi}_1 - \hat{\pi}_2} = \sqrt{\pi(1-\pi)\left(\dfrac{1}{n_1} + \dfrac{1}{n_2}\right)}$

and π is approximated by $\hat{\pi} = \dfrac{y_1 + y_2}{n_1 + n_2}$.

R.R.: For a given value of α,

1. reject H_0 if $z > z_\alpha$
2. reject H_0 if $z < -z_\alpha$
3. reject H_0 if $|z| > z_{\alpha/2}$.

Note: $n\hat{\pi}$ and $n(1-\hat{\pi})$ must be greater than or equal to 5 for both populations in order for the normal approximation (and hence for this test) to hold.

E X A M P L E   8.9

In a recent survey of county high school students ($n_1 = 100$ males and $n_2 = 100$ females), 58 of the males and 46 of the females sampled said they consume alcohol on a regular basis. Use the sample data to conduct a test $H_0: \pi_1 - \pi_2 = 0$ against the one-sided alternative $H_a: \pi_1 - \pi_2 > 0$, that a higher proportion of males than females consume alcohol on a regular basis. Use $\alpha = .05$.

Solution The four parts of the statistical test are shown here:

H_0: $\pi_1 - \pi_2 = 0$
H_a: $\pi_1 - \pi_2 > 0$

T.S.: $z = \dfrac{\hat{\pi}_1 - \hat{\pi}_2}{\sigma_{\hat{\pi}_1 - \hat{\pi}_2}}$, where $\sigma_{\hat{\pi}_1 - \hat{\pi}_2} = \sqrt{\pi(1-\pi)\left(\dfrac{1}{n_1} + \dfrac{1}{n_2}\right)}$

R.R.: For $\alpha = .05$, reject H_0 if $z > 1.645$.

From the sample data we find

$$\hat{\pi}_1 = \frac{58}{100} = .58, \qquad \hat{\pi}_2 = \frac{46}{100} = .46, \qquad \text{and} \qquad \hat{\pi} = \frac{58 + 46}{100 + 100} = .52.$$

Note also that $n\hat{\pi}$ and $n(1 - \hat{\pi})$ are 5 or more for both samples, validating the normal approximations to the binomial.

Substituting into the test statistic, we obtain

$$z = \frac{.58 - .46}{\sqrt{.52(.48)\left(\dfrac{1}{100} + \dfrac{1}{100}\right)}} = \frac{.12}{.071} = 1.69.$$

Conclusion: Since $z = 1.69$ exceeds 1.645, we reject $H_0: \pi_1 - \pi_2 = 0$; we have shown that a higher proportion of high school males than females in the county studied consumes alcohol on a regular basis. ▲

▼ EXERCISES

Basic Techniques

8.37 A random sample of $n_1 = 1,000$ observations was obtained from a binomial population with $\pi_1 = .4$. Another random sample, independent of the first sample, was selected from a binomial population with $\pi_2 = .2$. Does the normal approximation hold? Describe the sampling distribution for $\hat{\pi}_1 - \hat{\pi}_2$.

8.38 In a study to compare two binomial proportions, $n_1 = 50$, $n_2 = 40$, $y_1 = 20$, and $y_2 = 15$. Use these hypothetical data to construct a 90% confidence interval for $\pi_1 - \pi_2$.

8.39 Refer to Exercise 8.38. How large a sample should we take from each population in order to have a 90% confidence interval of the form $\hat{\pi}_1 - \hat{\pi}_2 \pm .01$? (Hint: Assuming that equal sample sizes will be taken from the two populations, solve the expression

$$z_{\alpha/2}\sigma_{\hat{\pi}_1 - \hat{\pi}_2} = .01$$

for n, the common sample size. Use $\hat{\pi}_1 = .40$ and $\hat{\pi}_2 = .375$ from Exercise 8.38.)

Applications

 8.40 A law student believes that the proportion of registered Republicans in favor of additional tax incentives is greater than the proportion of registered Democrats in favor of such incentives. The student acquired independent random samples of 200 Republicans and 200 Democrats and found 109 Republicans and 86 Democrats in favor of additional tax incentives. Use these data to test $H_0: \pi_1 - \pi_2 = 0$ versus H_a: $\pi_1 - \pi_2 > 0$. Give the level of significance for your test.

8.41 In a comparison of the incidence of tumor potential in two strains of rats, 100 rats (50 males, 50 females) were selected from each of two strains and were examined for a period of 1 year. All the rats were approximately the same age and were housed and fed under comparable conditions. Use the accompanying 1-year sample data to construct a 95% confidence interval for the difference in the proportions of rats exhibiting tumor potential for the two strains.

	Strain A	Strain B
Sample size	100	100
Number exhibiting tumor potential	25	15

8.42 There is a remedy for male pattern baldness—at least that's what millions of males hope since the FDA approved Upjohn's minoxidil for such a use. Minoxidil was investigated in a large, 27-center study where patients were randomly assigned to receive topical minoxidil or an identical-appearing placebo. Ignoring the center-to-center variation, suppose the preliminary results were as follows:

	Sample Size	% with New Hair Growth
Minoxidil group	310	32
Placebo	309	20

a. Use these data to test $H_0: \pi_1 - \pi_2 = 0$ versus $H_a: \pi_1 - \pi_2 \neq 0$. Give the p-value for your test.

b. If you were working for the FDA, what additional information might you want to examine in this study?

8.43 Is cocaine deadlier than heroin? A study reported in the *Journal of the American Medical Association* found that rats with unlimited access to cocaine had poorer health, had more behavior disturbances, and died at a higher rate than did a corresponding group of rats given unlimited access to heroin. The death rates after 30 days on the study were as follows:

	% Dead at 30 Days
Cocaine group	90
Heroin group	36

a. Suppose that 100 rats were used in each group. Conduct a test of $H_0: \pi_1 - \pi_2 = 0$ versus $H_a: \pi_1 - \pi_2 > 0$. Give the p-value for your test.

b. What implications are there for human use of the two drugs?

8.6 ▼ THE POISSON DISTRIBUTION

In Chapter 4 (and again in this chapter), we indicated that the normal distribution provides a good approximation to the binomial distribution provided $n \geq 5/\min (\pi, 1 - \pi)$. This requirement was needed to ensure that the binomial distribution was reasonably symmetric. However, there are many instances when the binomial probability distribution is sufficiently skewed so as to render the normal approximation inappropriate. For example, in observing patients administered a new drug product in a properly conducted clinical trial, the number of persons experiencing a particular side effect might be quite small. If π (the probability of observing a person with the side effect) is .001, $\min(\pi, 1 - \pi) = .001$. For this example, in order to approximate the binomial with a normal distribution, the sample size would have to be equal to or greater than $5/.001 = 5,000$.

Poisson distribution

In 1837, S. D. Poisson developed a discrete probability distribution, suitably called the **Poisson distribution**, which provides a good approximation to the binomial when π is small and n is large but $n\pi$ is less than 5. The probability of observing y successes in the n trials is given by the formula

$$P(y) = \frac{\mu^y e^{-\mu}}{y!},$$

e = 2.71828

where e is a constant approximately equal to 2.71828, and μ is the average value of y. Table 7 in the Appendix gives Poisson probabilities for various values of the parameter μ. For approximating binomial probabilities using the Poisson distribution, take

$$\mu = n\pi.$$

E X A M P L E   8.10

▼

Refer to the clinical trial alluded to at the beginning of this section, where $n = 1,000$ patients were treated with a new drug. Compute the probability that none of a sample of $n = 1,000$ patients experiences a particular side effect (such as nausea) when $\pi = .001$.

Solution   The mean of the binomial distribution is $\mu = n\pi = 1,000(.001) = 1$. Substituting into the Poisson probability distribution with $\mu = 1$, we have

$$P(y = 0) = \frac{(1)^0 e^{-1}}{0!} = e^{-1} = \frac{1}{2.71828} = .367879.$$

(Note also from Table 7 in the Appendix that the entry corresponding to $y = 0$ and $\mu = 1$ is 0.3679.)   ▲

E X A M P L E   8.11   ▼

Suppose that after a clinical trial of a new medication involving 1,000 patients, no patient experienced nausea. Would it be reasonable to infer that less than .001 of the entire population would experience this side effect while taking the drug?

Solution   Certainly not. We computed the probability of observing $y = 0$ in $n = 1,000$ trials assuming $\pi = .001$ (i.e., assuming .1% of the population would experience nausea) to be .368. Since this probability is quite large, it would not be wise to infer that $\pi < .001$.   ▲

Although the Poisson distribution provides a useful approximation to the binomial under certain conditions, the application of the Poisson distribution is not limited to these situations. In particular, the Poisson distribution has been useful in finding the probability of y occurrences of an event that *occurs randomly* over an interval of time, volume, space, and so on, provided certain assumptions are met.

1.   Events occur one at a time; two or more events do not occur precisely at the same time.
2.   The occurrence of an event in a given period is independent of the occurrence of the event in a nonoverlapping period; that is, the occurrence (or nonoccurrence) of an event during one period does not change the probability of an event occurring in some later period.

In many discussions of this topic, a third assumption is added:

3.   The expected number of events during any one period is the same as that during any other period.

Although the underlying mathematics is made easier with this third assumption, this appears to be irrelevent; the first two assumptions are sufficient for a Poisson distribution to apply.

While these assumptions seem to be somewhat restrictive, many situations appear to satisfy these conditions. For example, the number of arrivals of customers at a checkout counter, parking lot toll booth, inspection station, or garage repair shop during a specified time interval (such as 1 minute) could be approximated with a Poisson probability distribution. Similarly, the number of clumps of algae of a particular species observed in a unit volume of lake water visible under a microscope could be approximated by a Poisson probability distribution.

Confronted with a set of measurements, we may now wish to check the assumption that the data follow a Poisson probability distribution. To do this we make use of the goodness-of-fit test of Section 8.2, using the test statistic

tests using Poisson distribution

$$\chi^2 = \sum_i \left[\frac{(n_i - E_i)^2}{E_i} \right].$$

There are two types of null hypotheses. The first hypothesis is that the data arise from a Poisson distribution with $\mu = \mu_0$; that is, we wish to test $H_0: \mu = \mu_0$ (μ_0 is specified) against the alternative hypothesis $H_a: \mu \neq \mu_0$. The quantity n_i denotes the number of observations in cell i and E_i is the expected number of observations in cell i obtained from the probabilities for a Poisson distribution with mean μ_0. The computed value of the test statistic is then compared to the tabulated chi-square value in Table 5 in the Appendix with $a = \alpha$ and df $= k - 1$, where k is the number of cells.

The second null hypothesis we might be interested in is less specific. We test

$H_0:$ The observed cell counts all come from a common Poisson distribution with mean μ (unspecified).

The alternative is that not all cell counts come from a common Poisson distribution. The test statistic is

$$\chi^2 = \sum_i \left[\frac{(n_i - E_i)^2}{E_i} \right],$$

where for all cells E_i is the expected number of observations in cell i obtained from the probabilities for a Poisson distribution with a mean estimated from the sample data. The rejection region is then located for $a = \alpha$ and df $= k - 2$. Note the difference in the degrees of freedom for the two null hypotheses. In the latter test, we lose one degree of freedom because we must estimate the Poisson parameter μ.

E X A M P L E   8.12   ▼

Environmental engineers often utilize information contained in the number of different alga species and the number of cell clumps per species to measure the health of a lake. Those lakes exhibiting only a few species but many cell clumps are classified as oligotrophic. In one such investigation, a lake sample was analyzed under a microscope to determine the number of clumps of cells per microscope field. These data are summarized below for 150 fields examined under a microscope. Here y_i denotes the number of cell clumps per field and n_i denotes the number of fields with y_i cell clumps.

y_i	0	1	2	3	4	5	6	≥ 7
n_i	6	23	29	31	27	13	8	13

Use $\alpha = .05$ to test the null hypothesis that the sample data were drawn from a Poisson probability distribution.

Solution   Before we can compute the value of χ^2, first we must estimate the Poisson parameter μ and then compute the expected cell counts. The Poisson mean μ is estimated by using the sample mean $\bar{y}$. For these data,

$$\bar{y} = \frac{\sum_i n_i y_i}{n} = \frac{486}{150} \approx 3.3.$$

It should be noted that the sample mean was computed to be 3.3 by using all the sample data before the 13 largest values were collapsed into the final cell. This is why the sample mean computed here was rounded up to 3.3.

     The Poisson probabilities for $y = 0, 1, \ldots, 7$ or more can be found in Table 7 in the Appendix with $\mu = 3.3$. These probabilities are shown here.

y_i	0	1	2	3	4	5	6	≥ 7
$P(y_i)$ for $\mu = 3.3$	.037	.122	.201	.221	.182	.120	.066	.051

The expected cell count E_i can be computed for any cell using the formula $E_i = nP(y_i)$. Hence, for our data (with $n = 150$), the expected cell counts are as shown here.

y_i	0	1	2	3	4	5	6	≥ 7
E_i	5.55	18.30	30.15	33.15	27.30	18.00	9.90	7.65

Substituting these values into the test statistic, we have

$$\chi^2 = \sum_i \left[\frac{(n_i - E_i)^2}{E_i} \right]$$

$$= \frac{(6 - 5.55)^2}{5.55} + \frac{(23 - 18.30)^2}{18.30} + \cdots + \frac{(13 - 7.65)^2}{7.65} = 7.01.$$

The tabulated value of chi-square for $a = .05$ and $df = k - 2 = 6$ is 12.59. Since the computed value of chi-square does not exceed 12.59, we have insufficient evidence to reject the null hypothesis that the data were collected from a Poisson distribution.   ▲

A word of caution is given here for situations in which we are considering this test procedure. As we mentioned previously, when using a chi-square statistic, we should have all expected cell counts fairly large. In particular, we want all $E_i > 1$ and not more than 20% less than 5. In Example 8.12, if values of $y \geq 7$ had been considered individually, the Es would not have satisfied the criteria for the use of χ^2. That is why we combined all values of $y \geq 7$ into one category.

▼ EXERCISES

Basic Techniques

8.44 Use a Poisson approximation to the binomial to find $P(y \leq 2)$ for $n = 1,500$ and $\pi = .002$; also find for $\pi = .003$.

8.45 Compute the following Poisson probabilities using Table 7 in the Appendix.
  a. $P(y = 1)$ given $\mu = .5$, $\mu = 1.0$, and $\mu = 3.0$
  b. $P(y > 1)$ given $\mu = 1.7$, $\mu = 2.5$, and $\mu = 4.2$
  c. $P(y < 5)$ given $\mu = 0.2$, $\mu = 1.0$, and $\mu = 2.0$

8.46 Cars arrive at the exit gate for airport long-term parking at a rate of six per minute during rush hour. Find the following probabilities using Table 7 in the Appendix. (y is the number of cars arriving during any given minute in rush hour.)
  a. $P(y = 0)$
  b. $P(y > 1)$
  c. $P(y > .3)$

8.47 A firm is considering using telemarketing techniques to supplement traditional marketing methods. It is estimated that one of every 100 calls results in a sale. Suppose that 250 calls are made in a single day:
  a. Write an expression for the probability that there are less than six sales—do not do the mathematics.
  b. What assumptions are you making in part a?
  c. Use a normal approximation to compute $P(y < 6)$.
  d. Compute $P(y < 6)$ using the Poisson distribution.
  e. Which approximation (part c or part d) appears better? Why?

Applications

8.48 A certain birth defect occurs with probability .0001; that is, one of every 10,000 babies has this defect. If 5,000 babies are born at a particular hospital in a given year, what approximation should be used? What is the approximate probability that there is at least one baby with the defect?

8.49 One portion of a study to determine the effectiveness of an exclusive bus lane was directed at examining the number of conflicts (driving situations that could result in an accident) at a major intersection during a specified period of time. A previous study prior to the installation of the exclusive bus lane indicated that the number of conflicts per 5 minutes during the 7:00 to 9:00 A.M. peak period could be adequately approximated by a Poisson distribution with $\mu = 2$. The following data were based on a sample of 40 days; y_i denotes the number of conflicts and n_i denotes the number of 5-minute periods during which y was observed.

y_i	0	1	2	3	4	5	≥ 6
n_i	90	230	240	130	68	30	12

a. Does the Poisson assumption appear to hold?

b. Use these data to test the research hypothesis that the mean number of conflicts per 5 minutes differs from 2. (Hint: Use a chi-square test based on Poisson probabilities.)

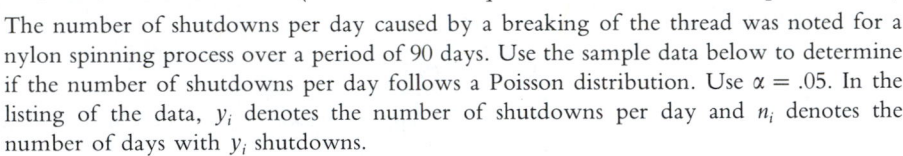 **8.50**   The number of shutdowns per day caused by a breaking of the thread was noted for a nylon spinning process over a period of 90 days. Use the sample data below to determine if the number of shutdowns per day follows a Poisson distribution. Use $\alpha = .05$. In the listing of the data, y_i denotes the number of shutdowns per day and n_i denotes the number of days with y_i shutdowns.

y_i	0	1	2	3	4	≥ 5
n_i	20	28	15	8	7	12

8.7 ▼ $r \times c$ CONTINGENCY TABLES: CHI-SQUARE TEST OF INDEPENDENCE

In all our calculations so far in this text, we have assumed that only one measurement is taken on each sampling unit. We might obtain the yield for an acre planted in wheat, a blood pressure reading on a patient who is being administered an anesthetic, or a measurement on the number of potential conflicts at a highway intersection during a 1-hour period. However, research problems in the sciences frequently involve more than one variable. If measurements are taken on two (or more) variables for each sampling unit, we say that we have **bivariate** (or **multivariate**) **data**.

bivariate and multivariate data

As with univariate count data, where the data may be summarized in a table, we frequently arrange bivariate data in a two-way table. For example, in a study of the public approval of a proposed high-speed bus lane for commuters, the interviewers might also ask individuals information about their occupations. We could then classify each person by his or her opinion concerning the new lane (favor, do not favor, undecided) and his or her occupation (white-collar worker, blue-collar worker, laborer).

What is the objective of such a classification? In most studies, either we wish to determine whether the two variables are related (dependent) or we wish to predict one variable based on knowledge of the other. This section deals with a test of independence for bivariate count data arranged in a two-way table. The two-way tables are sometimes called **contingency tables** because the alternative hypothesis in our test is that the two variables are dependent (i.e., there is a contingency between the two variables).

contingency tables

Consider the problem in which we would like to determine whether the following two variables are dependent: employee classification (staff, faculty, administrator) at a university and an employee's opinion about whether the local chapter of the teachers' union should be the sole collective bargaining agent for employee benefits. A random sample of 200 employees is taken from employee records and each employee is classified according to both variables. Suppose the results of the survey appear as shown in Table 8.6. Is there evidence to indicate that a person's opinion concerning collective bargaining depends on his or her employment status? That is, can we conclude that the two variables are dependent?

T A B L E 8.6

Classification of 200 Employees by Classification and Opinion on Collective Bargaining

| Employee Classification | Opinion on Collective Bargaining by Teachers' Union | | | |
	Favor	Do Not Favor	Undecided	Totals
Staff	30	15	15	60
Faculty	40	50	10	100
Administrator	10	25	5	40
Totals	80	90	30	200

independence

To answer this question we must define the concept of **independence**.

D E F I N I T I O N 8.2
Independence

▼

Two variables that have been categorized in a two-way table are **independent** if the probability that a measurement is classified into a given cell of the table is equal to the probability of being classified into that row times the probability of being classified into that column. This must be true for all cells of the table.

For example, suppose that the probability of selecting a person favoring the teachers' union in the university survey is π_1, the probability of selecting one who does not favor the union for collective bargaining is π_2, and the probability of selecting a person who is undecided is π_3 (note: $\pi_1 + \pi_2 + \pi_3 = 1$). Similarly, suppose that the probabilities of selecting a staff member, a faculty member, or an administrator are, respectively, π_A, π_B, and π_C (where $\pi_A + \pi_B + \pi_C = 1$). Then the two variables, employee classification and opinion concerning the teachers' union, are independent if the probability of classifying a person into a specific cell of the two-way table is obtained by multiplying the respective row and column probabilities. These ideas are illustrated in Table 8.7.

T A B L E 8.7

Cell Probabilities Showing Independence for the Collective Bargaining Survey

	Opinion		
Employee Classification	Favor, π_1	Do Not Favor, π_2	Undecided, π_3
Staff, π_A	$\pi_A\pi_1$	$\pi_A\pi_2$	$\pi_A\pi_3$
Faculty, π_B	$\pi_B\pi_1$	$\pi_B\pi_2$	$\pi_B\pi_3$
Administrator, π_C	$\pi_C\pi_1$	$\pi_C\pi_2$	$\pi_C\pi_3$

A test of the independence of two variables arranged in a two-way table makes use of the test statistic

test statistic

$$\chi^2 = \sum_{i,j} \left[\frac{(n_{ij} - E_{ij})^2}{E_{ij}} \right],$$

where n_{ij} and E_{ij} are, respectively, the observed and expected number of measurements falling in the cell for the ith row and the jth column.

D E F I N I T I O N 8.3

Expected Number of Measurements

> ▼
>
> The **expected number of measurements** E_{ij} falling in the i, j cell (cell of the ith row and jth column of the table) is taken to be
>
> $$E_{ij} = \frac{(\text{row } i \text{ total})(\text{column } j \text{ total})}{n}$$
>
> when the two variables are independent. (Note: If E_{ij} is not an integer, it should not be rounded to an integer value for the tests that follow.)

E X A M P L E 8.13

▼

Compute the expected number of measurements falling into each cell of Table 8.6.

Solution   The expected number of measurements falling in the 1, 1 cell (first row, first column) is

$$E_{11} = \frac{(\text{row 1 total})(\text{column 1 total})}{n} = \frac{(60)(80)}{200} = 24.$$

Similarly, the expected number of measurements in the 3, 2 cell is

$$E_{32} = \frac{(\text{row 3 total})(\text{column 2 total})}{n} = \frac{(40)(90)}{200} = 18.$$

These and the remaining cell counts appear in Table 8.8.

TABLE 8.8

Expected Cell Counts for the Collective Bargaining Survey

| | Opinion | | | |
Employee Classification	Favor	Do Not Favor	Undecided	Totals
Staff	24	27	9	60
Faculty	40	45	15	100
Administrator	16	18	6	40
Totals	80	90	30	200

Note that the expected counts in a row sum to the same row total as do the observed cell counts. The same applies for columns.   ▲

We can now summarize the *chi-square test of independence* for data arranged in a two-way table.

Chi-Square Test of Independence

▼

Null hypothesis: The two variables are independent.
Alternative hypothesis: The two variables are dependent.

Test statistic: $\chi^2 = \sum_{i,j} \left[\frac{(n_{ij} - E_{ij})^2}{E_{ij}} \right]$

Rejection region: Reject H_0 if χ^2 exceeds the tabulated value of chi-square (Table 5 in the Appendix) for $a = \alpha$ and $df = (r - 1)(c - 1)$, where

r = number of rows in the table
c = number of columns in the table.

The guidelines that Cochran (1954) proposed for the E_{ij}s (see Section 8.2) are still in effect when we use the chi-square test of independence. While agreeing with Cochran, Conover (1971) goes even further by stating that when the E_{ij}s are all about the same magnitude and both r and c are large, then even if the E_{ij}s are as small as 1, the approximation by a chi-square distribution will still be good. These guidelines give us a great deal of flexibility in applying the chi-square test without having to collapse some of the categories.

EXAMPLE 8.14

▼

Conduct a chi-square test of independence for the teachers' union data in Table 8.6. Use $\alpha = .05$.

Solution   Using the observed cell counts of Table 8.6 and the expected cell counts of Table 8.8, we can substitute these values into the test statistic.

$$\chi^2 = \sum_{i,j} \left[\frac{(n_{ij} - E_{ij})^2}{E_{ij}} \right]$$

$$= \frac{(30 - 24)^2}{24} + \frac{(15 - 27)^2}{27} + \frac{(15 - 9)^2}{9} + \frac{(40 - 40)^2}{40} + \frac{(50 - 45)^2}{45}$$

$$+ \frac{(10 - 15)^2}{15} + \frac{(10 - 16)^2}{16} + \frac{(25 - 18)^2}{18} + \frac{(5 - 6)^2}{6}$$

$$= 18.2.$$

The critical value of χ^2 for $a = .05$ and $\mathrm{df} = (r - 1)(c - 1) = 2(2) = 4$ is 9.488. Since the computed value, 18.2, exceeds 9.488, we reject H_0 and conclude that the two variables are dependent. In particular, we say that the proportion of persons favoring the teachers' union as the collective bargaining agent varies depending on the employee status. From Table 8.6 we see that a much higher proportion of the staff members favor the teachers' union as the collective bargaining agent than of either the faculty or administrators.

▲

The only function of this chi-square test is to determine whether the observed dependence is due to random fluctuations. When H_0: independence is rejected in favor of H_a: dependence, we conclude that the two variables that have been summarized in a contingency table are related; that is, an outcome on one variable affects or is affected by an outcome on the other variable. The rejection of H_0 does not, however, indicate the strength or type of the relation between the two variables. As was discussed in Chapter 3, a percentage comparison can help to identify how the variables are related, and certain "measures of associa-tion" to be presented in the next section can help to quantify the strength of the relation between the two variables.

The same chi-square test statistic applies to a slightly different sampling procedure. In our discussion of the chi-square goodness-of-fit test, an implicit assumption has been that the data summarized in the contingency table resulted from a single random sample from the population of interest. Often, separate random samples are taken from the *subpopulations* defined by the rows (or columns) of the contingency table. For example, the data of Table 8.6 might have resulted from separate random samples of sizes 60 (staff), 100 (faculty), and 40 (administrator) rather than from a single overall sample of 200 individuals. When random samples are taken for categories of the row (or column) variable, the test is called a *test of homogeneity* of the row (or column) distributions. So, when sampling by rows, the percentage distributions across columns is the same from row to row. This would be seen in a percentage comparison by rows for the sample data. Similarly, when sampling by columns, the percentage distribution across rows is the same from column to column.

Since the mechanics and conclusions are the same for the chi-square test of independence and for the chi-square test of homogeneity of the distributions, we will not worry about the distinction between the two tests.

▼ EXERCISES

Basic Techniques

8.51 Data from a random sample of 200 individuals are summarized here in a 2 × 2 contingency table.
  a. Give the null and alternative hypotheses for a test of independence.
  b. Compute the expected cell counts and the value of the χ^2 test statistic.
  c. Based on $\alpha = .05$, determine the rejection region and draw conclusions about the test of the null hypothesis.

	Column 1	Column 2
Row 1	20	70
Row 2	30	80

8.52 In a 2 × 3 contingency table, what is the minimum number of expected values that need to be computed if the remaining ones are computed by subtraction? What is the minimum number for a 3 × 3 table?

Applications

  8.53 A survey of student opinion concerning a proposed tuition increase was taken to determine whether student opinion was independent of gender. The results of 300

interviews are recorded in the accompanying table. Run a chi-square test of independence and give the level of significance for the test results. Draw conclusions.

	Opinion		
Gender of Student	Favor Increase	Oppose	Undecided
Female	91 ( )*	54 ( )*	13
Male	59	69	14

* Note: You need to use the formula for computing the expected cell count only for these cells; the remaining cell counts can be computed by subtraction from the appropriate row or column total.

 8.54 A scientist was interested in testing the effectiveness of a new drug product in controlling worms in the small intestine of sheep. A prestudy test was used to select 40 sheep with approximately the same level of infestation. These sheep were then randomly divided into two groups of 20. Those in the first group were given the drug product; those in the second group received no treatment and served as a control group. After 2 weeks, each of the 40 sheep was examined and classified as either "responder" or "nonresponder" depending on the observed worm count. The sample data are summarized here:

Classification	Group 1 (Drug-Tested)	Group 2 (Control)
Responder	15 ( )	7
Nonresponder	5	13

a. Compute the expected cell counts.
b. Run a chi-square test of independence with $\alpha = .05$. State the null hypothesis for this test and draw appropriate conclusions.

 8.55 A carcinogenicity study was conducted to examine the tumor potential of a drug product scheduled for initial testing in humans. A total of 300 rats (150 males and 150 females) were studied for a 6-month period. At the beginning of the study, 100 rats (50 males, 50 females) were randomly assigned to the control group, 100 to the low-dose group, and the remaining 100 (50 males, 50 females) to the high-dose group. On each day of the 6-month period, the rats in the control group received an injection of an inert solution, whereas those in the drug groups received an injection of the solution plus drug. The sample data are shown in the accompanying table.

	Number of Tumors	
Rat Group	One or More	None
Control	10	90
Low dose	14	86
High dose	19	81

a. Give the percentage of rats with one or more tumors for each of the three groups.

b. Conduct a chi-square test of independence with $\alpha = .05$.

c. Does there appear to be a drug-related problem regarding tumors for this drug product? That is, as the dose is increased, does there appear to be an increase in the proportion of rats with tumors?

8.56 SAS computer output for the data of Exercise 8.55 is shown here. Compare the output with your results in Exercise 8.55.

```
                        CATEGORICAL ANALYSIS

                  TABLE OF R_GROUP BY N_TUMORS

             R_GROUP      N_TUMORS

             FREQUENCY
             PERCENT
             ROW PCT
             COL PCT       > = 1     NONE    TOTAL


             CONTROL         10        90      100
                           3.33     30.00    33.33
                          10.00     90.00
                          23.26     35.02


             HIGHDOSE        19        81      100
                           6.33     27.00    33.33
                          19.00     81.00
                          44.19     31.52


             LOWDOSE         14        86      100
                           4.67     28.67    33.33
                          14.00     86.00
                          32.56     33.46


             TOTAL           43       257      300
                          14.33     85.67   100.00


          STATISTICS FOR TABLE OF R_GROUP BY N_TUMORS


     STATISTIC                          DF      VALUE    PROB


     CHI-SQUARE                          2      3.312    0.191
     LIKELIHOOD RATIO CHI-SQUARE         2      3.327    0.189
     MANTEL-HAENSZEL CHI-SQUARE          1      0.649    0.420
     PHI                                        0.105
     CONTINGENCY COEFFICIENT                    0.104
     CRAMER'S V                                 0.105

     SAMPLE SIZE = 300
```

 8.57 A total of 210 emphysema patients entering a clinic over a 1-year period were treated with one of two drugs (either the standard drug, A, or an experimental compound, B) for a period of 1 week. After this period, each patient's condition was rated as greatly improved, improved, or no change. The sample results are shown here.

	Patient's Condition		
Therapy	No Change	Improved	Greatly Improved
Standard, A	20	35	45
Experimental, B	15	45	50

a. Make a percentage comparison for the rows; does there appear to be a difference in the two therapies?

b. Run a chi-square test of independence and draw a conclusion. Use $\alpha = .05$. Does it agree with your speculation in part a?

8.58 Refer to Exercise 8.57. The sum of the expected values must equal $n = 210$, and the expected values will add to the appropriate row and column totals. Determine the minimum number of expected values that need to be computed. How is this number related to the degrees of freedom for the test?

8.59 The market research group of a particular firm conducted a survey in three cities to compare the sales potential of a new soft drink. Each person contacted was asked to try the new drink and to classify it as excellent, satisfactory, or unsatisfactory. The results of the survey are summarized in the accompanying table. Use the Minitab computer output shown here to conduct a chi-square test of independence. Give the approximate level of significance for your test and draw conclusions.

Classification	**City 1**	**City 2**	**City 3**
Excellent	62	51	45
Satisfactory	28	30	35
Unsatisfactory	10	19	20

```
MTB > NAME C1 'CITY1' C2 'CITY2' C3 'CITY3'
MTB > READ C1-C3
      3 ROWS READ
MTB > END

MTB > CHISQUARE C1 C2 C3

Expected counts are printed below observed counts

          CITY1    CITY2    CITY3    Total
    1        62       51       45      158
          52.67    52.67    52.67

    2        28       30       35      .93
          31.00    31.00    31.00

    3        10       19       20      49
          16.33    16.33    16.33

Total      100      100      100      300

ChiSq =  1.654 +  0.053 +  1.116 +
         0.290 +  0.032 +  0.516 +
         2.456 +  0.435 +  0.823 = 7.376

df = 4

MTB > STOP
```

8.60 A university conducted a self-study to satisfy the requirements for accreditation. One aspect of the self-study concerned faculty evaluations. Through the use of student evaluations of their instructors, each faculty member was classified both by rank and by ability as a teacher. Use the accompanying results to test the null hypothesis of independence of the two classifications. Use $\alpha = .05$, and draw a conclusion.

Teaching Evaluation	Rank			
	Instructor	Assistant Professor	Associate Professor	Professor
Above average	36	62	45	50
Average	48	50	35	43
Below average	30	13	20	35

8.61 A survey of admissions practices at a liberal arts college was conducted to determine whether there appeared to be a difference in the acceptance rates for white and minority (nonwhite) applicants. The results of this survey, which combine information from 4,000 applicants, are shown in the accompanying table. Use the accompanying computer printout to conduct a chi-square test of independence. Use Table 5 in the Appendix to obtain an approximate level of significance for the test and draw conclusions.

Applicant Accepted?	Applicant		Total
	Nonwhite	White	
Yes	38	126	164
No	362	3,474	3,836
Total	400	3,600	4,000

```
MTB > NAME C1 'NONWHITE' C2 'WHITE'
MTB > READ C1 C2
      2 ROWS READ
MTB > END
MTB > CHISQUARE C1 C2

Expected counts are printed below observed counts

        NONWHITE    WHITE     Total
    1         38      126       164
          16.40   147.60

    2        362     3474      3836
         383.60  3452.40

Total       400     3600      4000

ChiSq = 28.449 +  3.161 +
         1.216 +  0.135 = 32.961

df = 1

MTB > STOP
```

8.62 Computer output for the data of Exercise 8.55 is shown here. Compare the Minitab output with your results in Exercise 8.55 and the output in Exercise 8.56.

```
MTB > NAME C1 '>ONE' C2 'NONE'
MTB > READ C1 C2
       3 ROWS READ
MTB > END
MTB > CHISQUARE C1 C2

Expected counts are printed below observed counts

            >ONE      NONE      Total
     1        10        90        100
           14.33     85.67

     2        14        86        100
           14.33     85.67

     3        19        81        100
           14.33     85.67

Total        43       257        300

ChiSq =   1.310 +   0.219 +
          0.008 +   0.001 +
          1.519 +   0.254 = 3.312
df = 2

MTB > STOP
```

8.8 ▼ MEASURES OF ASSOCIATION

Cross-classification of data, presented in the previous sections, is performed to determine whether two variables, representing two systems of classification, are related. Hence, previous sections dealt solely with the question of whether the two variables are dependent and presented statistical tests for the null hypothesis that the two variables are independent. Our objective was to collect evidence to support the research hypothesis that the two variables are dependent and, therefore, related. In this section, we carry the study of the relationship between two variables one step further. It is not sufficient to know only that two variables are related; we also want measures of the strength of the relationship or association.

We discussed the use of the chi-square test of independence for contingency tables in Section 8.7. After we have performed a chi-square test of independence and found the variables to be dependent, we may be interested in the strength of the dependence between the two variables.

The contingency coefficient is a measure of association for cross-classification data. The contingency coefficient is computed as follows.

Contingency Coefficient

$$C = \sqrt{\frac{\chi^2}{n + \chi^2}},$$

where

$$\chi^2 = \sum \frac{(O - E)^2}{E} \quad \text{and} \quad n = \text{total sample size.}$$

The contingency coefficient is relatively easy to compute and satisfies the condition that it equals 0 when there is no association between the variables. However, it does have some disadvantages as a measure of association. First, the contingency coefficient is always less than 1, even when the two variables are completely dependent on one another. Second, C provides only an intuitive measure of the degree of association between two variables. It is used after a chi-square test of independence and hence is most frequently employed for data that satisfy the conditions required of a chi-square test. Third, contingency coefficients for two different sets of data can be compared only if the two-way tables are of the same size (the same number of rows and columns). This restriction is due to the fact that C can attain larger values for larger tables, so the range of possible values of C changes with the size of the table involved. Finally, it is difficult to compare the contingency coefficient with any of the other measures of association that will be presented later in this chapter.

Before we further discuss the contingency coefficient, we illustrate its use with an example.

E X A M P L E   8.15

▼

A study of migrants and their family relations was conducted to determine whether the variable "degree of kinship participation" in the extended family was independent of a family's socioeconomic status. Use the data in Table 8.9 to run a chi-square test of independence, and then compute the contingency coefficient C. Expected cell counts have been computed and are given in parentheses in the table.

Solution   Before computing C we should first run a test of independence to see if the data provide sufficient evidence to show that the two variables "socioeconomic status" and "degree of kinship participation" are dependent.

TABLE 8.9
Study of Newcomer Adaptations

Socioeconomic Status	Degree of Kinship Participation			Total
	Low	Medium	High	
Low	75 (88.69)	98 (102.48)	75 (56.83)	248
Medium	182 (184.53)	211 (213.23)	123 (118.24)	516
High	116 (99.78)	122 (115.29)	41 (63.93)	279
Total	373	431	239	1043

Source: Felix M. Berardo, "Internal Migrants and Extended Family Relations — A Study of Newcomer Adaptation" (Ph.D diss., Florida State University, 1965), P. 122. Used by permission.

Using the observed and expected cell frequencies from Table 8.9, we have

$$\chi^2 = \sum \frac{(O - E)^2}{E}$$

$$= \frac{(-13.69)^2}{88.69} + \frac{(-4.48)^2}{102.48} + \frac{(18.17)^2}{56.83} + \frac{(-2.53)^2}{184.53} + \frac{(-2.23)^2}{213.23}$$

$$+ \frac{(4.76)^2}{118.24} + \frac{(16.22)^2}{99.78} + \frac{(6.71)^2}{115.29} + \frac{(-22.93)^2}{63.93}$$

$$= 19.62.$$

Since the boundary of the rejection region for $\alpha = .05$ (and hence $a = .05$) with df $= (r - 1)(c - 1) = 2(2) = 4$ is 9.488, we conclude that the classifications are dependent.

Now let us compute a measure of the strength of this dependence, the contingency coefficient. Using $\chi^2 = 19.62$, we find the contingency coefficient to be

$$C = \sqrt{\frac{\chi^2}{n + \chi^2}} = \sqrt{\frac{19.62}{1043 + 19.62}} = \sqrt{.0185} = .136.$$

Unfortunately, very little can be said concerning the magnitude of C. However, we know from the chi-square test that the variables are dependent, and C gives us a measure of the strength of this relationship.

Some of the disadvantages of C can be alleviated by adjusting the computed value of C according to the maximum value C may attain. It can be shown that for any square two-way table (i.e., a table with the same number of rows and columns), the maximum value of C obtainable under perfect association is

$$C_{max} = \sqrt{\frac{r-1}{r}},$$

where r is the number of rows in the table. For $r = 2$, we have

$$C_{max} = \sqrt{\frac{1}{2}} = .707.$$

If we now form a modified (adjusted) version of the contingency coefficient,

$$C_{adj} = \frac{C}{C_{max}},$$

the largest value that C_{adj} can assume is 1.0, which occurs under perfect association when $C = C_{max}$. Applying this result to the data in Table 8.9, we see that, with $r = 3$ rows, we have

$$C_{max} = \sqrt{\frac{2}{3}} = .816.$$

Hence, the adjusted value of C is

$$C_{adj} = \frac{C}{C_{max}} = \frac{.136}{.816} = .17.$$

Using C_{adj} we now have a measure of association that equals 0 under no association and 1 when the two variables are perfectly associated. C_{adj} should be used in place of C for square two-way tables. ▲

Cramer's V is a second measure of association for comparing contingency table data but it avoids some of the problems associated with C. For a two-way table with r rows and c columns, we define V as follows.

Cramer's V

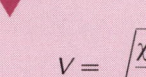

$$V = \sqrt{\frac{\chi^2}{nt}},$$

where n is the total sample size and t is the smaller of the two numbers $r - 1$ and $c - 1$; r and c are the number of rows and columns, respectively, for the two-way table.

It can be shown that V lies in the interval from 0 to 1.

EXAMPLE 8.16

Compute Cramer's V for the data of Example 8.15.

Solution   For $\chi^2 = 19.62$, $r = c = 3$, and $n = 1043$, we have

$$V = \sqrt{\frac{\chi^2}{nt}} = \sqrt{\frac{19.62}{1043(2)}} = .097.$$

We know from the chi-square test that the variables are related. V gives us another measure of the magnitude of this association.   ▲

Previously, we indicated that measures of association are distinguished by the scales of measurement for the two variables being studied and by the range of possible values for the measures of association. A third distinction between measures of association for two variables relates to their interpretations. The measures of association discussed in this section are based on the chi-square statistic and provide only a vague measure of the association between the two variables of interest. The interested reader can consult Ott et al. (1992) for details about other, more meaningful measures of association.

▼ EXERCISES

Basic Techniques

8.63   Use the data shown here to compute C and C_{adj}.

	Column 1	Column 2
Row 1	117	74
Row 2	83	126

8.64 It can be shown that $\chi^2 = 15.52$ for the data shown here.

	Column 1	Column 2
Row 1	11	15
Row 2	16	19
Row 3	44	11
Row 4	71	45

  a. Give the level of significance of the test for independence.
  b. Compute Cramer's V to measure the strength of the association between the two variables.

8.65 Refer to Exercise 8.64. Compute the contingency coefficient. Is C an appropriate measure for these data? Explain.

Application

 8.66 Social researchers conducted a study on the relationship between the type of book publisher of scholarly materials and the type of college represented. Using the data shown here,
  a. Run a chi-square test of independence based on $\alpha = .05$.
  b. Give a measure of the strength of the association.

	Type of Publisher		
Type of College	Secular	Religious	Total
Bible college	15	15	30
State college	25	0	25
Total	40	15	55

8.9 ▼ COMBINING SETS OF $r \times c$ CONTINGENCY TABLES (optional)

In the previous section, we discussed the chi-square test of independence for examining the dependence of two variables based on data arranged in a contingency table. Suppose a pharmaceutical company is developing a drug product for the treatment of epilepsy. In each of several clinics, patients are assigned at random to either a placebo or the new drug and treated for a period of 2 months.

At the end of the study, each patient is rated as either improved or not improved. If 100 patients (50 per treatment group) are to be enrolled in a particular clinic and we observe 40 and 15 patients improved in the new drug and placebo groups, respectively, the data could be displayed as shown in Table 8.10 and analyzed using the chi-square methods of the previous section. The null hypothesis of independence of the two classifications (treatment group and rating) could be restated in terms of the proportions, π_1 and π_2, of improved patients for the two populations. The new H_0 would be $H_0: \pi_1 - \pi_2 = 0$—namely, that there is no difference in the proportions of improved patients for the drug and placebo groups. Rejection of H_0 using the chi-square statistic from the test of independence test indicates that the population proportions are different for the two treatment groups.

TABLE 8.10

Number (%) of Patients Improved

	Improved	Not Improved	Total
New drug	40 (80%)	10	50
Placebo	15 (30%)	35	50

This same scenario can be extended to more than one clinic and we can extend our test procedure to deal with a set of q clinics ($q \geq 2$). For this situation, we would observe the sample percentage improved for the drug and placebo groups in each clinic; the data could be summarized using Table 8.11. The test for comparing the drug and placebo proportions combines sample information across the separate contingency tables to answer the question of whether, on the average, the improvement rates are the same for the two treatment groups. Before we do this, however, we need some additional notation, shown in Table 8.12.

Cochran (1954) proposed a test statistic for the hypothesis of no difference (on the average) for the improvement rates for a set of q 2×2 contingency tables. This same problem was addressed by Mantel and Haenszel (1959) and also extended to cover a set of q $2 \times c$ contingency tables. For 2×2 tables the

TABLE 8.11

Summary Table for a Set of 2 × 2 Contingency Tables

Clinic		Improved	Not Improved
1	Drug		
	Placebo		
2	Drug		
	Placebo		
⋮			
q	Drug		
	Placebo		

Table	Treatment	Response Category 1	2	Total
1	1	n_{111}	n_{112}	$n_{11.}$
	2	n_{121}	n_{122}	$n_{12.}$
	Total	$n_{1.1}$	$n_{1.2}$	$n_{1..}$
2	1	n_{211}	n_{212}	$n_{21.}$
	2	n_{221}	n_{222}	$n_{22.}$
	Total	$n_{2.1}$	$n_{2.2}$	$n_{2..}$
$\vdots$				
h	1	n_{h11}	n_{h12}	$n_{h1.}$
	2	n_{h21}	n_{h22}	$n_{h2.}$
	Total	$n_{h.1}$	$n_{h.2}$	$n_{h..}$
$\vdots$				

Mantel–Haenszel statistic for testing the equality of the improvement rates, on the average, can be written as

$$\chi^2_{\text{MH}} = \frac{\left\{ \sum_h \left(n_{h11} - \frac{n_{h1.} n_{h.1}}{n_{h..}} \right) \right\}^2}{\sum_h \frac{n_{h1.} n_{h2.} n_{h.1} n_{h.2}}{n_{h..}^2 (n_{h..} - 1)}},$$

which follows a chi-square distribution with df $= 1$. Let's see how this works for a set of sample data.

E X A M P L E 8.17 ▼

The pharmaceutical study discussed previously was extended to three clinics. In each clinic, as patients qualified for the study and gave their consent to participate, they were assigned to either the drug or placebo groups according to a predetermined random code. Each clinic was to treat 50 patients per group. The study results are summarized in Table 8.13. Use these data to test the null hypothesis of no difference in the improvement rates, on the average. Use the Mantel–Haenszel chi-square statistic and give the p-value for the test.

TABLE 8.13
Study Results

	Clinic	Improved	Not Improved	Total
1	Drug	40 (80%)	10	50
	Placebo	15 (30%)	35	50
	Total	55	45	100
2	Drug	35 (70%)	15	50
	Placebo	20 (40%)	30	50
	Total	55	45	100
3	Drug	43 (86%)	7	50
	Placebo	31 (62%)	19	50
	Total	74	26	100
Total		184	116	300

Solution   The necessary row and column totals in each clinic are given in Table 8.13. The numerator of the Mantel–Haenszel statistic is

$$\left\{\sum_h \left(n_{h11} - \frac{n_{h1.}n_{h.1}}{n_{h..}}\right)\right\}^2 = \left\{\left(40 - \frac{50(55)}{100}\right) + \left(35 - \frac{50(55)}{100}\right)\right.$$
$$\left. + \left(43 - \frac{50(74)}{100}\right)\right\}^2$$
$$= (12.5 + 7.5 + 6)^2 = 676,$$

whereas the denominator is

$$\sum_h \frac{n_{h1.}n_{h2.}n_{h.1}n_{h.2}}{n_{h.}^2(n_{h..} - 1)} = \frac{50(50)(55)(45)}{(100)^2(99)} + \frac{50(50)(55)(45)}{(100)^2(99)} + \frac{50(50)(74)(26)}{(100)^2(99)}$$
$$= 6.25 + 6.25 + 4.8586 = 17.3586.$$

Substituting, we obtain

$$\chi^2_{MH} = \frac{676}{17.3586} = 38.9432.$$

For df $= 1$, this result is significant at the $p < .001$ level. As can be seen from the sample data, the drug-treated groups have consistently higher improvement rates than the placebo groups.   ▲

EXAMPLE 8.18 ▼

Sample data are not always as obvious and conclusive as those given in Example 8.17. Use the revised sample data shown in Table 8.14 to conduct a Mantel–Haenszel test. Give the p-value for your test and interpret your findings.

TABLE 8.14

Revised Study Results

	Clinic	Improved	Not Improved	Total
1	Drug	35 (70%)	15	50
	Placebo	26 (52%)	24	50
	Total	61	39	100
2	Drug	28 (56%)	22	50
	Placebo	29 (58%)	21	50
	Total	57	43	100
3	Drug	37 (74%)	13	50
	Placebo	24 (48%)	26	50
	Total	61	39	100

Solution Using the row and column totals of Table 8.14, the numerator and denominator of χ^2_{MH} can be shown to be 110.25 and 18.21, respectively. The Mantel–Haenszel statistic is then

$$\chi^2_{MH} = 6.05.$$

Based on df $= 1$, this test result has a significance defined by $.01 < p < .025$. We conclude that although the drug product did not have a higher improvement rate in all three clinics, the data combined across clinics indicates that, on the average, the drug improvement rate is higher than the placebo rate ($.01 < p < .025$). ▲

Mantel and Haenszel also extended this test procedure to cover the situation in which we want a combined test based on sample data displayed in a set of q $2 \times c$ contingency tables. Returning to our example, suppose rather than having two response categories (e.g., improved, not improved) we have c different categories such as (worse, same, or better) or (none, slight, moderate, completely well). For these situations, it is possible to score the categories of the scale and run a Mantel–Haenszel test based on the difference in mean scores for the two treatment groups. Because the formulas become more involved, we will revert to available statistical software.

EXAMPLE 8.19

The data shown in Table 8.15 appeared in Sugiura and Otaka (1974). They summarize by age category and dose of radiation the numbers of deaths from leukemia (LD) observed at the Atomic Bombs Casualty Commission (ABCC) and the corresponding numbers of individuals who did not die from leukemia (NLD) during the period from 1950 to 1970. Run a Mantel–Haenszel mean score test (based on uniform scores) to compare the mean levels of radiation for the two groups (leukemia death and non–death from leukemia) across the five age categories. Interpret the results.

TABLE 8.15

Deaths from Leukemia Observed at ABCC (1950–1970)

Age (Years)	Survival Status	Not in City	0–9	10–49	Dose (Rads) 50–99	100–199	200+
0–9	LD	0	7	3	1	4	11
	NLD	5015	10,752	2989	694	418	387
10–19	LD	5	4	6	1	3	6
	NLD	5973	11,811	2620	771	792	820
20–34	LD	2	8	3	1	3	9
	NLD	5669	10,828	2798	797	596	624
35–49	LD	3	19	4	2	1	10
	NLD	6158	12,645	3566	972	694	608
50+	LD	3	7	3	2	2	6
	NLD	3695	9053	2415	655	393	289

```
           GENERALIZED COCHRAN-MANTEL-HAENSZEL TEST STATISTICS
         FOR AVERAGE PARTIAL ASSOCIATION IN THREE-WAY CONTINGENCY TABLES

        P.46 DEATHS FROM LEUKEMIA - UNIFORM SCORES FOR DOSE 00060015

        COLUMN SCORES: UNIFORM        1.00  2.00
        ROW SCORES: UNIFORM           1.00  2.00  3.00  4.00  5.00  6.00

                          SUMMARY ACROSS TABLES
**************************************************************************
                 A SUMMARY OF INDIVIDUAL TABLE STATISTICS
  TABLE  TABLE  SAMPLE      MULTIVARIATE          MEAN SCORE         CORRELATION
  NO.    FREQ.  SIZE     Q     D.F.  P        QMS    D.F.  P      QMA    D.F.  P
   1      1    20281  248.05   5   0.0     248.05    5   0.0    127.51   1   0.0
   2      1    22812   43.89   5   0.0      43.89    5   0.0     29.32   1   0.0
   3      1    21338  100.56   5   0.0     100.56    5   0.0     60.23   1   0.0
   4      1    24682   88.85   5   0.0      88.85    5   0.0     38.02   1   0.0
   5      1    16523   84.74   5   0.0      84.74    5   0.0     40.65   1   0.0

         TOTAL  105636  566.09  25   0.0     566.09   25   0.0    295.65   5   0.0

            B. GENERALIZED COCHRAN-MANTEL-HAENSZEL STATISTICS
         SAMPLE      MULTIVARIATE          MEAN SCORE         CORRELATION
         SIZE    Q(GMH)  D.F.  P       Q(CMMS)  D.F.  P    Q(CMMA)  D.F.  P
        105636   461.66   5   0.0      461.66    5  0.0    262.62    1   0.0
**************************************************************************
                                                          (continues)
```

```
                    GENERALIZED COCHRAN-MANTEL-HAENSZEL TEST STATISTICS
                   FOR AVERAGE PARTIAL ASSOCIATION IN THREE-WAY CONTINGENCY TABLES
                    P.46 DEATHS FROM LEUKEMIA - MIDPOINT SCORES FOR DOSE 00249515


           COLUMN SCORES: UNIFORM          1.00    2.00
           ROW SCORES: USER SPECIFIED      0.0     4.50    29.50   74.50   149.50   300.00

                                  SUMMARY ACROSS TABLES
    ************************************************************************************

                        A. SUMMARY OF INDIVIDUAL TABLE STATISTICS

    TABLE   TABLE    SAMPLE        MULTIVARIATE             MEAN SCORE            CORRELATION
     NO.    FREQ.     SIZE      Q       D.F.    P       QMS      D.F.   P      QMA    D.F.   P
      1       1      20281    248.05     5     0.0    248.05      5    0.0   226.32    1    0.0
      2       1      22812     43.89     5     0.0     43.89      5    0.0    39.01    1    0.0
      3       1      21338    100.56     5     0.0    100.56      5    0.0    94.29    1    0.0
      4       1      24682     88.85     5     0.0     88.85      5    0.0    65.87    1    0.0
      5       1      16523     84.74     5     0.0     84.74      5    0.0    76.04    1    0.0
            TOTAL   105636    566.09    25     0.0    566.09     25    0.0   501.52    5    0.0

                   B. GENERALIZED COCHRAN-MANTEL-HAENSZEL STATISTICS
            SAMPLE         MULTIVARIATE              MEAN SCORE            CORRELATION
             SIZE     Q(CMH)    D.F.    P      Q(CMMS)   D.F.   P     Q(CMMA)  D.F.   P
            105636   461.66      5     0.0     461.66      5   0.0   426.60     1    0.0
```

Solution   The test procedure based on mean scores can be summarized as follows:

H_0:   There is no difference, on the average, in mean radiation scores for the LD and NLD groups.

H_a: The mean radiation scores are different.

T.S.:   Generalized Mantel–Haenszel χ^2 statistic.

R.R.:   Reject H_0 if the observed chi-square value exceeds χ^2_α based on df = 1.

The generalized Mantel–Haenszel mean score test can be run using just about any set of scores. One possibility would be to use uniform scores (for example, 0, 1, 2, 3, 4, and 5) to the radiation dose categories listed in Table 8.15. Another possibility would be to use the midpoints of the intervals for the dose categories. This scoring system works fine for all but the first and last categories. Here, in order to compute the chi-square statistic, we will make the arbitrary assignment of 0 to the first (not in city) category and 300 to the 200+ category of dose. The results of the corresponding mean score tests for the two scoring systems are 262.6 and 426.6, respectively, shown in the preceding computer output under the label of Q(CMMA). Thus, we can conclude that for either scoring system, there is a highly significant difference in the mean radiation effect for the LD and NLD groups across the different age categories. In particular, since we are comparing two groups—those who died from leukemia (LD) and those who did not (NLD)—the higher mean associated with the LD group combined across the different categories indicates strong evidence of a connection between the level of radiation exposure and survival status. A more detailed analysis of these data is presented in Landis et al. (1978).   ▲

▼ EXERCISES

 8.67 A sample of 1,200 individuals arrested for driving under the influence of alcohol was obtained from police records. The research recorded the gender, socioeconomic status (from occupation information), and the number of previous alcohol-related arrests. These data are shown here:

Socioeconomic Status	Number of Previous Alcohol-Related Arrests	Male	Female
Low	0	110	130
	1–2	60	50
	>2	30	20
Medium	0	105	101
	1–2	75	55
	>2	20	44
High	0	90	80
	1–2	80	60
	>2	30	60

Use the output shown here to run a generalized Mantel–Haenszel test based on uniform scores to compare the average number of previous arrests for males and females. Give a *p*-value for your test and interpret your results. Would the scores 0, 1.5, and 2.5 give different results?

```
            GENERALIZED COCHRAN-MANTEL-HAENSZEL TEST STATISTICS
           FOR AVERAGE PARTIAL ASSOCIATION IN THREE-WAY CONTINGENCY TABLES
          P.48 NO. OF ALC. RELATED ARRESTS - UNIFORM SCORES FOR NO. 00280016

COLUMN SCORES: UNIFORM           1.00   2.00   3.00
ROW SCORES: UNIFORM              1.00   2.00

                              SUMMARY ACROSS TABLES
**************************************************************************************

                       A. SUMMARY OF INDIVIDUAL TABLE STATISTICS
 TABLE   TABLE   SAMPLE        MULTIVARIATE           MEAN SCORE          CORRELATION
  NO.    FREQ.    SIZE      Q     D.F.    P       QMS   D.F.    P      QMA   D.F.    P
   1       1      400     4.56     2   0.1021    4.49    1   0.0340   4.49    1   0.0340
   2       1      400    12.12     2   0.0023    3.56    1   0.0591   3.56    1   0.0591
   3       1      400    13.41     2   0.0012    6.54    1   0.0105   6.54    1   0.0105

          TOTAL   1200    30.10     6    0.0     14.60    3   0.0022  14.60    3   0.0022

                                                            (continues)
```

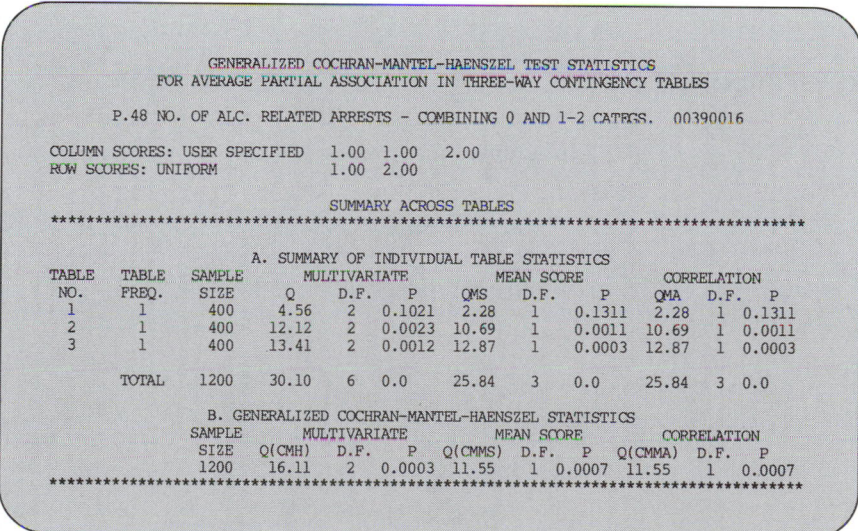

```
                B. GENERALIZED COCHRAN-MANTEL-HAENSZEL STATISTICS
        SAMPLE        MULTIVARIATE              MEAN SCORE              CORRELATION
         SIZE    Q(CMH)  D.F.    P     Q(CMMS)  D.F.    P     Q(CMMA)  D.F.    P
         1200    16.11    2   0.0003    2.17     1   0.1406    2.17     1   0.1406
     *********************************************************************************

                GENERALIZED COCHRAN-MANTEL-HAENSZEL TEST STATISTICS
            FOR AVERAGE PARTIAL ASSOCIATION IN THREE-WAY CONTINGENCY TABLES

            P.48 NO. OF ALC. RELATED ARRESTS - MID-PT. SCORES FOR NO.  00360016

     COLUMN SCORES: USER SPECIFIED     0.0    1.50    2.50
     ROW SCORES: UNIFORM               1.00   2.00


                            SUMMARY ACROSS TABLES
     *********************************************************************************

                    A. SUMMARY OF INDIVIDUAL TABLE STATISTICS
      TABLE    TABLE   SAMPLE        MULTIVARIATE         MEAN SCORE         CORRELATION
       NO.     FREQ.    SIZE    Q    D.F.    P     QMS   D.F.    P     QMA   D.F.    P
        1       1       400   4.56    2   0.1021  4.56   1   0.0327  4.56   1   0.0327
        2       1       400  12.12    2   0.0023  2.38   1   0.1230  2.38   1   0.1230
        3       1       400  13.41    2   0.0012  4.99   1   0.0254  4.99   1   0.0254

              TOTAL    1200   30.10   6   0.0    11.94   3   0.0076 11.94   3   0.0076


                B. GENERALIZED COCHRAN-MANTEL-HAENSZEL STATISTICS
        SAMPLE        MULTIVARIATE              MEAN SCORE              CORRELATION
         SIZE    Q(CMH)  D.F.    P     Q(CMMS)  D.F.    P     Q(CMMA)  D.F.    P
         1200    16.11    2   0.0003    1.08     1   0.2987    1.08     1   0.2987
     *********************************************************************************
```

8.68 Refer to Exercise 8.67. Combine the 0 and 1–2 categories and run the Mantel–Haenszel test for sets of 2×2 contingency tables. Interpret your findings.

```
                GENERALIZED COCHRAN-MANTEL-HAENSZEL TEST STATISTICS
            FOR AVERAGE PARTIAL ASSOCIATION IN THREE-WAY CONTINGENCY TABLES

            P.48 NO. OF ALC. RELATED ARRESTS - COMBINING 0 AND 1-2 CATEGS.  00390016

     COLUMN SCORES: USER SPECIFIED     1.00   1.00    2.00
     ROW SCORES: UNIFORM               1.00   2.00

                            SUMMARY ACROSS TABLES
     *********************************************************************************

                    A. SUMMARY OF INDIVIDUAL TABLE STATISTICS
      TABLE    TABLE   SAMPLE        MULTIVARIATE         MEAN SCORE         CORRELATION
       NO.     FREQ.    SIZE    Q    D.F.    P     QMS   D.F.    P     QMA   D.F.    P
        1       1       400   4.56    2   0.1021  2.28   1   0.1311  2.28   1   0.1311
        2       1       400  12.12    2   0.0023 10.69   1   0.0011 10.69   1   0.0011
        3       1       400  13.41    2   0.0012 12.87   1   0.0003 12.87   1   0.0003

              TOTAL    1200   30.10   6   0.0    25.84   3   0.0    25.84   3   0.0

                B. GENERALIZED COCHRAN-MANTEL-HAENSZEL STATISTICS
        SAMPLE        MULTIVARIATE              MEAN SCORE              CORRELATION
         SIZE    Q(CMH)  D.F.    P     Q(CMMS)  D.F.    P     Q(CMMA)  D.F.    P
         1200    16.11    2   0.0003   11.55     1   0.0007   11.55     1   0.0007
     *********************************************************************************
```

8.10 ▼ SUMMARY

This chapter has dealt with categorical data representing the number (frequency) of observations falling into each possible cell. For a single variable, we're interested in the cell counts of the separate categories of the variable. When observations are made on two or more variables, we're interested in the frequencies associated with the cells of the contingency table formed by cross-classifying observations according to the variables of interest.

Categorical data obtained from a single variable arise in a number of practical situations. We discussed a chi-square goodness-of-fit test that is used to test whether the sample frequencies (and percentages) associated with categories of a variable agree with what would be expected according to hypothesized cell percentages. We also examined estimation and test procedures for a binomial proportion π and for comparing two binomial proportions based on independent samples. Inferences related to a Poisson random variable were also discussed briefly.

Two-variable categorical data problems were introduced using a chi-square test of independence for data displayed in an $r \times c$ contingency table. We also presented two different measures of the strength of the relation between two variables in an $r \times c$ contingency table—the contingency coefficient and Cramer's V. The Mantel–Haenszel tests for combining information across sets of 2×2 contingency tables and for providing a mean score test on data combined across sets of $2 \times c$ contingency tables were also presented. Finally, it should be stressed that this chapter gave only a brief introduction to problems in the analysis of categorical data. Sequences of courses can be developed at the undergraduate (but more likely the graduate) level.

▼ KEY FORMULAS

1. Multinomial distribution

$$P(n_1, n_2, \ldots, n_k) = \frac{n!}{n_1! n_2! \ldots n_k!} \pi_1^{n_1} \pi_2^{n_2} \cdots \pi_k^{n_k}$$

2. Chi-square goodness-of-fit test

$$H_0: \quad \pi_i = \pi_{i0}$$

$$\text{T.S.:} \quad \chi^2 = \sum_i \left[\frac{(n_i - E_i)^2}{E_i} \right],$$

where

$$E_i = n\pi_{i0}$$

3. Confidence interval for π

$$\hat{\pi} = z_{\alpha/2}\sigma_{\hat{\pi}},$$

where

$$\hat{\pi} = \frac{y}{n}$$

and

$$\sigma_{\hat{\pi}} = \sqrt{\frac{\hat{\pi}(1 - \hat{\pi})}{n}}$$

4. Sample size required for a $100(1 - \alpha)\%$ confidence interval of the form $\hat{\pi} \pm E$

$$n = \frac{z_{\alpha/2}^2 \pi(1 - \pi)}{E^2}$$

(Hint: Use $\pi = .5$ if no estimate is available.)

5. Statistical test for π

$$H_0: \quad \pi = \pi_0$$

$$\text{T.S.:} \quad z = \frac{\hat{\pi} - \pi_0}{\sigma_{\hat{\pi}}},$$

where

$$\sigma_{\hat{\pi}} = \sqrt{\frac{\pi_0(1 - \pi_0)}{n}}$$

6. Confidence interval for $\pi_1 - \pi_2$

$$\hat{\pi}_1 - \hat{\pi}_2 \pm z_{\alpha/2}\sigma_{\hat{\pi}_1 - \hat{\pi}_2},$$

where

$$\sigma_{\hat{\pi}_1 - \hat{\pi}_2} = \sqrt{\frac{\hat{\pi}_1(1 - \hat{\pi}_1)}{n_1} + \frac{\hat{\pi}_2(1 - \hat{\pi}_2)}{n_2}}$$

7. Statistical test for $\pi_1 - \pi_2$

$$H_0: \quad \pi_1 - \pi_2 = 0$$

$$\text{T.S.:} \quad z = \frac{\hat{\pi}_1 - \hat{\pi}_2}{\sigma_{\hat{\pi}_1 - \hat{\pi}_2}},$$

where

$$\sigma_{\hat{\pi}_1 - \hat{\pi}_2} = \sqrt{\hat{\pi}(1 - \hat{\pi})\left(\frac{1}{n_1} + \frac{1}{n_2}\right)}$$

and

$$\hat{\pi} = \frac{y_1 + y_2}{n_1 + n_1}$$

8. Chi-square test of independence

$$\chi^2 = \sum_{i,j} \left[\frac{(n_{ij} - E_{ij})^2}{E_{ij}} \right],$$

where

$$E_{ij} = \frac{(\text{row } i \text{ total})(\text{column } j \text{ total})}{n}$$

9. Contingency coefficient

$$C = \sqrt{\frac{\chi^2}{n + \chi^2}}$$

10. Cramer's V

$$V = \sqrt{\frac{\chi^2}{nt}}$$

11. Mantel–Haenszel statistic

$$\chi^2_{MH} = \frac{\left\{ \sum_h \left(n_{h11} - \frac{n_{h1.} n_{h.1}}{n_{h..}} \right) \right\}^2}{\sum_h \frac{n_{h1.} n_{h2.} n_{h.1} n_{h.2}}{n_{h..}^2 (n_{h..} - 1)}}$$

▼ SUPPLEMENTARY EXERCISES

8.69 A sociologist studied the relationship between male skin color (light, medium, and dark) and job-mobility orientation (high, medium, and low). Use these data to run a chi-square test of independence. Interpret your results using $\alpha = .05$.

Job-Mobility Orientation	Male Skin Color		
	Light	Medium	Dark
High	35	84	51
Medium	49	78	23
Low	10	13	6

8.70 A study was conducted to determine the relationship between annual income and number of children per family. Compute percentages for each of the income categories; then run a chi-square test of independence and draw conclusions. Use $\alpha = .10$.

Number of Children per Family	Annual Income		
	< $20,000	$20,000–$40,000	> $40,000
≤ 2 children	38	45	22
> 2 children	220	95	30

8.71 Refer to Exercise 8.70. Assume that the data were obtained from the east. Data were also obtained from an additional 300 persons in the south and the west. Use these data and the accompanying computer output to determine whether there is a difference in annual income (on the average) for families with two or fewer than two children compared to families with more than two children. Combine data across the east, south, and west using uniform scores. Draw conclusions.

	Number of Children per Family	Annual Income		
		< $20,000	$20,000–$40,000	> $40,000
South	≤ 2 children	25	38	40
	> 2 children	120	50	27
West	≤ 2 children	36	39	27
	> 2 children	95	60	43

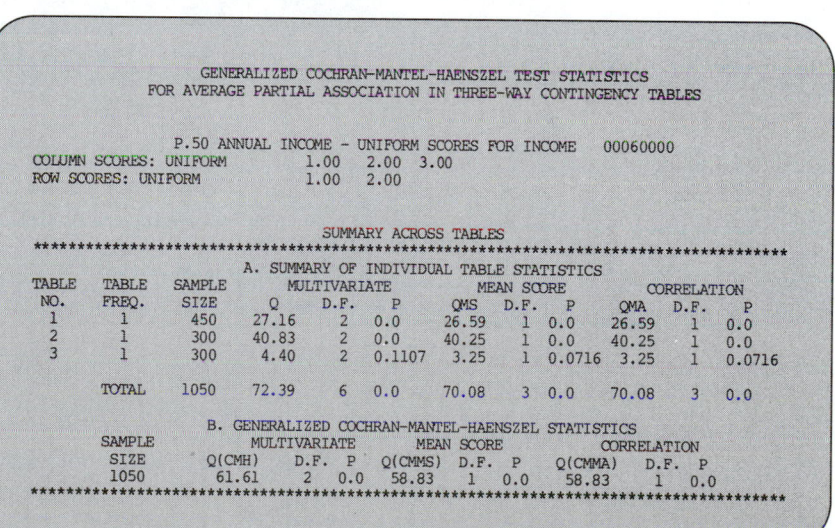

```
         GENERALIZED COCHRAN-MANTEL-HAENSZEL TEST STATISTICS
       FOR AVERAGE PARTIAL ASSOCIATION IN THREE-WAY CONTINGENCY TABLES


              P.50 ANNUAL INCOME - UNIFORM SCORES FOR INCOME    00060000
COLUMN SCORES: UNIFORM            1.00   2.00  3.00
ROW SCORES: UNIFORM               1.00   2.00


                          SUMMARY ACROSS TABLES
*************************************************************************
                  A. SUMMARY OF INDIVIDUAL TABLE STATISTICS
TABLE   TABLE   SAMPLE      MULTIVARIATE         MEAN SCORE          CORRELATION
NO.     FREQ.    SIZE     Q    D.F.   P       QMS   D.F.   P       QMA   D.F.   P
 1       1       450    27.16   2    0.0     26.59   1    0.0     26.59   1    0.0
 2       1       300    40.83   2    0.0     40.25   1    0.0     40.25   1    0.0
 3       1       300     4.40   2    0.1107   3.25   1    0.0716   3.25   1    0.0716

         TOTAL   1050    72.39   6    0.0     70.08   3    0.0     70.08   3    0.0

              B. GENERALIZED COCHRAN-MANTEL-HAENSZEL STATISTICS
         SAMPLE       MULTIVARIATE         MEAN SCORE          CORRELATION
          SIZE    Q(CMH)   D.F.  P    Q(CMMS)  D.F.  P    Q(CMMA)  D.F.  P
          1050    61.61     2   0.0    58.83    1   0.0    58.83    1   0.0
*************************************************************************
```

8.72 A random sample of 145 people of various occupations was taken to investigate the public opinion on police treatment. Each person was asked whether he or she would expect the police to treat him or her as good as, better than, or worse than a common criminal. The table below summarizes the results. Is there sufficient evidence to indicate that the expected treatment is independent of occupation? Use $\alpha = .10$.

	Expected Treatment			
Occupation	Better	As Good	Worse	Totals
Unemployed	6	23	11	40
Blue-collar worker	17	30	8	55
White-collar worker	16	28	6	50
Totals	39	81	25	145

 8.73 A sociological study was conducted to determine whether there is a relationship between the length of time blue-collar workers remain in their first job and the amount of their education. From union membership records, a random sample of persons was classified. The data are shown below.

a. Use the computer output that follows to identify the expected cell numbers.

```
                    CATEGORICAL ANALYSES

             TABLE OF YRS - JOB BY YRS - ED

  YRS - JOB            YRS - ED
  FREQUENCY
  PERCENT
  ROW PCT
  COL PCT   0-4.5    4.5-9    9-13.5    13.5      TOTAL

    0-2.5       5       21       30        33        89
             1.44     6.03     8.62      9.48      25.57
             5.62    23.60    33.71     37.08
             7.14    25.61    32.26     32.04

    2.5-5      15       35       40        30       120
             4.31    10.06    11.49      8.62      34.48
            12.50    29.17    33.33     25.00
            21.43    42.68    43.01     29.13

    5-7.5      22       16       15        30        83
             6.32     4.60     4.31      8.62      23.85
            26.51    19.28    18.07     36.14
            31.43    19.51    16.13     29.13

    7.5        28       10        8        10        56
             8.05     2.87     2.30      2.87      16.09
            50.00    17.86    14.29     17.86
            40.00    12.20     8.60      9.71

  TOTAL       70       82       93       103       348
            20.11    23.56    26.72     29.60    100.00

                                        (continues)
```

```
                    STATISTICS FOR 2-WAY TABLES

CHI-SQUARE                       57.830   DF=  9 PROB = 0.0001
PHI                               0.408
CONTINGENCY COEFFICIENT           0.377
CRAMER'S V                        0.235
LIKELIHOOD RATIO CHI-SQUARE      55.605   DF=  9 PROB = 0.0001

                    CATEGORICAL ANALYSES

          OBS    YRS-JOB    YRS-ED      FREQ
           1      0-2.5     0-4.5         5
           2      0-2.5     4.5-9        21
           3      0-2.5     9-13.5       30
           4      0-2.5     13.5         33
           5      2.5-5     0-4.5        15
           6      2.5-5     4.5-9        35
           7      2.5-5     9-13.5       40
           8      2.5-5     13.5         30
           9      5-7.5     0-4.5        22
          10      5-7.5     4.5-9        16
          11      5-7.5     9-13.5       15
          12      5-7.5     13.5         30
          13      7.5       0-4.5        28
          14      7.5       4.5-9        10
          15      7.5       9-13.5        8
          16      7.5       13.5         10

          N=16
```

b. Test the research hypothesis that the variable "length of time on first job" is related to the variable "amount of education."

Years on First Job	Years of Education			
	0–4.5	4.5–9	9–13.5	13.5
0–2.5	5	21	30	33
2.5–5	15	35	40	30
5–7.5	22	16	15	30
7.5	28	10	8	10

c. Give the level of significance for the test.

d. Draw your conclusions using $\alpha = .05$.

8.74 The personnel department of a large corporation was interested in determining the relationship between performance ratings of recently hired employees and the employees' college grade point averages. To do this, a random sample of 90 records was obtained, examined, and classified in the two-way table below. Is there evidence to indicate a relationship between the two variables "performace rating" and "college grade point average"? Use $\alpha = .05$.

Performance Rating	College Grade Point Average		
	A	B	C
Above average	19	8	3
Average	9	12	15
Below average	6	5	13

8.75 Television research to date suggest that "zipping" and "zapping" of commercials by VCR viewers is uncommon. Zipping is the use of the VCR remote control to fast-forward past commercials; zapping is the use of the remote control to change channels when commercials appear. Based on a random sample of 2,000 users of VCRs, 66% said that they did not skip the commercials. Obviously, with more than 30 million households with VCRs, widespread zipping and zapping would have a tremendous impact on the rates charged for TV advertising. Use the data to construct a 95% confidence interval for π, the proportion of VCR users that do not skip commercials.

8.76 Two researchers at Johns Hopkins University have studied the use of drug products in the elderly. Patients in a recent study were asked the extent to which physicians counseled them with regard to their drug therapies. The researchers found the following:

▼ 25.4% of the patients said their physicians did not explain what the drug was supposed to do.

▼ 91.6% indicated they were not told how the drug might "bother" them.

▼ 47.1% indicated their physicians did not ask how the drug "helped" or "bothered" them after therapy was started.

▼ 87.7% indicated the drug was not changed after discussion on how the therapy was helping or bothering them.

a. Assume that 500 patients were interviewed in this study. Summarize each of these results using a 95% confidence interval.

b. Do you have any comments about the validity of any of these results?

8.77 People over the age of 40 years tend to notice changes in their digestive systems that alter what and how much they can eat. A study was conducted to see whether this observation applies across different ethnic segments of our society. Random samples of Anglo-Saxons, Germans, Latin Americans, Italians, Spaniards, and African-Americans were obtained. The data from this survey are summarized here:

Ethnic Group	Sample Size Responding (60 of Each Group Were Contacted)	Number Reporting Altered Digestive System
Anglo-Saxon	55	7
German	58	6
Latin American	52	34
Italian	54	38
Spanish	30	20
African-American	49	31

a. Does it appear that there may be a bias due to the response rates?

b. Compare the rates (π_is) for the Anglo-Saxon and German groups using a 95% confidence interval.

8.78 Refer to Exercise 8.77. There seem to be two distinct rates—those around 12% and those around 70%. Combine the sample data for the first two groups and for the last four groups. Use these data to test the hypothesis—$H_0: \pi_1 - \pi_2 = 0$ versus $H_a: \pi_1 - \pi_2 < 0$. Here, π_1 corresponds to the population rate for the first combined group, and π_2 is the corresponding proportion for the second combined group. Give the p-value for your test.

8.79 Two sets of 60 ninth-graders were taught algebra I by different methods. The experimental group used self-paced modules developed for use at a computer with a display screen. The control group was given formal lectures by the teachers. At the end of the 4-month period, a comprehensive, standarized test was given to both groups. The experimental group had 65% scoring above 80 (out of 100), whereas the control group had just 47% above 80. Use these data to compare the percentages above 80 for the two groups. Give the p-value for your test.

8.80 Refer to Exercise 8.79. How might you have designed this study? What additional data might you want to collect?

8.81 A recent study conducted for a large university compared the effects of intensive behavioral training to change type A habits on the incidence of a second heart attack. A total of 290 patients who recently had a nonfatal heart attack were studied. Those randomized to group 1 received the intensive behavioral training in addition to routine medical care. The training was designed to help the patient to slow down and be more relaxed. Those assigned to group 2 received only the routine medical care. The data from this 5-year study are summarized below. Use these data to test $H_0: \pi_1 - \pi_2 = 0$ versus $H_a: \pi_1 - \pi_2 < 0$. Give a p-value for your test and draw a conclusion.

	Sample Size	Number of Second Heart Attacks Fatal and Nonfatal
Group 1	140	17
Group 2	150	29

8.82 A random sample of faculty members of a state university system was polled and classified by university and by which of the three collective bargaining agents (union 101, union 102, union 103) was preferred. The data appear here, and computer output follows.

	Bargaining Agent		
University	101	102	103
1	42	29	12
2	31	23	6
3	26	28	2
4	8	17	37

```
                        CATEGORICAL ANALYSES
               TABLE OF UNIV BY B_AGENT

        UNIV                   B_AGENT
    FREQUENCY
     PERCENT
     ROW PCT
     COL PCT      101        102        103       TOTAL
         1         42         29         12         83
                 16.09      11.11       4.60      31.80
                 50.60      34.94      14.46
                 39.25      29.90      21.05
         2         31         23          6         60
                 11.88       8.81       2.30      22.99
                 51.67      38.33      10.00
                 28.97      23.71      10.53
         3         26         28          2         56
                  9.96      10.73       0.77      21.46
                 46.43      50.00       3.57
                 24.30      28.87       3.51
         4          8         17         37         62
                  3.07       6.51      14.18      23.75
                 12.90      27.42      59.68
                  7.48      17.53      64.91
       TOTAL      107         97         57        261
                 41.00      37.16      21.84     100.00


                STATISTICS FOR 2-WAY TABLES

   CHI-SQUARE                        75.197     DF =   6   PROB=0.0001
   PHI                                0.537
   CONTINGENCY COEFFICIENT            0.473
   CRAMER'S V                         0.380
   LIKELIHOOD RATIO CHI-SQUARE       71.991     DF =   6   PROB=0.0001


                   CATEGORICAL ANALYSES

             OBS     UNIV    B_AGENT     FREQ
              1        1       101        42
              2        1       102        29
              3        1       103        12
              4        2       101        31
              5        2       102        23
              6        2       103         6
              7        3       101        26
              8        3       102        28
              9        3       103         2
             10        4       101         8
             11        4       102        17
             12        4       103        37

             N=12
```

a. Identify the expected cell numbers.
b. Use the computer output to determine whether there is evidence to indicate a difference in the distribution of preference across the four state universities.
c. Give the level of significance for the test.
d. Draw your conclusions.

$ 8.83 An advertising firm selected to conduct a market awareness study for a brand of house paint obtained information from a national survey of 1,500 randomly selected homeowners. Each homeowner selected for the survey was asked if he or she was familiar with a newly marketed line of interior latex paints. If 465 responded affirmatively, use the sample data to test the research hypothesis that the company has reached more than 30% of the homeowners with recent advertising. Use $\alpha = .05$.

8.84 Refer to Exercise 8.71. Use the computer output here to determine whether scores of 10, 30, and 60 make a difference in the conclusions. What about scores of 15, 30, and 100? What conclusions can be drawn and what reservations might you have?

```
            GENERALIZED COCHRAN-MANTEL-HAENSZEL TEST STATISTICS
          FOR AVERAGE PARTIAL ASSOCIATION IN THREE-WAY CONTINGENCY TABLES

            P.50 ANNUAL INCOME - SCORES OF 10,30,60 FOR INCOME    00140000

  COLUMN SCORES: USER SPECIFIED      10.00    30.00    60.00
  ROW SCORES: UNIFORM                 1.00     2.00

                       SUMMARY ACROSS TABLES
  **********************************************************************
                A. SUMMARY OF INDIVIDUAL TABLE STATISTICS

  TABLE  TABLE  SAMPLE      MULTIVARIATE          MEAN SCORE           CORRELATION
  NO.    FREQ.  SIZE      Q     D.F.   P      QMS    D.F.  P       QMA    D.F.  P
   1       1     450    27.16    2    0.0    25.20    1   0.0     25.20    1   0.0
   2       1     300    40.83    2    0.0    38.65    1   0.0     38.65    1   0.0
   3       1     300     4.40    2    0.1107  2.76    1   0.0966   2.76    1   0.0966

          TOTAL 1050    72.39    6    0.0    66.61    3   0.0     66.61    3   0.0

              B. GENERALIZED COCHRAN-MANTEL-HAENSZEL STATISTICS
          SAMPLE     MULTIVARIATE        MEAN SCORE           CORRELATION
          SIZE    Q(CMH)  D.F.  P  Q(CMMS)  D.F.  P   Q(CMMA)  D.F.  P
          1050    61.61    2   0.0  54.97    1   0.0   54.97    1    0.0
  **********************************************************************
            GENERALIZED COCHRAN-MANTEL-HAENSZEL TEST STATISTICS
          FOR AVERAGE PARTIAL ASSOCIATION IN THREE-WAY CONTINGENCY TABLES

            P.50 ANNUAL INCOME - SCORES OF 15,30,100 FOR INCOME    00170000

  COLUMN SCORES: USER SPECIFIED      15.00    30.00   100.00
  ROW SCORES: UNIFORM                 1.00     2.00

                       SUMMARY ACROSS TABLES
  **********************************************************************
                A. SUMMARY OF INDIVIDUAL TABLE STATISTICS

  TABLE  TABLE  SAMPLE      MULTIVARIATE          MEAN SCORE           CORRELATION
  NO.    FREQ.  SIZE      Q     D.F.   P      QMS    D.F.  P       QMA    D.F.  P
   1       1     450    27.16    2    0.0    18.69    1   0.0     18.69    1   0.0
   2       1     300    40.83    2    0.0    31.72    1   0.0     31.72    1   0.0
   3       1     300     4.40    2    0.1107  1.60    1   0.2057   1.60    1   0.2057

          TOTAL 1050    72.39    6    0.0    52.01    3   0.0     52.01    3   0.0

              B. GENERALIZED COCHRAN-MANTEL-HAENSZEL STATISTICS
          SAMPLE     MULTIVARIATE        MEAN SCORE           CORRELATION
          SIZE    Q(CMH)  D.F.  P  Q(CMMS)  D.F.  P   Q(CMMA)  D.F.  P
          1050    61.61    2   0.0  41.02    1   0.0   41.02    1    0.0
  **********************************************************************
```

 8.85 Faculty members at each of three universities were classified according to political ideology (left or right) and according to their academic tolerance (low, medium, or high).

University	Political Ideology	Academic Tolerance		
		Low	Medium	High
1	Left	11	16	44
	Right	15	19	11
2	Left	20	25	36
	Right	50	22	6
3	Left	5	3	4
	Right	30	33	25

a. Use descriptive statistics to characterize the three universities.

b. Conduct a Mantel–Haenszel mean score test based on uniform scores to compare the left and right ideologies across the three universities.

```
              GENERALIZED COCHRAN-MANTEL-HAENSZEL TEST STATISTICS
             FOR AVERAGE PARTIAL ASSOCIATION IN THREE-WAY CONTINGENCY TABLES

         P.54 POLITICAL IDEOLOGY - UNIFORM SCORES FOR ACADEMIC TOLERANCE   00510000


COLUMN SCORES: UNIFORM        1.00    2.00    3.00
ROW SCORES: UNIFORM           1.00    2.00


                          SUMMARY ACROSS TABLES
**************************************************************************************
                      A. SUMMARY OF INDIVIDUAL TABLE STATISTICS
  TABLE   TABLE   SAMPLE      MULTIVARIATE          MEAN SCORE          CORRELATION
  NO.     FREQ.    SIZE      Q     D.F.   P      QMS    D.F.  P      QMA   D.F.   P
   1        1      116     15.50    2   0.0004  13.17    1  0.0003  13.17   1   0.0003
   2        1      159     34.22    2   0.0     34.01    1  0.0     34.01   1   0.0
   3        1      100      0.71    2   0.7003   0.01    1  0.9144   0.01   1   0.9144

          TOTAL   375     50.42    6   0.0     47.19    3  0.0     47.19   3   0.0


                   B. GENERALIZED COCHRAN-MANTEL-HAENSZEL STATISTICS
          SAMPLE       MULTIVARIATE          MEAN SCORE          CORRELATION
           SIZE    Q(CMH)  D.F.   P     Q(CMMS)  D.F.  P     Q(CMMA)  D.F.   P
           375     41.90    2    0.0    39.82     1   0.0    39.82     1    0.0
**************************************************************************************
```

8.86 A study examining the effectiveness of a drug product for the treatment of arthritis was conducted at four different centers. At each center, 100 patients were treated, 50 with the test drug and 50 with an identical-appearing placebo. The data are shown here.

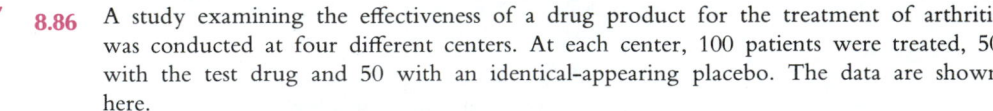

		Global Outcome			
Clinic	Treatment	Worse	Same	Better	Much Better/Completely Well
1	Placebo	10	15	17	8
	Test drug	12	14	10	14
2	Placebo	6	20	22	2
	Test drug	4	15	10	21
3	Placebo	7	25	12	6
	Test drug	5	22	12	11
4	Placebo	2	14	20	14
	Test drug	1	12	15	22

a. Compare the percentage distributions for the two treatment groups at the four universities.

b. One investigator wanted to collapse the global outcome categories into "improved" and "not improved." Comment on this.

c. Conduct a Mantel–Haenszel test on the collapsed data of part b using uniform scores. Draw conclusions.

8.87 Computer output for the full data of Exercise 8.86 is shown here. Compare your findings here to those in Exercise 8.86. Any lessons learned?

```
        GENERALIZED COCHRAN-MANTEL-HAENSZEL TEST STATISTICS
      FOR AVERAGE PARTIAL ASSOCIATION IN THREE-WAY CONTINGENCY TABLES

        P.55 GLOBAL OUTCOME - UNIFORM SCORES FOR OUTCOME   00610000

  COLUMN SCORES: UNIFORM      1.00    2.00    3.00    4.00
  ROW SCORES: UNIFORM         1.00    2.00

                          SUMMARY ACROSS TABLES
  ****************************************************************************
                A. SUMMARY OF INDIVIDUAL TABLE STATISTICS
  TABLE   TABLE   SAMPLE      MULTIVARIATE        MEAN SCORE        CORRELATION
   NO.    FREQ.    SIZE    Q    D.F.    P      QMS  D.F.    P     QMA  D.F.    P
    1       1      100   3.63    3   0.3042   0.08   1   0.7789  0.08   1   0.7789
    2       1      100  21.10    3   0.0001   8.84   1   0.0029  8.84   1   0.0029
    3       1      100   1.98    3   0.5775   1.72   1   0.1896  1.72   1   0.1896
    4       1      100   2.95    3   0.3995   1.93   1   0.1647  1.93   1   0.1647

          TOTAL   400  29.65   12   0.0031  12.57   4   0.0136 12.57   4   0.0136

               B. GENERALIZED COCHRAN-MANTEL-HAENSZEL STATISTICS
     SAMPLE        MULTIVARIATE        MEAN SCORE         CORRELATION
      SIZE    Q(CMH)  D.F.   P    Q(CMMS)  D.F.  P    Q(CMMA) D.F.   P
      400     20.80    3   0.0001   8.38    1  0.0038   8.38    1  0.0038
  ****************************************************************************
```

8.88 Legislators of a particular state were concerned that the enrollment (which affects budget allocations) at a particular university within the state system had been padded by allowing students to overenroll or to enroll for courses that required no academic work. To substantiate their initial findings, a random sample of 200 graduate students (from the 5,000 currently enrolled) was interviewed. If 20 students stated they had been allowed to pad their enrollments in the past quarter, use those data to construct a 99% confidence interval for π, the proportion of the entire student body with padded enrollments. Interpret your results.

 8.89 The relative sensitivities of two fuses were tested under controlled conditions by firing 40 rounds with Type I and 60 rounds with Type II, the firings being conducted in random order. Each round was classified according to whether the fuse functioned or not, with the following results:

Type of Fuse	Functioned	Did Not Function	Total
I	10	30	40
II	40	20	60
Total	50	50	100

a. Test the hypothesis that there is no difference between the sensitivities of the two fuses. Use $\alpha = .05$.

b. Use the previous data to demonstrate the relationship between the z test and the chi-square test of independence in a 2 × 2 table. (Hint: Compare z^2 and χ^2 for the two tests.)

 8.90 A study was conducted to investigate whether there is any relationship between voting record and education. One hundred five citizens selected at random were interviewed as to how often they vote and what level of formal education they had achieved. The results are shown in the table below. Is there evidence to indicate a relationship between level of education and frequency of voting? Use $\alpha = .05$.

	How Often Do You Vote?			
Education	Never	Some Elections	All Elections	Totals
Less than h.s. level	11	12	11	34
High school	7	15	13	35
College	2	20	14	36
Totals	20	47	38	105

 8.91 An extension of the traffic study for the implementation of a priority bus lane involved the sampling of public opinion concerning the bus lane during various phases of the study. Three different phases were to be studied. Phase 0 required the bus drivers to use the existing traffic lanes. In Phase 1, bus drivers made use of the exclusive bus lane with no preemption of the traffic signals at the intersection; Phase 2 allowed the bus drivers to extend the "green time" on a traffic signal to allow them to pass through before the light changed. Use the sample data below to determine whether the distribution of persons favoring, not favoring, or undecided changes from phase to phase. Use $\alpha = .05$.

	Opinion			
Phase	Favor	Do Not Favor	Undecided	Totals
0	80	90	30	200
1	60	112	28	200
2	50	125	25	200

 8.92 The quality-assurance unit of a large pharmaceutical company was engaged in comparing two new formulations of a drug product in tablet form. Both tablets contained the same amount of active ingredient, but varied in size, shape, and excipient (an inert substance that acts as a vehicle). A random sample of 100 tablets was obtained from a pilot batch for each formulation. The number of tablets classified as acceptable (or not acceptable) with regard to potency is shown below for each formulation. Use these data to place a 95% confidence interval about $\pi_1 - \pi_2$, the difference in the proportions of acceptable tablets.

Formulation	Number Acceptable	Number Not Acceptable	Sample Size
1	84	16	100
2	96	4	100

8.93 Refer to Exercise 8.92.
   a. Run a two-sided statistical test of $H_0: \pi_1 - \pi_2 = 0$. Draw conclusions.
   b. Run a chi-square test of independence for these data. Use $\alpha = .05$. Compare your results to part a.
   c. Suggest a relationship between the two tests (Hint: Compare z^2 and χ^2 for the two tests.)

 8.94 A matter of public concern is whether there is an association between the use of saccharin as a sweetener and the development of cancer. Suppose the 25 independent, controlled studies have been conducted at elevated doses (greater than or equal to 100 times the normal yearly human intake) in a particular strain of mice. In two of these studies, the treated group of mice had a higher incidence of cancer than in the corresponding control group of mice; no meaningful differences were observed between the treated and controlled mice in the other studies. What conclusions might you draw from these 25 studies? What additional data would you like to see? How do these results extrapolate to the human situation? What error rate should be controlled more carefully?

8.95 A survey of drivers was obtained to compare the proportions who use seat belts regularly for various age categories. These data are shown below. Analyze the data and draw conclusions. Use $\alpha = .05$.

Age (Years)	Regularity of Seat Belt Usage			
	Always	Regularly	Sometimes	Never
16–20	1	10	70	19
21–25	4	8	80	8
26–30	8	10	77	5
> 30	15	30	49	6

 8.96 Entry-level diastolic blood pressures for a group of 12 hypertensive females are shown in the next table, along with a corresponding diastolic blood pressure following 2 weeks of treatment with an antihypertensive medication. Note: The goal of the therapy is to reduce the diastolic blood pressure to 90 mm Hg.

Patient	Entry Level Diastolic BP	BP Following 2 Weeks
1	105	89
2	110	87
3	115	93
4	107	90
5	108	88
6	116	90
7	114	91
8	110	90
9	112	89
10	106	91
11	109	90
12	111	92

a. What percent of the patients were controlled at ≤ 90 mm Hg?

b. Give a p-value for the test $H_0: \pi = 0$ versus $H_a: \pi > 0$.

8.97 Refer to Exercise 8.96.

a. Consider using a t test for $H_0: \mu_d = 0$ versus $H_a: \mu_d > 0$. Might there be a problem with one (or more) of the assumptions? If so, which ones and why?

b. Suggest and perform a suitable test of location for the difference data. Give the p-value and interpret your findings.

8.98 A poll asked the following question of a random sample of 100 people in various countries: From what you have heard or read, which of these statements comes closest to the way you feel about the United States' involvement in the conflict in Nicaragua? (1) The United States should begin to withdraw its support of the contras. (2) The United States should carry on its present level of support. (3) The United States should increase its level of support.

Country	Withdraw Support	Maintain Present Level	Increase Support	Don't Know
Argentina	57	6	6	31
Australia	21	43	24	12
Brazil	76	5	5	14
Canada	41	16	23	20
Finland	81	4	5	10
France	72	8	5	15
Great Britain	45	15	15	25
India	66	4	8	22
Sweden	79	10	4	7
Uruguay	62	10	5	23
United States	31	10	53	6
West Germany	58	11	14	17

a. Would it be meaningful to do a percentage comparison of the data?
b. Suggest a statistical test to compare the responses across countries.
c. Run the test you selected in part b (using $\alpha = .05$) and draw conclusions.

8.99 A sample of 150 patients suffering from severe acquired immunodeficiency syndrome (AIDS) were treated with an experimental compound. Without treatment, these patients were given little chance for survival over the next 6 months. Baseline characteristics are shown here.

Baseline Characteristics	
Gender (% male)	92%
Age in years ($\bar{y} \pm$ SD)	32 ± 6
Weight in kg ($\bar{y} \pm$ SD)	68 ± 7

For the 120 patients who were treated for more than 1 week, the 6-month survival rate was 52%; the survival rate for the 30 not receiving more than 1 week of therapy was only 17%.

a. Use the survival data to give 95% confidence intervals on 6-month survival rates for those patients treated for more than one week and those not.

b. Would it be valid to make a comparison of these two survival rates for the subgroups of $n_1 = 120$ and $n_2 = 30$ patients? Why or why not?

c. Estimate the overall survival rate for the AIDS patients using a 95% confidence interval. Is this a more valid estimate of the survival rate for patients treated with the drug than is the estimate based on those treated for more than 1 week?

 8.100 As part of a reseach study conducted at a large behavioral treatment program, 160 subjects quitting smoking were randomized to one of four treatment programs.

> A: Reading and audiovisual (A/V) material only, no professional contact
> B: Reading and A/V material, low professional contact
> C: Reading and A/V material, high professional contact
> D: High professional contact only

Assessments were made at 4, 12, 26, and 52 weeks, with the results (abstinence rates) shown here:

	Time			
Program	4	12	26	52
A	50%	45%	35%	30%
B	70%	50%	47%	35%
C	95%	75%	60%	40%
D	90%	68%	56%	38%

a. Plot the sample data and draw some tentative conclusions.

b. At a given time point, what statistical test would be appropriate for comparing the abstinence rates?

c. Run these comparisons for each time point and draw conclusions. In general, what seems to be the pattern of abstinence?

8.101 A survey was conducted among regular shoppers at three different suburban shopping centers. Subjects were given a questionnaire to be completed in private and deposited in a centrally located, locked collection box. A total of 1,000 questionnaires were distributed, and 400 were returned. One of the items on the questionnaire had to do with family income levels.

	Family Income		
Shopping Center	< 50,000	50–100,000	> 100,000
A	60	25	15
B	66	50	9
C	127	40	8
Total	253	115	32

a. Do a percentage comparison of the family income distributions for the three shopping centers. Do the centers appear to have the same distributions?

b. Run a test of significance to compare the income distributions. Give a *p*-value for your test and draw a conclusion.

c. What effect might the response rate have on the conclusions drawn from this survey?

8.102 Three supermarkets have a policy of advertising specials on certain days of the week to attract customers. A customer count is maintained at these supermarkets from the hours of 11.00 A.M. to 2:00 P.M. for the five weekdays. The data collected was the following:

| | Day of the Week | | | | |
Supermarket	Mon.	Tues.	Wed.	Thurs.	Fri.
1	605	650	702	663	568
2	696	741	750	827	663
3	540	668	528	572	516

a. Do the data indicate that the choice of a supermarket depends on the day of the week? Use a procedure that has a 0.01 chance of being in error if this conclusion is drawn. If you decide that choice of a supermarket is related to day of the week, briefly describe the nature of that relationship based on the sample data.

b. Do the data support the hypothesis that the number of customers expected (for all three supermarkets together) is the same for the days Monday through Friday? Use a procedure that has a 0.05 chance of being in error if it is concluded that different days have different customer counts.

8.103 A survey was conducted among large corporations in the United States. Each was given a questionnaire to be completed and returned. One of the items of information sought had to do with the frequency of market offerings in the past year through three underwriters. The results compiled from returned questionnaires were as follows:

| | Frequency of Offerings | | |
Underwriter	0	1	2 or More
1	58	26	25
2	56	56	20
3	97	30	42

Use the output provided to determine whether percentage of market offerings differs among underwriters. Give a *p*-value for your test and draw a conclusion.

```
                        CATEGORICAL ANALYSIS
                      TABLE OF WRITER BY OFFERS

           WRITER                   OFFERS

           FREQUENCY
           PERCENT
           ROW PCT
           COL PCT      0        1        2       TOTAL
          ──────────────────────────────────────────────
              1          58       26       25       109
                       14.15     6.34     6.10     26.59
                       53.21    23.85    22.94
                       27.49    23.21    28.74
          ──────────────────────────────────────────────
              2          56       56       20       132
                       13.66    13.66     4.88     32.20
                       42.42    42.42    15.15
                       26.54    50.00    22.99
          ──────────────────────────────────────────────
              3          97       30       42       169
                       23.66     7.32    10.24     41.22
                       57.40    17.75    24.85
                       45.97    26.79    48.28
          ──────────────────────────────────────────────
           TOTAL        211      112       87       410
                       51.46    27.32    21.22    100.00

              STATISTICS FOR TABLE OF WRITER BY OFFERS

           STATISTIC                  DF       VALUE      PROB

      CHI-SQUARE                       4       23.977     0.000
      LIKELIHOOD RATIO CHI-SQUARE      4       23.488     0.000
      MANTEL-HAENSZEL CHI-SQUARE       1        0.087     0.769
      PHI                                       0.242
      CONTINGENCY COEFFICIENT                   0.235
      CRAMER'S V                                0.171

      SAMPLE SIZE=410
```

$ 8.104 The following data give the observed frequencies of errors per page of unread galley proof for a sample of 40 pages from a certain journal publisher.

Errors/Page	Observed Frequencies
0	5
1	9
2	5
3	7
4	4
5	2
6	3
7	2
8	1
9	0
10	2

Conduct a test to determine whether the errors per page follow a Poisson distribution with a mean rate of 3.2. Use $\alpha = .10$.

ANALYZING DATA: REGRESSION METHODS

LINEAR REGRESSION AND CORRELATION

9.1   INTRODUCTION

In Chapters 5 and 6, we considered estimation and the test of a hypothesis concerning a population mean or the difference between two population means. The problems we encountered were relatively simple and straightforward. Estimation of a population mean can become more involved, however, if the variable of interest, often called the *dependent variable*, is affected by one or more additional variables, called *independent variables*.

For example, suppose we are interested in estimating the mean weight gain per month for steers fed on a particular variety of feed. The dependent variable, weight gain, could be affected by many variables: initial weight of the steer, amount of feed offered per day, protein content of the feed, water content of the feed, and so on. The problem of estimating the mean weight gain per month must now take into account the *levels*, or *settings* of the independent variables.

Suppose we want to estimate the mean weight gain (in pounds per day) for steers fed a high-concentrate feed containing 15% protein, 10% water, and the rest carbohydrates. Here, 15% is a setting, or level, of the independent variable "protein content of feed" and 10% is a setting of the independent variable "water content of feed." Similarly, we might wish to estimate, or predict, the mean weight gain for other combinations of settings of the independent variables. Thus, estimating a population mean becomes a problem of estimating a population mean for each setting of the independent variables.

This estimation problem can be greatly simplified if we consider models relating a response (dependent variable) to a set of independent variables. In Chapters 9 and 10, we develop the concepts of linear regression. We will formulate a model and develop estimation and test procedures when the dependent variable is related to one independent variable.

In later chapters, we will expand the material of Chapters 9 and 10 to include models and inferences for multiple regression where the dependent variable is related to more than one independent variable. These same models will be useful in many other prediction or estimation problems. For example, a biologist may wish to predict an animal's pulse rate based on the amount of a particular drug administered and the length of time since the drug's administration. A political scientist may wish to predict the outcome of an election (in terms of numbers of votes cast for a particular candidate) based in various socioeconomic factors and previous voting records of the population under study. A sociologist may wish to relate or predict the average number of prison-free years between first and second offenses for persons characterized by various variables, such as age, gender, number of years of schooling, IQ, and so on. A market analyst might want to predict the year-end sales for a company based on various economic indices. Each of these problems bears certain similarities and can be attacked using the models to be proposed in this chapter.

equation of straight line

The simplest type of model relating a response y to a single quantitative independent variable x is given by the **equation of a straight line**.

$$y = \beta_0 + \beta_1 x,$$

y-intercept β_0
slope β_1

deterministic model

where β_0 is the **y-intercept** (value of y when $x = 0$), and β_1 is the **slope** of the straight line (change in y for a unit change in x). For a given equation, β_0 and β_1 are constants. An equation of this form is called a **deterministic model** because there is no error in reading y. That is, for a given value of independent variable x, we can predict y exactly using the deterministic equation $y = \beta_0 + \beta_1 x$.

Although deterministic models are simple to use, they are unrealistic in many situations, since a dependent variable y cannot always be adequately represented by a deterministic equation in one or more quantitative independent variables. Consider the data of Table 9.1, which gives hospital expenses covered by an insurance carrier and the number of days of confinement for a random sample of $n = 10$ patients. In the table, we let y be the dependent variable (expense) and x be the independent variable (number of days of confinement). Suppose the insurance carrier and hospital administrators are interested in estimating the average hospital expense for a given number of days of confinement. Can we use a deterministic model for this problem?

scatter diagram

First, we draw a picture of the data. The data of Table 9.1 can be plotted by using a **scatter diagram**. In a scatter diagram, we draw a vertical axis and a horizontal axis, labeled y and x, respectively. The $n = 10$ data points are then plotted as shown in Figure 9.1.

TABLE 9.1
Hospital Expense Data

Expense, y	Number of Days of Confinement, x
$ 50	1
175	3
180	6
200	7
60	2
140	4
420	12
540	15
170	5
300	9

FIGURE 9.1

Scatter Diagram of the Ten Data Points for the Hospital Expense Data of Table 9.1

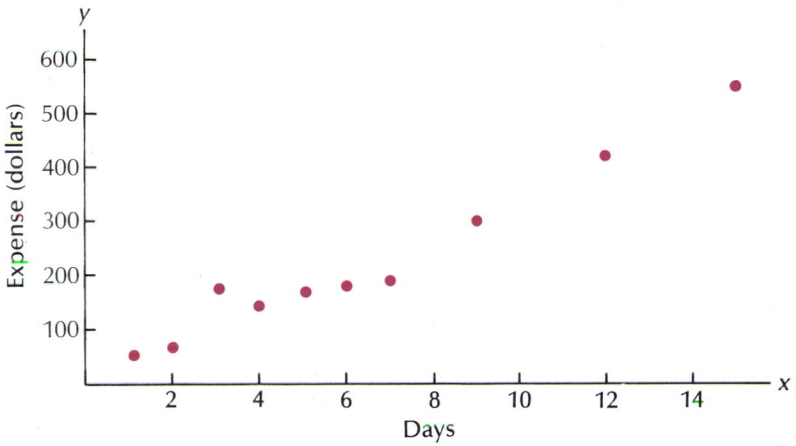

You can see from Figure 9.1 that a straight line would adequately describe the trend in the data. However, we cannot predict y *exactly* for a given value of x. Thus, for example, we could not predict the average value of y for $x = 8$ days of confinement using a deterministic model of the form $y = \beta_0 + \beta_1 x$.

A model that allows for the possibility that the observations do not lie on a straight line is the model

$$y = \beta_0 + \beta_1 x + \varepsilon,$$

random error ε

where ε is a **random error**. In this model, ε represents the difference between a measurement y and a point on the line $\beta_0 + \beta_1 x$. The random error term

ε takes into account all unpredictable and unknown factors that are not included in the model. For example, the amount of hospital expenses covered by the insurance carrier could be affected by such factors as the type of operation, complications following the operation, expenses previously submitted for coverage, and whether the insured's spouse has another insurance plan that could cover the expenses. The combined effects of all these and other factors not included in the model contribute to ε.

average value of ε is 0 for fixed x

expected value of y

One assumption made concerning the random error is that the **average value of ε for a given value of x is 0.** Thus, since β_0 and β_1 are constants, the *average of y* (often called the **expected value of y**) for a fixed value of x is $\beta_0 + \beta_1 x$. This line, denoted by

$$E(y) = \beta_0 + \beta_1 x,$$

is shown in Figure 9.2. A point on the line denotes the average value of y for the corresponding setting of x. The difference between a sample data point and the expected value of y (a point on the line $E(y) = \beta_0 + \beta_1 x$) is ε. Thus, the observed values of y deviate above or below the line by a random amount ε. The random errors associated with the ten data points listed in Table 9.1 are pictured in Figure 9.2.

FIGURE 9.2

A Plot of $E(y) = \beta_0 + \beta_1 x$ for the Hospital Expense Data of Table 9.1

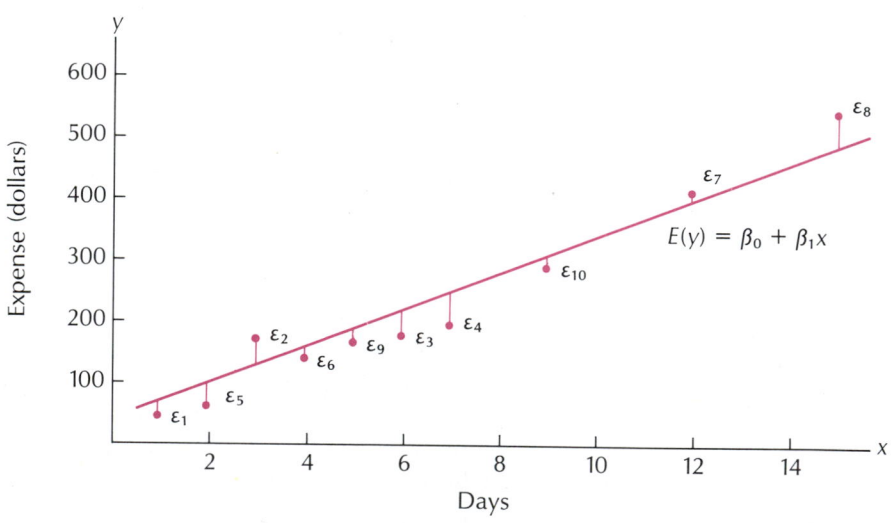

Unfortunately, since β_0 and β_1 are unknown parameters, we will never know the precise location of the line $E(y) = \beta_0 + \beta_1 x$. All we will have in a given experimental situation will be the n data points. In the next section, we show how to use the sample information to construct estimates, $\hat{\beta}_0$ and $\hat{\beta}_1$, of the parameters β_0 and β_1 to be used in formulating an estimate of the line $E(y) = \beta_0 + \beta_1 x$. Appropriate confidence intervals and tests of hypotheses are also discussed in Chapter 10.

9.2 ▼ LINEAR REGRESSION AND THE METHOD OF LEAST SQUARES

Consider the problem of obtaining estimates for parameters in the model

$$y = \beta_0 + \beta_1 x + \varepsilon$$

linear regression

for the **linear regression**

$$E(y) = \beta_0 + \beta_1 x.$$

There are many ways to determine an estimate of $E(y)$, which is represented by the equation

$$\hat{y} = \hat{\beta}_0 + \hat{\beta}_1 x.$$

eyeball fitting

One procedure, called the **eyeball fitting** technique, requires that we plot the data on a scatter diagram and then use a ruler to draw what we feel is the straight line that most accurately displays the linear trend of the data. Unfortunately, if each of us was given the same set of data, we might each come up with a different prediction equation.

method of least squares

residual, or error of prediction

The **method of least squares** is, in many respects, a formalization of the eyeball fitting routine just discussed. If we let $\hat{y}$ denote the predicted value of y for a given value of x, then the **error of prediction** (often called the **residual**) is $y - \hat{y}$, the difference between the actual value of y and what we predict it to be. *The method of least squares chooses the prediction line $\hat{y} = \hat{\beta}_0 + \hat{\beta}_1 x$ that minimizes the sum of the squared errors of prediction $\sum (y - \hat{y})^2$ for all sample points.* We can denote the sum of the squared errors of prediction for the linear model $y = \beta_0 + \beta_1 x + \varepsilon$ by

$$\sum (y - \hat{y})^2 = \sum (y - \hat{\beta}_0 - \hat{\beta}_1 x)^2.$$

Thus, the method of least squares consists of finding those estimates $\hat{\beta}_0$ and $\hat{\beta}_1$ that minimize $\sum (y - \hat{y})^2$.

While the procedure for deriving these estimates involves use of the calculus, we can summarize the results. The estimates, called **least squares estimates**, that minimize $\sum (y - \hat{y})^2$ are computed as shown here.

least squares estimates

Least Squares Estimates of β_1 and β_0

$$\hat{\beta}_1 = \frac{S_{xy}}{S_{xx}} \quad \text{and} \quad \hat{\beta}_0 = \bar{y} - \hat{\beta}_1\bar{x},$$

where

S_{xx}

$$S_{xx} = \sum (x - \bar{x})^2 = \sum x^2 - \frac{(\sum x)^2}{n}$$

S_{xy}

$$S_{xy} = \sum (x - \bar{x})(y - \bar{y}) = \sum xy - \frac{(\sum x)(\sum y)}{n}$$

These ideas can probably be best understood by working an example.

E X A M P L E   9.1

In a random sample of $n = 9$ steers, the live weights and dressed weights were recorded. In Table 9.2 we let y denote the dressed weight (in hundreds of pounds) and x denote the corresponding live weight (in hundreds of pounds). Use the sample data to obtain least squares estimates for the model

$$y = \beta_0 + \beta_1 x + \varepsilon.$$

T A B L E   9.2

Sample Data for Example 9.1; Live Weight (x) and Dressed Weight (y) of Steers

Live Weight (x)	Dressed Weight (y)
4.2	2.8
3.8	2.5
4.8	3.1
3.4	2.1
4.5	2.9
4.6	2.8
4.3	2.6
3.7	2.4
3.9	2.5

Solution   When we do not have the use of a calculator, the least squares estimates can be computed fairly easily if we construct a summary table, such as that shown in Table 9.3.

TABLE 9.3

Summary Table for the Data of Example 9.1

	x	x^2	y	y^2	xy
	4.2	17.64	2.8	7.84	11.76
	3.8	14.44	2.5	6.25	9.50
	4.8	23.04	3.1	9.61	14.88
	3.4	11.56	2.1	4.41	7.14
	4.5	20.25	2.9	8.41	13.05
	4.6	21.16	2.8	7.84	12.88
	4.3	18.49	2.6	6.76	11.18
	3.7	13.69	2.4	5.76	8.88
	3.9	15.21	2.5	6.25	9.75
Totals	37.2	155.48	23.7	63.13	99.02

Using the computational formulas for S_{xx} and S_{xy}, we have, from Table 9.3,

$$S_{xx} = \sum x^2 - \frac{(\sum x)^2}{n} = 155.48 - \frac{(37.2)^2}{9}$$

$$= 155.48 - 153.76 = 1.72$$

$$S_{xy} = \sum xy - \frac{(\sum x)(\sum y)}{n} = 99.02 - \frac{(37.2)(23.7)}{9}$$

$$= 99.02 - 97.96 = 1.06.$$

The least squares estimate for β_1 is

$$\hat{\beta}_1 = \frac{S_{xy}}{S_{xx}} = \frac{1.06}{1.72} = .616.$$

The sample means $\bar{x}$ and $\bar{y}$ are

$$\bar{x} = \frac{\sum x}{n} = \frac{37.2}{9} = 4.133$$

and

$$\bar{y} = \frac{\sum y}{n} = \frac{23.7}{9} = 2.633.$$

Substituting our calculated values into the formula for $\hat{\beta}_0$, we have

$$\hat{\beta}_0 = \bar{y} - \hat{\beta}_1 \bar{x} = 2.633 - .616(4.133) = .087.$$

The least squares equation for these data is

$$\hat{y} = .087 + .616x,$$

which is plotted in Figure 9.3 with the sample data superimposed.

FIGURE 9.3

Plot of the Least Squares Equation for the Data of Example 9.1

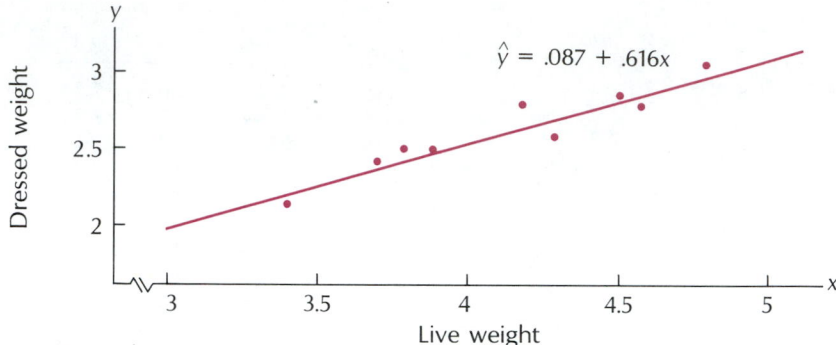

The slope $\hat{\beta}_1$ of this equation says that for every 1 unit increase in x (100 pounds live weight), we can expect a corresponding .616 unit increase in y (61.6 pounds in dressed weight). The predicted value of y for any value of x is given by a point on the line and can be computed by using the equation

$$\hat{y} = .087 + .616x. \qquad\qquad ▲$$

Algebraically, it can be shown that from a least squares fit of the model $y = \beta_0 + \beta_1 x + \varepsilon$,

$$y_i - \bar{y} = (y_i - \hat{y}_i) + (\hat{y}_i - \bar{y})$$

and that

$$\sum_i (y_i - \bar{y})^2 = \sum (y_i - \hat{y}_i)^2 + \sum (\hat{y}_i - \bar{y})^2.$$

Although the proof of this equality is beyond the scope of this text, we can obtain an intuitive understanding of this relationship by considering the following situation.

Suppose that we wish to use the model

$$y = \beta_0 + \varepsilon.$$

In this model, β_0 represents the population mean for the variable y, and, intuitively, we would estimate its value using the sample mean $\bar{y}$. (You can confirm this result by using the formula for the estimated intercept $\hat{\beta}_0$ in a linear model.) Since $\hat{y} = \bar{y}$ for this model, the sum of the squared errors of prediction is $\sum (y - \bar{y})^2$.

Now suppose the variable y is related to an independent variable x. From our previous work, we could fit the model $y = \beta_0 + \beta_1 x + \varepsilon$ to obtain

$$\hat{y} = \hat{\beta}_0 + \hat{\beta}_1 x.$$

For this model the sum of the squared prediction errors is

$$\sum (y - \hat{y})^2.$$

In Figure 9.4 we have presented two prediction equations: $\hat{y} = \bar{y}$ for the model $y = \beta_0 + \varepsilon$ and $\hat{y} = \hat{\beta}_0 + \hat{\beta}_1 x$ for the model $y = \beta_0 + \beta_1 x + \varepsilon$. Note that we can express the distance between an observation y and the sample mean $\bar{y}$ as the sum of two components, $(\hat{y} - \bar{y})$ and $(y - \hat{y})$. The quantity $(\hat{y} - \bar{y})$ represents that portion of the overall distance that can be attributed to the independent variable x (through the prediction equation $\hat{y} = \hat{\beta}_0 + \hat{\beta}_1 x$). The quantity $(y - \hat{y})$ represents that portion of the distance between y and $\bar{y}$ that cannot be accounted for by the independent variable x (and that we attribute to error). Combining this information for all sample observations, we can express the total variability in the sample measurements about the sample mean, $\sum (y - \bar{y})^2$, called the **sum of squares about the mean**, as the sum of the squared deviations of the predicted values from $\bar{y}$ called the **sum of squares due to regression**, $\sum (\hat{y} - \bar{y})^2$, and the sum of the squared errors of prediction, $\sum (y - \hat{y})^2$, called the **sum of squares for error**. Thus, we have

sum of squares about the mean

sum of squares due to regression

sum of squares for error

Sum of squares about the mean = sum of squares due to regression
+ sum of squares for error.

There is another way to view this equation. It can be shown that the sample mean $\bar{y}$ is also the average of the fitted values. Therefore, the sum of squares due to regression $\sum (\hat{y} - \bar{y})^2$ depicts variability in the fitted values. Similarly, $\sum (y - \hat{y})^2$ represents variability in the y-values about the fitted values. As a result,

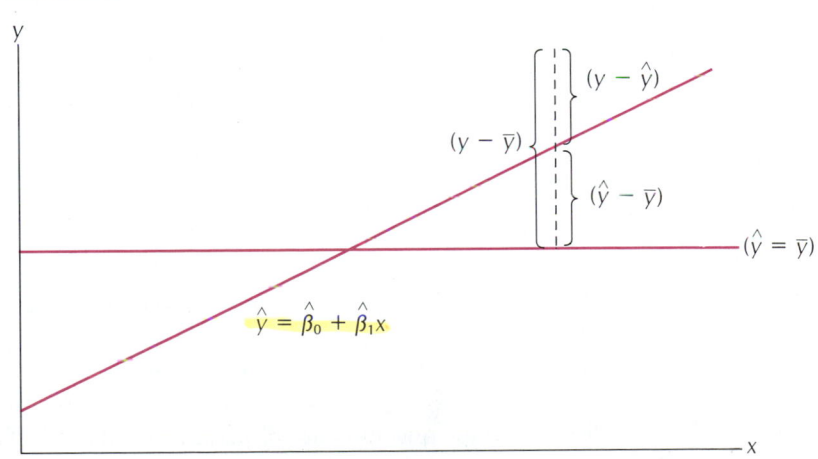

FIGURE 9.4
Relationship Between $\sum (y - \bar{y})^2$ and $\sum (y - \hat{y})^2$

the total variability in the y-values can be written as

$$\underbrace{\sum (y - \bar{y})^2}_{\substack{\text{total variability} \\ \text{in } y\text{-values}}} = \underbrace{\sum (\hat{y} - \bar{y})^2}_{\substack{\text{variability} \\ \text{explained by model}}} + \underbrace{\sum (y - \hat{y})^2}_{\substack{\text{unexplained} \\ \text{variability}}}.$$

Figure 9.4 shows how the total variability is partitioned into the two components.

Obviously, if we're interested in predicting y based on the independent variable x, the larger the explained variability is relative to the unexplained variability, the better the model "fits" the data, and this should lead to more precise prediction of y based on x.

E X A M P L E   9.2

▼

Consider the five data points listed in columns 1 and 2 of Table 9.4.

a. Fit the model

$$y = \beta_0 + \beta_1 x + \varepsilon.$$

b. Verify that

$$\sum (y - \bar{y})^2 = \sum (y - \hat{y})^2 + \sum (\hat{y} - \bar{y})^2.$$

Solution

a. For these data, it can be shown that

$$S_{xx} = \sum x^2 - \frac{(\sum x)^2}{n} = 66 - \frac{(16)^2}{5} = 14.8$$

$$S_{xy} = \sum xy - \frac{(\sum x)(\sum y)}{n} = 145 - \frac{(16)(38)}{5} = 23.4$$

$$\hat{\beta}_1 = 1.58$$

and

$$\hat{\beta}_0 = \bar{y} - \hat{\beta}_1 \bar{x} = 7.6 - 1.58(3.2) = 2.54.$$

b. For each x-value we then compute $\hat{y}$ from the least squares prediction equation. We also compute the quantities $(y - \bar{y})$, $(y - \hat{y})$, and $(\hat{y} - \bar{y})$. These quantities are displayed in Table 9.4. From columns 4, 5, and 6 in

T A B L E 9.4

Data and Computations for Example 9.2

x	y	$\hat{y}$	$y - \bar{y}$	$y - \hat{y}$	$\hat{y} - \bar{y}$
1	4	4.1216	−3.6000	−.1216	−3.4784
2	6	5.7027	−1.6000	.2973	−1.8973
3	7	7.2838	−.6000	−.2838	−.3162
4	9	8.8649	1.4000	.1351	1.2649
6	12	12.0271	4.4000	−.0271	4.4271

the table, we have

$\hat{y} = 1.58x + 2.54$

$$\sum (y - \bar{y})^2 = 37.2000$$
$$\sum (y - \hat{y})^2 = .2027$$
$$\sum (\hat{y} - \bar{y})^2 = 36.9982.$$

Note that, except for rounding errors,

$$\sum (y - \bar{y})^2 = \sum (y - \hat{y})^2 + \sum (\hat{y} - \bar{y})^2.$$

Intuitively, since the explained variability $\sum (\hat{y} - \bar{y})^2 = 36.9982$ accounts for almost all of the total variability $\sum (y - \bar{y})^2 = 37.2000$, the model appears to fit the data very well. A scatterplot of y versus x with the prediction equation superimposed shows how tightly the data group about the prediction equation.

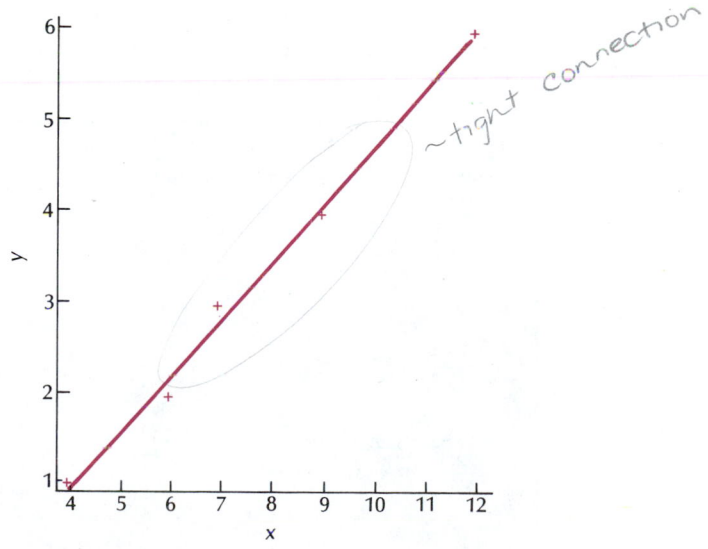

~tight connection

▼ EXERCISES

Basic Techniques

9.1   Plot the data shown here in a scatter diagram.

x	5	10	12	15	18	24
y	10	19	21	28	34	40

9.2   Use the equation $y = 1.8 + 2.0x$.
a. Predict y when $x = 3$.
b. Plot the equation on a graph with the horizontal axis scaled from 0 to 5 and the vertical axis scaled from 0 to 12.

9.3   Use the accompanying data to determine the least squares prediction equation.

x	1	2	3	4	5
y	2	4	6	7	9

9.4   Use the accompanying data to answer a and b.

x	1	3	5	7	9
y	1	4	8	9	12

a. Determine the least squares prediction equation.
b. Use the least squares prediction equation to predict y when $x = 6$.

9.5   Refer to the data of Exercise 9.1. Find the least squares prediction equation and compare it to the freehand regression line you found in Exercise 9.2.

9.6   A computer solution using SAS for the least squares prediction equation to the data is shown here.

```
OPTIONS NODATE NONUMBER PS=60 LS=78;
DATA ONE;
  INPUT X Y;
  CARDS;
   10      25
   18      55
   25      50
   40      75
   50     110
   63     138
   42      90
   30      60
    5      10
   55     100
  ;
PROC GLM;
  MODEL Y=X/P;
TITLE1 '      ';
RUN;
```

(continues)

```
Model: MODEL1
Dependent Variable: Y
                             Analysis of Variance

                              Sum of           Mean
      Source         DF       Squares         Square      F Value      Prob>F
      Model           1    13230.96994    13230.96994     162.560      0.0001
      Error           8      651.13006       81.39126
      C Total         9    13882.10000

           Root MSE        9.02171      R-square       0.9531
           Dep Mean       71.30000      Adj R-sq       0.9472
           C.V.           12.65317
                             Parameter Estimates
                       Parameter       Standard      T for H0:
      Variable   DF    Estimate         Error       Parameter=0    Prob > |T|
      INTERCEP    1    4.697852        5.95202071       0.789        0.4527
      X           1    1.970478        0.15454842      12.750        0.0001

                            Dep Var      Predict
                    Obs        Y          Value      Residual
                     1      25.0000      24.4026       0.5974
                     2      55.0000      40.1665      14.8335
                     3      50.0000      53.9598      -3.9598
                     4      75.0000      83.5170      -8.5170
                     5      110.0        103.2         6.7783
                     6      138.0        128.8         9.1620
                     7      90.0000      87.4579       2.5421
                     8      60.0000      63.8122      -3.8122
                     9      10.0000      14.5502      -4.5502
                    10      100.0        113.1       -13.0741

      Sum of Residuals             1.207923E-13
      Sum of Squared Residuals        651.1301
      Predicted Resid SS (Press)     1065.6517
```

a. Plot the data.

b. Determine the least squares prediction equation from the output here and draw the regression line in the data plot in part a.

c. Does the prediction equation seem to represent the data adequately?

d. Predict y (annual prescription volume) for $x = 35$.

Applications

 9.7 Family income and annual savings data are displayed here for a sample of nine families.

Annual Savings ($1,000)	Annual Income ($1,000)
1	36
2	39
2	42
5	45
5	48
6	51
7	54
8	56
7	59

a. Graph the data using a scatterplot.
b. Determine an eyeball fit to the data. Predict y (annual savings $1,000) based on an annual income of $x = \$45,000$.

9.8   Refer to Exercise 9.7.
a. Determine the least squares prediction equation.
b. Compute $\sum (y - \hat{y})^2$ for the least squares fit and for your eyeball fit. Note that SSE for the least squares fit will be less than or equal to that for the eyeball fit.

9.9   As one part of a study of commercial bank branches, data are obtained on the number of independent businesses, x, located in sample ZIP code areas and the number of bank branches, y, located in these areas. The commercial centers of cities are excluded.

x	92	116	124	210	216	267	306	378	415	502	615	703
y	3	2	3	5	4	5	5	6	7	7	9	9

$$\sum x = 3{,}944 \qquad \sum y = 65 \qquad \sum xy = 26{,}208$$
$$\sum x^2 = 1{,}732{,}524 \qquad \sum y^2 = 409 \qquad n = 12$$

a. Plot the data. Does a linear equation relating y to x appear plausible?
b. Calculate the regression equation (with y as the dependent variable).

9.10   The following data were obtained in a study of sales volume (per district) as a fraction of the number of client contacts per month.

Sales Volume ($1,000) y	Average Number of Client Contacts per Month x
15	10
26	15
28	17
30	20
32	23
86	46
109	53
95	48
130	59
160	65

a. Plot the data.
b. Eyeball a linear fit to the data and guess the value of the intercepts and slope.
c. Predict sales for $x = 50$.

9.11   Refer to Exercise 9.10.
a. Obtain the linear regression equation $\hat{y} = \hat{\beta}_0 + \hat{\beta}_1 x$ using the method of least squares. Compare this line to the one obtained in Exercise 9.10.
b. Predict the sales for $x = 50$ and compare to Exercise 9.10c.

9.12 An experiment was conducted to examine the effect of different concentrations of pectin on the firmness of canned sweet potatoes. Three concentrations were used: 0%, 1.5%, and 3% pectin by weight. Six number 303 × 406 cans were packed with sweet potatoes in a 25% (by weight) sugar solution. Two cans were randomly assigned to each of the pectin concentrations with the appropriate percentage of pectin added to the sugar syrup. The cans were then sealed and placed in a 25°C environment for 30 days. At the end of the storage time, the cans were opened and a firmness determination made for the contents of each can. These data appear below.

Pectin concentration	0%, 0%	1.5%, 1.5%	3.0%, 3.0%
Firmness reading	50.5, 46.8	62.3, 67.7	80.1, 79.2

a. Let x denote the pectin concentration of a can and y denote the firmness reading following the 30 days of storage at 25°C. Plot the sample data in a scatter diagram.
b. Obtain least squares estimates for the parameters in the model $y = \beta_0 + \beta_1 x + \varepsilon$.

9.13 Refer to Exercise 9.12. Predict the firmness for a can of sweet potatoes treated with a 1% concentration of pectin (by weight) after 30 days of storage at 25°C.

9.14 A study was conducted to examine the quality of fish after seven days in ice storage. Ten raw fish of the same kind and approximately the same size were caught and prepared for ice storage. Two of the fish were placed in storage immediately after being caught, two were placed in storage 3 hours after being caught, and two each were placed in storage at 6, 9, and 12 hours after being caught. Let y denote a measurement of fish quality (on a 10-point scale) after the seven days of storage, and x denote the time after being caught that the fish were placed in ice packing. The sample data appear below.

y	8.5	8.4	7.9	8.1	7.8	7.6	7.3	7.0	6.8	6.7
x	0	0	3	3	6	6	9	9	12	12

a. Plot the sample data in a scatter diagram.
b. Use the method of least squares to obtain estimates of the parameters in the model $y = \beta_0 + \beta_1 x + \varepsilon$.

9.15 Refer to Exercise 9.14. Predict the 7-day quality score of a fish placed in ice storage 10 hours after being caught. Would you be willing to predict a quality score for fish placed in storage 18 hours after being caught?

9.3 ▼ QUICK, PORTABLE STATISTICS (optional)

Sometimes it is necessary to get a quick, approximate fit to the model $y = \beta_0 + \beta_1 x + \varepsilon$. The procedure can be used provided we obtain three or more (x, y) observations. The rules for constructing the approximate linear regression are given here.

Procedure for Obtaining an Approximate Linear Regression Equation

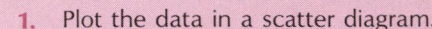

1. Plot the data in a scatter diagram.
2. Using two lines parallel to the vertical axis, divide the data into three groups of *roughly* the same number of observations.
3. For the lower-end group, find the median point; that is, the point corresponding to the medians for x and y based on measurements in that group.
4. Repeat (3) for the upper-end group.
5. Connect the two median points using a straight line. This line represents an approximate linear regression equation $\hat{y} = \hat{\beta}_0 + \hat{\beta}_1 x$.

EXAMPLE 9.3

Data were obtained on the reduction in cholesterol count (in mg per 100 ml of blood serum) for a random sample of 15 male volunteers participating in a study involving a low-cholesterol diet. Each volunteer participated in the study for 4 weeks. A pre-study cholesterol reading was obtained prior to beginning the diet, and the reduction in the count was observed for the 4-week period. In addition to cholesterol levels, ages of the volunteers were also recorded. These data are given in Table 9.5. Construct an approximate linear regression line.

TABLE 9.5

Age and Cholesterol-Reduction Data, Example 9.3

Age	Reduction in Cholesterol	Age	Reduction in Cholesterol
45	30	31	40
43	52	26	17
46	45	22	28
49	38	58	44
50	62	60	61
37	55	52	58
34	25	27	45
30	30		

Solution   We begin with a scatter diagram of the data (see Figure 9.5). The data are then divided into three approximately equal groups with two vertical lines. These have been inserted onto the diagram in Figure 9.6. The median points for the lower-end group can be found by drawing a vertical line that evenly divides the x-values and a horizontal line that evenly divides the y-values. The intersection of these two lines is the median point for the

FIGURE 9.5

Scatter Diagram of the Age and Cholesterol Count Data, Example 9.3

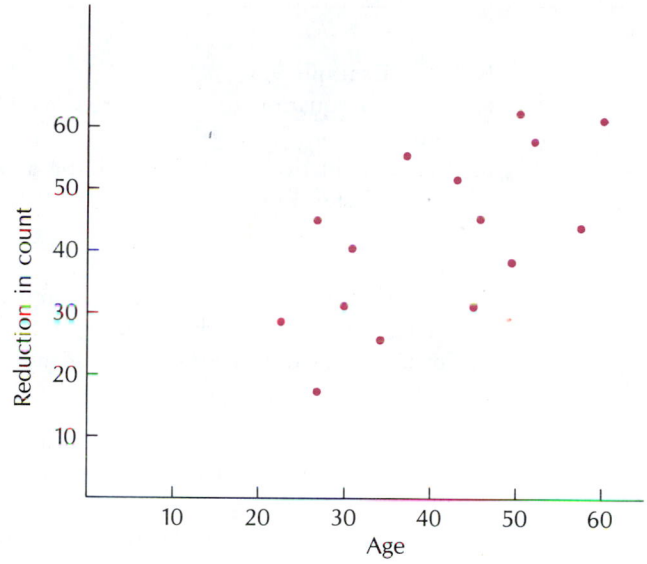

lower-end group (see Figure 9.6). Similarly, for the upper-end group, we find the median point indicated in Figure 9.6. The straight line joining the median points for the two end groups is the line that approximates the least squares equation for these data.

FIGURE 9.6

Approximate Linear Regression Line

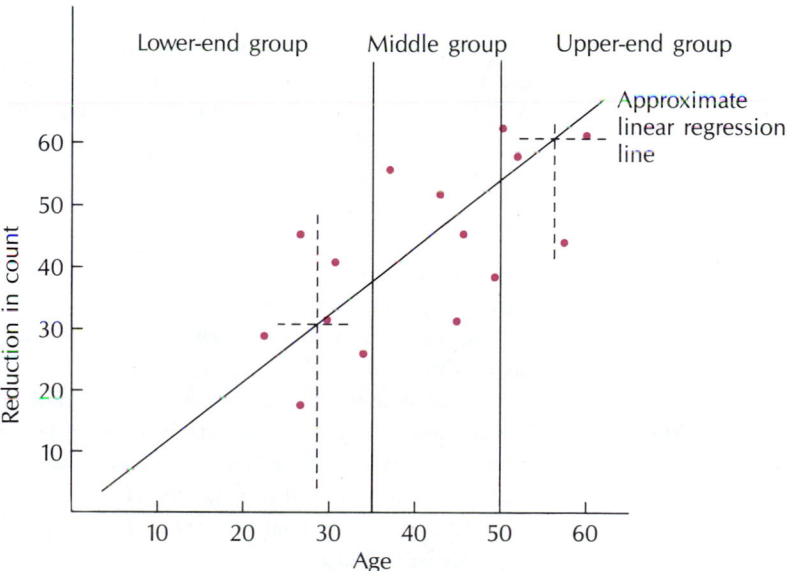

EXAMPLE 9.4 ▼

Refer to Example 9.3.

a. Write an equation for the approximate linear regression line shown in Figure 9.6.

b. Predict reduction in cholesterol for a 55-year-old male who uses the low-cholesterol diet.

Solution

a. From Figure 9.6 the y-intercept appears to be $\hat{\beta}_0 \approx -1$; the slope (change in y for unit change in x) is $\hat{\beta}_1 \approx 1$. Hence, $\hat{y} = -1 + x$.

b. For a 55-year-old male, $x = 55$ and $\hat{y} = -1 + 55 = 54$.   ▲

Although we've presented a good deal of material on linear regression so far in this chapter, we have not yet developed or discussed any formal estimation or test procedures; all inferences have been of an informal, seat-of-the-pants style. Inferences for linear regression are developed in detail in Chapter 10. But before we leave our introduction to regression, we will consider several additional topics: ways to linearize data, correlations, and a few extensions beyond the linear regression setting.

9.4 ▼ TRANSFORMATIONS TO LINEARIZE DATA

There are many reasons why we might want to re-express (transform) either the independent or dependent variable. One such reason is that we wish to simplify the underlying model; the analysis of the transformed data could be greatly simplified as well. For example, basic books on finance show that if a quantity y grows at a rate r per unit time, then the value y_t at time t is $y_t = y_0 e^{rt}$, a nonlinear model, where y_0 is the value at time 0. One way to deal with the model is to take natural logarithms of both sides to obtain $\log y_t = \log y_0 + rt$; this is a linear model. We will find that there will be many other situations in which we can use a transformation on either the dependent variable, the independent variable, or both variables to eliminate curvature in the data and hence simplify our analysis of the data.

How will we know which transformation will help to linearize the data? In situations that are characterized by the finance model, the appropriate transformation is obvious. In other situations in which we have no known model and only a scatter plot of the data, the choice of transformation to linearize the data is more a matter of trial and error. However, there are a few general guidelines that can be of help.

First, consider transformations of the dependent variable that will help linearize the sample data. For example, suppose a marketing research firm has observed the following trends in sales (y) as a function of mass media advertising expenses (x) for 10 different companies selling a similar product. These data are displayed in Table 9.6. A scatterplot of y versus x is shown in Figure 9.7.

TABLE 9.6
Sales and Expenditures (millions of dollars)

Company	Sales (y)	Expenditure (x)
1	2.5	1.0
2	2.6	1.6
3	2.7	2.5
4	5.0	3.0
5	5.3	4.0
6	9.1	4.6
7	14.8	5.0
8	17.5	5.7
9	23.0	6.0
10	28.0	7.0

FIGURE 9.7
Scatterplot of y versus x for Table 9.6

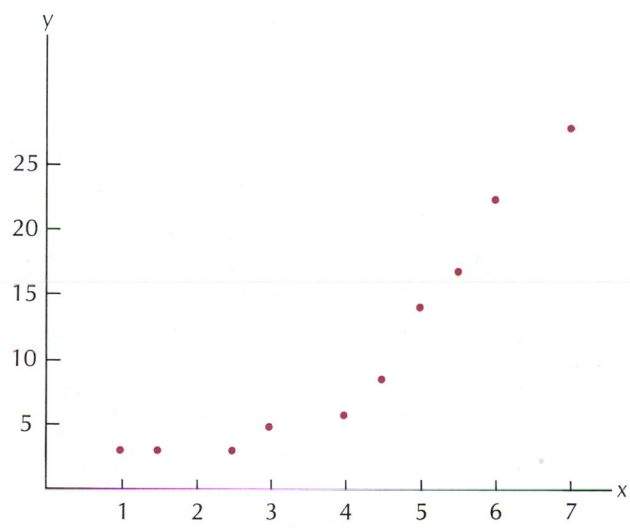

What re-expression of the dependent variable y will help to linearize the data? In order to answer this question, we make use of Figure 9.8, which displays the relationship between x and various transformations on the dependent variable y. This figure tells us that when a plot of y versus x is linear, successive plots of x versus y^2 and y^3 will curve upward whereas plots

FIGURE 9.8
Relationship Between x
and Various
Transformations on the
Dependent Variable y

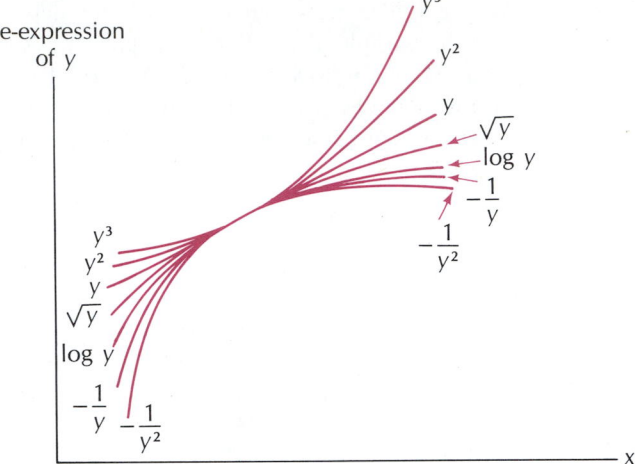

of x versus $\sqrt{y}$, log y, $-1/y$, and $-1/y^2$ will curve downward, with each successive plot having more severe curvature. The order of progression from curvature upward to curvature downward is

transformations on y

$$y^3$$
$$y^2$$
$$y$$
$$\sqrt{y}$$
$$\log y$$
$$-1/y$$
$$-1/y^2.$$

So if, in a scatterplot of the independent variable x versus the dependent variable y, the plot curves upward, we must proceed down on the scale to choose a transformation of the dependent variable that linearizes the data. For our data, we might consider either $\sqrt{y}$ or log y initially.

EXAMPLE 9.5

Re-express the y data from Table 9.6 using a log y transformation to linearize the data.

Solution   The x, y data from Table 9.6 have been reconstructed here with an additional column for $\log_{10} y$.

$\log_{10} y$	y	x
0.40	2.5	1
0.41	2.6	1.6
0.43	2.7	2.5
0.70	5	3
0.72	5.3	4
0.96	9.1	4.6
1.17	14.8	5
1.24	17.5	5.7
1.36	23	6
1.45	28	7

The scatterplot of log y versus x in Figure 9.9 shows that this transformation on the dependent variable does a reasonable job of linearizing the data.

FIGURE 9.9
Scatterplot of Log y versus x

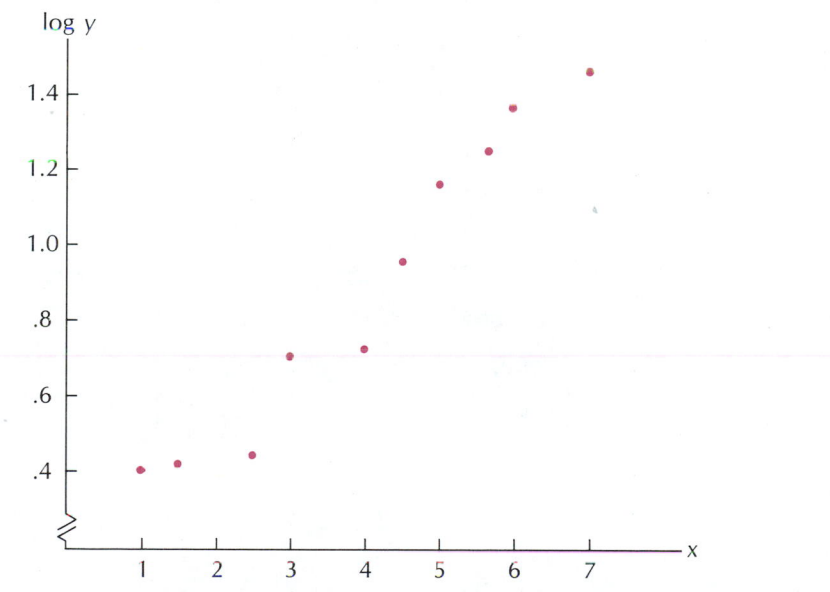

Sometimes it is convenient to consider a re-expression of the independent variable for linearizing the data. A similar progression of transformations exists for the independent variable as for the dependent variable with the direction of the re-expression depending on the shape of the plot. The order of transformation is the same as for transformations on the dependent variable, namely,

transformations on x

$$x^3$$
$$x^2$$
$$x$$
$$\sqrt{x}$$
$$\log x$$
$$-1/x$$
$$-1/x^2$$
etc.

There are differences, though, in which transformations help to linearize the sample data when a scatterplot has curvature. For example, when the plot of y versus x turns upward, choices of transformations on x to linearize the data are selected by *moving up rather than down* the list of transformations. This is just the opposite of what we would do using a transformation on y.

For the data displayed in Figure 9.7, we could consider an x^2 or x^3 transformation to linearize the data. Table 9.7 and Figure 9.10 display a re-expression of x using an x^2 transformation.

T A B L E 9.7

Re-Expression of x Using an x^2 Transformation

y	x	x^2
2.5	1	1
2.6	1.6	2.56
2.7	2.5	6.25
5	3	9
5.3	4	16
9.1	4.6	21.16
14.8	5	25
17.5	5.7	32.49
23	6	36
28	7	49

F I G U R E 9.10

Scatterplot for Data in Table 9.7

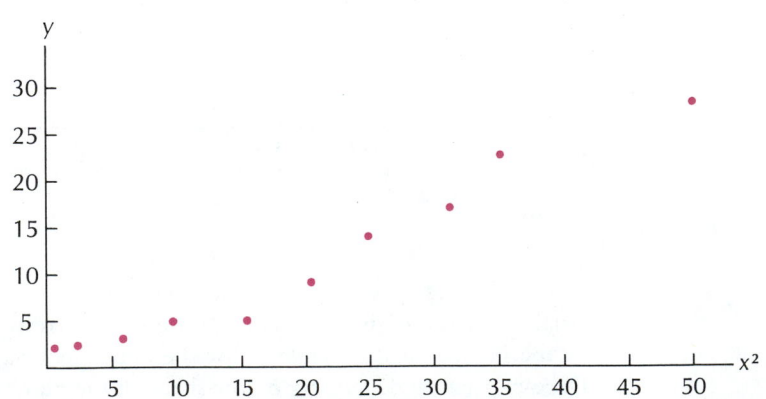

Since this transformation does a better job of linearizing the data than does the log y transformation of Example 9.5, we can use the linear regression model

$$y = \beta_0 + \beta_1 x^2 + \varepsilon$$

to analyze the sample data in Table 9.6. To analyze data for a model such as this, we merely define a new variable, say $z = x^2$, and proceed with a linear regression model using y and z.

To avoid some of the confusion associated with choosing an appropriate re-expression, we have summarized four general shapes of a scatterplot and the choices of transformations on either x and y for linearizing the data in Figure 9.11.

FIGURE 9.11

Transformations for *x* and *y* to Linearize Data

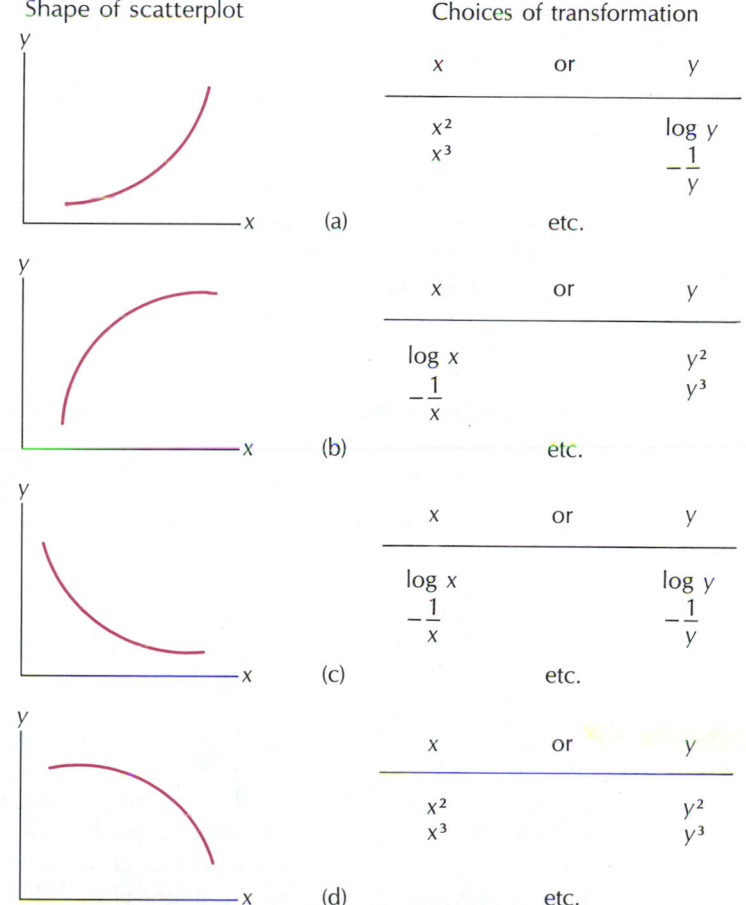

Shape of scatterplot

Choices of transformation

x	or	y
x^2		$\log y$
x^3		$-\dfrac{1}{y}$
	etc.	

(a)

x	or	y
$\log x$		y^2
$-\dfrac{1}{x}$		y^3
	etc.	

(b)

x	or	y
$\log x$		$\log y$
$-\dfrac{1}{x}$		$-\dfrac{1}{y}$
	etc.	

(c)

x	or	y
x^2		y^2
x^3		y^3
	etc.	

(d)

We have indicated that re-expressing the dependent or independent variable can transform a nonlinear model into the framework of a linear regression model for some variables. In addition, transformations on dependent variables can eliminate certain types of nonconstant variances. More discussion about this topic will be presented in Chapter 12.

▼ EXERCISES

Applications

$ 9.16 The sales, in thousands of units, of a small electronics firm for the last 10 years have been

Year	1	2	3	4	5	6	7	8	9	10
Sales	2.60	2.85	3.02	3.45	3.69	4.26	4.73	5.16	5.91	6.50

a. Plot sales against year.
b. Plot log sales against year. (Hint: It doesn't matter whether you use log base 10 or log base e.)
c. Which plot appears more nearly linear?

9.17 Refer to the data of Exercise 9.16.
a. Calculate the regression equation with sales as the dependent variable.
b. Calculate the regression equation with log sales as the dependent variable.
c. Which model appears to provide a better linear fit?

9.18 Use each of the regression equations of Exercise 9.17 to forecast sales in year 11. (Hint: You will have to convert log sales back to sales.) Which prediction appears more plausible?

9.19 Data relating advertising (x) to sales for a given product are shown here.

x	15	10	18	16	15	12	10	17	15	13	15	20	18	12
y	365	320	357	375	381	335	312	345	362	349	371	331	340	351

a. Plot the y versus x.
b. Is there evidence of nonlinearity in the plot?

9.5 ▼ CORRELATION

In this section, we will extend our study of the relationships between two variables. Not only might we like to predict the value of one variable (the dependent variable) based on information on an independent variable, as we have done in previous sections, but we might also wish to provide a measure of the strength of the relationship between these variables. This idea will be the topic of this section.

One measure of the strength of the relationship between two variables x and y is called the coefficient of linear correlation, or, simply, the **correlation coefficient**. Given n pairs of observations (x_i, y_i), we can compute the **sample correlation coefficient r** as

correlation coefficient

sample correlation coefficient r

$$r = \frac{S_{xy}}{\sqrt{S_{xx}S_{yy}}},$$

where

S_{yy}

$$S_{yy} = \sum y^2 - \frac{(\sum y)^2}{n}.$$

You will immediately note the similarity between r and the slope of the least squares equation

$$\hat{y} = \hat{\beta}_0 + \hat{\beta}_1 x$$

relating y to x:

$$\hat{\beta}_1 = \frac{S_{xy}}{S_{xx}}$$

$$r = \hat{\beta}_1 \sqrt{\frac{S_{xx}}{S_{yy}}}.$$

For experimental situations in which not all x-values and y-values are the same, both S_{xx} and S_{yy} are positive. Then r and $\hat{\beta}_1$ have the same sign. Because of the relationship between r and $\hat{\beta}_1$, the sample correlation coefficient r measures the strength of the linear relationship between x and y and is used to estimate the corresponding population coefficient of linear correlation ρ (Greek letter rho).

Properties of r and r^2

1. r lies between -1 and $+1$. $r > 0$ indicates a positive linear relationship and $r < 0$ a negative linear relationship between x and y. $r = 0$ indicates no linear relationship between x and y. (See Figure 9.12.)
2. r^2 gives the proportion of the total variability in the y-values that can be accounted for by the independent variable x.

FIGURE 9.12
Interpretation of *r*

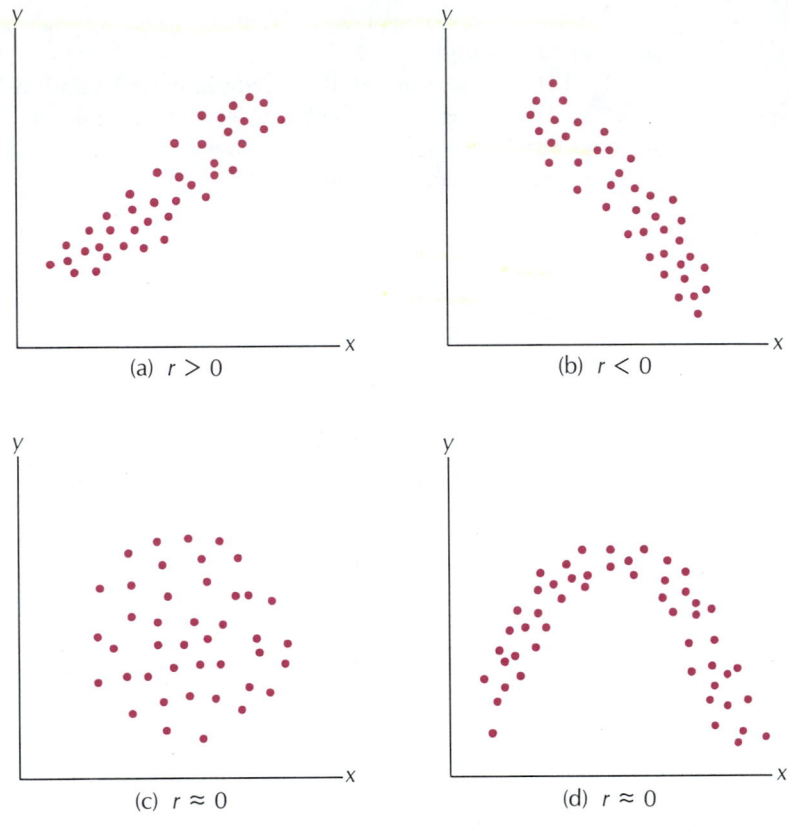

(a) $r > 0$

(b) $r < 0$

(c) $r \approx 0$

(d) $r \approx 0$

EXAMPLE 9.6

▼

An engineer is interested in calibrating a flow meter to be used on a liquid-soap production line. For the test, 10 different flow rates are fixed and the corresponding meter readings observed. These data are shown in Table 9.8.

TABLE 9.8
Data for Example 9.6

Observed Meter Flow Rate, *y*	Actual Flow Rate, *x*
1.4	1
2.3	2
3.1	3
4.2	4
5.1	5
5.8	6
6.8	7
7.6	8
8.7	9
9.5	10

a. Plot the sample data. Do the data look linear?

b. Compute the correlation coefficient between the observed and actual flow rates.

Solution

a. A plot of the data is shown here.

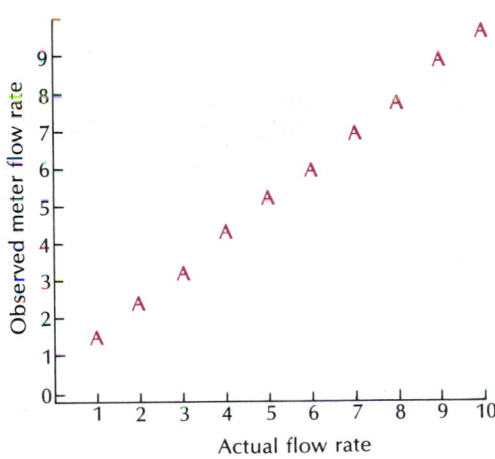

b. For these data, it can be shown that

$$S_{yy} = \sum y^2 - \frac{\left(\sum y\right)^2}{n} = 364.09 - \frac{(54.5)^2}{10} = 67.065$$

$$S_{xx} = \sum x^2 - \frac{\left(\sum x\right)^2}{n} = 385 - \frac{(55)^2}{10} = 82.5$$

and

$$S_{xy} = \sum xy - \frac{\left(\sum x\right)\left(\sum y\right)}{n} = 374.1 - \frac{55(54.5)}{10} = 74.35.$$

Combining, we have

$$r = \frac{S_{xy}}{\sqrt{S_{xx}S_{yy}}} = \frac{74.35}{\sqrt{82.5(67.065)}} = .9996.$$

> That is, there appears to be a very strong positive linear relationship between the flow rate x and the instrument reading y. Note, however, that this doesn't mean that the instrument is making accurate readings.   ▲

We can illustrate how r^2 measures the proportion of the total y variability accounted for by x for the linear model $y = \beta_0 + \beta_1 x + \varepsilon$. As noted in Section 9.2, the total variability of the y-values about their mean $\bar{y}$ can be expressed as

$$\sum (y - \bar{y})^2 = \sum (y - \hat{y})^2 + \sum (\hat{y} - \bar{y})^2,$$

where $\sum (\hat{y} - \bar{y})^2$ is that portion of the total variability that can be accounted for by the independent variable x and $\sum (y - \hat{y})^2$ is the sum of squares for error (SSE). Using a computational formula, we can rewrite the total sum of squares S_{yy} as

$$S_{yy} = \sum (y - \bar{y})^2 = \sum y^2 - \frac{(\sum y)^2}{n}.$$

Similarly, it can be shown that the sum of squares for error for a linear regression model can be written as

$$\sum (y - \hat{y})^2 = S_{yy} - \frac{S_{xy}^2}{S_{xx}}$$

and by subtraction, the sum of squares due to regression is

$$\sum (\hat{y} - \bar{y})^2 = \frac{S_{xy}^2}{S_{xx}}.$$

Then expressing both $\sum (y - \hat{y})^2$ and $\sum (\hat{y} - \bar{y})^2$ as a proportion of $S_{yy} = \sum (y - \bar{y})^2$, we have

$$\frac{\sum (y - \hat{y})^2}{S_{yy}} = 1 - \frac{S_{xy}^2}{S_{xx} S_{yy}} = 1 - r^2$$

and

$$\frac{\sum (\hat{y} - \bar{y})^2}{S_{yy}} = \frac{S_{xy}^2}{S_{xx} S_{yy}} = r^2.$$

Thus, r^2 represents that proportion of the total variability of the y-values that is accounted for by the independent variable x. Similarly, $1 - r^2$ represents that proportion of the total variability of the y-values that is not accounted for by the variable x.

EXAMPLE 9.7

▼

Refer to Example 9.6. What percent of the variability in flow meter readings (y) is accounted for by the actual flow rate (x)?

Solution For these data $r^2 = (.9996)^2 = .9992$. Thus, 99.92% of the variability in y is accounted for by x. ▲

The ordinary correlation coefficient r assesses the linear association between two variables x and y. In certain situations, the variable y may increase (or decrease) with increases in x but not necessarily in a linear fashion. When this happens, the correlation coefficient r will not depict the full extent of the relation between x and y. One way to handle this is to consider transforming the data; another way is to consider alternate models. These models will be discussed in multiple regression (Chapters 11 and 12). Another approach is to use the rank correlation coefficient, which measures the *monotonic* association between y and x. That is, the rank correlation coefficient measures whether y increases (or decreases) with x, even when the relation between y and x is not necessarily linear.

The rank order correlation coefficient is easy to calculate. We simply rank all x-values and all y-values separately and then calculate the ordinary correlation coefficient for the ranks. This correlation based on the ranks is called **Spearman's rank order correlation coefficient r_s**.

Spearman's rank order correlation coefficient r_s

EXAMPLE 9.8

▼

A corporation examined the relationship between profits ($1,000) and the percent of operating capacity being used for each of 12 plants.

Profits ($1,000), y	% of Operating Capacity, x
2.5	50
6.2	57
3.1	61
4.6	68
7.3	77
4.5	80
6.1	82
11.6	85
10.0	89
14.2	91
16.1	95
19.5	99

a. Plot the data. Are the data linear?
b. Compute the rank correlation coefficient.

Solution

a.

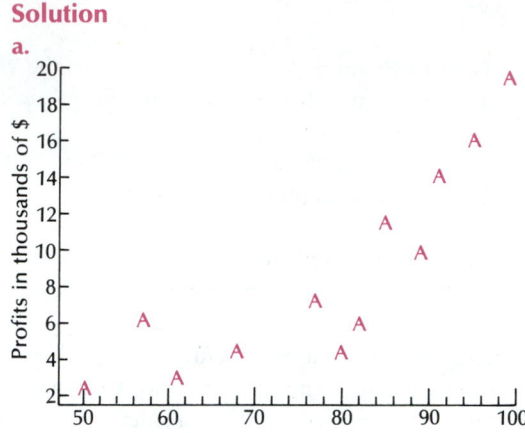

PLOT OF DATA

PLOT OF Y*X   LEGEND: A = 1 OBS, B = 2 OBS, ETC.

b. To compute r_s, first we need the ranks (from low to high) separately for each variable. These are shown here.

Rank on Profits	Rank on % Capacity
1	1
6	2
2	3
4	4
7	5
3	6
5	7
9	8
8	9
10	10
11	11
12	12

Then, if we let y denote the ranks on profits and x denote rank on % of operating capacity, we compute r_s as we would r. To do this, we need S_{yy}, S_{xx}, and S_{xy}.

For these data, it can be shown that

$$S_{yy} = 143$$
$$S_{xx} = 143$$

and

$$S_{xy} = 125.$$

Hence, the rank correlation coefficient is

$$r_s = \frac{125}{\sqrt{143(143)}} = 0.874.$$

For these data y increases with x, but as seen in the previous figure, the relation between y and x is not linear. ▲

If there are no ties in ranks for either of the two variables, there is a simpler formula for r_s that makes use of d_i, the difference between the y rank and x rank on observation i:

$$r_s = 1 - \frac{6 \sum d_i^2}{n(n^2 - 1)},$$

where n is the number of x_i, y_i observations.

E X A M P L E   9.9

▼

Compute r_s for the data of Example 9.8 using the simpler computational formula.

Solution The differences in ranks are shown here:

Rank on Profit	Rank on % Capacity	d_i
1	1	0
6	2	4
2	3	−1
4	4	0
7	5	2
3	6	−3
5	7	−2
9	8	1
8	9	−1
10	10	0
11	11	0
12	12	0

Then $\sum d_i^2 = 36$ and

$$r_s = 1 - \frac{6 \sum d_i^2}{n(n^2 - 1)}$$

$$= 1 - \frac{6(36)}{12(143)} = .874.$$

Note: This agrees with what we obtained in Example 9.8 except for rounding errors.   ▲

▼ EXERCISES

Basic Techniques

9.20   Plot the sample data shown here, compute the correlation coefficient, and interpret your findings.

x	1	2	3	4	6	9	10
y	2	4	5	7	8	12	13

9.21   Refer to Exercise 9.20. Suppose the first three y-values are 16, 12, and 10.
  a. Plot the data and guess a value for *r*.
  b. Compute the sample correlation coefficient, and compare the computed value to the guessed value.
  c. Why do the correlation coefficients differ for Exercises 9.20 and 9.21?

Applications

$ 9.22   The sales, in thousands of units, for the small electronics firm mentioned in Exercise 9.16 are repeated here.

Year	1	2	3	4	5	6	7	8	9	10
Sales	2.60	2.85	3.02	3.45	3.69	4.26	4.73	5.16	5.91	6.50

Compute the correlation coefficient between sales (*y*) and year (*x*).

9.23 An instructor believes that true–false tests are as effective as problem-type tests in judging a student's proficiency in mathematics. A test consisting of half true–false questions and half problems was given to ten calculus students selected at random. The test score results were as follows:

Student	1	2	3	4	5	6	7	8	9	10
T–F questions	48	40	25	10	16	21	23	19	35	32
Problems	45	47	20	12	12	15	25	16	30	32

a. Plot the data. Are the data linear?
b. Calculate the correlation coefficient to measure the strength of the linear relation for the two sets of test scores.

9.24 Compute the rank correlation coefficient for the data of Exercise 9.16. Which measure of association seems more appropriate, r or r_s? Refer to Exercise 9.22 for the value of r.

9.25 An experiment was conducted to investigate the amplitude of the shock wave recorded on sensors placed at different distances from an explosive charge. The charge is to be detonated underground, with three sensors placed at each of the three different distances from the charge, as illustrated in Figure 9.13. The shock-wave amplitudes are recorded and summarized according to the distance from the explosion. These data are given below.

Distance, x	5	5	5	10	10	10	15	15	15
Amplitude, y	8.6	8.2	8.1	5.8	6.2	6.1	5.2	4.8	4.7

FIGURE 9.13
Location of Sensors from the Charge for the Shock-Wave Experiment of Exercise 9.25

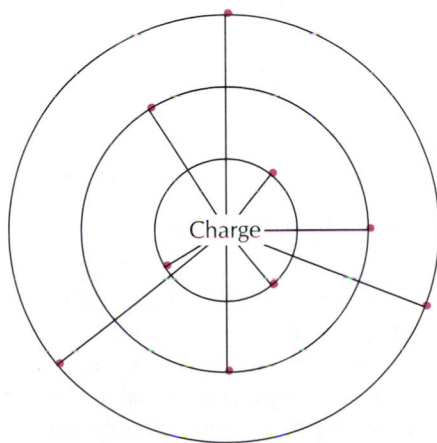

a. Plot the sample data. Are the data linear?

b. Choose between r and r_s as a measure of the strength of the relation between distance and amplitude. Compute its value.

9.26 A forester was interested in training an assistant to estimate the timber volume of a standing tree. Having trained her, the forester calibrated the assistant's estimates against known timber volumes. Perhaps a better way to quantify the assistant's estimate would be to base it on an objective reading, such as the basal area of the tree. If, indeed, volume is related to basal area, the assistant would have an objective way to estimate the timber volume of a tree. A random sample of 12 trees was obtained. For each tree included in the sample, the basal area x was recorded along with the cubic-foot volume after the tree was felled. These data appear here.

Tree	1	2	3	4	5	6	7	8	9	10	11	12
Basal area, x	.3	.5	.4	.9	.7	.2	.6	.5	.8	.4	.8	.6
Volume, y	6	9	7	19	15	5	12	9	20	9	18	13

a. Plot the data.

b. Obtain the linear regression equation $\hat{y} = \hat{\beta}_0 + \hat{\beta}_1 x$.

c. Compute and interpret the correlation coefficient between basal area and timber volume.

9.27 An equal number of families from eight different cities of various sizes were asked how much money they spend for food, clothing, and housing per year. The city sizes and average family responses are summarized below. (City size in 1,000s, expenditure in \$1,000s.)

City size	30	50	75	100	150	200	175	120
Expenditure	65	77	79	80	82	90	84	81

a. Plot the data.

b. Compute the correlation coefficient r.

9.28 Compute r_s for the data of Exercise 9.27. Is the shortcut formula appropriate? Why or why not?

9.6 ▼ A LOOK AHEAD: MULTIPLE REGRESSION

Multiple regression is discussed in detail in Chapters 11 and 12. In this section, we give you a brief idea of what lies ahead. Linear regression deals with situations in which a dependent variable can be expressed in a model as a linear function of a single quantitative independent variable. Multiple regression models, on the other hand, express the dependent variable using more than one independent variable or higher-degree terms than a single independent variable.* The adjective "multiple" refers to the fact that there is more than one term in the model (excluding the intercept).

Multiple regression models relating a response y to a single quantitative variable could be of the form.

multiple regression: single x

$$y = \beta_0 + \beta_1 x_1 + \beta_2 x_1^2 + \varepsilon$$
$$y = \beta_0 + \beta_1 x_1 + \beta_2 x_1^2 + \beta_3 x_1^3 + \varepsilon.$$

Plots of data for some of the previous examples and exercises of this chapter have indicated that a linear regression model may not be appropriate. For some of these situations, a model with higher-degree terms could represent a viable alternative.

For other situations, it is quite possible that more than one independent variable is related to the response and should be included in the model. For example, in studying the yield of a tomato crop, several independent variables—amount of fertilizer (x_1), amount of water (x_2), hours of sunlight on clear days (x_3)—could all have an effect on the yield. Hence, to formulate a model that adequately represents the yield of a tomato crop, we should include all these terms in a model. Typical multiple regression models relating a response y to two quantitative independent variables, x_1 and x_2, are shown here:

multiple regression: two xs

$$y = \beta_0 + \beta_1 x_1 + \beta_2 x_2 + \varepsilon$$
$$y = \beta_0 + \beta_1 x_1 + \beta_2 x_2 + \beta_3 x_1 x_2 + \varepsilon$$
$$y = \beta_0 + \beta_1 x_1 + \beta_2 x_1^2 + \beta_3 x_2 + \beta_4 x_2^2 + \beta_5 x_1 x_2 + \beta_6 x_1^2 x_2$$
$$+ \beta_7 x_1 x_2^2 + \beta_8 x_1^2 x_2^2 + \varepsilon.$$

Several of these models are illustrated in Figure 9.14.

Similarly, typical multiple regression models relating y to a set of three independent variables are as follows:

multiple regression: three xs

$$y = \beta_0 + \beta_1 x_1 + \beta_2 x_2 + \beta_3 x_3 + \varepsilon$$
$$y = \beta_0 + \beta_1 x_1 + \beta_2 x_2 + \beta_3 x_3 + \beta_4 x_1 x_2 + \beta_5 x_1 x_3 + \beta_6 x_2 x_3 + \varepsilon$$
$$y = \beta_0 + \beta_1 x_1 + \beta_2 x_1^2 + \beta_3 x_2 + \beta_4 x_2^2 + \beta_5 x_3 + \beta_6 x_3^2 + \beta_7 x_1 x_2 + \beta_8 x_1 x_3$$
$$+ \beta_9 x_2 x_3 + \varepsilon.$$

Again, the particular choice of model will depend on the actual experimental situation under study and the number of variables that affect the response. It suffices to say here that we will try to choose a model that best describes the relationship between the independent variables of interest and the response y.

---

* Some textbooks would not classify these models as multiple regression models.

FIGURE 9.14

Graphical Illustrations of Several Models Relating a Response y to Two Independent Variables x_1 and x_2

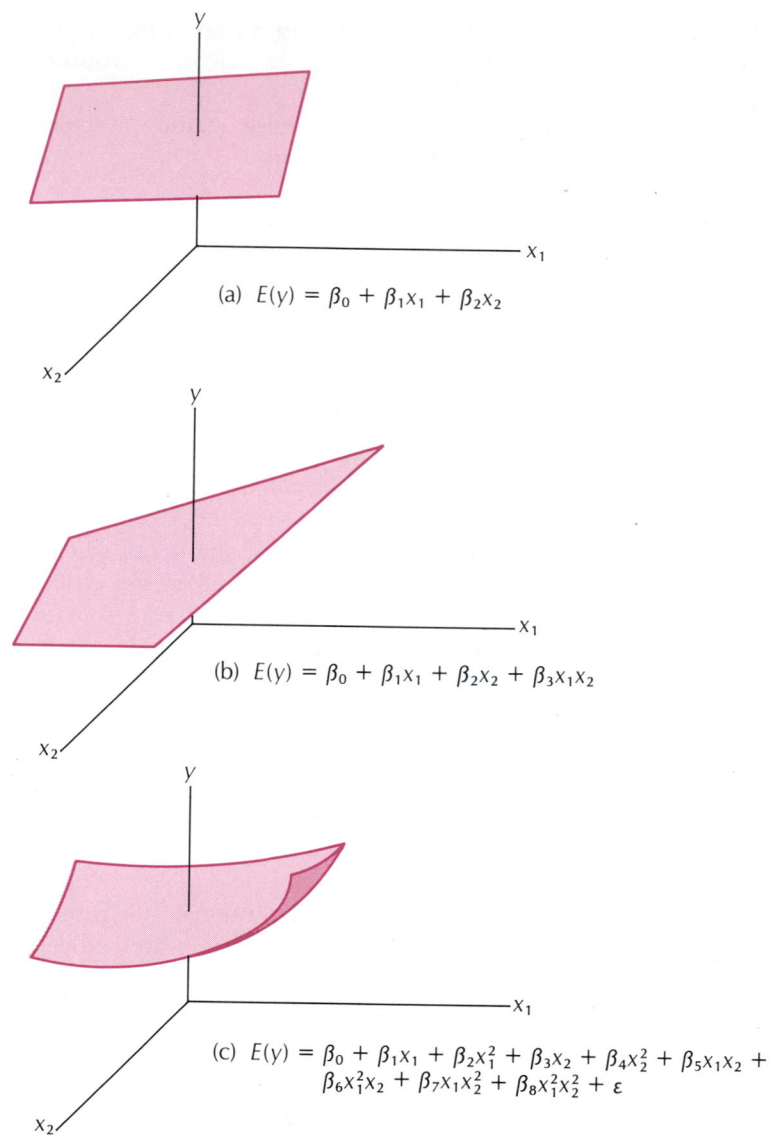

(a) $E(y) = \beta_0 + \beta_1 x_1 + \beta_2 x_2$

(b) $E(y) = \beta_0 + \beta_1 x_1 + \beta_2 x_2 + \beta_3 x_1 x_2$

(c) $E(y) = \beta_0 + \beta_1 x_1 + \beta_2 x_1^2 + \beta_3 x_2 + \beta_4 x_2^2 + \beta_5 x_1 x_2 + \beta_6 x_1^2 x_2 + \beta_7 x_1 x_2^2 + \beta_8 x_1^2 x_2^2 + \varepsilon$

degree

pth-degree term

Individual terms in a multiple regression model are classified by their exponents. The **degree** of a term is given by the sum of the exponents for the independent variables appearing in the term. Thus, any independent variable x_i that appears in the regression model as x_i^p is called a **pth-degree term**. Hence, x_i, x_i^2, x_i^3 are called first-, second-, and third-degree terms, respectively. Similarly, if two independent variables x_i and x_j appear together as $x_i^p x_j^q$, the term is called a $(p + q)$th-degree term. We would classify the terms $x_i x_j$, $x_i^2 x_j$, $x_i x_j^2$, $x_i^2 x_j^2$ as second-, third-, third-, and fourth-degree terms, respectively.

E X A M P L E   9.10

▼

An experimenter believes that the multiple regression model

$$y = \beta_0 + \beta_1 x_1 + \beta_2 x_1^2 + \beta_3 x_2 + \beta_4 x_1 x_2 + \varepsilon$$

adequately represents the relationship between a dependent variable y and two independent variables x_1 and x_2. Identify the degree of all terms in the model containing one or more independent variables.

Solution   Following the procedures just discussed for assigning degrees to terms, we have the following:

Term	Degree
$\beta_1 x_1$	first
$\beta_2 x_1^2$	second
$\beta_3 x_2$	first
$\beta_4 x_1 x_2$	second

▲

We just indicated that individual terms of a multiple regression model are classified by their exponents. We can also identify specific models by the types of terms that appear in the model. A **first-order model** is a multiple regression model that contains all possible first-degree terms in the independent variables.*

first-order model

E X A M P L E   9.11

▼

Write a first-order multiple regression model relating a response y to the independent variables x_1, x_2, and x_3.

Solution   A first-order model in x_1, x_2, and x_3 includes all first-degree terms in these variables. The appropriate model is

$$y = \beta_0 + \beta_1 x_1 + \beta_2 x_2 + \beta_3 x_3 + \varepsilon.$$

▲

second-order model

A **second-order model** is a multiple regression model that includes all possible first- and second-degree terms in the independent variables.

---

* Some textbooks refer to these models as multiple linear regression models because they have only first-degree (linear) terms in the independent variables.

EXAMPLE 9.12 ▼

Write a second-order model relating a dependent variable y to the independent variables x_1, x_2, and x_3.

Solution The appropriate second-order model would include the terms of the corresponding first-order model (Example 9.11) and all possible second-degree terms:

$$y = \beta_0 + \beta_1 x_1 + \beta_2 x_2 + \beta_3 x_3 + \beta_4 x_1^2 + \beta_5 x_2^2 + \beta_6 x_3^2 + \beta_7 x_1 x_2 \\ + \beta_8 x_1 x_3 + \beta_9 x_2 x_3 + \varepsilon.$$

▲

Higher-order models represent obvious extensions of the first- and second-order models. Several exercises at the end of this chapter offer you additional practice in formulating these models.

9.7 ▼ SUMMARY

This chapter begins to lay the foundation for linear and multiple regression, where a dependent variable y can be represented as a function of one or more independent variables. For the linear regression model $y = \beta_0 + \beta_1 x + \varepsilon$, we discussed a procedure for obtaining estimates of the parameters β_0 and β_1. The method of least squares chooses values for $\hat{\beta}_0$ and $\hat{\beta}_1$ in the linear regression line

$$\hat{y} = \hat{\beta}_0 + \hat{\beta}_1 x$$

so that the quantity $\sum (y - \hat{y})^2$ is minimized. For a given data set, these values are computed as

$$\hat{\beta}_1 = \frac{S_{xy}}{S_{xx}} \quad \text{and} \quad \hat{\beta}_0 = \bar{y} - \hat{\beta}_1 \bar{x},$$

where

$$S_{xx} = \sum x^2 - \frac{(\sum x)^2}{n}$$

and

$$S_{xy} = \sum xy - \frac{(\sum x)(\sum y)}{n}.$$

We also introduced a quick, portable way to approximate a linear regression equation and transformations for the independent and dependent variables, which may help to linearize data.

The correlation coefficient, which is closely related to the estimated slope $\hat{\beta}_1$, was presented as a measure of the strength of the linear relationship. When two variables are related in a positive or negative fashion, but not necessarily linearly related, the rank correlation coefficient r_s can be used as a measure of the strength of the monotonic relation between y and x. For r_s, ranks rather than original values for x and y are used in the calculations.

Finally, we presented a brief look at what's ahead beyond linear regression. After learning how to make statistical inferences (estimation and statistical tests) for linear regression in Chapter 10, we will be in a position to expand the model to include more than one independent variable. Model relations and methods of inferences for multiple regression are presented in Chapters 11 and 12.

▼ KEY FORMULAS

1. Formulas for least squares estimates, $\hat{\beta}_0$ and $\hat{\beta}_1$

$$\hat{\beta}_1 = \frac{S_{xy}}{S_{xx}} \quad \text{and} \quad \hat{\beta}_0 = \bar{y} - \hat{\beta}_1 \bar{x},$$

where

$$S_{xx} = \sum x^2 - \frac{(\sum x)^2}{n}$$

and

$$S_{xy} = \sum xy - \frac{(\sum x)(\sum y)}{n}$$

2. Correlation coefficient

$$r = \frac{S_{xy}}{\sqrt{S_{xx} S_{yy}}} \quad \text{or} \quad r = \hat{\beta}_1 \sqrt{\frac{S_{xx}}{S_{yy}}}$$

3. Rank order correlation coefficient (Spearman's)

$$r_s = 1 - \frac{6 \sum d_i^2}{n(n^2 - 1)}$$

when there are no tied ranks for either the x or y variables

▼ SUPPLEMENTARY EXERCISES

9.29 Suppose that a response y is related to two independent variables x_1 and x_2.
a. Write a first-order regression model.
b. Write three different regression models relating y to x_1 and x_2 using first- and second-degree terms.

9.30 Identify the degree for all terms in the model

$$y = \beta_0 + \beta_1 x_1 + \beta_2 x_1^2 + \beta_3 x_2 + \beta_4 x_1 x_2 + \beta_5 x_1^2 x_2 + \varepsilon$$

9.31 Sketch the deterministic model

$$y = 1.5 + 1.0x^2$$

for x in the range $-3 \le x \le 3$. (Hint: Substitute different values for x into the model to determine corresponding values for y. Plot the x and y points.)

9.32 Sketch the deterministic model

$$y = 1.5 + 2.5x + 1.0x^2$$

for x in the range $-3 \le x \le 3$.

9.33 Sketch the deterministic model

$$y = 1.5 - 2.5x + 1.0x^2$$

for x in the range $-3 \le x \le 3$. Compare this sketch to that for Exercise 9.32.

9.34 Refer to the data of Exercise 9.25.
a. Determine an approximate linear regression line using the method of Section 9.3.
b. Determine the linear regression line using least squares estimates for β_0 and β_1. Compare your results to part a.

9.35 Refer to Exercise 9.26. Use the method of Section 9.3 to obtain an approximate linear regression line. How does it compare to the linear regression line shown for the same data in the Minitab output displayed here?

```
MTB > NAME C1 'AREA'
MTB > NAME C2 'VOLUME'
MTB > SET C1
MTB > END
MTB > SET C2
MTB > END
MTB > PRINT C1 C2

 ROW    AREA   VOLUME

   1     0.3      6
   2     0.5      9
   3     0.4      7
   4     0.9     19
   5     0.7     15
   6     0.2      5
   7     0.6     12
   8     0.5      9
   9     0.8     20
  10     0.4      9
  11     0.8     18
  12     0.6     13
```

(continues)

```
MTB > REGRESS 'VOLUME' ON 1 PREDICTOR 'AREA'

The regression equation is
VOLUME = - 1.23 + 23.4 AREA

Predictor      Coef      Stdev     t-ratio        p
Constant     -1.234      1.080      -1.14      0.280
AREA         23.404      1.815      12.90      0.000

s = 1.295      R-sq = 94.3%      R-sq(adj) = 93.8%

Analysis of Variance

SOURCE        DF         SS         MS         F         p
Regression     1      278.90     278.90     166.35     0.000
Error         10       16.77       1.68
Total         11      295.67

Unusual Observations
Obs.    AREA     VOLUME      Fit   Stdev.Fit   Residual   St.Resid
  9    0.800     20.000   17.489       0.576      2.511      2.17R

R denotes an obs. with a large st. resid.
MTB > STOP
```

9.36 Refer to Example 9.8. Compute the correlation coefficient r for these data. Distinguish between the interpretation of r and r_s for these data.

9.37 Consider the regression model

$$y = \beta_0 + \beta_1 x_1 + \beta_2 x_1^2 + \beta_3 x_1^3 + \beta_4 x_1^4 + \varepsilon.$$

  a. Specify the degree of each term in the model.
  b. What is the order of the model?

9.38 Write a second-order model relating a response y to the independent variables x_1, x_2, and x_3.

9.39 Consider the multiple regression model

$$y = \beta_0 + \beta_1 x_1 + \beta_2 x_1^2 + \beta_3 x_2 + \beta_4 x_1 x_2 + \varepsilon.$$

  a. Specify the degree of each term in the general linear model.
  b. Is this a first- or second-order model? Explain.

9.40 Write a second-order model relating a response y to four independent variables (x_1, x_2, x_3 and x_4).

9.41 Sketch plots for these deterministic models for $-1 \leq x \leq 1$:
  a. $y = 1.2 + x$
  b. $y = 1.2 + x + .4x^2$
  c. $y = 1.2 + x + .4x^2 + .6x^3$

9.42 Sketch a graph using the five data points given here.

x	1	2	3	4	5
y	5	10	14	21	26

9.43 Refer to the sketch in Exercise 9.42. Write a regression model relating y to x.

9.44 Sketch a graph of the response y as a function of the independent variable x using the five data points given here.

x	0	-1	2	3	-2
y	15	12	31	50	11

9.45 Using the sketch drawn in Exercise 9.44, indicate the form (without values for the βs) of a multiple regression model relating the response y to the independent variable x.

9.46 If a second-order model relating y to x has one peak (or, equivalently, one valley) when sketched, and a third-order model has one peak and one valley when sketched, how many peaks and valleys do you think a fourth-order model has when sketched? Sketch a typical fourth-order model relating a response y to an independent variable x.

9.47 Earnings from a particular stock are listed below for the past 7 years. Sketch these seven data points.

Year	1992	1991	1990	1989	1988	1987	1986
Earnings per share	2.30	1.80	1.50	1.20	1.05	1.10	1.20

9.48 Refer to the sketch in Exercise 9.47.
  a. Suggest an appropriate multiple regression model relating earnings per share to the independent variable "year."
  b. Compute the rank order correlation coefficient r_s.

9.49 Yields in bushels of tomatoes are shown here for 12 equal-sized plots.
  a. Plot the data.
  b. Based on the plot, specify a linear or multiple regression model that may be appropriate.
  c. Compute either r or r_s depending on the model selected in part b.

Yield of 12 Equal-Sized Plots of Tomato Plantings for Different Amounts of Fertilizer

Plot	Yield, y (in bushels)	Amount of Fertilizer, x (in pounds per plot)
1	24	12
2	18	5
3	31	15
4	33	17
5	26	20
6	30	14
7	20	6
8	25	23
9	25	11
10	27	13
11	21	8
12	29	18

9.50 An experiment was conducted to examine the weight gain for chickens treated with various doses of a growth promotant. From a group of 15 chickens of approximately the same age and weight, a random assignment of three chickens was made to each dose group. The specified dose x (mg/kg) of growth promotant was added to the feed daily for a fixed period of time. Weight gains (in pounds) are shown here.

Dose x	Weight Gain y		
0 control	1.5	1.8	1.7
0.2	2.3	2.0	1.8
0.4	4.3	3.7	4.1
0.8	5.7	5.9	6.2
1.6	7.9	7.7	7.5

a. Plot the sample data.
b. Fit the data to obtain the linear regression equation $\hat{y} = \hat{\beta}_0 + \hat{\beta}_1 x$.

9.51 Refer to Exercise 9.50. Compute the correlation coefficient r as a measure of the strength of the linear relation between dose and weight gain. What economic implications do you see for this promotant?

9.52 The correlation coefficient can sometimes be used to measure the reliability (consistency or reproducibility) of a test. This can be done by examining the same individuals with the same test on two separate occasions or by examining the same individuals on two different, equivalent tests. Discuss the pros and cons of these two methods for measuring the reliability of a test. Is one method to be preferred to the other?

9.53 Each student in a group of 100 second-grade students was administered two equivalent forms of a word fluency test. The test required that a student write as many words as possible beginning with a given letter during a 5-minute period. The second test was the same as the first except that the letter was changed. The sample correlation coefficient was .75; interpret this result. What factors might contribute to the less-than-complete agreement between the two test results?

9.54 A sociologist working for the government of a large city collected data on the number of nonviolent crimes (in 1,000s) reported and the increase (or decrease) of all crimes over the previous reporting period. Quarterly data are shown here:

Quarter	Nonviolent Crimes	Increase (or decrease) in All Crimes
1	7.2	14.1
2	6.4	14.5
3	6.6	13.3
4	7.3	13.6
5	7.5	15.2
6	6.9	15.7
7	7.1	15.3
8	7.4	14.8
9	7.6	16.1
10	7.3	16.6
11	7.1	16.2
12	7.0	15.9

a. Plot the nonviolent crime data versus quarter. Also plot the increase versus quarter on the same graph.
b. Does there appear to be a relationship between the two crime variables?
c. Compute the correlation coefficient between nonviolent crimes and the increase in all crimes.

9.55 Refer to Exercise 9.54.

a. Lag the increase in crimes by one quarter and superimpose these data on the plot of nonviolent crimes by quarter.
b. Compute the correlation coefficient of the nonviolent crimes and lagged increase in crimes. (Note: You will only have 11 data points corresponding to quarters 1–11.)

9.56 A study was conducted to examine the efficiencies of various manufacturing sites of a large corporation. At each site, the average number of acceptable cartons of manu-factured goods per month was recorded, as was the average number of hours of assembly-line operation per month. These data are shown here.

Location	Average Number of Acceptable Cartons (1,000) y	Average Number of Hours of Line Operation x
1	12	20
2	11	38
3	15	40
4	16	45
5	20	57
6	18	68
7	22	74
8	26	79
9	20	81
10	21	86
11	27	93
12	32	104
13	33	110
14	34	120
15	31	138

a. Plot the sample data.
b. Obtain a least squares fit to these data, using a linear regression model.
c. Plot the least squares prediction equation on the graph of part a. Does this model appear to fit the data adequately?

9.57 Refer to the data of Exercise 9.56. Suppose the last four data points corresponding to locations 12, 13, 14, and 15 were as shown here, rather than as indicated in the previous exercise.

Location	y	x
12	25	104
13	23	110
14	20	120
15	15	138

a. Plot the entire new data set for the 15 locations.

b. Would a linear regression model fit these data well? If not, can you suggest a transformation of the x variable to linearize the plot? Plot the transformed data.

c. Refer to part b. Fit the transformed data using a linear regression model.

9.58 Refer to Exercise 9.57. What other way might you have linearized the plot using a transformation? Plot these transformed data as well. Which transformation seems more suitable for these data?

9.59 Refer to the data of Exercise 9.56. Obtain an appropriate linear regression equation using the median approach of Section 9.3. How close is this approximation to the least squares fit of Exercise 9.56?

9.60 An investigator was interested in examining the effect of different doses of a new drug on the pulse rates of human subjects. Four doses of the drug were used in the experiment (1.5, 2.0, 2.5, and 3.0 mL/kg of body weight). Three persons were randomly assigned to each of the four drug doses. After a prestudy pulse rate was recorded for each individual, subjects were injected with the appropriate drug dose. One hour later, pulse rates were again recorded. The changes in pulse rates are listed in the accompanying table.

Change in pulse rate, y	20, 21, 19	16, 17, 17	15, 13, 14	8, 10, 8
Drug dose, x	1.5, 1.5, 1.5	2.0, 2.0, 2.0	2.5, 2.5, 2.5	3.0, 3.0, 3.0

a. Plot the sample data.

b. Find the least squares line for these data using the accompanying SAS output.

```
OPTIONS NODATE NONUMBER LS=78 PS=60;
DATA RAW;
  INPUT DOSE CHANGE;
  LABEL DOSE   = 'DRUG DOSE'
        CHANGE= 'CHANGE IN PULSE RATE';
  CARDS;
  1.5   20
  1.5   21
  1.5   19
  2.0   16
  2.0   17
  2.0   17
  2.3
  2.5   15
  2.5   13
  2.5   14
  3.0    8
  3.0   10
  3.0    8
;
```
(continues)

```
PROC PRINT N;
TITLE1 'EFFECT OF DIFFERENT DOSES ON PULSE RATES';
TITLE2 'LISTING OF THE DATA';

PROC PLOT DATA=RAW;
  PLOT CHANGE*DOSE;
TITLE2 'PLOT OF THE DATA';

PROC REG DATA=RAW;
  MODEL CHANGE=DOSE / P;
TITLE2 'EXAMPLE OF PROC REG WITH THE P OPTION (PREDICTED VALUES)';
RUN;
```

```
EFFECT OF DIFFERENT DOSES ON PULSE RATES
      LISTING OF THE DATA

   OBS    DOSE    CHANGE

    1     1.5       20
    2     1.5       21
    3     1.5       19
    4     2.0       16
    5     2.0       17
    6     2.0       17
    7     2.3
    8     2.5       15
    9     2.5       13
   10     2.5       14
   11     3.0        8
   12     3.0       10
   13     3.0        8

         N = 13
```

```
       EFFECT OF DIFFERENT DOSES ON PULSE RATES
                  PLOT OF THE DATA

      Plot of CHANGE*DOSE.   Legend: A = 1 obs, B = 2 obs, etc.


   21 +   A
      |
   20 +   A
      |
 C 19 +   A
 H    |
 A 18 +
 N    |
 G 17 +              B
 E    |
   16 +              A
 I    |
 N 15 +                            A
      |
 P 14 +                            A
 U    |
 L 13 +                            A
 S    |
 E 12 +
      |
 R 11 +
 A    |
 T 10 +                                            A
 E    |
    9 +
      |
    8 +                                            B
      |
      ---+--------+--------+--------+--------+--------+--------+--------+--
        1.50     1.75     2.00     2.25     2.50     2.75     3.00
                            DRUG DOSE

NOTE: 1 obs had missing values.                    (continues)
```

```
                    EFFECT OF DIFFERENT DOSES ON PULSE RATES
              EXAMPLE OF PROC REG WITH THE P OPTION (PREDICTED VALUES)

  Model: MODEL1
  Dependent Variable: CHANGE      CHANGE IN PULSE RATE
                        Analysis of Variance

                             Sum of          Mean
  Source            DF      Squares         Square      F Value     Prob>F

  Model              1    201.66667      201.66667      168.056     0.0001
  Error             10     12.00000        1.20000
  C Total           11    213.66667

          Root MSE          1.09545      R-square      0.9438
          Dep Mean         14.83333      Adj R-sq      0.9382
          C.V.              7.38502

                         Parameter Estimates

                    Parameter      Standard     T for HO:
  Variable   DF      Estimate         Error     Parameter=0     Prob > |T|

  INTERCEP    1    31.333333    1.31148770          23.891         0.0001
  DOSE        1    -7.333333    0.56568542         -12.964         0.0001

                        Variable
  Variable   DF         Label

  INTERCEP    1    Intercept
  DOSE        1    DRUG DOSE


                    EFFECT OF DIFFERENT DOSES ON PULSE RATES
              EXAMPLE OF PROC REG WITH THE P OPTION (PREDICTED VALUES)
                          Dep Var     Predict
                   Obs    CHANGE       Value      Residual

                    1    20.0000     20.3333      -0.3333
                    2    21.0000     20.3333       0.6667
                    3    19.0000     20.3333      -1.3333
                    4    16.0000     16.6667      -0.6667
                    5    17.0000     16.6667       0.3333
                    6    17.0000     16.6667       0.3333
                    7                14.4667
                    8    15.0000     13.0000       2.0000
                    9    13.0000     13.0000     -178E-17
                   10    14.0000     13.0000       1.0000
                   11     8.0000      9.3333      -1.3333
                   12    10.0000      9.3333       0.6667
                   13     8.0000      9.3333      -1.3333

  Sum of Residuals            1.953993E-14
  Sum of Squared Residuals       12.0000
  Predicted Resid SS (Press)     17.7709
```

c. Predict the change in pulse rate that would accompany a drug dose of 2.3 mL/kg of body weight. (Note: A dose of $x = 2.3$ with no y-value was included as another observation so the software would compute the predicted value for $x = 2.3$.)

9.61 Refer to Exercise 9.60. Calculate the correlation coefficient, and then interpret your results.

9.62 A production supervisor was concerned about the quality of the outgoing product from his department. He felt very strongly that the percentage of defective items passing through his assembly line during a 30-minute period increased throughout the day. At nine 30-minute periods throughout the day, the assembly line was closely examined to determine the number of defectives being produced. For each of these 30-minute periods, the number of hours that workers had been working (from 8:00 A.M.) was also recorded. The data are given in the accompanying table.

Number of Defectives, y	Number of Hours, x, That Workers Are on the Job
13	1.0
14	1.5
16	2.5
14	2.0
15	3.5
20	4.5
18	4.0
18	5.5
20	6.0
Total $\sum y = 148$; $\sum y^2 = 2{,}490$	$\sum x = 30.5$; $\sum x^2 = 128.25$; $\sum xy = 535.5$

a. Plot the data.

b. Write a linear model relating the number of defectives, y, to the number of hours on the job, x, for these data.

c. Use the method of least squares to fit the model.

d. Predict the number of items that would be defective in a 30-minute period if the workers had just completed 5 hours of work.

9.63 Refer to Exercise 9.62. Compute the sample correlation coefficient to measure the strength of the linear relationship between x and y. Interpret your answer.

9.64 A chain of grocery stores conducted a study to determine the relationship between the amount of money, x, spent on advertising and the weekly volume, y, of sales. Six different levels of advertising expenditure were tried in a random order over a 6-week period. The accompanying data were observed (in units of $100).

Weekly sales volume, y	10.2	11.5	16.1	20.3	25.6	28.0
Amount spent on advertising, x	1.0	1.25	1.5	2.0	2.5	3.0

a. Plot these data on a scatter diagram.

b. Use the method of least squares to find the regression equation $\hat{y} = \hat{\beta}_0 + \hat{\beta}_1 x$.

c. Use the prediction equation of part b to estimate sales volume for an expenditure of $220 in advertising.

9.65 Refer to Exercise 9.64. Compute the correlation coefficient between the sales volume and advertising volume. Does there appear to be a strong linear relationship between x and y?

9.66 Suppose that the following data were collected on emphysema patients; the number of years, x, the patient smoked and inhaled and a physician's evaluation, y, of the patient's lung capacity (measured on a scale of 0 to 100). The results for a sample of ten patients appear in the accompanying table. (Note: $S_{xx} = 876.9$, $S_{yy} = 2{,}510$, and $S_{xy} = 1{,}148$.)

Patient	Years Smoking, x	Lung Capacity, y
1	25	55
2	36	60
3	22	50
4	15	30
5	48	75
6	39	70
7	42	70
8	31	55
9	28	30
10	33	35

a. Plot the data on a scatter diagram.

b. Use the method of least squares to find the regression line $\hat{y} = \hat{\beta}_0 + \hat{\beta}_1 x$.

c. Does there appear to be a positive linear relationship between x and y?

d. Calculate the correlation coefficient between the variables "lung capacity" and "number of years smoking."

e. Predict a person's lung capacity after 30 years of smoking.

9.67 An experiment was conducted to measure the strength of the linear relationship between two variables: a student's emotional stability (as measured by a guidance counselor's subjective judgment after an encounter session) and the student's score on an achievement test administered to children entering the first grade. The variable of emotional stability was measured on a scale of 0 to 40 (from low to high), and the achievement test was also measured from 0 to 40. Use the accompanying data from a random sample of 15 children to calculate the correlation coefficient. (Note: $S_{xx} = 485.33$, $S_{yy} = 522.93$, and $S_{xy} = 316.67$.)

Student	Emotional Stability, x	Achievement, y	Student	Emotional Stability, x	Achievement, y
1	23	31	9	32	33
2	21	23	10	29	35
3	31	34	11	16	21
4	34	29	12	29	22
5	26	29	13	23	24
6	22	27	14	27	28
7	14	21	15	25	15
8	18	17			

9.68 Conduct a study to determine whether there is a correlation between a social science major's performance in a math or a statistics course and his or her performance in a course in the social sciences. For example, you may wish to visit the department of sociology to obtain a random sample of 30 senior sociology majors. Contact these students to determine their (numerical) grades in a specific sociology course (such as

"Introductory Sociology") and a mathematics (or statistics) course required for graduation. Let x denote a student's sociology grade and y his or her mathematics (statistics) grade.

a. Identify the population from which the sample was drawn.

b. Find the least squares prediction equation, $\hat{y} = \hat{\beta}_0 + \hat{\beta}_1 x$.

c. Calculate the coefficient of linear correlation.

d. Describe the strength of the relationship between the two sets of scores.

9.69 Use the computer output for the data in Table 9.9 to answer the following.

a. Determine the least squares regression line

$$\hat{y} = \hat{\beta}_0 + \hat{\beta}_1 x_1 + \hat{\beta}_2 x_2 + \hat{\beta}_3 x_3 + \hat{\beta}_4 x_4.$$

b. Predict y (population increase) for birth rate $x_1 = 30$, death rate $x_2 = 15$, life expectancy $x_3 = 65$, and per capita GNP $x_4 = 1,100$.

c. Does the regression seem to fit the data? Explain.

TABLE 9.9

Demographic Characteristics of Ten Nations in 1980

Nation	Projected Population Increase for the Year 2000, by % (y)	Birth Rate (x_1)	Death Rate (x_2)	Life Expectancy (x_3)	Per Capita GNP (x_4)
Bolivia	67.9	44	19	47	$ 510
Cuba	27.0	18	6	72	810
Cyprus	16.7	19	8	73	2,110
Egypt	54.2	38	10	55	400
Ghana	81.2	48	17	48	390
Jamaica	27.3	29	7	70	1,110
Nigeria	93.1	50	18	42	560
South Africa	63.0	38	10	60	1,460
South Korea	33.8	23	7	62	1,160
Turkey	53.0	35	10	58	1,210

```
MTB > BRIEF 3
MTB > REGRESS C1 ON 4, C2 C3 C4 C5

The regression equation is
PROJ_POP = 81.9 + 1.26 BRTHRATE - 0.42 DEATHRTE - 1.19 LIFE_EXP + 0.00177 GNP

Predictor      Coef       Stdev      t-ratio        p
Constant      81.89       70.47        1.16      0.298
BRTHRATE     1.2553       0.6508       1.93      0.112
DEATHRTE      0.425       1.335       -0.32      0.763
LIFE_EXP    -1.1937       0.8479      -1.41      0.218
GNP        0.001773       0.006621     0.27      0.800

s = 7.729     R-sq = 94.8%     R-sq(adj) = 90.6%
                                              (continues)
```

```
Analysis of Variance

SOURCE        DF        SS          MS          F         p
Regression     4       5434.4      1358.6      22.74     0.002
Error          5        298.7        59.7
Total          9       5733.1

SOURCE        DF      SEQ SS
BRTHRATE       1       5295.4
DEATHRTE       1         15.2
LIFE_EXP       1        119.6
GNP            1          4.3

Obs.BRTHRATE    PROJ_POP      Fit  Stdev.Fit   Residual   St.Resid
  1     44.0      67.90     73.86      5.81      -5.96       -1.17
  2     18.0      27.00     17.43      6.15      -9.57        2.04R
  3     19.0      16.70     18.95      6.65      -2.25       -0.57
  4     38.0      54.20     60.40      5.00      -6.20       -1.05
  5     48.0      81.20     78.32      4.58       2.88        0.46
  6     29.0      27.30     33.73      5.60      -6.43       -1.21
  7     50.0      93.10     87.87      4.83       5.23        0.87
  8     38.0      63.00     56.31      5.35       6.69        1.20
  9     23.0      33.80     35.84      6.29      -2.04       -0.45
 10     35.0      53.00     54.49      3.77      -1.49       -0.22

R denotes an obs. with a large st. resid.

MTB > PRINT C1-C5

ROW   PROJ_POP   BRTHRATE   DEATHRTE   LIFE_EXP    GNP

  1     67.9        44         19         47       510
  2     27.0        18          6         72       810
  3     16.7        19          8         73      2110
  4     54.2        38         10         55       400
  5     81.2        48         17         48       390
  6     27.3        29          7         70      1110
  7     93.1        50         18         42       560
  8     63.0        38         10         60      1460
  9     33.8        23          7         62      1160
 10     53.0        35         10         58      1210

MTB > BRIEF 3
MTB > REGRESS C1 ON 1, C2

The regression equation is
PROJ_POP = - 19.8 + 2.09 BRTHRATE

Predictor       Coef       Stdev      t-ratio         p
Constant      -19.778      7.635       -2.59       0.032
BRTHRATE       2.0906      0.2125       9.84       0.000

s = 7.397      R-sq = 92.4%      R-sq(adj) = 91.4%

Analysis of Variance

SOURCE        DF        SS          MS          F         p
Regression     1       5295.4      5295.4      96.77     0.000
Error          8        437.8        54.7
Total          9       5733.1

Obs.BRTHRATE    PROJ_POP      Fit  Stdev.Fit   Residual   St.Resid
  1     44.0      67.90     72.21      3.13      -4.31       -0.64
  2     18.0      27.00     17.85      4.16       9.15        1.50
  3     19.0      16.70     19.94      3.99      -3.24       -0.52
  4     38.0      54.20     59.66      2.47      -5.46       -0.78
  5     48.0      81.20     80.57      3.75       0.63        0.10
  6     29.0      27.30     40.85      2.59     -13.55       -1.96
  7     50.0      93.10     84.75      4.09       8.35        1.35
  8     38.0      63.00     59.66      2.47       3.34        0.48
  9     23.0      33.80     28.31      3.34       5.49        0.83
 10     35.0      53.00     53.39      2.35      -0.39       -0.06
```

9.70   Refer to the data in Table 9.9. Use the Minitab output shown here to answer parts a and b.

```
MTB > REGRESS C1 ON 2, C2 C3

The regression equation is
PROJ_POP = - 18.8 + 1.88 BRTHRATE + 0.57 DEATHRTE

Predictor      Coef      Stdev     t-ratio        p
Constant     -18.828     8.241      -2.28      0.056
BRTHRATE      1.8754     0.4840      3.87      0.006
DEATHRTE      0.572      1.142       0.50      0.632

s = 7.770       R-sq = 92.6%     R-sq(adj) = 90.5%

Analysis of Variance

SOURCE        DF        SS        MS        F        p
Regression     2      5310.5    2655.3    43.98    0.000
Error          7       422.6      60.4
Total          9      5733.1

SOURCE        DF      SEQ SS
BRTHRATE       1      5295.4
DEATHRTE       1        15.2

Obs.BRTHRATE  PROJ_POP      Fit Stdev.Fit  Residual   St.Resid
  1      44.0    67.90    74.56     5.74     -6.66     -1.27
  2      18.0    27.00    18.36     4.49      8.64      1.36
  3      19.0    16.70    21.38     5.08     -4.68     -0.80
  4      38.0    54.20    58.16     3.97     -3.96     -0.59
  5      48.0    81.20    80.92     4.00      0.28      0.04
  6      29.0    27.30    39.56     3.74    -12.26     -1.80
  7      50.0    93.10    85.24     4.41      7.86      1.23
  8      38.0    63.00    58.16     3.97      4.84      0.72
  9      23.0    33.80    28.31     3.51      5.49      0.79
 10      35.0    53.00    52.53     3.00      0.47      0.07

MTB > REGRESS C1 ON 3, C2 C3 C4

The regression equation is
PROJ_POP = 77.7 + 1.26 BRTHRATE - 0.36 DEATHRTE - 1.11 LIFE_EXP

Predictor      Coef      Stdev     t-ratio        p
Constant      77.71     63.18       1.23      0.265
BRTHRATE      1.2560     0.5983      2.10      0.081
DEATHRTE     -0.363      1.208      -0.30      0.774
LIFE_EXP     -1.1053     0.7182     -1.54      0.174

s = 7.106       R-sq = 94.7%     R-sq(adj) = 92.1%

Analysis of Variance

SOURCE        DF        SS        MS        F        p
Regression     3      5430.1    1810.0    35.84    0.000
Error          6       303.0      50.5
Total          9      5733.1

SOURCE        DF      SEQ SS
BRTHRATE       1      5295.4
DEATHRTE       1        15.2
LIFE EXP       1       119.6

Obs.BRTHRATE  PROJ_POP      Fit Stdev.Fit  Residual   St.Resid
  1      44.0    67.90    74.13     5.25     -6.23     -1.30
  2      18.0    27.00    18.56     4.11      8.44      1.46
  3      19.0    16.70    17.98     5.14     -1.28     -0.26
  4      38.0    54.20    61.02     4.08     -6.82     -1.17
  5      48.0    81.20    78.78     3.92      2.42      0.41
  6      29.0    27.30    34.22     4.87     -6.92     -1.34
  7      50.0    93.10    87.56     4.30      5.54      0.98
  8      38.0    63.00    55.49     4.02      7.51      1.28
  9      23.0    33.80    35.53     5.68     -1.73     -0.41
 10      35.0    53.00    53.93     2.89     -0.93     -0.14
```

a. Find the following least squares regression equations:

$$\hat{y} = \hat{\beta}_0 + \hat{\beta}_1 x_1$$
$$\hat{y} = \hat{\beta}_0 + \hat{\beta}_1 x_1 + \hat{\beta}_2 x_2$$
$$\hat{y} = \hat{\beta}_0 + \hat{\beta}_1 x_1 + \hat{\beta}_2 x_2 + \hat{\beta}_3 x_3.$$

b. Refer to Exercise 9.69 and the output. Does it appear that all four variables are needed to predict y? Do we need the variables x_1, x_2, and x_3? The variables x_1 and x_2? Or might knowledge of x_1 (birth rate) be sufficient? Explain.

9.71 The fuel of a new four-cylinder diesel engine was studied under various external, controlled operating temperatures. For each setting, two different engines were studied and the fuel consumption recorded.

Observation	External Temperature (°F) x	Fuel Consumption (gallons) y
1	20	25
2	20	26
3	30	28
4	30	27
5	40	32
6	40	35
7	50	42
8	50	46
9	60	55
10	60	53
11	70	55
12	70	57
13	80	60
14	80	58
15	90	61
16	90	58

a. Plot the sample data. Do you think a linear regression line will be an adequate model?

b. Fit the least squares regression model $y = \beta_0 + \beta_1 x + \varepsilon$ and draw the prediction equation on the graph of part a.

c. What's the sample correlation coefficient for these data?

9.72 Refer to Exercise 9.71. Since there does not appear to be a simple, suitable transformation on one or both of the variables to linearize the data, rank the data separately and compute the rank order correlation coefficient measure.

9.73 The maximum volume of oxygen uptake (VO$_2$ max) has been used as a measure of cardiac status in healthy individuals as well as in persons suffering from cardiac-related illnesses (such as congestive heart failure). The VO$_2$ max readings for 12 healthy adult males following strenuous exercise are recorded here. In general, VO_2 max decreases with any increase in activity level.

Individual	VO_2 Max y	Duration of Exercise (minutes) x
1	82	10.0
2	73	9.5
3	68	10.2
4	74	10.5
5	66	11.0
6	63	11.3
7	58	11.6
8	54	12.0
9	56	12.1
10	51	12.5
11	55	12.8
12	44	13.0

a. Plot the data.
b. Does a linear regression equation seem to be appropriate for these data?
c. Fit the data using the model $y = \beta_0 + \beta_1 x + \varepsilon$.

9.74 Refer to Exercise 9.73.
a. Obtain the least squares prediction for the same data, except that the final observation is $x = 13$, $y = 30$.
b. How well does a linear regression equation fit these data?
c. What effect does the observation from part a have on the prediction equation, compared to the corresponding prediction equation of Exercise 9.73? (Hint: Plot both the regression equations.)

INFERENCES RELATED TO LINEAR REGRESSION AND CORRELATION

10.1 INTRODUCTION

In Chapter 9, we gave formulas for finding the least squares estimates for β_0 and β_1 in the linear regression model

$$y_i = \beta_0 + \beta_1 x_i + \varepsilon_i \qquad i = 1, 2, \ldots, n.$$

We would now like to use these estimates to make inferences about the relationship between y and x. For example, suppose that x represents the amount of force applied to a 1-foot section of steel and y denotes the corresponding increase in width of the steel sample. Applying the results of Chapter 9, we would obtain a random sample of n observations and compute the least squares estimates for β_0 and β_1 as

$$\hat{\beta}_1 = \frac{S_{xy}}{S_{xx}} \qquad \text{and} \qquad \hat{\beta}_0 = \bar{y} - \hat{\beta}_1 \bar{x}.$$

It might be of interest in this problem to test to see if there is a positive linear relationship between x and y. To do this, we could conduct a statistical test of the null hypothesis $H_0: \beta_1 = 0$ against the alternative hypothesis $H_a: \beta_1 > 0$. The estimate $\hat{\beta}_1$ will be used in this test.

10.2 ▼ INFERENCES ABOUT β_0 AND β_1

For $H_0 : \beta_1 = 0$
as x changes 1 unit,
y doesn't change.

Before we can make any inferences about parameters in the linear regression model, we need to expand on the assumptions we have for the model. Previously, we assumed that the random error term ε_i associated with observation y_i has expectation zero. In addition, we will assume the following:

1. $\varepsilon_1, \varepsilon_2, \ldots, \varepsilon_n$ are independent of each other.
2. ε for a given setting of the independent variable x is normally distributed with mean 0 and variance σ_ε^2. The variance σ_ε^2 is constant for all settings of x.

These two assumptions imply that y_i is normally distributed with mean $\beta_0 + \beta_1 x_i$ and constant variance σ_ε^2 and that y_i and y_j are independent. See Figure 10.1.

each $y_i \sim N(\beta_0 + \beta_1 x_1, \sigma_\varepsilon^2)$

FIGURE 10.1

Illustration of the Two Assumptions for Linear Regression

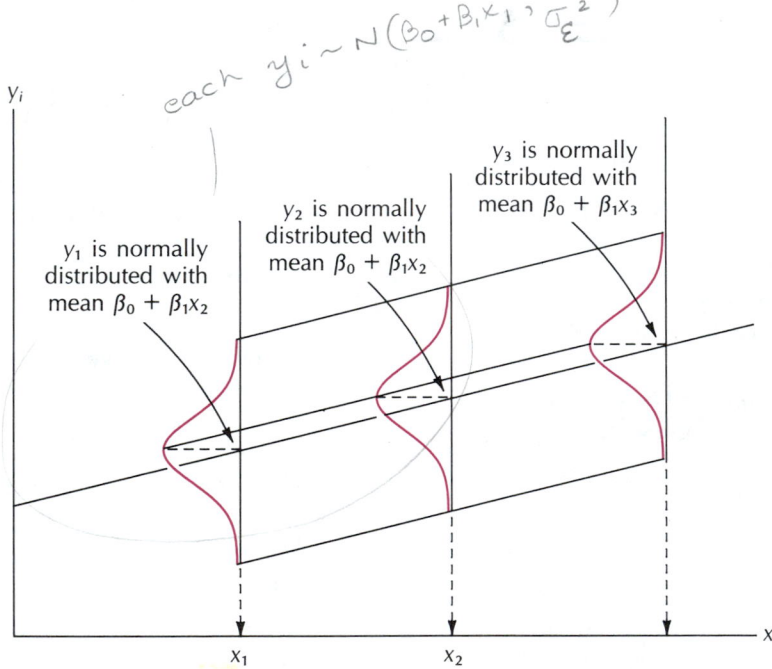

y_i

y_1 is normally distributed with mean $\beta_0 + \beta_1 x_2$

y_2 is normally distributed with mean $\beta_0 + \beta_1 x_2$

y_3 is normally distributed with mean $\beta_0 + \beta_1 x_3$

x_1 x_2 x

expected values

Under these assumptions, both $\hat{\beta}_0$ and $\hat{\beta}_1$ have sampling distributions that are normal with means (called **expected values**) and standard errors as shown here.

Expected Values and Standard Errors for $\hat{\beta}_0$ and $\hat{\beta}_1$ in Linear Regression

$$\mu_{\hat{\beta}_0} = \beta_0 \qquad\qquad \mu_{\hat{\beta}_1} = \beta_1$$

$$\sigma_{\hat{\beta}_0} = \sigma_\varepsilon \sqrt{\frac{\sum x^2}{nS_{xx}}} \qquad\qquad \sigma_{\hat{\beta}_1} = \frac{\sigma_\varepsilon}{\sqrt{S_{xx}}}$$

E X A M P L E   10.1 ▼

In examining the weight loss of a compound, a chemist hypothesized that weight loss y (in pounds) is linearly related to the relative humidity x of the room in which the process operates. From a sample of 12 observations, we find $\bar{y} = 4.8$, $\bar{x} = 6.5$, $S_{xy} = 138.2$, $S_{xx} = 101$, and $\sum x^2 = 608.1$. Compute $\hat{\beta}_0$, $\hat{\beta}_1$, and their standard errors.

Solution For these data,

$$\hat{\beta}_1 = \frac{S_{xy}}{S_{xx}} = 1.37$$

$$\sigma_{\hat{\beta}_1} = \frac{\sigma_\varepsilon}{\sqrt{S_{xx}}} = \frac{\sigma_\varepsilon}{10.05}$$

and

$$\hat{\beta}_0 = \bar{y} - \hat{\beta}_1 \bar{x} = 4.8 - 1.37(6.5) = -4.11$$

$$\sigma_{\hat{\beta}_0} = \sigma_\varepsilon \sqrt{\frac{\sum x^2}{nS_{xx}}} = .71\sigma_\varepsilon.$$

▲

The problem with our solution to Example 10.1 is that we still don't have the standard errors for $\hat{\beta}_0$ and $\hat{\beta}_1$ because we don't know σ_ε, the standard deviation of the εs, or we don't have an estimate of its value. In Chapter 9, we showed that

$$\underset{\substack{\text{sum of squares} \\ \text{about the mean}}}{\sum(y - \bar{y})^2} = \underset{\substack{\text{sum of squares} \\ \text{for regression}}}{\sum(\hat{y} - \bar{y})^2} + \underset{\substack{\text{sum of squares} \\ \text{for error}}}{\sum(y - \hat{y})^2.} \quad \text{(SSE)}$$

This last quantity is called by many different names: residual sum of squares, sum of squares about the regression, and sum of squares for error (SSE), to name just a few. For the remainder of this text, we will use the term SSE. Dividing SSE by $n - 2$ degrees of freedom, we can obtain an estimate of σ_ε^2 in linear regression. We designate this estimate as s_ε^2.

Estimate of σ_ε^2 and σ_ε for Linear Regression

$$s_\varepsilon^2 = \frac{\sum (y - \hat{y})^2}{n - 2} = \frac{SSE}{n - 2} \quad \sim n-2 \;\; b/c$$
$$\qquad\qquad\qquad\qquad\qquad we \; are \; estimating$$
$$s_\varepsilon = \sqrt{\frac{SSE}{n - 2}} \qquad\qquad 2 \; parameters$$

Note: If computer software is not available and calculations are done by hand or by using a calculator, use the shortcut formula SSE $= S_{yy} - \hat{\beta}_1 S_{xy}$.

E X A M P L E   10.2

▼

The yield per plot in bushels of corn was observed on $n = 10$ plots that had been fertilized in varying degrees. We let the independent variable x denote the amount of fertilizer applied. The data and the coded fertilizer values are recorded in Table 10.1. Use the sample data of Table 10.1 to obtain the least squares prediction equation $\hat{y}$ for the linear regression model $y = \beta_0 + \beta_1 x + \varepsilon$. Also, calculate estimates of σ_ε^2 and σ_ε.

T A B L E   10.1
Corn Yield Data for Example 10.2

Yield (in bushels)	Fertilizer (in pounds per plot)
12	2
13	2
13	3
14	3
15	4
15	4
14	5
16	5
17	6
18	6

Solution For these data (using a computer or hand computations) it is easily seen that

$$n = 10 \qquad\qquad \sum x = 40$$
$$\sum y = 147 \qquad\qquad \bar{x} = 4.0$$
$$\bar{y} = 14.7 \qquad\qquad \sum x^2 = 180$$
$$\sum y^2 = 2193 \qquad\qquad \sum xy = 611.$$

Substituting into appropriate linear regression formulas we find the least square estimates for β_1 and β_0 to be, respectively,

$$\hat{\beta}_1 = \frac{S_{xy}}{S_{xx}} = \frac{611 - \dfrac{40(147)}{10}}{180 - \dfrac{(40)^2}{10}} = \frac{23}{20} = 1.15$$

and

$$\hat{\beta}_0 = \bar{y} - \hat{\beta}_1 \bar{x} = 14.7 - 1.15(4.0) = 10.10 \text{ bushels.}$$

The estimate for σ_ε^2 (and σ_ε) requires that we find

$$S_{yy} = \sum y^2 - \frac{(\sum y)^2}{n} = 2193 - \frac{(147)^2}{10} = 32.10$$

calc similar to S_{xx}

and

$$SSE = S_{yy} - \hat{\beta}_1 S_{xy} = 32.10 - 1.15(23) = 5.65.$$

The estimate for σ_ε^2 from these data is

$$s_\varepsilon^2 = \frac{SSE}{n-2} = \frac{5.65}{8} = 0.71$$

and hence $s_\varepsilon = 0.84$.   ▲

E X A M P L E   10.3

▼

Refer to Example 10.2 to compute the *estimated* standard errors for $\hat{\beta}_0$ and $\hat{\beta}_1$.

Solution   The formulas for the standard errors are, respectively,

$$\sigma_{\hat{\beta}_0} = \sigma_\varepsilon \sqrt{\frac{\sum x^2}{n S_{xx}}} \quad \text{and} \quad \sigma_{\hat{\beta}_1} = \frac{\sigma_\varepsilon}{\sqrt{S_{xx}}}.$$

Using $s_\varepsilon = 0.84$ as the estimate of σ_ε and substituting into these formulas, we obtain the estimated standard errors for $\hat{\beta}_0$ and $\hat{\beta}_1$:

$$\hat{\sigma}_{\hat{\beta}_0} = s_\varepsilon \sqrt{\frac{\sum x^2}{nS_{xx}}} = .84 \sqrt{\frac{180}{10(20)}} = .80$$

and

$$\hat{\sigma}_{\hat{\beta}_1} = \frac{s_\varepsilon}{\sqrt{S_{xx}}} = \frac{.84}{4.47} = .19.$$

▲

estimated standard errors

Note that by substituting s_ε for σ_ε in the formulas for $\sigma_{\hat{\beta}_1}$ and $\sigma_{\hat{\beta}_0}$ we have obtained the **estimated standard errors** for $\hat{\beta}_1$ and $\hat{\beta}_0$. In practice, since we will never know σ_ε and hence will always substitute an estimate of its value in the standard error formulas, we will drop the word "estimated" and simply call $\hat{\sigma}_{\hat{\beta}_1}$ and $\hat{\sigma}_{\hat{\beta}_0}$ the standard errors for $\hat{\beta}_1$ and $\hat{\beta}_0$, respectively.

An additional assumption for the ε_i will allow us to construct confidence intervals and statistical tests for β_1 and β_0. If we assume that the ε_i from the linear regression model

$$y_i = \beta_0 + \beta_1 x_i + \varepsilon_i$$

are normally distributed, we can specify confidence intervals for β_0 and β_1 using the formula (estimate) $\pm t$(standard error). Fortunately, for large sample sizes, the confidence intervals are robust against modest departures from normality; other assumptions will be discussed in Section 10.6.

$100(1 - \alpha)\%$ Confidence Intervals for β_0 and β_1 in Linear Regression

$$\hat{\beta}_0 \pm t_{\alpha/2} s_\varepsilon \sqrt{\frac{\sum x^2}{nS_{xx}}}$$

and

$$\hat{\beta}_1 \pm t_{\alpha/2} \frac{s_\varepsilon}{\sqrt{S_{xx}}},$$

where

$$s_\varepsilon = \sqrt{\frac{SSE}{n-2}}$$

EXAMPLE 10.4 ▼

Use the data from Example 10.2 to develop 95% confidence intervals for β_0 and β_1.

Solution   The calculations from Example 10.2 yielded the linear regression equation $\hat{y} = 10.10 + 1.15x$. The $t_{.025}$ value for df $= 8$ is 2.306, $s_\varepsilon = 0.84$, $S_{xx} = 20$, and $\sum x^2 = 180$. Substituting these values into the appropriate formulas, we obtain the 95% confidence intervals shown here:

$n-2$

$$\beta_0: \quad 10.10 \pm 2.306(0.84)\sqrt{\frac{180}{10(20)}}, \quad \text{or} \quad 10.10 \pm 1.84$$

$$\beta_1: \quad 1.15 \pm 2.306 \frac{(0.84)}{\sqrt{20}}, \quad \text{or} \quad 1.15 \pm 0.43.$$

In other words, we are 95% confident that the true value of the intercept β_0 lies somewhere in the interval $8.26 \le \beta_0 \le 11.94$. Similarly, we are 95% confident that the true slope β_1 lies somewhere in the interval $0.72 \le \beta_1 \le 1.58$. ▲

EXAMPLE 10.5 ▼

A restaurant opening on a "reservations-only" basis would like to use the number of advance reservations x to predict the number of dinners y to be prepared. Data on reservations and number of dinners served for one day chosen at random from each week in a 100-week period gave the following results:

$$\bar{x} = 150 \qquad \bar{y} = 120$$
$$\sum (x - \bar{x})^2 = 90,000 \qquad \sum (y - \bar{y})^2 = 70,000.$$
$$\sum (x - \bar{x})(y - \bar{y}) = 60,000$$

a. Find the least squares estimates $\hat{\beta}_0$ and $\hat{\beta}_1$ for the linear regression line $\hat{y} = \hat{\beta}_0 + \hat{\beta}_1 x$.
b. Predict the number of meals to be prepared if the number of reservations is 135.
c. Construct a 90% confidence interval for the slope. Does information on x (number of advance reservations) help in predicting y (number of dinners prepared)?

Solution

a. The least squares estimates are given by

$$\hat{\beta}_1 = \frac{S_{xy}}{S_{xx}} = \frac{60{,}000}{90{,}000} = .67$$

and

$$\hat{\beta}_0 = \bar{y} - \hat{\beta}_1 \bar{x} = 120 - .67(150) = 19.50.$$

b. The predicted number of meals required for the number of advance reservations equal to 135 is

$$\hat{y} = 19.50 + .67(135) = 109.95, \qquad \text{or} \qquad 110.$$

c. The 90% confidence interval for β_1 uses the formula

$$\hat{\beta}_1 \pm t \,(\text{standard error}),$$

where the standard error is $s_\varepsilon / \sqrt{S_{xx}}$.

Although Table 4 in the Appendix does not list a t-value for $a = .05$ and df $= 98$, we'll use the t-value for the next higher df (df $= 120$); this value is 1.658.

The standard deviation s_ε can be computed using the summary sample data

$$s_\varepsilon^2 = \frac{\text{SSE}}{n - 2},$$

where

$$\begin{aligned}
\text{SSE} &= S_{yy} - \hat{\beta}_1 S_{xy} \\
&= 70{,}000 - 0.67(60{,}000) \\
&= 29{,}800.
\end{aligned}$$

Thus,

$$s_\varepsilon = \sqrt{\frac{29{,}800}{98}} = \sqrt{304.08} = 17.44$$

and the 90% confidence interval for β_1 is

$$0.67 \pm 1.658 \, \frac{(17.44)}{\sqrt{90{,}000}}$$

or

$$0.67 \pm .10.$$

Since we are 90% confident that the true value of β_1 lies somewhere in the interval $.57 \leq \beta_1 \leq .77$, we are thus confident the increase in y (number of dinners prepared) for every increase of one advance reservation is in the interval from .57 to .77. Also, since the interval for β_1 does not include 0 as a possible value for the slope, it appears that the number of advance reservations is a useful predictor of the number of meals to be prepared in the context of a linear regression model, $y = \beta_0 + \beta_1 x + \varepsilon$. ▲

Statistical tests for β_0 and β_1 use a t statistic of the form $t =$ estimate/standard error. Three different research hypotheses are presented along with the corresponding rejection regions. For a particular experimental situation, we must choose one of the specific alternatives shown here.

Statistical Tests for β_0 and β_1

▼

	Intercept, β_0	Slope, β_1
H_0:	$\beta_0 = 0$	H_0: $\beta_1 = 0$

H_a: 1. $\beta_0 > 0$ H_a: 1. $\beta_1 > 0$
     2. $\beta_0 < 0$ 2. $\beta_1 < 0$
     3. $\beta_0 \neq 0$ 3. $\beta_1 \neq 0$

T.S.: $t = \dfrac{\hat{\beta}_0}{s_\varepsilon \sqrt{\dfrac{\sum x^2}{n S_{xx}}}}$

T.S.: $t = \dfrac{\hat{\beta}_1}{\dfrac{s_\varepsilon}{\sqrt{S_{xx}}}}$

R.R.: For a given value of α and df $= n - 2$,

R.R.: Same as those shown for β_0

  1. reject H_0 if $t > t_\alpha$
  2. reject H_0 if $t < -t_\alpha$
  3. reject H_0 if $|t| > t_{\alpha/2}$.

E X A M P L E   10.6

▼

Refer to the data of Example 10.5. Confirm the conclusion we reached concerning β_1 by conducting a test of H_0: $\beta_1 = 0$ versus H_a: $\beta_1 \neq 0$. Use $\alpha = .10$.

Solution   The parts of the statistical test are given here:

$$H_0: \quad \beta_1 = 0$$
$$H_a: \quad \beta_1 \neq 0$$

$$\text{T.S.:} \quad t = \frac{\hat{\beta}_1}{\dfrac{s_\varepsilon}{\sqrt{S_{xx}}}} = \frac{0.67}{\dfrac{17.44}{\sqrt{90{,}000}}} = 11.53$$

R.R.:   For a two-tailed test with $\alpha = .10$ and df = 98, we will reject H_0 if $|t| > 1.645$.

Conclusion: Since $t = 11.53$ is greater than 1.645, we have sufficient evidence to reject H_0. It does appear that x is useful in predicting y.   ▲

E X A M P L E   10.7

▼

The laboratory of a hospital participating in the clinical trial of an antibiotic drug had to be validated to see that the laboratory personnel could accurately assay blood samples "spiked" with fixed amounts of the antibiotic. The validation consisted of the following experiment. Thirteen spiked samples (with amounts known only to the study investigator) were sent to the laboratory to be assayed for the amount of the antibiotic present. The results of the validation experiment are shown here. (Note: The spiked samples with known amounts added were supplied in a blinded fashion to the laboratory.) The amounts found are the assay results supplied by the hospital laboratory.

Amount Added (μg/ml) x	Amount Found (μg/ml) y
0	0
5	4.5
5	5.0
5	4.8
10	8.9
10	8.9
10	8.9
20	17.0
20	18.2
20	15.4
40	32.6
40	36.1
40	31.5

a. If the laboratory assay was completely accurate, the amount found would equal the amount added. Plotting data for such a situation would suggest a slope of 1 and an intercept of 0. Plot the sample data. Does it appear this assay is "accurate"?

b. Fix these data to a linear regression model.

c. Test the null hypothesis $H_0: \beta_1 = 1$ versus $H_a: \beta_1 \neq 1$. Give the p-value for your test.

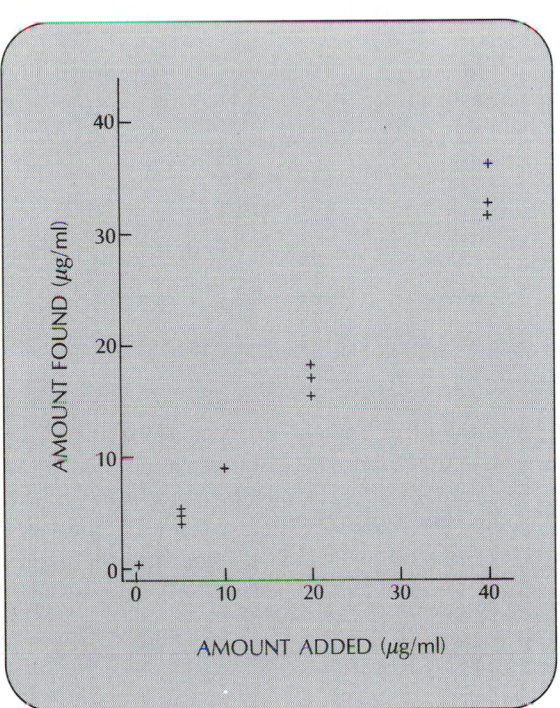

Solution

a. A plot of the sample data is shown here.

b. For the data, it can be shown that $\hat{\beta}_1 = .822$ with a standard error of $s_\varepsilon / \sqrt{S_{xx}} = .024$; the estimate of β_0 is $\hat{\beta}_0 = .529$ with a standard error of .537. The linear regression equation is then

$$\hat{y} = .529 + .822x.$$

c. The statistical test for H_0: $\beta_1 = 1$ is a slight variation of the test for H_0: $\beta_1 = 0$. The test statistic for this variation is

$$t = \frac{\hat{\beta}_1 - \beta_{1,0}}{s_\varepsilon / \sqrt{S_{xx}}},$$

where $\beta_{1,0}$ is the hypothesized value of β_1 under H_0. The test of H_0: $\beta_1 = 1$ is shown here.

$$H_0: \quad \beta_1 = 1$$
$$H_a: \quad \beta_1 \neq 1$$
$$\text{T.S.:} \quad t = \frac{.822 - 1}{.024} = -7.42$$

R.R.: Based on df $= 11$ and Table 4 of the Appendix, the p-value for the test result is $p < .001$.   ▲

A statistical test about β_1 can be restated in terms of an F statistic and put into the format of an "analysis of variance." (See Table 10.2.) Recall we showed in Chapter 9 that the total variability can be partitioned into two components: the sum of squares due to regression (SSREG) and the sum of squares for error (SSE). As indicated previously, the larger SSREG is relative to SSE, the better the model "fits" the data. It should come as no surprise, then, that we can form a test statistic based on SSREG and SSE. In particular, for

$$H_0: \quad \beta_1 = 0$$

and

$$H_a: \quad \beta_1 \neq 0,$$

TABLE 10.2
Analysis of Variance Table for Linear Regression

Source Due to	Sum of Squares (SS)	df	Mean Square (MS)	F
Regression	SSREG	1	MSREG = SSREG/1	F = MSREG/MSE
Error	SSE	$n - 2$	MSE = SSE/($n - 2$)	
Total	SSTotal	$n - 1$		

we can form an F statistic as

$$F = \frac{\text{SSREG}/1}{\text{SSE}/(n-2)} = \frac{\text{SSREG}}{s_\varepsilon^2},$$

which, under H_0, follows an F distribution with $df_1 = 1$ and $df_2 = n - 2$.

The F statistic shown here is the ratio of the explained variation to the unexplained variation divided by the respective degrees of freedom. So large values of F would provide evidence for rejection of H_0 in favor of H_a.

The resulting F test is sometimes summarized in an analysis of variance table (see Table 10.2), which lists the sources of variability, the sums of squares, their degrees of freedom, the mean squares (i.e., sums of squares divided by degrees of freedom), and the F statistic. An analysis of variance table for a linear regression problem is shown in Table 10.2.

Note that the F statistic is written as the ratio of the mean square regression (MSREG) to the mean square for error. This is equivalent to $F = \text{SSREG}/s_\varepsilon^2$, since $df_1 = 1$ and $s_\varepsilon^2 = \text{MSE}$. We will, however, use the more standard notation involving mean squares, to conform with notation to be used when we study analysis of variance techniques in more detail in Chapters 13–19.

EXAMPLE 10.8 ▼

Refer to the SAS output shown here for the data of Example 10.7.

a. Show that the square of the t statistic in Example 10.7 is equal to the computed F statistic shown in the output.

b. Locate the analysis of variance table for these data in the output and reconstruct it in the format shown in this section.

DEP VARIABLE: FOUND

ANALYSIS OF VARIANCE

SOURCE	DF	SUM OF SQUARES	MEAN SQUARE	F VALUE	PROB>F
MODEL	1	1675.71027	1675.71027	1149.032	0.0001
ERROR	11	16.04203876	1.45836716		
C TOTAL	12	1691.75231			

ROOT MSE	1.207629	R-SQUARE	0.9905
DEP MEAN	14.75385	ADJ R-SQ	0.9897
C.V.	8.185179		

PARAMETER ESTIMATES

VARIABLE	DF	PARAMETER ESTIMATE	STANDARD ERROR	T FOR HO: PARAMETER=0	PROB > |T|
INTERCEP	1	0.52906977	0.53691888	0.985	0.3456
ADDED	1	0.82187597	0.02424601	33.897	0.0001

TEST: B1EQ1	NUMERATOR	78.7103	DF:	1	F VALUE:	53.9715
	DENOMINATOR	1.45837	DF:	11	PROB >F:	0.0001

Solution

a. The computed value of t from Example 10.7 was $t = -7.42$ and hence $t^2 = 55.06$. Except for rounding errors, in the hand computation of t, this is equal to the value of the F statistic, 53.9715.

b. The only difference between the analysis of variance table shown in the output and the one that we have constructed here using the format of this section is that the regression source of variability is designated by MODEL and the degrees of freedom column precedes the SS column in the output. The reconstructed analysis of variance table is shown here.

Source	SS	df	MS	F	p
Regression	1,675.71027	1	1,675.71027	1,149.032	.0001
Error	16.04203876	11	1.45836716		
Total	1,691.75231	12			

▲

A variation on the inferences discussed so far arises when the experimenter knows the value of the intercept β_0. For example, in the calibration of an instrument, if known sample values x_i yield corresponding instrument readings y_i, then at least theoretically a value of $x_i = 0$ should give rise to an instrument reading of $y_i = 0$. In this situation, it might be logical to assume that the fitted regression goes through the origin $(0, 0)$; that is to say, $\beta_0 = 0$.

When β_0 is known, we have the following changes in our estimation and test procedure for β_1. The least squares estimate of β_1 is

$$\hat{\beta}_1 = \frac{\sum x_i y_i - \beta_0 \sum x_i}{\sum x_i^2} \quad \left(\text{Note: If } \beta_0 = 0, \ \hat{\beta}_1 = \frac{\sum x_i y_i}{\sum x_i^2} \right)$$

and the standard error of $\hat{\beta}_1$ is $\dfrac{s_\varepsilon}{\sum x_i^2}$, where s_ε^2 is now computed as

$$s_\varepsilon^2 = \frac{\sum (y_i - \hat{y}_i)^2}{n - 1}.$$

The test procedure for β_1 when β_0 is known is as shown here.

Statistical Test for β_1, Where Intercept (β_0) Is Known

▼

H_0: $\beta_1 = 0$

H_a: **1.** $\beta_1 > 0$
      2. $\beta_1 < 0$
      3. $\beta_1 \neq 0$

T.S.: $t = \dfrac{\hat{\beta}_1}{s_\varepsilon^2 / \sqrt{\sum x_i^2}}$,

where

$$\hat{\beta}_1 = \frac{\sum x_i y_i - \beta_0 \sum x_i}{\sum x_i^2}$$

and

$$s_\varepsilon^2 = \frac{\sum (y_i - \hat{y}_i)^2}{n - 1}$$

R.R.: For a given value of α and df $= n - 1$,

      1. reject H_0 if $t > t_\alpha$
      2. reject H_0 if $t < -t_\alpha$
      3. reject H_0 if $|t| > t_{\alpha/2}$.

In this section, we have developed inferences about β_0 and β_1 in parallel. However, in most situations the slope is the more important parameter. Some statistical software packages (e.g., SPSS) even omit the standard error for the intercept. Also we will not be presenting inferences concerning the correlation coefficient since r is a multiple of $\hat{\beta}_1$. (Recall that $r = \hat{\beta}_1 \sqrt{S_{xx}/S_{yy}}$.) Confidence limits for ρ can be obtained by multiplying the limits for β_1 by $\sqrt{S_{xx}/S_{yy}}$. Similarly, a test of H_0: $\rho = 0$ (no linear correlation between x and y) is equivalent to a test of H_0: $\beta_1 = 0$ (no linear relation between x and y).

▼ **EXERCISES**

Basic Techniques

10.1   Consider the data in the accompanying table.

    a. Compute $\sum x$, $\sum y$, $\sum x^2$, $\sum y^2$, and $\sum xy$.
    b. Use these computations to find estimates of β_0, β_1, σ_ε^2, and σ_ε.
    c. Give the standard errors for $\hat{\beta}_0$ and $\hat{\beta}_1$.

x	y
1	9
1	10
2	10
2	11
3	12
3	12
4	11
4	13
5	14
5	15

10.2 Refer to Exercise 10.1.
   a. Give all parts for a two-tailed statistical test of $H_0: \beta_1 = 0$ based on $\alpha = .05$.
   b. Conduct the test and draw a conclusion.
   c. Give the p-value for the test in part b.

10.3 Recall in Exercise 9.12 that we were interested in examining the relationship between the different concentrations of pectin (0%, 1.5%, and 3% by weight) on the firmness of canned sweet potatoes after storage in a controlled 25°C environment. The sample data for six cans are repeated here.

y (firmness)	50.5	46.8	62.3	67.7	80.1	79.2
x (concentration of pectin)	0	0	1.5	1.5	3.0	3.0

   a. Obtain the least squares estimates for the parameters in the model $y = \beta_0 + \beta_1 x + \varepsilon$.
   b. Obtain an estimate of σ_ε^2.
   c. Give the standard error of $\hat{\beta}_1$.

10.4 Refer to Exercise 10.3. Perform a statistical test of the null hypothesis that there is no linear relationship between the concentration of pectin and the firmness of canned sweet potatoes after thirty days of storage at 25°C. Give the p-value for this test and draw conclusions.

 10.5 The extent of disease transmission can be affected greatly by the viability of infectious organisms suspended in the air. Because of the infectious nature of the disease under study, the viability of these organisms must be studied in an airtight chamber. One way to do this is to disperse an aerosol cloud, prepared from a solution containing the organisms, into the chamber. The biological recovery at any particular time is the percentage of the total number of organisms suspended in the aerosol that are viable. The data in the accompanying table are the biological recovery percentages computed from 13 different aerosol clouds. For each of the clouds, recovery percentages were determined at different times.
   a. Plot the data.
   b. Since there is some curvature, try to linearize the data using the log of the biological recovery.

Cloud	Time, x (in minutes)	Biological Recovery (%)
1	0	70.6
2	5	52.0
3	10	33.4
4	15	22.0
5	20	18.3
6	25	15.1
7	30	13.0
8	35	10.0
9	40	9.1
10	45	8.3
11	50	7.9
12	55	7.7
13	60	7.7

10.6 Refer to Exercise 10.5.
 a. Fit the linear regression model $y = \beta_0 + \beta_1 x + \varepsilon$, where y is the log biological recovery.
 b. Compute an estimate of σ_ε.
 c. Identify the standard errors of $\hat{\beta}_0$ and $\hat{\beta}_1$.

10.7 Refer to Exercise 10.5. Conduct a test of the null hypothesis that $\beta_1 = 0$. Use $\alpha = .05$.

10.8 Refer to Exercise 10.5. Place a 95% confidence interval on β_0, the mean log biological recovery percentage at time zero. Interpret your findings. (Note: $E(y) = \beta_0$ when $x = 0$.)

10.9 An experiment was conducted to examine the relationship between the weight gain of chickens whose diets had been supplemented by different amounts of amino acid lysine and the amount of lysine ingested. Since the percentage of lysine is known, and we can monitor the amount of feed consumed, we can determine the amount of lysine eaten. A random sample of twelve 2-week-old chickens was selected for the study. Each was caged separately and was allowed to eat at will from feed composed of a base supplemented with lysine. The sample data summarizing weight gains and amounts of lysine eaten over the test period are given here. (In the data, y represents weight gain in grams, and x represents the amount of lysine ingested in grams.)
 a. Plot the data in a scatter diagram. Does a linear model seem appropriate?
 b. Fit the linear regression model $y = \beta_0 + \beta_1 x + \varepsilon$.

Chick	y	x	Chick	y	x
1	14.7	.09	7	17.2	.11
2	17.8	.14	8	18.7	.19
3	19.6	.18	9	20.2	.23
4	18.4	.15	10	16.0	.13
5	20.5	.16	11	17.8	.17
6	21.1	.23	12	19.4	.21

10.10 Refer to Exercise 10.9.
 a. Compute an estimate of σ_ε^2.
 b. Identify the standard error of $\hat{\beta}_1$.

c. Conduct a statistical test of the research hypothesis that for this diet preparation and length of study, there is a direct (positive) linear relationship between weight gain and the amount of lysine eaten.

10.11 Refer to Exercises 10.9 and 10.10.

a. For this example, would it make sense to give any physical interpretation to β_0? (Hint: The lysine was mixed in the feed.)

b. Consider an alternative model relating weight gain to amount of lysine ingested:

$$y = \beta_1 x + \varepsilon.$$

Distinguish between this model and the model $y = \beta_0 + \beta_1 x + \varepsilon$.

10.12 **a.** Refer to part b of Exercise 10.11. Compute $\hat{\beta}_1$ for the model $y = \beta_1 x + \varepsilon$, where

$$\hat{\beta}_1 = \frac{\sum xy}{\sum x^2}.$$

b. Which of the two models, $y = \beta_0 + \beta_1 x + \varepsilon$ or $y = \beta_1 x + \varepsilon$, appears to give a better fit to the sample data? (Hint: Plot the two prediction equations on a graph of the sample observations.)

10.13 Refer to Example 10.7. Use the sample data to construct a 95% confidence interval for the intercept β_0. Does the interval include 0 as a possible value? Does the validation experiment conform to theory in regard to the intercept?

10.14 Refer to Example 10.7. Compute the mean percent (of the added amount) received for each level of x. Plot these data. If there is a problem with the use of the assay at this laboratory, at what ranges of x does this occur?

10.15 Interest rates charged for home mortgages have, in general, declined over recent months. With the apparent favorable influence for new home building, the data shown here are the prevailing mortgage interest rates and the number of housing starts in a midwestern city over a period of 18 months.

Month	Interest Rate x	Number of Housing Starts y
1	10.5	360
2	10.3	340
3	10.6	370
4	11.4	360
5	11.8	330
6	11.3	300
7	11.0	290
8	10.5	340
9	10.2	360
10	10.0	370
11	9.8	380
12	9.8	390
13	9.9	375
14	10.0	350
15	10.0	345
16	9.9	360
17	9.8	380
18	9.7	395

a. Plot the data.

b. Use these data to obtain a linear regression equation.

c. Is the slope significantly different from 0?

10.16 Refer to Exercise 10.15. Predict the number of housing starts for interest rates of 10.2% and 9.5%. Do you predict that the prevailing interest rate will increase or decrease next month (month 19)?

10.17 Research in dentistry over the past 20 years has indicated that plaque from different locations in the mouth can differ in chemical composition. Since the quantity of plaque at a given site might be quite small, it is necessary to have a sensitive procedure in order to study the chemical composition of plaque. One such procedure relates the DNA content (one important chemical component) of plaque to the weight of plaque.

In order to study the relationship between weight of plaque and DNA content, ten male volunteers (ages 18–20) were selected at random from a group of volunteers. Over a 4-day period each person consumed his normal diet supplemented by 30 grams of sucrose per day. No tooth brushing was allowed. The 4-day accumulation of plaque for each person was weighed and analyzed for DNA content. These sample data are summarized in the accompanying table.

Person	Plaque Weight (mg) x	DNA (μg) y
1	42.7	260
2	52.3	303
3	24.6	175
4	33.4	214
5	41.8	226
6	36.7	246
7	27.0	181
8	47.3	251
9	31.4	154
10	33.9	247

a. Graph these data in a scatter plot.

b. Use the computer output shown here to determine the least squares regression line.

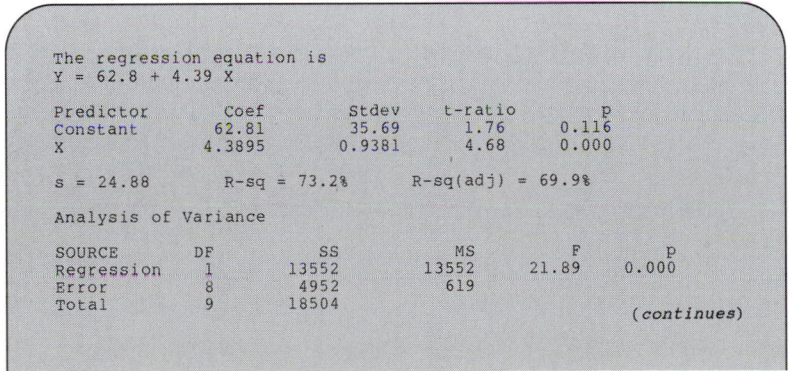

```
The regression equation is
Y = 62.8 + 4.39 X

Predictor      Coef      Stdev     t-ratio      p
Constant      62.81      35.69      1.76      0.116
X            4.3895     0.9381      4.68      0.000

s = 24.88      R-sq = 73.2%      R-sq(adj) = 69.9%

Analysis of Variance

SOURCE        DF        SS         MS        F         p
Regression     1       13552      13552     21.89     0.000
Error          8        4952        619
Total          9       18504
                                        (continues)
```

```
Obs.      X          Y        Fit Stdev.Fit   Residual   St.Resid
  1     42.7     260.00     250.24     9.46       9.76       0.42
  2     52.3     303.00     292.38    16.28      10.62       0.56
  3     24.6     175.00     170.79    14.13       4.21       0.21
  4     33.4     214.00     209.42     8.60       4.58       0.20
  5     41.8     226.00     246.29     9.01     -20.29      -0.87
  6     36.7     246.00     223.90     7.88      22.10       0.94
  7     27.0     181.00     181.32    12.32      -0.32      -0.01
  8     47.3     251.00     270.43    12.38     -19.43      -0.90
  9     31.4     154.00     200.64     9.52     -46.64      -2.03R
 10     33.9     247.00     211.61     8.42      35.39       1.51

R denotes an obs. with a large st. resid.
```

10.18 Refer to Exercise 10.17. If the least squares prediction equation is $\hat{y} = 62.81 + 4.39x$, test to see if there is a significant linear relationship between the DNA content and the plaque weight. That is, use the sample data to test whether the population slope β_1 differs from 0. Use $\alpha = .05$. (Note: $S_{yy} = 18,504.1$, $S_{xy} = 3,087.43$, and $S_{xx} = 703.37$.)

10.19 Use the output of Exercise 10.17 to compare the results of your test in Exercise 10.18.

10.3 ▼ QUICK, PORTABLE STATISTICS (optional)

One portable statistic for approximating a linear regression equation was discussed in Chapter 9. There is a portable statistical test that is useful for ascertaining whether there is an association between two variables.

Quadrant Sum Test for Association

H_0: The two variables are not correlated.
H_a: The two variables are correlated (a two-tailed test).

Test Procedure

1. Plot the data using a scatter diagram.
2. Draw a median line parallel to each axis. The median line parallel to the vertical axis will designate the median (midpoint) value of the variable plotted along the horizontal axis. Similarly, the median line parallel to the horizontal axis will correspond to the median of the variable plotted along the vertical axis (see Figure 10.2).
3. Beginning in the upper-right quadrant and moving counterclockwise, label the quadrants +, −, +, −, respectively (see Figure 10.2).

(continues)

4. Obtain the following four counts:
    a. Beginning from the right side of the scatter diagram and moving toward the left along the horizontal median line, count all observations (dots) that are on the same side of the horizontal median line. Stop counting when you encounter the first observation on the other side of the horizontal median line. Attach the sign of the quadrant to this count.
    b. Beginning from the top of the scatter diagram and moving down along the vertical median line, count all observations that are on the same side of the median line. Stop counting when you encounter the first observation on the other side of the vertical median line. Attach the sign of the quadrant to this count.
    c. Repeat step a, moving from left to right.
    d. Repeat step b, moving from bottom to top.
5. Let C denote the sum of the counts (with their appropriate signs) obtained in part 4.
6. For $\alpha = .10$, .05, or .01, reject H_0 if $|C| \geq 9$, 11, or 13, respectively.
7. If C is positive, the two variables are positively correlated; if C is negative, the two variables are negatively correlated.

FIGURE 10.2

Locating the Median Lines and Quadrants for the Quadrant Sum Test

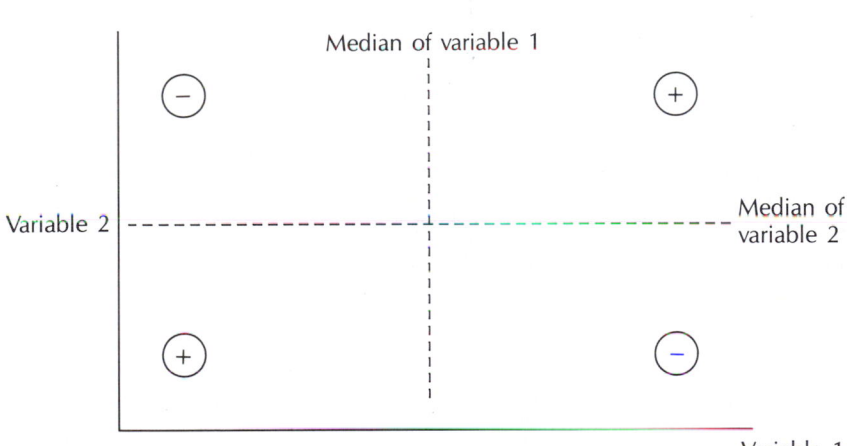

EXAMPLE 10.9 ▼

Recall that in Example 9.3 we had data on the age of volunteers and their reductions in cholesterol counts after 4 weeks on a low–cholesterol diet. For convenience, the data are shown in Table 10.3.

Age (years)	Reduction in Cholesterol	Age (years)	Reduction in Cholesterol
45	30	31	40
43	52	26	17
46	45	22	28
49	38	58	44
50	62	60	61
37	55	52	58
34	25	27	45
30	30		

Use the quadrant sum test with $\alpha = .05$ to determine whether there is a relationship between the age of a volunteer and the 4-week cholesterol reduction.

Solution   We must first construct a scatter diagram. This is shown in Figure 10.3. The four quadrants are labeled $+$, $-$, $+$, $-$, going counterclockwise from the upper right. Using the age and cholesterol count data in Table 10.3, we find the median age to be 43 and the median cholesterol count to be 44. The median lines have been drawn on the scatter diagram by using dotted lines.

1. Beginning at the extreme right, we count observations until we must cross the horizontal median to count the next observation (dot). There are four observations (see Figure 10.3). We attach a plus sign to this count since the four observations were in the upper-right quadrant.

F I G U R E 10.3

**Scatter Diagram for
Example 10.9**

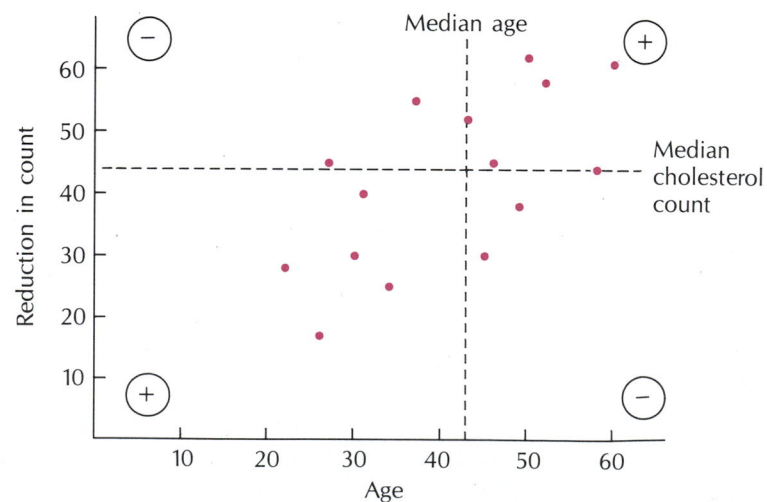

2. Beginning from the top, we count three observations before we must cross the vertical median line. Again we assign a plus sign to these three measurements.
3. Beginning from the extreme left, we count two observations before we must cross the horizontal median line. Since this observation was in the lower-left quadrant, it receives a plus sign.
4. Similarly, from the bottom we count four observations before we must cross the vertical median line. This count is assigned a plus sign because the measurements are in the lower-left quadrant.
5. The combined count (with appropriate sign) is

$$C = 4 + 3 + 2 + 4 = 13.$$

Since $|C| > 11$, we reject the null hypothesis, at the $\alpha = .05$ level, that the variables are uncorrelated. The sign of C indicates that there is a positive correlation between the age and cholesterol count reduction observed for persons treated with this diet. ▲

▼ EXERCISES

10.20 Refer to the data of Exercise 10.17. Use the quadrant sum test to determine if there is a significant association between plaque weight and DNA concentration.

10.21 Refer to Exercise 10.20. Do the data appear to be linear? How might you examine the linear relationship?

 10.22 The thermal pollution of automobiles was studied for 1987 and 1988 models. The data relating automobile weight to Btu (in 1,000s) per vehicle mile are shown in the accompanying table.
a. Plot the data in a scatter diagram.
b. Is there evidence for an association based on the quadrant sum test?

Weight, x (in 1,000 pounds)	Btu per Vehicle Mile, y (in 1,000s)
1.8	4
2.6	5.2
4.2	8.5
5.0	11.6
4.8	10.1
3.4	6.3

10.23 Refer to Exercise 10.22.
a. Use the approximate regression equation procedure of Section 9.3 to obtain an approximate fit to the data.

b. Compare the approximate prediction equation to the actual least squares prediction equation.

c. Plot both equations (the approximate prediction equation and the least squares line on your graph of part a of Exercise 10.22).

10.4 ▼ INFERENCES CONCERNING $E(y)$

The methods of previous sections can be expanded to include inferences concerning the average (expected) value of y for a given setting of the independent variable. For example, in evaluating the effects of different levels of advertising expenditure x on sales y, it may be of interest to estimate the average sales per month for a given level of expenditure x. The estimate of $E(y)$ for a specific setting of x can be obtained by evaluating the prediction equation

$$\hat{y} = \hat{\beta}_0 + \hat{\beta}_1 x$$

at that setting. It can be shown that in repeated sampling at a particular setting of x, the sampling distribution of $\hat{y}$ has a mean

$$E(y) = \beta_0 + \beta_1 x$$

and a variance given by

$$V(\hat{y}) = \sigma_\varepsilon^2 \left(\frac{1}{n} + \frac{(x - \bar{x})^2}{S_{xx}} \right).$$

Again assuming that the ε_is are normally distributed, a $100(1 - \alpha)\%$ confidence interval for $E(y)$ is given by the following formula.

$100(1 - \alpha)\%$ Confidence Interval for $E(y)$

$$\hat{y} \pm t_{\alpha/2} s_\varepsilon \sqrt{\frac{1}{n} + \frac{(x - \bar{x})^2}{S_{xx}}},$$

where

$$s_\varepsilon^2 = \frac{SSE}{n - 2}$$

and the t-value is based on df $= n - 2$.

E X A M P L E   10.10   ▼

Use the data of Example 10.2 to give a 90% confidence interval for the mean corn yield when 5 pounds of fertilizer are applied to a plot.

Solution   The prediction equation in Example 10.2 was

$$\hat{y} = 10.10 + 1.15x,$$

where x = fertilizer applied. For our example, we need $x = 5$ so that

$$\hat{y} = 10.10 + 1.15(5) = 15.85.$$

The variance of $\hat{y}$ can be computed by using $S_{xx} = 20$, $s_\varepsilon = 0.84$, $\bar{x} = 4$, and $n = 10$. The t-value in Table 4 in the Appendix for $a = .05$ and df $= n - 2 = 8$ is 1.86. Hence, the appropriate confidence interval for the average corn yield per plot when 5 pounds of fertilizer are applied is

$$15.85 \pm 1.86(.84)\sqrt{\frac{1}{10} + \frac{(5-4)^2}{20}}, \qquad \text{or} \qquad 15.85 \pm .61,$$

that is, 15.24 to 16.46 bushels.   ▲

E X A M P L E   10.11   ▼

In Example 10.10, we constructed a 90% confidence interval for the mean corn yield when 5 pounds of fertilizer are applied. Use the same sample data to construct a 90% confidence interval on $E(y)$ for any specific value of fertilizer in the range from 2 to 6. Graph your results.

Solution   Using the results from Example 10.10, $\hat{y} = 10.10 + 1.15x$, $s_\varepsilon = .84$, and

$$\sqrt{\frac{1}{n} + \frac{(x - \bar{x})^2}{S_{xx}}} = \sqrt{.1 + \frac{(x-4)^2}{20}}.$$

Our 90% confidence interval for $E(y)$, then, is of the form

$$\hat{y} \pm 1.86(.84)\sqrt{.1 + \frac{(x-4)^2}{20}}.$$

All we need to do is substitute a specific value of x in this form to determine a confidence interval. For fertilizer settings of 2, 3, 4, 5, and 6 the 90% confidence limits are given next.

x	90% Confidence Interval
2	11.544 to 13.256
3	12.945 to 14.155
4	14.206 to 15.194
5	15.245 to 16.455
6	16.144 to 17.856

confidence bands

Plotting the endpoints of the confidence intervals and connecting the points, we get the general 90% confidence interval on $E(y)$ for any value of x between 2 and 6. The graph in Figure 10.4 displays 90% **confidence bands** for $E(y)$. Notice how the width of the confidence interval (the vertical distance between the two curves of the graph) varies for different values of x. $E(y)$ can be estimated more precisely for values of x in the center of the experimental region. The widening of the gap between the bands at the extremities of the experimental region indicates that it would be unwise to extrapolate (try to estimate $E(y)$) beyond the region of experimentation.

FIGURE 10.4

90% Confidence Band for $E(y)$ Using the Data of Example 10.10

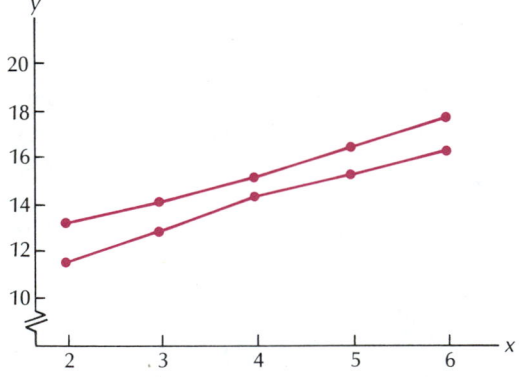

A statistical test concerning $E(y)$ for a given setting of the independent variable in linear regression can also be formulated using the test setup shown next. Thus, for example, we might wish to test that the mean corn yield per plot is 16 when 6 pounds of fertilizer are applied.

Test of a Hypothesis Concerning $E(y)$

H_0: $E(y) = \mu_0$

H_a: **1.** $E(y) > \mu_0$
          2. $E(y) < \mu_0$
          3. $E(y) \neq \mu_0$

T.S.: $t = \dfrac{\hat{y} - \mu_0}{s_\varepsilon \sqrt{\dfrac{1}{n} + \dfrac{(x - \bar{x})^2}{S_{xx}}}}$

R.R.:   For a general value of α and df $= n - 2$,

    1.   reject H_0 if $t > t_\alpha$
    2.   reject H_0 if $t < -t_\alpha$
    3.   reject H_0 if $|t| > t_{\alpha/2}$.

E X A M P L E   10.12   ▼

An experiment was run to examine the rate of growth of a particular type of bacteria. The growth y was determined for two different cultures at five equally spaced time intervals (1, 2, 3, 4, and 5 hours past culture seeding).

	Time (hours)				
Growth Rate, y	1	2	3	4	5
Culture 1	8.0	9.0	9.1	10.2	10.4
Culture 2	8.5	9.2	9.3	9.8	10.1

Use the computer output shown here to answer the following questions.
a. Determine the least squares fit to the linear regression model $y = \beta_0 + \beta_1 x + \varepsilon$.
b. Conduct a test of H_0: $\beta_1 = 0$. Give a p-value for the test.
c. Conduct a test of $E(y) = 9.5$ where $x = 3.5$. Use $\alpha = .05$.

Solution
a. The linear regression equation is $\hat{y} = 7.89 + 0.49x$.
b. The computed value of t for H_0: $\beta_1 = 0$ is shown as 8.42; the corresponding two-sided p-value is $p < .0001$.

c. Although the result of a statistical test of $H_0: E(y) = 9.5$ is not given directly, we can use the 95% confidence interval for $E(y)$ shown under observation eleven ($x = 3.5$) to reach a conclusion. Since the confidence limits $9.40 - 9.81$ include the value 9.5, we have insufficient evidence to reject $H_0: E(y) = 9.5$.   ▲

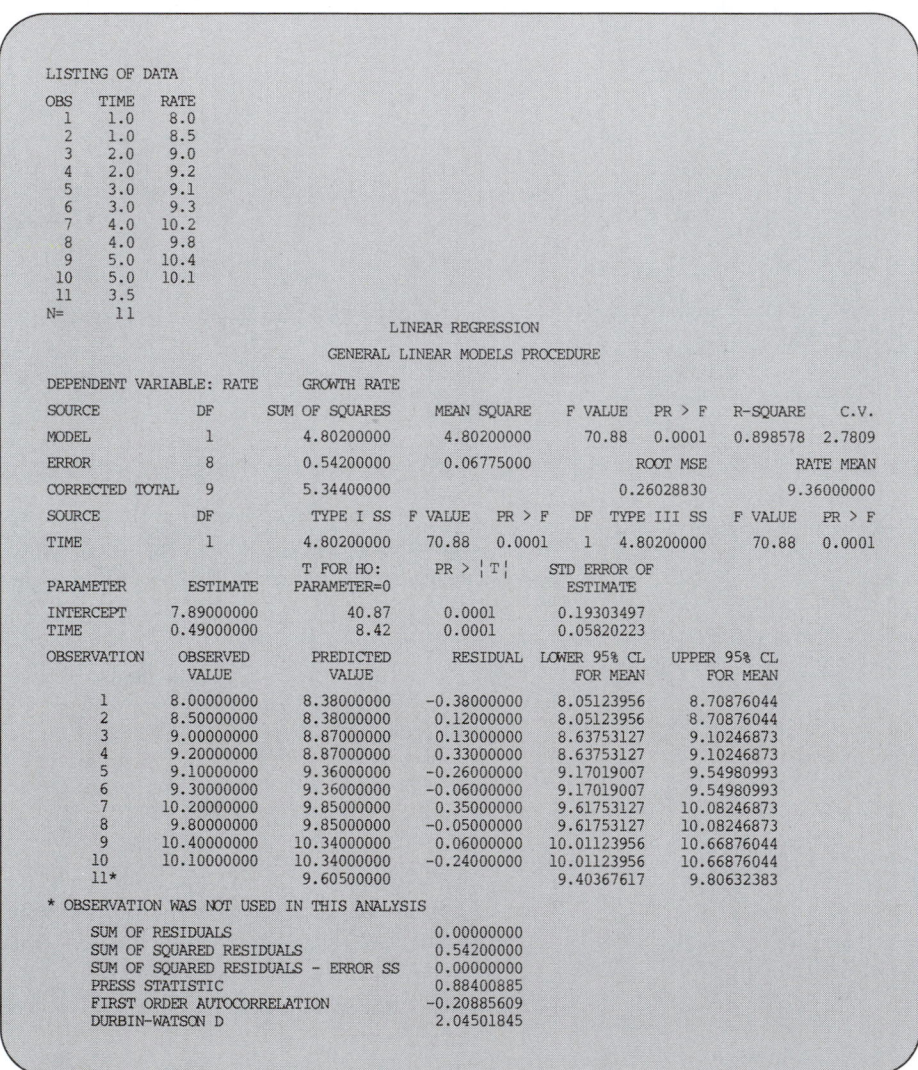

```
LISTING OF DATA

OBS    TIME    RATE
 1     1.0     8.0
 2     1.0     8.5
 3     2.0     9.0
 4     2.0     9.2
 5     3.0     9.1
 6     3.0     9.3
 7     4.0    10.2
 8     4.0     9.8
 9     5.0    10.4
10     5.0    10.1
11     3.5
N=     11
```

LINEAR REGRESSION

GENERAL LINEAR MODELS PROCEDURE

DEPENDENT VARIABLE: RATE   GROWTH RATE

SOURCE	DF	SUM OF SQUARES	MEAN SQUARE	F VALUE	PR > F	R-SQUARE	C.V.
MODEL	1	4.80200000	4.80200000	70.88	0.0001	0.898578	2.7809
ERROR	8	0.54200000	0.06775000		ROOT MSE		RATE MEAN
CORRECTED TOTAL	9	5.34400000			0.26028830		9.36000000

SOURCE	DF	TYPE I SS	F VALUE	PR > F	DF	TYPE III SS	F VALUE	PR > F
TIME	1	4.80200000	70.88	0.0001	1	4.80200000	70.88	0.0001

PARAMETER	ESTIMATE	T FOR H0: PARAMETER=0	PR > |T|	STD ERROR OF ESTIMATE
INTERCEPT	7.89000000	40.87	0.0001	0.19303497
TIME	0.49000000	8.42	0.0001	0.05820223

OBSERVATION	OBSERVED VALUE	PREDICTED VALUE	RESIDUAL	LOWER 95% CL FOR MEAN	UPPER 95% CL FOR MEAN
1	8.00000000	8.38000000	-0.38000000	8.05123956	8.70876044
2	8.50000000	8.38000000	0.12000000	8.05123956	8.70876044
3	9.00000000	8.87000000	0.13000000	8.63753127	9.10246873
4	9.20000000	8.87000000	0.33000000	8.63753127	9.10246873
5	9.10000000	9.36000000	-0.26000000	9.17019007	9.54980993
6	9.30000000	9.36000000	-0.06000000	9.17019007	9.54980993
7	10.20000000	9.85000000	0.35000000	9.61753127	10.08246873
8	9.80000000	9.85000000	-0.05000000	9.61753127	10.08246873
9	10.40000000	10.34000000	0.06000000	10.01123956	10.66876044
10	10.10000000	10.34000000	-0.24000000	10.01123956	10.66876044
11*		9.60500000		9.40367617	9.80632383

* OBSERVATION WAS NOT USED IN THIS ANALYSIS

```
SUM OF RESIDUALS                           0.00000000
SUM OF SQUARED RESIDUALS                    0.54200000
SUM OF SQUARED RESIDUALS - ERROR SS         0.00000000
PRESS STATISTIC                             0.88400885
FIRST ORDER AUTOCORRELATION               -0.20885609
DURBIN-WATSON D                             2.04501845
```

E X A M P L E   10.13   ▼

Refer to Example 10.12. Suppose we wanted to give a confidence interval for $E(y)$ when $x = 8.5$. What problem(s) might we encounter?

Solution   Since $x = 8.5$ is well outside the range of experimentation ($1 \leq x \leq 5$), one might have cause for concern due to extrapolation. The assumed linear model seems to fit the data well in the experimental region; it might be completely inappropriate near $x = 8.5$. Hence, any inferences based on the confidence interval for $E(y)$ at $x = 8.5$ would have to be viewed skeptically.   ▲

10.5 ▼ PREDICTING *y* FOR A GIVEN VALUE OF *x*

predict *y*

In Section 10.4, we were concerned with estimating the expected value of y for a given value of x. Suppose, however, that after obtaining a least squares prediction equation for the general linear model, an investigator would like to **predict** the actual value of y (say the next measurement) for a given value of the independent variable x. Note that this problem differs from the problem discussed in the previous section in that we do not want to estimate the average value of y for a given value of x; rather, we wish to predict what a particular observation will be for that same setting of x.

We still use the least squares equation $\hat{y}$ as our predictor, but the corresponding interval about the observation y is called a *prediction interval*. (Prediction intervals are constructed about variables, whereas confidence intervals are constructed about parameters.)

General $100(1 - \alpha)\%$ Prediction Interval

$$\hat{y} \pm t_{\alpha/2} s_\varepsilon \sqrt{1 + \frac{1}{n} + \frac{(x - \bar{x})^2}{S_{xx}}},$$

where

$$s_\varepsilon^2 = \frac{SSE}{n - 2}$$

and $t_{\alpha/2}$ is based on df $= n - 2$.

Note the similarity between the confidence interval for $E(y)$ and the prediction interval for the variable y. The only difference is that the above prediction interval has a 1 added to the quantity under the square root sign. This makes the interval wider to account for the fact that we're predicting a variable (future value of y) rather than a constant $E(y)$.

EXAMPLE 10.14

▼

Use the data of Example 10.2 to predict the actual crop yield for a plot fertilized with 5 pounds of fertilizer. Place a 90% prediction interval about the actual value of y.

Solution   Using our previous work from Example 10.10, the predicted value of y (using $\hat{y}$) at $x = 5$ is $\hat{y} = 15.85$. Also, for $x = 5$, $\bar{x} = 4$, $n = 10$, and $S_{xx} = 20$,

$$\frac{1}{n} + \frac{(x - \bar{x})^2}{S_{xx}} = .15.$$

The corresponding t-value for $a = .05$ and $df = n - 2 = 8$ is 1.86. Hence, the 90% prediction interval is

$$15.85 \pm 1.86(.84)\sqrt{1 + .15}, \quad \text{or} \quad 15.85 \pm 1.68,$$

that is, 14.17 to 17.53.

Note that the above interval is almost three times wider than the corresponding interval for $E(y)$ of Example 10.10. This is to be expected since here we are placing an interval about a quantity that may vary, whereas in Example 10.10, we were placing an interval about $E(y)$, which cannot vary. Since both intervals are called 90% intervals, the prediction interval must be wider to have the same fraction of intervals (.90) covering y in repeated sampling.   ▲

EXAMPLE 10.15

▼

a. Refer to the data of Example 10.2. Construct a general 90% prediction interval for y when x takes the values 2, 3, 4, 5, and 6.
b. Graph your results to show a 90% prediction band for y when $2 \leq x \leq 6$.
c. On the graph of part b superimpose the graph of the 90% confidence band for $E(y)$.

Solution
a. From previous calculations, $\hat{y} = 10.10 + 1.15x$, $s_\varepsilon = .84$, and

$$\sqrt{1 + \frac{1}{n} + \frac{(x - \bar{x})^2}{S_{xx}}} = \sqrt{1 + \frac{1}{10} + \frac{(x - 4)^2}{20}}.$$

Hence, a general 90% prediction interval is of the form

$$\hat{y} \pm 1.86(.84) \sqrt{1.1 + \frac{(x-4)^2}{20}}.$$

Substituting the values $x = 2, 3, 4, 5,$ and 6 into this form we have the intervals given here.

x	90% Prediction Interval
2	10.618 to 14.182
3	11.874 to 15.226
4	13.061 to 16.339
5	14.174 to 17.526
6	15.218 to 18.782

b,c. Plotting the endpoints of the prediction intervals and connecting the dots, we obtain the 90% prediction bands, shown by the solid lines in Figure 10.5. The dotted lines indicate the corresponding 90% confidence bands for $E(y)$. Notice that the prediction bands are wider than the corresponding confidence bands to allow for the fact that we are predicting the value of a random variable rather than estimating a parameter.

FIGURE 10.5

90% Prediction and Confidence Bands for y and $E(y)$, Example 10.15

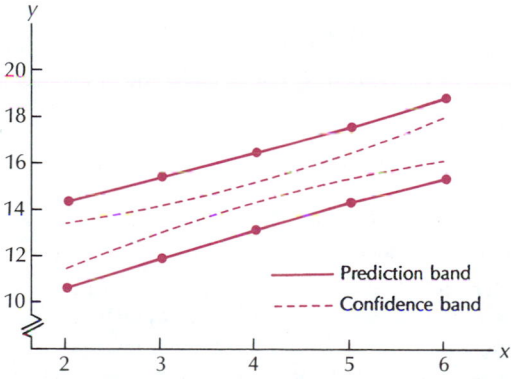

A study of Figure 10.5 and the formulas for these confidence and prediction intervals should suggest factors that may influence the precision of our confidence intervals for $E(y)$ and prediction intervals for y. First, the plus or minus terms in the confidence interval and prediction interval involve $t_{\alpha/2}$, s_ε, n, $(x - \bar{x})^2$, and S_{xx}. Forgetting about $t_{\alpha/2}$ and s_ε for the moment, the width of these intervals will decrease as n increases, $(x - \bar{x})^2$ decreases, and S_{xx} increases. Obviously, if we take more observations, n increases. The quantity $(x - \bar{x})^2$ can be made smaller by making predictions at values of x closer to the mean of the x-values ($\bar{x}$). This is seen in Figure 10.5, where the confidence (and prediction) interval is wider for values of x farther away from the center of the region $\bar{x} = 4$. Finally, we can increase $S_{xx} = \sum (x_i - \bar{x})^2$ and hence improve the confidence and prediction intervals based on our model by increasing the spread of the x-values in our sample. This is certainly an important point when the experimenter has control of the x-values. However, there is a point of diminishing returns. The width of the confidence interval for $E(y)$ and the prediction interval for y are adequate measures of precision *assuming* the model adequately fits the data. If the x-values are spread too far to make S_{xx} large (say, the values $x \neq 1$, 5, 6, 7, and 11 in Example 10.15), a linear regression model may no longer adequately describe the relation between x and y, thus rendering invalid the confidence interval for $E(y)$ and prediction interval for y discussed in this section.

The point is that you should understand the factors affecting the precision of the confidence and prediction intervals based on a linear regression model. However, you should not apply these methods blindly. Plot the data; see how well the model fits the data; plot the confidence and prediction intervals to see the penalty associated with predictions away from $\bar{x}$, and so on.

▼ EXERCISES

Basic Techniques

10.24 Refer to Exercise 10.5. For the least squares equation

$$\hat{y} = \hat{\beta}_0 + \hat{\beta}_1 x,$$

estimate the mean log biological recovery percentage at 30 minutes, using a 95% confidence interval.

10.25 Refer to Exercise 10.9. Estimate the mean weight gain for chickens fed on a diet supplemented with lysine if .19 grams of lysine were ingested over a study period of the same duration. Use a 95% confidence interval.

10.26 Refer to Exercise 10.25. Construct a 95% prediction interval for the weight gain of a chick chosen at random and observed to ingest .19 grams of lysine. Compare your results to the confidence interval of Exercise 10.25.

10.27 Using the data of Exercise 10.24, construct a 95% prediction interval for the log biological recovery percentage at 30 minutes. Compare your result to the confidence interval on $E(y)$ of Exercise 10.24.

Applications

⊕ **10.28** A chemist is interested in determining the weight loss y of a particular compound as a function of the amount of time the compound is exposed to the air. The data in the following table give the weight losses associated with $n = 12$ settings of the independent variable, exposure time.

Weight Loss and Exposure Time Data

Weight Loss, y (in pounds)	Exposure Time (in hours)
4.3	4
5.5	5
6.8	6
8.0	7
4.0	4
5.2	5
6.6	6
7.5	7
2.0	4
4.0	5
5.7	6
6.5	7

a. Find the least squares prediction equation for the model

$$y = \beta_0 + \beta_1 x + \varepsilon.$$

b. Test $H_0: \beta_1 = 0$; give the p-value for $H_a: \beta_1 > 0$ and draw conclusions.

10.29 Refer to Exercise 10.28 and the SAS computer output shown here.

```
        LISTING OF DATA

     OBS      WT_LOSS      TIME

      1        4.3          4
      2        5.5          5
      3        6.8          6
      4        8.0          7
      5        4.0          4
      6        5.2          5
      7        6.6          6
      8        7.5          7
      9        2.0          4
     10        4.0          5
     11        5.7          6
     12        6.5          7

     N=      12
```

(continues)

```
                        LINEAR REGRESSION
                 GENERAL LINEAR MODELS PROCEDURE
DEPENDENT VARIABLE: WT_LOSS    WEIGHT LOSS (LBS)

SOURCE          DF    SUM OF SQUARES    MEAN SQUARE   F VALUE  PR > F   R-SQUARE    C.V.

MODEL           1      26.00416667     26.00416667    40.22   0.0001   0.800888   14.5970

ERROR          10       6.46500000      0.64650000            ROOT MSE          WT_LOSS MEAN

CORRECTED TOTAL 11      32.46916667                          0.80405224          5.50833333

SOURCE          DF     TYPE I SS   F VALUE   PR > F    DF   TYPE III SS  F VALUE  PR > F

TIME            1      26.00416667   40.22   0.0001     1   26.00416667   40.22   0.0001

                                 T FOR HO:      PR > |T|     STD ERROR OF
PARAMETER        ESTIMATE       PARAMETER=0                    ESTIMATE

INTERCEPT       1.73333333         -1.49        0.1677       1.16518239
TIME            1.31666667          6.34        0.0001       0.20760539

OBSERVATION      OBSERVED        PREDICTED      RESIDUAL   LOWER 95% CL    UPPER 95% CL
                  VALUE           VALUE                    INDIVIDUAL      INDIVIDUAL

     1          4.30000000      3.53333333    0.76666667    1.54371634     5.52295033
     2          5.60000000      4.85000000    0.65000000    2.97100515     6.72899485
     3          6.80000000      6.16666667    0.63333333    4.28767181     8.04566152
     4          8.00000000      7.48333333    0.51666667    5.49371634     9.47295033
     5          4.00000000      3.53333333    0.46666667    1.54371634     5.52295033
     6          5.20000000      4.85000000    0.35000000    2.97100515     6.72899485
     7          6.60000000      6.16666667    0.43333333    4.28767181     8.04566152
     8          7.50000000      7.48333333    0.01666667    5.49371634     9.47295033
     9          2.00000000      3.53333333   -1.53333333    1.54371634     5.52295033
    10          4.00000000      4.85000000   -0.85000000    2.97100515     6.72899485
    11          5.70000000      6.16666667   -0.46666667    4.28767181     8.04566152
    12          6.50000000      7.48333333   -0.98333333    5.49371634     9.47295033

    SUM OF RESIDUALS                          0.00000000
    SUM OF SQUARED RESIDUALS                  6.46500000
    SUM OF SQUARED RESIDUALS - ERROR SS      -0.00000000
    PRESS STATISTIC                          10.03092643
    FIRST ORDER AUTOCORRELATION               0.60849016
    DURBIN-WATSON D                           0.54253674

                        LINEAR REGRESSION
                 GENERAL LINEAR MODELS PROCEDURE
DEPENDENT VARIABLE: WT_LOSS    WEIGHT LOSS (LBS)

SOURCE          DF   SUM OF SQUARES    MEAN SQUARE   F VALUE  PR > F   R-SQUARE    C.V.

MODEL           1     26.00416667     26.00416667    40.22   0.0001   0.800888   14.5970

ERROR          10      6.46500000      0.64650000            ROOT MSE          WT_LOSS MEAN

CORRECTED TOTAL 11     32.46916667                          0.80405224          5.50833333

SOURCE          DF      TYPE I SS  F VALUE   PR > F    DF   TYPE III SS  F VALUE  PR > F

TIME            1      26.00416667   40.22   0.0001     1   26.00416667   40.22   0.0001

                                 T FOR HO:      PR > |T|     STD ERROR OF
PARAMETER        ESTIMATE       PARAMETER=0                    ESTIMATE

INTERCEPT       1.73333333         -1.49        0.1677       1.16518239
TIME            1.31666667          6.34        0.0001       0.20760539

OBSERVATION      OBSERVED        PREDICTED      RESIDUAL   LOWER 95% CL    UPPER 95% CL
                  VALUE           VALUE                    FOR MEAN        FOR MEAN

     1          4.30000000      3.53333333    0.76666667    2.66793184     4.39873483
     2          5.50000000      4.85000000    0.65000000    4.28346174     5.41653826
     3          6.80000000      6.16666667    0.63333333    5.60012840     6.73320493
     4          6.00000000      7.48333333    0.51666667    6.61793184     6.34873483
     5          4.00000000      3.53333333    0.46666667    2.66793184     4.39873483
     6          5.20000000      4.85000000    0.35000000    4.28346174     5.41653826
     7          6.60000000      6.16666667    0.43333333    5.60012840     6.73320493
     8          7.50000000      7.48333333    0.01666667    6.61793184     6.34873483
     9          2.00000000      3.53333333   -1.53333333    2.66793184     4.39873483
    10          4.00000000      4.85000000   -0.85000000    4.28346174     5.41653826
    11          5.70000000      6.16666667   -0.46666667    5.60012840     6.73320493
    12          6.50000000      7.48333333   -0.98333333    6.61793184     8.34873483

    SUM OF RESIDUALS                          0.00000000
    SUM OF SQUARED RESIDUALS                  6.46500000
    SUM OF SQUARED RESIDUALS - ERROR SS      -0.00000000
    PRESS STATISTIC                          10.03092643
    FIRST ORDER AUTOCORRELATION               0.60849016
    DURBIN-WATSON D                           0.54253674
```

a. Identify the 95% confidence bands for $E(y)$ when $4 \leq x \leq 7$.
b. Identify the 95% prediction bands for y, $4 \leq x \leq 7$.
c. Distinguish between the meaning of the confidence bands and prediction bands in parts a and b.

10.6 ▼ EXAMINING LACK OF FIT IN LINEAR REGRESSION

In our study of linear regression, we have been concerned with how well a linear regression model $y = \beta_0 + \beta_1 x + \varepsilon$ fits but only from an intuitive standpoint. We could examine a scatterplot of the data to see if it looked linear and we could test whether the slope differed from 0; however, we had no way of testing to see if a higher-order model would be a more appropriate model for the relationship between y and x. This section will outline situations in which we can test for the validity of a linear regression model.

Pictures (or graphs) are always a good starting point for examining lack of fit. First use a scatterplot of y versus x. Second, a plot of residuals $y_i - \hat{y}_i$ versus predicted values $\hat{y}_i$ may give an indication of the following problems:

1. Outliers or erroneous observations. In examining the residual plot, your eye will naturally be drawn to data points with unusually high (in absolute value) residuals.
2. Violation of the assumptions. For the model $y = \beta_0 + \beta_1 x + \varepsilon$, we've assumed a linear relation between y and the dependent variable x, and independent, normally distributed errors with a constant variance.

The residual plot for a model and data set that has none of these apparent problems would look much like the plot in Figure 10.6. Note from this plot there are no extremely large residuals (and hence no apparent outliers) and that there is no trend in the residuals to indicate that the linear model is inappropriate. When a higher-order model is more appropriate, a residual plot more like that shown in Figure 10.7 would be observed.

A check of the constant variance assumption can be addressed in the y versus x scatterplot or with a plot of the residuals $(y_i - \hat{y}_i)$ versus x_i. For example, a pattern of residuals as shown in Figure 10.8 indicates homogeneous error variances across values of x; Figure 10.9 indicates the error variances increase with increasing values of x.

FIGURE 10.6
Residual Plot with No Apparent Pattern

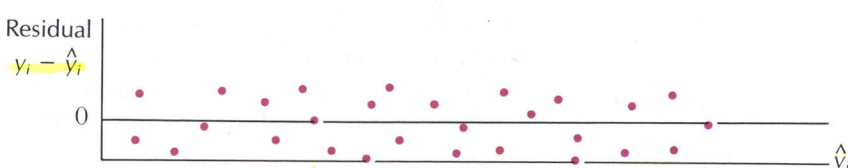

FIGURE 10.7
**Residual Plot Showing the
Need for a Higher-Order
Model**

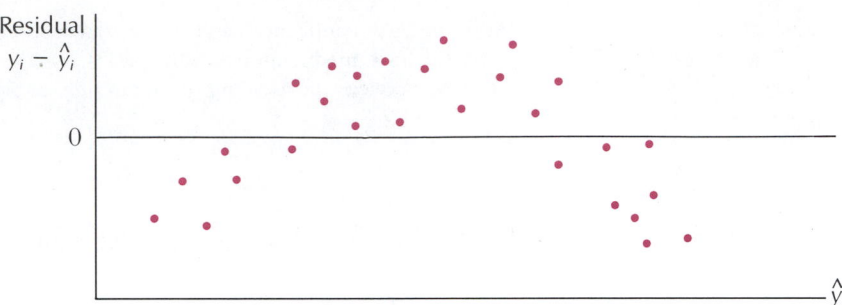

FIGURE 10.8
**Residual Plot Showing
Homogeneous Error
Variances**

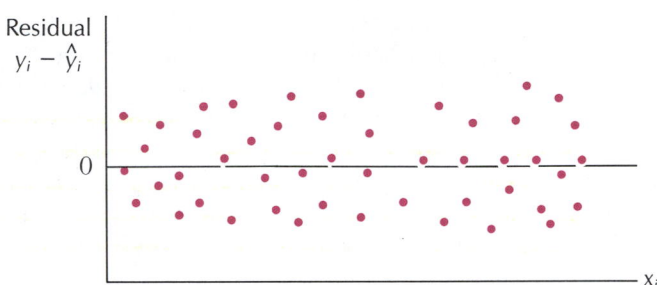

FIGURE 10.9
**Residual Plot Showing
Error Variances
Increasing with x**

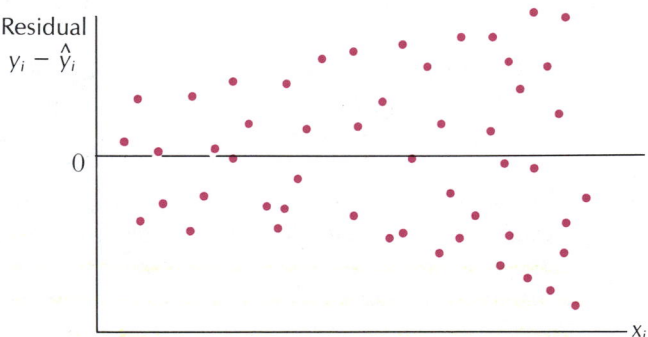

The question of independence of the errors and normality of the errors is addressed later in Chapter 12. We illustrate some of the points we've learned so far about residuals by way of an example.

EXAMPLE 10.16   ▼

The amount of heat loss was examined for a new brand of thermal panes. Random assignment of three different panes was made to each of the three outdoor temperature settings being considered. For each trial the window temperature was controlled at 68°F and 50% relative humidity.

Outdoor Temperature (°F)	Heat Loss
20	86, 80, 77
40	78, 84, 75
60	33, 38, 43

a. Plot the data.

b. Fit the linear regression model $y = \beta_0 + \beta_1 x + \varepsilon$ and test $H_0: \beta_1 = 0$ (give the p-value for your test).

c. Compute $\hat{y}_i$ and $y_i - \hat{y}_i$ for the nine observations. Plot $y_i - \hat{y}_i$ versus $\hat{y}_i$.

d. Does the constant variance assumption seem reasonable?

Solution   The computer output shown here can be used to address the four parts of this example. The student edition of EXECUSTAT produced the following analysis.

```
                    Simple Regression Analysis for HLOSS
..............................................................................
eeeeeeeeeeeeeeeeeeeeeeeeeeeeeeeeeeeeeeeeeeeeeeeeeeeeeeeeeeeeeeeeeeeeeeeeeeeeee
Linear model: loss = 109 - 1.075*temp

                          Table of Estimates
..............................................................................
eeeeeeeeeeeeeeeeeeeeeeeeeeeeeeeeeeeeeeeeeeeeeeeeeeeeeeeeeeeeeeeeeeeeeeeeeeeeee
                                      Standard         t          p
                      Estimate         Error        Value      Value
..............................................................................
eeeeeeeeeeeeeeeeeeeeeeeeeeeeeeeeeeeeeeeeeeeeeeeeeeeeeeeeeeeeeeeeeeeeeeeeeeeeee
Intercept               109           9.9694       10.93      0.0000
Slope                 -1.075          0.230747     -4.66      0.0023
..............................................................................
eeeeeeeeeeeeeeeeeeeeeeeeeeeeeeeeeeeeeeeeeeeeeeeeeeeeeeeeeeeeeeeeeeeeeeeeeeeeee
R-squared = 75.61%
Correlation coeff. = -0.870
Standard error of estimation = 11.3042
Durbin-Watson statistic = 1.27278
Mean absolute error = 8.66667
..............................................................................
eeeeeeeeeeeeeeeeeeeeeeeeeeeeeeeeeeeeeeeeeeeeeeeeeeeeeeeeeeeeeeeeeeeeeeeeeeeeee

                          Table of All Residuals
..............................................................................
eeeeeeeeeeeeeeeeeeeeeeeeeeeeeeeeeeeeeeeeeeeeeeeeeeeeeeeeeeeeeeeeeeeeeeeeeeeeee
                                  Predicted              Studentized
Row         temp        loss        loss      Residual     Residual
..............................................................................
eeeeee eeeeeeeeeeeeeeeeeeeeeeeeeeeeeeeeeeeeeeeeeeeeeeeeeeeeeeeeeeeeeeeeeeeeeeee
1            20          86         87.5        -1.5        -0.14
2            20          80         87.5        -7.5        -0.76
3            20          77         87.5       -10.5        -1.11
4            40          78         66          12          1.15
5            40          84         66          18          2.03
6            40          75         66           9          0.82
7            60          33         44.5       -11.5        -1.24
8            60          38         44.5        -6.5        -0.65
9            60          43         44.5        -1.5        -0.14
..............................................................................
eeeeee eeeeeeeeeeeeeeeeeeeeeeeeeeeeeeeeeeeeeeeeeeeeeeeeeeeeeeeeeeeeeeeeeeeeeeee

                                                            (continues)
```

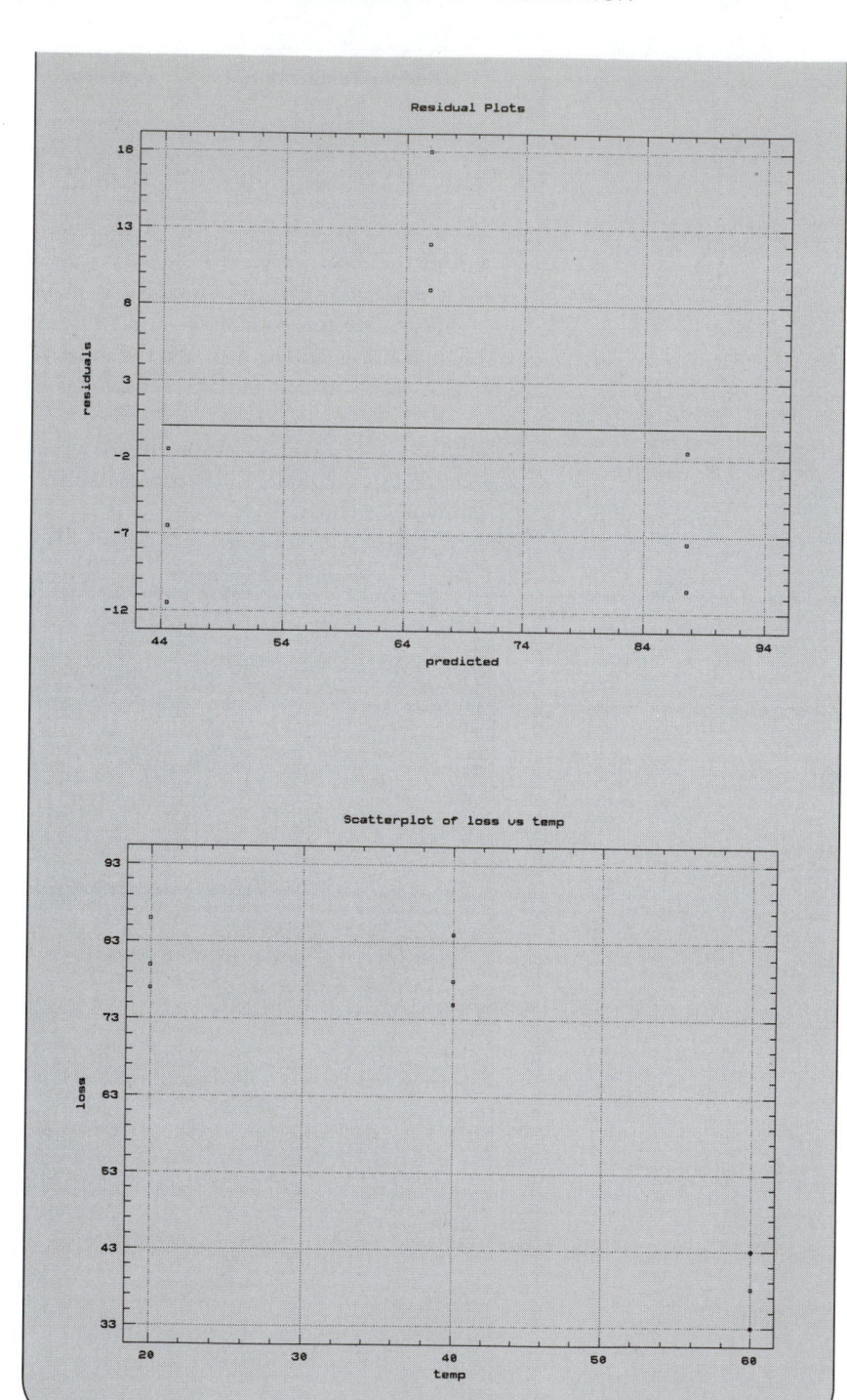

a. The scatterplot of y versus x certainly shows a downward linear trend, and there may be evidence of curvature as well.

b. The linear regression model seems to fit the data well, and the test of $H_0: \beta_1 = 0$ is significant at the $p = .0023$ level. But is this the best model for the data?

c. The plot of residuals $(y_i - \hat{y}_i)$ against the predicted values $\hat{y}_i$ is similar to Figure 10.7, suggesting that we may need additional terms in our model.

d. Since residuals associated with $x = 20$ (the first three), $x = 40$ (the second three), and $x = 60$ (the third three) are easily located, we really do not need a separate plot of residuals versus x to examine the constant variance assumption. It is clear from the original scatterplot and the residual plot shown that we do not have a problem. ▲

How can we test for the apparent lack of fit of the linear regression model in Example 10.16? When there is more than one observation per level of the independent variable, we can conduct a test for lack of fit of the fitted model by partitioning SSE into two parts, one **pure experimental error** and the other **lack of fit**. Let y_{ij} denote the response for the jth observation at the ith level of the independent variable. Then, if there are n_i observations at the ith level of the independent variable, the quantity

pure experimental error
lack of fit

$$\sum_j (y_{ij} - \bar{y}_i)^2$$

provides a measure of what we will call pure experimental error. This sum of squares has $n_i - 1$ degrees of freedom.

Similarly, for each of the other levels of x, we can compute a sum of squares due to pure experimental error. The pooled sum of squares

$$\text{SSP}_{\text{exp}} = \sum_{ij} (y_{ij} - \bar{y}_i)^2$$

called the sum of squares for pure experimental error, has $\sum_i (n_i - 1)$ degrees of freedom. With SS_{Lack} representing the remaining portion of SSE, we have

$$\text{SSE} = \underset{\substack{\text{due to pure} \\ \text{experimental} \\ \text{error}}}{\text{SSP}_{\text{exp}}} + \underset{\substack{\text{due to lack} \\ \text{to fit}}}{\text{SS}_{\text{Lack}}}$$

If SSE is based on $n - 2$ degrees of freedom in the linear regression model, then SS_{Lack} will have $\text{df} = n - 2 - \sum_i (n_i - 1)$.

Under the null hypothesis that our model is correct, we can form independent estimates of σ_ε^2, the model error variance, by dividing SSP_{exp} and SS_{Lack} by their respective degrees of freedom; these estimates are called **mean squares** and denoted by MSP_{exp} and MS_{Lack}, respectively.

mean squares

The test for lack of fit is summarized here.

A Test for Lack of Fit in Linear Regression

H_0: A linear regression model is appropriate.

H_a: A linear regression model is not appropriate.

T.S.: $F = \dfrac{MS_{Lack}}{MSP_{exp}}$,

where

$$MS_{exp} = \frac{SSP_{exp}}{\sum (n_i - 1)} = \frac{\sum_{ij} (y_{ij} - \bar{y}_i)^2}{\sum_i (n_i - 1)}$$

and

$$MS_{Lack} = \frac{SSE - SSP_{exp}}{n - 2 - \sum (n_i - 1)}$$

R.R.: For specified value of α, reject H_0 (the adequacy of the model) if the computed value of F exceeds the table value for $df_1 = n - 2 - \sum_i (n_i - 1)$ and $df_2 = \sum_i (n_i - 1)$.

Conclusion: If the F test is significant, this indicates that the linear regression model is inadequate. A nonsignificant result indicates that there is insufficient evidence to suggest that the linear regression model is inappropriate.

E X A M P L E   10.17

Refer to the data of Example 10.16. Conduct a test for lack of fit of the linear regression model.

Solution   Using a calculator, it is easy to show that the contributions to experimental error for the differential levels of x are as shown here.

Level of x	$\bar{y}_i$	Contribution to Pure Experimental Error $\sum_i (y_{ij} - \bar{y}_i)^2$	$n_i - 1$
20	81	42	2
40	79	42	2
60	38	50	2
Total		134	6

Summarizing these results, we have

$$SSP_{exp} = \sum_{ij} (y_{ij} - \bar{y}_i)^2 = 134.$$

The output shown for Example 10.16 gives SSE = 894.5; hence, by subtraction,

$$SS_{Lack} = SSE - SSP_{exp} = 894.5 - 134 = 760.5.$$

The sum of squares due to pure experimental error has $\sum_i (n_i - 1) = 6$ degrees of freedom; it therefore follows that with $n = 9$, SS_{Lack} has $n - 2 - \sum_i (n_i - 1) = 1$ degree of freedom. We find that

$$MSP_{exp} = \frac{SSP_{exp}}{6} = \frac{134}{6} = 22.33$$

and

$$MS_{Lack} = \frac{SS_{Lack}}{1} = 760.5.$$

The F statistic for the test of lack of fit is

$$F = \frac{MS_{Lack}}{MSP_{exp}} = \frac{760.5}{22.33} = 34.06.$$

Using $df_1 = 1$, $df_2 = 6$, and $\alpha = .05$ we will reject H_0 if $F \geq 5.99$.

Since the computed value of F exceeds 5.99, we reject H_0 and conclude that there is significant lack of fit for a linear regression model. The scatterplot shown in Example 10.16 confirms this nonlinearity.   ▲

To summarize: In situations where there is more than one y-value at one or more levels of x, it is possible to conduct a formal test for lack of fit of the linear regression model. This test should precede any inferences made using the fitted linear regression line. If the test for lack of fit is significant, some higher-order polynomial in x may be more appropriate. A scatterplot of the data and a residual plot from the linear regression line should help in selecting the appropriate model. More information on the selection of an appropriate model will be discussed along with multiple regression (Chapters 11 and 12).

If the F test for lack of fit is not significant, proceed with inferences based on the fitted linear regression line.

▼ EXERCISES

Applications

 10.30   A manufacturer of laundry detergent was interested in testing a new product prior to market release. One area of concern was the relationship between the height of the detergent suds in a washing machine as a function of the amount of detergent added in the wash cycle. For a standard size washing machine tub filled to the full level, random assignments of amounts of detergent were made and tested on the washing machine. The data appear next.

Height, y	Amount, x
28.1, 27.6	6
32.3, 33.2	7
34.8, 35.0	8
38.2, 39.4	9
43.5, 46.8	10

  a. Plot the data.
  b. Fit a linear regression model.
  c. Use a residual plot to investigate possible lack of fit.

10.31   Refer to Exercise 10.30.
  a. Conduct a test for lack of fit of the linear regression model.
  b. If the model is appropriate, give a 95% prediction band for y.

10.32   Refer to Exercise 9.19. Conduct a test for lack of fit and draw conclusions.

10.7 ▼ THE CALIBRATION PROBLEM: PREDICTING x FOR A GIVEN VALUE OF y

In experimental situations, we are often interested in estimating the value of the independent variable corresponding to a measured value of the dependent variable. This problem will be illustrated for the case in which the dependent variable y is linearly related to an independent variable x.

Consider the calibration of an instrument that measures the flow rate of a chemical process. Let x denote the actual flow rate and y denote a reading on the calibrating instrument. In the calibration experiment, the flow rate is controlled at n levels x_i, and the corresponding instrument readings y_i are observed. Suppose we assume a model of the form

$$y_i = \beta_0 + \beta_1 x_i + \varepsilon_i,$$

where the ε_is are independent identically distributed normal random variables with mean zero and variance σ_ε^2. Then, using the n data points (x_i, y_i), we can obtain the least squares estimates $\hat{\beta}_0$ and $\hat{\beta}_1$. Sometime in the future the experimenter will be interested in estimating the flow rate x from a particular instrument reading y.

The most commonly used estimate is found by replacing $\hat{y}$ by y and solving the least squares equation $\hat{y} = \hat{\beta}_0 + \hat{\beta}_1 x$ for x:

$$\hat{x} = \frac{y - \hat{\beta}_0}{\hat{\beta}_1}.$$

Two different inverse prediction problems will be discussed here. The first is for predicting x corresponding to an *observed* value of y; the second is for predicting x corresponding to the mean of $m > 1$ values of y that were obtained independent of the regression data. The solution to the first inverse problem is shown here.

Case 1: Predicting x Based on an Observed y-Value

Predictor of x: $\hat{x} = \dfrac{y - \hat{\beta}_0}{\hat{\beta}_1}$

$100(1 - \alpha)\%$ prediction limits for x:

$$\hat{x}_U = \bar{x} + \frac{1}{1 - c^2} [(\hat{x} - \bar{x}) + d]$$

$$\hat{x}_L = \bar{x} + \frac{1}{1 - c^2} [(\hat{x} - \bar{x}) - d],$$

(continues)

where

$$d = \frac{t_{\alpha/2} s_\varepsilon}{\hat{\beta}_1} \sqrt{\frac{n+1}{n}(1-c^2) + \frac{(\hat{x}-\bar{x})^2}{S_{xx}}},$$

$$s_\varepsilon^2 = \frac{SSE}{n-2}, \qquad c^2 = \frac{t_{\alpha/2}^2 s_\varepsilon^2}{\hat{\beta}_1^2 S_{xx}},$$

and $t_{\alpha/2}$ is based on df $= n - 2$.

It should be noted that since

$$t = \frac{\hat{\beta}_1}{s_\varepsilon / \sqrt{S_{xx}}}$$

is the test statistic for $H_0: \beta_1 = 0$, $c = t_{\alpha/2}/t$. We will require that $|t| > t_{\alpha/2}$; that is, β_1 must be significantly different from zero. Then $c^2 < 1$ and $0 < (1 - c^2) < 1$. The greater the strength of the linear relationship between x and y, the larger the quantity $(1 - c^2)$, making the width of the prediction interval narrower. Note also that we will get a better prediction of x when $\hat{x}$ is closer to the center of the experimental region, as measured by $\bar{x}$. Combining a prediction at an endpoint of the experimental region with a weak linear relationship between x and y ($t \approx t_{\alpha/2}$ and $c^2 < 1$) can create extremely wide limits for the prediction of x.

E X A M P L E   10.18   ▼

In Example 9.6, an engineer was interested in calibrating a flow-rate meter. The data are shown in Table 10.4.

Use these data to place a 95% prediction interval on x, the actual flow rate corresponding to an instrument reading of 4.0.

Solution   For these data, we found that $S_{xy} = 74.35$, $S_{xx} = 82.5$, and $S_{yy} = 67.065$. It follows that $\hat{\beta}_1 = 74.35/82.5 = .9012$, $\hat{\beta}_0 = \bar{y} - \hat{\beta}_1 \bar{x} = 5.45 - (.9012)(5.5) = .4934$, and SSE $= S_{yy} - \hat{\beta}_1 S_{xy} = 67.065 - (.9012)(74.35) = .0608$. The estimate of σ_ε^2 is based on $n - 2 = 8$ degrees of

TABLE 10.4

Data for the Calibration Problem of Example 10.18

Flow Rate, x	Instrument Reading, y
1	1.4
2	2.3
3	3.1
4	4.2
5	5.1
6	5.8
7	6.8
8	7.6
9	8.7
10	9.5

freedom.

$$s_\varepsilon^2 = \frac{\text{SSE}}{n-2} = \frac{.0608}{8} = .0076$$

$$s_\varepsilon = .0872.$$

For $\alpha = .05$, the t-value for df $= 8$ and $a = .025$ is 2.306.

$$c^2 = \frac{t_{\alpha/2}^2 s_\varepsilon^2}{\hat{\beta}_1^2 S_{xx}} = \frac{(2.306)^2(.0076)}{(.9012)^2(82.5)} = .0006$$

and $1 - c^2 = .9994$. Using $\hat{x} = 3.8910$, the upper and lower prediction limits for x when $y = 4.0$ are as follows:

$$\hat{x}_U = 5.5 + \frac{1}{.9994}\left[-1.6090 + \frac{2.306(.0872)}{.9012}\sqrt{\frac{11}{10}(.9994) + \frac{(-1.6090)^2}{82.5}}\right]$$

$$= 5.5 + \frac{1}{.9994}(-1.6090 + .2373) = 4.1274$$

$$\hat{x}_L = 5.5 + \frac{1}{.9994}(-1.6090 - .2373) = 3.6526.$$

Thus, the 95% prediction limits for x are 3.65 to 4.13. These limits are shown in Figure 10.10.

95% Prediction Interval for x When y = 4.0, Example 10.18

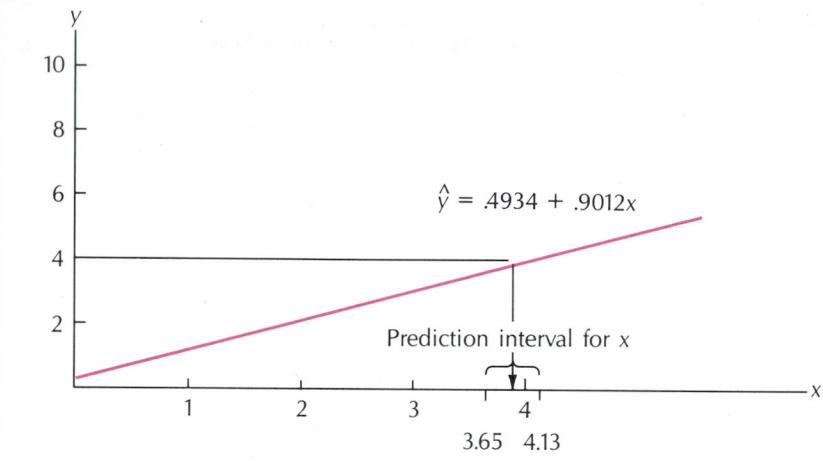

The solution to the second inverse prediction problem is summarized next.

Case 2: Predicting x Based on m y-Values

Predicting the value of x corresponding to $100P\%$ of the mean of m independent y-values. For $0 \le P \le 1$,

Predictor of x: $\hat{x} = \dfrac{P\bar{y}_m - \hat{\beta}_0}{\hat{\beta}_1}$

$$\hat{x}_U = \bar{x} + \frac{1}{1 - c^2}[(\hat{x} - \bar{x}) + g]$$

$$\hat{x}_L = \bar{x} + \frac{1}{1 - c^2}[(\hat{x} - \bar{x}) - g],$$

where

$$g = \frac{t_{\alpha/2}}{\hat{\beta}_1}\sqrt{\left(s_y^2 P^2 + \frac{s_\varepsilon^2}{n}\right)(1 - c^2) + \frac{(\hat{x} - \bar{x})^2 s_\varepsilon^2}{S_{xx}}}$$

and $\bar{y}_m$ and $s_{\bar{y}}$ are the mean and standard error, respectively, of m independent y-values.

▼ EXERCISES

Applications

10.33 A particular forester has become adept at estimating the volume (in cubic feet) of trees on a particular site prior to a timber sale. Since his operation has now expanded, he would like to train another person to assist in estimating the cubic-foot volume of trees. He decides to calibrate his assistant's estimations of actual tree volume. The forester selects a random sample of trees soon to be felled. For each tree, the assistant is to guess the cubic-foot volume y. In addition, the forester obtains the actual cubic-foot volume x after the tree has been chopped down. From these data the forester obtains the calibration curve for the model

$$y = \beta_0 + \beta_1 x + \varepsilon.$$

Then in the near future he can use the calibration curve to correct the assistant's estimates of tree volumes. The sample data are summarized below.

Tree	1	2	3	4	5	6	7	8	9	10
Estimated volume, y	12	14	8	12	17	16	14	14	15	17
Actual volume, x	13	14	9	15	19	20	16	15	17	18

Fit the calibration curve using the method of least squares. Does the evidence indicate that the slope is significantly greater than 0? Use $\alpha = .05$.

10.34 Refer to Exercise 10.33.
  a. Predict the actual tree volume for a tree the assistant estimates to have a cubic-foot volume of 13.
  b. Place a 95% prediction interval on x, the actual tree volume in part a.

10.35 Data from 24 patients were obtained to examine the relationship between dose (amount of drug) and cumulative urine volume for a drug product being studied as a diuretic. These data are shown here in the computer output.
  a. Locate the linear regression equation. Identify the independent and dependent variables.
  b. Use the output to predict dose based on individual y-values of 10, 14, and 19 cm^3. What are the corresponding 99% prediction limits for each of those cases?

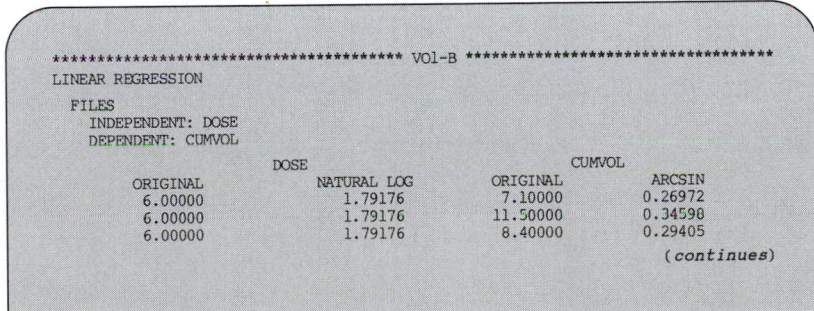

```
******************************************* VO1-B *******************************************
    LINEAR REGRESSION

       FILES
          INDEPENDENT: DOSE
          DEPENDENT: CUMVOL

                          DOSE                              CUMVOL
              ORIGINAL        NATURAL LOG        ORIGINAL            ARCSIN
              6.00000          1.79176           7.10000            0.26972
              6.00000          1.79176          11.50000            0.34598
              6.00000          1.79176           8.40000            0.29405
                                                                 (continues)
```

```
          6.00000                   1.79176              8.00000           0.28676
          6.00000                   1.79176              9.40000           0.31161
          6.00000                   1.79176             12.00000           0.35374
          9.00000                   2.19722             13.20000           0.37183
          9.00000                   2.19722             14.70000           0.39348
          9.00000                   2.19722             12.70000           0.36438
          9.00000                   2.19722             15.50000           0.40465
          9.00000                   2.19722             18.40000           0.44333
          9.00000                   2.19722             14.40000           0.38923
         13.50000                   2.60269             12.10000           0.35528
         13.50000                   2.60269             15.80000           0.40878
         13.50000                   2.60269             13.80000           0.38061
         13.50000                   2.60269             20.40000           0.46863
         13.50000                   2.60269             22.70000           0.49661
         13.50000                   2.60269             17.00000           0.42499
         20.25000                   3.00815             19.80000           0.46114
         20.25000                   3.00815             15.60000           0.40603
         20.25000                   3.00815             25.30000           0.52706
         20.25000                   3.00815             13.50000           0.37624
         20.25000                   3.00815             24.80000           0.52129
         20.25000                   3.00815             20.90000           0.47481
```

```
                      Y           Y            Y           Y          Y
              X      MEAN      VARIANCE       #        MINIMUM    MAXIMUM
         1.79176   0.31031     0.00113        6        0.26972    0.35374
         2.19722   0.39448     0.00079        6        0.36438    0.44333
         2.60269   0.42248     0.00282        6        0.35528    0.49661
         3.00815   0.46109     0.00368        6        0.37624    0.52706
TOTAL              0.39709     0.00210       24
```

ESTIMATED LINE

```
                       ESTIMATE            95% CONFIDENCE LIMITS
SLOPE:                  0.11847            [0.07553, 0.16140]
INTERCEPT:              0.11277            [0.00791, 0.21763]
```

INDEX OF SIGNIFICANCE OF SLOPE: 0.131

CORRELATION COEFFICIENT: 0.77342

TABLE OF RESIDUALS

```
                                                          STANDARDIZED
          X VALUE          OBSERVED Y      PREDICTED Y      RESIDUAL
          1.79176           0.26972          0.32504        -1.20333
          1.79176           0.34598          0.32504         0.45544
          1.79176           0.29405          0.32504        -0.67412
          1.79176           0.28676          0.32504        -0.83269
          1.79176           0.31161          0.32504        -0.29204
          1.79176           0.35374          0.32504         0.62432
          2.19722           0.37183          0.37307        -0.02713
          2.19722           0.39348          0.37307         0.44388
          2.19722           0.36438          0.37307        -0.18910
          2.19722           0.40465          0.37307         0.68689
          2.19722           0.44333          0.37307         1.52821
          2.19722           0.38923          0.37307         0.35134
          2.60269           0.35528          0.42111        -1.43193
          2.60269           0.40878          0.42111        -0.26814
          2.60269           0.38061          0.42111        -0.88100
          2.60269           0.46863          0.42111         1.03361
          2.60269           0.49661          0.42111         1.64217
          2.60269           0.42499          0.42111         0.08438
          3.00815           0.46114          0.46914        -0.17405
          3.00815           0.40603          0.46914        -1.37276
          3.00815           0.52706          0.46914         1.25964
          3.00815           0.37624          0.46914        -2.02086*
          3.00815           0.52129          0.46914         1.13415
          3.00815           0.47481          0.46914         0.12312
```

* RESIDUAL IS GREATER THAN OR EQUAL TO 2 TIMES STANDARD DEVIATION

(continues)

```
PREDICTION OF X
    99% PREDICTION LIMITS FOR INDIVIDUAL X VALUE

    INPUT VALUE        PREDICTION        PREDICTION LIMITS
      10.00000          5.83575*         [2.92557, 7.74469]
      14.00000          9.82768          [7.25149, 12.37486]
      19.00000          17.37828         [13.58904, 29.75220]

  * X VALUE LIES OUTSIDE RANGE OF VALUES IN THE INDEPENDENT FILE

                     --------------------------


PREDICTION OF X
  % OF CONTROL MEAN
  CONTROL VALUES:                             10.00000
                                             20.00000
                                             30.00000
                                             12.00000
     MEAN:                                    0.42969
     VARIANCE:                                0.01369

         %              PREDICTION        95% PREDICTION LIMITS
       75.0             5.86145*          [2.20339, 12.88173]
       50.0             2.36703*          [0.79139, 4.44615]

  * X VALUE LIES OUTSIDE RANGE OF VALUES IN THE INDEPENDENT FILE

                     --------------------------


  ********************************************************************************
```

▼ **10.36** Refer to the output of Exercise 10.35. Suppose the investigator wanted to predict the dose of the diuretic that would produce a response equivalent to 50% (and 75%) of the response obtained from four patients treated with a known diuretic. Predict x and give appropriate limits for each of these situations.

10.8 ▼ SUMMARY

This chapter followed closely the material presented in Chapter 9 for constructing a linear regression equation. Here we showed how to make inferences (using confidence intervals and statistical tests) related to a linear regression model. First, we considered confidence intervals and tests related to β_0 and β_1, the intercept and slope in the linear regression model. Second, we developed a confidence interval and statistical test for $E(y)$, the average (expected) value of y for a given setting of the independent variable x. By considering different values of x, we showed how to generate a confidence band for $E(y)$. Next, we developed a prediction interval and corresponding prediction bands for predicting the actual value (perhaps the next value) of y for a given value of x.

Finally, after these sections on inferences related to linear regression, we presented ways to examine lack of fit of a linear regression model to a data set, ways to linearize the relationship between y and x using transformations, and several quick, portable statistics.

This chapter has provided an important foundation for the further discussions of regression in Chapters 11 and 12.

▼ KEY FORMULAS

1. Means and standard errors for $\hat{\beta}_0$ and $\hat{\beta}_1$

$$\mu_{\hat{\beta}_0} = \beta_0 \qquad\qquad \mu_{\hat{\beta}_1} = \beta_1$$

$$\sigma_{\hat{\beta}_0} = \sigma_\varepsilon \sqrt{\frac{\sum x^2}{nS_{xx}}} \qquad \sigma_{\hat{\beta}_1} = \frac{\sigma_\varepsilon}{\sqrt{S_{xx}}}$$

Note: For all practical problems, σ_ε is unknown and must be estimated from the sample data.

2. Calculation of s_ε, an estimate of σ_ε for linear regression

$$s_\varepsilon = \sqrt{\frac{\text{SSE}}{n-2}} \qquad \text{where SSE} = S_{yy} - \hat{\beta}_1 S_{xy}$$

3. $100(1-\alpha)\%$ confidence intervals for β_0 and β_1

$$\beta_0: \quad \hat{\beta}_0 \pm t_{\alpha/2} s_\varepsilon \sqrt{\frac{\sum x^2}{nS_{xx}}}$$

$$\beta_1: \quad \hat{\beta}_1 \pm t_{\alpha/2} \frac{s_\varepsilon}{\sqrt{S_{xx}}}$$

4. Statistical tests for β_0 and β_1

$$H_0: \quad \beta_0 = 0 \qquad\qquad\qquad\qquad H_0: \quad \beta_1 = 0$$

$$\text{T.S.:} \quad t = \frac{\hat{\beta}_0}{s_\varepsilon \sqrt{\dfrac{\sum x^2}{nS_{xx}}}}, \quad df = n-2 \qquad \text{T.S.:} \quad t = \frac{\hat{\beta}_1}{\dfrac{s_\varepsilon}{\sqrt{S_{xx}}}}, \quad df = n-2$$

5. Alternative test for β_1

$$H_0: \quad \beta_1 = 0$$

$$\text{T.S.:} \quad F = \text{SSREG}/s_\varepsilon^2$$

$$df_1 = 1, \quad df_2 = n - 2$$

6. Test for β_1 (intercept known)

$$H_0: \quad \beta_1 = 0$$

$$\text{T.S.:} \quad t = \frac{\hat{\beta}_1}{s_\varepsilon / \sqrt{\sum x_i^2}}, \quad df = n - 1,$$

where

$$\hat{\beta}_1 = \frac{\sum x_i y_i - \beta_0 \sum x_i}{\sum x_i^2}$$

and

$$s_\varepsilon^2 = \sum (\hat{y}_i - y_i)^2 / (n - 1)$$

7. $100(1 - \alpha)\%$ confidence interval for $E(y)$

$$\hat{y} \pm t_{\alpha/2} s_\varepsilon \sqrt{\frac{1}{n} + \frac{(x - \bar{x})^2}{S_{xx}}}$$

8. Statistical test for $E(y)$

$$H_0: \quad E(y) = \mu_0$$

$$\text{T.S.:} \quad t = \frac{\hat{y} - \mu_0}{s_\varepsilon \sqrt{\frac{1}{n} + \frac{(x - \bar{x})^2}{S_{xx}}}}, \quad df = n - 2$$

9. $100(1 - \alpha)\%$ prediction interval for y

$$\hat{y} \pm t_{\alpha/2} s_\varepsilon \sqrt{1 + \frac{1}{n} + \frac{(x - \bar{x})^2}{S_{xx}}}$$

10.   Test for lack of fit

$$H_0: \quad \text{A linear regression model is appropriate}$$

$$\text{T.S.:} \quad F = \frac{\text{MS}_{\text{Lack}}}{\text{MSP}_{\text{exp}}},$$

where

$$\text{MSP}_{\text{exp}} = \frac{\text{SSP}_{\text{exp}}}{\sum (n_i - 1)} = \frac{\sum_{ij} (y_{ij} - \bar{y}_i)^2}{\sum_i (n_i - 1)}$$

and

$$\text{MS}_{\text{Lack}} = \frac{\text{SSE} - \text{SSP}_{\text{exp}}}{n - 2 - \sum (n_i - 1)}$$

11. $100(1 - \alpha)\%$ prediction interval for x based on a single y-value

$$\hat{x}_U = \bar{x} + \frac{1}{1 - c^2} [(\hat{x} - \bar{x}) + d]$$

$$\hat{x}_L = \bar{x} + \frac{1}{1 - c^2} [(\hat{x} - \bar{x}) - d],$$

where

$$c^2 = \frac{t_{\alpha/2}^2 s_\varepsilon^2}{\hat{\beta}_1^2 S_{xx}}$$

and

$$d = \frac{t_{\alpha/2} s_\varepsilon}{\hat{\beta}_1} \sqrt{\frac{n + 1}{n} (1 - c^2) + \frac{(\hat{x} - \bar{x})^2}{S_{xx}}}$$

12. $100(1 - \alpha)\%$ prediction interval for x based on m y-values

$$\hat{x}_U = \bar{x} + \frac{1}{1 - c^2} [(\hat{x} - \bar{x}) + g]$$

$$\hat{x}_L = \bar{x} + \frac{1}{1 - c^2} [(\hat{x} - \bar{x}) - g],$$

where

$$\hat{x} = \frac{P\bar{y}_m - \hat{\beta}_0}{\hat{\beta}_1}$$

and

$$g = \frac{t_{\alpha/2}}{\hat{\beta}_1} \sqrt{\left(s_y^2 P^2 + \frac{s_\varepsilon^2}{n}\right)(1 - c^2) + \frac{(\hat{x} - \bar{x})^2 s_\varepsilon^2}{S_{xx}}}$$

▼ SUPPLEMENTARY EXERCISES

10.37 In Example 9.1, we presented the live and dressed weights for a sample of nine steers. The data are shown here.

Live Weight, x	Dressed Weight, y
4.2	2.8
3.8	2.5
4.8	3.1
3.4	2.1
4.5	2.9
4.6	2.8
4.3	2.6
3.7	2.4
3.9	2.5

a. Plot the data.
b. In Example 9.1, we found $S_{xx} = 1.72$, $S_{xy} = 1.06$, and $\hat{y} = .087 + .616x$. Is there a positive linear relationship between x and y? Give a p-value associated with your conclusion.

10.38 Refer to Exercise 10.37. Use the data to predict the live weight for $y = 3.0$. Use a 95% prediction interval.

10.39 The thermal pollution data of Exercise 10.22 are shown here.

Weight, x (in 1,000 pounds)	Btu per Vehicle Mile, y (in 1,000s)
1.8	4.0
2.6	5.2
4.2	8.5
5.0	11.6
4.8	10.1
3.4	6.3

a. Compute s_ε^2.
b. Give a 95% confidence interval for β_1 and discuss the meaning of your finding.

10.40 Labor data (in terms of manhours) are presented here for the number of orders processed per month by a large manufacturing center.

Month	Orders Processed, x	Manhours Required to Process Orders, y
1	3,000	8,000
2	3,400	9,200
3	4,000	10,000
4	2,800	7,500
5	2,000	5,800
6	1,700	5,000
7	1,400	4,400
8	1,300	3,700
9	1,000	3,100
10	600	2,220
11	1,500	4,100
12	2,200	5,500
13	3,300	8,100
14	3,600	9,400
15	4,100	10,600
16	3,200	7,900

a. Examine the plot of orders versus months. Are there cyclical patterns?

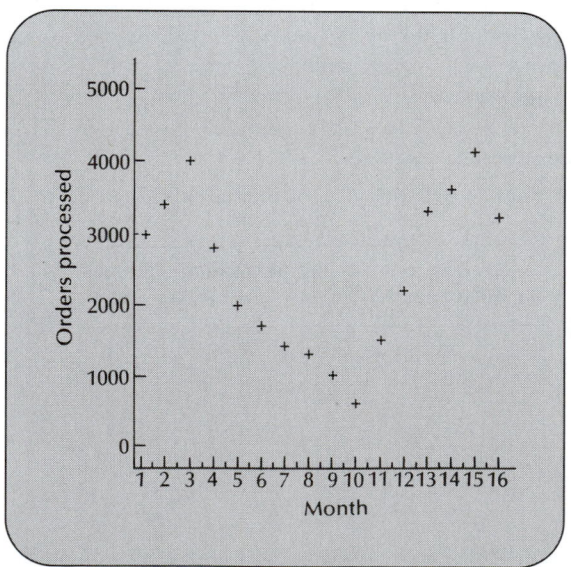

b. From the plot of orders (x) versus manhours (y) ignoring the apparent cyclical effect of part a, what regression model might adequately describe the data?

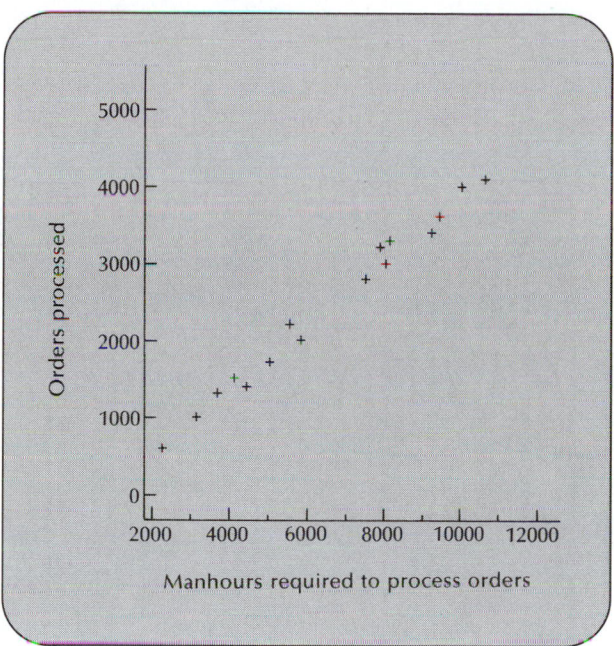

10.41 Refer to Exercise 10.40.
   a. Fit the linear regression model $y = \beta_0 + \beta_1 x + \varepsilon$, and draw conclusions about the slope and intercept.
   b. Show the results of your test for β_1 in an analysis of variance table.

10.42 Refer to Exercise 10.40. Redo your analysis without an intercept. Does this model provide a better fit to the data? Why or why not? Does it make practical sense to use such a model?

10.43 A company was interested in calibrating an instrument to measure the consistency of the liquid in a paper mill. In the making of paper, small particles of wood fiber are conveyed in a liquid flow to the screening process, where the pulp is separated from the water, dried, and made into paper. Controlling the consistency of the fibers in the liquid flow represents an important step in maintaining the final quality of the paper. Let us assume that for the calibration of the consistency meter, we are able to control the consistency of a simulated laboratory liquid flow at various levels x while monitoring the meter reading y. Assuming a model of the form

$$y = \beta_0 + \beta_1 x + \varepsilon,$$

use the following sample data to fit the model.

Meter reading, y	2.16	2.15	2.17	2.26	2.35	2.39	2.42	2.51
Actual consistency, x	2.16	2.17	2.18	2.20	2.22	2.24	2.26	2.28

Is there sufficient evidence to indicate that the slope is greater than 0? Use $\alpha = .05$.

10.44 Computer output for the data of Exercise 10.43 is shown here.

```
 2.16, 2.16
?2.17, 2.15
?2.18, 2.17
?2.20, 2.26
?2.22, 2.35
?2.24, 2.39
?2.26, 2.42
?2.28, 2.51
WANT A LISTING OF THIS FILE (Y OR N)? Y
WANT TRANSFORMATION OF INDEPENDENT VARIABLE
(0=NO, 1=NATURAL LOG, 2=COMMON LOG)? 0
WANT TRANSFORMATION ON DEPENDENT VARIABLE
(0=NO, 1=NATURAL LOG, 2=COMMON LOG, 3=ARCSIN, 4=SQUARE ROOT)?
0

SIMPLE LINEAR REGRESSION

                    FILES:            X              Y
              OBSERVATIONS:     INDEPENDENT       DEPENDENT
                                 2.1600           2.1600
                                 2.1700           2.1500
                                 2.1800           2.1700
                                 2.2000           2.2600
                                 2.2200           2.3500
                                 2.2400           2.3900
                                 2.2600           2.4200
                                 2.2800           2.5100

                                INDEPENDENT       DEPENDENT
      NO. OBSERVATIONS            8.0000           8.0000
      ARITHMETIC MEANS            2.2138           2.3013
      STANDARD DEVIATION          0.0437           0.1361

                                Y INTERCEPT        SLOPE
      ESTIMATE                    -4.5054          3.0747
      STANDARD DEVIATION           0.4367          0.1973
      T-STATISTIC-SIGNIFICANCE   -10.3159         15.5877
      DEGREES OF FREEDOM           6.0000          6.0000
      PROBABILITY                  0.0000          0.0000
      RESIDUAL SUM OF SQUARES               0.0031
      RESIDUAL MEAN SQUARE                  0.0005

WANT TO ESTIMATE EFFECTIVE DOSE (Y OR N)? Y
ENTER NO. OF ESTIMATES? 1
ENTER Y VALUES
?2.20

ESTIMATE OF X
    Y VALUE       ESTIMATE      95% CONFIDENCE INTERVAL
    2.2000        2.1808          [2.1597, 2.2002]
```

a. Compare the least squares estimates to those obtained in Exercise 10.43.
b. Give the level of significance for a test of $H_0: \beta_1 = 0$.
c. Identify 95% prediction limits for the independent variable (actual consistency of the flow) when the observed meter reading is 2.20.

10.45 Consider the following data.

y	x
1.0	−1
2.0	−1
1.0	−1
6.0	1
7.0	1
6.5	1
2.0	3
3.0	3

a. Fit the linear regression model $y = \beta_0 + \beta_1 x + \varepsilon$.

b. Plot the residuals versus $\hat{y}$. Is there evidence of lack of fit?

10.46 Refer to Exercise 10.45 and test for lack of fit. Draw conclusions and make recommendations for a model relating y to x.

10.47 An airline was interested in comparing the estimated time of arrival (ETA) and the actual time of arrival (ATA) for domestic flights. To do this a random sample of 20 flights was taken over the past month for flights with ETAs in the time frame from noon until 6:00 P.M. These data are shown here in order of ETA.

ETA	ATA
12:15	12:16
12:30	12:52
12:30	12:40
1:00	1:28
1:30	1:42
2:10	2:20
2:35	2:52
3:05	3:11
3:20	3:55
3:50	4:14
4:15	4:55
4:40	4:46
4:50	5:16
5:00	5:33
5:00	5:20
5:10	5:28
5:15	6:30
5:30	6:05
6:00	7:10
6:00	6:30

a. Plot the data.

b. Would a linear regression model appear appropriate for characterizing the relationship between ETA and ATA? If so, why? If not, why not?

c. Guess the value of the correlation coefficient.

10.48 **a.** Obtain the linear regression line for the data of Exercise 10.47.

b. Plot the residuals versus $\hat{y}$. Does any recognizable pattern emerge?

c. Can you test for lack of fit? If so, do it and interpret your results.

10.49 Refer to Exercises 10.47 and 10.48. Suppose that you were given a computer printout of the data for Exercise 10.47 and the eleventh observation was shown as (ETA = 4:15, ATA = 9.55).

a. What circumstances might have given rise to such a data point? Might such a value be an error? If so, how might it have occurred?

b. How might you check for errors and odd values in this or other data sets?

 10.50 A study was conducted to examine the effect of different levels of nitrogen on the yield of lettuce plants. Use the data shown here to fit a linear regression model. Test for possible lack of fit of the model.

Coded Nitrogen	Yield (Emergent Stalks per Plot)
1	21, 18, 17
2	24, 22, 26
3	34, 29, 32

10.51 The specific activity of the enzyme sucrase was measured by extracting a portion of the intestines of 24 patients who underwent an intestinal bypass. After the sections were extracted, they were homogenized and analyzed for enzyme activity (Carter (1981)). Two different methods can be used to measure the activity of sucrase: the homogenate method and the pellet method. Data for the 24 patients are shown here for the two methods:

Sucrase Activity as Measured by the Homogenate and Pellet Methods

Patient	Homogenate Method, y	Pellet Method, x
1	18.88	70.00
2	7.26	55.43
3	6.50	18.87
4	9.83	40.41
5	46.05	57.43
6	20.10	31.14
7	35.78	70.10
8	59.42	137.56
9	58.43	221.20
10	62.32	276.43
11	88.53	316.00
12	19.50	75.56
13	60.78	277.30
14	77.92	331.50
15	51.29	133.74
16	77.91	221.50
17	36.65	132.93
18	31.17	85.38
19	66.09	142.34
20	115.15	294.63
21	95.88	262.52
22	64.61	183.56
23	37.71	86.12
24	100.82	226.55

a. Examine the scatterplot of the data. Might a linear model adequately describe the relationship between the two methods?

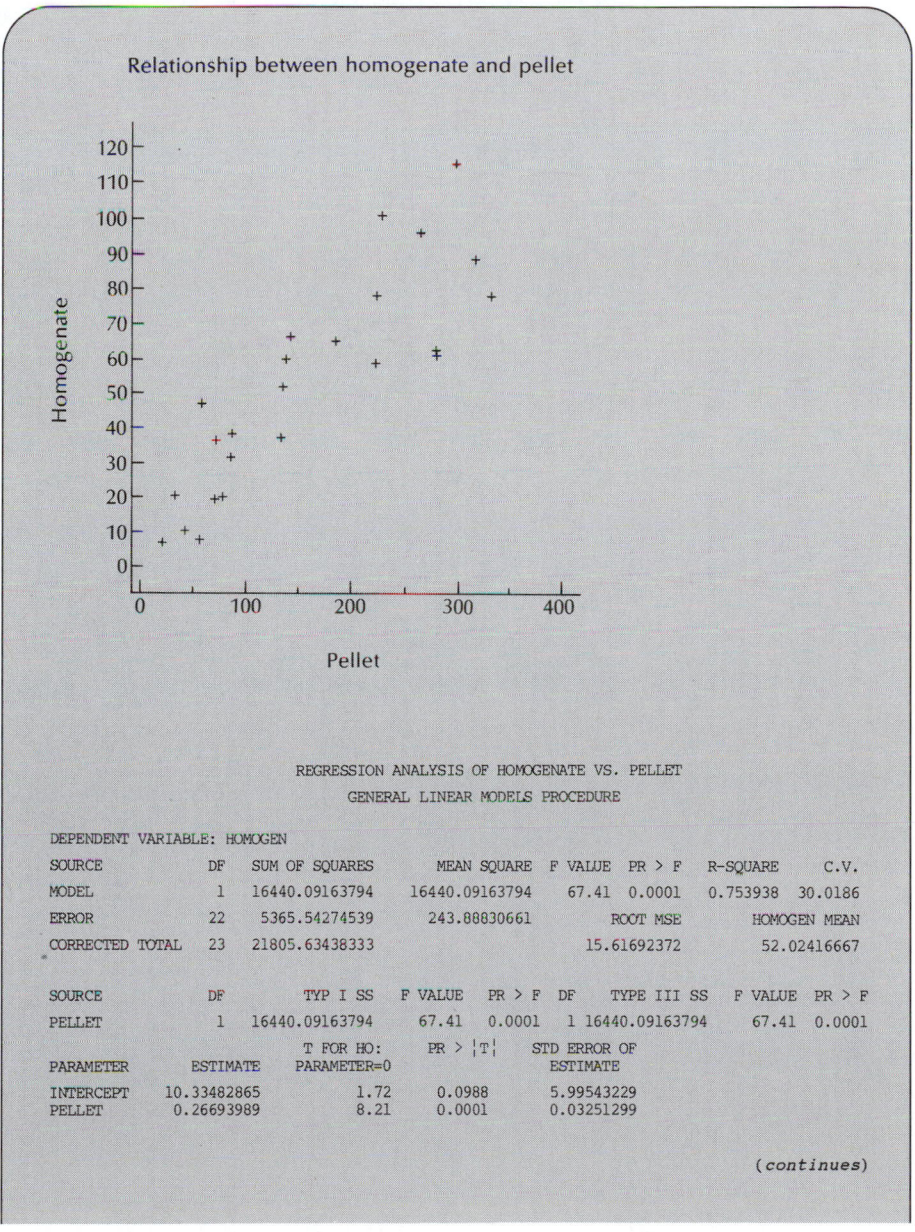

Relationship between homogenate and pellet

REGRESSION ANALYSIS OF HOMOGENATE VS. PELLET
GENERAL LINEAR MODELS PROCEDURE

DEPENDENT VARIABLE: HOMOGEN

SOURCE	DF	SUM OF SQUARES	MEAN SQUARE	F VALUE	PR > F	R-SQUARE	C.V.
MODEL	1	16440.09163794	16440.09163794	67.41	0.0001	0.753938	30.0186
ERROR	22	5365.54274539	243.88830661	ROOT MSE		HOMOGEN MEAN	
CORRECTED TOTAL	23	21805.63438333		15.61692372		52.02416667	

SOURCE	DF	TYP I SS	F VALUE	PR > F	DF	TYPE III SS	F VALUE	PR > F
PELLET	1	16440.09163794	67.41	0.0001	1	16440.09163794	67.41	0.0001

PARAMETER	ESTIMATE	T FOR H0: PARAMETER=0	PR > |T|	STD ERROR OF ESTIMATE
INTERCEPT	10.33482865	1.72	0.0988	5.99543229
PELLET	0.26693989	8.21	0.0001	0.03251299

(*continues*)

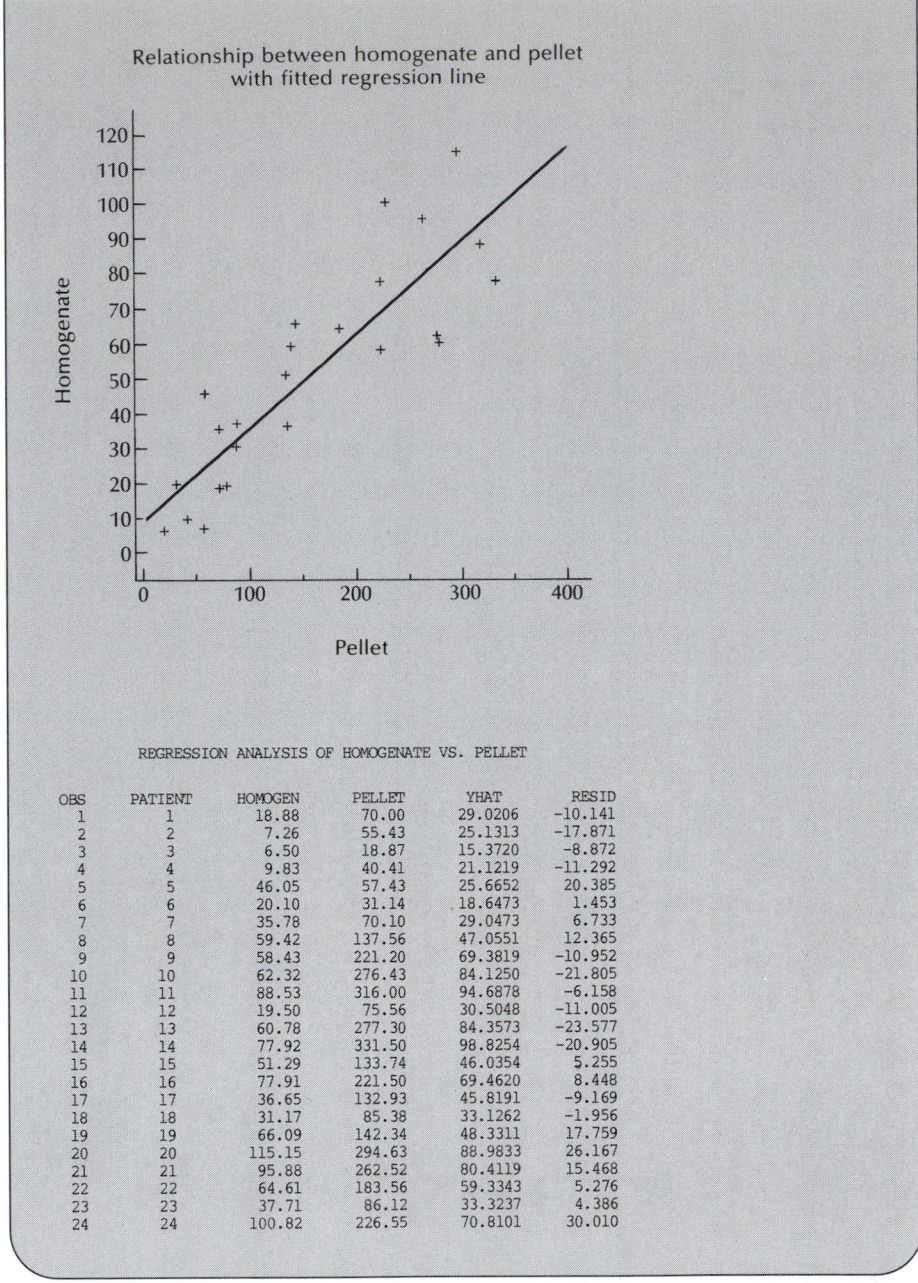

Relationship between homogenate and pellet
with fitted regression line

REGRESSION ANALYSIS OF HOMOGENATE VS. PELLET

OBS	PATIENT	HOMOGEN	PELLET	YHAT	RESID
1	1	18.88	70.00	29.0206	-10.141
2	2	7.26	55.43	25.1313	-17.871
3	3	6.50	18.87	15.3720	-8.872
4	4	9.83	40.41	21.1219	-11.292
5	5	46.05	57.43	25.6652	20.385
6	6	20.10	31.14	18.6473	1.453
7	7	35.78	70.10	29.0473	6.733
8	8	59.42	137.56	47.0551	12.365
9	9	58.43	221.20	69.3819	-10.952
10	10	62.32	276.43	84.1250	-21.805
11	11	88.53	316.00	94.6878	-6.158
12	12	19.50	75.56	30.5048	-11.005
13	13	60.78	277.30	84.3573	-23.577
14	14	77.92	331.50	98.8254	-20.905
15	15	51.29	133.74	46.0354	5.255
16	16	77.91	221.50	69.4620	8.448
17	17	36.65	132.93	45.8191	-9.169
18	18	31.17	85.38	33.1262	-1.956
19	19	66.09	142.34	48.3311	17.759
20	20	115.15	294.63	88.9833	26.167
21	21	95.88	262.52	80.4119	15.468
22	22	64.61	183.56	59.3343	5.276
23	23	37.71	86.12	33.3237	4.386
24	24	100.82	226.55	70.8101	30.010

b. Examine the residual plot; are there any potential problems uncovered by the plot?

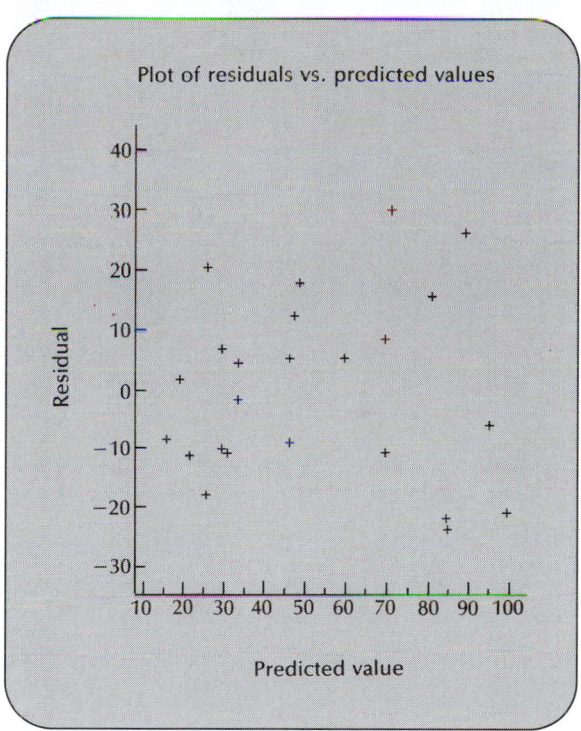

10.52 Refer to Exercise 10.51. In general, the pellet method is more time consuming than the homogenate method, yet it provides a more accurate measure of sucrase activity.
   a. How might you estimate the pellet reading based on a particular homogenate reading?
   b. How would you develop a confidence (prediction) interval about your point estimate?

10.53 A large chemical company examined the impact of immediate development expenditures on sales ($100,000) by collecting quarterly data for the preceding 6 years. A portion of the computer output for a linear regression analysis is shown here.
   a. Write the linear regression model and the least squares prediction equation.
   b. Give results and a conclusion for the test $H_0: \beta_1 = 0$.
   c. Can you suggest other variable costs that may be important?

```
DEPENDENT VARIABLE SALES

SOURCE              DF      SUM OF SQUARES      MEAN SQUARE     F VALUE
MODEL               1       35139.08936255      35139.0893255    24.5
ERROR               22      31582.24397078      1435.55654413
CORRECTED TOTAL     23      66721.33333333

SOURCE              DF        TYPE I SS        F VALUE       PR > F
DEVEL               1       35139.08936255       24.48       0.0001

                                T FOR HO:        | |      STD ERROR OF
PARAMETER       ESTIMATE      PARAMETER=0    PR > |T|       ESTIMATE
INTERCEPT     -663.89962895     -3.37        0.0028        197.0663320
DEVEL          144.48242659      4.95        0.0001         29.2031655

                  PR > F         R-SQUARE         C.V.
                  0.0001         0.526654        12.2090

                 STD DEV                        SALES MEAN
                37.88873901                     310.33333333
```

$ 10.54 A random sample of 14 pharmacies was used to examine the relation between sales volume and the profit before tax (PBT). These data are shown here:

Pharmacy	Sales Volume, x ($1,000)	PBT, y ($1,000)
1	38	1.3
2	20	2.1
3	48	2.2
4	44	2.6
5	56	3.3
6	39	4.0
7	65	4.1
8	84	4.2
9	82	5.5
10	105	5.7
11	126	7.0
12	52	7.5
13	80	7.7
14	101	7.9

a. Plot the sample data.
b. Calculate the sample correlation coefficient.
c. Is there a significant linear trend between sales (x) and PBT (y)?

10.55 Refer to Exercise 10.54. Refer to a residual plot for these data. Are there any obvious outliers? Do the assumptions for a linear regression model seem to hold?

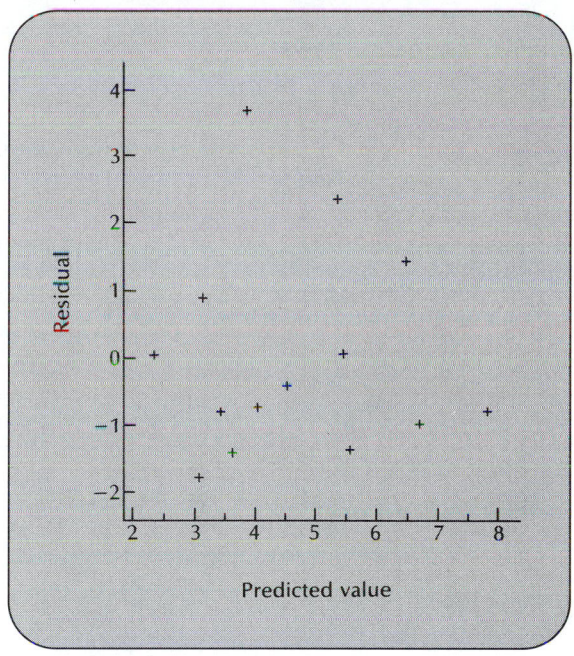

10.56 Refer to the following data:

x	1	1	1	2	2	2	4	4	4	8	8	8
y	13.5	15.4	16.1	18.2	19.6	20.2	21.8	22.2	23.1	23.6	24.7	24.9

a. Plot y versus x.
b. Compute $\log_{10} x$ and plot y versus $\log_{10} x$.
c. Which plot (parts a and b) looks more linear?

10.57 Refer to the data of Exercise 10.56.
a. Fit a linear regression model for y and x.
b. Compute s_ε^2 for this model.

10.58 Refer to the data of Exercise 10.56.
a. Fit a linear regression model for y and $\log_{10} x$.
b. Compute s_ε^2 for this model.
c. Compare the fit of this model to that for y versus x in Exercise 10.57 using s_ε^2. Which model provides the better fit? Does this agree with your opinion in Exercise 10.56, based on plots of the data?

10.59 Use the data in Exercise 10.56 and the linear regression of 10.58 to construct a 95% confidence band for $E(y)$. Within what limits would we expect $E(y)$ to be when $x = 5$? When $x = 9$?

10.60 Refer to Exercise 10.59. Construct a 95% prediction band on future values of y. Give the prediction limits for y when $x = 5$ and $x = 9$. Compare these to the confidence limits for $E(y)$ at these same values of x.

10.61 In screening for compounds useful in treating hypertension (high blood pressure), six rats are assigned to each of three groups. The rats in group 1 receive .1 mg/kg of a test compound; those in groups 2 and 3 receive .2 and .4 mg/kg, respectively. The response of interest is the decrease in blood pressure two hours postdose, compared to the corresponding predose blood pressure. The data are shown here:

	Dose, x	Blood Pressure Drop (mm Hg), y					
Group 1	.1 mg/kg	10	12	15	16	13	11
Group 2	.2 mg/kg	25	22	26	19	18	24
Group 3	.4 mg/kg	30	32	35	27	26	29

a. Use a software package to fit the model

$$y = \beta_0 + \beta_1 \log_{10} x + \varepsilon.$$

b. Use residual plots to examine the fit to the model in part a.

c. Conduct a statistical test of $H_0: \beta_1 = 0$ vs. $H_a: \beta_1 > 0$. Give the p-value for your test.

10.62 Population and area data are listed here by state. Plot the data. Compute the correlation coefficient. Can inferences be made based on these data?

State	Population (1,000)	Area (1,000 mi²)
Maine	1,125	33.3
New Hampshire	921	9.3
Vermont	511	9.6
Massachusetts	5,737	8.3
Rhode Island	947	1.2
Connecticut	3,108	5.0
New York	17,558	49.1
New Jersey	7,365	7.8
Pennsylvania	11,864	45.3
Ohio	10,798	41.3
Indiana	5,490	36.2
Illinois	11,427	56.3
Michigan	9,262	58.5
Wisconsin	4,706	56.2
Minnesota	4,076	84.4
Iowa	2,914	56.3

(*continues*)

State	Population (1,000)	Area (1,000 mi²)
Missouri	4,917	69.7
North Dakota	653	70.7
South Dakota	691	77.1
Nebraska	1,570	77.4
Kansas	2,364	82.2
Delaware	594	2.0
Maryland	4,217	10.5
Virginia	5,347	40.8
West Virginia	1,950	24.2
North Carolina	5,882	52.7
South Carolina	3,122	31.1
Georgia	5,463	58.9
Florida	9,746	58.7
Kentucky	3,661	40.4
Tennessee	4,591	42.1
Alabama	3,894	51.8
Mississippi	2,521	47.7
Arkansas	2,286	53.1
Louisiana	4,206	47.8
Oklahoma	3,025	70.0
Texas	14,229	266.8
Montana	787	147.0
Idaho	944	83.6
Wyoming	470	97.8
Colorado	2,890	104.0
New Mexico	1,303	121.6
Arizona	2,718	114.0
Utah	1,461	84.9
Nevada	800	110.6
Washington	4,132	68.1
Oregon	2,633	97.1
California	23,668	158.8
Alaska	402	591.0
Hawaii	965	6.5

Source: U.S. Bureau of Census 1980 Census of Population.

$ 10.63 The following data give the profit ($1,000) per trip of the space shuttle (y) and the dollar amount of items per payload for each trip (x).

Trip	1	2	3	4	5	6
Items	7,500	12,500	15,200	9,900	8,700	15,100
Profit	5	8	9	8	7	10

a. Plot the data.
b. Using the following output, give the least squares prediction equation for profit.

```
                      REGRESSION ANALYSIS OF PROFIT BY PAYLOAD ITEMS

                            GENERAL LINEAR MODELS PROCEDURE

DEPENDENT VARIABLE: PROFIT

SOURCE            DF  SUM OF SQUARES   MEAN SQUARE  F VALUE    PR > F   R-SQUARE        C.V.
MODEL              1     12.20451345   12.20451345    18.57    0.0126   0.822776     10.3491
ERROR              4      2.62881988    0.65720497              ROOT MSE          PROFIT MEAN
CORRECTED TOTAL    5     14.83333333                           0.81068179         7.83333333

SOURCE            DF       TYPE I SS   F VALUE   PR > F   DF   TYPE III SS  F VALUE  PR > F
ITEMS              1     12.20451345     18.57   0.0126    1   12.20451345    18.57  0.0126

                              T FOR HO:     PR > |T|    STD ERROR OF
PARAMETER     ESTIMATE     PARAMETER=0                    ESTIMATE

INTERCEPT   2.37654568           1.82      0.1436         1.30880887
ITEMS       0.00047519           4.31      0.0126         0.00011027
```

c. Is there a significant linear relationship between the number of items and the profit per trip? Give the value of the test statistic and the *p*-value.

10.64   Refer to Exercise 10.63. Use the output shown here to find the Spearman's rank order correlation. Is there a significant positive relationship between x and y? What's the value of the test statistic and the *p*-value for the test that lead you to this conclusion?

```
                        SPEARMANS RANK ORDER CORRELATION

VARIABLE   N          MEAN       STD DEV         MEDIAN         MINIMUM          MAXIMUM

PROFIT     6     7.83333333    1.72240142     8.00000000      5.00000000      10.00000000
ITEMS      6  11483.33333333  3287.80575258  11200.00000000  7500.00000000   15200.00000000


        SPEARMAN CORRELATION COEFFICIENTS / PROB > |R| UNDER HO:RHO=0 / N=6

                               PROFIT    ITEMS

                  PROFIT    1.00000    0.92763
                            0.0000     0.0077

                  ITEMS     0.92763    1.00000
                            0.0077     0.0000
```

10.65   A study was conducted to examine the list price of residential properties and the actual sale price as listed in the records of the County Court House. A random sample of 20 was taken from the list of residential sales recorded over the past 6 months. These data are shown here.

Sale	Sale Price ($1,000)	List Price ($1,000)	Sale	Sale Price ($1,000)	List Price ($1,000)
1	45.0	49.9	11	83.5	91.0
2	58.0	59.0	12	89.0	94.9
3	66.5	69.0	13	90.0	93.9
4	67.5	75.0	14	93.9	99.9
5	69.0	74.0	15	92.0	98.5
6	72.5	79.5	16	102.0	105.9
7	74.0	80.0	17	106.9	115.0
8	74.9	78.0	18	120.5	139.9
9	82.0	89.9	19	147.0	165.0
10	84.5	88.1	20	206.5	229.9

a. Examine a plot of the data.

b. Use the computer output shown here to give the least squares prediction equation relating sale price (y) to list price (x). Give the p-value for a test of $H_0: \beta_1 = 0$ versus $H_a: \beta_1 \neq 0$.

c. How well does the linear regression equation fit the data? Explain.

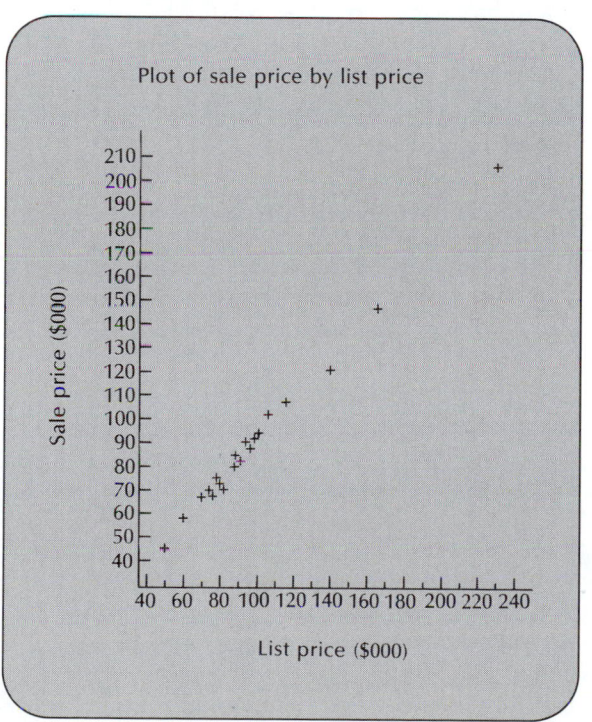

Plot of sale price by list price

```
      LIST OF DATA
   OBS    SALE      LIST

    1     45.0      49.9
    2     58.0      59.0
    3     66.5      69.0
    4     67.5      75.0
    5     69.0      74.0
    6     72.5      79.5
    7     74.0      80.0
    8     74.9      78.0
    9     82.0      89.9
   10     84.5      88.1
   11     83.5      91.0
   12     89.0      94.9
   13     90.0      93.9
   14     93.9      99.9
   15     92.0      98.5
   16    102.0     105.9
   17    106.9     115.0
   18    120.5     139.9
   19    147.0     165.0
   20    206.5     229.9
   21              150.0


  N=     21
```

REGRESSION ANALYSIS OF SALE PRICE BY LIST PRICE
GENERAL LINEAR MODELS PROCEDURE

DEPENDENT VARIABLE: SALE

SOURCE	DF	SUM OF SQUARES	MEAN SQUARE	F VALUE	PR > F	R-SQUARE	C.V.
MODEL	1	23515.88443941	23515.88443941	3282.72	0.0001	0.994547	2.9328
ERROR	18	128.94356059	7.16353114		ROOT MSE		SALE MEAN
CORRECTED TOTAL	19	23644.82800000			2.67647738		91.26000000

SOURCE	DF	TYPE I SS	F VALUE	PR > F	DF	TYPE III SS	F VALUE	PR > F
LIST	1	23515.88443941	3282.72	0.0001	1	23515.88443941	3282.72	0.0001

PARAMETER	ESTIMATE	T FOR HO: PARAMETER=0	PR > ¦T¦	STD ERROR OF ESTIMATE
INTERCEPT	5.29292116	3.28	0.0042	1.61538230
LIST	0.86998005	57.30	0.0001	0.01518421

OBSERVATION	OBSERVED VALUE	PREDICTED VALUE	RESIDUAL	LOWER 95% CL INDIVIDUAL	UPPER 95% CL INDIVIDUAL
1	45.00000000	48.70492575	-3.70492575	42.73547556	54.67437595
2	58.00000000	56.62174423	1.37825577	50.72152042	62.52196804
3	66.50000000	65.32154475	1.17845525	59.48167736	71.16141214
4	67.50000000	70.54142506	-3.04142506	64.72966185	76.35318827
5	69.00000000	69.67144501	-0.67144501	63.85542576	75.48746426
6	72.50000000	74.45633529	-1.95633529	68.66158932	80.25108127
7	74.00000000	74.89132532	-0.89132532	69.09825364	80.68439700
8	74.90000000	73.15136522	1.74863478	67.35133604	78.95139439
9	82.00000000	83.50412784	-1.50412784	77.73521919	89.27303648
10	84.50000000	81.93816374	2.56183626	76.16613943	87.71018805
11	83.50000000	84.46110589	-0.96110589	78.69382064	90.22839114
12	89.00000000	87.85402810	1.14597190	82.09078034	93.61727585
13	90.00000000	86.98404804	3.01595196	81.22002076	92.74807533
14	93.90000000	92.20392836	1.69607164	86.44193001	97.96592670
15	92.00000000	90.98595628	1.01404372	85.22405313	96.74785943
16	102.00000000	97.42380867	4.57619133	91.65748313	103.19013420
17	106.90000000	105.34062714	1.55937286	99.55564618	111.12560811
18	120.50000000	127.00313044	-6.50313044	121.09405289	132.91220799
19	147.00000000	148.83962974	-1.83962974	142.70308369	154.97617580
20	206.50000000	205.30133512	1.19866488	198.18192546	212.42074479
21*		135.78992896		129.80114167	141.77871626

* OBSERVATION WAS NOT USED IN THIS ANALYSIS

(continues)

```
                    REGRESSION ANALYSIS OF SALE PRICE BY LIST PRICE
                                        GENERAL LINEAR MODELS PROCEDURE
     DEPENDENT VARIABLE: SALE
              SUM OF RESIDUALS                           -0.00000000
              SUM OF SQUARED RESIDUALS                  128.94356059
              SUM OF SQUARED RESIDUALS - ERROR SS        -0.00000000
              PRESS STATISTIC                           161.70259371
              FIRST ORDER  AUTOCORRELATION                0.06291667
              DURBIN-WATSON D                             1.75657045
```

10.66 Refer to Exercise 10.65. Predict the sale price of a house to be listed at $150,000. Within what limits should the sale price be? (Hint: Use 95% limits.)

10.67 A supermarket chain conducted an experiment to investigate the effect of price (P) on the weekly demand y (in pounds) for a house brand of coffee. Eight supermarket stores that had nearly equal past records of demand for the product were used in the experiment. Eight prices were randomly assigned to the stores and were advertised using the same procedures. The number of pounds of coffee sold during the following week was recorded for each of the stores and is shown below:

Demand (y, pounds)	Price (P, dollars)
1,120	3.00
999	3.10
932	3.20
884	3.30
807	3.40
760	3.50
701	3.60
688	3.70

The data were analyzed using SAS with the results shown in the following computer output.

a. Suppose that a supermarket that had been selling coffee for $3.70/pound is considering a raise in price to $3.80/pound. Give a 90% confidence interval for the expected change in sales per week.

b. The supermarket in part a that is considering a raise in price has decided that the price increase will be desirable as long as average sales will not drop below 550 pounds/week. Compute a point estimate for the estimated average sales in subsequent weeks if the price is raised to $3.80/pound. Can the supermarket be 95% certain that sales will be at least 550 pounds? (In answering these questions, assume that a linear regression model is a correct one.)

c. Comment on the assumption of linearity for this model. Are there any (1) theoretical or (2) empirical reasons to question this assumption?

d. What is the interpretation of the intercept for this analysis? Is the intercept a meaningful number here? Comment.

```
         LISTING OF DATA

   OBS    DEMAND    PRICE

    1      1120      3.0
    2       999      3.1
    3       932      3.2
    4       884      3.3
    5       807      3.4
    6       760      3.5
    7       701      3.6
    8       688      3.7
    9                3.8

               REGRESSION ANALYSIS OF DEMAND BY PRICE
                  GENERAL LINEAR MODELS PROCEDURE

 DEPENDENT VARIABLE: DEMAND

 SOURCE       DF    SUM OF SQUARES    MEAN SQUARE   VALUE    PR > F   R-SQUARE           C.V.

 MODEL         1   155246.72023808  155246.72023808 182.89  0.0001   0.968235        3.3824

 ERROR         6     5093.15476192     848.85912699                 ROOT MSE      DEMAND MEAN

 CORRECTED TOTAL 7 160339.87500000                               29.13518709     861.37500000

 SOURCE       DF      TYPE I SS    F VALUE   PR > F   DF      TYPE III SS   F VALUE   PR > F

 PRICE         1   155246.72023808  182.89   0.0001    1   155246.72023808  182.89   0.0001

                                   T FOR HO:    PR > |T|   STD ERROR OF
 PARAMETER          ESTIMATE      PARAMETER=0               ESTIMATE

 INTERCEPT     2898.09523810         19.20      0.0001   150.95636910
 PRICE          607.97619048        -13.52      0.0001    44.95656970

 OBSERVATION    OBSERVED       PREDICTED       RESIDUAL    LOWER 95% CL    UPPER 95% CL
                 VALUE          VALUE                       FOR MEAN        FOR MEAN

     1        1120.00000000  1074.16666667    45.83333333  1028.14833941  1120.18499393
     2         999.00000000  1013.36904762   -14.36904762   976.06459051  1050.67350473
     3         932.00000000   952.57142857   -20.57142857   922.44536171   982.69749543
     4         884.00000000   891.77380952    -7.77380952   865.97538716   917.57223189
     5         807.00000000   830.97619048   -23.97619048   805.17776811   856.77461284
     6         760.00000000   770.17857143   -10.17857143   740.05250457   800.30463829
     7         701.00000000   709.38095238    -8.38095238   672.07649527   746.68540949
     8         688.00000000   648.58333333    39.41666667   602.56500607   694.60166059
     9*                       587.78571429                  532.23599174   643.33543684

 * OBSERVATION WAS NOT USED IN THIS ANALYSIS

     SUM OF RESIDUALS                              0.00000000
     SUM OF SQUARED RESIDUALS                    5093.15476190
     SUM OF SQUARED RESIDUALS - ERROR SS           -0.00000002
     PRESS STATISTIC                            12885.97459378
     FIRST ORDER AUTOCORRELATION                   -0.00347221
     DURBIN-WATSON D                               1.28943863
```

§ **10.68** A computer–equipment outlet sells an imported personal computer (PC) on a franchise basis and performs preventive maintenance and repair service on this PC. The following data have been collected from 16 recent calls on users to perform routine preventive maintenance service. For each call, x is the number of machines serviced and y is the total number of minutes spent by the service person.

x	6	5	1	5	4	7	4	4
y	86	95	18	69	62	101	39	53

x	2	8	5	2	7	1	4	5
y	33	102	65	25	105	17	55	68

a. Based on the data plot and a look at the residuals, does the linear regression equation provide a good fit to the data? Are there any apparent outliers? What could be done to evaluate this further?

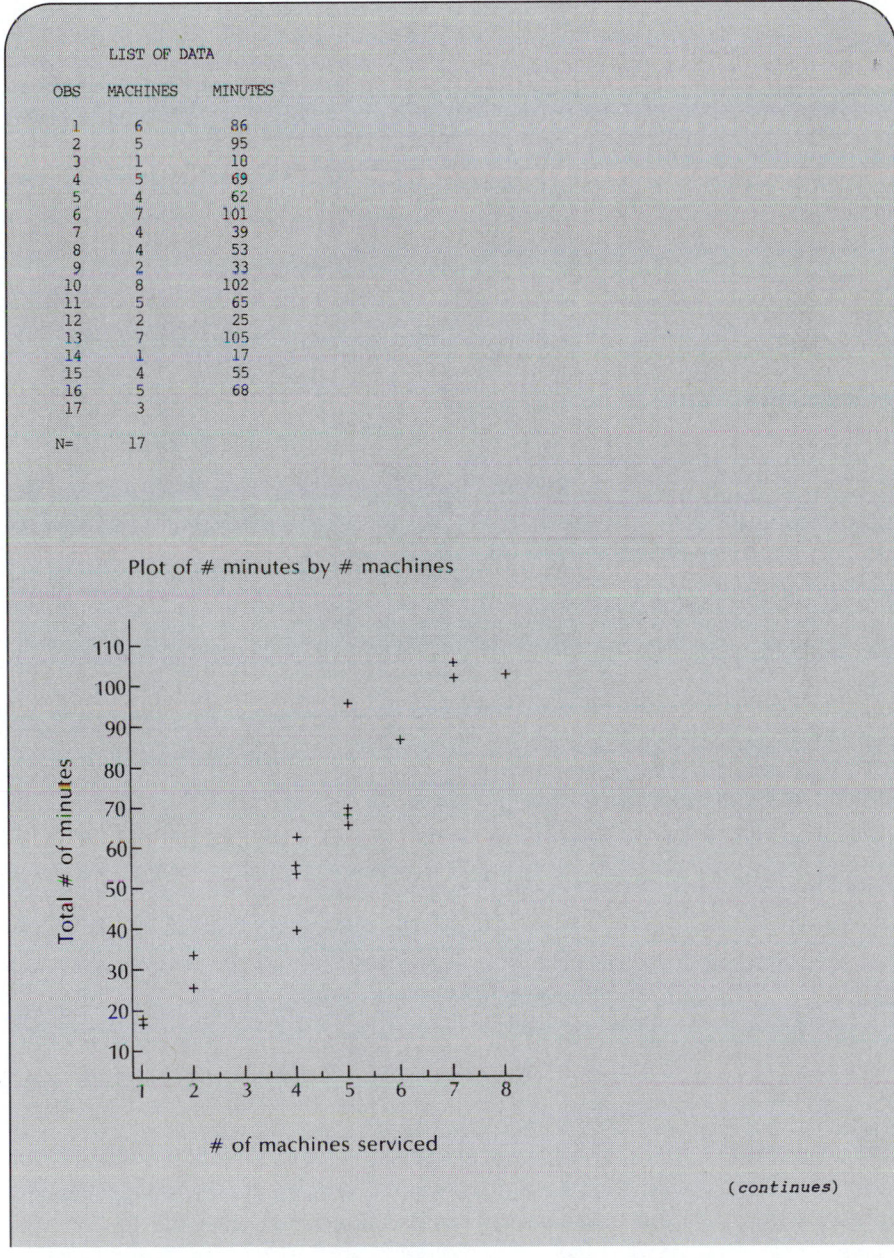

```
        LIST OF DATA

 OBS   MACHINES   MINUTES

   1       6         86
   2       5         95
   3       1         10
   4       5         69
   5       4         62
   6       7        101
   7       4         39
   8       4         53
   9       2         33
  10       8        102
  11       5         65
  12       2         25
  13       7        105
  14       1         17
  15       4         55
  16       5         68
  17       3

 N=       17
```

Plot of # minutes by # machines

Total # of minutes

# of machines serviced

(continues)

```
                          GENERAL LINEAR MODELS PROCEDURE
DEPENDENT VARIABLE: MINUTES        TOTAL NUMBER OF MINUTES

SOURCE             DF      SUM OF SQUARES      MEAN SQUARE   F VALUE  PR > F  R-SQUARE    C.V.

MODEL              1      12363.91848859    12363.91848859   144.85  0.0001  0.911865  14.8866

ERROR              14      1195.01901141       85.35850081            ROOT MSE     MINUTES MEAN

CORRECTED TOTAL    15     13558.93750000                            9.23896644     62.06250000

SOURCE             DF         TYPE I SS   F VALUE  PR > F DF   TYPE III SS  F VALUE  PR > F

MACHINES           1     12363.91848859   144.85  0.0001  1 12363.91848859  144.85  0.0001

                                T FOR HO:      PR > |T|     STD ERROR OF
PARAMETER           ESTIMATE    PARAMETER=0                   ESTIMATE

INTERCEPT          2.06844106        0.38      0.7122       5.49397893
MACHINES          13.71292776       12.04      0.0001       1.13939815

OBSERVATION        OBSERVED        PREDICTED       RESIDUAL     LOWER 95% CL    UPPER 95% CL
                    VALUE           VALUE                       INDIVIDUAL      INDIVIDUAL
      1           86.00000000     84.34600760     1.65399240    63.53813898    105.15387622
      2           95.00000000     70.63307985    24.36692015    50.45063670     91.11552300
      3           18.00000000     15.78136882     2.21863118    -6.24638827     37.80912591
      4           69.00000000     70.63307985    -1.63307985    50.15063670     91.11552300
      5           62.00000000     56.92015209     5.07984791    36.47418705     77.36611713
      6          101.00000000     98.05893536     2.94106464    76.64986616    119.46800456
      7           39.00000000     56.92015209   -17.92015209    36.47418705     77.36611713
      8           53.00000000     56.92015209    -3.92015209    36.47418705     77.36611713
      9           33.00000000     29.49429658     3.50570342     8.26028403     50.72830912
     10          102.00000000    111.77186312    -9.77186312    89.50814781    134.03557843
     11           65.00000000     70.63307985    -5.63307985    50.15063670     91.11552300
     12           25.00000000     29.49429658    -4.49429658     8.26028403     50.72830912
     13          105.00000000     98.05893536     6.94106464    76.64986616    119.46800456
     14           17.00000000     15.78136882     1.21863118    -6.24638827     37.80912591
     15           55.00000000     56.92015209    -1.92015209    36.47418705     77.36611713
     16           68.00000000     70.63307985    -2.63307985    50.15063670     91.11552300
     17 *                         43.20722433                   22.50726242     63.90718625

* OBSERVATION WAS NOT USED IN THIS ANALYSIS

     SUM OF RESIDUALS                       -0.00000000
     SUM OF SQUARED RESIDUALS             1195.01901141
     SUM OF SQUARED RESIDUALS - ERROR SS    -0.00000000
     PRESS STATISTIC                      1466.75264743
     FIRST ORDER AUTOCORRELATION             0.10650215
     DURBIN-WATSON D                         1.77890479
```

b. Based on the original prediction equation, if you were to send a service person on a call to perform preventive maintenance on three PCs, is it likely that she will spend less than 45 minutes? Explain how you reached your conclusion.

$ **10.69** The following data give the number of pounds of meat sold in a week by a grocery store (y) and the number of meat items advertised (x) that week.

Week	1	2	3	4	5	6
Pounds sold	7,700	11,025	8,150	8,500	8,750	7,920
Number of items	5	8	9	8	7	10

Give a measure of the strength of relationship between x and y, based on the computer output shown here. Interpret this finding.

```
                        SPEARMANS RANK CORRELATION

VARIABLE   N       MEAN        STD DEV        MEDIAN       MINIMUM        MAXIMUM

ITEMS      6     7.83333333    1.72240142    8.00000000    5.00000000    10.00000000
POUNDS     6  8674.16666667 1212.84960596 8325.00000000 7700.00000000 11025.00000000

       SPEARMAN CORRELATION COEFFICIENTS / PROB > |R| UNDER HO:RHO=0 / N = 6

                           ITEMS         POUNDS

                ITEMS     1.00000       -0.02899
                          0.0000         0.9565

                POUNDS   -0.02899        1.00000
                          0.9565         0.0000
```

MULTIPLE REGRESSION AND THE GENERAL LINEAR MODEL

11.1 ▾ INTRODUCTION

The simplest type of regression model relating the dependent variable y to a quantitative independent variable x is the one discussed in Chapter 9.

$$y = \beta_0 + \beta_1 x + \varepsilon.$$

Under the assumption that the average value of ε (also called the **expected value of ε**) for a given value of x is $E(\varepsilon) = 0$, this model indicates that the expected value of y for a given value of x is described by the straight line

$$E(y) = \beta_0 + \beta_1 x.$$

Not all data sets are adequately described by a model for which the expectation is a straight line. For example, consider the data of Table 11.1, which gives the yields (in bushels) for 14 equal-sized plots planted in tomatoes for different levels of fertilization. It is evident from the scatterplot in Figure 11.1 that a linear equation will not adequately represent the relationship between yield and the amount of fertilizer applied to the plot. The reason for this is that, whereas a modest amount of fertilizer may well enhance the crop yield, too much fertilizer can be destructive.

TABLE 11.1
Yield of 14 Equal-Sized Plots of Tomato Plantings for Different Amounts of Fertilizer

Plot	Yield, y (in bushels)	Amount of Fertilizer, x (in pounds per plot)
1	24	12
2	18	5
3	31	15
4	33	17
5	26	20
6	30	14
7	20	6
8	25	23
9	25	11
10	27	13
11	21	8
12	29	18
13	29	22
14	26	25

FIGURE 11.1
Scatterplot of the Yield Versus Fertilizer Data in Table 11.1

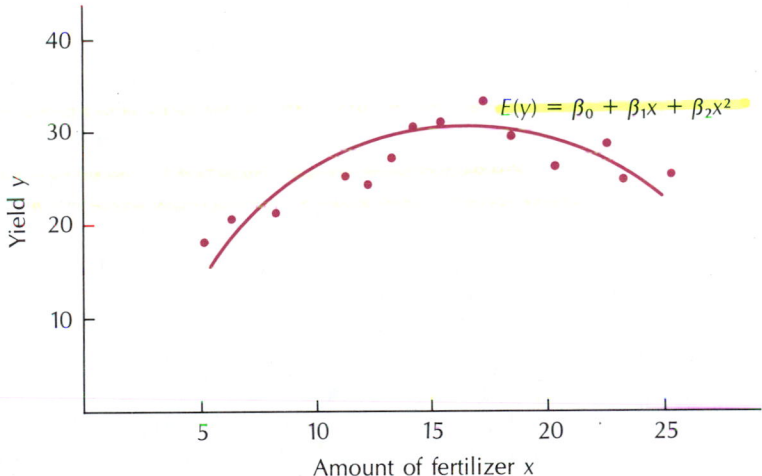

A model for this physical situation might be

$$y = \beta_0 + \beta_1 x + \beta_2 x^2 + \varepsilon.$$

Again with the assumption that $E(\varepsilon) = 0$, the expected value of y for a given value of x is

$$E(y) = \beta_0 + \beta_1 x + \beta_2 x^2.$$

One such line is plotted in Figure 11.1, superimposed on the data of Table 11.1.

A general polynomial regression model relating a dependent variable y to a single quantitative independent variable x is given by

$$y = \beta_0 + \beta_1 x + \beta_2 x^2 + \cdots + \beta_p x^p + \varepsilon$$

with

$$E(y) = \beta_0 + \beta_1 x + \beta_2 x^2 + \cdots + \beta_p x^p.$$

The choice of p and hence the choice of an appropriate regression model will depend on the experimental situation.

multiple regression model

The **multiple regression model** that relates a dependent variable y to a set of quantitative independent variables is a direct extension of a polynomial regression model in one independent variable. We write the multiple regression model as

$$y = \beta_0 + \beta_1 x_1 + \beta_2 x_2 + \cdots + \beta_k x_k + \varepsilon.$$

Any of the independent variables may be powers of other independent variables; for example, x_2 might be x_1^2. In fact, there are many other possibilities; x_3 might

cross-product term

be a **cross-product term** equal to $x_1 x_2$, x_4 might be $\log x_1$, etc. The only restriction is that no x is a perfect linear function of any other x.

first-order model

The simplest type of multiple regression equation is a **first-order model**, in which each of the independent variables appears, but there are no cross-product terms or terms in powers of the independent variables. For example, when three quantitative independent variables are involved, the first-order multiple regression model would be

$$y = \beta_0 + \beta_1 x_1 + \beta_2 x_2 + \beta_3 x_3 + \varepsilon.$$

For these first-order models we can attach some meaning to the βs. The parameter β_0 is the y-intercept, which represents the expected value of y when each x is zero. For cases in which it does not make sense to have each x be zero, β_0 (or its estimate) should be used only as part of the prediction equation, and not given an interpretation by itself.

partial slopes

The other parameters $(\beta_1, \beta_2, \ldots, \beta_k)$ in the multiple regression equation are sometimes called **partial slopes**. In linear regression, the parameter β_1 is the slope of the regression line and it represents the expected change in y for a unit increase in x. In a first-order multiple regression model, β_1 represents the expected change in y for a unit increase in x_1 when all other xs are held constant. In general then, $\beta_j (j \neq 0)$ represents the expected change in y for a unit increase in x_j while holding all other xs constant.

Besides the usual assumptions for a multiple regression model (see Chapter 9) there is an additional assumption that is implied when we use a first-order multiple regression model. Since the expected change in y for a unit change in x_j is constant and does not depend on the value of any other x, we are in fact assuming that the effects of the independent variables are **additive**.

additive effects

When might this additional assumption of additivity be warranted? Figure 11.2(a) shows a scatterplot of y versus x_1; Figure 11.2(b) shows the same plot with an ID attached to the different levels of a second independent variable x_2 (x_2 takes on the values of 1, 2, or 3). From Figure 11.2(a) we see that y is approximately linear in x_1. The parallel lines of Figure 11.2(b) corresponding to the three levels of the independent variable x_2 indicate that the expected change in y for a unit change in x_1 remains the same no matter which level of x_2 is used. These data suggest that the effects of x_1 and x_2 are additive; hence, a first-order model of the form $y = \beta_0 + \beta_1 x_1 + \beta_2 x_2 + \varepsilon$ is appropriate.

FIGURE 11.2
(a) Scatterplot of y Versus x_1; (b) Scatterplot of y Versus x_1, Indicating Additivity of Effects for x_1 and x_2

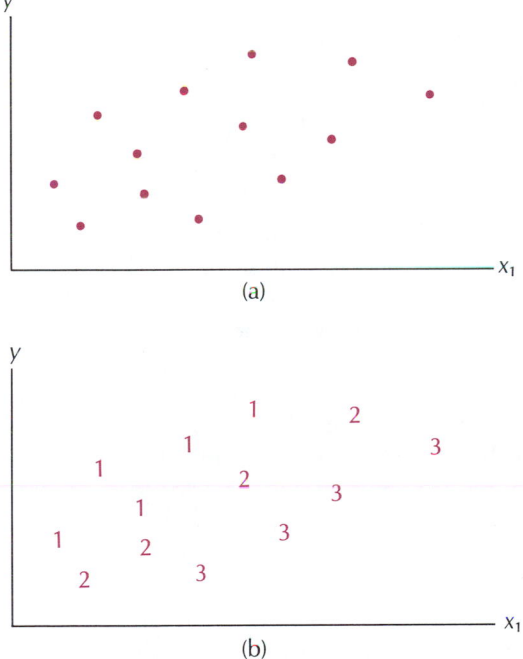

(a)

(b)

interaction

Figure 11.3 displays a situation in which **interaction** is present between the variables x_1 and x_2. Even though a scatterplot of y versus x_1 is as shown in Figure 11.2(a), the nonparallel lines of Figure 11.3 indicate that the expected change in y for a unit change in x_1 now depends on the level of x_2. When this occurs, the independent variables x_1 and x_2 are said to interact. A first-order model, which assumes additivity of the effects, would not be appropriate here. At the very least, we would include a cross-product term ($x_1 x_2$) in the model.

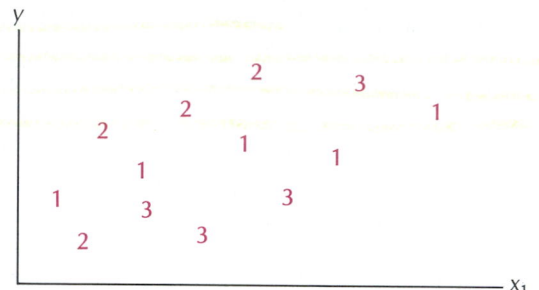

The simplest model allowing for interaction between x_1 and x_2 is

$$y = \beta_0 + \beta_1 x_1 + \beta_2 x_2 + \beta_3 x_1 x_2 + \varepsilon.$$

Note that for a given value of x_2 (say $x_2 = 2$), the expected value of y is

$$E(y) = \beta_0 + \beta_1 x_1 + \beta_2(2) + \beta_3 x_1(2)$$
$$= (\beta_0 + 2\beta_2) + (\beta_1 + 2\beta_3)x_1.$$

Here the intercept and slope are $(\beta_0 + 2\beta_2)$ and $(\beta_1 + 2\beta_3)$, respectively. The corresponding intercept and slope for $x_2 = 3$ can be shown to be $(\beta_0 + 3\beta_2)$ and $(\beta_1 + 3\beta_3)$. Clearly, the slopes of the two regression lines are not the same, and hence we have nonparallel lines.

Not all experiments can be modeled using a first-order multiple regression model. For these situations where a higher-order multiple regression model may be appropriate, it will be more difficult to assign a literal interpretation to the βs because of the presence of terms that contain cross-products or powers of the independent variables. Our focus will be on finding a multiple regression model that provides a good fit to the sample data, not on interpreting individual βs, except as they relate to the overall model.

The models that we have described briefly have been for regression problems where the experimenter is interested in developing a model to relate a response to one or more *quantitative* independent variables. The problem of modeling an experimental situation is not restricted to the quantitative independent-variable case.

Consider the problem of writing a model for an experimental situation in which a response y is related to a set of *qualitative* independent variables or to both quantitative and qualitative independent variables. For the first situation (relating y to one or more qualitative independent variables), let's suppose that we want to compare the average number of lightning discharges per minute for a storm, as measured from two different tracking posts located 30 miles apart. If we let y denote the number of discharges recorded on an oscilloscope during a 1-minute period, we could write the following two models:

For tracking post 1: $y = \mu_1 + \varepsilon$

For tracking post 2: $y = \mu_2 + \varepsilon$.

Thus, we assume that observations at tracking post 1 randomly "bob" about a population mean μ_1. Similarly, at tracking post 2, observations differ from a population mean μ_2 by a random amount ε. These two models are not new and could have been used to describe observations when comparing two population means in Chapter 6. What is new is that we can combine these two models into a single model of the form

$$y = \beta_0 + \beta_1 x_1 + \varepsilon,$$

dummy variable

where β_0 and β_1 are unknown parameters, ε is a random error term, and x_1 is a **dummy variable** with the following interpretation. We let

$x_1 = 1$ if an observation is obtained from tracking post 2

$x_1 = 0$ if an observation is obtained from tracking post 1.

For observations obtained from tracking post 1, we substitute $x_1 = 0$ into our model to obtain

$$y = \beta_0 + \beta_1(0) + \varepsilon = \beta_0 + \varepsilon.$$

Hence, $\beta_0 = \mu_1$, the population mean for observations from tracking post 1. Similarly, by substituting $x_1 = 1$ in our model, the equation for observations from tracking post 2 is

$$y = \beta_0 + \beta_1(1) + \varepsilon = \beta_0 + \beta_1 + \varepsilon.$$

Since $\beta_0 = \mu_1$ and $\beta_0 + \beta_1$ must equal μ_2, we have $\beta_1 = \mu_2 - \mu_1$, the difference in means between observations from tracking posts 2 and 1.

This model, $y = \beta_0 + \beta_1 x_1 + \varepsilon$, which relates y to the qualitative independent variable tracking post, can be extended to a situation in which the qualitative variable has more than two levels. We do this by using more than one dummy variable. Consider an experiment in which we're interested in four levels of qualitative variables. We call these levels **treatments**. We could write the model

treatments

$$y = \beta_0 + \beta_1 x_1 + \beta_2 x_2 + \beta_3 x_3 + \varepsilon,$$

where

$x_1 = 1$ if treatment 2, $x_1 = 0$ otherwise

$x_2 = 1$ if treatment 3, $x_2 = 0$ otherwise

$x_3 = 1$ if treatment 4, $x_3 = 0$ otherwise.

To interpret the βs in this equation, it is convenient to construct a table of the expected values. Since ε has expectation zero, the general expression for the expected value of y is

$$E(y) = \beta_0 + \beta_1 x_1 + \beta_2 x_2 + \beta_3 x_3.$$

The expected value for observations on treatment 1 is found by substituting $x_1 = 0$, $x_2 = 0$, and $x_3 = 0$; after this substitution, we find $E(y) = \beta_0$. The expected value for observations on treatment 2 is found by substituting $x_1 = 1$, $x_2 = 0$, and $x_3 = 0$ into the $E(y)$ formula; this substitution yields $E(y) = \beta_0 + \beta_1$. Substitutions of $x_1 = 0$, $x_2 = 1$, $x_3 = 0$ and $x_1 = 0$, $x_2 = 0$, $x_3 = 1$ yield expected values for treatments 3 and 4, respectively. These expected values are summarized in Table 11.2.

TABLE 11.2

Expected Values for an Experiment with Four Treatments

	Treatment		
1	2	3	4
$E(y) = \beta_0$	$E(y) = \beta_0 + \beta_1$	$E(y) = \beta_0 + \beta_2$	$E(y) = \beta_0 + \beta_3$

If we identify the mean of treatment 1 as μ_1, the mean of treatment 2 as μ_2, and so on, then from Table 11.2 we have

$$\mu_1 = \beta_0, \qquad \mu_2 = \beta_0 + \beta_1, \qquad \mu_3 = \beta_0 + \beta_2, \qquad \text{and} \qquad \mu_4 = \beta_0 + \beta_3.$$

Solving these equations for the βs, we have

$$\beta_0 = \mu_1 \qquad\qquad \beta_2 = \mu_3 - \mu_1$$
$$\beta_1 = \mu_2 - \mu_1 \qquad\qquad \beta_3 = \mu_4 - \mu_1.$$

Any comparison among the treatment means can be phrased in terms of the βs. For example, the comparison $\mu_4 - \mu_3$ could be written as $\beta_3 - \beta_2$. Likewise, $\mu_3 - \mu_2$ could be written as $\beta_2 - \beta_1$.

EXAMPLE 11.1 ▼

Consider a hypothetical situation for an experiment with four treatments ($t = 4$) in which we know the means for the four treatments. If $\mu_1 = 7$, $\mu_2 = 9$, $\mu_3 = 6$, and $\mu_4 = 15$, determine values for β_0, β_1, β_2, and β_3 in the model

$$y = \beta_0 + \beta_1 x_1 + \beta_2 x_2 + \beta_3 x_3 + \varepsilon,$$

where

$$x_1 = 1 \text{ if treatment 2}, \qquad x_1 = 0 \text{ otherwise}$$
$$x_2 = 1 \text{ if treatment 3}, \qquad x_2 = 0 \text{ otherwise}$$
$$x_3 = 1 \text{ if treatment 4}, \qquad x_3 = 0 \text{ otherwise.}$$

Solution   Based on what was presented in Table 11.2, we know that

$$\beta_0 = \mu_1$$
$$\beta_1 = \mu_2 - \mu_1$$
$$\beta_2 = \mu_3 - \mu_1$$

and

$$\beta_3 = \mu_4 - \mu_1.$$

Using the known values for μ_1, μ_2, μ_3, and μ_4, it follows that

$$\beta_0 = 7$$
$$\beta_1 = 9 - 7 = 2$$
$$\beta_2 = 6 - 7 = -1$$

and

$$\beta_3 = 15 - 7 = 8.$$ ▲

EXAMPLE 11.2   ▼

Refer to Example 11.1. Express $\mu_3 - \mu_2$ and $\mu_3 - \mu_4$ in terms of the βs. Check your findings by substituting values for the βs.

Solution   Using the relationship between the βs and the μs, it can be seen that

$$\beta_2 - \beta_1 = (\mu_3 - \mu_1) - (\mu_2 - \mu_1) = \mu_3 - \mu_2$$

and

$$\beta_2 - \beta_3 = (\mu_3 - \mu_1) - (\mu_4 - \mu_1) = \mu_3 - \mu_4.$$

Substituting computed values for the βs, we have

$$\beta_2 - \beta_1 = -1 - (2) = -3$$

and

$$\beta_2 - \beta_3 = -1 - (8) = -9.$$

These computed values are identical to the "known" differences for $\mu_3 - \mu_2$ and $\mu_3 - \mu_4$, respectively.   ▲

EXAMPLE 11.3   ▼

Use dummy variables to write the model for an experiment with t treatments. Identify the βs.

Solution   The model could be written in the form

$$y = \beta_0 + \beta_1 x_1 + \beta_2 x_2 + \cdots + \beta_{t-1} x_{t-1} + \varepsilon,$$

where

$$x_1 = 1 \text{ if treatment 2,} \qquad x_1 = 0 \text{ otherwise}$$
$$x_2 = 1 \text{ if treatment 3,} \qquad x_2 = 0 \text{ otherwise}$$
$$\vdots \qquad\qquad\qquad \vdots$$
$$x_{t-1} = 1 \text{ if treatment } t, \qquad x_{t-1} = 0 \text{ otherwise.}$$

The table of expected values would be

	Treatment		
1	2	$\cdots$	t
$E(y) = \beta_0$	$E(y) = \beta_0 + \beta_1$	$\cdots$	$E(y) = \beta_0 + \beta_{t-1}$

from which we obtain

$$\beta_0 = \mu_1$$
$$\beta_1 = \mu_2 - \mu_1$$
$$\vdots$$
$$\beta_{t-1} = \mu_t - \mu_1.$$

▲

In the procedure just described, we have a response related to the qualitative variable "treatments," and for t levels of the treatments, we enter $(t - 1)$ βs into our model, using dummy variables. More will be said about the use of the models for more than one qualitative independent variable in Chapters 15 and 16, where we consider the analysis of variance for several different experimental designs.

11.2 ▼ THE GENERAL LINEAR MODEL

It is important at this point to recognize that a single general model can be used for multiple regression models in which a response is related to a set of quantitative independent variables, and for models that relate y to a set of qualitative independent variables. This model, called the **general linear model** has the form

general linear model

$$y = \beta_0 + \beta_1 x_1 + \beta_2 x_2 + \cdots + \beta_k x_k + \varepsilon.$$

For multiple regression models, the xs represent independent variables (such as weight or amount of water), independent variables raised to powers, and cross-product terms involving the independent variables. A few regression models were discussed in Section 11.1; more about the use of the general linear model in regression will be discussed in the remainder of this chapter and in Chapter 12.

The xs of the general linear model represent dummy variables (coded 0 and 1) or products of dummy variables when y is related to a set of qualitative independent variables. We discussed how to use dummy variables for representing y in terms of a single qualitative variable in Section 11.1; the same approach can be used to relate y to more than one qualitative independent variable. This will be discussed in Chapter 15, where we present more analysis of variance techniques.

The general linear model can also be used for the case in which y is related to both qualitative and quantitative independent variables. A particular example of this is discussed in Section 12.5, and other applications are presented in Chapter 19.

Why is this model called the general *linear* model, especially since it can be used for polynomial models? The word "linear" in the general linear model refers to how the βs are entered in the model, not to how the independent variables appear in the model. A general linear model is linear (used in the usual algebraic sense) in the βs.

Why are we discussing the general linear model now? The techniques that we will develop in this chapter for making inferences about a single β, a set of βs, and $E(y)$ in multiple regression are those that apply to any general linear model. Thus, using general linear model techniques we have a common thread to inferences about multiple regression (Chapters 11 and 12) and the analysis of variance (Chapters 15, 16, 17, and 18). As you study these six chapters, try whenever possible to make the connection back to a general linear model; we'll help you with this connection.

▼ EXERCISES

Basic Techniques

11.1 **a.** Write a first-order multiple regression model relating a response y to three qualitative independent variables.

b. Show how this model can be written as a general linear model.

11.2 Write a second-order multiple regression model relating a response y to three quantitative independent variables. Include all possible terms. (Hint: A first-order model contains terms in the x_j; a second-order model includes these terms as well as squares and cross-products.)

11.3 Refer to Exercise 11.2. Show that the model you wrote can be written in the form of a general linear model. Identify the terms.

11.4 Consider the model

$$y = \beta_0 + \beta_1 x_1 + \beta_2 x_2 + \varepsilon,$$

where

$$x_1 = \begin{cases} 1 & \text{if treatment 2} \\ 0 & \text{otherwise} \end{cases}$$

$$x_2 = \begin{cases} 1 & \text{if treatment 3} \\ 0 & \text{otherwise.} \end{cases}$$

a. Interpret the βs in the model.

b. Identify the difference in mean responses for treatments 2 and 3 using the model.

11.5 (Optional) Refer to Exercise 11.4. Suppose that the model is expanded to include the term $\beta_3 x_3$, where x_3 is a dummy variable for the qualitative variable "location."

$$x_3 = \begin{cases} 1 & \text{if location 2} \\ 0 & \text{otherwise.} \end{cases}$$

a. Interpret the βs for this model. (Hint: Consider all combinations of the three treatments and two locations.)

b. Write the difference in mean response for treatments 2 and 3 for location 2. Is it the same for location 1?

c. Identify an experimental situation in which this model might be a reasonable approximation.

11.6 (Optional) A study was done to examine the effect of a quantitative independent variable (age) on reaction time (as measured by braking time). The experiment included males and females. Two models were proposed:

$$y = \beta_0 + \beta_1 x_1 + \beta_2 x_2 + \varepsilon$$

and

$$y = \beta_0 + \beta_1 x_1 + \beta_2 x_2 + \beta_3 x_1 x_2,$$

where

$$x_1 = \text{age (in years)}$$

and

$$x_2 = \begin{cases} 1 & \text{female} \\ 0 & \text{if male.} \end{cases}$$

Interpret the βs for the two models and explain a practical difference between the two models.

11.3 ▼ LEAST SQUARES SOLUTION TO THE GENERAL LINEAR MODEL

The general linear model relates a response y to a set of independent variables (qualitative or quantitative). For a random sample of n measurements we can write the ith observation as

$$y_i = \beta_0 + \beta_1 x_{i1} + \beta_2 x_{i2} + \cdots + \beta_k x_{ik} + \varepsilon_i \qquad (i = 1, 2, \ldots, n; n > k),$$

where $x_{i1}, x_{i2}, \ldots, x_{ik}$ are the settings of the independent variables corresponding to the observation y_i.

In order to find least squares estimates for $\beta_0, \beta_1, \ldots,$ and β_k in a general linear model, we follow the same procedure that we did for a linear regression model in Chapter 9. A random sample of n observations is obtained; the least squares prediction equation

$$\hat{y} = \hat{\beta}_0 + \hat{\beta}_1 x_1 + \cdots + \hat{\beta}_k x_k$$

is found by choosing $\hat{\beta}_0, \hat{\beta}_1, \ldots, \hat{\beta}_k$ to minimize the expression $\text{SSE} = \sum_i (y_i - \hat{y}_i)^2$. But, although it was easy to write down the solutions to $\hat{\beta}_0$ and $\hat{\beta}_1$ for the linear regression model,

$$y = \beta_0 + \beta_1 x + \varepsilon,$$

the estimates for $\beta_0, \beta_1, \ldots, \beta_k$ must be found by solving a set of simultaneous equations, called the *normal equations*, shown here.

	y_i	$\hat{\beta}_0$	$x_{i1}\hat{\beta}_1$	$\cdots$	$x_{ik}\hat{\beta}_k$
1	$\sum y_i$	$= n\hat{\beta}_0$	$+ \sum x_{i1}\hat{\beta}_1$	$+ \cdots +$	$\sum x_{ik}\hat{\beta}_k$
x_{i1}	$\sum x_{i1}y_i$	$= \sum x_{i1}\hat{\beta}_0$	$+ \sum x_{i1}^2\hat{\beta}_1$	$+ \cdots +$	$\sum x_{i1}x_{ik}\hat{\beta}_k$
$\vdots$	$\vdots$				
x_{ik}	$\sum x_{ik}y_i$	$= \sum x_{ik}\hat{\beta}_0$	$+ \sum x_{ik}x_{i1}\hat{\beta}_1$	$+ \cdots +$	$\sum x_{ik}^2\hat{\beta}_k$

Note the pattern associated with these equations. By labeling the rows and columns as we have done, we can obtain any term in the normal equations by multiplying the row and column elements and summing. For example, the last term in the second equation is found by multiplying the row element (x_{i1}) by the column element $(x_{ik}\hat{\beta}_k)$ and summing; the resulting term is $\sum x_{i1}x_{ik}\hat{\beta}_k$. Because all terms in the normal equations can be formed in this way, it is fairly simple to write down the equations to be solved in order to obtain the least squares estimates $\hat{\beta}_0, \hat{\beta}_1, \ldots, \hat{\beta}_k$. The solution to these equations is not necessarily trivial; that's why we'll enlist the help of various statistical software packages for their solution.

EXAMPLE 11.4 ▼

In Exercise 10.28, we presented data for the weight loss of a compound for different amounts of time the compound was exposed to the air. Additional information was also available on the humidity of the environment during exposure. The complete data are presented in Table 11.3.

TABLE 11.3

Weight Loss, Exposure Time, and Relative Humidity Data for Example 11.4

Weight Loss, y (pounds)	Exposure Time, x_1 (hours)	Relative Humidity, x_2
4.3	4	.20
5.5	5	.20
6.8	6	.20
8.0	7	.20
4.0	4	.30
5.2	5	.30
6.6	6	.30
7.5	7	.30
2.0	4	.40
4.0	5	.40
5.7	6	.40
6.5	7	.40

a. Set up the normal equations for this regression problem if the assumed model is

$$y = \beta_0 + \beta_1 x_1 + \beta_2 x_2 + \varepsilon,$$

where x_1 is exposure time and x_2 is relative humidity.

b. Use the computer output shown here to determine the least squares estimates of β_0, β_1, and β_2. Predict weight loss for 6.5 hours of exposure and a relative humidity of .35.

```
        LISTING OF DATA

 OBS   WT_LOSS    TIME    HUMID

  1      4.3       4.0     0.20
  2      5.5       5.0     0.20
  3      6.8       6.0     0.20
  4      8.0       7.0     0.20
  5      4.0       4.0     0.30
  6      5.2       5.0     0.30
  7      6.6       6.0     0.30
  8      7.5       7.0     0.30
  9      2.0       4.0     0.40
 10      4.0       5.0     0.40
 11      5.7       6.0     0.40
 12      6.5       7.0     0.40
 13                6.5     0.35

 N = 13
```

LEAST SQUARES ANALYSIS

GENERAL LINEAR MODELS PROCEDURE

DEPENDENT VARIABLE: WT_LOSS WEIGHT-LOSS (POUNDS)

SOURCE	DF	SUM OF SQUARES	MEAN SQUARE	F VALUE	PR > F	R-SQUARE	C.V.
MODEL	2	31.12416667	15.56208333	104.13	0.0001	0.958576	7.0181
ERROR	9	1.34500000	0.14944444				
					STD DEV		WT_LOSS MEAN
CORRECTED TOTAL	11	32.46916667			0.38658045		5.50833333

SOURCE	DF	TYPE I SS	F VALUE	PR > F	DF	TYPE IV SS	F VALUE	PR > F
TIME	1	26.00416667	174.01	0.0001	1	26.00416667	174.01	0.0001
HUMID	1	5.12000000	34.26	0.0002	1	5.12000000	34.26	0.0002

PARAMETER	ESTIMATE	T FOR H0: PARAMETER = 0	PR > |T|	STD ERROR OF ESTIMATE
INTERCEPT	0.66666667	0.96	0.3620	0.69423219
TIME	1.31666667	13.19	0.0001	0.09981464
HUMID	-8.00000000	-5.85	0.0002	1.36676829

OBSERVATION	OBSERVED VALUE	PREDICTED VALUE	RESIDUAL	LOWER 95% CL FOR MEAN	UPPER 95% CL FOR MEAN
1	4.30000000	4.33333333	-0.03333333	3.80984168	4.85682499
2	5.50000000	5.65000000	-0.15000000	5.23518217	6.06481783
3	6.80000000	6.96666667	-0.16666667	6.55184883	7.38148450
4	8.00000000	8.28333333	-0.28333333	7.75984168	8.80682499
5	4.00000000	3.53333333	0.46666667	3.11090353	3.95576314
6	5.20000000	4.85000000	0.35000000	4.57345478	5.12654522
7	6.60000000	6.16666667	0.43333333	5.89012144	6.44321189
8	7.50000000	7.48333333	0.01666667	7.06090353	7.90576314
9	2.00000000	2.73333333	-0.73333333	2.20984168	3.25682499
10	4.00000000	4.05000000	-0.05000000	3.63518217	4.46481783
11	5.70000000	5.36666667	0.33333333	4.95184883	5.78148450
12	6.50000000	6.68333333	-0.18333333	6.15984168	7.20682499
13*		6.42500000		6.05268960	6.79731040

*OBSERVATION WAS NOT USED IN THIS ANALYSIS

```
SUM OF RESIDUALS                          0.00000000
SUM OF SQUARED RESIDUALS                  1.34500000
SUM OF SQUARED RESIDUALS - ERROR SS      -0.00000000
PRESS STATISTIC                           2.61233345
FIRST ORDER AUTOCORRELATION               0.15902520
DURBIN-WATSON D                           1.65613383
```

Solution

a. The three normal equations for this model are shown here.

	y_i	$\hat{\beta}_0$	$x_{i1}\hat{\beta}_1$	$x_{i2}\hat{\beta}_2$
1	$\sum y_i =$	$n\hat{\beta}_0$ +	$\sum x_{i1}\hat{\beta}_1$ +	$\sum x_{i2}\hat{\beta}_2$
x_{i1}	$\sum x_{i1}y_i =$	$\sum x_{i1}\hat{\beta}_0$ +	$\sum x_{i1}^2\hat{\beta}_1$ +	$\sum x_{i1}x_{i2}\hat{\beta}_2$
x_{i2}	$\sum x_{i2}y_i =$	$\sum x_{i2}\hat{\beta}_0$ +	$\sum x_{i2}x_{i1}\hat{\beta}_1$ +	$\sum x_{i2}^2\hat{\beta}_2$

For these data, we have

$$\sum y_i = 66.10, \qquad \sum x_{i1} = 66, \qquad \sum x_{i2} = 3.60,$$
$$\sum x_{i1}y_i = 383.3, \qquad \sum x_{i2}y_i = 19.19, \qquad \sum x_{i1}x_{i2} = 19.8.$$
$$\sum x_{i1}^2 = 378, \qquad \sum x_{i2}^2 = 1.16,$$

Substituting these values into the normal equation yields the result shown here:

$$66.1 = 12\hat{\beta}_0 + 66\hat{\beta}_1 + 3.6\hat{\beta}_2$$
$$383.3 = 66\hat{\beta}_0 + 378\hat{\beta}_1 + 19.8\hat{\beta}_2$$
$$19.19 = 3.6\hat{\beta}_0 + 19.8\hat{\beta}_1 + 1.16\hat{\beta}_2.$$

b. The normal equations of part a could be solved to determine $\hat{\beta}_0$, $\hat{\beta}_1$, and $\hat{\beta}_2$. The solution would agree with that shown here in the output. The least squares prediction equation is

$$\hat{y} = 0.667 + 1.317x_1 - 8.000x_2,$$

where x_1 is exposure time and x_2 is relative humidity. Substituting $x_1 = 6.5$ and $x_2 = .35$, we have

$$\hat{y} = 0.667 + 1.317(6.5) - 8.000(.35) = 6.428.$$

This value agrees with the predicted value shown as observation 13 in the output, except for rounding errors. ▲

▼ EXERCISES

Applications

11.7 A pharmaceutical firm would like to obtain information on the relationship between the dose level and potency of a drug product. To do this, each of 15 test tubes is inoculated with a virus culture and incubated for 5 days at 30°C. Three test tubes are randomly assigned to each of the 5 different dose levels to be investigated (2, 4, 8, 16, and 32 mg). Each tube is injected with only one dose level and the response of interest (a measure of the protective strength of the product against the virus culture) is obtained. The data are given here.

Dose Level	Response
2	5, 7, 3
4	10, 12, 14
8	15, 17, 18
16	20, 21, 19
32	23, 24, 29

a. Plot the data.
b. Fit a linear regression model to these data.
c. What other regression model might be appropriate?
d. SAS computer output is shown for both a linear and quadratic regression equation. Which regression equation appears to fit the data better? Why?

OBS	DOSE	RESPONSE
1	2	5
2	2	7
3	2	3
4	4	10
5	4	12
6	4	14
7	8	15
8	8	17
9	8	18
10	16	20
11	16	21
12	16	19
13	32	23
14	32	24
15	32	29

N = 15

LINEAR REGRESSION ANALYSIS

GENERAL LINEAR MODELS PROCEDURE

DEPENDENT VARIABLE: RESPONSE

SOURCE	DF	SUM OF SQUARES	MEAN SQUARE	F VALUE	PR > F	R-SQUARE	C.V.
MODEL	1	590.91612903	590.91612903	44.28	0.0001	0.773046	23.1207
ERROR	13	173.48387097	13.34491315		ROOT MSE		RESPONSE MEAN
CORRECTED TOTAL	14	764.40000000			3.65306900		15.80000000

SOURCE	DF	TYPE I SS	F VALUE	PR > F	DF	TYPE III SS	F VALUE	PR > F
DOSE	1	590.91612903	44.28	0.0001	1	590.91612903	44.28	0.0001

PARAMETER	ESTIMATE	T FOR H0: PARAMETER=0	PR > |T|	STD ERROR OF ESTIMATE
INTERCEPT	8.66666667	6.07	0.0001	1.42786770
DOSE	0.57526882	6.65	0.0001	0.08645016

(continues)

```
OBSERVATION          OBSERVED          PREDICTED          RESIDUAL
                      VALUE             VALUE
     1              5.00000000         9.81720430        -4.81720430
     2              7.00000000         9.81720430        -2.81720430
     3              3.00000000         9.81720430        -6.81720430
     4             10.00000000        10.96774194        -0.96774194
     5             12.00000000        10.96774194         1.03225806
     6             14.00000000        10.96774194         3.03225806
     7             15.00000000        13.26881720         1.73118280
     8             17.00000000        13.26881720         3.73118280
     9             18.00000000        13.26881720         4.73118280
    10             20.00000000        17.87096774         2.12903226
    11             21.00000000        17.87096774         3.12903226
    12             19.00000000        17.87096774         1.12903226
    13             23.00000000        27.07526882        -4.07526882
    14             24.00000000        27.07526882        -3.07526882
    15             29.00000000        27.07526882         1.92473118

         SUM OF RESIDUALS                             0.00000000
         SUM OF SQUARED RESIDUALS                   173.48387097
         SUM OF SQUARED RESIDUALS - ERROR SS         -0.00000000
         FIRST ORDER AUTOCORRELATION                  0.53691663
         DURBIN-WATSON D                              0.77105118
```

Linear regression analysis
plot of residuals vs. predicted values

X-axis: Predicted value
Y-axis: Residual

```
                        QUADRATIC REGRESSION ANALYSIS

                       GENERAL LINEAR MODELS PROCEDURE

DEPENDENT VARIABLE:  RESPONSE
```

SOURCE	DF	SUM OF SQUARES	MEAN SQUARE	F VALUE	PR > F	R-SQUARE	C.V.
MODEL	2	673.82061986	336.91030993	44.63	0.0001	0.881503	17.3887
ERROR	12	90.57938014	7.54828168		ROOT MSE		RESPONSE MEAN
CORRECTED TOTAL	14	764.40000000			2.74741363		15.80000000

(continues)

SOURCE	DF	TYPE I SS	F VALUE	PR > F	DF	TYPE III SS	F VALUE	PR > F
DOSE	1	590.91612903	78.28	0.0001	1	205.97013882	27.29	0.0002
DOSE*DOSE	1	82.90449083	10.98	0.0062	1	82.90449083	10.98	0.0062

PARAMETER	ESTIMATE	T FOR H0: PARAMETER=0	PR > |T|	STD ERROR OF ESTIMATE
INTERCEPT	4.48366013	2.71	0.0191	1.65720368
DOSE	1.50632511	5.22	0.0002	0.28836373
DOSE*DOSE	-0.02698714	-3.31	0.0062	0.00814314

OBSERVATION	OBSERVED VALUE	PREDICTED VALUE	RESIDUAL
1	5.00000000	7.38836180	-2.38836180
2	7.00000000	7.38836180	-0.38836180
3	3.00000000	7.38836180	-4.38836180
4	10.00000000	10.07716635	-0.07716635
5	12.00000000	10.07716635	1.92283365
6	14.00000000	10.07716635	3.92283365
7	15.00000000	14.80708412	0.19291588
8	17.00000000	14.80708412	2.19291588
9	18.00000000	14.80708412	3.19291588
10	20.00000000	21.67615433	-1.67615433
11	21.00000000	21.67615433	-0.67615433
12	19.00000000	21.67615433	-2.67615433
13	23.00000000	25.05123340	-2.05123340
14	24.00000000	25.05123340	-1.05123340
15	29.00000000	25.05123340	3.94876660

SUM OF RESIDUALS	0.00000000
SUM OF SQUARED RESIDUALS	90.57938014
SUM OF SQUARED RESIDUALS - ERROR SS	0.00000000
FIRST ORDER AUTOCORRELATION	0.21674175
DURBIN-WATSON D	1.33139643

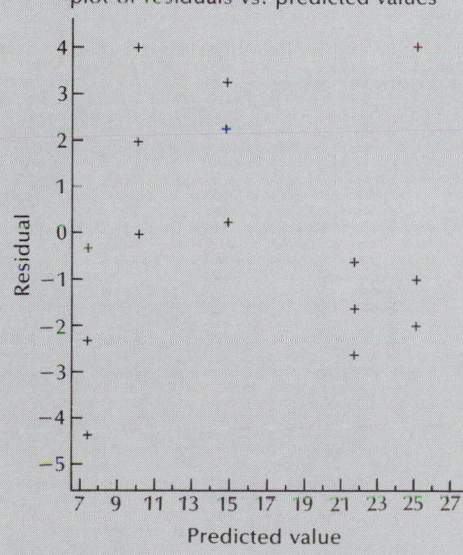

Quadratic regression analysis
plot of residuals vs. predicted values

11.8 Refer to the data of Exercise 11.7. Many times a logarithmic transformation can be used on the dose levels to linearize the response with respect to the independent variable.

a. Refer to a set of log tables (see, for example, the Chemical Rubber Company tables) or use a calculator to obtain the logarithms of the five dose levels.

b. If x_1 denotes the log dose, fit the model

$$y = \beta_0 + \beta_1 x_1 + \varepsilon.$$

A residual plot is shown here in the output.

c. Compare your results in part b to those for Exercise 11.7. Does the logarithmic transformation provide a better linear fit than that in Exercise 11.7?

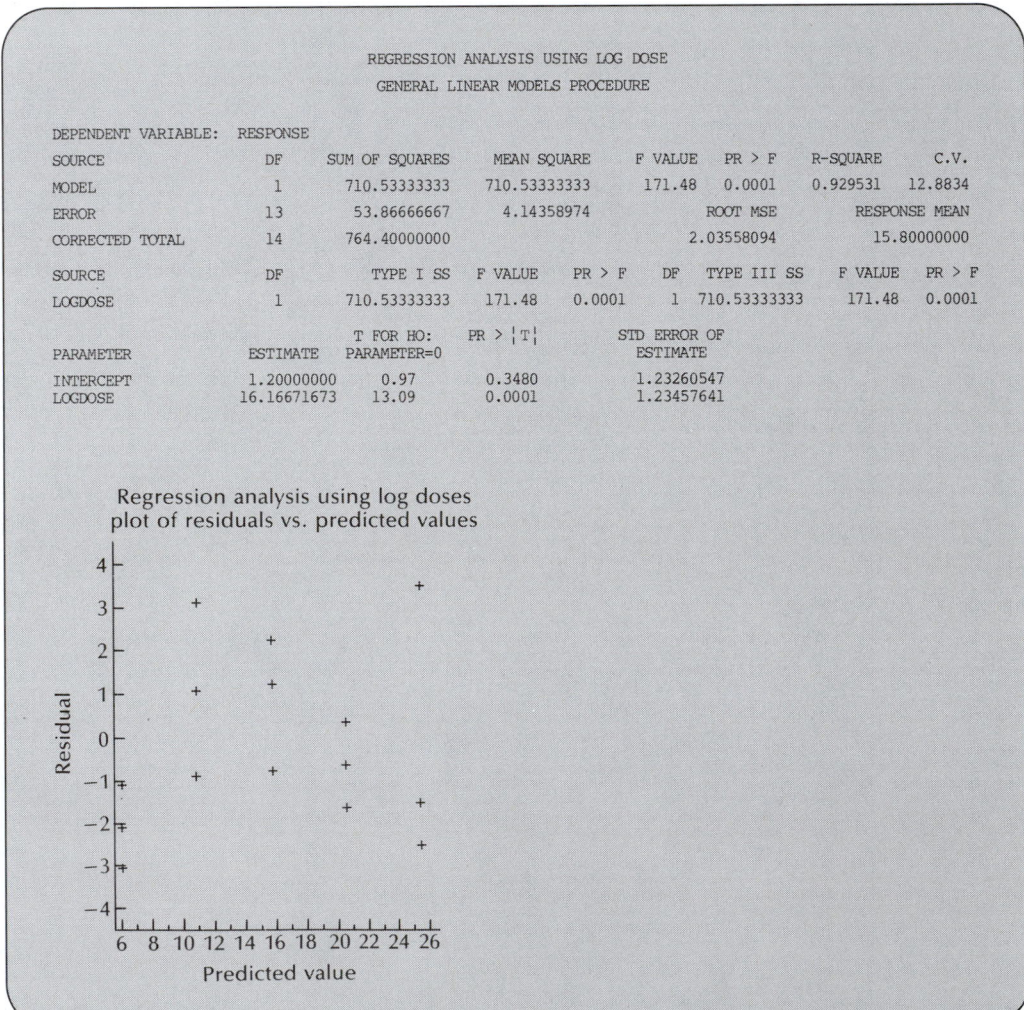

REGRESSION ANALYSIS USING LOG DOSE
GENERAL LINEAR MODELS PROCEDURE

DEPENDENT VARIABLE: RESPONSE

SOURCE	DF	SUM OF SQUARES	MEAN SQUARE	F VALUE	PR > F	R-SQUARE	C.V.
MODEL	1	710.53333333	710.53333333	171.48	0.0001	0.929531	12.8834
ERROR	13	53.86666667	4.14358974		ROOT MSE		RESPONSE MEAN
CORRECTED TOTAL	14	764.40000000			2.03558094		15.80000000

SOURCE	DF	TYPE I SS	F VALUE	PR > F	DF	TYPE III SS	F VALUE	PR > F
LOGDOSE	1	710.53333333	171.48	0.0001	1	710.53333333	171.48	0.0001

PARAMETER	ESTIMATE	T FOR H0: PARAMETER=0	PR > |T|	STD ERROR OF ESTIMATE
INTERCEPT	1.20000000	0.97	0.3480	1.23260547
LOGDOSE	16.16671673	13.09	0.0001	1.23457641

Regression analysis using log doses
plot of residuals vs. predicted values

11.9 The abrasive effect of a wear tester for experimental fabrics was tested on a particular fabric while run at six different machine speeds. Forty-eight identical 5-inch-square pieces of fabric were cut, with eight squares randomly assigned to each of the six machine speeds 100, 120, 140, 160, 180, and 200 revolutions per minute (rev/min). The order of assignment of the squares to the machine was random, with each square tested for a 3-minute period at the appropriate machine setting. The amount of wear was measured and recorded for each square. The data appear here.

Machine Speed (in rev/min)	Wear
100	23.0, 23.5, 24.4, 25.2, 25.6, 26.1, 24.8, 25.6
120	26.7, 26.1, 25.8, 26.3, 27.2, 27.9, 28.3, 27.4
140	28.0, 28.4, 27.0, 28.8, 29.8, 29.4, 28.7, 29.3
160	32.7, 32.1, 31.9, 33.0, 33.5, 33.7, 34.0, 32.5
180	43.1, 41.7, 42.4, 42.1, 43.5, 43.8, 44.2, 43.6
200	54.2, 43.7, 53.1, 53.8, 55.6, 55.9, 54.7, 54.5

a. Generate a graph of the data (since the variability within a speed is about the same for all speeds, you can save time while still maintaining the trend by plotting the sample mean for each speed).
b. What type of regression model appears appropriate?
c. Output for linear, quadratic, and cubic regression models is shown on the following pages. Which regression equation gives a better fit? Why?
d. Is there anything peculiar about the data? What might have happened?

```
LISTING OF DATA

OBS    SPEED    WEAR

 1      100     23.0
 2      100     23.5
 3      100     24.4
 4      100     25.2
 5      100     25.6
 6      100     26.1
 7      100     24.8
 8      100     25.6
 9      120     26.7
10      120     26.1
11      120     25.8
12      120     26.3
13      120     27.2
14      120     27.9
15      120     28.3
16      120     27.4
17      140     28.0
18      140     28.4
19      140     27.0
20      140     28.8
21      140     29.8
22      140     29.4
23      140     28.7
24      140     29.3
25      160     32.7
```

(continues)

```
    LISTING OF DATA
   OBS    SPEED    WEAR
   26     160      32.1
   27     160      31.9
   28     160      33.0
   29     160      33.5
   30     160      33.7
   31     160      34.0
   32     160      32.5
   33     180      43.1
   34     180      41.7
   35     180      42.4
   36     180      42.1
   37     180      43.5
   38     180      43.8
   39     180      44.2
   40     180      43.6
   41     200      54.2
   42     200      43.7
   43     200      53.1
   44     200      53.8
   45     200      55.6
   46     200      55.9
   47     200      54.7
   48     200      54.5
 N =    48
```

LINEAR REGRESSION ANALYSIS

GENERAL LINEAR MODELS PROCEDURE

DEPENDENT VARIABLE: WEAR

SOURCE	DF	SUM OF SQUARES	MEAN SQUARE	F VALUE	PR > F	R-SQUARE	C.V.
MODEL	1	4326.79207143	4326.79207143	291.47	0.0001	0.863693	11.0305
ERROR	46	682.84709524	14.84450207		ROOT MSE		WEAR MEAN
CORRECTED TOTAL	47	5009.63916667			3.85285635		34.92916667

SOURCE	DF	TYPE I SS	F VALUE	PR > F	DF	TYPE III SS	F VALUE	PR > F
SPEED	1	4326.79207143	291.47	0.0001	1	4326.79207143	291.47	0.0001

PARAMETER	ESTIMATE	T FOR H0: PARAMETER=0	PR > |T|	STD ERROR OF ESTIMATE
INTERCEPT	-6.76547619	-2.70	0.0096	2.50470943
SPEED	0.27796429	17.07	0.0001	0.01628129

OBSERVATION	OBSERVED VALUE	PREDICTED VALUE	RESIDUAL
1	23.00000000	21.03095238	1.96904762
2	23.50000000	21.03095238	2.46904762
3	24.40000000	21.03095238	3.36904762
4	25.20000000	21.03095238	4.16904762
5	25.60000000	21.03095238	4.56904762
6	26.10000000	21.03095238	5.06904762
7	24.80000000	21.03095238	3.76904762
8	25.60000000	21.03095238	4.56904762
9	26.70000000	26.59023810	0.10976190
10	26.10000000	26.59023810	-0.49023810
11	25.80000000	26.59023810	-0.79023810
12	26.30000000	26.59023810	-0.29023810
13	27.20000000	26.59023810	0.60976190
14	27.90000000	26.59023810	1.30976190
15	28.30000000	26.59023810	1.70976190
16	27.40000000	26.59023810	0.80976190
17	28.00000000	32.14952381	-4.14952381
18	28.40000000	32.14952381	-3.74952381
19	27.00000000	32.14952381	-5.14952381
20	28.80000000	32.14952381	-3.34952381
21	29.80000000	32.14952381	-2.34952381
22	29.40000000	32.14952381	-2.74952381
23	28.70000000	32.14952381	-3.44952381
24	29.30000000	32.14952381	-2.84952381
25	32.70000000	37.70880952	-5.00880952
26	32.10000000	37.70880952	-5.60880952

(continues)

```
                              LINEAR REGRESSION ANALYSIS

                           GENERAL LINEAR MODELS PROCEDURE

  DEPENDENT VARIABLE:  WEAR

  OBSERVATION          OBSERVED          PREDICTED          RESIDUAL
                        VALUE             VALUE

       27             31.90000000       37.70880952       -5.60880952
       28             33.00000000       37.70880952       -4.70880952
       29             33.50000000       37.70880952       -4.20880952
       30             33.70000000       37.70880952       -4.00880952
       31             34.00000000       37.70880952       -3.70880952
       32             32.50000000       37.70880952       -5.20880952
       33             43.10000000       43.26809524       -0.16809524
       34             41.70000000       43.26809524       -1.56809524
       35             42.40000000       43.26809524       -0.86809524
       36             42.10000000       43.26809524       -1.16809524
       37             43.50000000       43.26809524        0.23190476
       38             43.80000000       43.26809524        0.53190476
       39             44.20000000       43.26809524        0.93190476
       40             43.60000000       43.26809524        0.33190476
       41             54.20000000       48.82738095        5.37261905
       42             43.70000000       48.82738095       -5.12738095
       43             53.10000000       48.82738095        4.27261905
       44             53.80000000       48.82738095        4.97261905
       45             55.60000000       48.82738095        6.77261905
       46             55.90000000       48.82738095        7.07261905
       47             54.70000000       48.82738095        5.87261905
       48             54.50000000       48.82738095        5.67261905


        SUM OF RESIDUALS                          0.00000000
        SUM OF SQUARED RESIDUALS                682.84709524
        SUM OF SQUARED RESIDUALS - ERROR SS       -0.00000000
        FIRST ORDER AUTOCORRELATION               0.73342816
        DURBIN-WATSON D                           0.48034159
```

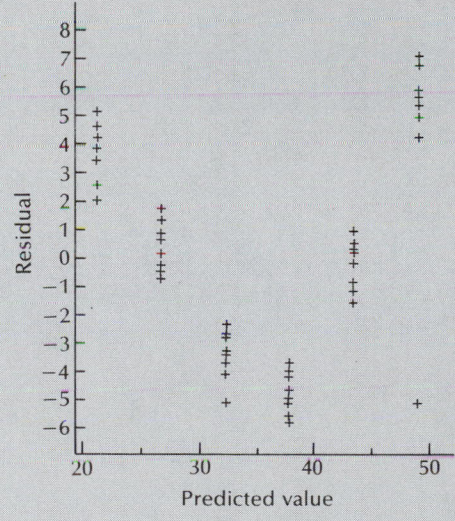

Linear regression analysis
plot of residuals vs. predicted values

(continues)

QUADRATIC REGRESSION ANALYSIS

GENERAL LINEAR MODELS PROCEDURE

DEPENDENT VARIABLE: WEAR

SOURCE	DF	SUM OF SQUARES	MEAN SQUARE	F VALUE	PR > F	R-SQUARE	C.V.
MODEL	2	4839.89302381	2419.94651190	641.53	0.0001	0.966116	5.5604
ERROR	45	169.74614286	3.77213651		ROOT MSE		WEAR MEAN
CORRECTED TOTAL	47	5009.63916667			1.94219888		34.92916667

SOURCE	DF	TYPE I SS	F VALUE	PR > F	DF	TYPE III SS	F VALUE	PR > F
SPEED	1	4326.79207143	1147.04	0.0001	1	261.47616353	69.32	0.0001
SPEED*SPEED	1	513.10095238	136.02	0.0001	1	513.10095238	136.02	0.0001

PARAMETER	ESTIMATE	T FOR HO: PARAMETER=0	PR > |T|	STD ERROR OF ESTIMATE
INTERCEPT	63.13928571	10.31	0.0001	6.12529888
SPEED	-0.70507143	-8.33	0.0001	0.08468583
SPEED*SPEED	0.00327679	11.66	0.0001	0.00028096

OBSERVATION	OBSERVED VALUE	PREDICTED VALUE	RESIDUAL
1	23.00000000	25.40000000	-2.40000000
2	23.50000000	25.40000000	-1.90000000
3	24.40000000	25.40000000	-1.00000000
4	25.20000000	25.40000000	-0.20000000
5	25.60000000	25.40000000	0.20000000
6	26.10000000	25.40000000	0.70000000
7	24.80000000	25.40000000	-0.60000000
8	25.60000000	25.40000000	0.20000000
9	26.70000000	25.71642857	0.98357143
10	26.10000000	25.71642857	0.38357143
11	25.80000000	25.71642857	0.08357143
12	26.30000000	25.71642857	0.58357143
13	27.20000000	25.71642857	1.48357143
14	27.90000000	25.71642857	2.18357143
15	28.30000000	25.71642857	2.58357143
16	27.40000000	25.71642857	1.68357143
17	28.00000000	28.65428571	-0.65428571
18	28.40000000	28.65428571	-0.25428571
19	27.00000000	28.65428571	-1.65428571
20	28.80000000	28.65428571	0.14571429
21	29.80000000	28.65428571	1.14571429
22	29.40000000	28.65428571	0.74571429
23	28.70000000	28.65428571	0.04571429
24	29.30000000	28.65428571	0.64571429
25	32.70000000	34.21357143	-1.51357143
26	32.10000000	34.21357143	-2.11357143
27	31.90000000	34.21357143	-2.31357143
28	33.00000000	34.21357143	-1.21357143
29	33.50000000	34.21357143	-0.71357143
30	33.70000000	34.21357143	-0.51357143
31	34.00000000	34.21357143	-0.21357143
32	32.50000000	34.21357143	-1.71357143
33	43.10000000	42.39428571	0.70571429
34	41.70000000	42.39428571	-0.69428571
35	42.40000000	42.39428571	0.00571429
36	42.10000000	42.39428571	-0.29428571
37	43.50000000	42.39428571	1.10571429
38	43.80000000	42.39428571	1.40571429
39	44.20000000	42.39428571	1.80571429
40	43.60000000	42.39428571	1.20571429
41	54.20000000	53.19642857	1.00357143
42	43.70000000	53.19642857	-9.49642857
43	53.10000000	53.19642857	-0.09642857
44	53.80000000	53.19642857	0.60357143
45	55.60000000	53.19642857	2.40357143
46	55.90000000	53.19642857	2.70357143
47	54.70000000	53.19642857	1.50357143
48	54.50000000	53.19642857	1.30357143

SUM OF RESIDUALS	0.00000000
SUM OF SQUARED RESIDUALS	169.74614286
SUM OF SQUARED RESIDUALS - ERROR SS	0.00000000
FIRST ORDER AUTOCORRELATION	0.25731691
DURBIN-WATSON D	1.44142233

(continues)

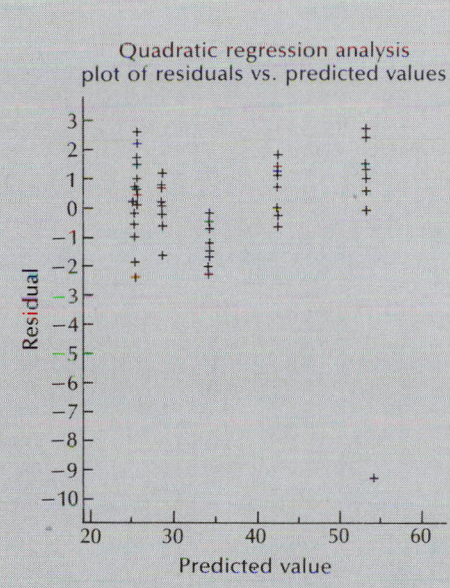

Quadratic regression analysis
plot of residuals vs. predicted values

CUBIC REGRESSION ANALYSIS

GENERAL LINEAR MODELS PROCEDURE

DEPENDENT VARIABLE: WEAR

SOURCE	DF	SUM OF SQUARES	MEAN SQUARE	F VALUE	PR > F	R-SQUARE	C.V.
MODEL	3	4846.78202381	1615.59400794	436.49	0.0001	0.967491	5.5079
ERROR	44	162.85714286	3.70129870		ROOT MSE		WEAR MEAN
CORRECTED TOTAL	47	5009.63916667			1.92387596		34.92916667

SOURCE	DF	TYPE I SS	F VALUE	PR > F	DF	TYPE III SS	F VALUE	PR > F
SPEED	1	4326.79207143	1168.99	0.0001	1	0.43368923	0.12	0.7338
SPEED*SPEED	1	513.10095238	138.63	0.0001	1	1.67992922	0.45	0.5040
SPEED*SPEED*SPEED	1	6.88900000	1.86	0.1794	1	6.88900000	1.86	0.1794

PARAMETER	ESTIMATE	T FOR H0: PARAMETER=0	PR > |T|	STD ERROR OF ESTIMATE
INTERCEPT	18.87261905	0.57	0.5704	33.00952220
SPEED	0.23847718	0.34	0.7338	0.69668199
SPEED*SPEED	-0.00320759	-0.67	0.5040	0.00476113
SPEED*SPEED*SPEED	0.0000144	1.36	0.1794	0.00001056

OBSERVATION	OBSERVED VALUE	PREDICTED VALUE	RESIDUAL
1	23.00000000	25.05416667	-2.05416667
2	23.50000000	25.05416667	-1.55416667
3	24.40000000	25.05416667	-0.65416667
4	25.20000000	25.05416667	0.14583333
5	25.60000000	25.05416667	0.54583333
6	26.10000000	25.05416667	1.04583333
7	24.80000000	25.05416667	-0.25416667
8	25.60000000	25.05416667	0.54583333
9	26.70000000	26.20059524	0.49940476
10	26.10000000	26.20059524	-0.10059524
11	25.80000000	26.20059524	-0.40059524
12	26.30000000	26.20059524	0.09940476
13	27.20000000	26.20059524	0.99940476
14	27.90000000	26.20059524	1.69940476

(*continues*)

OBSERVATION	OBSERVED VALUE	PREDICTED VALUE	RESIDUAL
15	28.30000000	26.20059524	2.09940476
16	27.40000000	26.20059524	1.19940476
17	28.00000000	28.93095238	-0.93095238
18	28.40000000	28.93095238	-0.53095238
19	27.00000000	28.93095238	-0.93095238
20	28.80000000	28.93095238	-0.13095238
21	29.80000000	28.93095238	0.86904762
22	29.40000000	28.93095238	0.46904762
23	28.70000000	28.93095238	-0.23095238
24	29.30000000	28.93095238	0.36904762
25	32.70000000	33.93690476	-1.23690476
26	32.10000000	33.93690476	-1.83690476
27	31.90000000	33.93690476	-2.03690476
28	33.00000000	33.93690476	-0.93690476
29	33.50000000	33.93690476	-0.43690476
30	33.70000000	33.93690476	-0.23690476
31	34.00000000	33.93690476	0.06309524
32	32.50000000	33.93690476	-1.43690476
33	43.10000000	41.91011905	1.18988095
34	41.70000000	41.91011905	-0.21011905
35	42.40000000	41.91011905	0.48988095
36	42.10000000	41.91011905	0.18988095
37	43.50000000	41.91011905	1.58988095
38	43.80000000	41.91011905	1.88988095
39	44.20000000	41.91011905	2.28988095
40	43.60000000	41.91011905	1.68988095
41	54.20000000	53.54226190	0.65773810
42	43.70000000	53.54226190	-9.84226190
43	53.10000000	53.54226190	-0.44226190
44	53.80000000	53.54226190	0.25773810
45	55.60000000	53.54226190	2.05773810
46	55.90000000	53.54226190	2.35773810
47	54.70000000	53.54226190	1.15773810
48	54.50000000	53.54226190	0.95773810

SUM OF RESIDUALS	0.00000000
SUM OF SQUARED RESIDUALS	162.85714286
SUM OF SQUARED RESIDUALS - ERROR SS	-0.00000000
FIRST ORDER AUTOCORRELATION	0.23779258
DURBIN-WATSON D	1.49287270

Cubic regression analysis
plot of residuals vs. predicted values

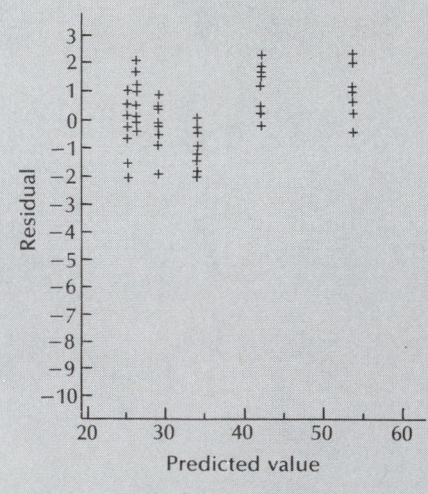

11.10 Refer to the data of Exercise 11.9. Suppose that another variable was controlled and that the first four squares at each speed were treated with a .2 concentration of protective coating, whereas the second four squares were treated with a .4 concentration of the same coating. x_1 denotes the machine speed and x_2 denotes the concentration of the protective coating. Fit these models using available statistical software. Which model seems to provide a better fit to the data? Why?

$$y = \beta_0 + \beta_1 x_1 + \beta_2 x_1^2 + \beta_3 x_2 + \varepsilon$$
$$y = \beta_0 + \beta_1 x_1 + \beta_2 x_1^2 + \beta_3 x_2 + \beta_4 x_1 x_2 + \beta_5 x_1^2 x_2 + \varepsilon$$

11.11 In Exercise 10.30, we compared the height of detergent suds to the amount of detergent added. Another variable that could affect the height of the suds is the degree of agitation in the wash cycle (measured in minutes). The complete data are presented here.

Height, y	Agitation, x_1	Amount, x_2
28.1	1	6
32.3	1	7
34.8	1	8
38.2	1	9
43.5	1	10
60.3	2	6
63.7	2	7
65.4	2	8
69.2	2	9
72.9	2	10
88.2	3	6
89.3	3	7
94.1	3	8
95.7	3	9
100.6	3	10

Fit the following two models using available statistical software.

$$y = \beta_0 + \beta_1 x_1 + \beta_2 x_2 + \varepsilon$$
$$y = \beta_0 + \beta_1 x_1 + \beta_2 x_2 + \beta_3 x_1 x_2 + \varepsilon$$

11.4 ▼ INFERENCES ABOUT A SINGLE PARAMETER IN THE GENERAL LINEAR MODEL

We make inferences about any of the parameters in the general linear model just as we did for β_0 and β_1 in the linear regression model, $y = \beta_0 + \beta_1 x + \varepsilon$.

Before we do this, however, we must introduce the *coefficient of determination, R^2*. The coefficient of determination for the model $y = \beta_0 + \beta_1 x_1 + \cdots + \beta_k x_k + \varepsilon$ is defined as the proportion of the variability in the dependent variable

y that is accounted for by the independent variables $x_1, x_2, \ldots, x_k$ of the model. When there is only one independent variable in the regression equation, the coefficient of determination is r^2, the square of the simple correlation coefficient between y and the independent variable, presented in Chapter 9. The coefficient of determination is designated by the symbol $R^2_{y \cdot x_1 x_2 \cdots x_k}$ and is computed as follows:

$$R^2_{y \cdot x_1 x_2 \cdots x_k} = \frac{S_{yy} - \text{SSE}}{S_{yy}}, \qquad 0 \le R^2 \le 1,$$

where

$$S_{yy} = \sum y_i^2 - \frac{(\sum y_i)^2}{n}$$

and

$$\text{SSE} = \sum (y_i - \hat{y}_i)^2$$
$$= \sum y_i^2 - \hat{\beta}_0 \sum y_i - \hat{\beta}_1 \sum x_{i1} y_i - \cdots - \hat{\beta}_k \sum x_{ik} y_i.$$

EXAMPLE 11.5

▼

Refer to the data for Example 11.4 and compute $R^2_{y \cdot x_1 x_2}$. Compare this to the value shown in the computer output.

Solution

a. For these data we have

$$S_{yy} = \sum y_i^2 - (\sum y_i)^2/n$$
$$= 396.57 - (66.1)^2/12$$
$$= 396.57 - 364.10 = 32.47$$

and

$$\text{SSE} = \sum y_i^2 - \hat{\beta}_0 \sum y_i - \hat{\beta}_1 \sum x_{i1} y_i - \hat{\beta}_2 \sum x_{i2} y_i$$
$$= 396.57 - .667(66.1) - 1.317(383.3) - (-8.000)(19.19)$$
$$= 1.195*.$$

---

*It is important to carry as many decimals as possible to avoid rounding errors. Had we used the eight places to the right of the decimal for the $\hat{\beta}$s, the value of SSE would be 1.345, as shown in the output.

Substituting into the formula for R^2 we find

$$R^2_{y \cdot x_1 x_2} = \frac{32.47 - 1.195}{32.47} = .963.$$

b. The output for Example 11.4 lists the coefficient of determination as .958576. Our result in part a agrees with this value except for rounding error. ▲

There is no general relationship between the coefficient of determination R^2 and squares of the individual correlation coefficients r_{yx_j}. If the independent variables are uncorrelated, then

$$R^2 = r^2_{yx_1} + r^2_{yx_2} + \cdots + r^2_{yx_k}.$$

multicollinearity

But when the independent variables are themselves correlated, it is difficult to separate R^2 into the predictive contribution of each independent variable. Most problems that have a model with more than one independent variable are affected (to a greater or lesser degree) by *collinearity* or **multicollinearity**, where the independent variables are themselves correlated. For these situations, where the xs account for overlapping pieces of the variability in the y-values, we often find that

$$R^2_{y \cdot x_1 x_2 \cdots x_k} < r^2_{yx_1} + r^2_{yx_2} + \cdots + r^2_{yx_k}.$$

We can now write down the expression for $s_{\hat{\beta}_j}$, the *estimated standard error for* $\hat{\beta}_j$ in the general linear model.

Estimated Standard Error for $\hat{\beta}_j$ in the General Linear Model

$$s_{\hat{\beta}_j} = s_\varepsilon \sqrt{\frac{1}{S_{x_j x_j}(1 - R^2_{x_j \cdot x_1 \cdots x_{j-1} x_{j+1} \cdots x_k})}},$$

where

$$s_\varepsilon = \sqrt{\frac{SSE}{n - (k + 1)}}$$

(continues)

is the standard deviation of the fitted line,

$$S_{x_j x_j} = \sum_i x_{ij}^2 - \frac{\left(\sum_i x_{ij}\right)^2}{n}$$

is the sum of squares, and

$$\sum_i (x_{ij} - \bar{x}_j)^2$$

for the variable x_j, and $R^2_{x_j \cdot x_1 \cdots x_{j-1} x_{j+1} \cdots x_k}$ is the coefficient of determination for the model with x_j as the *dependent* variable and all other xs in the model.

This formula for the estimated standard error of $\hat{\beta}_j$ is *not* recommended for use in computing $s_{\hat{\beta}_j}$; we will rely on software packages for that. The reason we choose to present this formula is for its interpretive value, especially in the presence of multicollinearity. If the independent variable x_j is highly correlated with one or more of the other independent variables, $R^2_{x_j \cdot x_1 \cdots x_{j-1} x_{j+1} \cdots x_k}$ will be large and $1 - R^2$ will be small, making $s_{\hat{\beta}_j}$ large. This makes sense. By definition, $\hat{\beta}_j$ is the estimated change in y for a 1-unit increase in x_j, *while holding the other xs constant*. In the presence of correlation between x_j and the other xs, it will be difficult to estimate the change in y for a unit increase x_j while the other xs remain constant. This difficulty is reflected in the large value of $s_{\hat{\beta}_j}$.

EXAMPLE 11.6   ▼

a. Compute estimated standard errors for $\hat{\beta}_1$ and $\hat{\beta}_2$ in the data of Example 11.4.
b. Compare these standard errors to those shown in the output for Example 11.4.

Solution

a. The quantities we need for $s_{\hat{\beta}_1}$ and $s_{\hat{\beta}_2}$ are $S_{x_1 x_1}$, $S_{x_2 x_2}$, s_ε, $R^2_{x_1 \cdot x_2}$ and $R^2_{x_2 \cdot x_1}$. From Example 11.4 we have

$$S_{x_1 x_1} = \sum x_{i1}^2 - \left(\sum x_{i1}\right)^2/n = 378 - (66)^2/12 = 15$$

and

$$S_{x_2 x_2} = \sum x_{i2}^2 - \left(\sum x_{i2}\right)^2/n = 1.16 - (3.6)^2/12 = .08.$$

Using SSE $= 1.345$, the value obtained in Example 11.5 after carrying sufficient decimal places, it follows that

$$s_\varepsilon = \sqrt{\frac{SSE}{n - (k + 1)}} = \sqrt{\frac{1.345}{9}} = .387.$$

The quantity $R^2_{x_1 \cdot x_2}$ obtained by fitting the linear regression model

$$x_1 = \beta_0 + \beta_1 x_2 + \varepsilon$$

is simply $r^2_{x_1 x_2}$, the square of the correlation coefficient for x_1 and x_2:

$$r_{x_1 x_2} = \frac{S_{x_1 x_2}}{\sqrt{S_{x_1 x_1} S_{x_2 x_2}}}.$$

For these data,

$$S_{x_1 x_2} = \sum x_{i1} x_{i2} - \frac{\sum x_{i1} \sum x_{i2}}{n} = 19.8 - \frac{66(3.6)}{12} = 0.$$

Hence, $r_{x_1 x_2} = 0$; x_1 and x_2 are uncorrelated. It follows that

$$s_{\hat{\beta}_1} = .387 \sqrt{\frac{1}{15(1 - 0)}} = .387 \sqrt{\frac{1}{15}} = .100,$$

and since $R^2_{x_1 \cdot x_2} = R^2_{x_2 \cdot x_1} = 0$,

$$s_{\hat{\beta}_2} = .387 \sqrt{\frac{1}{.08}} = 1.368.$$

b. Except for slight rounding errors, the estimated standard errors computed for $\hat{\beta}_1$ and $\hat{\beta}_2$ agree with those shown in the output of Example 11.4. ▲

Knowing the estimated standard error for $\hat{\beta}_j$, we can write down formulas for interval estimation of β_j and a statistical test for β_j. These are summarized here.

$100(1 - \alpha)$% Confidence Interval for β_j

▼

$$\hat{\beta}_j \pm t_{\alpha/2} s_{\hat{\beta}_j},$$

where $t_{\alpha/2}$ is the tabulated t-value for df $= n - (k + 1)$ and $a = \alpha/2$.

Statistical Test for H_0: $\beta_j = 0$

▼

H_0: $\beta_j = 0$

H_a: **1.** $\beta_j > 0$
     2. $\beta_j < 0$
     3. $\beta_j \neq 0$

T.S.: $t = \dfrac{\hat{\beta}_j}{s_{\hat{\beta}_j}}$

R.R.: For df $= n - (k + 1)$ and specified value α, reject H_0 if

     1. $t > t_\alpha$
     2. $t < -t_\alpha$
     3. $|t| > t_{\alpha/2}$.

E X A M P L E   11.7

▼

An experiment was conducted to examine the potential for deterioration of a new commercial paint when exposed to the atmosphere. Under controlled conditions (settings of the independent variables, temperature x_1 and exposure time x_2) the deterioration y of the paint was measured. These data are shown here.

y	120	101	110	105	92	130
x_1 (°C)	-10	-10	0	0	10	10
x_2 (months)	1	3	2	2	1	3

a. Consider the model $y = \beta_0 + \beta_1 x_1 + \beta_2 x_2 + \varepsilon$. Give an interpretation to the test H_0: $\beta_2 = 0$.

b. SAS was used to fit the model of part a. A portion of the output is shown here. Test H_0: $\beta_2 = 0$ and draw a conclusion.

c. Distinguish between the model shown in part a and the model $y = \beta_0 + \beta_1 x_1 + \beta_2 x_2 + \beta_3 x_1 x_2 + \varepsilon$. Which model is more appropriate?

```
      LISTING OF DATA

   OBS     Y     X1    X2

    1     120    -10    1
    2     101    -10    3
    3     110     0     2
    4     105     0     2
    5      92    10     1
    6     130    10     3

   N =      6
```

MULTIPLE LINEAR REGRESSION ANALYSIS

GENERAL LINEAR MODELS PROCEDURE

DEPENDENT VARIABLE: Y

SOURCE	DF	SUM OF SQUARES	MEAN SQUARE	F VALUE	PR > F	R-SQUARE	C.V.
MODEL	2	90.50000000	45.25000000	0.16	0.8575	0.097382	15.2476
ERROR	3	838.83333333	279.61111111			ROOT MSE	Y MEAN
CORRECTED TOTAL	5	929.33333333				16.72157621	109.66666667

SOURCE	DF	TYPE I SS	F VALUE	PR > F	DF	TYPE III SS	F VALUE	PR > F
X1	1	0.25000000	0.00	0.9780	1	0.25000000	0.00	0.9780
X2	1	90.25000000	0.32	0.6097	1	90.25000000	0.32	0.6097

PARAMETER	ESTIMATE	T FOR H0: PARAMETER=0	PR > |T|	STD ERROR OF ESTIMATE
INTERCEPT	100.16666667	5.55	0.0116	18.06136659
X1	0.02500000	0.03	0.9780	0.83607881
X2	4.75000000	0.57	0.6097	8.36078811

Solution

a. In general, H_0: $\beta_j = 0$ states that x_j has no predictive power over and above that provided by the other independent variables. It does not say that x_j has no predictive power by itself. For our model, $y = \beta_0 + \beta_1 x_1 + \beta_2 x_2 + \varepsilon$, H_0: $\beta_2 = 0$ states that x_2 has no predictive power over and above x_1.

b. The t-value for H_0: $\beta_2 = 0$ is $t = .57$; since this value is less than $t_{.05,3} = 2.353$ (and $t_{.10,3} = 1.638$), we have insufficient evidence to reject H_0. The variable x_2 does not appear to have any predictive power over and above x_1. In fact, since $R^2 = .097$ (i.e., 9.7%), neither x_1 nor x_2 seems to provide much predictive power for y.

c. The model $y = \beta_0 + \beta_1 x_1 + \beta_2 x_2 + \beta_3 x_1 x_2 + \varepsilon$ allows for an "interaction" between the independent variables x_1 and x_2; that is, it allows for the fact that the expected change in y for a unit change in x_1 *does* depend on the level of x_2. A scatterplot of y vs. x_1 with levels of x_2 marked is shown in Figure 11.4.

   Figure 11.4 shows that the expected change in y for a unit change in x_1 depends on the level of x_2 being considered. For example, when $x_2 = 1$, y decreases from 120 to 92 when x_1 changes from -10 to $+10$. However,

FIGURE   11.4

Scatterplot of y Versus x_1 with the Levels of x_2 (1, 2, or 3) Marked

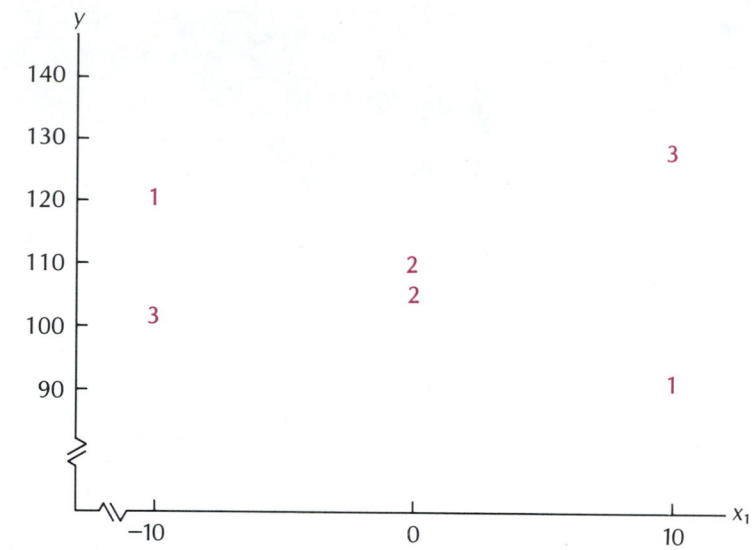

when $x_2 = 3$, y increases from 101 to 130 when x_1 changes from -10 to $+10$. It is clear then that a model such as

$$y = \beta_0 + \beta_1 x_1 + \beta_2 x_2 + \beta_3 x_1 x_2 + \varepsilon,$$

which allows for nonadditivity (interaction), may be more appropriate.   ▲

▼ EXERCISES

Applications

11.12   In Exercise 10.5, we presented data on the biological recovery of organisms suspended in aerosol clouds formed in an airtight chamber and the times at which the recoveries were determined. These data are shown here for your convenience.

Cloud	Time, x (in minutes)	Biological Recovery (%)
1	0	70.6
2	5	52.0
3	10	33.4
4	15	22.0
5	20	18.3
6	25	15.1
7	30	13.0
8	35	10.0
9	40	9.1
10	45	8.3
11	50	7.9
12	55	7.7
13	60	7.7

a. Because the assumption of equal variances at different settings of x (time) would probably not be satisfied, logarithms (to the base 10) of the biological recoveries were used. Give logs of the biological recoveries.

b. Plot the transformed biological recoveries versus time and suggest a possible model.

11.13 Refer to the output shown here for fitting the regression model $y = \beta_0 + \beta_1 x + \beta_2 x^2 + \varepsilon$ to the data of Exercise 11.12.

a. Determine the prediction equation.

b. Determine s_ε and the standard errors for $\hat{\beta}_1$ and $\hat{\beta}_2$.

```
                  LISTING OF DATA

    OBS    TIME    RECOVERY    LOG_REC    TIME_2

     1       0       70.6       1.84880        0
     2       5       52.0       1.71600       25
     3      10       33.4       1.52375      100
     4      15       22.0       1.34242      225
     5      20       18.3       1.26245      400
     6      25       15.1       1.17898      625
     7      30       13.0       1.11394      900
     8      35       10.0       1.00000     1225
     9      40        9.1       0.95904     1600
    10      45        8.3       0.91908     2025
    11      50        7.9       0.89763     2500
    12      55        7.7       0.86649     3025
    13      60        7.7       0.88649     3600

   N =     13
```

REGRESSION ANALYSIS

GENERAL LINEAR MODELS PROCEDURE

DEPENDENT VARIABLE: LOG_REC   LOG BASE 10 OF RECOVERY

SOURCE	DF	SUM OF SQUARES	MEAN SQUARE	F VALUE	PR > F	R-SQUARE	C.V.
MODEL	2	1.28344124	0.64172062	1217.91	0.0001	0.995911	1.9209
ERROR	10	0.00526904	0.00052690		ROOT MSE		LOG_REC MEAN
CORRECTED TOTAL	12	1.28871028			0.02295439		1.19500589

SOURCE	DF	TYPE I SS	F VALUE	PR > F	DF	TYPE III SS	F VALUE	PR > F
TIME	1	1.15231437	2186.95	0.0001	1	0.40321970	765.26	0.0001
TIME_2	1	0.13112687	248.86	0.0001	1	0.13112687	248.86	0.0001

PARAMETER	ESTIMATE	T FOR H0: PARAMETER=0	PR > |T|	STD ERROR OF ESTIMATE
INTERCEPT	1.85047412	113.17	0.0001	0.01649659
TIME	-0.03533741	-27.66	0.0001	0.00127741
TIME_2	0.00032372	15.78	0.0001	0.00002052

11.14 Refer to Exercise 11.13. Conduct a test of $H_0: \beta_2 = 0$ and interpret your findings.

11.15 Refer to Exercise 11.12.

a. Give an interpretation to β_0 for the model

$$y = \beta_0 + \beta_1 x + \varepsilon.$$

b. Would the model $y = \beta_1 x + \varepsilon$ be an alternate model? Why?

11.16 Refer to Example 11.7. Use the output here to test $H_0: \beta_3 = 0$ for the model $y = \beta_0 + \beta_1 x_1 + \beta_2 x_2 + \beta_3 x_1 x_2 + \varepsilon$. Give a p-value for this test and draw a conclusion. Does the test of $H_0: \beta_3 = 0$ reflect what was seen in the scatterplot of Example 11.7?

```
            LISTING OF DATA
   OBS    Y    X1   X2    X1X2
    1    120   -10   1    -10
    2    101   -10   3    -30
    3    110    0    2     0
    4    105    0    2     0
    5     92   10    1     10
    6    130   10    3     30
   N =    6
```

REGRESSION ANALYSIS

GENERAL LINEAR MODELS PROCEDURE

DEPENDENT VARIABLE: Y

SOURCE	DF	SUM OF SQUARES	MEAN SQUARE	F VALUE	PR > F	R-SQUARE	C.V.
MODEL	3	902.75000000	300.91666667	22.64	0.0426	0.971395	3.3244
ERROR	2	26.58333333	13.29166667		ROOT MSE		Y MEAN
CORRECTED TOTAL	5	929.33333333			3.64577381		109.66666667

SOURCE	DF	TYPE I SS	F VALUE	PR > F	DF	TYPE III SS	F VALUE	PR > F
X1	1	0.25000000	0.02	0.9035	1	638.45000000	48.03	0.0202
X2	1	90.25000000	6.79	0.1211	1	90.25000000	6.79	0.1211
X1X2	1	812.25000000	61.11	0.0160	1	812.25000000	61.11	0.0160

PARAMETER	ESTIMATE	T FOR H0: PARAMETER=0	PR > |T|	STD ERROR OF ESTIMATE
INTERCEPT	100.16666667	25.44	0.0015	3.93788578
X1	-2.82500000	-6.93	0.0202	0.40760990
X2	4.75000000	2.61	0.1211	1.82288690
X1X2	1.42500000	7.82	0.0160	0.18228869

11.17 Consider the data shown here.

y	x_1	x_2
3	-2	-1
10	-1	0
12	0	0
12	1	0
13	2	1

a. Compute r_{yx_1}, r_{yx_2} and $r_{x_1 x_2}$.

b. Does $R^2_{y \cdot x_1 x_2} = r^2_{yx_1} + r^2_{yx_2}$? Interpret your finding.

 11.18 An experiment was conducted to examine the effect of temperature (x_1) and humidity (x_2) on the yield of a production process. The sample data are shown here.

Yield, y	Temperature, x_1	Humidity, x_2
65	70	50
78	100	50
52	130	50
70	70	80
77	100	80
83	130	80

a. Use the computer output shown here to determine the least squares fit to the model

$$y = \beta_0 + \beta_1 x_1 + \beta_2 x_2 + \beta_3 x_1^2 + \beta_4 x_1 x_2 + \varepsilon$$

b. Identify and interpret R^2.

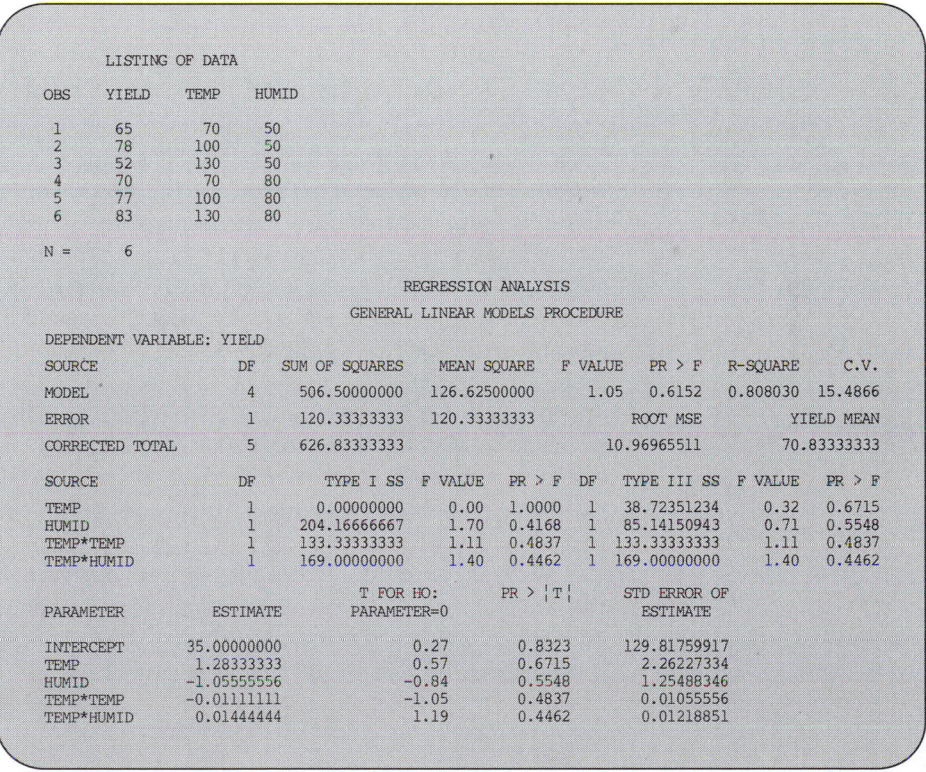

```
      LISTING OF DATA

OBS    YIELD    TEMP    HUMID

 1       65       70      50
 2       78      100      50
 3       52      130      50
 4       70       70      80
 5       77      100      80
 6       83      130      80

N =      6

                           REGRESSION ANALYSIS
                      GENERAL LINEAR MODELS PROCEDURE

DEPENDENT VARIABLE: YIELD

SOURCE              DF   SUM OF SQUARES   MEAN SQUARE   F VALUE    PR > F   R-SQUARE     C.V.

MODEL                4     506.50000000   126.62500000    1.05     0.6152  0.808030  15.4866

ERROR                1     120.33333333   120.33333333            ROOT MSE         YIELD MEAN

CORRECTED TOTAL      5     626.83333333                          10.96965511      70.83333333

SOURCE              DF       TYPE I SS  F VALUE   PR > F  DF    TYPE III SS  F VALUE   PR > F

TEMP                 1     0.00000000     0.00    1.0000   1    38.72351234    0.32    0.6715
HUMID                1   204.16666667     1.70    0.4168   1    85.14150943    0.71    0.5548
TEMP*TEMP            1   133.33333333     1.11    0.4837   1   133.33333333    1.11    0.4837
TEMP*HUMID           1   169.00000000     1.40    0.4462   1   169.00000000    1.40    0.4462

                                      T FOR HO:     PR > |T|     STD ERROR OF
PARAMETER           ESTIMATE         PARAMETER=0                  ESTIMATE

INTERCEPT          35.00000000          0.27        0.8323      129.81759917
TEMP                1.28333333          0.57        0.6715        2.26227334
HUMID              -1.05555556         -0.84        0.5548        1.25488346
TEMP*TEMP          -0.01111111         -1.05        0.4837        0.01055556
TEMP*HUMID          0.01444444          1.19        0.4462        0.01218851
```

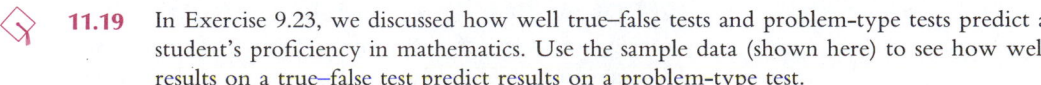

11.19 In Exercise 9.23, we discussed how well true–false tests and problem-type tests predict a student's proficiency in mathematics. Use the sample data (shown here) to see how well results on a true–false test predict results on a problem-type test.

Student	1	2	3	4	5	6	7	8	9	10
T–F	48	40	25	10	16	21	23	19	35	32
Problems	45	47	20	12	12	15	25	16	30	32

a. Fit the linear regression model

$$y = \beta_0 + \beta_1 x + \varepsilon,$$

where y is the problem test score and x is the corresponding true–false test score for a student.

b. Test for the significance of β_1; give the p-value for your test and draw a conclusion. What does this say about p?

c. Compute R and R^2; interpret your findings. What is the value of r?

11.5 ▼ INFERENCES CONCERNING $E(y)$ AND y

The methods of previous sections can be expanded to include inferences concerning the average value of y for a given setting of the independent variables. For example, suppose that in a regression study we obtain the prediction equation $\hat{y} = 22.6 + 1.2x$. What does it mean to predict y when $x = 10$? It could mean predicting the average (expected) y-value for all cases having $x = 10$; we would predict $\hat{y} = 22.6 + 1.2(10) = 34.6$. Or it could mean predicting the y-value for one particular case having $x = 10$; again we would use $\hat{y} = 22.6 + 1.2(10) = 34.6$. The difference in the two solutions is in the standard errors. Because the standard errors are more complicated to write down, we will make use of software packages to develop inferences about $E(y)$ and y. Those readers interested in the actual formulas are referred to Section 10.5 for the single independent variable case and to Section 11.7 for a discussion of the matrix formulas for the k independent variable case.

We will illustrate the output for data analyzed by hand in Chapter 10 so that the output can be compared to previous work when necessary. However, keep in mind that the software packages are completely general and can be used for any number of independent variables.

EXAMPLE 11.8 ▼

In Example 10.2, we recorded the corn yield (in bushels) for 10 plots with varying degrees of fertilization. These data are repeated here for your convenience.

Yield (bushels)	Fertilizer (pounds per plot)
12	2
13	2
13	3
14	3
15	4
15	4
14	5
16	5
17	6
18	6

a. Use the output shown here to obtain a fit to the linear regression equation $y = \beta_0 + \beta_1 x + \varepsilon$.

b. Find a 90% confidence interval for $E(y)$ when $x = 5$.

c. Construct 90% confidence bands for $E(y)$ when $2 \le x \le 6$.

```
         LISTING OF DATA
   OBS      YIELD   FERTILIZ

    1        12        2
    2        13        2
    3        13        3
    4        14        3
    5        15        4
    6        15        4
    7        14        5
    8        16        5
    9        17        6
   10        18        6

  N = 10


                            LINEAR REGRESSION ANALYSIS
                         GENERAL LINEAR MODELS PROCEDURE

  DEPENDENT VARIABLE: YIELD      YIELD (BUSHELS)

                             SUM OF        MEAN
  SOURCE              DF      SQUARES       SQUARE      F VALUE    PR > F    R-SQUARE    C.V.
  MODEL                1    26.45000000   26.45000000    37.45     0.0003   0.823988   5.7169
  ERROR                8     5.65000000    0.70625000              STD DEV            YIELD MEAN
  CORRECTED TOTAL      9    32.10000000                            0.84038682         14.70000000


  SOURCE       DF      TYPE I SS      F VALUE    PR > F    DF    TYPE IV SS     F VALUE   PR > F
  FERTILIZ      1    26.45000000       37.45     0.0003     1   26.45000000     37.45    0.0003

                                       T FOR HO:                      STD ERROR OF
  PARAMETER         ESTIMATE          PARAMETER=0      PR > |T|        ESTIMATE
  INTERCEPT       10.00000000            12.67          0.0001         0.79726094
  FERTILIZ         1.15000000             6.12          0.0003         0.18791620

                                                                        (continues)
```

OBSERVATION	OBSERVED VALUE	PREDICTED VALUE	RESIDUAL	LOWER 90% CL FOR MEAN	UPPER 90% CL FOR MEAN
1	12.00000000	12.40000000	-0.40000000	11.54404094	13.25595906
2	13.00000000	12.40000000	0.60000000	11.54404094	13.25595906
3	13.00000000	13.55000000	-0.55000000	12.94474555	14.15525445
4	14.00000000	13.55000000	0.45000000	12.94474555	14.15525445
5	15.00000000	14.70000000	0.30000000	14.20581181	15.19418819
6	15.00000000	14.70000000	0.30000000	14.20581181	15.19418819
7	14.00000000	15.85000000	-1.85000000	15.24474555	16.45525445
8	16.00000000	15.85000000	0.15000000	15.24474555	16.45525445
9	17.00000000	17.00000000	-0.00000000	16.14404094	17.85595906
10	18.00000000	17.00000000	1.00000000	16.14404094	17.85595906

SUM OF RESIDUALS	0.00000000
SUM OF SQUARED RESIDUALS	5.65000000
SUM OF SQUARED RESIDUALS - ERROR SS	0.00000000
PRESS STATISTIC	8.79139107
FIRST ORDER AUTOCORRELATION	-0.25221239
DURBIN-WATSON D	2.29911504

Solution

a. The output gives $\hat{y} = 10.1 + 1.15x$.

b. The 90% confidence interval for $E(y)$ when $x = 5$ is 15.245 to 16.455. Our hand calculations of Example 10.10 agree with this result except for rounding errors.

c. The 90% confidence bands for $E(y)$ can be constructed using the 90% confidence intervals shown in the output for $x = 2, 3, 4, 5,$ and 6. These results agree with our hand calculations in Example 10.11. First plot the endpoints of the confidence intervals for $x = 2, 3, 4, 5,$ and 6. Then if we connect all the lower endpoints and all the upper endpoints, we obtain the 90% confidence bands for $E(y)$. (See Figure 11.5.)

FIGURE 11.5
90% Confidence Bands for $E(y)$, Example 11.8

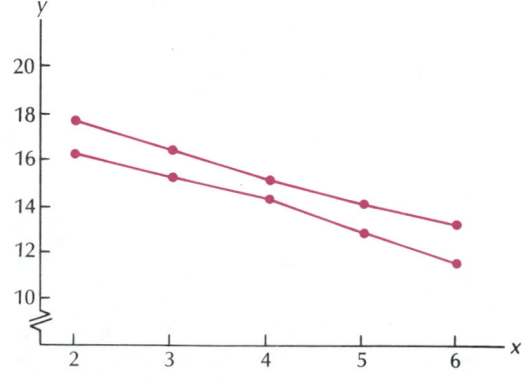

There is another problem closely related to that of estimating $E(y)$ for a given setting of the independent variables. Suppose that after obtaining least squares estimates for the βs of a multiple regression equation, we want to *predict* the actual value of y (say the next observation) for a given setting of the independent variables. Note that this problem differs from the problem discussed previously in that we do not want to estimate the average value of y for a given value of x, but rather we wish to predict what a particular observation will be for that same setting of x.

prediction interval

We still use the least squares equation y as our predictor, but the corresponding interval about the observation y is called a **prediction interval**. (Prediction intervals are constructed about variables, whereas confidence intervals are constructed about parameters.) Again, we will use computer output to illustrate the solution to such problems since the algebraic formulas can be quite involved without the use of matrices.

EXAMPLE 11.9

▼

Refer to the data for Example 11.8 and the computer output shown here.
a. Place a 90% prediction interval about the value of y when $x = 5$.
b. Develop 90% prediction bands for y when $2 \le x \le 6$.

```
                        LINEAR REGRESSION ANALYSIS
                      GENERAL LINEAR MODELS PROCEDURE

DEPENDENT VARIABLE YIELD    YIELD (BUSHELS)
                        SUM OF        MEAN
SOURCE           DF     SQUARES       SQUARE      F VALUE   PR > F   R-SQUARE      C.V.
MODEL            1      26.45000000   26.45000000  37.45    0.0003   0.823988     5.7169
ERROR            8      5.65000000    0.70625000            STD DEV           YIELD MEAN
CORRECTED TOTAL  9      32.10000000                         0.84038682         14.70000000

SOURCE     DF    TYPE I SS    F VALUE   PR > F    DF    TYPE IV SS   F VALUE  PR > F
FERTILIZ   1     26.45000000  37.45     0.0003    1     26.45000000  37.45    0.0003

                              T FOR HO:             STD ERROR OF
PARAMETER       ESTIMATE      PARAMETER = 0  PR > |T|    ESTIMATE
INTERCEPT       10.10000000   12.67          0.0001      0.79726094
FERTILIZ        1.15000000    6.12           0.0003      0.18791620
                OBSERVED      PREDICTED               LOWER 90% CL   UPPER 90% CL
OBSERVATION     VALUE         VALUE       RESIDUAL    FOR MEAN       FOR MEAN
1               12.40000000   12.40000000 -0.40000000 10.61817913    14.18182087
2               13.00000000   12.40000000  0.60000000 10.61817913    14.18182087
3               13.00000000   13.55000000 -0.55000000 11.87412630    15.22587370
4               14.00000000   13.55000000  0.45000000 11.87412630    15.22587370
5               15.00000000   14.70000000  0.30000000 13.06096319    16.33903681
6               15.00000000   14.70000000  0.30000000 13.06096319    16.33903681
7               14.00000000   15.85000000 -1.85000000 14.17412630    17.52587370
8               16.00000000   15.85000000  0.15000000 14.17412630    17.52587370
9               17.00000000   17.00000000 -0.00000000 15.21817913    18.78182087
10              18.00000000   17.00000000  1.00000000 15.21817913    18.78182087
SUM OF RESIDUALS                                       0.00000000
SUM OF SQUARED RESIDUALS                               5.65000000
SUM OF SQUARED RESIDUALS - ERROR SS                    0.00000000
PRESS STATISTIC                                        8.79139107
FIRST ORDER AUTOCORRELATION                           -0.25221239
DURBIN-WATSON D                                        2.29911504
```

c. Graph both the 90% confidence bands of Example 11.8 and the 90% prediction bands of part b. Comment on your findings.

Solution

a. From the output shown here, the 90% prediction interval corresponding to $x = 5$ is 14.174 to 17.526. This result agrees with our hand calculations in Example 10.14.

b. The additional intervals necessary to construct a 90% prediction band for y when $2 \leq x \leq 6$ can be extracted from the output. These intervals agree with those obtained by hand in Example 10.14.

c. Plotting the endpoints of the intervals shown for $x = 2, 3, 4, 5,$ and 6 and then connecting all the lower endpoints and all the upper endpoints, we obtain the 90% prediction bands shown in Figure 11.6.

   As we have seen previously, the prediction bands are wider than the corresponding confidence bands to allow for the fact that we are predicting the value of a random variable rather than estimating a parameter.

FIGURE 11.6
90% Prediction and Confidence Bands for y **and** $E(y)$**, Example 11.9**

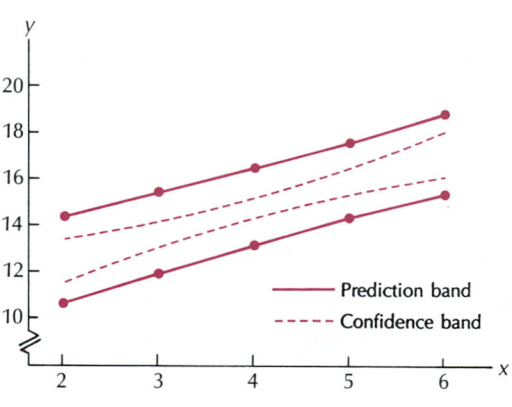

▼ EXERCISES

Applications

11.20   Refer to Exercise 11.12. For the regression equation

$$\hat{y} = \hat{\beta}_0 + \hat{\beta}_1 x + \hat{\beta}_2 x^2,$$

estimate the mean log biological recovery percentage at 30 minutes using a 95% confidence interval.

11.21 Refer to Example 11.9. In general, how would the prediction bands change if you used a higher degree of certainty (say 99%)? What if you use a lower degree of certainty (say 80%)?

11.22 A portion of a computer output for the weight loss data of Example 11.4 is shown here. Locate 95% confidence bands for $E(y)$.

```
                    LISTING OF DATA
        OBS    WT_LOSS    TIME    HUMIDITY
         1       4.3        4        0.2
         2       5.5        5        0.2
         3       6.8        6        0.2
         4       8.0        7        0.2
         5       4.0        4        0.3
         6       5.2        5        0.3
         7       6.6        6        0.3
         8       7.5        7        0.3
         9       2.0        4        0.4
        10       4.0        5        0.4
        11       5.7        6        0.4
        12       6.5        7        0.4

     N =     12
```

REGRESSION ANALYSIS

GENERAL LINEAR MODELS PROCEDURE

DEPENDENT VARIABLE: WT_LOSS

SOURCE	DF	SUM OF SQUARES	MEAN SQUARE	F VALUE	PR > F	R-SQUARE	C.V.
MODEL	2	31.12416667	15.56208333	104.13	0.0001	0.958576	7.0181
ERROR	9	1.34500000	0.14944444		ROOT MSE		WT_LOSS MEAN
CORRECTED TOTAL	11	32.46916667			0.38658045		5.50833333

SOURCE	DF	TYPE I SS	F VALUE	PR > F	DF	TYPE III SS	F VALUE	PR > F
TIME	1	26.00416667	174.01	0.0001	1	26.00416667	174.01	0.0001
HUMIDITY	1	5.12000000	34.26	0.0002	1	5.12000000	34.26	0.0002

PARAMETER	ESTIMATE	T FOR H0: PARAMETER=0	PR > |T|	STD ERROR OF ESTIMATE
INTERCEPT	0.66666667	0.96	0.3620	0.69423219
TIME	1.31666667	13.19	0.0001	0.09981464
HUMIDITY	-8.00000000	-5.85	0.0002	1.36676829

OBSERVATION	OBSERVED VALUE	PREDICTED VALUE	RESIDUAL	LOWER 95% CL FOR MEAN	UPPER 95% CL FOR MEAN
1	4.30000000	4.33333333	-0.03333333	3.80984144	4.85682523
2	5.50000000	5.65000000	-0.15000000	5.23518198	6.06481802
3	6.80000000	6.96666667	-0.16666667	6.55184865	7.38148469
4	8.00000000	8.28333333	-0.28333333	7.75984144	8.80682523
5	4.00000000	3.53333333	0.46666667	3.11090334	3.95576333
6	5.20000000	4.85000000	0.35000000	4.57345465	5.12654535
7	6.60000000	6.16666667	0.43333333	5.89012132	6.44321201
8	7.50000000	7.48333333	0.01666667	7.06090334	7.90576333
9	2.00000000	2.73333333	-0.73333333	2.20984144	3.25682523
10	4.00000000	4.05000000	-0.05000000	3.63518198	4.46481802
11	5.70000000	5.36666667	0.33333333	4.95184865	5.78148469
12	6.50000000	6.68333333	-0.18333333	6.15984144	7.20682523

```
     SUM OF RESIDUALS                        0.00000000
     SUM OF SQUARED RESIDUALS                1.34500000
     SUM OF SQUARED RESIDUALS -ERROR SS     -0.00000000
     PRESS STATISTIC                         2.61233345
     FIRST ORDER AUTOCORRELATION             0.15902520
     DURBIN-WATSON D                         1.65613383
```

11.23   Refer to Exercise 10.9. Construct a 95% prediction interval for the weight gain of a chick chosen at random and observed to ingest .19 grams of lysine. Compare your results to the confidence interval of Exercise 10.25.

11.24   Using the data of Exercise 11.20, construct a 95% prediction interval for the log biological recovery percentage at 30 minutes. Compare your result to the confidence interval on $E(y)$ of Exercise 11.20.

11.6 ▼ INFERENCES CONCERNING A SET OF βs IN A GENERAL LINEAR MODEL (optional)

The general linear model that we have discussed has the form

$$y = \beta_0 + \beta_1 x_1 + \cdots + \beta_k x_k + \varepsilon,$$

where the xs associated with qualitative independent variables are dummy variables and the xs associated with quantitative independent variables have a quantitative interpretation (such as initial weight, amount of water added, and stirring time). The ability to write both qualitative and quantitative independent variables in the framework of a general linear model will enable us to develop one procedure for simultaneously testing the significance of one or more βs of the model.

Consider a model for relating the sales volume y to advertising expenditure x_1, with the form

$$y = \beta_0 + \beta_1 x_1 + \beta_2 x_1^2 + \varepsilon.$$

In Section 11.4, we considered the problem of testing the significance of a single β in the general linear model. Thus, a test of H_0: $\beta_2 = 0$ would be a test of the null hypothesis that y is linearly related to x_1 (sales expenditure). However, an experimenter might wish to test

$$H_0: \quad \beta_1 = \beta_2 = 0$$

which hypothesizes that y is *not related* in a linear or quadratic way to the independent variable x_1.

The procedure for simultaneously testing that a set of βs is equal to 0 in the general linear model will have the following null and research hypotheses:

$$H_0: \quad \beta_{g+1} = \beta_{g+2} = \cdots = \beta_k = 0 \qquad (k > g)$$
$$H_a: \quad \text{At least one of the } \beta\text{s is nonzero.}$$

complete model
reduced model

To formulate a test statistic, we must specify two models, model 1—often referred to as the **complete model**—and model 2—referred to as the **reduced model**:

$$\text{model 1:} \quad y = \beta_0 + \beta_1 x_1 + \beta_2 x_2 + \cdots + \beta_g x_g + \beta_{g+1} x_{g+1} + \cdots + \beta_k x_k + \varepsilon$$
$$(k > g)$$

$$\text{model 2:} \quad y = \beta_0 + \beta_1 x_1 + \beta_2 x_2 + \cdots + \beta_g x_g + \varepsilon$$

You will note that model 1 represents the general linear model and model 2 is a general linear model under the assumption that H_0 is true. Thus, model 1 contains k independent variables, model 2 contains g independent variables, and we are testing that a set of $(k - g)$ βs is equal to 0. The xs in the two models represent either quantitative independent variables or dummy variables associated with qualitative independent variables.

Using the method of least squares, we fit both models separately, and for each model we calculate the sum of squares for error. Letting SSE_1 and SSE_2 denote the sum of squares for error for models 1 and 2, respectively, we can

$\text{SSE}_2 - \text{SSE}_1$

examine the difference **$\text{SSE}_2 - \text{SSE}_1$**. For n sample observations, the degrees of freedom for a sum of squares for error can be computed as $n - $ (the number of parameters in the model). Hence, SSE_1 and SSE_2 have, respectively, $n - (k + 1)$ and $n - (g + 1)$ degrees of freedom. When H_0 is true, SSE_2 and SSE_1 will be of approximately the same magnitude, although SSE_1 will be less than SSE_2 owing to the fact that the complete model has more terms in it. When H_a is true, and at least one of the βs under test is different from 0, SSE_2 will be much larger than SSE_1. Since $\text{SSE}_2 - \text{SSE}_1$ will be greater than 0 under either H_0 or H_a, we

SS_{drop}

call the difference $\text{SSE}_2 - \text{SSE}_1$ the **drop in the sum of squares** for error attributable to the variables $x_{g+1}, x_{g+2}, \ldots, x_k$. If the sum of squares drop is large, this implies that the sum of squares for error has been greatly reduced by including the variables $x_{g+1}, x_{g+2}, \ldots, x_k$ in the model, and intuitively we would reject H_0.

How large must $\text{SSE}_2 - \text{SSE}_1$ be in order to reject the hypothesis that $x_{g+1}, x_{g+2}, \ldots, x_k$ are unrelated to the response y? When H_0 is true, the quantity $\text{SS}_{\text{drop}} = \text{SSE}_2 - \text{SSE}_1$ divided by the number of parameters under test in H_0 $(k - g)$ provides an unbiased estimate of σ_ε^2, the variance associated with an observation in the general linear model. We designate this estimate as

MS_{drop}

the **mean square drop**:

$$\text{MS}_{\text{drop}} = \frac{\text{SSE}_2 - \text{SSE}_1}{k - g}.$$

The mean square error for model 1,

$$\text{MSE}_1 = \frac{\text{SSE}_1}{n - (k + 1)},$$

also provides an unbiased estimate of σ_ε^2. It can be shown that when H_0 is true, the ratio

F statistic

$$\frac{\text{MS}_{\text{drop}}}{\text{MSE}_1}$$

follows an F distribution (**the F statistic**), with $k - g$ and $n - (k + 1)$ degrees of freedom, respectively.

When H_a is true, MSE_1 is still an unbiased estimate of σ_ε^2, whereas MS_{drop} is an unbiased estimate of $\sigma_\varepsilon^2 +$ (a positive function of $\beta_{g+1}, \beta_{g+2}, \ldots, \beta_k$). Thus, large values of $F = \text{MS}_{\text{drop}}/\text{MSE}_1$ will indicate rejection of the null hypothesis.

E X A M P L E   11.10   ▼

In Example 11.4, we considered a situation in which a chemist was interested in determining the weight loss y of a compound as a function of the length of time the compound was exposed to the air and the relative humidity of the environment during exposure. An experiment was performed using three relative humidities and four exposure times, with weight loss (in pounds) recorded for each factor–level combination. We obtained a least squares fit to a first-order model, relating the response to the two variables.

Now let us consider a model of the form

$$y = \beta_0 + \beta_1 x_1 + \beta_2 x_2 + \beta_3 x_2^2 + \beta_4 x_1 x_2 + \beta_5 x_1 x_2^2 + \varepsilon,$$

where $x_1 =$ exposure time and $x_2 =$ relative humidity. The computer output shows the least squares fit for the complete and reduced model corresponding to $H_0: \beta_4 = \beta_5 = 0$. Give all parts of the F test of $H_0: \beta_4 = \beta_5 = 0$ and interpret the results of your test. Use $\alpha = .05$.

```
          LISTING OF DATA
   OBS     WT_LOSS    TIME    HUMIDITY
    1        4.3       4        0.2
    2        5.5       5        0.2
    3        6.8       6        0.2
    4        8.0       7        0.2
    5        4.0       4        0.3
    6        5.2       5        0.3
    7        6.6       6        0.3
    8        7.5       7        0.3
    9        2.0       4        0.4
   10        4.0       5        0.4
   11        5.7       6        0.4
   12        6.5       7        0.4

  N =       12                          (continues)
```

```
                        COMPLETE MODEL
               GENERAL LINEAR MODELS PROCEDURE

DEPENDENT VARIABLE: WT_LOSS

SOURCE       DF    SUM OF SQUARES    MEAN SQUARE    F VALUE    PR > F    R-SQUARE    C.V.
MODEL         5      32.04216667      6.40843333     90.05     0.0001   0.986849   4.8430

ERROR         6       0.42700000      0.07116667              ROOT MSE        WT_LOSS MEAN
CORRECTED TOTAL 11    32.46916667                             0.26677081      5.50833333


SOURCE              DF      TYPE I SS   F VALUE   PR > F    DF   TYPE III SS  F VALUE   PR > F
TIME                 1     26.00416667   365.40   0.0001     1    0.28212844    3.96    0.0936
HUMIDITY             1      5.12000000    71.94   0.0001     1    0.16601864    2.33    0.1775
HUMIDITY*HUMIDITY    1      0.60166667     8.45   0.0271     1    0.24448677    3.44    0.1132
TIME*HUMIDITY        1      0.19600000     2.75   0.1481     1    0.09174312    1.29    0.2995
TIME*HUMIDI*HUMIDI   1      0.12033333     1.69   0.2412     1    0.12033333    1.69    0.2412


                            T FOR HO:   PR > |T|    STD ERROR OF
PARAMETER            ESTIMATE    PARAMETER=0            ESTIMATE

INTERCEPT            -9.69000000     -1.39    0.2150      6.99071885
TIME                  2.48000000      1.99    0.0936      1.24556547
HUMIDITY             75.50000000      1.53    0.1775     49.43184702
HUMIDITY*HUMIDITY  -152.00000000     -1.85    0.1132     82.00762160
TIME*HUMIDITY       -10.00000000     -1.14    0.2995      8.80747788
TIME*HUMIDI*HUMIDI   19.00000000      1.30    0.2412     14.61163920



                        REDUCED MODEL
               GENERAL LINEAR MODELS PROCEDURE

DEPENDENT VARIABLE: WT_LOSS

SOURCE       DF    SUM OF SQUARES    MEAN SQUARE    F VALUE    PR > F    R-SQUARE    C.V.
MODEL         3      31.72583333     10.57527778    113.81     0.0001   0.977106   5.5339

ERROR         8       0.74333333      0.09291667              ROOT MSE        WT_LOSS MEAN
CORRECTED TOTAL 11    32.46916667                             0.30482235      5.50833333


SOURCE              DF      TYPE I SS   F VALUE   PR > F    DF   TYPE III SS  F VALUE   PR > F
TIME                 1     26.00416667   279.87   0.0001     1   26.00416667  279.87    0.0001
HUMIDITY             1      5.12000000    55.10   0.0001     1    0.30844037    3.32    0.1059
HUMIDITY*HUMIDITY    1      0.60166667     6.48   0.0345     1    0.60166667    6.48    0.0345

                            T FOR HO:   PR > |T|    STD ERROR OF
PARAMETER            ESTIMATE    PARAMETER=0            ESTIMATE

INTERCEPT            -3.29166667     -2.00    0.0810      1.64904855
TIME                  1.31666667     16.73    0.0001      0.07870479
HUMIDITY             20.50000000      1.82    0.1059     11.25162025
HUMIDITY*HUMIDITY   -47.50000000     -2.54    0.0345     18.66648065
```

Solution The null hypothesis $H_0: \beta_4 = \beta_5 = 0$ states there is no interaction (linear or quadratic) between the independent variables exposure time x_1 and relative humidity x_2. The complete and reduced models for this null hypothesis are

$$\text{complete model:} \quad y = \beta_0 + \beta_1 x_1 + \beta_2 x_2 + \beta_3 x_2^2 + \beta_4 x_1 x_2 + \beta_5 x_1 x_2^2 + \varepsilon$$
$$\text{reduced model:} \quad y = \beta_0 + \beta_1 x_1 + \beta_2 x_2 + \beta_3 x_2^2 + \varepsilon.$$

From the computer output, we have

$$SSE_1 = .427, \qquad df_1 = 6$$

and from the reduced model,

$$SSE_2 = .743, \qquad df_2 = 8.$$

It follows that

$$SS_{drop} = SSE_2 - SSE_1 = .743 - .427 = .316$$

$$MS_{drop} = \frac{SSE_2 - SSE_1}{2} = .158$$

and

$$MSE_1 = \frac{SSE_1}{6} = .071.$$

The parts of the test are summarized here.

$$H_0: \quad \beta_4 = \beta_5 = 0$$
$$H_a: \quad \text{At least one different from } 0$$

$$\text{T.S.:} \quad F = \frac{MS_{drop}}{MSE_1} = \frac{.158}{.071} = 2.23$$

R.R.: For $\alpha = .05$ the critical F value based on 2 and 6 degrees of freedom is 5.14. Since the observed value of F does not exceed the critical value, we have insufficient evidence to reject H_0. Practically, we are unable to detect any linear or quadratic interaction between time of exposure and relative humidity.

▲

EXAMPLE 11.11 ▼

Consider an experiment to investigate the effects of four different pesticides on the yield of fruit from three different varieties of citrus trees. Write a model that contains main effects for both qualitative variables but no interaction. Use the data, reproduced in Table 11.4, to test for no difference

TABLE 11.4

Data for the Fruit Yield Experiment of Example 11.11

	Pesticide			
Variety	**1**	**2**	**3**	**4**
1	29	50	43	53
2	41	58	42	73
3	66	85	69	85

among mean yields for pesticides by fitting complete and reduced models. Use $\alpha = .05$.

a. Write a model that contains dummy variables for pesticides and for varieties.

b. Interpret the βs.

c. Fit complete and reduced models to test the hypothesis of no differences among mean yields for pesticides. Use $\alpha = .05$.

Solution

a. We could use the model

$$y = \beta_0 + \overbrace{\beta_1 x_1 + \beta_2 x_2}^{\text{varieties}} + \overbrace{\beta_3 x_3 + \beta_4 x_4 + \beta_5 x_5}^{\text{pesticides}} + \varepsilon,$$

where $x_1, x_2, \ldots, x_5$ are dummy variables defined in the following way:

$x_1 = 1$ if variety 2 $\qquad x_1 = 0$ otherwise
$x_2 = 1$ if variety 3 $\qquad x_2 = 0$ otherwise
$x_3 = 1$ if pesticide 2 $\qquad x_3 = 0$ otherwise
$x_4 = 1$ if pesticide 3 $\qquad x_4 = 0$ otherwise
and
$x_5 = 1$ if pesticide 4 $\qquad x_5 = 0$ otherwise

b. We can form a table of expected values by substituting appropriate values for the dummy variables. For example, an observation on variety 1, pesticide 1 has $x_1 = x_2 = \cdots = x_5 = 0$; hence, $E(y) = \beta_0$. Similarly, an observation on variety 2, pesticide 3 has $x_1 = 1$, $x_2 = 0$, $x_3 = 0$, $x_4 = 1$, and $x_5 = 0$; so $E(y) = \beta_0 + \beta_1 + \beta_4$. The remaining entries for the expected value table shown here were computed in the same way.

	Pesticide			
Variety	1	2	3	4
1	β_0	$\beta_0 + \beta_3$	$\beta_0 + \beta_4$	$\beta_0 + \beta_5$
2	$\beta_0 + \beta_1$	$\beta_0 + \beta_1 + \beta_3$	$\beta_0 + \beta_1 + \beta_4$	$\beta_0 + \beta_1 + \beta_5$
3	$\beta_0 + \beta_2$	$\beta_0 + \beta_2 + \beta_3$	$\beta_0 + \beta_2 + \beta_4$	$\beta_0 + \beta_2 + \beta_5$

It is clear from this table that

β_0 is the expected value (mean yield) for an observation on variety 1, pesticide 1.

β_1 is the difference in expected values (mean yields) for varieties 2 and 1 (for a given pesticide).

β_2 is the difference in expected values for varieties 3 and 1 (for a given pesticide).

β_3 is the difference in expected values for pesticides 2 and 1 (for a given variety).

$$\vdots$$

β_5 is the difference in expected values for pesticides 4 and 1 (for a given variety).

c. The null hypothesis for testing no difference among mean yields for pesticides is

$$H_0: \quad \beta_3 = \beta_4 = \beta_5 = 0$$

For this test, the complete and reduced models are

$$\text{complete model:} \quad y = \beta_0 + \beta_1 x_1 + \beta_2 x_2 + \cdots + \beta_5 x_5 + \varepsilon$$
$$\text{reduced model:} \quad y = \beta_0 + \beta_1 x_1 + \beta_2 x_2 + \varepsilon$$

From the computer output shown here,

$$SSE_1 = 151.50, \qquad df_1 = 6, \qquad MSE_1 = 25.25$$
$$SSE_2 = 1342.50, \qquad df_2 = 9, \qquad MSE_2 = 149.17$$

```
                         LISTING OF DATA
   OBS    VARIETY    PEST_CDE    YIELD    X1    X2    X3    X4    X5
    1        1          1          29      0     0     0     0     0
    2        1          2          50      0     0     1     0     0
    3        1          3          43      0     0     0     1     0
    4        1          4          53      0     0     0     0     1
    5        2          1          41      1     0     0     0     0
    6        2          2          58      1     0     1     0     0
    7        2          3          42      1     0     0     1     0
    8        2          4          73      1     0     0     0     1
    9        3          1          66      0     1     0     0     0
   10        3          2          85      0     1     1     0     0
   11        3          3          69      0     1     0     1     0
   12        3          4          85      0     1     0     0     1
  N =    12

                         COMPLETE MODEL
                  GENERAL LINEAR MODELS PROCEDURE

DEPENDENT VARIABLE: YIELD
SOURCE              DF      SUM OF SQUARES     MEAN SQUARE    F VALUE    PR > F   R-SQUARE    C.V.
MODEL                5      3416.16666667    683.23333333      27.06    0.0005   0.957535   8.6887
ERROR                6       151.50000000     25.25000000               ROOT MSE        YIELD MEAN
CORRECTED TOTAL     11      3567.66666667                              5.02493781      57.833333333
                                                                                      (continues)
```

SOURCE	DF	TYPE I SS	F VALUE	PR > F	DF	TYPE III SS	F-VALUE	PR > F
X1	1	112.66666667	4.46	0.0791	1	190.12500000	7.53	0.0336
X2	1	2112.50000000	83.66	0.0001	1	2112.50000000	83.66	0.0001
X3	1	169.00000000	6.69	0.0414	1	541.50000000	21.45	0.0036
X4	1	84.50000000	3.35	0.1171	1	54.00000000	2.14	0.1940
X5	1	937.50000000	37.13	0.0009	1	937.50000000	37.13	0.0009

PARAMETER	ESTIMATE	TO FOR HO: PARAMETER=0	PR > |T|	STD ERROR OF ESTIMATE
INTERCEPT	31.25000000	8.79	0.0001	3.55316760
X1	9.75000000	2.74	0.0336	3.55316760
X2	32.50000000	9.15	0.0001	3.55316760
X3	19.00000000	4.63	0.0036	4.10284454
X4	6.00000000	1.46	0.1940	4.10284454
X5	25.00000000	6.09	0.0009	4.10284454

REDUCED MODEL

GENERAL LINEAR MODELS PROCEDURE

DEPENDENT VARIABLE: YIELD

SOURCE	DF	SUM OF SQUARES	MEAN SQUARE	F VALUE	PR > F	R-SQUARE	C.V.
MODEL	2	2225.16666667	1112.58333333	7.46	0.0123	0.623704	21.1182
ERROR	9	1342.50000000	149.16666667		ROOT MSE		YIELD MEAN
CORRECTED TOTAL	11	3567.66666667			12.21338064		57.8333333

SOURCE	DF	TYPE I SS	F VALUE	PR > F	DF	TYPE III SS	F VALUE	PR > F
X1	1	112.66666667	0.76	0.4074	1	190.12500000	1.27	0.2881
X3	1	2112.50000000	14.16	0.0045	1	2112.50000000	14.16	0.0045

PARAMETER	ESTIMATE	T FOR HO: PARAMETER=0	PR > |T|	STD ERROR OF ESTIMATE
INTERCEPT	43.75000000	7.16	0.0001	6.10669032
X1	9.75000000	1.13	0.2881	8.63616427
X2	32.50000000	3.76	0.0045	8.63616427

Hence, $SS_{drop} = 1342.50 - 151.50 = 1191.00$ and $MS_{drop} = SS_{drop}/3 = 397.00$. The entire test is shown here.

H_0: $\beta_3 = \beta_4 = \beta_5 = 0$
H_a: At least one of the βs is different from 0.

T.S.: $F = \dfrac{MS_{drop}}{MSE_1} = \dfrac{397.00}{25.25} = 15.72$

R.R.: For $\alpha = .05$, reject H_0 if $F > F_{.05}$ with 3 and 6 degrees of freedom. This value is 4.76. Since $F > 4.76$, we reject H_0 and conclude that the pesticide means are not all equal. ▲

▼ EXERCISES

Applications

 11.25 Refer to Example 11.10. Under the assumption that there is no interaction between the variables "exposure time" and "relative humidity," start with the complete model

$$y = \beta_0 + \beta_1 x_1 + \beta_2 x_2 + \beta_3 x_2^2 + \varepsilon.$$

Use an appropriate reduced model to test the hypothesis of no linear or quadratic effect due to x_2. Use $\alpha = .05$.

11.26   Refer to Exercise 11.25.

a. Fit a complete and reduced model to test for no linear effect due to x_1. Use $\alpha = .05$.

b. How else might you test this same hypothesis without fitting complete and reduced models?

11.7 ▼ MATRIX NOTATION FOR THE GENERAL LINEAR MODEL (optional)

Recall that a model relating a response y to a set of k independent variables of the form

$$y = \beta_0 + \beta_1 x_1 + \beta_2 x_2 + \cdots + \beta_k x_k + \varepsilon$$

has been called the general linear model. If a sample of $n(n > k)$ measurements is obtained for n settings of the independent variables $x_1, x_2, \ldots, x_k$, we can write an individual observation as

$$y_i = \beta_0 + \beta_1 x_{i1} + \beta_2 x_{i2} + \cdots + \beta_k x_{ik} + \varepsilon_i \qquad (i = 1, 2, \ldots, n),$$

where $x_{i1}, x_{i2}, \ldots, x_{ik}$ are the settings of the independent variables for the response y_i and ε_i is the random error for the ith response.

The entire set of n observations can be expressed in the general linear model using matrix notation. (Refer to the chapter appendix for a summary of basic matrix operations.) Let the $n \times 1$ matrix $\mathbf{Y}$

$$\mathbf{Y} = \begin{bmatrix} y_1 \\ y_2 \\ \vdots \\ y_n \end{bmatrix}$$

be the matrix of observations, and let the $n \times (k + 1)$ matrix $\mathbf{X}$

$$\mathbf{X} = \begin{bmatrix} 1 & x_{11} & x_{12} & \cdots & x_{1k} \\ 1 & x_{21} & x_{22} & \cdots & x_{2k} \\ \vdots & \vdots & \vdots & & \vdots \\ 1 & x_{n1} & x_{n2} & \cdots & x_{nk} \end{bmatrix}$$

be a matrix of settings for the independent variables augmented with a column of 1s. The first row of $\mathbf{X}$ contains a 1 and the settings on the k independent variables for the first observation. Row 2 contains a 1 and corresponding settings on the independent variables for y_2. Similarly, the other rows contain settings for the remaining observations. Let

$$\boldsymbol{\beta} = \begin{bmatrix} \beta_0 \\ \beta_1 \\ \beta_2 \\ \vdots \\ \beta_k \end{bmatrix}$$

be a $(k + 1) \times 1$ matrix containing the unknown parameters for the general linear model and let

$$\boldsymbol{\varepsilon} = \begin{bmatrix} \varepsilon_1 \\ \varepsilon_2 \\ \vdots \\ \varepsilon_n \end{bmatrix}$$

general linear model, matrix notation

be an $n \times 1$ matrix of errors associated with the n observations. Then we can write the **general linear model** in **matrix notation** as

$$\begin{bmatrix} y_1 \\ y_2 \\ \vdots \\ y_n \end{bmatrix} = \begin{bmatrix} 1 & x_{11} & x_{12} & \cdots & x_{1k} \\ 1 & x_{21} & x_{22} & \cdots & x_{2k} \\ \vdots & \vdots & \vdots & & \vdots \\ 1 & x_{n1} & x_{n2} & \cdots & x_{nk} \end{bmatrix} \begin{bmatrix} \beta_0 \\ \beta_1 \\ \vdots \\ \beta_k \end{bmatrix} + \begin{bmatrix} \varepsilon_1 \\ \varepsilon_2 \\ \vdots \\ \varepsilon_n \end{bmatrix}$$

or simply

$$\mathbf{Y} = \mathbf{X}\boldsymbol{\beta} + \boldsymbol{\varepsilon}.$$

Note that to obtain the equation for y_1, we multiply row one of $\mathbf{X}$ times the matrix of βs and then add ε_1 to obtain

$$\begin{bmatrix} y_1 \end{bmatrix} = \begin{bmatrix} 1 & x_{11} & x_{12} & \cdots & x_{1k} \end{bmatrix} \begin{bmatrix} \beta_0 \\ \beta_1 \\ \vdots \\ \beta_k \end{bmatrix} + \begin{bmatrix} \varepsilon_1 \end{bmatrix}$$

or

$$y_1 = \beta_0 + \beta_1 x_{11} + \beta_2 x_{12} + \cdots + \beta_k x_{1k} + \varepsilon_1,$$

which is precisely as it was defined previously. In fact, any observation y_i is obtained by multiplying the ith row of $\mathbf{X}$ times the matrix of βs and then adding ε_i.

If we let the matrix

$$\hat{\boldsymbol{\beta}} = \begin{bmatrix} \hat{\beta}_0 \\ \hat{\beta}_1 \\ \hat{\beta}_2 \\ \vdots \\ \hat{\beta}_k \end{bmatrix}$$

represent the matrix of least squares estimates for the parameters of the general linear model, then, provided the matrix $\mathbf{X'X}$ has an inverse, we can find these estimates using the matrix equation given here.

Matrix Equation for Least Squares Estimates

$$\hat{\boldsymbol{\beta}} = (\mathbf{X'X})^{-1}\mathbf{X'Y}$$

This matrix solution gives the set of parameter estimates $\hat{\beta}_0, \hat{\beta}_1, \ldots, \hat{\beta}_k$ in the general linear model that minimizes $\sum_i (y_i - \hat{y}_i)^2$ for the data collected.

E X A M P L E   11.12   ▼

Refer to the data of Example 9.1, where we related the dressed weight y of a steer to its live weight x using the linear regression model

$$y = \beta_0 + \beta_1 x + \varepsilon.$$

Use matrices to find the least squares estimates of β_0 and β_1. Compare your results to those obtained in Example 9.1.

Solution   The model $y = \beta_0 + \beta_1 x + \varepsilon$ can be considered a general linear model with $k = 1$ independent variable. Using the $n = 9$ sample observations, we can specify the following matrices:

$$\mathbf{Y} = \begin{bmatrix} 2.8 \\ 2.5 \\ 3.1 \\ 2.1 \\ 2.9 \\ 2.8 \\ 2.6 \\ 2.4 \\ 2.5 \end{bmatrix} \qquad \mathbf{X} = \begin{bmatrix} 1 & 4.2 \\ 1 & 3.8 \\ 1 & 4.8 \\ 1 & 3.4 \\ 1 & 4.5 \\ 1 & 4.6 \\ 1 & 4.3 \\ 1 & 3.7 \\ 1 & 3.9 \end{bmatrix}$$

Note that the second column of $\mathbf{X}$ gives the settings (live weights) corresponding to the observed dressed weights (y).

The transpose of $\mathbf{X}$ is

$$\mathbf{X}' = \begin{bmatrix} 1 & 1 & 1 & 1 & 1 & 1 & 1 & 1 & 1 \\ 4.2 & 3.8 & 4.8 & 3.4 & 4.5 & 4.6 & 4.3 & 3.7 & 3.9 \end{bmatrix}.$$

Thus,

$$\mathbf{X'X} = \begin{bmatrix} 1 & 1 & 1 & 1 & 1 & 1 & 1 & 1 & 1 \\ 4.2 & 3.8 & 4.8 & 3.4 & 4.5 & 4.6 & 4.3 & 3.7 & 3.9 \end{bmatrix} \begin{bmatrix} 1 & 4.2 \\ 1 & 3.8 \\ 1 & 4.8 \\ 1 & 3.4 \\ 1 & 4.5 \\ 1 & 4.6 \\ 1 & 4.3 \\ 1 & 3.7 \\ 1 & 3.9 \end{bmatrix}$$

$$= \begin{bmatrix} 9 & 37.2 \\ 37.2 & 155.48 \end{bmatrix}$$

and

$$\mathbf{X'Y} = \begin{bmatrix} 1 & 1 & 1 & 1 & 1 & 1 & 1 & 1 & 1 \\ 4.2 & 3.8 & 4.8 & 3.4 & 4.5 & 4.6 & 4.3 & 3.7 & 3.9 \end{bmatrix} \begin{bmatrix} 2.8 \\ 2.5 \\ 3.1 \\ 2.1 \\ 2.9 \\ 2.8 \\ 2.6 \\ 2.4 \\ 2.5 \end{bmatrix}$$

$$= \begin{bmatrix} 23.7 \\ 99.02 \end{bmatrix}.$$

The inverse of $\mathbf{X'X}$ can be shown to be

$$(\mathbf{X'X})^{-1} = \begin{bmatrix} 10.0439 & -2.4031 \\ -2.4031 & .5814 \end{bmatrix}$$

so the least squares estimates of β_0 and β_1 are

$$\hat{\boldsymbol{\beta}} = (\mathbf{X}'\mathbf{X})^{-1}\mathbf{X}'\mathbf{Y}$$

$$= \begin{bmatrix} 10.0439 & -2.4031 \\ -2.4031 & .5814 \end{bmatrix} \begin{bmatrix} 23.7 \\ 99.02 \end{bmatrix} = \begin{bmatrix} .085 \leftarrow \\ .617 \leftarrow \end{bmatrix} \begin{matrix} \hat{\beta}_0 \\ \hat{\beta}_1 \end{matrix}$$

and the linear regression line is

$$\hat{y} = .085 + .617x.$$

This result is identical to that obtained for Example 9.1, except for rounding errors. ▲

EXAMPLE 11.13 ▼

Refer to the data of Example 11.4, in which a chemist was interested in determining the weight loss y of a compound as a function of the exposure time and relative humidity. Set up all the matrices for fitting the regression model $y = \beta_0 + \beta_1 x_1 + \beta_2 x_2 + \varepsilon$.

Solution   The matrices required to obtain $\hat{\boldsymbol{\beta}}$ are

$$\mathbf{Y} = \begin{bmatrix} 4.3 \\ 5.5 \\ \vdots \\ 6.5 \end{bmatrix} \qquad \mathbf{X} = \begin{bmatrix} 1 & 4 & .20 \\ 1 & 5 & .20 \\ \vdots & \vdots & \vdots \\ 1 & 7 & .40 \end{bmatrix}$$

$$\mathbf{X}'\mathbf{X} = \begin{bmatrix} 12 & 66 & 3.6 \\ 66 & 378 & 19.8 \\ 3.6 & 19.8 & 1.16 \end{bmatrix} \qquad \text{and} \qquad \mathbf{X}'\mathbf{Y} = \begin{bmatrix} 66.1 \\ 383.3 \\ 19.19 \end{bmatrix}.$$

From these matrices one could compute

$$\hat{\boldsymbol{\beta}} = (\mathbf{X}'\mathbf{X})^{-1}\mathbf{X}'\mathbf{Y}.$$

The important thing is not the computation, but the understanding that all general model problems (multiple regression and others) have a common format and are solved via matrices using the same formulas. ▲

In a similar way, all the inferential procedures about general linear models that were discussed earlier in this chapter can be set up using matrices and solved simply using available statistical software packages.

We begin as follows. Under the assumptions that the ε_i are independent and each normally distributed with mean 0 and variance σ_ε^2, the least squares estimator $\hat{\beta}_i$ has mean β_i (the parameter estimated) and standard error $\sqrt{v_{ii}}\sigma_\varepsilon$, where v_{ii} is the ith diagonal element of $(\mathbf{X}'\mathbf{X})^{-1}$

$$(\mathbf{X}'\mathbf{X})^{-1} = \begin{bmatrix} v_{00} & v_{01} & \cdots & v_{0k} \\ v_{10} & v_{11} & \cdots & v_{1k} \\ \vdots & \vdots & & \vdots \\ v_{k0} & v_{k1} & \cdots & v_{kk} \end{bmatrix}.$$

Note that the first diagonal element is labeled v_{00} to correspond to $\hat{\beta}_0$, so the standard error of $\hat{\beta}_0$ is $\sqrt{v_{00}}\sigma_\varepsilon$.

EXAMPLE 11.14

▼

Refer to Example 11.4. For the regression model $y = \beta_0 + \beta_1 x_1 + \beta_2 x_2 + \varepsilon$, it can be shown that

$$(\mathbf{X}'\mathbf{X})^{-1} = \begin{bmatrix} 3.2250 & -.3667 & -3.7500 \\ -.3667 & .0667 & .0000 \\ -3.7500 & .0000 & 12.5000 \end{bmatrix}.$$

a. Determine the standard errors for $\hat{\beta}_0$, $\hat{\beta}_1$, and $\hat{\beta}_2$.
b. If MSE $= .149$, estimate the standard errors for $\hat{\beta}_0$, $\hat{\beta}_1$, and $\hat{\beta}_2$. Compare your values to those given in the output for Example 11.4.

Solution
a. Using the standard error formula based on matrices,

$$\sigma_{\hat{\beta}_0} = \sqrt{v_{00}}\sigma_\varepsilon = \sqrt{3.2250}\sigma_\varepsilon$$
$$\sigma_{\hat{\beta}_1} = \sqrt{v_{11}}\sigma_\varepsilon = \sqrt{0.0667}\sigma_\varepsilon$$

and

$$\sigma_{\hat{\beta}_2} = \sqrt{v_{22}}\sigma_\varepsilon = \sqrt{12.5000}\sigma_\varepsilon.$$

b. Substituting $\sqrt{\text{MSE}} = \sqrt{.149} = .386$ for σ_ε, the (estimated) standard errors for $\hat{\beta}_0$, $\hat{\beta}_1$, and $\hat{\beta}_2$ are, respectively, .693, .100, and 1.365. These agree (except for rounding errors) with the standard errors shown in the output for Example 11.4. ▲

Using the previous result on standard errors and the fact that SSE = $\mathbf{Y'Y} - \hat{\boldsymbol{\beta}}'\mathbf{X'Y}$, it is possible to make inferences about a single β_i or a set of βs.

Inferences About a Single β_i and a Set of βs

100(1 − α)% confidence interval for β_i: $\hat{\beta}_i \pm t_{\alpha/2} s_\varepsilon \sqrt{v_{ii}}$.

Statistical test for H_0: $\beta_i = 0$ T.S.: $t = \dfrac{\hat{\beta}_i}{s_\varepsilon \sqrt{v_{ii}}}$,

where

$$s_\varepsilon = \sqrt{\text{MSE}} = \sqrt{\frac{\text{SSE}}{n - (k + 1)}}.$$

Test for H_0: $\beta_{g+1} = \beta_{g+2} = \cdots = \beta_k = 0$ makes use of SSE = $\mathbf{Y'Y} - \hat{\boldsymbol{\beta}}'\mathbf{X'Y}$ computed for both the complete and reduced models.

A confidence interval for $E(y)$ and the corresponding prediction interval for y are as follows:

100(1 − α)% Confidence Interval for $E(y)$

100(1 − α)% confidence interval for $E(y)$

$$\hat{y} \pm t_{\alpha/2} s_\varepsilon \sqrt{\boldsymbol{\ell}'(\mathbf{X'X})^{-1}\boldsymbol{\ell}}.$$

100(1 − α)% prediction interval for y

$$\hat{y} \pm t_{\alpha/2} s_\varepsilon \sqrt{1 + \boldsymbol{\ell}'(\mathbf{X'X})^{-1}\boldsymbol{\ell}},$$

where the matrix

$$\boldsymbol{\ell} = \begin{bmatrix} 1 \\ x_1 \\ x_2 \\ \vdots \\ x_k \end{bmatrix}$$

displays the desired settings of the independent variable and

$$s_\varepsilon = \sqrt{\text{MSE}} = \sqrt{\frac{\text{SSE}}{n - (k + 1)}}.$$

The results of this section indicate that with a basic understanding of a few matrix operations, it is possible to use a common format for all the estimation and test procedures discussed to date for models in the form of a general linear model. It is this common format that has enabled computer software vendors to develop statistical software packages with broad applicability rather than creating separate programs for each type of problem.

▼ EXERCISES

Applications

11.27 The data from Exercise 9.12 related to the firmness of canned sweet potatoes treated with various concentrations of pectin and stored at 25°C are shown here.

y (firmness)	50.5	46.8	62.3	67.7	80.1	79.2
x (concentration of pectin)	0	0	1.5	1 .5	3.0	3.0

a. Obtain the least squares estimates for the parameters in the model $y = \beta_0 + \beta_1 x + \varepsilon$ by using the matrix approach.
b. Give the standard error of $\hat{\beta}_1$.
c. Obtain an estimate of σ_ε.

11.28 Refer to Exercise 11.27.
a. Give an estimate of the standard error of $\hat{\beta}_1$.
b. Perform a statistical test of the null hypothesis that there is no linear relationship between the concentration of pectin and the firmness of canned sweet potatoes after 30 days of storage at 25°C. Use $\alpha = .05$.

11.29 Refer to Exercise 11.12.
a. Using a matrix approach, fit the general linear model

$$y = \beta_0 + \beta_1 x + \beta_2 x^2 + \varepsilon$$

to obtain $\hat{y}$.
b. Compute an estimate of σ_ε^2.
c. Identify the standard errors of $\hat{\beta}_0$, $\hat{\beta}_1$, and $\hat{\beta}_2$.

11.30 Refer to Exercise 11.12. Conduct a test of the null hypothesis that $\beta_2 = 0$, that is, the log of the biological recovery percentage (y) is linearly related to time (x). Use $\alpha = .05$.

11.31 Refer to Exercise 11.12. Place a 95% confidence interval on β_0, the mean log biological recovery percentage at time zero. (Note: $E(y) = \beta_0$ when $x = 0$.)

11.32 Refer to Example 11.8. Set up all matrices for developing a 90% confidence interval on $E(y)$, the mean corn yield, when 5 pounds of fertilizer are applied.

11.8 ▼ SUMMARY

This chapter consolidates the material for expressing a response y as a function of one or more independent variables. Multiple regression models (where all the independent variables are quantitative) and models that incorporate information on qualitative variables were discussed and can be represented in the form of a general linear model

$$y = \beta_0 + \beta_1 x_1 + \beta_2 x_2 + \cdots + \beta_k x_k + \varepsilon.$$

After discussing various models and the interpretation of βs in these models, we presented the normal equations used in obtaining the least squares estimates $\hat{\beta}$.

A confidence interval and statistical test about an individual parameter β_j were developed using $\hat{\beta}_j$ and the standard error of $\hat{\beta}_j$. We also considered a statistical test about a set of βs, a confidence interval for $E(y)$ based on a set of xs, and a prediction interval for a given set of xs.

All of these inferences involve a fair to moderate amount of numerical calculation unless statistical software programs or packages are available. Sometimes these calculations can be done by hand if one is familiar with matrix operations (see Section 11.7 and the appendix to this chapter). However, even these methods become unmanageable as the number of independent variables increases. So, the message should be very clear. Inferences about general linear models should be done using available computer software to facilitate the analysis and to minimize computational errors. Our job in these situations is to review and interpret the output.

Aside from a few exercises that will probe your understanding of the mechanics involved with these calculations, most of the exercises in the remainder of this chapter and in the regression problems of the next chapter will make extensive use of computer output.

▼ KEY FORMULAS

1.
$$R^2_{y \cdot x_1 x_2 \cdots x_k} = \frac{S_{yy} - \text{SSE}}{S_{yy}},$$

where

$$S_{yy} = \sum y_i^2 - \frac{(\sum y_i)^2}{n}$$

and

$$\text{SSE} = \sum (y_i - \hat{y}_i)^2$$
$$= \sum y_i^2 - \hat{\beta}_0 \sum y_i - \hat{\beta}_1 \sum x_{i1} y_i - \cdots - \hat{\beta}_k \sum x_{ik} y_i$$

2.
$$s_{\hat{\beta}_j} = s_\varepsilon \sqrt{\frac{1}{S_{x_j x_j}(1 - R^2_{x_j \cdot x_1 \cdots x_{j-1} x_{j+1} \cdots x_k})}},$$

where

$$s_\varepsilon = \sqrt{\frac{SSE}{n - (k - 1)}}$$

and

$$S_{x_j x_j} = \sum_i (x_{ij} - \bar{x}_j)^2$$

3.

$$MS_{drop} = \frac{SSE_2 - SSE_1}{k - g}$$

4.

$$\hat{\beta} = (\mathbf{X'X})^{-1}\mathbf{X'Y}$$

▼ SUPPLEMENTARY EXERCISES

11.33 The data presented here illustrate the effects of correlated and uncorrelated independent variables.

y	7	10	16	12	16	15	19	22	20	23	22	24
x_1	1	2	3	4	1	2	3	4	1	2	3	4
x_2	1	1	1	1	2	2	2	2	3	3	3	3
x_3	1	3	5	7	9	11	13	15	17	19	21	23

a. Plot x_1 versus x_2, x_1 versus x_3, and x_2 versus x_3.
b. Which of the plots in part a indicate uncorrelated independent variables?

11.34 Computer output for the data from Exercise 11.33 is shown for the following models:

1. $y = \beta_0 + \beta_1 x_1 + \beta_2 x_2 + \varepsilon$
2. $y = \beta_0 + \beta_1 x_1 + \beta_3 x_3 + \varepsilon$
3. $y = \beta_0 + \beta_2 x_2 + \beta_3 x_3 + \varepsilon$
4. $y = \beta_0 + \beta_1 x_1 + \beta_2 x_2 + \beta_3 x_3 + \varepsilon$

```
    LISTING OF DATA

 OBS    Y    X1   X2   X3
  1     7    1    1    1
  2    10    2    1    3
  3    16    3    1    5
  4    12    4    1    7
  5    16    1    2    9
  6    15    2    2   11
  7    19    3    2   13
  8    22    4    2   15
  9    20    1    3   17
 10    23    2    3   19
 11    22    3    3   21
 12    24    4    3   23

 N =   12
```

(continues)

REGRESSION ANALYSIS, MODEL 1

GENERAL LINEAR MODELS PROCEDURE

DEPENDENT VARIABLE: Y

SOURCE	DF	SUM OF SQUARES	MEAN SQUARE	F VALUE	PR > F	R-SQUARE	C.V.
MODEL	2	290.60000000	145.30000000	35.28	0.0001	0.886877	11.8218
ERROR	9	37.06666667	4.11851852		ROOT MSE		Y MEAN
CORRECTED TOTAL	11	327.66666667			2.02941334		17.1666667

SOURCE	DF	TYPE I SS	F VALUE	PR > F	DF	TYPE III SS	F VALUE	PR > F
X1	1	48.60000000	11.80	0.0074	1	48.60000000	11.80	0.0074
X2	1	242.00000000	58.76	0.0001	1	242.00000000	58.76	0.0001

PARAMETER	ESTIMATE	T FOR H0: PARAMETER=0	PR > |T|	STD ERROR OF ESTIMATE
INTERCEPT	1.66666667	0.82	0.4327	2.02941334
X1	1.80000000	3.44	0.0074	0.52399227
X2	5.50000000	7.67	0.0001	0.71750597

REGRESSION ANALYSIS, MODEL 2

GENERAL LINEAR MODELS PROCEDURE

DEPENDENT VARIABLE: Y

SOURCE	DF	SUM OF SQUARES	MEAN SQUARE	F VALUE	PR > F	R-SQUARE	C.V.
MODEL	2	290.60000000	145.30000000	35.28	0.001	0.886877	11.8218
ERROR	9	37.06666667	4.11851852		ROOT MSE		Y MEAN
CORRECTED TOTAL	11	327.66666667			2.02941334		17.16666667

SOURCE	DF	TYPE I SS	F VALUE	PR > F	DF	TYPE III SS	F VALUE	PR > F
X1	1	48.60000000	11.80	0.0074	1	2.42517483	0.59	0.4625
X3	1	242.00000000	58.76	0.0001	1	242.00000000	58.76	0.0001

PARAMETER	ESTIMATE	T FOR H0: PARAMETER=0	PR > |T|	STD ERROR OF ESTIMATE
INTERCEPT	7.85416667	5.01	0.0007	1.56633788
X1	0.42500000	0.77	0.4625	0.55384459
X3	0.68750000	7.67	0.0001	0.08968825

REGRESSION ANALYSIS, MODEL 3

GENERAL LINEAR MODELS PROCEDURE

DEPENDENT VARIABLE: Y

SOURCE	DF	SUM OF SQUARES	MEAN SQUARE	F VALUE	PR > F	R-SQUARE	C.V.
MODEL	2	290.60000000	145.30000000	35.28	0.0001	0.886877	11.8218
ERROR	9	37.06666667	4.11851852		ROOT MSE		Y MEAN
CORRECTED TOTAL	11	327.66666667			2.02941334		17.16666667

SOURCE	DF	TYPE I SS	F VALUE	PR > F	DF	TYPE III SS	F VALUE	PR > F
X2	1	242.00000000	58.76	0.0001	1	2.42517483	0.59	0.4625
X3	1	48.60000000	11.80	0.0074	1	48.60000000	11.80	0.0074

PARAMETER	ESTIMATE	T FOR H0: PARAMETER=0	PR > |T|	STD ERROR OF ESTIMATE
INTERCEPT	9.76666667	5.22	0.0005	1.87102665
X2	-1.70000000	-0.77	0.4625	2.21537835
X3	0.90000000	3.44	0.0074	0.26199614

(continues)

```
                          REGRESSION ANALYSIS, MODEL 4
                          GENERAL LINEAR MODELS PROCEDURE

DEPENDENT VARIABLE: Y
SOURCE           DF      SUM OF SQUARES    MEAN SQUARE    F VALUE    PR > F    R-SQUARE      C.V.
MODEL            2        290.60000000    145.30000000     35.28    0.0001    0.886877    11.8218
ERROR            9         37.06666667      4.11851852                ROOT MSE            Y MEAN
CORRECTED TOTAL  11       327.66666667                               2.02941334       17.16666667

SOURCE           DF        TYPE I SS    F VALUE   PR > F    DF    TYPE III SS   F VALUE   PR > F
X1               1        48.60000000    11.80    0.0074     0     0.00000000
X2               1       242.00000000    58.76    0.0001     0     0.00000000
X3               0         0.00000000                        0     0.00000000

                            T FOR HO:     PR > |T|     STD ERROR OF
PARAMETER        ESTIMATE    PARAMETER=0                 ESTIMATE
INTERCEPT       1.66666667 B     0.82      0.4327        2.02941334
X1              1.80000000 B     3.44      0.0074        0.52399227
X2              5.50000000 B     7.67      0.0001        0.71750597
X3              0.00000000 B

NOTE: THE X'X MATRIX HAS BEEN DEEMED SINGULAR AND A GENERALIZED INVERSE HAS BEEN EMPLOYED TO SOLVE THE NORMAL EQUATIONS.
      THE ABOVE ESTIMATES REPRESENT ONLY ONE OF MANY POSSIBLE SOLUTIONS TO THE NORMAL EQUATIONS.  ESTIMATES FOLLOWED BY
      THE LETTER B ARE BIASED AND DO NOT ESTIMATE THE PARAMETER BUT ARE BLUE FOR SOME LINEAR COMBINATION OF PARAMETERS
      (OR ARE ZERO).  THE EXPECTED VALUE OF THE BIASED ESTIMATORS MAY BE OBTAINED FROM THE GENERAL FORM OF ESTIMABLE
      FUNCTIONS.  FOR THE BIASED ESTIMATORS, THE STD ERR IS THAT OF THE BIASED ESTIMATOR AND THE T VALUE TESTS
      HO: E(BIASED ESTIMATOR) = 0.  ESTIMATES NOT FOLLOWED BY THE LETTER B ARE BLUE FOR THE PARAMETER.
```

What do you observe based on the output for these four models? Does it confirm what you saw from the plots in Exercise 11.33?

11.35   Refer to Exercise 11.34. Might there be a problem with correlated independent variables for one or more of these models? How would this be detected?

11.36   An experiment was conducted to determine the effects of two bonding agents (starch and calcium phosphate) on the dissolution properties of the tablet form of a drug product. Sample data were collected and then used to determine the prediction equation $\hat{y} = 77.5 + 0.2x_1 + 10x_1^2 + 5.8x_2 + 6.5x_1x_2$. If $n = 6$, $\sum y = 425$, $\sum y^2 = 30{,}731$, and SSE $= 120.45$, determine R^2.

11.37   In Exercise 9.25, we presented an experiment to investigate the shock wave amplitudes recorded from sensors placed at fixed distances from an explosive charge. These data are presented here for your convenience.

Distance, x	5	5	5	10	10	10	15	15	15
Amplitude, y	8.6	8.2	8.1	5.8	6.2	6.1	5.2	4.8	4.7

a. Plot the data.
b. Determine the quadratic regression equation $y = \beta_0 + \beta_1 x + \beta_2 x^2$ from the output shown.
c. Compute and interpret the coefficient of determination.

```
        LISTING OF DATA
   OBS    DISTANCE    AMPTUDE
    1        5         8.6
    2        5         8.2
    3        5         8.1
    4       10         5.8
    5       10         6.2
    6       10         6.1
    7       15         5.2
    8       15         4.8
    9       15         4.7

   N =     9
```

COMPLETE MODEL

GENERAL LINEAR MODELS PROCEDURE

DEPENDENT VARIABLE: AMPTUDE

SOURCE	DF	SUM OF SQUARES	MEAN SQUARE	F VALUE	PR > F	R-SQUARE	C.V.
MODEL	2	17.98222222	8.99111111	147.13	0.0001	0.980017	3.8559
ERROR	6	0.36666667	0.06111111		ROOT MSE		AMPTUDE MEAN
CORRECTED TOTAL	8	18.34888889			0.24720662		6.41111111

SOURCE	DF	TYPE I SS	F VALUE	PR > F	DF	TYPE III SS	F VALUE	PR > F
DISTANCE	1	17.34000000	283.75	0.0001	1	1.92666667	31.53	0.0014
DISTANCE*DISTANCE	1	0.64222222	10.51	0.0176	1	0.64222222	10.51	0.0176

PARAMETER	ESTIMATE	T FOR HO: PARAMETER=0	PR > |T|	STD ERROR OF ESTIMATE
INTERCEPT	11.70000000	18.81	0.0001	0.62212301
DISTANCE	-0.79333333	-5.61	0.0014	0.14129035
DISTANCE*DISTANCE	0.02266667	3.24	0.0176	0.00699206

11.38   Refer to Exercise 11.37. Give the *p*-value for a test of $H_0: \beta_2 = 0$. Interpret your findings. Does this agree with what you saw in the data for Exercise 11.37?

11.39   Refer to Example 11.4. Determine 95% prediction bands for the data using the output shown here.

LEAST SQUARES ANALYSIS

GENERAL LINEAR MODELS PROCEDURE

DEPENDENT VARIABLE: WT_LOSS   WEIGHT LOSS (POUNDS)

SOURCE	DF	SUM OF SQUARES	MEAN SQUARE	F VALUE	PR > F	R-SQUARE	C.V.
MODEL	2	31.12416667	15.56208333	104.13	0.0001	0.958576	7.0181
ERROR	9	1.34500000	0.14944444		STD DEV		WT-LOSS
							MEAN
CORRECTED TOTAL	11	32.46916667			0.38658045		5.50833333

SOURCE	DF	TYPE I SS	F VALUE	PR > F	DF	TYPE IV SS	F VALUE	PR > F
TIME	1	26.00416667	174.01	0.0001	1	26.00416667	174.01	0.0001
HUMID	1	5.12000000	34.26	0.0002	1	5.12000000	34.26	0.0002

PARAMETER	ESTIMATE	T FOR HO: PARAMETER=0	PR > |T|	STD ERROR OF ESTIMATE
INTERCEPT	0.66666667	0.96	0.3620	0.69423219
TIME	1.31666667	13.19	0.0001	0.09981464
HUMID	-8.00000000	-5.85	0.0002	1.36676829

(continues)

OBSERVATION	OBSERVED VALUE	PREDICTED VALUE	RESIDUAL	LOWER 95% CL INDIVIDUAL	UPPER 95% CL INDIVIDUAL
1	4.30000000	4.33333333	-0.03333333	3.31411004	5.35255662
2	5.50000000	5.65000000	-0.15000000	4.68209172	6.61790828
3	6.80000000	6.96666667	-0.16666667	5.99875839	7.93457495
4	8.00000000	8.28333333	-0.28333333	7.26411004	9.30255662
5	4.00000000	3.53333333	0.46666667	2.56213843	4.50452824
6	5.20000000	4.85000000	0.35000000	3.93280326	5.76719674
7	6.60000000	6.16666667	0.43333333	5.24946993	7.08386341
8	7.50000000	7.48333333	0.01666667	6.51213843	8.45452824
9	2.00000000	2.73333333	-0.67333333	1.71411004	3.75255662
10	4.00000000	4.05000000	0.05000000	3.08209172	5.01790828
11	5.70000000	5.36666667	0.33333333	4.39875839	6.33457495
12	6.50000000	6.68333333	-0.18333333	5.66411004	7.70255662
13*		6.42500000		5.47453294	7.37546706

*OBSERVATION WAS NOT USED IN THIS ANALYSIS

SUM OF RESIDUALS	0.00000000
SUM OF SQUARED RESIDUALS	1.34500000
SUM OF SQUARED RESIDUALS - ERROR SS	-0.00000000
PRESS STATISTIC	2.61233345
FIRST ORDER AUTOCORRELATION	0.15902520
DURBIN-WATSON D	1.65613383

11.40 A study of demand for imported subcompact cars consisted of data from 12 metropolitan areas. The variables were:

DEMAND	imported subcompact car sales as a percentage of total sales
EDUC	average number of years of schooling completed by adults
INCOME	per capita income
POPN	area population
FAMSIZE	average size of intact families

BMDP output is shown here.

MULTIPLE R	0.9813		STD. ERROR OF EST.	1.7777	
MULTIPLE R-SQUARE	0.9629				

ANALYSIS OF VARIANCE

	SUM OF SQUARES	DF	MEAN SQUARE	F RATIO	P(TAIL)
REGRESSION	574.073	4	143.518	45.412	0.00004
RESIDUAL	22.123	7	3.160		

VARIABLE		COEFFICIENT	STD. ERROR	STD. REG COEFF	T	P(2 TAIL)
INTERCEPT		44.01318				
EDUC	2	5.08694	1.970	0.471	2.582	0.036
INCOME	3	3.01868	2.204	0.351	1.370	0.213
POPN	4	0.20527	0.373	0.045	0.550	0.600
FAMSIZE	5	-5.40869	3.850	-0.228	-1.405	0.203

a. Write down the regression equation. Place the standard error of each coefficient below the coefficient, perhaps in parentheses.

b. Locate R^2 and the residual standard deviation.

11.41   Refer to Exercise 11.40.
  a. Draw conclusions from the statistical tests summarized in the output.
  b. Would you consider an alternate model? If so, what would it be?

11.42   A study was conducted to examine the effects of temperature and lighting intensity on office productivity. Let y denote a measure of productivity, x_1 denote office temperature in °F, and x_2 denote lighting intensity.
  a. Write a first-order regression model.
  b. Write a second-order regression model allowing for all possible terms in x_1 and x_2.

11.43   Refer to Exercise 11.42. Here are the data:

y	45	49	47	57	48	53	51	54	56	64
x_1	64	64	66	66	68	68	70	70	72	72
x_2	60	65	60	65	60	65	60	65	60	65

  a. Suggest plots that may help to formulate a regression model.
  b. Fit the model that you think is appropriate and draw some conclusions.
  c. Do you have any recommendations for further experimentation?

11.44   The following data were recorded for a regression study:

y	15	12	14	18	19	16	17	26	20	22	24
x_1	6	7	7	8	8	9	9	10	10	11	12
x_2	10	12	13	14	15	15	16	17	18	19	19

  a. Write a first-order regression model.
  b. Use the sample data to fit the model.
  c. Calculate the residual standard deviation.

11.45   Refer to Exercise 11.44.
  a. Calculate the estimated standard errors for $\hat{\beta}_1$ and $\hat{\beta}_2$.
  b. Compute 90% confidence intervals for β_1 and β_2.

11.46   Computer output is shown here for a study of consumer product sales (Plan y) for a company as a function of advertising expense (Plan x).

```
CORRELATIONS

           ADV    ADVSQ   SALES
            1       2       3
ADV      1  1.000
ADVSQ    2  0.994   1.000
SALES    3  0.363   0.274   1.000

SQUARED MULTIPLE CORRELATIONS OF EACH INDEPENDENT VARIABLE WITH ALL OTHER
INDEPENDENT VARIABLES
                                                    (continues)
```

```
(MEASURES OF MULTICOLLINEARITY OF PREDICTOR VARIABLES)
AND TESTS OF SIGNIFICANCE OF MULTIPLE REGRESSION

DEGREES OF FREEDOM FOR F-STATISTICS ARE    1 AND    12

    VARIABLE                                    SIGNIFICANCE
  NO.    NAME         COEF      F-STATISTIC    (P LESS THAN)
   1     ADV          25.746      1074.10         0.00000
   2     ADVSQ        -0.686         8.89         0.0197
         INTERCEPT  -144.312
```

a. Identify the regression equation.

b. Give the p-value for a test of $H_0: \beta_2 = 0$. Does this suggest that a quadratic regression model is to be preferred to a linear regression model?

11.47 The manager of documentation for a computer software firm wants to forecast the time required to document moderate-sized computer programs. Records are available for 26 programs. The variables are $y =$ number of writer-days needed, $x_1 =$ number of subprograms, $x_2 =$ average number of lines per subprogram, $x_3 = x_1 x_2$, $x_4 = x_2^2$, and $x_5 = x_1 x_2^2$. A portion of the output from a regression analysis (SAS) of the data is shown here:

```
DEPENDENT VARIABLE: Y
SOURCE            DF   SUM OF SQUARES   MEAN SQUARE   F VALUE   PR > F    R SQUARE      C.V.
MODEL              5    2546.02735209   509.20547042    44.31   0.0001   0.917195    11.9597
ERROR             20     229.85726330    11.49286316             ROOT MSE            Y MEAN
CORRECTED TOTAL   25    2775.88461538                            3.39011256        28.34615385

                                 T FOR H0:     PR > |T|   STD. ERROR OF
PARAMETER      ESTIMATE        PARAMETER=0                   ESTIMATE
 INTERCEPT   -16.81979712         -1.45        0.1636      11.63104920
 X1            1.47018752          4.02        0.0007       0.36594367
 X2            0.99477822          1.63        0.1194       0.61144114
 X1X2         -0.02400705         -1.01        0.3243       0.02375645
 X2SQ         -0.01031004         -1.40        0.1774       0.00737400
 X1X2SQ        0.00024957          0.71        0.4862       0.00035178
```

a. Write the multiple regression model and locate the residual standard deviation.

b. What does the variable x_3 represent in terms of the problem?

c. Does x_3 have a statistically significant predictive value as "last predictor in"?

11.48 The model $y = \beta_0 + \beta_1 x_1 + \beta_2 x_2 + \varepsilon$ was fit to the data of Exercise 11.47. Selected output is shown here:

```
DEPENDENT VARIABLE: Y
SOURCE            DF   SUM OF SQUARES   MEAN SQUARE   F VALUE   PR > F    R SQUARE      C.V.
MODEL              2    2516.12160091  1258.06080045   111.39   0.0001   0.906422    11.8558
ERROR             23     259.76301448    11.29404411             ROOT MSE            Y MEAN
CORRECTED TOTAL   25    2775.88461538                            3.36066126        28.34615385
```
(continues)

PARAMETER	ESTIMATE	T FOR HO: PARAMETER=0	PR > |T|	STD. ERROR OF ESTIMATE
INTERCEPT	0.84008527	0.24	0.8089	3.43374955
X1	1.01583472	12.81	0.0001	0.07929252
X2	0.05582624	1.08	0.2897	0.05150660

a. Write the complete and reduced-form estimated models.

b. Is the improvement in R^2 obtained by adding x_3, x_4, and x_5 statistically significant at $\alpha = .05$? What is the p-value for this test?

11.49 A producer of various feed additives for cattle conducted a study of the number of days of feedlot time required to bring beef cattle to market weight. Eighteen steers of essentially identical age and weight were purchased and brought to a feedlot. Each steer was fed a diet with a specific combination of protein content, antibiotic concentration, and percentage of feed supplement. The data were:

STEER:	1	2	3	4	5	6	7	8	9
PROTEIN:	10	10	10	10	10	10	15	15	15
ANTIBIO:	1	1	1	2	2	2	1	1	1
SUPPLEM:	3	5	7	3	5	7	3	5	7
TIME:	88	82	81	82	83	75	80	80	75
STEER:	10	11	12	13	14	15	16	17	18
PROTEIN:	15	15	15	20	20	20	20	20	20
ANTIBIO:	2	2	2	1	1	1	·2	2	2
SUPPLEM:	3	5	7	3	5	7	3	5	7
TIME:	77	76	72	79	74	75	74	70	69

Computer output from a regression analysis follows:

MULTIPLE R	0.9490	STD. ERROR OF EST.	1.7096
MULTIPLE R-SQUARE	0.9007		

ANALYSIS OF VARIANCE

	SUM OF SQUARES	DF	MEAN SQUARE	F RATIO	P(TAIL)
REGRESSION	371.0832	3	123.6944	42.323	0.0000
RESIDUAL	40.9166	14	2.9226		

VARIABLE		COEFFICIENT	STD. ERROR	STD. REG COEFF	T	P(2 TAIL)	TOLERANCE
INTERCEPT		102.70834					
PROTEIN	2	-0.83333	0.09870	-0.711	-8.443	0.0000	1.00000
ANTIBIO	3	-4.00000	0.80590	-0.418	-4.963	0.0002	1.00000
SUPPLEM	4	-1.37500	0.24675	-0.469	-5.572	0.0001	1.00000

a. Write down the regression equation.

b. Find the standard deviation.

c. Find the R^2 value.

11.50 Refer to Exercise 11.49.

a. Predict the feedlot time required for a steer fed 15% protein, 1.5% antibiotic concentration, and 5% supplement.

b. Do these values of the independent variables represent a major extrapolation from the data?

c. Give a 95% confidence interval for the mean time predicted in part a.

Note: This requires access to computer software for multiple regression.

11.51 The data of Exercise 11.49 were also analyzed by a regression model using only protein content as an independent variable, yielding the following output:

```
MULTIPLE R               0.7111    STD. ERROR OF EST.            3.5678
MULTIPLE R-SQUARE        0.5057

ANALYSIS OF VARIANCE
                  SUM OF SQUARES    DF      MEAN SQUARE   F RATIO    P(TAIL)
REGRESSION           208.3332        1       208.3332     16.367     0.0009
RESIDUAL             203.6667       16        12.7292

                                             STD. REG
VARIABLE          COEFFICIENT   STD. ERROR    COEFF      T     P(2 TAIL)   TOLERANCE
INTERCEPT           89.83334
PROTEIN 2           -0.83333      0.20599    -0.711    -4.046    0.0009     1.00000
```

a. Write the regression equation.

b. Find the R^2 value.

c. Test the null hypothesis that the coefficients of ANTIBIO and SUPPLEM are 0 at $\alpha = .05$.

 11.52 The accompanying table gives demographic data for 12 male patients with congestive heart failure enrolled in a study of an experimental compound.

Demographic Data for Patients with Heart Failure (NYHA Class III or IV)

Patient	Age (yrs)	Disease Duration	Height (cm)	Weight (kg)	Baseline Cardiac Index (L/min/m²)	Baseline Pulmonary Capillary Wedge Pressure (mm Hg)
01	67	5 yr	172.0	57.0	1.6	40
02	45	2 yr	170.0	67.0	2.4	25
03	59	8 yr	172.7	102.0	2.2	39
04	63	1 yr	175.3	74.9	1.7	39
05	55	1 yr	172.7	92.0	2.3	34
06	65	1 yr	178.0	90.0	1.6	36
07	62	2 yr	163.0	67.0	1.4	36
08	60	1 yr	182.5	72.0	2.2	17
09	72	2 yr	168.0	71.0	1.3	37
10	44	3 mo	163.0	68.0	2.4	28
11	63	5 yr	172.0	82.0	2.1	38
12	63	1 yr	163.0	64.0	1.1	36

a. Summarize these data using a box plot for each variable.

b. Construct scatterplots to display (1) age by cardiac index (CI) and by pulmonary capillary wedge pressure (PCWP) and (2) disease duration by CI and by PCWP.

Is there evidence of a correlation between age and CI or PCWP? What about correlation between duration of disease and CI or PCWP?

11.53 The data of Exercise 11.52 were used to fit several multiple regression models. y_1 = CI, y_2 = PCWP, x_1 = age, x_2 = disease duration.

a. $y_1 = \beta_0 + \beta_1 x_1 + \beta_2 x_2 + \varepsilon$

b. $y_1 = \beta_0 + \beta_1 x_1 + \beta_2 x_2 + \beta_3 x_1 x_2 + \varepsilon$

c. $y_2 = \beta_0 + \beta_1 x_1 + \beta_2 x_2 + \varepsilon$

d. $y_2 = \beta_0 + \beta_1 x_1 + \beta_2 x_2 + \beta_3 x_1 x_2 + \varepsilon$

```
                    LISTING OF DATA
    OBS    PATIENT    AGE    DURATION    CI    PCWP
     1        1        67      5.00      1.6    40
     2        2        45      2.00      2.4    25
     3        3        59      8.00      2.2    39
     4        4        63      1.00      1.7    39
     5        5        55      1.00      2.3    34
     6        6        65      1.00      1.6    36
     7        7        62      2.00      1.4    36
     8        8        60      1.00      2.2    17
     9        9        72      2.00      1.3    37
    10       10        44      0.25      2.4    28
    11       11        63      5.00      2.1    38
    12       12        63      1.00      1.1    36

  N =    12

                    REGRESSION ANALYSIS, MODEL 1
                  GENERAL LINEAR MODELS PROCEDURE

DEPENDENT VARIABLE: CI

SOURCE              DF    SUM OF SQUARES    MEAN SQUARE    F VALUE    PR > F    R-SQUARE      C.V.

MODEL               2       1.56955279      0.78477639       9.30     0.0065   0.673869    15.6333

ERROR               9       0.75961388      0.08440154              ROOT MSE              CI MEAN

CORRECTED TOTAL    11       2.32916667                            0.229051943          1.85833333


SOURCE              DF       TYPE I SS     F VALUE    PR > F    DF    TYPE III SS    F VALUE    PR > F

AGE                 1       1.36216181      16.14     0.0030     1    1.53467632      18.18     0.0021
DURATION            1       0.20739097       2.46     0.1514     1    0.20739097       2.46     0.1514

                                           T FOR HO:     PR > |T|    STD. ERROR OF
PARAMETER             ESTIMATE           PARAMETER=0                   ESTIMATE

INTERCEPT           4.47562208               7.00       0.0001      0.63976685
AGE                -0.04620336              -4.26       0.0021      0.01083529
DURATION            0.06039479               1.57       0.1514      0.03852829


                    REGRESSION ANALYSIS, MODEL 2
                  GENERAL LINEAR MODELS PROCEDURE

DEPENDENT VARIABLE: CI

SOURCE              DF    SUM OF SQUARES    MEAN SQUARE    F VALUE    PR > F    R-SQUARE      C.V.

MODEL               3       1.57161289      0.52387096       5.53     0.0237   0.674753    16.5592

ERROR               8       0.75755378      0.09469422              ROOT MSE              CI MEAN

CORRECTED TOTAL    11       2.32916667                            0.30772426           1.85833333
```

(*continues*)

SOURCE	DF	TYPE I SS	F VALUE	PR > F	DF	TYPE III SS	F VALUE	PR > F
AGE	1	1.36216181	14.38	0.0053	1	0.64786628	6.84	0.0309
DURATION	1	0.20739097	2.19	0.1772	1	0.00015009	0.00	0.9692
AGE*DURATION	1	0.00206010	0.02	0.8864	1	0.00206010	0.02	0.8864

PARAMETER	ESTIMATE	T FOR H0: PARAMETER=0	PR > |T|	STD. ERROR OF ESTIMATE
INTERCEPT	4.59930709	4.27	0.0027	1.07814691
AGE	−0.04833990	−2.62	0.0309	0.01848097
DURATION	−0.02240952	−0.04	0.9692	0.56287924
AGE*DURATION	0.00137554	0.15	0.8864	0.00932590

REGRESSION ANALYSIS, MODEL 3

GENERAL LINEAR MODELS PROCEDURE

DEPENDENT VARIABLE: PCWP

SOURCE	DF	SUM OF SQUARES	MEAN SQUARE	F VALUE	PR > F	R-SQUARE	C.V.
MODEL	2	221.88101159	110.94050579	3.26	0.0862	0.420030	17.2873
ERROR	9	306.36898841	34.04099871				
CORRECTED TOTAL	11	528.25000000			ROOT MSE		PCWP MEAN
					5.83446645		33.75000000

SOURCE	DF	TYPE I SS	F VALUE	PR > F	DF	TYPE III SS	F VALUE	PR > F
AGE	1	162.57201147	4.78	0.0567	1	115.29741682	3.39	0.0989
DURATION	1	59.30900012	1.74	0.2194	1	59.30900012	1.74	0.2194

PARAMETER	ESTIMATE	T FOR H0: PARAMETER=0	PR > |T|	STD. ERROR OF ESTIMATE
INTERCEPT	7.29878609	0.57	0.5839	12.84835977
AGE	0.40047458	1.84	0.0989	0.21760372
DURATION	1.02132713	1.32	0.2194	0.77375900

REGRESSION ANALYSIS, MODEL 4

GENERAL LINEAR MODELS PROCEDURE

DEPENDENT VARIABLE: PCWP

SOURCE	DF	SUM OF SQUARES	MEAN SQUARE	F VALUE	PR > F	R-SQUARE	C.V.
MODEL	3	228.56514750	76.18838250	2.03	0.1878	0.432684	18.1348
ERROR	8	299.68485250	37.46060656				
CORRECTED TOTAL	11	528.25000000			ROOT MSE		PCWP MEAN
					6.12050705		33.75000000

SOURCE	DF	TYPE I SS	F VALUE	PR > F	DF	TYPE III SS	F VALUE	PR > F
AGE	1	162.57201147	4.34	0.0708	1	21.54668440	0.58	0.4700
DURATION	1	59.30900012	1.58	0.2438	1	4.08124565	0.11	0.7498
AGE*DURATION	1	6.68413591	0.18	0.6838	1	6.68413591	0.18	0.6838

PARAMETER	ESTIMATE	T FOR H0: PARAMETER=0	PR > |T|	STD. ERROR OF ESTIMATE
INTERCEPT	14.34402569	0.67	0.5224	21.44389171
AGE	0.27877467	0.76	0.4700	0.36757883
DURATION	−3.69530133	−0.33	0.7498	11.19543293
AGE*DURATION	0.07835228	0.42	0.6838	0.18548824

Which model provides the best fit to the cardiac index data? To the pulmonary capillary wedge pressure data? Do these analyses confirm what you concluded in Exercise 11.52? Explain.

APPENDIX ▼ MATRIX OPERATIONS (optional)

A *matrix* is defined to be a rectangular array of numbers. We indicate a particular matrix by a boldface capital letter. The numbers of a matrix, called *elements*, appear in rows and columns, as indicated in Figure 11.7.

FIGURE 11.7
2 × 2, 3 × 3, and 4 × 4 Identity Matrices

$$\mathbf{I} = \begin{bmatrix} 1 & 0 \\ 0 & 1 \end{bmatrix}$$

$$\mathbf{I} = \begin{bmatrix} 1 & 0 & 0 \\ 0 & 1 & 0 \\ 0 & 0 & 1 \end{bmatrix}$$

$$\mathbf{I} = \begin{bmatrix} 1 & 0 & 0 & 0 \\ 0 & 1 & 0 & 0 \\ 0 & 0 & 1 & 0 \\ 0 & 0 & 0 & 1 \end{bmatrix}$$

dimension of matrix

Note that in addition to identifying a matrix by a capital boldface letter, we can also indicate the **dimension of a matrix** by specifying the number of rows and columns in the matrix. Thus, a 3 × 3 (read "3 by 3") matrix contains 3 rows and 3 columns, a 2 × 3 matrix contains 2 rows and 3 columns, and a 1 × 4 matrix contains 1 row and 4 columns.

identity matrix

One important type of matrix is the **identity matrix**. An identity matrix, denoted by **I**, is a square matrix (same number of rows and columns) the diagonal elements of which, proceeding from the upper left to the lower right of the matrix, are 1, with all off-diagonal elements 0. Three identity matrices are shown in Figure 11.7.

As with other quantities used in statistics and mathematics, we will want to perform operations with matrices, such as addition, multiplication, and so on. Thus, in the following discussions, we define the matrix operations we will need in our statistical work.

addition of matrices

Two matrices, **A** *and* **B**, *can be added only if they have the same dimensions*. For example, we could add two 3 × 3 matrices because both matrices have the same dimensions, but we could not add a 2 × 3 matrix and a 3 × 3 matrix. *The sum of two matrices*, **A** *and* **B**, *the dimensions of which are the same forms a new matrix by adding the corresponding elements of* **A** *and* **B**. This new matrix, **A** + **B**, has the same dimensions as **A** and **B**.

We illustrate addition with some examples.

EXAMPLE 11.15 ▼

Suppose a 2 × 2 matrix **A** and a 2 × 2 matrix **B** are as shown below.

$$\mathbf{A} = \begin{bmatrix} 1 & 3 \\ 2 & 5 \end{bmatrix} \qquad \mathbf{B} = \begin{bmatrix} 4 & 1 \\ 3 & 7 \end{bmatrix}$$

Find the sum of the two matrices.

Solution Since the two matrices **A** and **B** have the same dimensions, we can add them. The sum of the two matrices, denoted by **A** + **B**, is

$$\mathbf{A} + \mathbf{B} = \begin{bmatrix} 1 & 3 \\ 2 & 5 \end{bmatrix} + \begin{bmatrix} 4 & 1 \\ 3 & 7 \end{bmatrix} = \begin{bmatrix} (1+4) & (3+1) \\ (2+3) & (5+7) \end{bmatrix} = \begin{bmatrix} 5 & 4 \\ 5 & 12 \end{bmatrix}.$$

▲

EXAMPLE 11.16 ▼

For the 2 × 3 matrix **A** and 2 × 3 matrix **B** shown here, find **A** + **B**.

$$\mathbf{A} = \begin{bmatrix} 3 & 4 & -6 \\ 9 & 1 & 1 \end{bmatrix} \qquad \mathbf{B} = \begin{bmatrix} 1 & 7 & 4 \\ 0 & -9 & 5 \end{bmatrix}$$

Solution The sum of these two matrices is

$$\mathbf{A} + \mathbf{B} = \begin{bmatrix} (3+1) & (4+7) & (-6+4) \\ (9+0) & (1-9) & (1+5) \end{bmatrix} = \begin{bmatrix} 4 & 11 & -2 \\ 9 & -8 & 6 \end{bmatrix}.$$

▲

Using Example 11.16, you can easily verify that the addition of two matrices **A** and **B** is commutative; that is,

$$\mathbf{A} + \mathbf{B} = \mathbf{B} + \mathbf{A}.$$

multiplication of matrices

*We can multiply a matrix **A** times a matrix **B** if the number of columns in **A** is equal to the number of rows in **B**.* For example, we could multiply a 2 × 3 matrix **A** times a 3 × 2 matrix **B** because the number of columns in **A** and the number of rows in **B** is the same, namely, 3. Similarly, we could multiply a 1 × 5 matrix **A** times a 5 × 4 matrix **B**.

The multiplication of matrices is somewhat complex. Basically, an element in the product matrix **AB** is found by multiplying each element in a row in **A** times each element in the corresponding column in **B** and adding the results. This procedure is best illustrated and understood by working some examples.

E X A M P L E   11.17 ▼

Let the 2×2 matrix **A** and the 2×1 matrix **B** be given as follows:

$$\mathbf{A} = \begin{bmatrix} 3 & 1 \\ 2 & 4 \end{bmatrix} \qquad \mathbf{B} = \begin{bmatrix} 1 \\ 2 \end{bmatrix}$$

Find the product **AB**.

Solution   The first thing to note in the multiplication of two matrices **A** and **B** is that the resulting product **AB** will be a new matrix with dimensions given by

number of rows of **AB** = number of rows of **A**
number of columns of **AB** = number of columns of **B**.

In our example,

$$\begin{array}{cc} \mathbf{A} & \mathbf{B} \\ \underline{2 \times 2} & \underline{2 \times 1} \end{array}$$

will be a new 2×1 matrix. There will be two elements in the new matrix, one in the first row and first column and one in the second row and first column.

The element in the first row, first column, is found by multiplying the elements of the first *row* of **A** times the corresponding elements of the first *column* of **B** and adding the result:

$$\begin{bmatrix} 3 & 1 \\ 2 & 4 \end{bmatrix} \begin{bmatrix} 1 \\ 2 \end{bmatrix} = \begin{bmatrix} 3 \cdot 1 + 1 \cdot 2 \end{bmatrix} = \begin{bmatrix} 5 \end{bmatrix}.$$

The elements in the second row, first column, of the new matrix is found by multiplying the elements of the *second* row of **A** times the elements of the *first* column **B** and adding the result.

$$\begin{bmatrix} 3 & 1 \\ 2 & 4 \end{bmatrix} \begin{bmatrix} 1 \\ 2 \end{bmatrix} = \begin{bmatrix} 2 \cdot 1 + 4 \cdot 2 \end{bmatrix} = \begin{bmatrix} 10 \end{bmatrix}$$

The product **AB** is then

$$\mathbf{AB} = \begin{bmatrix} 3 & 1 \\ 2 & 4 \end{bmatrix} \begin{bmatrix} 1 \\ 2 \end{bmatrix} = \begin{bmatrix} 5 \\ 10 \end{bmatrix}.$$

▲

E X A M P L E   11.8

Let a 2×2 matrix **A** and a 2×3 matrix **B** be given as follows:

$$\mathbf{A} = \begin{bmatrix} 1 & 2 \\ 0 & 3 \end{bmatrix} \qquad \mathbf{B} = \begin{bmatrix} 4 & 4 & 0 \\ 8 & 3 & 2 \end{bmatrix}.$$

a. Find **AB**.
b. Find **BA**.

Solution

a. Again, we will illustrate this matrix multiplication in separate parts. First we know that the new matrix will be a 2×3 matrix.

$$\begin{array}{cc} \mathbf{A} & \mathbf{B} \\ \underline{2} \times 2 & 2 \times \underline{3} \end{array}$$

The element in the first row, first column, of the new matrix (called the (1,1) element) is obtained by multiplying the elements in the first row of **A** times corresponding elements in the first column of **B** and adding:

$$(1, 1)\text{ element} = \begin{bmatrix} 1 & 2 \\ 0 & 3 \end{bmatrix}\begin{bmatrix} 4 & 4 & 0 \\ 8 & 3 & 2 \end{bmatrix} = \begin{bmatrix} 1 \cdot 4 + 2 \cdot 8 & & \\ & & \end{bmatrix}.$$

The element in the first row, second column (called the (1, 2) element), of the new matrix is formed by multiplying the elements of the first row of **A** and the second column of **B** and adding:

$$(1, 2)\text{ element} = \begin{bmatrix} 1 & 2 \\ 0 & 3 \end{bmatrix}\begin{bmatrix} 4 & 4 & 0 \\ 8 & 3 & 2 \end{bmatrix} = \begin{bmatrix} & 1 \cdot 4 + 2 \cdot 3 & \\ & & \end{bmatrix}.$$

The remaining elements are found in a similar manner:

$$(1, 3)\text{ element} = \begin{bmatrix} 1 & 2 \\ 0 & 3 \end{bmatrix}\begin{bmatrix} 4 & 4 & 0 \\ 8 & 3 & 2 \end{bmatrix} = \begin{bmatrix} & & 1 \cdot 0 + 2 \cdot 2 \\ & & \end{bmatrix}$$

$$(2, 1)\text{ element} = \begin{bmatrix} 1 & 2 \\ 0 & 3 \end{bmatrix}\begin{bmatrix} 4 & 4 & 0 \\ 8 & 3 & 2 \end{bmatrix} = \begin{bmatrix} & & \\ 0 \cdot 4 + 3 \cdot 8 & & \end{bmatrix}$$

$$(2, 2)\text{ element} = \begin{bmatrix} 1 & 2 \\ 0 & 3 \end{bmatrix}\begin{bmatrix} 4 & 4 & 0 \\ 8 & 3 & 2 \end{bmatrix} = \begin{bmatrix} & & \\ & 0 \cdot 4 + 3 \cdot 3 & \end{bmatrix}$$

$$(2, 3)\text{ element} = \begin{bmatrix} 1 & 2 \\ 0 & 3 \end{bmatrix}\begin{bmatrix} 4 & 4 & 0 \\ 8 & 3 & 2 \end{bmatrix} = \begin{bmatrix} & & \\ & & 0 \cdot 0 + 3 \cdot 2 \end{bmatrix}.$$

Combining our results we have

$$\mathbf{AB} = \begin{bmatrix} 20 & 10 & 4 \\ 24 & 9 & 6 \end{bmatrix}.$$

b. To multiply

$$\underset{2 \times 3}{\mathbf{B}} \quad \text{times} \quad \underset{2 \times 2}{\mathbf{A}},$$

the number of columns in **B** must equal the number of rows in **A**. Since this does not hold in this example, we cannot multiply **B** times **A**. Note: *This also implies that* **BA** *is not necessarily equal to* **AB**.   ▲

transpose A′

The **transpose** of a matrix **A** is a new matrix, denoted by **A′** (called "A prime"), formed by interchanging the correspondence rows and columns of the original matrix **A**. Thus, the first row of **A** becomes the first column of **A′**, and so on.

E X A M P L E   11.19   ▼

Determine the transpose of the matrix

$$\mathbf{A} = \begin{bmatrix} 1 & 5 & 6 \\ 3 & 2 & 4 \end{bmatrix}.$$

Solution   We obtain the transpose of **A** by forming a new matrix, where the first and second rows of **A** become the first and second columns of the new matrix.

$$\mathbf{A}' = \begin{bmatrix} 1 & 3 \\ 5 & 2 \\ 6 & 4 \end{bmatrix}$$

Note that **A** had dimensions 2 × 3, whereas **A′** has the reverse dimensions, 3 × 2.   ▲

determinant

We can associate a number with every square matrix (that is, the number of rows equals the number of columns), which we call the **determinant** of a matrix. The determinant of a 1 × 1 matrix, that is, a matrix with only one element, is the value of that element. Thus, the determinant of the matrix **A** = [4] is 4. The computation of determinants will be illustrated for 2 × 2 and 3 × 3 matrices.

Let **A** be a 2×2 matrix with elements

$$\mathbf{A} = \begin{bmatrix} a_{11} & a_{12} \\ a_{21} & a_{22} \end{bmatrix}.$$

The determinant of the matrix **A**, denoted by $|\mathbf{A}|$, is the quantity

$$|\mathbf{A}| = a_{11}a_{22} - a_{21}a_{12}.$$

Note that the determinant of a 2×2 matrix is formed by multiplying elements along the diagonal of the matrix from upper left to lower right and then subtracting the product of the diagonal elements from lower left to upper right (see Figure 11.8).

FIGURE 11.8

Calculations for the Determinant of a 2 × 2 Matrix

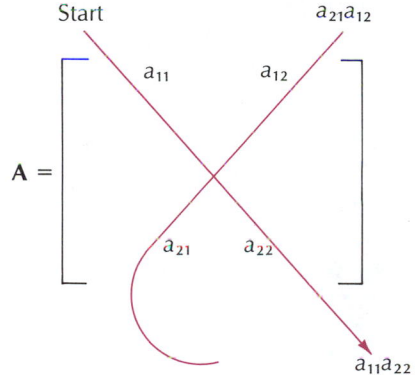

EXAMPLE 11.20

Find the determinant of the matrix

$$\mathbf{A} = \begin{bmatrix} 2 & 3 \\ 1 & 5 \end{bmatrix}.$$

Solution Using the computational formula, we have

$$|\mathbf{A}| = 2(5) - 1(3) = 10 - 3 = 7.$$

The determinant of a 3×3 matrix can be computed in a similar way. Let

$$\mathbf{A} = \begin{bmatrix} a_{11} & a_{12} & a_{13} \\ a_{21} & a_{22} & a_{23} \\ a_{31} & a_{32} & a_{33} \end{bmatrix}.$$

Then the determinant of this matrix is

$$|\mathbf{A}| = a_{11}a_{22}a_{33} + a_{21}a_{32}a_{13} + a_{31}a_{23}a_{12} - a_{13}a_{22}a_{31} - a_{23}a_{32}a_{11} - a_{33}a_{21}a_{12}.$$

Although this computation is more difficult than that for the 2×2 matrix, it can be remembered easily using the procedure shown in Figure 11.9.

FIGURE 11.9

Computation of |A| for a 3 × 3 Matrix

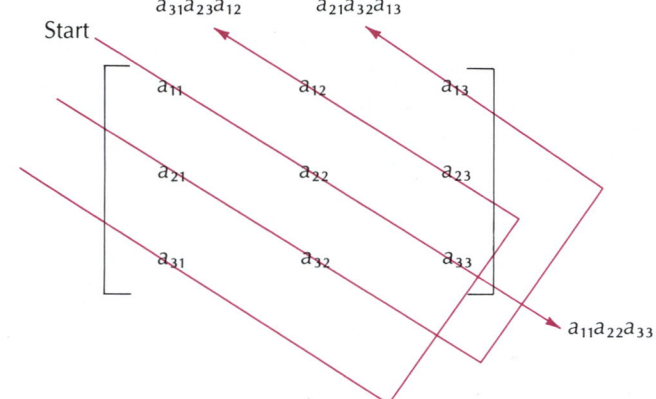

(a) Computation of first three terms of $|\mathbf{A}|$

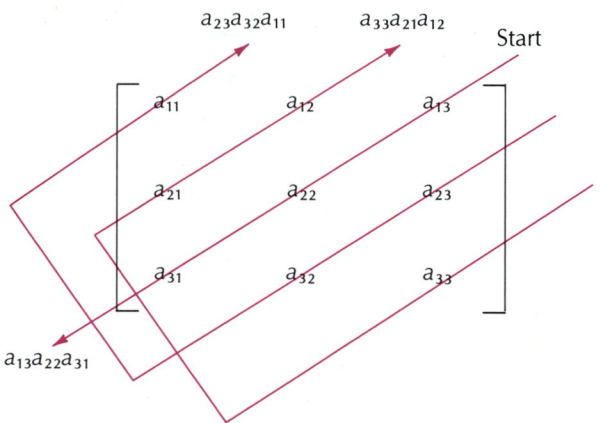

(b) Computation of last three terms of $|\mathbf{A}|$

EXAMPLE 11.21 ▼

Find the determinant of the matrix

$$\mathbf{A} = \begin{bmatrix} 1 & 3 & 1 \\ 2 & 1 & 1 \\ 1 & 3 & 3 \end{bmatrix}.$$

Solution   The first three terms of $|\mathbf{A}|$, starting in the upper–left corner, are

$$1(1)(3) = 3 \qquad 2(3)(1) = 6 \qquad 1(1)(3) = 3.$$

The last three terms are

$$1(1)(1) = 1 \qquad 1(3)(1) = 3 \qquad 3(2)(3) = 18.$$

Combining, we have

$$|\mathbf{A}| = 3 + 6 + 3 - 1 - 3 - 18 = -10. \qquad\qquad ▲$$

Although there are computational formulas for calculating the determinant of a square matrix with more than three rows (and columns), the computational tricks for 2×2 and 3×3 matrices do not extend to higher-dimensional matrices. *When working with larger matrices, we will use a software package to obtain a solution.*

We can now use the definition of a determinant to define a new concept. The **cofactor** associated with the (i, j) element of a square matrix $\mathbf{A}$ is defined to be $(-1)^{i+j}$ times the determinant of the matrix formed by deleting all the elements in row i and column j of the matrix $\mathbf{A}$. We illustrate this idea with an example.

cofactor of element

E X A M P L E   11.22   ▼

Find the cofactors of the $(1, 2)$ element and the $(2, 2)$ element of the matrix

$$\mathbf{A} = \begin{bmatrix} 1 & 3 & 1 \\ 2 & 1 & 1 \\ 1 & 3 & 3 \end{bmatrix}.$$

Solution   To find the cofactor of the $(1, 2)$ element of $\mathbf{A}$, we delete the first row and second column of $\mathbf{A}$.

$$\begin{bmatrix} 1 & 3 & 1 \\ 2 & 1 & 1 \\ 1 & 3 & 3 \end{bmatrix}$$

The determinant of the matrix that remains is $2(3) - 1(1) = 6 - 1 = 5$, so the cofactor of the $(1, 2)$ element is $(-1)^{1+2}5 = -5$.

To find the cofactor of the (2, 2) element of **A**, we first delete the second row and second column of the matrix.

$$\begin{bmatrix} 1 & 3 & 1 \\ 2 & 1 & 1 \\ 1 & 3 & 3 \end{bmatrix}$$

The determinant of the remaining matrix is $1(3) - 1(1) = 2$. Hence, the cofactor of the (2, 2) element is $(-1)^{2+2}2 = 2$.   ▲

cofactor matrix

A **cofactor matrix** associated with a square matrix **A** is a new matrix formed by replacing each element by its cofactor.

E X A M P L E   11.23   ▼

Find the cofactor matrix associated with the matrix

$$\mathbf{A} = \begin{bmatrix} 1 & 3 & 1 \\ 2 & 1 & 1 \\ 1 & 3 & 3 \end{bmatrix}.$$

Solution   We have already found the cofactors associated with the (1, 2) and (2, 2) elements in Example 11.22. In a similar way, we can find the cofactors of the other elements. Table 11.5 summarizes these computations.

T A B L E   11.5
Computations for the Cofactor Matrix A in Example 11.23

Element (i, j)	Determinant after Deleting ith Row and jth Column	Cofactor of Element (i, j)
(1, 1)	$1(3) - 3(1) = 0$	0
(1, 2)	See Example 11.22	-5
(1, 3)	$2(3) - 1(1) = 5$	$(-1)^{1+3}5 = 5$
(2, 1)	$3(3) - 3(1) = 6$	$(-1)^{2+1}6 = -6$
(2, 2)	See Example 11.22	2
(2, 3)	$1(3) - 1(3) = 0$	0
(3, 1)	$3(1) - 1(1) = 2$	$(-1)^{3+1}2 = 2$
(3, 2)	$1(1) - 2(1) = -1$	$(-1)^{3+2}(-1) = 1$
(3, 3)	$1(1) - 2(3) = -5$	$(-1)^{3+3}(-5) = -5$

Hence, the cofactor matrix is

$$\begin{bmatrix} 0 & -5 & 5 \\ -6 & 2 & 0 \\ 2 & 1 & -5 \end{bmatrix}.$$

▲

inverse A^{-1}

The final matrix concept that we present makes use of many of the previous operations and results we have discussed. The **inverse** of a square matrix **A**, denoted by A^{-1} (read "*A* inverse"), is a new matrix with the property that both

$$AA^{-1} = I \quad \text{and} \quad A^{-1}A = I.$$

It should be noted that not all square matrices have an inverse; only those with a nonzero determinant have inverses.

Now we will show how we can apply our previous results to obtain the inverse of a 2×2 matrix and a 3×3 matrix.

A Procedure for Obtaining the Inverse of a 2×2 or 3×3 Square Matrix A

1. Find the determinant of **A**. If the determinant is nonzero, proceed; otherwise, the inverse does not exist.
2. Find the cofactor matrix associated with **A**.
3. Find the transpose of the cofactor matrix.
4. Divide each element of this transposed matrix by $|A|$. The resulting matrix is A^{-1}.

Note: For matrices larger than 3×3, we will find the inverse of a matrix by using a computer software package.

E X A M P L E   11.24 ▼

Find the inverse of the matrix

$$A = \begin{bmatrix} 1 & 3 & 1 \\ 2 & 1 & 1 \\ 1 & 3 & 3 \end{bmatrix}.$$

Solution   Referring to Examples 11.21 and 11.23 and following the steps for obtaining A^{-1}, we have the following.

1. $|A| = -10.$
2. The cofactor matrix of **A** is

$$\begin{bmatrix} 0 & -5 & 5 \\ -6 & 2 & 0 \\ 2 & 1 & -5 \end{bmatrix}.$$

3. The transpose of the cofactor matrix is

$$
\begin{bmatrix}
0 & -6 & 2 \\
-5 & 2 & 1 \\
5 & 0 & -5
\end{bmatrix}.
$$

4. Dividing each element by $|\mathbf{A}| = -10$, we find the inverse of $\mathbf{A}$ to be

$$
\mathbf{A}^{-1} =
\begin{bmatrix}
0 & .6 & -.2 \\
.5 & -.2 & -.1 \\
-.5 & 0 & .5
\end{bmatrix}.
$$

Note that

$$
\mathbf{A}\mathbf{A}^{-1} =
\begin{bmatrix}
1 & 3 & 1 \\
2 & 1 & 1 \\
1 & 3 & 3
\end{bmatrix}
\begin{bmatrix}
0 & .6 & -.2 \\
.5 & -.2 & -.1 \\
-.5 & 0 & .5
\end{bmatrix}
=
\begin{bmatrix}
1 & 0 & 0 \\
0 & 1 & 0 \\
0 & 0 & 1
\end{bmatrix}
$$

$$
\mathbf{A}^{-1}\mathbf{A} =
\begin{bmatrix}
0 & .6 & -.2 \\
.5 & -.2 & -.1 \\
-.5 & 0 & .5
\end{bmatrix}
\begin{bmatrix}
1 & 3 & 1 \\
2 & 1 & 1 \\
1 & 3 & 3
\end{bmatrix}
=
\begin{bmatrix}
1 & 0 & 0 \\
0 & 1 & 0 \\
0 & 0 & 1
\end{bmatrix}.
$$

Hence, our calculations are correct.   ▲

EXAMPLE 11.25   ▼

Find the inverse of the matrix

$$
\mathbf{A} =
\begin{bmatrix}
5 & 0 & 0 \\
0 & 6 & 0 \\
0 & 0 & 2
\end{bmatrix}.
$$

diagonal matrix

Solution   Although we could follow the general procedure for the inverse of a matrix, we can simplify our work when we have a **diagonal matrix**, that is, a matrix with nonzero elements on the diagonal from the upper left to the lower right and with all other elements being 0. The inverse of any diagonal matrix $\mathbf{A}$ is a matrix with each diagonal element equal to the inverse of the corresponding diagonal element of $\mathbf{A}$. All other elements are 0. Thus, we can immediately write $\mathbf{A}^{-1}$ as

$$\mathbf{A}^{-1} = \begin{bmatrix} \frac{1}{5} & 0 & 0 \\ 0 & \frac{1}{6} & 0 \\ 0 & 0 & \frac{1}{2} \end{bmatrix}.$$

▲

▼ APPENDIX EXERCISES

11.54 Consider the two matrices

$$\mathbf{A} = \begin{bmatrix} 2 & 1 \\ 3 & 2 \end{bmatrix} \qquad \mathbf{B} = \begin{bmatrix} 2 & 1 \\ 1 & 1 \end{bmatrix}.$$

a. Compute $\mathbf{A} + \mathbf{B}$.
b. Verify that $\mathbf{A} + \mathbf{B} = \mathbf{B} + \mathbf{A}$.
c. Compute $\mathbf{AB}$.

11.55 Refer to Exercise 11.54.
a. Find the matrices $\mathbf{A}'$ and $\mathbf{B}'$.
b. Compute $\mathbf{A}^{-1}$.

11.56 Refer to Exercise 11.54. Compute $(\mathbf{AB})^{-1}$.

11.57 Consider the 3×3 matrix

$$\mathbf{A} = \begin{bmatrix} 3 & 0 & 2 \\ 0 & 2 & 0 \\ 2 & 0 & 2 \end{bmatrix}.$$

a. Find $\mathbf{A}'$.
b. Compute $|\mathbf{A}|$.

11.58 Refer to Exercise 11.57. Find $\mathbf{A}^{-1}$. Verify that $\mathbf{AA}^{-1} = \mathbf{A}^{-1}\mathbf{A} = \mathbf{1}$.

11.59 Consider the matrix

$$\mathbf{A} = \begin{bmatrix} 1 & 1 \\ 1 & 2 \\ 1 & 3 \end{bmatrix}.$$

a. Find $\mathbf{A}'$.
b. Compute $\mathbf{A}'\mathbf{A}$.

11.60 Refer to Exercise 11.59. Compute $(\mathbf{A}'\mathbf{A})^{-1}$.

11.61 Refer to Exercises 11.59 and 11.60. Let

$$\mathbf{B} = \begin{bmatrix} 10 \\ 22 \end{bmatrix}.$$

Compute $(\mathbf{A}'\mathbf{A})^{-1}\mathbf{B}$.

11.62 Consider the two matrices

$$\mathbf{A} = \begin{bmatrix} 1 & -1 & 1 \\ 2 & 0 & 1 \\ 3 & 1 & 3 \end{bmatrix} \quad \mathbf{B} = \begin{bmatrix} 1 & 0 & 0 \\ 0 & 1 & 0 \\ 0 & 0 & 1 \end{bmatrix}.$$

a. Compute $\mathbf{A} + \mathbf{B}$.
b. Compute $\mathbf{AB}$.

11.63 Refer to Exercise 11.62.
a. Find $|\mathbf{A}|$.
b. Compute $\mathbf{A}^{-1}$.

MORE ON MULTIPLE REGRESSION

12.1 ▾ INTRODUCTION

In Chapter 11, we presented the background information needed in order to use multiple regression. We discussed the general linear model and its use in multiple regression and introduced the normal equations, a set of simultaneous equations used in obtaining least squares estimates for the βs of a multiple regression equation. Next, we presented standard errors associated with the $\hat{\beta}_j$ and their use in inferences about a single parameter β_j, a set of βs, $E(y)$ and a future value of y. Finally, we condensed all of these inferential techniques using matrices (in an optional appendix).

This chapter is devoted to putting multiple regression into practice. How does one begin to develop an appropriate multiple regression for a given problem? While there are no hard and fast rules, we can offer a few hints to enable you to put multiple regression into practice.

First, for each problem you must decide upon the dependent variable and candidate independent variables for the regression equation. This selection process will be discussed in Section 12.2. Next, in Section 12.3, we consider how one selects the form of the multiple regression equation. The final step in the process of developing a multiple regression is to check for violation of the underlying assumptions. Tools for assessing the validity of the assumptions will be discussed in Section 12.4.

Following these steps *once* for a given problem will not ensure that you have an appropriate model. Rather, the regression equation seems to evolve as these steps are applied repeatedly, depending on the particular problem. For example, having considered candidate independent variables (step 1) and selected the form for a regression model involving some of these variables (step 2), we may find that certain assumptions have been violated (step 3). This will mean that we may have to return to either step 1 or step 2, but, hopefully, we have learned from our previous deliberations and can modify the variables under consideration and/or the model(s) selected for consideration. Eventually, a regression model will emerge that meets the needs of the experimenter. Then the analysis techniques of Chapter 11 can be used to draw inferences about model parameters $E(y)$ and y.

12.2 ▼ SELECTING THE VARIABLES (STEP 1)

The problem of trying to select reasonable candidate independent variables for inclusion in a multiple regression model can be a difficult task, but the time and attention paid to this selection process will reap benefits later in the form of better predictions. The input of a person knowledgeable in the subject matter field is a valuable source of advice on reasonable (independent) variables that could influence the response (dependent variable) of interest.

For example, in trying to predict annual sales for a recently released personal computer, it would be wise to consult an electrical engineer and perhaps other technical people to learn about the unique software and hardware features of the computer relative to the competition. Is it a new generation computer or merely a minor modification of what's presently available? In addition, it would be wise to consult with a marketing person who has more than passing knowledge of factors that have affected the marketability of other microcomputers. With information obtained from these two sources, one could piece together a list of candidate independent variables. Ideally, this list would contain variables that are closely related to the dependent variable, but not to one another.

E X A M P L E   12.1   ▼

Construct a list of independent variables that might be useful in developing a multiple regression model for predicting sales of a new microcomputer.

Solution   The list of candidate variables is quite long. For convenience, we have divided the list into those variables related to the attributes of the microprocessor and those related to the competition and the general market. The list, however, is not exhaustive.

	Microprocessor		Competition/Market
	1. Purchase price		1. Number of competitors
	2. Ease of use (Is it user friendly?)		2. Purchase price of market leader
	3. Specificity		3. Maintenance fee of market leader
	4. Annual maintenance fee		4. Volume of the market
	5. Number of possible applications		5. GNP for last three quarters
	6. Number of unique features		
	7. Can it be copied?		

One way to sort out which independent variables should be included in the regression model from the list of variables generated from discussions with experts is to resort to any one of a number of selection procedures. We will consider several of these in this text; for further details, the reader can consult Draper and Smith (1981).

The first selection procedure involves performing *all possible regressions* with the dependent variable and one or more of the independent variables from the list of candidate variables. Obviously, this approach should not be attempted unless the analyst has access to a computer with suitable software and sufficient core to run a large number of regression models relatively efficiently.

For purposes of illustration, we will use hypothetical data on prescription sales data (volume per month) obtained for a random sample of 20 independent pharmacies. These data, along with data on the total floor space, percent of floor

T A B L E   12.1

Data on 20 Independent Pharmacies

OBS	PHARMACY	VOLUME	FLOOR—SP	PRESC—RX	PARKING	SHOPCNTR	INCOME
1	1	22	4900	9	40	1	18
2	2	19	5800	10	50	1	20
3	3	24	5000	11	55	1	17
4	4	28	4400	12	30	0	19
5	5	18	3850	13	42	0	10
6	6	21	5300	15	20	1	22
7	7	29	4100	20	25	0	8
8	8	15	4700	22	60	1	15
9	9	12	5600	24	45	1	16
10	10	14	4900	27	82	1	14
11	11	18	3700	28	56	0	12
12	12	19	3800	31	38	0	8
13	13	15	2400	36	35	0	6
14	14	22	1800	37	28	0	4
15	15	13	3100	40	43	0	6
16	16	16	2300	41	20	0	5
17	17	8	4400	42	46	1	7
18	18	6	3300	42	15	0	4
19	19	7	2900	45	30	1	9
20	20	17	2400	46	16	0	3
N = 20							

space allocated to the prescription department, the number of parking spaces available for the store, whether the pharmacy is in a shopping center, and the per capita income for the surrounding community are recorded in Table 12.1.

Before running all possible regressions for the data of Table 12.1, we need to consider what criterion should be used to select the best-fitting equation from all possible regressions. The first and perhaps simplest criterion for selecting the best regression equation from the set of all possible regression equations is to compute an estimate of the error variance σ_ε^2 using $s_\varepsilon^2 = \text{MSE} = \text{SSE}/[n - (k + 1)]$. Since this quantity is used in most inferences (statistical tests and confidence intervals) about model parameters and $E(y)$, it would seem reasonable to choose the model that has the smallest value of s_ε^2. A second criterion makes use of the *coefficient of determination* R^2 for each model; by examining in detail the models that have the highest R^2 values, we can see whether there is some consistent pattern that suggests the number and identity of the variables to include in the model.

E X A M P L E   12.2

▼

Refer to the data of Table 12.1. Use the R^2 criterion to determine the best-fitting regression equation for 1, 2, 3, and 4 independent variables.

Solution   SAS output is provided here, and the regression equations with the highest R^2 values are summarized in Table 12.2.

```
                          REGRESSION ANALYSES
                PROC RSQUARE - ALL POSSIBLE SUBSETS ANALYSIS

   N = 20                        REGRESSION MODELS FOR DEPENDENT VARIABLE VOLUME

   NUMBER IN
    MODEL       R-SQUARE           C(P)                  VARIABLES IN MODEL
      1        0.00480421       30.45388047      PARKING
      1        0.03353172       29.11293360      FLOOR_SP
      1        0.04105340       28.76183600      SHOPCNTR
      1        0.14798995       23.77023759      INCOME
      1        0.43933184       10.17094219      PRESC_RX

      2        0.04210776       30.71262010      PARKING SHOPCNTR
      2        0.06855667       29.47803470      FLOOR_SP PARKING
      2        0.20543099       23.08899693      PARKING INCOME
      2        0.23487329       21.71468547      FLOOR_SP INCOME
      2        0.25653635       20.70349407      FLOOR_SP SHOPCNTR
      2        0.49576794        9.53661080      SHOPCNTR INCOME
      2        0.53142435        7.87223587      PRESC_RX PARKING
      2        0.54748785        7.12242198      PRESC_RX INCOME
      2        0.64706473        2.47435928      PRESC_RX SHOPCNTR
      2        0.66566267        1.60624219      FLOOR_SP PRESC_RX

      3        0.25569607       22.74271718      FLOOR_SP PARKING INCOME
      3        0.26507110       22.30510820      FLOOR_SP PARKING SHOPCNTR
      3        0.49828073       11.41931841      PARKING SHOPCNTR INCOME
      3        0.50012580       11.33319388      FLOOR_SP SHOPCNTR INCOME
      3        0.60243233        6.55771633      PRESC_RX PARKING INCOME
      3        0.64711563        4.47198330      PRESC_RX SHOPCNTR INCOME
      3        0.66259120        3.74961255      PRESC_RX PARKING SHOPCNTR
      3        0.66641145        3.57129027      FLOOR_SP PRESC_RX INCOME

                                                    (continues)
```

3	0.67943313	2.96346249	FLOOR_SP PRESC_RX PARKING
3	0.69072432	2.43641080	FLOOR_SP PRESC_RX SHOPCNTR
4	0.50128901	13.27889728	FLOOR_SP PARKING SHOPCNTR INCOME
4	0.66300855	5.73013127	PRESC_RX PARKING SHOPCNTR INCOME
4	0.68058567	4.90966443	FLOOR_SP PRESC_RX PARKING INCOME
4	0.69326657	4.31774327	FLOOR_SP PRESC_RX SHOPCNTR INCOME
4	0.69873953	4.06227626	FLOOR_SP PRESC_RX PARKING SHOPCNTR
5	0.70007369	6.00000000	FLOOR_SP PRESC_RX PARKING SHOPCNTR INCOME

T A B L E   12.2
Best-Fitting Models, R^2 Criterion, Example 12.2

Number of Independent Variables	R^2	Variables
1	.439	Prescription sales
2	.666	Floor space, prescription sales
3	.691	Floor space, prescription sales, shopping center
4	.694	All except per capita income

Although there is a good jump in R^2 going from one to two independent variables, very little improvement is seen thereafter. Hence, the best-fitting model based on the R^2 criterion involves the independent variables floor space and prescription sales.

One problem with using R^2 as a criterion for the best-fitting regression equation is that R^2 increases for each independent variable, even when the new x has very little predictive power. Other possible criteria for selecting the best regression that do not increase with the addition of each are presented here.

We should keep in mind that the object of our search is to choose the subset of independent variables that generates the best prediction equation for *future* values of y; unfortunately, however, since we do not know these future values, we focus on criteria that choose the best fitting regression equations to the known sample y-values. One possible bridge between this emphasis on the best fit to the known sample y-values and that on choosing the best predictor of future y-values is to split the sample data into two parts—one part used for fitting the various regression equations and the other part for validating how well the prediction equations can predict "future" values. Although there is no universally accepted rule for deciding how many of the data should be included in the "fitting" portion of the sample and how many go into the "validating" portion of the sample, it's reasonable to split the total sample in half, provided the total sample size n is greater than $2p + 20$, where p is the number of parameters in the largest potential regression model. A possible criterion for the best prediction equation would be to minimize $\sum (y_i - \hat{y}_i)^2$ for the validating portion of the total sample.

Once the regression model is selected from the data-splitting approach, the entire set of sample data is used to obtain the final prediction equation. So, even though it appears we would only use part of the data, the entire data set is used to obtain the final prediction equation.

Observations do cost money, however, and it may be impractical to obtain enough observations to apply the data-splitting approach for choosing the best-fitting regression equation. In these situations, a form of validation can be accomplished using the PRESS statistic. For a sample of y-values and a proposed regression model relating y to a set of xs, we first remove the first observation and fit the model using the remaining $n - 1$ observations. Based on the fitted equation, we estimate the first observation (denoted by $\hat{y}_1^*$) and compute the residual $y_1 - \hat{y}_1^*$. This process is repeated $n - 1$ times, successively removing the second, third, . . . , nth observation, each time computing the residual for the removed observation. The PRESS statistic is defined as

$$\text{PRESS} = \sum_{i=1}^{n} (y_i - y_i^*)^2.$$

The model that gives the smallest value for the PRESS statistic is chosen as the best-fitting model.

EXAMPLE 12.3

▼

Compute the PRESS statistic for the data of Table 12.1 to determine the best-fitting regression equation.

Solution   SAS output is provided here. The best-fitting model based on the lowest value of the PRESS statistic involves the independent variables floor space and prescription sales.

```
                                    LISTING OF DATA

  OBS    PHARMACY    VOLUME    FLOOR_SP    PRESC_RX    PARKING    SHOPCNTR    INCOME

   1         1          22        4900         9          40         1          18
   2         2          19        5800        10          50         1          20
   3         3          24        5000        11          55         1          17
   4         4          28        4400        12          30         0          19
   5         5          18        3850        13          42         0          10
   6         6          21        5300        15          20         1          22
   7         7          29        4100        20          25         0           8
   8         8          15        4700        22          60         1          15
   9         9          12        5600        24          45         1          16
  10        10          14        4900        27          82         1          14
  11        11          18        3700        28          56         0          12
  12        12          19        3800        31          38         0           8
  13        13          15        2400        36          35         0           6
  14        14          22        1800        37          28         0           4
  15        15          13        3100        40          43         0           6
  16        16          16        2300        41          20         0           5
  17        17           8        4400        42          46         1           7
  18        18           6        3300        42          15         0           4
  19        19           7        2900        45          30         1           9
  20        20          17        2400        46          16         0           3

  N =   20                                                            (continues)
```

```
                        REGRESSION ANALYSIS
      PRESS STATISTIC FOR REGRESSION MODELS, DEPENDENT VARIABLE VOLUME

      NUMBER     PRESS STATISTIC      VARIABLES
        IN                           IN MODEL
      MODEL
         1           907.636         PARKING
                     887.545         SHOPCNTR
                     869.668         FLOOR_SP
                     772.163         INCOME
                     516.391         PRESC_RX

         2           975.912         PARKING SHOPCNTR
                     916.644         PARKING FLOOR_SP
                     797.404         FLOOR_SP INCOME
                     787.578         PARKING INCOME
                     762.507         FLOOR_SP SHOPCNTR
                     547.150         INCOME PRESC_RX
                     485.820         SHOPCNTR INCOME
                     479.976         PARKING PRESC_RX
                     368.757         SHOPCNTR PRESC_RX
                     347.007         FLOOR_SP PRESC_RX

         3           890.550         PARKING FLOOR_SP INCOME
                     819.792         PARKING FLOOR_SP SHOPCNTR
                     602.214         FLOOR_SP SHOPCNTR INCOME
                     523.006         PARKING SHOPCNTR INCOME
                     513.246         PARKING INCOME PRESC_RX
                     482.387         FLOOR_SP INCOME PRESC_RX
                     455.424         SHOPCNTR INCOME PRESC_RX
                     378.166         PARKING SHOPCNTR PRESC_RX
                     371.671         PARKING FLOOR_SP PRESC_RX
                     370.843         FLOOR_SP SHOPCNTR PRESC_RX

         4           684.190         PARKING FLOOR_SP SHOPCNTR INCOME
                     513.468         PARKING FLOOR_SP INCOME PRESC_RX
                     471.086         FLOOR_SP SHOPCNTR INCOME PRESC_RX
                     458.014         PARKING SHOPCNTR INCOME PRESC_RX
                     405.832         PARKING FLOOR_SP SHOPCNTR PRESC_RX

         5           513.915         PARKING FLOOR_SP SHOPCNTR INCOME PRESC_RX
```

To this point, we've considered criteria for selecting the best-fitting regression model from a subset of independent variables. In general, if we choose a model that leaves out one or more "important" predictor variables, our model is *underspecified* and the additional variability in the *y*-values that would be accounted for with these variables becomes part of the estimated error variance. At the other end of the spectrum, if we choose a model that contains one or more "extraneous" predictor variables, our model is *overspecified* and we stand the chance of having a *multicollinearity* problem. This problem will be dealt with later. The point is that a final criterion, based on the C_p statistic, seems to balance some pros and cons of previously presented selection criteria, along with the problems of over- and underspecification, to arrive at a choice of the best-fitting subset regression equation. The C_p statistic (see Mallows (1973)) is

$$C_p = \frac{\text{SSE}_p}{s_\varepsilon^2} - (n - 2p),$$

where SSE_p is the sum of squares for error from a model with p parameters (including $\hat{\beta}_0$) and s_ε^2 is the mean square error from the regression equation with the largest number of independent variables. For a given selection problem, compute C_p for every regression equation that is fit. Theory suggests that the best-fitting model should have $C_p \approx p$.

E X A M P L E   12.4

▼

Refer to the output of Example 12.2. Determine the value of C_p for all possible regressions with 1, 2, 3, 4, and 5 independent variables. Select the best-fitting equation for 1, 2, 3, and 4 independent variables. Which regression equation seems to give the best overall fit, based on the C_p statistic?

Solution   The best-fitting models are summarized in Table 12.3. Based on the C_p criterion, there would be very little difference between the best-fitting models for 2, 3, or 4 independent variables in the model. The most "important" predictive variables seem to be floor space and prescription sales since they appear in the best-fitting models for 2, 3, and 4 independent variables. Note that these are the same important independent variables found in Example 12.2.

T A B L E   12.3
Best-Fitting Models, C_p
Criterion

Number of Independent Variables	p	C_p	Variables
1	2	10.17	Prescription sales
2	3	1.61	Floor space, prescription sales
		2.47	Prescription sales, shopping center
3	4	2.96	Floor space, prescription sales, parking spaces
4	5	4.06	Floor space, prescription sales, parking spaces, shopping center

▲

Best subset regression provides another procedure for finding the best-fitting regression equation from a set of candidate independent variables. The beauty of this procedure is that it uses an algorithm to avoid running all possible regressions. The user indicates the number (k) of best subset regressions desired. Some programs also allow the user to specify the criterion (for example, C_p or maximum R^2), but others fix the criterion. For instance in the Biomedical System of programs, BMDP9R allows for the best subset regression using the C_p statistic. Having indicated a value for k, the program computes the best k subset regressions on the basis of the C_p statistic and then identifies the best of the best. We will illustrate this with the data of Table 12.1.

E X A M P L E   12.5

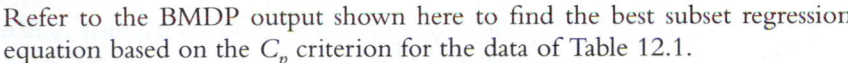

Refer to the BMDP output shown here to find the best subset regression equation based on the C_p criterion for the data of Table 12.1.

```
NUMBER OF CASES READ..............20
SUMMARY STATISTICS FOR EACH VARIABLE

                                                                 SMALLEST   LARGEST
                         STANDARD   COEFFICIENT  SMALLEST  LARGEST  STANDARD   STANDARD
          VARIABLE   MEAN  DEVIATION  OF VARIATION  VALUE    VALUE    SCORE      SCORE   SKEWNESS  KURTOSIS
  3 FLOORSP   3932.50000  1177.67333  0.299472  1800.00000  5800.00000  -1.81      1.59     -0.18    -1.25
  4 PRESC RX    27.55000    12.99585  0.471719     9.00000    46.00000  -1.43      1.42     -0.05    -1.62
  5 PARKING     38.80000    16.84168  0.434064    15.00000    82.00000  -1.41      2.57      0.60    -0.04
  6 SHOPCNTR     0.45000     0.51042  1.134262     0.0         1.00000  -0.88      1.08      0.19    -2.06
  7 INCOME      11.15000     6.01992  0.539903     3.00000    22.00000  -1.35      1.80      0.29    -1.44
  2 VOLUME      17.15000     6.28511  0.366479     6.00000    29.00000  -1.77      1.89      0.02    -0.71

VALUES FOR KURTOSIS GREATER THAN ZERO INDICATE DISTRIBUTIONS
WITH HEAVIER TAILS THAN THE NORMAL DISTRIBUTION
```

Solution The output is shown here. The best-fitting equation involves floor space and the prescription sales as independent variables as we obtained in Example 12.2. The C_p value is 1.61 and the prediction equation is

$$\hat{y} = 48.291 - .004(\text{floor space}) - .582(\text{prescription sales}).$$

```
STATISTICS FOR "BEST" SUBSET

MALLOWS´ CP                                              1.61
SQUARED MULTIPLE CORRELATION                             0.66566
MULTIPLE CORRELATION                                     0.81588
ADJUSTED SQUARED MULT. CORR.                             0.62633
RESIDUAL MEAN SQUARE                                    14.760993
STANDARD ERROR OF EST.                                   3.842004
F-STATISTIC                                             16.92
NUMERATOR DEGREES OF FREEDOM                             2
DENOMINATOR DEGEES OF FREEDOM                           17
SIGNIFICANCE (TAIL PROB.)                                0.0001

NOTE THAT THE ABOVE F-STATISTIC AND
ASSOCIATED SIGNIFICANCE TEND TO BE
LIBERAL WHENEVER A SUBSET OF VARIABLES
IS SELECTED BY THE CP OR ADJUSTED
R-SQUARE CRITERIA.

       VARIABLE    REGRESSION   STANDARD    STAND           2TAIL               CONTRIBUTION
  NO.   NAME       COEFFICIENT  ERROR      COEF.  T-STAT.   SIG.   TOLERANCE    TO R-SQ
        INTERCEPT   48.2909      6.89043    7.683   7.01    0.000
   3    FLOOR_SP    -0.00384228  0.00113262 -0.720  -3.39   0.003  0.436658     0.22633
   4    PRESC_RX    -0.581890    0.102637   -1.203  -5.67   0.000  0.436658     0.63213

THE CONTRIBUTION TO R-SQUARED FOR EACH VARIABLE IS THE AMOUNT
BY WHICH R-SQUARED WOULD BE REDUCED IF THAT VARIABLE WERE
REMOVED FROM THE REGRESSION EQUATION.
```

backward elimination
stepwise regression

There are a number of other procedures that can be used to select the best regression and, although we won't spend a great deal more time on this subject, we will mention briefly the **backward elimination** method and **stepwise regression** procedure.

The backward elimination method begins with fitting the regression model, which contains all the candidate independent variables. For each independent variable x_j, we compute

$$F_j = \frac{\text{SSdrop}_j}{\text{MSE}} \qquad j = 1, 2, \ldots,$$

where SSdrop_j is the drop in the sum of squares error obtained for the complete model, which contains all xs except x_j. MSE is the mean square error for the complete model. Let min F_j denote the smallest F_j value. If min $F_j < F_\alpha$, where α is the preselected significance level, remove the independent variable corresponding to min F_j from the regression equation. The backward elimination process then begins all over again with one variable removed from the list of candidate independent variables.

Backward elimination starts with the complete model with all independent variables entered and eliminates variables one at a time until a reasonable candidate regression model is found. This occurs when, in a particular step, min $F_j > F_\alpha$; the resulting complete model is the best-fitting regression equation. Stepwise regression, on the other hand, works in the other direction starting with the model $y = \beta_0 + \varepsilon$ and adding variables one at a time until a stopping criterion is satisfied. At the initial stage of the process, the first variable entered into the equation is the one with the largest F test for regression. At the second stage, the two variables to be included in the model are the variables with the largest F test for regression of two variables. Note that the variable entered in the first step might not be included in the second step; that is, the best single variable might not be one of the best two variables. Because of this, some people use a simplified stepwise regression (sometimes called *forward selection*) whereby, once a variable is entered, it cannot be eliminated from the regression equation at a later stage.

EXAMPLE 12.6   ▼

Use the data of Example 12.2 to find the variables to be included in a regression equation based on backward elimination. Comment on your findings.

Solution   SAS output is shown for a backward elimination procedure applied to the data of Table 12.1. As indicated, backward elimination begins with all (five) candidate variables in the regression equation. This is designated as step 0 in the backward elimination process. Then one by one, independent variables are eliminated until min $F_j > F_\alpha$. Note that in step 1, the variable income is removed and in step 2, the variable parking is removed from the

REGRESSION ANALYSIS, USING BACKWARD ELIMINATION

BACKWARD ELIMINATION PROCEDURE FOR DEPENDENT VARIABLE VOLUME

STEP 0   ALL VARIABLES ENTERED   R SQUARE = 0.70007369   C(P) =   6.00000000

	DF	SUM OF SQUARES	MEAN SQUARE	F	PROB>F
REGRESSION	5	525.44030541	105.08806108	6.54	0.0025
ERROR	14	225.10969459	16.07926390		
TOTAL	19	750.55000000			

	B VALUE	STD ERROR	TYPE II SS	F	PROB>F
INTERCEPT	42.08710826				
FLOOR_SP	-0.00241878	0.00183889	27.81923726	1.73	0.2095
PRESC_RX	-0.50046955	0.16429694	149.19783807	9.28	0.0087
PARKING	-0.03690284	0.06546687	5.10907792	0.32	0.5819
SHOPCNTR	-3.09957355	3.24983522	14.62673442	0.91	0.3564
INCOME	0.10666360	0.42742012	1.00135642	0.06	0.8066

BOUNDS ON CONDITION NUMBER: 7.823107,   117.1991

- -

STEP 1   VARIABLE INCOME REMOVED   R SQUARE = 0.69873952   C(P) =   4.06227626

	DF	SUM OF SQUARES	MEAN SQUARE	F	PROB>F
REGRESSION	4	524.43894899	131.10973725	8.70	0.0008
ERROR	15	226.11105101	15.07407007		
TOTAL	19	750.55000000			

	B VALUE	STD ERROR	TYPE II SS	F	PROB>F
INTERCEPT	43.46782063				
FLOOR_SP	-0.00228513	0.00170330	27.13112543	1.80	0.1997
PRESC_RX	-0.52910174	0.11386382	325.48983690	21.59	0.0003
PARKING	-0.03952477	0.06256589	6.01580808	0.40	0.5371
SHOPCNTR	-2.71387948	2.76799605	14.49041122	0.96	0.3424

BOUNDS ON CONDITION NUMBER: 5.071729,   46.98862

- -

STEP 2   VARIABLE PARKING REMOVED   R SQUARE = 0.69072432   C(P) = 2.43641080

	DF	SUM OF SQUARES	MEAN SQUARE	F	PROB>F
REGRESSION	3	518.42314091	172.80771364	11.91	0.0002
ERROR	16	232.12685909	14.50792869		
TOTAL	19	750.55000000			

	B VALUE	STD ERROR	TYPE II SS	F	PROB>F
INTERCEPT	42.82702645				
FLOOR_SP	-0.00247284	0.00164539	32.76871130	2.26	0.1523
PRESC_RX	-0.52941361	0.11170410	325.87978038	22.46	0.0002
SHOPCNTR	-3.03834296	2.66836223	18.81002755	1.30	0.2716

BOUNDS ON CONDITION NUMBER: 4.917388,   30.31995

- -

REGRESSION ANALYSIS, USING BACKWARD ELIMINATION

BACKWARD ELIMINATION PROCEDURE FOR DEPENDENT VARIABLE VOLUME

STEP 3   VARIABLE SHOPCNTR REMOVED   R SQUARE = 0.66566267   C(P) = 1.60624219

	DF	SUM OF SQUARES	MEAN SQUARE	F	PROB>F
REGRESSION	2	499.61311336	249.80655668	16.92	0.0001
ERROR	17	250.93688664	14.76099333		
TOTAL	19	750.55000000			

	B VALUE	STD ERROR	TYPE II SS	F	PROB>F
INTERCEPT	48.29085530				
FLOOR_SP	-0.00384228	0.00113262	169.87259933	11.51	0.0035
PRESC_RX	-0.58189034	0.10263739	474.44587802	32.14	0.0001

BOUNDS ON CONDITION NUMBER: 2.290122,   9.160487

- -

ALL VARIABLES IN THE MODEL ARE SIGNIFICANT AT THE 0.1000 LEVEL.

SUMMARY OF BACKWARD ELIMINATION PROCEDURE FOR DEPENDENT VARIABLE VOLUME

STEP	VARIABLE REMOVED	NUMBER IN	PARTIAL R**2	MODEL R**2	C(P)	F	PROB>F
1	INCOME	4	0.0013	0.6987	4.06228	0.0623	0.8066
2	PARKING	3	0.0080	0.6907	2.43641	0.3991	0.5371
3	SHOPCNTR	2	0.0251	0.6657	1.60624	1.2965	0.2716

regression equation. Step 3 is the final step in the process for this example; the variable shopping center is removed. As indicated in the output, the remaining variables comprise the best-fitting regression equation based on backward elimination. That equation is

$$\hat{y} = 48.291 - .004(\text{floor space}) - .582(\text{prescription sales}).$$

This is identical to the result we obtained from the other variable selection procedures.   ▲

EXAMPLE 12.7   ▼

Note the results of stepwise regression applied to the data of Table 12.1.

Solution   The SAS output for the data of Table 12.1 is shown here. Stepwise regression begins with the model $y = \beta_0 + \varepsilon$ and adds variables one at a time. For these data, the variable prescription sales was entered in step 1 of the stepwise procedure, the variable floor space was added to the regression model in step 2, and the variable shopping center was added in step 3. No other variables met the entrance criterion of $p = .5$ for inclusion in the model. If the

```
                    REGRESSION ANALYSIS, USING FORWARD SELECTION
                 FORWARD SELECTION PROCEDURE FOR DEPENDENT VARIABLE VOLUME
    STEP  1    VARIABLE PRESC_RX ENTERED    R SQUARE = 0.43933184      C(P) =    10.17094219
                        DF              SUM OF SQUARES        MEAN SQUARE          F       PROB>F
    REGRESSION           1               329.74051403        329.74051403       14.10      0.0014
    ERROR               18               420.80948597         23.37830478
    TOTAL               19               750.55000000

                       B VALUE            STD ERROR          TYPE II SS           F       PROB>F
    INTERCEPT       25.98133346
    PRESC_RX        -0.32055657          0.08535423         329.74051403        14.10      0.0014
    BOUNDS ON CONDITION NUMBER:              1.            1
    -----------------------------------------------------------------------------------------------
    STEP  2    VARIABLE FLOOR_SP ENTERED    R SQUARE = 0.66566267     C(P) =    1.60624219
                        DF              SUM OF SQUARES        MEAN SQUARE          F       PROB>F
    REGRESSION           2               499.61311336        249.80655668       16.92      0.0001
    ERROR               17               250.93688664         14.76099333
    TOTAL               19               750.55000000
                       B VALUE            STD ERROR          TYPE II SS           F       PROB>F
    INTERCEPT       48.29085530
    FLOOR_SP        -0.00384228          0.00113262         169.87259933        11.51      0.0035
    PRESC_RX        -0.58189034          0.10263739         474.44587802        32.14      0.0001
    BOUNDS ON CONDITION NUMBER:        2.290122      9.160487
    -----------------------------------------------------------------------------------------------
                                                                                     (continues)
```

```
STEP  3    VARIABLE SHOPCNTR ENTERED      R SQUARE = 0.69072432       C(P) =    2.43641080

                DF              SUM OF SQUARES         MEAN SQUARE          F        PROB>F

REGRESSION       3              518.42314091          172.80771364       11.91      0.0002
ERROR           16              232.12685909           14.50792869
TOTAL           19              750.55000000

                B VALUE         STD ERROR             TYPE II SS           F        PROB>F

INTERCEPT      42.82702645
FLOOR_SP       -0.00247284      0.00164539             32.76871130        2.26       0.1523
PRESC_RX       -0.52941361      0.11170410            325.87978038       22.46       0.0002
SHOPCNTR        3.03834296      2.66836223             18.81002755        1.30       0.2716

BOUNDS ON CONDITION NUMBER:      4.917388.      30.31995
---------------------------------------------------------------------------------------
       NO OTHER VARIABLES MET THE 0.5000 SIGNIFICANCE LEVEL FOR ENTRY INTO THE MODEL.

                    REGRESSION ANALYSIS, USING FORWARD SELECTION

         SUMMARY OF FORWARD SELECTION PROCEDURE FOR DEPENDENT VARIABLE VOLUME

                  VARIABLE    NUMBER    PARTIAL     MODEL
         STEP     ENTERED       IN       R**2        R**2       C(P)         F        PROB>F

           1      PRESC_RX       1       0.4393     0.4393     10.1709    14.1046      0.0014
           2      FLOOR_SP       2       0.2263     0.6657      1.6062    11.5082      0.0035
           3      SHOPCNTR       3       0.0251     0.6907      2.4364     1.2965      0.2716
```

criterion was more selective, requiring a relatively small *p*–value (say .15 or less) for each new independent variable, the stepwise regression procedure would not include the variable shopping center in step 3 (with a *p*–value of .2716) and we would arrive at the same best-fitting regression equation as we obtained previously with other methods.   ▲

In a typical regression problem, you ascertain which variables are potential candidates for inclusion in a regression model (step 1) by discussions with experts and/or by using any one of a number of possible selection procedures. For example, we could run all possible regressions, apply a best-subset regression approach, or follow a stepwise regression (a backward elimination) procedure. This list is by no means exhaustive. Sometimes the various criteria do single out the same model as best (or near best, as seen with the data of Table 12.1). Other times you may get different models from the different criteria. Which approach is best? Which one should we believe and use?

The most important response to these questions is that with the availability and accessibility of large-scale computer and major software systems, it is possible to work effectively with any of these selection procedures; no one procedure is universally accepted as better than the others. Hence, rather than attempting to use some or all of the procedures, you should begin to use one method (perhaps because of the availability of particular software in your computer facility) and learn as much as you can about it by continued use. Then you will be well equipped to solve almost any regression problem to which you are exposed.

▼ EXERCISES

Applications

 12.1 **Class Project**   The director of admissions at your college or university is interested in developing a regression model that will be useful in predicting a student's end-of-the-year grade point average (GPA) based on his or her high school record. Discuss this project among yourselves and seek out additional experts in order to develop a list of candidate independent variables for inclusion in the regression model. Should only one model be developed or should you consider more than one regression model? Might dummy variables be useful?

12.2 **Class Project**   See Exercise 12.1. Obtain data from the admissions office and apply one of the selection procedures to identify a possible regression model.

12.3 A random sample of 45 students at a state university was asked to decide whether each of the following acts should be considered a crime. The acts presented were aggravated assault, armed robbery, arson, atheism, auto theft, burglary, civil disobedience, communism, drug addiction, embezzlement, forcible rape, gambling, homosexuality, land fraud, nazism, payola, price fixing, prostitution, sexual abuse of children, sexual discrimination, shoplifting, strikes, strip mining, treason, and vandalism. For each student the interviewer determined the number of acts considered a crime and other information concerning the interviewee (years of college education, age, income of parents, and gender). The data are shown here. Use the output of a best-subset regression program to ascertain which variables should be included in the model. Can you suggest other variables that should have been addressed in the interview?

```
              LISTING OF DATA
  OBS    CRIME    AGE    COLLEGE    INCOME    SEX
    1      23      16       2         63       1
    2      25      18       2         72       1
    3      22      18       2         75       1
    4      16      18       2         61       0
    5      19      19       2         65       1
    6      19      19       2         70       1
    7      18      20       2         78       1
    8      16      19       2         76       0
    9      12      18       2         53       0
   10      13      19       2         56       0
   11      16      19       2         59       1
   12      13      20       2         55       0
   13      13      21       2         60       0
   14      14      20       2         52       0
   15      14      24       3         54       0
   16      13      25       3         55       0
   17      16      25       3         55       0
   18      16      27       4         56       1
   19      14      28       4         52       1
   20      20      38       4         59       0
   21      25      29       4         63       1
   22      19      30       4         55       1
   23      23      31       4         59       0
   24      25      32       4         52       1
```

(continues)

```
25      22      32      4       55      1
26      25      31      4       57      0
27      17      30      4       46      1
28      14      29      4       35      0
29      12      29      4       32      0
30      10      28      4       30      0
31       8      27      4       29      0
32       7      26      4       28      0
33       5      25      4       25      0
34       9      24      3       33      0
35       7      23      3       26      0
36       9      23      3       28      1
37      10      22      3       38      0
38       4      22      3       24      0
39       6      22      3       28      0
40       8      21      3       29      1
41      11      21      2       35      1
42      10      20      2       33      0
43       6      19      2       27      0
44       7      21      3       24      0
45      15      21      2       53      1

N =     45
```

```
                            REGRESSION ANALYSIS
          BACKWARD  ELIMINATION PROCEDURE FOR INDEPENDENT VARIABLE CRIMES

          BACKWARD ELIMINATION PROCEDURE FOR DEPENDENT VARIABLE CRIME

STEP 0    ALL VARIABLES ENTERED        R SQUARE = 0.82783940      C(P) =    5.00000000

                  DF            SUM OF SQUARES      MEAN SQUARE          F      PROB>F

REGRESSION         4            1301.62108953      325.40527238      48.09     0.0001
ERROR             40             270.69002158        6.76725054
TOTAL             44            1572.31111111

                B VALUE           STD ERROR          TYPE II SS          F      PROB>F

INTERCEPT      -10.82338752
INCOME           0.29025487       0.03141812        577.57817022      85.35     0.0001
AGE              0.43238152       0.20236447         30.89427247       4.57     0.0388
COLLEGE         -0.02399594       1.22148794          0.00261162       0.00     0.9844
SEX              2.45416550       0.87466592         53.27648156       7.87     0.0077

BOUNDS ON CONDITION NUMBER:      7.476669,     68.21544

---------------------------------------------------------------------------------------

STEP 1    VARIABLE COLLEGE REMOVED        R SQUARE = 0.82783774    C(P) =    3.00038592

                  DF            SUM OF SQUARES      MEAN SQUARE          F      PROB>F

REGRESSION         3            1301.61847791      433.87282597      65.72     0.0001
ERROR             41             270.69263320        6.60225935
TOTAL             44            1572.31111111

                B VALUE           STD ERROR          TYPE II SS          F      PROB>F

INTERCEPT      -10.82193315
INCOME           0.29058236       0.02630415        805.71727230     122.04     0.0001
AGE              0.42872187       0.07806990        199.10244384      30.16     0.0001
SEX              2.45108843       0.84997169         54.90370062       8.32     0.0062

BOUNDS ON CONDITION NUMBER:      1.202437,     10.21103

---------------------------------------------------------------------------------------

ALL VARIABLES IN THE MODEL ARE SIGNIFICANT AT THE 0.1000 LEVEL.
```

```
                        SUMMARY OF BACKWARD ELIMINATION PROCEDURE FOR DEPENDENT VARIABLE CRIME

                        VARIABLE    NUMBER     PARTIAL      MODEL
                STEP    REMOVED       IN       R**2          R**2        C(P)            F       PROB>F
                  1     COLLEGE        3       0.0000       0.8278      3.00039        0.0004     0.9844
```

(continues)

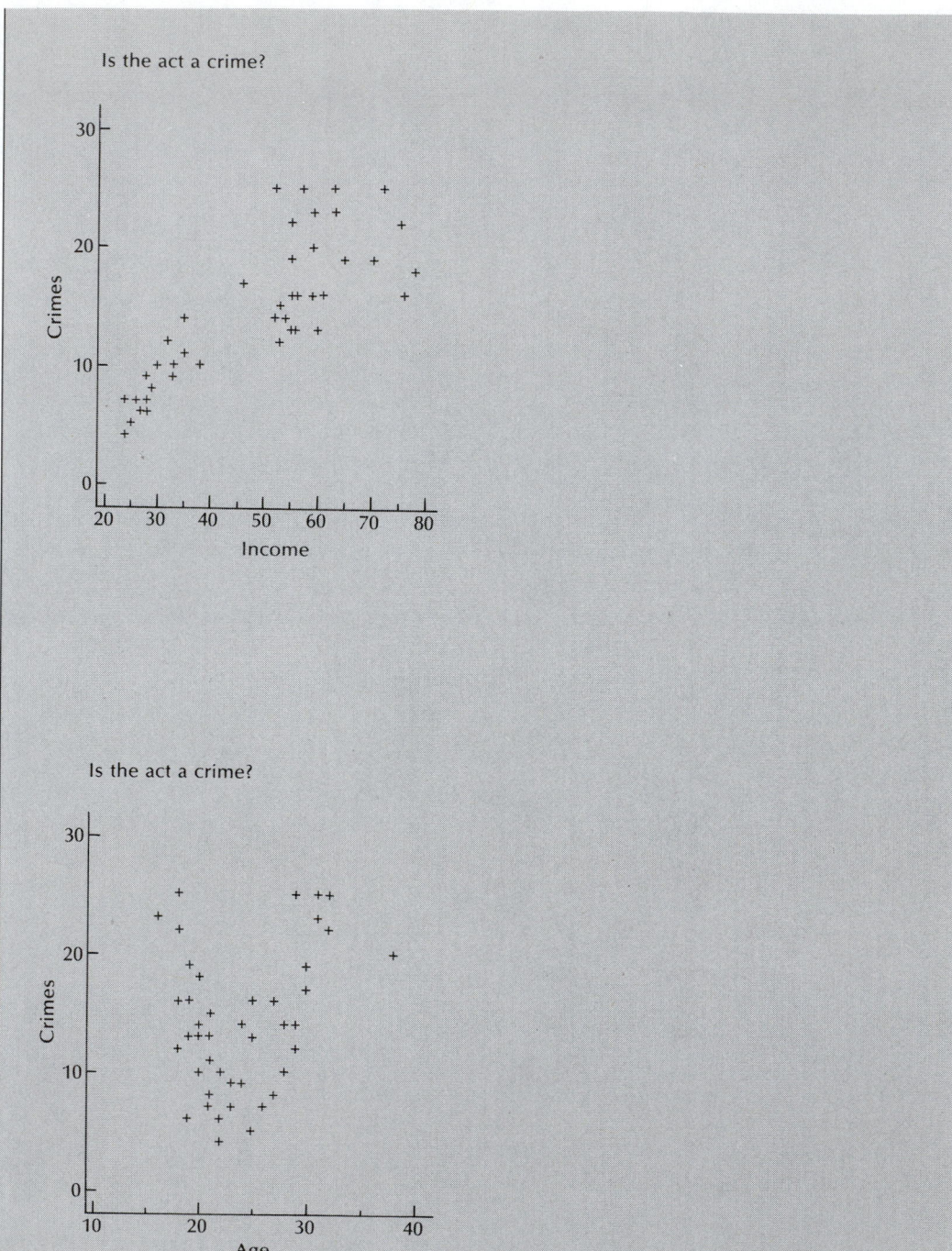

(continues)

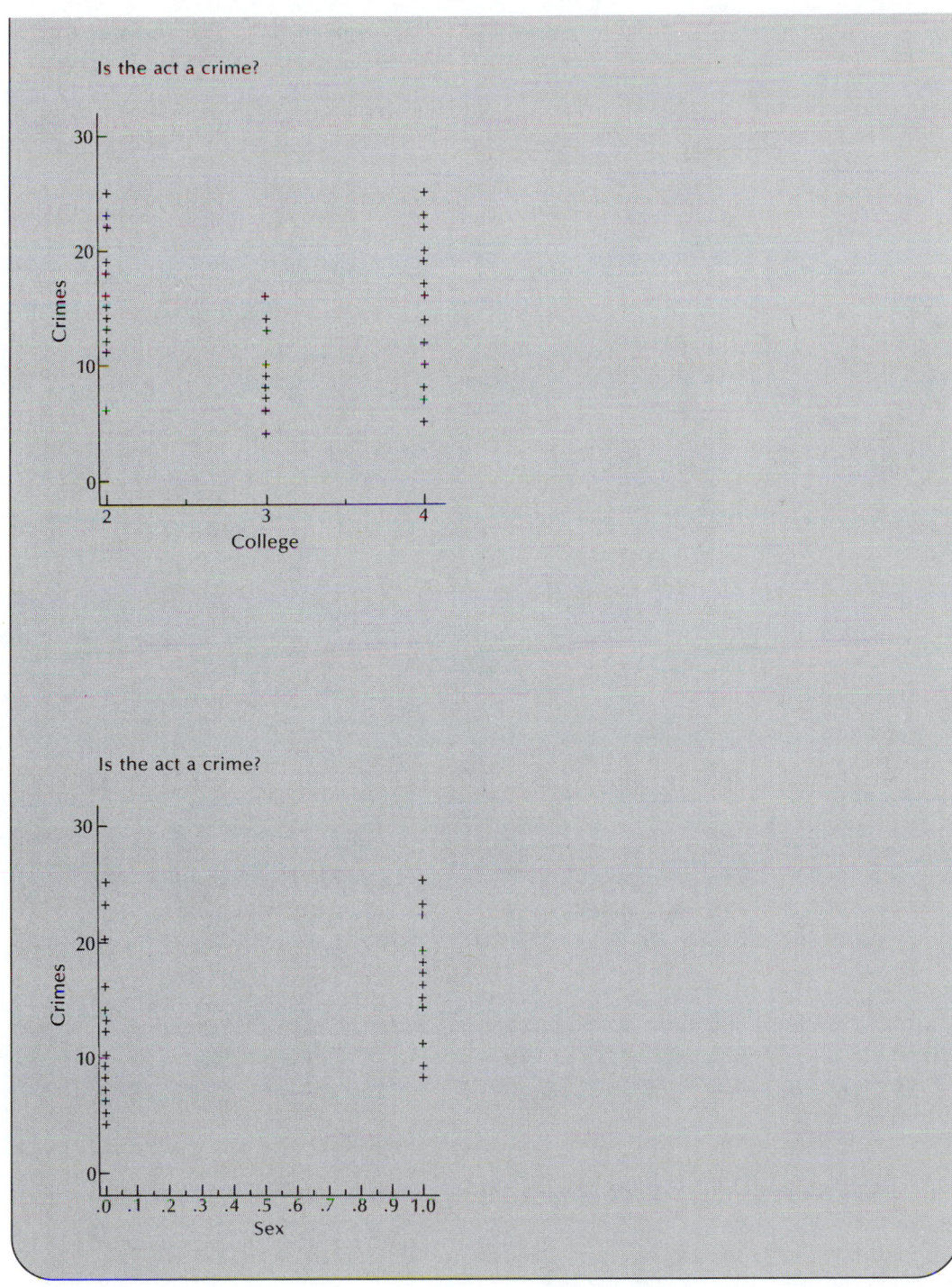

12.4   Refer to Exercise 12.3. Computer output from a stepwise regression program is shown here. Comment on the results of this analysis compared to that done in Exercise 12.3.

```
                              REGRESSION ANALYSIS
                FORWARD SELECTION PROCEDURE FOR INDEPENDENT VARIABLE CRIMES
               FORWARD SELECTION PROCEDURE FOR DEPENDENT VARIABLE CRIME

STEP 1    VARIABLE INCOME ENTERED      R SQUARE = 0.66453936      C(P) =    36.94132731

                 DF          SUM OF SQUARES         MEAN SQUARE          F        PROB>F
REGRESSION        1          1044.86262180        1044.86262180       85.18      0.0001
ERROR            43           527.44848931          12.26624394
TOTAL            44          1572.31111111

                B VALUE       STD ERROR          TYPE II SS            F        PROB>F
INTERCEPT     -0.19647505
INCOME         0.30177022      0.03269660        1044.86262180       85.18      0.0001
BOUNDS ON CONDITION NUMBER:        1,            1
------------------------------------------------------------------------------------------
STEP 2    VARIABLE AGE ENTERED     R SQUARE = 0.79291863          C(P) =    0.11353325

                 DF          SUM OF SQUARES         MEAN SQUARE          F        PROB>F
REGRESSION        2          1246.71477730         623.35738865       80.41      0.0001
ERROR            42           325.59633381           7.75229366
TOTAL            44          1572.31111111

                B VALUE       STD ERROR          TYPE II SS            F        PROB>F
INTERCEPT    -11.33832496
INCOME         0.32018698      0.02624270        1154.03879316      148.86      0.0001
AGE            0.43163600      0.08458942         201.85215549       26.04      0.0001
BOUNDS ON CONDITION NUMBER:  1.01928,     4.077119

------------------------------------------------------------------------------------------

STEP 3    VARIABLE SEX ENTERED        R SQUARE = 0.82783774     C(P) = 3.00038592

                 DF          SUM OF SQUARES         MEAN SQUARE          F        PROB>F
REGRESSION        3          1301.61847791         433.87282597       65.72      0.0001
ERROR            41           270.69263320           6.60225935
TOTAL            44          1572.31111111

                B VALUE       STD ERROR          TYPE II SS            F        PROB>F
INTERCEPT    -10.82193315
INCOME         0.29058236      0.02630415         805.71727230      122.04      0.0001
AGE            0.42872187      0.07806990         199.10244384       30.16      0.0001
SEX            2.45108843      0.84997169          54.90370062        8.32      0.0062
BOUNDS ON CONDITION NUMBER:  1.202437,     10.21103

------------------------------------------------------------------------------------------

NO OTHER VARIABLES MET THE 0.5000 SIGNIFICANCE LEVEL FOR ENTRY INTO THE MODEL.


                              REGRESSION ANALYSIS
                FORWARD SELECTION PROCEDURE FOR INDEPENDENT VARIABLE CRIMES
             SUMMARY OF FORWARD SELECTION PROCEDURE FOR DEPENDENT VARIABLE CRIME

          VARIABLE    NUMBER    PARTIAL      MODEL
STEP      ENTERED       IN       R**2        R**2        C(P)        F        PROB>F
  1       INCOME         1       0.6645      0.6645      36.9413    85.1820    0.0001
  2       AGE            2       0.1284      0.7929       9.1135    26.0377    0.0001
  3       SEX            3       0.0349      0.8278       3.0004     8.3159    0.0062
```

12.5   A company is interested in the effects of various food additives (protein and antibiotics) on the amount of time it takes to bring cattle to a desired market weight. Discuss what variable should be examined in arriving at a multiple regression equation for predicting the time to market weight.

12.3 ▼ MODEL FORMATION (STEP 2)

In Section 12.2, we suggested several ways to develop a list of candidate independent variables for a given regression problem. We can and should seek the advice of experts in the subject matter area to provide a starting point and we can employ any one of several selection procedures to come up with a possible regression model. This section involves refining the information gleaned from step 1 in order to develop a useful multiple regression model.

Having chosen a subset of k independent variables to be candidates for inclusion in the multiple regression and the dependent variable y, we still may not know the actual relationship between the dependent and independent variables. Suppose the assumed regression model is of a lower order than is the actual model relating y to $x_1, x_2, \ldots, x_k$. Then provided there is more than one observation per factor–level combination of the independent variables, we can conduct a test of the inadequacy of a fitted polynomial model using the equation $F = \text{MS}_{\text{Lack}}/\text{MSE}$ as discussed in Chapter 10.

Another way to examine an assumed (fitted) model for lack of fit is to examine scatterplots of residuals $(y_i - \hat{y}_i)$ versus x_j. For example, suppose that step 1 has indicated that the variables x_1, x_2, and x_3 constitute a reasonable subset of independent variables to be related to a response y using a multiple regression equation. Not knowing which polynomial function of the independent variables to use, we could start by fitting the multiple linear regression model

$$y = \beta_0 + \beta_1 x_1 + \beta_2 x_2 + \beta_3 x_3 + \varepsilon$$

to obtain the least squares prediction equation $y = \hat{\beta}_0 + \hat{\beta}_1 x_1 + \hat{\beta}_2 x_2 + \hat{\beta}_3 x_3$. A plot of the residuals $(y_i - \hat{y}_i)$ versus each one of the xs would shed some light as to which higher-degree terms may be appropriate.

E X A M P L E   12.8   ▼

In a radioimmunoassay, a hormone with a radioactive trace is added to a test tube containing an antibody that is specific to that hormone. The two will combine to form an antigen–antibody complex. In order to measure the extent of the reaction of the hormone with the antibody, we measure the amount of hormone that is bound to the antibody relative to the amount remaining free. Typically, experimenters measure the ratio of the bound/free radioactive count (y) for each dose of hormone (x) added to a test tube. Frequently, the relation between y and x is nearly linear. Data from 11 test tubes in a radioimmunoassay experiment are shown in Table 12.4.

a. Plot the sample data and fit the linear regression model

$$y = \beta_0 + \beta_1 x + \varepsilon.$$

T A B L E 12.4
Radioimmunoassay Data,
Example 12.8

Bound/Free Count	Dose (concentration)
9.900	0.00
10.465	0.25
10.312	0.50
13.633	0.75
20.784	1.00
36.164	1.25
62.045	1.50
78.327	1.75
90.307	2.00
97.348	2.25
102.686	2.50

b. Plot the residuals versus count and versus $\hat{y}$. Does a linear model adequately fit the data?

c. Suggest an alternative (if appropriate).

Solution   Computer output is shown here.

```
      LISTING OF DATA

   OBS    COUNT    DOSE
    1     9.900    0.00
    2    10.465    0.25
    3    10.312    0.50
    4    13.633    0.75
    5    20.784    1.00
    6    36.164    1.25
    7    62.045    1.50
    8    78.327    1.75
    9    90.307    2.00
   10    97.348    2.25
   11   102.686    2.50

   N =     11




                          SIMPLE LINEAR REGRESSION

                       GENERAL LINEAR MODELS PROCEDURE

 DEPENDENT VARIABLE: COUNT

 SOURCE           DF    SUM OF SQUARES      MEAN SQUARE    F VALUE    PR > F    R-SQUARE    C.V.
 MODEL             1    13577.44212015    13577.44212015    111.44    0.0002    0.925273    22.8242
 ERROR             9     1096.54306185      121.83811798                        ROOT MSE       COUNT MEAN
 CORRECTED TOTAL  10    14673.98518200                                         11.03803053      48.36100000

 SOURCE           DF       TYPE I SS      F VALUE    PR > F    DF     TYPE III SS    F VALUE    PR > F
 DOSE              1    13577.44212015    111.44    0.0001    1    13577.44212015    111.4    0.0001

                                    T FOR HO:      PR > |T|      STD ERROR OF
 PARAMETER          ESTIMATE       PARAMETER=0                    ESTIMATE
 INTERCEPT         7.18881818         1.15          0.2780       6.22628894
 DOSE             44.43985455        10.56          0.0001       4.20973967

                                                                       (continues)
```

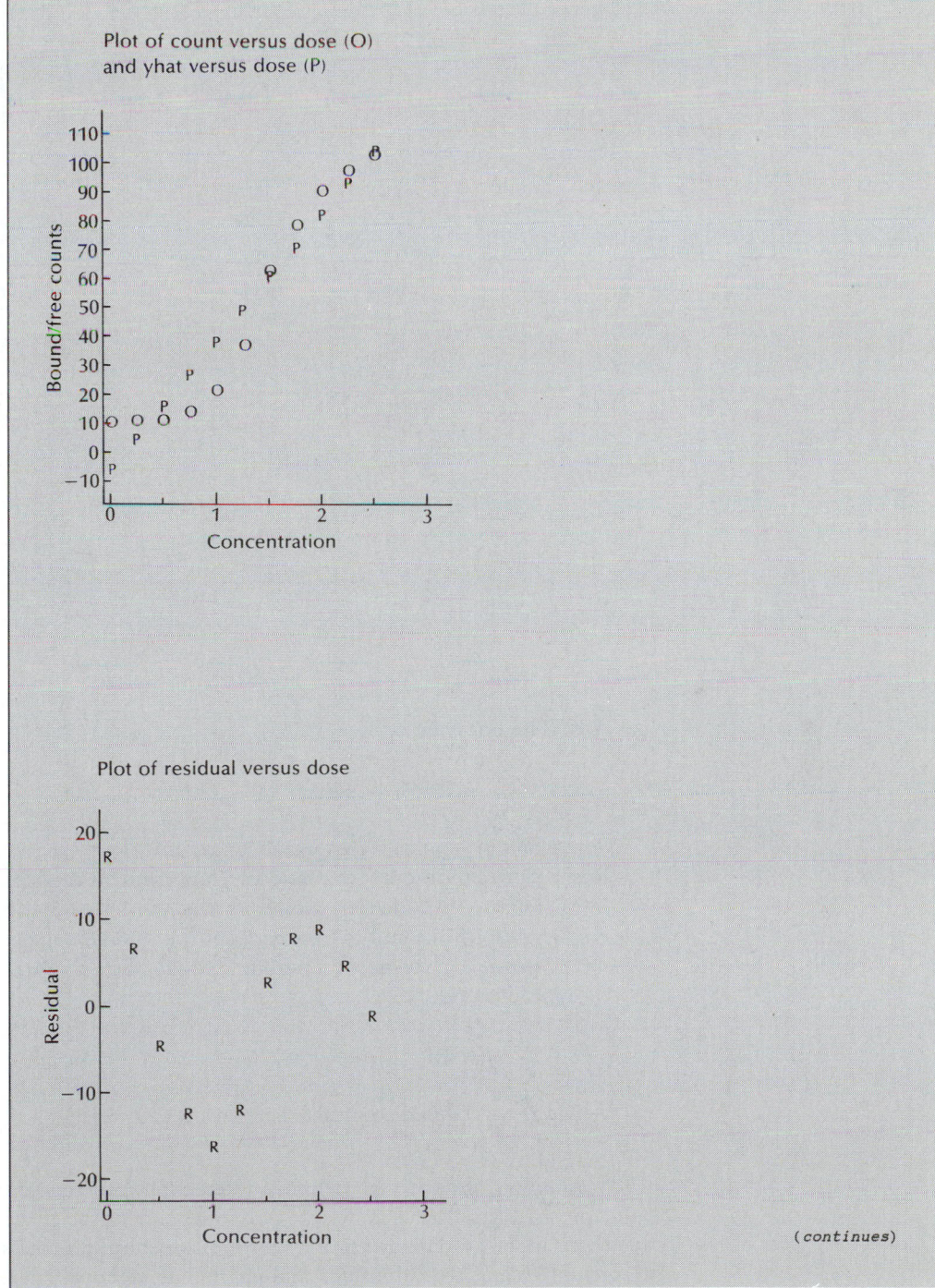

Plot of count versus dose (O)
and yhat versus dose (P)

Plot of residual versus dose

(continues)

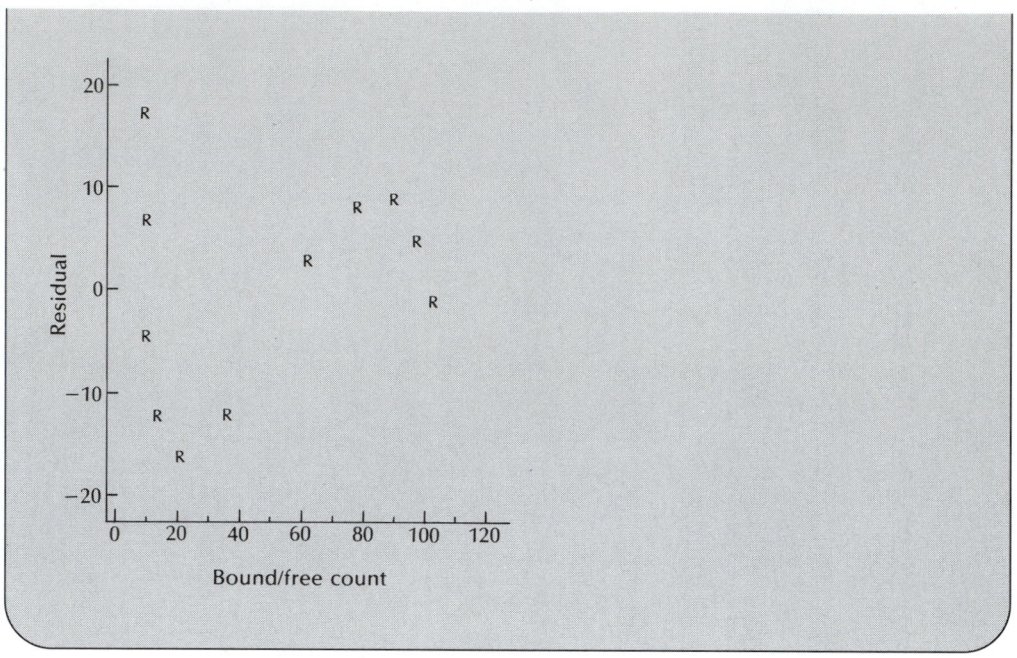

a, b. The linear fit is

$$\hat{y} = -7.189 + 44.440x.$$

The plot of y (count) versus x (concentration) clearly shows a lack of fit of the linear regression model; the residual plots confirm this same lack of fit. The linear regression underestimates counts at the lower and upper ends of the concentration scale and overestimates at the middle concentrations.

c. A possible alternative model would be a quadratic model in concentration,

$$y = \beta_0 + \beta_1 x + \beta_2 x^2 + \varepsilon.$$

More will be said about this later in the chapter.   ▲

Scatterplots are not very helpful in detecting interactions among the independent variables, other than for the two independent variable case. The reason is that there are just too many variables for most practical problems and it is difficult to present the interrelationships among independent variables and their joint effects on the response y using two-dimensional scatterplots. Perhaps the most reasonable suggestion is to use one of the best subset regression methods of the previous section, some trial-and-error fitting of models using the candidate independent variables, and a bit of common sense to determine which interaction terms should be used in the multiple regression model.

The presence of dummy variables (for qualitative independent variables) presents no major problem for ascertaining the adequacy of the fit of a polynomial model. The important thing to remember is that when quantitative and dummy variables are included in the same regression model, for each setting of the dummy variables, we obtain a regression in the quantitative variables. Hence, plotting methods for detecting an inadequate fit should be applied separately for each setting of the dummy variables. By examining these plots carefully, we can also detect potential differences in the forms of the polynomial models for different settings of the dummy variables.

E X A M P L E   12.9

▼

A company analyst is interested in developing a regression model for predicting automobile sales for standard and luxury models of a particular make in a given territory. Empirical discussions and some substantive research into previous sales patterns for the company in that territory tend to indicate that the prevailing interest rate for car loans and the price per gallon of gasoline are the key predictive variables. The number of cars sold per month (in 1,000s) for the previous 18 months is shown here for gasoline-powered standard and luxury models. Fit a linear regression model and use residual plots to determine what (if any) higher-order terms are required. Do the same conclusions hold for standard and luxury models? Make suggestions for additional terms in the multiple regression equation.

Solution   A multiple regression model of the form

$$y = \beta_0 + \beta_1 x_1 + \beta_2 x_2 + \beta_3 x_3 + \varepsilon,$$

where

y = number of sales per month (in 1,000s)

x_1 = price per gallon

x_2 = interest rate

$x_3 = \begin{cases} 1 & \text{if standard} \\ 0 & \text{if luxury} \end{cases}$

was fit to the data. From the output, the regression equation is

$$\hat{y} = 56.074 - 16.144x_1 - 2.332x_2 + 14.422x_3.$$

Substituting $x_3 = 0$ and 1 into this equation, we obtain the separate regression equations for the luxury and standard cars, respectively:

$x_3 = 0$ (luxury cars)
$$\hat{y} = 56.074 - 16.144x_1 - 2.332x_2$$
$x_3 = 1$ (standard cars)
$$\hat{y} = 56.074 - 16.144x_1 - 2.332x_2 + 14.422$$
$$= 70.496 - 16.144x_1 - 2.332x_2.$$

Plots of y versus x_1 and x_2 for the two model types show clear negative linear relationships between sales and price per gallon of gasoline or interest rates. However, the slopes appear to be greater for the standard model than for the luxury model. This is borne out in the residual plots for the two models.

```
              LISTING OF DATA
 OBS    MONTH      Y       X1      X2      X3
  1       1      22.1     1.39    12.1      1
  2       1       7.2     1.39    12.1      0
  3       2      15.4     1.44    12.2      1
  4       2       5.4     1.44    12.2      0
  5       3      11.7     1.45    12.3      1
  6       3       7.6     1.45    12.3      0
  7       4      10.3     1.32    14.2      1
  8       4       2.5     1.32    14.2      0
  9       5      11.4     1.35    15.8      1
 10       5       2.4     1.35    15.8      0
 11       6       7.5     1.28    16.3      1
 12       6       1.7     1.28    16.3      0
 13       7      13.0     1.26    16.5      1
 14       7       4.3     1.26    16.5      0
 15       8      12.8     1.26    14.7      1
 16       8       3.7     1.26    14.7      0
 17       9      14.6     1.25    13.4      1
 18       9       3.9     1.25    13.4      0
 19      10      18.9     1.24    12.9      1
 20      10       7.0     1.24    12.9      0
 21      11      19.3     1.20    11.2      1
 22      11       6.8     1.20    11.2      0
 23      12      30.1     1.20    10.9      1
 24      12      10.1     1.20    10.9      0
 25      13      28.2     1.18    10.3      1
 26      13       9.4     1.18    10.3      0
 27      14      25.6     1.10     9.7      1
 28      14       7.9     1.10     9.7      0
 29      15      37.5     1.11     9.6      1
 30      15      14.1     1.11     9.6      0
 31      16      36.1     1.14     9.1      1
 32      16      14.5     1.14     9.1      0
 33      17      39.8     1.17     7.8      1
 34      17      14.9     1.17     7.8      0
 35      18      44.3     1.18     8.3      1
 36      18      15.6     1.18     8.3      0

 N =     36
```

(continues)

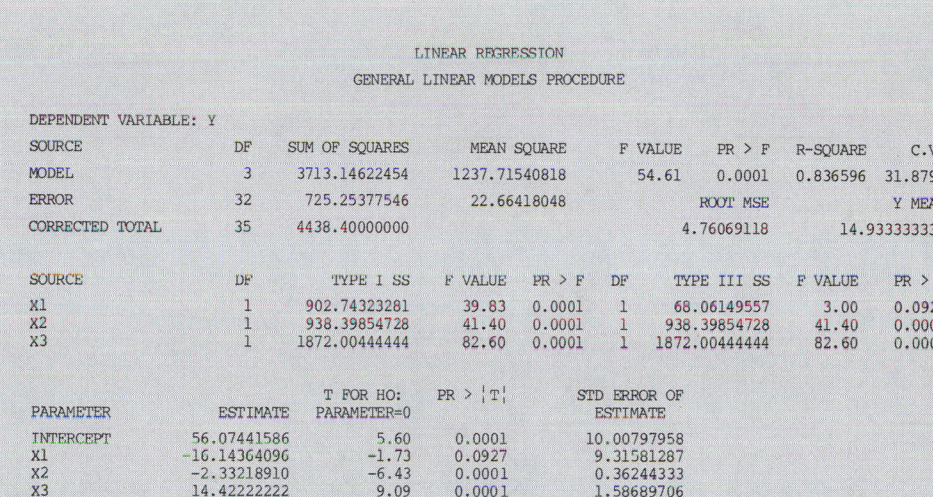

LINEAR REGRESSION

GENERAL LINEAR MODELS PROCEDURE

DEPENDENT VARIABLE: Y

SOURCE	DF	SUM OF SQUARES	MEAN SQUARE	F VALUE	PR > F	R-SQUARE	C.V.
MODEL	3	3713.14622454	1237.71540818	54.61	0.0001	0.836596	31.8796
ERROR	32	725.25377546	22.66418048		ROOT MSE		Y MEAN
CORRECTED TOTAL	35	4438.40000000			4.76069118		14.933333333

SOURCE	DF	TYPE I SS	F VALUE	PR > F	DF	TYPE III SS	F VALUE	PR > F
X1	1	902.74323281	39.83	0.0001	1	68.06149557	3.00	0.0927
X2	1	938.39854728	41.40	0.0001	1	938.39854728	41.40	0.0001
X3	1	1872.00444444	82.60	0.0001	1	1872.00444444	82.60	0.0001

PARAMETER	ESTIMATE	T FOR HO: PARAMETER=0	PR > |T|	STD ERROR OF ESTIMATE
INTERCEPT	56.07441586	5.60	0.0001	10.00797958
X1	-16.14364096	-1.73	0.0927	9.31581287
X2	-2.33218910	-6.43	0.0001	0.36244333
X3	14.42222222	9.09	0.0001	1.58689706

Plot of monthly sales
versus price per gallon

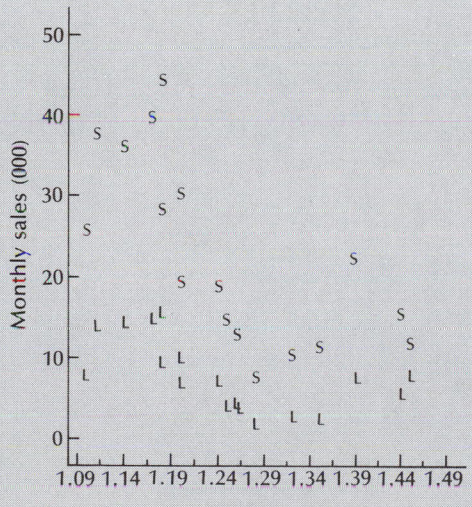

(*continues*)

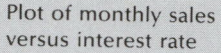

Plot of monthly sales
versus interest rate

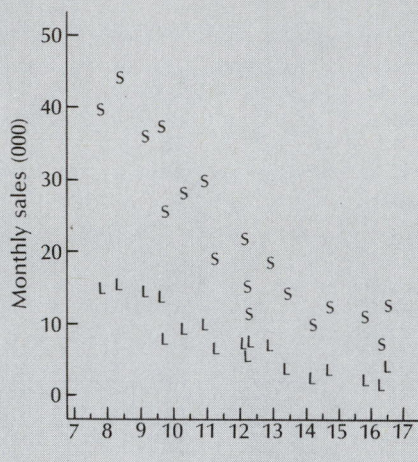

Plot of residual values
versus price per gallon

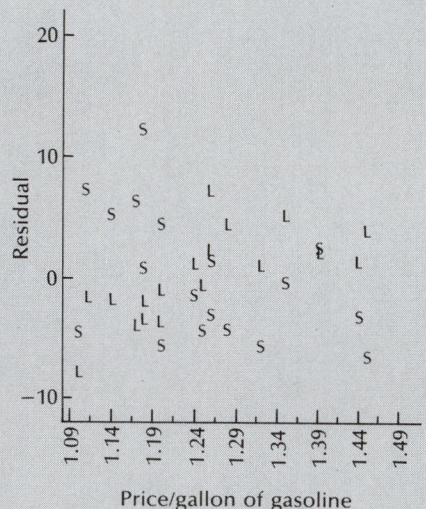

(*continues*)

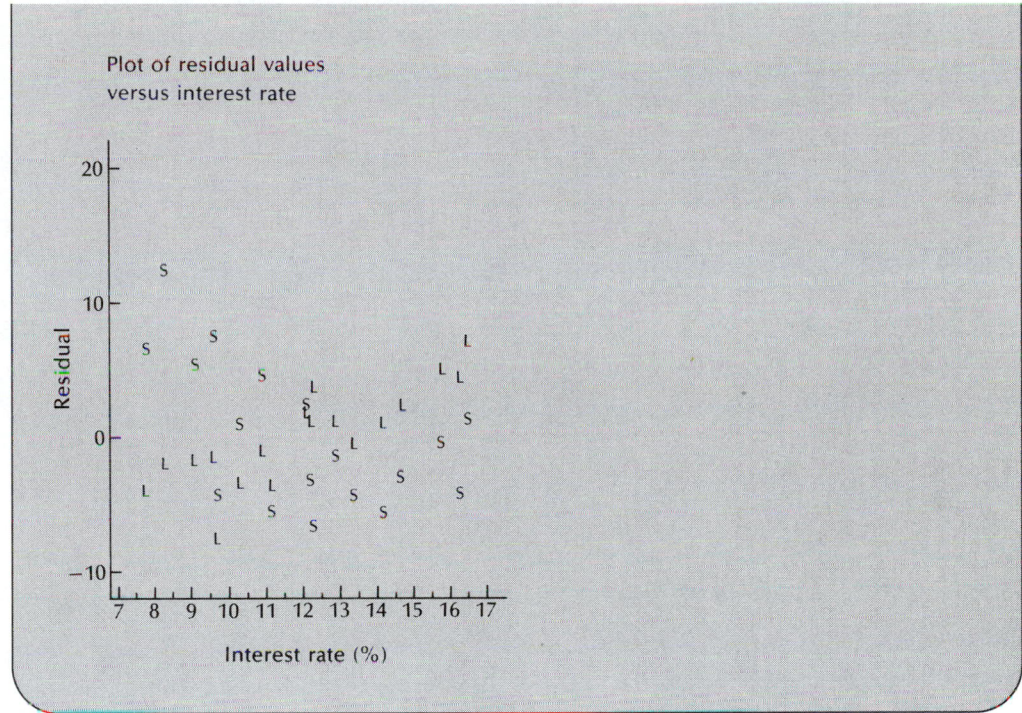

Plot of residual values versus interest rate

Plots of residuals versus price per gallon and versus interest rates for luxury models show underestimation for smaller values of x_1 and x_2 and over-estimation for the larger values of x_1 and x_2. Corresponding residual plots for the standard models show fairly good fits to the data, although there may be some curvature that could be accounted for by including higher-order terms in x_1 and x_2 in the regression model.

A regression model of the form

$$y = \beta_0 + \beta_1 x_1 + \beta_2 x_1^2 + \beta_3 x_2 + \beta_4 x_2^2 + \beta_5 x_3 + \beta_6 x_1 x_3 + \beta_7 x_1^2 x_3 + \beta_8 x_2 x_3 + \beta_9 x_2^2 x_3 + \varepsilon$$

would allow for curvature in y (sales) due to x_1 (price per gallon) and to x_2 (interest rate); the model also allows for different regression coefficients for the two car models. One might also consider adding interaction terms between the two quantitative independent variables. Some output for this model follows.

OBS	MONTH	Y	X1	X2	X3	X1_2	X2_2
1	1	22.1	1.39	12.1	1	1.9321	146.41
2	1	7.2	1.39	12.1	0	1.9321	146.41
3	2	15.4	1.44	12.2	1	2.0736	148.84
4	2	5.4	1.44	12.2	0	2.0736	148.84
5	3	11.7	1.45	12.3	1	2.1025	151.29
6	3	7.6	1.45	12.3	0	2.1025	151.29
7	4	10.3	1.32	14.2	1	1.7424	201.64
8	4	2.5	1.32	14.2	0	1.7424	201.64
9	5	11.4	1.35	15.8	1	1.8225	249.64
10	5	2.4	1.35	15.8	0	1.8225	249.64
11	6	7.5	1.28	16.3	1	1.6384	265.69
12	6	1.7	1.28	16.3	0	1.6384	265.69
13	7	13.0	1.26	16.5	1	1.5876	272.25
14	7	4.3	1.26	16.5	0	1.5876	272.25
15	8	12.8	1.26	14.7	1	1.5876	216.09
16	8	3.7	1.26	14.7	0	1.5876	216.09
17	9	14.6	1.25	13.4	1	1.5625	179.56
18	9	3.9	1.25	13.4	0	1.5625	179.56
19	10	18.9	1.24	12.9	1	1.5376	166.41
20	10	7.0	1.24	12.9	0	1.5376	166.41
21	11	19.3	1.20	11.2	1	1.4400	125.44
22	11	6.8	1.20	11.2	0	1.4400	125.44
23	12	30.1	1.20	10.9	1	1.4400	118.81
24	12	10.1	1.20	10.9	0	1.4400	118.81
25	13	28.2	1.18	10.3	1	1.3924	106.09
26	13	9.4	1.18	10.3	0	1.3924	106.09
27	14	25.6	1.10	9.7	1	1.2100	94.09
28	14	7.9	1.10	9.7	0	1.2100	94.09
29	15	37.5	1.11	9.6	1	1.2321	92.16
30	15	14.1	1.11	9.6	0	1.2321	92.16
31	16	36.1	1.14	9.1	1	1.2996	82.81
32	16	14.5	1.14	9.1	0	1.2996	82.81
33	17	39.8	1.17	7.8	1	1.3689	60.84
34	17	14.9	1.17	7.8	0	1.3689	60.84
35	18	44.3	1.18	8.3	1	1.3924	68.89
36	18	15.6	1.18	8.3	0	1.3924	68.89

N = 36

LINEAR REGRESSION

GENERAL LINEAR MODELS PROCEDURE

DEPENDENT VARIABLE: Y

SOURCE	DF	SUM OF SQUARES	MEAN SQUARE	F VALUE	PR > F	R-SQUARE	C.V.
MODEL	9	4203.08834550	467.00981617	51.60	0.0001	0.946983	20.1455
ERROR	26	235.31165450	9.05044825		ROOT MSE		Y MEAN
CORRECTED TOTAL	35	4438.40000000			3.00839629		14.93333333

SOURCE	DF	TYPE I SS	F VALUE	PR > F	DF	TYPE III SS	F VALUE	PR > F
X1	1	902.74323281	99.75	0.0001	1	0.02708386	0.00	0.9568
X1*X1	1	270.27042800	29.86	0.0001	1	0.02587043	0.00	0.9578
X2	1	684.25426357	75.60	0.0001	1	29.10270106	3.22	0.0846
X2*X2	1	90.22789329	9.97	0.0040	1	15.45933401	1.71	0.2027
X3	1	1872.00444444	206.84	0.0001	1	2.24286224	0.25	0.6228
X1*X3	1	218.90085296	24.19	0.0001	1	7.39894543	0.82	0.3742
X1*X1*X3	1	24.95517426	2.76	0.1088	1	8.24522060	0.91	0.3486
X2*X3	1	124.22121378	13.73	0.0010	1	29.98826625	3.31	0.0802
X2*X2*X3	1	15.51084238	1.71	0.2019	1	15.51084238	1.71	0.2019

PARAMETER	ESTIMATE	T FOR H0: PARAMETER=0	PR > |T|	STD ERROR OF ESTIMATE
INTERCEPT	40.56734222	0.28	0.7785	142.71008115
X1	12.20097631	0.05	0.9568	223.03567864
X1*X1	-4.59508961	-0.05	0.9578	85.94630524
X2	-5.32756126	-1.79	0.0846	2.97096029
X2*X2	0.15363504	1.31	0.2027	0.11755197
X3	-100.46988523	-0.50	0.6228	201.82253225
X1*X3	285.19322700	0.90	0.3742	315.42008163
X1*X1*X3	-116.01343994	-0.95	0.3486	121.54643051
X2*X3	-7.64808102	-1.82	0.0802	4.20157234
X2*X2*X3	0.21763441	1.31	0.2019	0.16624360

(continues)

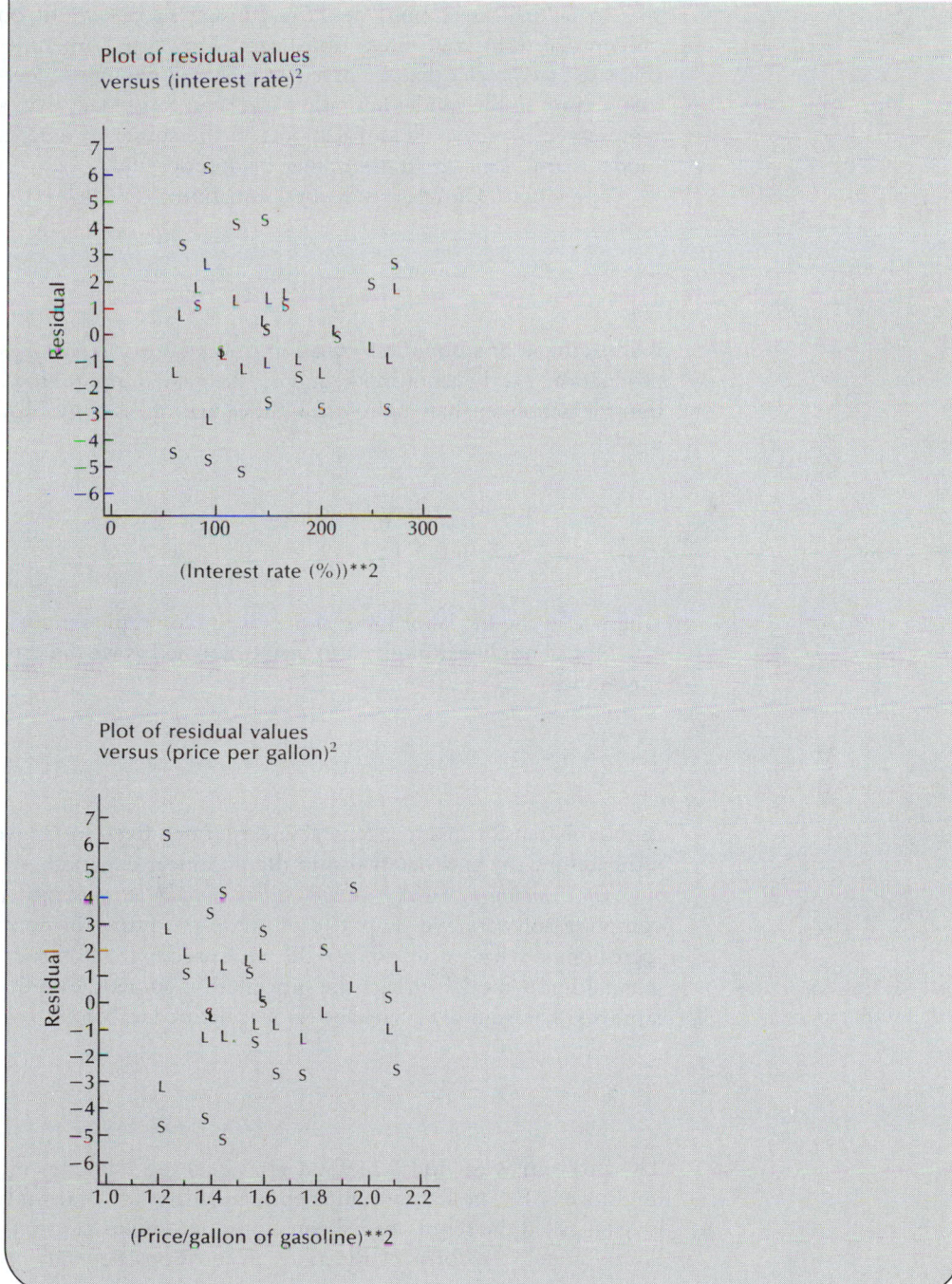

Plot of residual values
versus (interest rate)^2

(Interest rate (%))**2

Plot of residual values
versus (price per gallon)^2

(Price/gallon of gasoline)**2

So far in this section, we have considered lack of fit only as it relates to polynomial terms and interaction terms. However, sometimes the lack of fit is unrelated to the fact that we have not included enough higher-degree terms and interactions in the model but rather is related to the fact that y is not adequately represented by any polynomial model in the subset of independent variables. A model that is *nonlinear* in the βs may be appropriate.

The Cobbs–Douglas production equation

$$y = \alpha_1 l^{\alpha_2} c^{\alpha_3}$$

is an example of a nonlinear equation in the constants, α_1, α_2, and α_3. Here y is production, l is the labor input, and c is the capital input. However, a logarithmic transformation enables us to treat the equation as if we had a general linear model:

$$\begin{aligned} \log y &= \log \alpha_1 + \alpha_2 \log l + \alpha_3 \log c \\ &= \beta_0 + \beta_1 x_1 + \beta_2 x_2, \end{aligned}$$

where x_1 is the log labor input and x_2 is the log capital input.

Not all nonlinear models can be transformed as we did this one. For example, the model

$$y = \alpha_0 e^{-\alpha_1 x_1} + \alpha_2 e^{-\alpha_3 x_2}$$

cannot be transformed to a general linear model. Even so, we would like to obtain estimates for the parameters (αs) in the nonlinear equation.

The remaining material in this section should be considered optional. We'll use computer software and output to illustrate the fitting of nonlinear models. The logic behind what we are doing is the same used in the least squares method for the general linear model; in fact the procedure is sometimes called **nonlinear least squares**. The sum of squares for error is defined as before,

nonlinear least squares

$$\text{SSE} = \sum_i (y_i - \hat{y}_i)^2.$$

The problem is to find a method for obtaining estimates $\hat{\alpha}_1$, $\hat{\alpha}_2$, ... that will minimize SSE. The set of simultaneous equations used for finding these estimates is again called the set of normal equations, but unlike least squares for the general linear model, the form of the normal equations depends on the form of the nonlinear model being used. Also, since the normal equations involve nonlinear functions of the parameters, their solutions can be quite complicated. Because of this technical difficulty, a number of iterative methods have been developed for obtaining a solution to the normal equations.

For those of you with a background in calculus, the normal equations for a nonlinear model involve partial derivatives of the nonlinear function with respect to each of the parameters α_i. Fortunately most of the computer software packages currently marketed (for example, SAS, NONLIN, BMDP) approximate the derivative and do not require one to give the form of the normal equations; only the form of the nonlinear equation is needed. We will illustrate this with the data from a previous example.

Recall that in Example 12.8 we fit a linear regression model to the radio–immunoassay data; a residual plot for that model suggested that a quadratic model might be more appropriate:

$$y = \beta_0 + \beta_1 x + \beta_2 x^2 + \varepsilon.$$

Computer output for the revised model is shown here. Note that the cyclical pattern is still apparent in the residual plot and hence the quadratic model is still inadequate.

A nonlinear model that may help to flatten the S–shape of the data plot shown in the output has the following form:

$$y = \frac{\beta_0 - \beta_3}{1 + (x/\beta_2)^{\beta_1}} + \beta_3$$

```
                              QUADRATIC REGRESSION
                         GENERAL LINEAR MODELS PROCEDURE

DEPENDENT VARIABLE: COUNT

SOURCE                 DF      SUM OF SQUARES      MEAN SQUARE    F VALUE    PR > F   R-SQUARE       C.V.
MODEL                   2      13964.37985960    6982.18994780     78.72    0.0001   0.951642    19.4746
ERROR                   8        709.60528640      88.70066080                ROOT MSE         COUNT MEAN
CORRECTED TOTAL        10      14673.98518200                              9.41810282       48.36100000

SOURCE                 DF         TYPE I SS   F VALUE    PR > F   DF    TYPE III SS   F VALUE   PR > F
DOSE                    1      13577.44212015    153.07    0.0001    1   153.70381752      1.73   0.2245
DOSE*DOSE               1        386.93777546      4.36    0.0702    1   386.93777546      4.36   0.0702

                                  T FOR HO:    PR > |T|   STD ERROR OF
PARAMETER              ESTIMATE   PARAMETER=0             ESTIMATE
INTERCEPT            2.88439860          0.40    0.6982    7.17520734
DOSE               17.57794312          1.32    0.2245   13.35331689
DOSE*DOSE          10.74476457          2.09    0.0702    5.14445966

                                                                   (continues)
```

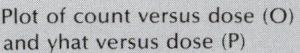

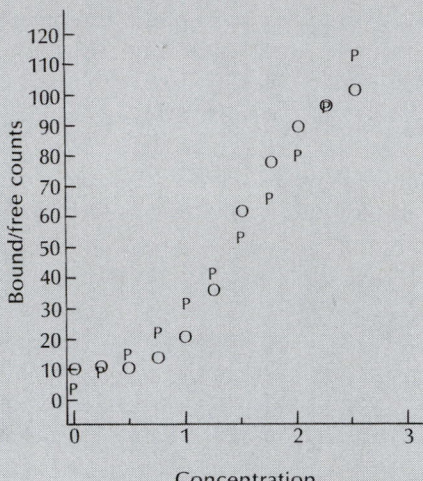

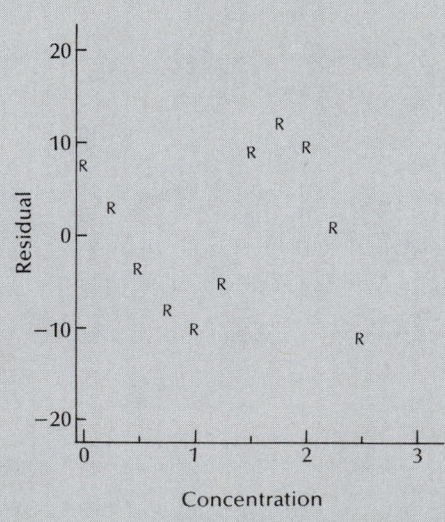

(*continues*)

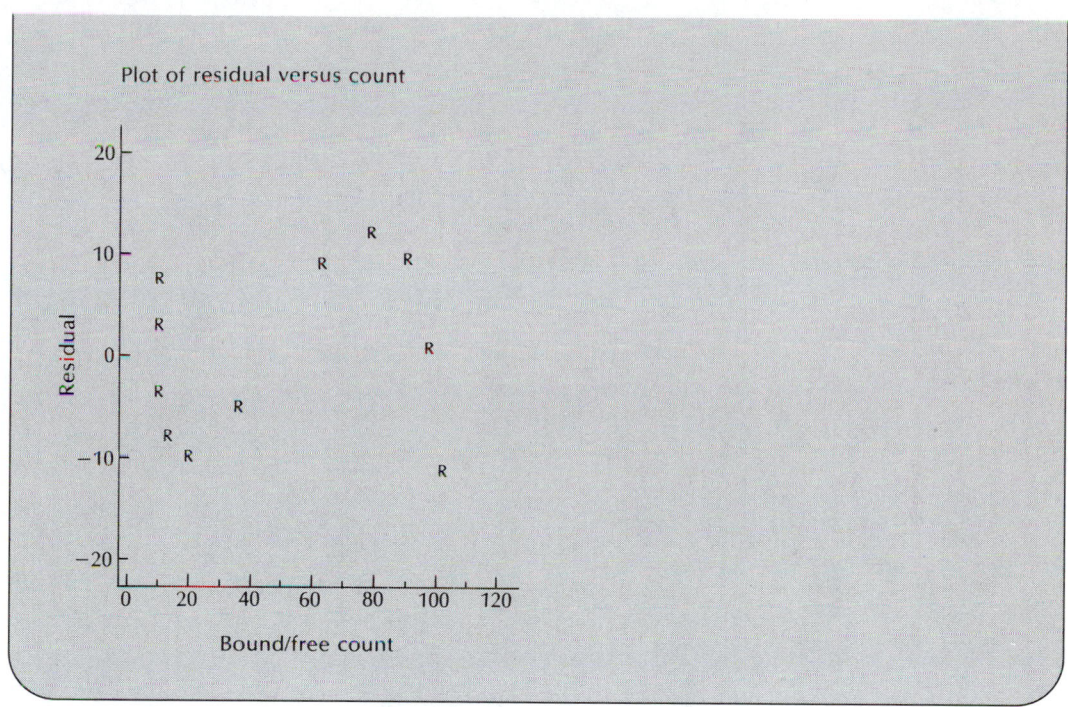

Plot of residual versus count

E X A M P L E 12.10 ▼

Use a nonlinear estimation program to fit the radioimmunoassay data to the preceding model.

Solution SAS was used to fit this model to the sample data. As can be seen from the residual plot, the nonlinear model provides a much better fit to the sample data than either the linear or quadratic model.

By way of explanation, the parameters have the following interpretations:

β_0: value of y at the lower end of the curve
β_3: value of y at the upper end of the curve
β_1: concentration (x) corresponding to the value of y midway between β_0 and β_3
β_2: a measure of the slope

```
     LISTING OF DATA
  OBS      COUNT     DOSE
    1       9.900    0.00
    2      10.465    0.25
    3      10.312    0.50
    4      13.633    0.75
    5      20.784    1.00
    6      36.164    1.25
    7      62.045    1.50
    8      78.327    1.75
    9      90.307    2.00
   10      97.348    2.25
   11     102.686    2.50
N =    11
```

NON-LINEAR REGRESSION

NON-LINEAR LEAST SQUARES SUMMARY STATISTICS   DEPENDENT VARIABLE COUNT

SOURCE	DF	SUM OF SQUARES	MEAN SQUARE
REGRESSION	4	40390.959650	10097.739913
RESIDUAL	7	9.675063	1.382152
UNCORRECTED TOTAL	11	40400.634713	
(CORRECTED TOTAL)	10	14673.985182	

PARAMETER	ESTIMATE	ASYMPTOTIC STD. ERROR	ASYMPTOTIC 95 % CONFIDENCE INTERVAL LOWER	UPPER
B0	10.3172000	0.6302686688	8.82683942	11.80756051
B1	5.3701025	0.2558487401	4.76511153	5.97509351
B2	1.4863325	0.0154082328	1.44989755	1.52276749
B3	107.3776302	1.7275763169	103.29252858	111.46273188

ASYMPTOTIC CORRELATION MATRIX OF THE PARAMETERS

CORR	B0	B1	B2	B3
B0	1.0000	0.4316	0.1141	-0.2553
B1	0.4316	1.0000	-0.5150	-0.8088
B2	0.1141	-0.5150	1.0000	0.7938
B3	-0.2553	-0.8088	0.7938	1.0000

Plot of count versus dose (O)
and yhat versus dose (P)

(continues)

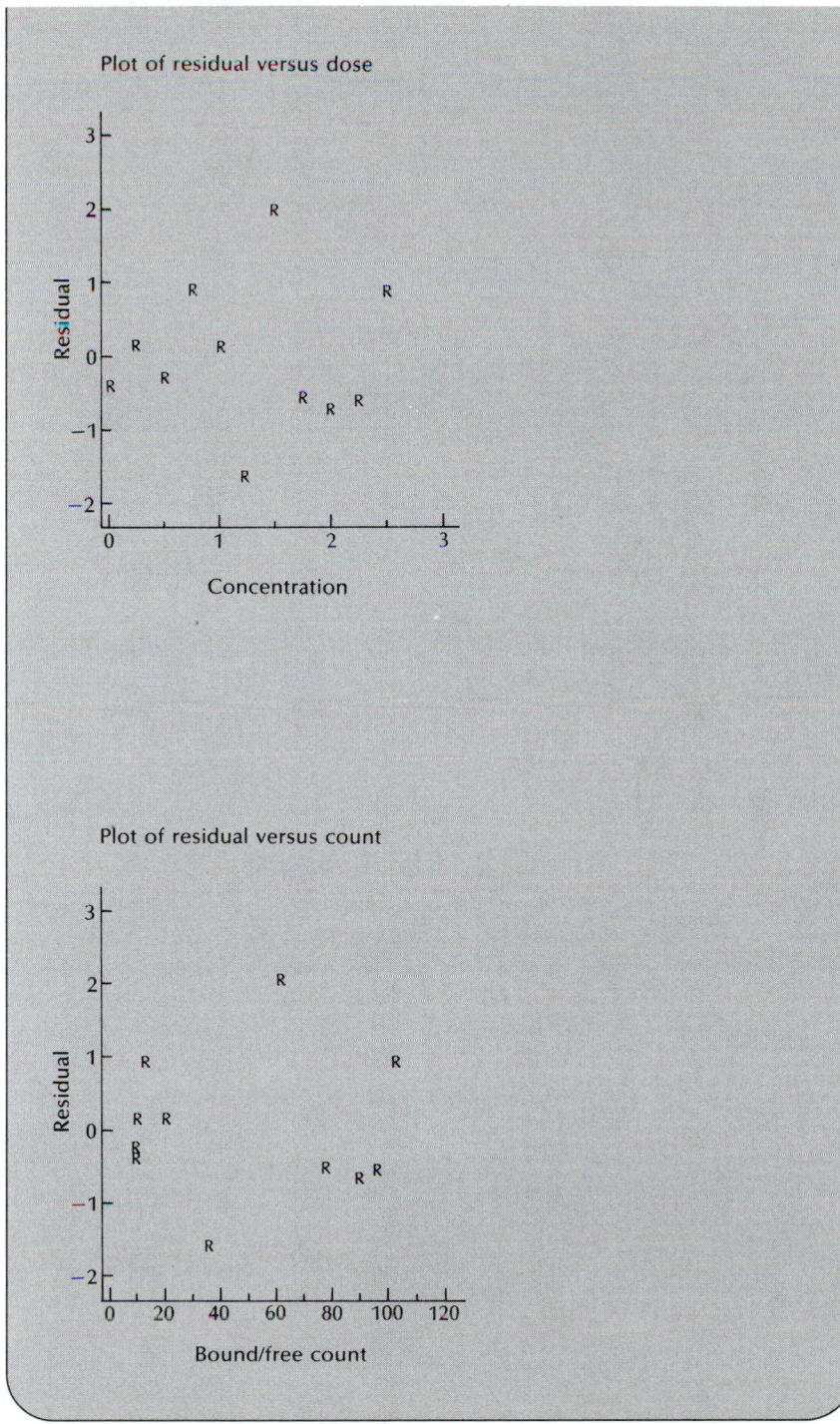

We can also use the fitted equation to predict y (count ratio) based on concentration.   ▲

▼ EXERCISES

Applications

 **12.6** Peak blood level data (in mg/ml) were obtained for 20 patients for a single dose of a drug product. In addition to the peak blood level, the patient's weight (lb) and the amount of drug (mg) were recorded. Use the output shown here to fit a linear regression line and use residual plots to identify possible additional terms to be included in the regression model.

```
OBS     BLOOD    DOSE    WEIGHT
 1       300      1       120
 2       250      1       135
 3       210      1       150
 4       150      1       128
 5       210      2       150
 6       230      2       160
 7       350      2       135
 8       270      2       180
 9       380      4       132
10       330      4       148
11       270      4       190
12       240      4       195
13       340      8       150
14       330      8       160
15       180      8       200
16       320      8       140
17       270     16       195
18       290     16       170
19       315     16       161
20       350     16       145
```

GENERAL LINEAR MODELS PROCEDURE

DEPENDENT VARIABLE: BLOOD   PEAK BLOOD LEVEL

SOURCE	DF	SUM OF SQUARES	MEAN SQUARE	F VALUE	PR > F	R-SQUARE	C.V.
MODEL	2	22290.44079456	11145.22039728	3.68	0.0468	0.302392	19.6953
ERROR	17	51423.30920544	3024.90054150		STD DEV		BLOOD MEAN
CORRECTED TOTAL	19	73713.75000000			54.99909582		279.25000000

SOURCE	DF	TYPE I SS	F VALUE	PR > F	DF	TYPE IV SS	F VALUE	PR > F
DOSE	1	8452.70329301	2.79	0.1129	1	16119.33338339	5.33	0.0338
WEIGHT	1	13837.73750155	4.57	0.0473	1	13837.73750155	4.57	0.0473

PARAMETER	ESTIMATE	T FOR H0: PARAMETER=0	PR > |T|	STD ERROR OF ESTIMATE
INTERCEPT	432.60229380	5.11	0.0001	84.69454320
DOSE	5.54666579	2.31	0.0338	2.40278001
WEIGHT	-1.19428513	-2.14	0.0473	0.55838151

(continues)

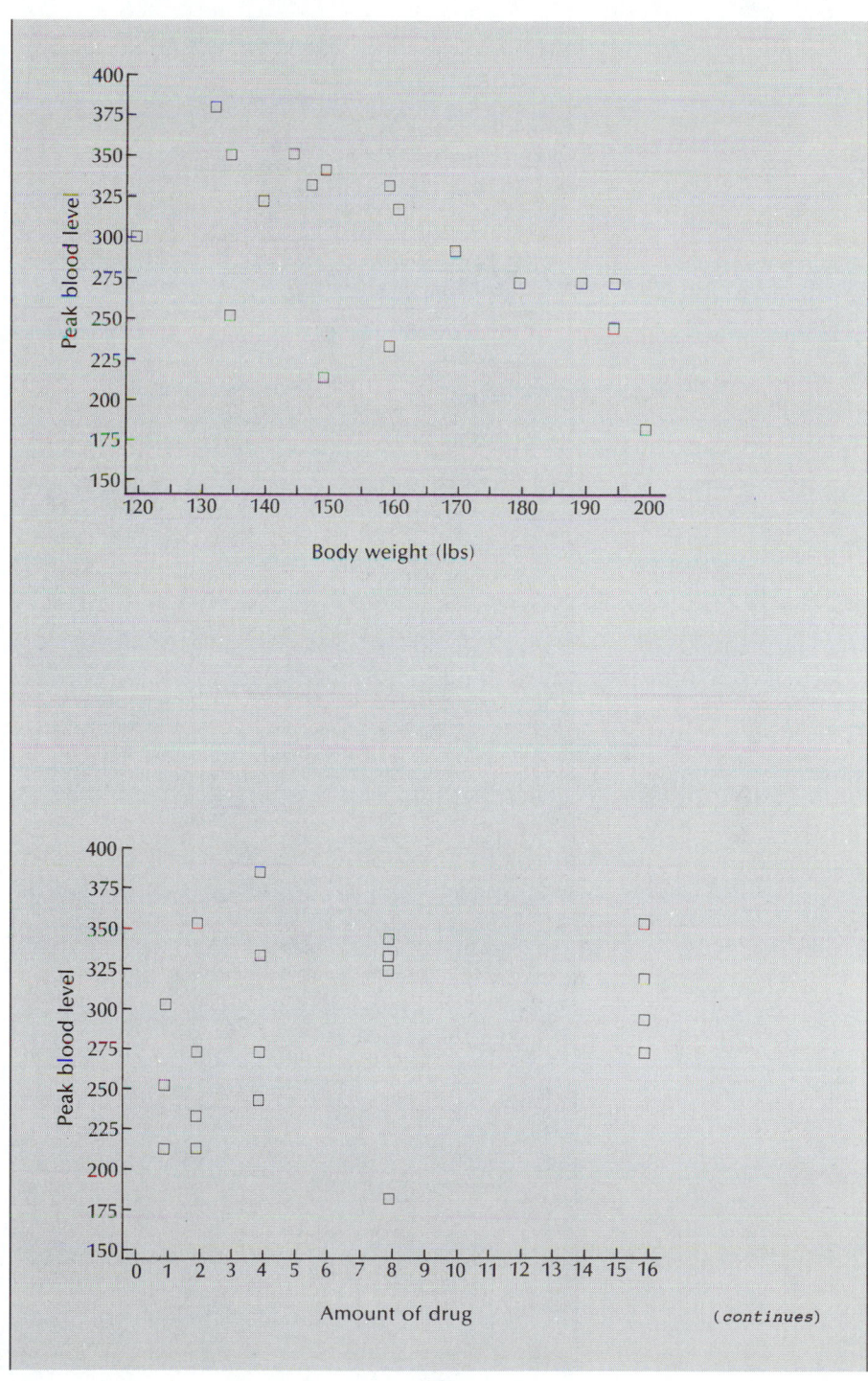

(*continues*)

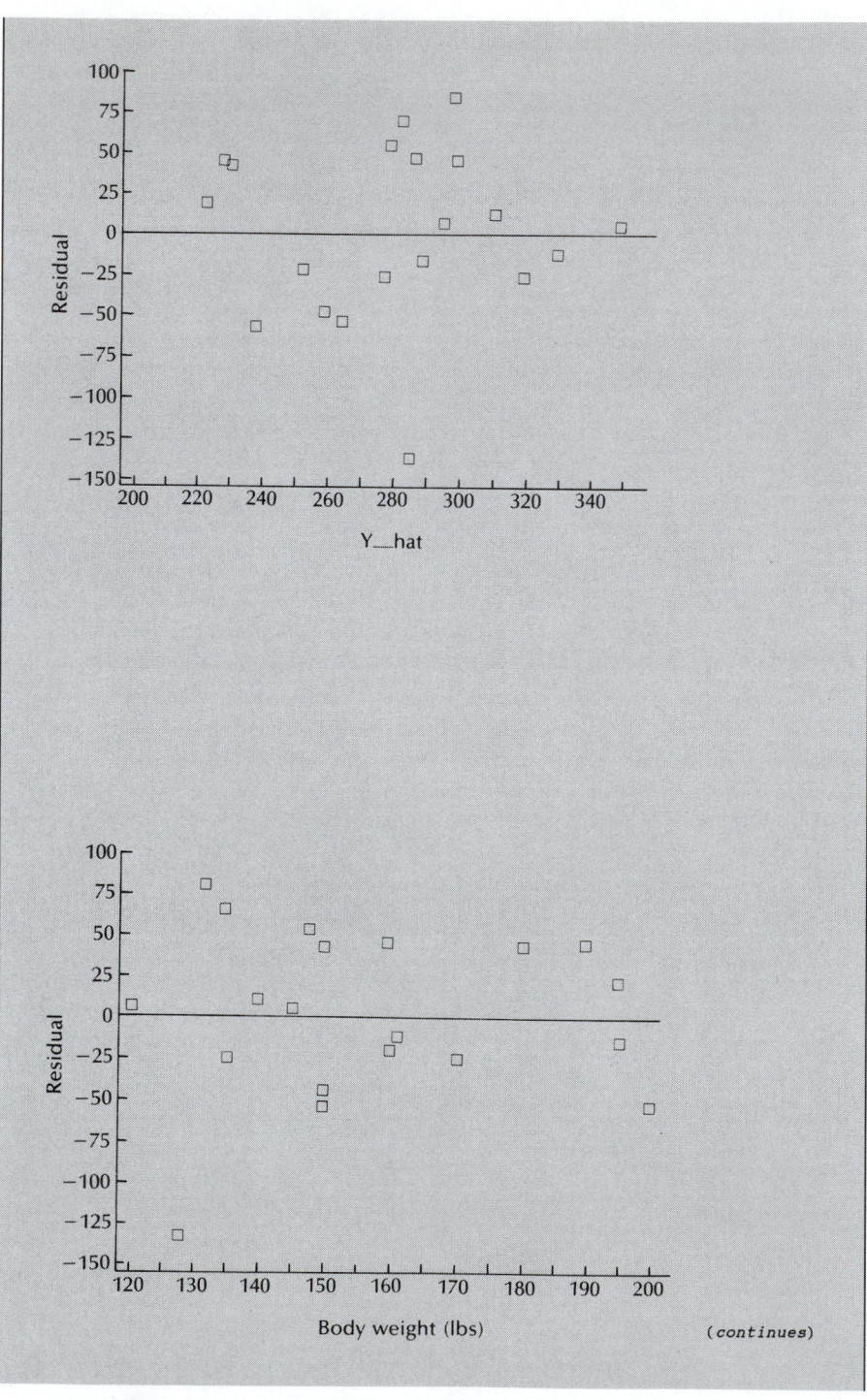

(*continues*)

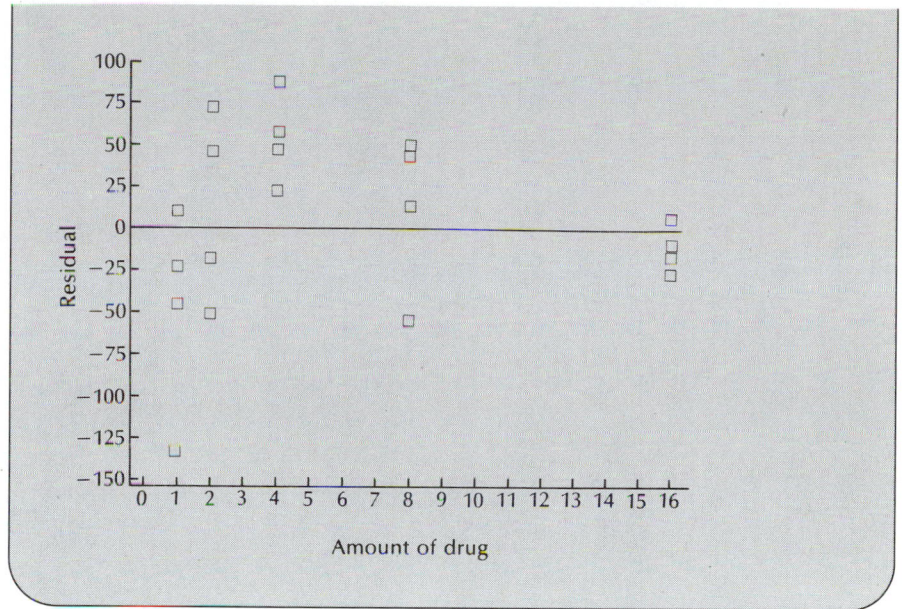

12.7 Refer to Exercise 12.6. Identify and discuss the fit of the model for the output shown here.

```
                      GENERAL LINEAR MODELS PROCEDURE

DEPENDENT VARIABLE: BLOOD    PEAK BLOOD LEVEL

SOURCE          DF   SUM OF SQUARES    MEAN SQUARE    F VALUE      PR > F      R-SQUARE      C.V.
MODEL            3   41167.19623371   13722.39874457    6.75       0.0038      0.558474     16.1510
ERROR           16   32546.55376629    2034.15961039                STD DEV                BLOOD MEAN
CORRECTED TOTAL 19   73713.75000000                               45.10165862            279.25000000

SOURCE          DF     TYPE I SS      F VALUE    PR > F   DF      TYPE IV SS    F VALUE    PR > F
LOG_DOSE         1   13690.00000000    6.73      0.0196    1    9899.32954861    4.87      0.0423
WEIGHT           1   21798.10707872   10.72      0.0048    1     309.92533215    0.15      0.7014
LOG_DOSE*WEIGHT  1    5679.08915499    2.79      0.1142    1    5679.08915499    2.79      0.1142


                                     T FOR HO:                        STD ERROR OF
PARAMETER            ESTIMATE       PARAMETER=0      PR > |T|           ESTIMATE
INTERCEPT          288.06239408        2.25          0.0390          128.09498236
LOG_DOSE           402.52746916        2.21          0.0423          182.46734172
WEIGHT              -0.34416249       -0.39          0.7014            0.88171355
LOG_DOSE*WEIGHT     -1.98696306       -1.67          0.1142            1.18916731

                                                                      (continues)
```

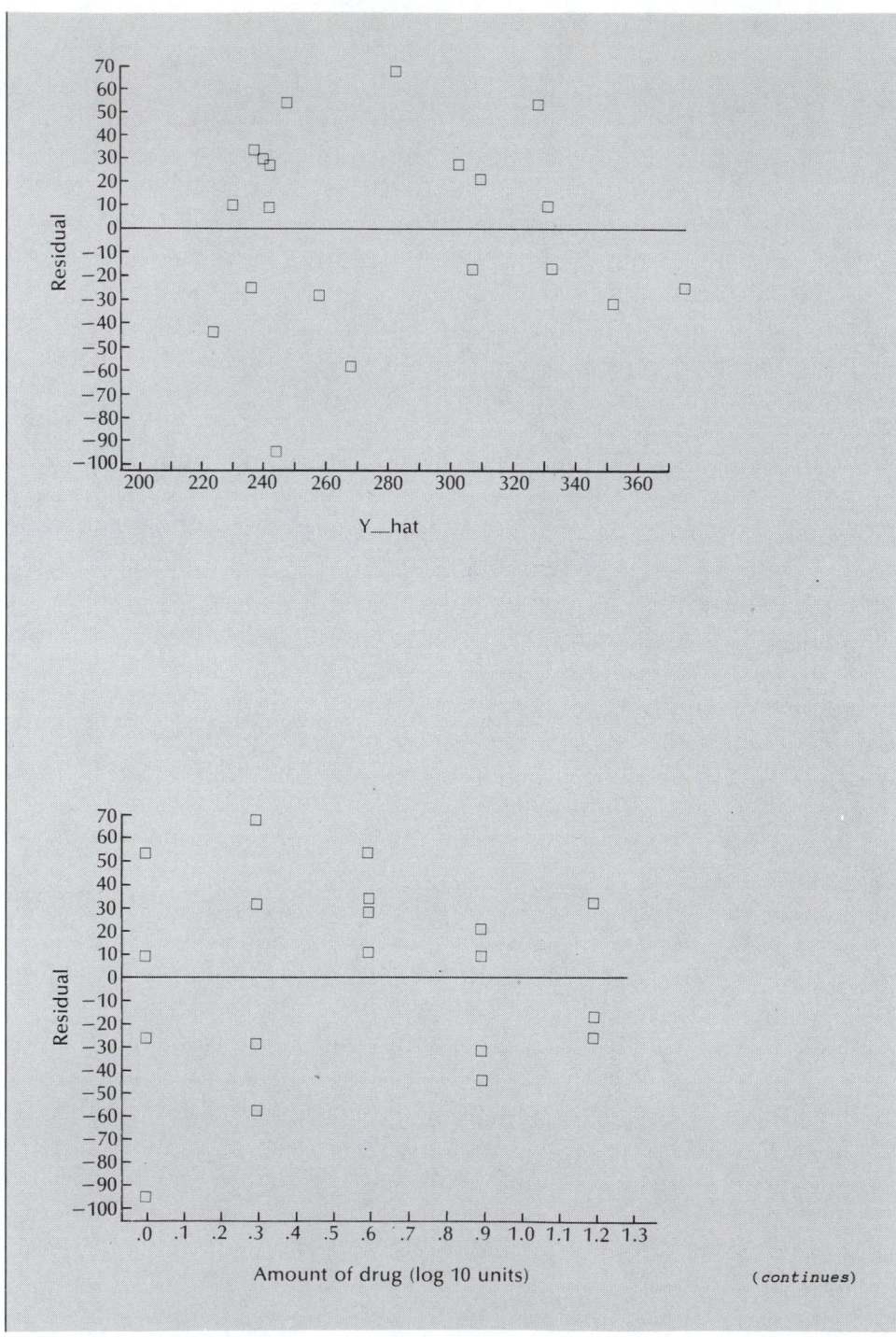

(continues)

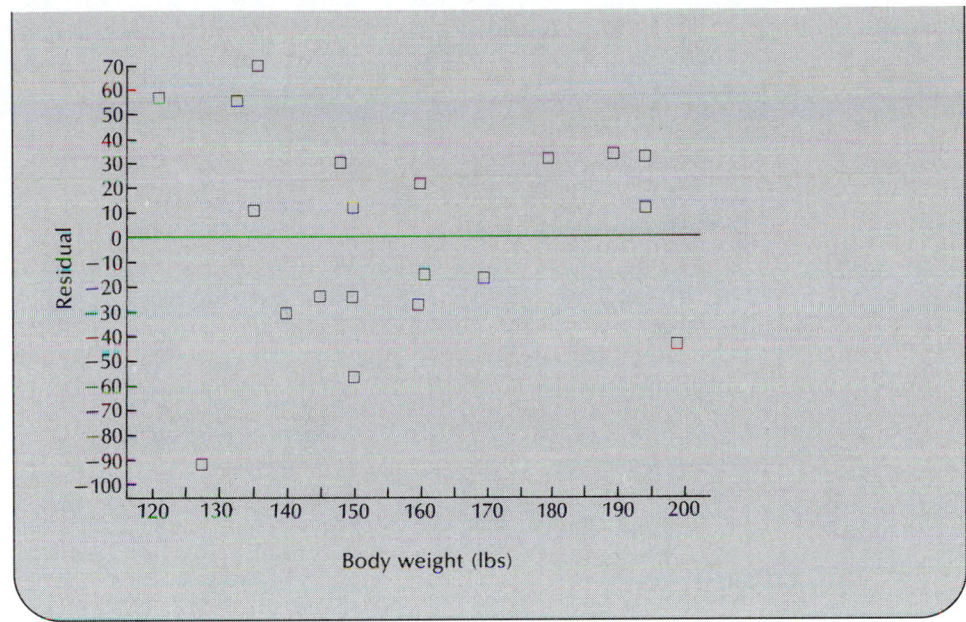

12.8 (Optional) Several different forms of nonlinear models are useful in predicting the plasma concentration from single or multiple doses of a drug product. The simplest model is

$$C(t) = \frac{k_a D}{V(k_a - k_e)} (e^{-k_e t} - e^{-k_a t}),$$

where $C(t)$ is the concentration of drug in the body at time t, D/V is the ratio of the amount of drug absorbed to the volume of blood, k_a is the absorption rate of the drug, and k_e is the elimination rate. The data for an experiment consist of concentrations obtained from blood samples at various points in time. The problem is to fit the data to the model to obtain estimates of k_a, k_e, and D/V. The concentration data shown here were fit using NONLIN.

```
                    FUNCTION 1

        X        OBSERVED Y       X       OBSERVED Y
    0.170000      5.10000      5.00000      18.5000
    0.250000     12.1000       6.00000      14.5000
    0.500000    133.000        7.00000      10.8000
    0.750000    127.000        8.00000      10.1000
    1.00000     109.600       10.0000        7.30000
    1.50000      77.0000      12.0000        3.60000
    2.00000      59.1000      16.0000        1.80000
    3.00000      41.3000      24.0000        1.00000
    4.00000      25.2000
                                              (continues)
```

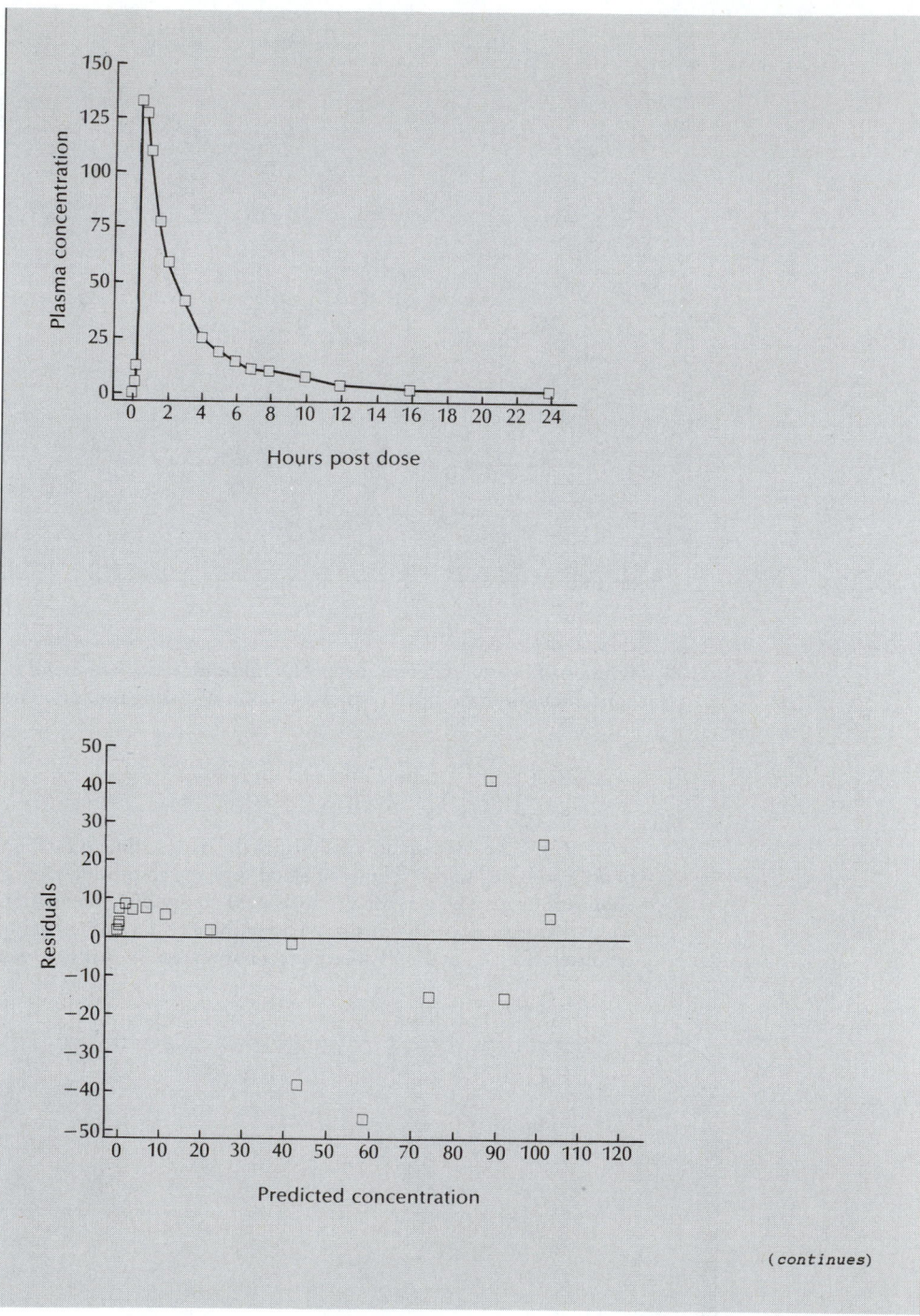

(continues)

```
          FUNCTION 1        ... ARE PREDICTED POINTS,      *** ARE OBSERVED POINTS

133.0
              *
130.3
127.7
125.0
              *
122.4
119.7          .
117.0
114.4
111.7
109.1
              *
106.4
103.7          .
101.1          .
98.42
95.76          .   .
93.10
90.44
87.78          .
85.12
82.46
79.80       .     .
77.14
74.48
              *
71.82  .
69.16          .
66.50
63.84
61.18          .
58.52
              *
55.86
53.20          .
50.54
47.88          .
45.22
42.56       .
39.90          .
                 *
37.24
34.58          .
31.92
29.26
26.60
23.94          .
                 *
21.28          .
18.62            .
15.96             .

13.30              *
                    .
                      *
10.64                 .
                          *
7.980        *            ..
                              *
5.320                  ...    .
                                 *
2.660                   ....
0.3993E-04                .....................................................
      0.1700   2.553   4.936   7.319   9.702   12.08   14.47   16.85   19.23   21.62   24.00

                                                                    (continues)
```

```
                        SUMMARY OF NONLINEAR ESTIMATION

AFTER 25 ITERATIONS THE ESTIMATES AND THEIR VARIABILITY ARE:

          NO.      ESTIMATE      STD. DEV.              95% CONFIDENCE LIMITS
 ^
 KA        1        1.60860       1.26013        -1.09411        4.31131     UNIVAR
                                                 -2.44040        5.65760     S PLANE
 ^
 KE        2        0.663720      0.465020       -0.333646       1.66109     UNIVAR
                                                 -0.830461       2.15790     S PLANE
 ^
 D/V       3        194.207       109.909        -41.5237        429.938     UNIVAR
                                                 -158.947        547.362     S PLANE


PREDICTED CONCENTRATION AT T = 9 IS 0.8414
HALF-LIFE KA = 0.4309
HALF-LIFE K = 1.0443
```

Give estimates of the model parameters. Predict concentration when $t = 3.0$. Does this model seem to fit the data? (Note: Estimates of the model parameter are given in the order $\hat{k}_a$, $\hat{k}_e$, and D/V.)

12.9 (Optional) Refer to Exercise 12.8. A second, more complicated model may be appropriate for the data of the previous exercise. The model is

$$C(t) = k_a \frac{D}{V} \left[\frac{\alpha - k_{21}}{(\alpha - \beta)(k_a - \alpha)} e^{-\alpha t} + \frac{k_{21} - \beta}{(\alpha - \beta)(k_a - \beta)} e^{-\beta t} + \frac{k_a - k_{21}}{(k_a - \alpha)(k_a - \beta)} e^{-k_a t} \right],$$

where $\alpha + \beta = k_e + k_{12} + k_{21}$ and $\alpha\beta = k_e k_{21}$. Use the output shown here to determine whether this model provides a better fit to the sample concentration data. (Note: The estimates are shown in the order $\hat{k}_a$, $\hat{\alpha}$, $\hat{\beta}$, $\hat{k}_{21}$, and D/V.)

```
                        SUMMARY OF NONLINEAR ESTIMATION

AFTER 13 ITERATIONS THE ESTIMATES AND THEIR VARIABILITY ARE:

    NO.      ESTIMATE      STD. DEV.              95% CONFIDENCE LIMITS
    1        6.83567       26.5429        -50.9965         64.6679      UNIVAR
                                          -98.5367         112.208      S PLANE
    2        4.19859       3.06471        -2.47885         10.8760      UNIVAR
                                          -7.96796         16.3651      S PLANE
    3        0.556948      0.140483       0.250861         0.863036     UNIVAR
                                          -0.754321E - 03  1.11465      S PLANE
    4        11.5141       150.713        -316.862         339.890      UNIVAR
                                          -586.799         609.827      S PLANE
    5        62.8093       878.265        -1850.77         1976.39      UNIVAR
                                          -3423.80         3549.42      S PLANE

PREDICTED AT T = 9 IS 1.3690
HALF-LIFE KA = 0.1014
HALF-LIFE B = 1.2445
K1 = -0.283 K2 = -0.762 K3 = 0.479

                                                        (continues)
```

FUNCTION 1 ... ARE PREDICTED POINTS, ***ARE OBSERVED POINTS

(continues)

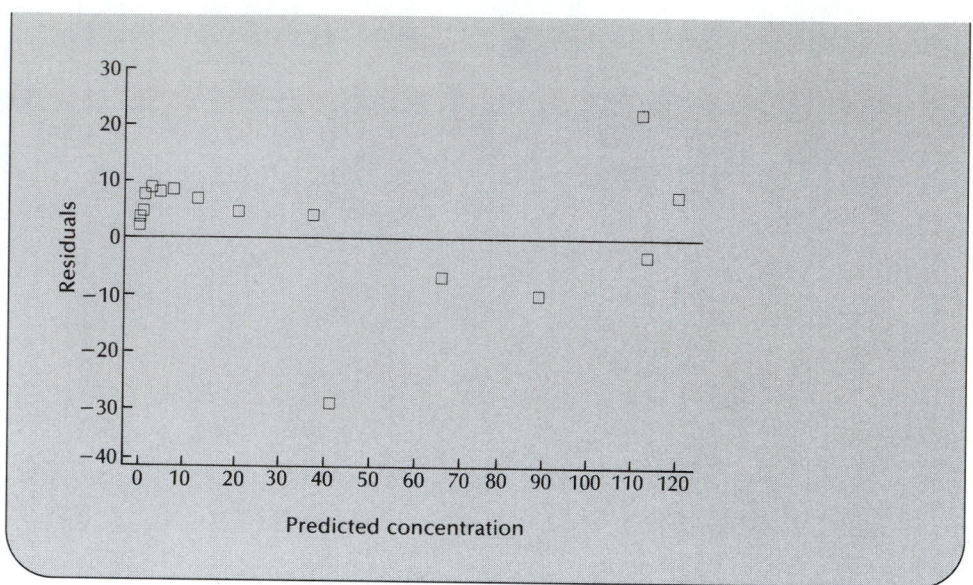

12.4 ▼ RESIDUAL ANALYSIS: CHECKING MODEL ASSUMPTIONS (STEP 3)

Now that we have identified possible independent variables (step 1) and considered the form of the multiple regression model (step 2), we should check whether the assumptions underlying the chosen model are valid. Recall that in Chapter 11 we indicated that the basic assumptions for a regression model of the form

$$y_i = \beta_0 + \beta_1 x_{i1} + \beta_2 x_{i2} + \cdots + \beta_k x_{ik} + \varepsilon_i$$

are

1. Zero expectation: $E(\varepsilon_i) = 0$ for all i.
2. Constant variance: $V(\varepsilon_i) = \sigma_\varepsilon^2$ for all i.
3. Normality: ε_i is normally distributed.
4. Independence: The ε_i are independent.

Note that since the assumptions for multiple regression are written in terms of the random errors ε_i, it would seem reasonable to check the assumptions by using the residuals $y_i - \hat{y}_i$, which are *estimates* of the ε_i.

The first assumption, zero expectation, deals with model selection and whether additional independent variables need to be included in the model. If we have done our job in steps 1 and 2, Assumption 1 should hold. The use of residual plots to check for inadequacy (lack of fit) of the model was discussed briefly in Chapter 9 and again in Section 12.3.

The assumptions of constant variance can be examined using residual plots. One of the simplest residual plots for detecting nonconstant variance is a plot of the residuals versus the predicted values, $\hat{y}_i$. Most of the available statistical software systems can provide these plots as part of the regression analysis.

E X A M P L E   12.11　▼

The data that are shown in Table 12.5 were fit to the model $y = \beta_0 + \beta_1 x + \beta_2 x^2 + \varepsilon$ using SAS. Examine the plot residuals versus $\hat{y}_i$ to detect possible nonconstant variance. Can you identify a pattern of non-constant variance?

T A B L E   12.5
Data for Example 12.11

y	11	10	2	14	22	10	20	19	32	23	40	37
x	.5	1	1.2	1.4	1.7	1.8	2	2.3	2.5	2.8	3	3.1
y	30	43	55	29	45	60	53	30	42	25	63	51
x	3.5	3.6	3.8	4.2	4.4	5.1	5.2	5.4	5.5	6	6.2	6.3

Solution　As can be seen from the SAS residual plot, the magnitudes of the residuals are generally increasing with the magnitudes of the predicted values of y, suggesting possible nonconstant variance. And, since y_i is directly related to x via the regression model (i.e., y increases with x), the residuals are increasing with the magnitude of the xs. This pattern in the residuals suggests that the variance of ε_i (and hence $V(y_i)$) is increasing with x. The accompanying plot of y versus x tends to bear this out.

```
LISTING OF DATA
OBS    Y      X
  1    11    0.5
  2    10    1.0
  3     2    1.2
  4    14    1.4
  5    22    1.7
  6    10    1.8
  7    20    2.0
  8    19    2.3
```

(*continues*)

```
 9    32    2.5
10    23    2.8
11    40    3.0
12    37    3.1
13    30    3.5
14    43    3.6
15    55    3.8
16    29    4.2
17    45    4.4
18    60    5.1
19    53    5.2
20    30    5.4
21    42    5.5
22    25    6.0
23    63    6.2
24    51    6.3
N =   24
```

QUADRATIC REGRESSION

GENERAL LINEAR MODELS PROCEDURE

DEPENDENT VARIABLE: Y

SOURCE	DF	SUM OF SQUARES	MEAN SQUARE	VALUE	PR > F	R-SQUARE	C.V.
MODEL	2	4458.40551726	2229.20275863	20.45	0.001	0.660717	32.7142
ERROR	21	2289.42781607	109.02037219		ROOT MSE		Y MEAN
CORRECTED TOTAL	23	6747.83333333			10.44128211		31.91666667

SOURCE	DF	TYPE I SS	F VALUE	PR > F	DF	TYPE III SS	F VALUE	PR> F
X	1	4141.99721230	37.99	0.0001	1	970.14118949	8.90	0.0071
X*X	1	316.40830496	2.90	0.1032	1	316.40830496	2.90	0.1032

PARAMETER	ESTIMATE	T FOR HO: PARAMETER=0	PR > |T|	STD ERROR OF ESTIMATE
INTERCEPT	-6.87174737	-0.77	0.4472	8.87156858
X	17.10536096	2.98	0.0071	5.73414408
X*X	-1.34903610	-1.70	0.1032	0.79186923

Plot of Y*X

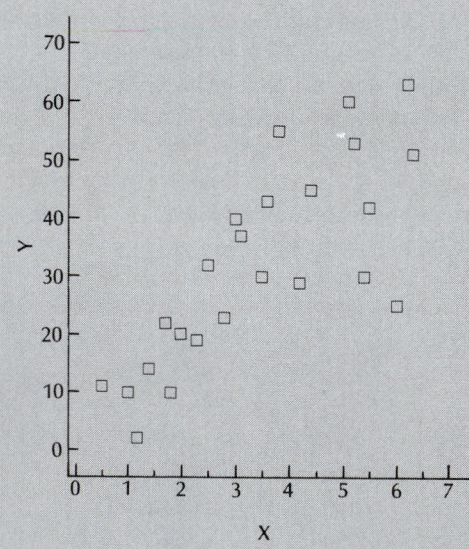

(continues)

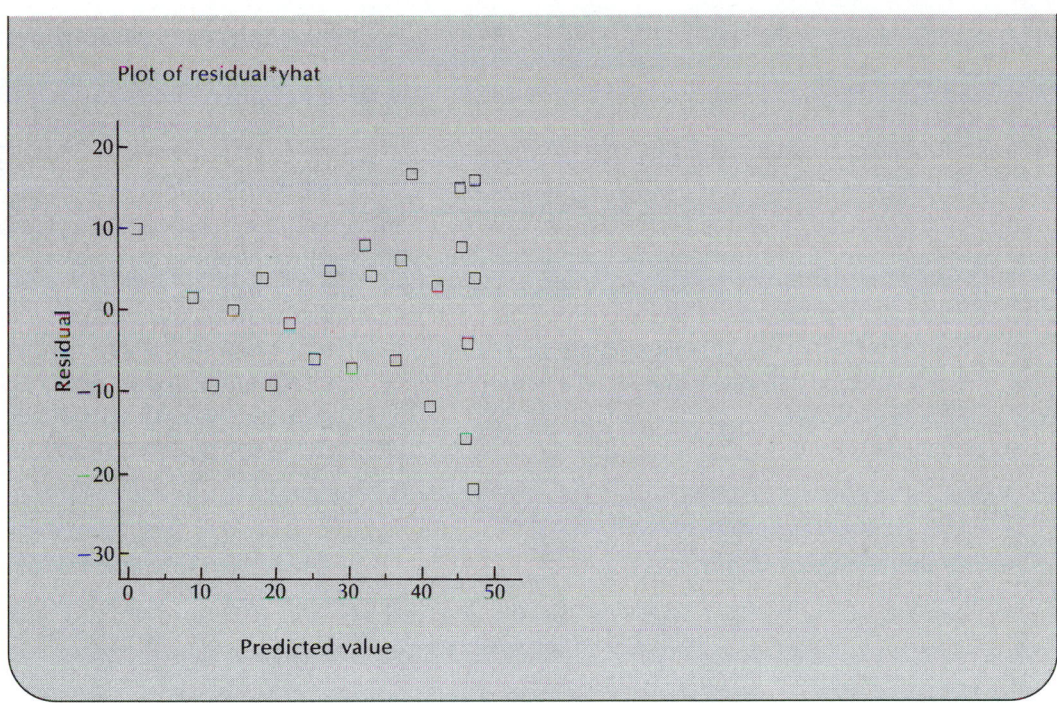

Plot of residual*yhat

What are the consequences of having a nonconstant variance problem in a regression model? First, if the variance about the regression line is not constant, the least squares estimates may not be as accurate as possible. A technique called **weighted least squares** (see Draper and Smith (1981)) will give more accuracy. but, perhaps more important, the weighted least squares technique improves the statistical tests (F and t tests) on model parameters and the interval estimates for parameter because they are, in general, based on smaller standard errors.

The more serious pitfall involved with inferences in the presence of nonconstant variance seems to be for estimates $E(y)$ and predictions of y. For these inferences, the point estimate y is sound but the width of the interval may be too large or too small depending on whether we're predicting in a low or high variance section of the experimental region.

The remedy for nonconstant variance seems to be weighted least squares. We will not cover this technique in the text. However, when the nonconstant variance possesses a pattern related to y, a reexpression (transformation) of y may resolve the problem, accomplishing the same end as weighted least squares. Several transformations for y were discussed in Chapter 9; ones that help to stabilize the variance when there is a pattern to the nonconstant variance will be discussed in

weighted least squares

Chapter 13 for the analysis of variance. They can also be applied in certain regression situations, as well.

EXAMPLE 12.12   ▼

Refer to the data of Example 12.11, where we detected a nonconstant variance problem. Since the variance about the regression line seemed to increase with x, a square root transformation on y was tried to stabilize the variance. Examine the computer output and residual plot shown here to determine whether the nonconstant variance problem has been eliminated. The student edition of EXECUSTAT produced the following analysis.

```
Row    y         x          sqrty
1      11        0.5        3.31662
2      10        1          3.16228
3      2         1.2        1.41421
4      14        1.4        3.74166
5      22        1.7        4.69042
6      10        1.8        3.16228
7      20        2          4.47214
8      19        2.3        4.3589
9      32        2.5        5.65685
10     23        2.8        4.79583
11     40        3          6.32456
12     37        3.1        6.08276
13     30        3.5        5.47723
14     43        3.6        6.55744
15     55        3.8        7.4162
16     29        4.2        5.38516
17     45        4.4        6.7082
18     60        5.1        7.74597
19     53        5.2        7.28011
20     30        5.4        5.47723
21     42        5.5        6.48074
22     25        6          5
23     63        6.2        7.93725
24     51        6.3        7.14143
```

```
                Polynomial Regression Analysis for SQRTY
.....................................................................
eeeeeeeeeeeeeeeeeeeeeeeeeeeeeeeeeeeeeeeeeeeeeeeeeeeeeeeeeeeeeeeeeeeeeee
Dependent variable: sqrty

                            Table of Estimates
.....................................................................
eeeeeeeeeeeeeeeeeeeeeeeeeeeeeeeeeeeeeeeeeeeeeeeeeeeeeeeeeeeeeeeeeeeeeee
                              Standard            t           P
                 Estimate     Error          Value       Value
.....................................................................
eeeeeeeeeeeeeeeeeeeeeeeeeeeeeeeeeeeeeeeeeeeeeeeeeeeeeeeeeeeeeeeeeeeeeee
Intercept        1.18979      0.811301          1.47        0.157
x                1.99022      0.524385          3.80        0.001
x 2             -0.176856     0.0724161        -2.44        0.023
.....................................................................
eeeeeeeeeeeeeeeeeeeeeeeeeeeeeeeeeeeeeeeeeeeeeeeeeeeeeeeeeeeeeeeeeeeeeee
R-squared = 70.16%
Adjusted R-squared = 67.31%
Standard error of  estimation = 0.95485
Durbin-Watson statistic = 2.29858
Mean absolute error = 0.7439
.....................................................................
eeeeeeeeeeeeeeeeeeeeeeeeeeeeeeeeeeeeeeeeeeeeeeeeeeeeeeeeeeeeeeeeeeeeeee
                                                      (continues)
```

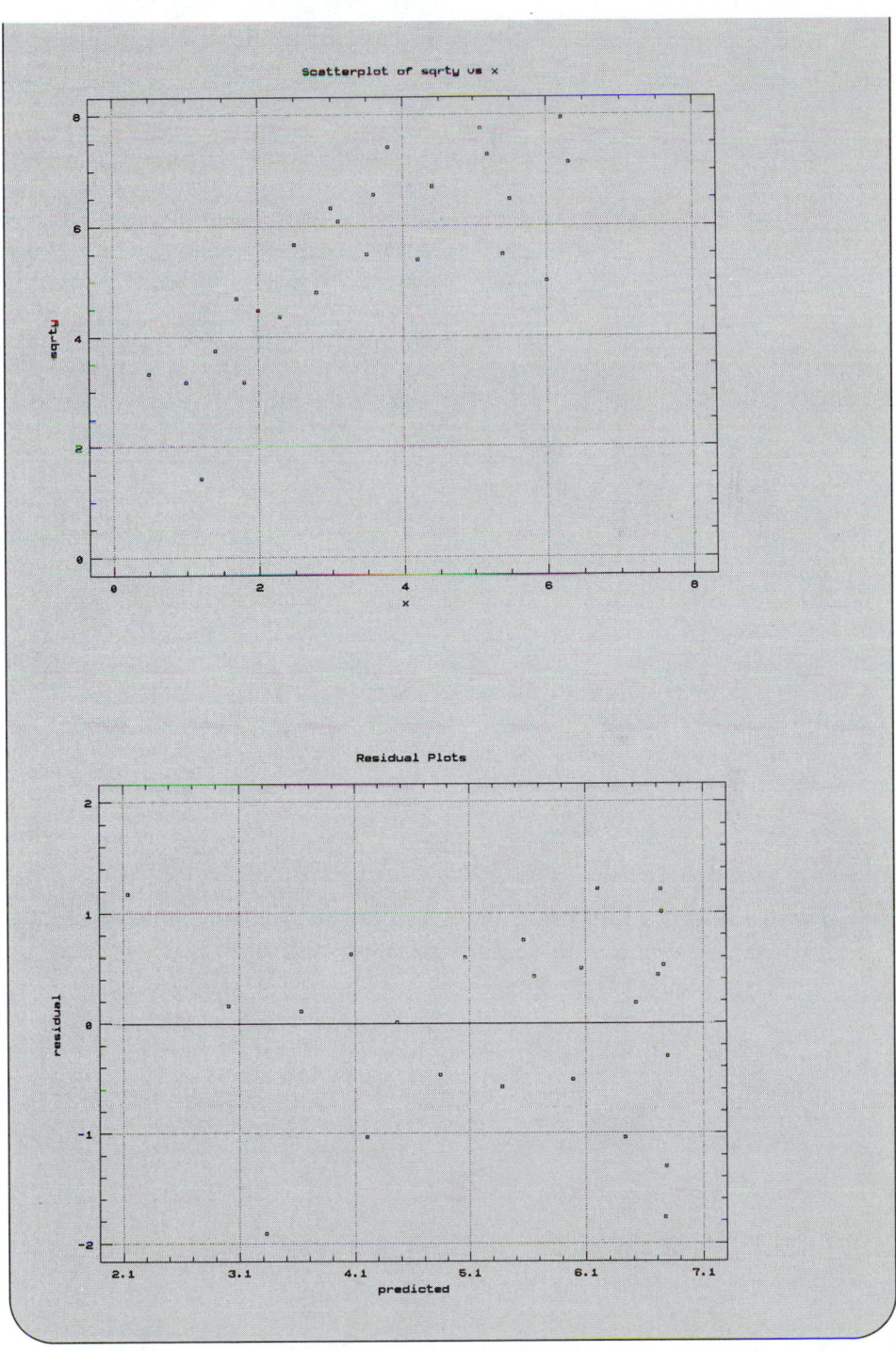

Solution The output shown here documents that this model provides a much better fit to the sample data; note, especially, the residual plot.   ▲

The third assumption for multiple regression is that of normality of the ε_i. Skewness and/or outliers are examples of forms of nonnormality that may be detected through the use of certain scatterplots and residual plots.

A plot of the residuals in the form of a histogram or a stem-and-leaf plot will help to detect skewness. By assumption, the ε_i are normally distributed with mean 0. If a histogram of the residuals is not symmetrical about 0, some skewness is present. For example, the residual plot in Figure 12.1(a) is symmetrical on 0 and suggests no skewness. In contrast, the residual plot in Figure 12.1(b) is skewed to the right.

probability plot

Another way to detect nonnormality is through the use of a normal **probability plot** of the residuals. The idea behind the plot is that if the residuals are normally distributed, the normal probability plot will be approximately a straight line.

F I G U R E 12.1
Top: Residuals Centered on Zero; Bottom: Residuals Skewed to Right

(a) Middle of
interval   Number of observations

−2.0	3	× × ×
−1.5	10	× × × × × × × × × ×
−1.0	16	× × × × × × × × × × × × × × × ×
−0.5	15	× × × × × × × × × × × × × × ×
0.0	20	× × × × × × × × × × × × × × × × × × × ×
0.5	15	× × × × × × × × × × × × × × ×
1.0	11	× × × × × × × × × × ×
1.5	6	× × × × × ×
2.0	3	× × ×
2.5	0	
3.0	1	×

(b) Middle of
interval   Number of observations

−2.0	3	× × ×
−1.5	10	× × × × × × × × × ×
−1.0	16	× × × × × × × × × × × × × × × ×
−0.5	15	× × × × × × × × × × × × × × ×
0.0	18	× × × × × × × × × × × × × × × × × ×
0.5	12	× × × × × × × × × × × ×
1.0	7	× × × × × × ×
1.5	5	× × × × ×
2.0	3	× × ×
2.5	3	× × ×
3.0	2	× ×
3.5	0	
4.0	1	×

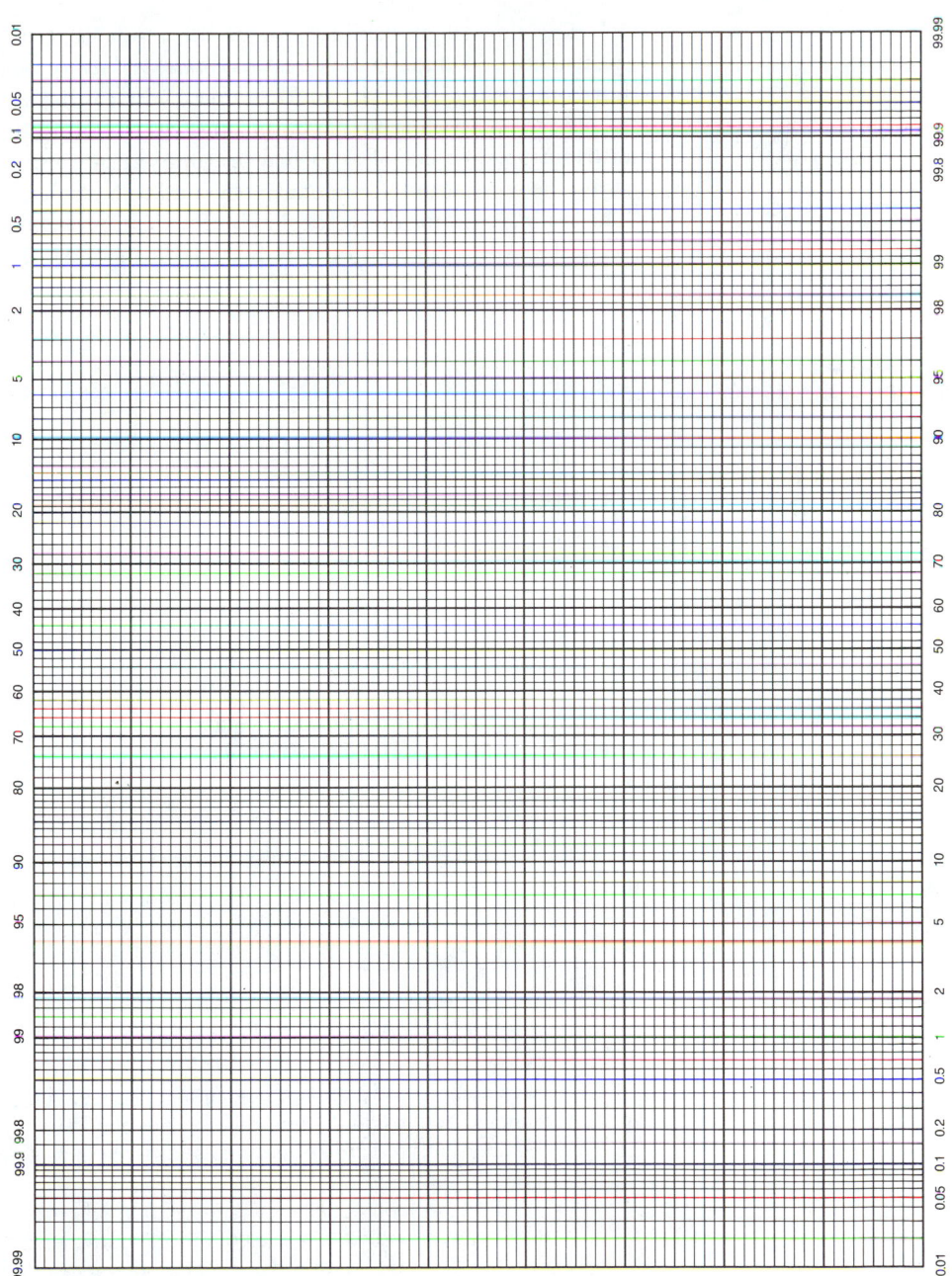

FIGURE 12.2
Normal Probability Paper

The normal probability plot of the residuals is simple to construct using specially designed graph paper called *normal probability paper* (see Figure 12.2). The horizontal axis is scaled in uniform units, whereas the vertical axis goes from .01 to .99, corresponding to the cumulative distribution function (cdf) of a normal distribution. The cdf of a normal distribution plot using uniformly scaled units along the vertical and horizontal axes would be S-shaped, as shown in Figure 12.3. The scaling used for the vertical axis of normal probability paper "stretches" the vertical axis to straighten out the S-shape of the normal cdf.

FIGURE 12.3
S-Shape of the Normal cdf

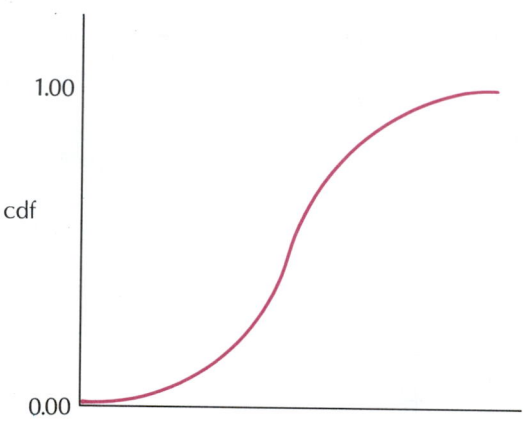

The steps we follow in constructing a normal probability plot are listed next.

**Constructing a Normal
Probability Plot**

1. Compute the residuals; divide each one by $s_\varepsilon = \sqrt{MSE}$ to standardize it. (Note: The standardized residuals from a normal distribution should have mean 0 and standard deviation 1; hence, almost all of the standardized residuals should lie between -3 and $+3$.)
2. Number the standardized residuals from the largest negative (1) to the largest positive (n).
3. Label the horizontal axis of a piece of normal probability paper from -5 to $+5$ corresponding to values of the standardized residuals.
4. Plot the value of the ith ordered standardized residual on the x axis and $(i - .5)/n$ along the y axis.
5. If the plot is nearly linear, it suggests the normality assumption is not violated.

Most regression software packages offer an option for generating probability plots. This is illustrated in Example 12.13.

E X A M P L E   12.13   ▼

Refer to the data of Example 12.11. Use the computer output shown here to determine whether there is evidence of nonnormality.

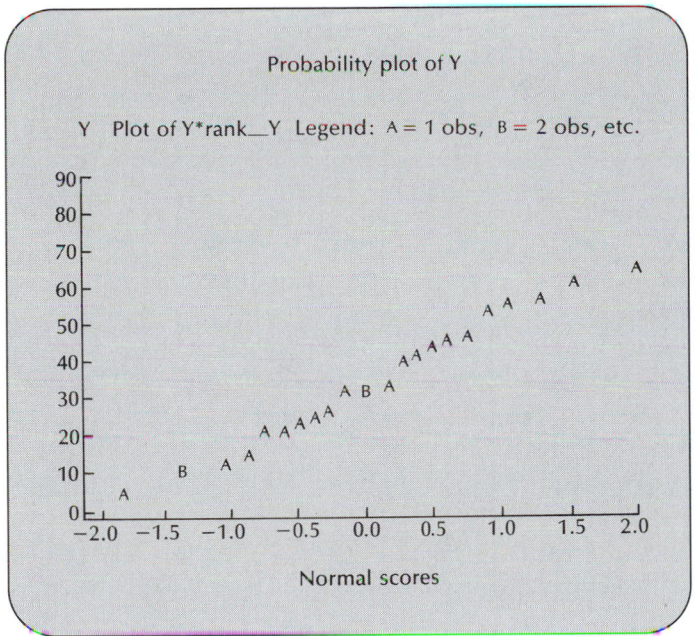

Solution   The fact that the probability plot is nearly linear suggests that the normality assumption has not been violated for the data of Example 12.11.

▲

The presence of one or more outliers is perhaps a more subtle form of nonnormality that may be detected by using a scatterplot and one or more residual plots. For the linear regression model $y = \beta_0 + \beta_1 x + \varepsilon$, a scatterplot of y versus x will help detect the presence of an outlier. This is shown in Table 12.6 and Figure 12.4. It certainly appears that the circled data point is an outlier.

TABLE 12.6
Listing of Data

Obs	x	y
1	10	120
2	20	115
3	21	250
4	27	210
5	29	300
6	33	330
7	40	295
8	44	400
9	52	380
10	56	460
11	62	125
12	68	510
N = 12		

FIGURE 12.4
Scatterplot of the Data in Table 12.6

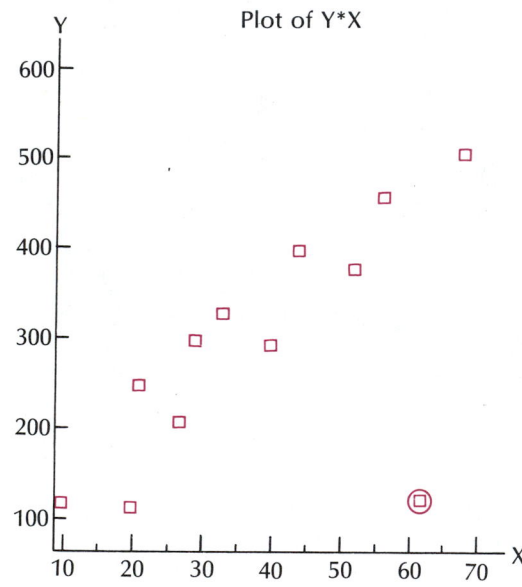

Computer output for a linear fit to the data of Table 12.6 is shown here, along with a residual plot and a normal probability plot. Again the data point corresponding to the suspected outlier (62, 125) is circled in each plot. The student edition of EXECUSTAT produced the following analysis.

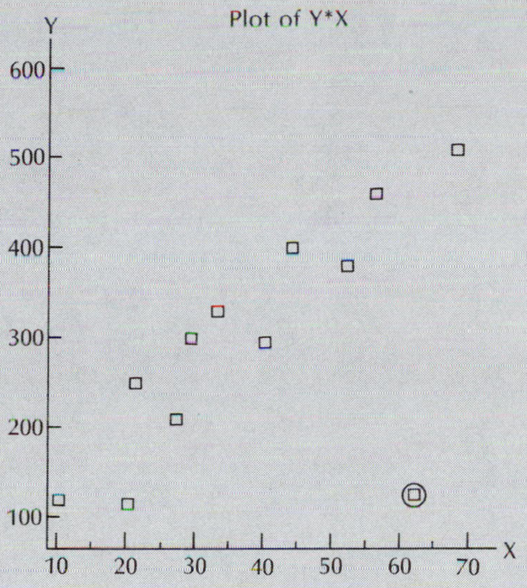

```
        Y              Plot of Y*X
      600

      500

      400

      300

      200

      100

           10  20  30  40  50  60  70      X
```

```
              Simple Regression Analysis for OTT
eeeeeeeeeeeeeeeeeeeeeeeeeeeeeeeeeeeeeeeeeeeeeeeeeeeeeeeeeeeeeeeeeeeeeeeeee
Linear model: y = 114.357 + 4.59461*x

                        Table of Estimates
eeeeeeeeeeeeeeeeeeeeeeeeeeeeeeeeeeeeeeeeeeeeeeeeeeeeeeeeeeeeeeeeeeeeeeeeee
                             Standard          t            P
              Estimate        Error          Value        Value
eeeeeeeeeeeeeeeeeeeeeeeeeeeeeeeeeeeeeeeeeeeeeeeeeeeeeeeeeeeeeeeeeeeeeeeeee
Intercept      114.357        75.5332         1.51         0.1610
Slope          4.59461        1.7868          2.57         0.0278

eeeeeeeeeeeeeeeeeeeeeeeeeeeeeeeeeeeeeeeeeeeeeeeeeeeeeeeeeeeeeeeeeeeeeeeeee
R-squared = 39.80%
Correlation coeff. = 0.631
Standard error of estimation = 108.053
Durbin-Watson statistic = 2.64723
Mean absolute error = 72.8884

eeeeeeeeeeeeeeeeeeeeeeeeeeeeeeeeeeeeeeeeeeeeeeeeeeeeeeeeeeeeeeeeeeeeeeeeee

                       Table of All Residuals
eeeeeeeeeeeeeeeeeeeeeeeeeeeeeeeeeeeeeeeeeeeeeeeeeeeeeeeeeeeeeeeeeeeeeeeeee
                               Predicted                 Studentized
Row        x          y            y        Residual      Residual
eeeeeeeeeeeeeeeeeeeeeeeeeeeeeeeeeeeeeeeeeeeeeeeeeeeeeeeeeeeeeeeeeeeeeeeeee
1          10        120        160.304     -40.3035       -0.43
2          20        115        206.25      -91.2497       -0.92
3          21        250        210.844      39.1557        0.38
4          27        210        238.412     -28.4119       -0.27
5          29        300        247.601      52.3988        0.49
6          33        330        265.98       64.0204        0.60
7          40        295        298.142      -3.14192      -0.03
8          44        400        316.52       83.4796        0.80
9          52        380        353.277      26.7227        0.25
10         56        460        371.656      88.3443        0.89
11         62        125        399.223     -274.223       -6.90
12         68        510        426.791      83.2089        0.93

eeeeeeeeeeeeeeeeeeeeeeeeeeeeeeeeeeeeeeeeeeeeeeeeeeeeeeeeeeeeeeeeeeeeeeeeee
```

(continues)

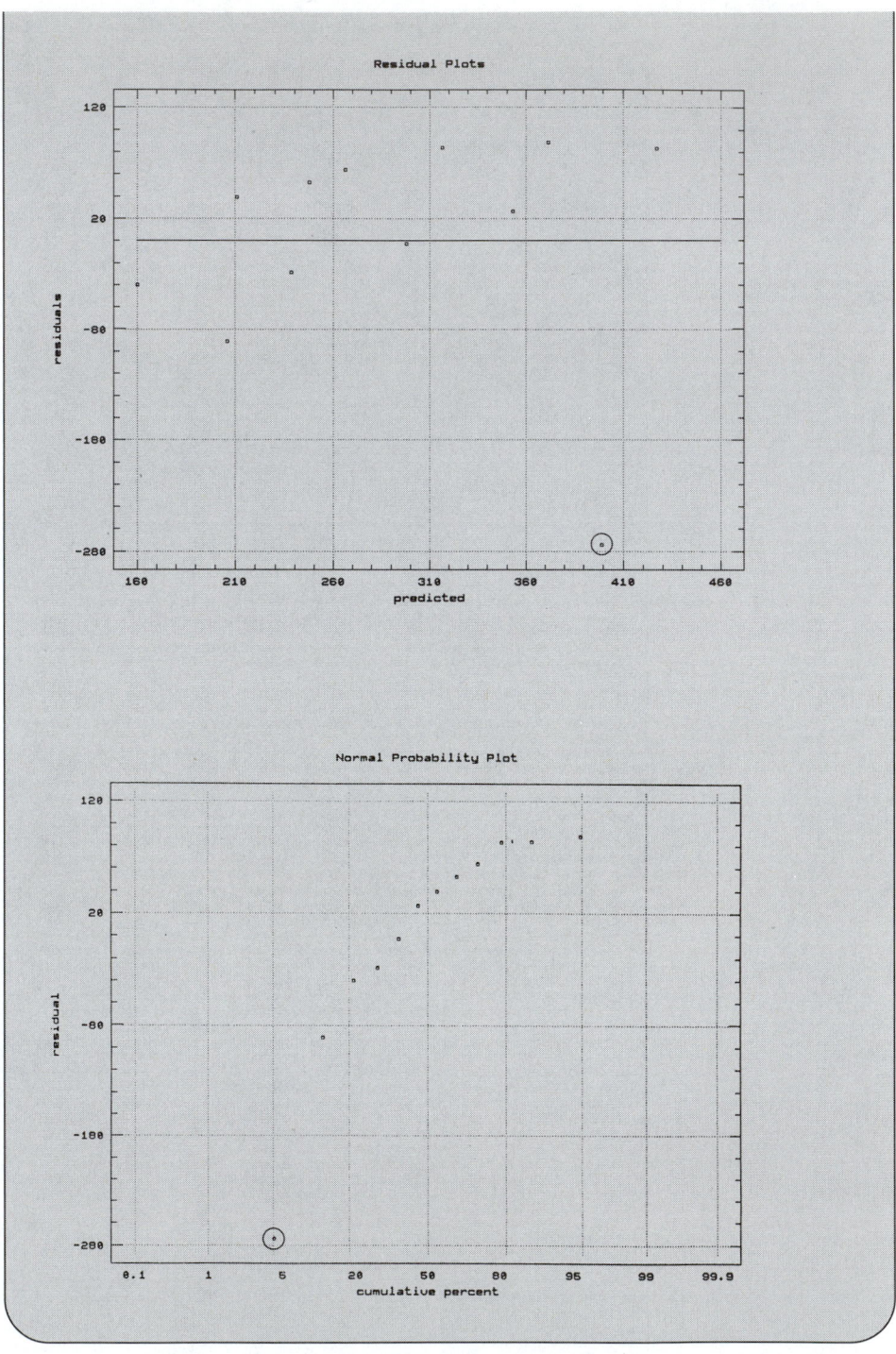

This data set helps to illustrate one of the problems in trying to identify outliers. Sometimes a single plot is not sufficient. For this example, the scatterplot and the probability plot clearly identify the outlier, whereas the residual plot is less conclusive since the outlier adversely affects the linear fit to the data by pulling the fitted line toward the outlier. This makes some of the other residuals larger than they should be. The message is clear: *Don't jump to conclusions without examining the data in several different ways.* The problem becomes even more difficult with multiple regression, where simple scatterplots are not possible.

The final assumption is that the ε_i are statistically independent, and hence uncorrelated. When the time sequence of the observations is known, as is the case with **time series** data, where observations are taken at successive points in time, it is possible to construct a plot of the residuals versus time to observe where the

time series

**FIGURE 12.5
(a) Positive Serial
Correlation; (b) Negative
Serial Correlation; (c) No
Apparent Serial
Correlation**

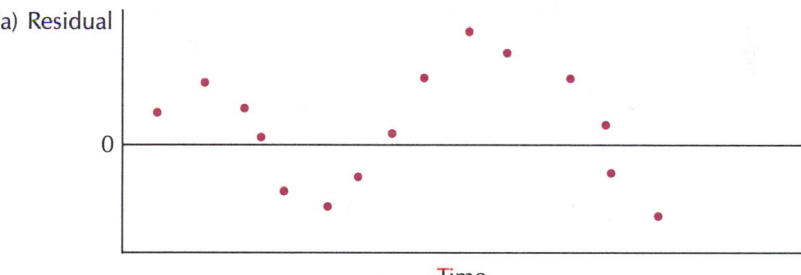

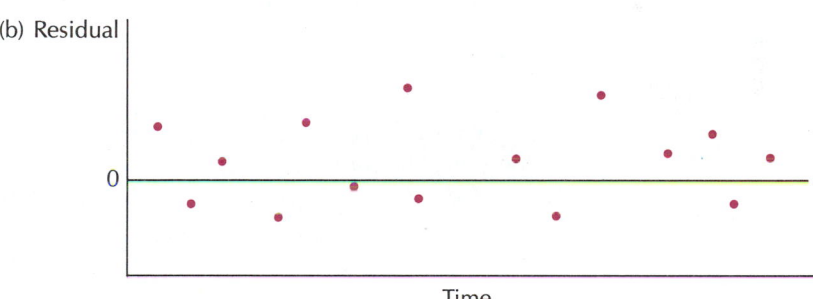

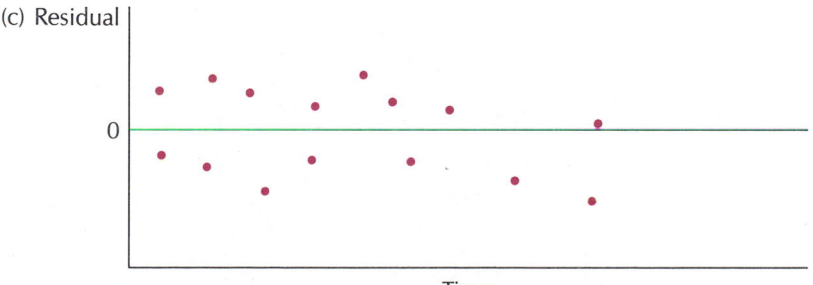

serial correlation

residuals are **serially correlated**. If, for example, there is a positive serial correlation, adjacent residuals (in time) tend to be similar; negative serial correlation implies that adjacent residuals are dissimilar. These patterns of positive and negative serial correlation are displayed in Figures 12.5(a) and 12.5(b), respectively. Figure 12.5(c) shows a residual plot with no apparent serial correlation.

A formal statistical test for serial correlation is based on the *Durbin–Watson statistic*. Let $\hat{\varepsilon}_t$ denote the residual at time t and n the total number of time points.

Durbin–Watson test statistic

Then the **Durbin–Watson test statistic** is

$$d = \frac{\sum_{t=1}^{n-1} (\hat{\varepsilon}_{t+1} - \hat{\varepsilon}_t)^2}{\sum_t \hat{\varepsilon}_t^2}.$$

The logic behind this statistic is as follows: If there is a positive serial correlation, then successive residuals will be similar and their squared difference $(\hat{\varepsilon}_{t+1} - \hat{\varepsilon}_t)^2$ will tend to be smaller than it would be if the residuals were uncorrelated. Similarly, if there is a negative serial correlation among the residuals, the squared difference of successive residuals will tend to be larger than when no correlation exists.

positive and negative serial correlation

When there is no serial correlation, the expected value of the Durbin–Watson test statistic d is approximately 2.0; **positive serial correlation** makes $d < 2.0$ and **negative serial correlation** makes $d > 2.0$. Although critical values of d have been tabulated by J. Durbin and G. S. Watson (1951), values of d less than approximately 1.5 (or greater than approximately 2.5) lead one to suspect positive (or negative) serial correlation.

E X A M P L E   12.14   ▼

Sample data corresponding to retail sales for a particular line of personalized computers by month are shown here.

Month, x	Sales (millions of dollars), y
1	6.0
2	6.3
3	6.1
4	6.8
5	7.5
6	8.0
7	8.1
8	8.5
9	9.0
10	8.7
11	7.9
12	8.2
13	8.4
14	9.0

Plot the data. Also plot the residuals by time based on a linear regression equation. Does there appear to be serial correlation?

Solution   It is clear from the scatterplot of the sample data and from the residual plot of the linear regression that there is serial correlation present in the data.

LISTING OF DATA

OBS	MONTH	SALES
1	1	6.0
2	2	6.3
3	3	6.1
4	4	6.8
5	5	7.5
6	6	8.0
7	7	8.1
8	8	8.5
9	9	9.0
10	10	8.7
11	11	7.9
12	12	8.2
13	13	8.4
14	14	9.0

N =   14

LINEAR REGRESSION

GENERAL LINEAR MODELS PROCEDURE

DEPENDENT VARIABLE: SALES

SOURCE	DF	SUM OF SQUARES	MEAN SQUARE	F VALUE	PR > F	R-SQUARE	C.V.
MODEL	1	10.57539560	10.57539560	34.30	0.0001	0.740833	7.1645
ERROR	12	3.69960440	0.30830037		ROOT MSE		SALES MEAN
CORRECTED TOTAL	13	14.27500000			0.55524802		7.75000000

SOURCE	DF	TYPE I SS	F VALUE	PR > F	DF	TYPE III SS	F VALUE	PR > F
MONTH	1	10.57539560	34.30	0.0001	1	10.57539560	34.30	0.0001

PARAMETER	ESTIMATE	T FOR HO: PARAMETER=0	PR > |T|	STD ERROR OF ESTIMATE
INTERCEPT	6.13296703	19.57	0.0001	0.31344787
MONTH	0.21560440	5.86	0.0001	0.03681259

OBSERVATION	OBSERVED VALUE	PREDICTED VALUE	RESIDUAL
1	6.00000000	6.34857143	-0.34857143
2	6.30000000	6.56417582	-0.26417582
3	6.10000000	6.77978022	-0.67978022
4	6.80000000	6.99538462	-0.19538462
5	7.50000000	7.21098901	0.28901099
6	8.00000000	7.42659341	0.57340659
7	8.10000000	7.64219780	0.45780220
8	8.50000000	7.85780220	0.64219780
9	9.00000000	8.07340659	0.92659341
10	8.70000000	8.28901099	0.41098901
11	7.90000000	8.50461538	-0.60461538
12	8.20000000	8.72021978	-0.52021978
13	8.40000000	8.93582418	-0.53582418
14	9.00000000	9.15142857	-0.15142857

SUM OF RESIDUALS	0.00000000
SUM OF SQUARED RESIDUALS	3.69960440
SUM OF SQUARED RESIDUALS - ERROR SS	-0.00000000
FIRST ORDER AUTOCORRELATION	0.66819228
DURBIN-WATSON D	0.62457541

(continues)

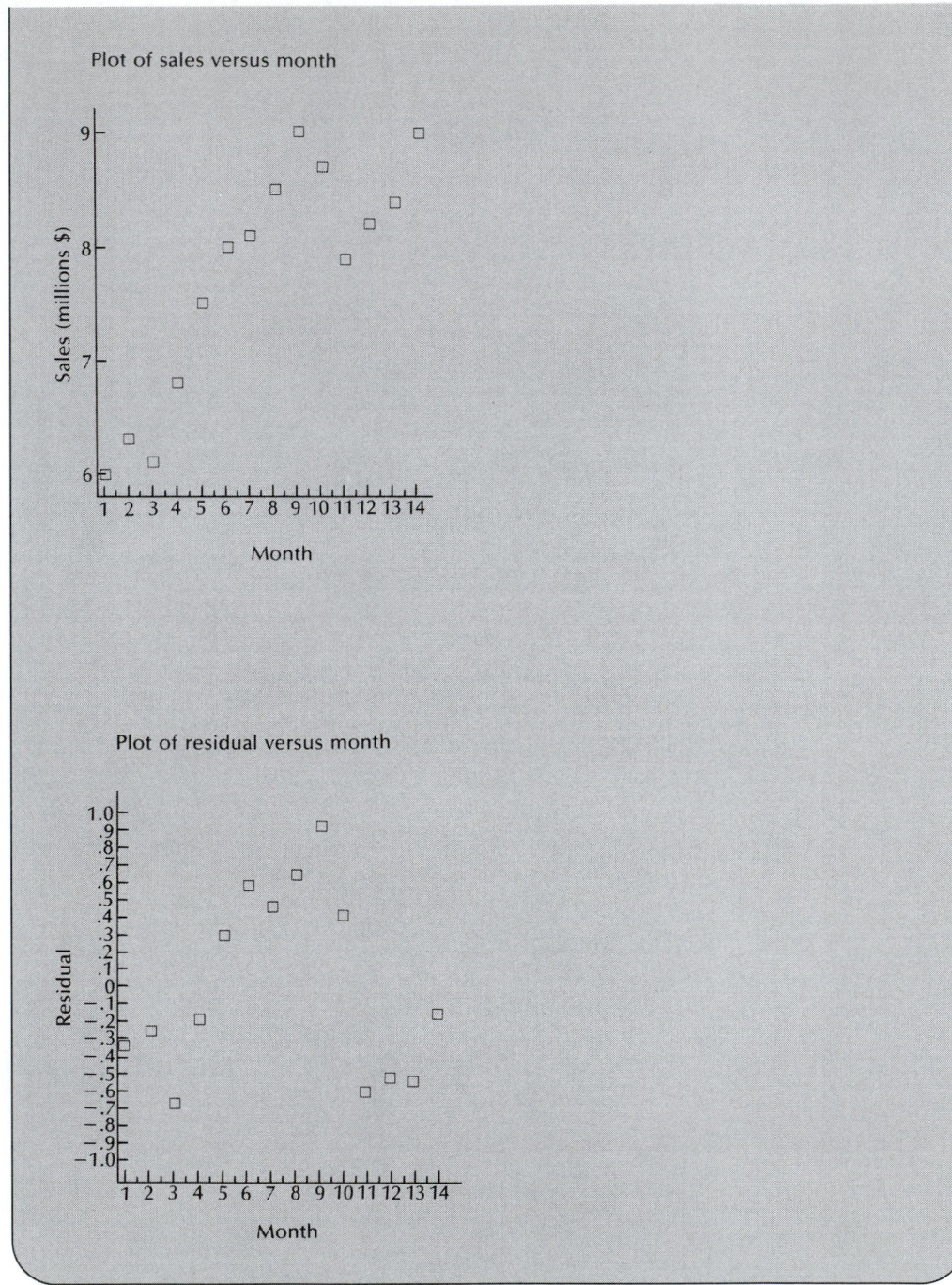

EXAMPLE 12.15 ▼

Determine the value of the Durbin–Watson statistic for the data of Example 12.14. Does it confirm the impressions you obtained from the plots?

Solution   Based on the output of Example 12.14, we find $d = .62457541$. Since this value is much less than 1.5, we have evidence of positive serial correlation; the residual plot bears this out.   ▲

If serial correlation is suspected, then the proposed multiple regression model is inappropriate and some alternative must be sought. A study of the many approaches to analyzing time series data where the errors are not independent can consume many years; hence, we cannot expect to solve many of these problems within the confines of this text. We will, however, suggest a simplified regression approach, based on *first differences*, which may alleviate the problem.

Regression based on first differences is simple to use and, as might be expected, is only a crude approach to the problem of serial correlation. For a simple linear regression of y on x, we compute the differences $y_t - y_{t-1}$ and $x_t - x_{t-1}$. A regression of the $n - 1$ y differences on the corresponding $n - 1$ x differences may eliminate the serial correlation. If not, you should consult someone more familiar with analyzing time series data.

The residual plots that we have discussed can be useful in diagnosing problems in fitting regression models to data. Unfortunately however, they, too, can be misleading because the residuals are subject to random variation. Some researchers have suggested that it is better to use "standardized" residuals in order to detect problems with a fitted regression model. One particular type of standardized residual, called the Studentized residual, has become part of the output for some of the major software packages such as SAS.

If the software package you use works with standardized residuals, you can replace plots of the ordinary residuals with plots of the standardized residuals to perform the diagnostic evaluation of the fit of a regression model. In theory, these standardized residuals have a mean of 0 and a standard deviation of 1. So large residuals would be ones with an absolute value of, say, 3 or more.

▼ EXERCISES

Basic Techniques

12.10   Several different patterns of residuals are shown in the plots of Figure 12.6. Indicate whether the plot suggests a problem, and, if so, indicate the potential problem and a possible solution.

FIGURE 12.6
**Residual Plots for
Exercise 12.10**

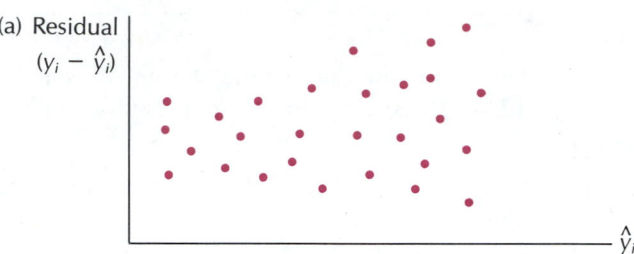

(a) Residual $(y_i - \hat{y}_i)$

$\hat{y}_i$

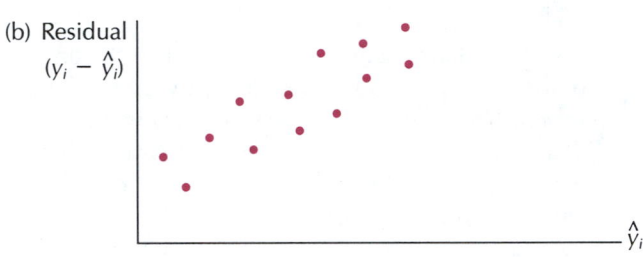

(b) Residual $(y_i - \hat{y}_i)$

$\hat{y}_i$

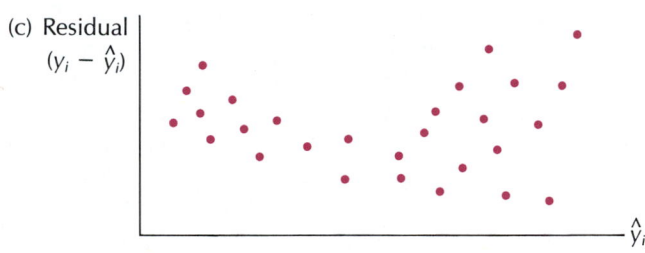

(c) Residual $(y_i - \hat{y}_i)$

$\hat{y}_i$

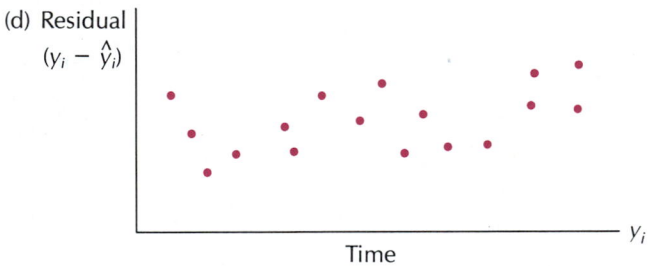

(d) Residual $(y_i - \hat{y}_i)$

Time y_i

12.11 Refer to the data of Example 12.14. Form first differences and regress the y differences on the x differences. Is there evidence of serial correlation for the difference model? What plot(s) did you use to reach a conclusion?

12.12 What is the value of the Durbin–Watson statistic for the data of Exercise 12.11? Does it agree with your previous conclusion?

Applications

12.13 A researcher in the social sciences examined the relationship between the rate (per 1,000) of nonviolent crimes y based on the rate 5 years ago x_1, the present unemployment rate x_2 for cities. Data from 20 different cities are shown here.

OBS	RATE	RATE 5	UNEMPLOY
1	20	14	14.3
2	10	10	7.4
3	14	16	12.0
4	15	10	10.7
5	13	16	8.1
6	4	12	6.5
7	11	8	8.6
8	16	7	10.4
9	13	12	10.5
10	13	20	12.2
11	15	14	8.3
12	8	10	8.0
13	15	10	12.2
14	15	20	14.1
15	6	13	8.3
16	3	2	8.7
17	5	10	9.5
18	13	14	14.4
19	14	16	13.8
20	7	8	10.0

N = 20

Use the output shown here to:

a. Determine the fit to the model

$$y = \beta_0 + \beta_1 x_1 + \beta_2 x_2 + \beta_3 x_1 x_2 + \varepsilon.$$

b. Examine the assumptions underlying the regression model. Discuss whether the assumptions appear to hold. If they don't, suggest possible remedies.

LISTING OF DATA

OBS	RATE	RATE_5	UNEMPLOY
1	20	14	14.3
2	10	10	7.4
3	14	16	12.0
4	15	10	10.7
5	13	16	8.1
6	4	12	6.5
7	11	8	8.6
8	16	7	10.4
9	13	12	10.5

(continues)

```
OBS     RATE    RATE_5    UNEMPLOY

10      13      20        12.2
11      15      14         8.3
12       8      10         8.0
13      15      10        12.2
14      15      20        14.1
15       6      13         8.3
16       3       2         8.7
17       5      10         9.5
18      13      14        14.4
19      14      16        13.8
20       7       8        10.0

N =     20
```

MULTIPLE REGRESSION

GENERAL LINEAR MODELS PROCEDURE

DEPENDENT VARIABLE: RATE

SOURCE	DF	SUM OF SQUARES	MEAN SQUARE	F VALUE	PR > F	R-SQUARE	C.V.
MODEL	3	227.86512995	75.95504332	7.10	0.0030	0.571091	28.4388
ERROR	16	171.13487005	10.69592938		ROOT MSE		RATE MEAN
CORRECTED TOTAL	19	399.00000000			3.27046317		11.50000000

SOURCE	DF	TYPE I SS	F VALUE	PR > F	DF	TYPE III SS	F VALUE	PR > F
RATE_5	1	87.59158010	8.19	0.0113	1	53.18992084	4.97	0.0404
UNEMPLOY	1	97.79946979	8.86	0.0089	1	79.79956840	7.46	0.0148
RATE_5*UNEMPLOY	1	45.47408006	4.25	0.0558	1	45.47408006	4.25	0.0558

PARAMETER	ESTIMATE	T FOR H0: PARAMETER=0	PR > |T|	STD ERROR OF ESTIMATE
INTERCEPT	-29.87534146	-2.13	0.0492	14.03734719
RATE_5	2.33165740	2.23	0.0404	1.04558496
UNEMPLOY	3.81909377	2.73	0.0148	1.39820005
RATE_5*UNEMPLOY	-0.20308407	-2.06	0.0558	0.09849250

OBSERVATION	OBSERVED VALUE	PREDICTED VALUE	RESIDUAL
1	20.00000000	16.72347242	3.27652758
2	10.00000000	6.67430534	3.32569466
3	14.00000000	14.26816091	-0.26816091
4	15.00000000	12.57554050	2.42445950
5	13.00000000	12.04614112	0.95385888
6	4.00000000	7.08809946	-3.08809946
7	11.00000000	7.64994023	3.35005977
8	16.00000000	11.38031534	4.61968466
9	13.00000000	12.61643923	0.38356077
10	13.00000000	13.79823765	-0.79823765
11	15.00000000	10.86797161	4.13202839
12	8.00000000	7.74725719	0.25274281
13	15.00000000	15.25792012	-0.25792012
14	15.00000000	13.33732118	1.66267882
15	6.00000000	10.22191198	-4.22191198
16	3.00000000	4.48042637	-1.48042637
17	5.00000000	10.42963681	-5.42963681
18	13.00000000	16.82106410	-3.82106410
19	14.00000000	15.29370850	-1.29370850
20	7.00000000	10.72212993	-3.72212993

MULTIPLE REGRESSION

GENERAL LINEAR MODELS PROCEDURE

DEPENDENT VARIABLE: RATE

```
        SUM OF RESIDUALS                        -0.00000000
        SUM OF SQUARED RESIDUALS                171.13487005
        SUM OF SQUARED RESIDUALS - ERROR SS     -0.00000000
        FIRST ORDER AUTOCORRELATION              0.29418193
        DURBIN-WATSON D                          1.26794898
```

(continues)

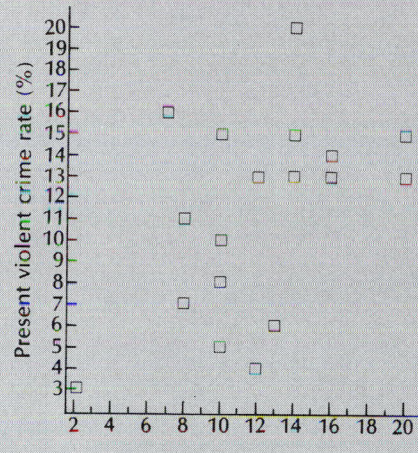

Plot of present violent crime rate versus rate 5 years ago

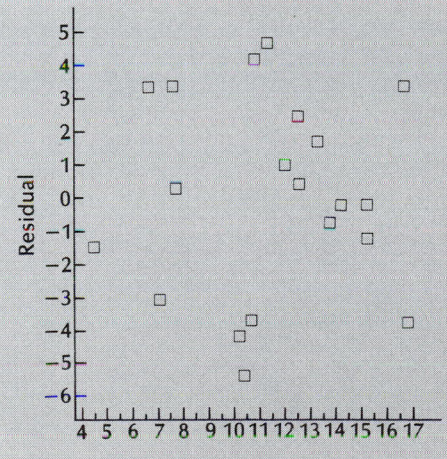

Plot of residual versus predicted value

(*continues*)

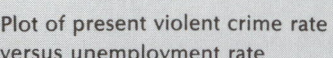

Plot of present violent crime rate
versus unemployment rate

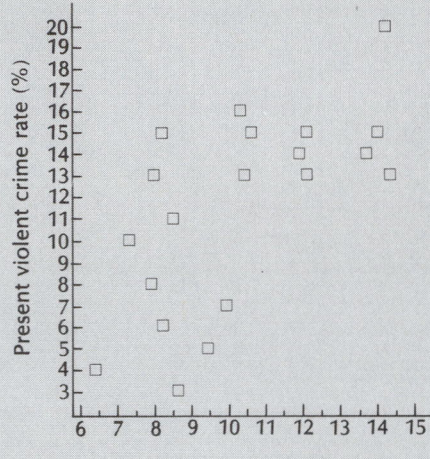

Plot of residual versus
violent crime rate 5 years ago

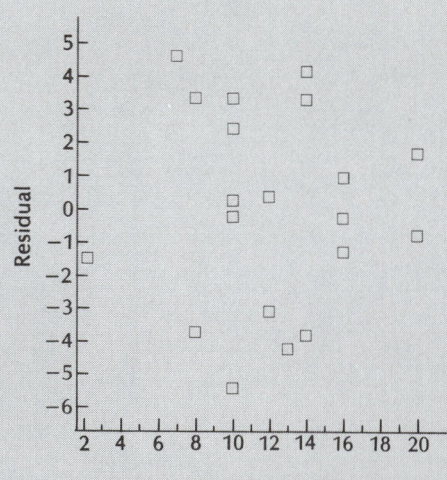

(*continues*)

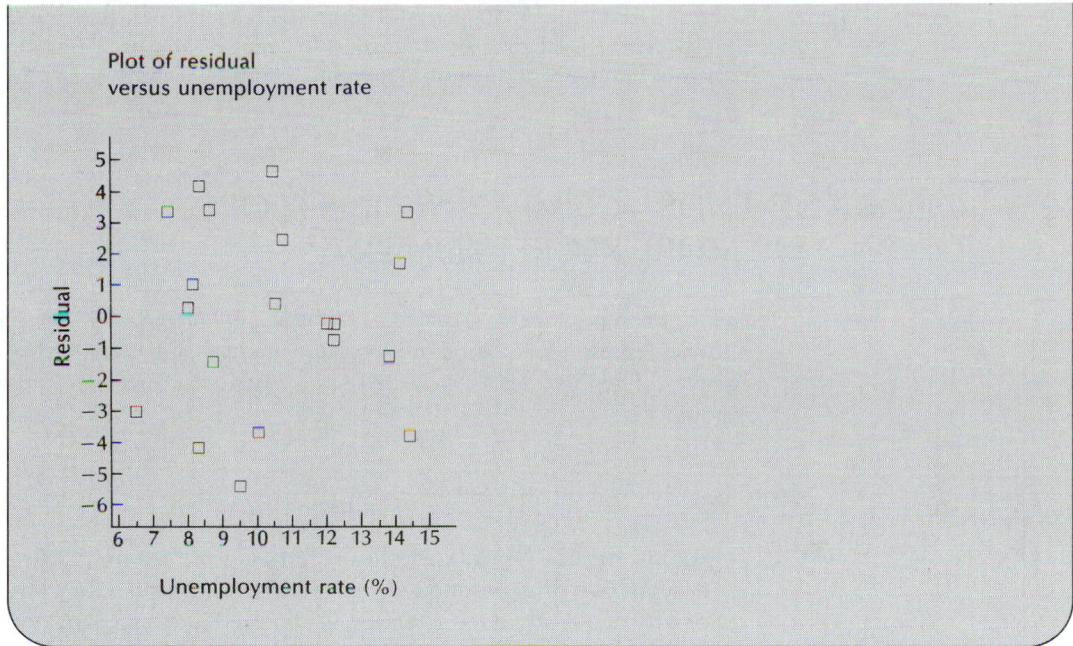

Plot of residual versus unemployment rate

12.14 Refer to Exercise 12.13. Predict y for $x_1 = 9$ and $x_2 = 16$. Might there be a problem with this prediction? If so, why?

12.15 Estimates ($\hat{y}$s) and residuals from a securities firm's regression model for the prediction of earnings per share (per quarter) are shown here for 25 different high-technology companies. Is there any evidence that the assumptions have been violated? Are any additional tests or plots warranted?

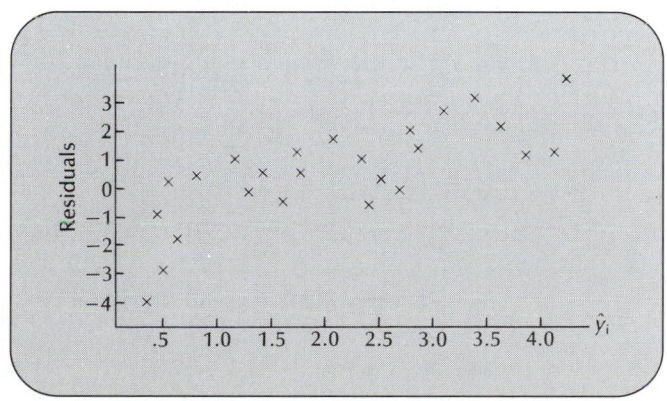

12.16   Refer to Exercise 12.15. Suppose that these data represent estimates and residuals for the earnings per share for the past 25 quarters of a single company. Assess the possibility of serial correlation. Are any adjustments required? If so, which?

12.5 ▼ ODDS AND ENDS—COMPARING THE SLOPES OF TWO OR MORE REGRESSION LINES

This topic represents a special case of the general problem of constructing a multiple regression equation for both qualitative and quantitative independent variables. The best way to illustrate this particular problem is by way of an example.

E X A M P L E   12.16   ▼

An investigator was interested in comparing the responses of rats to different doses of two drug products (A and B). The study called for a sample of 60 rats of a particular strain to be randomly allocated into two equal groups. The first group of rats was to receive drug A, with 10 rats randomly assigned to each of three doses (5, 10, and 20 mg). Similarly, the 30 rats in group 2 were to receive drug B, with 10 rats randomly assigned to the 5-, 10-, and 20-mg doses. In the study, each rat received its assigned dose, and after a 30-minute observation period, it was scored for signs of anxiety on a 0-to-30-point scale. Assume a rat's anxiety score is a linear function of the dosage of the drug. Write a model relating a rat's scores to the two independent variables "drug product" and "drug dose." Interpret the βs.

Solution   For this experimental situation, we have one qualitative variable (drug product) and one quantitative variable (drug dose). Letting x_1 denote the drug dose, we have the model

$$y = \beta_0 + \beta_1 x_1 + \beta_2 x_2 + \beta_3 x_1 x_2 + \varepsilon,$$

where

$x_1 = $ drug dose
$x_2 = 1$ if product B $x_2 = 0$ otherwise.

The expected value for y in our model is

$$E(y) = \beta_0 + \beta_1 x_1 + \beta_2 x_2 + \beta_3 x_1 x_2.$$

Substituting $x_2 = 0$ and $x_2 = 1$, respectively, for drugs A and B, we obtain the expected rat anxiety score for a given dose:

$$\text{drug A:} \quad E(y) = \beta_0 + \beta_1 x_1$$
$$\text{drug B:} \quad E(y) = \beta_0 + \beta_1 x_1 + \beta_2 + \beta_3 x_1 = (\beta_0 + \beta_2) + (\beta_1 + \beta_3)x_1.$$

linear regression lines

These two expected values represent **linear regression lines**. The parameters in the model can be interpreted in terms of the slopes and intercepts associated with these regression lines. In particular,

y-intercept

slope

β_0:   **y-intercept** for product A regression line

β_1:   **slope** of product A regression line

β_2:   difference in y-intercepts of regression lines for products B and A

β_3:   difference in slopes of regression lines for products B and A

intersecting lines

Figure 12.7(a) indicates a situation where $\beta_3 \neq 0$ (that is, there is an interaction between the two variables "drug product" and "drug dose"). Thus, the regression lines are not parallel. Figure 12.7(b) indicates a case in

parallel lines

which $\beta_3 = 0$ (no inter-action), which results in parallel regression lines.

▲

Indeed, many other experimental situations are possible, depending on the signs and magnitudes of the parameters β_0, β_1, β_2, and β_3.

FIGURE 12.7
Comparing Two Regression Lines

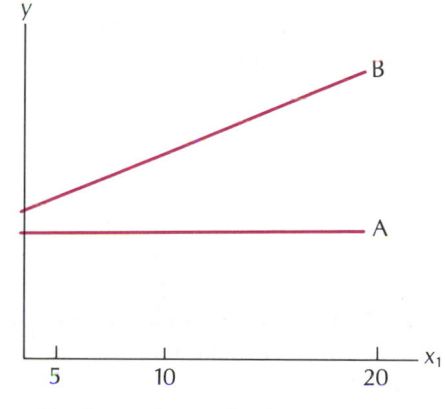

(a) $\beta_3 \neq 0$; interaction is present;
intersecting lines

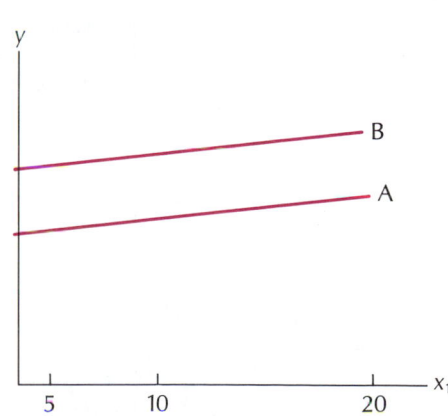

(b) $\beta_3 = 0$; interaction is not present;
parallel lines

E X A M P L E   12.17

▼

Sample data for the experiment discussed in Example 12.16 are listed in Table 12.7. The response of interest is an anxiety score obtained from trained investigators. Use these data to fit the general linear model

$$y = \beta_0 + \beta_1 x_1 + \beta_2 x_2 + \beta_3 x_1 x_2 + \varepsilon.$$

**TABLE   12.7
Rat Anxiety Scores,
Example 12.17**

Drug Product	Drug Dose (mg)		
	5	10	20
A	15 16	18 16	20 17
	16 15	17 15	19 18
	18 16	18 19	21 21
	13 17	19 18	18 20
	19 15	20 16	19 17
	av = 16	av = 17.6	av = 19.0
B	16 15	19 18	24 23
	17 15	21 20	25 24
	18 18	22 21	23 22
	17 17	23 22	25 26
	15 16	20 19	25 24
	av = 16.4	av = 20.5	av = 24.1

Of particular interest to the experimenter is a comparison between the slopes of the regression lines. A difference in slopes would indicate that the drug products have different effects on the anxiety of the rats. Conduct a statistical test of the equality of the two slopes. Use $\alpha = .05$.

Solution   Using the complete model

$$y = \beta_0 + \beta_1 x_1 + \beta_2 x_2 + \beta_3 x_1 x_2 + \varepsilon,$$

we obtain a least squares fit of

$$\hat{y} = 15.30 + .19 x_1 - .70 x_2 + .30 x_1 x_2$$

with $SSE_1 = 133.63$ (see the computer output that follows).

The reduced model corresponding to the null hypothesis $H_0: \beta_3 = 0$ (that is, the slopes are the same) is

$$y = \beta_0 + \beta_1 x_1 + \beta_2 x_2 + \varepsilon$$

```
      LISTING OF DATA

OBS      PRODUCT      DOSE      SCORE
  1         0           5        15
  2         0           5        16
  3         0           5        16
  4         0           5        15
  5         0           5        18
  6         0           5        16
  7         0           5        13
  8         0           5        17
  9         0           5        19
 10         0           5        15
 11         0          10        18
 12         0          10        16
 13         0          10        17
 14         0          10        15
 15         0          10        18
 16         0          10        19
 17         0          10        19
 18         0          10        18
 19         0          10        20
 20         0          10        16
 21         0          20        20
 22         0          20        17
 23         0          20        19
 24         0          20        18
 25         0          20        21
 26         0          20        21
 27         0          20        18
 28         0          20        20
 29         0          20        19
 30         0          20        17
 31         1           5        16
 32         1           5        15
 33         1           5        17
 34         1           5        15
 35         1           5        10
 36         1           5        18
 37         1           5        17
 38         1           5        17
 39         1           5        15
 40         1           5        16
 41         1          10        19
 42         1          10        18
 43         1          10        21
 44         1          10        20
 45         1          10        22
 46         1          10        21
 47         1          10        23
 48         1          10        22
 49         1          10        20
 50         1          10        19
 51         1          20        24
 52         1          20        23
 53         1          20        25
 54         1          20        24
 55         1          20        23
 56         1          20        22
 57         1          20        25
 58         1          20        26
 59         1          20        25
 60         1          20        24

N =     60
```

FULL MODEL

GENERAL LINEAR MODELS PROCEDURE

DEPENDENT VARIABLE: SCORE

SOURCE	DF	SUM OF SQUARES	MEAN SQUARE	F VALUE	PR > F	R-SQUARE	C.V.
MODEL	3	442.10476190	147.36825397	61.76	0.0001	0.767898	8.1588
ERROR	56	133.62857143	2.38622449		ROOT MSE		SCORE MEAN
CORRECTED TOTAL	59	575.73333333			1.54474091		18.93333333

(continues)

SOURCE	DF	TYPE I SS	F VALUE	PR > F	DF	TYPE III SS	F VALUE	PR > F
DOSE	1	272.00476190	113.99	0.0001	1	42.75238095	17.92	0.0001
PRODUCT	1	117.60000000	49.28	0.0001	1	1.63333333	0.68	0.4116
DOSE*PRODUCT	1	52.50000000	22.00	0.0001	1	52.50000000	22.00	0.0001

PARAMETER	ESTIMATE	T FOR HO: PARAMETER=0	PR > |T|	STD ERROR OF ESTIMATE
INTERCEPT	15.30000000	25.57	0.0001	0.59827558
DOSE	0.19142857	4.23	0.0001	0.04522538
PRODUCT	-0.70000000	-0.83	0.4116	0.84608944
DOSE*PRODUCT	0.30000000	4.69	0.0001	0.06395835

REDUCED MODEL

GENERAL LINEAR MODELS PROCEDURE

DEPENDENT VARIABLE: SCORE

SOURCE	DF	SUM OF SQUARES	MEAN SQUARE	F VALUE	PR > F	R-SQUARE	C.V.
MODEL	2	389.60476190	194.80238095	59.66	0.0001	0.676710	9.5443
ERROR	57	186.12857143	3.26541353		ROOT MSE		SCORE MEAN
CORRECTED TOTAL	59	575.7333333			1.80704553		18.93333333

SOURCE	DF	TYPE I SS	F VALUE	PR > F	DF	TYPE III SS	F VALUE	PR > F
DOSE	1	272.00476190	83.30	0.0001	1	272.00476190	83.30	0.0001
PRODUCT	1	117.60000000	36.01	0.0001	1	117.60000000	36.01	0.0001

PARAMETER	ESTIMATE	T FOR HO: PARAMETER=0	PR > |T|	STD ERROR OF ESTIMATE
INTERCEPT	13.55000000	24.77	0.0001	0.54711020
DOSE	0.34142857	9.13	0.0001	0.03740940
PRODUCT	2.80000000	6.00	0.0001	0.46657715

for which we obtain

$$\hat{y} = 13.55 + .34x_1 + 2.80x_2$$

and $SSE_2 = 186.13$. The reduction in the sum of squares for error attributed to $x_1 x_2$ is

$$SS_{drop} = SSE_2 - SSE_1 = 186.13 - 133.63 = 52.50.$$

Using $MSE_1 = SSE_1/56 = 133.63/56 = 2.39$ and $MS_{drop} = 52.50$ (since we are testing only one β),

$$F = \frac{MS_{drop}}{MSE_1} = \frac{52.50}{2.39} = 22.00.$$

Since the observed value of F exceeds 4.00, the table value for $df_1 = 1$, $df_2 = 56$ (actually, 60), and $a = .05$, we reject H_0 and conclude that the slopes for the two groups are different. It should be noted that we could have obtained the same result by testing $H_0: \beta_3 = 0$ using a t test. From the computer output, the t statistic is 4.69, which is significant at the .0001 level. For this type of test, the t statistic and F statistic are related; namely, $t^2 = F$ (here $4.69^2 \approx 22$).

The results presented here for comparing the slope of two regression lines can be readily extended to the comparison of three or more regression lines by including additional dummy variables and all possible interaction terms between the quantitative variable x_1 and the dummy variables. Thus, for example, in comparing the slopes of three regression lines, the model would contain the quantitative variable x_1, two dummy variables x_2 and x_3, and two interaction terms $x_1 x_2$ and $x_1 x_3$.   ▲

▼ EXERCISES

Applications

12.17   An experimenter wished to compare the potencies of three different drug products. To do this, 12 test tubes were inoculated with a culture of the virus under study and incubated for 2 days at 35°C. Four dosage levels (.2, .4, .8, and 1.6 μg per tube) were to be used from each of the three drug products, with only one dose–drug product combination for each of the 12 test tube cultures. One means of comparing the drug products would be to examine their slopes (with respect to dose).

  a. Write a general linear model relating the response y to the independent variables "dose" and "drug product." Make the expected response a linear function of log dose (x_1). Identify the parameters in the model.

  b. It would seem reasonable to assume that the three separate response lines have a common intercept β_0, since this would correspond to a 0 dosage level of any of the drug products. Change the model of part a to reflect this change.

12.18   Refer to Exercise 12.17.

  a. Use the following data to make a comparison among the three slopes. Fit a complete and a reduced model for your test. Use $\alpha = .05$.

	Drug Product		
Dose	A	B	C
.2	2.0	1.8	1.3
.4	4.3	4.1	2.0
.8	6.5	4.9	2.8
1.6	8.9	5.7	3.4

b. Is there evidence to indicate that the slopes are equal?

c. Suggest how you might test the null hypothesis that the intercepts are all equal to 0.

12.6 ▼ ODDS AND ENDS—GENCAT: A GENERAL LINEAR MODEL PROGRAM FOR CATEGORICAL DATA (optional)

In this section, we will present an example of a procedure that can be used to compare s different multinomial populations, where each population has r possible responses. The procedure represents an extension of our study of two-dimensional contingency tables (Chapter 14). We present the material in this chapter to take advantage of what we have learned about writing models for experimental situations. The example we will discuss makes use of methodology developed by Grizzle, Starmer, and Koch (1969), with the format explained in more detail by Forthofer, Starmer, and Grizzle (1971).

The subject of analyses for multidimensional contingency tables has been widely studied, and we hope only to scratch the surface in this section. However, by following the example, you can see the utility of the approach and then refer to the references for more details of additional applications.

π_{ij}

Assume there are s different multinomial populations, each with r categories of response. Let π_{ij} (Greek letter pi) denote the cell probability for the ith row and jth column. Then the cell probabilities for the s populations are as listed in Table 12.8. Note that for any row we require

$$\sum_j \pi_{ij} = 1.$$

The procedure we will consider for analyzing categorical data is similar to the general linear model approach of Chapters 11 and 12. In particular, we construct

TABLE 12.8
Cell Probabilities for the s Populations

Population	\multicolumn{5}{c}{Category}
	1	2	3	$\cdots$	r
1	π_{11}	π_{12}	π_{13}	$\cdots$	π_{1r}
2	π_{21}	π_{22}	π_{23}	$\cdots$	π_{2r}
3	π_{31}	π_{32}	π_{33}	$\cdots$	π_{3r}
$\vdots$	$\vdots$	$\vdots$	$\vdots$		$\vdots$
s	π_{s1}	π_{s2}	π_{s3}	$\cdots$	π_{sr}

u different functions of the multinomial population probabilities by fitting the model

model

$$\mathbf{A}\boldsymbol{\pi} = \mathbf{X}\boldsymbol{\beta}.$$

(Note that this model is similar to the general linear model $\mathbf{Y} = \mathbf{X}\boldsymbol{\beta}$.) The matrix $\mathbf{A}$ in our model is a matrix of constants that defines the u different functions of the πs and is used to specify the response. The matrix $\boldsymbol{\pi}$ is given by

$$\boldsymbol{\pi} = \begin{bmatrix} \pi_{11} \\ \pi_{12} \\ \vdots \\ \pi_{sr} \end{bmatrix}.$$

The design matrix $\mathbf{X}$ is similar to those used in the general linear model, and $\boldsymbol{\beta}$ is a matrix of v unknown parameters:

$$\boldsymbol{\beta} = \begin{bmatrix} \beta_1 \\ \beta_2 \\ \vdots \\ \beta_v \end{bmatrix} \qquad v \le u.$$

To bring this procedure to life, consider the following example. Suppose that a multi-clinic investigation was conducted to compare four different treatments (nasal sprays) for their effectiveness in relieving the symptoms of ragweed allergy. Each investigator was to obtain a random sample of patients who responded positively to a skin test for ragweed allergy and who volunteered to participate in the study. A double-blind procedure was used, in which each physician was supplied with a random code to assign one of the four sprays to patients, but neither the investigator nor the patient knew which medication would be received. On the appointed day, volunteers were asked to report to the physician's office, where they were examined for allergic symptoms due to ragweed. Then each person was shown how to self-administer one dose of the assigned nasal spray. After a 4-hour period, each patient was rated as to whether he or she had improved while on the drug (yes or no).

The data appear in Table 12.9. An entry is the number of yes or no responses for a treatment–investigator combination.

Note that this example differs from the two-dimensional contingency tables of Section 8.7. In particular, we have $s = 24$ (six investigators times four treatments)

TABLE 12.9
Yes and No Responses for the Multiclinic Investigation

Investigator	Treatment	Response No	Response Yes	Sample Size
1	A	1	2	3
	B	0	1	1
	C	2	1	3
	D	1	2	3
2	A	2	3	5
	B	1	2	3
	C	0	3	3
	D	0	4	4
3	A	10	11	21
	B	13	5	18
	C	12	7	19
	D	7	11	18
4	A	1	0	1
	B	3	0	3
	C	0	1	1
	D	2	0	2
5	A	2	4	6
	B	4	2	6
	C	7	6	13
	D	4	4	8
6	A	15	4	19
	B	14	4	18
	C	13	5	18
	D	14	8	22
				218

different multinomial populations with each population having $r = 2$ possible categories (yes or no).

The **π matrix** of unknown probabilities associated with our data is shown here. Several of the individual πs have been identified to show you how the matrix has been constructed using a single subscript.

π matrix

$$
\boldsymbol{\pi} = \begin{bmatrix} \pi_1 \\ \pi_2 \\ \pi_3 \\ \vdots \\ \pi_8 \\ \pi_9 \\ \vdots \\ \pi_{48} \end{bmatrix}
$$

probability of a no response on treatment A, investigator 1
probability of a yes response on treatment A, investigator 1
probability of a no response on treatment B, investigator 1

probability of a yes response on treatment D, investigator 1
probability of a no response on treatment A, investigator 2

probability of a yes response on treatment D, investigator 6

tests

There are a number of different hypotheses that we might wish to pose concerning the cell probabilities for the 24 multinomial populations.

For example, we might hypothesize that the proportion of no responses is linearly related to a treatment and an investigator effect. To do this, we define the following parameters:

μ: overall mean effect
α_1:   effect due to investigator 1
α_2:   effect due to investigator 2
$\vdots$
α_5:   effect due to investigator 5
β_1:   effect due to treatment A
β_2:   effect due to treatment B
β_3:   effect due to treatment C.

Note that we have not defined an effect for treatment D or for investigator 6, since these effects can be stated in terms of the previous effects. In particular, we take

investigator 6 effect: $\alpha_6 = -\alpha_1 - \alpha_2 - \alpha_3 - \alpha_4 - \alpha_5$
treatment D effect: $\beta_4 = -\beta_1 - \beta_2 - \beta_3.$

We now define the following matrix model relating the theoretical cell probabilities (the πs) to a linear function of the αs, βs, and μ. This is given by our general model $\mathbf{A}\pi = \mathbf{X}\beta$, where $\mathbf{A}$ is a 24×48 matrix of the form

A matrix

$$\mathbf{A} = \begin{bmatrix} 1 & 0 & & & & & \\ & & 1 & 0 & & & \\ & & & & 1 & 0 & \\ & & & & & & \ddots \\ & & & & & & 1 & 0 \end{bmatrix}.$$

In general, we can write the matrix $\mathbf{A}$ as

A*

$$\mathbf{A} = \begin{bmatrix} \mathbf{A}^* & & \\ & \mathbf{A}^* & \\ & & \mathbf{A}^* \end{bmatrix},$$

where, for our case, $\mathbf{A}^* = \begin{bmatrix} 1 & 0 \end{bmatrix}$.

The **X matrix** is the design matrix corresponding to the π matrix. For our example,

X matrix

$$
\mathbf{X} =
\begin{bmatrix}
\mu & \alpha_1 & \alpha_2 & \alpha_3 & \alpha_4 & \alpha_5 & \beta_1 & \beta_2 & \beta_3 \\
1 & 1 & 0 & 0 & 0 & 0 & 1 & 0 & 0 \\
1 & 1 & 0 & 0 & 0 & 0 & 0 & 1 & 0 \\
1 & 1 & 0 & 0 & 0 & 0 & 0 & 0 & 1 \\
1 & 1 & 0 & 0 & 0 & 0 & -1 & -1 & -1 \\
1 & 0 & 1 & 0 & 0 & 0 & 1 & 0 & 0 \\
1 & 0 & 1 & 0 & 0 & 0 & 0 & 1 & 0 \\
1 & 0 & 1 & 0 & 0 & 0 & 0 & 0 & 1 \\
1 & 0 & 1 & 0 & 0 & 0 & -1 & -1 & -1 \\
1 & 0 & 0 & 1 & 0 & 0 & 1 & 0 & 0 \\
1 & 0 & 0 & 1 & 0 & 0 & 0 & 1 & 0 \\
1 & 0 & 0 & 1 & 0 & 0 & 0 & 0 & 1 \\
1 & 0 & 0 & 1 & 0 & 0 & -1 & -1 & -1 \\
1 & 0 & 0 & 0 & 1 & 0 & 1 & 0 & 0 \\
1 & 0 & 0 & 0 & 1 & 0 & 0 & 1 & 0 \\
1 & 0 & 0 & 0 & 1 & 0 & 0 & 0 & 1 \\
1 & 0 & 0 & 0 & 1 & 0 & -1 & -1 & -1 \\
1 & 0 & 0 & 0 & 0 & 1 & 1 & 0 & 0 \\
1 & 0 & 0 & 0 & 0 & 1 & 0 & 1 & 0 \\
1 & 0 & 0 & 0 & 0 & 1 & 0 & 0 & 1 \\
1 & 0 & 0 & 0 & 0 & 1 & -1 & -1 & -1 \\
1 & -1 & -1 & -1 & -1 & -1 & 1 & 0 & 0 \\
1 & -1 & -1 & -1 & -1 & -1 & 0 & 1 & 0 \\
1 & -1 & -1 & -1 & -1 & -1 & 0 & 0 & 1 \\
1 & -1 & -1 & -1 & -1 & -1 & -1 & -1 & -1 \\
\end{bmatrix}.
$$

For our data, the matrix of parameters is

β matrix

$$
\boldsymbol{\beta} =
\begin{bmatrix}
\mu \\
\alpha_1 \\
\alpha_2 \\
\alpha_3 \\
\alpha_4 \\
\alpha_5 \\
\beta_1 \\
\beta_2 \\
\beta_3
\end{bmatrix}
$$

With these three matrices, we see that the model

$$\mathbf{A}\boldsymbol{\pi} = \mathbf{X}\boldsymbol{\beta}$$

indicates that the probability of a no response for a particular multinomial population is linearly related to the treatment and investigator effect for that cell. For example, from the first row of both sides of the equation, we have π_1, the probability of a no response on investigator 1 and treatment A, equal to $\mu + \alpha_1 + \beta_1$.

test for fit

Having specified a model, we can perform a number of different tests similar to those indicated by an analysis of variance table for the general linear model. The first test we consider is a test of the adequacy of the linear model $\mathbf{A}\boldsymbol{\pi} = \mathbf{X}\boldsymbol{\beta}$. Some people refer to this as a test for the treatment-by-investigator interaction. Since this test will be performed automatically by the computer program we will use, we will not go into any details here. Then, provided there is no significant interaction, we proceed to examine the main effects due to treatments (sprays) and investigators.

Unlike the analysis of variance tests for the general linear model, our tests will utilize chi-square rather than F statistics. The degrees of freedom for the chi-square test will be the same as the numerator degrees of freedom in a comparable F test. The details of the calculations are left to the computer solution (Gencat).

As with the general linear model, we wish to test that sets of parameters equal 0. Main effects null hypotheses can be stated in matrix notation as

$$H_0: \quad \mathbf{C}\boldsymbol{\beta} = \mathbf{0}.$$

investigator effects

For testing that there is no effect due to investigators ($\alpha_1 = \alpha_2 = \cdots = \alpha_5 = 0$), we can use the **C** matrix

$$\mathbf{C} = \begin{bmatrix} 0 & 1 & 0 & 0 & 0 & 0 & 0 & 0 & 0 \\ 0 & 0 & 1 & 0 & 0 & 0 & 0 & 0 & 0 \\ 0 & 0 & 0 & 1 & 0 & 0 & 0 & 0 & 0 \\ 0 & 0 & 0 & 0 & 1 & 0 & 0 & 0 & 0 \\ 0 & 0 & 0 & 0 & 0 & 1 & 0 & 0 & 0 \end{bmatrix}.$$

treatment effects

Similarly, for testing $H_0: \beta_1 = \beta_2 = \beta_3 = 0$ (no treatment effects), we use

$$\mathbf{C} = \begin{bmatrix} 0 & 0 & 0 & 0 & 0 & 0 & 1 & 0 & 0 \\ 0 & 0 & 0 & 0 & 0 & 0 & 0 & 1 & 0 \\ 0 & 0 & 0 & 0 & 0 & 0 & 0 & 0 & 1 \end{bmatrix}.$$

No **C** matrix need be specified for the interaction test; it is performed automatically.

E X A M P L E   12.18   ▼

Use the previous data to fit a model relating the probability of a no response to a linear function of a treatment and an investigator effect. Test for interaction and main effects. Use $\alpha = .05$ for each test.

Solution   The computer program we will use (Gencat) is an improved version of the one described in detail in Forthofer, Starmer, and Grizzle (1971). Similar analyses can be done using SAS.

We will identify the control fields (without the appropriate job-control language that is required by your computer center) necessary to run this job.

Columns	Description
1–5	number of categories of response (enter r)
6–10	number of multinomial populations (enter s)
11–15	1
16–20	1
21–25	0
26–30	number of **C** matrices
31–35	0
36–40	0

Parameter specifications

For our data, the parameter specifications are as shown here.

2   24   1   1   0   2   0   0

Data.   The data are entered by populations, beginning with a new line for each population. The number of responses falling into the first category of a population is recorded anywhere in the first ten columns, with the decimal recorded. The first 10 columns are called a *10-column field*. The number of responses falling into the second category of a population is recorded in the next 10-column field (columns 11–20), and so on. For cells with no responses, we enter $1/r$ (in our case, .5) rather than a 0. Our data are presented here.

1.0	2.0
.5	1.0
2.0	1.0
1.0	2.0
2.0	3.0
1.0	2.0
.5	3.0
.5	4.0
10.0	11.0

13.0	5.0
12.0	7.0
7.0	11.0
1.0	.5
3.0	.5
.5	1.0
2.0	.5
2.0	4.0
4.0	2.0
7.0	6.0
4.0	4.0
15.0	4.0
14.0	4.0
13.0	5.0
14.0	8.0

A* Matrix. The **A*** matrix is entered by rows in 10-column fields with the decimals entered

```
1.0    0.0
```

X Matrix. The first line for the **X** matrix contains the number of parameters in our model entered on the right side (right-justified) of the first 5-column field (columns 1–5). For example, if the **X** matrix contains 6 parameters, the number 6 would be entered in column 5. For an **X** matrix containing 12 parameters, the number 12 would be entered in columns 4 and 5. Then with the second line, we enter the **X** matrix *by columns* in fields of 5, with the decimal entered. Each column starts a new line. These lines are shown for our example.

```
9
1.0  1.0  1.0  1.0  1.0  1.0  1.0  1.0  1.0  1.0  1.0  1.0  1.0  1.0  1.0  1.0
1.0  1.0  1.0  1.0  1.0  1.0  1.0  1.0
1.0  1.0  1.0  1.0  0.0  0.0  0.0  0.0  0.0  0.0  0.0  0.0  0.0  0.0  0.0  0.0
0.0  0.0  0.0  0.0  −1.0  −1.0  −1.0  −1.0
0.0  0.0  0.0  0.0  1.0  1.0  1.0  1.0  0.0  0.0  0.0  0.0  0.0  0.0  0.0  0.0
0.0  0.0  0.0  0.0  −1.0  −1.0  −1.0  −1.0
0.0  0.0  0.0  0.0  0.0  0.0  0.0  0.0  1.0  1.0  1.0  1.0  0.0  0.0  0.0  0.0
0.0  0.0  0.0  0.0  −1.0  −1.0  −1.0  −1.0
0.0  0.0  0.0  0.0  0.0  0.0  0.0  0.0  0.0  0.0  0.0  0.0  1.0  1.0  1.0  1.0
0.0  0.0  0.0  0.0  −1.0  −1.0  −1.0  −1.0
0.0  0.0  0.0  0.0  0.0  0.0  0.0  0.0  0.0  0.0  0.0  0.0  0.0  0.0  0.0  0.0
1.0  1.0  1.0  1.0  −1.0  −1.0  −1.0  −1.0
1.0  0.0  0.0  −1.0  1.0  0.0  0.0  −1.0  1.0  0.0  0.0  −1.0  1.0  0.0  0.0  −1.0
1.0  0.0  0.0  −1.0  1.0  0.0  0.0  −1.0
0.0  1.0  0.0  −1.0  0.0  1.0  0.0  −1.0  0.0  1.0  0.0  −1.0  0.0  1.0  0.0  −1.0
0.0  1.0  0.0  −1.0  0.0  1.0  0.0  −1.0
0.0  0.0  1.0  −1.0  0.0  0.0  1.0  −1.0  0.0  0.0  1.0  −1.0  0.0  0.0  1.0  −1.0
0.0  0.0  1.0  −1.0  0.0  0.0  1.0  −1.0
```

C Matrices.   For each **C** matrix needed for a specific null hypothesis, we first enter a line giving the number of rows of the **C** matrix in a 5-column field, right-justified. Succeeding lines contain the **C** matrix entered *by rows* in 5-column fields, with the decimal entered. Each row of **C** starts a new line.

```
5
0.0  1.0  0.0  0.0  0.0  0.0  0.0  0.0  0.0
0.0  0.0  1.0  0.0  0.0  0.0  0.0  0.0  0.0
0.0  0.0  0.0  1.0  0.0  0.0  0.0  0.0  0.0
0.0  0.0  0.0  0.0  1.0  0.0  0.0  0.0  0.0
0.0  0.0  0.0  0.0  0.0  1.0  0.0  0.0  0.0
3
0.0  0.0  0.0  0.0  0.0  0.0  1.0  0.0  0.0
0.0  0.0  0.0  0.0  0.0  0.0  0.0  1.0  0.0
0.0  0.0  0.0  0.0  0.0  0.0  0.0  0.0  1.0
```

A copy of the output follows. Comments have been made within the output to explain the parts of the output in which we are interested.

```
                         GENERALIZED CHI-SQUARE ANALYSIS
R   =     2   R IS THE NUMBER OF CATEGORIES OF RESPONSE
S   =    24   S IS THE NUMBER OF POPULATIONS
U   =     1   U IS THE RANK OF THE A MATRIX
MM  =     1   MM = 1 IF LEAST SQUARES ANALYSIS IS USED AND ZERO OTHERWISE
ML  =     0   ML = 0 IF TESTING A LINEAR HYPOTHESIS AND ML = 1 IF TESTING A LOGARITHMIC HYPOTHESIS
NC  =     2   NC IS THE NUMBER OF SETS OF CONTRASTS TO BE TESTED BY LEAST SQUARES ANALYSIS
IK  =     0   IK = 1 IF K IS THE IDENTITY MATRIX AND ZERO OTHERWISE
ISW =     0   ISW = 1 IF YOU WISH TO REANALYZE THE DATA ENTERED IN THE PRECEDING PROBLEM AND
                  ZERO OTHERWISE


          FREQUENCY TABLES (S × R)
                  1.            2.
                  1.            1.
                  2.            1.
                  1.            2.
                  2.            3.
                  1.            2.
                  1.            3.
                  1.            4.
                 10.           11.        This table gives the number of responses falling into the
                 13.            5.        no and yes categories of each population. (Note: When
                 12.            7.        a zero response was observed for a particular cell, we
                  7.           11.        entered 0.5, which the computer then rounded to 1. for
                  1.            1.        this frequency table.)
                  3.            1.
                  1.            1.
                  2.            1.
                  2.            4.
                  4.            2.
                  7.            6.
                  4.            4.
                 15.            4.
                  4.            4.
                 13.            5.
                 14.            8.
                                                                         (continues)
```

```
          PROBABILITY TABLES (S × R)
              0.33333        0.66667
              0.33333        0.66667
              0.66667        0.33333
              0.33333        0.66667
              0.40000        0.60000
              0.33333        0.66667
              0.14286        0.85714
              0.11111        0.88889
              0.47619        0.52381
              0.72222        0.27778
              0.63158        0.36842
              0.38889        0.61111
              0.66667        0.33333
              0.85714        0.14286
              0.33333        0.66667
              0.80000        0.20000
              0.33333        0.66667
              0.66667        0.33333
              0.53846        0.46154
              0.50000        0.50000
              0.78947        0.21053
              0.77778        0.22222
              0.72222        0.27778
              0.63636        0.36364
```

This table gives the sample proportions of no and yes for each population, based on the frequency table entered into the computer.

```
      CHI-SQUARE = 225.1973    DF = 24  P = 0.0

                    DESIGN MATRIX    This is our X matrix
        1.     1.     0.     0.     0.     0.     1.     0.     0.
        1.     1.     0.     0.     0.     0.     0.     1.     0.
        1.     1.     0.     0.     0.     0.     0.     0.     1.
        1.     1.     0.     0.     0.     0.    -1.    -1.    -1.
        1.     0.     1.     0.     0.     0.     1.     0.     0.
        1.     0.     1.     0.     0.     0.     0.     1.     0.
        1.     0.     1.     0.     0.     0.     0.     0.     1.
        1.     0.     1.     0.     0.     0.    -1.    -1.    -1.
        1.     0.     0.     1.     0.     0.     1.     0.     0.
        1.     0.     0.     1.     0.     0.     0.     1.     0.
        1.     0.     0.     1.     0.     0.     0.     0.     1.
        1.     0.     0.     1.     0.     0.    -1.    -1.    -1.
        1.     0.     0.     0.     1.     0.     1.     0.     0.
        1.     0.     0.     0.     1.     0.     0.     1.     0.
        1.     0.     0.     0.     1.     0.     0.     0.     1.
        1.     0.     0.     0.     1.     0.    -1.    -1.    -1.
        1.     0.     0.     0.     0.     1.     1.     0.     0.
        1.     0.     0.     0.     0.     1.     0.     1.     0.
        1.     0.     0.     0.     0.     1.     0.     0.     1.
        1.     0.     0.     0.     0.     1.    -1.    -1.    -1.
        1.    -1.    -1.    -1.    -1.    -1.     1.     0.     0.
        1.    -1.    -1.    -1.    -1.    -1.     0.     1.     0.
        1.    -1.    -1.    -1.    -1.    -1.     0.     0.     1.
        1.    -1.    -1.    -1.    -1.    -1.    -1.    -1.    -1.
```

ESTIMATED MODEL 3 PARAMETERS These are least squares estimates of the model parameters.
0.53613D 00 -0.92943D-01 -0.30425D 00 0.20820D-01 0.19908D 00 -0.20553D-01 -0.14319D-01 0.10228D 00
0.14669D 01

```
              VARIANCE COVARIANCE MATRIX OF THE ESTIMATED MODEL PARAMETERS
    0.16851D 02  0.18434D-02 -0.12320D-03 -0.11849D-02  0.12164D-02 -0.47443D-03  0.10812D-03  0.78970D-05
   -0.96186D-05
    0.18434D-02  0.15851D-01 -0.33372D-02 -0.23548D-02 -0.48790D-02 -0.30238D-02 -0.24452D-03  0.40121D-03
   -0.12907D-03
   -0.12320D-03 -0.33372D-02  0.79601D-02 -0.40146D-03 -0.28617D-02 -0.10533D-02  0.10747D-03  0.46891D-03
   -0.52128D-04
   -0.11849D-02 -0.23548D-02 -0.40146D-03  0.37066D-02 -0.16988D-02 -0.31286D-04 -0.12321D-03 -0.87075D-04
    0.24444D-04
    0.12164D-02 -0.48790D-02 -0.28617D-02 -0.16988D-02  0.13577D-01 -0.25169D-02  0.39683D-03 -0.94708D-03
    0.51481D-03
   -0.47443D-03 -0.30238D-02 -0.10533D-02 -0.31286D-04 -0.25169D-02  0.65797D-02  0.97526D-04  0.20367D-03
   -0.46156D-03
```

(continues)

```
    0.10812D-03 -0.24452D-03  0.10747D-03 -0.12321D-03  0.39683D-03  0.97526D-04  0.28285D-02 -0.95478D-03
   -0.93921D-03
    0.78970D-05  0.40121D-03  0.46891D-03 -0.87075D-04 -0.94708D-03  0.20367D-03 -0.95478D-03  0.28637D-02
   -0.97081D-03
   -0.96186D-05 -0.12907D-03 -0.52128D-04  0.24444D-04  0.51481D-03 -0.47156D-03 -0.93921D-03 -0.97081D-03
    0.28066D-02
```

CHI-SQUARE DUE TO ERROR = 7.0677   DF = 15   P = 0.9557   This is our chi-square test for interaction. The p-value is the probability of x^2 being greater than the observed value.

```
                               F(P) LINEAR MODEL
    0.33333D 00  0.33333D 00  0.66667D 00  0.33333D 00  0.40000D 00  0.33333D 00  0.14286D 00  0.11111D 00
    0.47619D 00  0.72222D 00  0.63158D 00  0.38889D 00  0.66667D 00  0.85714D 00  0.33333D 00  0.80000D 00
    0.33333D 00  0.66667D 00  0.53846D 00  0.50000D 00  0.78947D 00  0.77778D 00  0.72222D 00  0.63636D 00

                          F(P) PREDICTED FROM FITTED MODEL
    0.42886D 00  0.54546D 00  0.45785D 00  0.34055D 00  0.21755D 00  0.33415D 00  0.24654D 00  0.12924D 00
    0.54263D 00  0.65922D 00  0.57161D 00  0.45432D 00  0.72089D 00  0.83749D 00  0.74988D 00  0.63258D 00
    0.50125D 00  0.61785D 00  0.53024D 00  0.41294D 00  0.71965D 00  0.83625D 00  0.74864D 00  0.63134D 00

                       F(P)-F(P) PREDICTED = RESIDUAL
   -0.95531D-01 -0.21213D 00  0.20882D 00 -0.72208D-02  0.18245D 00 -0.81888D-03 -0.10368D 00 -0.18134D-01
   -0.66436D-01  0.62998D-01  0.59965D-01 -0.65428D-01 -0.54225D-01  0.19654D-01 -0.41655D 00  0.16742D 00
   -0.16792D 00  0.48816D-01  0.82207D-02  0.87056D-01  0.69823D-01 -0.58471D-01 -0.26416D-01  0.50226D-02
```

```
                 C MATRIX    The C matrix for investigators
0.    1.    0.    0.    0.    0.    0.    0.    0.
0.    0.    1.    0.    0.    0.    0.    0.    0.
0.    0.    0.    1.    0.    0.    0.    0.    0.
0.    0.    0.    0.    1.    0.    0.    0.    0.
0.    0.    0.    0.    0.    1.    0.    0.    0.
```

```
                    ESTIMATED MODEL CONTRASTS
-0.92943D-01   -0.30425D 00    0.20820D-01    0.19908D 00    -0.20553D-01

          STANDARD DEVIATIONS OF THE ESTIMATED MODEL CONTRASTS
0.12590D 00     0.89219D-01     0.60882D-01     0.11652D 00     0.81115D-01
```

CHI-SQUARE = 25.0614   DF = 5   P = 0.0001   This is our chi-square test for investigators.

```
                 C MATRIX
0.    0.    0.    0.    0.    0.    1.    0.    0.
0.    0.    0.    0.    0.    0.    0.    1.    0.       The C matrix for treatments
0.    0.    0.    0.    0.    0.    0.    0.    1.
```

```
             ESTIMATED MODEL CONTRASTS
  -0.14319D-01    0.10228D 00    0.14669D-01

STANDARD DEVIATIONS OF THE ESTIMATED MODEL CONTRASTS
     0.53184D-01     0.53514D-01     0.52978D-01
```

CHI-SQUARE = 5.7173   DF = 3   P = 0.1262   This is our chi-square test for treatments.

To summarize the results of this analysis, the test for interaction between investigators and treatments (sprays) was nonsignificant ($p = .9557$). The tests for main effects showed a highly significant effect ($p = .0001$) due to differences among investigators, but the treatment effect achieved a level of significance of only .1262. Thus, although there were significant differences in the proportions of no responses for different investigators, there did not appear to be differences in the proportions of no responses for the different sprays.

▲

12.7 ▼ SUMMARY

This chapter is one of the key chapters in the text because it presents some of the practical problems associated with multiple regression problems. Step 1 of the process is to decide upon the dependent variable and a set of candidate independent variables for inclusion in the model. We discussed the invaluable nature of information from an expert in the subject matter field and the utility of some of the best subset regression techniques for choosing which variables to include in the model.

Step 2 is involved with the actual polynomial form of the particular multiple regression equation. In particular, attention should be paid to lack of fit of a proposed model to data collected on the dependent and independent variables of interest. A formal test for lack of fit of a polynomial model is possible where there are repetitions of observations at one or more than one settings of the independent variables. Lack of fit can also be examined using residual plots.

Following steps 1 and 2 as we've discussed them can sometimes be a problem, depending on the data that are available. For example, if data are available on many variables at the time when the multiple regression model is being formulated, then consultation with experts and application of one (or more) of the best subset regression techniques can be useful in culling the list of potential independent variables (step 1). The regression model is then modified in step 2 based on the discussions and analyses of step 1. Sometimes, however, data are not available on many possible independent variables. For these situations, step 1 consists of discussions with experts to determine which variables may be important predictors; data are then gathered on these variables. After the data are obtained on these candidate independent variables, the subset regression techniques and the model formulation techniques of step 2 can be applied to refine the model.

The final step of the multiple regression problem is to check the underlying assumptions of multiple regression: zero expectation, constant variance, normality, and independence. Although some formal tests were presented, violation of the assumption is checked best by closely examining the data using scatterplots and various residual plots. The more experience one gains in examining and interpreting data with these plots, the better will be the resulting regression equations.

▼ SUPPLEMENTARY EXERCISES

12.19 Use the following data to fit a model. Plot the data and suggest a polynomial model.

y	7	8	6	12	15	13	7	10	11	14	16	17
x	10	10	10	15	15	15	20	20	20	25	25	25

12.20 Refer to the data of Exercise 12.19.
  a. Fit the model $y = \beta_0 + \beta_1 x + \beta_2 x^2 + \beta_3 x^3 + \varepsilon$.
  b. Test for lack of fit using $\alpha = .05$.
  c. Examine a residual plot for violation of the regression assumptions.

12.21 Refer to Exercise 12.19. Suppose that the third, fifth, sixth, and tenth observations are missing.
  a. Fit a cubic model.
  b. Examine the residuals and compare the fits for the models of Exercises 12.20 and 12.21.

12.22 A pharmaceutical firm wanted to obtain information on the relationship between the dose level of a drug product and its potency. To do this, each of 15 test tubes were inoculated with a virus culture and incubated for 5 days at 30°C. Three test tubes were randomly assigned to each of the five different dose levels to be investigated (2, 4, 8, 16, and 32 mg). Each tube was injected with only one dose level and the response of interest (a measure of the protective strength of the product against the virus culture) was obtained. The data are given below.

Dose Level	Response
2	5, 7, 3
4	10, 12, 14
8	15, 17, 18
16	20, 21, 19
32	23, 24, 29

  a. Plot the data.
  b. Fit both a linear and a quadratic model to these data.
  c. Which model seems more appropriate?
  d. Compare your results in part b to those obtained in the SPSS computer output that follows. (Note: VAR01 = dose, VAR02 = response, VAR03 = dose^2.)

```
EXAMPLE OF REGRESSION MODELS

FILE   NONAME

**********************************************MULTIPLE REGRESSION****************************VARIABLE LIST 1
                                                                                            REGRESSION LIST 1

DEPENDENT VARIABLE      VAR02

VARIABLE(S) ENTERED ON STEP NUMBER 1      VAR01

MULTIPLE R           0.87923    ANALYSIS OF VARIANCE    DF    SUM OF SQUARES    MEAN SQUARE      F
R SQUARE             0.77305    REGRESSION               1       590.91613       590.91613   44.28025
ADJUSTED R SQUARE    0.75559    RESIDUAL                13       173.48387        13.34491
STANDARD ERROR       3.65307

---------- VARIABLES IN THE EQUATION ---------------------- VARIABLES NOT IN THE EQUATION --------
  VARIABLE     B       BETA     STD ERROR B      F      VARIABLE   BETA IN   PARTIAL   TOLERANCE    F
VAR01       0.57527  0.87923    0.08645       44.280
(CONSTANT)  8.66667

MAXIMUM STEP REACHED                                                              (continues)
```

```
EXAMPLE OF REGRESSION MODELS

FILE   NONAME

**********************************************MULTIPLE REGRESSION**********************************************

DEPENDENT VARIABLE:     VAR02      FROM      VARIABLE LIST 1
                                             REGRESSION LIST 1

              OBSERVED        PREDICTED                              PLOT OF STANDARDIZED RESIDUAL
   SEQNUM      VAR02            VAR02        RESIDUAL          -2.0    -1.0    0.0     1.0     2.0
     1        5.000000         9.817204     -4.817204                  *
     2        7.000000         9.817204     -2.817204                      *      |
     3        3.000000         9.817204     -6.817204                  *         |
     4        10.00000        10.86774      -.9677419                      *   |
     5        12.00000        10.96774       1.032258                       | *
     6        14.00000        10.96774       3.032258                       |   *
     7        15.00000        13.26882       1.731182                       |  *
     8        17.00000        13.26882       3.731182                       |    *
     9        18.00000        13.26882       4.731182                       |     *
    10        20.00000        17.87096       2.129032                       |  *
    11        21.00000        17.87096       3.129032                       |   *
    12        19.00000        17.87096       1.129032                       | *
    13        23.00000        27.07526      -4.075269                  *     |
    14        24.00000        27.07526      -3.075269                    *   |
    15        29.00000        27.07526       1.924730                       | *

DURBIN-WATSON TEST OF RESIDUAL DIFFERENCES COMPARED BY CASE ORDER (SEQNUM).

VARIABLE LIST 1.  REGRESSION LIST 1.  DURBIN-WATSON TEST   0.77105

**********************************MULTIPLE REGRESSION**************************VARIABLE LIST 1
                                                                         REGRESSION LIST 1
DEPENDENT VARIABLE..   VAR02
VARIABLE(S) ENTERED ON STEP NUMBER 1..   VAR01
                                         VAR03

MULTIPLE R        0.93888    ANALYSIS OF VARIANCE    DF   SUM OF SQUARES   MEAN SQUARE      F
R SQUARE          0.88150    REGRESSION               2.    673.82062      336.91031    44.63404
ADJUSTED R SQUARE 0.86175    RESIDUAL                12.     90.57938        7.54828
STANDARD ERROR    2.74741

----------------VARIABLES IN THE EQUATION--------------   --------VARIABLES NOT IN THE EQUATION----------

VARIABLE      B         BETA      STD ERROR B      F       VARIABLE   BETA IN   PARTIAL   TOLERANCE    F
VAR01      1.50633    2.30224     0.28836       27.287
VAR03     -0.02699   -1.46062     0.00814       10.983
(CONSTANT) 4.48366

ALL VARIABLES ARE IN THE EQUATION

EXAMPLE OF REGRESSION MODELS

FILE   NONAME

**********************************MULTIPLE REGRESSION***********************************************************

DEPENDENT VARIABLE:     VAR02      FROM      VARIABLE LIST 1
                                             REGRESSION LIST 1

              OBSERVED        PREDICTED                              PLOT OF STANDARDIZED RESIDUAL
   SEQNUM      VAR02            VAR02        RESIDUAL          -2.0    -1.0    0.0     1.0     2.0
     1        5.000000         7.388367     -2.388361                     *     |
     2        7.000000         7.388367     -.3883618                        *  |
     3        3.000000         7.388367     -4.388361                  *        |
     4        10.00000        10.07717      -.7716632E-01                       |
     5        12.00000        10.07717       1.922833                       | *
     6        14.00000        10.07717       3.922833                       |   *
     7        15.00000        14.80708       .1929159                       | *
     8        17.00000        14.80708       2.192915                       |*
     9        18.00000        14.80708       3.192915                       |  *
    10        20.00000        21.67615      -1.676154                    *   |
    11        21.00000        21.67615      -.6761543                      *  |
    12        19.00000        21.67615      -2.676154                    *    |
    13        23.00000        25.05122      -2.051233                     *   |
    14        24.00000        25.05122      -1.051233                      * |
    15        29.00000        25.05122       3.948766                       | *

DURBIN-WATSON TEST OF RESIDUAL DIFFERENCES COMPARED BY CASE ORDER (SEQNUM).

VARIABLE LIST 1.  REGRESSION LIST 1.  DURBIN-WATSON TEST 1.33140
```

12.23 Refer to the data of Exercise 12.22. Many times, a logarithmic transformation can be used on the dose levels to linearize the response with respect to the independent variable.

a. Refer to a set of log tables or an electronic calculator to obtain the logarithms of the five dose levels.

b. Where x_1 denotes the log dose, fit the model

$$y = \beta_0 + \beta_1 x_1 + \varepsilon.$$

c. Compare your results in part b to those shown in the computer printout that follows. (Note: VAR04 = log dose.)

d. Which of the three models seems more appropriate? Why?

```
EXAMPLE OF REGRESSION MODELS

FILE   NONAME

*******************************************MULTIPLE REGRESSION****************************VARIABLE LIST 1
                                                                                         REGRESSION LIST 1

DEPENDENT VARIABLE       VAR02

VARIABLE(S) ENTERED ON STEP NUMBER 1      VAR04

MULTIPLE R          0.96412    ANALYSIS OF VARIANCE    DF    SUM OF SQUARES    MEAN SQUARE        F
R SQUARE            0.92953    REGRESSION               1.      710.53334      710.53334     171.47775
ADJUSTED R SQUARE   0.92411    RESIDUAL                13.       53.86666        4.14359
STANDARD ERROR      2.03558

--------- VARIABLES IN THE EQUATION----------------------VARIABLES NOT IN THE EQUATION -------

VARIABLE       B        BETA      STD ERROR E      F      VARIABLE   BETA IN  PARTIAL   TOLERANCE     F
VAR04       16.16672   0.96412    1.23458      171.478
(CONSTANT)   1.20000

MAXIMUM STEP REACHED

EXAMPLE OF REGRESSION MODELS

FILE   NONAME

*******************************************MULTIPLE REGRESSION**************************************************
DEPENDENT VARIABLE:      VAR02      FROM       VARIABLE LIST 1
                                               REGRESSION LIST 1

             OBSERVED         PREDICTED                          PLOT OF STANDARDIZED RESIDUAL
  SEQNUM       VAR02            VAR02          RESIDUAL      -2.0    -1.0    0.0    1.0    2.0
     1        5.000000        6.066669        -1.066667                       *  |
     2        7.000000        6.066669         .9333332                       | *
     3        3.000000        6.066669        -3.066667               *       |
     4       10.00000        10.93333         -.9333344                     * |
     5       12.00000        10.93333         1.066665                       |  *
     6       14.00000        10.93333         3.066665                       |    *
     7       15.00000        15.80000         -.7999990                     * |
     8       17.00000        15.80000         1.200001                       |*
     9       18.00000        15.80000         2.200001                       |  *
    10       20.00000        20.66666         -.6666647                     * |
    11       21.00000        20.66666          .3333353                       |*
    12       19.00000        20.66666        -1.666664                   *   |
    13       23.00000        25.53333        -2.533335                *      |
    14       24.00000        25.53333        -1.533335                    *  |
    15       29.00000        25.53333         3.466664                       |   *

DURBIN-WATSON TEST OF RESIDUAL DIFFERENCES COMPARED BY CASE ORDER (SEQNUM).

VARIABLE LIST 1.  REGRESSION LIST 1.   DURBIN-WATSON TEST 1.71667
```

12.24 An experiment was conducted to examine the weather resistance of a new commercial paint as a function of two independent variables, temperature x_1 and exposure time x_2. The sample data are listed here.

y	120	101	110	105	92	130
x_1 (°C)	−10	−10	0	0	10	10
x_2 (months)	1	3	2	2	1	3

a. Fit the model

$$y = \beta_0 + \beta_1 x_1 + \beta_2 x_2 + \beta_3 x_1 x_2 + \varepsilon.$$

b. Examine the residuals and comment on your findings.

12.25 Refer to Exercise 12.24.
a. Could we fit the following model?

$$y = \beta_0 + \beta_1 x_1 + \beta_2 x_1^2 + \beta_3 x_2 + \beta_4 x_2^2 + \beta_5 x_1 x_2 + \beta_6 x_1 x_2^2 + \beta_7 x_1^2 x_2 + \beta_8 x_1^2 x_2^2 + \varepsilon.$$

b. Test for lack of fit of the model in Exercise 12.24. Make a recommendation.

12.26 (Optional) Refer to Example 12.18. If we wish to make pairwise comparisons among the four treatments, we can do so by identifying different **C** matrices for a null hypothesis of the form $\mathbf{C}\boldsymbol{\beta} = \mathbf{0}$. For example, to test for no significant difference between treatments A and B, we would use the **C** matrix

$$\mathbf{C} = [0 \quad 0 \quad 0 \quad 0 \quad 0 \quad 0 \quad 1 \quad -1 \quad 0].$$

Identify the **C** matrices for the following comparisons. (Hint: For comparisons with treatment D, recall that $\beta_4 = -\beta_1 - \beta_2 - \beta_3$.)
a. A versus C **b.** A versus D **c.** B versus D **d.** C versus D

12.27 (Optional) Refer to Exercise 12.26. Run the Gencat procedure to make the five pairwise treatment comparisons of the nasal sprays: A versus B, A versus C, A versus D, B versus D, C versus D. Note that the number of **C** matrices to be used will change the parameter card of Example 12.18.

12.28 The abrasive effect of a wear tester for experimental fabrics was tested on a particular fabric while run at six different machine speeds. Forty-eight identical 5-inch-square pieces of fabric were cut, with eight squares randomly assigned to each of the six machine speeds 100, 120, 140, 160, 180, and 200 revolutions per minute (rev/min). The order of assignment of the squares to the machine was random, with each square tested for a 3-minute period at the appropriate machine setting. The amount of wear was measured and recorded for each square. The data appear in the accompanying table.
a. Plot the mean data per revolutions per minute level and suggest a model.
b. Fit the suggested model to the data.
c. Suggest which residual plots might be useful in checking the assumptions underlying the model.

Machine Speed (rev/min)	Wear
100	23.0, 23.5, 24.4, 25.2, 25.6, 26.1, 24.8, 25.6
120	26.7, 26.1, 25.8, 26.3, 27.2, 27.9, 28.3, 27.4
140	28.0, 28.4, 27.0, 28.8, 29.8, 29.4, 28.7, 29.3
160	32.7, 32.1, 31.9, 33.0, 33.5, 33.7, 34.0, 32.5
180	43.1, 41.7, 42.4, 42.1, 43.5, 43.8, 44.2, 43.6
200	54.2, 43.7, 53.1, 53.8, 55.6, 55.9, 54.7, 54.5

12.29 Refer to the data of Exercise 12.28. Suppose that another variable was controlled and that the first four squares at each speed were treated with a .2 concentration of protective coating, while the second four squares were treated with a .4 concentration of the same coating. Given that x_1 denotes the machine speed and x_2 denotes the concentration of the protective coating, fit these models:

$$y = \beta_0 + \beta_1 x_1 + \beta_2 x_1^2 + \beta_3 x_2 + \varepsilon$$
$$y = \beta_0 + \beta_1 x_1 + \beta_2 x_1^2 + \beta_3 x_2 + \beta_4 x_1 x_2 + \beta_5 x_1^2 x_2 + \varepsilon.$$

12.30 A laundry detergent manufacturer was interested in testing a new product prior to market release. One area of concern was the relationship between the height of the detergent suds in a washing machine as a function of the amount of detergent added and the degree of agitation in the wash cycle. For a standard size washing machine tub filled to the full level, random assignments of different agitation levels (measured in minutes) and amounts of detergent were made and tested on the washing machine. The data are as shown in the accompanying table.

a. Plot the data and suggest a model.

b. Does the assumption of normality appear to hold?

c. Fit an appropriate model.

d. Use residual plots to detect possible violations of the assumptions.

Height, y	Agitation, x_1	Amount, x_2
28.1	1	6
32.3	1	7
34.8	1	8
38.2	1	9
43.5	1	10
60.3	2	6
63.7	2	7
65.4	2	8
69.2	2	9
72.9	2	10
88.2	3	6
89.3	3	7
94.1	3	8
95.7	3	9
100.6	3	10

12.31 Refer to Exercise 12.30. Would the following model be more appropriate? Why or why not?

$$y = \beta_0 + \beta_1 x_1 + \beta_2 x_1^2 + \beta_3 x_2 + \beta_4 x_2^2 + \beta_5 x_1 x_2 + \beta_6 x_1 x_2^2 + \beta_7 x_1^2 x_2 + \beta_8 x_1^2 x_2^2 + \varepsilon.$$

12.32 Refer to the data of Exercise 12.30.
a. Can we test for lack of fit for the following model?

$$y = \beta_0 + \beta_1 x_1 + \beta_2 x_1^2 + \beta_3 x_2 + \beta_4 x_2^2 + \beta_5 x_1 x_2 + \beta_6 x_1 x_2^2 + \beta_7 x_1^2 x_2 + \beta_8 x_1^2 x_2^2 + \varepsilon.$$

b. Write the complete model for the sample data. Note that if there were replication at one or more design points, the number of degrees of freedom for SS_{Lack} would be identical to the difference between the number of parameters in the complete model and the number of parameters in the model of part a.

12.33 Refer to Example 12.9.
a. Identify the parameters in the model.
b. Fit the "complete" model.
c. Draw conclusions relative to the standard and luxury models.

12.34 (Optional) The time to failure of a particular jet engine component has a probability distribution given by

$$f(t) = \frac{\beta t^{\beta - 1}}{\theta^\beta} e^{-(t/\theta)^\beta} \qquad t, \theta, \beta > 0.$$

For this same component type, we can represent the rate of failure for a population of components as

$$h(t) = \frac{\beta t^{\beta - 1}}{\theta^\beta}.$$

Data were collected for the first 24 months of engine service, and component failure rates were obtained for each month. SAS was used to fit the nonlinear $h(t)$ function, sometimes referred to as the *hazard* function.
a. Identify estimates of the scale parameter (θ) and the shape parameter (β).

```
           LISTING OF DATA
    OBS      MONTH      RATE
     1         1       0.17977
     2         2       0.00000
     3         3       0.00000
     4         4       0.32022
     5         5       0.00000
     6         6       0.00000
     7         7       0.00000
     8         8       0.00000
     9         9       0.48333
    10        10       0.35694
    11        11       1.59485
    12        12       1.10666
    13        13       1.23159
    14        14       1.48853        (continues)
```

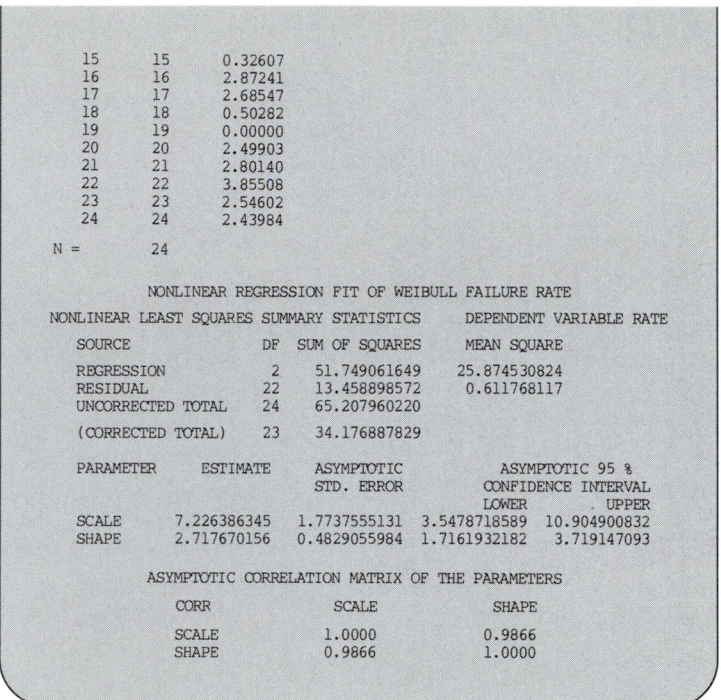

b. Comment on the fit of the data based on the plots.

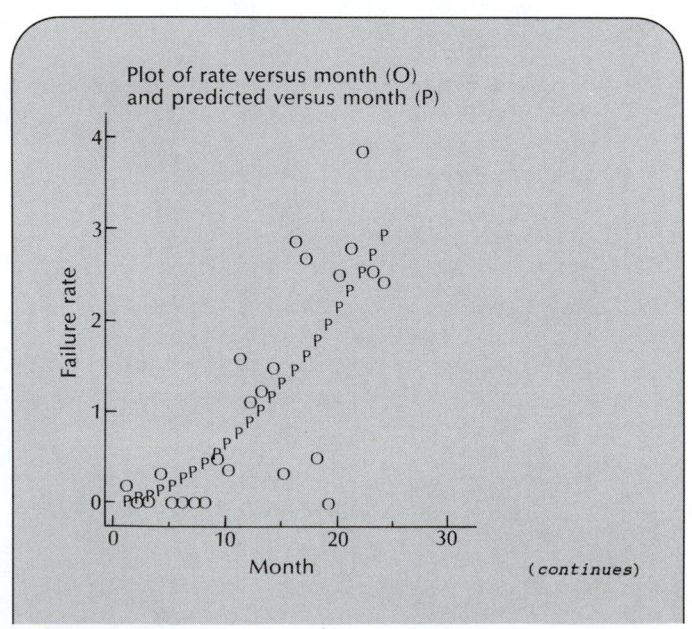

(*continues*)

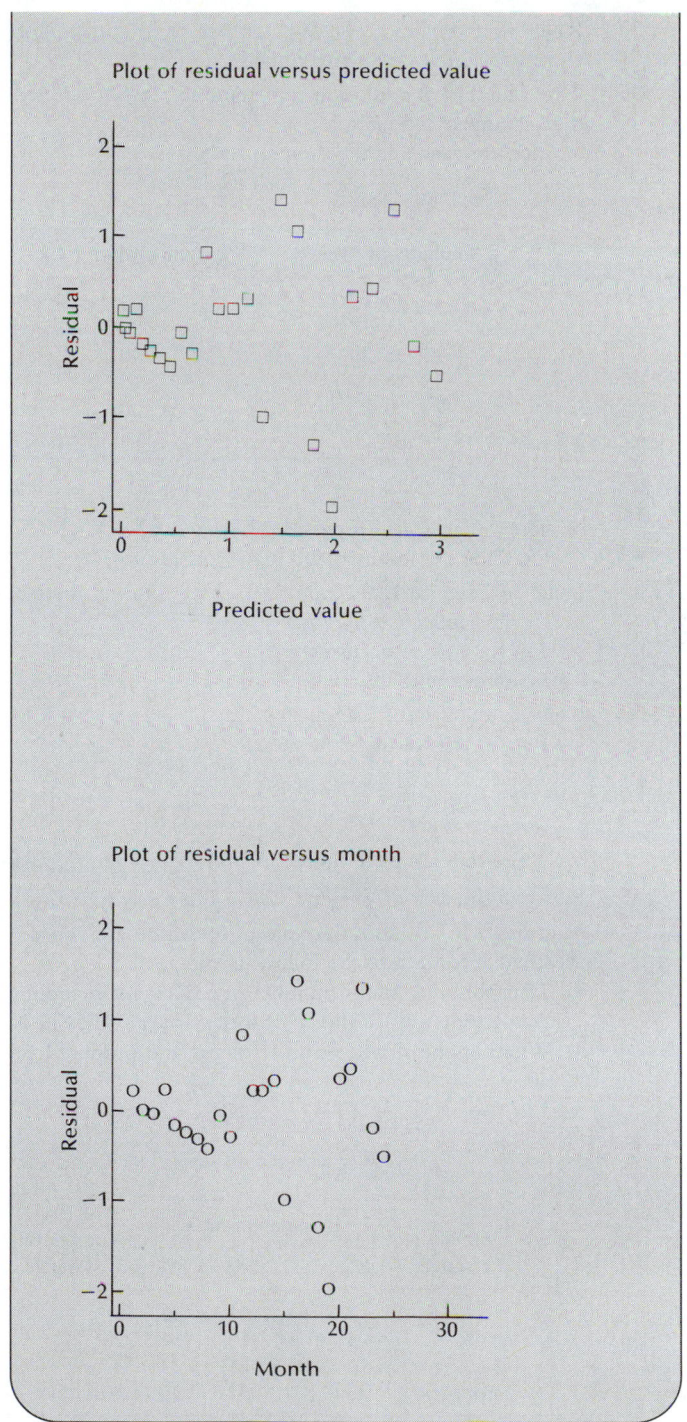

12.35 Refer to Exercise 12.34. What other technique might we use to analyze these data? Comment on any potential pitfalls.

12.36 The solubility of a solution was examined for six different temperature settings, shown in the accompanying table.
  a. Plot the data, and fit as appropriate.

y, Solubility by Weight	x, Temperature (°C)
43, 45, 42	0
32, 33, 37	25
21, 28, 29	50
15, 14, 9	75
12, 10, 8	100
7, 6, 2	125

  b. Test for lack of fit if possible. Use $\alpha = .05$.
  c. Examine the residuals and draw conclusions.

12.37 Refer to Exercise 12.36. Suppose we are missing observations 5, 8, and 14.
  a. Fit the model $y = \beta_0 + \beta_1 x + \beta_2 x^2 + \varepsilon$.
  b. Test for lack of fit, using $\alpha = .05$.
  c. Again examine the residuals.

12.38 Refer to the data of Exercise 12.29.
  a. Test for lack of fit of the model

$$y = \beta_0 + \beta_1 x_1 + \beta_2 x_1^2 + \beta_3 x_2 + \beta_4 x_1 x_2 + \beta_5 x_1^2 x_2 + \varepsilon.$$

  b. Write the complete model for this experimental situation.

12.39 Refer to the data of Exercise 12.22. Test for lack of fit of a quadratic model.

12.40 (Optional) In a random sample of 498 motorists on a major toll road, each driver was classified according to the following categories.
  1. Transportation status (number of passengers including the driver): 1, 2, 3, 4 or more.
  2. Destination area of the city for the driver: NE, SE, NW, SW.
  3. Whether the driver planned to use the proposed public transportation system: yes, no.

	Yes					No			
	Destination Area					Destination Area			
Number of Passengers	NE	SE	NW	SW	Number of Passengers	NE	SE	NW	SW
1	7	10	15	22	1	31	27	15	7
2	4	8	13	14	2	18	16	9	4
3	13	10	21	29	3	12	13	4	3
4 or more	28	17	32	35	4 or more	15	18	16	12

These data, the number of motorists in each classification, are summarized in two two-way tables. Analyze these data using Gencat or some other multidimensional contingency table analysis program. If you use Gencat, follow the format of Example 12.18. Draw your conclusions.

12.41 Refer to Example 12.17. Suggest another way to test for parallelism. Use the computer output given in Example 12.17.

12.42 A psychologist is interested in examining the effects of sleep deprivation on a person's ability to perform simple arithmetic tasks. To do this, prospective subjects are screened to obtain individuals whose daily sleep patterns were closely matched. From this group, 20 subjects are chosen. Each individual selected is randomly assigned to one of five groups, four individuals per group.

Group 1:   0 hours of sleep
Group 2:   2 hours of sleep
Group 3:   4 hours of sleep
Group 4:   6 hours of sleep
Group 5:   8 hours of sleep

All subjects are then placed on a standard routine for the next 24 hours.

The following day after breakfast, each individual is tested to determine the number of arithmetic additions done correctly in a 10-minute period. That evening the amount of sleep each person is allowed depends on the group to which he or she had been assigned. The following morning after breakfast, each person is again tested using a different but equally difficult set of additions.

Let the response of interest be the difference in the number of correct responses on the first test day minus the number correct on the second test day. The data are presented here.

Group	Response, y
1	39, 33, 41, 40
2	25, 29, 34, 26
3	10, 18, 14, 17
4	4, 6, −1, 9
5	−5, 0, −3, −8

a. Plot the sample data and use the plot to suggest a model.
b. Fit the suggested model.
c. Examine the fitted model for possible violation of assumptions.

12.43 An experiment was conducted to determine the relationship between the amount of warping y for a particular alloy and the temperature (in °C) under which the experiment was conducted. The sample data appear in the table below. Note that three observations were taken at each temperature setting. Use the computer output that follows to complete parts a through d.

Amount of Warping	Temperature (°C)
10, 13, 12	15
14, 12, 11	20
14, 12, 16	25
18, 19, 22	30
25, 21, 20	35
23, 25, 26	40
30, 31, 34	45
35, 33, 38	50

a. Plot the data to determine whether a linear or quadratic model appears more appropriate.

b. If a linear model is fit, indicate the prediction equation. Superimpose the prediction equation over the scatter diagram of y versus x.

c. If a quadratic model is fit, identify the prediction equation. Superimpose the quadratic prediction equation on the scatter diagram. Which fit looks better, the linear or the quadratic?

d. Predict the amount of warping at a temperature of 27°C, using both the linear and the quadratic prediction equations.

```
    LISTING OF DATA
  OBS     WARPING     TEMP
   1         10        15
   2         13        15
   3         12        15
   4         14        20
   5         12        20
   6         11        20
   7         14        25
   8         12        25
   9         16        25
  10         18        30
  11         19        30
  12         22        30
  13         25        35
  14         21        35
  15         20        35
  16         23        40
  17         25        40
  18         26        40
  19         30        45
  20         31        45
  21         34        45
  22         35        50
  23         33        50
  24         38        50

 N =    24
                            QUADRATIC REGRESSION
                      GENERAL LINEAR MODELS PROCEDURE

DEPENDENT VARIABLE: WARPING

SOURCE                DF    SUM OF SQUARES    MEAN SQUARE    F VALUE   PR > F   R-SQUARE     C.V.
MODEL                  1     1571.62698413   1571.62698413   265.55   0.0001   0.923491   11.3593
ERROR                 22      130.20634921      5.91847042            ROOT MSE        WARPING MEAN
CORRECTED TOTAL       23     1701.83333333                          2.43279066       21.416666667
                                                                              (continues)
```

SOURCE	DF	TYPE I SS	F VALUE	PR > F	DF	TYPE III SS	F VALUE	PR > F
TEMP	1	1571.62698413	265.55	0.0001	1	1571.62698413	265.55	0.0001

PARAMETER	ESTIMATE	T FOR HO: PARAMETER=0	PR > |T|	STD ERROR OF ESTIMATE
INTERCEPT	-1.53968254	-1.03	0.3138	1.49370995
TEMP	0.70634921	16.30	0.0001	0.04334604

OBSERVATION	OBSERVED VALUE	PREDICTED VALUE	RESIDUAL
1	10.00000000	9.05555556	0.94444444
2	13.00000000	9.05555556	3.94444444
3	12.00000000	9.05555556	2.94444444
4	14.00000000	12.58730159	1.41269841
5	12.00000000	12.58730159	-0.58730159
6	11.00000000	12.58730159	-1.58730159
7	14.00000000	16.11904762	-2.11904762
8	12.00000000	16.11904762	-4.11904762

QUADRATIC REGRESSION

GENERAL LINEAR MODELS PROCEDURE

DEPENDENT VARIABLE: WARPING

9	16.00000000	16.11904762	-0.11904762
10	18.00000000	19.65079365	-1.65079365
11	19.00000000	19.65079365	-0.65079365
12	22.00000000	19.65079365	2.34920635
13	25.00000000	23.18253968	1.81746032
14	21.00000000	23.18253968	-2.18253968
15	20.00000000	23.18253968	-3.18253968
16	23.00000000	26.71428571	-3.71428571
17	25.00000000	26.71428571	-1.71428571
18	26.00000000	26.71428571	-0.71428571
19	30.00000000	30.24603175	-0.24603175
20	31.00000000	30.24603175	0.75396825
21	34.00000000	30.24603175	3.75396825
22	35.00000000	33.77777778	1.22222222
23	33.00000000	33.77777778	-0.77777778
24	38.00000000	33.77777778	4.22222222

SUM OF RESIDUALS	0.00000000
SUM OF SQUARED RESIDUALS	130.20634921
SUM OF SQUARED RESIDUALS - ERROR SS	0.00000000
FIRST ORDER AUTOCORRELATION	0.47433866
DURBIN-WATSON D	0.90755753

QUADRATIC REGRESSION

GENERAL LINEAR MODELS PROCEDURE

DEPENDENT VARIABLE: WARPING

SOURCE	DF	SUM OF SQUARES	MEAN SQUARE	F VALUE	PR > F	R-SQUARE	C.V.
MODEL	2	1613.92063492	806.96031746	192.76	0.0001	0.948342	9.5535
ERROR	21	87.91269841	4.18631897			ROOT MSE	WARPING MEAN
CORRECTED TOTAL	23	1701.83333333				2.04604960	21.41666667

SOURCE	DF	TYPE I SS	F VALUE	PR > F	DF	TYPE III SS	F VALUE	PR > F
TEMP	1	1571.62698413	375.42	0.0001	1	0.15969355	0.04	0.8470
TEMP*TEMP	1	42.29365079	10.10	0.0045	1	42.29365079	10.10	0.0045

PARAMETER	ESTIMATE	T FOR HO: PARAMETER=0	PR > |T|	STD ERROR OF ESTIMATE
INTERCEPT	9.17857143	2.55	0.0186	3.59852022
TEMP	-0.04682540	-0.20	0.8470	0.23974742
TEMP*TEMP	0.01158730	3.18	0.0045	0.00364553

12.44 One use of multiple regression is in the setting of performance standards. In other words, a regression equation can be used to predict how well an individual ought to perform when certain conditions are met. In a study of this type, designed to identify an equation that could be used to predict the sales of individual salespeople, data from a random sample of 50 sales territories from four sections of the country (northeast, southeast, midwest, and west) were collected. Data on individual sales performances, as well as on several potential predictor variables, were collected. The variables were as follows.

y = sales-territory performance measured by aggregate sales, in units credited to territory salesperson

x_1 = time with company (months)

x_2 = advertising, or company effort (dollar expenditures in ads in territory)

x_3 = market share (the weighted average of past market share magnitudes for four previous years)

x_4 = indicator variable for section of country (1 = northeast, 0 = otherwise)

x_5 = indicator variable for section of country (1 = southeast, 0 = otherwise)

x_6 = indicator variable for section of country (1 = midwest, 0 = otherwise)

x_7 = indicator variable (1 = male salesperson, 0 = female salesperson)

These data were analyzed using Minitab, with the following results:

```
MTB > DESCRIBE C1-C10

        N      MEAN     MEDIAN    TRMEAN     STDEV    SEMEAN
Y      50      3335       3396      3277      1579       223
X1     50     96.62      85.00     93.86     66.33      9.38
X2     50      5002       5069      4915      2370       335
X3     50     7.335      7.305     7.297     1.668     0.236
C5     50     2.460      2.000     2.455     1.129     0.160
X4     50    0.8200     1.0000    0.8636    0.3881    0.0549
X5     50    0.2600     0.0000    0.2273    0.4431    0.0627
X6     50    0.2600     0.0000    0.2273    0.4431    0.0627
X7     50    0.2400     0.0000    0.2045    0.4314    0.0610

        MIN       MAX        Q1        Q3
Y       131      7205      2033      4367
X1      000    237.00     40.00    144.25
X2      222     10832      3038      6564
X3    4.131    11.205     5.987     8.569
C5    1.000     4.000     1.000     3.250
X4   0.0000    1.0000    1.0000    1.0000
X5   0.0000    1.0000    0.0000    1.0000
X6   0.0000    1.0000    0.0000    1.0000
X7   0.0000    1.0000    0.0000    0.2500
MTB > REGRESS 'Y' ON 7 'X1' 'X2' 'X3' 'X4' 'X5' 'X6' 'X7'

The regression equation is
Y = 16.4 - 0.000546X1 + 0.667X2 + 0.0302X3 - 0.116X4 - 0.041X5 -33.3X6 - 33.6X7

Predictor      Coef        Stdev      t-ratio
Constant     16.3944      0.2931        55.94
    X1     -0.0005463    0.0007607      -0.72
    X2      0.666689     0.000047    14315.675
    X3      0.03024      0.06467         0.47
    X4      -0.1163      0.1128         -1.03
    X5      -0.0412      0.1201         -0.34
    X6     -33.3155      0.1204       -276.81
    X7     -33.6118      0.1185       -283.70
S = 0.2864    R-sq =100.0%    R-sq(adj) = 100.0%       (continues)
```

```
Analysis of Variance
SOURCE         DF        SS          MS
Regression      7    122189056     17455576
Error          42            3            0
Total          49    122189056


SOURCE    DF     SEQ SS
  X1       1    33243924
  X2       1    88931584
  X3       1           1
  X4       1          80
  X5       1        4972
  X6       1        1880
  X7       1        6602
```

Obs.	X1	Y	Fit	Stdev. Fit	Residual	St.Resid.
1	62	3407.00	3406.54	0.09	0.46	1.68
2	70	131.00	131.17	0.14	-0.17	-0.69
3	186	4650.00	4649.93	0.09	0.07	0.27
4	13	1971.00	1970.91	0.11	0.09	0.35
5	20	4168.00	4167.94	0.11	0.06	0.21
6	0	3047.00	3047.28	0.10	-0.28	-1.03
7	31	1196.00	1195.91	0.13	0.09	0.36
8	61	2415.00	2414.91	0.10	0.09	0.34
9	48	1987.00	1987.12	0.09	-0.12	-0.46
10	101	2214.00	2213.84	0.10	0.16	0.61
11	145	4333.00	4333.14	0.27	-0.14	-1.36X
12	200	6253.00	6253.08	0.12	-0.08	-0.29
13	81	1714.00	1713.87	0.12	0.13	0.49
14	124	5146.00	5146.01	0.09	-0.01	-0.04
15	24	3469.00	3469.27	0.11	-0.27	-1.04
16	216	4124.00	4123.60	0.11	0.40	1.53
17	232	3851.00	3851.17	0.14	-0.17	-0.69
18	109	2172.00	2171.83	0.10	0.17	0.64
19	75	1743.00	1743.25	0.12	-0.25	-0.97
20	5	2269.00	2268.93	0.11	0.07	0.27
21	12	3429.00	3429.24	0.10	-0.24	-0.88
22	90	1986.00	1985.83	0.10	0.17	0.64
23	209	3623.00	3623.21	0.12	-0.21	-0.82
24	167	5429.00	5429.16	0.15	-0.16	-0.64
25	170	4511.00	4511.22	0.10	-0.22	-0.81
26	42	1478.00	1477.94	0.12	0.06	0.24
27	167	3385.00	3385.22	0.11	-0.22	-0.84
28	98	1660.00	1660.84	0.11	-0.84	-3.16R
29	144	1212.00	1211.69	0.12	0.31	1.20
30	78	4592.00	4592.00	0.09	0.00	0.00
31	116	2876.00	2875.85	0.09	0.15	0.55
32	89	4349.00	4349.02	0.09	-0.02	-0.06
33	37	2096.00	2095.80	0.09	0.20	0.72
34	34	5308.00	5308.07	0.11	-0.07	0.26
35	165	5731.00	5730.01	0.10	0.99	3.70R
36	41	1121.00	1120.84	0.11	0.16	0.62
37	80	2356.00	2355.91	0.12	0.09	0.34
38	140	7205.00	7204.80	0.13	0.20	0.79
39	48	3562.00	3561.96	0.13	0.04	0.15
40	203	4133.00	4132.94	0.11	0.06	0.23
41	71	2049.00	2049.12	0.09	-0.12	-0.42
42	13	2512.00	2511.90	0.09	0.10	0.36
43	144	3722.00	3721.89	0.09	0.11	0.40
44	11	2806.00	2805.74	0.13	0.26	1.01
45	34	1477.00	1477.10	0.09	-0.10	-0.37
46	94	4040.00	4039.96	0.08	0.04	0.16
47	237	6633.00	6633.36	0.12	-0.36	-1.37
48	115	3203.00	3203.04	0.12	-0.04	-0.17
49	66	4423.00	4423.27	0.10	-0.27	-1.00
50	113	5563.00	5563.38	0.10	-0.38	-1.40

R denotes an obs. with a large st. resid.
X denotes an obs. whose X value gives it large influence.

Conduct a test to determine whether salespersons in the west make more (other things being equal) than salespersons in the northeast. Give the null and alternative hypotheses, the computed and the critical values of the test statistic, and your conclusion. Use $\alpha = .05$.

12.45 Refer to Exercise 12.44. What is the estimated average increase in sales territory performance of a salesperson when advertising in the territory increases by $1,000?

12.46 Refer to Exercise 12.44. Conduct a test to determine whether males sell (on average) 200 units more than females (other things being equal). Use $\alpha = .05$.

12.47 Refer to Exercise 12.44. A particular concern of one company sales manager is that different regional attitudes may well affect the performance of males and females unequally.

  a. Suggest a new regression model that allows for the possibility of an interaction effect between the four regions of the country and the gender of the salesperson.

  b. Interpret the "new" βs in this model.

$ **12.48** A random sample of 22 residential properties was used in a regression of price on nine different independent variables. The variables used in this study were:

> PRICE = selling price (dollars)
> BATHS = number of baths (powder room = 1/2 bath)
> BEDA = dummy variable for number of bedrooms (1 = 2 bedrooms, 0 = otherwise)
> BEDB = dummy variable for number of bedrooms (1 = 3 bedrooms, 0 = otherwise)
> BEDC = dummy variable for number of bedrooms (1 = 4 bedrooms, 0 = otherwise)
> CARA = dummy variable for type of garage (1 = no garage, 0 = otherwise)
> CARB = dummy variable for type of garage (1 = one-car garage, 0 = otherwise)
>   AGE = age in years
>   LOT = lot size in square yards
>   DOM = days on the market

In this study, homes had two, three, four, or five bedrooms and either no garage or one- or two-car garages. Hence, we are using two dummy variables to code for the three categories of garage.

The data were analyzed using Minitab, with the results that follow. Using the full regression model (nine independent variables), estimate the average difference in selling price between:

  a. Properties with no garage and properties with a one-car garage.
  b. Properties with a one-car and properties with a two-car garage.
  c. Properties with no garage and properties with a two-car garage.

```
MTB > NAME C1 'PRICE' C2 'BATHS' C3 'BEDA' C4 'BEDB' C5 'BEDC'
MTB > NAME C6 'CARA' C7 'CARB' C8 'AGE' C9 'LOT' C10 'DOM'
MTB > READ C1-C10
      22 ROWS READ
MTB > END
MTB > SAVE 'HOUSES'

Worksheet saved into file: HOUSES.MTW
MTB > PRINT C1-C10
                                                          (continues)
```

```
    ROW    PRICE   BATHS   BEDA    BEDB    BEDC    CARA    CARB    AGE    LOT    DOM

      1    25750    1.0     1       0       0       1       0      23    9680   164
      2    37950    1.0     0       1       0       0       1       7    1889    67
      3    46450    2.5     0       1       0       0       0       9    1941   315
      4    46550    2.5     0       0       1       1       0      18    1813    61
      5    47950    1.5     1       0       0       0       1       2    1583   234
      6    49950    1.5     0       1       0       0       0      10    1533   116
      7    52450    2.5     0       0       1       0       0       4    1667   162
      8    54050    2.0     0       1       0       0       1       5    3450    80
      9    54850    2.0     0       1       0       0       0       5    1733    63
     10    52050    2.5     0       1       0       0       0       5    3727   102
     11    54392    2.5     0       1       0       0       0       7    1725    48
     12    53450    2.5     0       1       0       0       0       3    2811   423
     13    59510    2.5     0       1       0       0       1      11    5653   130
     14    60102    2.5     0       1       0       0       0       7    2333   159
     15    63850    2.5     0       0       1       0       0       6    2022   314
     16    62050    2.5     0       0       0       0       0       5    2166   135
     17    69450    2.0     0       1       0       0       0      15    1836    71
     18    82304    2.5     0       0       1       0       0       8    5066   338
     19    81850    2.0     0       1       0       0       0       0    2333   147
     20    70050    2.0     0       1       0       0       0       4    2904   115
     21   112450    2.5     0       0       1       0       0       1    2930    11
     22   127050    3.0     0       0       1       0       0       9    2904    36

MTB > DESCRIBE C1-C10


              N      MEAN    MEDIAN   TRMEAN    STDEV    SEMEAN
PRICE        22     62023    54621    60585    22749     4850
BATHS        22     2.182    2.500    2.200    0.524    0.112
BEDA         22    0.0909   0.0000   0.0500   0.2942   0.0627
BEDB         22     0.591    1.000    0.600    0.503    0.107
BEDC         22    0.2727   0.0000   0.2500   0.4558   0.0972
CARA         22    0.0909   0.0000   0.0500   0.2942   0.0627
CARB         22    0.1818   0.0000   0.1500   0.3948   0.0842
AGE          22      7.45     6.50     7.05     5.48     1.17
LOT          22      2895     2250     2624     1868      398
DOM          22     149.6    123.0    142.9    109.8     23.4


              MIN      MAX       Q1       Q3
PRICE        25750   127050    49450    69600
BATHS        1.000    3.000    2.000    2.500
BEDA        0.0000   1.0000   0.0000   0.0000
BEDB         0.000    1.000    0.000    1.000
BEDC        0.0000   1.0000   0.0000   1.0000
CARA        0.0000   1.0000   0.0000   0.0000
CARB        0.0000   1.0000   0.0000   0.0000
AGE          0.00    23.00     4.00     9.25
LOT          1533     9680     1793     3060
DOM          11.0    423.0     66.0    181.5


MTB > REGRESS 'PRICE' ON 9 PREDICTORS C2-C10


The regression equation is
PRICE = 39617 + 11686 BATHS + 15128 BEDA + 2477 BEDB + 26114 BEDC - 44023 CARA
        - 12375 CARB - 506 AGE + 3.40 LOT - 86.0 DOM

Predictor      Coef       Stdev     t-ratio        p
Constant      39617       30942        1.28     0.225
BATHS         11686       10428        1.12     0.284
BEDA          15128       26254        0.58     0.575
BEDB           2477       17783        0.14     0.892
BEDC          26114       18118        1.44     0.175
CARA         -44023       22775       -1.93     0.077
CARB         -12375       10759       -1.15     0.272
AGE            -506        1111       -0.46     0.657
LOT           3.399       2.504        1.36     0.200
DOM          -86.05       35.72       -2.41     0.033

s = 16531      R-sq = 69.8%     R-sq(adj) = 47.2%
```

(continues)

```
Analysis of Variance

SOURCE       DF          SS          MS        F       p
Regression    9  7588195840   843132864     3.09    0.036
Error        12  3279394048   273282848
Total        21 10867590144


SOURCE       DF      SEQ SS
BATHS         1  3352323072
BEDA          1    24291496
BEDB          1   668205888
BEDC          1   261898224
CARA          1  1261090304
CARB          1   133807624
AGE           1        5848
LOT           1   300736096
DOM           1  1585837312


Unusual Observations
Obs.    BATHS      PRICE      Fit Stdev.Fit  Residual   St.Resid
   7     2.50      52450    84651     7506     -32201      -2.19R
  16     2.50      62050    62050    16531          0       * X

R denotes an obs. with a large st. resid.
X denotes an obs. whose X value gives it large influence.

MTB > REGRESS 'PRICE' ON 7 PREDICTORS C2,C3,C5,C6,C7.C9,C10

The regression equation is
PRICE = 39091 + 11712 BATHS + 14183 BEDA + 24531 BEDC - 50962 CARA - 12121 CARB
          + 3.08 LOT - 84.8 DOM

Predictor       Coef       Stdev     t-ratio        p
Constant       39091       21445       1.82      0.090
BATHS          11712        9531       1.23      0.239
BEDA           14183       16759       0.85      0.412
BEDC           24531        9021       2.72      0.017
CARA          -50962       15878      -3.21      0.006
CARB          -12121       10010      -1.21      0.246
LOT            3.082       2.231       1.38      0.189
DOM           -84.81       33.24      -2.55      0.023

s = 15443      R-sq = 69.3%    R-sq(adj) = 53.9%

Analysis of Variance

SOURCE       DF          SS          MS         F        p
Regression    7  7528777728  1075539712     4.51    0.008
Error        14  3338812416   238486608
Total        21 10867590144


SOURCE       DF      SEQ SS
BATHS         1  3352323072
BEDA          1    24291496
BEDC          1   929454592
CARA          1  1261501440
CARB          1   133856232
LOT           1   274448000
DOM           1  1552902528

Unusual Observations
Obs.    BATHS      PRICE      Fit Stdev.Fit  Residual   St.Resid
   7     2.50      52450    84299     6973     -31849      -2.31R

R denotes an obs. with a large st. resid.

MTB > REGRESS 'PRICE' ON 6 PREDICTORS C2,C5,C6,C7,C9,C10

The regression equation is
PRICE = 44534 + 8336 BATHS + 24649 BEDC - 47007 CARA - 10588 CARB + 3.54 LOT
          - 76.7 DOM
```

(continues)

```
Predictor        Coef       Stdev     t-ratio        p
Constant        44534       20264        2.20     0.044
BATHS            8336        8574        0.97     0.346
BEDC            24649        8934        2.76     0.015
CARA           -47007       15030       -3.13     0.007
CARB           -10588        9751       -1.09     0.295
LOT             3.539        2.144        1.65     0.120
DOM            -76.67        31.51       -2.43     0.028

s = 15296       R-sq = 67.7%      R-sq(adj) = 54.8%

Analysis of Variance

SOURCE        DF          SS           MS          F        p
Regression     6  7357974528   1226329088       5.24    0.004
Error         15  3509615104    233974336
Total         21 10867589120
SOURCE        DF      SEQ SS
BATHS          1  3352323072
BEDC           1   883193344
CARA           1  1307168128
CARB           1   111305152
LOT            1   318872864
DOM            1  1385112064

Unusual Observations
Obs.   BATHS      PRICE      Fit Stdev.Fit  Residual   St.Resid
  7     2.50      52450    83502      6843    -31052     -2.27R

R denotes an obs. with a large st. resid.

MTB > REGRESS 'PRICE' ON 5 PREDICTORS C5,C6,C7,C9,C10


The regression equation is
PRICE = 62606 + 28939 BEDC - 52659 CARA - 14153 CARB + 3.52 LOT - 75.6 DOM

Predictor        Coef       Stdev     t-ratio        p
Constant        62606        8056        7.77     0.000
BEDC            28939        7755        3.73     0.002
CARA           -52659       13837       -3.81     0.002
CARB           -14153        9019       -1.57     0.136
LOT             3.523        2.140        1.65     0.119
DOM            -75.64        31.44       -2.41     0.029

s = 15270       R-sq = 65.7%      R-sq(adj) = 54.9%

Analysis of Variance

SOURCE        DF          SS           MS          F        p
Regression     5  7136792576   1427358464       6.12    0.002
Error         16  3730797312    233174832
Total         21 10867590144

SOURCE        DF      SEQ SS
BEDC           1  2901187584
CARA           1  2274636288
CARB           1   292810432
LOT            1   318495200
DOM            1  1349662976


Unusual Observations
Obs.   BEDC       PRICE      Fit Stdev.Fit  Residual   St.Resid
  1     0.00      25750    31641     13849     -5891     -0.92 X
  4     1.00      46550    40659     13849      5891      0.92 X
  7     1.00      52450    85164      6614    -32714     -2.38R
 22     1.00     127050    99052      7948     27998      2.15R

R denotes an obs. with a large st. resid.
X denotes an obs. whose X value gives it large influence.

MTB > REGRESS 'PRICE' ON 4 PREDICTORS C5,C6,C9,C10
```

(continues)

```
The regression equation is
PRICE = 59313 + 31921 BEDC - 48742 CARA + 3.02 LOT - 69.0 DOM

Predictor       Coef        Stdev      t-ratio        p
Constant       59313         8105        7.32      0.000
BEDC           31921         7836        4.07      0.001
CARA          -48742        14183       -3.44      0.003
LOT            3.025         2.206        1.37      0.188
DOM           -69.00        32.46       -2.13      0.049

s = 15913      R-sq = 60.4%      R-sq(adj) = 51.1%

Analysis of Variance

SOURCE        DF          SS            MS          F        p
Regression     4   6562672128    1640668032      6.48    0.002
Error         17   4304917504     253230448
Total         21  10867589120

SOURCE        DF      SEQ SS
BEDC           1   2901187584
CARA           1   2274636288
LOT            1    242949280
DOM            1   1143899008

Unusual Observations
Obs.    BEDC       PRICE        Fit  Stdev.Fit  Residual   St.Resid
   1    0.00       25750       28533     14284     -2783      -0.40 X
   4    1.00       46550       43767     14284      2783       0.40 X
   7    1.00       52450       85098      6893    -32648      -2.28R
  22    1.00      127050       97533      8221     29517       2.17R

R denotes an obs. with a large st. resid.
X denotes an obs. whose X value gives it large influence.

MTB > REGRESS 'PRICE' ON 3 PREDICTORS C5,C6,C10

The regression equation is
PRICE = 66338 + 30129 BEDC - 38457 CARA - 60.4 DOM

Predictor       Coef        Stdev      t-ratio        p
Constant       66338         6433       10.31      0.000
BEDC           30129         7913        3.81      0.001
CARA          -38457        12329       -3.12      0.006
DOM           -60.41        32.62       -1.85      0.081

s = 16298      R-sq = 56.0%      R-sq(adj) = 48.7%

Analysis of Variance

SOURCE        DF          SS            MS          F        p
Regression     3   6086432256    2028810752      7.64    0.002
Error         18   4781157888     265619888
Total         21  10867590144

SOURCE        DF      SEQ SS
BEDC           1   2901187584
CARA           1   2274636288
DOM            1    910608192

Unusual Observations
Obs.    BEDC       PRICE        Fit  Stdev.Fit  Residual   St.Resid
   1    0.00       25750       17975     12322      7775       0.73 X
   4    1.00       46550       54326     12322     -7775      -0.73 X
   7    1.00       52450       86682      6960    -34232      -2.32R
  22    1.00      127050       94293      8065     32757       2.31R

R denotes an obs. with a large st. resid.
X denotes an obs. whose X value gives it large influence.

MTB > REGRESS 'PRICE' ON 2 PREDICTORS C5,C6
```

(continues)

```
The regression equation is
PRICE = 57231 + 29518 BEDC - 35840 CARA

Predictor        Coef       Stdev      t-ratio        p
Constant        57231        4403        13.00    0.000
BEDC            29518        8396         3.52    0.002
CARA           -35840       13006        -2.76    0.013

s = 17308        R-sq = 47.6%      R-sq(adj) = 42.1%

Analysis of Variance

SOURCE          DF          SS            MS         F         p
Regression       2    5175823872    2587911936      8.64    0.002
Error           19    5691765760     299566624
Total           21   10867589120

SOURCE          DF       SEQ SS
BEDC             1    2901187584
CARA             1    2274636288

Unusual Observations
Obs.    BEDC      PRICE      Fit  Stdev.Fit  Residual   St.Resid
  1     0.00      25750     21391    12939      4359       0.38 X
  4     1.00      46550     50909    12939     -4359      -0.38 X
  7     1.00      52450     86749     7391    -34299      -2.19R
 22     1.00     127050     86749     7391     40301       2.58R

R denotes an obs. with a large st. resid.
X denotes an obs. whose X value gives it large influence.

MTB > REGRESS 'PRICE' ON 1 PREDICTOR C5

The regression equation is
PRICE = 54991 + 25785 BEDC

Predictor        Coef       Stdev      t-ratio        p
Constant        54991        4989        11.02    0.000
BEDC            25785        9554         2.70    0.014

s = 19958        R-sq = 26.7%      R-sq(adj) = 23.0%

Analysis of Variance

SOURCE          DF          SS            MS         F         p
Regression       1    2901187584    2901187584      7.28    0.014
Error           20    7966402048     398320096
Total           21   10867589120

Unusual Observations
Obs.    BEDC      PRICE      Fit  Stdev.Fit  Residual   St.Resid
 22     1.00     127050     80776     8148     46274       2.54R

R denotes an obs. with a large st. resid.
MTB > STOP
```

12.49 Refer to Exercise 12.48. Conduct a test using the full regression model to determine whether the depreciation (decrease) in house price per year of age is less than \$2,500. Give the null hypothesis for your test and the p-value. Draw a conclusion. Use $\alpha = .05$.

12.50 Refer to Exercise 12.48. Suppose that we wished to modify our nine-variable model to allow for the possibility that the relationship between "price" and "age" differs depending on the number of bedrooms.

    a. Formulate such a model.

    b. What combination of model parameters represents the difference between a five-bedroom, one-garage home and a two-bedroom, two-garage home?

12.51   Refer to Exercise 12.48. What is your choice of a "best" model (at the .05 level) from the original set of nine variables? Why did you choose this model?

12.52   Refer to Exercise 12.48. In another study involving the same 22 properties, "price" was regressed on a single independent variable, "list," which was the listing price of the property in thousands of dollars.

```
MTB > PRINT 'PRICE' 'LIST'

 ROW     PRICE        LIST

   1     25750       29900
   2     37950       39900
   3     46450       44900
   4     46550       47500
   5     47950       49900
   6     49950       49900
   7     52450       53000
   8     54050       54900
   9     54850       54900
  10     52050       55900
  11     54392       55900
  12     53450       56000
  13     59510       62000
  14     60102       62500
  15     63850       63900
  16     62050       66900
  17     69450       72500
  18     82304       82254
  19     81850       82900
  20     70050       99900
  21    112450      117000
  22    127050      139000

MTB > DESCRIBE 'PRICE' 'LIST'

                 N      MEAN    MEDIAN    TRMEAN     STDEV    SEMEAN
PRICE           22     62023     54621     60585     22749      4850
LIST            22     65521     55950     63628     25551      5447

               MIN       MAX        Q1        Q3
PRICE        25750    127050     49450     69600
LIST         29900    139000     49900     74939

MTB > REGRESS 'PRICE' ON 1 PREDICTOR 'LIST'

The regression equation is
PRICE = 5406 + 0.864 LIST

Predictor        Coef       Stdev     t-ratio         p
Constant         5406        3363        1.61     0.124
LIST          0.86411     0.04797       18.01     0.000

s = 5616       R-sq = 94.2%     R-sq(adj) = 93.9%

Analysis of Variance

SOURCE       DF          SS             MS          F         p
Regression    1 10236690432   10236690432     324.51     0.000
Error        20   630899840      31544992
Total        21 10867590144

Unusual Observations
Obs.     LIST      PRICE        Fit  Stdev.Fit   Residual   St.Resid
  20    99900      70050      91731       2038     -21681     -4.14R
  22   139000     127050     125518       3723       1532      0.36 X

R denotes an obs. with a large st. resid.
X denotes an obs. whose X value gives it large influence.

MTB > STOP
```

a. Using the regression results, predict the selling price of a home that is listed at $70,000.

b. What is the chance that your prediction is off by more than $3,000?

12.53 A study was conducted involving the relationship between the selling price (in thousands of dollars) of a home and two independent variables, the number of rooms and the number of square feet. The following data were collected on 22 properties sold in a particular residential area.

Row	Price	Rooms	Sq ft
1	25.75	5	986
2	37.95	5	998
3	46.45	7	1690
4	46.55	8	1829
5	47.95	6	1186
6	49.95	6	1734
7	52.45	7	1684
8	54.05	7	1846
9	54.85	7	1690
10	52.05	7	1910
11	54.39	7	1784
12	53.45	6	1690
13	59.51	7	1590
14	60.10	8	1855
15	63.85	8	2212
16	62.05	10	2784
17	69.45	7	2190
18	82.30	8	2259
19	81.85	7	1919
20	70.05	7	1685
21	112.45	10	2654
22	127.05	10	2756

Use the computer output shown here to address parts a, b, and c.

```
        LISTING OF DATA

  OBS   PRICE   ROOMS   SQ_FT

    1   25.75     5       986
    2   37.95     5       998
    3   46.45     7      1690
    4   46.55     8      1829
    5   47.95     6      1186
    6   49.95     6      1734
    7   52.45     7      1684
    8   54.05     7      1846
    9   54.85     7      1690
   10   52.05     7      1910
   11   54.39     7      1784
   12   53.45     6      1690
```

(continues)

```
          LISTING OF DATA

     OBS    PRICE    ROOMS    SQ_FT
     13     59.51      7      1590
     14     60.10      8      1855
     15     63.85      8      2212
     16     62.05     10      2784
     17     69.45      7      2190
     18     82.30      8      2259
     19     81.85      7      1919
     20     70.05      7      1685
     21    112.45     10      2654
     22    127.05     10      2756

 N  =    22
```

MULTIPLE REGRESSION

GENERAL LINEAR MODELS PROCEDURE

DEPENDENT VARIABLE: PRICE

SOURCE	DF	SUM OF SQUARES	MEAN SQUARE	F VALUE	PR > F	R-SQUARE	C.V.
MODEL	2	6816.77693315	3408.38846658	15.99	0.0001	0.627265	23.5416
ERROR	19	4050.68890321	213.19415280			ROOT MSE	PRICE MEAN
CORRECTED TOTAL	21	10867.46583636				14.60116957	62.02272727

SOURCE	DF	TYPE I SS	F VALUE	PR > F	DF	TYPE III SS	F VALUE	PR > F
ROOMS	1	6357.37363636	29.82	0.0001	1	109.54214216	0.51	0.4822
SQ_FT	1	459.40329679	2.15	0.1585	1	459.40329679	2.15	0.1585

PARAMETER	ESTIMATE	T FOR HO: PARAMETER=0	PR > |T|	STD ERROR OF ESTIMATE
INTERCEPT	-16.97597859	-0.90	0.3815	18.94658431
ROOMS	4.33606202	0.72	0.4822	6.04912439
SQ_FT	0.02551127	1.47	0.1585	0.01737891

OBSERVATION	OBSERVED VALUE	PREDICTED VALUE	RESIDUAL
1	25.7500000	29.85843923	-4.10843923
2	37.9500000	30.16457441	7.78542559
3	46.4500000	56.49049414	-10.04049414
4	46.5500000	64.37262206	-17.82262206
5	47.9500000	39.29675434	8.65324566
6	49.9500000	53.27692781	-3.32692781
7	52.4500000	56.33742655	-3.88742655
8	54.0500000	60.47025156	-6.42025156
9	54.8500000	56.49049414	-1.64049414
10	52.0500000	62.10297254	-10.05297254
11	54.3900000	58.88855310	-4.49855310
12	53.4500000	52.15443213	1.29556787
13	59.5100000	53.93936760	5.57063240
14	60.1000000	65.03591496	-4.93591496
15	63.8500000	74.14343673	-10.29343673
16	62.0500000	97.40800460	-35.35800460
17	69.4500000	69.24612687	0.20387313
18	82.3000000	75.34246621	6.95753379
19	81.8500000	62.33257393	19.51742607
20	70.0500000	56.36293782	13.68706218
21	112.4500000	94.09154009	18.35845991
22	127.0500000	96.69368917	30.35631083

MULTIPLE REGRESSION

GENERAL LINEAR MODELS PROCEDURE

DEPENDENT VARIABLE: PRICE

```
     SUM OF RESIDUALS                            0.00000000
     SUM OF SQUARED RESIDUALS                 4050.68890321
     SUM OF SQUARED RESIDUALS - ERROR SS       -0.00000000
     FIRST ORDER AUTOCORRELATION                0.39259574
     DURBIN-WATSON D                            0.98314796
```

a. Conduct a test to see whether the two variables "rooms" and "sq ft," taken together, contain information about "price." Use $\alpha = .05$.

b. Conduct a test to see whether the coefficient of "rooms" is equal to 0. Use $\alpha = .05$.

c. Conduct a test to see whether the coefficient of "sq ft" is equal to 0. Use $\alpha = .05$.

12.54 Refer to Exercise 12.53.

a. Explain the apparent inconsistency between the result of part a and the results of parts b and c.

b. What do you think would happen to the T-ratio of "sq ft" if "rooms" were dropped from the model?

12.55 A study was conducted to determine whether infection surveillance and control programs have reduced the rates of hospital-acquired infection in U.S. hospitals. This data set consists of a random sample of 28 hospitals selected from 338 hospitals participating in a larger study.

Each line of the data set provides information on variables for a single hospital. The variables are as follows:

RISK = output variable, average estimated probability of acquiring infection in hospital (in percent)

STAY = input variable, average length of stay of all patients in hospital (in days)

AGE = input variable, average age of patients (in years)

RCR = input variable, ratio of number of cultures performed to number of patients without signs or symptoms of hospital-acquired infection (times 100)

SCHOOL = dummy input variable for medical school affiliation, 1 = yes, 0 = no

DV_1 = dummy input variable for region of country, 1 = northeast, 0 = other

DV_2 = dummy input variable for region of country, 1 = north central, 0 = other

DV_3 = dummy input variable for region of country, 1 = south, 0 = other

(Note that there are four geographic regions of the country—northeast, north central, south, and west. These four regions of the country require only three dummy variables to code for them.)

The data were analyzed using SAS with the following results.

```
                          LISTING OF DATA
OBS    RISK    STAY    AGE     RCR    SCHOOL    DV1    DV2    DV3
 1     4.1     7.13    55.7    9.0      0        0      0      1
 2     1.6     8.82    58.2    3.8      0        1      0      0
 3     2.7     8.34    56.9    8.1      0        0      1      0
 4     5.6     8.95    53.7   18.9      0        0      0      1
 5     5.7    11.20    56.5   34.5      0        0      0      0
 6     5.1     9.76    50.9   21.9      0        1      0      0
 7     4.6     9.68    57.8   16.7      0        0      1      0
 8     5.4    11.18    45.7   60.5      1        1      0      0
```

(*continues*)

```
                              LISTING OF DATA

    OBS    RISK    STAY    AGE    RCR    SCHOOL    DV1    DV2    DV3

     9     4.3     8.67    48.2   24.4      0        0      1      0
    10     6.3     8.84    56.3   29.6      0        0      0      0
    11     4.9    11.07    53.2   28.5      1        0      0      0
    12     4.3     8.30    57.2    6.8      0        0      1      0
    13     7.7    12.78    56.8   46.0      1        0      0      0
    14     3.7     7.58    56.7   20.8      0        1      0      0
    15     4.2     9.00    56.3   14.6      0        0      1      0
    16     5.6    10.12    51.7   14.9      1        0      1      0
    17     5.5     8.37    50.7   15.1      0        1      0      0
    18     4.6    10.16    54.2    8.4      1        0      0      1
    19     6.5    19.56    59.9   17.2      0        0      0      0
    20     5.5    10.90    57.2   10.6      0        1      0      0
    21     1.8     7.67    51.7    2.5      0        0      1      0
    22     4.2     8.88    51.5   10.1      0        0      1      0
    23     5.6    11.48    57.6   20.3      0        0      0      0
    24     4.3     9.23    51.6   11.6      0        1      0      0
    25     7.6    11.41    61.1   16.6      0        0      0      0
    26     7.8    12.07    43.7   52.4      0        1      0      0
    27     3.1     8.63    54.0    8.4      0        0      0      0
    28     3.9    11.15    56.5    7.7      0        0      0      0

    N =    28
```

```
                                         CORRELATION

VARIABLE    N          MEAN          STD DEV           SUM          MINIMUM        MAXIMUM

STAY       28     10.03321429      2.37286052     280.93000000     7.13000000    19.56000000
AGE        28     54.33928571      4.08024503    1521.50000000    43.70000000    61.10000000
RCR        28     19.28214286     14.32881204     539.90000000     2.50000000    60.50000000
SCHOOL     28      0.17857143      0.39002103       5.00000000     0.00000000     1.00000000
DV1        28      0.28571429      0.46004371       8.00000000     0.00000000     1.00000000
DV2        28      0.28571429      0.46004371       8.00000000     0.00000000     1.00000000
DV3        28      0.10714286      0.31497039       3.00000000     0.00000000     1.00000000
```

```
            PEARSON CORRELATION COEFFICIENTS / PROB > |R| UNDER HO:RHO=0 / N = 28

                    STAY       AGE       RCR      SCHOOL     DV1       DV2       DV3

        STAY      1.00000    0.18019   0.35014   0.20586   0.07993  -0.32591  -0.19127
                  0.0000     0.3589    0.0678    0.2933    0.6860    0.0906    0.3296

        AGE       0.18019    1.00000  -0.47243  -0.23498  -0.39490  -0.06737   0.01678
                  0.3589    -0.0000    0.0111    0.2287    0.0375    0.7334    0.9325

        RCR       0.35014   -0.47243   1.00000   0.41016   0.23847  -0.31552  -0.17682
                  0.0678     0.0111    0.0000    0.0302    0.2217    0.1019    0.3681

        SCHOOL    0.20586   -0.23498   0.41016   1.00000  -0.08847  -0.08847   0.13998
                  0.2933     0.2287    0.0302    0.0000    0.6544    0.6544    0.4774

        DV1      -0.07993   -0.39490   0.23847  -0.08847   1.00000  -0.40000  -0.21909
                  0.6860     0.0375    0.2217    0.6544    0.0000    0.0349    0.2627

        DV2      -0.32591   -0.06737  -0.31552  -0.08847  -0.40000   1.00000  -0.21909
                  0.0906     0.7334    0.1019    0.6544    0.0349    0.0000    0.2627

        DV3      -0.19127    0.01678  -0.17682   0.13998  -0.21909  -0.21909   1.00000
                  0.3296     0.9325    0.3681    0.4774    0.2627    0.2627    0.0000
```

```
                             STEPWISE REGRESSION
              BACKWARD ELIMINATION, DEPENDENT VARIABLE RISK

        BACKWARD ELIMINATION PROCEDURE FOR DEPENDENT VARIABLE RISK

  STEP 0   ALL VARIABLES ENTERED    R SQUARE = 0.60724861    C(P) =    8.00000000

                  DF        SUM OF SQUARES        MEAN SQUARE         F      PROB>F

  REGRESSION       7         39.49805177         5.64257882         4.42    0.0041
  ERROR           20         25.54623394         1.27731170
  TOTAL           27         65.04428571
```

(continues)

```
                            STEPWISE REGRESSION
                  BACKWARD ELIMINATION, DEPENDENT VARIABLE RISK

            BACKWARD ELIMINATION PROCEDURE FOR DEPENDENT VARIABLE RISK
```

	B VALUE	STD ERROR	TYPE II SS	F	PROB>F
INTERCEPT	-1.07800774				
STAY	0.23613428	0.11569116	5.32126218	4.17	0.0547
AGE	0.04359681	0.07810854	0.39793239	0.31	0.5829
RCR	0.06923673	0.02278287	11.79650358	9.24	0.0065
SCHOOL	-0.41516871	0.64822732	0.52395194	0.41	0.5291
DV1	-0.26955673	0.68941266	0.19527144	0.15	0.6999
DV2	-0.19268071	0.71943459	0.09162010	0.07	0.7916
DV3	0.70243224	0.88962481	0.79632801	0.62	0.4390

BOUNDS ON CONDITION NUMBER: 2.315515.   94.11721

STEP 1   VARIABLE DV2 REMOVED   R SQUARE = 0.60584002   C(P) =   6.07172885

	DF	SUM OF SQUARES	MEAN SQUARE	F	PROB>F
REGRESSION	6	39.40643167	6.56773861	5.38	0.0017
ERROR	21	25.63785404	1.22085019		
TOTAL	27	65.04428571			

	B VALUE	STD ERROR	TYPE II SS	F	PROB>F
INTERCEPT	-1.81224950				
STAY	0.24597088	0.10725430	6.42096620	5.26	0.0322
AGE	0.05262498	0.06888511	0.71251762	0.58	0.4534
RCR	0.07154787	0.02061408	14.70713325	12.05	0.0023
SCHOOL	-0.42280540	0.63312506	0.54445805	0.45	0.5115
DV1	-0.15497958	0.52853481	0.10496975	0.09	0.7722
DV3	0.83288104	0.72780215	1.59882767	1.31	0.2653

BOUNDS ON CONDITION NUMBER: 1.1929521.   53.56369

STEP 2   VARIABLE DV1 REMOVED   R SQUARE = 0.60422621   C(P) =   4.15390986

	DF	SUM OF SQUARES	MEAN SQUARE	F	PROB>F
REGRESSION	5	39.30146193	7.86029239	6.72	0.0006
ERROR	22	25.74282379	1.17012835		
TOTAL	27	65.04428571			

	B VALUE	STD ERROR	TYPE II SS	F	PROB>F
INTERCEPT	-2.21637907				
STAY	0.24760767	0.10486035	6.52437780	5.58	0.0275
AGE	0.05898907	0.06400415	0.99394033	0.65	0.3667
RCR	0.07097867	0.02005725	14.61240661	12.49	0.0019
SCHOOL	-0.38736862	0.60843670	0.47429829	0.41	0.5309
DV3	0.87192445	0.70049715	1.81291925	1.55	0.2263

BOUNDS ON CONDITION NUMBER: 1.905871,   36.65382

STEP 3   VARIABLE SCHOOL REMOVED   R SQUARE = 0.59693428   C(P) =   2.52523447

	DF	SUM OF SQUARES	MEAN SQUARE	F	PROB>F
REGRESSION	4	38.82716364	9.70679091	8.52	0.0002
ERROR	23	26.21712207	1.13987487		
TOTAL	27	65.04428571			

	B VALUE	STD ERROR	TYPE II SS	F	PROB>F
INTERCEPT	-2.30479519				
STAY	0.23848508	0.10252510	6.16764346	5.41	0.0292
AGE	0.06257589	0.06292612	1.12722159	0.99	0.3304
RCR	0.06713326	0.01892561	14.34276871	12.58	0.0017
DV3	0.76072793	0.66954727	1.47147677	1.29	0.2676

BOUNDS ON CONDITION NUMBER: 1.741914,   23.03492

(continues)

```
                            STEPWISE REGRESSION
                 BACKWARD ELIMINATION, DEPENDENT VARIABLE RISK
            BACKWARD ELIMINATION PROCEDURE FOR DEPENDENT VARIABLE RISK

STEP 4    VARIABLE AGE REMOVED       R SQUARE = 0.57960421      C(P) = 1.40772979

                  DF        SUM OF SQUARES        MEAN SQUARE        F        PROB>F
REGRESSION         3        37.69994205          12.56664735       11.03      0.0001
ERROR             24        27.34434367           1.13934765
TOTAL             27        65.04428571

                B VALUE       STD ERROR          TYPE II SS          F        PROB>F
INTERCEPT      0.88480344
STAY           0.28060533      0.09334523         10.29588785        9.04      0.0061
RCR            0.05622030      0.01541554         15.15391450       13.30      0.0013
DV3            0.74723631      0.66925498          1.42032908        1.25      0.2753
BOUNDS ON CONDITION NUMBER:   1.162616,     10.11556

-----------------------------------------------------------------------------------

STEP 5    VARIABLE DV3 REMOVED      R SQUARE = 0.55776787      C(P) =    0.51969728

                  DF        SUM OF SQUARES        MEAN SQUARE        F        PROB>F
REGRESSION         2        36.27961297          18.13980648       15.77      0.0001
ERROR             25        28.76467275           1.15058691
TOTAL             27        65.04428571

                B VALUE       STD ERROR          TYPE II SS          F        PROB>F
INTERCEPT      1.15123509
STAY           0.26598212      0.09287658          9.43651980        8.20      0.0084
RCR            0.05416385      0.01538042         14.26927648       12.40      0.0017
BOUNDS ON CONDITION NUMBER:   1.139728,     4.558912

-----------------------------------------------------------------------------------

ALL VARIABLES IN THE MODEL ARE SIGNIFICANT AT THE 0.1000 LEVEL.

           SUMMARY OF BACKWARD ELIMINATION PROCEDURE FOR DEPENDENT VARIABLE RISK

                 VARIABLE     NUMBER    PARTIAL    MODEL
          STEP   REMOVED       IN        R**2      R**2       C(P)         F       PROB>F
           1     DV2           6        0.0014    0.6058     6.07173     0.0717    0.7916
           2     DV1           5        0.0016    0.6042     4.15391     0.0860    0.7722
           3     SCHOOL        4        0.0073    0.5969     2.52523     0.4053    0.5309
           4     AGE           3        0.0173    0.5796     1.40773     0.9889    0.3304
           5     DV3           2        0.0218    0.5578     0.51970     1.2446    0.2753
```

Does the set of seven input variables contain information about the output variable, "risk"? Give a p-value for your test.

Based on the full regression model (seven input variables), can we be at least 95% certain that hospitals in the south have at least .5% higher risk of infection than hospitals in the west, all other things being equal?

Refer to Exercise 12.55.

Consider the following two statements:

There is multicollinearity between region of the country and whether a hospital has a medical school.

There is an interaction effect between region of the country and whether a hospital has a medical school.

What is the difference between these two statements? What evidence is needed to ascertain the truth or falsity of the statements? Is this evidence present in the accompanying output? If it is, do you think the statements are true or false?

b. Construct a model that allows for the possibility of an interaction effect between region of the country and medical school affiliation. For this model, what is the difference in intercept between a hospital in the northeast affiliated with a medical school and a hospital in the west not affiliated with one?

12.57 Refer to Exercise 12.55. Suppose that we decide to eliminate from the full model some variables that we think contribute little to explaining the output variable. What would your final choice of a model be? Why would you choose this model?

12.58 Refer to Exercise 12.55. Predict the infection risk of a patient in a medical school–affiliated hospital in the northeast, where the average stay of patients is 10 days, the average age is 64, and the routine culturing ratio is 20%. Is this prediction an interpolation or an extrapolation? How do you know?

12.59 Thirty volunteers participated in the following experiment. The subjects took their own pulse rates (which is easiest to do by holding the thumb and forefinger of one hand on the pair of arteries on the side of the neck). They were then asked to flip a coin. If their coin came up heads, they ran in place for 1 minute. Then all subjects took their own pulse rates again. The difference in the before and after pulse rates was recorded, as well as other data on student characteristics. A regression was run to "explain" the pulse rate differences using the other variables as independent variables. The variables were:

$$PULSE = \text{difference between the before and after pulse rates}$$
$$RUN = \text{dummy variable, } 1 = \text{did not run in place, } 0 = \text{ran in place}$$
$$SMOKE = \text{dummy variable, } 1 = \text{does not smoke, } 0 = \text{smokes}$$
$$HEIGHT = \text{height in inches}$$
$$WEIGHT = \text{weight in pounds}$$
$$PHYS1 = \text{dummy variable, } 1 = \text{a lot of physical exercise, } 0 = \text{otherwise}$$
$$PHYS2 = \text{dummy variable, } 1 = \text{moderate physical exercise, } 0 = \text{otherwise}$$

a. Perform an appropriate test to determine whether the entire set of independent variables explains a significant amount of the variability of "pulse." Draw a conclusion based on $\alpha = .01$.

b. Does multicollinearity seem to be a problem here? What is your evidence? What effect does multicollinearity have on your ability to make predictions using regression?

```
                   LISTING OF DATA
  OBS    PULSE    RUN    SMOKE    HEIGHT    WEIGHT    PHYS1    PHYS2
   1      -29      0       1        66       140        0        1
   2      -17      0       1        72       145        0        1
   3      -14      0       0        73       160        1        0
   4      -22      0       0        73       190        0        0
   5      -21      0       1        69       155        0        1
   6      -25      0       1        73       165        0        0
   7       -5      0       1        72       150        1        0
   8       -9      0       1        74       190        0        1
   9      -18      0       1        72       195        0        1
  10      -23      0       1        71       138        0        1
  11      -14      0       0        74       160        0        0
  12      -21      0       1        72       155        0        1
  13        8      0       0        70       153        1        0
```

(continues)

```
                         LISTING OF DATA
     OBS   PULSE   RUN   SMOKE   HEIGHT   WEIGHT   PHYS1   PHYS2

      14    -13     0      1       67      145       0       1
      15    -21     0      1       71      170       1       0
      16     -1     0      1       72      175       1       0
      17    -16     0      0       69      175       0       1
      18    -15     1      1       68      145       0       0
      19      4     1      0       75      190       0       1
      20     -3     1      1       72      180       1       0
      21      2     1      0       67      140       0       1
      22     -5     1      1       70      150       0       1
      23     -1     1      1       73      155       0       1
      24     -5     1      1       74      148       1       0
      25     -6     1      0       68      150       0       1
      26     -6     1      0       73      155       0       1
      27      8     1      0       66      130       0       1
      28     -1     1      1       69      160       0       1
      29     -5     1      1       66      135       1       0
      30     -3     1      1       75      160       1       0

     N =    30
```

```
                                  CORRELATION

 VARIABLE    N        MEAN           STD DEV           SUM           MINIMUM          MAXIMUM

 RUN        30     0.43333333      0.50400693      13.00000000      0.00000000      1.00000000
 SMOKE      30     0.66666667      0.47946330      20.00000000      0.00000000      1.00000000
 HEIGHT     30    70.86666667      2.77592275    2126.00000000     66.00000000     75.00000000
 WEIGHT     30   158.63333333     17.53908607    4759.00000000    130.00000000    195.00000000
 PHYS1      30     0.30000000      0.46609160       9.00000000      0.00000000      1.00000000
 PHYS2      30     0.56666667      0.50400693      17.00000000      0.00000000      1.00000000
```

```
        PEARSON CORRELATION COEFFICIENTS / PROB > |R| UNDER HO:RHO=0 / N = 30

                        RUN       SMOKE      HEIGHT     WEIGHT     PHYS1      PHYS2

        RUN          1.00000    -0.09513   -0.12981   -0.25056    0.01468    0.08597
                     0.00000     0.6170     0.4924     0.1817     0.9386     0.6515

        SMOKE       -0.09513     1.00000    0.01727   -0.06834    0.15430   -0.04757
                     0.6170      0.0000     0.9278     0.7197     0.4156     0.8029

        HEIGHT      -0.12981     0.01727    1.00000    0.59885    0.19189   -0.28919
                     0.4942      0.9278     0.0000     0.0005     0.3097     0.1211

        WEIGHT      -0.25056    -0.06834    0.59885    1.00000    0.01392   -0.11221
                     0.1817      0.7197     0.0005     0.0000     0.9418     0.5549

        PHYS1        0.01468     0.15430    0.19189    0.01392    1.00000   -0.74863
                     0.9386      0.4156     0.3097     0.9418     0.0000     0.0001

        PHYS2        0.08597    -0.04757   -0.28919   -0.11221   -0.74863    1.00000
                     0.6515      0.8029     0.1211     0.5549     0.0001     0.0000
```

```
                            STEPWISE REGRESSION
           BACKWARD ELIMINATION, DEPENDENT VARIABLE PULSE

        BACKWARD ELIMINATION PROCEDURE FOR DEPENDENT VARIABLE PULSE

 STEP 0    ALL VARIABLES ENTERED       R SQUARE = 0.62973045     C(P) = 7.00000000

                     DF        SUM OF SQUARES        MEAN SQUARE        F       PROB>F

 REGRESSION           6        1850.58887109        308.43147852      6.52      0.0004
 ERROR               23        1088.11112891         47.30917952
 TOTAL               29        2938.70000000

                  B VALUE          STD ERROR        TYPE II SS         F       PROB>F

 INTERCEPT       -31.68830679
 RUN              11.40166481       2.66171908      868.07553823     18.35      0.0003
 SMOKE            -6.89029281       2.74454278      298.18154585      6.30      0.0195
 HEIGHT            0.13169561       0.60021947        2.27754970      0.05      0.8283
 WEIGHT            0.02303608       0.09440380        2.81697901      0.06      0.8094
 PHYS1            13.43465041       4.25117641      472.47616161      9.99      0.0044
 PHYS2             7.80635269       3.97815470      182.17065424      3.85      0.0619

 BOUNDS ON CONDITION NUMBER:     2.464274,     62.50691
```

(continues)

```
                          STEPWISE REGRESSION
                 BACKWARD ELIMINATION, DEPENDENT VARIABLE PULSE

             BACKWARD ELIMINATION PROCEDURE FOR DEPENDENT VARIABLE PULSE

 STEP  1    VARIABLE HEIGHT REMOVED      R SQUARE = 0.62895543    C(P) =    5.04814181

                     DF          SUM OF SQUARES      MEAN SQUARE         F     PROB>F

 REGRESSION           5          1848.31132139      369.66226428      89.14   0.0001
 ERROR               24          1090.38867861       45.43286161
 TOTAL               29          2938.70000000

                   B VALUE         STD ERROR        TYPE II SS          F     PROB>F

 INTERCEPT        -24.25519127
 RUN               11.43076116     2.60516294       874.68284765      19.25   0.0002
 SMOKE             -6.85327902     2.68448142       296.10525519       6.52   0.0175
 WEIGHT             0.03529782     0.07456145        10.18209732       0.22   0.6402
 PHYS1             13.44838310     4.16556957       473.54521380      10.42   0.0036
 PHYS2              7.65315557     3.83795325       180.65576063       3.98   0.0576

 BOUNDS ON CONDITION NUMBER:   2.406131,        40.22006

 ---------------------------------------------------------------------------------

 STEP  2    VARIABLE WEIGHT REMOVED    R SQUARE = 0.62549060     C(P) =    3.26336637

                     DF          SUM OF SQUARES      MEAN SQUARE         F     PROB>F

 REGRESSION           4          1838.12922407      459.53230602      10.44   0.0001
 ERROR               25          1100.57077593       44.02283104
 TOTAL               29          2938.70000000

                   B VALUE         STD ERROR        TYPE II SS          F     PROB>F

 INTERCEPT        -18.30152045
 RUN               11.13212935     2.48810400       881.24648295      20.02   0.0001
 SMOKE             -6.96392377     2.63262467       307.96107626       7.00   0.0139
 PHYS1             13.32514812     4.09240540       466.72897076      10.60   0.0032
 PHYS2              7.45071026     3.75440264       173.37705597       3.94   0.0583

 BOUNDS ON CONDITION NUMBER:  2.396734,       27.36375

 ---------------------------------------------------------------------------------

 ALL VARIABLES IN THE MODEL ARE SIGNIFICANT AT THE 0.1000 LEVEL.

             SUMMARY OF BACKWARD ELIMINATION PROCEDURE FOR DEPENDENT VARIABLE PULSE

              VARIABLE    NUMBER    PARTIAL     MODEL
       STEP   REMOVED      IN       R**2        R**2        C(P)          F       PROB>F

         1    HEIGHT        5       0.0008      0.6290     5.04814      0.0481    0.8283
         2    WEIGHT        4       0.0035      0.6255     3.26337      0.2241    0.6402
```

c. Based on the full regression model (six dependent variables), compute a point estimate of the average increase in "pulse" for individuals who engaged in a lot of physical activity compared to those who engaged in little physical activity. Can we be 95% certain that the actual average increase is greater than 0?

12.60 Refer to Exercise 12.59.

a. Give the implied regression line of pulse-rate difference on height and weight for a smoker who did not run in place and who has engaged in little physical activity.

b. Consider the following two statements:

1. There is multicollinearity between the "smoke" variable and the physical activity dummy variables.

2. There is an interaction effect between the "smoke" variable and the physical activity dummy variables.

Is there any difference between these two statements? Explain the relationships that would exist in the data set if each of these two statements were correct.

12.61 Refer to Exercise 12.59.

a. What is your choice of a good predictive equation? Why did you choose that particular equation?

b. The model as constructed does not contain any interaction effects. Construct a model that allows for the possibility of an interaction effect between each pair of qualitative variables.

 **12.62** The data for this exercise were taken from a chemical assay of calcium discussed in Brown, Healy, and Kearns (1981). A set of standard solutions is prepared and these and the unknowns are read on a spectrophotometer in arbitrary units (y). A linear regression model is fit to the standards and the values of the unknowns (x) are read off from this. The preparation of the standard and unknown solutions involves a fair amount of laboratory manipulation, and the actual concentrations of the standards may differ slightly from their target values, the very precise instrumentation being capable of detecting this. The target values are 2.0, 2.0, 2.5, 3.0, 3.0 mmol per liter; the "duplicates" are made up independently. The sequence of reading the standards and unknowns is repeated four times. Two specimens of each unknown are included in each assay and the four sequences of readings are done twice, first with the flame conditions in the instrument optimized, and then with a slightly weaker flame.

The data in the following table relate to assays on the above pattern of a set of six unknowns performed by four laboratories.

The standards are identified as 2.0A, 2.0B, 2.5, 3.0A, 3.0B; the unknowns are identified as U1, U2, W1, W2, Y1, Y2.

y: spectrophotometer reading

x: actual mmol per liter

Laboratory/Solution		Measurements			
1 W1	1206	1202	1202	1201
1 2.0A	1068	1071	1067	1066
1 W2	1194	1193	1189	1185
1 2.0B	1072	1068	1064	1067
1 U1	1387	1387	1384	1380
1 2.5	1333	1321	1326	1317
1 U2	1394	1390	1383	1376
1 3.0A	1579	1576	1578	1572
1 Y1	1478	1480	1473	1466
1 3.0B	1579	1571	1579	1567
1 Y2	1483	1477	1482	1472
2 W1	1017	1017	1012	1020
2 2.0A	910	916	915	915
2 W2	1012	1018	1015	1023
2 2.0B	913	923	914	921
2 U1	1188	1199	1197	1202
2 2.5	1129	1148	1136	1147

(continues)

Laboratory/Solution	Measurements			
2 U2	1186	1196	1193	1199
2 3.0A	1359	1378	1370	1373
2 Y1	1263	1280	1280	1279
2 3.0B	1349	1361	1359	1363
2 Y2	1259	1269	1259	1265
3 W1	1090	1098	1090	1100
3 2.0A	969	975	969	972
3 U2	1088	1092	1087	1085
3 2.0B	969	960	960	966
3 U1	1270	1261	1261	1269
3 2.5	1196	1196	1209	1200
3 W2	1261	1268	1270	1273
3 3.0A	1451	1440	1439	1449
3 Y1	1352	1349	1353	1343
3 3.0B	1439	1433	1433	1445
3 Y2	1349	1353	1349	1355
4 2.0A	1122	1117	1119	1120
4 W2	1256	1254	1256	1263
4 W1	1260	1251	1252	1264
4 2.0B	1122	1110	1111	1116
4 U2	1453	1447	1451	1455
4 2.5	1386	1381	1381	1387
4 U1	1450	1446	1448	1457
4 3.0A	1656	1663	1659	1665
4 Y2	1543	1548	1543	1545
4 3.0B	1658	1658	1661	1660
4 Y1	1545	1546	1548	1544

a. Plot y versus x for the standards, one graph for each laboratory.

b. Fit the linear regression equation $y = \beta_0 + \beta_1 x + \varepsilon$ for each laboratory and predict the value of x corresponding to the y for each of the unknowns. Compute the standard deviation of predicted values of x based on the four predicted x-values for each of the unknowns.

c. Which laboratory appears to make better predictions of x, mmol of calcium per liter? Why?

12.63 Refer to Exercise 12.62. Suppose you average the y-values for each of the unknowns and fit the ys in the linear regression model of Exercise 12.62.

a. Do your linear regression lines change for each of the laboratories?

b. Will predictions of x change based on these new regression lines for the four laboratories? Explain.

12.64 Refer to Exercise 12.62. Using the independent variable x, suggest a single general linear model that could be used to fit the data from all four laboratories. Identify the parameters in this general linear model.

12.65   Refer to Exercise 12.64.
   a. Fit the data to the model of Exercise 12.64.
   b. Give separate regression models for each of the laboratories.
   c. How do these regression models compare to the previous regression equations for the laboratories?
   d. What advantage(s) might there be to fitting a single model rather than separate models for the laboratories?

INTRODUCTION TO THE ANALYSIS OF VARIANCE

13.1 ▼ INTRODUCTION

In Chapter 6, we presented methods for comparing two population means, based on independent random samples. Very often the two-sample problem is a simplification of what we encounter in practical situations. For example, suppose we wish to compare the mean hourly wage for nonunion farm laborers from three different ethnic groups (African-American, Anglo-American, and Hispanic) employed by a large produce company. Independent random samples of farm laborers would be selected from each of the three ethnic groups (populations). Then using the information from the three sample means, we would try to make an inference about the corresponding population mean hourly wages. Most likely, the sample means would differ, but this does not necessarily imply a difference among the population means for the three ethnic groups. How do you decide whether the differences among the sample means are large enough to imply that the corresponding population means are different? We will answer this question using a statistical testing procedure called an *analysis of variance*.

13.2 ▼ THE LOGIC BEHIND AN ANALYSIS OF VARIANCE

The reason we call the testing procedure an analysis of variance can be seen by using the example in Section 13.1. Assume that we wish to compare the three ethnic mean hourly wages based on samples of five workers selected from each of the ethnic groups. Although a sample of size five from each of the populations seems pitifully small, it illustrates the basic ideas.

Suppose the sample data (hourly wages, in dollars) are as shown in Table 13.1. Do these data present sufficient evidence to indicate differences among the three population means? A brief visual inspection of the data indicates very little variation with a sample, whereas the variability among the sample means is much larger. Since the variability among the sample means is so large *in comparison to the* **within–sample variation**, we might conclude intuitively that the corresponding population means are different.

within-sample variation

TABLE 13.1
A Comparison of Three Sample Means (small amount of within-sample variation)

Sample from Population		
1	2	3
5.90	5.51	5.01
5.92	5.50	5.00
5.91	5.50	4.99
5.89	5.49	4.98
5.88	5.50	5.02
$\bar{y}_1 = 5.90$	$\bar{y}_2 = 5.50$	$\bar{y}_3 = 5.00$

within

TABLE 13.2
A Comparison of Three Sample Means (large amount of within-sample variation)

Sample from Population		
1	2	3
5.90	6.31	4.52
4.42	3.54	6.93
7.51	4.73	4.48
7.89	7.20	5.55
3.78	5.72	3.52
$\bar{y}_1 = 5.90$	$\bar{y}_2 = 5.50$	$\bar{y}_3 = 5.00$

between

Table 13.2 illustrates a situation in which the sample means are the same as given in Table 13.1 but the variability within a sample is much larger. In contrast to the data in Table 13.1, the **between-sample variation** is small relative to the within-sample variability. We would be less likely to conclude that the corresponding population means differ based on these data.

between-sample variation

The variations in the two sets of data, Tables 13.1 and 13.2, are shown graphically in Figure 13.1. The strong evidence to indicate a difference in population means for the data of Table 13.1 is apparent in Figure 13.1(a). The lack of evidence to indicate a difference in population means for the data of Table 13.2 is indicated by the overlapping of data points for the samples in Figure 13.1(b).

FIGURE 13.1
Dot Diagrams for the Data of Table 13.1 and Table 13.2: ○, **Measurement from Sample 1;** ●, **Measurement from Sample 2;** □, **Measurement from Sample 3**

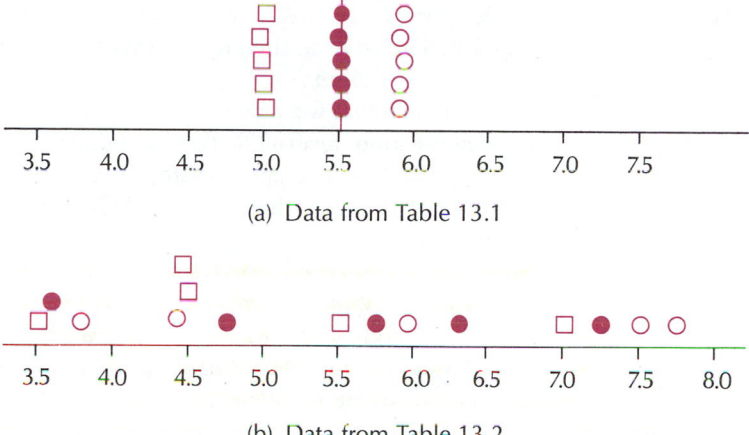

(a) Data from Table 13.1

(b) Data from Table 13.2

analysis of variance

The preceding discussion, with the aid of Figure 13.1, should indicate what we mean by an **analysis of variance**. All differences in sample means are judged statistically significant (or not) by comparing them to the variation within samples. The details of the testing procedure will be presented in the next section.

13.3 A STATISTICAL TEST ABOUT MORE THAN TWO POPULATION MEANS: AN ANALYSIS OF VARIANCE

In Chapter 6, we presented a method for testing the equality of two population means. We hypothesized two normal populations (1 and 2) with means denoted by μ_1 and μ_2, respectively, and a common variance σ^2. To test the null hypothesis that $\mu_1 = \mu_2$, independent random samples of sizes n_1 and n_2 were drawn from the two populations. The sample data were then used to compute the value of the test statistic

$$t = \frac{\bar{y}_1 - \bar{y}_2}{s_p\sqrt{(1/n_1) + (1/n_2)}},$$

where

$$s_p^2 = \frac{(n_1 - 1)s_1^2 + (n_2 - 1)s_2^2}{(n_1 - 1) + (n_2 - 1)} = \frac{(n_1 - 1)s_1^2 + (n_2 - 1)s_2^2}{n_1 + n_2 - 2}$$

pooled estimate of σ^2

is a pooled estimate of the common population variance σ^2. The rejection region for a specified value of α, the probability of a Type I error, was then found using Table 4 in the Appendix.

Suppose now that we wish to extend this method to test the equality of more than two population means. The test procedure described here applies to only two means and therefore is inappropriate. Hence, we will employ a more general method of data analysis, the analysis of variance. We illustrate its use with the following example.

Students from five different campuses throughout the country were surveyed to determine their attitudes toward industrial pollution. Each student sampled was asked a specific number of questions and then given a total score for the interview. Suppose that nine students are surveyed at each of the five campuses and we wish to examine the average student score for each of the five campuses.

We label the set of all test scores that could have been obtained from campus I as population I, and we will assume that this population possesses a mean μ_1. A random sample of $n_1 = 9$ measurements (scores) is obtained from this population to monitor student attitudes toward pollution. The set of all scores that could have been obtained from students on campus II is labeled population II (which has a mean μ_2). The data from a random sample of $n_2 = 9$ scores are obtained from this population. Similarly μ_3, μ_4, and μ_5 represent the means of the populations for scores from campuses III, IV, and V, respectively. We also obtain random samples of nine student scores from each of these populations.

From each of these five samples, we calculate a sample mean and variance. The sample results can then be summarized as shown in Table 13.3.

TABLE   13.3
Summary of the Sample Results for Five Populations

	Population				
	I	II	III	IV	V
Sample mean	$\bar{y}_1$	$\bar{y}_2$	$\bar{y}_3$	$\bar{y}_4$	$\bar{y}_5$
Sample variance	s_1^2	s_2^2	s_3^2	s_4^2	s_5^2

If we are interested in testing the equality of the population means (i.e., $\mu_1 = \mu_2 = \mu_3 = \mu_4 = \mu_5$), we might be tempted to run all possible pairwise comparisons of two population means. Hence, if we assume that the five distributions are approximately normal with the same variance σ^2, we could run 10 t tests comparing two means, as listed here (see Section 6.2).

multiple t tests

Null Hypotheses

$\mu_1 = \mu_2$	$\mu_1 = \mu_4$	$\mu_2 = \mu_3$	$\mu_2 = \mu_5$	$\mu_3 = \mu_5$
$\mu_1 = \mu_3$	$\mu_1 = \mu_5$	$\mu_2 = \mu_4$	$\mu_3 = \mu_4$	$\mu_4 = \mu_5$

One obvious disadvantage to this test procedure is that it is tedious and time consuming. But a more important and less apparent disadvantage of running multiple t tests to compare means is that the probability of falsely rejecting at least one of the hypotheses increases as the number of t tests increases. Thus, although we may have the probability of a Type I error fixed at $\alpha = .05$ for each individual test, the probability of falsely rejecting *at least one* of those tests is larger than .05. In other words, the combined probability of a Type I error for the set of 10 hypotheses would be larger than the value .05 set for each individual test. Indeed, it can be proved that the combined probability could be as large as .40.

What we need is a single test of the hypothesis "all five population means are equal," which will be less tedious than the individual t tests and can be performed with a specified probability of a Type I error (say, .05). This test is the analysis of variance.

First we assume that the five sets of measurements are normally distributed, with means given by μ_1, μ_2, μ_3, μ_4, and μ_5 and with a common variance σ^2. Next we consider the quantity

$\sim N(\mu_1 \rightarrow \mu_5, \sigma^2)$

$$s_W^2 = \frac{(n_1 - 1)s_1^2 + (n_2 - 1)s_2^2 + (n_3 - 1)s_3^2 + (n_4 - 1)s_4^2 + (n_5 - 1)s_5^2}{(n_1 - 1) + (n_2 - 1) + (n_3 - 1) + (n_4 - 1) + (n_5 - 1)}$$

$$= \frac{(n_1 - 1)s_1^2 + (n_2 - 1)s_2^2 + (n_3 - 1)s_3^2 + (n_4 - 1)s_4^2 + (n_5 - 1)s_5^2}{n_1 + n_2 + n_3 + n_4 + n_5 - 5}.$$

Note that this quantity is merely an extension of

$$s_p^2 = \frac{(n_1 - 1)s_1^2 + (n_2 - 1)s_2^2}{n_1 + n_2 - 2},$$

which is used as an estimate of the common variance for two populations for a test of the hypothesis $\mu_1 = \mu_2$ (Section 6.2). Thus, s_W^2 represents a combined estimate of the common variance σ^2, and it measures the variability of the observations within the five populations. (The subscript W refers to the within-sample variability.)

Next we consider a quantity that measures the variability between or among the population means. If the null hypothesis $\mu_1 = \mu_2 = \mu_3 = \mu_4 = \mu_5$ is true, then the populations are identical, with mean μ and variance σ^2. Drawing single samples from the five populations is then equivalent to drawing five different samples from the same population. What kind of variation might be expected for these sample means? If the variation is too great, we would reject the hypothesis that $\mu_1 = \mu_2 = \mu_3 = \mu_4 = \mu_5$.

To discuss the variation from sample mean to sample mean, we need to know the distribution of the mean of a sample of nine observations in repeated sampling. From Chapter 4 we know that the sampling distribution for $\bar{y}$ based on $n = 9$ measurements will have the same mean μ and variance $\sigma^2/9$. Since we have drawn five samples of nine observations each, we can estimate the variance of the distribution of sample means, $\sigma^2/9$, using the formula

$$\text{sample variance (of the means)} = \frac{\sum \bar{y}^2 - [(\sum \bar{y})^2/5]}{5-1}.$$

Note that we merely consider the $\bar{y}$s as a sample of five observations and calculate the "sample variance." This quantity estimates $\sigma^2/9$ and hence $9 \times$ (sample variance of the means) estimates σ^2. We designate this quantity as s_B^2; the subscript B designates a measure of the variability among the sample means for the five populations. For this problem $s_B^2 = $ (9 times the sample variance of the means).

Under the null hypothesis that all five population means are identical, we have two estimates of σ^2, namely, s_W^2 and s_B^2. Suppose the ratio

$$\frac{s_B^2}{s_W^2}$$

is used as the test statistic to test the hypothesis that $\mu_1 = \mu_2 = \mu_3 = \mu_4 = \mu_5$. What is the distribution of this quantity if we were to repeat the experiment over and over again, each time calculating s_B^2 and s_W^2?

For our example s_B^2/s_W^2 follows an F distribution, with degrees of freedom that can be shown to be $df_1 = 4$ for s_B^2 and $df_2 = 40$ for s_W^2. The proof of these remarks is beyond the scope of this text. However, we make use of this result for testing the null hypothesis $\mu_1 = \mu_2 = \mu_3 = \mu_4 = \mu_5$.

The test statistic used to test equality of the population means is

test statistic

$$F = \frac{s_B^2}{s_W^2}.$$

When the null hypothesis is true, both s_B^2 and s_W^2 estimate σ^2, and F would be expected to assume a value near $F = 1$. When the hypothesis of equality is false, s_B^2 will tend to be larger than s_W^2 due to the differences among the population means. Hence, we will reject the null hypothesis in the upper tail of the distribution of $F = s_B^2/s_W^2$; for $\alpha = .05$, the critical value of $F = s_B^2/s_W^2$ is 2.61. (See Figure 13.2.) If the calculated value of F falls in the rejection region, we conclude that not all five population means are identical.

This procedure can be generalized (and simplified) with only slight modifications in the formulas to test the equality of t (where t is an integer equal to or greater than 2) population means from normal populations with a common variance σ^2. Random samples of sizes $n_1, n_2, \ldots, n_t$ are drawn from the respective populations. We then compute the sample means and variances. The null hypothesis $\mu_1 = \mu_2 = \cdots = \mu_t$ is tested against the alternative that at least one of the population means is different from the others.

Before presenting the generalized test procedure, it is convenient to introduce the notation to be used in the shortcut computational formulas for s_B^2 and s_W^2.

FIGURE 13.2
Critical Value of *F* for
$\alpha = .05$, $df_1 = 4$, and
$df_2 = 40$

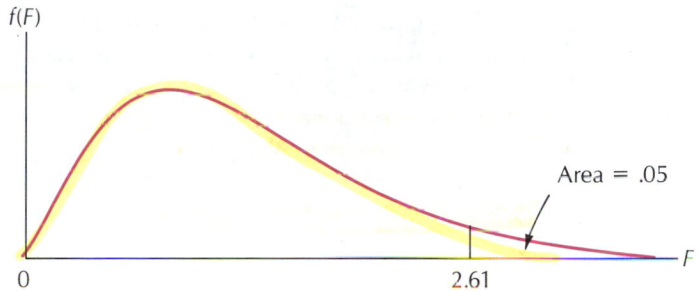

completely randomized design

analysis of variance

The experimental setting where a random sample of observations is taken from each of *t* different populations is called a **completely randomized design**. Consider a completely randomized design where four observations are obtained from each of the five populations. If we let y_{ij} denote the *j*th observation from population *i*, we could display the sample data for this completely randomized design as shown in Table 13.4. Using Table 13.4, we can introduce notation that is helpful when performing an **analysis of variance (AOV)** for a completely randomized design.

TABLE 13.4
**Summary of Sample Data
for a Completely
Randomized Design**

Population	Data				Total	Mean
1	y_{11}	y_{12}	y_{13}	y_{14}	$y_1.$	$\bar{y}_1.$
2	y_{21}	y_{22}	y_{23}	y_{24}	$y_2.$	$\bar{y}_2.$
3	y_{31}	y_{32}	y_{33}	y_{34}	$y_3.$	$\bar{y}_3.$
4	y_{41}	y_{42}	y_{43}	y_{44}	$y_4.$	$\bar{y}_4.$
5	y_{51}	y_{52}	y_{53}	y_{54}	$y_5.$	$\bar{y}_5.$

**Notation Needed for the
AOV of a Completely
Randomized Design**

y_{ij}: The *j*th sample observation selected from population *i*. For example, y_{23} denotes the third sample observation drawn from polulation 2.

n_i: The number of sample observations selected from population *i*. In our data set, n_1, the number of observations obtained from population 1, is 4. Similarly, $n_2 = n_3 = n_4 = n_5 = 4$. However, it should be noted that the sample sizes need not be the same. Thus, we might have $n_1 = 12$, $n_2 = 3$, $n_3 = 6$, $n_4 = 10$, and so forth.

n_T: The total sample size; $n_T = \sum n_i$. For the data given in Table 13.4, $n_T = n_1 + n_2 + n_3 + n_4 + n_5 = 20$.

$y_i.$: The sum (total) of the sample measurements obtained from population *i*.

$y..$: The sum (grand total) of *all* sample observations: $y.. = \sum y_i.$

$\bar{y}_i.$: The average of the n_i sample observations drawn from population *i*, $\bar{y}_i. = y_i./n_i$.

$\bar{y}..$: The average of all sample observations; $\bar{y}.. = y../n_T$.

With this notation it is possible to establish the following algebraic identities. (Although we will use these results in later calculations for s_W^2 and s_B^2, the proofs of these identities are beyond the scope of this text.) The variability of the n_T sample measurements about their mean $\bar{y}_{..}$ can be measured using the sum of the squared deviations $(y_{ij} - y_{..})^2$. This quantity,

$$TSS = \sum_{i,j} (y_{ij} - \bar{y}_{..})^2 = (n_T - 1)s^2,$$

total sum of squares

is called the **total sum of squares** of the measurements about their mean. The double summation in TSS means that we must sum the squared deviations for all rows (i) and columns (j) of the one-way classification.

It is possible to partition the total sum of squares as follows:

$$\sum_{i,j} (y_{ij} - \bar{y}_{..})^2 = \sum_{i,j} (y_{ij} - \bar{y}_{i.})^2 + \sum_{i} n_i(\bar{y}_{i.} - \bar{y}_{..})^2.$$

The first quantity on the right side of the equation measures the variability of an observation y_{ij} about its sample mean $\bar{y}_{i.}$. Thus,

$$SSW = \sum_{i,j} (y_{ij} - \bar{y}_{i.})^2 = (n_1 - 1)s_1^2 + (n_2 - 1)s_2^2 + \cdots + (n_t - 1)s_t^2$$

avg in pop i

within-sample sum of squares

is a measure of the *within-sample* variability. SSW is referred to as the **within-sample sum of squares** and is used to compute s_W^2.

The second expression in the total sum of squares equation measures the variability of the sample means $\bar{y}_{i.}$ about the overall mean $\bar{y}_{..}$. This quantity, which measures the variability *between* (or among) the sample means, is referred to as the **sum of squares between samples** (SSB) and is used to compute s_B^2.

between-sample sum of squares

 # obs. in i

$$SSB = \sum_{i} n_i(\bar{y}_{i.} - \bar{y}_{..})^2$$

sum i   avg all

Although the formulas for TSS, SSW, and SSB are easily interpreted, they are not easy to use for calculations. Instead, we use the shortcut formulas shown here.

Shortcut Sum of Squares Formulas for a Completely Randomized Design

$$TSS = \sum_{i,j} y_{ij}^2 - \frac{y_{..}^2}{n_T}$$

→ sum total / total obs.

$$SSB = \sum_{i} \frac{y_{i.}^2}{n_i} - \frac{y_{..}^2}{n_T}$$

$$SSW = TSS - SSB$$

An analysis of variance for a completely randomized design with t populations has the following null and alternative hypotheses:

H_0: $\mu_1 = \mu_2 = \mu_3 = \cdots = \mu_t$ (i.e., the t population means are equal)

H_a: At least one of the t population means differs from the rest.

The quantities s_B^2 and s_W^2 can be computed using the shortcut formulas

$$s_B^2 = \frac{\text{SSB}}{t-1} \qquad s_W^2 = \frac{\text{SSW}}{n_T - t},$$

where $t-1$ and $n_T - t$ are the degrees of freedom for s_B^2 and s_W^2, respectively.

mean square

Historically, people have referred to a sum of squares divided by its degrees of freedom as a **mean square**. Hence, s_B^2 is often called the *mean square between samples* and s_W^2, the *mean square within samples*.

The null hypothesis of equality of the t population means is rejected if

$$F = \frac{s_B^2}{s_W^2},$$

exceeds the tabulated value of F for $a = \alpha$, $\text{df}_1 = t - 1$, and $\text{df}_2 = n_T - t$.

AOV table

After completing the F test, the results of a study are then summarized in an *analysis of variance table*. The format of an **AOV table** is shown in Table 13.5. The AOV table lists the sources of variability in the first column. The second column lists the sums of squares associated with each source of variability. Since we showed that the total sum of squares (TSS) can be partitioned into two parts, then SSB and SSW must add up to TSS in the AOV table. The third column of the table gives the degrees of freedom associated with the sources of variability. Again, we have a check; $(t - 1) + (n_T - t)$ must add to $n_T - 1$. The mean squares are found in the fourth column of Table 13.5, and the F test for the equality of the t population means is given in the fifth column.

TABLE 13.5

An Example of an AOV Table for a Completely Randomized Design

Source	Sum of Squares	Degrees of Freedom	Mean Square	F Test
Between samples	SSB	$t-1$	$s_B^2 = \text{SSB}/(t-1)$	s_B^2/s_W^2
Within samples	SSW	$n_T - t$	$s_W^2 = \text{SSW}/(n_T - t)$	
Totals	TSS	$n_T - 1$		

EXAMPLE 13.1 ▼

A horticulturist was investigating the phosphorus content of tree leaves from three different varieties of apple trees (1, 2, and 3). Random samples of five leaves from each of the three varieties were analyzed for phosphorus content.

TABLE 13.6
Phosphorus Content of Leaves from Three Different Trees, Example 13.1

Variety	Phosphorus Content					Totals	Means
1	.35	.40	.58	.50	.47	2.30	0.46
2	.65	.70	.90	.84	.79	3.88	0.78
3	.60	.80	.75	.73	.66	3.54	0.71
Total						9.72	0.65

The data are given in Table 13.6. Use these data to test the hypothesis of equality of the mean phosphorus levels for the three varieties. Use $\alpha = .05$.

Solution   The null and alternative hypotheses for this example are

$$H_0: \quad \mu_1 = \mu_2 = \mu_3$$
$$H_a: \quad \text{At least one of the population means differs from the rest.}$$

The sample sizes are $n_1 = n_2 = n_3 = 5$, for which $n_T = 15$. From the sample data we see that the total (sum) for all observations on variety 1 is $y_{1.} = 2.30$. Similarly, the totals for varieties 2 and 3 are $y_{2.} = 3.88$, and $y_{3.} = 3.54$. The sum of all sample measurements is then

$$y_{..} = 9.72.$$

Using the sample measurements, the total sum of squares, TSS, is

$$\text{TSS} = \sum_{i,j} y_{ij}^2 - \frac{y_{..}^2}{n_T} = (.35)^2 + (.40)^2 + \cdots + (.66)^2 - \frac{(9.72)^2}{15}$$
$$= 6.673 - 6.299 = .374.$$

The sample totals can then be used to compute the sum of squares between samples, SSB.

$$\text{SSB} = \sum_i \frac{y_{i.}^2}{n_i} - \frac{y_{..}^2}{n_T} = \frac{(2.30)^2 + (3.88)^2 + (3.54)^2}{5} - \frac{(9.72)^2}{15}$$
$$= 6.575 - 6.299 = .276.$$

Then the sum of squares within samples, SSW, is

$$\text{SSW} = \text{TSS} - \text{SSB} = .374 - .276 = .098.$$

The AOV table for these data is shown in Table 13.7. The critical value of $F = s_B^2/s_W^2$ is 3.89, which is obtained from Table 6 in the Appendix for $a = .05$, $df_1 = 2$, and $df_2 = 12$. Since the computed value of F, 17.25, exceeds 3.89, we

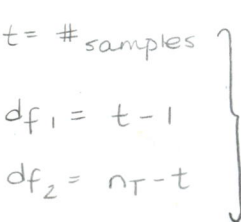

TABLE 13.7
AOV Table for the Data for Example 13.1

$t - 1$ $F = \dfrac{s_B^2}{s_w^2}$

Source	Sum of Squares	Degrees of Freedom	Mean Square	F Test
s_B Between samples	.276	2	.276/2 = .138	.138/.008 = 17.25
s_w Within samples	.098	12	.098/12 = .008	
Totals	.374	14		

$n_T - t$

reject the null hypothesis of equality of the mean phosphorus content for the three varieties. It appears from the data that the mean for variety 1 is smaller than the means for varieties 2 and 3. ▲

EXAMPLE 13.2

▼

A clinical psychologist wished to compare three methods for reducing hostility levels in university students. A certain test (HLT) was used to measure the degree of hostility. A high score on the test indicated great hostility. Eleven students obtaining high and nearly equal scores were used in the experiment. Four were selected at random from among the 11 problem cases and treated with method 1. Four of the remaining seven students were selected at random and treated with method 2. The remaining three students were treated with method 3. All treatments were continued for a one-semester period. Each student was given the HLT test at the end of the semester, with the results shown in Table 13.8. Use these data to perform an analysis of variance to determine if there are differences among mean scores for the three methods. Use $\alpha = .05$.

TABLE 13.8
HLT Test Scores, Example 13.2

$n_1 = 4$
$n_2 = 4$
$n_3 = 3$

Method	Test Scores				Totals
1	80	92	87	83	342
2	70	81	78	74	303
3	63	76	70		209
Total					854

Solution The null and alternative hypotheses are

$H_0: \quad \mu_1 = \mu_2 = \mu_3$
$H_a: \quad$ At least one of the population means differs from the rest.

For $n_1 = 4$, $n_2 = 4$, and $n_3 = 3$, we have a total sample size of $n_T = 11$. The totals from Table 13.8 are

$y_{1.} = 342, \qquad y_{2.} = 303, \qquad y_{3.} = 209, \qquad y_{..} = 854.$

Substituting into the computational formulas for TSS and SSB, we have

$$\text{TSS} = \sum_{i,j} y_{ij}^2 - \frac{y_{..}^2}{n_T} = (80)^2 + (92)^2 + \cdots + (70)^2 - \frac{(854)^2}{11}$$

$$= 66{,}988 - 66{,}301.45 = 686.55$$

$$\text{SSB} = \sum \frac{y_{i.}^2}{n_i} - \frac{y_{..}^2}{n_T} = \frac{(342)^2}{4} + \frac{(303)^2}{4} + \frac{(209)^2}{3} - 66{,}301.45$$

$$= 66{,}753.58 - 66{,}301.45 = 452.13.$$

Then

$$\text{SSW} = 686.55 - 452.13 = 234.42.$$

The AOV table for these data is shown in Table 13.9.

TABLE 13.9
**AOV Table for the Data of
Example 13.2**

Source	SS	df	MS	F
Between samples	452.13	2	226.07	226.07/29.3 = 7.72
Within samples	234.42	8	29.30	
Totals	686.55	10		

The critical value of F is obtained from Table 6 in the Appendix for $a = .05$, $df_1 = 2$, and $df_2 = 8$; this value is 4.46. Since the computed value of F, 7.72, exceeds the tabulated value, 4.46, we reject the null hypothesis of equality of the mean scores for the three groups. Computer output shown here verifies the results we obtained by hand.

```
        LISTING OF DATA
   OBS     METHOD     SCORE
    1        1          80
    2        1          92
    3        1          87
    4        1          83
    5        2          70
    6        2          81
    7        2          78
    8        2          74
    9        3          63
   10        3          76
   11        3          70
   N =      11
                              (continues)
```

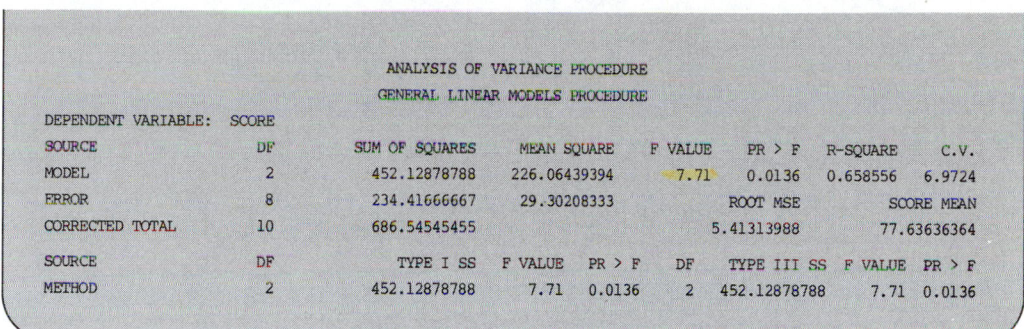

ANALYSIS OF VARIANCE PROCEDURE
GENERAL LINEAR MODELS PROCEDURE

DEPENDENT VARIABLE: SCORE

SOURCE	DF	SUM OF SQUARES	MEAN SQUARE	F VALUE	PR > F	R-SQUARE	C.V.
MODEL	2	452.12878788	226.06439394	7.71	0.0136	0.658556	6.9724
ERROR	8	234.41666667	29.30208333		ROOT MSE		SCORE MEAN
CORRECTED TOTAL	10	686.54545455			5.41313988		77.63636364

SOURCE	DF	TYPE I SS	F VALUE	PR > F	DF	TYPE III SS	F VALUE	PR > F
METHOD	2	452.12878788	7.71	0.0136	2	452.12878788	7.71	0.0136

▲

▼ EXERCISES

Applications

 13.1 Sample data from an experiment aimed at comparing the tar content (milligrams) of five different brands of cigarettes gave the following results:

Brand	$\bar{y}_{i\cdot}$	s_i	n_i
1	9.6	1.3	10
2	10.2	1.4	10
3	10.8	1.1	10
4	11.5	1.2	10
5	13.6	1.5	10

a. Based on your intuition, is there evidence to indicate any differences among the mean contents of the five brands?

b. Run an analysis of variance to confirm or reject your conclusion of part a. Use $\alpha = .05$.

 13.2 The number of units of production was recorded for a random sample of ten hourly periods from the three bottling assembly lines of a plant. These data are shown here:

	Assembly Line	
1	2	3
290	258	249
265	276	257
286	277	264
275	243	266
288	248	278
250	259	273
279	265	281
294	282	254
285	275	261
293	268	265

a. Plot the data separately for each line. Are there any obvious differences?

b. Identify the means and standard deviations for the three lines using the output shown here.

```
          LISTING OF DATA
    OBS     LINE     UNITS
     1       1        290
     2       1        265
     3       1        286
     4       1        275
     5       1        288
     6       1        250
     7       1        279
     8       1        294
     9       1        285
    10       1        293
    11       2        258
    12       2        276
    13       2        277
    14       2        243
    15       2        248
    16       2        259
    17       2        265
    18       2        282
    19       2        275
    20       2        268
    21       3        249
    22       3        257
    23       3        264
    24       3        266
    25       3        278
    26       3        273
    27       3        281
    28       3        254
    29       3        261
    30       3        265

N =     30
```

```
          MEANS BY LINE
VARIABLE        MEAN        STANDARD
                            DEVIATION

--------------- LINE=1 ---------------

UNITS      280.50000000    13.89844116

--------------- LINE=2 ---------------

UNITS      265.10000000    12.99957264

--------------- LINE=3 ---------------

UNITS      264.80000000    10.26103742
```

```
                    ANALYSIS OF VARIANCE PROCEDURE
                    GENERAL LINEAR MODELS PROCEDURE
DEPENDENT VARIABLE: UNITS
```

SOURCE	DF	SUM OF SQUARES	MEAN SQUARE	F VALUE	PR > F	R-SQUARE	C.V.
MODEL	2	1612.46666667	806.23333333	5.17	0.0125	0.277082	4.6209
ERROR	27	4207.00000000	155.81481481		ROOT MSE		UNITS MEAN
CORRECTED TOTAL	29	5819.46666667			12.48258045		270.1333333

SOURCE	DF	TYPE I SS	F VALUE	PR > F	DF	TYPE III SS	F VALUE	PR > F
LINE	2	1612.46666667	5.17	0.0125	2	1612.46666667	5.17	0.0125

13.4 ▼ THE MODEL FOR OBSERVATIONS IN A COMPLETELY RANDOMIZED DESIGN

We formulated a model (equation) to relate a response y to a set of quantitative independent variables in Chapter 9. In this section, we will consider a model for the completely randomized design (sometimes referred to as a one-way classification). While the model at first may appear to be quite different from those of Chapter 9, we will see later that it is very similar.

assumptions

We make the following **assumptions** concerning the sample measurements and the populations from which they were drawn:

1. The samples are independent random samples. Results from one sample in no way affect the measurements observed in another sample.
2. Each sample is selected from a normal population.
3. The mean and variance for population i are, respectively, μ_i and $\sigma^2 (i = 1, 2, \ldots, t)$.

To summarize, we assume that the t populations are independently normally distributed with different means but a common variance σ^2.

We can now formulate a model (equation) that encompasses the assumptions listed above. Recall that we previously let y_{ij} denote the jth sample observation from population i.

model

$$y_{ij} = \mu + \alpha_i + \varepsilon_{ij}$$

terms

This model states that y_{ij}, the jth sample measurement selected from population i, is the sum of three **terms**. The term μ denotes an overall mean that is an unknown constant. The term α_i denotes an effect due to population i; α_i is an unknown constant. The term ε_{ij} denotes a random error associated with the jth observation from population i. We assume that ε_{ij} is normally distributed, with a mean of 0 and a variance σ_ε^2. In addition, the errors are independent; that is, the error associated with one observation in no way affects the error associated with another observation.

Since the εs are normally distributed with mean 0, the mean or expected value of y_{ij}, denoted by $E(y_{ij})$, is

$$E(y_{ij}) = \mu + \alpha_i.$$

That is, y_{ij} has been selected from a population with mean $\mu + \alpha_i$. The effect α_i, representing the deviation of the ith population from the overall mean μ, may assume a positive, zero, or negative value. Hence, the mean for population i can be

TABLE 13.10
Summary of Some of the Assumptions for a Completely Randomized Design

Population	Population Mean	Population Variance	Sample Measurements
1	$\mu + \alpha_1$	σ_ε^2	$y_{11}, y_{12}, \ldots, y_{1n_1}$
2	$\mu + \alpha_2$	σ_ε^2	$y_{21}, y_{22}, \ldots, y_{2n_2}$
$\vdots$	$\vdots$	$\vdots$	$\vdots$
t	$\mu + \alpha_t$	σ_ε^2	$y_{t1}, y_{t2}, \ldots, y_{tn_t}$

greater than, equal to, or less than μ, the overall mean. The variance for each of the t populations can be shown to be σ_ε^2. Finally, because the εs are normally distributed, each of the t populations is normal. A summary of the assumptions for a one-way classification is shown in Table 13.10.

The null hypothesis for a one-way analysis of variance is that $\mu_1 = \mu_2 = \cdots = \mu_t$. Using our model, this would be equivalent to the null hypothesis

$$H_0: \quad \alpha_1 = \alpha_2 = \cdots = \alpha_t = 0.$$

If H_0 is true, then all populations have the same unknown mean μ. Indeed, many textbooks use this latter null hypothesis for the analysis of variance in a completely randomized design. The corresponding alternative hypothesis is

$$H_a: \quad \text{At least one of the } \alpha_i \text{s differs from 0.}$$

In this section, we have presented a brief description of the model associated with the analysis of variance for a completely randomized design. Although some authors bypass an examination of the model, we believe it is a necessary part of an analysis of variance discussion.

You may be concerned with checking the validity of the underlying assumptions in an analysis of variance. In practice, you should make at least a rough check before proceeding. In the next section, we will discuss how to test the "equality of variance" assumption. The assumption of normality is not too critical since we are basing the analysis of variance test on means (and hence the Central Limit Theorem applies). Nonnormality in the form of skewed distributions will not affect conclusions drawn from an analysis of variance unless the skewness is severe and the sample sizes are small. However, to guard against gross violations of the normality assumption, the data for each sample should be plotted separately. If the data for one or more of the samples appear nonnormal, the Kruskal–Wallis test of Section 13.6 (also referred to as the Kruskal–Wallis one-way analysis of variance by ranks) can be used. The null hypothesis for the Kruskal–Wallis test is that the t populations are identical.

13.5 ▼ CHECKING ON THE EQUAL VARIANCE ASSUMPTION

The assumption of equal population variances, like the assumption of normality of the populations, has been made in several places in the text, such as for the t test when comparing two population means and now for the analysis of variance F test in a completely randomized design.

Let us consider first an experiment where we wish to compare t population means based on independent random samples from each of the populations. Recall that we assume we are dealing with normal populations with a common variance σ_{ε}^2 and possibly different means. If there were just two populations of interest, we could verify the assumption of equality of the two population variances using the F test of Chapter 7. However, with $t > 2$, rather than making all pairwise F tests, we seek a single test that can be used to verify the assumption of equality of the population variances.

The one test we will use in this text for the null hypothesis

$$H_0: \quad \sigma_1^2 = \sigma_2^2 = \cdots = \sigma_t^2$$

Hartley's test

was proposed by H. O. Hartley (**Hartley's test**) (1940 and 1950) and represents a logical extension to the F test for $t = 2$. If s_i^2 denotes the sample variance computed from the ith sample, the test statistic is

F_{max}

$$F_{max} = \frac{s_{max}^2}{s_{min}^2},$$

where s_{max}^2 and s_{min}^2 are the largest and smallest of the s_i^2s respectively. The test procedure is summarized here.

Hartley's Test for Homogeneity of Population Variances

▼

$H_0: \quad \sigma_1^2 = \sigma_2^2 = \cdots = \sigma_t^2$, i.e., homogeneity of variances

$H_a:$ Not all population variances are the same

T.S.: $F_{max} = \dfrac{s_{max}^2}{s_{min}^2}$

R.R.: For a specified value of α, reject H_0 if F_{max} exceeds the tabulated F value (Table 14) for $a = \alpha$, t, and $df_2 = n - 1$, where n is the number of observations in each sample.

It should be noted that, theoretically, we required the sample sizes to be all the same. In practice, if the sample sizes are nearly equal, the largest n_i can be used for running the test of homogeneity. This procedure will result in the probability of a Type I error being slightly more than the nominal value α.

Several comments should be made. Most practitioners do not routinely run Hartley's test. One reason is that the test is extremely sensitive to departures from normality. So, in checking one assumption (constant variance), the practitioner would have to be very careful about departures from another analysis of variance assumption (normality of the populations). Fortunately, as we mentioned in Chapter 6, the assumption of homogeneity (equality) of population variances is less critical when the sample sizes are nearly equal. When the sample sizes are nearly equal, the variances can be markedly different and the p-values for an analysis of variance will still be only mildly distorted. Thus, we recommend that Hartley's test be used only for the more extreme cases. In these extreme situations where homogeneity of the population variances is a problem, a transformation of the data may help to stabilize the variances. Then inferences can be made from an analysis of variance.

transformation of data

A **transformation of the sample data** is defined to be a process in which the measurements on the original scale are systematically converted to a new scale of measurement. For example, if the original variable is y and the variances associated with the variable across the treatments are not equal (heterogeneous), it may be necessary to work with a new variable such as $\sqrt{y}$, log y, or some other transformed variable.

How can we select the appropriate transformation? This is no easy task and often takes a great deal of experience in the experimenter's area of application. In spite of these difficulties, we can consider several guidelines for choosing an appropriate transformation.

guidelines for selecting y_T

Many times the variances across the populations of interest are heterogeneous and seem to vary with the magnitude of the population mean. For example, it may be that the larger the population mean, the larger the population variance. When we are able to identify how the variance varies with the population mean, we can define a suitable transformation from the variable y to a new variable y_T. Three specific situations are presented in Table 13.11.

TABLE 13.11
Transformation to Achieve Uniform Variance

Relationship Between μ and σ^2	y_T	Variance of y_T (for a Given k)
$\sigma^2 = k\mu$ (when $k = 1$, y is a Poisson variable)	$y_T = \sqrt{y}$ or $\sqrt{y + .375}$	$1/4$; $(k = 1)$
$\sigma^2 = k\mu^2$	$y_T = \log y$ or $\log (y + 1)$	1; $(k = 1)$
$\sigma^2 = k\pi(1 - \pi)$ (when $k = 1/n$, y is a binomial variable)	$y_T = \arcsin \sqrt{y}$	$1/4n$; $(k = 1/n)$

The first row of Table 13.11 suggests that if y is a Poisson* random variable the variance of y is equal to the mean of y. Thus, if the different populations correspond to different Poisson populations, the variances will be heterogeneous provided the means are different. The transformation that will stabilize the variances is $y_T = \sqrt{y}$; or, if the Poisson means are small (under 5), the transformation $y_T = \sqrt{y + .375}$ is better.

E X A M P L E   13.3

▼

The mean dissolved oxygen contents (in ppm) of three different lakes were to be compared based on independent random samples of 10 observations taken from the center of each lake at a depth of 1 foot. The sample data are given in Table 13.12.

T A B L E   13.12

Mean Dissolved Oxygen Contents (in ppm) of Three Lakes, Example 13.3

	Lake	
1	2	3
0	1	14
2	3	26
1	4	25
3	6	18
1	8	19
2	7	22
3	5	21
4	3	16
1	4	20
5	5	30
$\bar{y} = 2.2$	$\bar{y} = 4.6$	$\bar{y} = 21.1$
$s = 1.55$	$s = 2.07$	$s = 4.84$

a. Run a test of the equality of the population variances. Use Hartley's test, with $\alpha = .05$.

b. Transform the data using $y_T = \sqrt{y + .375}$.

c. Compute the sample means and sample standard deviations for the transformed data.

Solution

a. The F test for the equality of population variances has

$$F_{\max} = \frac{(4.84)^2}{(1.55)^2} = 9.75.$$

---

* The Poisson random variable is a useful discrete random variable with applications as an approximation for the binomial (when n is large but $n\pi$ is small) and as a model for events occurring randomly in time. For additional information see Hildebrand and Ott (1991) and Mendenhall (1987).

The critical value of F_{max} for $a = .05$, $t = 3$, and $df_2 = 9$ is 5.34. Since F_{max} is greater than 5.34, we reject the hypothesis of homogeneity of the population variances.

b. The square root data appear in Table 13.13.

c. The sample means and standard deviations for the transformed data are shown in Table 13.14. Although the original data had heterogeneous

T A B L E 13.13
Square Root Transformations ($\sqrt{y} = .375$) of the Data of Table 13.12

| | Lake | |
1	2	3
0.612	1.173	3.791
1.541	1.837	5.136
1.173	2.092	5.037
1.837	2.525	4.287
1.173	2.894	4.402
1.541	2.716	4.730
1.837	2.318	4.623
2.092	1.837	4.047
1.173	2.092	4.514
2.318	2.318	5.511

T A B L E 13.14
Sample Means and Standard Deviations for the Data in Table 13.13

| | Lake | | |
	1	2	3
Sample mean	1.53	2.18	4.61
Sample standard deviation	.51	.50	.52

variances, the sample variances are all approximately .25, as indicated in Table 13.11. ▲

$y_T = \log y$

The second transformation indicated in Table 13.11 is for an experimental situation where the population variance is approximately equal to the square of the population mean, or equivalently, where $\sigma = \mu$. Actually, the logarithmic trans-

coefficient of variation

formation is appropriate any time the **coefficient of variation** σ_i/μ_i is constant across the populations of interest.

E X A M P L E 13.4 ▼

Irritable bowel syndrome (IBS) is a nonspecific intestinal disorder characterized by abdominal pain and irregular bowel habits. Each person in a random sample of 24 patients having periodic attacks of IBS was randomly assigned to one of three treatment groups, A, B, and C. The number of hours of relief while on therapy is recorded for each patient in Table 13.15.

T A B L E 13.15
Data for Hours of Relief While on Therapy, Example 13.4

	Treatment	
A	B	C
4.2	4.1	38.7
2.3	10.7	26.3
6.6	14.3	5.4
6.1	10.4	10.3
10.2	15.3	16.9
11.7	11.5	43.1
7.0	19.8	48.6
3.6	12.6	29.5
$\bar{y} = 6.46$	$\bar{y} = 12.34$	$\bar{y} = 27.35$
$s = 3.22$	$s = 4.53$	$s = 15.66$

a. Test for differences among the population variances. Use $\alpha = .05$.
b. Since there are no 0 y values, use the transformation $y_T = \ln y$ ("ln" denotes logarithms to the base e) to try to stabilize the variances.
c. Compute the sample means and the sample standard deviations for the transformed data.

Solution
a. The F test for a test of the null hypothesis $H_0: \sigma_1^2 = \sigma_2^2 = \sigma_3^2$ is

$$F_{max} = \frac{(15.66)^2}{(3.22)^2} = \frac{245.24}{10.37} = 23.65.$$

Since the computed value of F_{max} exceeds 6.94, the tabulated value (Table 14) for $a = .05$, $t = 3$, and $df_2 = 7$, we reject H_0 and conclude that the population variances are different.
b. The transformed data are shown in Table 13.16. Note: Natural logs have been tabulated [*CRC Standard Mathematical Tables* (1961)] but can also be computer generated.

T A B L E 13.16
Natural Logarithms of the Data in Table 13.15

	Treatment	
A	B	C
1.435	1.411	3.656
.833	2.370	3.270
1.887	2.660	1.686
1.808	2.342	2.332
2.322	2.728	2.827
2.460	2.442	3.764
1.946	2.986	3.884
1.281	2.534	3.384

.c. The sample means and standard deviations for the transformed data are given in Table 13.17. Although the sample variances are not exactly the same, they certainly do not indicate that the corresponding population variances are different.

TABLE 13.17
Sample Means and Standard Deviations for the Data of Table 13.16

	Treatment		
	A	B	C
Sample mean	1.75	2.43	3.10
Sample standard deviation	.54	.46	.77

▲

$$y_T = \arcsin \sqrt{y}$$

The third transformation listed in Table 13.11 is particularly appropriate for data recorded as percentages or proportions. You will recall that in Chapter 4 we introduced the binomial distribution, where y designates the number of successes in n identical trials and $\hat{\pi} = y/n$ provides an estimate of π, the proportion of experimental units in the population possessing the characteristic. Although we may not have mentioned this while studying the binomial, the variance of $\hat{\pi}$ is given by $\pi(1 - \pi)/n$. Thus, if the response variable is $\hat{\pi}$, the proportion of successes in a random sample of n observations, then the variance of $\hat{\pi}$ will vary, depending on the values of π for the populations from which the samples were drawn. See Table 13.18.

TABLE 13.18
Variance of $\hat{\pi}$, the Sample Proportion, for Several Values of π and $n = 20$

Values of π	$\pi(1 - \pi)/n$
.01	.0005
.05	.0024
.1	.0045
.2	.0080
.3	.0105
.4	.0120
.5	.0125

Since the variance of $\hat{\pi}$ is symmetrical about $\pi = .5$, the variance of $\hat{\pi}$ for $\pi = .7$ and $n = 20$ would be .0105. Similarly, we can determine $\pi(1 - \pi)/n$ for other values of $\pi > .5$. The important thing to note is that if the populations have values of π in the vicinity of approximately .3 to .5, there is very little difference in the variances for $\hat{\pi}$. However, the variance of $\hat{\pi}$ is quite variable for either large or small values of π, and for these situations we should consider the possibility of transforming the sample proportions to stabilize the variances.

The transformation we recommend is $\arcsin \sqrt{\hat{\pi}}$ (sometimes written as $\sin^{-1} \sqrt{\hat{\pi}}$). That is, we are transforming the sample proportion into the angle whose sine is $\sqrt{\hat{\pi}}$. Some experimenters express these angles in degrees, others in radians.

For consistency we will always express our angles in radians. Table 15* of the Appendix provides arcsin computations for various values of $\hat{\pi}$.

E X A M P L E 13.5 ▼

In a national public opinion poll, a random sample of 30 registered voters was obtained from each of 24 different standard metropolitan statistical areas (SMSA). Each of the 30 voters in a sample was asked whether he or she favored limiting the FBI director to a fixed term in office (such as 10 years). The following data are the sample proportions for the 24 SMSAs. Transform the data by using $y_T = \arcsin \sqrt{\hat{\pi}}$. Calculate the sample mean and standard deviation for the transformed data.

.13	.60	.33	.03	.43	.43
.17	.70	.47	.10	.60	.60
.30	.10	.57	.20	.20	.67
.53	.20	.70	.33	.30	.77

Solution Using a calculator or Table 15 in the Appendix, the transformed data are

.37	.89	.61	.17	.72	.72
.42	.99	.76	.32	.89	.89
.58	.32	.86	.46	.46	.96
.82	.46	.99	.61	.58	1.07

The sample mean and standard deviation are, respectively, .66 and .25. ▲

when $\pi = 0, 1$

One comment should be made concerning the situation in which a **sample proportion of 0 or 1** is observed. For these cases we recommend substituting $1/4n$ and $1 - (1/4n)$, respectively, as the corresponding sample proportions to be used in the calculations.

In this section, we discussed Hartley's test for checking the equality of variance assumption and the analysis of variance, and transformations of data that can alleviate the problem of nonconstant variances. As an added benefit, the transformations presented in this section also (sometimes) decrease the nonnormality of the data. Still there will be times when the presence of severe skewness or outliers causes nonnormality that could not be eliminated by these transformations. Wilcoxon's rank sum test (Chapter 6) can be used for comparing two populations in the presence of nonnormality when working with two independent samples.

---

* Table 15 in the Appendix gives 2 arcsin $\sqrt{\hat{\pi}}$.

For data based on more than two independent samples we can address non-normality using the Kruskal–Wallis test (Section 13.6). It should be noted that these tests are also based on a transformation (the rank transformation) of the sample data.

▼ EXERCISES

Applications

 13.3 The data of Example 13.5 are shown here. Suppose that the four columns represent four geographic locations of the country (NE, SE, NW, SW) and that a random sample of 100 voters was obtained from six selected SMSAs within each geographic location. Analyze the sample data by using the arcsin transformation to determine if there are differences among the four geographic locations. Use $\alpha = .05$.

NE	SE	NW	SW
.13	.10	.03	.20
.17	.20	.10	.30
.30	.33	.20	.43
.53	.47	.33	.60
.60	.57	.43	.67
.70	.70	.60	.77

13.4 Refer to Exercise 13.3. Suppose that the rows correspond to different socioeconomic levels, so that one SMSA was selected from each socioeconomic level in each of the four geographic locations. The sample data then represent the proportion of favorable responses based on independent samples of 100 people for each socioeconomic–geographic location combination.

a. Analyze the transformed data and draw conclusions.

b. Comment on the proposal to take two random samples of size 50 for each socio-economic–geographic location combination, rather than taking one sample of 100 voters.

13.6 ▼ A NONPARAMETRIC ALTERNATIVE: THE KRUSKAL–WALLIS TEST

The concept of a rank sum test can be extended to a comparison of more than two populations. In particular, suppose that n_1 observations are drawn at random from population 1, n_2 from population 2, . . . , and n_k from population k. We may wish to test the hypothesis that the k samples were drawn from identical distributions.

The following test procedure, sometimes called the Kruskal–Wallis test, is then appropriate.

Extension of the Rank Sum Test for More Than Two Populations

H_0:   The k distributions are identical

H_a: Not all the distributions are the same

T.S.: $H = \dfrac{12}{n_T(n_T + 1)} \sum_i \dfrac{T_i^2}{n_i} - 3(n_T + 1)$,

where n_i is the number of observations from sample i ($i = 1, 2, \ldots, k$), n_T is the combined (total) sample size; that is, $n_T = \sum_i n_i$ and T_i denotes the sum of the ranks for the measurements in sample i after the combined sample measurements have been ranked.

R.R.:   For a specified value of α, reject H_0 if H exceeds the critical value of χ^2 for $a = \alpha$ and df $= k - 1$.

Note: When there are a large number of ties in the ranks of the sample measurements, use

H'

$$H' = \frac{H}{1 - [\sum_j (t_j^3 - t_j)/(n_T^3 - n_T)]}$$

where t_j is the number of observations in the jth group of tied ranks.

E X A M P L E   13.6

Three random samples of clerics were drawn, one containing ten Methodist ministers, the second containing ten Catholic priests, and the third containing ten Pentecostal ministers. Each of the clerics was then examined, using a test to measure his or her knowledge about causes of mental illness. These test scores are listed in Table 13.19.

T A B L E   13.19
Scores for Knowledge of Mental Illness for the Clerics, Example 13.6

Methodist	Catholic	Pentecostal
32	32	28
30	32	21
30	26	15
29	26	15
26	22	14
23	20	14
20	19	14
19	16	11
18	14	9
12	14	8

Use the data to determine if the three groups of clerics differ with respect to their knowledge about the causes of mental illness. Use $\alpha = .05$.

Solution   The research and null hypotheses for this example can be stated as follows:

H_a: At least one of the three groups of clerics differs from the others with respect to knowledge about causes of mental illness.

H_0:   There is no difference among the three groups with respect to knowledge about the causes of mental illness (i.e., the samples of scores were drawn from identical populations).

Before computing H we must first jointly rank the 30 test scores from lowest to highest. From Table 13.19 we see that 8 is the lowest test score, and this cleric is assigned the rank of 1. Similarly, the scores 9, 11, and 12 receive the ranks 2, 3, and 4, respectively. Five clerics have a test score of 14, and since these 5 scores occupy the ranks 5, 6, 7, 8, and 9, we assign each one a rank of 7, the average of the occupied ranks. In a similar way, we can assign the remaining ranks to test scores. Table 13.20 lists the 30 test scores and associated ranks (in parentheses).

T A B L E   13.20
Test Scores and Ranks for the Clerics Study

Methodist		Catholic		Pentecostal	
32	(29)	32	(29)	28	(24)
30	(26.5)	32	(29)	21	(18)
30	(26.5)	26	(22)	15	(10.5)
29	(25)	26	(22)	15	(10.5)
26	(22)	22	(19)	14	(7)
23	(20)	20	(16.5)	14	(7)
20	(16.5)	19	(14.5)	14	(7)
19	(14.5)	16	(12)	11	(3)
18	(13)	14	(7)	9	(2)
12	(4)	14	(7)	8	(1)

Note from Table 13.20 that the sums of the ranks for the three groups are 197, 178, and 90. Hence, the computed value of H is

$$H = \frac{12}{30(30+1)} \left[\frac{(197)^2}{10} + \frac{(178)^2}{10} + \frac{(90)^2}{10} \right] - 3(30+1)$$

$$= \frac{12}{930} (3880.9 + 3168.4 + 810) - 93$$

$$= 8.4.$$

Since there are groups of tied ranks, we will use H' and compare its value to H. To do this we form the g groups composed of identical ranks, shown in the accompanying table.

From this information, we calculate the quantity

$$\frac{\sum_i (t_j^3 - t_i)}{n_T^3 - n_T} = \frac{1}{30^3 - 30} [(5^3 - 5) + (2^3 - 2) + (2^3 - 2) + (2^3 - 2) + (3^3 - 3) + (2^3 - 2) + (3^3 - 3)]$$

$$= \frac{192}{26{,}970} = .0071.$$

Rank	Group	t_j
1	1	1
2	2	1
3	3	1
4	4	1
7, 7, 7, 7, 7	5	5
10.5, 10.5	6	2
12	7	1
13	8	1
14.5, 14.5	9	2
16.5, 16.5	10	2
18	11	1
19	12	1
20	13	1
22, 22, 22	14	3
24	15	1
25	16	1
26.5, 26.5	17	2
29, 29, 29	18	3

Substituting this value into the formula for H', we have

$$H' = \frac{H}{1 - .0071} = \frac{8.4}{.9929} = 8.46.$$

So, even with more than half of the measurements involved in ties, H' and H are nearly the same. The critical value of chi-square with $a = .05$ and $df = k - 1 = 2$ can be found using Table 5 in the Appendix. This value is 5.991. Since the observed value of H' is greater than 5.991, we reject the null hypothesis and conclude that at least one of the clergy groups has more knowledge about the causes of mental illness than the other two groups. ▲

▼ EXERCISES

13.5   The yields (in pounds) of five different varieties (A, B, C, D, E) of 4-year-old orange trees in one orchard were to be compared. A random sample of seven trees of each variety was obtained from the orchard. The yields for these trees are presented here.

A	B	C	D	E
13	27	40	17	36
19	31	44	28	32
39	36	41	41	34
38	29	37	45	29
22	45	36	15	25
25	32	38	13	31
10	44	35	20	30

Conduct a test of the null hypothesis that the five varieties have the same yield distributions. Use $\alpha = .01$. Draw conclusions.

13.6   In Exercise 6.22 we discussed an experiment to compare the weights of the combs of roosters fed two different vitamin-supplemented diets. Twenty-eight healthy roosters were randomly divided into two groups, with one group receiving diet I and the other receiving diet II. After the study period, the comb weight (in mg) was recorded for each rooster. These data are given here.

Diet I	73	130	115	144	127	126	112	76	68	101	126	49	110	123
Diet II	80	72	73	60	55	74	67	89	75	66	93	75	68	76

a. Plot the sample data separately for the two diets.
b. Compute the sample means and variances.
c. Based on parts a and b, is there reason to suspect nonconstant variances or non-normality?
d. Run an analysis of variance (if appropriate) and draw conclusions. Use $\alpha = .05$.

13.7   Refer to Exercise 13.6. Suggest a nonparametric alternative and conduct the test. Compare your results to those obtained from Exercise 13.6, and reach a conclusion based on the results of both tests.

13.7 ▼ SUMMARY

In this chapter, we presented methods for extending the results of Chapter 6 to include a comparison among t population means. An independent random sample is drawn from each of the t populations. A measure of the within-sample variability is computed as $s_W^2 = SSW/(n_T - t)$. Similarly, a measure of the between-sample variability is obtained as $s_B^2 = SSB/(t - 1)$.

The decision to accept or reject the null hypothesis of equality of the t population means depends on the computed value of $F = s_B^2/s_W^2$. Under H_0, both s_B^2 and s_W^2 estimate σ_ε^2, the variance common to all t populations. Under the alternative hypothesis, s_B^2 estimates $\sigma_\varepsilon^2 + \theta$, where θ is a positive quantity, whereas s_W^2 still estimates σ_ε^2. Thus, large values of F indicate a rejection of H_0. Critical values for F are obtained from Table 6 in the Appendix for $df_1 = t - 1$ and $df_2 = n_T - t$. This test procedure, called an analysis of variance, is usually summarized in an analysis of variance (AOV) table.

You might be puzzled at this point by the following question: Suppose we reject H_0 and conclude that at least one of the means differs from the rest; which ones differ from the others? This chapter has not answered this question; Chapter 14 attacks this problem through procedures based on multiple comparisons.

In this chapter, we also discussed the assumptions underlying an analysis of variance for a completely randomized design. Independent random samples are absolutely necessary. The assumption of normality is least critical because we are dealing with means and the Central Limit Theorem holds for reasonable sample sizes. The equal variance assumption is critical only when the sample sizes are markedly different; this is a good argument for equal (or nearly equal) sample sizes. A test for equality of variances makes use of the $F_{\max}$ statistic, $s_{\max}^2/s_{\min}^2$.

Sometimes the sample data indicate that the population variances are different. Then, when the relationship between the population mean and the population standard deviation is either known or suspected, it is convenient to transform the sample measurements y to new values y_T in order to stabilize the population variances, using the transformations suggested in Table 13.11. These transformations include the square root, logarithmic, arcsin, and many others.

The topics in this chapter are certainly not covered in exhaustive detail. However, the material is sufficient for training the beginning researcher to be aware of the assumptions underlying his or her project and to consider either running an alternative analysis (such as using a nonparametric statistical method, the Kruskal–Wallis test) when appropriate or applying a transformation to the sample data.

▼ KEY FORMULAS

1. Analysis of variance for a completely randomized design

$$\text{TSS} = \sum_{i,j} (y_{ij} - \bar{y}_{..})^2$$

$$\text{SSB} = \sum_{i} n_i(\bar{y}_{i\cdot} - \bar{y}_{..})^2$$

$$\text{SSW} = \sum_{i,j} (y_{ij} - \bar{y}_{i\cdot})^2$$

2. Analysis of variance for a completely randomized design (shortcut formulas)

$$\text{TSS} = \sum_{i,j} y_{ij}^2 - \frac{y_{..}^2}{n_T}$$

$$\text{SSB} = \sum_{i} \frac{y_{i.}^2}{n_i} - \frac{y_{..}^2}{n_T}$$

$$\text{SSW} = \text{TSS} - \text{SSB}$$

3. Model for a completely randomized design

$$y_{ij} = \mu + \alpha_i + \varepsilon_{ij}$$

4. Hartley's test for homogeneity of variances

$$H_0: \quad \sigma_1^2 = \sigma_2^2 = \cdots = \sigma_t^2$$

$$\text{T.S.:} \quad F_{\text{max}} = \frac{s_{\text{max}}^2}{s_{\text{min}}^2}$$

5. Kruskal–Wallis test

$$H_0: \quad \text{The } k \text{ distributions are identical}$$

$$\text{T.S.:} \quad H = \frac{12}{n_T(n_T + 1)} \sum \frac{T_i^2}{n_i} - 3(n_T + 1)$$

▼ SUPPLEMENTARY EXERCISES

13.8 An experiment was conducted to compare the number of major defectives observed along each of five production lines in which changes were being instituted. Production was monitored continuously during the period of changes, and the number of major defectives was recorded per day for each line. These data are shown here.

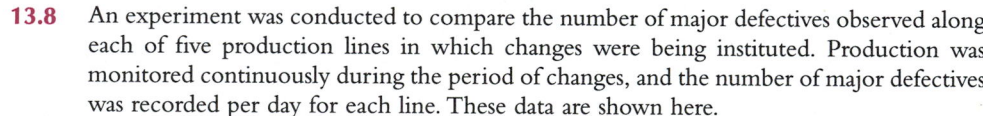

		Production Line		
1	2	3	4	5
34	54	75	44	80
44	41	62	43	52
32	38	45	30	41
36	32	10	32	35
51	56	68	55	58

a. Compute $\bar{y}$ and s^2 for each sample. Does there appear to be a problem with nonconstant variances? Use Hartley's test based on $\alpha = .05$.

b. Use a square root transformation on the data and conduct an analysis on the transformed data.

c. Draw your conclusions concerning differences among production lines.

13.9 Do a Kruskal–Wallis test on the data represented in Exercise 13.8. Does this test confirm the conclusions drawn in Exercise 13.8? If the results differ, which analysis do you believe? Use $\alpha = .05$.

13.10 The Agricultural Experiment Station of a university tested two different herbicides and their effects on crop yield. From 90 acres set aside for the experiment, herbicide 1 was used on a random sample of 30 acres, herbicide 2 was used on a second random sample of 30 acres, and the remaining 30 acres were used as a control. At the end of the growing season, the yields (in bushels per acre) were

	Sample Mean	Sample Standard Deviation
Herbicide 1	90.2	6.5
Herbicide 2	89.3	7.8
Control 3	85.0	7.4

Use these data to conduct a one-way analysis of variance. Use $\alpha = .05$. Are any of the yields different? If so, which ones?

13.11 Research from the Department of Fruit Crops at a university compared four different preservatives to be used in freezing strawberries. The yield from a strawberry patch was prepared for freezing and randomly divided into four equal groups. Within each group the strawberries were treated with the appropriate preservative and packaged into 8 small plastic bags for freezing at 0°C. Those in group I served as a control group, while those in groups II, III, and IV were assigned one of three newly developed preservatives. After all 32 bags of strawberries were prepared, they were stored at 0°C for a period of 6 months. At the end of this time, the contents of each bag were allowed to thaw and then rated on a scale of 1 to 10 points for discoloration. (Note that a low score indicates little discoloration.) These ratings are given below.

Group I	10	8	7.5	8	9.5	9	7.5	7
Group II	6	7.5	8	7	6.5	6	5	5.5
Group III	3	5.5	4	4.5	3	3.5	4	4.5
Group IV	2	1	2.5	3	4	3.5	2	2

a. We might be concerned with the normality of the data. To avoid any problems, refer to Section 13.6 to run a Kruskal–Wallis one-way analysis of variance by ranks. Use $\alpha = .05$.

b. Use the output here to compare the results of a one-way analysis of variance to the results for the Kruskal–Wallis test of part a. Draw conclusions based on the results of both tests.

```
       LISTING OF DATA
   OBS    GROUP    DISCOLOR
    1      1        10.0
    2      1         8.0
    3      1         7.5
    4      1         8.0
    5      1         9.5
    6      1         9.0
    7      1         7.5
    8      1         7.0
    9      2         6.0
   10      2         7.5
   11      2         8.0
   12      2         7.0
   13      2         6.5
   14      2         6.0
   15      2         5.0
   16      2         5.5
   17      3         3.0
   18      3         5.5
   19      3         4.0
   20      3         4.5
   21      3         3.0
   22      3         3.5
   23      3         4.0
   24      3         4.5
   25      4         2.0
   26      4         1.0
   27      4         2.5
   28      4         3.0
   29      4         4.0
   30      4         3.5
   31      4         2.0
   32      4         2.0
N =    32
```

```
        ANALYSIS OF VARIANCE PROCEDURE
       GENERAL LINEAR MODELS PROCEDURE
           CLASS LEVEL INFORMATION
       CLASS    LEVELS    VALUES
       GROUP       4      1 2 3 4

NUMBER OF OBSERVATIONS IN DATA SET = 32
```

```
                    ANALYSIS OF VARIANCE PROCEDURE
                   GENERAL LINEAR MODELS PROCEDURE
DEPENDENT VARIABLE: RANK1    RANK FOR VARIABLE DISCOLOR
```

SOURCE	DF	SUM OF SQUARES	MEAN SQUARE	F VALUE	PR > F	R-SQUARE	C.V.
MODEL	3	2331.93750000	777.31250000	56.74	0.0001	0.858751	22.4313
ERROR	28	383.56250000	13.69866071		ROOT MSE		RANK1 MEAN
CORRECTED TOTAL	31	2715.50000000			3.70117018		16.50000000

SOURCE	DF	TYPE I SS	F VALUE	PR > F	DF	TYPE III SS	F VALUE	PR > F
GROUP	3	2331.93750000	56.74	0.0001	3	2331.93750000	56.74	0.0001

```
NUMBER OF OBSERVATIONS IN DATA SET = 32
                    ANALYSIS OF VARIANCE PROCEDURE
                   GENERAL LINEAR MODELS PROCEDURE
DEPENDENT VARIABLE: DISCOLOR
```

SOURCE	DF	SUM OF SQUARES	MEAN SQUARE	F VALUE	PR > F	R-SQUARE	C.V.
MODEL	3	159.18750000	53.06250000	55.67	0.0001	0.856422	18.3771
ERROR	28	26.68750000	0.95312500		ROOT MSE		DISCOLOR MEAN
CORRECTED TOTAL	31	185.87500000			0.97628121		5.31250000

SOURCE	DF	TYPE I SS	F VALUE	PR > F	DF	TYPE III SS	F VALUE	PR > F
GROUP	3	159.18750000	55.67	0.0001	3	159.18750000	55.67	0.0001

13.12 Refer to Exercise 13.11. Suppose that some of the 32 bags were stored improperly and could not be analyzed at the end of the 6-month period. In particular, the sample sizes in the four groups were, respectively, 8, 6, 5, and 7. These data appear here.

Group I	10	8	7.5	8	9.5	9	7.5	7
Group II	6	7.5	8	7	6.5	6		
Group III	3	5.5	4	4.5	3			
Group IV	2	1	2.5	3	4	3.5	2	

Rerun an analysis of variance and/or a Kruskal–Wallis test and draw conclusions. Use $\alpha = .05$.

 13.13 An experiment was conducted to compare the starch content of tomato plants grown in sandy soil supplemented by one of three different nutrients, A, B, or C. Eighteen tomato seedlings of one particular variety were selected for the study, with six assigned to each of the nutrient groups. All seedlings were planted in a sand culture and maintained under a controlled environment. Those seedlings assigned to nutrient A served as the control group (receiving distilled water only). Plants assigned to nutrient B were fed a weak concentration of Hoagland nutrient, while those assigned to nutrient C received the Hoagland nutrient at full strength. The stem starch contents were determined 25 days after planting and are recorded below, in micrograms per milligram.

Nutrient A	22	20	21	18	16	14
Nutrient B	12	14	15	10	9	6
Nutrient C	7	9	7	6	5	3

a. Run an analysis of variance to test for differences in starch content for the three nutrient groups. Use $\alpha = .05$.
b. Draw your conclusions.

13.14 Although we often have well-planned experiments with equal numbers of observations per treatment, we still end up with unequal numbers at the end of a study. Suppose that although six plants were allocated to each of the nutrient groups of Exercise 13.13, only five survived in group B and four in group C. The data for the stem starch contents are given here.

Nutrient A	22	20	21	18	16	14
Nutrient B	12	14	15	10	9	
Nutrient C	7	9	7	6		

a. Write an appropriate model for this experimental situation. Define all terms.
b. Assuming that nutrients B and C did not cause the plants to die, perform an analysis of variance to compare the treatment means. Use $\alpha = .05$.

13.15 Salary disputes and their eventual resolutions often leave both employers and employees embittered by the entire ordeal. To assess employee reactions to a recently devised salary and fringe benefits plan, the personnel department obtained random samples of 15 employees from each of three divisions: manufacturing, marketing, and research. Each employee sampled was asked to respond (in confidence) to a series of questions. Several employees refused to cooperate, as reflected in the unequal sample sizes. The data are given here.

	Manufacturing	Marketing	Research
Sample size	12	14	11
Sample mean	25.2	32.6	28.1
Sample variance	3.6	4.8	5.3

a. Write a model for this experimental situation.

b. Use the summary of the scored responses to compare the means for the three divisions (the higher a score, the higher the employee acceptance). Use $\alpha = .01$.

13.16 The yields of corn, in bushels per plot, were recorded for four different varieties of corn, A, B, C, and D. In a controlled greenhouse experiment, each variety was randomly assigned to eight of 32 plots available for the study. The yields are listed here.

A	2.5	3.6	2.8	2.7	3.1	3.4	2.9	3.5
B	3.6	3.9	4.1	4.3	2.9	3.5	3.8	3.7
C	4.3	4.4	4.5	4.1	3.5	3.4	3.2	4.6
D	2.8	2.9	3.1	2.4	3.2	2.5	3.6	2.7

a. Write an appropriate statistical model.

b. Perform an analysis of variance on these data and draw your conclusions. Use $\alpha = .05$.

13.17 Refer to Exercise 13.16. Perform a Kruskal–Wallis analysis of variance by ranks (with $\alpha = .05$) and compare your results to those in Exercise 13.16.

13.18 Many corporations make use of the Wide Area Telephone System (WATS), where, for a fixed rent per month, the corporation can make as many long distance calls as it likes. Depending on the area of the country in which the corporation is located, it can rent a WATS line for certain geographic bands. For example, in Ohio, these bands might include the following states:

Band I:	Ohio	
Band II:	Indiana	Pennsylvania
	Kentucky	Tennessee
	Maryland	Virginia
	Michigan	West Virginia
	North Carolina	Washington, D.C.
Band III:	32 states shown in Figure 13.3, plus Washington, D.C.	

To monitor the use of the WATS lines, a corporation selected a random sample of 12 calls from each of the following areas in a given month. The length of the conversation (in minutes) was recorded for each call. (Band III excludes states in Band II and Ohio.)

Ohio	2	3	5	8	4	6	18	19	9	6	7	5
Band II	6	8	10	15	19	21	10	12	13	2	5	7
Band III	12	14	13	20	25	30	5	6	21	22	28	11

FIGURE 13.3
WATS Line Coverage

Band II

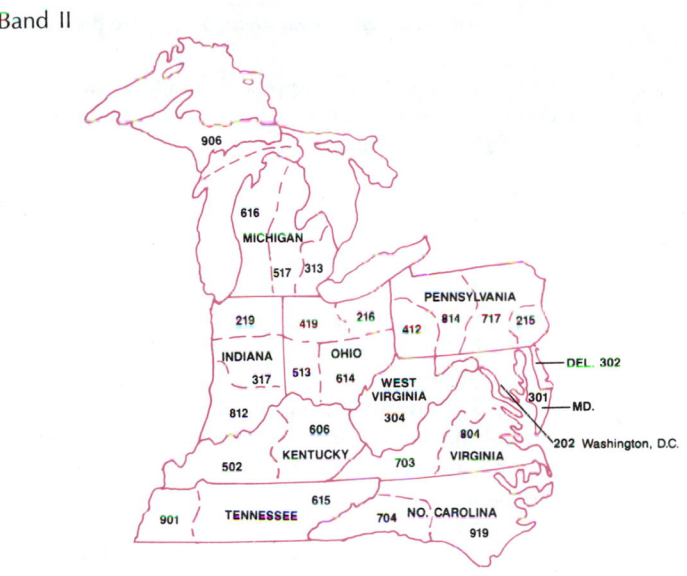

Band III

Perform an analysis of variance to compare the mean lengths of calls for the three areas. Use $\alpha = .05$.

13.19 Refer to Exercise 13.18. Suppose that rather than 12 calls from each area, we obtained a random sample of 15 calls. Use the additional three measurements recorded here for the total of 15 observations per area.

Ohio	10	11	4
Band II	8	28	31
Band III	29	50	120

Analyze the sample data to compare the mean number of call times for the three areas. Use $\alpha = .05$.

13.20 Refer to Exercise 13.18. Some researchers would argue that durations of telephone calls may not be normally distributed. Perform a Kruskal–Wallis one-way analysis of variance by ranks and compare your results to those of Exercise 13.18. Use $\alpha = .05$.

13.21 Examine the SAS computer output that follows for the analysis of variance performed in Example 13.2.
   a. Identify the sums of squares and degrees of freedom for methods, error, and total.
   b. Give the mean squares and F test for the equality of method means.
   c. Give the level of significance for the test in part b.
   d. What is the coefficient of determination?

```
                              ANALYSIS OF VARIANCE PROCEDURE
                              GENERAL LINEAR MODELS PROCEDURE

DEPENDENT VARIABLE: SCORE
SOURCE              DF     SUM OF SQUARES    MEAN SQUARE    F VALUE    PR > F    R-SQUARE     C.V.
MODEL               2        452.12878788    226.06439394       771    0.0136   0.658556   6.9724
ERROR               8        234.41666667     29.30208333              ROOT MSE          SCORE MEAN
CORRECTED TOTAL    10        686.54545455                            5.41313988          77.63636364

SOURCE              DF      TYPE I SS    F VALUE    PR > F    DF    TYPE III SS    F VALUE    PR> F
METHOD              2     452.12878788       7.71    0.0136     2   452.12878788       7.71   0.0136
```

13.22 Four different types of pillows were tested by a panel of consumers. Each panelist examined only one pillow and rated it on a scale from 1 (inferior) to 7 (superior). The data are summarized here.

Pillow Type	Rating	n_i
1	1, 3, 5, 7, 2, 3, 4	7
2	7, 6, 7, 7, 6	5
3	1, 2, 3, 2, 3, 2, 1	7
4	4, 3, 4, 1, 5	5

a. Perform a one-way analysis of variance using $\alpha = .05$.

b. Draw conclusions.

13.23 **a.** Do the data of Exercise 13.22 suggest violations of the AOV assumptions?

b. Suggest an alternate analysis and compare your results to those of Exercise 13.22.

c. Draw final conclusions.

13.24 Use data from either a laboratory course or from some other source to make comparisons of three or more methods, varieties, or plans. Use an analysis of variance with $\alpha = .05$.

13.25 An experiment was conducted to test the effects of five different diets in turkeys. Six turkeys were randomly assigned to each of the five diet groups and were fed for a fixed period of time.

Group	Weight Gained (pounds)
Control diet	4.1, 3.3, 3.1, 4.2, 3.6, 4.4
Control diet + level 1 of additive A	5.2, 4.8, 4.5, 6.8, 5.5, 6.2
Control diet + level 2 of additive A	6.3, 6.5, 7.2, 7.4, 7.8, 6.7
Control diet + level 1 of additive B	6.5, 6.8, 7.3, 7.5, 6.9, 7.0
Control diet + level 2 of additive B	9.5, 9.6, 9.2, 9.1, 9.8, 9.1

a. Plot the data separately for each sample.

b. Compute $\bar{y}$ and s^2 for each sample.

c. Is there any evidence of unequal variances or nonnormality? Explain.

d. Assuming that the five groups were comparable with respect to initial weights of the turkeys, use the weight-gained data to draw conclusions concerning the different diets. Use $\alpha = .05$.

13.26 Run a Kruskal–Wallis test for the data of Exercise 13.25. Do these results confirm what you concluded from an analysis of variance? What overall conclusions can be drawn? Use $\alpha = .05$.

13.27 Some researchers have conjectured that stem-pitting disease in peach tree seedlings might be related to the presence or absence of nematodes in the soil. Hence, weed and soil treatment using herbicides might be effective in promoting seedling growth. An experiment was conducted to compare peach tree seedling growth with soil and weeds treated with one of three herbicides:

A. control (no herbicide)
B: herbicide with Nemagone
C: herbicide without Nemagone

Of the 18 seedlings chosen for the study, 6 were randomly assigned to each treatment group. Soil and weeds in the growing areas for the three groups were treated with the appropriate herbicide. At the end of the study period, the height (in centimeters) was recorded for each seedling. Use the following sample data to run an analysis of variance for detecting differences among the seedling heights for the three groups. Use $\alpha = .05$. Draw your conclusions.

Herbicide A	66	67	74	73	75	64
Herbicide B	85	84	76	82	79	86
Herbicide C	91	93	88	87	90	86

13.28 Suppose that the data of Exercise 13.8 represented data from five different weeks by day of the week.

Day of Week	Data
Monday	34, 54, 75, 44, 80
Tuesday	44, 41, 62, 43, 52
Wednesday	32, 38, 45, 30, 41
Thursday	36, 32, 10, 32, 35
Friday	51, 56, 68, 55, 58

a. Do a one-way analysis of variance to assess the variability due to days.
b. Draw conclusions based on $\alpha = .01$.

13.29 Refer to the data of Exercise 13.25. To illustrate the effect that an extreme value can have on conclusions from an analysis of variance, suppose that the weight gained by the fifth turkey in the level 2, additive B group was 12.8 (or 15.8) rather than 9.8.
a. What effect does this have on the assumptions for an analysis of variance?
b. With 9.8 replaced by 15.8, if someone unknowingly ran an analysis of variance, what conclusions would he or she draw?

13.30 Refer to Exercise 13.29. What happens to the Kruskal–Wallis test if the value 9.8 is replaced by 12.8 (or 15.8)? Might there be a reason to run both a Kruskal–Wallis test and an analysis of variance? Why?

13.31 Is the Kruskal–Wallis test more powerful than an analysis of variance, in certain situations, for detecting differences among treatment means? Explain.

 13.32 A small corporation makes wire coating (for insulation) on three different machines. The machines are identical with respect to make, model, and age but seem to differ with respect to variation of the inside diameter dimension (millimeters). Management wishes to test this variation and collected data from each machine.

Machine A	Machine B	Machine C
105	56	183
3	43	144
90	1	219
217	37	86
22	14	39

Conduct a test for the homogeneity of the population variances. Would it be appropriate to proceed with an analysis of variance based on the results of this test? Explain. Use $\alpha = .05$.

CHAPTER 14

MULTIPLE COMPARISONS

14.1 INTRODUCTION

In Chapter 13, we introduced a procedure for testing the equality of t population means. The test statistic $F = s_B^2/s_W^2$ was used to determine whether the between–sample variability was large relative to the within-sample variability. If the computed value of F for the sample data exceeded the critical value obtained from Table 6 in the Appendix, the null hypothesis $H_0: \mu_1 = \mu_2 = \cdots = \mu_t$ was rejected in favor of the alternative hypothesis

H_a: At least one of the t population means differs from the rest.

While rejection of the null hypothesis does give us some information concerning the population means, we do not know which means differ from each other. For example, does μ_1 differ from μ_2 or μ_3? Does μ_3 differ from μ_4, μ_5, and μ_6? **Multiple-comparison procedures** have been developed to answer questions such as these. Although many multiple-comparison procedures have been proposed, we will focus on just a few of the more common methods. After studying these few procedures, you should be able to evaluate the results of most published material using multiple comparisons or to suggest an appropriate multiple-comparison procedure in an experimental situation.

A word of caution should be voiced at this time. Whenever we are analyzing data, it is tempting to analyze those comparisons that appear to be interesting after seeing the sample data. This practice has sometimes been called **data dredging** or **data snooping**, and the confidence coefficient for a single comparison does not

data dredging
data snooping

807

reflect the after-the-fact nature of the comparison. For example, we know from previous work that the interval estimate for the difference between two population means using the formula

$$(\bar{y}_1 - \bar{y}_2) \pm t_{\alpha/2} s_p \sqrt{\frac{1}{n_1} + \frac{1}{n_2}}$$

has a confidence coefficient of $1 - \alpha$. Suppose we had run an analysis of variance to test the hypothesis

$$H_0: \quad \mu_1 = \mu_2 = \mu_3 = \mu_4 = \mu_5 = \mu_6$$

for six populations, but decided to compute a confidence interval for μ_1 and μ_2 only after we saw that the largest sample mean was $\bar{y}_1.$ and the smallest was $\bar{y}_2.$. In this situation, the confidence coefficient would not be $1 - \alpha$ as originally thought; that value applies only to a preplanned comparison, one planned before looking at the sample data.

One way to allow for data snooping after observing the sample data is to use a multiple-comparison procedure that has a confidence coefficient to cover all comparisons that could be done after observing the sample data. Some of these procedures are discussed in this chapter.

The other possibility is to use data-snooping comparisons as a basis for generating hypotheses that must be confirmed in future testing. Here, the data-snooping comparisons serve an exploratory, or hypothesis-generating, role, and inferences would not be made based on the data snoop. Further testing would be done to confirm (or not) the hypothesis generated in the data snoop.

exploratory hypothesis generation

confirmation

14.2 ▼ LINEAR CONTRASTS

Before developing several different multiple-comparison procedures, we need the following notation and definitions. Consider a one-way classification where we wish to make comparisons among the t population means $\mu_1, \mu_2, \ldots, \mu_t$. These comparisons among t population means can be written in the form

$$l = a_1 \mu_1 + a_2 \mu_2 + \cdots + a_t \mu_t = \sum_{i=1}^{t} a_i \mu_i,$$

where the a_is are constants satisfying the property that $\sum a_i = 0$. For example, if we wanted to compare μ_1 to μ_2, we would write the linear form

$$l = \mu_1 - \mu_2.$$

Note here that $a_1 = 1$, $a_2 = -1$, $a_3 = a_4 = \cdots = a_t = 0$, and $\sum_i a_i = 0$. Similarly, we could compare the mean for population 1 to the average of the means for populations 2 and 3. Then l would be of the form

$$l = \mu_1 - \frac{(\mu_2 + \mu_3)}{2},$$

where $a_1 = 1$, $a_2 = a_3 = -\frac{1}{2}$, $a_4 = a_5 = \cdots = a_t = 0$, and $\sum_i a_i = 0$.

$\hat{l}$

linear contrast

An estimate of the linear form l, designated by $\hat{l}$, is formed by replacing the μ_is in l with their corresponding sample means $\bar{y}_{i\cdot}$. The estimate $\hat{l}$ is called a **linear contrast**.

DEFINITION 14.1
Linear Contrast

> ▼
>
> $\hat{l} = a_1 \bar{y}_{1\cdot} + a_2 \bar{y}_{2\cdot} + \cdots + a_t \bar{y}_{t\cdot} = \sum_i a_i \bar{y}_{i\cdot}$ is called a **linear contrast** among the t sample means and can be used to estimate $l = \sum_i a_i \mu_i$. The a_is are constants satisfying the constraint $\sum_i a_i = 0$.

The variance of the linear contrast $\hat{l}$ can be estimated as follows:

$\hat{V}(\hat{l})$

$$\hat{V}(\hat{l}) = s_W^2 \left[\frac{a_1^2}{n_1} + \frac{a_2^2}{n_2} + \cdots + \frac{a_t^2}{n_t} \right] = s_W^2 \sum_i \frac{a_i^2}{n_i},$$

where n_i is the number of sample observations selected from population i and s_W^2 is the mean square within samples obtained from the analysis of variance table for the one-way classification. If all the sample sizes are the same (i.e., all $n_i = n$), then

$$\hat{V}(\hat{l}) = \frac{s_W^2}{n} \sum_i a_i^2.$$

There are many different contrasts that can be formed among the t sample means. However, if each of the sample means is based on the same number of observations (i.e., $n_i = n$), we have the following definition.

DEFINITION 14.2
Orthogonal

> ▼
>
> Two contrasts l_1 and l_2, where
>
> $$\hat{l}_1 = \sum_i a_i \bar{y}_{i\cdot} \quad \text{and} \quad \hat{l}_2 = \sum_i b_i \bar{y}_{i\cdot}$$

(continues)

are said to be **orthogonal** if

$$a_1 b_1 + a_2 b_2 + \cdots + a_t b_t = \sum_i a_i b_i = 0.$$

Note: The sample sizes must be the same.

mutually orthogonal

A set of contrasts is said to be **mutually orthogonal** if all pairs of contrasts in the set are orthogonal.

E X A M P L E   14.1   ▼

Consider a one-way classification for comparing $t = 4$ population means. Are the following contrasts orthogonal?

$$\hat{l}_1 = \bar{y}_{1.} - \bar{y}_{2.} \qquad \hat{l}_2 = \bar{y}_{3.} - \bar{y}_{4.}$$

$\hat{l}_1 = a_1 = 1 \quad a_2 = -1 \qquad \hat{l}_2 \quad b_1 = 1 \quad b_2 = -1$

Solution   We can rewrite the contrasts in the following form:

$$\hat{l}_1 = \bar{y}_{1.} - \bar{y}_{2.} + 0(\bar{y}_{3.}) + 0(\bar{y}_{4.})$$
$$\hat{l}_2 = 0(\bar{y}_{1.}) + 0(\bar{y}_{2.}) + \bar{y}_{3.} - \bar{y}_{4.},$$

where we see that $a_1 = 1$, $a_2 = -1$, $a_3 = 0$, $a_4 = 0$, and $b_1 = 0$, $b_2 = 0$, $b_3 = 1$, $b_4 = -1$. It is then apparent that

$$\sum_i a_i b_i = a_1 b_1 + a_2 b_2 + a_3 b_3 + a_4 b_4 = 0$$

and hence the contrasts are orthogonal.   ▲

E X A M P L E   14.2   ▼

Refer to Example 14.1. Are the given contrasts orthogonal?

$$\hat{l}_1 = \bar{y}_{1.} - \bar{y}_{2.} \qquad \text{and} \qquad \hat{l}_2 = \bar{y}_{1.} - \bar{y}_{3.}$$

Solution   Rewriting the contrasts as

$$\hat{l}_1 = \bar{y}_{1.} - \bar{y}_{2.} + 0(\bar{y}_{3.}) + 0(\bar{y}_{4.})$$
$$\hat{l}_2 = \bar{y}_{1.} + 0(\bar{y}_{2.}) - \bar{y}_{3.} + 0(\bar{y}_{4.}),$$

we see that

$$\sum_i a_i b_i = (1)(1) + (-1)(0) + (0)(-1) + (0)(0) = 1$$

which indicates that the two contrasts are not orthogonal.   ▲

The concept of orthogonality between linear contrasts is important in the study of multiple-comparison procedures. Recall that prior to running an analysis of variance among the t population means in a one-way classification, we assumed that

1. The t populations were normally distributed with a common variance σ_ε^2 but different means (under H_0 we assume that the means are equal).
2. Independent random samples were obtained from the t populations.

t − 1 contrasts

If we assume that each of the sample means is based on the same number of observations, then it can be shown that $t - 1$ orthogonal contrasts can be formed using the t sample means. These $t - 1$ **contrasts** form a set of mutually orthogonal contrasts. (An easy way to remember $t - 1$ is to refer to the number of degrees of freedom associated with the between-sample source of variability in the AOV table.) In addition, it can be shown that the sums of squares for the $t - 1$ contrasts will add up to the treatment sum of squares. Mutual orthogonality is desirable because it leads to independence of the $t - 1$ sums of squares associated with the orthogonal contrasts. As we will see later in the chapter, methods have also been developed that use nonorthogonal contrasts among the sample means. These methods will be particularly appropriate when the experimenter is making all pairwise comparisons among t treatment means.

▼ EXERCISES

Basic Techniques

14.1   Consider the expressions

$$\hat{l}_1 = \bar{y}_{1.} + \bar{y}_{2.} - 2\bar{y}_{3.}$$
$$\hat{l}_2 = \bar{y}_{1.} + \bar{y}_{2.} - 2\bar{y}_{4.}$$

a. Are $\hat{l}_1$ and $\hat{l}_2$ linear contrasts?
b. Are $\hat{l}_1$ and $\hat{l}_2$ orthogonal?

14.2   Refer to Exercise 14.1. Construct a set of three mutually orthogonal contrasts for comparing four population means.

14.3   Write down a set of four mutually orthogonal contrasts to be used in comparing five population means.

14.3 ▼ WHICH ERROR RATE IS CONTROLLED?

Let us suppose that an experimenter wishes to compare t population means using c independent (orthogonal) contrasts. Each comparison among the t population means can be tested using a t test of the following form:

t test

$$H_0: \quad l = 0$$

$$H_a: \quad l > 0 \text{ (for a one-tailed test)}$$

$$\text{T.S.:} \quad t = \frac{\hat{l}}{\sqrt{\hat{V}(\hat{l})}} = \frac{\hat{l}}{\sqrt{s_W^2 \sum_i a_i^2/n_i}}.$$

The rejection region for the computed value of the test statistic can be obtained from Table 4 in the Appendix with $a = \alpha/2$ and $df = n_T - t$.

If each of the comparisons is tested with the same value of α, and if we assume that s_W^2 has an infinite number of degrees of freedom (so the tests are independent), then when all the null hypotheses are true, the probability of falsely rejecting H_0 on at least one of the t tests can be shown to be $1 - (1 - \alpha)^c$. This quantity is sometimes called an **overall error rate** for the c comparisons. We can see from Table 14.1 that as c increases for a given value of α, the probability of falsely rejecting H_0 on at least one of the t tests becomes quite large. Hence, if an experimenter wished to compare $t = 20$ population means by using $c = 10$ orthogonal contrasts, the probability of falsely rejecting H_0 on at least one of the t tests could be as high as .401 when each individual test was performed with $\alpha = .05$.

$1 - (1 - \alpha)^c$ overall error rate

TABLE 14.1
A Comparison of the Overall Error Rate for c Independent Contrasts Among $t(t > c)$ Sample Means

c, Number of Contrasts	α, Probability of a Type I Error on an Individual Test		
	.10	.05	.01
1	.100	.050	.010
2	.190	.097	.020
3	.271	.143	.030
4	.344	.185	.039
5	.410	.226	.049
⋮	⋮	⋮	⋮
10	.651	.401	.096

The results of Table 14.1 are disturbing and may lead us to question significant results when they appear. The problem can be alleviated somewhat by **controlling the overall error rate** rather than the error rate (Type I error) for the individual t test. Suppose, for example, that we wished the overall error rate for $c = 10$ orthogonal contrasts among $t = 20$ population means to be .10. What value of α

controlling overall error rate

must we use on the individual t tests to achieve an overall error rate of .10? Assuming s_W^2 is based on a large number of degrees of freedom, this problem can be solved by determining the value of α for which

$$1 - (1 - \alpha)^{10} = .10.$$

error rate for nonorthogonal contrasts

The method of solution for this equation is not important now; we can see from Table 14.1 that by using $\alpha = .01$ for all 10 tests, the overall error rate would be approximately .10.

Although controlling the overall error rate for comparisons using orthogonal contrasts is fairly simple, it is difficult to obtain an expression equivalent to $1 - (1 - \alpha)^c$ for comparisons made with nonorthogonal contrasts. For example, suppose we wish to make all pairwise comparisons among $t = 4$ population means. Previous results indicate that we could make $t - 1 = 3$ orthogonal (independent) contrasts, but there are six possible pairwise comparisons among the population means (1 and 2, 1 and 3, 1 and 4, 2 and 3, 2 and 4, and 3 and 4). If each of these six comparisons is made using a t test with $\alpha = .05$, what is the overall error rate? Pearson and Hartley (1942, 1943) and Harter (1957) attacked the problem of determining the probability of falsely rejecting H_0 on at least one of the t tests for nonindependent contrasts. The solution is not easy, however, and is beyond the scope of this text. One alternative is to redefine the overall error rate and to determine a testing procedure that controls the overall error rate at a desired level. Indeed, a major difference among the multiple-comparison procedures we will discuss in the following sections is the error rate that each procedure controls.

14.4 ▼ FISHER'S LEAST SIGNIFICANT DIFFERENCE

Recall that we are interested in determining which population means differ after we have rejected the hypothesis of equality of t population means in an analysis of variance. R. A. Fisher (1949) developed a procedure for making pairwise comparisons among a set of t population means. The procedure is called Fisher's least significant difference (LSD).

The α-level of Fisher's LSD is valid for a given comparison only if the LSD is used for independent (orthogonal) comparisons or for preplanned comparisons. However, since many people find Fisher's LSD easy to compute and hence use it for making all possible pairwise comparisons (particularly those that look "interesting" following the completion of the experiment), researchers recommend applying Fisher's LSD only after the F test for treatments has been shown to be significant. This revised approach is sometimes referred to as **Fisher's protected LSD**. Simulation studies (Cramer and Swanson (1973)) suggest that the error rate for the protected LSD is controlled on an experimentwise basis at a level approximately equal to the α-level for the F test.

Fisher's protected LSD

We'll illustrate Fisher's protected procedure, but continue to call it Fisher's LSD. This procedure is summarized here.

Fisher's Least Significant Difference Procedure

1. Perform an analysis of variance to test H_0; $\mu_1 = \mu_2 = \cdots = \mu_t$ against the alternative hypothesis that at least one of the means differs from the rest.
2. If there is insufficient evidence to reject H_0 using $F = s_B^2/s_W^2$, we proceed no further.
3. If H_0 is rejected, define the **least significant difference (LSD)** to be the observed difference between two sample means necessary to declare the corresponding population means different.
4. For a specified value of α, the least significant difference for comparing μ_i to μ_j is

$$LSD = t_{\alpha/2} \sqrt{s_W^2\left(\frac{1}{n_i} = \frac{1}{n_j}\right)},$$

where n_i and n_j are the respective sample sizes from population i and j and t is the critical t value (Table 4 of the Appendix) for $a = \alpha/2$ and df denoting the degrees of freedom for s_W^2. Note that for $n_i = n_j = n$

$$LSD = t_{\alpha/2} \sqrt{\frac{2s_W^2}{n}}$$

5. All pairs of sample means are then compared. If $|\bar{y}_{i.} - \bar{y}_{j.}| \geq LSD$, we declare the corresponding population means μ_i and μ_j different.
6. For each pairwise comparison of population means, the probability of a Type I error is fixed at a specified value of α.

Note: The LSD procedure is analogous to a two-sample for any two population means μ_i and μ_j; except that we use the error term s_W^2 from the analysis of variance rather than the pooled sample variance for samples i and j.

Fischer's

$n_i = n_j$

s_W^2, $t_{\alpha/2}$

EXAMPLE 14.3

Hydrochloric acid (HCl) is used in the preparation of certain dyes. Six different batches of HCl were used to produce a particular dye. Five measurements on the yield (in grams of dye) were obtained from each batch. A summary of the sample data is given in Tables 14.2 and 14.3. Use Fisher's LSD procedure to make all pairwise comparisons among the six population (batch) mean yields. Use $\alpha = .05$.

T A B L E 14.2
Summary of the Dye Yields for Example 14.3

Batch	Sample Mean
1	505
2	528
3	564
4	498
5	600
6	470

T A B L E 14.3
AOV Table for the Data of Example 14.3

Source	df	SS	MS	F
Between batches	5	56,360	11,272	4.60
Within batches	24	58,824	2,451	
Total	29			

Solution   We can solve this problem by following the five steps listed for the LSD procedure.

steps for LSD procedure

> *Step 1.*   We use the AOV table in Table 14.3. The F test of H_0: $\mu_1 = \mu_2 = \cdots = \mu_6$ is based on

$$F = \frac{s_B^2}{s_W^2} = 4.60.$$

For $\alpha = .05$ with $df_1 = 5$ and $df_2 = 24$, we reject H_0 if F exceeds 2.62 (see Table 6 in the Appendix).

> *Steps 2, 3.*   Since 4.60 is greater than 2.62, we reject H_0 and conclude that at least one of the population means differs from the rest.

> *Step 4.*   The least significant difference for comparing two means based on samples of size 5 is then

$$\text{LSD} = t_{\alpha/2} \sqrt{\frac{2s_W^2}{5}} = 2.064 \sqrt{\frac{2(2451)}{5}} = 64.63.$$

Note that the appropriate t value (2.064) was obtained from Table 4 with $a = \alpha/2 = .025$ and $df = 24$.

> *Step 5.*   When we have equal sample sizes, it is convenient to use the following procedures rather than make all pairwise comparisons among the sample means, because the same LSD is to be used for all comparisons.

a. We rank the sample means from lowest to highest.

Population	6	4	1	2	3	5
Sample mean	470	498	505	528	564	600

b. We compute the sample difference

$$\bar{y}_{\text{largest}} - \bar{y}_{\text{smallest}}.$$

If this difference is greater than the LSD, we declare the corresponding population means significantly different from each other. Next we compute the sample difference

$$\bar{y}_{\text{2nd largest}} - \bar{y}_{\text{smallest}}$$

and compare the result to the LSD. We continue to make comparisons with $\bar{y}_{\text{smallest}}$:

$$\bar{y}_{\text{3rd largest}} - \bar{y}_{\text{smallest}}$$

and so on, until we find either that all sample differences involving $\bar{y}_{\text{smallest}}$ exceed the LSD (and hence the corresponding population means are different) or that a sample difference involving $\bar{y}_{\text{smallest}}$ is less than the LSD. In the latter case we stop and make no further comparisons with $\bar{y}_{\text{smallest}}$. For our data, comparisons with $\bar{y}_{\text{smallest}}$, $\bar{y}_{6.}$ give the following results:

Comparison	Conclusion
$\bar{y}_{\text{largest}} - \bar{y}_{\text{smallest}} = \bar{y}_{5.} - \bar{y}_{6.} = 130$	$>$ LSD; proceed
$\bar{y}_{\text{2nd largest}} - \bar{y}_{\text{smallest}} = \bar{y}_{3.} - \bar{y}_{6.} = 94$	$>$ LSD; proceed
$\bar{y}_{\text{3rd largest}} - \bar{y}_{\text{smallest}} = \bar{y}_{2.} - \bar{y}_{6.} = 58$	$<$ LSD; stop

summary diagram

To summarize our results we make the following diagram:

Population   <u>6   4   1   2</u>   3   5

Those populations joined by the underline have means that are not significantly different from $\bar{y}_{6.}$. Note that populations 3 and 5 have sample differences with population 6 that exceed the LSD and hence are not underlined.

c. We now make similar comparisons with $\bar{y}_{\text{2nd smallest}}$, $\bar{y}_{4.}$ in this case, using the procedures of part b.

Comparison	Conclusion
$\bar{y}_{5.} - \bar{y}_{4.} = 102$	$>$ LSD; proceed
$\bar{y}_{3.} - \bar{y}_{4.} = 66$	$>$ LSD; proceed
$\bar{y}_{2.} - \bar{y}_{4.} = 30$	$<$ LSD; stop

Population   6   <u>4   1   2</u>   3   5

d. Continue with $\bar{y}_{3\text{rd smallest}}$, or $\bar{y}_{1.}$ in our example.

Comparison	Conclusion
$\bar{y}_{5.} - \bar{y}_{1.} = 95$	> LSD; proceed
$\bar{y}_{3.} - \bar{y}_{1.} = 59$	< LSD; stop

Population   6   4   <u>1   2   3</u>   5

e. Continue with $\bar{y}_{4\text{th smallest}}$, or $\bar{y}_{2.}$ in our example.

Comparison	Conclusion
$\bar{y}_{5.} - \bar{y}_{2.} = 72$	> LSD; proceed
$\bar{y}_{3.} - \bar{y}_{2.} = 36$	< LSD; stop

Population   6   4   1   <u>2   3</u>   5

f. Continue with $\bar{y}_{5\text{th smallest}}$, or $\bar{y}_{3.}$ in our example.

Comparison	Conclusion
$\bar{y}_{5.} - \bar{y}_{3.} = 36$	< LSD; stop

Population   6   4   1   2   <u>3   5</u>

g. We can summarize steps a through f as follows:

Population   <u>6   4   1   2</u>   3   5

Those populations not underlined by a common line are declared to have means that are significantly different according to the least significant difference criterion. Note that we can eliminate the second and fourth lines from the top of part g since they are part of the first and third lines, respectively. The revised summary of significant and nonsignificant results is

Population   <u>6   4   1   2</u>   3   5

In conclusion, we have μ_6, μ_4, μ_1, and μ_2 significantly less than μ_5. Also, μ_6 and μ_4 are significantly less than μ_3.   ▲

While the LSD procedure described in Example 14.3 may seem quite laborious, its application is quite simple. First, we run an analysis of variance. If we reject the null hypothesis of equality of the population means, we compute the LSD for all pairs of sample means. When the sample sizes are the same, this difference is a single number for all pairs. We can use the stepwise procedure described in steps 5a through 5g of Example 14.3. We need not write all those steps down, only the summary lines. The final summary (as given in step 5g) gives a handy visual display of the pairwise comparisons using Fisher's LSD.

Several remarks should be made concerning the LSD method for pairwise comparisons.

First, there is the possibility that the overall F test in our analysis of variance is significant but that no pairwise differences are significant using the LSD procedure. This apparent anomaly can occur because the null hypothesis H_0: $\mu_1 = \mu_2 = \cdots = \mu_t$ for the F test is equivalent to the hypothesis that all possible comparisons (paired or otherwise) among the population means are zero. For a given set of data, the comparisons that are significant might not be of the form $\mu_i - \mu_j$, the form we are using in our paired comparisons.

Fisher's confidence interval

Second, Fisher's LSD procedure can also be used to form a confidence interval for $\mu_i - \mu_j$. A $100(1 - \alpha)\%$ confidence interval has the form

$$(\bar{y}_{i\cdot} - \bar{y}_{j\cdot}) \pm LSD.$$

LSD for equal sample sizes

Third, when all the sample sizes are the same, the LSD for all pairs is

$$t_{\alpha/2}\sqrt{\frac{2s_W^2}{n}}.$$

14.5 ▼ TUKEY'S W PROCEDURE

We are aware of the major drawback of a multiple-comparison procedure with a controlled per-comparison error rate. Even when $\mu_1 = \mu_2 = \cdots = \mu_t$, unless α, the per-comparison error rate (such as with Fisher's unprotected LSD) is quite small, there is a high probability of declaring at least one pair of means significantly different when running multiple comparisons. To avoid this, other multiple-comparison procedures have been developed that control different error rates.

Studentized range distribution

Tukey (1953) proposed a procedure that makes use of the **Studentized range distribution.** When more than two sample means are being compared, to test the largest and smallest sample means, we could use the test statistic

$$\frac{\bar{y}_{largest} - \bar{y}_{smallest}}{s_p\sqrt{1/n}},$$

s_p = pooled
n = # of observations

where n is the number of observations in each sample and s_p is a pooled estimate of the common population standard deviation σ. This test statistic is very similar to that for comparing two means (Section 6.2), but it does not possess a t distribution. One reason it does not is that we have waited to determine which two sample means (and hence population means) we would compare until we observed the largest and smallest sample means. This procedure is quite different from that of specifying $H_0 : \mu_1 - \mu_2 = 0$, observing $\bar{y}_1$. and $\bar{y}_2$., and forming a t statistic.

The quantity

$$\frac{\bar{y}_{largest} - \bar{y}_{smallest}}{s_p\sqrt{1/n}}$$

follows a Studentized range distribution. We will not discuss the properties of this distribution, but will illustrate its use in Tukey's multiple-comparison procedure.

Tukey's *W* Procedure

W

$q_\alpha(t, v)$

experimentwise error rate

1. Rank the t sample means.
2. Two population means μ_i and μ_j are declared different if

$$|\bar{y}_{i.} - \bar{y}_{j.}| \geq W,$$

where

$$W = q_\alpha(t, v)\sqrt{\frac{s_W^2}{n}}.$$

v = degrees of freedom
t = # popn's.

s_W^2 is the mean square within samples based on v degrees of freedom, $q_\alpha(t, v)$ is the **upper-tail critical value of the Studentized range** for comparing t different populations, and n is the number of observations in each sample. A discussion follows showing how to obtain values of $q_\alpha(t, v)$ from Table 10 in the Appendix.

3. The error rate that is controlled is an **experimentwise error rate**. Thus, the probability of observing an experiment with one or more pairwise comparisons falsely declared significant is specified at α.

We can obtain values of $q_\alpha(t, v)$ from Table 10 in the Appendix. Values of v are listed along the left column of the table with values of t across the top row. Upper-tail values for the Studentized range are then presented for $a = .05$ and .01. For example, in comparing 10 population means based on 9 degrees of freedom for s_W^2, the .05 upper-tail critical value of the Studentized range is $q_{.05}(10, 9) = 5.74$.

EXAMPLE 14.4 ▼

Refer to the data of Example 14.3. Use Tukey's W procedure with $\alpha = .05$ to make pairwise comparisons among the six population means.

Solution   Step 1 was performed in Example 14.3. For

$t = 6$ (we are making pairwise comparisons among six means)
$v = 24$ (s_W^2 had 24 degrees of freedom in the AOV)
$\alpha = .05$ (we specified the experimentwise error rate at .05)
$n = 5$ (there were five sample observations selected from each population)

we find

$$q_{.05}(6, 24) = 4.37.$$

The absolute value of each difference in sample means must then be compared to

$$W = q_\alpha(t, v)\sqrt{\frac{s_W^2}{n}} = 4.37\sqrt{\frac{2451}{5}} = 96.75.$$

By substituting W for the LSD, we can use the same stepwise procedure for comparing sample means that we used in step 5 of the solution to Example 14.3.

Having ranked the sample means from low to high, comparisons against $\bar{y}_{\text{smallest}}$, which is $\bar{y}_{6\cdot}$, yield

Population	6	4	1	2	3	5

Comparisons with $\bar{y}_{\text{2nd smallest}}$ ($\bar{y}_{4\cdot}$) yield

Population	6	4	1	2	3	5

Similarly, comparisons with $\bar{y}_{1\cdot}$, $\bar{y}_{2\cdot}$, and $\bar{y}_{3\cdot}$ yield

Population	6	4	1	2	3	5

Combining our results we obtain

Population   6   4   1   2   3   5

which simplifies to

Population   6   4   1   2   3   5

All populations not underlined by a common line have population means that are significantly different from each other. That is, μ_6 and μ_4 are significantly less than μ_5.   ▲

By examining the multiple-comparison summaries using the least significant difference (Example 14.3) and Tukey's *W* procedure (Example 14.4), we see that Tukey's procedure is more conservative (declares fewer significant differences) than the LSD procedure. For example, applying Tukey's procedure to the data of Table 14.2 shows that μ_3 is no longer significantly larger than μ_6 and μ_4. Similarly, μ_5 is no longer significantly larger than μ_1 and μ_2. The explanation for this is that although both procedures have an experimentwise error rate, the per-comparison error rate of the protected LSD method has been shown to be larger than that for Tukey's *W* procedure.

Tukey's procedure can be modified to account for unequal samples using a method proposed originally by Tukey in 1953 and then independently by Kramer in 1956. The procedure, sometimes referred to as the Tukey–Kramer procedure, makes use of the quantity

$$W_{ij} = q_\alpha(t, v) \sqrt{\frac{s_W^2}{2}\left(\frac{1}{n_i} + \frac{1}{n_j}\right)}$$

for determining whether μ_i and μ_j are different. Note that W_{ij} becomes the familiar *W* from Tukey's procedure when $n_i = n_j$. The rest of the procedure remains the same.

Tukey's confidence interval   Tukey's procedure can also be used to construct confidence intervals for comparing two means. However, unlike the confidence intervals that can be formed from Fisher's LSD, Tukey's procedure enables us to construct simultaneous confidence intervals for all pairs of treatment differences. For a specified α level from which we compute *W*, the overall probability is $1 - \alpha$ that all differences $\mu_i - \mu_j$ will be included in an interval of the form

$$(\bar{y}_{i\cdot} - \bar{y}_{j\cdot}) \pm W.$$

That is, the probability is $1 - \alpha$ that all the intervals $(\bar{y}_{i\cdot} - \bar{y}_{j\cdot}) \pm W$ include the corresponding population differences $\mu_i - \mu_j$. For unequal sample sizes, W_{ij} replaces W in the confidence interval.

14.6 ▼ STUDENT–NEWMAN–KEULS PROCEDURE

The Student–Newman–Keuls (SNK) procedure provides a modification of the Tukey W procedure. Although the SNK procedure also makes use of the Studentized range statistic, different critical values are used depending on the number of steps separating the means being tested. To compare the two procedures, let's refer to Example 14.3. Ranked in order from lowest to highest, the sample means are

Sample mean	470	498	505	528	564	600
Sample i	6	4	1	2	3	5

and the critical value of the Studentized range for Tukey's W procedure is

$$q_\alpha(t, v) = q_{.05}(6, 24) = 4.37.$$

This same value of q is used for all pairwise comparisons of the six treatment means.

The SNK procedure makes use of a critical value

$$W_r = q_\alpha(r, v) \sqrt{\frac{s_W^2}{n}}$$

for means that are r steps apart when the t sample means are ranked from lowest to highest. For our example, $\bar{y}_{\text{largest}}$ and $\bar{y}_{\text{smallest}}$ are six "steps" apart and they would be compared using

$$W_6 = q_\alpha(6, v) \sqrt{\frac{s_W^2}{n}}$$

$$= 4.37 \sqrt{\frac{2451}{5}} = 96.75.$$

(Note: This is W for Tukey's W procedure.) However, $\bar{y}_{\text{largest}}$ and $\bar{y}_{\text{2nd smallest}}$ are five "steps" apart and they would be compared to

$$W_5 = q_\alpha(5, v) \sqrt{\frac{s_W^2}{n}}$$

$$= 4.17 \sqrt{\frac{2451}{5}} = 92.33.$$

TABLE 14.4
Values of r, $q_\alpha(r, v)$ and W_r for Example 14.3

r	2	3	4	5	6
$q_\alpha(r, v)$	2.92	3.53	3.90	4.17	4.37
W_r	64.65	78.16	86.35	92.33	96.75

The complete set of critical values W_r needed for the data of Example 14.3 is shown in Table 14.4. Values of $q_\alpha(r, v)$ are obtained from Table 10 in the Appendix by replacing t by r.

The Student–Newman–Keuls procedure, which relies on the number of ordered steps between two sample means when determining the significance of an observed sample difference, has neither an experimentwise nor a per-comparison error rate. Rather, the error rate is defined for means the same number of ordered steps apart. Since the critical value W_r decreases as the number of steps between the means being compared decreases, the SNK proceure is less conservative and hence will generally declare more significant differences than will Tukey's W procedure, which utilizes the largest value for W no matter how many steps separate the means being compared.

The SNK procedure is summarized here.

SNK Procedure

1. Rank the t sample means from lowest to highest.
2. For two means $\bar{y}_{i.}$ and $\bar{y}_{j.}$ that are r steps apart, we declare μ_i and μ_j different if

$$|\bar{y}_{i.} - \bar{y}_{j.}| \geq W_r,$$

where $W_r = q_\alpha(r, v)\sqrt{s_W^2/n}$, n is the number of observations per sample, s_W^2 is the mean square within samples from the AOV table, v is the degrees of freedom for s_W^2, and $q_\alpha(r, v)$ is the critical value of the Studentized range. Values of $q_\alpha(r, v)$ are given in Table 10 in the Appendix for $\alpha = .05$ and $.01$.

(Note: Use the column labeled t to locate the desired value for r.)

EXAMPLE 14.5

Refer to the data of Example 14.3. Run the SNK procedure to make all pairwise comparisons based on $\alpha = .05$.

Solution The critical values of W_r to be used were shown previously.

1. Beginning with $\bar{y}_{\text{largest}}$ ($\bar{y}_{5.}$) every sample mean is compared to $\bar{y}_{\text{smallest}}$ ($\bar{y}_{6.}$) using the appropriate value of W_r. These results are summarized here.

Comparison	W_r	Conclusion
$\bar{y}_{5.} - \bar{y}_{6.} = 130$	96.75	> 96.75; proceed
$\bar{y}_{3.} - \bar{y}_{6.} = 94$	92.33	> 92.33; proceed
$\bar{y}_{2.} - \bar{y}_{6.} = 58$	86.35	< 86.35; stop

2. Similarly, we can make comparisons with $\bar{y}_{2\text{nd smallest}}$ $(\bar{y}_{4.})$, $\bar{y}_{1.}$, $\bar{y}_{2.}$, and $\bar{y}_{3.}$:

Comparison	W_r	Conclusion
$\bar{y}_{5.} - \bar{y}_{4.} = 102$	92.33	> 92.33; proceed
$\bar{y}_{3.} - \bar{y}_{4.} = 66$	86.35	< 86.35; stop
$\bar{y}_{5.} - \bar{y}_{1.} = 95$	86.35	> 86.35; proceed
$\bar{y}_{3.} - \bar{y}_{1.} = 59$	78.16	< 78.16; stop
$\bar{y}_{5.} - \bar{y}_{2.} = 72$	78.16	< 78.16; stop
$\bar{y}_{5.} - \bar{y}_{3.} = 36$	64.65	< 64.65; stop

The results of these multiple comparisons made using the SNK procedure are shown here:

Population   6   4   1   2   3   5

▲

All populations not underlined by a common line have population means that are different from each other. These results using the SNK procedure are slightly different from those shown for Tukey's W procedure. In particular, μ_6 and μ_3 are declared different as well as μ_5 and μ_1 for the SNK procedure. This example illustrates the fact that the SNK procedure will tend to declare more differences (and hence be less conservative) than Tukey's W procedure.

The SNK procedure can be modified to account for different sample sizes. When $n_i \neq n_j$, the value of W_r for means r steps apart is modified in the same way as Tukey's procedure. Here we would use

$$q_\alpha(r, v) \sqrt{\frac{s_W^2}{2}\left(\frac{1}{n_i} + \frac{1}{n_j}\right)}$$

to assess whether μ_i differs from μ_j.

14.7 ▼ DUNCAN'S NEW MULTIPLE RANGE TEST

Duncan (1955) developed a multiple-comparison procedure for obtaining all pairwise comparisons among t sample means. Although his procedure makes use of the Studentized range, his error rate is neither on an experimentwise basis (as with Tukey's) nor on a per-comparison basis. When the sample means have been ranked from lowest to highest, the error rate is designated in the following way. In general, if two sample means are r steps apart, Duncan defines the **protection level** as

protection level

$$(1 - \alpha)^{r-1}.$$

The probability of falsely rejecting the equality of two population means when the sample means are r steps apart is then taken to be

error rate

$$1 - (1 - \alpha)^{r-1}.$$

For $\alpha = .05$, we illustrate the concept of a protection level in Table 14.5.

T A B L E   14.5

Duncan Protection Level Using $\alpha = .05$ When the Sample Means Are r Steps Apart

Number of Steps Apart, r	Protection Level, $(1 - .05)^{r-1}$	Probability of Falsely Rejecting H_0, $1 - (1 - .05)^{r-1}$
2	.950	.050
3	.903	.097
4	.857	.143
5	.815	.185
6	.774	.226
7	.735	.265

Duncan's reasons for allowing the protection level to decrease for increasing values of r have their basis in results presented in Section 14.2. As we indicated there, it is possible to form $t - 1$ orthogonal contrasts for comparing t treatment means. Using those contrasts, we can partition the treatment sum of squares into $t - 1$ single degree-of-freedom sums of squares. (We will discuss this partitioning later.) If we assume the degrees of freedom for s_W^2 are quite large, then the $(t - 1)$ F statistics are nearly independent. Then when each F test is conducted at a preset α value, and we assume $\mu_1 = \mu_2 = \cdots = \mu_t$, the probability of rejecting H_0 for one or more contrasts is approximately

$$1 - (1 - \alpha)^{t-1}.$$

Duncan argued that since experimenters have little or no reservations in performing these multiple F tests for orthogonal contrasts even though the overall α level increases with t, it is reasonable to construct a multiple-comparison test for which the protection level decreases with the number of sample means included in a comparison. Thus, Duncan uses a α-value equal to the quantity $1 - (1 - \alpha)^{r-1}$ when a pair of sample means are r steps apart ($r = 2, \ldots, t$).

Because the protection level decreases with increasing r, *Duncan's multiple range test is very powerful*. That is, there is a high probability of declaring a difference when there is actually a difference between the population means. This has been one of the reasons Duncan's procedure has been extremely popular among researchers.

We summarize Duncan's new multiple range test for pairwise comparisons of t population means.

Duncan's New Multiple Range Test

▼

1. Rank the t sample means.
2. Two population means are declared significantly different if the absolute value of their sample differences exceeds

W_r'

$$W_r' = q_\alpha'(r, v) \sqrt{\frac{s_W^2}{n}},$$

where n is the number of observations in each sample mean, s_W^2 is the mean square within samples obtained from the analysis of variance table, v is the number of degrees of freedom for s_W^2, and $q_\alpha'(r, v)$ is the critical value of the Studentized range required for Duncan's procedure when the means being compared are r steps apart. Values of $q_\alpha'(r, v)$ are given in Table 11 in the Appendix for $\alpha = .05$ or $.01$ and various combinations of r and v.

$q_\alpha'(r, v)$

We illustrate the use of Duncan's procedure with the data of Example 14.3.

EXAMPLE 14.6   ▼

Refer to the data of Example 14.3. Run Duncan's multiple range test with $\alpha = .05$ to make all pairwise comparisons among the six population means.

Solution   Recall from Example 14.3 that $n = 5$, $s_W^2 = 2451$, and $v = 24$. Using this information we can set up the following table for r, $q_\alpha'(r, v)$, and W_r'.

r	2	3	4	5	6
$q_\alpha'(r, v)$	2.92	3.07	3.15	3.22	3.28
W_r'	64.65	67.97	69.74	71.29	72.62

For example, when two means are $r = 2$ steps apart, $q'_{.05}(2, 24)$ is 2.92. Then

$$W'_2 = q'_{.05}(2, 24) \sqrt{\frac{s^2_W}{n}} = 2.92 \sqrt{\frac{2451}{5}} = 64.65.$$

Thus, two sample means $r = 2$ steps apart will be declared significantly different if the absolute value of their difference exceeds 64.7. The remainder of the entries in the table for different values of r were computed in a similar manner.

The sample means, ranked in order from lowest to highest, are

Population	6	4	1	2	3	5
Sample mean	470	498	505	528	564	600

Beginning with the largest mean, $\bar{y}_{5.}$, each sample mean is compared to the smallest mean, $\bar{y}_{6.}$, using the appropriate value of W'_r. For example, $\bar{y}_{5.}$ and $\bar{y}_{6.}$ are $r = 6$ steps apart, and their difference must be compared to $W'_6 = 72.6$. The comparisons with $\bar{y}_{\text{smallest}}$ are shown in the following table.

Comparison	Conclusion
$\bar{y}_{5.} - \bar{y}_{6.} = 130$	> 72.62; proceed
$\bar{y}_{3.} - \bar{y}_{6.} = 94$	> 71.29; proceed
$\bar{y}_{2.} - \bar{y}_{6.} = 58$	< 69.74; stop

Similarly, comparisons are made with $\bar{y}_{4.}$, $\bar{y}_{1.}$, $\bar{y}_{2.}$, and $\bar{y}_{3.}$ as follows:

Comparison	Conclusion
$\bar{y}_{5.} - \bar{y}_{4.} = 102$	> 71.29; proceed
$\bar{y}_{3.} - \bar{y}_{4.} = 66$	< 69.35; stop
$\bar{y}_{5.} - \bar{y}_{1.} = 95$	> 69.74; proceed
$\bar{y}_{3.} - \bar{y}_{1.} = 59$	< 67.97; stop
$\bar{y}_{5.} - \bar{y}_{2.} = 72$	< 67.97; proceed
$\bar{y}_{3.} - \bar{y}_{2.} = 36$	< 64.65; stop
$\bar{y}_{5.} - \bar{y}_{3.} = 36$	< 64.65; stop

The results of these pairwise comparisons can be summarized as usual.

Population   6 4 1 2 3 5

In conclusion, μ_6, μ_4, μ_1, and μ_2 are significantly less than μ_5; μ_6 is significantly less than μ_3.   ▲

We can compare the results obtained by using Duncan's new multiple range test to those obtained for Fisher's LSD, Tukey's procedure, and the SNK procedure. Based on the summary lines for the four procedures, we see that Duncan's procedure for the data of Table 14.2 results in conclusions more like the LSD and SNK procedures than Tukey's. The only difference in the conclusions for Fisher's LSD and Duncan's procedure is that in using the LSD, the means for populations 3 and 4 were found to be different—for Duncan's, they were not. The difference between the SNK and Duncan's procedures is with μ_2 and μ_5.

Unlike the other multiple-comparison procedures discussed, Duncan's new multiple range test cannot be used to form confidence intervals for the pairwise differences $\mu_i - \mu_j$. We can, however, adopt Duncan's procedure for unequal sample sizes if the n_is are nearly equal by replacing n by

$$\tilde{n} = \frac{t}{\dfrac{1}{n_1} + \dfrac{1}{n_2} + \cdots + \dfrac{1}{n_t}}$$

in the formula for W'_r. Duncan (1957) and Kramer (1957) have also presented extensions for situations where the population variances are different and where the sample means are correlated.

14.8 ▼ THE k RATIO RULE FOR MAKING PAIRWISE COMPARISONS AMONG TREATMENT MEANS

In this chapter, we have spent considerable time discussing the merits of different multiple-comparison procedures. Fisher's LSD procedure can be used to make all pairwise comparisons among t population means while the controlled error rate for each comparison is at a specified level α. We found that even when $\mu_1 = \mu_2 = \cdots = \mu_t$, unless the per-comparison error rate α is quite small, the probability of falsely declaring at least one pair of means significantly different is quite large for Fisher's LSD.

To avoid this difficulty, Tukey proposed a multiple-comparison procedure utilizing an experimentwise error rate. Numerous other authors have offered solutions to this same problem, but none of these has adequately answered the question of which error rate to control.

The dilemma (see Duncan (1975)) of multiple comparisons can be illustrated with the following simplistic example. Suppose we wish to make pairwise comparisons among t (t very large) population means. In addition, suppose that null and alternative hypotheses for each of the pairwise comparisons is of the form

$$H_0: \quad \mu_i - \mu_j = 0$$
$$H_a: \quad \mu_i - \mu_j \neq 0$$

and that for each test the LSD, based on $\alpha = .05$, is used to determine whether two population means are significantly different.

Case I. Suppose that only 5% of all the differences in population means were declared significant. Intuitively, since we set $\alpha = .05$, we would be inclined to think we erred by declaring 5% of the differences significant when in fact all the population means are identical. To guard against declaring too many differences significant, we would certainly want to increase the magnitude of the absolute difference (LSD) in sample means required for significance. This is precisely the approach taken by Tukey (and others, such as Scheffé), who required that the absolute difference in sample means exceed W (where $W > \text{LSD}$) before declaring significance. Certainly the use of Tukey's procedure would be preferred to Fisher's LSD, to protect against a situation as described in case I.

Case II. Suppose, however, that only 5% of the population differences were declared to be nonsignificant. Then, intuitively, we might think we had erred on the opposite side and declared 5% of the population differences nonsignificant when in fact all t population means are different. To protect against such an occurrence, we would decrease the magnitude of the absolute difference in sample means required to declare significance. Clearly in situations such as this, Fisher's LSD procedure would be preferable to Tukey's W procedure.

Waller and Duncan (1969) proposed a multiple-comparison procedure that uses the sample data to help in determining whether we need to use a conservative rule (much like Tukey's) or a nonconservative rule (much like Fisher's). The procedure makes use of the computed value of $F = s_B^2 / s_W^2$, the test statistic for the null hypothesis $H_0: \mu_1 = \mu_2 = \cdots = \mu_t$. If the computed value of F is small, then the sample data tend to indicate that the population means are rather homogenous. For this situation, to protect against case I, we would require a large absolute difference in sample means to declare significance. In contrast, if the computed value of F is large, the sample data would tend to indicate or confirm that the population means are heterogeneous. For this situation, to protect against case II, we would require a small absolute difference in sample means to declare significance.

Although a complete explanation of this procedure is well beyond the scope of this text, adequate tables have been developed to enable us to apply it to multiple comparisons resulting from the analysis of variance for a completely randomized design (Chapter 13), and from the other standard experimental designs (to be discussed in Chapter 15). There are two restrictions that are placed on the Waller–Duncan procedure. First, we require all sample sizes to be the same, and, second, the procedure should not be used when a priori we would expect certain of the population means to differ more than others.

error weight ratio k

Fisher's LSD required us to choose the comparisonwise error rate α. Now we must specify the **error weight ratio k**, which designates the seriousness of a Type I error relative to a Type II error. While we cannot go into a detailed discussion of the rationale for choosing k, we can provide several guidelines. Whereas typical values of α for Fisher's LSD may be .10, .05, or .01, corresponding values of k, the error weight ratio, are 50, 100, or 500.

The Waller–Duncan test procedure is summarized here.

Waller–Duncan k Ratio Procedure

▼

1. Choose k, the error weight ratio.
2. Perform an analysis of variance to obtain the computed value of F for

$$H_0: \quad \mu_1 = \mu_2 = \cdots = \mu_t.$$

3. Compute

$$LSD = t_c \sqrt{s_W^2 \left(\frac{2}{n}\right)},$$

where t_c is obtained from Tables 12 or 13 in the Appendix and is based on k, df_1, df_2, and F; s_W^2 is the mean square error from the AOV table; and n is the number of observations selected from each population. (Note: This procedure requires that we have the same number of observations from each population.)

4. Follow a stepwise procedure similar to that for Fisher's LSD to declare

$$\mu_i - \mu_j \neq 0 \quad \text{if} \quad |\bar{y}_{i\cdot} - \bar{y}_{j\cdot}| > LSD.$$

Before we proceed with an example, let us consider the format of Tables 12 and 13. Because of the extensiveness of the tables, only two values of k (100 and 500) have been included. (Tables for $k = 50$ are presented in Waller and Duncan (1972).) The table entry is the t_c-value corresponding to specific values of k, df_1 (df for MST), df_2 (df for MSE), and $F = MST/MSE$. Thus, for $k = 100$, $df_1 = 6$, $df_2 = 8$, and $F = 3.0$, the t-value for Waller–Duncan's least significant difference is 2.61.

Because of the complexity of the tables, not all values of df_1, df_2, and F can be given for a fixed error weight ratio. Fortunately, it is possible to interpolate to obtain the appropriate value of t_c.

Interpolation will often be required for values of F. To obtain the appropriate value of t_c when the computed value of F falls between two tabulated F values, interpolate linearly using one of two quantities

$$a = \frac{1}{\sqrt{F}} \quad \text{or} \quad b = \sqrt{\frac{F}{F-1}}.$$

Table 14.6 indicates under what situations we use a and under what conditions we interpolate using b.

T A B L E 14.6
Situations in Which to Use *a* or *b* for Interpolation in the Waller–Duncan Procedure

		df₂	
F	**df₁**	**≤100**	**>100**
≤2.4	≤60	a	a
	>60	a	b

		df₂	
		≤20	**>20**
>2.4	≤20	a	b
	>20	b	b

E X A M P L E 14.7 ▼

Refer to Example 14.3. We were interested in comparisons among 6 batches used to produce a particular dye. The computed value of F was 4.60, based on $df_1 = 5$ and $df_2 = 24$. Find the value of t_c to be used in a k ratio procedure for pairwise comparisons among the 6 batch means. Use an error weight ratio of 100.

Solution From Table 12 in the Appendix we note that there are no t_c-values listed for $F = 4.60$; we must interpolate between the values listed for $F = 4.0$ and $F = 6.0$. Note also that there are no t_c-values for $df_1 = 5$. However, very little difference is observed whether we use $df_1 = 4$ or 6. For $k = 100$, $df_1 = 4$, $df_2 = 24$, and $F = 4.0$, the table entry is 2.18. For the same settings except $df_1 = 6$, $t_c = 2.19$. There is no change in t_c for $k = 100$, $df_2 = 24$, and $F = 6.0$ when $df_1 = 4$ or 6. In both cases, $t_c = 2.06$.

Because df_1 has very little effect on t_c in the range of 4 to 6 for this problem, we will work with $df_1 = 4$. To obtain that t_c-value corresponding to 4.60, we see from Table 14.6 that we must interpolate using

$$b = \sqrt{\frac{F}{F-1}}.$$

From Table 12 in the Appendix, for $F = 4.0$, then $b = 1.155$, and for $F = 6.0$, then $b = 1.095$. The distance between 1.155 and 1.095 is .060. Computing b for $F = 4.60$, we have

$$b = \sqrt{\frac{4.60}{3.60}} = 1.13.$$

We now interpolate linearly with respect to b. First we draw a graph with the horizontal axis labeled with b and the vertical axis with t_c. We plot the values of t_c corresponding to $b = 1.155$ and $b = 1.095$. Then we connect these two points with a straight line. We use the straight line to determine the interpolated value of t_c corresponding to $b = 1.13$. (See Figure 14.1.) As can be seen from Figure 14.1, the interpolated value of t_c corresponding to $b = 1.13$ is $t_c = 2.13$.

FIGURE 14.1

Interpolating Linearly with Respect to b to Find t_c: Example 14.7

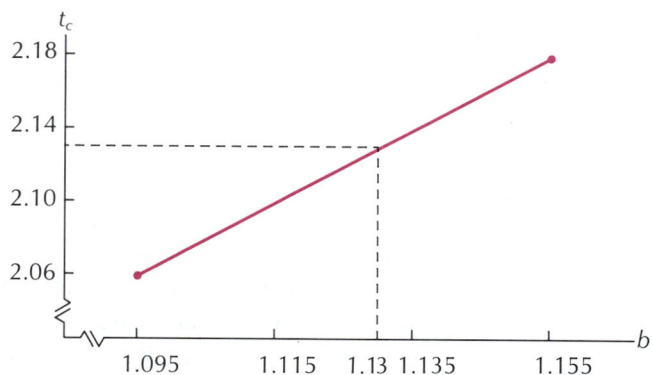

EXAMPLE 14.8 ▼

Refer to Example 14.3 and perform all pairwise comparisons between population means using the k ratio procedure with $k = 100$.

Solution   In Example 14.7, we found the appropriate value of t_c to be 2.13. Recalling that $s_W^2 = 2451$ and that there were five observations per batch, the least significant difference using the Waller–Duncan approach is

$$LSD = t_c \sqrt{s_W^2 \left(\frac{2}{n}\right)} = 2.13 \sqrt{2451 \left(\frac{2}{5}\right)} = 66.69.$$

The stepwise procedure of Fisher's LSD can be employed to summarize our sample findings. The lines summarizing our results are seen to be

Population   6   4   1   2   3   5

It is interesting to compare our results in Example 14.8 with those of previous multiple-comparison procedures for the same set of data. We see that our results are identical to the results from Duncan's multiple range test and very similar to those from Fisher's LSD and the SNK procedure. The reason that the k ratio results are similar to the results for the less conservative multiple-comparison procedures is due to the fact that the value of $F = s_B^2/s_W^2$ is fairly large, indicating that the population means are more heterogeneous. This situation requires a nonconservative rule. If the computed value of F was small, we would have found the results of the k ratio procedure similar to one of the more conservative rules. ▲

14.9 ▼ SCHEFFÉ'S S METHOD

The five multiple-comparison procedures discussed so far have been developed for pairwise comparisons among t population means. A more general procedure, proposed by Scheffé (1953), can be used to make all possible comparisons among the t population means. Although Scheffé's procedure can be applied to pairwise comparisons among the t population means, it is more conservative (less sensitive) than any of the other three multiple comparison procedures for detecting significant differences among pairs of population means because the "family" of comparisons it is trying to protect is larger than that for pairwise comparisons.

Scheffé's S Method for Multiple Comparisons

1. Consider any linear comparison among the t population means of the form

$$l = a_1\mu_1 + a_2\mu_2 + \cdots + a_t\mu_t.$$

We wish to test the null hypothesis

H_0: $l = 0$

against the alternative

H_a: $l \neq 0$.

(continues)

2. The test statistic is

$$\hat{l} = a_1 \bar{y}_1. + a_2 \bar{y}_2. + \cdots + a_t \bar{y}_t.$$

3. Let

$$S = \sqrt{\hat{V}(\hat{l})} \sqrt{(t-1)F_{\alpha, df_1, df_2}}$$

where, from Section 14.2,

$$\hat{V}(\hat{l}) = s_W^2 \sum_i \frac{a_i^2}{n_i}$$

t is the total number of population means, F_{α, df_1, df_2} is the upper-tail critical value of the F distribution with $a = \alpha$, $df_1 = t - 1$, and df_2 is the degrees of freedom for s_W^2.

4. For a specified value of α, we reject H_0 if $|\hat{l}| > S$.

5. The error rate that is controlled is an *experimentwise error rate*. If we consider all imaginable contrasts, the probability of observing an experiment with one or more contrasts falsely declared significant is designated by α.

E X A M P L E   14.9   ▼

Refer to the data of Table 14.2. Suppose that three of the batches (6, 4, and 2) were prepared from one concentration of HCl, and the other three batches (1, 3, and 5) were prepared from another concentration of HCl. Use the sample data and Scheffé's procedure to compare the mean dye yields for the two different concentrations of HCl. Let $\alpha = .05$.

Solution   Assume the even-numbered batches are from concentration I of HCl and the odd-numbered batches from concentration II. A contrast of particular importance is

$$\hat{l} = \bar{y}_1. + \bar{y}_3. + \bar{y}_5. - \bar{y}_2. - \bar{y}_4. - \bar{y}_6.$$

which compares the means for batches from concentration II to those for concentration I. In particular, we would like to test

$H_0: \quad l = 0.$
$H_a: \quad l \neq 0.$

The estimated value of l is

$$\hat{l} = \bar{y}_{1.} + \bar{y}_{3.} + \bar{y}_{5.} - \bar{y}_{2.} - \bar{y}_{4.} - \bar{y}_{6.}$$

$$= 505 + 564 + 600 - 528 - 498 - 470 = 173.$$

To compute

$$S = \sqrt{\hat{V}(\hat{l})} \; \sqrt{(t-1)F_{\alpha, df_1, df_2}},$$

we must first calculate $\hat{V}(\hat{l})$. Using the formula

$$\hat{V}(\hat{l}) = s_W^2 \sum_i \frac{a_i^2}{n_i}$$

with all sample sizes equal to 5 and $s_W^2 = 2451$, we have

$$\hat{V}(\hat{l}) = 2451 \left[\frac{1}{5} + \frac{1}{5} + \frac{1}{5} + \frac{1}{5} + \frac{1}{5} + \frac{1}{5} \right] = 2941.2.$$

From Table 6 for $\alpha = .05$, $df_1 = t - 1 = 5$, and $df_2 = 24$ (the degrees of freedom for s_W^2),

$$F_{.05, 5, 24} = 2.62.$$

The computed value of S is then

$$S = \sqrt{2941.2} \, \sqrt{5(2.62)} = (54.23)(3.62) = 196.31.$$

Since the absolute value of $\hat{l}$, 173, does not exceed $S = 196.31$, we have insufficient evidence to indicate that the means for batches from concentration II differ from those for concentration I. ▲

Scheffé's confidence interval

Scheffé's method can also be used for constructing a simultaneous confidence interval for all possible (not necessarily pairwise) contrasts using the t treatment means. In particular, there is a probability equal to $1 - \alpha$ that all possible comparisons of the form $l = \sum a_i \mu_i$, where $\sum a_i = 0$, will be encompassed by intervals of the form

$$\hat{l} - S < l < \hat{l} + S$$

14.10 ▼ SUMMARY

Four different multiple-comparison procedures (Fisher's, Tukey's, SNK, and Duncan's) were presented for making pairwise comparisons of t population means. Another procedure, Scheffé's, can be applied to any linear combination (including pairwise comparisons) of the means. For each procedure, we have tried to indicate which error rate is controlled and how conservative the procedure is relative to the others presented. Since all the pairwise, multiple-comparison procedures compute the magnitude of the difference $|\bar{y}_{i\cdot} - \bar{y}_{j\cdot}|$ that is needed to declare μ_i and μ_j different, we can get some feel for how conservative one procedure is relative to another by comparing the magnitudes of the differences required for significance using the data of Example 14.3. This information is shown in Table 14.7.

TABLE 14.7
Critical Difference
$|\bar{y}_{i\cdot} - \bar{y}_{j\cdot}|$ **for Sample Means**
r Steps Apart

Multiple-Comparison Procedure	Number of Steps Separating Means				
	2	3	4	5	6
Fisher's	64.63	64.63	64.63	64.63	64.63
SNK	64.65	78.16	86.35	92.33	96.75
Tukey's	96.75	96.75	96.75	96.75	96.75
Duncan's	64.65	67.97	69.74	71.29	72.62
Scheffé's	113.33	113.33	113.33	113.33	113.33

As can be seen, Scheffé's procedure is *very* conservative and should not be used for pairwise comparisons.

A fifth procedure for pairwise multiple comparisons was the Waller–Duncan k ratio test. Since critical differences depend on the computed value of the F test for treatments, it can be either a conservative or a nonconservative rule, depending on the weight of sample evidence for rejection of the null hypothesis; the smaller the value of F, the more conservative the rule.

Which procedure should you use? We generally prefer the SNK procedure for efficacy (effectiveness) comparisons. But our reasons for this choice have a great deal to do with our work setting and the regulations surrounding our decision. Since our environment may be entirely different from yours, the decision regarding which procedure to use, and when to use it, is up to the individual. For a given problem, determine whether your decisions regarding differences should, in general, be more (or less) conservative. Then choose a procedure that exhibits the desired characteristic.

▼ KEY FORMULAS

1. Fisher's LSD procedure

$$LSD = t_{\alpha/2} \sqrt{s_W^2 \left(\frac{1}{n_i} + \frac{1}{n_j} \right)}$$

2.   Tukey's W procedure

$$W = q_\alpha(t, v)\sqrt{\frac{s_W^2}{2}\left(\frac{1}{n_i} + \frac{1}{n_j}\right)}; \text{ when } n_i = n_j = n,\ W = q_\alpha(t, v)\sqrt{\frac{s_W^2}{n}}$$

3.   SNK procedure

$$W_r = q_\alpha(r, v)\sqrt{\frac{s_W^2}{2}\left(\frac{1}{n_i} + \frac{1}{n_j}\right)}; \text{ when } n_i = n_j = n,\ W_r = q_\alpha(r, v)\sqrt{\frac{s_W^2}{n}}$$

4.   Duncan's multiple range test

$$W'_r = q'_\alpha(r, v)\sqrt{\frac{s_W^2}{n}} \qquad (\text{Note: } n_i = n_j = n)$$

5.   Scheffé's method

$$S = \sqrt{\hat{V}(\hat{l})}\ \sqrt{(t - 1)F_{\alpha, df_1, df_2}},$$

where

$$\hat{V}(\hat{l}) = s_W^2 \sum \frac{a_i^2}{n_i}$$

▼ SUPPLEMENTARY EXERCISES

14.4   Refer to the data of Example 13.1. There a horticulturist was investigating the phosphorus content of the leaves from three different varieties of apple trees.
  a. Perform an analysis of variance.
  b. Use Duncan's multiple range test procedure to run all pairwise comparisons. Use $\alpha = .05$.
  c. Compare your conclusions in part b to those in the SAS computer output shown here.

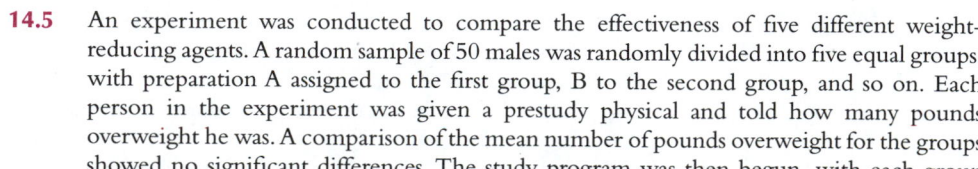

```
        LISTING OF DATA
   OBS     VARIETY    PHOSPHOR
    1         1          0.35
    2         1          0.40
    3         1          0.58
    4         1          0.50
    5         1          0.47
    6         2          0.65
    7         2          0.70
    8         2          0.90
    9         2          0.84
   10         2          0.79
   11         3          0.60
   12         3          0.80
   13         3          0.75
   14         3          0.73
   15         3          0.66
  N =      15
```

ANALYSIS OF VARIANCE PROCEDURE

GENERAL LINEAR MODELS PROCEDURE

DEPENDENT VARIABLE: PHOSPHOR

SOURCE	DF	SUM OF SQUARES	MEAN SQUARE	F VALUE	PR > F	R-SQUARE	C.V.
MODEL	2	0.27664000	0.13832000	16.97	0.0003	0.738810	13.9317
ERROR	12	0.09780000	0.00815000		ROOT MSE		PHOSPHOR MEAN
CORRECTED TOTAL	14	0.37444000			0.09027735		0.64800000

SOURCE	DF	TYPE I SS	F VALUE	PR > F	DF	TYPE III SS	F VALUE	PR > F
VARIETY	2	0.27664000	16.97	0.0003	2	0.27664000	16.97	0.0003

ANALYSIS OF VARIANCE PROCEDURE

GENERAL LINEAR MODELS PROCEDURE

DUNCAN'S MULTIPLE RANGE TEST FOR VARIABLE: PHOSPHOR
NOTE: THIS TEST CONTROLS THE TYPE I COMPARISONWISE ERROR RATE,
        NOT THE EXPERIMENTWISE ERROR RATE

ALPHA = 0.05   DF = 12   MSE = 0.00815

NUMBER OF MEANS	2	3
CRITICAL RANGE	0.124165	0.130067

MEANS WITH THE SAME LETTER ARE NOT SIGNIFICANTLY DIFFERENT

DUNCAN GROUPING	MEAN	N	VARIETY
A	0.77600	5	2
A			
A	0.70800	5	3
B	0.46000	5	1

▼ **14.5** An experiment was conducted to compare the effectiveness of five different weight-reducing agents. A random sample of 50 males was randomly divided into five equal groups, with preparation A assigned to the first group, B to the second group, and so on. Each person in the experiment was given a prestudy physical and told how many pounds overweight he was. A comparison of the mean number of pounds overweight for the groups showed no significant differences. The study program was then begun, with each group taking the prescribed preparation for a fixed period of time. At the end of the study period, weight losses were recorded. The data are given here.

A	12.4	10.7	11.9	11.0	12.4	12.3	13.0	12.5	11.2	13.1
B	9.1	11.5	11.3	9.7	13.2	10.7	10.6	11.3	11.1	11.7
C	8.5	11.6	10.2	10.9	9.0	9.6	9.9	11.3	10.5	11.2
D	8.7	9.3	8.2	8.3	9.0	9.4	9.2	12.2	8.5	9.9
E	12.7	13.2	11.8	11.9	12.2	11.2	13.7	11.8	11.5	11.7

Run an analysis of variance to determine if there are any significant differences among the weight losses for the five diet preparations. Use $\alpha = .05$.

14.6 Refer to Exercise 14.5. Run all pairwise comparisons of population means using the following procedures.
  a. Fisher's LSD, $\alpha = .05$.
  b. Tukey's W, $\alpha = .05$.
  c. Duncan's multiple range test, $\alpha = .05$
  d. Compare the conclusions for the three procedures.

14.7 Use a computer program to run an analysis of variance for the data of Exercise 14.5. Draw conclusions from the output. Compare your results to the conclusions for Exercise 14.5.

14.8 Refer to Exercise 14.5. Examine the widths of 95% confidence intervals for pairwise comparisons of diets A, B, ..., E, using Fisher's LSD, Tukey's W, and Scheffé's S method.

14.9 Refer to Exercise 14.5. Suppose that preparation D was a placebo. Use Scheffé's procedure to make the following comparisons. Set $\alpha = .05$.
  a. $\mu_D - \frac{1}{4}(\mu_A + \mu_B + \mu_C + \mu_E)$
  b. $\mu_A - \mu_E$
  c. $\mu_A - \frac{1}{2}(\mu_B + \mu_E)$
  d. $\mu_A - \frac{1}{3}(\mu_B + \mu_C + \mu_E)$

14.10 Refer to Exercise 14.9. Give the corresponding Scheffé 95% confidence intervals for these four comparisons.

14.11 Refer to Exercise 13.27.
  a. Use Fisher's LSD procedure with $\alpha = .05$ to declare significant differences.
  b. Use Tukey's W method with $\alpha = .05$ to make all pairwise comparisons of population means for the data of Exercise 13.27. Compare your conclusions to those of part a.
  c. How would your conclusions differ using the SNK procedure?

14.12 Refer to Exercise 13.27.
  a. Use the Waller–Duncan method, with $k = 100$.
  b. Compare your results for parts a and b of Exercise 14.11 and with part a of this exercise.

14.13 Refer to Exercise 13.13. Use Duncan's multiple range procedure to declare significant differences among the population means. Set $\alpha = .05$.

14.14 Refer to Exercise 14.5. Determine the appropriate value of t_c for the Waller–Duncan k ratio procedure when $k = 100$ and $k = 500$.

14.15   Make all pairwise comparisons among the 5 diet preparations of Exercise 14.5 using the Waller–Duncan method. Use an error weight ratio of $k = 100$. Compare your results to those of Exercise 14.6.

14.16   Use Duncan's multiple range procedure to compare the four population means of Exercise 13.16.

14.17   Refer to Exercise 13.12. Perform all pairwise comparisons of the four population means. Be sure to identify the procedure you use and the error rate controlled. Use $\alpha = .05$.

14.18   Set up confidence intervals for all pairwise comparisons of treatment means in Exercise 14.4, using Tukey's W procedure. Interpret your findings.

14.19   Refer to the data of Exercise 13.19. Make all pairwise comparisons of the three areas using the Waller–Duncan procedure with $k = 100$.

14.20   The nitrogen contents of red clover plants inoculated with three strains of *Rhizobium* are given here.

3DOK1	3DOK5	3DOK7
19.4	18.2	20.7
32.6	24.6	21.0
27.0	25.5	20.5
32.1	19.4	18.8
33.0	21.7	18.6
	20.8	20.1
		21.3

Is there evidence of a difference in the effects of the three treatments on nitrogen content? Analyze the data completely and draw conclusions. Use $\alpha = .01$.

14.21   An experiment was conducted to compare three methods of grass-seed preparation: mechanical scarification (ms), acid dip (ad), and hot water dip (hw). One hundred grass seeds were placed in each of 150 petri dishes. Among the 150 dishes, 50 were randomly assigned ms, 50 ad, and 50 hw. After a period of 2 weeks, the germination rates were checked for each dish.

Method	Mean Germination	Standard Deviation
ms	65.3	7.2
ad	82.1	5.4
hw	73.8	6.5

Analyze these data using a one-way analysis and draw conclusions. Use $\alpha = .05$.

14.22   Refer to Exercise 14.21. Use the SNK procedure to identify method differences. Summarize your results.

14.23 An experiment was conducted to assess the relative merits of four different gasoline blends. Twenty automobiles of the same type, model, and engine size were used with five randomly assigned to each of the blends. Summary test data for the blends are shown here.

Blend	Mean (mi/gal)	Standard Deviation
1 (control)	26.2	4.3
2 (control − additive x)	28.1	5.6
3 (control − additive y)	29.6	5.1
4 (control − additives x and y)	38.2	7.3

Run an analysis and draw conclusions. Use $\alpha = .05$.

14.24 Refer to Exercise 14.23 and consider the following linear contrasts. Describe what each contrast is measuring.
   a. $\hat{l}_1 = \bar{y}_{1.} + \bar{y}_{2.} - \bar{y}_{3.} - \bar{y}_{4.}$
   b. $\hat{l}_2 = \bar{y}_{1.} + \bar{y}_{3.} - \bar{y}_{2.} - \bar{y}_{4.}$
   c. $\hat{l}_3 = \bar{y}_{1.} - \bar{y}_{2.} - \bar{y}_{3.} + \bar{y}_{4.}$

14.25 Use Scheffé's method to conduct a test of significance on $\hat{l}_3$ of Exercise 14.24, based on $\alpha = .05$. What do you conclude?

14.26 Construct a confidence interval for l_1 and l_2 of Exercise 14.24 using Scheffé's method. Draw conclusions.

14.27 Refer to Exercise 13.25. Use Tukey's procedure to run all pairwise comparisons with $\alpha = .05$. Summarize your results.

14.28 How would the SNK procedure change your conclusions to Exercise 14.27?

14.29 Refer to Exercise 14.27. Suppose that levels 1 and 2 for additives A and B are 1.0 and 3.0 mg/kg, respectively.
   a. Plot the mean weight gained versus $x =$ mg/kg for the two additives (include the control data, $x = 0$).
   b. Fit separate linear regression models for additives A and B.

14.30 Refer to Exercise 14.29. Estimate the slopes for the two lines using 95% confidence intervals. Does it appear that additives A and B have different regression lines? (A method for comparing slopes is presented in Chapter 11.)

ANALYSIS OF VARIANCE FOR SOME STANDARD EXPERIMENTAL DESIGNS

15.1 ▾ INTRODUCTION

The design of an experiment is the process of planning an experiment. A large part of scientific reasoning consists of drawing conclusions from experiments (studies) that have been carefully designed, appropriately conducted, and properly analyzed. In this chapter, we will present a brief preview of some standard experimental designs and their analyses.

Section 15.2 reviews the analysis of variance for a completely randomized design discussed in Chapter 13. Here the focus of interest is the comparison of treatment means. Sections 15.3 and 15.4 deal with extensions of the completely randomized design, where the focus remains the same—namely, treatment mean comparisons—but where other "nuisance" variables must be controlled. For each of these designs, we will consider the arrangement of treatments, the advantages and disadvantages of the design, a model, and an analysis of variance for data from such a design. Section 15.5 introduces factorial experiments that focus on the evaluation of the effects of two or more independent variables (factors) on a response rather than on comparisons of treatment means as in the designs of Sections 15.2 through 15.4. Particular attention is given to measuring the effects of each factor alone or in combination with the other factors. Not all designs focus on either comparison of treatment means or examination of the effects of factors on a response. In Section 15.6, we discuss designs that combine the attributes of the "block" designs of Sections 15.3 and 15.4 with those of factorial experiments in Section 15.5. The remaining sections of this chapter deal with multiple-comparison procedures for the different experimental designs and a useful unifying thread showing the relationship between regression and the analysis of variance.

15.2 ▼ COMPLETELY RANDOMIZED DESIGN

Recall that the completely randomized design is concerned with the comparison of t population (treatment) means μ_1, μ_2, ..., μ_t. We assume that there are t different populations from which we are to draw independent random samples of sizes n_1, n_2, ..., n_t, respectively. Or, in the terminology of the design of experiments, we assume that there are $n_1 + n_2 + \cdots + n_t$ homogeneous *experimental units* (people or objects on which a measurement is made). Then treatments are randomly allocated to the experimental units in such a way that n_1 units receive treatment 1, n_2 receive treatment 2, and so on. The objective of the experiment is to make inferences about the corresponding treatment (population) means.

Consider the following example. A horticultural laboratory is interested in examining the leaves of apple trees to detect nutritional deficiencies using three different laboratory procedures. In particular, the laboratory would like to determine whether there is a difference in mean assay readings for apple leaves utilizing three different laboratory procedures (A, B, C). The experimental units in this investigation are apple tree leaves and the treatments are the three levels of the qualitative variable "laboratory procedure." If a single analyst takes a random sample of nine leaves from the same tree, randomly assigns three leaves to each of the three procedures, and assays the leaves using the assigned treatment, we could use the three sample means to estimate the corresponding mean leaf nutritional deficiency for the three laboratory test procedures, or use the analysis of variance methods of Chapter 13 to run a statistical test of the hypothesis that all three treatment means are identical. The design used for this investigation is a completely randomized design with three observations for each treatment.

The completely randomized design has several advantages and disadvantages when used as an experimental design for comparing t treatment means.

Advantages and Disadvantages of a Completely Randomized Design

▼

Advantages

1. It is extremely easy to construct the design.
2. The design is easy to analyze even though the sample sizes might not be the same for each treatment.
3. The design can be used for any number of treatments.

Disadvantages

1. Although the completely randomized design can be used for any number of treatments, it is best suited for situations in which there are relatively few treatments.
2. The experimental units to which treatments are applied must be as homogeneous as possible. Any extraneous sources of variability will tend to inflate the error term, making it more difficult to detect differences among the treatment means.

15.3 ▼ RANDOMIZED BLOCK DESIGN

Let us now change the horticultural problem slightly and see how well the completely randomized design suits our needs. Suppose that, rather than relying upon one analyst, we use three analysts for the leaf assays. If we randomly assigned three apple leaves to each of the analysts, we might end up with a randomization scheme like the one listed in Table 15.1.

TABLE 15.1

Random Assignment of the Nine Leaves to the Three Analysts

Each analyst does same procedure.

	Analyst	
1	2	3
A	B	C
A	B	C
A	B	C

Even though we still have three observations for each treatment in this scheme, any differences that we may observe among the leaf determinations for the three laboratory procedures may be due entirely to differences among the analysts who assayed the leaves. For example, if we tested the hypothesis H_0: $\mu_A - \mu_B = 0$ against H_a: $\mu_A - \mu_B \neq 0$ and were led to reject H_0, we would not be able to tell whether μ_A differs from μ_B because assays from analyst 1 are different from those for analyst 2 or because the properties of determinations by procedure A differ markedly from those for procedure B. This example illustrates a situation in which the nine experimental units (tree leaves) are affected by an extraneous source of variability: the analyst. In this case, the units differ markedly and would not be a homogeneous set on which we could base an evaluation of the effects of the three treatments.

The completely randomized design just described can be modified to gain additional information concerning the means μ_A, μ_B, and μ_C. We can block out the undesirable variability among analysts by using the following experimental design. We *restrict* our randomization of treatments to experimental units to ensure that each analyst performs a determination using each of the three procedures. The order of these determinations for each analyst is randomized. One such randomization is listed in Table 15.2. Note that each analyst will assay three leaves, one leaf for each of the three procedures. Hence, pairwise comparisons among the laboratory procedures that utilize the sample means will be free of any variability among analysts. For example, if we ran the test

$$H_0: \quad \mu_A - \mu_B = 0$$
$$H_a: \quad \mu_A - \mu_B \neq 0$$

and rejected H_0, the difference between μ_A and μ_B would be due to a difference between the nutritional deficiencies detected by procedures A and B and not due to a difference among the analysts, since each analyst would have assayed one leaf for each of the three procedures.

TABLE 15.2
A Different Assignment of Leaves to Analysts

Analyst		
1	2	3
A	B	A
C	A	B
B	C	C

Each analyst conducts all 3 proced.

randomized block design
blocks

This design, which represents an extension to the completely randomized design, is called a **randomized block design**; the analysts in our experiment are called **blocks**. By using this design, we have effectively filtered out any variability among the analysts, enabling us to make more precise comparisons among the treatments means μ_A, μ_B, and μ_C.

In general, we can use a randomized block design to compare t different treatment means when an extraneous source of variability (blocks) is present. If there are b different blocks, we would run each of the t treatments in each block to filter out the block-to-block variability. In our example, we had $t = 3$ treatment means (laboratory procedures) and $b = 3$ blocks (analysts).

The randomized block design has certain advantages and disadvantages, as shown here.

Advantages and Disadvantages of the Randomized Block Design

▼

Advantages

1. It is a useful design for comparing t treatment means in the presence of a single extraneous source of variability.
2. The statistical analysis is simple.
3. The design is easy to construct.
4. It can be used to accommodate any number of treatments in any number of blocks.

Disadvantages

1. Since the experimental units within a block must be homogeneous, the design is best suited for a relatively small number of treatments.
2. This design controls for only one extraneous source of variability (due to blocks). Additional extraneous sources of variability tend to increase the error term, making it more difficult to detect treatment differences.
3. The effect of each treatment on the response must be approximately the same from block to block.

The definition of a randomized block design is given next.

DEFINITION 15.1
Randomized Block Design

▼

A **randomized block design** is an experimental design for comparing t treatments in b blocks. Treatments are randomly assigned to experimental units within a block, with each treatment appearing exactly once in every block.

Note that while these data look similar to the data presentation for a completely randomized design (see Table 13.4), the difference is in the way treatments were assigned to the experimental units.

Consider the data for a randomized block design as arranged in Table 15.3.

TABLE 15.3
A Randomized Block Design

Treatment	Block 1	2	$\cdots$	b	Totals	Mean
1	y_{11}	y_{12}	$\cdots$	y_{1b}	$y_{1\cdot}$	$\bar{y}_{1\cdot}$
2	y_{21}	y_{22}	$\cdots$	y_{2b}	$y_{2\cdot}$	$\bar{y}_{2\cdot}$
$\vdots$	$\vdots$	$\vdots$		$\vdots$	$\vdots$	$\vdots$
t	y_{t1}	y_{t2}	$\cdots$	y_{tb}	$y_{t\cdot}$	$\bar{y}_{t\cdot}$
Totals	$y_{\cdot1}$	$y_{\cdot2}$	$\cdots$	$y_{\cdot b}$	$y_{\cdot\cdot}$	
Mean	$\bar{y}_{\cdot1}$	$\bar{y}_{\cdot2}$	$\cdots$	$\bar{y}_{\cdot b}$		$\bar{y}_{\cdot\cdot}$

Using Table 15.3 we can introduce notation that is helpful in performing an analysis of variance. This notation is presented here.

Notation Needed for the AOV of a Randomized Block Design

▼

y_{ij}: observation for treatment i in block j
t: number of treatments
b: number of blocks

$y_{i\cdot}$: total for all observations receiving treatment i
$y_{\cdot j}$: total for all observations in block j
$y_{\cdot\cdot}$: total for all sample observations
$\bar{y}_{i\cdot}$: sample mean for treatment i; $\bar{y}_{i\cdot} = y_{i\cdot}/b$
$\bar{y}_{\cdot j}$: sample mean for block j; $\bar{y}_{\cdot j} = y_{\cdot j}/t$
$\bar{y}_{\cdot\cdot}$: overall sample mean; $\bar{y}_{\cdot\cdot} = y_{\cdot\cdot}/bt$

total sum of squares

The **total sum of squares** of the measurements about their mean $\bar{y}$ is defined as before:

$$TSS = \sum_{ij} (y_{ij} - \bar{y}_{\cdot\cdot})^2.$$

It is possible to partition the total sum of squares into three separate sources of variability: one due to the variability among treatments, one due to the variability among blocks, and one due to the variability among the y_{ij}s that is not accounted for by either treatments or blocks. We call this final source of variability "error." The **partition of TSS** can be shown to take the following form:

partition of TSS

$$\sum_{ij} (y_{ij} - \bar{y}_{\cdot\cdot})^2 = \underbrace{b \sum_i (\bar{y}_{i\cdot} - \bar{y}_{\cdot\cdot})^2}_{SST} + \underbrace{t \sum_j (\bar{y}_{\cdot j} - \bar{y}_{\cdot\cdot})^2}_{SSB} + \underbrace{\sum_{ij} (y_{ij} - \bar{y}_{i\cdot} - \bar{y}_{\cdot j} + \bar{y}_{\cdot\cdot})^2}_{SSE}.$$

The first quantity on the right side of the equation measures the variability of the treatment means $\bar{y}_{i\cdot}$ from the overall mean. Thus,

$$SST = b \sum_i (\bar{y}_{i\cdot} - \bar{y}_{\cdot\cdot})^2,$$

between-treatment sum of squares

called the **between-treatment sum of squares**, is a measure of the between-treatment variability. Similarly, the second quantity,

$$SSB = t \sum_j (\bar{y}_{\cdot j} - \bar{y}_{\cdot\cdot})^2,$$

between-block sum of squares

measures the variability between the block means $\bar{y}_{\cdot j}$ and the overall mean. It is called the **between-block sum of squares**. The third source of variability,

sum of squares for error   referred to as the **sum of squares for error** (SSE), represents the variability in the y_{ij}s not accounted for by the block and treatment sources.

Although the sum of squares formulas just discussed are instructive, they are not convenient to use in calculations. The shortcut formulas given below are more convenient tools for calculations.

Shortcut Sums of Squares Formulas for a Randomized Block Design

$$TSS = \sum_{ij} y_{ij}^2 - \frac{y_{..}^2}{bt}$$

$$SST = \sum_{i} \frac{y_{i\cdot}^2}{b} - \frac{y_{..}^2}{bt}$$

$$SSB = \sum_{j} \frac{y_{\cdot j}^2}{t} - \frac{y_{..}^2}{bt}$$

$$SSE = TSS - SST - SSB$$

The model for an observation in a randomized block design can be written in the form

model

$$y_{ij} = \mu + \alpha_i + \beta_j + \varepsilon_{ij},$$

where the terms of the model are defined as follows:

μ: an overall mean, which is an unknown constant
α_i: an effect due to treatment i; α_i is an unknown constant
β_j: an effect due to block j; β_j is an unknown constant
ε_{ij}: a random error associated with the response on treatment i, block j. We assume that the ε_{ij}s are normally distributed with mean 0 and unknown variance σ_ε^2. In addition, the errors are assumed to be independent. (Technically speaking, we need not assume that the ε_{ij}s are normally distributed at this point, but we must make this assumption prior to running an analysis of variance.)

$\varepsilon_{ij} \sim N(\mu, \sigma^2)$
$\sim N(0, \sigma_\varepsilon^2)$

The assumptions given above for our model can be shown to imply that y_{ij}, the response on treatment i in block j, is normally distributed with mean $\mu + \alpha_i + \beta_j$ and variance σ_ε^2. A table of population means (expected values) for the data of Table 15.3 is shown in Table 15.4.

Several comments should be made concerning the table of expected values. First, two observations that receive the same treatment (appear in the same row of Table 15.4) have population means that differ by block effects only. For

	Block			
Treatment	1	2	$\cdots$	b
1	$E(y_{11}) = \mu + \alpha_1 + \beta_1$	$E(y_{12}) = \mu + \alpha_1 + \beta_2$	$\cdots$	$E(y_{1b}) = \mu + \alpha_1 + \beta_b$
2	$E(y_{21}) = \mu + \alpha_2 + \beta_1$	$E(y_{22}) = \mu + \alpha_2 + \beta_2$	$\cdots$	$E(y_{2b}) = \mu + \alpha_2 + \beta_b$
$\vdots$	$\vdots$			$\vdots$
t	$E(y_{t1}) = \mu + \alpha_t + \beta_1$	$E(y_{t2}) = \mu + \alpha_t + \beta_2$	$\cdots$	$E(y_{tb}) = \mu + \alpha_t + \beta_b$

example, the expected values associated with y_{11} and y_{12} (two observations receiving treatment 1) are, respectively, $\mu + \alpha_1 + \beta_1$ and $\mu + \alpha_1 + \beta_2$. Thus, the difference in their means is $\beta_1 - \beta_2$, which accounts for the fact that y_{11} appeared in block 1 and y_{12} in block 2. Second, two observations appearing in the same block (in the same column of Table 15.4) have means that differ by a treatment effect only. For example, y_{11} and y_{21} both appear in block 1. The difference in their means is, from Table 15.4,

$$(\mu + \alpha_1 + \beta_1) - (\mu + \alpha_2 + \beta_1) = \alpha_1 - \alpha_2,$$

which accounts for the fact that the observations received different treatments. Finally, when two observations receive a different treatment *and* appear in different blocks, their expected values differ by effects due to treatments and to blocks. Thus, y_{11} and y_{22} have expectations that differ by

$$(\mu + \alpha_1 + \beta_1) - (\mu + \alpha_2 + \beta_2) = (\alpha_1 - \alpha_2) + (\beta_1 - \beta_2).$$

filtering

Using the information we have learned concerning the model for a randomized block design, we can illustrate the concept of **filtering** and show how the randomized block design filters out the variability due to blocks. Consider a randomized block design with $t = 3$ treatments (1, 2, and 3) laid out in $b = 3$ blocks as shown in Table 15.5.

TABLE 15.5
Randomized Block Design with $t = 3$ Treatments and $b = 3$ Blocks

Block	Treatment		
1	1	2	3
2	1	3	2
3	3	1	2

The model for this randomized block design is

$$y_{ij} = \mu + \alpha_i + \beta_j + \varepsilon_{ij} \qquad (i = 1, 2, 3; j = 1, 2, 3).$$

Suppose we wish to estimate the difference in mean response for treatments 2 and 1, namely, $\alpha_2 - \alpha_1$. The difference in sample means, $\bar{y}_{2\cdot} - \bar{y}_{1\cdot}$, would represent a point estimate of $\alpha_2 - \alpha_1$. By substituting into our model, we have

$$\bar{y}_{1.} = \sum_j \frac{y_{ij}}{3}$$

$$= \frac{1}{3}[(\mu + \alpha_1 + \beta_1 + \varepsilon_{11}) + (\mu + \alpha_1 + \beta_2 + \varepsilon_{12}) + (\mu + \alpha_1 + \beta_3 + \varepsilon_{13})]$$

$$= \mu + \alpha_1 + \bar{\beta} + \bar{\varepsilon}_1,$$

where $\bar{\beta}$ represents the mean of the three block effects β_1, β_2, and β_3, and $\bar{\varepsilon}_1$ represents the mean of the three random errors ε_{11}, ε_{12}, and ε_{13}. Similarly, it is easy to show that

$$\bar{y}_{2.} = \mu + \alpha_2 + \bar{\beta} + \bar{\varepsilon}_2$$

and hence

$$\bar{y}_{2.} - \bar{y}_{1.} = (\alpha_2 - \alpha_1) + (\bar{\varepsilon}_2 - \bar{\varepsilon}_1).$$

Note how the block effects cancel, leaving the quantity $(\bar{\varepsilon}_2 - \bar{\varepsilon}_1)$ as the error of estimation using $\bar{y}_{2.} - \bar{y}_{1.}$ to estimate $\alpha_2 - \alpha_1$.

If a completely randomized design had been employed instead of a randomized block design, treatments would have been assigned to experimental units at random and it is quite unlikely that each treatment would have appeared in each block. When the same treatment appears more than once in a block and we calculate an estimate of $\alpha_2 - \alpha_1$ using $\bar{y}_{2.} - \bar{y}_{1 \cdot j}$ all block effects would not cancel out as they did previously. Then the error of estimation would include not only $\bar{\varepsilon}_2 - \bar{\varepsilon}_1$ but also the block effects that do not cancel; that is,

$$\bar{y}_{2.} - \bar{y}_{1.} = \alpha_2 - \alpha_1 + [(\bar{\varepsilon}_2 - \bar{\varepsilon}_1) + (\text{block effects that do not cancel})].$$

Hence, the randomized block design filters out variability due to blocks by decreasing the error of estimation for a comparison of treatment means.

Having indicated the shortcut formulas for calculating sums of squares, and having specified the model associated with a randomized block design, we can now formulate the analysis of variance. The null hypothesis of no **difference among the treatment means** is equivalent to testing

test for treatment effects

$$H_0: \quad \alpha_1 = \alpha_2 = \cdots = \alpha_t = 0.$$

As we observed in Table 15.4, any time we compare the mean response of two treatments (say i and i') in the same block, the difference in their mean response is $\alpha_i - \alpha_{i'}$. Thus, under H_0, we are assuming that treatments have the same mean response within a block.

The alternative hypothesis is

H_a: At least one α_i is different from 0 (i.e., at least one of the treatment means differs from the rest).

The test statistic is

$$F = \frac{MST}{MSE},$$

where MST and MSE are mean squares computed from the appropriate sums of squares in the AOV table of Table 15.6.

TABLE 15.6
Analysis of Variance Table for a Randomized Block Design

Source	SS	df	MS	F
Treatments	SST	$t-1$	MST $= $ SST$/(t-1)$	MST/MSE
Blocks	SSB	$b-1$	MSB $=$ SSB$/(b-1)$	MSB/MSE
Error	SSE	$(b-1)(t-1)$	MSE $=$ SSE$/(b-1)(t-1)$	
Totals	TSS	$bt-1$		

unbiased estimates

When $H_0: \alpha_1 = \alpha_2 = \cdots = \alpha_t = 0$ is true, both MST and MSE are **unbiased estimates** of σ_ε^2, the variance associated with the observations in our model. That is, when H_0 is true, both MST and MSE have a mean value in repeated sampling equal to σ_ε^2, and we would expect $F = $ MST/MSE to be near 1. When H_a is true, the mean of MSE, called the **expected mean square for error**, is still

expected mean square for error

$$E(MSE) = \sigma_\varepsilon^2.$$

However, MST is no longer unbiased for σ_ε^2. In fact, the expected mean square for treatments can be shown to be

$$E(MST) = \sigma_\varepsilon^2 + b\theta_T,$$

where θ_T is a positive function of the α_is. Because MST will tend to overestimate σ_ε^2 under H_a, the ratio $F = $ MST/MSE will be greater than 1, and we will reject H_0 in the upper tail of the distribution of F.

For a specified probability of a Type 1 error, the F test for $H_0: \alpha_1 = \alpha_2 = \cdots = \alpha_t = 0$ will reject H_0 if the computed value of F exceeds the critical value of F for $a = \alpha$, $df_1 = t - 1$, and $df_2 = (b-1)(t-1)$. Note that df_1 and df_2

correspond to the degrees of freedom for MST and MSE, respectively, in the AOV table.

We may also be interested in testing whether it was advantageous to block. **test for effects of blocks** That is, is there an **effect due to blocks** that we have effectively filtered out? Recall that the expected values for two observations from different blocks receiving the same treatment differed by an effect due to blocks only (see Table 15.4). The hypothesis of no effect due to blocks can be written in the form

$$H_0: \quad \beta_1 = \beta_2 = \cdots = \beta_b = 0.$$

The alternative hypothesis and test statistic are then

$$H_a: \quad \text{At least one of the } \beta\text{s differs from } 0$$

$$\text{T.S.:} \quad F = \frac{\text{MSB}}{\text{MSE}}.$$

Under H_0, both MSB and MSE are unbiased estimates of σ_ε^2 (i.e., $E(\text{MSB}) = E(\text{MSE}) = \sigma_\varepsilon^2$). But under H_a, the expected mean squares are

$$E(\text{MSB}) = \sigma_\varepsilon^2 + t\theta_B \quad \text{and} \quad E(\text{MSE}) = \sigma_\varepsilon^2,$$

where θ_B is a positive function of the βs. Since MSB will tend to overestimate σ_ε^2 under H_a, we will reject H_0 for large values of F. For a specified value of α, we can locate the rejection region in the F table of the Appendix for $a = \alpha$, $\text{df}_1 = b - 1$, and $\text{df}_2 = (b - 1)(t - 1)$.

E X A M P L E   15.1   ▼

An experiment was conducted to compare the effects of three different insecticides on a particular variety of string beans. Four different plots were prepared, with each plot subdivided into three rows. A suitable distance was maintained between the rows within a plot. Each row was planted with 100 seeds and then maintained under the insecticide assigned to the row. The insecticides were randomly assigned to the rows within a plot so that each insecticide appeared in one row in all four plots. The response of interest was the number of seedlings that emerged per row. These data are given in Table 15.7.

a. Set up an appropriate statistical model for this experimental situation.
b. Run an analysis of variance to compare the three insecticides.
c. Summarize your results in an AOV table. Use $\alpha = .05$.

TABLE 15.7

Number of Seedlings by Insecticide and Plot, Example 15.1

Insecticide	Plot			
	1	2	3	4
1	56	49	65	60
2	84	78	94	93
3	80	72	83	85

Solution We recognize this experimental design as a randomized block design with $b = 4$ blocks (plots) and $t = 3$ treatments (insecticides) per block. The appropriate statistical model is

$$y_{ij} = \mu + \alpha_i + \beta_j + \varepsilon_{ij} \qquad (i = 1, 2, 3; j = 1, 2, 3, 4).$$

From the sample data, the treatment and block totals can be shown to be

Insecticides: $y_{1\cdot} = 230$ Plots: $y_{\cdot 1} = 220$
$y_{2\cdot} = 349$ $y_{\cdot 2} = 199$
$y_{3\cdot} = 320$ $y_{\cdot 3} = 242$
$y_{\cdot\cdot} = 899$ $y_{\cdot 4} = 238$
            $y_{\cdot\cdot} = 899$

Substituting into the corresponding shortcut formulas for the sums of squares, we have

$$\text{SST} = \sum_i \frac{y_{i\cdot}^2}{b} - \frac{y_{\cdot\cdot}^2}{bt} = \frac{(230)^2 + (349)^2 + (320)^2}{4} - \frac{(899)^2}{12}$$

$$= 69{,}275.25 - 67{,}350.08 = 1{,}925.17$$

$$\text{SSB} = \sum_j \frac{y_{\cdot j}^2}{t} - \frac{y_{\cdot\cdot}^2}{bt} = \frac{(220)^2 + (199)^2 + (242)^2 + (238)^2}{3} - \frac{(899)^2}{12}$$

$$= 67{,}736.33 - 67{,}350.08 = 386.25.$$

Then

$$\text{TSS} = \sum_{ij} y_{ij}^2 - \frac{y_{\cdot\cdot}^2}{bt} = (56)^2 + (49)^2 + \cdots + (85)^2 - 67{,}350.08$$

$$= 2334.92.$$

By subtraction,

$$\text{SSE} = \text{TSS} - \text{SST} - \text{SSB} = 2{,}334.92 - 1{,}925.17 - 386.25 = 23.50.$$

TABLE 15.8
AOV Table for the Data
of Example 15.1

Source	SS	df	MS	F
Treatments	1,925.17	2	962.59	962.59/3.92 = 245.56
Blocks	386.25	3	128.75	128.75/3.92 = 32.84
Error	23.50	6	3.92	
Totals	2,334.92	11		

The analysis of variance table in Table 15.8 summarizes our results. Note that the mean square for a source in the AOV table is computed by dividing the sum of squares for that source by its degrees of freedom.

The F test for treatments, namely,

$$H_0: \quad \alpha_1 = \alpha_2 = \alpha_3 = 0 \text{ (no differences among treatment means)}$$

makes use of the F statistic MST/MSE. Since the computed value of F, 245.56, is greater than the tabulated F-value, 5.14, based on $df_1 = 2$, $df_2 = 6$, and $a = .05$, we reject H_0 and conclude that there are differences among the three treatment (insecticide) means.

The test for blocks

$$H_0: \quad \beta_1 = \beta_2 = \beta_3 = \beta_4 = 0 \text{ (no differences among plot means)}$$

utilizes the computed value of $F = \text{MSB/MSE}$ as the test statistic. Since the computed value of F, 32.84, is greater than the tabulated value, 4.76, based on $df_1 = 3$, $df_2 = 6$, and $a = .05$, we reject the null hypothesis and conclude that there are differences among the plot means. Further, since the $F_{.01}$ value (9.78) is also greatly exceeded, the level of significance for the test is much less than .01. ▲

EXAMPLE 15.2 ▼

Compare the computer output shown here to the hand calculations for the data of Example 15.1.

```
ANALYSIS OF VARIANCE PROCEDURE

DEPENDENT VARIABLE: NO_SEED
    SOURCE          DF   SUM OF SQUARES   MEAN SQUARE  F VALUE    PR > F   R-SQUARE      C.V.
    MODEL            5   2311.41666667    462.28333333  118.03    0.0001   0.989935     2.6417
    ERROR            6     23.50000000      3.91666667                     ROOT MSE   NO_SEED MEAN
    CORRECTED TOTAL  11  2334.91666667                                     1.97905701   74.91666667

    SOURCE          DF      ANOVA SS     F VALUE    PR > F
    INSECT           2   1925.16666667    245.77    0.0001
    PLOT             3    386.25000000     32.87    0.0004
```

Solution   Except for minor rounding errors in our hand calculations, the results from the output agree with the analysis of variance of Example 15.1. Note that the p-value for treatment (insecticide) differences is .0001.   ▲

We have discussed randomized block designs briefly in Chapter 13 and in more detail here. But we might still ask whether blocking has increased our precision for comparing treatment means in a given experiment. Let MSE_{RB} and MSE_{CR} denote the mean square errors for a randomized block design and a completely randomized design, respectively. One measure of precision for the two designs is the variance of $\bar{y}_{i\cdot}$, the ith treatment mean $(i = 1, 2, \ldots, t)$. For a randomized block design, the estimated variance for $\bar{y}_{i\cdot}$, is MSE_{RB}/b. A similar expression for a completely randomized design is MSE_{CR}/r, where r is the number of observations (replications) of each treatment required to satisfy the relationship

$$\frac{MSE_{CR}}{r} = \frac{MSE_{RB}}{b} \quad \text{or} \quad \frac{MSE_{CR}}{MSE_{RB}} = \frac{r}{b}.$$

relative efficiency

The quantity r/b is called the **relative efficiency** of the randomized block design. The larger the value of MSE_{CR} to MSE_{RB}, the larger r must be to obtain the same precision for a treatment mean as obtained with the randomized block design.

Because the analysis of data must be appropriate to the experimental design used, we don't perform an analysis of variance for a completely randomized design when we employ a randomized block design. However, we can use the mean squares from the analysis of variance for the randomized block design to obtain the relative efficiency by using the formula

$$\frac{MSE_{CR}}{MSE_{RB}} = \frac{(b-1)MSB + b(t-1)MSE}{(bt-1)MSE}.$$

E X A M P L E   15.3   ▼

Refer to Example 15.1. Compute the relative efficiency of the randomized block design relative to a completely randomized design.

Solution   From the AOV table in Example 15.1, $MSB = 128.75$ and $MSE = 3.92$. Hence, the relative efficiency of this randomized block design relative to a completely randomized design is

$$\frac{MSE_{CR}}{MSE_{RB}} = \frac{3(128.75) + 4(2)(3.92)}{11(3.92)} = 9.68.$$

That is, approximately ten times as many observations of each treatment would be required in a completely randomized design to obtain the same precision for treatment comparisons as with this randomized block design.

▲

▼ EXERCISES

Basic Techniques

15.1 **a.** Analyze the hypothetical data shown here for a randomized block design with $b = 6$ blocks and $t = 2$ treatments. Use the analysis of variance methods of this section with $\alpha = .05$.

	Treatment	
Block	A	B
1	58	47
2	324	331
3	206	163
4	94	75
5	39	30
6	418	397

b. Give the efficiency of the randomized block design relative to the completely randomized design. Explain.

15.2 Refer to Exercise 15.1. Analyze these same data using the paired t-methods of Chapter 6. Compare your results. Use $\alpha = .05$. (Hint: $t^2 = F$ for the test on treatments.)

Applications

 15.3 An experiment is conducted to compare four different mixtures of the components oxidizer, binder, and fuel used in the manufacturing of rocket propellant. The four mixtures under test, corresponding to settings of the mixture proportions for oxide, are shown here.

Mixture	Oxidizer	Binder	Fuel
1	.4	.4	.2
2	.4	.2	.4
3	.6	.2	.2
4	.5	.3	.2

To compare the four mixtures, five different samples of propellant are prepared from each mixture and readied for testing. Each of five investigators is randomly assigned one sample of each of the four mixtures and asked to measure the propellant thrust. These data are summarized next.

| | **Investigator** | | | | |
Mixture	1	2	3	4	5
1	2,340	2,355	2,362	2,350	2,348
2	2,658	2,650	2,665	2,640	2,653
3	2,449	2,458	2,432	2,437	2,445
4	2,403	2,410	2,418	2,397	2,405

a. Identify the blocks and treatments for this experimental design.
b. Indicate the method of randomization.
c. Why would this design be preferable to a completely randomized design?

15.4 Refer to Exercise 15.3.
   a. Use the computer output shown here to conduct an analysis of variance. Use $\alpha = .05$.
   b. What conclusions can you draw concerning the best mixture from the four tested? (Note: The higher the response value, the better the rocket propellant thrust.)
   c. Compute the relative efficiency of the randomized block design.

```
ANALYSIS OF VARIANCE PROCEDURE

DEPENDENT VARIABLE: THRUST
     SOURCE        DF    SUM OF SQUARES    MEAN SQUARE      F VALUE    PR > F    R-SQUARE       C.V.
   MODEL            7    261713.45000000   37387.63571429   542.96    0.0001    0.996853      0.3368
   ERROR           12       826.30000000      68.85833333              ROOT   MSE       THRUST  MEAN
   CORRECTED TOTAL 19    262539.75000000                               8.29809215       2463.75000000
     SOURCE        DF      ANOVA  SS        F VALUE    PR > F
   MIXTURE          3    261260.95000000   1264.73    0.0001
   INVEST           4       452.50000000      1.64    0.2273
```

 15.5 A study was undertaken to compare the starting salaries of bachelor's degree candidates at a particular university for the academic years 1990–1991 and 1989–1990. Since there could be a great deal of variability in salaries from one discipline to another, the investigator blocked on curricula within the university. Then the median salary for bachelor's candidates for 1989–1990 and for 1990–1991 was obtained from the most recent lists of graduates to ascertain the starting salary. It should be noted that only those students who had accepted a job were considered in this study. Use the following sample data of starting salaries (in $1,000) to conduct an analysis of variance, testing separately for an effect due to treatments and blocks. Use $\alpha = .05$ for both tests.

Curriculum	1990–1991	1989–1990
Accounting	22.0	21.5
Agricultural sciences	20.6	20.3
Biological sciences	20.0	19.8
Business (general)	24.8	24.6
Chemistry	22.4	22.2
Computer science	26.9	26.5
Engineering (civil)	24.3	24.9
Engineering (chemical)	23.0	22.6
Humanities	16.5	16.6
Mathematics	20.7	21.1
Social sciences	16.9	16.8

15.6   The computer output shown here gives the analysis of variance for a taste-test experiment where each of 12 persons was asked to sample three new formulations of a widely used bulk laxative.

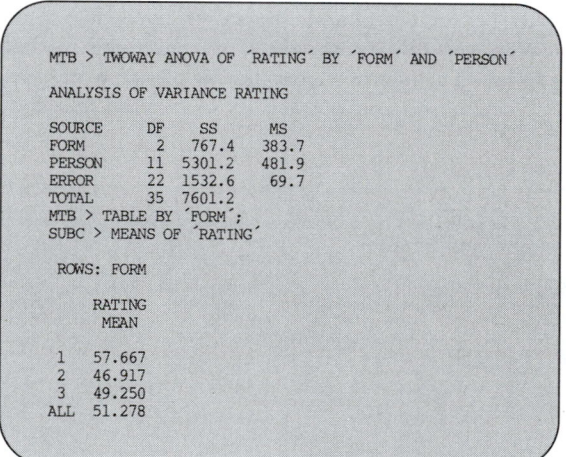

```
MTB > TWOWAY ANOVA OF 'RATING' BY 'FORM' AND 'PERSON'

ANALYSIS OF VARIANCE RATING

SOURCE     DF    SS      MS
FORM        2   767.4   383.7
PERSON     11  5301.2   481.9
ERROR      22  1532.6    69.7
TOTAL      35  7601.2
MTB > TABLE BY 'FORM';
SUBC > MEANS OF 'RATING'

 ROWS: FORM

     RATING
      MEAN

 1   57.667
 2   46.917
 3   49.250
ALL  51.278
```

a. Run an F test to compare the formulation means. Give the p-value for your test.

b. Also run an F test comparing the people. Do you think it was appropriate to block on people? Explain

15.7   Refer to Exercise 15.6.

a. Which formulation means appear to be different?

b. Calculate 95% confidence intervals for all pairwise differences in formulation means. Which means appear different based on these intervals? Explain.

15.4 ▼ LATIN SQUARE DESIGN

The apple leaf problem of Sections 15.2 and 15.3 can be complicated further in the following way. Suppose that each leaf assay takes a long time and only one can be done by each analyst per day. If we used the randomized block design of Table 15.2, letting the first row denote day 1, the second row denote day 2, and the third row denote day 3, the design could be listed as shown in Table 15.9.

TABLE 15.9

A Randomized Block Design for the Leaf Assay in the Presence of a Day Effect

	Analyst		
Day	1	2	3
1	A	B	A
2	C	A	B
3	B	C	C

Suppose now that we tested H_0: $\mu_A - \mu_B = 0$ against H_a: $\mu_A - \mu_B \neq 0$. Two procedure A determinations were done on day 1 and one was done on day 2, whereas procedure B was used on each of the 3 days. Thus, if we reject H_0, we would not be certain whether μ_A differed from μ_B because of a difference in the laboratory procedures or because of a difference among the 3 days. Sometimes laboratory equipment must be calibrated daily and new chemical solutions must be prepared. Differences in determinations from day to day could be due to differences among the solutions or to differences in calibration accuracy.

This example illustrates a situation in which the experimental units (leaves) are affected by a second extraneous source of variability, days. We can modify the randomized block design to filter out this second source of variability, the variability among days, in addition to filtering out the first source, variability among analysts. To do this, we restrict our randomization to ensure that each treatment appears in each row (day) and in each column (analyst). One such randomization is shown in Table 15.10. Note that the test procedures have been assigned to analysts and to days so that each procedure is performed once a day

TABLE 15.10

Assignment of Leaves to Analysts and Days

	Analyst		
Day	1	2	3
1	A	B	C
2	B	C	A
3	C	A	B

and once by each analyst. Hence, pairwise comparisons among treatment procedures that involve the sample means are free of variability among days and analysts.

Latin square design

This experimental design is called a **Latin square design**. In general, a Latin square design can be used to compare t treatment means in the presence of two extraneous sources of variability, which we block off into t rows and t columns. The t treatments are then randomly assigned to the rows and columns so that each treatment appears in every row and every column of the design (see Table 15.10).

The advantages and disadvantages of the Latin square design are listed here.

Advantages and Disadvantages of the Latin Square Design

▼

Advantages
_____

1. The design is particularly appropriate for comparing t treatment means in the presence of two sources of extraneous variation, each measured at t levels.
2. The analysis is still quite simple.

Disadvantages
_____

1. Although a Latin square can be constructed for any value of t, it is best suited for comparing t treatments when $5 \leq t \leq 10$.
2. Any additional extraneous sources of variability tend to inflate the error term, making it more difficult to detect differences among the treatment means.
3. The effect of each treatment on the response must be approximately the same across rows and columns.

The definition of a Latin square design is given here.

DEFINITION 15.2
$t \times t$ Latin Square Design

▼

A $t \times t$ **Latin square design** contains t rows and t columns. The t treatments are randomly assigned to experimental units within the rows and columns so that each treatment appears in every row and in every column.

A typical assignment of treatments for a 4×4 Latin square comparing the treatments I, II, III, and IV is shown in Table 15.11. Note that each treatment appears in all four rows and all four columns.

T A B L E   15.11
A 4 × 4 Latin Square Design

Row	Column 1	2	3	4
1	1	2	3	4
2	2	3	4	1
3	3	4	1	2
4	4	1	2	3

The notation for a Latin square design is only slightly more complicated than that for a randomized block design.

Notation Needed for the AOV of a $t \times t$ Latin Square Design

▼

y_{ijk}: response on treatment i in row j and column k

t: number of treatments; also the number of rows and the number of columns

$y_{i\cdot\cdot}$: total for all observations receiving treatment i

$\bar{y}_{i\cdot\cdot}$: sample mean for treatment i; $\bar{y}_{i\cdot\cdot} = y_{i\cdot\cdot}/t$

$y_{\cdot j\cdot}$: total for all observations in row j

$\bar{y}_{\cdot j\cdot}$: sample mean for row j; $\bar{y}_{\cdot j\cdot} = y_{\cdot j\cdot}/t$

$y_{\cdot\cdot k}$: total for all observations in column k

$\bar{y}_{\cdot\cdot k}$: sample mean for column k; $\bar{y}_{\cdot\cdot k} = y_{\cdot\cdot k}/t$

$y_{\cdots}$: total for all sample measurements

$\bar{y}_{\cdots}$: overall sample mean; $\bar{y}_{\cdots} = y_{\cdots}/t^2$

partitioning TSS

With this notation, we can show a partition of the total sum of squares into four components. The first three components measure variability among the treatments, rows, and columns, respectively. The other source is due to random error.

$$\sum_{ij} (y_{ijk} - \bar{y}_{\cdots})^2 = t \sum_i (\bar{y}_{i\cdot\cdot} - \bar{y}_{\cdots})^2 + t \sum_j (\bar{y}_{\cdot j\cdot} - \bar{y}_{\cdots})^2 + t \sum_k (\bar{y}_{\cdot\cdot k} - \bar{y}_{\cdots})^2$$

$$+ \sum_{ij} (y_{ijk} - \bar{y}_{i\cdot\cdot} - \bar{y}_{\cdot j\cdot} - \bar{y}_{\cdot\cdot k} + 2\bar{y}_{\cdots})^2$$

Note that the sum of squares is obtained by summing over only two of the three subscripts. Even though observations are identified by treatment, row, and column, by summing over treatments (i) and rows (j), we have also summed over columns (k).

The algebraic verification of the partitioning TSS into the four components is not important here and is beyond the scope of this text. We will concentrate instead on the interpretation of the partitioning.

The first quantity on the right side of the equation for TSS measures the variability of the treatment means $\bar{y}_{i\cdot\cdot}$ about the overall mean $\bar{y}_{\cdots}$. As before, we call this source of variability the sum of squares between treatments:

$$SST = t \sum_i (\bar{y}_{i\cdot\cdot} - \bar{y}_{\cdots})^2.$$

Similarly, the second and third terms of the equation measure, respectively, the variability between rows and the variability between columns. These are designated by

$$SSR = t \sum_j (\bar{y}_{\cdot j\cdot} - \bar{y}_{\cdots})^2$$

$$SSC = t \sum_k (\bar{y}_{\cdot\cdot k} - \bar{y}_{\cdots})^2.$$

The final source of variability, designated as the sum of squares for error (SSE), represents all additional variability in the measurements not accounted for by rows, columns, or treatments.

The simplified computational formulas useful in the AOV of a Latin square are given here.

Shortcut Sums of Squares Formulas for a Latin Square Design

$$TSS = \sum_{ij} y_{ijk}^2 - \frac{y_{\cdots}^2}{t^2}$$

$$SST = \sum_i \frac{y_{i\cdot\cdot}^2}{t} - \frac{y_{\cdots}^2}{t^2}$$

$$SSR = \sum_j \frac{y_{\cdot j\cdot}^2}{t} - \frac{y_{\cdots}^2}{t^2}$$

$$SSC = \sum_k \frac{y_{\cdot\cdot k}^2}{t} - \frac{y_{\cdots}^2}{t^2}$$

$$SSE = TSS - SST - SSR - SSC$$

The model for a response in a Latin square design is the same as that for a randomized block design, with the addition of one more term to account for the second blocking variable. Thus,

model

$$y_{ijk} = \mu + \alpha_i + \beta_j + \gamma_k + \varepsilon_{ijk},$$

where the terms are defined as follows:

y_{ijk}:   the response on treatment i in row j and column k

μ: an overall mean; μ is a constant

α_i: an effect due to treatment i; α_i is a constant

β_j: an effect due to row j; β_j is a constant

γ_k: an effect due to column k; γ_k is a constant

ε_{ijk}:   a random error associated with the response for treatment i in row j and column k. We assume the ε_{ijk}s are normally distributed with mean 0 and unknown variance σ_ε^2. As before, the εs are assumed to be independent.

These assumptions for this model imply that y_{ijk}, the response for treatment i in row j and column k, is normally distributed with mean

$$E(y_{ijk}) = \mu + \alpha_i + \beta_j + \gamma_k$$

and variance of σ_ε^2.

filtering
    We can use the model to illustrate how a Latin square design **filters** out extraneous variability due to rows and columns. For purposes of illustration, we will consider the Latin square design shown in Table 15.11. If we wish to estimate $\alpha_3 - \alpha_1$, the difference in mean response for treatments 3 and 1, using the difference in sample means $\bar{y}_{3..} - \bar{y}_{1..}$, we can substitute into our model to obtain expressions for $\bar{y}_{3..}$ and $\bar{y}_{1..}$. If y_{ijk} denotes the observation in treatment i in row j and column k, we have, from Table 15.11,

$$\bar{y}_{1..} = \frac{1}{4}(y_{111} + y_{142} + y_{133} + y_{124})$$

$$= \mu + \alpha_1 + \frac{1}{4}(\beta_1 + \beta_2 + \beta_3 + \beta_4) + \frac{1}{4}(\gamma_1 + \gamma_2 + \gamma_3 + \gamma_4) + \bar{\varepsilon}_1,$$

where $\bar{\varepsilon}_1$ is the mean of the random errors for the 4 observations on treatment 1. Similarly,

$$\bar{y}_{3..} = \frac{1}{4}(y_{331} + y_{322} + y_{313} + y_{344})$$

$$= \mu + \alpha_3 + \frac{1}{4}(\beta_1 + \beta_2 + \beta_3 + \beta_4) + \frac{1}{4}(\gamma_1 + \gamma_2 + \gamma_3 + \gamma_4) + \bar{\varepsilon}_3.$$

Then the sample difference is

$$\bar{y}_{3..} - \bar{y}_{1..} = \alpha_3 - \alpha_1 + (\bar{\varepsilon}_3 - \bar{\varepsilon}_1)$$

and the error of estimation for $\alpha_3 - \alpha_1$ is $\bar{\varepsilon}_3 - \bar{\varepsilon}_1$.

If a randomized block design had been used with blocks representing rows, treatments would be randomized within the rows only. It is quite possible for the same treatment to appear more than once in the same column. Then the sample difference would be

$$\bar{y}_{3..} - \bar{y}_{1..} = \alpha_3 - \alpha_1 + [(\bar{\varepsilon}_3 - \bar{\varepsilon}_1) + \text{(column effects that do not cancel)}].$$

Thus, the error of estimation would be inflated by the column effects that do not cancel out. Following the same reasoning, if a completely randomized design was used when a Latin square design was appropriate, the error of estimation would be inflated by both row and column effects that do not cancel out.

test for treatment effects

We can **test specific hypotheses concerning the parameters in our model**. In particular, we may wish to test the hypothesis of no difference among the t treatment means. This hypothesis can be stated in the form

$$H_0: \quad \alpha_1 = \alpha_2 = \cdots = \alpha_t = 0 \text{ (i.e., the } t \text{ treatment means are identical)}.$$

The alternative hypothesis would be

$$H_a: \quad \text{At least one of the } \alpha_i\text{s differs from the rest (i.e., at least one treatment mean is different from the others).}$$

The test statistic for our test would be

$$F = \frac{\text{MST}}{\text{MSE}}.$$

For our model,

$$E(\text{MSE}) = \sigma_\varepsilon^2 \quad \text{and} \quad E(\text{MST}) = \sigma_\varepsilon^2 + t\theta_T.$$

Since it can be shown that θ_T is 0 under H_0 and is a positive function of the αs under H_a, we will reject H_0 in the upper tail of the F distribution. The appropriate degrees of freedom are obtained from the AOV table shown in Table 15.12.

SSE

We should note that we compute **SSE** and the degrees of freedom for SSE by subtraction. Thus, knowing the degrees of freedom for treatments, rows, and

TABLE 15.12

AOV Table for a $t \times t$ Latin Square Design

Source	SS	df	MS	F
Treatments	SST	$t-1$	MST $=$ SST/$(t-1)$	MST/MSE
Rows	SSR	$t-1$	MSR $=$ SSR/$(t-1)$	MSR/MSE
Columns	SSC	$t-1$	MSC $=$ SSC/$(t-1)$	MSC/MSE
Error	SSE	t^2-3t+2	MSE $=$ SSE/(t^2-3t+2)	
Totals	TSS	t^2-1		

columns, we can subtract this sum from $t^2 - 1$ to obtain the degrees of freedom for error.

Tests similar to that for treatments can be formulated for rows and columns. **test for rows** The **test for rows** is as follows:

$$H_0: \quad \beta_1 = \beta_2 = \cdots = \beta_t = 0 \text{ (i.e., no effect due to rows)}$$
$$H_a: \quad \text{at least one of the } \beta\text{s differs from } 0$$

$$\text{T.S.:} \quad F = \frac{\text{MSR}}{\text{MSE}}.$$

test for columns The **test for columns** is as follows:

$$H_0: \quad \gamma_1 = \gamma_2 = \cdots = \gamma_t = 0 \text{ (i.e., no effect due to columns)}$$
$$H_a: \quad \text{at least one of the } \gamma\text{s differs from } 0$$

$$\text{T.S.:} \quad F = \frac{\text{MSC}}{\text{MSE}}.$$

EXAMPLE 15.4

▼

A petroleum company was interested in comparing the miles per gallon achieved by four different gasoline blends (1, 2, 3, and 4). Because there can be considerable variability due to drivers and due to car models, these two extraneous sources of variability were included as "blocking" variables in the following Latin square design.

Each operator drove each car model over a standard course with the assigned gasoline blend by the Latin square design of Table 15.13. The miles-per-gallon data are shown in Table 15.14. Use the sample data in Table 15.14 to perform an analysis of variance. Make all appropriate tests using $\alpha = .05$.

Solution   Using Table 15.14, we find the treatment totals to be

$$y_{1..} = 94.3 \qquad y_{2..} = 100.2 \qquad y_{3..} = 72.2 \qquad y_{4..} = 89.4.$$

T A B L E   15.13
4 × 4 Latin Square Assignment for the Gasoline Blend Experiment of Example 15.4

Driver	Car Model			
	1	2	3	4
1	4	2	3	1
2	2	3	1	4
3	3	1	4	2
4	1	4	2	3

T A B L E   15.14

Sample Data for the Gasoline Blend Study of Example 15.4

Driver	Car Model								Totals
	1		2		3		4		
1	4	15.5	2	33.9	3	13.2	1	29.1	91.7
2	2	16.3	3	26.6	1	19.4	4	22.8	85.1
3	3	10.8	1	31.1	4	17.1	2	30.3	89.3
4	1	14.7	4	34.0	2	19.7	3	21.6	90.0
Totals		57.3		125.6		69.4		103.8	356.1

Thus,

$$\text{SST} = \sum_i \frac{y_{i\cdot\cdot}^2}{t} - \frac{y_{\cdot\cdot\cdot}^2}{t^2} = \frac{(94.3)^2 + (100.2)^2 + (72.2)^2 + (89.4)^2}{4} - \frac{(356.1)^2}{16}$$

$$= \frac{32{,}137.73}{4} - 7{,}925.45 = 108.98.$$

Similarly,

$$\text{SSR} = \sum_j \frac{y_{\cdot j\cdot}^2}{t} - \frac{y_{\cdot\cdot\cdot}^2}{t^2} = \frac{(91.7)^2 + \cdots + (90.0)^2}{4} - 7{,}925.45$$

$$= 7{,}931.35 - 7{,}925.45 = 5.9$$

$$\text{SSC} = \sum_k \frac{y_{\cdot\cdot k}^2}{t} - \frac{y_{\cdot\cdot\cdot}^2}{t^2} = \frac{(57.3)^2 + \cdots + (103.8)^2}{4} - 7{,}925.45$$

$$= 8{,}662.36 - 7{,}925.45 = 736.91$$

$$\text{TSS} = \sum_{ij} y_{ijk}^2 - \frac{y_{\cdot\cdot\cdot}^2}{t^2} = (15.5)^2 + (33.9)^2 + \cdots + (21.6)^2 - 7{,}925.45$$

$$= 8{,}801.05 - 7{,}925.45 = 875.6.$$

The sum of squares for error can be found by subtraction:

$$SSE = TSS - SST - SSR - SSC$$
$$= 875.6 - 108.98 - 5.9 - 736.91 = 23.81$$

The results of these calculations and F tests for treatments, rows, and columns can be summarized in an analysis of variance table, as shown in Table 15.15.

TABLE 15.15

AOV Table for the Data of Example 15.4

Source	SS	df	MS	F
Treatments (blends)	108.98	3	36.33	9.15
Rows (drivers)	5.90	3	1.97	0.50
Columns (car models)	736.91	3	245.64	61.87
Error	23.81	6	3.97	
Totals	875.6	15		

The F-value for $a = .05$, $df_1 = 3$, and $df_2 = 6$ is 4.76. Since the computed value of F for treatments and for columns exceeds 4.76, we conclude that there are significant differences among the blends and among the car models.

▲

As with the randomized block design, we can compare the efficiency of the Latin square design to that of the completely randomized design. Let MSE_{LS} and MSE_{CR} denote the mean square errors, respectively, for a Latin square design and a completely randomized design. The **relative efficiency** is

relative efficiency

$$\frac{MSE_{CR}}{MSE_{LS}} = \frac{MSR + MSC + (t-1)MSE}{(t+1)MSE}.$$

EXAMPLE 15.5

▼

Computer output is shown for the analysis of variance of Example 15.4. Compare the output to our hand calculations.

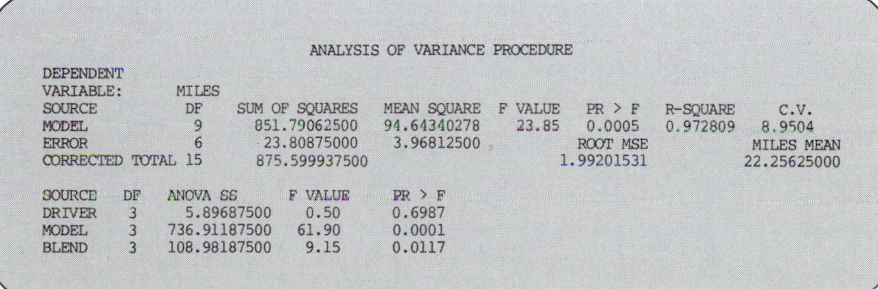

Solution   Except for minor rounding errors that we may have made in our hand calculations, the results of Example 15.4 agree with the output.   ▲

E X A M P L E   15.6   ▼

Refer to the data of Example 15.4. Compute the efficiency of the Latin square design relative to a completely randomized design.

Solution   For these data, $t = 4$, MSR = 1.97, MSC = 245.64, and MSE = 3.97. Substituting into the formula for relative efficiency, we have

$$\frac{\text{MSE}_{CR}}{\text{MSE}_{LS}} = \frac{1.97 + 245.64 + 3(3.97)}{5(3.97)} = 13.07.$$

That is, it would take approximately 13 times as many observations in using a completely randomized design to gather the same amount of information on the treatment means as it would take when using the Latin square design.

▲

▼ EXERCISES

Applications

🚜   **15.8**   An experiment was planned to compare two different fertilizer placements (broadcast, band) and two different rates of fertilizer flow on watermelon yields. Recent research has shown that broadcast application (scattering over the outer area) of fertilizer is superior to bands of fertilizer applied near the seed for watermelon yields. For this experiment the investigators wished to compare two nitrogen–phosphorus–potassium (broadcast and band) fertilizers applied at a rate of 160–70–135 pound/acre and two brands of micro-nutrients (A and B). These four combinations were to be studied in a Latin square field plot.

The treatments were randomly assigned according to a Latin square design conducted over a large farm plot, which was divided into rows and columns. A watermelon plant dry weight was obtained for each row–column combination 30 days after the emergence of the plants. These data are shown next.

			Column					
Row	1		2		3		4	
1	1	1.75	3	1.43	4	1.28	2	1.66
2	2	1.70	1	1.78	3	1.40	4	1.31
3	4	1.35	2	1.73	1	1.69	3	1.41
4	3	1.45	4	1.36	2	1.65	1	1.73

Treatment 1 broadcast, A   Treatment 3 band, A
Treatment 2 broadcast, B   Treatment 4 band, B

a. Write an appropriate statistical model for this experiment.

b. Use the data to run an analysis of variance. Give the *p*-value for each test and draw conclusions.

15.9 Refer to Exercise 15.8. In addition to obtaining 30-day-emergence dry weights, the watermelon yields (in tons per acre) were also recorded after the growing season. Use the data below to conduct an analysis of variance ($\alpha = .05$). Draw conclusions.

		Column		
Row	1	2	3	4
1	9.5	6.8	4.9	7.1
2	7.9	9.1	6.6	5.3
3	5.6	7.6	8.7	6.7
4	7.1	5.4	6.9	8.8

15.10 Refer to the data of Exercise 15.10.

a. Locate the analysis of variance in the output shown here.

b. Compare your results to those of Exercise 15.9.

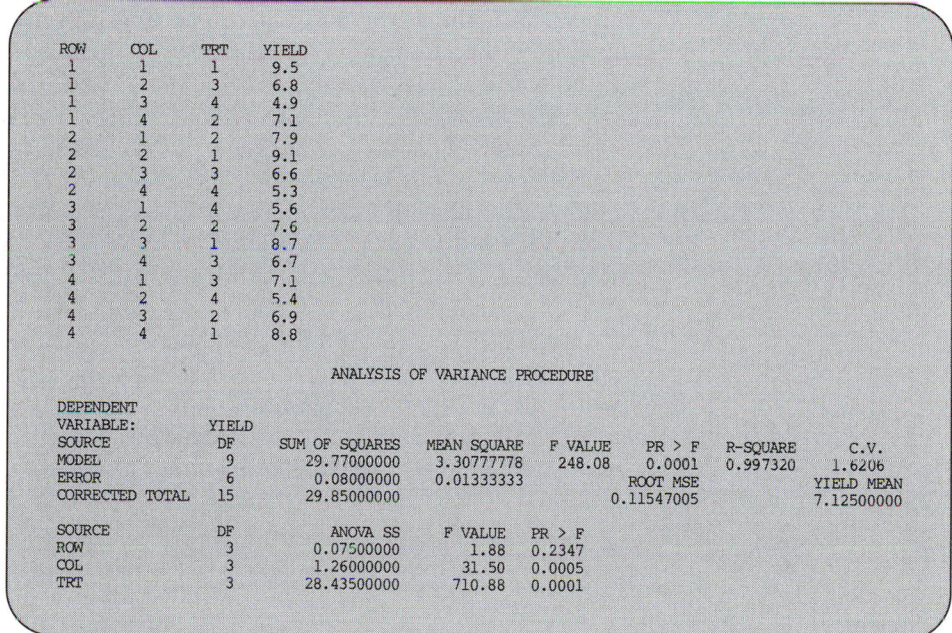

```
ROW     COL     TRT     YIELD
1       1       1       9.5
1       2       3       6.8
1       3       4       4.9
1       4       2       7.1
2       1       2       7.9
2       2       1       9.1
2       3       3       6.6
2       4       4       5.3
3       1       4       5.6
3       2       2       7.6
3       3       1       8.7
3       4       3       6.7
4       1       3       7.1
4       2       4       5.4
4       3       2       6.9
4       4       1       8.8
```

ANALYSIS OF VARIANCE PROCEDURE

DEPENDENT
VARIABLE: YIELD

SOURCE	DF	SUM OF SQUARES	MEAN SQUARE	F VALUE	PR > F	R-SQUARE	C.V.
MODEL	9	29.77000000	3.30777778	248.08	0.0001	0.997320	1.6206
ERROR	6	0.08000000	0.01333333			ROOT MSE	YIELD MEAN
CORRECTED TOTAL	15	29.85000000				0.11547005	7.12500000

SOURCE	DF	ANOVA SS	F VALUE	PR > F
ROW	3	0.07500000	1.88	0.2347
COL	3	1.26000000	31.50	0.0005
TRT	3	28.43500000	710.88	0.0001

15.11 Refer to Table 15.13. Derive the error of estimation for estimating $\alpha_4 - \alpha_2$ using $\bar{y}_{4\cdot\cdot} - \bar{y}_{2\cdot\cdot}$.

15.5 ▼ FACTORIAL EXPERIMENTS

In Section 15.2, we reviewed the completely randomized design for comparing $t \geq 2$ treatment means. Sections 15.3 and 15.4 were devoted to a discussion of the randomized block design and the Latin square design. Both designs are extensions of the completely randomized design; the randomized block design allows one to compare t treatment means in the presence of one extraneous (nuisance) source of variability, whereas the Latin square design allows the experimenter to control two such sources in order to compare treatment means. Suppose now that, rather than comparing t levels of a single factor, we wish to examine the effects of two or more independent variables on a response y. For example, suppose that we want to examine the effects of temperature x_1 and pressure x_2 on the bond strength y of a new adhesive product. Two major problems arise. First, we must consider the number of levels and the actual settings of these levels for each independent variable (factor). Second, having chosen the levels for each factor, we must choose the factor–level combinations (treatments) that will be applied to the experimental units.

The ability to choose appropriate settings for the independent variables depends a great deal on the experimenter's knowledge of the physical situation under study. Then, assuming the experimenter has chosen the levels of each independent variable, he or she faces the task of deciding which factor–level combinations should be assigned to the experimental units. For purposes of illustration, suppose that an experimenter is interested in examining the effects of two independent variables, nitrogen and phosphorus, on the yield of a crop. For simplicity we will assume that two levels have been selected for the study of each factor: 40 and 60 pounds per plot for nitrogen, 10 and 20 pounds per plot for phosphorus. For this study the experimental units are small, relatively homogeneous plots that have been partitioned from the acreage of a farm.

one-at-a-time approach

As discussed in Chapter 2, one approach for examining the effects of two or more factors on a response is the **one-at-a-time approach**. To examine the effect of a single variable, an experimenter varies the levels of this variable while holding the levels of the other independent variables fixed. This process is continued until the effect of each variable on the response has been examined while holding the other independent variables constant. For our experiment the factor–level combinations chosen might be as shown in Table 15.16. These factor–level combinations are illustrated in Figure 15.1.

T A B L E   15.16

Factor–Level Combinations for a One-at-a-Time Approach

Combination	Nitrogen	Phosphorus
1	60	10
2	40	10
3	40	20

FIGURE 15.1

Factor–Level Combinations for a One-at-a-Time Approach

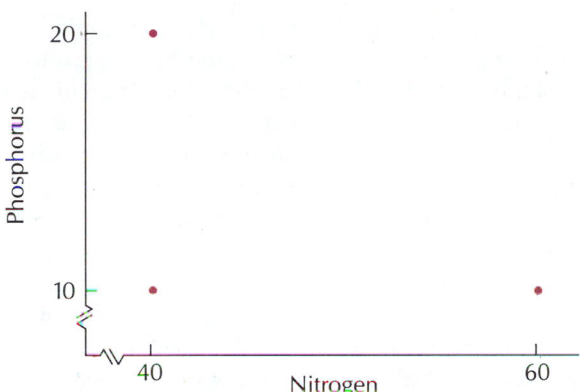

From the graph in Figure 15.1, we see that there is one difference that can be used to measure the effects of nitrogen and phosphorus separately. The difference in response for combinations 1 and 2 would estimate the effect of nitrogen; the difference in response for combinations 2 and 3 would estimate the effect of phosphorus.

Hypothetical yields corresponding to the three factor–level combinations of our experiment are given in Table 15.17. Suppose the experimenter is interested in using the sample information to determine the factor–level combination that will give the maximum yield. From the table, we see that crop yield increases when the nitrogen application is increased from 40 to 60 (holding phosphorus at 10). Yield also increases when the phosphorus setting is changed from 10 to 20 (at a fixed nitrogen setting of 40). Thus, it might seem logical to predict that increasing both the nitrogen and phosphorus applications to the soil will result in a larger crop yield. The fallacy in this argument is that our prediction is based on the assumption that the effect of one factor is the same for both levels of the other factor.

TABLE 15.17

Yields for the Three Factor–Level Combinations

Observation (yield)	Nitrogen	Phosphorus
145	60	10
125	40	10
160	40	20
?	60	20

We know from our investigation what happens to yield when the nitrogen application is increased from 40 to 60 for a phosphorus setting of 10. But will the yield also increase by approximately 20 units when the nitrogen application is changed from 40 to 60 at a setting of 20 for phosphorus?

To answer this question, we could apply the factor–level combination of 60 nitrogen–20 phosphorus to another experimental plot and observe the crop yield. If the yield is 180, then the information obtained from the three factor–level combinations would be correct and would have been useful in predicting the factor–level combination that produces the greatest yield. However, suppose the yield obtained from the high settings of nitrogen and phosphorus turns out to be 110. If this happens, the two factors nitrogen and phosphorus are said to **interaction**   **interact**. That is, the effect of one factor on the response does not remain the same for different levels of the second factor, and the information obtained from the one-at-a-time approach would lead to a faulty prediction.

The two outcomes just discussed for the crop yield at 60–20 setting are displayed in Figure 15.2, along with the yields at the three initial design points. Figure 15.2 (top) illustrates a situation with no interaction between the two factors. The effect of nitrogen on yield is the same for both levels of phosphorus. In contrast, Figure 15.2 (bottom) illustrates a case in which the two factors nitrogen and phosphorus do interact.

FIGURE 15.2

Yields of the Three Design Points and Possible Yield at a Fourth Design Point

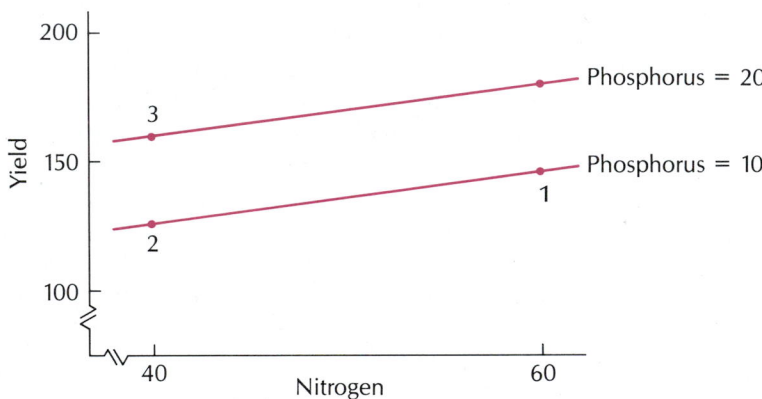

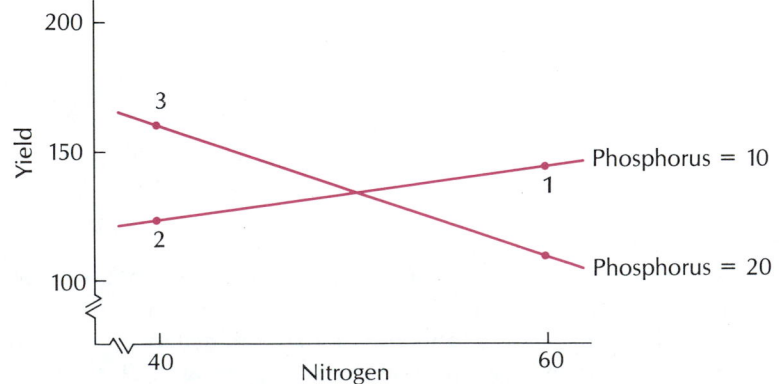

We have seen that the one-at-a-time approach to investigating the effect of two factors on a response is suitable only for situations in which the two factors do not interact. Although this was illustrated for the simple case in which two factors were to be investigated at each of two levels, the inadequacies of a one-at-a-time approach are even more salient when trying to investigate the effects of more than two factors on a response.

factorial experiment

Factorial experiments are useful for examining the effects of two or more factors on a response y, whether or not interaction exists. As before, the choice of the number of levels of each variable and the actual settings of these variables is important. But assuming we have made these selections with help from an investigator knowledgeable in the area being examined, we must decide at what factor–level combinations we will observe y.

Classically, factorial experiments have not been referred to as designs because they deal with the choice of levels and the selection of factor–level combinations (treatments) rather than with how the treatments are assigned to experimental units. Unless otherwise specified, we will assume that treatments are assigned to experimental units at random. The factorial–level combinations will then correspond to the "treatments" of a completely randomized design.

DEFINITION 15.3
Factorial Experiment

▼

A **factorial experiment** is an experiment in which the response y is observed at all factor–level combinations of the independent variables.

Using our previous example, if we are interested in examining the effect of two levels of nitrogen x_1 at 40 and 60 pounds per plot and two levels of phosphorus x_2 at 10 and 20 pounds per plot on the yield of a crop, we could use a completely randomized design where the four factor–level combinations (treatments) of Table 15.18 are assigned at random to the experimental units.

Similarly, if we wished to examine x_1 at the two levels 40 and 60 and x_2 at the three levels 10, 15, and 20, we could use the six factor–level combinations of Table 15.19 as treatments in a completely randomized design.

TABLE 15.18
2 × 2 Factorial Experiment for Crop Yield

Factor–Level x_1	Combinations x_2
40	10
40	20
60	10
60	20

TABLE 15.19
2 × 3 Factorial
Experiment for Crop Yield

Factor–Level	Combinations
x_1	x_2
40	10
40	15
40	20
60	10
60	15
60	20

EXAMPLE 15.7 ▼

An auto manufacturer is interested in examining the effect of engine speed x_1, measured in revolutions per minute, and ground speed x_2, measured in miles per hour, on gasoline mileage. The investigators, in consultation with company mechanics and other personnel, decided to consider settings of x_1 at 800, 1,000, and 1,200 and settings of x_2 at 30, 50, and 70. Give the factor–level combinations to be used in a 3 × 3 factorial experiment.

Solution Using the definition of factorial experiment, we would observe gasoline mileage at the following settings of x_1 and x_2:

x_1	800	800	800	1,000	1,000	1,000	1,200	1,200	1,200
x_2	30	50	70	30	50	70	30	50	70

▲

The examples of factorial experiments presented in this section have concerned two independent variables. However, the procedure applies to any number of factors and levels per factor. Thus, if we had four different factors x_1, x_2, x_3, and x_4 at two, three, three, and four levels, respectively, we could formulate a 2 × 3 × 3 × 4 factorial experiment by considering all $2 \cdot 3 \cdot 3 \cdot 4 = 72$ factor–level combinations.

One final comparison should be made between the one-at-a-time approach and a factorial experiment. Not only do we get information concerning factor interactions using a factorial experiment, but also, when there are no interactions, we get at least the same amount of information about the effects of each individual factor using fewer observations. To illustrate this idea, let us consider the 2 × 2 factorial experiment with nitrogen and phosphorus. If there is no interaction between the two factors, the data appear as shown in Figure 15.3(a). For convenience, the data are reproduced in Table 15.20, with the four sample combinations designated by the numbers 1 through 4. If a 2 × 2 factorial experiment is used and no interaction exists between the two factors, we can obtain two independent differences to examine the effects of each of the factors on the response. Thus, from Table 15.20, the differences between observations 1

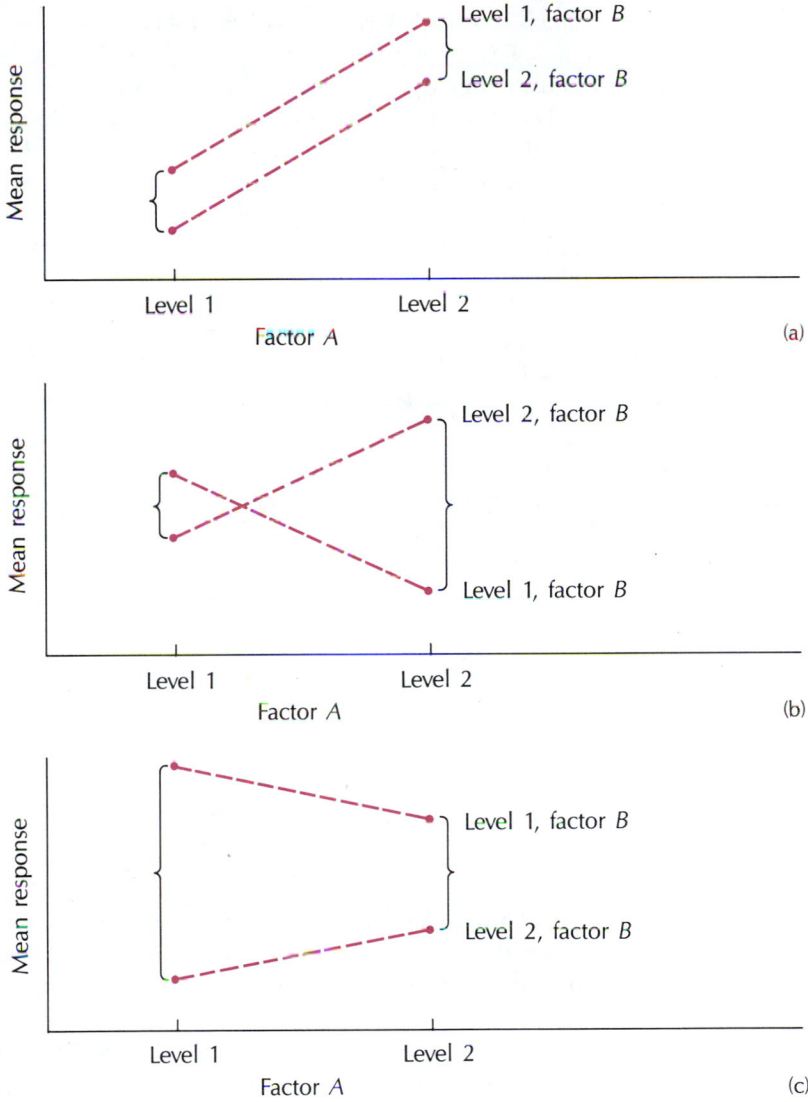

FIGURE 15.3
Illustrations of the Absence and Presence of Interaction in a 2 × 2 Factorial Experiment: (a) Factors *A* and *B* do not interact; (b) factors *A* and *B* interact; (c) factors *A* and *B* interact

TABLE 15.20
Factor–Level Combinations for a 2 × 2 Factorial Experiment

Combination	Yield	Nitrogen	Phosphorus
1	145	60	10
2	125	40	10
3	165	40	20
4	180	60	20

and 4 and the difference between observations 2 and 3 would be used to measure the effect of phosphorus. Similarly, the difference between observations 4 and 3 and the difference between observations 2 and 1 would be used to measure the effect of the two levels of nitrogen on plot yield.

If we employed a one-at-a-time approach for the same experimental situation, it would take six observations (two observations at each of the three initial factor–level combinations shown in Table 15.20) to obtain the same number of independent differences for examining the separate effects of nitrogen and phosphorus when no interaction is present.

The model for a two-factor factorial experiment with no interaction, such as the 2×2 factorial experiment with nitrogen and phosphorus, is

$$y_{ij} = \mu + \alpha_i + \beta_j + \varepsilon_{ij},$$

where the α_i and β_j are constants and the random errors ε_{ij} are independent, normally distributed with mean 0 and variance σ_ε^2. Here, y_{ij} denotes an observation taken at the ith level of factor A and the jth level of factor B. Since terms included in the model are added to one another, the model is sometimes referred **additive model**   to as an **additive model**. Expected values for a 2×2 factorial experiment are shown in Table 15.21.

T A B L E   15.21
Expected Values for a
2×2 Factorial Experiment

	Factor B	
Factor A	Level 1	Level 2
Level 1	$\mu + \alpha_1 + \beta_1$	$\mu + \alpha_1 + \beta_2$
Level 2	$\mu + \alpha_2 + \beta_1$	$\mu + \alpha_2 + \beta_2$

This model assumes that difference in population means (expected values) for any two levels of factor A is the same no matter what level of B we're considering. The same property holds when comparing two levels of factor B. For example, the difference in mean response for levels 1 and 2 of factor A is the same value, $\alpha_1 - \alpha_2$, no matter what level of factor B we are considering. Thus, a test for no differences among the two levels of factor A would be of the form H_0: $\alpha_1 - \alpha_2 = 0$. Similarly, the difference between levels of factor B is $\beta_1 - \beta_2$ for either level of factor A, and a test of no difference between the factor B means is H_0: $\beta_1 - \beta_2 = 0$. This phenomenon was also noted for the randomized block design.

interaction   If the assumption of additivity of terms in the model does not hold, then we may need another model that employs terms to account for **interaction**. Consider a two-factor factorial experiment with n observations per factor–level combination (treatment). For this experimental situation, the model is

$$y_{ijk} = \mu + \alpha_i + \beta_j + \alpha\beta_{ij} + \varepsilon_{ijk},$$

where y_{ijk} is the response obtained for the kth observation at the ith level of factor A and jth level of factor B and $\alpha\beta_{ij}$ is an effect due to the ith level of A and jth level of B*. The expected values for a 2×2 factorial experiment with n observations per cell are presented in Table 15.22.

TABLE 15.22
Expected Values for a 2×2 Factorial Experiment, with Replications

Factor A	Factor B	
	Level 1	Level 2
Level 1	$\mu + \alpha_1 + \beta_1 + \alpha\beta_{11}$	$\mu + \alpha_1 + \beta_2 + \alpha\beta_{12}$
Level 2	$\mu + \alpha_2 + \beta_1 + \alpha\beta_{21}$	$\mu + \alpha_2 + \beta_2 + \alpha\beta_{22}$

As can be seen from Table 15.22, the difference in mean response for levels 1 and 2 of factor A on level 1 of factor B is

$$(\alpha_1 - \alpha_2) + (\alpha\beta_{11} - \alpha\beta_{21}),$$

but for level 2 of factor B this difference is

$$(\alpha_1 - \alpha_2) + (\alpha\beta_{12} - \alpha\beta_{22}).$$

Since the difference in mean response for levels 1 and 2 of factor A is *not* the same for different levels of factor B, the model is no longer additive, and we say that the two factors interact.

DEFINITION 15.4
Interact

Two factors A and B are said to **interact** if the difference in mean responses for two levels of one factor is not constant across levels of the second factor.

In measuring the octane rating of gasoline, interaction can occur when two components of the blend are combined to form a gasoline mixture. The octane properties of the blended mixture may be quite different than would be expected by examining each component of the mixture. Interaction in this situation could have a positive or negative effect on the performance of the blend, in which case the components are said to potentiate, or antagonize, one another.

---

* It is convenient to use the notation $\alpha\beta_{ij}$ to denote a new term for the model. It does not, however, represent the multiplication of two terms α_i and β_j.

profile plot

We can amplify the notion of an interaction with the **profile plots** shown in Figure 15.3. As we see from Figure 15.3(a), when no interaction is present, the difference in the mean response between levels 1 and 2 of factor B (as indicated by the braces) is the same for both levels of factor A. However, for the two illustrations in Figure 15.3(b) and (c), we see that the difference between the levels of factor B changes from level 1 to level 2 of factor A. For these cases, we have an interaction between the two factors.

It should be noted that an interaction is not restricted to two factors. With three factors A, B, and C, we might have an interaction between factors A and B, A and C, and B and C, and the two-factor interactions would have interpretations that follow immediately from Definition 15.4. Thus, the presence of an AC interaction indicates that the difference in mean response for levels of factor A varies across levels of factor C. A three-way interaction between factors A, B, and C might indicate that the difference in mean response for levels of C changes across combinations of levels for factors A and B.

In general, the analysis of variance table for a three-factor factorial experiment depends on whether we have $n > 1$ observations per cell. Consider a three-factor experiment with a levels of factor A, b levels of factor B, c levels of factor C, and n observations per cell. Before presenting these tables, we need the notation defined here.

Notation Needed for the AOV of a Three-Factor Factorial Experiment with $n > 1$ Observations per Cell

▼

y_{ijkm}: mth observation at the ith level of factor A, jth level of factor B, and kth level of factor C

$y_{ijk\cdot}$ $(\bar{y}_{ijk\cdot})$:   sum (mean) of the observations at the ith level of factor A, jth level of factor B, and kth level of factor C; $y_{ijk\cdot} = \sum_m y_{ijkm}$ and $\bar{y}_{ijk\cdot} = y_{ijk\cdot}/n$

$y_{ij\cdot\cdot}$ $(\bar{y}_{ij\cdot\cdot})$:   sum (mean) of the observations at the ith level of factor A and jth level of factor B; $y_{ij\cdot\cdot} = \sum_{k,m} y_{ijkm}$ and $\bar{y}_{ij\cdot\cdot} = y_{ij\cdot\cdot}/cn$

$y_{i\cdot k\cdot}$ $(\bar{y}_{i\cdot k\cdot})$:   sum (mean) of the observations at the ith level of factor A and kth level of factor C; $y_{i\cdot k\cdot} = \sum_{j,m} y_{ijkm}$ and $\bar{y}_{i\cdot k\cdot} = y_{i\cdot k\cdot}/bn$

$y_{\cdot jk\cdot}$ $(\bar{y}_{\cdot jk\cdot})$:   sum (mean) of the observations at the jth level of factor B and kth level of factor C; $y_{\cdot jk\cdot} = \sum_{i,m} y_{ijkm}$ and $\bar{y}_{\cdot jk\cdot} = y_{\cdot jk\cdot}/an$

$y_{i\cdots}$ $(\bar{y}_{i\cdots})$:   sum (mean) of the observations at the ith level of factor A; $y_{i\cdots} = \sum_{jkm} y_{ijkm}$ and $\bar{y}_{i\cdots} = y_{i\cdots}/bcn$

(continues)

$y_{\cdot j \cdot \cdot}$ ($\bar{y}_{\cdot j \cdot \cdot}$):   sum (mean) of the observations at the jth level of factor B;

$$y_{\cdot j \cdot \cdot} = \sum_{ikm} y_{ijkm} \text{ and } \bar{y}_{\cdot j \cdot \cdot} = y_{\cdot j \cdot \cdot}/acn$$

$y_{\cdot \cdot k \cdot}$ ($\bar{y}_{\cdot \cdot k \cdot}$):   sum (mean) of the observations at the kth level of factor C;

$$y_{\cdot \cdot k \cdot} = \sum_{ijm} y_{ijkm} \text{ and } \bar{y}_{\cdot \cdot k \cdot} = y_{\cdot \cdot k \cdot}/abn$$

$y_{\cdot \cdot \cdot \cdot}$ ($\bar{y}_{\cdot \cdot \cdot \cdot}$):   sum (mean) of all the observations; $y_{\cdot \cdot \cdot \cdot} = \sum_{ijkm} y_{ijkm}$ and

$$\bar{y}_{\cdot \cdot \cdot \cdot} = y_{\cdot \cdot \cdot \cdot}/abcn$$

partitioning sums of squares

The appropriate AOV formulas for a three-factor factorial experiment with n observations per cell can be subdivided into sums of squares for main effects (variability between levels of a single factor), two-way interactions, and three-way interactions.

main effects

The sums of squares for **main effects** are

$$SSA = nbc \sum_i (\bar{y}_{i \cdot \cdot \cdot} - \bar{y}_{\cdot \cdot \cdot \cdot})^2 = \sum_i \frac{y_{i \cdot \cdot \cdot}^2}{nbc} - \frac{y_{\cdot \cdot \cdot \cdot}^2}{nabc}$$

$$SSB = nac \sum_j (\bar{y}_{\cdot j \cdot \cdot} - \bar{y}_{\cdot \cdot \cdot \cdot})^2 = \sum_j \frac{y_{\cdot j \cdot \cdot}^2}{nac} - \frac{y_{\cdot \cdot \cdot \cdot}^2}{nabc}$$

$$SSC = nab \sum_k (\bar{y}_{\cdot \cdot k \cdot} - \bar{y}_{\cdot \cdot \cdot \cdot})^2 = \sum_k \frac{y_{\cdot \cdot k \cdot}^2}{nab} - \frac{y_{\cdot \cdot \cdot \cdot}^2}{nabc}.$$

two-way interactions

The sums of squares for **two-way interactions** are

$$SSAB = nc \sum_{ij} (\bar{y}_{ij \cdot \cdot} - \bar{y}_{i \cdot \cdot \cdot} - \bar{y}_{\cdot j \cdot \cdot} + \bar{y}_{\cdot \cdot \cdot \cdot})^2 = \sum_{ij} \frac{y_{ij \cdot \cdot}^2}{nc} - \frac{y_{\cdot \cdot \cdot \cdot}^2}{nabc} - SSA - SSB$$

$$SSAC = nb \sum_{ik} (\bar{y}_{i \cdot k \cdot} - \bar{y}_{i \cdot \cdot \cdot} - \bar{y}_{\cdot \cdot k \cdot} + \bar{y}_{\cdot \cdot \cdot \cdot})^2 = \sum_{ik} \frac{y_{i \cdot k \cdot}^2}{nb} - \frac{y_{\cdot \cdot \cdot \cdot}^2}{nabc} - SSA - SSC$$

$$SSBC = na \sum_{jk} (\bar{y}_{\cdot jk \cdot} - \bar{y}_{\cdot j \cdot \cdot} - \bar{y}_{\cdot \cdot k \cdot} + \bar{y}_{\cdot \cdot \cdot \cdot})^2 = \sum_{jk} \frac{y_{\cdot jk \cdot}^2}{na} - \frac{y_{\cdot \cdot \cdot \cdot}^2}{nabc} - SSB - SSC.$$

three-way interaction

The sum of squares for the **three-way interaction** is

$$SSABC = n \sum_{ijk} (\bar{y}_{ijk\cdot} - \bar{y}_{ij\cdot\cdot} - \bar{y}_{i\cdot k\cdot} - \bar{y}_{\cdot jk\cdot} + \bar{y}_{i\cdot\cdot\cdot} + \bar{y}_{\cdot j\cdot\cdot} + \bar{y}_{\cdot\cdot k\cdot} - \bar{y}_{\cdot\cdot\cdot\cdot})^2$$

$$= \sum_{ijk} \frac{y_{ijk\cdot}^2}{n} - \frac{y_{\cdot\cdot\cdot\cdot}^2}{nabc} - SSAB - SSAC - SSBC - SSA - SSB - SSC.$$

These same formulas can be modified for a two-factor experiment with n observations per cell. For this situation, we eliminate factor C from all consideration and obtain the following formulas:

$$TSS = \sum_{ijk} y_{ijk}^2 - \frac{y_{\cdot\cdot\cdot}^2}{nab}$$

$$SSA = nb \sum_{i} (\bar{y}_{i\cdot\cdot} - \bar{y}_{\cdot\cdot\cdot})^2 = \sum_{i} \frac{y_{i\cdot\cdot}^2}{nb} - \frac{y_{\cdot\cdot\cdot}^2}{nab}$$

$$SSB = na \sum_{j} (\bar{y}_{\cdot j\cdot} - \bar{y}_{\cdot\cdot\cdot})^2 = \sum_{j} \frac{y_{\cdot j\cdot}^2}{na} - \frac{y_{\cdot\cdot\cdot}^2}{nab}$$

$$SSAB = n \sum_{ij} (\bar{y}_{ij\cdot} - \bar{y}_{i\cdot\cdot} - \bar{y}_{\cdot j\cdot} + \bar{y}_{\cdot\cdot\cdot})^2 = \sum_{ij} \frac{y_{ij\cdot}^2}{n} - \frac{y_{\cdot\cdot\cdot}^2}{nab} - SSA - SSB.$$

Note the similarities and patterns for the two-factor and three-factor sums of squares formulas. These same patterns and similarities apply for more than three-factor experiments. Fortunately, however, we can use the output from computer software packages to obtain these sums of squares, when needed.

We illustrate the AOV table for a three-factor factorial experiment with a levels of A, b levels of B, c levels of C, and $n = 1$ observation per cell. The AOV table is given in Table 15.23.

T A B L E   15.23
AOV for a Three-Factor Factorial Experiment Without Replication

Source	SS	df	MS
Main effects			
A	SSA	$a - 1$	MSA = SSA/$(a - 1)$
B	SSB	$b - 1$	MSB = SSB/$(b - 1)$
C	SSC	$c - 1$	MSC = SSC/$(c - 1)$
Interactions			
AB	SSAB	$(a - 1)(b - 1)$	MSAB = SSAB/$(a - 1)(b - 1)$
AC	SSAC	$(a - 1)(c - 1)$	MSAC = SSAC/$(a - 1)(c - 1)$
BC	SSBC	$(b - 1)(c - 1)$	MSBC = SSBC/$(b - 1)(c - 1)$
ABC	SSABC	$(a - 1)(b - 1)(c - 1)$	MSABC = SSABC/$(a - 1)(b - 1)(c - 1)$
Totals	TSS	$abc - 1$	

The total sum of squares formulas is as before:

$$\text{TSS} = \sum_{ijk} y_{ijk}^2 - \frac{y_{...}^2}{abc}.$$

You will note, however, that there is no source of variability designated as error since there are no more degrees of freedom. This is true for any factorial with $n = 1$ observation per cell.

F tests

To construct **F tests** for the sources of variability listed in the AOV table, we must do one of two things:

1. Assume that one or more of the higher-order interactions are negligible. The affected sums of squares are then combined to form an error sum of squares. (Note that this is why we illustrated the 2×2 factorial [with $n = 1$] using an additive model.)
2. Replicate the experiment to generate additional degrees of freedom for error.

E X A M P L E 15.8

▼

An experiment was conducted to determine the effects of four different pesticides on the yield of fruit from three different varieties (B_1, B_2, B_3) of a citrus tree. Four trees from each variety were randomly selected from an orchard. The four pesticides were then randomly assigned to trees of a particular variety and applications were made according to recommended levels. Yields of fruit, in bushels per tree, were obtained after the test period. These data appear in Table 15.24. Set up an analysis of variance table, computing all the sums of squares and the mean squares. Use $\alpha = .05$ for all F tests.

Solution The experiment just described is a 3×4 factorial with $n = 1$ observation per cell. Factor A, pesticides, is investigated at $a = 4$ levels and

T A B L E 15.24
Data for Example 15.8; Yield of Fruit (in bushels per tree)

Variety, *B*	Pesticide, *A*				Totals
	1	2	3	4	
1	29	50	43	53	175
2	41	58	42	73	214
3	66	85	69	85	305
Totals	136	193	154	211	694

factor B, varieties, at $b = 3$ levels. The sources of variability and the corresponding degrees of freedom are shown next.

Source	df
Pesticides, A	3
Varieties, B	2
AB	6
Total	11

Using the AOV formulas, we have, from Table 15.24:

$$\frac{y_{...}^2}{nab} = \frac{(694)^2}{12} = 40,136.33$$

$$SSA = \sum_i \frac{y_{i..}^2}{nb} - \frac{y_{...}^2}{nab}$$

$$= \frac{(136)^2 + (193)^2 + (154)^2 + (211)^2}{3} - 40,136.33$$

$$= 41,327.33 - 40,136.33 = 1,191$$

$$SSB = \sum_j \frac{y_{.j.}^2}{na} - \frac{y_{...}^2}{nab} = \frac{(175)^2 + (214)^2 + (305)^2}{4} - 40,136.33$$

$$= 42,361.5 - 40,136.33 = 2,225.17$$

$$TSS = \sum_{ijk} y_{ijk}^2 - \frac{y_{...}^2}{nab} = (29)^2 + (50)^2 + \cdots + (85)^2 - 40,136.33$$

$$= 43,704 - 40,136.33 = 3,567.67.$$

The formula for the AB interaction is

$$SSAB = \sum_{ij} \frac{y_{ij.}^2}{n} - \frac{y_{...}^2}{nab} - SSA - SSB.$$

For $n = 1$ observation at each AB combination, we have

$$\sum_{ij} \frac{y_{ij.}^2}{n} = (29)^2 + (50)^2 + \cdots + (85)^2 = 43,704.$$

Then

$$SSAB = 43{,}704 - 40{,}136.33 - 1{,}191 - 2{,}225.17 = 151.50.$$

This same result could be obtained by subtraction as

$$SSAB = TSS - SSA - SSB = 3{,}567.67 - 1{,}191 - 2{,}225.17 = 151.50.$$

The analysis of variance table is then as shown in Table 15.25.

T A B L E 15.25

AOV Table for the Fruit Yield Experiment of Example 15.8

Source	SS	df	MS
Pesticides, A	1,191.00	3	397
Varieties, B	2,225.17	2	1,112.59
AB	151.50	6	25.25
Totals	3,567.67	11	

As can be seen from this AOV table, we have no degrees of freedom remaining for error.

If we are willing to assume that the AB interaction is negligible, we can designate the AB interaction source as error and use SSAB as SSE. For our data, if we are willing to assume that the difference in mean yield for the two pesticides remains constant for all varieties, the AOV table would be as shown in Table 15.26.

T A B L E 15.26

AOV Table for the Fruit Yield Experiment of Example 15.8, with the AB Interaction Designated as Error

Source	SS	df	MS	F
A	1,191.00	3	397.00	$397/25.25 = 15.72$
B	2,225.17	2	1,112.59	$1{,}112.59/25.25 = 44.06$
Error	151.50	6	25.25	
Totals	3,567.67	11		

To test for no difference among pesticides, with $\alpha = .05$, we use the test statistic

$$F = \frac{MSA}{MSE} = \frac{397.00}{25.25} = 15.72.$$

Since the computed value of F exceeds 4.76, the tabulated value for $a = .05$, $df_1 = 3$, and $df_2 = 6$, we reject H_0 and conclude that there are differences among the mean yields for the four pesticides.

Similarly, to test for no difference among varieties, we use

$$F = \frac{\text{MSB}}{\text{MSE}} = 44.06.$$

Based on $\alpha = .05$, the computed value of F exceeds the tabulated value of 5.14 for $a = .05$, $df_1 = 2$, and $df_2 = 6$, and we conclude that there are differences among the mean yields for varieties.   ▲

EXAMPLE 15.9   ▼

Refer to Example 15.8. Suppose that before beginning the study, the experimenter realized that it cannot be assumed that the AB interaction was negligible in order to perform F tests in an analysis of variance; so the experimenter decided to use 24 citrus trees (8 of each variety) and randomly assigned 2 trees to each factor–level combination.

a. Construct a plot of the data, sometimes called a profile plot.
b. Write an appropriate model.
c. Use the yield data of Table 15.27 to conduct an analysis of variance. Use $\alpha = .05$ for all F tests.

TABLE 15.27
Data for the 3 × 4 Factorial Experiment of Fruit Yield, Example 15.9, with $n = 2$ Observations per Cell

Variety, B	Pesticide, A				Totals
	1	2	3	4	
1	49	50	43	53	375
	39	55	38	48	
2	55	67	53	85	474
	41	58	42	73	
3	66	85	69	85	626
	68	92	62	99	
Totals	318	407	307	443	1,475

Solution

a. To construct a profile plot we need the $\bar{y}_{ij\cdot}$, the sample means for the factor–level combinations of A and B. These are given in Table 15.28. Using these sample means it is easy to form the profile plot shown in Figure 15.4.
b. The profile plot of part a does not indicate much evidence of an interaction between factors A and B since the lines are essentially parallel. However, since we have two observations per factor–level combination, we'll

TABLE 15.28
Sample Means for Factor–Level Combinations of A and B

	Pesticide, A			
Variety, B	1	2	3	4
1	44	52.5	40.5	50.5
2	48	62.5	47.5	79
3	67	88.5	65.5	92

FIGURE 15.4
Profile Plot for Example 15.9

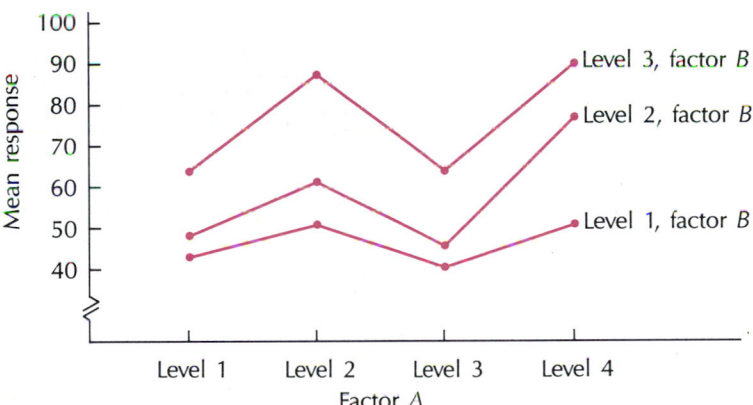

include interaction terms in the model and run the test for interaction as part of the analysis of variance. From the profile plot, there is a suggestion of an effect due to factor B (since the factor B line segments have different heights) and of an effect due to factor A (since the line segments are not horizontal). The model is

$$y_{ijk} = \mu + \alpha_i + \beta_j + \alpha\beta_{ij} + \varepsilon_{ijk}.$$

c. The sums of squares for an analysis of variance make use of the data totals shown in Tables 15.27 and 15.29.

$$SSA = \sum_i \frac{y_{i\cdot\cdot}^2}{nb} - \frac{y_{\cdots}^2}{nab} = \frac{(318)^2 + (407)^2 + (307)^2 + (443)^2}{6} - \frac{(1{,}475)^2}{24}$$

$$= 92{,}878.5 - 90{,}651.04 = 2{,}227.46$$

$$SSB = \sum_j \frac{y_{\cdot j\cdot}^2}{na} - \frac{y_{\cdots}^2}{nab} = \frac{(375)^2 + (474)^2 + (626)^2}{8} - 90{,}651.04$$

$$= 94{,}647.13 - 90{,}651.04 = 3{,}996.09.$$

T A B L E 15.29

y_{ij} **Totals for the Data of Table 15.27**

	Pesticide, A				
Variety, B	1	2	3	4	Totals
1	88	105	81	101	375
2	96	125	95	158	474
3	134	177	131	184	626
Totals	318	407	307	443	1,475

To compute the AB interaction, we must calculate the totals y_{ij} for all cells of the table. These are given in Table 15.29.

Using the data of Table 15.29 we compute SSAB.

$$
\begin{aligned}
\text{SSAB} &= \sum_{ij} \frac{y_{ij}^2}{n} - \frac{y_{...}^2}{nab} - \text{SSA} - \text{SSB} \\
&= \frac{(88)^2 + (105)^2 + \cdots + (184)^2}{2} - 90{,}651.04 \\
&\quad - 2{,}227.46 - 3{,}996.09 \\
&= 97{,}331.5 - 90{,}651.04 - 2{,}227.46 - 3{,}996.09 = 456.91.
\end{aligned}
$$

From Table 15.27, the total sum of squares is computed as

$$
\begin{aligned}
\text{TSS} &= \sum_{ijk} y_{ijk}^2 - \frac{y_{...}^2}{nab} = (49)^2 + (39)^2 + (50)^2 + \cdots + (85)^2 \\
&\quad + (99)^2 - 90{,}651.04 \\
&= 97{,}839 - 90{,}651.04 = 7{,}187.96.
\end{aligned}
$$

The sum of squares and degrees of freedom for error can be obtained by subtraction:

$$
\begin{aligned}
\text{SSE} &= \text{TSS} - \text{SSA} - \text{SSB} - \text{SSAB} \\
&= 7{,}187.96 - 2{,}227.46 - 3{,}996.09 - 456.91 = 507.50 \\
\text{df} &= 23 - 3 - 2 - 6 = 12
\end{aligned}
$$

The analysis of variance table for this 3×4 factorial experiment with $n = 2$ observations per cell is given in Table 15.30.

TABLE 15.30

AOV Table for the Fruit Yield Experiment of Example 15.9

Source	SS	df	MS	F
A	2,227.46	3	742.49	742.49/42.29 = 17.56
B	3,996.09	2	1,998.05	1,998.05/42.29 = 47.25
AB	456.91	6	76.15	76.15/42.29 = 1.80
Error	507.50	12	42.29	

The first test of significance would be to test for no interaction between factors A and B. The F statistic is

$$F = \frac{\text{MSAB}}{\text{MSE}} = \frac{76.15}{42.29} = 1.80.$$

The computed value of F does not exceed the tabulated value of 3.00 for $a = .05$, $df_1 = 6$, and $df_2 = 12$. Hence, we have insufficient evidence to indicate an interaction between A and B; this confirms what we saw in the profile plot. We then proceed as in the AOV table and test separately for differences among pesticides and among varieties. Comparing the computed F-values to the appropriate critical value from the tables, we find a significant effect due to both factors. The patterns of responses are displayed in the profile plot. ▲

significant interaction

The results of an F test for main effects for A or B must be interpreted very carefully in the presence of a **significant interaction**. The first thing we would do is to construct a profile plot using the sample means. If the profile plot for Example 15.9 had appeared as shown in Figure 15.5, there would have been an indication of an interaction between factors A and B (pesticides and varieties). Indeed, the F test for interaction would undoubtedly have been significant.

FIGURE 15.5

Significant Interaction Present; Tests on Main Effects Are Appropriate

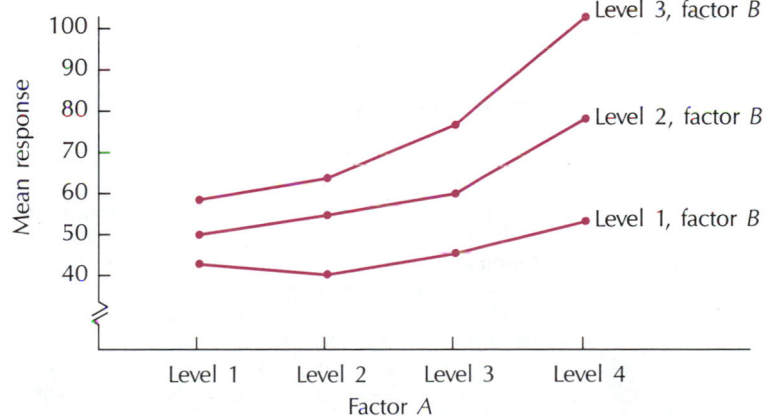

Would F tests for main effects have been appropriate for the profile plot of Figure 15.5? The answer is yes because this is an *orderly* interaction; the *order* of the means for levels of factor B is always the same even though the *magnitude* of the differences between levels of factor B may change from level to level of factor A. Clearly, the profile plot in Figure 15.5 shows that the level 3 mean of factor B is always larger than the means for levels 2 and 1. Similarly, the level 2 mean for B is always larger than the mean for level 1, factor B.

When the interaction is orderly, a test on main effects can be meaningful; however, not all interactions are so well behaved. The profile in Figure 15.6 shows a situation in which a test of main effects in the presence of interaction might be misleading. A *disorderly* interaction, such as in Figure 15.6, can obscure main effects. It's not that tests on main effects are inappropriate, it's that they might not show the effects when present.

FIGURE 15.6

Significant Interaction Present; Tests on Main Effects Are Inappropriate

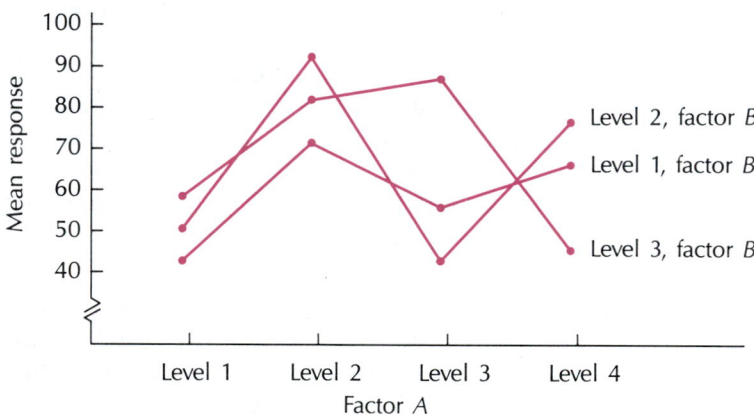

▼ **EXERCISES**

Applications

$ 15.12 A national trade association asked relocation experts from four different companies to estimate the total transfer costs for the same ten hypothetical employees in order to generate data for input into proposed revisions of corporate tax structures. These data ($1,000) are shown here:

	Employee									
Company	1	2	3	4	5	6	7	8	9	10
1	76	93	109	74	54	95	64	92	141	87
2	75	95	102	68	52	90	65	96	120	76
3	80	98	110	82	67	88	80	101	133	91
4	76	95	108	75	64	98	76	97	136	93

a. Identify the design.
b. Write the model, identifying terms.
c. Give the analysis of variance table without performing the calculations.

15.13 Compute the sums of squares for the analysis of variance table of Exercise 15.12. Perform appropriate F tests and draw conclusions. Use $\alpha = .05$. Can a test be made for the interaction effect?

15.14 SAS output for the data for Exercise 15.12 is shown here. Compare the AOV from SAS to the one you obtained in Exercise 15.13.

```
                              RANDOMIZED BLOCK DESIGN
                            ANALYSIS OF VARIANCE PROCEDURE

DEPENDENT VARIABLE: DOLLAR
SOURCE              DF     SUM OF SQUARES      MEAN SQUARE     F VALUE   PR > F   R-SQUARE    C.V.
MODEL               12     16065.80000000    1338.81666667     67.11    0.0001   0.967563   5.0015
ERROR               27       538.60000000      19.94814815               ROOT MSE       DOLLAR MEAN
CORRECTED TOTAL     39     16604.40000000                               4.46633498       89.30000000


SOURCE              DF            ANOVA SS     F VALUE   PR > F
COMPANY              3      497.40000000        8.31    0.0004
EMPLOYEE             9    15568.40000000       86.72    0.0001
```

15.6 FACTORIAL EXPERIMENTS COMBINED WITH BLOCKING DESIGNS

In the previous section, we defined a factorial experiment to be an experiment where the response y is observed at all factor–level combinations of the independent variables. The factor–level combinations of the independent variables (treatments) were randomly assigned to the experimental units; hence, we were employing a completely randomized design to investigate the effects of the factors on the response.

Sometimes the objectives of a study are such that we wish to investigate the effects of certain factors on a response while blocking out certain other extraneous sources of variability. Such situations require a block design with treatments from a factorial experiment. We will draw on our knowledge of block designs (randomized block designs and Latin square designs) to effectively block out the extraneous sources of variability in order to focus on the effects of the factors on the response of interest. This can be illustrated with the following example.

EXAMPLE 15.10 ▼

An investigator wants to examine the effects of two factors (each measured at three levels) on a response y. It is determined that two observations are desired at each factor–level combination, but only nine observations can be done each day. Propose an appropriate experimental design.

Solution   Since nine observations can be obtained each day, it would be possible to run a complete replication of the 3×3 factorial experiment on two different days to get the desired number of observations. The design is shown here.

	Day 1, Factor *B*		
Factor *A*	1	2	3
1			
2			
3			

	Day 2, Factor *B*		
Factor *A*	1	2	3
1			
2			
3			

It should be noted that this design is really a randomized block design where the blocks are days and the treatments are the nine factor–level combinations of the 3×3 factorial experiment. So, with the randomized block design, we are able to block or filter out the variability due to the nuisance variable days while comparing the treatments. Since the treatments are factor–level combinations from a factorial experiment, we can examine the effects of the two factors (A and B) on the response while filtering out the day-to-day variability.

The analysis of variance for this design follows from our discussions in Sections 15.2 and 15.4.   ▲

EXAMPLE 15.11 ▼

Construct an analysis of variance table identifying the sources of variability and the degrees of freedom for the 3×3 factorial experiment laid off in a randomized block design with $b = 2$ discussed in Example 15.10.

Solution   The analysis of variance table for a randomized block design with $t = 9$ and $b = 2$ is shown here:

Source	SS	df
Treatments	SST	8
Blocks	SSB	1
Error	SSE	8
Total	TSS	17

Since the treatments of this randomized block are the nine factor–level combinations of a 3×3 factorial experiment, we can subdivide the sum of squares treatment (SST) into the sources of variability for a 3×3 factorial experiment from Section 15.5. The revised AOV table is shown here.

Source	SS	df
Treatments	SST	8
A	SSA	2
B	SSB	2
AB	SSAB	4
Blocks	SSB	1
Error	SSE	8
Total	TSS	17

So, rather then running an overall test comparing the treatment means using $F = \text{MST/MSE}$, we could conduct the analysis of variance for a factorial experiment in order to examine the interaction and main effects. These F tests would use the appropriate numerator mean squares (MSAB, MSA, and MSB) and MSE from this analysis. ▲

▼ EXERCISES

Basic Techniques

15.15 Diagram a design that has a 3×5 factorial experiment laid off in a randomized block design with $b = 3$ blocks. Give the complete analysis of variance table (sources, SSs, dfs).

15.16 Diagram a design which has a $2 \times 4 \times 3$ factorial experiment laid off in a randomized block design with $b = 2$ blocks. Give the complete analysis of variance for this experimental design.

15.7 ▼ THE ESTIMATION OF TREATMENT DIFFERENCES AND MULTIPLE COMPARISONS

We have emphasized the analysis of variance associated with a randomized block design, a Latin square design, and factorial experiments. There are times, however, when we might be more interested in estimating the difference in mean response for two treatments (different levels of the same factor or different combinations of levels). For example, an environmental engineer might be more interested in estimating the difference in the mean dissolved oxygen content for a lake before and after rehabilitative work than in testing to see whether there is a difference. Thus, the engineer is asking the question, "What is the difference in mean dissolved oxygen content?" instead of the question, "Is there a difference between the mean content before and after the cleanup project?"

Fisher's LSD procedure can be used to estimate the difference in treatment means for a randomized block design, a Latin square design, and k-factor factorial experiments with various designs. Let $\bar{y}_i$ denote the mean response for treatment i, $\bar{y}_{i'}$ denote the mean response for treatment i', and n_t denote the number of observations in each treatment. A $100(1 - \alpha)\%$ confidence interval on $\mu_i - \mu_{i'}$, the difference in mean response for the two treatments, is defined as shown here.

100(1 − α)% Confidence Interval for the Difference in Treatment Means

$$(\bar{y}_i - \bar{y}_{i'}) \pm t_{\alpha/2} s_\varepsilon \sqrt{\frac{2}{n_t}},$$

where s_ε is the square root of MSE in the AOV table and $t_{\alpha/2}$ can be obtained from Table 4 in the Appendix for $a = \alpha/2$ and the degrees of freedom for MSE.

EXAMPLE 15.12 ▼

A company was interested in comparing three different display panels for use by air traffic controllers. Each display panel was to be examined under five different simulated emergency conditions. Thirty highly trained air traffic controllers with similar work experience were enlisted for the study. A random assignment of controllers to display-panel–emergency conditions was made, with two controllers assigned to each factor–level combination. The time (in seconds) required to stabilize the emergency situation was recorded for each controller at a panel-emergency condition. These data appear in Table 15.31.

TABLE 15.31

Display Panel Data for Example 15.12 (time in seconds)

Display Panel, *B*	Emergency Condition, *A*					Totals
	1	2	3	4	5	
1	18	31	22	39	15	251
	16	35	27	36	12	
2	13	33	24	35	10	235
	15	30	21	38	16	
3	24	42	40	52	28	378
	28	46	37	57	24	
Totals	114	217	171	257	105	864

a. Construct a profile plot.
b. Run an analysis of variance that includes a test for interaction.

Solution

a. The sample means are shown in Figure 15.7.

Display Panel, B	Emergency Condition, A				
	1	2	3	4	5
1	17	33	24.5	37.5	13.5
2	14	31.5	22.5	36.5	13
3	26	44	38.5	54.5	26

FIGURE 15.7

Plot of Panel Means for Each Emergency Condition

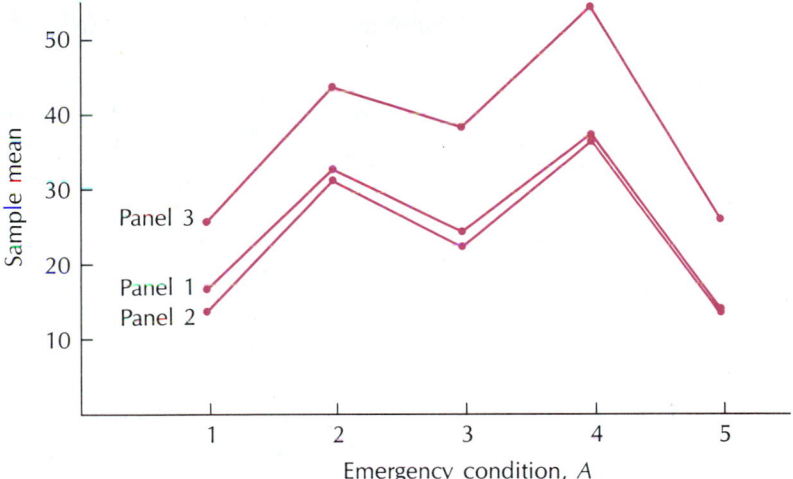

b. Substituting into the appropriate sums of squares formulas for a two–factor experiment with $n = 2$ observations per cell and using the data from Table 15.31, we have

$$\frac{y^2_{...}}{nab} = \frac{(864)^2}{30} = 24{,}883.2$$

$$SSA = \sum_i \frac{y^2_{i..}}{nb} - \frac{y^2_{...}}{nab} = \frac{(114)^2 + (217)^2 + \cdots + (105)^2}{6} - 24{,}883.2$$

$$= 27{,}733.33 - 24{,}883.2 = 2850.13$$

$$SSB = \sum_j \frac{y^2_{.j.}}{na} - \frac{y^2_{...}}{nab} = \frac{(251)^2 + (235)^2 + (378)^2}{10} - 24{,}883.2$$

$$= 26{,}111 - 24{,}883.2 = 1227.8.$$

TABLE 15.32
y_{ij} Totals for the Data of
Table 15.31

Display Panel, B	Emergency Condition, A					Totals
	1	2	3	4	5	
1	34	66	49	75	27	251
2	28	63	45	73	26	235
3	52	88	77	109	52	378
Totals	114	217	171	257	105	864

Combining the two observations in each cell, we obtain a table of y_{ij}. totals, as shown in Table 15.32.

Substituting into the formula for SSAB, we have

$$SSAB = \sum_{ij} \frac{y_{ij}^2}{n} - \frac{y_{...}^2}{nab} - SSA - SSB$$

$$= \frac{(34)^2 + (66)^2 + \cdots + (52)^2}{2} - 24,883.2 - 2,850.13 - 1,227.8$$

$$= 29,006 - 24,883.2 - 2,850.13 - 1,227.8 = 44.87.$$

The total sum of squares can be computed using the original observations in Table 15.31.

$$TSS = \sum_{ijk} y_{ijk}^2 - \frac{y_{...}^2}{nab}$$

$$= (18)^2 + (16)^2 + (31)^2 + \cdots + (28)^2 + (24)^2 - 24,883.2$$
$$= 29,112 - 24,883.2 = 4,228.8.$$

By subtraction, we find SSE to be

$$SSE = TSS - SSA - SSB - SSAB$$
$$= 4,228.8 - 2,850.13 - 1,227.8 - 44.87 = 106.$$

We can summarize our computations in an AOV table, Table 15.33.

The first test of significance should be a test for interaction between factors A and B. Since the computed value of F, .79, is less than the critical value of F, 2.64, for $a = .05$, $df_1 = 8$, and $df_2 = 15$, we have insufficient evidence to indicate an interaction between emergency conditions and display panels. As indicated by the profile plot, the difference in mean reaction time for controllers on two different display panels appears constant for all five emergency conditions.

TABLE 15.33
AOV Table for the Data
of Table 15.31

Source	SS	df	MS	F
Emergency conditions A	2,850.13	4	712.53	$712.53/7.07 = 100.78$
Display panel B	1,227.8	2	613.9	$613.9/7.07 = 86.83$
AB	44.87	8	5.61	$5.61/7.07 = .79$
Error	106	15	7.07	
Totals	4,228.8	29		

EXAMPLE 5.13 ▼

Refer to Example 15.12. Since there is no interaction, estimate the difference in mean response (reaction time) for display panels 2 and 3 using a 95% confidence interval.

Solution For these data,

$$\bar{y}_{.3.} = \frac{378}{10} = 37.8 \quad \text{and} \quad \bar{y}_{.2.} = \frac{235}{10} = 23.5.$$

The t-value for $a = .025$ and df $= 23$ is 2.069; the estimate of σ_ε is

$$s_\varepsilon = \sqrt{\text{MSE}} = 2.66.$$

The appropriate confidence interval is based on the formula from page 892:

$$\bar{y}_{.3.} - \bar{y}_{.2.} \pm t_{\alpha/2} s_\varepsilon \sqrt{\frac{2}{n_t}}.$$

For this experiment, with $n_t = 10$, we have

$$37.8 - 23.5 \pm 2.069(2.66) \sqrt{\frac{2}{10}}$$

$$14.3 \pm 2.46,$$

or 11.84 to 16.76.

Assume now that we have chosen analysis of variance rather than estimation as the inferential technique for answering our practical questions. We compute the appropriate sums of squares and perform F tests on the sources of variability. Having rejected a hypothesis of the form "no difference in mean response for levels of factor A," do we stop and draw no further conclusion? No, we do not, because the F test is only a preliminary test to determine if there are any differences among the treatment means. We must then try to determine which levels of the factor differ from the rest.

As mentioned in Chapter 14, we could perform multiple t tests among the factor–level means, but the overall error rate could be quite large. Presumably, we would proceed with one of the **multiple comparison procedures**, such as Tukey's, Duncan's, or Scheffé's, with a controlled error rate. All these procedures can be performed following an analysis of variance for a randomized block design, a Latin square design, or a k-factorial experiment. The quantity s_w^2 is replaced by MSE in the various formulas, and the sample size n refers to the number of observations per mean in the comparison. The degrees of freedom for MSE are obtained from the AOV table.   ▲

using multiple comparison procedure

E X A M P L E   15.14   ▼

Refer to Example 15.12 and the data in Table 15.31. Use Tukey's W procedure to locate significant differences among display panels.

Solution   For Tukey's W procedure we use the formula presented in Chapter 14:

$$W = q_\alpha(t, v) \sqrt{\frac{s_w^2}{n}},$$

where s_w^2 is MSE from the AOV table, based on $v = 15$ degrees of freedom, and $q_\alpha(t, v)$ is the upper-tail critical value of the Studentized range (with $a = \alpha$) for comparing t different population means. The value of $q_\alpha(t, v)$ from Table 10 in the Appendix for comparing the three display panel means, each of which has ten observations per sample mean, is

$$q_{.05}(3, 15) = 3.67.$$

For ten observations per mean, the value of W is

$$W = q_\alpha(t, v) \sqrt{\frac{s_w^2}{n}} = 3.67 \sqrt{\frac{7.07}{10}} = 3.09.$$

The display panel means are, from Table 15.31,

$$\bar{y}_{.1.} = \frac{251}{10} = 25.1 \qquad \bar{y}_{.2.} = \frac{235}{10} = 23.5 \qquad \bar{y}_{.3.} = \frac{378}{10} = 37.8.$$

First we rank the sample means from lowest to highest:

Display panel	2	1	3
Means	23.5	25.1	37.8

For two means that differ (in absolute value) by more than $W = 3.09$, we declare them to be significantly different from each other. The results of our multiple comparison procedure are summarized here:

Display panel   2   1   3

Thus, display panels 1 and 2 both have a mean reaction time significantly lower than display panel 3, but we are unable to detect any difference between panels 1 and 2.   ▲

15.8 ▼ RELATIONSHIP BETWEEN REGRESSION AND ANALYSIS OF VARIANCE (optional)

The link between regression methods and analysis of variance procedures is provided by the general linear model. In Section 11.1, we showed how dummy variables could be used to describe a response y as a function of a qualitative independent variable using a general linear model. Recall that the model for a completely randomized design (with t treatments) could be written as

$$y = \beta_0 + \beta_1 x_1 + \beta_2 x_2 + \cdots + \beta_{t-1} x_{t-1} + \varepsilon,$$

where

$$x_1 = 1 \text{ if treatment 2, } x_1 = 0 \text{ otherwise}$$
$$x_2 = 1 \text{ if treatment 3, } x_2 = 0 \text{ otherwise}$$
$$\vdots$$
$$x_{t-1} = 1 \text{ if treatment } t, x_{t-1} = 0 \text{ otherwise.}$$

For this model, we have

$$\beta_0 = \mu_1$$
$$\beta_1 = \mu_2 - \mu_1$$
$$\beta_2 = \mu_3 - \mu_1$$
$$\vdots$$
$$\beta_{t-1} = \mu_t - \mu_1.$$

We also indicated in Chapter 11 that the general linear model could be used to relate y to more than one qualitative independent variable. For a randomized block design, we have a response related to two qualitative independent variables: blocks and treatments. Applying our previous results, if there are b blocks and t treatments, we will enter $(b - 1)$ βs for blocks and $(t - 1)$ βs for treatments into our model.

EXAMPLE 15.15

▼

Write a model for a randomized block design to compare $t = 4$ treatments in $b = 3$ blocks. The design is shown in Table 15.34.

TABLE 15.34

Design for $b = 3$ Blocks and $t = 4$ Treatments, Example 15.15

Blocks	Treatments			
1	I	III	II	IV
2	II	IV	III	I
3	IV	I	II	III

Solution   The model can be written in the form

$$
y = \beta_0 + \overbrace{\beta_1 x_1 + \beta_2 x_2}^{\text{blocks}} + \overbrace{\beta_3 x_3 + \beta_4 x_4 + \beta_5 x_5}^{\text{treatments}} + \varepsilon,
$$

where

$x_1 = 1$ if block 2 $\qquad$ $x_1 = 0$ otherwise
$x_2 = 1$ if block 3 $\qquad$ $x_2 = 0$ otherwise
$x_3 = 1$ if treatment II $\qquad$ $x_3 = 0$ otherwise
$x_4 = 1$ if treatment III $\qquad$ $x_4 = 0$ otherwise
$x_5 = 1$ if treatment IV $\qquad$ $x_5 = 0$ otherwise.

We can easily interpret the βs using a table of expected values. To obtain the expected value of an observation for a given block-treatment combination, we substitute appropriate values for $x_1, x_2, \ldots, x_5$ in the formula

$$
E(y) = \beta_0 + \beta_1 x_1 + \beta_2 x_2 + \cdots + \beta_5 x_5.
$$

For example, the expected value for an observation in block 2 on treatment II would have $x_1 = 1$, $x_2 = 0$, $x_3 = 1$, $x_4 = 0$, and $x_5 = 0$, giving

$$
E(y) = \beta_0 + \beta_1(1) + \beta_2(0) + \beta_3(1) + \beta_4(0) + \beta_5(0) = \beta_0 + \beta_1 + \beta_3.
$$

The table of expected values is given in Table 15.35.

T A B L E 15.35

Table of Expected Values for a Randomized Block Design, with $b = 3$ and $t = 4$, Example 15.15

	Treatment			
Block	I	II	III	IV
1	β_0	$\beta_0 + \beta_3$	$\beta_0 + \beta_4$	$\beta_0 + \beta_5$
2	$\beta_0 + \beta_1$	$\beta_0 + \beta_1 + \beta_3$	$\beta_0 + \beta_1 + \beta_4$	$\beta_0 + \beta_1 + \beta_5$
3	$\beta_0 + \beta_2$	$\beta_0 + \beta_2 + \beta_3$	$\beta_0 + \beta_2 + \beta_4$	$\beta_0 + \beta_2 + \beta_5$

The mean response for block 1 and treatment 1 is β_0. If we consider any treatment and compare blocks 2 and 1, the difference in the mean response is β_1. For example, when using treatment II, the difference in the mean response for blocks 2 and 1 is, from Table 15.35,

$$(\beta_0 + \beta_1 + \beta_3) - (\beta_0 + \beta_3) = \beta_1.$$

Note that this is true for any treatment. Hence, $\beta_1 = \mu_2 - \mu_1$, the difference in the mean response for blocks 2 and 1. Similarly, in comparing blocks 3 and 1 for a given treatment, the difference is $\beta_2 = \mu_3 - \mu_1$.

In the same way, we can compare two treatments for a given block. For example, it follows immediately that $\beta_3 = \mu_{II} - \mu_I$, the difference in mean response for treatments II and I. Similarly, in block 3 the difference in mean response for treatments II and I is

$$(\beta_0 + \beta_2 + \beta_3) - (\beta_0 + \beta_2) = \beta_3.$$

In the same fashion, we can show that

$$\beta_4 = \mu_{III} - \mu_I$$
$$\beta_5 = \mu_{IV} - \mu_I.$$

These results can easily be extended to a Latin square design.   ▲

E X A M P L E 15.16   ▼

Nylon is spun on a series of machines. When a break in the nylon thread occurs during the spinning process, the machine operator must stop the machine and rethread the nylon prior to continuing. Investigators would like to compare the output of three different spinning machines (I, II, III) using three different operators (A, B, C). To control day-to-day variability in addition to operator variability, it was decided to use a 3 × 3 Latin square design, as shown in Table 15.36. Write an appropriate model for this Latin square design, using dummy variables. Interpret the βs for the model.

TABLE 15.36

Latin Square Design for the Spinning Machines Experiment, Example 15.16

	Day		
Operator	1	2	3
A	I	II	III
B	II	III	I
C	III	I	II

Solution   The model for this Latin square design must relate a response to three qualitative independent variables, rows (operators), columns (days), and treatments (machines). For each qualitative variable, we will include two β's.

$$y = \beta_0 + \overbrace{\beta_1 x_1 + \beta_2 x_2}^{\text{operators}} + \overbrace{\beta_3 x_3 + \beta_4 x_4}^{\text{days}} + \overbrace{\beta_5 x_5 + \beta_6 x_6}^{\text{machines}} + \varepsilon,$$

where

$x_1 = 1$ if operator B $x_1 = 0$ otherwise
$x_2 = 1$ if operator C $x_2 = 0$ otherwise
$x_3 = 1$ if day 2 $x_3 = 0$ otherwise
$x_4 = 1$ if day 3 $x_4 = 0$ otherwise
$x_5 = 1$ if machine II $x_5 = 0$ otherwise
$x_6 = 1$ if machine III $x_6 = 0$ otherwise.

For this model, with $E(\varepsilon) = 0$, we can write the expected value of y as

$$E(y) = \beta_0 + \beta_1 x_1 + \beta_2 x_2 + \beta_3 x_3 + \beta_4 x_4 + \beta_5 x_5 + \beta_6 x_6.$$

By substituting appropriate values for the dummy variables $x_1, x_2, \ldots, x_6$, we obtain the table of expected values, Table 15.37.

Although it is readily apparent that β_0 represents the mean response for machine I run by operator A during day 1, it is more difficult to identify the other βs. Since each machine appears in each row (operator) and each column (day), the βs due to rows and columns will be eliminated when

TABLE 15.37

Expected Values for the 3 × 3 Latin Square Design of the Spinning Machines Experiment, Example 15.16

	Day		
Operator	1	2	3
A	I β_0	II $\beta_0 + \beta_3 + \beta_5$	III $\beta_0 + \beta_4 + \beta_6$
B	II $\beta_0 + \beta_1 + \beta_5$	III $\beta_0 + \beta_1 + \beta_3 + \beta_6$	I $\beta_0 + \beta_1 + \beta_4$
C	III $\beta_0 + \beta_2 + \beta_6$	I $\beta_0 + \beta_2 + \beta_3$	II $\beta_0 + \beta_2 + \beta_4 + \beta_5$

comparing two machines. If we average the three expected values for observations on machine I and the three expected values from Table 15.37 for machine II, we have

$$\text{average for machine I} = \frac{1}{3} [\beta_0 + (\beta_0 + \beta_2 + \beta_3) + (\beta_0 + \beta_1 + \beta_4)]$$

$$= \beta_0 + \frac{1}{3} (\beta_1 + \beta_2 + \beta_3 + \beta_4)$$

$$\text{average for machine II} = \frac{1}{3} [(\beta_0 + \beta_1 + \beta_5) + (\beta_0 + \beta_3 + \beta_5)$$

$$+ (\beta_0 + \beta_2 + \beta_4 + \beta_5)]$$

$$= \beta_0 + \beta_5 + \frac{1}{3} (\beta_1 + \beta_2 + \beta_3 + \beta_4).$$

The difference in these averages represents $\mu_{II} - \mu_{I}$. Subtracting, we have

$$\beta_5 = \mu_{II} - \mu_{I}.$$

Similarly, by comparing machine III to machine I, we can show that

$$\beta_6 = \mu_{III} - \mu_{I}.$$

The same reasoning can be used to obtain

$$\beta_1 = \mu_B - \mu_A$$
$$\beta_2 = \mu_C - \mu_A$$
$$\beta_3 = \mu_2 - \mu_1$$
$$\beta_4 = \mu_3 - \mu_1.$$

Now that the βs have been identified, these interpretations are exactly what we might have imagined, having examined the completely randomized design and a randomized block design. ▲

main effects terms

interaction terms

factorial experiment

The models that we have written in this section have had only **main effects terms** (terms involving only one x) for each of the qualitative independent variables. But it is also possible to write models containing **interaction terms**: terms involving the product of xs between two or more variables.

Consider a 2×3 **factorial experiment** in which an experimenter would like to compare three diet preparations (A, B, C) under two diet plans using a completely randomized design. A fixed number of overweight persons would be assigned to each of the six factor–level combinations (treatments). Since it is possible that the difference in mean response (weight loss) for two different diet preparations is not the same for each diet plan, we must include interaction terms as well as main effects in our models. An appropriate model is given by

model

$$y = \beta_0 + \overbrace{\beta_1 x_1 + \beta_2 x_2 + \beta_3 x_3}^{\text{main effects}} + \overbrace{\beta_4 x_1 x_2 + \beta_5 x_1 x_3}^{\text{interaction}} + \varepsilon,$$

where

$x_1 = 1$ if plan 2 $x_1 = 0$ otherwise
$x_2 = 1$ if diet B $x_2 = 0$ otherwise
$x_3 = 1$ if diet C $x_3 = 0$ otherwise.

The main effects terms in our model are

$\beta_1 x_1$ for diet plans
$\beta_2 x_2$ and $\beta_3 x_3$ for diet preparations.

Interaction terms are formed from cross-products of the xs involved in main effects. Thus, from the product of x_1 and x_2, we have the interaction term $\beta_4 x_1 x_2$. Similarly, we have $\beta_5 x_1 x_3$ from the product of x_1 and x_3. It might also appear that we should have an interaction term involving the product of x_2 with x_3. However, *interaction terms are always formed by products of x values from different independent variables*. No term involving levels of the same variable could contribute toward measuring this effect.

To interpret the βs associated with our model, we again form a table of expected values by substituting appropriate combinations for x_1, x_2, and x_3 into the general formula

$$E(y) = \beta_0 + \beta_1 x_1 + \beta_2 x_2 + \beta_3 x_3 + \beta_4 x_1 x_2 + \beta_5 x_1 x_3.$$

For example, the expected response on plan 2 and diet B can be found by substituting $x_1 = 1$, $x_2 = 1$, and $x_3 = 0$ into $E(y)$.

$$E(y) = \beta_0 + \beta_1(1) + \beta_2(1) + \beta_3(0) + \beta_4(1)(1) + \beta_5(1)(0)$$
$$= \beta_0 + \beta_1 + \beta_2 + \beta_4$$

The table of expected values is given in Table 15.38.

T A B L E 15.38
**Expected Values of a
2 × 3 Factorial Experiment**

		Diet Preparation	
Diet Plan	A	B	C
1	β_0	$\beta_0 + \beta_2$	$\beta_0 + \beta_3$
2	$\beta_0 + \beta_1$	$\beta_0 + \beta_1 + \beta_2 + \beta_4$	$\beta_0 + \beta_1 + \beta_3 + \beta_5$

From Table 15.38 we have the following interpretations for main effects:

β_0: mean response for plan 1, diet A
β_1: difference in mean response for plans 2 and 1 on diet A
β_2: difference in mean response for diets B and A on plan 1
β_3: difference in mean response for diets C and A on plan 1.

The interaction βs are slightly more complicated to interpret since they measure the failure of diet preparations to have the same effect across different diet plans. Using the first two columns of Table 15.38, we can find β_4 as the sum of the expected values in the diagonal left-to-right direction minus the sum of the expected values in the diagonal right-to-left direction.

	A	B
1	β_0	$\beta_0 + \beta_2$
2	$\beta_0 + \beta_1$	$\beta_0 + \beta_1 + \beta_2 + \beta_4$

$$[\beta_0 + (\beta_0 + \beta_1 + \beta_2 + \beta_4)] - [(\beta_0 + \beta_2) + (\beta_0 + \beta_1)] = \beta_4$$

Similarly, taking the first and third columns of Table 15.38 and subtracting the right-to-left sum from the left-to-right sum, we have

$$[\beta_0 + (\beta_0 + \beta_1 + \beta_3 + \beta_5)] - [(\beta_0 + \beta_3) + (\beta_0 + \beta_1)] = \beta_5.$$

no interaction   We should note that if there were **no interaction** between the variables "diet plan" and "diet preparation," the parameters β_4 and β_5 would both equal 0. Using Table 15.38, with $\beta_4 = \beta_5 = 0$, we have the following interpretations for the parameters in the reduced model:

$y = \beta_0 + \beta_1 x_1 + \beta_2 x_2 + \beta_3 x_3 + \varepsilon$
β_0: mean response for diet A, plan 1
$\beta_1 = \mu_2 - \mu_1$: the difference in mean response for plans 2 and 1
$\beta_2 = \mu_B - \mu_A$: the difference in mean response for diets B and A
$\beta_3 = \mu_C - \mu_A$: the difference in mean response for diets C and A.

Note that in the absence of interaction, the interpretation of a main effect term for one variable does not depend on the level of another variable. For example, with interaction present,

$$\beta_1 = \mu_2 - \mu_1 \text{ on diet A (that is, the difference in mean response for plans 2}$$
$$\text{and 1 while on diet A)}$$

and without interaction,

$\beta_1 = \mu_2 - \mu_1$ (that is, the difference in mean response for plans 2 and 1 for any diet preparation).

EXAMPLE   15.17

▼

Refer to the experiment just discussed, which involved the investigation of diet plans and diet preparations. Write a complete model (indicate main effects and interactions) for an experiment to compare four diet preparations using three diet plans.

Solution   The model must relate a response to two qualitative independent variables: diet plans and diet preparations. The variable "diet plan" will have $3 - 1 = 2$ main effects terms; the variable "diet preparation" will have $4 - 1 = 3$ main effects terms, and there will be $2(3) = 6$ interaction terms formed from products of x values, one from each of the two variables. The model can be written as follows:

$$y = \beta_0 + \overbrace{\beta_1 x_1 + \beta_2 x_2}^{\text{diet plans}} + \overbrace{\beta_3 x_3 + \beta_4 x_4 + \beta_5 x_5}^{\text{diet preparations}}$$

$$+ \overbrace{\beta_6 x_1 x_3 + \beta_7 x_1 x_4 + \beta_8 x_1 x_5 + \beta_9 x_2 x_3 + \beta_{10} x_2 x_4 + \beta_{11} x_2 x_5}^{\text{interaction terms}} + \varepsilon,$$

where

$x_1 = 1$ if plan 2	$x_1 = 0$ otherwise
$x_2 = 1$ if plan 3	$x_2 = 0$ otherwise
$x_3 = 1$ if diet B	$x_3 = 0$ otherwise
$x_4 = 1$ if diet C	$x_4 = 0$ otherwise
$x_5 = 1$ if diet D	$x_5 = 0$ otherwise.

▲

Being able to write models for regression problems *and* models for analysis of variance problems using a general linear model means that there is a link between regression and analysis of variance methods. We have already seen in previous chapters that we can make tests concerning a set of βs in a regression model by fitting complete and reduced models (see Chapter 11). The same procedure can also be done to make the required F tests for an analysis of variance if the model is written as a general linear model. The study described in Example 15.8 can be used to illustrate this fact.

The study design described in Example 15.8 is a 3×4 factorial involving three varieties and four pesticides. The no-interaction model, written in the form of a general linear model, is

$$y = \beta_0 + \beta_1 x_1 + \beta_2 x_2 + \beta_3 x_3 + \beta_4 x_4 + \beta_5 x_5 + \varepsilon,$$

where

$$\text{varieties} \begin{cases} x_1 = 1 & \text{if variety 2} \\ = 0 & \text{otherwise} \\ x_2 = 1 & \text{if variety 3} \\ = 0 & \text{otherwise} \end{cases} \qquad \text{pesticides} \begin{cases} x_3 = 1 & \text{if pesticide 2} \\ = 0 & \text{otherwise} \\ x_4 = 1 & \text{if pesticide 3} \\ = 0 & \text{otherwise} \\ x_5 = 1 & \text{if pesticide 4} \\ = 0 & \text{otherwise.} \end{cases}$$

The hypothesis of no difference among the pesticide means, $H_0: \beta_3 = \beta_4 = \beta_5 = 0$, is the same as that specified for an analysis of variance F test on pesticide. In addition, the sum of squares drop (SS_{drop}) obtained by fitting models 1 and 2 is identical to the sum of squares due to pesticides of Example 15.8.

$$\text{model 1:} \qquad y = \beta_0 + \overbrace{\beta_1 x_1 + \beta_2 x_2}^{\text{varieties}} + \overbrace{\beta_3 x_3 + \beta_4 x_4 + \beta_5 x_5}^{\text{pesticides}} + \varepsilon$$
$$SSE_1 = 151.50$$
$$\text{model 2:} \qquad y = \beta_0 + \beta_1 x_1 + \beta_2 x_2 + \varepsilon$$
$$SSE_2 = 1{,}342.50$$
$$SS_{drop}(\text{due to pesticides}) = SSE_2 - SSE_1 = 1{,}191.00.$$

In the same way, the sum of squares drop obtained for testing no difference among the variety means ($H_0: \beta_1 = \beta_2 = 0$) is identical to the sum of squares due to varieties in the analysis of variance of Example 15.8. Here,

$$\text{model 1:} \qquad y = \beta_0 + \overbrace{\beta_1 x_1 + \beta_2 x_2}^{\text{varieties}} + \overbrace{\beta_3 x_3 + \beta_4 x_4 + \beta_5 x_5}^{\text{pesticides}} + \varepsilon$$
$$SSE_1 = 151.50$$
$$\text{model 2:} \qquad y = \beta_0 + \beta_3 x_3 + \beta_4 x_4 + \beta_5 x_5 + \varepsilon$$
$$SSE_2 = 2{,}376.67.$$

Although we have not fit model 2, it can be shown that the sum of squares for error is as given here. Also,

$$SS_{drop}(\text{due to varieties}) = SSE_2 - SSE_1 = 2{,}225.17.$$

Note that SSE_1, the sum of squares for error for the complete model, is SSE from the AOV table of Example 15.8.

The method of obtaining sums of squares (SS_{drop}) due to various sources of variability by fitting complete and reduced models can be used for *any* experimental design. All we need to do is to specify an appropriate general linear model. Then, by hypothesizing that various parameters in the model are 0, we can obtain SS_{drop} and run an appropriate F test. The F test is identical to that used in a standard analysis of variance. *In fact, as we have developed this section, we see that an analysis of variance consists of testing hypotheses concerning parameters in the general linear model.*

Why don't we fit complete and reduced models every time we conduct an analysis of variance? While this would certainly be possible, it is not practical. The reason is that many designs are **balanced designs**.

balanced design

DEFINITION 15.5
Balanced Design

▼

A **balanced design** has each level of one independent variable appearing the same number of times with each level of another independent variable, and this is true for all pairs of independent variables.

For many balanced designs it is possible to obtain shortcut computational formulas for calculating the sum of squares drop for a particular source of variability. For example, by Definition 15.5, it is clear that Latin square, randomized block, and completely randomized designs are balanced designs, as are factorial experiments laid off in one of these designs. Balanced designs have shortcut formulas for calculating the sum of squares associated with the various sources of variability in an AOV table. These formulas are simplified expressions for calculating SS_{drop} by fitting a complete and reduced model.

At this point, you might be concerned with identifying an unbalanced design. An example of an unbalanced design would be any balanced design with one or more missing observations. Suppose an experimenter was interested in comparing the drop in potency for three different concentrations of a drug product stored at three different temperatures. Assume further that two bottles were to be stored for 6 weeks at each factor–level combination. Since final potency determinations must be made after the 6-week period in order to compute the drop in potency, it is possible that one or more bottles would be broken. If such an accident did occur, we would have an unequal number of observations at the different factor–level combinations, making the design *unbalanced*. The shortcut formulas developed earlier in this chapter would no longer be appropriate for computing sums of squares for sources in the AOV table. While the general procedure for fitting complete and reduced models can be used, we will discuss the analysis of unbalanced designs in Chapter 16.

15.9 ▼ SUMMARY

In this chapter, we have extended our discussion of the analysis of variance presented in Chapter 13. We did this by considering several basic experimental designs. The completely randomized design, the randomized block design, and the Latin square design illustrate how we can block out undesirable background variability to obtain more precise comparisons among treatment means. The factorial experiment is useful in investigating the effect of one or more factors on a response; it can be used with either a completely randomized design, a randomized block design, or a Latin square design. Thus, an experimenter may wish to examine the effects of two or more factors on a response while blocking out one or more extraneous sources of variability.

For each design discussed in this chapter, we presented a description of the design layout (including arrangement of treatments), potential advantages and disadvantages, a model, and the analysis of variance. Finally, we discussed how one could make comparisons between pairs of treatment means for these designs.

The reader should recognize that the designs presented are only the most basic ones, and so far we have only dealt with the situation in which we had a *balanced design*; that is, an equal number of observations per treatment or factor–level combination. Chapter 16 extends the results of this chapter to some unbalanced (and perhaps more practical) situations.

▼ KEY FORMULAS

1. Randomized block design

$$TSS = \sum_{ij} (y_{ij} - \bar{y}_{..})^2 = \sum_{ij} y_{ij}^2 - \frac{y_{..}^2}{bt}$$

$$SST = b \sum_{i} (\bar{y}_{i.} - \bar{y}_{..})^2 = \sum_{i} \frac{y_{i.}^2}{b} - \frac{y_{..}^2}{bt}$$

$$SSB = t \sum_{j} (\bar{y}_{.j} - \bar{y}_{..})^2 = \sum_{j} \frac{y_{.j}^2}{t} - \frac{y_{..}^2}{bt}$$

$$SSE = TSS - SST - SSB$$

2.   Latin square design

$$\text{TSS} = \sum_{ij} (y_{ijk} - \bar{y}_{...})^2 = \sum_{ij} y_{ijk}^2 - \frac{y_{...}^2}{t^2}$$

$$\text{SST} = t \sum_i (\bar{y}_{i..} - \bar{y}_{...})^2 = \sum_i \frac{y_{i..}^2}{t} - \frac{y_{...}^2}{t^2}$$

$$\text{SSR} = t \sum_j (\bar{y}_{.j.} - \bar{y}_{...})^2 = \sum_j \frac{y_{.j.}^2}{t} - \frac{y_{...}^2}{t^2}$$

$$\text{SSC} = t \sum_k (\bar{y}_{..k} - \bar{y}_{...})^2 = \sum_k \frac{y_{..k}^2}{t} - \frac{y_{...}^2}{t^2}$$

$$\text{SSE} = \text{TSS} - \text{SST} - \text{SSR} - \text{SSC}$$

3.   Three-factor factorial equipment in a completely randomized design

$$\text{SSA} = \sum_i \frac{y_{i...}^2}{nbc} - \frac{y_{....}^2}{nabc}$$

$$\text{SSB} = \sum_j \frac{y_{.j..}^2}{nac} - \frac{y_{....}^2}{nabc}$$

$$\text{SSC} = \sum_k \frac{y_{..k.}^2}{nab} - \frac{y_{....}^2}{nabc}$$

$$\text{SSAB} = \sum_{ij} \frac{y_{ij..}^2}{nc} - \frac{y_{....}^2}{nabc} - \text{SSA} - \text{SSB}$$

$$\text{SSAC} = \sum_{ik} \frac{y_{i.k.}^2}{nb} - \frac{y_{....}^2}{nabc} - \text{SSA} - \text{SSC}$$

$$\text{SSBC} = \sum_{jk} \frac{y_{.jk.}^2}{na} - \frac{y_{....}^2}{nabc} - \text{SSB} - \text{SSC}$$

$$\text{SSABC} = \sum_{ijk} \frac{y_{ijk.}^2}{n} - \frac{y_{....}^2}{nabc} - \text{SSAB} - \text{SSAC}$$

$$- \text{SSBC} - \text{SSA} - \text{SSB} - \text{SSC}$$

4. $100(1 - \alpha)\%$ confidence interval for difference in treatment means

$$\bar{y}_i - \bar{y}_{i'} \pm t_{\alpha/2}s_\varepsilon \sqrt{\frac{2}{n_t}}$$

▼ SUPPLEMENTARY EXERCISES

15.17 An experimenter was interested in examining the bond strength of a new adhesive product prepared under three different temperature settings (280°F, 300°F, and 320°F) and four different pressure settings (100, 150, 200, and 250 psi). A single fixed amount of adhesive is to be prepared and tested at each temperature–pressure setting combination. Identify the design.

15.18 An oil company has been experimenting with a new gasoline additive. As part of the testing program, the company examined the effect on miles per gallon (mi/gal) of four additive concentrations and five different octane levels for the gasoline. If one gasoline mixture is to be made and tested at each concentration–octane combination, identify the experimental design.

15.19 A company executive was interested in comparing the cost per mile for all cars of a particular brand with V-8 engines (351 horsepower) and all cars of the same make with six-cylinder engines (250 horsepower). Since the entire fleet of company cars for salespeople consisted of approximately 600 automobiles, it was decided to obtain a random sample of data from 16 cars, 8 from each engine type. To avoid geographic variability, a random sample of 1 car of each engine type was selected from each of 8 geographic areas throughout the country. Cost per mile was determined for each car sampled during the period of January 1986 to March 1987. These data appear here.

Geographic area	1	2	3	4	5	6	7	8
351 hp, V-8	.0837	.0564	.0703	.0502	.0638	.0483	.0746	.0694
250 hp, 6-cylinder	.0523	.0371	.0464	.0481	.0535	.0335	.0444	.0528

a. Do the data suggest a difference in the mean cost per mile for the two engine types?
b. Give the level of significance for your test. Draw some conclusions.
c. Can you identify additional sources of variability that could be blocked?

15.20 Write the model and complete an analysis of variance table for a 3×5 factorial experiment.
a. How many degrees of freedom do you have for the error term when the two-way interaction is included?
b. How many degrees of freedom do you have for the error term if you assume there is no two-way interaction?

15.21 Examine the analysis of variance for the data of Example 15.8 shown in the BMDP computer output that follows.

```
ANALYSIS OF VARIANCE FOR 1-ST DEPENDENT VARIABLE

                 SUM OF      DEGREES OF     MEAN                        PROB. F
   SOURCE        SQUARES      FREEDOM       SQUARE         F           EXCEEDED
   MEAN        90650.93750       1        90650.93750  2143.45068       0.000
   VARIETY      3896.07349       2         1998.03662    47.24377       0.000
   PESTICIDE    2227.44238       3          742.48071    17.55603       0.000
   VAR × PEST    456.91479       6           76.15247     1.80063       0.182
   ERROR         507.50464      12           42.29205
```

a. Construct an analysis of variance table from the computer output.

b. Give the p-value and conclusion for the F test on the variety-by-treatment interaction. Is it appropriate to proceed with tests for main effects?

c. Draw conclusions from the p-values listed for varieties and pesticides.

d. Compare your results to those of Example 15.9.

15.22 A study was conducted to compare the yield of soybeans in a factorial experiment consisting of four manganese rates (from $MnSO_4$) and four copper rates (from $CuSO_45H_2O$). A large plot was subdivided into 16 separate subplots, to which the 16 factor–level combinations (treatments) were applied. Soybeans were then planted over the entire plot in rows 3 feet apart. The sample data are given here (in kilograms/hectare).

| | **Mn** | | | |
Cu	20	50	80	110
1	1558	2003	2490	2830
3	1590	2020	2620	2860
5	1550	2010	2490	2750
7	1328	1760	2280	2630

Construct a profile plot and write an appropriate statistical model for this experimental situation. Identify the design.

15.23 Refer to Exercise 15.22.

a. Perform an analysis of variance. Can you test for an interaction? Use $\alpha = .05$.

b. Assuming that there is no interaction between the two variables Cu and Mn, write a regression model expressing the soybean yield in terms of two quantitative variables (Cu and Mn).

15.24 Write the model and complete an analysis of variance table for a $4 \times 3 \times 6$ factorial experiment laid off in a completely randomized design. (Assume that the three-way interaction is nonsignificant.)

15.25 An experiment was set up to compare the effect of different soil pH and calcium additives on the increase in trunk diameters for orange trees. Annual applications of elemental sulfur, gypsum, soda ash, and other ingredients were applied to provide pH value levels of 4, 5, 6, and 7. Three levels of a calcium supplement (100, 200, and 300 pounds per acre) were also applied. All factor–level combinations of these two variables were used in the experiment. At the end of a two-year period, three diameters were examined at each factor–level combination. These data appear next.

pH Value	Calcium		
	100	200	300
4.0	5.2	7.4	6.3
	5.9	7.0	6.7
	6.3	7.6	6.1
5.0	7.1	7.4	7.3
	7.4	7.3	7.5
	7.5	7.1	7.2
6.0	7.6	7.6	7.2
	7.2	7.5	7.3
	7.4	7.8	7.0
7.0	7.2	7.4	6.8
	7.5	7.0	6.6
	7.2	6.9	6.4

a. Construct a profile plot. What do the data suggest?

b. Write an appropriate statistical model.

c. Perform an analysis of variance and identify the experimental design. Use $\alpha = .05$.

15.26 Refer to Exercise 15.8. Use Fisher's LSD to determine significant differences among the four treatments (broadcast and band methods of fertilizer applications).

15.27 Refer to Exercise 15.22. Use Tukey's W procedure to determine differences among the four manganese rates. Use $\alpha = .05$.

15.28 Refer to Exercise 15.25. Use Duncan's multiple range test (for $\alpha = .05$) to declare significant differences.

15.29 An experiment was conducted to compare the average oral body temperature for persons taking one of nine different medications often prescribed for a specific disorder. To do this, each of three investigators with prior clinical experience of this nature was to obtain a random sample of patients from his or her practice who satisfy the study entrance criteria. Then the investigator was to allocate the medications randomly, one to each person. Each patient in the study was given the assigned medication at 6:00 A.M. of the assigned study day. Temperatures were taken at hourly intervals beginning at 8:00 A.M. and continuing for 10 hours. During this time, the patients were not allowed to do any physical activity and had to lie in bed. To eliminate the variability of temperature readings within a day, the average of the hourly determinations was the response of interest. These data are given in the accompanying table.

a. Write an appropriate statistical model and identify the parameters of the model.

b. Perform an analysis of variance to test for a difference in mean temperatures for medications and then for investigators. Summarize your results in an AOV table. Use $\alpha = .05$. (Hint: To simplify some of your calculations, subtract a constant, such as 97, from each of the sample measurements.)

| | Medication | | | | | | | | |
Investigator	A	B	C	D	E	F	G	H	I
1	97.8	98.1	98.0	97.3	97.9	97.9	97.1	98.0	97.8
	97.2	98.1	97.8	97.3	97.8	97.9	97.6	97.8	98.0
	97.6	98.0	98.1	97.5	97.8	97.8	97.3	98.0	97.7
	97.2	97.7	97.8	97.5	97.7	97.8	97.7	97.9	97.9
	97.6	97.7	97.9	97.6	97.8	97.6	97.5	98.0	97.8
2	97.6	97.8	97.9	97.5	97.8	98.0	97.6	97.9	98.0
	97.4	97.7	98.1	97.4	97.8	97.7	97.5	98.0	97.6
	97.3	97.6	97.8	97.5	97.7	97.8	97.6	97.9	98.0
	97.5	97.7	97.8	97.6	97.7	97.9	97.5	97.9	97.9
	97.5	97.7	97.6	97.7	97.8	97.8	97.3	97.8	97.9
3	97.5	97.6	98.0	97.9	97.7	97.9	97.4	97.8	98.0
	97.9	97.7	97.8	97.8	97.8	98.0	97.8	97.8	98.1
	97.6	97.9	98.1	97.8	97.9	97.7	97.4	98.0	97.9
	97.6	97.9	97.7	97.8	98.0	97.9	97.6	97.9	98.1
	97.7	97.8	98.7	97.6	98.1	97.9	97.6	97.8	97.9

15.30   Refer to Exercise 15.29. Use a computer program to run an analysis of variance. Compare your results to those of Exercise 15.29, in which we coded the measurements for ease of calculations.

 15.31   A physician was interested in examining the relationship between work performed by individuals in an exercise tolerance test and the excess weight (as determined by standard weight–height tables) they carried. To do this, a random sample of 28 healthy adult females, ranging in age from 25 to 40, was selected from the community clinic during routine visits for physical examinations. The selection process was restricted so that seven persons were selected from each of the following weight classifications.

Normal weight (less than 10% underweight)
1%–10% overweight
11%–20% overweight
More than 20% overweight.

As part of the physical examination, each person was required to exercise on a bicycle ergometer until the onset of fatigue. The time to fatigue (in minutes) was recorded for each person. These data are given next.

Classification	Fatigue Time
Normal	25, 28, 19, 27, 23, 30, 35
1%–10% overweight	24, 26, 18, 16, 14, 12, 17
11%–20% overweight	15, 18, 17, 25, 12, 10, 23
More than 20% overweight	10, 9, 18, 14, 6, 4, 15

a. Identify the experimental design and write an appropriate statistical model.

b. Use $\alpha = .05$ and perform an analysis of variance.

15.32 Refer to Exercise 15.31.

a. How would you design an experiment to investigate the effects of age, gender, and excess weight on fatigue time?

b. Suppose the physician wanted to investigate the relationship among the quantitative variables percentage overweight, age, and fatigue time. Write a possible model.

15.33 An experiment was conducted to compare the heat loss for five different designs for commercial thermal panes. To do this, a sample of five panes of each design was obtained. The panes were then randomly assigned to the five exterior temperature settings (in °F) listed below so that each design appeared in each temperature setting. The interior temperature of the test was controlled at 68°F. Use the sample data here to compare the heat loss associated with the five pane designs.

			Pane Design		
Exterior Temperature Setting (°F)	A	B	C	D	E
80	8.4	8.6	9.2	9.1	10.3
60	8.4	8.7	9.3	9.4	10.7
40	8.9	9.1	9.7	9.9	10.9
20	10.4	10.7	10.6	10.5	11.3
0	10.8	11.2	11.1	11.3	11.6

a. Identify the experimental design and write an appropriate statistical model.

b. Use $\alpha = .05$ to run an analysis of variance.

15.34 Refer to Exercise 15.33. Use Tukey's W procedure to compare the treatment means. Set $\alpha = .05$.

15.35 Refer to Exercise 15.25. Suppose that rather than running three observations in each cell of the 4×3 factorial experiment at the same orange grove, a separate factorial experiment with one observation per cell is run at each of three different orange groves, as shown below.

Grove 1   Grove 2   Grove 3

4×3 factorial 4×3 factorial 4×3 factorial

Assume that the first, second, and third observations at each factor–level combination of Exercise 15.25 represent observations from groves 1, 2, and 3, respectively.

a. Identify the experimental design.

b. Write an appropriate linear statistical model, assuming there is no three-way interaction.

c. Perform an analysis of variance and give the level of significance for each test.

15.36 An experiment was conducted to examine the effects of different levels of reinforcement and different levels of isolation on children's ability to recall. A single analyst was to work with a random sample of 36 children selected from a relatively homogeneous group of fourth-grade students. Two levels of reinforcement (none and verbal) and three levels of isolation (20, 40, and 60 minutes) were to be used. Students were randomly assigned to the six treatment groups, with a total of six students being assigned to each group.

Each student was to spend a 30-minute session with the analyst. During this time, the student was to memorize a specific passage, with reinforcement provided as dictated by the group to which the student was assigned. Following the 30-minute session, the student was isolated for the time specified for his or her group and then tested for recall of the memorized passage. These data appear in the accompanying table.

Level of Reinforcement	Time of Isolation (Minutes)					
	20		40		60	
None	26	19	30	36	6	10
	23	18	25	28	11	14
	28	25	27	24	17	19
Verbal	15	16	24	26	31	38
	24	22	29	27	29	34
	25	21	23	21	35	30

Use the computer output shown here to draw your conclusions.

```
EXAMPLE OF TWO-WAY ANOVA WITH INTERACTION
FILE    NONAME
********************** ANALYSIS OF VARIANCE ****************************
     VAR03
     VAR01
     VAR02
**********************************************************************

     SOURCE OF          SUM OF              MEAN              SIGNIF
     VARIATION          SQUARES      DF     SQUARE      F     OF F
MAIN EFFECTS           352.223        3    117.408    7.441   0.001
  VAR01                196.000        1    196.000   12.423   0.001
  VAR02                156.223        2     78.112    4.951   0.014
2-WAY INTERACTIONS    1058.665        2    529.333   33.549   0.001
  VAR01   VAR02       1058.665        2    529.333   33.549   0.001
EXPLAINED             1410.888        5    282.177   17.885   0.001
RESIDUAL               473.330       30     15.778
TOTAL                 1884.218       35     53.835

36 CASES WERE PROCESSED.
 0 CASES (0.0 PCT) WERE MISSING.
```

 15.37 A food-processing plant has tested several different formulations of a new breakfast drink. Each of six panels rated the 12 different formulations obtained from combining one of three levels of sweetness, one of two levels of caloric content, and one of two colors.
a. Identify the design.
b. Write an appropriate model.
c. Give the analysis of variance table for this design.

15.38 Critique the SAS output shown here for the design described in Exercise 15.37. Has anything been omitted?

SOURCE	SUM OF SQUARES	DEGREES OF FREEDOM	MEAN SQUARE	F	TAIL PROBABILITY
MEAN	171307.55556	1	171307.55556	6234.40	0.0000
A	4149.52778	2	2074.76389	75.51	0.0000
B	624.22222	1	624.22222	22.72	0.0000
C	3200.00000	1	3200.00000	116.46	0.0000
AB	488.52778	2	244.26389	8.89	0.0004
AC	203.08333	2	101.54167	3.70	0.0307
BC	80.22222	1	80.22222	2.92	0.0927
ABC	24.19444	2	12.09722	0.44	0.6459
ERROR	1648.66667	60	27.47778		

15.39 Refer to Exercise 15.38. Assume there was no panel-to-panel variability (and hence MSE was an appropriate measure of error), and draw conclusions about the formulations. Based on the cell means shown here, which ones appear different? Would a series of profile plots help to explain what is happening? Explain.

	Color			
	1		2	
Sweetness Level	Caloric Level		Caloric Level	
	1	2	1	2
1	59.5	42.5	54.5	40.1
2	66.8	49.6	64.7	50.1
3	52.0	39.3	35.1	30.2

15.40 Job performance reviews were based on a numerical rating scale for random samples of 12, 9, and 18 employees from three divisions of a corporation.

Summary data are shown here:

Division	n	$\bar{y}$	s
Research	12	21.2	8.3
Development	9	15.4	7.3
Commercial	18	27.4	8.2

a. Identify the design.
b. Write an appropriate model.

15.41 Refer to Exercise 15.40. Perform an analysis of variance and draw conclusions. (Note: A high score is good.) Use $\alpha = .10$.

15.42 Stability data were generated for 2-mL vials manufactured with 30 mg/mL of active ingredient of a drug product. These data are shown here. Note that triplicate measurements were taken at each laboratory and time point.

Time (in months at 30°C)	Laboratory	mg/mL of Active Ingredient	pH
1	1	30.03	3.61
1	1	30.10	3.60
1	1	30.14	3.57
3	1	30.10	3.50
3	1	30.18	3.45
3	1	30.23	3.48
6	1	30.03	3.56
6	1	30.03	3.74
6	1	29.96	3.81
9	1	29.81	3.60
9	1	29.79	3.55
9	1	29.82	3.59
1	2	30.12	3.87
1	2	30.10	3.80
1	2	30.02	3.84
3	2	29.90	3.70
3	2	29.95	3.80
3	2	29.85	3.75
6	2	29.75	3.90
6	2	29.85	3.90
6	2	29.80	3.90
9	2	29.75	3.77
9	2	29.85	3.74
9	2	29.80	3.76

a. Write a model that could be used to relate y (either mg/mL of active ingredient or pH) to the independent variables, time (in months) and laboratory. (Hint: Use a dummy variable and allow for interaction.)

b. Interpret the βs in the model for part a.

c. Give an analysis of variance table for the model of part a without computing the necessary sums of squares.

15.43 Refer to Exercise 15.42. Computer output is shown for an analysis of variance for both dependent variables (i.e., $y_1 = $ mg/mL of active ingredient and $y_2 = $ pH). Draw conclusions about the stability of these 2-mL vials based on these analyses. Use $\alpha = .05$.

ANALYSIS OF VARIANCE FOR DEPENDENT VARIABLE MG_ML OF ACTIVE

GENERAL LINEAR MODELS PROCEDURE

DEPENDENT VARIABLE: MG_ML

SOURCE	DF	SUM OF SQUARES	MEAN SQUARE	F VALUE	PR > F	R-SQUARE	C.V.
MODEL	3	0.38500499	0.12833500	21.12	0.0001	0.760078	0.2602
ERROR	20	0.12152834	0.00607642		ROOT MSE		MG_ML MEAN
CORRECTED TOTAL	23	0.50653333			0.07795138		29.95666667

SOURCE	DF	TYPE I SS	F VALUE	PR > F	DF	TYPE III SS	F VALUE	PR > F
TIME	1	0.29170249	48.01	0.0001	1	0.17123832	28.18	0.0001
LAB	1	0.09126667	15.02	0.0009	1	0.04021903	6.62	0.0182
TIME*LAB	1	0.00203583	0.34	0.5692	1	0.00203583	0.34	0.5692

PARAMETER	ESTIMATE	T FOR HO: PARAMETER=0	PR > |T|	STD ERROR OF ESTIMATE
INTERCEPT	30.20553288	722.07	0.0001	0.04183178
TIME	-0.03941043	-5.31	0.0001	0.00742394
LAB	-0.15219955	-2.57	0.0182	0.05915907
TIME*LAB	0.00607710	0.58	0.5692	0.01049904

ANALYSIS OF VARIANCE FOR DEPENDENT VARIABLE PH

GENERAL LINEAR MODELS PROCEDURE

DEPENDENT VARIABLE: PH

SOURCE	DF	SUM OF SQUARES	MEAN SQUARE	F VALUE	PR > F	R-SQUARE	C.V.
MODEL	3	0.30464096	0.10154699	12.67	0.0001	0.655289	2.4196
ERROR	20	0.16025488	0.00801274		ROOT MSE		PH MEAN
CORRECTED TOTAL	23	0.46489583			0.08951393		3.69958333

SOURCE	DF	TYPE I SS	F VALUE	PR > F	DF	TYPE III SS	F VALUE	PR > F
TIME	1	0.00126193	0.16	0.6957	1	0.00663061	0.83	0.3738
LAB	1	0.29703750	37.07	0.0001	1	0.12982259	16.20	0.0007
TIME*LAB	1	0.00634152	0.79	0.3442	1	0.00634152	0.79	0.3842

PARAMETER	ESTIMATE	T FOR HO: PARAMETER=0	PR > |T|	STD ERROR OF ESTIMATE
INTERCEPT	3.55149660	73.93	0.0001	0.04803670
TIME	0.00775510	0.91	0.3738	0.00852514
LAB	0.27344671	4.03	0.0007	0.06793416
TIME*LAB	-0.01072562	-0.89	0.3842	0.01205636

15.44 The following models were fit to the data of Exercise 15.42.

$$y_i = \beta_0 + \beta_1 x_1 + \beta_2 x_2 + \beta_3 x_1 x_2 + \varepsilon,$$

where $y_1 = $ mg/mL and $y_2 = $ pH

$x_1 = $ time in months
$x_2 = 1$ if laboratory 2
  0 otherwise.

a. How might you graph these data and the fitted lines? Does a linear relationship between y_i and x_1 seem to fit these data? For both laboratories?

b. For both y_1 and y_2, estimate the difference in the slopes of the two regression lines using a 95% confidence interval.

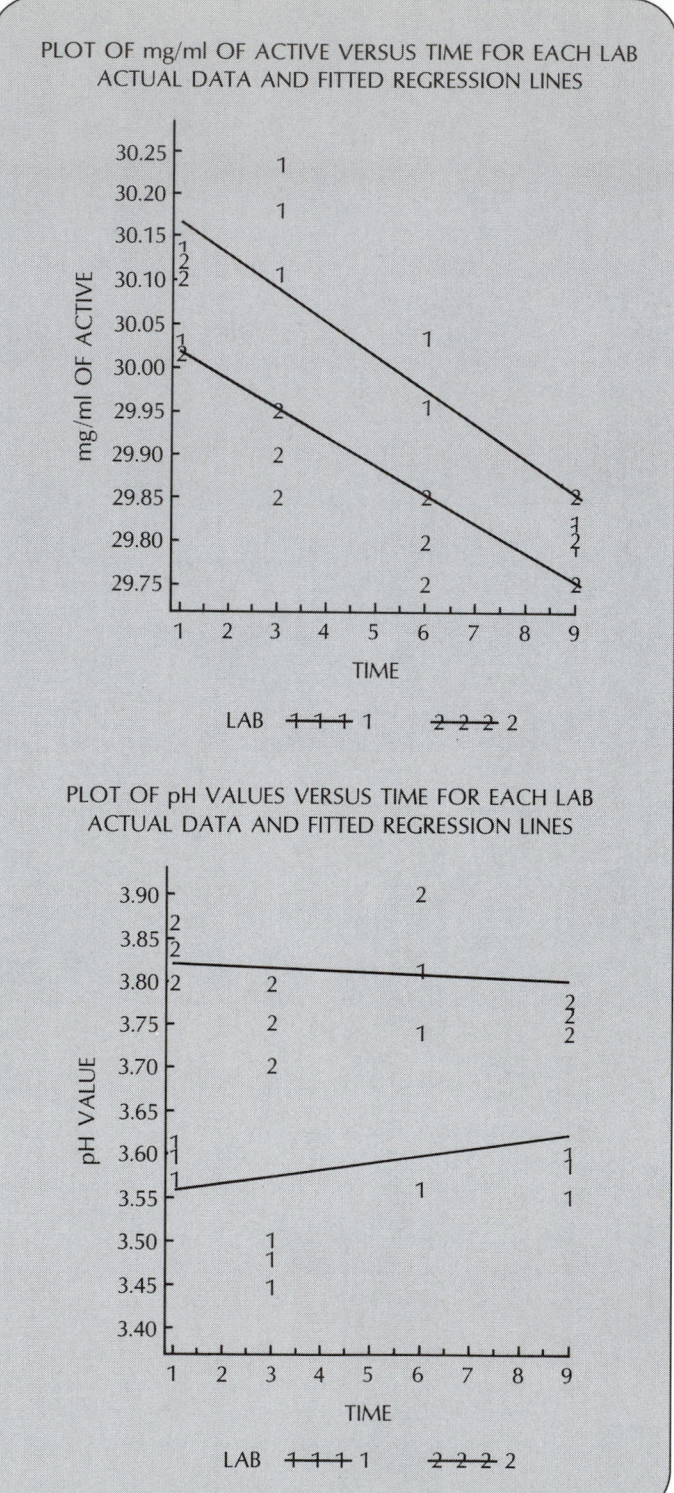

15.45   Refer to Exercise 15.42. The same type of data on mg/mL and pH were generated at 40°C as were obtained at 30°C in Exercise 15.42. These data are shown here.

Time (in months at 40°C)	Laboratory	mg/mL of Active Ingredient	pH
1	1	30.08	3.61
1	1	30.10	3.60
1	1	30.14	3.59
3	1	30.03	3.39
3	1	30.18	3.45
3	1	30.26	3.29
6	1	29.90	3.63
6	1	29.90	3.71
6	1	29.96	3.65
9	1	29.81	3.51
9	1	29.85	3.38
9	1	29.72	3.32
1	2	30.12	3.80
1	2	30.10	3.70
1	2	30.02	3.81
3	2	29.90	3.70
3	2	29.85	3.80
3	2	29.80	3.75
6	2	29.75	3.80
6	2	29.70	3.70
6	2	29.75	3.70
9	2	29.65	3.64
9	2	29.75	3.68
9	2	29.70	3.60

a. Without computing the necessary sums of squares, give an analysis of variance table for this three-factor experiment. Include all possible interactions.

b. Computer output is shown here for the analysis of variance of part a for both dependent variables (i.e., $y_1 =$ mg/mL and $y_2 =$ pH). Interpret the results of these analyses.

```
                    ANALYSIS OF VARIANCE FOR DEPENDENT VARIABLE MG_ML OF ACTIVE
                              GENERAL LINEAR MODELS PROCEDURE

DEPENDENT VARIABLE: MG_ML

SOURCE              DF      SUM OF SQUARES      MEAN SQUARE    F VALUE    PR > F    R-SQUARE      C.V.
MODEL                7         0.98852273       -0.14121753     22.08     0.0001   0.794382     0.2672
ERROR               40         0.25586893        0.00639672               ROOT MSE          MG_ML MEAN
CORRECTED TOTAL     47         1.24439167                                 0.07997952        29.93708333
                                                                                          (continues)
```

SOURCE	DF	TYPE I SS	F VALUE	PR > F	DF	TYPE III SS	F VALUE	PR > F
TIME	1	0.72777149	113.77	0.0001	1	0.39245581	61.35	0.0001
LAB	1	0.22963333	35.90	0.0001	1	0.08149613	12.74	0.0009
TIME*LAB	1	0.00107959	0.17	0.6834	1	0.00107959	0.17	0.6834
TEMP	1	0.01840833	2.88	0.0976	1	0.00000857	0.00	0.9710
TIME*TEMP	1	0.00797194	1.25	0.2709	1	0.00170139	0.27	0.6089
LAB*TEMP	1	0.00270000	0.42	0.5196	1	0.00000346	0.00	0.9816
TIME*LAB*TEMP	1	0.00095805	0.15	0.7008	1	0.00095805	0.15	0.7008

ANALYSIS OF VARIANCE FOR DEPENDENT VARIABLE PH

GENERAL LINEAR MODELS PROCEDURE

DEPENDENT VARIABLE: PH

SOURCE	DF	SUM OF SQUARES	MEAN SQUARE	F VALUE	PR > F	R-SQUARE	C.V.
MODEL	7	0.69408435	0.09915491	10.08	0.0001	0.638258	2.7108
ERROR	40	0.39338231	0.00983456		ROOT MSE		PH MEAN
CORRECTED TOTAL	47	1.08746667			0.09916934		3.65833333

SOURCE	DF	TYPE I SS	F VALUE	PR > F	DF	TYPE III SS	F VALUE	PR > F
TIME	1	0.01158186	1.18	0.2843	1	0.00026670	0.03	0.8700
LAB	1	0.56767500	57.72	0.0001	1	0.22710248	23.09	0.0001
TIME*LAB	1	0.00714427	0.73	0.3991	1	0.00714427	0.73	0.3991
TEMP	1	0.08167500	8.30	0.0063	1	0.00007612	0.01	0.9303
TIME*TEMP	1	0.02491888	2.53	0.1193	1	0.01728914	1.76	0.1924
LAB*TEMP	1	0.00030000	0.03	0.8622	1	0.00108908	0.11	0.7410
TIME*LAB*TEMP	1	0.00078934	0.08	0.7784	1	0.00078934	0.08	0.7784

15.46   Shelf life data for a given product (% of theory) are shown here for each of eight batches of product tested at 0, 3, 6, 12, and 24 weeks.

LISTING OF DATA

BATCH	MONTH	THEORY
1	0	100.7
1	3	100.7
1	6	101.3
1	12	101.0
1	24	102.0
2	0	98.7
2	3	102.3
2	6	100.0
2	12	101.7
2	24	99.3
3	0	102.0
3	3	101.3
3	6	101.7
3	12	101.3
3	24	101.2
4	0	99.7
4	3	100.3
4	6	100.7
4	12	100.7
4	24	100.2
5	0	100.0
5	3	100.7
5	6	101.3
5	12	100.3
5	24	100.7

(*continues*)

```
          LISTING OF DATA
     BATCH    MONTH    THEORY

       6        0       98.5
       6        3       99.4
       6        6       99.3
       6       12      100.3
       6       24      100.0
       7        0       99.7
       7        3      100.2
       7        6      100.7
       7       12       99.3
       7       24      100.2
       8        0      100.3
       8        3       99.0
       8        6      101.7
       8       12      101.0
       8       24      100.0
```

Three different analyses are shown here. Identify the objective(s) of each analysis and draw conclusions. Use $\alpha = .05$.

ANALYSIS OF VARIANCE
FULL MODEL
GENERAL LINEAR MODELS PROCEDURE

DEPENDENT VARIABLE: THEORY

SOURCE	DF	SUM OF SQUARES	MEAN SQUARE	F VALUE	PR > F	R-SQUARE	C.V.
MODEL	15	16.04425000	1.06961667	1.46	0.1977	0.477351	0.8514
ERROR	24	17.56675000	0.73194792		ROOT MSE		THEORY MEAN
CORRECTED TOTAL	39	33.61100000			0.85553955		100.48500000

SOURCE	DF	TYPE I SS	F VALUE	PR > F	DF	TYPE III SS	F VALUE	PR > F
MONTH	1	0.33153125	0.45	0.5074	1	0.33153125	0.45	0.5074
BATCH	7	13.50300000	2.64	0.0360	7	9.47932169	1.85	0.1234
MONTH*BATCH	7	2.20971875	0.43	0.8729	7	2.20971875	0.43	0.8729

ANALYSIS OF VARIANCE
REDUCED MODEL
GENERAL LINEAR MODELS PROCEDURE

DEPENDENT VARIABLE: THEORY

SOURCE	DF	SUM OF SQUARES	MEAN SQUARE	F VALUE	PR > F	R-SQUARE	C.V.
MODEL	8	13.83453125	1.72931641	2.71	0.0218	0.411607	0.7949
ERROR	31	19.77646875	0.63795060		ROOT MSE		THEORY MEAN
CORRECTED TOTAL	39	33.61100000			0.79871810		100.48500000

SOURCE	DF	TYPE I SS	F VALUE	PR > F	DF	TYPE III SS	F VALUE	PR > F
MONTH	1	0.33153125	0.52	0.4764	1	0.33153125	0.52	0.4764
BATCH	7	13.50300000	3.02	0.0153	7	13.50300000	3.02	0.0153

(continues)

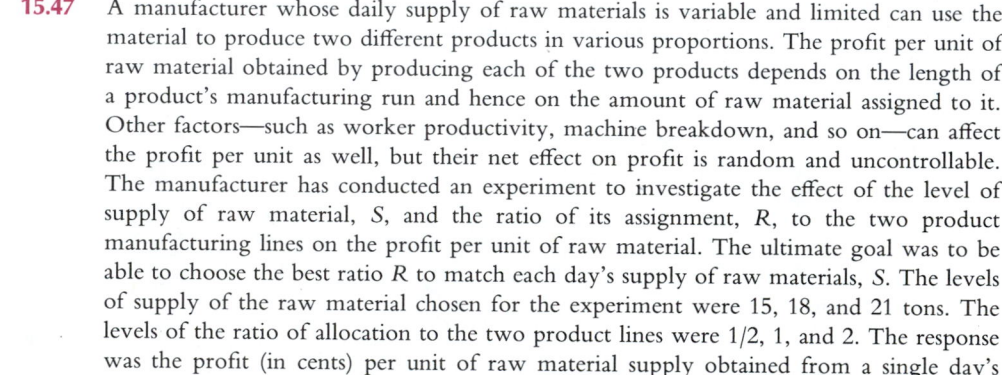

ANALYSIS OF VARIANCE
SIMPLE REGRESSION

GENERAL LINEAR MODELS PROCEDURE

DEPENDENT VARIABLE: THEORY

SOURCE	DF	SUM OF SQUARES	MEAN SQUARE	F VALUE	PR > F	R-SQUARE	C.V.
MODEL	1	0.33153125	0.33153125	0.38	0.5420	0.009864	0.9313
ERROR	38	33.27946875	0.87577549		ROOT MSE		THEORY MEAN
CORRECTED TOTAL	39	33.61100000			0.93582877		100.48500000

SOURCE	DF	TYPE I SS	F VALUE	PR > F	DF	TYPE III SS	F VALUE	PR > F
MONTH	1	0.33153125	0.38	0.5420	1	0.33153125	0.38	0.5420

PARAMETER	ESTIMATE	T FOR HO: PARAMETER=0	PR > |T|	STD ERROR OF ESTIMATE
INTERCEPT	100.38843750	465.41	0.0001	0.21569787
MONTH	0.01072917	0.62	0.5420	0.01743814

15.47 A manufacturer whose daily supply of raw materials is variable and limited can use the material to produce two different products in various proportions. The profit per unit of raw material obtained by producing each of the two products depends on the length of a product's manufacturing run and hence on the amount of raw material assigned to it. Other factors—such as worker productivity, machine breakdown, and so on—can affect the profit per unit as well, but their net effect on profit is random and uncontrollable. The manufacturer has conducted an experiment to investigate the effect of the level of supply of raw material, S, and the ratio of its assignment, R, to the two product manufacturing lines on the profit per unit of raw material. The ultimate goal was to be able to choose the best ratio R to match each day's supply of raw materials, S. The levels of supply of the raw material chosen for the experiment were 15, 18, and 21 tons. The levels of the ratio of allocation to the two product lines were 1/2, 1, and 2. The response was the profit (in cents) per unit of raw material supply obtained from a single day's production. Three replications of each combination were conducted in a random sequence. The data for the 27 days are shown in the following table.

Ratio of Raw Material Allocation (R)	Raw Material Supply (tons)		
	15	18	21
1/2	22, 20, 21	21, 19, 20	19, 18, 20
1	21, 20, 19	23, 24, 22	20, 19, 21
2	17, 18, 16	21, 11, 20	20, 22, 24

a. Draw conclusions based on the analysis of variance shown here. Use $\alpha = .05$.

b. Identify the two best combinations of R and S. Are these two combinations significantly different? Use a procedure that limits the error rate of all pairwise comparisons of combinations to be no more than 0.05.

```
     LISTING OF DATA
 RATIO    SUPPLY    PROFIT
  0.5      15        22
  0.5      15        20
  0.5      15        21
  0.5      18        21
  0.5      18        19
  0.5      18        20
  0.5      21        19
  0.5      21        18
  0.5      21        20
  1.0      15        21
  1.0      15        20
  1.0      15        19
  1.0      18        23
  1.0      18        24
  1.0      18        22
  1.0      21        20
  1.0      21        19
  1.0      21        21
  2.0      15        17
  2.0      15        18
  2.0      15        16
  2.0      18        21
  2.0      18        11
  2.0      18        20
  2.0      21        20
  2.0      21        22
  2.0      21        24
```

ANALYSIS OF VARIANCE
2 FACTOR FACTORIAL

GENERAL LINEAR MODELS PROCEDURE

DEPENDENT VARIABLE: PROFIT

SOURCE	DF	SUM OF SQUARES	MEAN SQUARE	F VALUE	PR > F	R-SQUARE	C.V.
MODEL	8	93.18518519	11.64814815	2.54	0.0482	0.529907	10.7550
ERROR	18	82.66666667	4.59259259		ROOT MSE		PROFIT MEAN
CORRECTED TOTAL	26	175.85185185			2.14303350		19.92592593

SOURCE	DF	TYPE I SS	F VALUE	PR > F	DF	TYPE III SS	F VALUE	PR > F
RATIO	2	22.29629630	2.43	0.1166	2	22.29629630	2.43	0.1166
SUPPLY	2	4.96296296	0.54	0.5917	2	4.96296296	0.54	0.5917
RATIO*SUPPLY	4	65.92592593	3.59	0.0255	4	65.92592593	3.59	0.0255

ANALYSIS OF VARIANCE
2 FACTOR FACTORIAL

GENERAL LINEAR MODELS PROCEDURE

MEANS

RATIO	SUPPLY	N	PROFIT
1	15	3	20.0000000
1	18	3	23.0000000
1	21	3	20.0000000
2	15	3	17.0000000
2	18	3	17.3333333
2	21	3	22.0000000
0.5	15	3	21.0000000
0.5	18	3	20.0000000
0.5	21	3	19.0000000

15.48 A manufacturer frequently sends small packages to a customer in another city via air freight, and, in many cases, it is important for a package to reach the customer as soon as possible. Three different firms offer air freight service, including pickup and delivery, on a 24-hour basis. The head of the manufacturer's shipping department would like to know whether the firms differ in speed of service and whether the time of day makes any difference. An experiment is designed to investigate these issues. Packages are sent at random times, and the air freight firm used for each package also is randomly chosen. The customer records the time that each package arrives, so that the time elapsed during shipment can be determined. These times are rounded to the nearest hour. The experimental results for a total of 54 packages are shown in the following table.

	Firm		
Time	1	2	3
Morning	8, 6, 6, 12, 7, 8	11, 11, 9, 10, 8, 11	7, 4, 6, 4, 9, 7
Afternoon	7, 10, 8, 11, 9, 11	10, 13, 10, 12, 11, 10	10, 8, 6, 5, 8, 6
Night	13, 11, 14, 11, 9, 12	12, 6, 9, 9, 10, 6	8, 11, 9, 9, 10, 12

a. Suppose that the preceding analysis were to be done using the dummy variable approach instead of the AOV approach. How many dummy variables would be needed to include both main effects and interaction effects in the model? What would the R^2 for this regression be?

b. What evidence is relevant for deciding whether the choice of best firm will be different at different times of the day? What conclusion would you draw using a 5% level of significance? Construct a graph that depicts the nature of any differences in firm as a function of the time of day.

c. Does any firm appear to be better than the other two firms? How could you compare the best firm and the second-best firm using a confidence interval?

```
        LISTING OF DATA
   TIME        FIRM    SPEED
   MORNING       1       8
   MORNING       1       6
   MORNING       1       6
   MORNING       1      12
   MORNING       1       7
   MORNING       1       8
   MORNING       2      11
   MORNING       2      11
   MORNING       2       9
   MORNING       2      10
   MORNING       2       8
   MORNING       2      11
   MORNING       3       7
   MORNING       3       4
   MORNING       3       6
   MORNING       3       4
   MORNING       3       9
```

(*continues*)

```
        LISTING OF DATA

TIME       FIRM    SPEED
MORNING      3        7
AFTRNOON     1        7
AFTRNOON     1       10
AFTRNOON     1        8
AFTRNOON     1       11
AFTRNOON     1        9
AFTRNOON     1       11
AFTRNOON     2       10
AFTRNOON     2       13
AFTRNOON     2       10
AFTRNOON     2       12
AFTRNOON     2       11
AFTRNOON     2       10
AFTRNOON     3       10
AFTRNOON     3        8
AFTRNOON     3        6
AFTRNOON     3        5
AFTRNOON     3        8
AFTRNOON     3        6
NIGHT        1       13
NIGHT        1       11
NIGHT        1       14
NIGHT        1       11
NIGHT        1        9
NIGHT        1       12
NIGHT        2       12
NIGHT        2        6
NIGHT        2        9
NIGHT        2        9
NIGHT        2       10
NIGHT        2        6
NIGHT        3        8
NIGHT        3       11
NIGHT        3        9
NIGHT        3        9
NIGHT        3       10
NIGHT        3       12
```

ANALYSIS OF VARIANCE
2 FACTOR FACTORIAL

GENERAL LINEAR MODELS PROCEDURE

DEPENDENT VARIABLE: SPEED

SOURCE	DF	SUM OF SQUARES	MEAN SQUARE	F VALUE	PR > F	R-SQUARE	C.V.
MODEL	8	154.37037037	19.29629630	6.06	0.0001	0.518537	19.6682
ERROR	45	143.33333333	3.18518519		ROOT MSE		SPEED MEAN
CORRECTED TOTAL	53	297.70370370			1.78470871		9.07407407

SOURCE	DF	TYPE I SS	F VALUE	PR > F	DF	TYPE III SS	F VALUE	PR > F
TIME	2	38.25925926	6.01	0.0049	2	38.25925926	6.01	0.0049
FIRM	2	50.03703704	7.85	0.0012	2	50.03703704	7.85	0.0012
TIME*FIRM	4	66.07407407	5.19	0.0016	4	66.07407407	5.19	0.0016

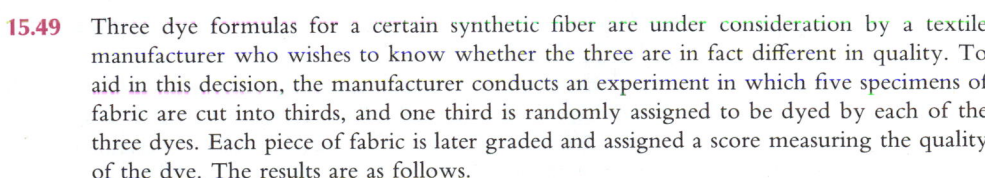

15.49 Three dye formulas for a certain synthetic fiber are under consideration by a textile manufacturer who wishes to know whether the three are in fact different in quality. To aid in this decision, the manufacturer conducts an experiment in which five specimens of fabric are cut into thirds, and one third is randomly assigned to be dyed by each of the three dyes. Each piece of fabric is later graded and assigned a score measuring the quality of the dye. The results are as follows.

	Fabric Specimen				
Dye	1	2	3	4	5
A	74	78	76	82	77
B	81	86	90	93	73
C	95	99	90	87	93

a. Identify the design.

b. Run an analysis of variance and draw conclusions about the dyes. Use $\alpha = .05$.

c. Give a measure of the efficiency of this design to one not blocking on fabric specimens.

```
LISTING OF DATA

DYE    SPECIMEN    QUALITY
 A        1          74
 A        2          78
 A        3          76
 A        4          82
 A        5          77
 B        1          81
 B        2          86
 B        3          90
 B        4          93
 B        5          73
 C        1          95
 C        2          99
 C        3          90
 C        4          87
 C        5          93


              ANALYSIS OF VARIANCE
            RANDOMIZED BLOCK DESIGN

         GENERAL LINEAR MODELS PROCEDURE

DEPENDENT VARIABLE: QUALITY
```

SOURCE	DF	SUM OF SQUARES	MEAN SQUARE	F VALUE	PR > F	R-SQUARE	C.V.
MODEL	6	688.00000000	114.66666667	3.34	0.0596	0.714484	6.9022
ERROR	8	274.93333333	34.36666667		ROOT MSE		QUALITY MEAN
CORRECTED TOTAL	14	962.93333333			5.86230899		84.93333333

SOURCE	DF	TYPE I SS	F VALUE	PR > F	DF	TYPE III SS	F VALUE	PR > F
DYE	2	593.73333333	8.64	0.0100	2	593.73333333	8.64	0.0100
SPECIMEN	4	94.26666667	0.69	0.6216	4	94.26666667	0.69	0.6216

15.50 An experiment tested the effect of factory music on workers' production. Four music programs (A, B, C, D) were compared with no music (E). Each program was played for one entire day, and five replications for each program were desired. The length of the experiment was thus 5 weeks. To control for variation in week and day of week, a Latin square design was adopted for the 25 days of the experiment. Each program was played once on each day of the week and once each week.

Week	Monday		Tuesday		Wednesday		Thursday		Friday	
1	A	133	B	139	C	140	D	140	E	145
2	B	136	C	141	D	143	E	146	A	139
3	C	140	A	138	E	142	B	139	D	139
4	D	129	E	132	A	137	C	136	B	140
5	E	132	D	144	B	143	A	142	C	142

Use the output shown here to analyze these data, and draw conclusions.

```
              LISTING OF DATA
WEEK    DAY          MUSIC      OUTPUT
 1      MONDAY        A          133
 1      TUESDAY       B          139
 1      WEDNDAY       C          140
 1      THURSDAY      D          140
 1      FRIDAY        E          145
 2      MONDAY        B          136
 2      TUESDAY       C          141
 2      WEDNDAY       D          143
 2      THURSDAY      E          146
 2      FRIDAY        A          139
 3      MONDAY        C          140
 3      TUESDAY       A          138
 3      WEDNDAY       E          142
 3      THURSDAY      B          139
 3      FRIDAY        D          139
 4      MONDAY        D          129
 4      TUESDAY       E          132
 4      WEDNDAY       A          137
 4      THURSDAY      C          136
 4      FRIDAY        B          140
 5      MONDAY        E          132
 5      TUESDAY       D          144
 5      WEDNDAY       B          143
 5      THURSDAY      A          142
 5      FRIDAY        C          142
```

```
                     ANALYSIS OF VARIANCE
                     LATIN SQUARE DESIGN
                 GENERAL LINEAR MODELS PROCEDURE

DEPENDENT VARIABLE: OUTPUT
```

SOURCE	DF	SUM OF SQUARES	MEAN SQUARE	F VALUE	PR > F	R-SQUARE	C.V.
MODEL	12	313.12000000	26.09333333	2.59	0.0561	0.721741	2.2805
ERROR	12	120.72000000	10.06000000		ROOT MSE		OUTPUT MEAN
CORRECTED TOTAL	24	433.84000000			3.17175031		139.08000000

SOURCE	DF	TYPE I SS	F VALUE	PR > F	DF	TYPE III SS	F VALUE	PR > F
MUSIC	4	11.84000000	0.29	0.8761	4	11.84000000	0.29	0.8761
WEEK	4	123.44000000	3.07	0.0589	4	123.44000000	3.07	0.0589
DAY	4	177.84000000	4.42	0.0200	4	177.84000000	4.42	0.0200

15.51 The yields of wheat (in pounds) are shown here for 25 plots (5 farms, 5 plots per farm). The treatments (fertilizers) applied to each plot are shown in parentheses.

Farm		Plot								
	1		2		3		4		5	
1	(D)	10.3	(E)	8.6	(A)	6.7	(C)	7.6	(B)	5.8
2	(E)	8.8	(B)	6.7	(C)	6.7	(A)	4.8	(D)	6.0
3	(A)	6.3	(C)	8.3	(B)	6.8	(D)	8.0	(E)	8.8
4	(C)	8.9	(D)	7.4	(E)	8.2	(B)	6.2	(A)	4.4
5	(B)	7.3	(A)	4.4	(D)	7.7	(E)	6.8	(C)	6.7

a. Identify the designs.

b. Do an analysis of variance and draw conclusions concerning the five fertilizers. Use $\alpha = .01$.

15.52   Refer to Exercise 15.51. Run a multiple-comparison procedure to make all pairwise comparisons of the treatment means. Identify which error rate was controlled.

ANALYSIS OF VARIANCE FOR SOME UNBALANCED DESIGNS

16.1   INTRODUCTION

We examined the analysis of variance for balanced designs in Chapters 13 and 15, where we used appropriate shortcut formulas (and corresponding computer solutions) to construct AOV tables and set up hypothesis tests. In Chapter 15, we also considered another way of performing an analysis of variance. We found that the null hypothesis under test in an analysis of variance can be expressed in terms of one or more βs in the general linear model. We also saw that the sum of squares associated with a source of variability in the analysis of variance table can be found as the drop in the sum of squares for error obtained from fitting reduced and complete models. Although we did not advocate the use of complete and reduced models for obtaining the sums of squares for sources of variability in balanced designs, we did indicate that the procedure was completely general and could be used for any experimental design. In particular, in this chapter, we will make use of complete and reduced models for obtaining the sums of squares in the analysis for *unbalanced designs*, where shortcut formulas are no longer readily available and easy to apply.

You might ask why an experimenter would run a study using an unbalanced design, especially since unbalanced designs seem to be more difficult to analyze. In point of fact, most studies do begin by using a balanced design, but for any one of many different reasons, the experimenter is unable to obtain the same number of observations per cell as dictated by the balanced design being employed. Consider a study of three different weight-reducing agents in which five different clinics (blocks) are employed and patients are to be randomly assigned to the three treatment groups according to a randomized block design. Even if the experimenter plans to have five overweight persons assigned to each treatment at each clinic, the final count will almost certainly show an imbalance of persons assigned to each treatment group. Almost every clinic could be expected to have a few people who would not complete the study. Some people might move from the community, others might drop out due to a lack of efficacy in the program, and so on. In addition, the experimenter might find it impossible to locate 15 overweight people at each clinic who are willing to participate in the study. Because an unbalanced design at the end of a study occurs quite often, we must learn how to analyze data arising from unbalanced designs.

16.2 ▼ A RANDOMIZED BLOCK DESIGN WITH ONE OR MORE MISSING OBSERVATIONS

unbalanced design

Any time the number of observations is not the same for all factor–level combinations, we call the design **unbalanced**. Thus, a randomized block design or a Latin square design with one or more missing observations is an unbalanced design. We will begin our examination by considering a simple case, a randomized block design with one missing observation.

value of missing observation

The analysis of variance for a randomized block design with one missing observation can be performed rather easily by using the shortcut formulas for a balanced design, after we have estimated the **value of the missing observation**. The formula for the missing observation M is given by

$$M = \frac{ty_{i\cdot} + by_{\cdot j} - y_{\cdot\cdot}}{(t-1)(b-1)},$$

where t is the number of treatments, b is the number of blocks, $y_{i\cdot}$ is the sum of all the observations on the treatment with the missing observation, $y_{\cdot j}$ is the sum of all measurements in the block with the missing observation, and $y_{\cdot\cdot}$ is the sum of all the measurements.

We illustrate the analysis of variance for this design with an example.

E X A M P L E   16.1   ▼

An experiment was conducted to determine the nutritional value of diets for cows that are supplemented by whey. Five dairies were involved in the study. Each cow in a sample of four cows from a dairy was randomly assigned to one of the four treatment groups, so that a total of five cows were in each treatment group.

Treatment 1:   water only
Treatment 2:   whey plus 30.2 L of water/day
Treatment 3:   whey plus 15.1 L of water/day
Treatment 4:   whey only.

In addition to the liquid portion of the diet listed for each treatment group, each cow was fed 7.5 kg of grain per day.

One response of interest was the amount of hay consumed per day. These data (in kilograms per animal) are listed in Table 16.1. Unfortunately, as can be seen from the data, the cow on diet 4 from dairy 2 was dropped from the study and no replacement was made. The cow developed an infection (unrelated to the treatment) and was dropped from the study for safety reasons.

T A B L E   16.1

Consumption of Hay for Cows, Example 16.1

	Treatment			
Dairy	1	2	3	4
1	15.4	9.6	9.5	8.4
2	14.8	9.3	9.4	—
3	15.9	9.8	9.7	9.3
4	15.5	9.4	9.2	8.1
5	14.7	9.2	9.0	7.9

Estimate the missing value and then perform an analysis of variance. Use $\alpha = .01$.

Solution   For this randomized block design with $b = 5$ and $t = 4$, the quantities $y_{i\cdot}$, $y_{\cdot j}$, and $y_{\cdot\cdot}$ are defined as follows:

$y_{i\cdot}$ = sum of all observations on treatment 4
$\quad = 8.4 + 9.3 + 8.1 + 7.9 = 33.7$
$y_{\cdot j}$ = sum of all observations in block 2
$\quad = 14.8 + 9.3 + 9.4 = 33.5$
$y_{\cdot\cdot}$ = sum of all measurements
$\quad = 15.4 + 9.6 + \cdots + 7.9 = 204.1.$

The estimate of the missing value is

$$M = \frac{ty_{i\cdot} + by_{\cdot j} - y_{\cdot\cdot}}{(t-1)(b-1)} = \frac{4(33.7) + 5(33.5) - 204.1}{3(4)}$$

$$= \frac{98.2}{12} = 8.2.$$

Having estimated the missing value, we can compute sums of squares for our analysis of variance by using the shortcut formulas of Chapter 15. The treatment and block totals are given by

$y_{1\cdot} = 76.3$	$y_{\cdot 1} = 42.9$
$y_{2\cdot} = 47.3$	$y_{\cdot 2} = 41.7$
$y_{3\cdot} = 46.8$	$y_{\cdot 3} = 44.7$
$y_{4\cdot} = 41.9$	$y_{\cdot 4} = 42.2$
	$y_{\cdot 5} = \underline{40.8}$
Totals 212.3	212.3

Note that the new totals for treatment 4 and for block 2 incorporate the estimated missing observation. Similarly, the sum of all measurements includes the estimated missing value.

$$\text{SST} = \frac{(76.3)^2 + \cdots + (41.9)^2}{5} - \frac{(212.3)^2}{20} = 2{,}400.97 - 2{,}253.56 = 147.41$$

$$\text{SSB} = \frac{(42.9)^2 + \cdots + (40.8)^2}{4} - 2{,}253.56 = 2.16$$

$$\text{TSS} = \sum_{ij} y_{ij}^2 - \frac{y_{\cdot\cdot}^2}{bt} = (15.4)^2 + (9.6)^2 + \cdots + (7.9)^2 - 2{,}253.56$$

$$= 2{,}403.69 - 2{,}253.56 = 150.13$$

By subtraction, SSE = .56.

The only difference in the analysis of variance table for unbalanced and balanced randomized block designs is that since n refers to the number of actual observations, the error for an unbalanced design loses one degree of freedom for each missing observation when compared to the corresponding balanced design. The AOV table for our example is shown in Table 16.2.

The F tests for treatments and blocks are both significant, using $\alpha = .01$ (the critical values of F are 6.22 and 5.67, respectively). As can be seen from the data, those cows on treatment 1 (water only) consumed much more hay than cows on any of the diets supplemented with whey.

T A B L E 16.2
AOV Table for the Data of Example 16.1

Source	SS	df	MS	F
Treatments	147.41	3	49.14	982.80
Blocks	2.16	4	.54	10.80
Error	.56	11	.05	
Totals	150.13	18		

comparisons among treatment means

Having seen an analysis of variance, we may wish to make certain **comparisons among the treatment means.** We'll run pairwise comparisons using Fisher's least significant difference. The least significant difference between the treatment with a missing observation and any other treatment mean is

LSD

$$LSD = t_{\alpha/2} \sqrt{MSE\left(\frac{2}{b} + \frac{t}{b(b-1)(t-1)}\right)}.$$

For any pair of treatments with no missing value, the least significant difference is as before; namely,

$$LSD = t_{\alpha/2} \sqrt{\frac{2MSE}{b}}.$$

fitting complete and reduced models

The formulas for estimating missing observations in a randomized block design become more complicated with more missing data, as do the formulas for least significant differences. Because of this, we will consider **fitting complete and reduced models** to analyze unbalanced designs. We will illustrate the procedure first by examining an unbalanced randomized block design.

Because it would require more data input for a computer solution using the general linear model format with dummy variables presented in Chapters 11 and 12, we will represent the complete and reduced models for testing treatments as follows:

models

complete model: $y_{ij} = \mu + \beta_j + \alpha_i + \varepsilon_{ij}$
(model 1)
reduced model: $y_{ij} = \mu + \beta_j + \varepsilon_{ij}$
(model 2),

where β_j is the jth block effect and α_i is the ith treatment effect.

By fitting model 1 (using SAS or other computer software), we obtain SSE_1. Similarly, a fit of model 2 yields SSE_2. The difference in the two sums of squares for error, $SSE_2 - SSE_1$, gives the drop in the sum of squares due to treatments. Since this is an unbalanced design, the block effects do not cancel out

SST$_{adj}$

when comparing treatment means as they do in a balanced randomized block design (see Chapter 15). The difference in the sums of squares, $SSE_2 - SSE_1$, has been adjusted for any effects due to blocks caused by the imbalance in the design. This difference is called the **sum of squares due to treatments adjusted for blocks**.

$$SSE_2 - SSE_1 = SST_{adj}$$

The sum of squares due to blocks **unadjusted for any treatment differences** is obtained by subtraction:

$$SSB = TSS - SST_{adj} - SSE,$$

where SSE and TSS are sums of squares from the complete model. (Note: We could also obtain SSB, the uncorrected sum of squares for blocks, using the shortcut formula of Section 15.3).

AOV table, treatments

The **analysis of variance table for testing the effect of treatments** is shown in Table 16.3. In the table, n is the number of actual observations.

T A B L E   16.3
AOV Table for Testing the Effects of Treatments, Unbalanced Randomized Block Design

Source	SS	df	MS	F
Blocks	SSB	$b - 1$	—	—
Treatments$_{adj}$	SST$_{adj}$	$t - 1$	MST$_{adj}$	MST$_{adj}$/MSE
Error	SSE	by subtraction	MSE	
Totals	TSS	$n - 1$		

The corresponding sum of squares for testing the effect of blocks has the same complete model (model 1) as before, and

$$y_{ij} = \mu + \alpha_i + \varepsilon_{ij}$$

SSB$_{adj}$

is the reduced model (model 2). The sum of squares drop, $SSE_2 - SSE_1$, is the **sum of squares due to blocks after adjusting for the effects of treatments**. By subtraction, we obtain

$$SST = TSS - SSB_{adj} - SSE.$$

AOV table, blocks

The **AOV table** is shown in Table 16.4.

Note that SST and SST$_{adj}$ are not the same quantity in an unbalanced design; they will be the same only for a balanced design. Similarly, SSB and SSB$_{adj}$ are

T A B L E 16.4

AOV Table for Testing Effects of Blocks, Unbalanced Randomized Block Design

Source	SS	df	MS	F
Blocks$_{adj}$	SSB$_{adj}$	$b - 1$	MSB$_{adj}$	MSB$_{adj}$/MSE
Treatments	SST	$t - 1$	—	—
Error	SSE	by subtraction	MSE	—
Totals	TSS	$n - 1$		

different quantities in an unbalanced design. For an unbalanced design, we have the following identities:

$$\text{TSS} = \text{SST}_{adj} + \text{SSB} + \text{SSE}$$
$$= \text{SST} + \text{SSB}_{adj} + \text{SSE},$$

but

$$\text{TSS} \neq \text{SST}_{adj} + \text{SSB}_{adj} + \text{SSE}.$$

▼ EXERCISES

Basic Techniques

16.1 Refer to the data of Example 16.1 and the SAS computer output shown here. Some items are notated to help you identify quantities in the output.

```
SAS
DATA COWS;
INPUT Y BLOCKS TREATS;
CARDS;

   19 OBSERVATIONS IN DATA SET COWS    3 VARIABLES
PROC PRINT;

   OBS      Y      BLOCKS     TREATS
    1     15.4       1          1
    2      9.6       1          2
    3      9.5       1          3
    4      8.4       1          4
    5     14.8       2          1
    6      9.3       2          2
    7      9.4       2          3
    8     15.9       3          1
    9      9.8       3          2
   10      9.7       3          3
   11      9.3       3          4
   12     15.5       4          1
   13      9.4       4          2
   14      9.2       4          3
   15      8.1       4          4
   16     14.7       5          1
   17      9.2       5          2
   18      9.0       5          3
   19      7.9       5          4
```

(continues)

```
PROC GLM;
CLASSES BLOCKS TREATS;
MODEL Y = BLOCKS TREATS;
TITLE "EXERCISE 15.1";
```

GENERAL LINEAR MODELS PROCEDURE

SOURCE	DF	SUM OF SQUARES	MEAN SQUARE	F VALUE	PROB>F	R-SQUARE	C.V.
REGRESSION	7	143.41548246	20.48792607	394.80383	0.0001	0.99603550	2.12065%
ERROR	11	0.57083333	0.05189394			STD DEV	Y MEAN
CORRECTED TOTAL	18	143.98631579				0.22780241	10.74211

SOURCE	DF	TYPE I SS	F VALUE	PROB>F	TYPE III SS	F VALUE	PROB>F
BLOCKS	4	2.61464912	12.59612	0.0007	2.11266667	10.17781	0.0014
TREATS	3	140.80083333	904.41411	0.0001	140.80083333	904.41411	0.0001

SSB   F test for treatments (adj)   SSB$_{adj}$ F test for blocks (adj)

SST$_{adj}$ F test for Treat (adj)

a. Indicate the complete and reduced models for testing treatments.

b. Construct an analysis of variance table for testing treatments. Give the level of significance for your test and draw conclusions.

c. Indicate the complete and reduced models for testing blocks.

d. Construct an analysis of variance table for testing blocks. Give the levels of significance for your test.

16.2 Refer to Example 16.1. Use the least significant difference criterion for identifying which treatments differ from the others. Use $\alpha = .05$.

16.3 Refer to Example 15.1. Suppose that the first observation in block 1 (plot 1) is missing. Analyze the data by estimating the missing value and then performing an analysis of variance. Use $\alpha = .05$.

16.4 Refer to Exercise 16.3. Perform the corresponding analysis by fitting complete and reduced models. Compare your conclusions to those in Exercise 16.3.

16.5 Refer to Exercise 16.1. Fit the reduced model $y_{ij} = \mu + \alpha_i + \varepsilon_{ij}$ to obtain SSE$_2$. The sum of squares drop will be the sum of squares due to blocks, adjusted for treatments. Verify that this computer value for SSB$_{adj}$ is the same as that shown in TYPE III SS column of the computer output in Exercise 16.1.

16.6 Refer to the data of Exercise 15.3. Suppose that in the rocket propellant test for the second mixture to be analyzed by investigator 3, a piece of equipment malfunctioned. Instead of going back to the laboratories to prepare a duplicate mixture, the investigators proceeded to obtain the remaining propellent thrust data.

a. Estimate the missing value.

b. Perform an analysis of variance, using $\alpha = .05$.

16.7 Refer to Exercise 16.6.

a. Use complete and reduced models to obtain an analysis of variance. Compare your results to those in Exercise 16.6.

b. How would you analyze the data if the response for mixture 4 and investigator 1 was also missing?

16.3 ▼ A LATIN SQUARE DESIGN WITH MISSING DATA

Recall that a $t \times t$ Latin square design can be used to compare t treatment means while filtering out two additional sources of variability (rows and columns). The treatments are randomly assigned in such a way that each treatment appears in every row and in every column. In this section, we will illustrate the method for performing an analysis of variance in a Latin square design when one observation is missing. Then we will use the general method of fitting complete and reduced models with missing observations, described for the randomized block design in Section 16.2, for more complicated designs.

estimating missing value

The formula for **estimating a single missing value** in a Latin square design is

$$M = \frac{t(y_{i..} + y_{.j.} + y_{..k}) - 2y_{...}}{(t-1)(t-2)},$$

where $y_{i..}$, $y_{.j.}$, and $y_{..k}$ represent the treatment, row, and column totals, respectively, corresponding to the missing observation, and t is the number of treatments in the Latin square design.

E X A M P L E   16.2   ▼

A company has considered the properties (such as strength, elongation, and so on) of many different variations of nylon stocking in trying to select the experimental stockings to be placed in extensive consumer acceptance surveys.

Five versions (A, B, C, D, and E) of the stockings have passed the preliminary screening and are scheduled for more extensive testing. As part of the testing, five samples of each type are to be examined for elongation under constant stress by each of five investigators on five separate days. The analyses are to be performed following the random assignment of a Latin square. The elongation data (in centimeters) are displayed in Table 16.5.

T A B L E   16.5
Elongation Data for Example 16.2

Investigator	Day 1		Day 2		Day 3		Day 4		Day 5	
1	B	22.1	A	18.6	C	23.0	E	24.3	D	17.1
2	C	23.5	D	16.5	A	18.7	B	22.0	E	—
3	D	17.4	E	23.8	B	22.8	C	23.9	A	20.0
4	A	20.3	B	23.4	E	25.9	D	18.7	C	24.2
5	E	25.7	C	24.8	D	18.9	A	20.6	B	24.6

Note that the measurement on variety E stockings for investigator 2 is missing and that the experiment was not rerun to obtain an observation. Use the methods of this section to estimate the missing value.

Solution   For our data the treatment, row, and column totals corresponding to the missing observations are

$$y_{5..} = 99.70 \qquad y_{.2.} = 80.70 \qquad y_{..5} = 85.90.$$

Then with $t = r = c = 5$ and $y_{...} = 520.80$, we find

$$M = \frac{5(99.7 + 80.7 + 85.9) - 2(520.8)}{4(3)} = 24.2. \qquad \blacktriangle$$

The analysis could now proceed as for a balanced Latin square design, using the shortcut formulas.

Having located a significant effect due to treatments, we can make pairwise treatment comparisons using the following formulas. The least significant difference between the treatment with the missing value and any other treatment is

LSD

$$\text{LSD} = t_{\alpha/2} \sqrt{\text{MSE}\left(\frac{2}{t} + \frac{1}{(t-1)(t-2)}\right)}.$$

For any other pair of treatments, the LSD is as before:

$$\text{LSD} = t_{\alpha/2} \sqrt{\frac{2\text{MSE}}{t}}.$$

fitting complete and reduced models

For Latin squares with more than one missing observation, it might be easier to use the method of **fitting complete and reduced models** to adjust for imbalances caused by the missing values. In general, using the complete model

$$y_{ijk} = \mu + \alpha_i + \beta_j + \gamma_k + \varepsilon_{ijk}$$

and a computer solution, we would get the analysis of variance table shown in Table 16.6. Note that the sum of squares due to rows is unadjusted, the sum of squares for columns is adjusted for rows, and the sum of squares for treatments is adjusted for both rows and columns.

Even though we do not have all the information for the analysis of variance tables, the corresponding tests for either rows or columns can be obtained from the computer output for this same model by using the TYPE III sums of squares

Source	SS	df	MS	F
Rows	SSR	$t-1$	—	—
Columns (adjusted for rows)	$SSC_{adj}R$	$t-1$	—	—
Treatments (adjusted for rows, columns)	$SST_{adj}R,C$	$t-1$	$MST_{adj}R,C$	$MST_{adj}R,C/MSE$
Error	SSE	by subtraction	MSE	
Totals	TSS	$n-1$		

column. Here we obtain $SSR_{adj}C,T$ and $SSC_{adj}R,T$. In the computer output, the F test and level of significance for these tests are given in the adjacent columns to the right of the partial sums of squares.

▼ EXERCISES

Basic Techniques

16.8 Refer to Example 16.2. Perform an analysis of variance, using the estimated value 24.2. Use $\alpha = .05$ to draw your conclusions.

16.9 Use the SAS computer output shown here to give an analysis of variance table for testing the effect of treatments adjusted for rows (investigators) and columns (days). Indicate the results of testing separately for effects of rows and columns. Use $\alpha = .05$. Compare your results to those of Exercise 16.8.

```
SAS

DATA NYLON;
INPUT Y:INV DAY TRT;
CARDS;

24 OBSERVATIONS IN DATA SET NYLON     4 VARIABLES
PROC PRINT;

OBS     Y      INV    DAY    TRT
  1    22.1     1      1      2
  2    23.5     2      1      3
  3    17.4     3      1      4
  4    20.3     4      1      1
  5    25.7     5      1      5
  6    18.6     1      2      1
  7    16.5     2      2      4
  8    23.8     3      2      5
  9    23.4     4      2      2
 10    24.8     5      2      3
 11    23.0     1      3      3
```

(*continues*)

```
12    18.7   2    3    1
13    22.8   3    3    2
14    25.9   4    3    5
15    18.9   5    3    4
16    24.3   1    4    5
17    22.0   2    4    2
18    23.9   3    4    3
19    18.7   4    4    4
20    20.6   5    4    1
21    17.1   1    5    4
22    20.0   3    5    1
23    24.2   4    5    3
24    24.6   5    5    2

PROC GLM;
CLASSES INV DAY TRT;
MODEL Y=INV DAY TRT;
TITLE "EXERCISE 15.9";


                              GENERAL LINEAR MODELS PROCEDURE

SOURCE            DF      SUM OF SQUARES    MEAN SQUARE    F VALUE   PROB>F   R-SQUARE        C.V.
REGRESSION        12      189.95683333      15.82973611   120.65626 0.0001   0.99245994   1.66918%
ERROR             11        1.44316667       0.13119697
                                                                     STD DEV      Y MEAN
CORRECTED TOTAL   23      191.40000000
                                                                    0.35221122    21.70000


SOURCE    DF    TYPE I SS SSR    F VALUE    PROB>F    TYPE III SS    F VALUE    PROB>F
INV        4     22.32850000     42.54767   0.0001    14.36883333    27.38027   0.0001
DAY        4      2.13400000      4.06640   0.0289     0.94283333     1.79660   0.1994
TRT        4    165.49433333    315.35472   0.0001   165.49433333   315.35472   0.0001

        SSC_adj R                         SSC_adj R,T      F-tests for
        SST_adj R,C                       SST_adj R,C      Rows adj
                                          SSR_adj CT       Columns adj
                                                           Treat adj
```

⚙ **16.10** The data of Example 15.4 have been reproduced here. Suppose that a car malfunction invalidated the data for driver 3, model 4, and blend 2. Rather than rent the speedway for another day, the investigators decided to analyze the data without replacing the missing value.

Driver		Car Model						
		1		2		3		4
1	IV	15.5	II	33.9	III	13.2	I	29.1
2	II	16.3	III	26.6	I	19.4	IV	22.8
3	III	10.8	I	31.1	IV	17.1	II	—
4	I	14.7	IV	34.0	II	19.7	III	21.6

a. Run an analysis of variance by estimating the missing value. Use $\alpha = .05$.

b. Make treatment comparisons by using Fisher's least significant difference, with $\alpha = .05$.

16.11 Use the method of fitting complete and reduced models to obtain an analysis of variance for the data in Exercise 16.10.

16.4 ▼ INCOMPLETE BLOCK DESIGNS

So far, we have discussed the analysis for unbalanced designs where the imbalance was not planned but rather was caused by some accident during the collection of the sample data. Sometimes, however, we may be forced to design an experiment in which we must sacrifice some balance to perform the experiment. To illustrate, suppose that a regulatory agency would like to compare the mean potencies for three different batches (A, B, C) of the same drug product. Assume for the sake of simplicity that each analyst can do just two analyses per day and that there are three analysts available on a given day. Thus, it would be possible to complete a comparison of the three batches on a single day if each analyst examines just two of the three possible batches. One possible design would be the arrangement listed here.

Analyst		
1	2	3
A	C	B
B	A	C

incomplete block design

balanced incomplete block design

We can think of this design as a partial (incomplete) randomized block design, where the number of treatments (batches) per block (analyst) is less than t, the number of treatments. In fact, a design such as this has become known as an **incomplete block design.**

There are many different types of incomplete block designs. The one we have constructed belongs to a class of designs that statisticians have called **balanced incomplete block designs.** Although these designs are not balanced as we have defined the term in Definition 15.5, the designs do retain some balance. For example, even though all treatments do not appear in the same block, each pair of treatments appears together in a block the same number of times. The pairs of treatments AB, AC, and BC in our design appear together once in a block. We achieved this ''balance'' by taking all possible combinations of two of the three treatments for blocks.

Consider now an extension to the balanced incomplete block design, with three treatments per block. Suppose we want to compare nine different batches in sets of three each. If each analyst can run three analyses, how many analysts would be required to maintain a balance similar to that obtained with our previous design?

Following similar logic, we could consider all possible combinations of three treatments with one analyst assigned to each different combination. Without too much trouble, we can show that this would require 84 analysts. Obviously,

Block	Treatments			Block	Treatments		
1	A	B	C	7	A	E	I
2	D	E	F	8	B	F	G
3	G	H	I	9	C	D	H
4	A	D	G	10	A	F	H
5	B	E	H	11	B	D	I
6	C	F	I	12	C	E	G

it would be prohibitive for most companies to employ 84 different analysts to compare the nine treatments in sets of three. Fortunately, there are other balanced incomplete designs that accomplish the experimental objective. One such design is shown in Table 16.7. By employing 12 analysts (blocks), we can compare the nine treatments.

EXAMPLE 16.3   ▼

Identify the values for the design constants listed here for the balanced incomplete block design described in the previous discussion.

- t: number of treatments
- k: number of treatments per block
- b: number of blocks
- r: number of repetitions of each treatment
- λ: number of times each pair of treatments appears together in a block

Solution   From Table 16.7 we see that

$$t = 9 \qquad k = 3 \qquad b = 12 \qquad r = 4.$$

In addition, after a cursory check we see that $\lambda = 1$ (i.e., each pair of treatments appears once in a block).   ▲

Before considering the analysis of variance for balanced incomplete blocks, we should note that a balanced incomplete block design does not exist for all possible values of t, k, b, and r. Although some researchers in statistics have been concerned with methods for constructing balanced incomplete block designs and other more complicated incomplete block designs, we encourage you to refer to tables of incomplete block designs (see Cochran and Cox (1957)) when searching for a design to satisfy certain experimental objectives.

The analysis of variance for a balanced incomplete block design can be performed either by using specifically developed shortcut formulas or by using the method of fitting complete and reduced models as discussed for unbalanced

Source	SS	df	MS	F
Blocks	SSB	$b-1$	—	—
Treatments$_{adj}$	SST$_{adj}$	$t-1$	MST$_{adj}$	MST$_{adj}$/MSE
Error	SSE	by subtraction	MSE	
Totals	TSS	$n-1$		

T A B L E 16.8
Analysis of Variance Table for a Balanced Incomplete Block Design

designs. We will present the shortcut formulas for the analysis of variance table shown in Table 16.8.

shortcut formulas

The quantities SSB (the sum of square unadjusted for treatments) and the total sum of squares are computed as we have done previously:

$$\text{TSS} = \sum_{ij} y_{ij}^2 - \frac{y_{..}^2}{n}, \qquad \text{where } n \text{ is the actual number of measurements}$$

and

$$\text{SSB} = \sum_{j} \frac{y_{.j}^2}{k} - \frac{y_{..}^2}{n},$$

where $y_{.j}$ is the sum of all observations in block j. Then if we define

$y_{i.} = $ sum of all observations on treatment i
$B_{(i)} = $ sum of all measurements for blocks that contain treatment i,

the sum of squares for treatments adjusted for blocks is

$$\text{SST}_{adj} = \frac{t-1}{nk(k-1)} \sum_{i} (ky_{i.} - B_{(i)})^2.$$

The sum of squares for error is found by subtraction:

$$\text{SSE} = \text{TSS} - \text{SSB} - \text{SST}_{adj}.$$

As indicated in the analysis of variance table, the test statistic for testing the hypothesis of no difference among the treatment means is MST$_{adj}$/MSE.

E X A M P L E 16.4

▼

A large company enlisted the help of a random sample of 20 potential consumers in a given geographical location to compare the physical characteristics (such as firmness and rebound) of eight experimental pillows and one presently marketed pillow. Since it was assumed that people would have a difficult time in distinguishing differences among the pillows when presented with all nine at once, it was decided to employ the balanced incomplete block design shown in Table 16.7.

After the pillow types were randomly assigned the letters from A to I, tables were prepared with the appropriate pillow types assigned to each table. For example, pillows G, H, and I were placed on table 3. Each pillow was sealed in an identical white pillowcase and hence could not be distinguished from the others by color. The only marking on the pillowcase was a four-digit number, which provided the investigators with an identification code. With all tables in place, the 20 potential consumers were asked to proceed one at a time through the design from table 1 to table 12, stopping at each table to compare the three pillows. All persons were to record a firmness score for each pillow at each table, based on a 1 to 5 point scale (higher score indicates more firmness). The sums of scores for each pillow at the 12 tables are recorded in Table 16.9 (with letters identifying the pillow type).

T A B L E 16.9
Sum of Firmness Scores, Example 16.4

Block (table)	Treatment (pillow)						Block Totals
1	A	59	B	26	C	38	123
2	D	85	E	92	F	69	246
3	G	74	H	52	I	27	153
4	A	62	D	70	G	68	200
5	B	27	E	98	H	59	184
6	C	31	F	60	I	35	126
7	A	63	E	85	I	30	178
8	B	22	F	73	G	75	170
9	C	45	D	74	H	51	170
10	A	52	F	76	H	43	171
11	B	18	D	79	I	41	138
12	C	41	E	84	G	81	206
							2,065

Use the shortcut formulas of this section to perform an analysis of variance. Use $\alpha = .05$ to test the null hypothesis of no differences among treatment (pillow) means.

Solution   For an analysis using the shortcut formulas, it is convenient to construct a table of totals, as shown in Table 16.10.

TABLE 16.10
Totals for the Data of
Table 16.9

Treatment	y_i	$B_{(i)}$	$ky_i - B_{(i)}$
A	236	672	36
B	93	615	−336
C	155	625	−160
D	308	754	170
E	359	814	263
F	278	713	121
G	298	729	165
H	205	678	−63
I	133	595	−196
Total	2,065		

From Table 16.10, we find that

$$SST_{adj} = \frac{(t-1)\sum_i (ky_i - B_{(i)})^2}{n(k)(k-1)} = \frac{8(322,112.00)}{36(3)(2)} = 11,930.07.$$

Similarly, using the block totals from the raw data table, we obtain

$$SSB = \sum_j \frac{y_{\cdot j}^2}{k} - \frac{y_{\cdot \cdot}^2}{n} = \frac{368,991.00}{3} - \frac{(2,065)^2}{36}$$

$$= 122,997.00 - 118,450.69 = 4,546.31.$$

The total sum of squares is

$$TSS = \sum_{ij} y_{ij}^2 - \frac{y_{\cdot \cdot}^2}{n} = 135,435.00 - 118,450.69 = 16,984.31.$$

The analysis of variance table for testing the hypothesis of no differences among the treatment means is shown in Table 16.11. Since the computed value of F, 46.97, exceeds the table value, 2.59, for $df_1 = 8$, $df_2 = 16$, and $a = .05$, we conclude that there are differences among the nine treatment means.

TABLE 16.11
AOV Table for the Data
of Example 16.4

Source	SS	df	MS	F
Blocks	4,546.31	11	—	—
Treatments_adj	11,930.07	8	1,491.26	46.97
Error	507.93	16	31.75	
Totals	16,984.31	35		

comparison among
treatment means

Following the observation of a significant F test concerning differences among treatment means, we naturally might like to determine which treatment means are significantly different from others. To do this, we make use of the following notation: $\hat{\mu}_i$, an estimate of the mean for treatment i, given by

$$\hat{\mu}_i = \bar{y}_{..} + \frac{ky_{i.} - B_{(i)}}{t\lambda},$$

where $\bar{y}_{..}$ is the overall sample mean. An estimate of the difference between two treatment means i and i' is then

$$\hat{\mu}_i - \hat{\mu}_{i'.} = \frac{[ky_{i.} - B_{(i)}] - [ky_{i'.} - B_{(i')}]}{t\lambda}.$$

The least significant difference between any pair of treatment means is

LSD

$$\text{LSD} = t_{\alpha/2}\sqrt{\frac{2k\text{MSE}}{t\lambda}}.$$

E X A M P L E   16.5   ▼

Compute the least significant difference for the nine treatment means of Example 16.4. Determine all pairwise differences, using $\alpha = .05$.

Solution   For these data the overall sample mean is $\bar{y}_{..} = 2{,}065/36 = 57.36$. Then using the column $ky_{i.} - B_{(i)}$, we have the following estimated treatment means:

Treatment	$\bar{y}_{..} + \dfrac{ky_{i.} - B_{(i)}}{t\lambda}$
A	61.36
B	20.03
C	39.58
D	76.25
E	86.58
F	70.80
G	75.69
H	50.36
I	35.58

Using MSE $= 31.75$, based on df $= 16$, we obtain

$$\text{LSD} = t_{\alpha/2} \sqrt{\frac{2k\text{MSE}}{t\lambda}} = 2.12 \sqrt{\frac{2(3)31.75}{9(1)}} = 9.75.$$

The nine treatment means are arranged next in ascending order, with a summary of the significant results. Those treatments underlined by a common line are not significantly different from each other, using the least significant difference criterion (see Chapter 14).

B	I	C	H	A	F	G	D	E
20.03	35.58	39.58	50.36	61.36	70.80	75.69	76.25	86.58

16.5 ▼ SUMMARY

In this chapter, we have discussed the analysis of variance for some unbalanced designs, beginning with a discussion of the analysis for a randomized block design with one missing observation. Two possible analyses were proposed. The first required that we estimate the missing value and then proceed with the usual shortcut formulas developed in Chapter 15. While estimating a single missing value is quite easy to do, the procedure becomes more difficult when there is more than one missing value. The second procedure, that of fitting complete and reduced models to obtain adjusted sums of squares, can be used for one or more missing observations.

With the Latin square design, we again showed how to estimate a single missing observation and proceed with the usual analysis. But, as with the randomized block design, the method of analysis by fitting complete and reduced models is more appropriate when there is more than one missing value.

Finally, we considered another class of unbalanced designs, incomplete block designs. The particular designs that we discussed were incomplete randomized block designs in which not all treatments appear in each block. These incomplete block designs retain a certain amount of balance, since all pairs of treatments appear together in a block the same number of times. The analysis for balanced incomplete block designs was illustrated using appropriate shortcut formulas. While no example was given in the chapter to show a computer solution for a balanced incomplete block design, we can obtain the analysis of variance for testing treatment differences following the procedure outlined for a randomized block design with missing observations.

▼ KEY FORMULAS

1. Missing observation, randomized block design

$$M = \frac{ty_{i\cdot} + by_{\cdot j} - y_{\cdot\cdot}}{(t-1)(b-1)}$$

2. Fisher's LSD for a randomized block design
   a. For any pair of treatments with no missing value

$$LSD = t_{\alpha/2} \sqrt{\frac{2MSE}{b}}$$

   b. Between the treatment with a missing value and any other treatment

$$LSD = t_{\alpha/2} \sqrt{MSE\left(\frac{2}{b} + \frac{t}{b(b-1)(t-1)}\right)}$$

3. Equalities for randomized block design

$$SSB = TSS - SST_{adj} - SSE$$
$$SST = TSS - SSB_{adj} - SSE$$

4. Missing observation, Latin square design

$$M = \frac{t(y_{i\cdot\cdot} + y_{\cdot j\cdot} + y_{\cdot\cdot k}) - 2y_{\cdot\cdot\cdot}}{(t-1)(t-2)}$$

5. Fisher's LSD for a Latin square design
   a. For any pair of treatments with no missing value

$$LSD = t_{\alpha/2} \sqrt{\frac{2MSE}{t}}$$

   b. Between the treatment with the missing value and any other treatment

$$LSD = t_{\alpha/2} \sqrt{MSE\left(\frac{2}{t} + \frac{1}{(t-1)(t-2)}\right)}$$

6. Sums of squares for an incomplete block design

$$SST_{adj} = \frac{t-1}{n(k)(k-1)} \sum_i (k_{y_{i\cdot}} - B_{(i)})^2$$

$$SSE = TSS - SSB - SST_{adj}$$

7. Pairwise comparisons of treatment means, incomplete block design

$$\hat{\mu}_i - \hat{\mu}_{i'} = \frac{[ky_{i\cdot} - B_{(i)}] - [ky_{i'\cdot} - B_{(i')}]}{t\lambda}$$

$$LSD = t_{\alpha/2}\sqrt{\frac{2k\text{MSE}}{t\lambda}}$$

▼ SUPPLEMENTARY EXERCISES

16.12 A physician was interested in comparing the effects of six different antihistamines in persons extremely sensitive to a ragweed skin allergy test. To do this, a random sample of ten allergy patients was selected from the physician's private practice, with treatments (antihistamines) assigned to each patient according to the experimental design shown in the following table. Each person then received injections of the assigned antihistamines in different sections of the right arm. The area of redness surrounding the point of injection was measured after a fixed period of time. The data are shown in the table.

Person	Treatments					
1	B	25	A	41	F	40
2	E	37	B	46	A	42
3	C	45	D	33	B	37
4	E	34	D	35	A	46
5	B	31	F	42	D	34
6	C	56	E	36	F	65
7	D	33	A	42	C	67
8	F	49	D	37	E	30
9	C	59	A	40	F	55
10	B	36	C	57	E	34

a. Identify the design.
b. Identify the characteristics of the design.

16.13 Refer to the data of Exercise 16.12. Do the data indicate differences among the treatment means? Use $\alpha = .05$.

16.14 Refer to Exercise 16.13. Use the least significant difference criterion for determining treatment differences, with $\alpha = .05$.

16.15 Use a computer program to perform the same analysis as in Exercise 16.13. Compare the results of both exercises.

16.16 Refer to Example 16.4. Use a computer program to perform an analysis of variance. Are your results the same as those found in the example?

16.17 Indicate how you would test for a significant effect due to blocks in a balanced incomplete block design.

16.18 The marketing research group of a corporation examined the public response to the introduction of a new TV game module by comparing weekly sales ($1,000) volumes for three different store chains in each of four geographic locations.

		Chain		
Geographic Area		1	2	3
N	W1	35	17	7
	W2	30	22	12
S	W1	42	30	22
	W2	48	28	19
E	W1	35	35	15
	W2	38	40	20
W	W1	22	43	28
	W2	26	48	23

a. Write an appropriate model (including an effect for weeks) and the sources of variability in an analysis of variance table.

b. How would your model change if we analyze the total 2-week sales data?

c. Run an analysis of variance on the 2-week sales data using shortcut formulas from Chapter 15. Use $\alpha = .05$.

16.19 Refer to Exercise 16.18. Use Tukey's procedure to compare the different geographic areas by chain means. Use $\alpha = .05$.

16.20 Refer to Exercise 16.18. Suppose that the week 1 data were not available in the north and east for chain 1, due to logistics problems that slowed the introduction of the product by a week.

a. Write an appropriate model.

b. Suggest a method for analyzing the data using available software.

c. Write model(s) for the procedure described in part b.

16.21 A foreign automobile manufacturer is spending hundreds of millions of dollars to construct a large manufacturing plant (about 70 acres under one roof) here in the United States. One of their objectives is to produce cars of high quality in the United States using U.S. workers. One part of the massive orientation program for new employees is to send about 20% of them to the home country for additional training. One measure of the worth of this additional training is whether the product quality is better on assembly lines where 20% of the employees have had the homeland orientation and been able to share it with their fellow employees. Data from six assembly lines (three with the additional orientation) are shown here. Two different inspectors examined each of two cars chosen at random for defects from the assembly lines. Use these data to answer the following questions.

	Additional Training			No Additional Training		
		Inspector				Inspector
Assembly Line	1	2	Assembly Line	1	2
1	6	6	4	8	7
	3	4		5	5
2	4	3	5	10	9
	2	2		4	4
3	2	3	6	15	13
	1	1		7	6

a. Suggest an appropriate dependent variable.
b. Write a model for this experimental situation and identify all terms.
c. Fill out the sources and degrees of freedom for an AOV table.

16.22 Refer to the conditions of Exercise 16.21.
a. Suggest a means to analyze these data.
b. Use the output shown here to draw conclusions.
c. Can you suggest any plots that might be helpful in interpreting the data?

```
            LISTING OF DATA

OBS   DEFECTS   LINE  INSPECT   TRAIN
 1       6       1       1       1
 2       6       1       2       1
 3       3       1       1       1
 4       4       1       2       1
 5       4       2       1       1
 6       3       2       2       1
 7       2       2       1       1
 8       2       2       2       1
 9       2       3       1       1
10       3       3       2       1
11       1       3       1       1
12      11       3       2       1
13       8       4       1       0
14       7       4       2       0
15       5       4       1       0
16       5       4       2       0
17      10       5       1       0
18       9       5       2       0
19       4       5       1       0
20       4       5       2       0
21      15       6       1       0
22      13       6       2       0
23       7       6       1       0
24       6       6       2       0

N = 24
```

(continues)

```
                    FIRST MODEL

        GENERAL LINEAR MODELS PROCEDURE
           CLASS LEVEL INFORMATION

        CLASS       LEVELS     VALUES
        INSPECT       2        1 2
        LINE          6        1 2 3 4 5 6
        TRAIN         2        0 1

        NUMBER OF OBSERVATIONS IN DATA SET = 24


                             FIRST MODEL

                  GENERAL LINEAR MODELS PROCEDURE
DEPENDENT VARIABLE: DEFECTS  NUMBER OF DEFECTS
SOURCE              DF   SUM OF SQUARES   MEAN SQUARE  F VALUE   PR>F    R-SQUARE    C.V.
MODEL               11     190.83333333    17.34848485    1.98   0.1275  0.645070  54.6100
ERROR               12     105.00000000     8.75000000            STD DEV          DEFECTS MEAN
CORRECTED TOTAL     23     295.83333333                          2.95803989        5.41666667


SOURCE                  DF     TYPE I SS   F VALUE    PR>F    DF    TYPE III SS   F VALUE   PR>F
TRAIN                    1   130.66666667    14.93   0.0023   1   130.66666667    14.93   0.0023
LINE (TRAIN)             4    56.66666667     1.62   0.2329   4    56.66666667     1.62   0.2329
INSPECT                  1     0.66666667     0.08   0.7872   1     0.66666667     0.08   0.7872
INSPECT*TRAIN            1     1.50000000     0.17   0.6861   1     1.50000000     0.17   0.6861
INSPECT*LINE(TRAIN)      4     1.33333333     0.04   0.9968   4     1.33333333     0.04   0.9968


                             SECOND MODEL

                  GENERAL LINEAR MODELS PROCEDURE
DEPENDENT VARIABLE: DEFECTS  NUMBER OF DEFECTS

SOURCE              DF   SUM OF SQUARES   MEAN SQUARE  F VALUE  PR>F   R-SQUARE    C.V.
MODEL                7     189.50000000    27.07142857    4.07  0.0095  0.640563  47.5929
ERROR               16     106.33333333     6.64583333           STD DEV          DEFECTS MEAN
CORRECTED TOTAL     23     295.83333333                         2.57795138        5.41666667


SOURCE              DF    TYPE I SS   F VALUE    PR>F    DF    TYPE III SS   F VALUE   PR>F
TRAIN                1  130.66666667   19.66   0.0004   1   130.66666667   19.66   0.0004
LINE (TRAIN)         4   56.66666667    2.13   0.1240   4    56.66666667    2.13   0.1240
INSPECT              1    0.66666667    0.10   0.7555   1     0.66666667    0.10   0.7555
INSPECT*TRAIN        1    1.50000000    0.23   0.6411   1     1.50000000    0.23   0.6411


                             THIRD MODEL

                  GENERAL LINEAR MODELS PROCEDURE
DEPENDENT VARIABLE: DEFECTS    NUMBER OF DEFECTS

SOURCE              DF   SUM OF SQUARES   MEAN SQUARE  F VALUE   PR>F   R-SQUARE    C.V.
MODEL                6     188.00000000    31.33333333    4.94   0.0043  0.635493  46.4965
ERROR               17     107.83333333     6.34313725           STD DEV          DEFECTS MEAN
CORRECTED TOTAL     23     295.83333333                         2.51855857        5.41666667

SOURCE          DF    TYPE I SS   F VALUE    PR>F   DF    TYPE III SS   F VALUE   PR>F
RAIN             1  130.66666667   20.60   0.0003   1   130.66666667   20.60   0.0003
LINE (TRAIN)     4   56.66666667    2.23   0.1084   4    56.66666667    2.23   0.1084
INSPECT          1    0.66666667    0.11   0.7497   1     0.66666667    0.11   0.7497
```

16.23   Refer to Exercise 16.21. Suppose that inspector 2 was unable to evaluate the second car from assembly line 4 and that inspector 1 missed car 1 from assembly line 3.

a. Does the model change? Suggest a method for analyzing the data.

b. Use the computer output shown here to draw conclusions.

LISTING OF ADJUSTED DATA

OBS	DEFECTS	LINE	INSPECT	TRAIN
1	6	1	1	1
2	6	1	2	1
3	3	1	1	1
4	4	1	2	1
5	4	2	1	1
6	3	2	2	1
7	2	2	1	1
8	2	2	2	1
9	3	3	2	1
10	1	3	1	1
11	1	3	2	1
12	8	4	1	0
13	7	4	2	0
14	5	4	1	0
15	10	5	1	0
16	9	5	2	0
17	4	5	1	0
18	4	5	2	0
19	15	6	1	0
20	13	6	2	0
21	7	6	1	0
22	6	6	2	0

N = 22

FIRST MODEL

GENERAL LINEAR MODELS PROCEDURE
    CLASS LEVEL INFORMATION

CLASS	LEVELS	VALUES
INSPECT	2	1 2
LINE	6	1 2 3 4 5 6
TRAIN	2	0 1

NUMBER OF OBSERVATIONS IN DATA SET = 22

FIRST MODEL
GENERAL LINEAR MODELS PROCEDURE
DEPENDENT VARIABLE: DEFECTS   NUMBER OF DEFECTS

SOURCE	DF	SUM OF SQUARES	MEAN SQUARE	F VALUE	PR>F	R-SQUARE	C.V.
MODEL	11	180.81818182	16.43801653	1.60	0.2326	0.638216	57.2637
ERROR	10	102.50000000	10.25000000		STD DEV		DEFECTS MEAN
CORRECTED TOTAL	21	283.31818182			3.20156212		5.59090909

SOURCE	DF	TYPE I SS	F VALUE	PR>F	DF	TYPE III SS	F VALUE	PR>F
TRAIN	1	127.68181818	12.46	0.0055	1	124.32142857	12.13	0.0059
LINE (TRAIN)	4	49.30303030	1.20	0.3683	4	48.75000000	1.19	0.3733
INSPECT	1	0.18750000	0.02	0.8951	1	0.03571429	0.00	0.9541
INSPECT*TRAIN	1	1.02083333	0.10	0.7588	1	0.89285714	0.09	0.7739
INSPECT*LINE (TRAIN)	4	2.62500000	0.06	0.9913	4	2.62500000	0.06	0.9913

SECOND MODEL

GENERAL LINEAR MODELS PROCEDURE
DEPENDENT VARIABLE: DEFECTS   NUMBER OF DEFECTS

SOURCE	DF	SUM OF SQUARES	MEAN SQUARE	F VALUE	PR>F	R-SQUARE	C.V.
MODEL	7	178.19318182	25.45616883	3.39	0.0247	0.628951	49.0125
ERROR	14	105.12500000	7.50892857		STD DEV		DEFECTS MEAN
CORRECTED TOTAL	21	283.31818182			2.274024243		5.59090909

SOURCE	DF	TYPE I SS	F VALUE	PR>F	DF	TYPE III SS	F VALUE	PR>F
TRAIN	1	127.68181818	17.00	0.0010	1	123.52083333	16.45	0.0012
LINE (TRAIN)	4	49.30303030	1.64	0.2191	4	50.14166667	1.67	0.2127
INSPECT	1	0.18750000	0.02	0.8767	1	0.18750000	0.02	0.8767
INSPECT*TRAIN	1	1.02083333	0.14	0.7179	1	1.0283333	0.14	0.7179

(continues)

```
                                    THIRD MODEL

                          GENERAL LINEAR MODELS PROCEDURE
     DEPENDENT VARIABLE: DEFECTS   NUMBER OF DEFECTS

     SOURCE           DF    SUM OF SQUARES   MEAN SQUARE   F VALUE     PR>F     R-SQUARE     C.V.
     MODEL             6    177.17234848     292.52872475    4.17      0.0115   0.625348   47.5799
     ERROR            15    106.14583333       7.07638889              STD DEV             DEFECTS MEAN
     CORRECTED TOTAL  21    283.31818182                               2.66014828          5.59090909

     SOURCE           DF    TYPE I SS     F VALUE    PR>F     DF    TYPE III SS    F VALUE    PR>F
     TRAIN             1    127.68181818    18.04    0.0007    1    123.52083333    17.46    0.0008
     LINE (TRAIN)      4     49.30303030     1.74    0.1933    4     49.27234848     1.74    0.1935
     INSPECT           1      0.18750000     0.03    0.8729    1      0.18750000     0.03    0.8729
```

$ **16.24**   Refer to Exercise 16.21. In addition to examining the total number of defects, each defect was classified as major or minor. These data are shown here.

	Inspector			
	1		2	
Assembly Line	Major	Minor	Major	Minor
1	1	5	1	5
	0	3	0	4
2	0	4	0	3
	0	2	0	2
3	0	2	0	3
	0	1	0	1
4	3	7	2	5
	0	5	0	5
5	3	7	3	6
	0	4	0	4
6	5	10	5	8
	1	6	1	5

   a. How consistent are the inspectors at evaluating major defects? Minor defects?
   b. Choose a dependent variable for analyzing these data.
   c. Write an appropriate model, identifying all terms.
   d. Fill out the sources and degrees of freedom for an AOV table.

16.25   Refer to Exercise 16.24. Suggest a dependent variable that accounts for both major and minor defects but does not obscure their identity. Would any transformations of the data be appropriate?

ANALYSIS OF VARIANCE FOR SOME FIXED-, RANDOM-, AND MIXED-EFFECTS MODELS

17.1 ▼ INTRODUCTION

In previous chapters, we have been able to write the model for a response in terms of k independent variables, using the **general linear model**

$$y_i = \beta_0 + \beta_1 x_{i1} + \beta_2 x_{i2} + \cdots + \beta_k x_{ik} + \varepsilon_i.$$

Initially, we assumed $x_1, x_2, \ldots, x_k$ to be independent variables measured without error; $\beta_0, \beta_1, \ldots, \beta_k$ to be unknown parameters; and the random error ε_i associated with observation i to have $E(\varepsilon_i) = 0$. Then we expanded these assumptions to include the following:

1. $\varepsilon_1, \varepsilon_2, \ldots, \varepsilon_n$ are independent of one another.
2. For a given setting of the independent variables $x_1, x_2, \ldots, x_k$, the variance of ε_i is σ_ε^2.

Thus, although we had many terms in the model, there was only one source of random variation.

DEFINITION 17.1
Fixed-effects Model

A model that can be written in the form of a general linear model with $k > 0$ independent variables and one random component is called a **fixed-effects model**.

All models discussed so far in this text relating a response y to one or more independent variables have fallen into the category of fixed-effects models. Inferences for fixed-effects models are stated in terms of one or more parameters in the general linear model.

Sometimes, however, we must account for more than one source of random variation in an experimental situation, and the model must then be expanded **components of variation** to accommodate these additional **components of variation**. For example, the blocks in a randomized block design might represent a random sample of b blocks taken from a population of all possible blocks. Then the effects due to blocks are considered to be random effects rather than fixed effects. Appropriate changes would be made in the model to reflect this difference in interpretation.

Before we give examples of situations that might warrant the inclusion of more than one random source of variability in a model, we define two different types of models.

DEFINITION 17.2
Random-effects Model

A model that can be written in the form of a general linear model with $k = 0$ independent variables and more than one random component is called a **random-effects model**.

DEFINITION 17.3
Mixed-effects Model

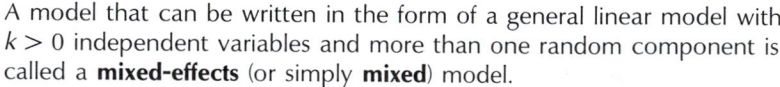

A model that can be written in the form of a general linear model with $k > 0$ independent variables and more than one random component is called a **mixed-effects** (or simply **mixed**) model.

In this chapter, we will consider various random–effects and mixed models. For each model we will indicate the appropriate analysis of variance and also relate the analysis discussed to those that we would obtain with the corresponding fixed-effects model.

17.2 ▼ A ONE-FACTOR EXPERIMENT WITH TREATMENT EFFECTS RANDOM: A RANDOM-EFFECTS MODEL

fixed-effects model

The best way to illustrate the difference between the fixed- and random-effects models for a one-factor experiment is by an example. Suppose we want to compare readings made on the intensities of the electrostatic discharges of lightning at three different tracking stations within a 20-mile radius of the central computing facilities of a university. If these three tracking stations are the only feasible tracking stations for such an operation and inferences are to be about these stations only, then we could write the **fixed-effects model** in the form of a general linear model (see Section 11.2):

$$y = \beta_0 + \beta_1 x_1 + \beta_2 x_2 + \varepsilon,$$

where

$x_1 = 1$ if tracking station 2 $x_1 = 0$ otherwise
$x_2 = 1$ if tracking station 3 $x_2 = 0$ otherwise.

Also, β_0, β_1, and β_2 are unknown parameters representing mean intensities or differences in mean intensities. Equivalently, using the results of Section 13.4, we could write the fixed-effects model as

$$y_{ij} = \mu + \alpha_i + \varepsilon_{ij},$$

where y_{ij} is the jth observation at tracking station i ($i = 1, 2, 3$), μ is an overall mean, and α_i is a fixed effect due to tracking station i. For both of these models, ε is assumed to be normally distributed, with mean 0 and variance σ^2.

Suppose, however, that rather than being concerned about only these three tracking stations, we consider these stations as a random sample of three taken from the many possible locations for tracking stations. Inferences would now relate not just to what happened at the sampled locations but also to what might happen at other possible locations for tracking stations. A model that can account for this difference in interpretation is the **random-effects model**

random-effects model

$$y_{ij} = \mu + \alpha_i + \varepsilon_{ij}.$$

While the model looks the same as the previous fixed-effects model, some of the assumptions are different.

assumptions

1. μ is still an overall mean, which is an unknown constant.
2. α_i is a random effect due to the ith tracking station. We assume that α_i is normally distributed, with mean 0 and variance σ_α^2.

3. The α_is are independent.
4. As before, ε_{ij} is normally distributed, with mean 0 and variance σ_ε^2.
5. The ε_{ij}s are independent.
6. The random components α_i and ε_{ij} are independent.

The difference between the fixed-effects model and the random-effects model can be illustrated by supposing we were to repeat the experiment. For the fixed-effects model, we would use the same three tracking stations, so it would make sense to make inferences about the mean intensities or differences in mean intensities at these three locations. However, for the random-effects model, we would take another random sample of three tracking stations (i.e., take another sample of three αs). Now rather than concentrating on the effect of a particular group of three αs from one experiment, we would examine the variability of the population of all possible α values. This will be illustrated using the analysis of variance table given in Table 17.1.

T A B L E 17.1

An AOV Table for a One-Factor Experiment: Fixed or Random Model

Source	SS	df	MS	EMS Fixed Effects	EMS Random Effects
Treatments	SST	$t-1$	MST	$\sigma_\varepsilon^2 + n\theta_T$	$\sigma_\varepsilon^2 + n\sigma_\alpha^2$
Error	SSE	$t(n-1)$	MSE	σ_ε^2	σ_ε^2
Totals	TSS	$tn-1$			

EMS

AOV table

test for means

The analysis of variance table is the same for a fixed- or random-effects model, with the exception that the **expected mean squares (EMS)** columns are different. You will recall that this column was not used in our tables in Chapters 13 and 15, since all mean squares except MSE had an expectation under the alternative hypothesis equal to σ_ε^2 plus a positive constant, which depended on the parameters under test. In general, with t treatments (tracking stations) and r observations per treatment, the **AOV table** would appear as shown in Table 17.1. For the fixed-effects model, θ_T is a positive function of the constants α_i, whereas σ_α^2 represents the variance of the population of α_i values for the random-effects model. Referring to our example, a **test for the equality of the mean** intensities at the three tracking stations in the fixed-effects model is (from Chapter 13)

H_0: $\alpha_1 = \alpha_2 = \alpha_3 = 0$ (i.e., the 3 means are identical)
H_a: At least one α is different from 0.
T.S.: $F = \text{MST}/\text{MSE}$, based on $\text{df}_1 = t - 1$ and $\text{df}_2 = t(n - 1)$.

test for σ_α^2

A **test concerning the variability for the population of α values** in the random-effects model makes use of the same test statistic. The null hypothesis and alternative hypothesis are

$$H_0: \quad \sigma_\alpha^2 = 0$$
$$H_a: \quad \sigma_\alpha^2 > 0$$

Since we assumed that the αs sampled were selected from a normal population with mean 0 and variance σ_α^2, the null hypothesis states that the αs were drawn from a normal population with mean 0 and variance 0; that is, all α values in the population are equal to 0.

Thus, although the forms of the null hypotheses are different for the two models, the meanings attached to them are very similar. For the fixed-effects model, we are assuming that the sampled αs (which are the only αs) are identically 0, whereas in the random-effects model, the null hypothesis leads us to assume that the sample αs, as well as all other αs in the population, are 0.

The alternative hypotheses are also similar. In the fixed-effects model, we are assuming that at least one of the αs is different from the rest; that is, there is some variability among the set of αs. For the random-effects model, the alternative hypothesis is that $\sigma_\alpha^2 > 0$; that is, not all α values are the same in the population.

EXAMPLE 17.1

▼

Consider the problem we have used to illustrate a one-factor experiment with random treatment effects. Two graduate students working for a professor in electrical engineering have been funded to record lightning discharge intensities (intensities of the electrical field) at three tracking stations. Because of the high frequency of thunderstorms in the summer months (in Florida, storms occur on 80 or more days per year), the graduate students were to choose a point at random on a map of the 20-mile-radius region and assemble their tracking equipment (provided they could get permission of the property owners). Each day during the hours from 8 A.M. to 5 P.M., they were to monitor their instruments until the maximum intensity had been recorded for five separate storms. The process was then repeated separately at the two other locations chosen at random. The sample data (in volts per meter) appear in Table 17.2.

TABLE 17.2

Lightning Discharge Intensities (in volts per meter), Example 17.1

Tracking Station	Intensities					Totals
1	20	1,050	3,200	5,600	50	9,920
2	4,300	70	2,560	3,650	80	10,660
3	100	7,700	8,500	2,960	3,340	22,600
Total						43,180

a. Write an appropriate statistical model, defining all terms.
b. Perform an analysis of variance and interpret your results. Use $\alpha = .05$.

Solution   A model appropriate for this one-factor experiment, with tracking stations chosen at random, is

$$y_{ij} = \mu + \alpha_i + \varepsilon_{ij} \qquad (i = 1, 2, 3; j = 1, 2, \ldots, 5),$$

where the terms in the model were as defined in this section.

The computational formulas for sums of squares are identical to those for a fixed–effects model.

$$\frac{y_{..}^2}{nt} = \frac{(43,180)^2}{15} = 124,300,826.7$$

$$SST = \sum_i \frac{y_{i.}^2}{n} - \frac{y_{..}^2}{nt} = \frac{(9,920)^2 + (10,660)^2 + (22,600)^2}{5} - \frac{(43,180)^2}{15}$$

$$= 144,560,400 - 124,300,826.7 = 20,259,573.3$$

$$TSS = \sum_{ij} y_{ij}^2 - \frac{y_{..}^2}{nt}$$

$$= 232,550,000 - 124,300,826.7 = 108,249,173.3$$

By subtraction,

$$SSE = TSS - SST = 87,989,600.$$

We can use these calculations to construct an AOV table, as shown in Table 17.3.

The F test for $H_0 \colon \sigma_\alpha^2 = 0$ is based on $df_1 = 2$ and $df_2 = 12$ degrees of freedom. Since the computed value of F, 1.38, does not exceed 3.89, the value in Table 6 for $a = .05$, $df_1 = 2$, and $df_2 = 12$, we have insufficient evidence to indicate that there is a significant random component due to variability in intensities from tracking station to tracking station. Rather, as an electrical engineer postulated, it is probably best to work with a single

TABLE 17.3
AOV Table for the Data of Example 17.1

Source	SS	df	MS	EMS	F
Tracking stations	20,259,573.3	2	10,129,786.65	$\sigma_\varepsilon^2 + 5\sigma_\alpha^2$	1.38
Error	87,989,600.0	12	7,332,466.67	σ_ε^2	
Totals	108,249,173.3	14			

tracking station, since most of the variability in intensities is related to the distance of the tracking station from the point of discharge, and we have no control of this source.   ▲

▼ EXERCISES

Applications

    17.1   A pharmaceutical company would like to examine the potency of a liquid medication mixed in large vats. To do this, a random sample of five vats from a month's production was obtained, and four separate samples were selected from each vat.

   a. Write a random-effects model for this experimental situation, identifying all terms in the model.

   b. Run an analysis of variance for the sample data given here. Use $\alpha = .05$.

Vat 1	Vat 2	Vat 3	Vat 4	Vat 5
3.2	2.6	3.4	4.2	1.8
3.8	2.9	3.9	4.4	2.3
3.5	2.8	3.3	4.3	1.9
3.0	2.0	3.1	4.2	2.1

    17.2   Suppose that the pharmaceutical company of Exercise 17.1 wishes to estimate μ, the expected potency for a measurement made on a vat selected at random. We have not discussed this topic in this section, but as might be expected, we can estimate μ using $\bar{y}$, the mean of the sample data. However, it is not so obvious, but nonetheless true, that the variance of $\bar{y}$ is

$$\frac{\sigma_\varepsilon^2}{20} + \frac{\sigma_\alpha^2}{5},$$

where σ_α^2 is the variance of the random vat effect α_i. Since we do not know σ_ε^2 or σ_α^2, we can form estimates by using the MS and EMS columns of the AOV table shown below.

Source	MS	EMS
Vats	MSA	$\sigma_\varepsilon^2 + 4\sigma_\alpha^2$
Error	MSE	σ_ε^2

From the abbreviated table, we see that MSA has expected mean square of $\sigma_\varepsilon^2 + 4\sigma_\alpha^2$, and hence MSA/20 provides an estimate of the variance of $\bar{y}$, $(\sigma_\varepsilon^2/20) + (\sigma_\alpha^2/5)$. In general, for a random-effects model in a one-factor experiment with the same number of observations per treatment, $\bar{y}$ provides an estimate of μ and $\sqrt{\text{MSA}/nt}$ provides an estimate of the standard error of $\bar{y}$.

a. Using the sample data of Exercise 17.1, form a point estimate of μ, the average potency for a measurement made on a vat selected at random.

b. Give a 95% confidence interval for μ of part a. (Hint: The formula is: point estimate ± 2.131 standard error.)

17.3 ▼ EXTENSIONS OF RANDOM-EFFECTS MODELS

The ideas presented for a random-effects model in a one-factor experiment can be extended to any of the block designs and factorial experiments covered in Chapters 13 and 15. Although we will not have time to cover all such situations, we will consider first a randomized block design in which the block effects and the treatment effects are random.

Consider an experiment to examine the effects of different analysts and subjects in chemical analyses for the DNA content of plaque. Three female subjects (ages 18–20 years) were chosen for the study. Each subject was allowed to maintain her usual diet, supplemented with 30 mg (15 tablets) of sucrose per day. No toothbrushing or mouthwashing was allowed during the study. At the end of the week, plaque was scraped from the entire dentition of each subject and divided into three samples. Each of three analysts chosen at random was then given an unmarked sample of plaque from each of the subjects and asked to perform an analysis for the DNA content (in micrograms). The two-factor experiment of sample data could then be organized as shown in Table 17.4.

T A B L E   17.4

DNA Concentrations for Samples of Plaque

		Subject		
Analyst	1	2	3	**Totals**
1				
2				
3				
Totals				

randomized block design   This experimental design is recognized as a **randomized block design**, with subjects representing blocks and analysts being the treatments. The experimental units are samples of plaque scraped from the dentition of subjects. If we assume that the three subjects represent a random sample from a large population of possible subjects, and, similarly, that the three analysts represent a random sample from a large population of possible analysts, we can write the following random-effects model relating DNA concentration to the two factors "analysts" and "subjects":

random-effects model

$$y_{ij} = \mu + \alpha_i + \beta_j + \varepsilon_{ij}.$$

We assume the following:

assumptions

1. μ is an overall unknown concentration mean.
2. α_i is a random effect due to the ith analyst. α_i is normally distributed, with mean 0 and variance σ_α^2.
3. The α_is are independent.
4. β_j is a random effect due to the jth subject. β_j is a normally distributed random variable, with mean 0 and variance σ_β^2.
5. The β_js are independent.
6. The α_is and the β_js are independent.

Again note the difference between assuming that the treatments and blocks are random rather than fixed effects. If, for example, the three analysts chosen for the study were the only analysts of interest, we would be concerned with differences in mean DNA concentrations for these specific analysts. Now, however, treating the effect due to an analyst as a random variable, our inference will be about the population of analysts' effects. Since the mean of this normal population is assumed to be 0, we want to determine whether the variance σ_α^2 is greater than 0.

The AOV table for a general two-factor completely randomized design of a levels of factor A and b levels of factor B and no replication is shown in Table 17.5. This same AOV can apply to a randomized block design where A denotes treatment and B denotes blocks. As with the random-effects model for a one-factor experiment, the analysis of variance tables for a fixed- and random-effects models of a two-factor experiment are identical, except for the expected mean squares.

The computation of sums of squares and mean squares would proceed exactly as shown in Chapter 15, using the appropriate shortcut formulas. The difference

TABLE 17.5

AOV Table for a Two-Factor Experiment, a Levels of Factor A and b Levels of Factor B

Source	SS	df	MS	EMS Fixed Effects	EMS Random Effects
A	SSA	$a - 1$	MSA	$\sigma_\varepsilon^2 + b\theta_A$	$\sigma_\varepsilon^2 + b\sigma_\alpha^2$
B	SSB	$b - 1$	MSB	$\sigma_\varepsilon^2 + a\theta_B$	$\sigma_\varepsilon^2 + a\sigma_\beta^2$
Error	SSE	$(a - 1)(b - 1)$	MSE	σ_ε^2	σ_ε^2
Totals	TSS	$ab - 1$			

T A B L E   17.6

Difference in Test Procedures for Factor A

	Fixed-Effects Model	Random-Effects Model
H_0:	$\alpha_1 = \alpha_2 = \cdots = \alpha_a = 0$	H_0: $\sigma_\alpha^2 = 0$
H_a:	At least one of the αs differs from the rest.	H_a: $\sigma_\alpha^2 > 0$
T.S.:	$F = \dfrac{\text{MSA}}{\text{MSE}}$	T.S.: $F = \dfrac{\text{MSA}}{\text{MSE}}$
R.R.:	Based on $df_1 = a - 1$, $df_2 = (a - 1)(b - 1)$	R.R.: Same

tests

in test procedures is illustrated in Table 17.6 for factor A. Similar results would also apply to factor B.

Rather than proceed with an example at this point, we will discuss a random–effects model for a factorial experiment with $n > 1$ observations at each factor–level combination. Then we will illustrate the test procedure.

In Chapter 15, we considered the fixed-effects model for a 2×2 factorial experiment in a completely randomized design with $n > 1$ observations per cell. The random-effects model for an **$a \times b$ factorial** experiment would be of the same form as the corresponding fixed-effects experiment, but with different assumptions.

$a \times b$ factorial, $n > 1$

model

$$y_{ijk} = \mu + \alpha_i + \beta_j + \alpha\beta_{ij} + \varepsilon_{ijk},$$

where y_{ijk} is the response for the kth observation at the ith level of factor A and the jth level of factor B; μ, α_i, β_j, and ε_{ijk} are defined as before for the random-effects model without replication. In addition, we assume the following:

assumptions

1. $\alpha\beta_{ij}$ is a random effect due to the ith level of factor A and the jth level of factor B. $\alpha\beta_{ij}$ is normally distributed, with mean 0 and variance $\sigma_{\alpha\beta}^2$.
2. The $\alpha\beta_{ij}$s are independent.
3. The α_is, β_js, and $\alpha\beta_{ij}$s are independent.

AOV tables

The appropriate **AOV tables** for fixed- and random-effects models are shown in Table 17.7.

T A B L E   17.7

AOV Table for an $a \times b$ Factorial Experiment with n Observations per Cell

				EMS	
Source	SS	df	MS	Fixed Effects	Random Effects
A	SSA	$a - 1$	MSA	$\sigma_\varepsilon^2 + nb\theta_A$	$\sigma_\varepsilon^2 + n\sigma_{\alpha\beta}^2 + bn\sigma_\alpha^2$
B	SSB	$b - 1$	MSB	$\sigma_\varepsilon^2 + na\theta_B$	$\sigma_\varepsilon^2 + n\sigma_{\alpha\beta}^2 + an\sigma_\beta^2$
AB	SSAB	$(a - 1)(b - 1)$	MSAB	$\sigma_\varepsilon^2 + n\theta_{AB}$	$\sigma_\varepsilon^2 + n\sigma_{\alpha\beta}^2$
Error	SSE	$ab(n - 1)$	MSE	σ_ε^2	σ_ε^2
Totals	TSS	$abn - 1$			

TABLE 17.8

A Comparison of Appropriate Interaction Tests for Fixed- and Random-Effects Models

	Fixed-Effects Model	Random-Effects Model
	H_0: $\alpha\beta_{11} = \alpha\beta_{12} = \cdots = \alpha\beta_{ab} = 0$	H_0: $\sigma^2_{\alpha\beta} = 0$
	H_a: At least one $\alpha\beta_{ij}$ differs from the rest.	H_a: $\sigma^2_{\alpha\beta} > 0$
T.S.:	$F = \dfrac{\text{MSAB}}{\text{MSE}}$	T.S.: $F = \dfrac{\text{MSAB}}{\text{MSE}}$
R.R.:	Based on $df_1 = (a-1)(b-1)$, $df_2 = ab(n-1)$	R.R.: Same

The appropriate tests using the AB interaction sum of squares are illustrated in Table 17.8 for the two models.

Now, unlike the one-factor experiment and the two-factor experiment without replication, the test statistic for main effects are different for the fixed- and random-effects models. In addition, for the random-effects model, the tests for σ^2_α and σ^2_β can proceed even when the test on the AB interaction ($\sigma^2_{\alpha\beta}$) is significant. We've seen previously that for fixed-effects models, a test for main effects in the presence of a significant interaction only seems to make sense when the profile plot suggest that the interaction is "orderly." For random-effects models, we're interested in identifying the various sources of variability (e.g., $\sigma^2_{\alpha\beta}$, σ^2_α, and σ^2_β) that affect the response y. Tests for σ^2_α and σ^2_β do make sense even when $\sigma^2_{\alpha\beta}$ has been shown to be greater than zero.

For the fixed-effects model following a nonsignificant test on the AB interaction, we can test for main effects due to factors A and B by using

$$F = \frac{\text{MSA}}{\text{MSE}} \quad \text{and} \quad F = \frac{\text{MSB}}{\text{MSE}},$$

respectively. As we see from the expected mean squares column of Table 17.7, no matter what the results of the test H_0: $\sigma^2_{\alpha\beta} = 0$, we can form an F test for the components σ^2_α and σ^2_β using the test procedures shown in Table 17.9. Note that the test statistics differ from those used in the fixed-effects case, where the denominator of all F statistics is MSE.

tests, main effects

TABLE 17.9

Tests for an $a \times b$ Factorial Experiment with Replication: Random-Effects Model

	Factor A		Factor B
	H_0: $\sigma^2_\alpha = 0$		H_0: $\sigma^2_\beta = 0$
	H_a: $\sigma^2_\alpha > 0$		H_a: $\sigma^2_\beta > 0$
T.S.:	$F = \dfrac{\text{MSA}}{\text{MSAB}}$	T.S.:	$F = \dfrac{\text{MSB}}{\text{MSAB}}$
R.R.:	Based on $df_1 = (a-1)$, $df_2 = (a-1)(b-1)$	R.R.:	Based on $df_1 = (b-1)$, $df_2 = (a-1)(b-1)$

EXAMPLE 17.2

▼

Consider the experimental data shown in Table 17.10 for a 3 × 3 factorial experiment in a completely randomized design with $n = 2$ observations per cell.

TABLE 17.10

3 × 3 Factorial Experiment with $n = 2$ Observations per Cell

Factor B	Factor A 1	2	3	Totals
1	13.2	10.6	8.5	63.3
	12.3	9.8	8.9	
	25.5	20.4	17.4	
2	12.5	9.6	7.9	62.0
	12.9	10.7	8.4	
	25.4	20.3	16.3	
3	13.0	9.9	8.3	62.5
	12.4	10.3	8.6	
	25.4	20.2	16.9	
Totals	76.3	60.9	50.6	187.8

Perform an analysis of variance for this experiment. Conduct all tests with $\alpha = .05$, and draw your conclusions.

Solution   Using the shortcut formulas of Chapter 15, we obtain the sums of squares as follows:

$$\text{TSS} = \sum_{ijk} y_{ijk}^2 - \frac{y_{...}^2}{nab} = 2,017.38 - \frac{(187.8)^2}{18} = 2,017.38 - 1,959.38 = 58$$

$$\text{SSA} = \sum_{i} \frac{y_{i..}^2}{nb} - \frac{y_{...}^2}{nab} = \frac{(76.3)^2 + (60.9)^2 + (50.6)^2}{6} - \frac{(187.8)^2}{18}$$

$$= 2,015.14 - 1,959.38 = 55.76$$

$$\text{SSB} = \sum_{j} \frac{y_{.j.}^2}{na} - \frac{y_{...}^2}{nab} = \frac{(63.3)^2 + (62.0)^2 + (62.5)^2}{6} - \frac{(187.8)^2}{18}$$

$$= 1,959.52 - 1,959.38 = .14$$

$$\text{SSAB} = \sum_{ij} \frac{y_{ij.}^2}{n} - \frac{y_{...}^2}{nab} - \text{SSA} - \text{SSB}$$

$$= 2,015.46 - 1,959.38 - 55.76 - .14 = .18$$

$$\text{SSE} = \text{TSS} - \text{SSA} - \text{SSB} - \text{SSAB} = 1.92.$$

Our results are summarized in an analysis of variance table, Table 17.11.

TABLE 17.11
AOV Table for the Data of Example 17.2

Source	SS	df	MS	EMS
A	55.76	2	27.88	$\sigma_\varepsilon^2 + 2\sigma_{\alpha\beta}^2 + 6\sigma_\alpha^2$
B	.14	2	.07	$\sigma_\varepsilon^2 + 2\sigma_{\alpha\beta}^2 + 6\sigma_\beta^2$
AB	.18	4	.05	$\sigma_\varepsilon^2 + 2\sigma_{\alpha\beta}^2$
Error	1.92	9	.21	σ_ε^2
Totals	58.00	17		

We can proceed with appropriate statistical tests, using the results presented in the AOV table. For the AB interaction we have

$$H_0: \quad \sigma_{\alpha\beta}^2 = 0$$
$$H_a: \quad \sigma_{\alpha\beta}^2 > 0$$

T.S.: $\quad F = \dfrac{\text{MSAB}}{\text{MSE}} = \dfrac{.05}{.21} = .24$

R.R.: For $\alpha = .05$, we will reject H_0 if F exceeds 3.63, the critical value for $a = .05$, $df_1 = 4$, and $df_2 = 9$.

Conclusion: There is insufficient evidence to reject H_0. There does not appear to be a significant variability in the response y for combinations of levels of factors A and B.

For factor B we have

$$H_0: \quad \sigma_\beta^2 = 0$$
$$H_a: \quad \sigma_\beta^2 > 0$$

T.S.: $\quad F = \dfrac{\text{MSB}}{\text{MSAB}} = \dfrac{.07}{.05} = 1.4$

R.R.: For $\alpha = .05$, we will reject H_0 if F exceeds 6.94, the critical value for $a = .05$, $df_1 = 2$, and $df_2 = 4$.

Conclusion: There is insufficient evidence to indicate a significant variability in DNA determinations from analyst to analyst. ▲

Note that if $\sigma_{\alpha\beta}^2 = 0$, both MSAB and MSE are unbiased estimates of σ_ε^2, and MSB has expectation equal to $\sigma_\varepsilon^2 + 6\sigma_\beta^2$. It can be argued that when MSAB is small relative to MSE, it would be wise to pool the sum of squares for AB

and for error to form a combined estimate of σ^2. We can illustrate this procedure for our data. The pooled mean square based on 13 degrees of freedom is

MS_{pooled}

$$MS_{pooled} = \frac{SSAB + SSE}{4 + 9} = \frac{2.10}{13} = .16$$

and can be used to test for main effects. The F test for H_0: $\sigma_\beta^2 = 0$ would be

$$F = \frac{MSB}{MS_{pooled}} = \frac{.07}{.16} = .44.$$

Comparing the computed value of F to the critical value 3.81 for $a = .05$, $df_1 = 2$, and $df_2 = 13$, we see that there is insufficient evidence to reject H_0. Since this is the same conclusion we reached by using $F = MSB/MSAB$, we recommend pooling only in cases in which MSAB is considerably less than MSE.

The test for factor A follows:

H_0: $\sigma_\alpha^2 = 0$
H_a: $\sigma_\alpha^2 > 0$

T.S.: $F = \dfrac{MSA}{MSAB} = \dfrac{27.88}{.05} = 557.6$

R.R.: For $\alpha = .05$, we will reject H_0 if F exceeds 6.94, the critical value based on $a = .05$, $df_1 = 2$, and $df_2 = 4$.

Conclusion: Since the observed value of F is much larger than 6.94, we reject H_0 and conclude that there is a significant variability in DNA concentrations in plaque collections from females between the ages of 18 and 20 years.

In this section, we have compared a random-effects model to a fixed-effects model for the completely randomized design and for the $a \times b$ factorial experiment with n observations per cell. This study has been in no way exhaustive, but it has shown that there are alternatives to a fixed-effects model. A more detailed study of the random-effects model would certainly include factorial experiments with more than two factors and the **nested sampling experiment** of Section

nested sampling experiment

17.6. For the latter design, levels of factor B are nested (rather than cross-classified) within levels of factor A. For example, in considering the potency of a chemical, we could sample different manufacturing plants, batches of chemicals within a plant, and determinations within a batch. Note that the factor "batches" is not cross-classified with the factor "plants" since, for example, batch 1 for plant 1 is different from batch 1 for plant 2.

In Section 17.4 we will consider extending the results of this section to include a mixed model for an $a \times b$ factorial experiment.

▼ EXERCISES

Applications

▼ **17.3** Refer to Example 17.2. Suppose that only one observation was made by an analyst on a plaque sample from each subject. Taking the first observation for each factor–level combination, we have the following sample data.

Analyst	Subject			Totals
	1	2	3	
1	13.2	10.6	8.5	32.3
2	12.5	9.6	7.9	30.0
3	13.0	9.9	8.3	31.2
Totals	38.7	30.1	24.7	93.5

a. Write an appropriate linear statistical model identifying all terms in the model.
b. Write down the expected mean squares.

17.4 Refer to Exercise 17.3. Perform an analysis of variance. Use $\alpha = .05$ for all tests.

$ **17.5** Officials of a marketing research corporation were interested in studying the effect of a new promotional campaign for an improved brand of D-cell batteries. The study was conducted in a random sample of four standard metropolitan statistical areas (SMSAs), which had outlet stores for a random sample of three chain stores (selected from a large list of grocery, drug, and department stores). Sales volumes (in dollars) were recorded for a random sample of two weeks following the promotional campaign in the designated areas. These data are shown next.

Chain Store	SMSA			
	1	2	3	4
1	98	149	79	340
	112	126	61	302
2	87	96	119	125
	75	138	104	133
3	140	159	169	460
	190	185	150	420

a. Write an appropriate linear statistical model. List the assumptions and identify terms.
b. Perform an analysis of variance, showing expected mean squares. Use $\alpha = .05$.

17.4 ▼ MIXED-EFFECTS MODELS

mixed-effects model

In Section 17.3, we compared the analysis of variance tables for fixed- and random–effects models for a randomized block design and for a general $a \times b$ factorial laid out in a completely randomized design. Suppose, however, that we have a **mixed–effects model** for these same experimental designs where one effect is fixed and the other is random. For example, in Section 17.3, we considered an experiment to examine the effects of different subjects and different analysts on the DNA content of plaque. If the three subjects were selected at random and if the three analysts chosen were the only analysts of interest, we would have a mixed model for a randomized block design with fixed analysts and random subjects.

Let's consider a mixed model for a general $a \times b$ factorial in a completely randomized design. The model is the same as that given in Section 17.3 except that there are different assumptions.

$$y_{ijk} = \mu + \alpha_i + \beta_j + \alpha\beta_{ij} + \varepsilon_{ijk},$$

where we use the following assumptions:

assumptions

1. μ is an overall unknown mean.
2. α_i is a fixed effect corresponding to the ith level of factor A.
3. β_j is a random effect due to the jth level of factor B. β_j is normally distributed, with mean 0 and variance σ_β^2.
4. $\alpha\beta_{ij}$ is a random effect due to the ith level of A and jth level of B. $\alpha\beta_{ij}$ is normally distributed, with mean 0 and variance $\sigma_{\alpha\beta}^2$.
5. The β_js and $\alpha\beta_{ij}$s are all independent.

Using these assumptions, the analysis of variance table (incorporating Table 17.7) for a fixed, random, or mixed model in a two-factor experiment with replication is as shown in Table 17.12.

T A B L E   17.12
AOV for an $a \times b$ Factorial Experiment with n Observations per Cell

				EMS		
Source	SS	df	MS	Fixed Effects	Random Effects	Mixed Effects A Fixed, B Random
A	SSA	$a-1$	MSA	$\sigma_\varepsilon^2 + bn\theta_A$	$\sigma_\varepsilon^2 + n\sigma_{\alpha\beta}^2 + bn\sigma_\alpha^2$	$\sigma_\varepsilon^2 + n\sigma_{\alpha\beta}^2 + bn\theta_A$
B	SSB	$b-1$	MSB	$\sigma_\varepsilon^2 + an\theta_B$	$\sigma_\varepsilon^2 + n\sigma_{\alpha\beta}^2 + an\sigma_\beta^2$	$\sigma_\varepsilon^2 + an\sigma_\beta^2$
AB	SSAB	$(a-1)(b-1)$	MSAB	$\sigma_\varepsilon^2 + n\theta_{AB}$	$\sigma_\varepsilon^2 + n\sigma_{\alpha\beta}^2$	$\sigma_\varepsilon^2 + n\sigma_{\alpha\beta}^2$
Error	SSE	$ab(n-1)$	MSE	σ_ε^2	σ_ε^2	σ_ε^2
Totals	TSS	$nab-1$				

The expected mean squares column of Table 17.12 can be helpful in determining appropriate tests of significance. The test for $\sigma_{\alpha\beta}^2$ is the same in the mixed model as in the random-effects model.

test for $\sigma_{\alpha\beta}^2$

$$H_0: \quad \sigma_{\alpha\beta}^2 = 0$$
$$H_a: \quad \sigma_{\alpha\beta}^2 > 0$$

$$\text{T.S.:} \quad F = \frac{\text{MSAB}}{\text{MSE}}$$

R.R.: Based on $df_1 = (a-1)(b-1)$ and $df_2 = ab(n-1)$.

No matter what the results of our tests for $\sigma_{\alpha\beta}^2$, we could proceed to use the following tests for factors A and B, which follow from entries in the expected mean squares column of Table 17.12. For factor A we have

test, factor A

$$H_0: \quad \alpha_1 = \alpha_2 = \cdots = \alpha_a = 0$$
$$H_a: \quad \text{At least one of the } \alpha\text{s differs from the rest.}$$

$$\text{T.S.:} \quad F = \frac{\text{MSA}}{\text{MSAB}}$$

R.R.: Based on $df_1 = (a-1)$ and $df_2 = (a-1)(b-1)$.

For factor B we have

test, factor B

$$H_0: \quad \sigma_{\beta}^2 = 0$$
$$H_a: \quad \sigma_{\beta}^2 > 0$$

$$\text{T.S.:} \quad F = \frac{\text{MSB}}{\text{MSE}}$$

R.R.: Based on $df_1 = (b-1)$ and $df_2 = ab(n-1)$.

The analysis of variance procedure outlined for a mixed-effects model from an $a \times b$ factorial experiment can be used as well for a randomized block design, where treatments are fixed, blocks are assumed to be random, and there are n observations for each block and treatment. We'll illustrate this latter situation with the following example.

EXAMPLE 17.3 ▼

A corporation is interested in comparing two different sunscreens (s_1 and s_2) for protecting the skin of persons who want to avoid burning or additional tanning while exposed to the sun. A random sample of subjects (ages 20–25 years) agreed to participate in the study. For each person two 1-inch × 1-inch squares were marked off on either side of the back, under the shoulder but above the small of the back. Sunscreen s_1 was then randomly assigned to the two squares on one side of the back, with s_2 assigned to the other two squares. A reading based on the color of skin in a square was made prior to the application of a fixed amount of the assigned sunscreen, and then again after application and exposure to the sun for a 2-hour period. The data recorded in Table 17.13 are differences (postexposure minus preexposure) for the persons in the study. A high response indicates more burning.

TABLE 17.13

Data (differences) for the Sunscreen Experiment, Example 17.3

Sunscreen, A	Subjects, B										Totals
	1	2	3	4	5	6	7	8	9	10	
1	8.2	3.6	10.7	3.9	12.9	5.5	9.1	13.7	8.1	2.5	
	7.6	3.5	10.3	4.4	12.1	5.9	9.7	13.2	8.7	2.8	156.4
	15.8	7.1	21.0	8.3	25.0	11.4	18.8	26.9	16.8	5.3	
2	6.1	4.3	9.6	2.3	12.4	4.8	8.3	12.9	8.0	2.1	
	6.8	4.7	9.2	2.5	12.8	4.0	8.6	13.6	7.5	2.5	143.0
	12.9	9.0	18.8	4.8	25.2	8.8	16.9	26.5	15.5	4.6	
Totals	28.7	16.1	39.8	13.1	50.2	20.2	35.7	53.4	32.3	9.9	299.4

You will recognize this experiment as a randomized block design with two treatments as a fixed effect (A), ten subjects as a random effect (B), and $n = 2$ observations for each treatment and subject. Use these data to run an analysis of variance and draw conclusions.

Solution   We can compute the sums of squares for the sources of variability in the AOV table using the usual shortcut formulas.

$$TSS = \sum_{ijk} y_{ijk}^2 - \frac{y_{...}^2}{nab}$$

$$= 2{,}771.6 - \frac{(299.4)^2}{40} = 2{,}771.6 - 2{,}241.01 = 530.59$$

$$SSA = \sum_i \frac{y_{i..}^2}{20} - \frac{y_{...}^2}{40} = 2{,}245.50 - 2{,}241.01 = 4.49$$

$$SSB = \sum_j \frac{y_{.j.}^2}{4} - \frac{y_{...}^2}{40} = 2{,}758.50 - 2{,}241.01 = 517.49$$

$$SSAB = \sum_{ij} \frac{y_{ij.}^2}{2} - \frac{y_{...}^2}{40} - SSA - SSB$$

$$= 2{,}768.96 - 4.49 - 517.49 - 2{,}241.01 = 5.97$$

Then by subtraction, SSE = TSS − SSA − SSB − SSAB = 2.64.

TABLE 17.14
AOV Table for the Data of Example 17.3

Source	SS	df	MS	EMS Mixed Model
A	4.49	1	4.49	$\sigma_\varepsilon^2 + 2\sigma_{\alpha\beta}^2 + 20\theta_A$
B	517.49	9	57.50	$\sigma_\varepsilon^2 + 4\sigma_\beta^2$
AB	5.97	9	.66	$\sigma_\varepsilon^2 + 2\sigma_{\alpha\beta}^2$
Error	2.64	20	.13	σ_ε^2
Totals	530.59	39		

Substituting $a = 2$, $b = 10$, and $n = 2$ into an AOV table similar to that shown in Table 17.12, we have the results shown in Table 17.14.

A test for the random component $\alpha\beta_{ij}$ is as follows:

$$H_0: \quad \sigma_{\alpha\beta}^2 = 0$$
$$H_a: \quad \sigma_{\alpha\beta}^2 > 0$$

$$\text{T.S.:} \quad F = \frac{\text{MSAB}}{\text{MSE}} = \frac{.66}{.13} = 5.08$$

R.R.: For $\alpha = .05$, we will reject H_0 if the computed value of F exceeds 2.39, the value in Table 6 for $a = .05$, $\text{df}_1 = 9$, and $\text{df}_2 = 20$.

Conclusion: Since 5.08 exceeds 2.39, we reject H_0 and conclude that $\sigma_{\alpha\beta}^2 > 0$; that is, there is a significant source of random variation due to the combination of the ith level of A (sunscreens) and the jth level of B (persons). We would infer from this that one sunscreen may not necessarily be better than the other for all persons.

Even with this significant F test for $\sigma_{\alpha\beta}^2$, we can proceed to test for effects due to blocks and treatments separately. For blocks we have

$$H_0: \quad \sigma_\beta^2 = 0$$
$$H_a: \quad \sigma_\beta^2 > 0$$

$$\text{T.S.:} \quad F = \frac{\text{MSB}}{\text{MSE}} = \frac{57.50}{.13} = 442.31$$

R.R.: For $\alpha = .05$, we will reject H_0 if F exceeds 2.39, the value in Table 6 for $a = .05$, $\text{df}_1 = 9$, and $\text{df}_2 = 20$.

Conclusion: Since 442.31 exceeds 2.39, we reject H_0 and conclude that $\sigma_\beta^2 > 0$. Thus there is a significant source of random variation due to variability from person to person.

For factor A we have

$$H_0: \quad \alpha_1 = \alpha_2 = 0$$
$$H_a: \quad \alpha_1 \neq \alpha_2$$

T.S.: $\quad F = \dfrac{\text{MSA}}{\text{MSAB}} = \dfrac{4.49}{.66} = 6.80$

R.R.: For $\alpha = .05$, we will reject H_0 if F exceeds 5.12, the value in Table 6 for $a = .05$, $df_1 = 1$, and $df_2 = 9$.

Conclusion: Since $6.80 > 5.12$, we reject H_0 and conclude that the mean response (post minus pre) differs for the two sunscreens. Since $\bar{y}_{s1} = 156.4/20 = 7.82$ and $\bar{y}_{s2} = 143/20 = 7.15$, we would conclude that s_2 offers more protection on the average than s_1. However, as noted previously, there are significant sources of variability due to persons and the combination of persons with sunscreens. ▲

This discussion of mixed models, which has been illustrated for a randomized block design with n observations per cell, provides only a brief introduction to the study of mixed models. Indeed, we could spend one or more quarters of study at the graduate level covering topics appropriate for mixed models. For more advanced work, we could examine factorial experiments with three or more factors (some random, others fixed). In addition, when examining the effect of two factors (both fixed effects) on a response while blocking on a third factor (which is random), a **split-plot design** becomes an important alternative to a factorial experiment that is laid off in a randomized block design. The difference between a split-plot design and the factorial experiment set off in a randomized block design lies in the method of applying treatments (factor–level combinations) to experimental units. For each block, levels of factor 1 are randomly assigned to experimental units. Then levels of the second factor are randomly assigned to subunits within each level of factor 1. This randomization is quite different from the randomization used in a factorial experiment that is laid off in a randomized block design. A discussion of this topic is presented in Section 17.6.

split-plot design

▼ EXERCISES

Applications

17.6 Prior to conducting a clinical trial that involves a subjective evaluation of a patient's progress, the participating physicians are asked to agree on certain criteria for reaching an evaluation. To examine the consistency in their evaluations before the initiation of a particular clinical trial, a pilot study was conducted on four patients who had been treated with a drug that was to be included in the trial. Each of the five physicians who were to participate in the study was asked to evaluate (on a 0-to-10 point scale) the degree of cure after a 2-week treatment period. Since the clinical evaluations of a patient's cure were to be based on the results of a bacterial culture analysis, each physician analyzed two cultures from each patient. This feature was unknown to the physicians, who were merely told they would be analyzing eight separable bacterial cultures. The evaluations based on these cultures are recorded here.

a. Treating physicians as fixed and patients as random, write an appropriate linear statistical model. Identify all terms in the model.

b. Show the expected mean squares column in the AOV table.

| | **Patient** | | | |
Physician	1	2	3	4
1	7.2	4.2	9.5	5.4
	9.6	3.5	9.3	3.9
2	8.5	2.9	8.8	6.3
	9.6	3.3	9.2	6.0
3	9.1	1.8	7.6	6.1
	8.6	2.4	7.1	5.6
4	8.2	3.6	7.3	5.0
	9.0	4.4	7.0	5.4
5	7.8	3.7	9.2	6.5
	8.0	3.9	8.3	6.9

17.7 Refer to Exercise 17.6. Perform an analysis of variance. Draw your conclusions, using $\alpha = .05$.

17.5 RULES FOR OBTAINING EXPECTED MEAN SQUARES

We discussed the AOVs for one- and two-factor experiments for fixed–effects models in Chapter 15 and for random or mixed models earlier in this chapter. We will see in this section that for any k-factors experiment of data, with n observations per factor–level combination, it is possible to write expected mean squares for all main effects and interactions for fixed, random, or mixed models using some rather simple rules. *The importance of these rules is that, having written down the expected mean squares for an unfamiliar experimental design, we often can construct appropriate F tests.* The assumptions for the fixed and random models will be the same as we have used in describing fixed, random, and mixed models in previous sections.

classifying interactions

Two rules for **classifying interactions** as fixed or random effects are needed before we can proceed with the rules for obtaining expected mean squares.

Rules for the Classification of Interactions

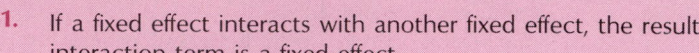

1. If a fixed effect interacts with another fixed effect, the resulting interaction term is a fixed effect.
2. If a random effect interacts with another effect (fixed or random), the resulting interaction term is a random component.

E X A M P L E   17.4

Consider a 3×6 factorial with two observations per factor–level combination. Classify the AB interaction as fixed or random for the following situations:

a. A and B are both fixed effects.
b. A is fixed and B is random.
c. A and B are both random

Solution   We apply the rules for classifying interactions.
a. AB is a fixed effect since A (fixed) interacts with B (fixed).
b. AB is a random component since A (fixed) interacts with B (random).
c. AB is random since A (random) interacts with B (random).   ▲

E X A M P L E   17.5

Consider a factorial experiment in the factors A, B, and C. Classify the AB, AC, BC, and ABC interactions as fixed or random when A and B are fixed effects and C is random.

Solution   We apply the classification rules.

AB is fixed; A (fixed) interacts with B (fixed).
AC is random; A (fixed) interacts with C (random).
BC is random; B (fixed) interacts with C (random).
ABC is random; A (fixed) interacts with BC (random).   ▲

mean square table

Before we state the rules for determining expected mean squares, it is convenient to construct a **mean square table**. These steps are summarized here and illustrated for an $a \times b$ factorial experiment with factor A random, B fixed, and n observations for each factor–level combination of A and B.

Steps in Constructing a Mean Square Table

▼

1. Write the model for the experiment. For an $a \times b$ factorial experiment, the model is

$$y_{ijk} = \mu + \alpha_i + \beta_j + \alpha\beta_{ij} + \varepsilon_{k(ij)}.$$

Note: We use brackets in the ε-term to indicate that there are $k = 1, 2, \ldots, n$ observations for each factor–level combination of factors A and B (i.e., for each choice of i, j).

2. Construct a table with each term in the model (except μ) forming a row heading. This table takes the following form for our example.

α_i
β_j
$\alpha\beta_{ij}$
$\varepsilon_{k(ij)}$

3. Form a column in the table for each subscript in the model.

	i	j	k
α_i			
β_j			
$\alpha\beta_{ij}$			
$\varepsilon_{k(ij)}$			

4. Above each column heading indicate whether the subscript corresponds to a fixed (F) or random (R) effect.

5. Also indicate the number levels for each subscript. Additions to the table from steps 4 and 5 are shown here for our example.

	a	b	n
	R	F	R
	i	j	k
α_i			
β_j			
$\alpha\beta_{ij}$			
$\varepsilon_{k(ij)}$			

(continues)

6. For each row in a given column, enter the number of levels associated with the column subscript, unless the row term contains the column subscript.

	a R i	b F j	n R k
α_i		b	n
β_j	a		n
$\alpha\beta_{ij}$			n
$\varepsilon_{k(ij)}$			

7. Examine the terms of the model, which are listed in the first column of the table. For each term with brackets in the subscript, place a 1 under the column(s) with a subscript included in the brackets.

	a R i	b F j	n R k
α_i		b	n
β_j	a		n
$\alpha\beta_{ij}$			n
$\varepsilon_{k(ij)}$	1	1	

8. Fill in the remaining cells of a column with a 0 if the column is headed by an F; fill in the remaining cells of a column headed by an R with a 1.

	a R i	b F j	n R k
α_i	1	b	n
β_j	a	0	n
$\alpha\beta_{ij}$	1	0	n
$\varepsilon_{k(ij)}$	1	1	1

This is the *mean square table* used for computing the expected mean squares for a two-factor experiment with factor A random, factor B fixed, and n observations per factor–level combination of A and B.

It's easy to compute expected mean squares once you have the mean square table. These rules are listed here.

Rules for Obtaining an EMS Using a Mean Square Table

1. Examine the subscript(s) of the term.
2. Eliminate any row in the mean square table that doesn't have the subscript(s).
3. Cover each column of the table headed by a nonbracketed subscript of the term.
4. Multiply the remaining, uncovered entries in each row to obtain the coefficients of terms in the expected mean square.

E X A M P L E   17.6

Compute $E(MSA)$ for a two-factor experiment with a levels of factor A (random), b levels of factor B (fixed), and n observations per factor–level combination.

Solution   Refer to the mean square table just given. We note that α_i has the subscript i; thus, we eliminate the second row of the table. Covering column 1 (which is headed by i), we multiply the remaining entries. Table 17.15 shows these steps, the multiplication, and the terms of the EMS.

T A B L E   17.15

Mean Square Table for Computing $E(MSA)$

	a R i	b F j	n R k	Product of Remaining Entries	Term EMS
α_i	1	b	n	bn	$bn\sigma_\alpha^2$
β_j	a	0	n	—	
$\alpha\beta_{ij}$	1	0	n	0	
$\varepsilon_{k(ij)}$	1	1	1	1	σ_ε^2

Using the last column of Table 17.15, we have

$$E(MSA) = \sigma_\varepsilon^2 + bn\sigma_\alpha^2.$$

The computation $E(MSB)$ follows in a similar manner. The term β_j has a subscript j, so we eliminate the second column (which is headed by j) and the first row (which contains no j) of the mean square table. The remaining entries are multiplied to obtain the coefficients of the expected mean square (see Table 17.16).

Hence,

$$E(MSB) = \sigma_\varepsilon^2 + n\sigma_{\alpha\beta}^2 + an\theta_B,$$

TABLE 17.16
Mean Square Table for Computing $E(MSB)$

	a R i	b F j	n R k	Product of Remaining Entries	Term EMS
α_i	1	b	n	—	
β_j	a	0	n	an	$an\theta_\beta$
$\alpha\beta_{ij}$	1	0	n	n	$n\sigma^2_{\alpha\beta}$
$\varepsilon_{k[ij]}$	1	1	1	1	σ^2_ε

where θ_B is a constant of the form

$$\theta_B = \frac{\sum_j \beta_j^2}{b-1}.$$

▲

EXAMPLE 17.7 ▼

a. Set up the mean square table for computing expected mean squares for a two-factor experiment with factor A fixed, factor B random, and n observations for each factor–level combination of A and B.
b. Compute $E(MSA)$.

Solution
a. The model for this experiment situation is

$$y_{ijk} = \mu + \alpha_i + \beta_j + \alpha\beta_{ij} + \varepsilon_{k[ij]}.$$

The corresponding mean square table is shown in Table 17.17.
b. $E(MSA)$ is found as shown in Table 17.18.

TABLE 17.17
Mean Square Table for Example 17.7

	a F i	b R j	n R k
α_i	0	b	n
β_j	a	1	n
$\alpha\beta_{ij}$	0	1	n
$\varepsilon_{k[ij]}$	1	1	1

TABLE 17.18
Computations for $E(MSA)$

	a F i	b R j	n R k	Product of Remaining Entries	Term EMS
α_i	0	b	n	bn	$bn\theta_\alpha$
β_j	a	1	n	—	
$\alpha\beta_{ij}$	0	1	n	n	$n\sigma^2_{\alpha\beta}$
$\varepsilon_{k[ij]}$	1	1	1	1	σ^2_ε

Thus, $E(\text{MSA}) = \sigma_\varepsilon^2 + n\sigma_{\alpha\beta}^2 + bn\theta_A$, where $\theta_A = \sum_i \alpha_i^2/(a-1)$. This agrees with what we obtained in the previous example for the fixed factor (B) with appropriate changes in notation. ▲

Previously, we have been concerned with only fixed–effects models. For these models the test statistics are always formed using the affected mean square in the numerator divided by MSE. But for random and mixed models the test statistics are not always the same. The test statistic for interaction, F equals MSAB/MSE, is the same for the fixed, random, and mixed models; but the F tests for factors A and B change depending on the assumptions for α_i and β_j. For example, the F test for factor A is MSA/MSE for A fixed, B fixed, and for A random, B fixed. In contrast, the F test for factor A is MSA/MSAB for A fixed, B random, and for A random, B random. So you can see the importance of knowing the expected mean squares for random and mixed models.

There's a special case of the two-factor experiment that should be mentioned: When there is only one observation ($n = 1$) per factor–level combination of A and B, and we have the model

$$y_{ij} = \mu + \alpha_i + \beta_j + \alpha\beta_{ij} + \varepsilon_{ij}.$$

You can see from the abbreviated AOV table of Table 17.19 that there are no degrees of freedom for error, so there is no test for interaction and depending on the model there may not be a valid test for the main effects. The only possible remedy is for the situation in which one can assume that there is no interaction between factors A and B, in which case all main effects can be tested using the mean square error from the model

$$y_{ij} = \mu + \alpha_i + \beta_j + \varepsilon_{ij}$$

regardless of whether we have a fixed, random, or mixed model. When the no-interaction assumption is not reasonable, the experiment must allow for replication ($n > 1$) at the factor–level combinations of A and B in order to obtain valid tests of the interaction and main effects.

T A B L E   17.19
Abbreviated AOV Table, Two-Factor Experiment ($n = 1$)

Source	df
A	$a - 1$
B	$b - 1$
AB	$(a - 1)(b - 1)$
Error	—
Total	$ab - 1$

These same rules that we used for the two-factor experiment can also be used for more complicated experiments and although the rules may seem a bit cumbersome, with practice they are quite easy to use. We'll give one more example using a three-factor experiment. For additional details regarding assumptions, derivations, and more complicated applications, see Hicks (1973).

E X A M P L E   17.8

▼

Give the expected mean squares for a $3 \times 5 \times 2$ factorial experiment with $n = 4$ observations per factor–level combination. Treat factors A and B as fixed and factor C as random.

Solution   The complete model for this experiment is given here along with the corresponding mean square table:

$$y_{ijkl} = \mu + \alpha_i + \beta_j + \gamma_k + \alpha\beta_{ij} + \alpha\gamma_{ik} + \beta\gamma_{ik} + \alpha\beta\gamma_{ijk} + \varepsilon_{l[ijk]}.$$

We'll set up the mean square table for general values of a, b, c, and n, and then substitute later. The mean square table using the rules discussed previously is shown in Table 17.20. The expected mean squares for the model terms can be obtained by applying the EMS rules to this mean square table. For example, for $E(MSA)$, we have the uncovered entries in Table 17.21.

From the last column of this table we have

$$E(MSA) = \sigma_\varepsilon^2 + bn\sigma_{\alpha\gamma}^2 + bcn\theta_A.$$

Substituting $a = 3$, $b = 5$, $c = 2$, and $n = 4$, this becomes

$$E(MSA) = \sigma_\varepsilon^2 + 20\sigma_{\alpha\gamma}^2 + 40\theta_A,$$

T A B L E   17.20
Mean Square Table for Example 17.8

	a F i	b F j	c R k	n R l
α_i	0	b	c	n
β_j	a	0	c	n
γ_k	a	b	1	n
$\alpha\beta_{ij}$	0	0	c	n
$\alpha\gamma_{ik}$	0	b	1	n
$\beta\gamma_{jk}$	a	0	1	n
$\alpha\beta\gamma_{ijk}$	0	0	1	n
$\varepsilon_{l[ijk]}$	1	1	1	1

T A B L E 17.21

Computations for $E(\text{MSA})$, Example 17.8

	a F i	b F j	c R k	n R l	Product of Remaining Entries	Term EMS
α_i	0	b	c	n	bcn	$bcn\theta_A$
β_j	a	0	c	n	—	
γ_k	a	b	1	n	—	
$\alpha\beta_{ij}$	0	0	c	n	0	
$\alpha\gamma_{ik}$	0	b	1	n	bn	$bn\sigma^2_{\alpha\gamma}$
$\beta\gamma_{jk}$	a	0	1	n	—	
$\alpha\beta\gamma_{ijk}$	0	0	1	n	0	
$\varepsilon_{l(ijk)}$	1	1	1	1	1	σ^2_ε

where

$$\theta_A = \sum_i \alpha_i^2/2.$$

Similarly, the expected mean squares for factors B and C can be shown to be

$$E(\text{MSB}) = \sigma^2_\varepsilon + an\sigma^2_{\beta\gamma} + acn\theta_B$$
$$= \sigma^2_\varepsilon + 12\sigma^2_{\beta\gamma} + 24\theta_B$$

and

$$E(\text{MSC}) = \sigma^2_\varepsilon + abn\sigma^2_\gamma$$
$$= \sigma^2_\varepsilon + 60\sigma^2_\gamma.$$

The table for computing the expected mean square for the AB interaction is shown in Table 17.22. The expected mean square for MSAB is

$$E(\text{MSAB}) = \sigma^2_\varepsilon + n\sigma^2_{\alpha\beta\gamma} + cn\theta_{AB}$$
$$= \sigma^2_\varepsilon + 4\sigma^2_{\alpha\beta\gamma} + 8\theta_{AB},$$

T A B L E 17.22

Computations for $E(\text{MSAB})$, Example 17.8

	a F i	b F j	c R k	n R l	Product of Remaining Entries	Term EMS
α_i	0	b	c	n	—	
β_j	a	0	c	n	—	
γ_k	a	b	1	n	—	
$\alpha\beta_{ij}$	0	0	c	n	cn	$cn\theta_{AB}$
$\alpha\gamma_{ik}$	0	b	1	n	—	
$\beta\gamma_{jk}$	a	0	1	n	—	
$\alpha\beta\gamma_{ijk}$	0	0	1	n	n	$n\sigma^2_{\alpha\beta\gamma}$
$\varepsilon_{l(ijk)}$	1	1	1	1	1	σ^2_ε

where

$$\theta_{AB} = \sum_{ij} \frac{\alpha\beta_{ij}^2}{8}.$$

After application of the EMS rules to the AC and BC interactions, we obtain

$$\begin{aligned} E(\text{MSAC}) &= \sigma_\varepsilon^2 + bn\sigma_{\alpha\gamma}^2 \\ &= \sigma_\varepsilon^2 + 20\sigma_{\alpha\gamma}^2 \end{aligned}$$

and

$$\begin{aligned} E(\text{MSBC}) &= \sigma_\varepsilon^2 + an\sigma_{\beta\gamma}^2 \\ &= \sigma_\varepsilon^2 + 12\sigma_{\beta\gamma}^2. \end{aligned}$$

Similarly, it can be shown that MSABC and MSE have expectations

$$\begin{aligned} E(\text{MSABC}) &= \sigma_\varepsilon^2 + n\sigma_{\alpha\beta\gamma}^2 \\ &= \sigma_\varepsilon^2 + 4\sigma_{\alpha\beta\gamma}^2 \end{aligned}$$

and

$$E(\text{MSE}) = \sigma_\varepsilon^2.$$

A summary of the expected mean squares, which we have computed using the EMS rules of this chapter, for the $3 \times 5 \times 2$ factorial with $n = 4$ observations per cell is shown in Table 17.23.

T A B L E   17.23

Partial AOV for Example 17.8

Source	df	EMS
A	$a - 1$	$\sigma_\varepsilon^2 + bn\sigma_{\alpha\gamma}^2 + bcn\theta_A$
B	$b - 1$	$\sigma_\varepsilon^2 + an\sigma_{\beta\gamma}^2 + acn\theta_B$
C	$c - 1$	$\sigma_\varepsilon^2 + abn\sigma_\gamma^2$
AB	$(a - 1)(b - 1)$	$\sigma_\varepsilon^2 + n\sigma_{\alpha\beta\gamma}^2 + cn\theta_{AB}$
AC	$(a - 1)(c - 1)$	$\sigma_\varepsilon^2 + bn\sigma_{\alpha\gamma}^2$
BC	$(b - 1)(c - 1)$	$\sigma_\varepsilon^2 + an\sigma_{\beta\gamma}^2$
ABC	$(a - 1)(b - 1)(c - 1)$	$\sigma_\varepsilon^2 + n\sigma_{\alpha\beta\gamma}^2$
Error	$abc(n - 1)$	σ_ε^2

EXAMPLE 17.9

▼

Refer to Example 17.8. Give an appropriate F statistic for

$$H_0: \quad \theta_A = 0$$

and

$$H_0: \quad \sigma_{\beta\gamma}^2 = 0$$

Solution   Using the expected mean squares listed in Table 17.23, it is clear that the test statistic for $H_0: \theta_A = 0$ is $F = \text{MSA/MSAC}$; the test statistic for $H_0: \sigma_{\beta\gamma}^2 = 0$ is $F = \text{MSBC/MSE}$.   ▲

We can always obtain valid tests for all sources of variability in fixed-effects models, but this is not true for some random- and mixed-effects models. Some researchers have developed approximate F tests that can be constructed for sources of variability in random- or mixed-effects models where no valid F test is available (see, for example, Satterthwaite (1946) and Welch (1956)). We will not cover this topic in this text since these tests can become quite involved and are open to some controversy.

estimation and prediction   For some random and mixed models, the objective of the researcher might include **estimation** of the variances for random effects and **prediction** of the average value of y. We briefly discussed estimation of the expected value of y and how we use the expected mean squares to obtain the variance of our estimates for a random-effects model. Estimation of the average value of y for a mixed model is more difficult and beyond the scope of this text.

estimates of variance components   Another use of expected mean squares is in the **estimation of the variances** associated with random effects in the model. For example, in a random-effects model for a one-factor experiment of t treatments and n observations per treatment, the expected mean squares for treatments and error are $\sigma_\varepsilon^2 + n\sigma_\alpha^2$ and σ_ε^2, respectively. As before, MSE is our estimate of σ_ε^2. Similarly, since MST estimates $\sigma_\varepsilon^2 + n\sigma_\alpha^2$, by substituting MSE for σ_ε^2, we can equate MST to the expected mean square for treatments and solve for σ_α^2. The solution, $(\text{MST} - \text{MSE})/n$, is an unbiased estimate of the variance of the treatments' source of variability in a one-factor experiment. This procedure of equating mean squares to expected mean squares can be used for obtaining estimates of variance components (variances of random effects) in random- and mixed-effects models for balanced designs. The problem of variance component estimation for unbalanced designs is a difficult one and is beyond the scope of this text. If you are interested in this topic, we refer you to three references for additional reading on the subject of variance component estimation: Hicks (1973), Mendenhall (1968), and Searle (1971).

17.6 ▼ NESTED SAMPLING AND THE SPLIT-PLOT DESIGN

Sometimes in an experiment one factor is "nested" within another. This can be illustrated with the following example. A pharmaceutical company conducted tests to determine the stability of its product (under room–temperature conditions) at a specific point in time. Two manufacturing sites were used. At each site, a random sample of three batches of the product was obtained and additional random samples of ten different tablets were obtained from each batch. The design can be represented as shown in Figure 17.1.

FIGURE 17.1

Two-Factor Experiment with Batches Nested in Sites

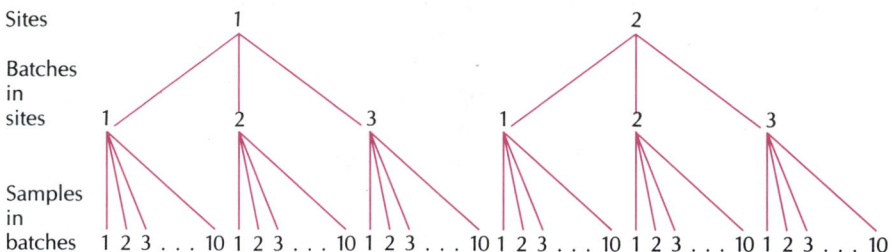

Although this might look like the usual two–factor experiment with sites (factor A) and batches (factor B), it should be noted that the three batches taken from site 1 are different from the three batches taken from site 2. In this sense, factor B (batches) is said to be *nested* in factor A (sites). For this experimental situation, it will be impossible to evaluate the effect of the interaction of factor B with factor A, because each level of factor B does not appear with each level of factor A, as would happen with a factorial arrangement of factors A and B. Here, the three batches within a site are unique to that site.

The general model for a two–factor experiment (n observations per cell) where factor B is nested in factor A can be written as

$$y_{ijk} = \mu + \alpha_i + \beta_{j(i)} + \varepsilon_{ijk} \qquad \begin{aligned} i &= 1, 2, \ldots, a \\ j &= 1, 2, \ldots, b \\ k &= 1, 2, \ldots, n. \end{aligned}$$

Note that this model is similar to the model for the two–factor experiment of Section 17.3, except that there is no interaction term $\alpha\beta_{ij}$ and the term for factor B, $\beta_{j(i)}$, is subscripted to denote the jth level of factor B is nested in the ith level of factor A. The analysis of variance table for this design is shown in Table 17.24.

TABLE 17.24

AOV Table for a Two-Factor Experiment (n observations per cell) with Factor B Nested in Factor A

					EMS	
Source	SS	df	MS	Fixed	Mixed (A Fixed)	Random
A	SSA	$a - 1$	MSA	$\sigma_\varepsilon^2 + bn\theta_A$	$\sigma_\varepsilon^2 + n\sigma_\beta^2 + bn\theta_A$	$\sigma_\varepsilon^2 + n\sigma_\beta^2 + bn\sigma_\alpha^2$
$B(A)$	SSB(A)	$a(b - 1)$	MSB(A)	$\sigma_\varepsilon^2 + n\theta_B$	$\sigma_\varepsilon^2 + n\sigma_\beta^2$	$\sigma_\varepsilon^2 + n\sigma_\beta^2$
Error	SSE	$ab(n - 1)$	MSE	σ_ε^2	σ_ε^2	σ_ε^2
Total	TSS	$abn - 1$				

Three of the more common situations are shown in Table 17.24 with the expected mean squares. Note the following in particular:

1. The F test for factor B is always

$$F = \frac{\text{MSB(A)}}{\text{MSE}}.$$

2. The F test for factor A in the fixed-effects model is

$$F = \frac{\text{MSA}}{\text{MSE}}.$$

For the random- and mixed-effects model, however, the corresponding test for factor A is

$$F = \frac{\text{MSA}}{\text{MSB(A)}}.$$

3. When $n = 1$, there is no test for factor B, but we can test for factor A in the random- and mixed-effects model using

$$F = \frac{\text{MSA}}{\text{MSB(A)}}.$$

E X A M P L E 17.10 ▼

An experiment was conducted to determine the content uniformity of film-coated tablets produced for a cardiovascular drug used to lower blood pressure. A random sample of three batches was obtained from each of two blending sites; within each batch a random sample of five tablets was assayed to determine content uniformity. These data are shown here:

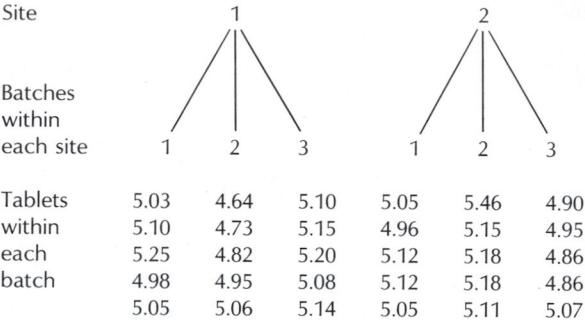

Site	1			2		
Batches within each site	1	2	3	1	2	3
Tablets	5.03	4.64	5.10	5.05	5.46	4.90
within	5.10	4.73	5.15	4.96	5.15	4.95
each	5.25	4.82	5.20	5.12	5.18	4.86
batch	4.98	4.95	5.08	5.12	5.18	4.86
	5.05	5.06	5.14	5.05	5.11	5.07

a. Run an analysis of variance. Use $\alpha = .05$.
b. Is there evidence to indicate batch-to-batch variability in content uniformity? Does the F test run depend on whether we assume batches are fixed or random?
c. Draw conclusions about batch.

Solution

a. For these data we have $a = 2$ sites, $b = 3$ batches within each blender, and $n = 5$ tablets per batch. For these data, the analysis of variance table is shown here. (See following output.)

Source	SS	df	MS	F
A	.01825	1	.01825	.16
$B(A)$	.45401	4	.11350	9.39
Error	.29020	24	.01209	
Total	.76246	29		

```
CONTENT UNIFORMITY OF FILM-COATED TABLETS
          INPUT DATA

   OBS     SITE    BATCH    CONTENT

    1       1        1       5.03
    2       1        1       5.10
    3       1        1       5.25
    4       1        1       4.98
    5       1        1       5.05
    6       1        2       4.64
    7       1        2       4.73
    8       1        2       4.82
```
(continues)

```
OBS     SITE    BATCH    CONTENT
 9        1        2       4.95
10        1        2       5.06
11        1        3       5.10
12        1        3       5.15
13        1        3       5.20
14        1        3       5.08
15        1        3       5.14
16        2        1       5.05
17        2        1       4.96
18        2        1       5.12
19        2        1       5.12
20        2        1       5.05
21        2        2       5.46
22        2        2       5.15
23        2        2       5.18
24        2        2       5.18
25        2        2       5.11
26        2        3       4.90
27        2        3       4.95
28        2        3       4.86
29        2        3       4.86
30        2        3       5.07

N=       30
```

CONTENT UNIFORMITY OF FILM-COATED TABLETS

ANALYSIS OF VARIANCE PROCEDURE

CLASS LEVEL INFORMATION

CLASS	LEVELS	VALUES
SITE	2	1 2
BATCH	3	1 2 3

NUMBER OF OBSERVATIONS IN DATA SET = 30

CONTENT UNIFORMITY OF FILM-COATED TABLETS

ANALYSIS OF VARIANCE PROCEDURE

DEPENDENT VARIABLE: CONTENT

SOURCE	DF	SUM OF SQUARES	MEAN SQUARE	F VALUE	PR>F	R-SQUARE	C.V.
MODEL	5	0.47226667	0.09445333	7.81	0.0002	0.619393	2.1803
ERROR	24	0.29020000	0.01209167		ROOT MSE		CONTENT MEAN
CORRECTED TOTAL	29	0.76246667			0.10996211		5.04333333

SOURCE	DF	ANOVA SS	F VALUE	PR > F
SITE	1	0.01825333	1.51	0.2311
BATCH(SITE)	4	0.45401333	9.39	0.0001

TESTS OF HYPOTHESES USING THE ANOVA MS FOR BATCH(SITE) AS AN ERROR TERM

SOURCE	DF	ANOVA SS	F VALUE	PR > F
SITE	1	0.01825333	0.16	0.7089

b., c. The F test for batches is

$$F = \frac{\text{MSB(A)}}{\text{MSE}} = 9.39$$

based on $df_1 = 4$ and $df_2 = 24$ degrees of freedom. Since the observed value of F, 9.39, exceeds the tabled value of F for $\alpha = .05$, we conclude that there is considerable batch–to–batch variability in content uniformity of tablets. This test does not depend on whether the batches are random. ▲

By now you may have realized that a whole new series of experimental designs have opened up with the introduction of nested effects. Thinking beyond the two-factor design, one could imagine a general multifactor design with factor A, factor B nested in levels of factor A, factor C nested in levels of A and B, and so on. The analysis of variance table for a three-factor nested design with all factors random is shown in Table 17.25.

T A B L E   17.25

AOV for a Three-Factor Nested Design—All Factors Random (n observations per cell)

Source	SS	df	MS	EMS
A	SSA	$a - 1$	MSA	$\sigma_\varepsilon^2 + n\sigma_\gamma^2 + cn\sigma_\beta^2 + bcn\sigma_\alpha^2$
$B(A)$	SSB(A)	$a(b - 1)$	MSB(A)	$\sigma_\varepsilon^2 + n\sigma_\gamma^2 + cn\sigma_\beta^2$
$C(A, B)$	SSC(A, B)	$ab(c - 1)$	MSC(A, B)	$\sigma_\varepsilon^2 + n\sigma_\gamma^2$
Error	SSE	$abc(n - 1)$	MSE	σ_ε^2
Total	TSS	$abcn - 1$		

Other extensions of these designs are possible as well. For example, one could have a three-factor experiment, where factors A and B are cross–classified but factor C is nested within levels of factor A and B. This would be an example of a *partially nested design*.

Suppose that a marketing research firm is responsible for sampling potential customers to obtain their opinions on two products (A_1 and A_2) in four geographic areas of the country ($B_1, \ldots, B_4$). A random sample of six stores selling product A_i is obtained in each geographic area. For each store selected for product A_i in geographic area B_j, ten people are interviewed concerning product i. For this design, factor C (stores) would be nested in levels of factors A (products) and B (geographic areas) and there would be $n = 10$ observations (opinions) for each level of factor C (stores) nested in levels of factors A and B.

The possibilities of nested and partial nested designs are seemingly endless, but, unfortunately, we will not have an opportunity to examine them here. The

split-plot design

interested reader should refer to Winer (1971) and Mendenhall (1968) for a more extensive treatment of this topic. We will, however, consider one very popular design that is similar to a partially nested design. It is called the **split-plot design** because it had its origin in agriculture experimentation. We will illustrate its use with an example.

The yields of three different varieties of soybeans are to be compared under two different levels of fertilizer application. If we were interested in getting (say) $n = 2$ observations at each combination of fertilizer and variety of soybeans, we would need 12 equal-sized plots. Taking fertilizers as factor A and varieties as a treatment factor T, one possible design would be the standard 2×3 factorial experiment with $n = 2$ observations per factor–level combination. But, since the application of fertilizer to a plot occurs when the soil is being prepared for planting, it would be difficult (logistically) to first apply fertilizer A_1 to six of the plots dictated by the factorial arrangement of factors A and T and then fertilizer A_2 to the other six plots before planting the required varieties of soybeans in each plot.

A design that would be easier to execute would have each fertilizer applied to two larger "wholeplots" and then the varieties of soybeans planted in three "subplots" (equal in size to the plots of the previous design) within each wholeplot. A design of this type appears in Figure 17.2.

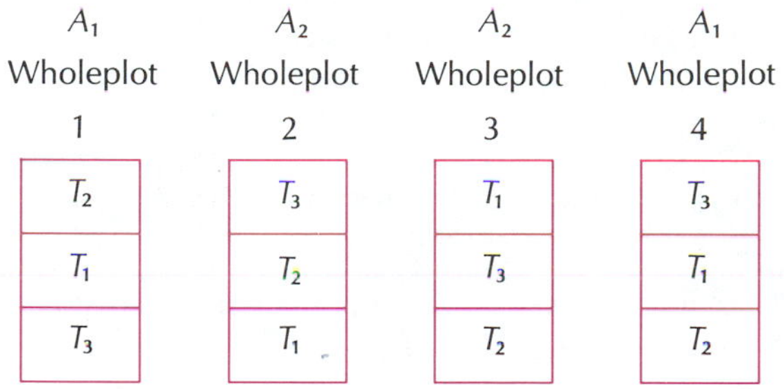

FIGURE 17.2
Split-Plot Design

A_1 Wholeplot 1	A_2 Wholeplot 2	A_2 Wholeplot 3	A_1 Wholeplot 4
T_2	T_3	T_1	T_3
T_1	T_2	T_3	T_1
T_3	T_1	T_2	T_2

This design is called a split-plot design, and with this design there is a two-stage randomization. First, levels of factor A (fertilizers) are randomly assigned to the wholeplots; second, the levels of factor T (soybeans) are randomly assigned to the subplots within a wholeplot (see Figure 17.3). Using this design, it would be much easier to prepare the soil and to apply the appropriate fertilizer to the larger wholeplots and then to plant varieties of soybeans in the subplots, rather than to prepare the soil and to apply fertilizer to the subplots and then to plant soybeans in the subplots, as would be the case for a standard 2×3 factorial experiment.

FIGURE 17.3

Two-Stage Randomization for a Split-Plot Design

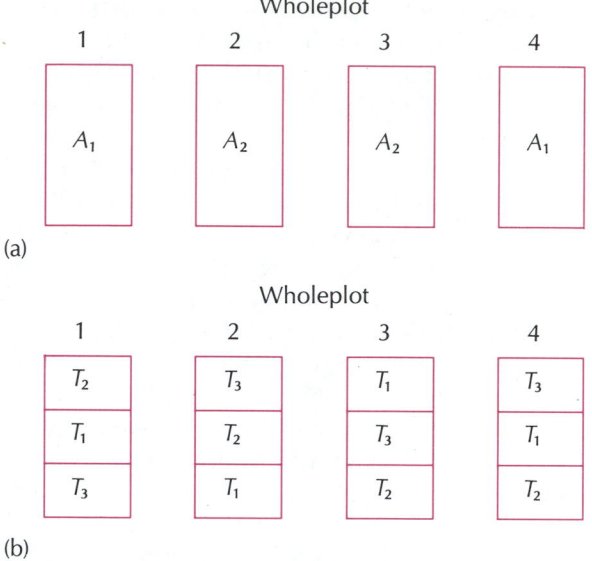

A variation on this design introduces a *blocking factor* (such as farms). So for our example, there may be $b = 2$ farms with $a = 2$ wholeplots per farm and $t = 3$ subplots per wholeplot. This design is shown in Figure 17.4.

The model for this more general two-factor split-plot design laid off in b blocks is as follows:

$$y_{ijk} = \mu + \alpha_i + \beta_j + \alpha\beta_{ij} + \tau_k + \alpha\tau_{ik} + \varepsilon_{ijk},$$

where y_{ijk} denotes the measurement receiving the ith level of factor A and the kth level of factor T in the jth block. The parameters α_i, τ_k, and $\alpha\tau_{ik}$ are the usual main effects and interaction parameters for a two-factor experiment, whereas

FIGURE 17.4

A Two-Factor Split-Plot Design Laid Out in Blocks

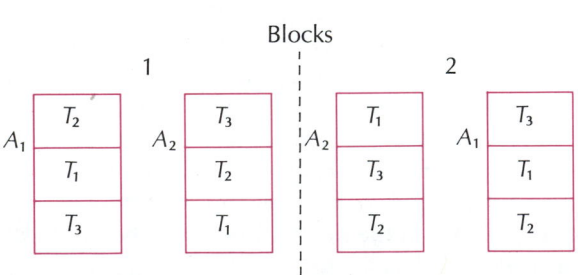

β_j is the effect due to block j and $\alpha\beta_{ij}$ is the interaction between the ith level of factor A and the jth block. The analysis corresponding to this model is shown in Table 17.26. Here we assume factors A and T are fixed effects, whereas blocks are random.

TABLE 17.26

AOV for a Two-Factor Split-Plot Design Laid Off in Blocks (A, T fixed; blocks random)

Source	SS	df	EMS
Between wholeplots			
Blocks	SSB	$b - 1$	$\sigma_\varepsilon^2 + at\sigma_\beta^2$
A	SSA	$a - 1$	$\sigma_\varepsilon^2 + t\sigma_{\alpha\beta}^2 + bt\theta_A$
AB (wholeplot error)	SSAB	$(a - 1)(b - 1)$	$\sigma_\varepsilon^2 + t\sigma_{\alpha\beta}^2$
Within wholeplots			
T	SST	$(t - 1)$	$\sigma_\varepsilon^2 + ab\theta_T$
AT	SSAT	$(a - 1)(t - 1)$	$\sigma_\varepsilon^2 + b\theta_{AT}$
Subplot error	SSE	$a(b - 1)(t - 1)$	σ_ε^2
Totals	TSS	$abt - 1$	

The sums of squares for the sources of variability listed in Table 17.26 can be obtained using the general shortcut formulas for main effects and interactions in a factorial experiment. Using these expected mean squares, we can obtain a valid F test for factor A in the wholeplot portion of the analysis and for factor T and the AT interaction in the subplot portion. These are shown here. Note that no test is made for the variability due to blocks.

Wholeplot Analysis

$$H_0: \quad \theta_A = 0 \text{ (or, equivalently, } H_0\text{: all } \alpha_i = 0\text{)}, \quad F = \frac{\text{MSA}}{\text{MSAB}}$$

Subplot Analysis

$$H_0: \quad \theta_{AT} = 0 \text{ (or, equivalently, } H_0\text{: all } \alpha\tau_{ik} = 0\text{)}, \quad F = \frac{\text{MSAT}}{\text{MSE}}$$

$$H_0: \quad \theta_T = 0 \text{ (or, equivalently, } H_0\text{: all } \tau_k = 0\text{)}, \quad F = \frac{\text{MST}}{\text{MSE}}$$

EXAMPLE 17.11 ▼

Soybean yields (in bushels per subplot unit) are shown here for a two–factor split–plot design laid off in $b = 3$ blocks. Fertilizers (factor A) were applied at random to the wholeplot units within each farm. Soybean varieties (factor

T) were then randomly allocated to the subplots within each wholeplot. Conduct an analysis of variance using these sample data. Give an approximate p-value for each test.

	1			2			3	
	Fertilizers			Fertilizers			Fertilizers	
Varieties	1	2	Varieties	2	1	Varieties	1	2
1	10.6	10.9	2	11.9	11.5	3	9.5	9.8
2	11.4	11.7	3	12.6	12.1	1	8.1	8.2
3	11.8	12.4	1	11.6	10.8	2	8.7	9.3

Solution   For these data with $a = 2$, $b = 3$, and $t = 3$, the sums of squares are as shown (see TYPE III SS column in the following computer output):

$$SSA = 0.841$$
$$SSB = 29.934$$
$$SSAB = 0.0597$$
$$SST = 4.955$$
$$SSAT = 0.021$$
$$SSE = 0.598$$
$$TSS = 35.325$$

```
                        GENERAL LINEAR MODELS PROCEDURE

DEPENDENT VARIABLE: RESPONSE

SOURCE              DF      SUM OF SQUARES    MEAN SQUARE    F VALUE    PR > F    R-SQUARE    C.V.

MODEL                9      34.72666667       3.85851852     51.59     0.0001    0.983062    2.5519

ERROR                8       0.59833333       0.07479167               ROOT MSE          RESPONSE MEAN

CORRECTED TOTAL     17      35.32500000                                0.27348065         10.71666667

SOURCE              DF      TYPE I SS     F VALUE  PR > F    DF    TYPE III SS    F VALUE   PR > F

FERT                 1      0.84500000     11.30   0.0099     1    0.84072727     11.24    0.0100
BLOCK                2     28.86333333    192.96   0.0001     2   29.93366667    200.11    0.0001
BLOCK*FERT           2      0.04333333      0.29   0.7560     2    0.05966667      0.40    0.6837
VARIETY              2      4.95450000     33.12   0.0001     2    4.95450000     33.12    0.0001
VARIETY*FERT         2      0.02050000      0.14   0.8739     2    0.02050000      0.14    0.8739

TESTS OF HYPOTHESES USING THE TYPE III MS FOR BLOCK*FERT AS AN ERROR TERM

SOURCE              DF      TYPE III SS   F VALUE  PR > F

FERT                 1      0.84072727     28.18   0.0337
```

The analysis of variance table using Type III sums of squares is shown here.

Source	SS	df	MS	F	p-value
Between wholeplots					
Blocks	29.934	2	14.967	—	—
A	.841	1	.841	28.18	.0337
AB (wholeplots)	.060	2	.030	—	—
Within wholeplots					
T	4.955	2	2.477	33.12	.0001
AT	.021	2	.010	.14	.8739
Subplot error	.598	8	.075	—	—
Totals	35.325	17			

▲

The distinction between this two-factor split-plot design and the standard two-factor experiments discussed in Chapter 15 lies in the randomization. In a split-plot design, there are two stages to the randomization process; first, levels of factor A are randomized to the wholeplots within each block, and then levels of factor B are randomized to the subplot units within each wholeplot of every block. In contrast, for a two-factor experiment laid off in a randomized block design (see Section 15.3), the randomization is a one-step procedure; treatments (factor–level combinations of the two factors) are randomized to the experimental units in each block. Much more has been done with split-plot designs than we will be able to cover in this text. The interested reader is referred to Steel and Torrie (1980), Snedecor and Cochran (1980), Winer (1971), and Mendenhall (1968), where the analyses for the basic split-plot design and more complicated variations on this design are discussed.

17.7 ▼ SUMMARY

Fixed, random, and mixed models are easily distinguished if we think in terms of the general linear model. The fixed-effects model relates a response to $k \geq 1$ independent variables and one random component, while a random-effects model is a general linear model with $k = 0$ and more than one random component. The mixed model, a combination of the fixed- and the random-effects models, relates a response to $k \geq 1$ independent variables and more than one random component.

The application of random-effects models to experimental situations was illustrated for the completely randomized design and for the $a \times b$ factorial experiment laid off in a completely randomized design. Similarities were noted between tests of significance in an analysis of variance for a random-effects model and for the corresponding fixed-effects model. Inferences resulting from an analysis of variance for a mixed model were illustrated using the $a \times b$ factorial experiment.

Unfortunately, in an introductory course, only a limited amount of time can be devoted to a discussion of random- and mixed–effects models. To expand our discussion in the text, the results of Section 17.5 are useful in developing the expected mean squares for sources of variability in the analysis of variance table for balanced designs. Using these expectations we can then attempt to construct appropriate test statistics for evaluating the significance of any of the fixed or random effects in the model.

The hardest part in any of these problems involving random- or mixed–effects models arises from trying to estimate $E(y)$, with an appropriate confidence interval for a random–effects model and the average value of y at some level or combination of levels for fixed effects in a mixed model. In Exercise 17.2, we illustrated how to obtain an estimate of $E(y)$ for a random–effects model and how to construct an approximate confidence interval. The problem becomes even more complicated for mixed models and hence is discussed in more advanced studies.

The final topics covered in this chapter were nested designs and split–plot designs. A brief introduction showed several variations on the basic factorial experiments discussed in Chapter 15 and in earlier sections of this chapter. The designs presented represent only a few of the more common designs possible when considering nested effects in a multifactor experimental setting. The interested reader should consult the references at the end of this book to pursue these topics in more detail.

▼ SUPPLEMENTARY EXERCISES

17.8 Distinguish between inferences related to θ_A (when factor A is fixed) and σ_α^2 (when factor A is random).

17.9 Refer to Example 17.8. Can valid test statistics be formulated to test for each of the sources of variability? List the valid tests and the corresponding test statistics.

17.10 Consider a factorial experiment with a levels of factor A, b levels of factor B, c levels of factor C, and n observations per factor–level combination.
   a. Write down the expected mean squares for all sources of variability in an analysis of variance table, treating A, B, and C as fixed effects.
   b. Note appropriate tests for all sources.

17.11 Refer to Exercise 17.10.
   a. Repeat part a, but now consider factors A, B, and C as random effects.
   b. Write down the test statistics for appropriate F tests for sources of variability.

17.12 Refer to Exercise 17.10. Suppose now that factor A is fixed and factors B and C are random.
   a. Write down expected mean squares for all sources in an AOV table.

b. Indicate the test statistics for those sources of variability that can be tested using an exact F test.

 17.13 The civil engineering department at a university was awarded a large grant to study the campus traffic problems and to recommend alternative solutions. One small phase of the study involved obtaining daily counts on the number of cars crossing, but not making use of, the campus facilities. To do this, a team of volunteers was stationed at each entrance to monitor simultaneously the license number and the time of entrance or exit for each car passing through the checkpoint. By comparing lists for all checkpoints and allowing a reasonable time for cars to traverse the campus, the teams were able to determine the number of cars crossing but not using the campus facilities during the 8:00 A.M. to 5:00 P.M. time period. A random sample of 6 weeks throughout the academic year was used, with 2 midweek days selected for study in the weeks sampled. The traffic volume data appear next.

Week 1	Week 2	Week 3	Week 4	Week 5	Week 6
680	438	539	264	693	530
618	520	600	198	646	575

a. Write an appropriate linear statistical model. Identify all terms in the model.
b. Perform an analysis of variance, indicating expected mean squares. Use $\alpha = .05$.

17.14 Refer to Exercise 17.13. Estimate the average number of cars crossing but not using the campus facilities for a midweek day of a randomly selected week and give an approximate confidence interval. (Hint: Refer to Exercise 17.2.)

17.15 Refer to Exercise 17.6. Suppose the five physicians chosen for the pilot study were considered to be a random sample from many possible physicians.
a. Write an appropriate model. Indicate how the assumptions for this model differ from those of part a in Exercise 17.6.
b. Compare the AOV table and conclusions drawn here to those of Exercise 17.7.

17.16 Refer to Exercise 17.15.
a. Which model and analysis seem to be more appropriate?
b. Might you also consider a fixed-effects model? Why or why not?

17.17 Refer to Exercise 15.3. Suppose that we consider the five investigators as a random sample from a population of all possible investigators for the rocket propellant experiment.
a. Write an appropriate linear statistical model, identifying all terms and listing your assumptions.
b. Perform an analysis of variance. Include an expected mean squares column in the analysis of variance table.

17.18 Refer to Exercise 17.17. Indicate the differences in the hypothesis under test and differences in the conclusions drawn for the fixed and random effects.

17.19 Obtain the expected mean squares for the experiment described in Example 15.4 if we assume that the drivers were selected at random from the many possible drivers in a large city.

17.20 Obtain expected mean squares for the experiment described in Exercise 15.9, assuming that both rows and columns are random effects.

17.21   Refer to the data of Exercise 15.30.
  a. Give the expected mean squares under the assumption that investigators were selected at random from a large group of similarly qualified persons throughout the country.
  b. Indicate how your analysis and conclusions would change from those in Exercise 15.31 under the assumption of part a.

17.22   Refer to the data of Example 17.2.
  a. Using the expected mean squares, give formulas for estimates of σ_ε^2, σ_γ^2, σ_β^2, and σ_α^2.
  b. Using the data of Example 17.2 and the formulas of part a, find estimates for all the variance components.

17.23   Refer to Exercise 17.5. Use the expected mean squares to obtain estimates of all the variance components.

17.24   Core soil samples are taken in each of six locations within a territory being investigated for surface mining of bituminous coal. Each of the core samples is divided into four sub-samples for separate analyses of the sulfur content of the sample.
  a. Identify the design and give a model for this experimental setting.
  b. Give the sources of variability and degrees of freedom for an AOV.

17.25   The sample data for Exercise 17.24 are shown here. Run an AOV and draw conclusions. Use $\alpha = .05$.

	Analyses			
Location	1	2	3	4
1	15.2	16.8	17.5	16.2
2	13.1	13.8	12.6	12.9
3	17.5	17.1	16.7	16.5
4	18.3	18.4	18.6	17.9
5	12.8	13.6	14.2	14.0
6	13.5	13.9	13.6	14.1

17.26   Tablet hardness is one comparative measure for different formulations of the same drug product; some combinations of ingredients (in addition to the active drug) in a formulation give rise to harder tablets than do other combinations. Suppose that three batches of a formulation are to be examined; 10 different 1-pound samples of tablets are obtained from each batch, and six tablets are tested from each of the 1-pound samples.
  a. Identify the design.
  b. Give an appropriate model with assumptions.
  c. Give the sources of variability and degrees of freedom for an AOV.

17.27   Refer to Exercise 17.26. Given that the sums of squares are as listed here, perform an analysis of variance and draw conclusions about the tablet hardness data for the formulation under study. Use $\alpha = .05$.

Source	SS
Between batches	2,200.8
Between samples within a batch	1,650.4
Between tablets within a sample	90.3

EXPERIMENTS WITH REPEATED MEASURES (optional)

18.1   INTRODUCTION

In all the experimental situations discussed so far in this text (except for the paired difference experiment), we have assumed that only one observation is taken on each experimental unit. For example, in an experiment to compare the effects of three different cardiovascular compounds on blood pressure, we could use a single-factor design where n_1 patients are assigned to compound 1, n_2 to compound 2, and n_3 to compound 3. Then the model would be

$$y_{ij} = \mu + \alpha_i + \varepsilon_{ij},$$

where α_i is the (fixed or random) effect due to compound i and ε_{ij} is the random effect associated with patient j treated with compound i. For this design, we would get one measurement (y_{ij}) for each patient.

The practicalities of many applied research settings make it mandatory from a cost and efficiency standpoint to obtain more than one observation per experiment unit. For example, in conducting clinical research, it is often difficult to find patients who have the condition to be studied *and* who are willing to participate in a clinical trial. Hence, it is important to obtain as much information as possible once a suitable number of patients have been located. In this chapter, we will consider several different experimental settings involving one or more factors and repeated measures. Rather than obtaining just one observation per patient, as suggested in the design with n_i patients treated with drug product i, we could obtain t different measurements corresponding to t different time points

TABLE 18.1
Repeated Time Points for Each Patient

Compound	Time Period			
	1	2	$\cdots$	t
1	y_{111} $\vdots$ y_{1n_11}	y_{112} $\vdots$ y_{1n_12}	$\cdots$ $\cdots$	y_{11t} $\vdots$ y_{1n_1t}
2	y_{211} $\vdots$ y_{2n_21}	y_{212} $\vdots$ y_{2n_22}	$\cdots$ $\cdots$	y_{21t} $\vdots$ y_{2n_2t}
3	y_{311} $\vdots$ y_{3n_31}	y_{312} $\vdots$ y_{3n_32}	$\cdots$ $\cdots$	y_{31t} $\vdots$ y_{3n_3t}

following administration of the assigned treatment. This experimental setting is shown in Table 18.1.

In Table 18.1, y_{ijk} denotes the observation at the time k for the jth patient on compound i. Note that we are getting $t > 1$ observations per patient, rather than only 1. The methods of this chapter can be used to analyze these data, as well as those in several other experimental situations. In this chapter, we will focus on experimental settings in the pharmaceutical industry. Similar, comparable applications can be found for these designs in the R & D and manufacturing operations of most major industries, such as the automotive, chemical, and aerospace industries.

18.2 ▼ SINGLE-FACTOR EXPERIMENTS WITH REPEATED MEASURES

In the previous section, we discussed some reasons why one might want to get more than one observation per patient. Another reason for obtaining more than one observation per patient is that frequently the variability *among* or *between* patients is much greater than the variability *within* a patient. We observed this in the paired t-test example of Section 6.5. If this is the case, it might be better to block on patients and to give each patient each treatment. Then the comparison among compounds is a within–patient comparison, rather than a comparison between patients, as would be the case with the single-factor experiment with n_i different patients assigned to compound i. A single-factor design that reflects this within-patient emphasis is shown in Table 18.2.

With this design, n patients would be treated separately with each of the three compounds. Presumably, the order of treatment would be randomized and there would be a sufficiently long "washout" period between the treatments, so that the results from one compound would not affect the results for another compound.

TABLE 18.2

A Within-Patient
Comparison of
Compounds 1, 2, and 3

	Patient			
Compound	1	2	...	n
1	y_{11}	y_{12}	...	y_{1n}
2	y_{21}	y_{22}	...	y_{2n}
3	y_{31}	y_{32}	...	y_{3n}

Here again, we are obtaining more than one observation per patient and presumably getting more useful information about the three drug products in question. One model for this experimental setting is

$$y_{ij} = \mu + \alpha_i + \pi_j + \varepsilon_{ij}.$$

Note that this model looks like any other single-factor experimental setting with a compounds and n patients. However, the assumptions are different because we are obtaining more than one observation per patient. For this model, we make the following assumptions.

1. α_i is a constant and $\sum \alpha_i = 0$.
2. The π_j are independent and normally distributed $(0, \sigma_\pi^2)$.
3. The ε_{ij} are independent of the π_j.
4. The ε_{ij} are independent and normally distributed $(0, \sigma_\varepsilon^2)$.

From these assumptions it can be shown that the variance of y_{ij} is $\sigma_\pi^2 + \sigma_\varepsilon^2$ and the covariance for any two observations from patient j is constant. These assumptions give rise to a variance–covariance matrix for the y_{ij}, which exhibits *compound symmetry*. For example, with $a = 3$, and $n = 2$, the variance–covariance for the y_{ij} would appear as

$$\mathbf{V}_{6 \times 6} = \begin{bmatrix} E & F \\ F & E \end{bmatrix},$$

where

$$\mathbf{E}_{3 \times 3} = (\sigma_\pi^2 + \sigma_\varepsilon^2) \begin{bmatrix} 1 & \rho & \rho \\ \rho & 1 & \rho \\ \rho & \rho & 1 \end{bmatrix}$$

and

$$\mathbf{F}_{3 \times 3} = \begin{bmatrix} 0 & 0 & 0 \\ 0 & 0 & 0 \\ 0 & 0 & 0 \end{bmatrix}.$$

Note that with compound symmetry, two observations on the same patient are correlated, whereas observations on two different patients are not. The analysis of variance for the experimental design being discussed and this set of assumptions is shown in Table 18.3. This AOV looks all too familiar. When the assumptions hold, and hence when compound symmetry holds, the statistical test on factor A ($F = \text{MSA/MSE}$) is appropriate. But there are some other more general conditions* that also lead to a valid F test for factor A using $F = \text{MSA/MSE}$. How restrictive are these assumptions and how can we tell when the test is appropriate?

T A B L E   18.3

AOV for the Experimental Setting Depicted in Table 18.2

Source	SS	df	EMS (*A* fixed, patients random)
Between patients	SSP	$n - 1$	$\sigma_\varepsilon^2 + a\sigma_\pi^2$
Within patients			
A	SSA	$a - 1$	$\sigma_\varepsilon^2 + n\theta_A$
Error	SSE	$(a - 1)(n - 1)$	σ_ε^2
Totals			

There are no easy answers to these questions since there are no simple tests to check for compound symmetry. The general conditions (called the Huynh–Feldt conditions) under which the F test for factor A is valid are often not met because observations on the same patient taken closely in time are more highly correlated than are observations taken farther apart in time. So be careful about this. In general, when the variance–covariance matrix does not follow a pattern of compound symmetry, the F test for factor A has a positive bias, which allows rejection of H_0: all $\alpha_i = 0$ more often than is indicated by the critical F-values.

From a practical standpoint, the best thing to do in a given experimental setting is to make certain that there is sufficient time between applications of the treatment to allow washout (or elimination) of the previous treatment and to make certain that the design is applied in only those situations where the disease is relatively stable, so that following treatment and washout, each patient (or experimental unit) is essentially the same as prior to receiving treatment. For example, even when studying the effect of blood-pressure–lowering drugs, we would expect the hypertension to be stable enough that the patients would return to their predrug level blood pressures after washout of the first assigned compound before receiving the second assigned compound, and so on.

In Section 18.3, more will be said about how to judge whether the underlying assumptions for the test hold, and if they do not, how to proceed. For further information on this topic, refer to higher-level textbooks covering repeated-measures experiments in detail (for example, Winer (1962)).

---

* *JASA* 1970 and *SAS Statistics User's Guide*.

18.3 ▼ TWO-FACTOR EXPERIMENTS WITH REPEATED MEASURES ON ONE OF THE FACTORS

We can extend our discussion of repeated measures experiments to two-factor settings. For example, in comparing the blood–pressure–lowering effects of cardiovascular compounds, we could randomize the patients so that n different patients receive each of the three compounds. But rather than having each patient receive each compound, we could take multiple measurements across time for each patient. For example, we might be interested in obtaining blood pressure readings immediately prior to receiving a single dose of the assigned compound and then every 15 minutes for the first hour and hourly thereafter for the next 6 hours. The data for this type of setting are depicted in Table 18.1. Note that this is a two-factor experiment (compounds and time) with repeated measures taken over one of the two factors (time).

The model that we will use for a two-factor experiment comparing a levels of factor A (compounds), having n patients per level of factor A, and b levels of factor B (time) is

$$y_{ijk} = \mu + \alpha_i + \pi_{j(i)} + \beta_k + \alpha\beta_{ik} + \varepsilon_{ijk},$$

where α_i, β_k, and $\alpha\beta_{ik}$ are fixed effects corresponding to main effects for factor A (compounds), factor B (times), and their interaction, respectively. The term $\pi_{j(i)}$ denotes a random effect due to the jth patient in the ith level of factor A. We assume that the $\pi_{j(i)}$ are independent and normally distributed $(0, \sigma_\pi^2)$.

Based on these assumptions, we have the analysis of variance shown in Table 18.4. Based on Table 18.4, it is clear that the following tests can be performed:

1. H_0: $\theta_{AB} = 0$

$$F = \frac{\text{MSAB}}{\text{MSE}}$$

2. H_0: $\theta_B = 0$

$$F = \frac{\text{MSB}}{\text{MSE}}$$

3. H_0: $\theta_A = 0$

$$F = \frac{\text{MSA}}{\text{MSP(A)}}.$$

TABLE 18.4
Analysis of Variance for a Two-Factor Experiment, Repeated Measures on One Factor

Source	SS	df	EMS (A, B fixed; patients random)
Between patients			
A	SSA	$a - 1$	$\sigma_\varepsilon^2 + b\sigma_\pi^2 + bn\theta_A$
Patients in A	SSP(A)	$a(n - 1)$	$\sigma_\varepsilon^2 + b\sigma_\pi^2$
Within patients			
B	SSB	$b - 1$	$\sigma_\varepsilon^2 + an\theta_B$
AB	SSAB	$(a - 1)(b - 1)$	$\sigma_\varepsilon^2 + n\theta_{AB}$
Error	SSE	$a(b - 1)(n - 1)$	σ_ε^2
Totals	TSS	$abn - 1$	

EXAMPLE 18.1 ▼

Ten subjects agreed to participate in a study to examine the concentration of drug in the bloodstream for two different dosage forms (capsule and tablet) of the same product following a single dose. Presumably, within limits, the higher the concentration, the more effective the drug product. Five subjects were allocated at random to the capsule form and the other five to the tablet form. Subjects fasted from 8:00 P.M. of the night prior to starting the study until four hours following ingestion of the assigned dose form (at 8:00 A.M. of the next day). Blood samples (15 mL) were obtained immediately preceding the assigned dose and at .5, 1, 2, 3, and 4 hours after dosing, and were analyzed for the concentration of the drug product in the bloodstream. These data (in ng/mL) are shown here.

	Tablet Time							Capsule Time					
Subject	0	.5	1	2	3	4	Subject	0	.5	1	2	3	4
1	0	50	75	120	60	30	1	0	30	55	80	130	65
2	0	40	80	135	70	40	2	0	25	50	75	125	60
3	0	55	75	125	85	50	3	0	35	65	85	140	85
4	0	70	85	140	90	40	4	0	45	70	90	145	80
5	0	60	90	150	95	50	5	0	50	75	95	160	90

a. Plot the mean sample data (response versus time) for each of the dose forms. Do the dose forms seem to have different availability patterns across time?

b. Run a repeated measures analysis of variance.

Solution

a.

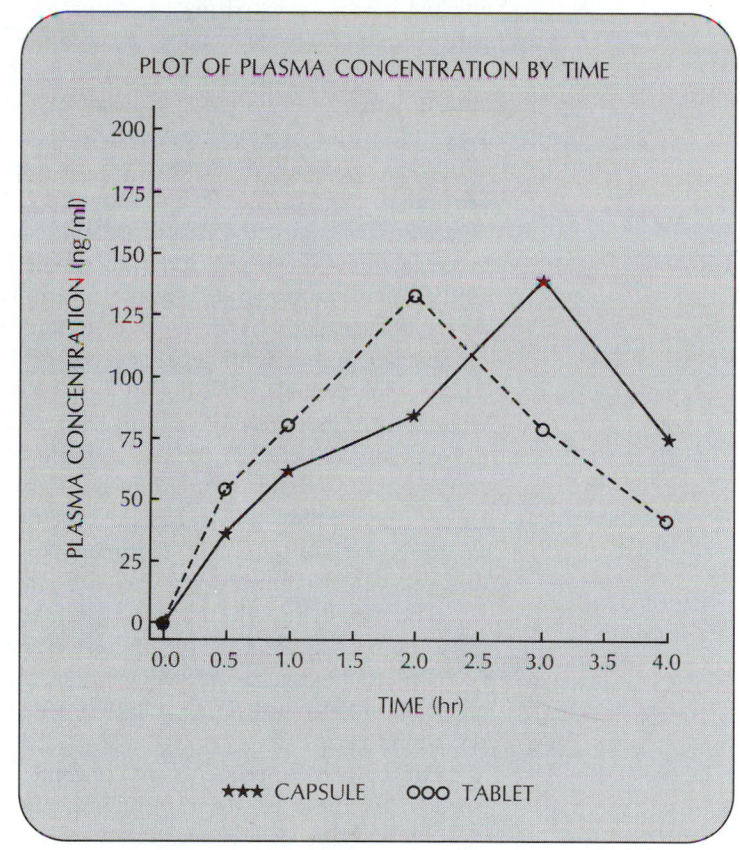

PLOT OF PLASMA CONCENTRATION BY TIME

★★★ CAPSULE   ooo TABLET

b.

Source	SS	df	MS	F	p-value
Between patients					
Formulation	33.75	1	33.75	.08	.7810
Patients in formulation	3,266.67	8	408.33		
Within patients					
Time	86,692.08	5	17,338.42	424.61	.0001
Time × formulation	19,478.75	5	3,895.75	95.41	.0001
Error	1,633.33	40	40.83		

Conclusion: There is evidence (based on the significant time × formulation interaction) that the two formulations have different availability (concentration) patterns across time.   ▲

The F test for factor A is based on between-subject effects and hence is *not* affected by the repeated measures on factor B. But, the F-ratios for the within-patients effects are affected and, as with the one-factor experiment with repeated measures, we must worry about the conditions under which these F tests are appropriate. If compound symmetry of the variance–covariance matrix for the y_{ijk}s holds, then we can apply these tests; also if the Huynh–Feldt conditions alluded to previously hold, then we can apply these F tests. Some have suggested (Greenhouse and Geisser (1959); Huynh and Feldt (1970)) that "adjusted" F-values be used to determine the statistical significance of a repeated measures F test when there is some departure from the underlying conditions for that test. The adjustments recommended by the various authors follow the same pattern. A quantity epsilon is defined as a multiplicative adjustment factor for the numerator and denominator degrees of freedom for the F test in question. This epsilon (which we'll denote by e) is not to be confused with the random error term ε in our models. For most of these adjustments, the multiplicative factor e ranges between 0 and 1, taking on a value of 1 when the underlying conditions for a valid F test are met and smaller values as the degree of departure from those conditions increases. A value of e having been determined for a given situation, the computed F statistic is compared to the critical value for an F distribution with numerator and denominator degrees of freedom multiplied by e.

The ideas behind the adjustment can be seen if we use the experimental setting for Table 18.4 as the basis for discussion. Here we have a two-factor experiment with repeated measures on the second factor (B). The F tests for the within-patient effects, B and AB shown in Table 18.4, are valid provided the Huynh–Feldt conditions hold.

For a given experiment, we compute a value of e and adjust the degrees of freedom for the F test by multiplying df_1 and df_2 by e. So, to run a test of H_0: $\theta_{AB} = 0$, a value of e is computed from the sample data and the computed F statistic

$$F = \frac{\text{MSAB}}{\text{MSE}}$$

is compared to a critical value of F_α based on $\text{df}_1 = e(a-1)(b-1)$ and $\text{df}_2 = ea(b-1)(n-1)$. Note that when $e = 1$, the underlying conditions hold and we have the original, recommended degrees of freedom, $\text{df}_1 = (a-1)(b-1)$ and $\text{df}_2 = a(b-1)(n-1)$.

In experimental situations where repeated measures data are to be analyzed and where you have access to SAS, you can use PROC GLM to compute revised p-values for two different adjustments to the degrees of freedom. The first adjustment, proposed by Greenhouse and Geisser (1959), uses a sample estimate of e. This adjustment, labeled "G-G" in the SAS output, has been shown, in simulation studies, to be ultraconservative, because the actual p-value may be

```
                           GENERAL LINEAR MODELS PROCEDURE

DEPENDENT VARIABLE: CONC

SOURCE              DF      SUM OF SQUARES      MEAN SQUARE     F VALUE    PR > F    R-SQUARE    C.V.

MODEL               19     109471.25000000    5761.64473684     141.10    0.0001    0.985299   9.6698

ERROR               40       1633.33333333      40.83333333               ROOT MSE          CONC MEAN

CORRECTED TOTAL     59     111104.58333333                               6.39009650        66.0833333333

SOURCE              DF        TYPE I SS    F VALUE    PR > F    DF      TYPE III SS   F VALUE    PR > F
FORM                 1      33.75000000       0.83    0.3687     1     33.75000000       0.83    0.3687
PATIENT(FORM)        8    3266.66666667      10.00    0.0001     8   3266.66666667      10.00    0.0001
TIME                 5   86692.08333333     424.61    0.0001     5  86692.08333333     424.61    0.0001
FORM*TIME            5   19478.75000000      95.41    0.0001     5  19478.75000000      95.41    0.0001


TESTS OF HYPOTHESES USING THE TYPE III MS FOR PATIENT(FORM) AS AN ERROR TERM
SOURCE              DF       TYPE III SS    F VALUE    PR > F
FORM                 1      33.75000000       0.08    0.7810




                           GENERAL LINEAR MODELS PROCEDURE
              TESTS OF HYPOTHESES FOR BETWEEN SUBJECTS EFFECTS

SOURCE              DF      TYPE III SS     MEAN SQUARE     F VALUE    PR > F
FORM                 1     33.75000000     33.75000000        0.08    0.7810

ERROR                8   3266.66666667    408.33333333




                           GENERAL LINEAR MODELS PROCEDURE
           UNIVARIATE TESTS OF HYPOTHESES FOR WITHIN SUBJECT EFFECTS

                                                                                ADJUSTED   PR > F
SOURCE              DF      TYPE III SS     MEAN SQUARE     F VALUE   PR > F    G     G     H    F
TIME                 5   86692.08333333   17338.41686667     424.61   0.0001   0.0001      0.0001
TIME*FORM            5   19478.75000000    3895.75000000      95.41   0.0001   0.0001      0.0001

ERROR(TIME)         10    1633.33333333      40.83333333

                      GREENHOUSE-GEISSER EPSILON = 0.5571
                      HUYNH-FELDT EPSILON = 0.9916
```

much smaller than that indicated by the p-value using the G-G adjustment. The second adjustment factor (proposed by Huynh and Feldt (1970)) is based on a different formula for e. Once again, however, an estimate of this adjustment factor is computed from the sample data. The degrees of freedom for critical values of the F statistics are then adjusted using the estimate of e. This adjustment is labeled "H-F" in the PROC GLM output. Although the Greenhouse–Geisser e and Huynh–Feldt e both must be in the interval $0 < e \leq 1$, the H-F estimate of e can sometimes be greater than 1. In these situations, a value of $e = 1$ is used in determining the appropriate degrees of freedom for the F test.

EXAMPLE 18.2 ▼

Refer to the output for Example 18.1.

a. Locate the estimated values for the Greenhouse–Geisser adjustment factor and the Huynh–Feldt adjustment factor.

b. Are the conclusions for the tests on time effects and the time formulation interaction affected by these adjustments?

Solution

a. The Greenhouse-Geisser estimate of e is .5571 and the Huynh–Feldt estimate of e is .9916.

b. Time effects: F tests based on the G-G adjustment and on the H-F adjustment yield p-values of .0001 and .0001, respectively, the same as the original F tests; hence, the adjustments do not change the original conclusion.

Time × formulation interactions: As with the F test on time, the adjustments did not change the p-values or the conclusions drawn from the original tests.

In conclusion, if you have access to SAS when doing an analysis of variance for a repeated measures experiment, it would be wise to check the effects of adjustments to F tests on the factors affected by repeated measures. If the conclusions based on the original ($e = 1$) test differ from those based on the H-F or G-G adjustment, we recommend adhering to the conclusions based on the less conservative H-F adjustment.   ▲

▼ EXERCISES

Applications

  18.1 A study was run to compare the effects of group therapy over time on patients who had been tested for depression and categorized into one of four levels of depression based on the Hamilton Depression Scale. There were 24 patients, with six patients per group, and each patient participated in the group therapy and was tested daily for a period of two weeks (14 times). Give the sources of variability and degrees of freedom for a repeated measures AOV.

18.2 Draw conclusions for Exercise 18.1 based on the sums of squares listed in the following table.

	SS
Between patients	
Depression group	962
Patients in group	582
Within patients	
Time	165
Time × group	120
Error	65

18.3 An antihistamine is frequently studied using a model to examine its effectiveness (compared to a placebo) in inhibiting a positive skin reaction to a known allergen. Consider the following situation. Individuals are screened to find 20 subjects who demonstrate sensitivity to the allergen to be used in the study. The 20 subjects are then randomly assigned to one of two treatment groups (the known antihistamine and an identical-appearing placebo), with 10 subjects per group. At the start of the study, a baseline (predrug) sensitivity reading is obtained, and then each patient begins taking the assigned medication for 3 days. Then skin sensitivity readings are taken at 1, 2, 3, 4, and 8 hours following the first dose. The percent inhibition of the skin sensitivity reaction (reduction in swelling area where the allergen is applied, compared to baseline) is shown here for each of the 20 patients.

		Percent Inhibition Time (hours)				
Treatment	Patient	1	2	3	4	8
1	1	10.5	28.2	15.3	43.0	29.0
	2	41.2	25.3	27.8	28.0	53.2
	3	43.0	20.8	29.3	5.2	26.5
	4	61.4	61.6	62.8	43.8	19.6
	5	5.0	28.2	31.6	19.5	2.3
	6	−10.2	27.2	38.1	35.5	18.0
	7	−12.9	22.1	34.0	43.4	34.2
	8	27.1	26.5	38.8	28.5	17.4
	9	13.0	19.7	23.5	29.4	39.6
	10	28.9	26.1	11.2	18.1	16.5
2	1	3.0	9.3	1.0	15.0	3.0
	2	−1.5	−10.1	20.2	18.3	13.5
	3	10.8	20.6	28.3	25.2	15.8
	4	15.3	19.8	25.4	31.3	21.7
	5	8.7	8.0	17.5	26.6	16.4
	6	−4.6	5.8	12.7	15.6	29.6
	7	−16.6	28.4	32.7	34.4	15.8
	8	9.4	15.7	22.7	29.8	23.2
	9	−19.3	15.7	21.7	30.4	26.1
	10	−12.8	12.3	0.1	21.3	10.6

(A negative value means there was an increase in swelling, compared to baseline.)
a. Compare means and standard deviations by time period for each group.
b. Plot these data showing mean percent inhibition by time for each treatment group. Does the antihistamine group appear to differ from the placebo group?

18.4 Refer to the data from Exercise 18.3. Give a model for this design and run a repeated measures analysis of variance to compare the two treatment groups. Do the analysis of variance results agree with your intuition based on the plot of Exercise 18.3?

18.5 Another question that may be asked relates to the onset of antihistaminic activity. How might you define onset? For each of the treatment groups, use a t test to determine the test time at which there is a significant reduction from baseline. What do these results suggest?

18.4 ▼ CROSSOVER DESIGNS

We will now consider an extension to the single-factor experiment discussed in Section 18.2. Recall that in Table 18.2 we presented data for an experimental situation where each of n patients received the same three compounds in a random order. A Latin square arrangement of the compounds is an experimental design that provides the same advantages as the single-factor experiment with repeated measures (namely, multiple observations per patient and a within-patient comparison of the treatments) while offering some protection that patients didn't change with time.

A 3×3 Latin square design for this experimental situation is shown in Table 18.5. The design itself is called a three-period crossover design.

T A B L E   18.5
A 3 × 3 Latin Square Design

		Factor B (periods)		
Sequence	**Patient**	1	2	3
1	n	A_1	A_2	A_3
2	n	A_2	A_3	A_1
3	n	A_3	A_1	A_2

With this design, $3n$ patients are randomly assigned to the sequences (rows) of the design, n to each sequence. The periods correspond to the order in which the compounds are taken. The model for this design is

$$y_{ijkl} = \mu + \delta_k + \pi_{l(k)} + \alpha_i + \beta_j + \alpha\beta_{ij}^* + \varepsilon_{ijkl},$$

where δ_k is the fixed effect for the kth sequence and $\pi_{l(k)}$ is the random effect of the lth patient in sequence k; α_i and β_j are the fixed effects for compounds (factor A) and periods (factor B). The reason for the asterisk on the interaction term will be discussed later. The analysis of variance for this design (three-period crossover design) is shown in Table 18.6.

The sums of squares, degrees of freedom, and expected mean squares for the between-patient effects for factors A and B in the within-patient portion of the analysis of variance are straightforward, but you will notice that there is an asterisk on the AB interaction term and that this interaction term has only two, rather than four, degrees of freedom. Actually, the missing two degrees of freedom are in the sum of squares due to sequences. In fact, it can be shown that the AB interaction sums of squares is equal to

$$SSAB = SSSeq + SSAB^*.$$

We will use this identity to compute SSAB*.

TABLE 18.6
Analysis of Variance for a Three-Period Crossover Design

Source	SS	df	EMS (A, B fixed; patients random)
Between patients			
Sequence	SSSeq	2	$\sigma_\varepsilon^2 + 3\sigma_\pi^2 + 3n\theta_{Seq}$
Patients in sequences	SSP(Seq)	$3(n-1)$	$\sigma_\varepsilon^2 + 3\sigma_\pi^2$
Within patients			
A (compounds)	SSA	2	$\sigma_\varepsilon^2 + 3n\theta_A$
B (periods)	SSB	2	$\sigma_\varepsilon^2 + 3n\theta_B$
AB*	SSAB*	2	$\sigma_\varepsilon^2 + n\theta_{AB}$
Error	SSE	$3(2)(n-1)$	σ_ε^2
Totals	TSS	$9n-1$	

The test that we run in the analysis of variance will give us partial information about the AB interaction; actually, we are testing the within–patient portion of that interaction.

EXAMPLE 18.3 ▼

Twelve males volunteered to participate in a study to compare the durations of effect of three different formulations of a drug product. Formulation 1 was a 50-mg tablet, formulation 2 was a 100-mg tablet, and formulation 3 was a sustained–release formulation capsule. A three-period crossover design was used, with four volunteers assigned to each of the three treatment sequences. On each treatment day, volunteers were given their assigned formulation and were observed to determine the duration of effect (blood pressure lowering). There was a 1-week washout between each treatment period of the study. The sample data are shown here.

		Period		
Sequence	Patient	1	2	3
1	$n = 4$	A_1	A_2	A_3
2	$n = 4$	A_2	A_3	A_1
3	$n = 4$	A_3	A_1	A_2

		Period		
Sequence	Patient	1	2	3
1	1	1.5	2.2	3.4
	2	2.0	2.6	3.1
	3	1.6	2.7	3.2
	4	1.1	2.3	2.9
2	1	2.5	3.5	1.9
	2	2.8	3.1	1.5
	3	2.7	2.9	2.4
	4	2.4	2.6	2.3
3	1	3.3	1.9	2.7
	2	3.1	1.6	2.5
	3	3.6	2.3	2.2
	4	3.0	2.5	2.0

Solution   Based on the analysis of variance shown in the accompanying computer output, there is a hint of a period by treatment interaction ($p = .0853$), which appears negligible in the presence of a highly significant treatment effect ($p = .0001$). This is borne out in the plot of mean durations

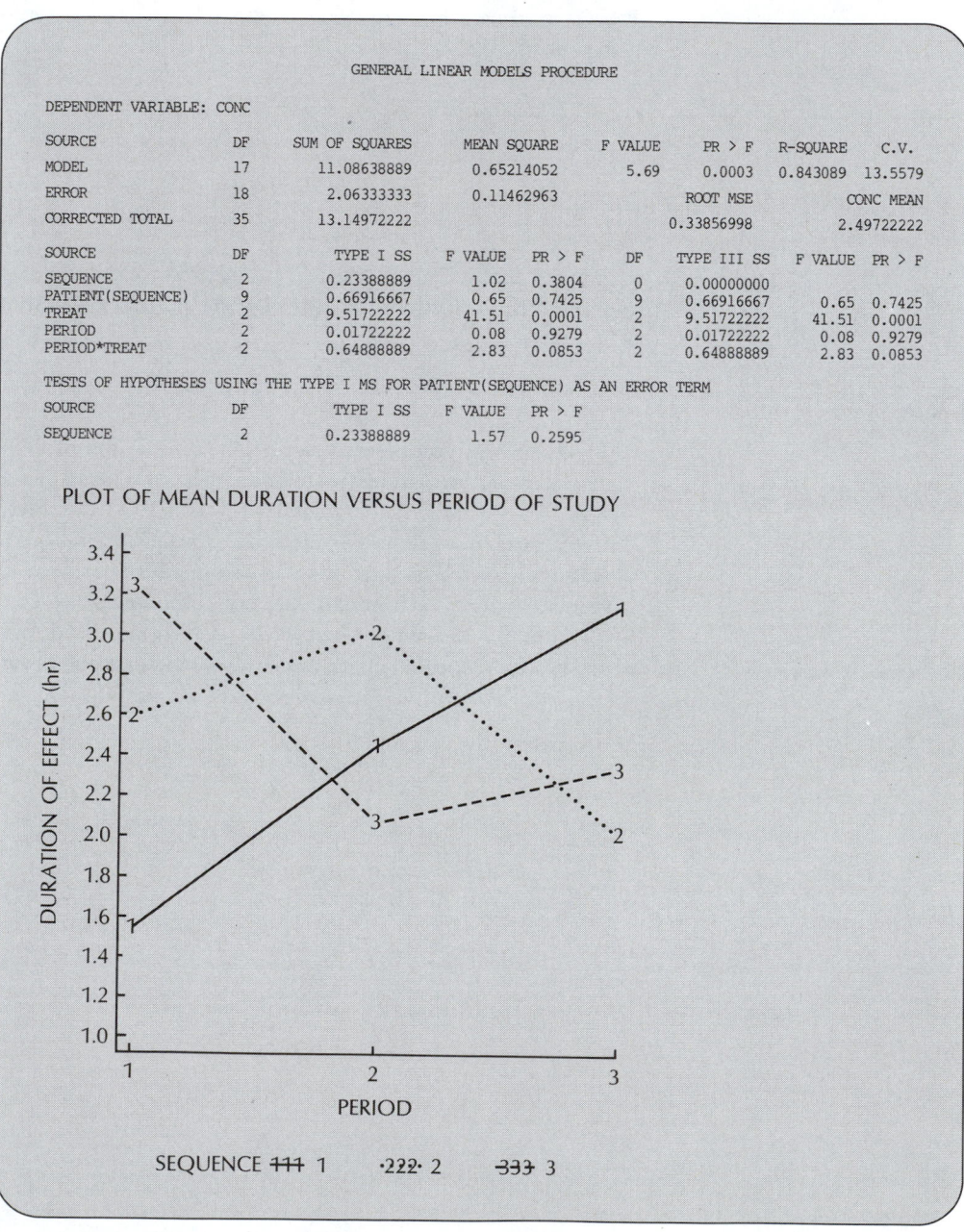

GENERAL LINEAR MODELS PROCEDURE

DEPENDENT VARIABLE: CONC

SOURCE	DF	SUM OF SQUARES	MEAN SQUARE	F VALUE	PR > F	R-SQUARE	C.V.
MODEL	17	11.08638889	0.65214052	5.69	0.0003	0.843089	13.5579
ERROR	18	2.06333333	0.11462963		ROOT MSE		CONC MEAN
CORRECTED TOTAL	35	13.14972222			0.33856998		2.49722222

SOURCE	DF	TYPE I SS	F VALUE	PR > F	DF	TYPE III SS	F VALUE	PR > F
SEQUENCE	2	0.23388889	1.02	0.3804	0	0.00000000		
PATIENT(SEQUENCE)	9	0.66916667	0.65	0.7425	9	0.66916667	0.65	0.7425
TREAT	2	9.51722222	41.51	0.0001	2	9.51722222	41.51	0.0001
PERIOD	2	0.01722222	0.08	0.9279	2	0.01722222	0.08	0.9279
PERIOD*TREAT	2	0.64888889	2.83	0.0853	2	0.64888889	2.83	0.0853

TESTS OF HYPOTHESES USING THE TYPE I MS FOR PATIENT(SEQUENCE) AS AN ERROR TERM

SOURCE	DF	TYPE I SS	F VALUE	PR > F
SEQUENCE	2	0.23388889	1.57	0.2595

PLOT OF MEAN DURATION VERSUS PERIOD OF STUDY

SEQUENCE ⁺⁺⁺ 1   ·222· 2   ₃₃₃ 3

versus period for the three sequences shown here. As can be seen, the longest durations on the average were observed with formulation 3 followed by formulation 2 and then 1. ▲

When there are only two compounds to be examined, the Latin square arrangement, called a two-period crossover design, would have $2n$ patients randomly assigned to the two sequences, n to each sequence. The two-period crossover design is shown in Table 18.7.

TABLE 18.7
Layout for a Two-Period Crossover Design

		Factor B (periods)	
Sequence	Patient	1	2
1	n	A_1	A_2
2	n	A_2	A_1

The corresponding analysis of variance for the model is

$$y_{ijkl} = \mu + \delta_k + \pi_{l(k)} + \alpha_i + \beta_j + \varepsilon_{ijkl},$$

where δ_k is the fixed effect due to sequence k, and α_i and β_j are the fixed effects due to treatment i and period j. As before, $\pi_{l(k)}$ represents the lth person in sequence k.

Note there is no AB interaction term in this model. We must assume this interaction is negligible; otherwise the design is inappropriate because there are no degrees of freedom available for testing the significance of the AB interaction. The AOV table for a two-period crossover design is shown in Table 18.8.

There are many other extensions to the repeated measures designs discussed in this chapter. For example, one could combine the concept of repeated measures

TABLE 18.8
AOV Table for a Two-Period Crossover Design

Source	SS	df	EMS (A, B fixed; patients random)
Between patients			
Sequences	SSSeq	1	$\sigma_\varepsilon^2 + 2\sigma_\pi^2 + 2n\theta_{Seq}$
Patients in sequences	SSP(Seq)	$2(n-1)$	$\sigma_\varepsilon^2 + 2\sigma_\pi^2$
Within patients			
A	SSA	1	$\sigma_\varepsilon^2 + 2n\theta_A$
B	SSB	1	$\sigma_\varepsilon^2 + 2n\theta_B$
Error	SSE	$2(n-1)$	σ_ε^2
Totals	TSS	$4n-1$	

T A B L E   18.9
Two-Period Crossover Design with Repeated Measures

Sequence	Period	
	1 Time 1 2···t	2 Time 1 2···t
1	A_1	A_2
2	A_2	A_1

on the same factor illustrated in Table 18.4 with the crossover design. Such a plan is illustrated in Table 18.9. Thus, rather than taking one observation per patient within each period, we would take observations at t different time points. For example, we could measure blood pressure every 15 minutes for the first hour following treatment with compound i, and then hourly for the next 7 hours. This would be done in each of the periods for a total of 10 blood pressure measurements on each patient in each time period.

Although we won't give the analysis of variance for this extension to the repeated measures experiments discussed in this chapter, and will not cover other more complicated repeated measures designs, we want you to be aware of the wealth of possible designs that are available if you are willing to take more than one observation per experimental unit. The interested reader is referred to Winer (1971, Chapters 5, 7, and 10) and Buncher (1981, Chapter 7).

18.5 ▼ SUMMARY

In this chapter, we have discussed some of the initial concepts and designs associated with repeated measures experiments. We introduced single- and two-factor experiments, analyses for these experiments, and the special topics of two and three-period crossover designs. These methods are only a beginning, however. Rather than presenting an exhaustive, detailed account of the subject, we have looked at these few situations to see the applicability and utility of some of the repeated measures designs and procedures. Facility in designing and analyzing such experiments can be gained only after more detailed coverage of repeated measures topics through additional reading and course work.

▼ SUPPLEMENTARY EXERCISES

18.6 An investigational drug product was studied under sleep laboratory conditions to determine its effect on duration of sleep. A group of 16 patients willing to participate in the study were randomly assigned to one of two drug sequences; 8 were to receive the investigational drug in period 1 and an identical-appearing placebo in period 2, and the remaining 8 patients were to receive the treatment in the reverse order.

a. Identify the design.

b. Give a model for this design.

c. State the assumptions that might affect the appropriateness of this design.

18.7   Sleep duration data (in hours/night) are shown for the patients of Exercise 18.6.

		Period	
Sequence	**Patient**	**1**	**2**
1	1	8.6	8.0
	2	7.5	7.1
	3	8.3	7.4
	4	8.4	7.3
	5	6.4	6.4
	6	6.9	6.8
	7	6.5	6.1
	8	6.0	5.7
2	9	7.3	7.9
	10	7.5	7.6
	11	6.4	6.3
	12	6.8	7.5
	13	7.1	7.7
	14	8.2	8.6
	15	7.2	7.8
	16	6.7	6.9

Sequence 1 received the investigational drug first and placebo second; the reverse order applied to sequence 2.

a. Compute means and standard errors per sequence, per period.

b. Plot these data to show what happened during the study. Does the investigational drug appear to affect sleep duration? In what way? Use $\alpha = .05$.

c. Run a repeated measures analysis of variance for this design. Draw conclusions. Does the analysis of variance confirm your impressions in part b?

18.8   Refer to Exercise 18.6. Suppose we ignore the order in which the patients received the treatments. Count the number of patients who had higher sleep duration on the investigational drug than on placebo.

a. Suggest another simple test for assessing the effectiveness of the investigational drug.

b. Give a p-value for the test of part a.

18.9   Refer to Exercise 18.6. Suppose the sleep durations for period 2 of sequence 1 were as follows:

8.5   7.6   8.5   8.3   7.2   7.0   6.4   6.1

a. Plot the study data for both sequences.

b. Does the design still seem to be appropriate? Is there a possible explanation for what happened?

18.10   Refer to Exercise 18.9. In spite of the results from period 2, we can still get a between-patient comparison of the treatment groups if we use the period 1 results only. Suggest an appropriate test, run the test, and give the *p*-value for your test. Draw a conclusion.

▼ **18.11**   Many of us have been exposed to advertising related to the "bioavailability" of generic and brand-name formulations of the same drug product. One way to compare the bioavailability of two formulations of a drug product is to compare areas under the concentration curve (AUC) for subjects treated with both formulations. For example, the shaded area in Figure 18.1 represents the AUC for a patient treated with a single dose of

FIGURE   18.1

AUC for a Patient Treated with a Single Dose of Drug, Exercise 18.11

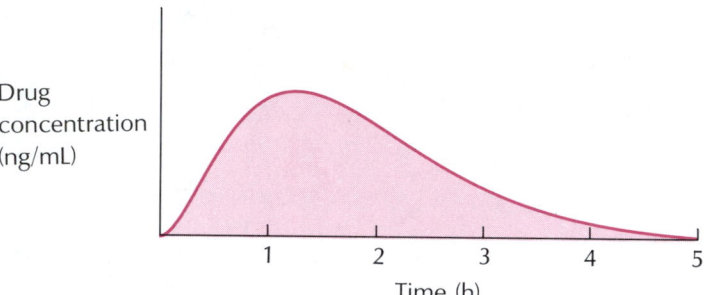

a drug. A two-period crossover design was used to compare the bioavailability of a brand-name (A_1) and generic version (A_2) of a weight-reducing agent. A random sample of six subjects was assigned to sequence 1 (A_2, then A_1), and another six subjects to sequence 2 (A_1, then A_2). The AUCs for these patients are shown here.

Sequence	Patient	Period 1	Period 2
1	1	80.2	40.4
	2	79.1	38.5
	3	85.1	54.3
	4	108.4	96.6
	5	41.2	66.5
	6	72.7	78.2
2	1	74.6	35.6
	2	125.3	67.3
	3	145.5	90.8
	4	86.7	86.5
	5	107.8	103.1
	6	79.7	83.9

a. Plot the formulation means (AUC) by period for each sequence.
b. Is there evidence of a period effect?
c. Do the formulations appear to differ relative to AUC?

18.12   Refer to Exercise 18.11. Run an analysis of variance for a two-period crossover design. Does your analysis confirm the intuition you expressed in Exercise 18.11? Use $\alpha = .05$.

18.13   Refer to Exercise 18.11. Compare the mean AUCs for the two formulations using *only* the period 1 data. Does this analysis confirm the analysis of Exercise 18.12? Why or why not might the analysis of Exercise 18.12 be more suitable than the "parallel" analysis of this exercise?

18.14   A study was conducted to demonstrate the effectiveness of an investigational drug product in reducing the number of epileptic seizures in patients who have not been helped by standard therapy. Thirty patients participated in the study, with 15 randomized to the drug treatment group and 15 to the placebo group. Patient demographic data are displayed here.

		Group	
		Investigational Drug ($n_1 = 15$)	Placebo ($n_2 = 15$)
Age (yr)	Mean ($\pm$ SD)	37.2 ($\pm$ 10.5)	39.5 ($\pm$ 9.6)
	Range	19–68	21–65
Gender	M	20	16
	F	10	14
Duration of illness (yr)	Mean ($\pm$ SD)	10.7 ($\pm$ 6.5)	11.5 ($\pm$ 7.3)
	Range	1–18	1–26

a. Do the groups appear to be comparable related to these demographic variables?
b. Are the mean ages or durations of illness different? How would you make this comparison?
c. How might you compare the sex distributions of the two groups?

18.15   The seizure data for the study of Exercise 18.14 are shown here. Note that we have baseline seizure rates as well as seizure rates for five months while on therapy.
a. Plot the mean seizure rates by month for the two groups. Does the investigational drug appear to work?
b. Run a repeated measures AOV and draw conclusions based on $\alpha = .01$.

			Time (months)				
Group	Patient	Baseline	1	2	3	4	5
Drug	1	15	11	10	6	5	3
	2	13	6	5	1	2	1
	3	12	8	3	0	3	0
	4	18	4	2	3	1	2
	5	30	15	14	10	8	20
	6	14	7	9	3	4	1
	7	25	12	18	13	10	6
	8	22	21	18	16	17	25
	9	23	17	14	10	7	1
	10	14	2	1	0	0	0
	11	15	4	5	6	3	2
	12	17	8	7	8	2	6
	13	26	13	10	9	7	4

(*continues*)

Group	Patient	Baseline	Time (months)				
			1	2	3	4	5
	14	28	2	1	3	1	3
	15	29	27	29	25	24	22
Placebo	1	16	15	18	14	13	12
	2	18	14	13	12	10	15
	3	14	10	5	4	6	7
	4	19	15	16	9	12	15
	5	12	10	14	16	17	12
	6	11	13	8	7	6	11
	7	31	32	30	21	24	20
	8	32	35	34	31	20	24
	9	21	20	18	15	16	18
	10	26	22	23	21	15	14
	11	13	10	14	12	8	6
	12	17	15	10	3	2	3
	13	18	16	12	14	13	11
	14	23	15	14	18	19	20
	15	10	8	11	10	9	6

18.16 Refer to the data of Exercise 18.15.
  a. Consider the change in seizure rate from baseline to the 5-month reading. Compare the two groups using these data. Do you reach a similar conclusion?
  b. Since seizure rates can be quite variable, some people might compare the maximum change for patients in the two groups. Do these data support your previous conclusions?

18.17 Gasoline efficiency ratings were obtained on a random sample of 12 automobiles, 6 each of two different models. These ratings were taken at five different times for each of the 12 automobiles.
  a. Compute the mean efficiencies for each model at each time point, and plot these data.
  b. Draw conclusions from the analysis of variance. Use $\alpha = .05$.
  c. What effects, if any, do the correction factors have on the within-model comparisons in the analysis of variance shown here?

```
                    LISTING OF DATA
  OBS    MODEL    CAR    TIME1    TIME2    TIME3    TIME4    TIME5
   1       1       1     1.43     1.47     1.39     1.40     1.44
   2       1       2     1.50     1.41     1.51     1.53     1.41
   3       1       3     1.79     1.88     1.89     2.00     1.90
   4       1       4     1.87     1.78     2.00     2.00     2.11
   5       1       5     1.85     1.89     1.93     1.86     1.81
   6       1       6     1.89     1.66     1.78     1.77     1.67
   7       2       7     1.63     1.62     1.64     1.63     1.53
   8       2       8     1.81     1.83     1.84     1.83     1.86
   9       2       9     2.25     2.10     2.34     2.27     2.32
  10       2      10     1.79     1.80     1.92     2.03     2.02
  11       2      11     2.11     2.00     2.33     2.46     2.35
  12       2      12     2.10     2.03     2.00     2.09     1.87
```
(continues)

REPEATED MEASURES ANALYSIS OF VARIANCE
GENERAL LINEAR MODELS PROCEDURE
TESTS OF HYPOTHESES FOR BETWEEN SUBJECTS EFFECTS

SOURCE	DF	TYPE III SS	MEAN SQUARE	F VALUE	PR > F
MODEL	1	0.95760667	0.95760667	3.38	0.0960
ERROR	10	2.83722667	0.28372267		

REPEATED MEASURES ANALYSIS OF VARIANCE
GENERAL LINEAR MODELS PROCEDURE
UNIVARIATE TESTS OF HYPOTHESES FOR WITHIN SUBJECT EFFECTS

						ADJUSTED PR > F	
SOURCE	DF	TYPE III SS	MEAN SQUARE	F VALUE	PR > F	G - G	H - F
TIME	4	0.09579333	0.02394833	3.03	0.0285	0.0719	0.0512
TIME*MODEL	4	0.01182667	0.00295667	0.37	0.8260	0.6906	0.7528
ERROR(TIME)	40	0.31654000	0.00791350				

GREENHOUSE-GEISSER EPSILON = 0.4943
HUYNH-FELDT EPSILON = 0.6770

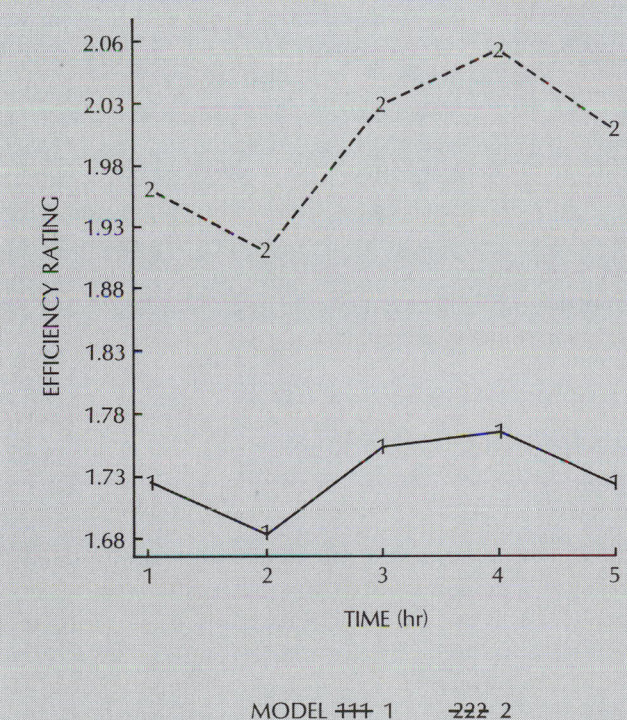

PLOT OF EFFICIENCY RATING VERSUS TIME

MODEL 111 1   222 2

THE ANALYSIS OF COVARIANCE

19.1 INTRODUCTION

The analysis of covariance is a procedure for comparing treatment means that incorporates information on a quantitative variable x. This topic appears in most statistical methods textbooks, but it has been our experience that many students become so engrossed in the computational formulas that they rarely understand the problems involved. We will approach the subject in a different way and try to avoid becoming too involved with formulas. Since the analysis of covariance combines features of the analysis of variance and regression, we will make use of a general linear model. By referring to and building on our work with general linear models in preceding chapters, we can more easily understand the topic of covariance analysis. We begin our presentation with a completely randomized design.

19.2 A COMPLETELY RANDOMIZED DESIGN WITH ONE COVARIATE

A completely randomized design is used to compare t population means. To do this, we obtain a random sample of n_i observations on the variable y in the ith population ($i = 1, 2, \ldots, t$). Now, in addition to measuring the response variable y on each experimental unit, we measure a second variable x, often called a *covariable*, or a **covariate**. For example, in studying the effects of different methods of reinforcement on the reading achievement levels of 8-year-old children, we could measure not only the final achievement level y for each child but also the

covariate

prestudy reading performance level x. Ultimately, we would want to make comparisons among the different methods while taking into account information on both y and x.

It should be noted that x can be thought of as an independent variable, but unlike most situations discussed in previous chapters, we cannot control the value of x (as we controlled settings of temperature or pressure) prior to observing the variable. In spite of this, we may still write a model for the completely randomized design treating the covariate as an independent variable.

For purposes of illustration, we will consider the problem of comparing $t = 2$ treatments from a completely randomized design with one covariable. Later, we'll show how these results can be extended to include $t > 2$ treatments and one or more covariables.

E X A M P L E   19.1   ▼

An investigator is interested in comparing two drug products (A and B) in overweight female volunteers. The experiment calls for 20 randomly selected subjects who are at least 25% overweight. Ten of these women are to be randomly assigned to product 1 and the remaining 10 to product 2. The response of interest is a score on a rating scale used to measure the mood of a subject. To obtain a score, a subject must complete a checklist indicating how each of 50 adjectives describes her mood at that time. From this checklist, we can obtain an anxiety–tension score, a danger–hostility score, an active score (measuring such factors as alertness and energy), and many others.

On the study day, all 20 volunteers are required to complete the checklist at 8 A.M. Then each subject is given the prescribed medication (1 or 2). Each subject is required to complete the checklist again at 10 A.M.

Write a model relating a subject's 10 A.M. checklist score y to the two independent variables "drug product" and "8 A.M. (predrug) checklist score." Interpret the βs.

Solution   For this experimental situation, we have one qualitative independent variable (drug) and one quantitative independent variable. Letting x_1 denote the checklist score at 8 A.M., the model is

$$y = \beta_0 + \beta_1 x_1 + \beta_2 x_2 + \beta_3 x_1 x_2 + \varepsilon,$$

where

$x_1 =$ checklist score at 8 A.M.
$x_2 = 1$ if product 2 $x_2 = 0$ otherwise.

The expected value of y for our model is

$$E(y) = \beta_0 + \beta_1 x_1 + \beta_2 x_2 + \beta_3 x_1 x_2.$$

Substituting $x_2 = 0$ and $x_2 = 1$, respectively, for drug products 1 and 2, we have the expected value of y, the adjective checklist score at 10 A.M.

product 1: $E(y) = \beta_0 + \beta_1 x_1$
product 2: $E(y) = (\beta_0 + \beta_2) + (\beta_1 + \beta_3)x_1$.

Thus, β_0 and β_1 are the intercept and slope, respectively, defining the linear relationship between y and x_1 for product 1. Since $\beta_0 + \beta_2$ and $\beta_1 + \beta_3$ represent the corresponding intercept and slope, respectively, for product 2, β_2 is the difference between the intercepts for lines 1 and 2, whereas β_3 is the difference between the slopes for the two lines. ▲

The major assumptions in an analysis of covariance are (1) the treatment's regression equations are linear in the covariable, and (2) the linear regressions for the different treatments are parallel. Assumption (2) indicates that β_3, the difference in slope, is 0. Taking the two assumptions for our example, we use the following expected values:

product 1: $E(y) = \beta_0 + \beta_1 x$
product 2: $E(y) = (\beta_0 + \beta_2) + \beta_1 x$.

adjusted treatment means

The objective of an analysis of covariance is to compare the treatment means after adjusting for differences among the treatments due to differences in the covariable levels for the treatment groups. The **adjusted treatment means** are found by predicting y for each treatment group corresponding to $x = \bar{x}$, the mean value of the covariable across all treatment groups.

E X A M P L E   19.2   ▼

The data for Example 19.1 are shown here.

Drug 1		Drug 2	
8 A.M. x_1	10 A.M. y	8 A.M. x_1	10 A.M. y
5	20	7	19
10	23	12	26
12	30	27	33
9	25	24	35
23	34	18	30
21	40	22	31
14	27	26	34
18	38	21	28
6	24	14	23
13	31	9	22

a. Write the model for an analysis of covariance.

b. Use the computer output shown here to give the linear regression equations for both treatment groups.

c. Compute the sample mean anxiety scores for the two products; also compute the adjusted treatment means.

d. Does there appear to be a difference between the adjusted treatment means?

```
                            ANALYSIS OF COVARIANCE
                        GENERAL LINEAR MODELS PROCEDURE

DEPENDENT VARIABLE: Y

SOURCE              DF    SUM OF SQUARES    MEAN SQUARE    F VALUE    PR > F    R-SQUARE    C.V.

MODEL               2     546.22979465     273.11489733    37.96     0.0001    0.817037    9.3627

ERROR               17    122.32020535     7.19530620            ROOT MSE             Y MEAN

CORRECTED TOTAL     19    668.55000000                           2.68240679          28.65000000
```

SOURCE	DF	TYPE I SS	F VALUE	PR > F	DF	TYPE III SS	F VALUE	PR > F
X1	1	430.92383794	59.89	0.0001	1	540.17979465	75.07	0.0001
X2	1	115.30595671	16.03	0.0009	1	115.30595671	16.03	0.0009

PARAMETER	ESTIMATE	T FOR HO: PARAMETER=0	PR > T	STD ERROR OF ESTIMATE
INTERCEPT	18.35999493	12.15	0.0001	1.51153263
X1	0.82748130	8.66	0.0001	0.09550226
X2	-5.15465839	4.00	0.0009	1.28765245

OBSERVATION	OBSERVED VALUE	PREDICTED VALUE	RESIDUAL
1	20.00000000	22.49740145	-2.49740145
2	24.00000000	23.32488275	0.67511725
3	25.00000000	25.80732666	-0.80732666
4	23.00000000	26.63480796	-3.63480796
5	30.00000000	28.28977057	1.71022943
6	31.00000000	29.11725187	1.88274813
7	27.00000000	29.94473317	-2.94473317
8	38.00000000	33.25465839	4.74534161
9	40.00000000	35.73710229	4.26289771
10	34.00000000	37.39206490	-3.39206490
11	19.00000000	18.99770567	0.00229433
12	22.00000000	20.65266827	1.34733173
13	26.00000000	23.13511218	2.86488782
14	23.00000000	24.79007479	-1.79007479
15	30.00000000	28.10000000	1.90000000
16	28.00000000	30.58244391	-2.58244391
17	31.00000000	31.40992521	-0.40992521
18	35.00000000	33.06488782	1.93511218
19	34.00000000	34.71985042	-0.71985042
20	33.00000000	35.54733173	-2.54733173

```
    SUM OF RESIDUALS                         0.00000000
    SUM OF SQUARED RESIDUALS               122.32020535
    SUM OF SQUARED RESIDUALS - ERROR SS    -0.00000000
    FIRST ORDER AUTOCORRELATION            -0.20370243
    DURBIN-WATSON D                         2.30336716
```

Solution

a. The analysis of covariance model for this situation is

$$y = \beta_0 + \beta_1 x_1 + \beta_2 x_2 + \varepsilon,$$

where

$$x_1 = \text{prestudy checklist score}$$

and

$$x_2 = \begin{cases} 1 & \text{if product 2} \\ 0 & \text{if product 1.} \end{cases}$$

The separate regression equations are:

product 1: $y = \beta_0 + \beta_1 x_1 + \varepsilon$
product 2: $y = (\beta_0 + \beta_2) + \beta_1 x_1 + \varepsilon.$

Hence β_1 is the common slope and β_2 is the difference in intercepts.

b. The least square prediction equations for the two products can be obtained from the parameter estimates shown in the output:

product 1: $\hat{y} = 18.360 + 0.827 x_1$
product 2: $\hat{y} = (18.360 - 5.155) + 0.827 x_1$
            $= 13.205 + 0.827 x_1.$

These prediction equations are shown in Figure 19.1.

c. From the output shown here we have the following sample means.

	Product 1	Product 2	Overall
y	29.2	28.1	28.65
x_1	13.1	18.0	15.55

In order to get the adjusted treatment means, we substitute the mean pretreatment value ($\bar{x}_1 = 15.55$) into the separate prediction equations.

Adjusted treatment mean, product 1:

$$\hat{y} = 18.360 + 0.827(15.55) = 31.220$$

Adjusted treatment mean, product 2:

$$\hat{y} = 13.205 + 0.827(15.55) = 26.065.$$

FIGURE 19.1

**Analysis of Covariance
Predicted Line**

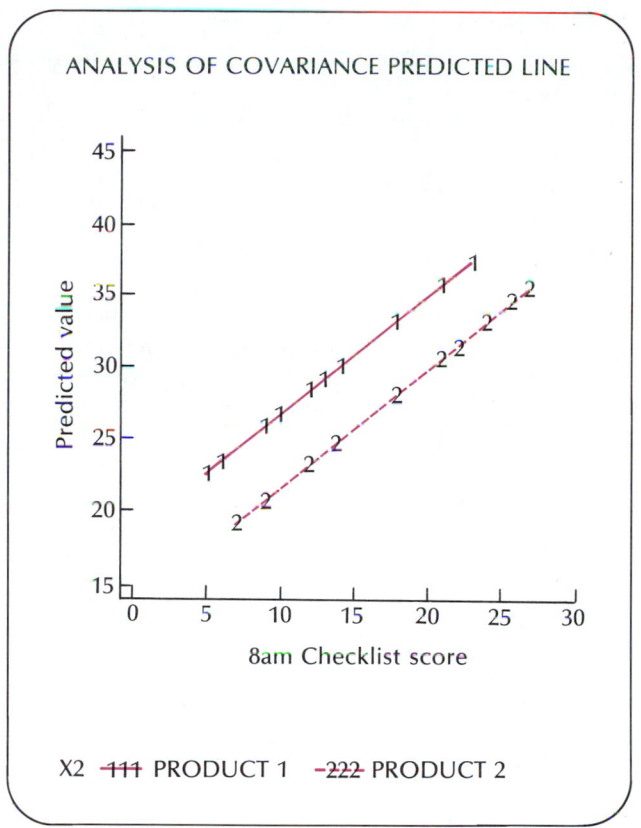

ANALYSIS OF COVARIANCE PREDICTED LINE

X2 ~~111~~ PRODUCT 1   ~~222~~ PRODUCT 2

Note that although the difference in treatment means for products 1 and 2 is only $29.2 - 28.1 = 1.1$, when we adjust for the fact that the mean pretreatment (covariable) score is 13.1 for product 1 and 18.0 for product 2, the difference between A and B is

$$31.220 - 26.065 = 5.155.$$

d. It does appear that the treatment 1 response (adjusted) for the covariable is higher than the corresponding response for treatment 2.   ▲

EXAMPLE 19.3

▼

Refer to the computer output of Example 19.2.
a. Compare the difference in adjusted treatment means based on your calculations to the difference in intercepts from the output.
b. Suggest a test procedure for assessing the equality of the adjusted treatment means for products 1 and 2. Give the p-value for this test.

Solution

a. The difference in adjusted treatment means is

$$31.220 - 26.065 = 5.155$$

whereas the difference in intercepts is

$$18.360 - 13.205 = 5.155.$$

b. Because the regression lines are parallel, the *difference* in adjusted treatment means is also equal to the difference in the intercepts for the two regressions. Thus, a formal test of the equality of adjusted treatments would be identical to a test of the equality of the intercepts of the linear regressions. From the computer output, the p-value for $H_0: \beta_2 = 0$ (difference in intercepts) is .0009; there is a difference in response to treatment for products 1 and 2.   ▲

Now that we have worked through a simple analysis of covariance, we should stand back and assess whether some of the initial assumptions can be supported. We'll review this for the data of Example 19.2.

EXAMPLE 19.4   ▼

a. Plot the data of Example 19.2 by treatment group. Is there evidence of lack of fit of a linear regression model? Can we test for lack of fit?
b. Is there evidence to indicate nonparallel linear regression? Can we test for this? How?

Solution   The sample data are displayed in Figure 19.2 by treatment group.
a. Recall that in order to test for lack of fit, we must be able to calculate a sum of squares due to pure experimental error. Since there are no replications of y values at any of the settings of the independent variables x_1 and x_2, we cannot test for lack of fit of the linear model.
b. Under the assumption that a linear regression model adequately fits the data for the two drugs, we could test for parallelism of the two regression lines using the model

$$y = \beta_0 + \beta_1 x_1 + \beta_2 x_2 + \beta_3 x_1 x_2 + \varepsilon,$$

where

$$x_1 = \text{covariable}$$

FIGURE 19.2

Analysis of Covariance, Example 19.4

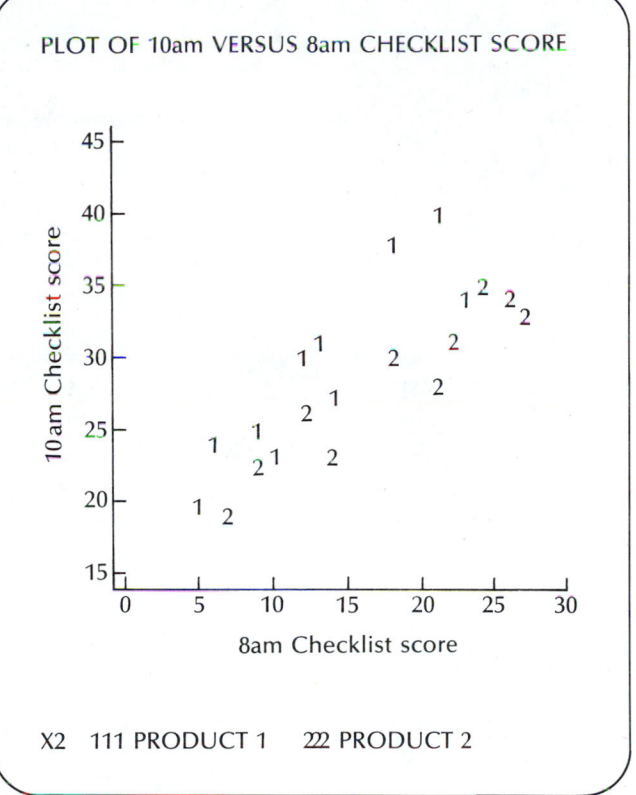

PLOT OF 10am VERSUS 8am CHECKLIST SCORE

X2   111 PRODUCT 1   222 PRODUCT 2

and

$$x_2 = \begin{cases} 0 & \text{if product 1} \\ 1 & \text{if product 2.} \end{cases}$$

For this model (substituting $x_2 = 0$ and 1), it is easy to see that β_3 is the difference in slope for the two regression lines

product 1 ($x_2 = 0$): $y = \beta_0 + \beta_1 x_1 + \varepsilon$
product 2 ($x_2 = 1$): $y = (\beta_0 + \beta_2) + (\beta_1 + \beta_3)x_1.$

Hence, a test for parallelism of the linear regression lines can be done using $H_0: \beta_3 = 0$. Based on the output shown here, the t-value for this test is -1.34, with probability greater than .1991. (These values are read from the computer output.) Since this probability is greater than .05, we have insufficient evidence to reject the null hypothesis of parallelism for the two lines.

▲

```
                            ANALYSIS OF COVARIANCE
                         GENERAL LINEAR MODELS PROCEDURE

DEPENDENT VARIABLE: Y
SOURCE                  DF    SUM OF SQUARES   MEAN SQUARE   F VALUE   PR > F   R-SQUARE   C.V.
MODEL                   3      558.56687442   186.48895814    27.09   0.0001   0.8935490  9.1512
ERROR                  16      109.98312557     6.87394535             ROOT MSE          Y MEAN
CORRECTED TOTAL        19      668.55000000                            2.62182100      28.65000000

SOURCE           DF      TYPE I SS    VALUE   PR > F   DF    TYPE III SS   F VALUE  PR > F
X1                1    130.92383791   62.69   0.0001    1    112.89948313    45.52  0.0001
X2                1    115.30595671   16.77   0.0008    1      1.21055917     0.18  0.6803
X1*X2             1     12.33707978    1.79   0.1991    1     12.33707978     1.79  0.1991


                             T FOR HO:     PR > T      STD ERROR OF
PARAMETER        ESTIMATE   PARAMETER=0                  ESTIMATE

INTERCEPT      16.42262086     7.94        0.0001       2.06736784
X1              0.97537245     6.75        0.0001       0.14456764
X2              1.31392521     0.42        0.6803       3.13098272
X1*X2           0.25363332     1.34        0.1991       0.18932291


OBSERVATION       OBSERVED          PREDICTED           RESIDUAL
                    VALUE             VALUE
      1          20.00000000       21.29948313        -1.29948313
      2          24.00000000       22.27485558         1.72514442
      3          25.00000000       25.20097294        -0.20097294
      4          23.00000000       26.17634539        -3.17634539
      5          30.00000000       28.12709030         1.87290970
      6          31.00000000       29.10246275         1.89753725
      7          27.00000000       30.07783521        -3.07783521
      8          38.00000000       33.97932502         4.02067498
      9          40.00000000       36.90544238         3.09455762
     10          34.00000000       38.85618729        -4.85618729
     11          19.00000000       20.16086957        -1.16086957
     12          22.00000000       21.60434783         0.39565217
     13          26.00000000       23.76956522         2.23043478
     14          23.00000000       25.21304348        -2.21304348
     15          30.00000000       28.10000000         1.90000000
     16          28.00000000       30.26521739        -2.26521739
     17          31.00000000       30.98695652         0.01304348
     18          35.00000000       32.43043478         2.56956522
     19          34.00000000       33.87391304         0.12608696
     20          33.00000000       34.59565217        -1.59565217


                            ANALYSIS OF COVARIANCE
                         GENERAL LINEAR MODELS PROCEDURE


DEPENDENT VARIABLE: Y
          SUM OF RESIDUALS                         0.00000000
          SUM OF SQUARED RESIDUALS               109.98312557
          SUM OF SQUARED RESIDUALS - ERROR SS     -0.00000000
          FIRST ORDER AUTOCORRELATION             -0.29462211
          DURBIN-WATSON D                          2.55074048
```

Suppose we are interested in comparing $t = 3$ treatments with only one covariable. Using the notation of the general linear model, we let x_1 denote the covariate. Then we can write the model

$$y = \beta_0 + \beta_1 x_1 + \beta_2 x_2 + \beta_3 x_3 + \beta_4 x_1 x_2 + \beta_5 x_1 x_3 + \varepsilon,$$

where

$$x_2 = 1 \text{ if treatment 2} \qquad x_2 = 0 \text{ otherwise}$$
$$x_3 = 1 \text{ if treatment 3} \qquad x_3 = 0 \text{ otherwise.}$$

From our discussions in Chapters 11 and 15, we recognize this model as a general linear model relating a response y to a quantitative variable (the covariate x_1) and a qualitative variable (treatments). The βs of the model can be interpreted using Table 19.1. As can be seen from the table of expected values, the model defines a straight line for each of the three treatments. The intercepts and slopes for the three lines are as indicated in Table 19.1.

T A B L E   19.1

Expected Values for the Model $y = \beta_0 + \beta_1 x_1 + \beta_2 x_2 + \beta_3 x_3 + \beta_4 x_1 x_2 + \beta_5 x_1 x_3 + \varepsilon$

Treatment	Expected Values of y
1	$\beta_0 + \beta_1 x_1$
2	$(\beta_0 + \beta_2) + (\beta_1 + \beta_4)x_1$
3	$(\beta_0 + \beta_3) + (\beta_1 + \beta_5)x_1$

assumptions

Typically, for a completely randomized design with a single covariate, the response y is assumed to be linearly related to the covariate, and the slope of the straight-line relationship is assumed to be the same for all treatments. While we have already written the model to relate y linearly to x_1, we need not be bound by these assumptions, for it is possible to write the model with higher-order terms in x_1 (provided, of course, there are enough different values of x_1 recorded in the sample data). At present, however, we will illustrate a covariance analysis with a model that is linear in the covariate x_1.

tests

The assumption of constant slope of the three treatment groups can be tested using the null hypothesis

$$H_0: \quad \beta_4 = \beta_5 = 0 \text{ (the slopes are identical)}$$

against the alternative hypothesis that at least one of the slopes differs from the rest. With our original model as the complete model and the model

reduced model

$$y = \beta_0 + \beta_1 x_1 + \beta_2 x_2 + \beta_3 x_3 + \varepsilon$$

as the reduced model, the mean square drop based on two degrees of freedom would be the numerator in an F test on the null hypothesis

$$H_0: \quad \beta_4 = \beta_5 = 0.$$

If there is insufficient evidence to reject the hypothesis of equality of the slopes $(\beta_4 = \beta_5 = 0)$, we use the reduced model to describe the experimental situation.

Under this model, the straight–line relationship between y and the covariate x_1 would have expectations

treatment 1: $\beta_0 + \beta_1 x_1$
treatment 2: $(\beta_0 + \beta_2) + \beta_1 x_1$
treatment 3: $(\beta_0 + \beta_3) + \beta_1 x_1$.

The model for a covariance analysis in a completely randomized design is not usually written in terms of the parameters in a general linear model. In other sources, you might see the model written as

$$y_{ij} = \alpha_i + \beta x_{ij} + \varepsilon_{ij} \qquad (i = 1, 2, \ldots, t; j = 1, 2, \ldots, n_i).$$

Note that we assume that the relationship between the response y and the covariate x is linear with the same slope but different intercepts across treatment groups. The reduced model illustrated for our $t = 3$ example,

$$y = \beta_0 + \beta_1 x_1 + \beta_2 x_2 + \beta_3 x_3 + \varepsilon,$$

is the general linear model analogy to this same situation.

Referring to our example, when all three treatments have the same slope (i.e., $\beta_4 = \beta_5 = 0$ in the complete model), a test of the equality of treatment means can be made using the sum of squares due to treatments adjusted for the covariate. We do this by fitting a complete and a reduced model, using the model

$$y = \beta_0 + \beta_1 x_1 + \beta_2 x_2 + \beta_3 x_3 + \varepsilon$$

for the null hypothesis $H_0: \beta_2 = \beta_3 = 0$. The sum of squares drop ($\text{SSE}_2 - \text{SSE}_1$) will be the sum of squares due to treatments adjusted for the covariate. The test statistic $F = \text{MS}_{\text{drop}}/\text{MSE}$ is based on $\text{df}_1 = $ two degrees of freedom (the number of βs set equal to 0 under H_0).

▼ EXERCISES

Basic Techniques

19.1   Consider a completely randomized design for $t = 5$ treatments, with a single covariate x_1 and six observations per treatment. Write the complete general linear model under the assumption that the response y is linearly related to the covariate x_1 for each treatment. Identify the parameters in your model.

19.2 Refer to Exercise 19.1. Indicate the relationships among the parameters of the model for the following cases; show a graph for each case.
  a. The lines are not parallel.
  b. The lines are parallel, but do not coincide.
  c. The lines are coincident.

19.3 Refer to Exercise 19.1. How would you test for parallelism among the straight lines for the five treatment groups? Identify how you would obtain the test statistic. What are the degrees of freedom associated with the test statistics?

19.4 Refer to Exercise 19.1. Assume the lines are parallel. Give the test for adjusted treatment means. How would you estimate the mean response for treatment 1 with $x_1 = 5$?

19.5 Perform an analysis of covariance for the data shown. (Hint: When people refer to performing an analysis of covariance, they usually mean to assume a linear relationship, then test for parallelism, and then test for adjusted treatment means assuming parallelism.) The data are given here. Use $\alpha = .05$.

	Treatment				
1		2		3	
x	y	x	y	x	y
26	100	24	118	37	124
35	150	28	134	31	95
28	106	29	138	14	60
21	95	32	147	27	86
29	113	36	165	18	68
34	144	35	159	25	81

19.3 THE EXTRAPOLATION PROBLEM

In the previous section, we discussed how to compare two (or more) treatments from a completely randomized design with one covariable. If the regression equations for the treatments are linear in the covariable and parallel, we said we could compare the treatments using the adjusted treatment means. However, as with most methods, the analysis of covariance methods should not be used blindly. Even if the linearity and parallelism assumptions hold, we can have problems if the values of the covariable do not have considerable overlap for the treatment groups. We will illustrate this with an example.

Suppose that we are interested in comparing self-esteem scores for alcoholics and drug addicts. A sample of nine alcoholics and a sample of nine drug addicts were obtained and for each individual, we obtained his or her self-esteem score and age. These data are shown in Table 19.2.

T A B L E   19.2
**Self-Esteem Scores and
Ages for a Sample of
Alcoholics and Drug
Addicts**

Alcoholics		Drug Addicts	
Self-Esteem	Age	Self-Esteem	Age
25	15	20	30
22	17	17	31
24	18	18	33
20	19	15	35
21	21	14	36
17	22	15	37
14	23	12	38
16	24	10	40
15	25	11	41

If we blindly followed the analysis of covariance procedures without looking at the data, we would likely find the regression equations for alcoholics and addicts to be reasonably linear and parallel and the test for adjusted treatments to be significant. This is all shown in the accompanying output.

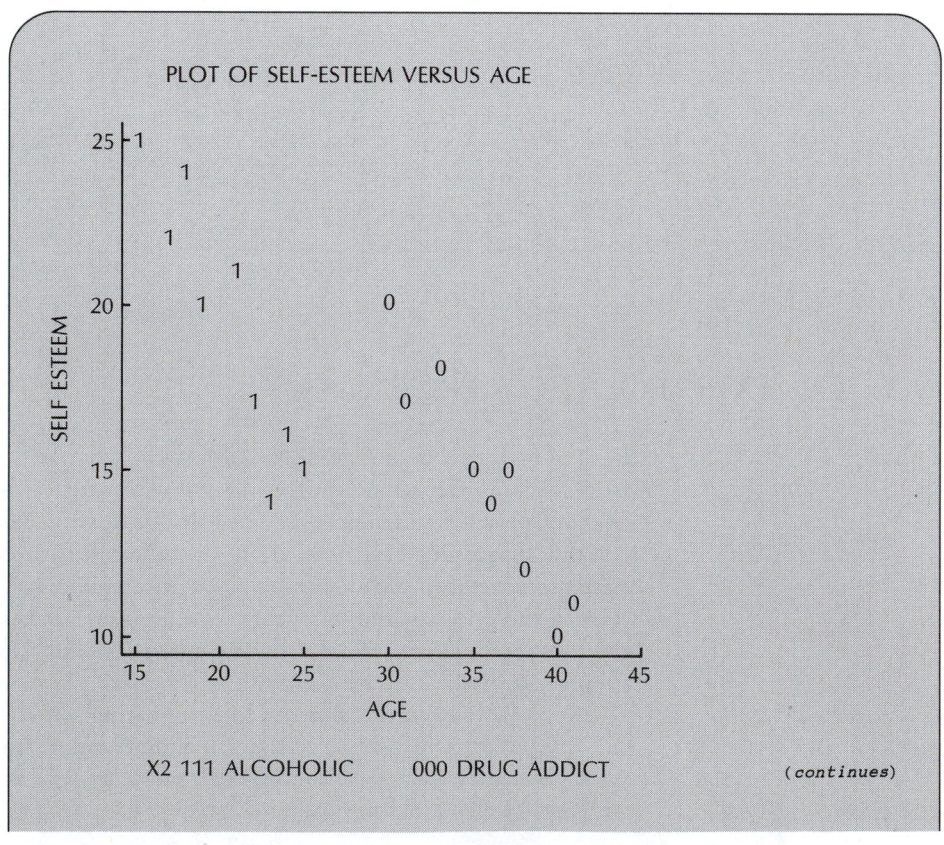

(continues)

```
                    ANALYSIS OF COVARIANCE
                     SEPARATE SLOPE MODEL
                 GENERAL LINEAR MODELS PROCEDURE

DEPENDENT VARIABLE: Y

SOURCE            DF    SUM OF SQUARES   MEAN SQUARE   F VALUE   PR > F   R-SQUARE   C.V.

MODEL             3     286.60610719     95.53536906    18.82    0.0001   0.912758  8.2284

ERROR            14      27.39389281      1.95670663            ROOT MSE            Y MEAN

CORRECTED TOTAL  17     314.00000000                            1.39882330        17.00000000

SOURCE            DF      TYPE I SS    F VALUE  PR > F   DF   TYPE III SS   F VALUE  PR > F

X1                1     212.84398455   108.78   0.0001    1   34.96467690    17.87   0.0008
X2                1      70.27928120    35.92   0.0001    1    0.43265088     0.22   0.6454
X1*X2             1       3.48284145     1.78   0.2035    1    3.48284145     1.78   0.2035

                              T FOR HO:    PR > T    STD ERROR OF
PARAMETER         ESTIMATE    PARAMETER=0            ESTIMATE

INTERCEPT       38.96789918        5.12    0.0002    7.61698531
X1              -1.34831741       -4.23    0.0008    0.31896295
X2               2.60800443        0.47    0.6454    5.54628759
X1*X2            0.26036560        1.33    0.2035    0.19515497


                    ANALYSIS OF COVARIANCE
                      COMMON SLOPE MODEL
                 GENERAL LINEAR MODELS PROCEDURE

DEPENDENT VARIABLE:

SOURCE            DF    SUM OF SQUARES   MEAN SQUARE   F VALUE   PR > F   R-SQUARE   C.V.

MODEL             2     283.12326574    141.56163287    68.77    0.0001   0.901666  8.4396

ERROR            15      30.87673428      2.05844895            ROOT MSE            Y MEAN

CORRECTED TOTAL  17     314.00000000                            1.43472957        17.0000000

SOURCE            DF      TYPE I SS    F VALUE  PR > F   DF   TYPE III SS   F VALUE  PR > F

X1                1     212.84398455   103.40   0.0001    1  185.12326574    89.93   0.0001
X2                1      70.27928120    34.14   0.0001    1   70.27928120    34.14   0.0001

                              T FOR HO;    PR > T    STD ERROR OF
PARAMETER         ESTIMATE    PARAMETER=0            ESTIMATE

INTERCEPT       28.92404838       24.83    0.0001    1.18877463
X1               0.94290288        9.48    0.0001    0.09942750
X2               9.68641053        5.84    0.0001    1.65775088
```

Do alcoholics and drug addicts really have different self-esteem scores? One possible explanation for the difference in scores is that we're dealing with two different age groups; the alcoholics sampled ranged in age from 15 to 25 years, whereas the drug addicts were between the ages of 30 and 41. This difference in ages for the two groups is borne out in the scatterplot shown in Figure 19.3.

The mean ages for the two groups are 20.44 and 35.67 years, respectively, while the combined mean age is 28.06 years. Note that the combined mean is outside the age range for each of the separate samples. We have no information about self-esteem scores for drug addicts less than 30 years of age and no information about self-esteem scores for alcoholics above the age of 25. Hence,

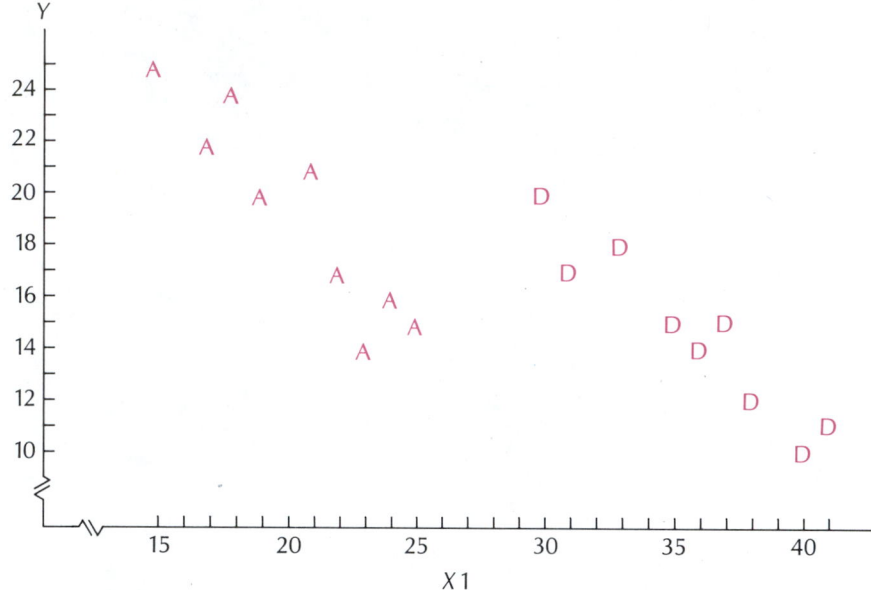

it would be inappropriate to compare the predicted self-esteem scores at the "adjusted" age (28.06) since this involves an extrapolation beyond the ages observed for the separate samples. For this example, it would be difficult to make any comparison between the alcoholics and drug addicts because of the age differences and other possible (unmeasured) differences between the two groups.

So, don't forget to look at your data. The potential for extrapolation, although not as obvious as for our example, should become apparent with plots of the data. Then you can avoid using an analysis of covariance to make comparisons of adjusted treatment means when the adjustment (or, in fact, any comparison) may be inappropriate. These same problems can occur with the extensions of these methods to include more than one covariable and more complicated experimental designs; it's just more difficult to detect the problem.

19.4 ▼ MULTIPLE COVARIATES AND MORE COMPLICATED DESIGNS

The sample procedures discussed in Section 19.2 can also be applied to completely randomized designs with one or more covariates. Including more than one covariate in the model merely means that we have more than one quantitative independent variable in our model. For example, we might wish to compare the social status y of several different occupational groups while incorporating information on the number of years x_1 of formal education beyond high school and the income level x_2 of each individual in a group. As mentioned previously, we need not restrict ourselves to linear terms in the covariate(s). Thus, we might have a response related to two covariates (x_1 and x_2) and $t = 3$ treatments using the model

$$y = \beta_0 + \beta_1 x_1 + \beta_2 x_1^2 + \beta_3 x_2 + \beta_4 x_3 + \beta_5 x_4 + \beta_6 x_1 x_3$$
$$+ \beta_7 x_1 x_4 + \beta_8 x_1^2 x_3 + \beta_9 x_1^2 x_4 + \beta_{10} x_2 x_3 + \beta_{11} x_2 x_4 + \varepsilon,$$

where

$x_3 = 1$ if treatment 2 $x_3 = 0$ otherwise
$x_4 = 1$ if treatment 3 $x_4 = 0$ otherwise.

We can readily obtain an interpretation of the βs by using a table of expected values similar to Table 19.1.

An analysis of covariance for more complicated designs can also be obtained using general linear model methodology. Consider an analysis of covariance for a randomized block design. For simplicity, we will assume that there are two blocks, three treatments, one covariate x_1, and more than one observation per cell.

E X A M P L E   19.5   ▼

Write the model for the experimental situation described in the previous paragraph, assuming the response is linearly related to x_1 for each treatment. Identify the parameters in your model.

Solution   The model is

$$y = \beta_0 + \beta_1 x_1 + \beta_2 x_2 + \beta_3 x_3 + \beta_4 x_4 + \beta_5 x_1 x_2$$
$$+ \beta_6 x_1 x_3 + \beta_7 x_1 x_4 + \varepsilon,$$

where

$x_2 = 1$ if block 2 $x_2 = 0$ otherwise
$x_3 = 1$ if treatment 2 $x_3 = 0$ otherwise
$x_4 = 1$ if treatment 3 $x_4 = 0$ otherwise.

We immediately recognize this as a model relating a response y to a quantitative variable x_1 and two qualitative variables: blocks and treatments. An interpretation of the βs in the model is readily obtained from the table of expected values shown in Table 19.3.

T A B L E   19.3

Expected Values for the Randomized Block Design with One Covariate, Example 19.5

Block	Treatment	Expected Values
1	1	$\beta_0 + \beta_1 x_1$
	2	$(\beta_0 + \beta_3) + (\beta_1 + \beta_6)x_1$
	3	$(\beta_0 + \beta_4) + (\beta_1 + \beta_7)x_1$
2	1	$(\beta_0 + \beta_2) + (\beta_1 + \beta_5)x_1$
	2	$(\beta_0 + \beta_2 + \beta_3) + (\beta_1 + \beta_5 + \beta_6)x_1$
	3	$(\beta_0 + \beta_2 + \beta_4) + (\beta_1 + \beta_5 + \beta_7)x_1$

▲

The model we formulated in Example 19.5 not only provides for a linear relationship between y and x_1 for each of the treatments in each block, but it also allows for differences among intercepts and slopes. If we wanted to test for the equality of the slopes across treatments and blocks, we would use the null hypothesis

$$H_0: \quad \beta_5 = \beta_6 = \beta_7 = 0.$$

If there is insufficient evidence to reject H_0, we would proceed with the reduced model (obtained by setting $\beta_5 = \beta_6 = \beta_7 = 0$ in our model)

$$y = \beta_0 + \beta_1 x_1 + \beta_2 x_2 + \beta_3 x_3 + \beta_4 x_4 + \varepsilon.$$

A test for differences among treatments adjusted for the covariate could be obtained by fitting a complete and a reduced model for the null hypothesis

$$H_0: \quad \beta_3 = \beta_4 = 0.$$

▼ EXERCISES

Basic Techniques

19.6 Write a model for a 4×4 Latin square design with one covariate x_1. Assume the response is linearly related to the covariate. Identify the parameters in the model.

19.7 Refer to Exercise 19.6.
   a. Indicate how you would test for parallelism among the different straight lines. How many degrees of freedom would the F test have?
   b. Indicate how you would perform a test for the effects of treatments adjusted for the covariate.

19.8 Refer to Exercise 19.6. Write a complete model assuming that the response is a second-order function of the covariate x_1. Can you identify parameters in the model? How would you test for parallelism of the second-order model?

19.5 ▼ SUMMARY

In this chapter, we presented a procedure called the analysis of covariance. Here, for each value of y, we also observe a value of a concomitant variable x. This second variable, called a covariate, is recognized as an uncontrolled quantitative independent variable. Because of this fact, we can formulate models using the general linear model methodology of previous chapters.

In most situations when reference is made to an analysis of covariance, it is assumed that the response is linearly related to the covariate x, with the slope of the line the same for all treatment groups. Then a test for treatments adjusted for the covariate is performed. Actually, many people run analyses of covariance without checking the assumptions of parallelism. Rather than trying to force a particular model onto an experimental situation, it would be much better to postulate a reasonable (not necessarily linear) model relating the response y to the covariate x through the design used. Then by knowing the meanings of the parameters in the model, we can postulate hypotheses concerning the parameters and test these hypotheses by fitting complete and reduced models.

▼ SUPPLEMENTARY EXERCISES

 19.9 An investigator studied the effects of three different antidepressants (A, B, and C) on patient ratings of depression. To do this, patients were stratified into six age–gender combinations. From a random sample of three patients from each stratum, the experimenter randomly allocated the three antidepressants. On the day the study was to be initiated, a baseline (pretreatment) depression scale rating was obtained from each patient. The assigned therapy was then administered and maintained for one week. At this time, a second rating (posttreatment) was obtained from each patient. The pre- and posttreatment ratings appear below (higher score indicates more depression).

Block	Gender	Age (years)	Pretreatment A	B	C	Posttreatment A	B	C
1	F	<20	48	36	31	21	25	17
2	F	20–40	43	31	28	22	21	19
3	F	>40	44	35	29	18	24	18
4	M	<20	42	38	29	26	20	17
5	M	20–40	37	34	28	21	24	15
6	M	>40	41	36	26	18	24	19

a. Identify the experimental design.

b. Write a first-order model relating the posttreatment response y to the pretreatment rating x_1 for each treatment.

19.10   Refer to Exercise 19.9.

    a. Use a computer program to fit the model of part b of Exercise 19.8. Use $\alpha = .05$.

    b. Test for parallelism of the lines.

    c. Assuming that the lines are parallel, test for differences in treatment means adjusted for the covariate. Use $\alpha = .05$.

19.11   Refer to Exercises 19.9 and 19.10.

    a. Assuming parallelism of the response lines, perform a test for block differences adjusted for the covariate. Use $\alpha = .05$.

    b. How might you partition the block sum of squares into five meaningful single-degree-of-freedom sums of squares?

    c. Write a model and perform the tests suggested in part b. Use $\alpha = .05$.

19.12   A random sample of 10 measurements was selected from each of two populations. In addition to measuring a response variable y on each experimental unit, a second variable x was also measured. The sample data appear in the accompanying table.

Population 1		**Population 2**	
x	y	x	y
30	165	24	180
27	170	31	169
20	130	20	171
21	156	26	161
33	167	20	180
29	151	25	170
27	165	22	169
25	162	30	160
28	169	24	178
32	173	21	182

    a. Plot the sample data.

    b. Write a first-order model relating the response y to the covariate x.

19.13   **a.** Fit the model of Exercise 19.12b.

    b. Test for parallelism. Use $\alpha = .05$.

    c. What other tests are appropriate?

COMMUNICATING AND DOCUMENTING THE RESULTS OF ANALYSES

COMMUNICATING AND DOCUMENTING THE RESULTS OF ANALYSES

20.1   INTRODUCTION

In Chapter 1, we introduced the subject of statistics as a study of making sense of data and identified the four major components of making sense of data, namely, data gathering, data summarization, data analysis, and communicating results. In this chapter, we deal with methods or ways of communicating the results of statistical analyses. But rather than tell you what to write—which, of course, depends on the particular problem being discussed, the intended audience, and the form of the communication—we consider some important elements of a statistical report. We also discuss some of the potential pitfalls of effective communication. Finally, we discuss how one might document analyses to ensure reproducibility of results at a later date.

20.2   THE DIFFICULTY OF GOOD COMMUNICATION

We've spent time throughout the text making sense of data; the final step in this process is the communication of results. How might one communicate the results of a study or survey? The list of possibilities is almost endless, including all forms of verbal and written communication. There is quite a range of possibilities for verbal and written communication. For example, written communication within a company can vary from an informal short note or memo to a formal project report (Figure 20.1).

FIGURE 20.1
Forms of Written and Verbal Communication

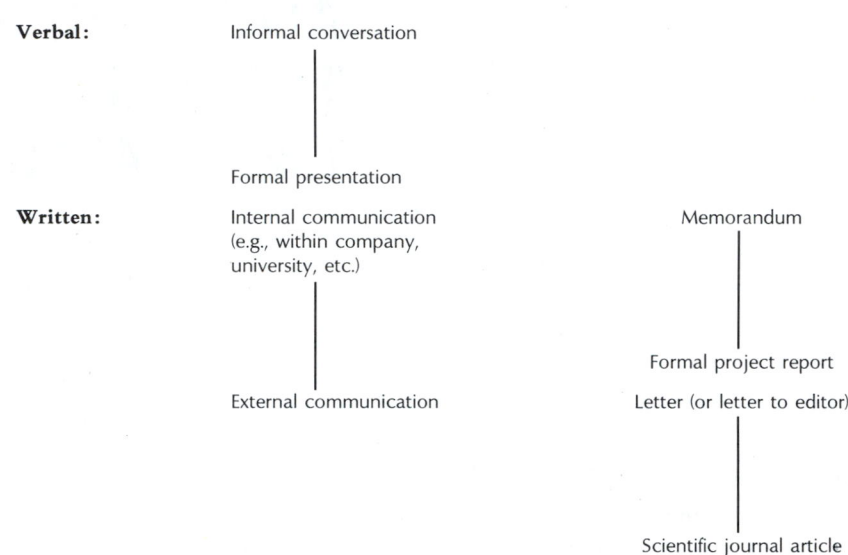

Communicating the results of a statistical analysis in concise, unambiguous terms is difficult. In fact, descriptions of most things are difficult. Try, for example, to describe the person sitting next to you so precisely that a stranger could select the individual from a group of others having similar physical characteristics. It is not an easy task. Fingerprints, voiceprints, and photographs—all pictorial descriptions—are the most precise methods of human identification. The description of a set of measurements is also a difficult task. But, like the description of a person, it can be accomplished more easily by using graphics or pictorial methods.

Cave drawings convey to us scattered bits of information about the life of prehistoric people. Similarly, vast quantities of knowledge about the ancient lives and cultures of the Babylonians, Egyptians, Greeks, and Romans are brought to life by means of drawings and sculpture. Art has been used to convey a picture of various lifestyles, history, and culture in all ages. Not surprisingly, use of graphs and tables in conjunction with a written description can help to convey the meaning of a statistical analysis.

In reading the results of a statistical analysis and in communicating the results of our own analyses, we must be careful not to distort them because of the way the data and results are presented. You have all heard the expression, "It is easy to lie with statistics." The idea is *not* new. The famous British statesman Disraeli is quoted as saying, "There are three kinds of lies: lies, damned lies, and statistics." Where do things go wrong?

First of all, distortion of truth can occur only when we communicate. And since communication can be accomplished with graphs, pictures, sound, aroma, taste, words, numbers, or any other means devised to reach our senses, distortions can occur using any one or any combination of these methods of communication.

In this respect, statements that we make could be misleading to others because we might have omitted something in the explanation of the data-gathering stage or with the analyses done. For example, we might unintentionally fail to clearly explain the meaning of a numerical statement. Or, we might omit some background information that is necessary for a clear interpretation of the results. Even a correct statement may appear to be distorted if the reader lacks knowledge of elementary statistics. Thus a very clear expression of an inference using a 95% confidence interval is meaningless to a person who has not been exposed to the introductory concepts of statistics.

Now we will look at some potential hurdles to effective communication that must be carefully considered when we present the results of a statistical analysis—or when we try to interpret what someone else has presented.

20.3 COMMUNICATION HURDLES: GRAPHICAL DISTORTIONS

Pictures can easily distort the truth. The marketing of many products, including soft drinks, beers, cosmetics, clothing, automobiles, and many more, involves the use of attractive, youthful models. The not-so-subtle impression we are left with is that (somehow) by using the product, we, too, will look like these models. Have you ever "stepped back" from one of these commercials and wondered how the commercial message relates to the quality and usefulness of the product? Or how you are being misled by a commercial? Try it sometime. Mail-order catalog sketches of products are frequently more attractive than the real thing, but we usually take this type of distortion for granted.

Statistical pictures are the histograms, frequency polygons, pie charts, and bar graphs of Chapter 3. These drawings or displays of numerical results are difficult to combine with sketches of lovely women or handsome men and hence are secure from the most common form of graphic distortion. But other distortions are possible. One could shrink or stretch the axes, thus distorting the actual results. The idea behind these distortions is that shallow and steep slopes are commonly associated with small and large increases, respectively.

For example, suppose that the values of a leading consumer price index over the first 6 months of the year were 160, 165, 178, 189, 196, and 210. We might show the upward movement of this consumer price index by using the frequency polygon of Figure 20.2. In this graph, the increase in the index is apparent, but it does not appear to be very great. On the other hand, we could present the sample data in a much different light, as shown in Figure 20.3. For this graph, the vertical axis is stretched and does not include 0. Note the impression of a substantial rise that is indicated by the steeper slope. Another way to achieve the same effect—to decrease or include a slope—is to stretch or shrink the horizontal axis.

FIGURE 20.2

Changes in a Consumer Price Index

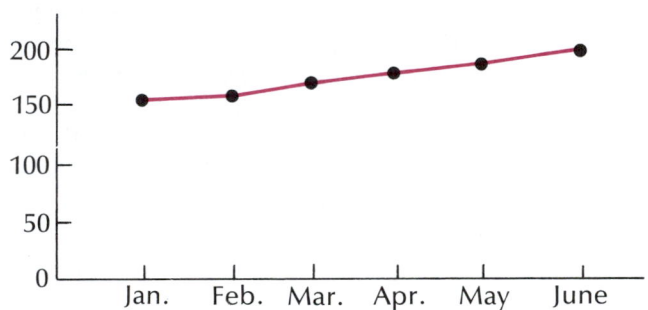

FIGURE 20.3

Changes in a Consumer Price Index

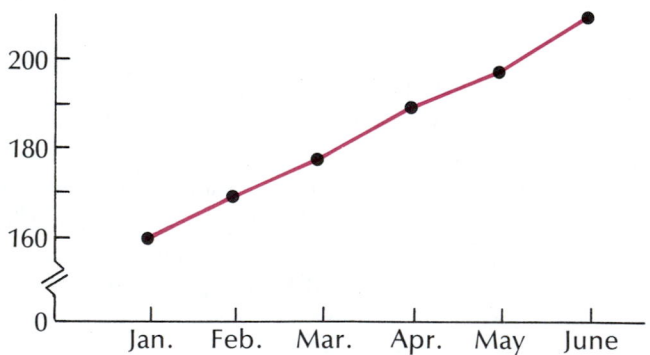

When we present data in the form of bar graphs, histograms, frequency polygons, or other figures, we must be careful not to shrink or stretch axes because it will catch most readers off guard. Increases or decreases in responses should be judged large or small depending on the arbitrary importance to the observer of the change, not on the slopes shown in graphic representations. In reality, most people look only at the slopes in the "pictures."

20.4 ▼ COMMUNICATION HURDLES: BIASED SAMPLES

One of the most common statistical distortions occurs because the experimenter unwittingly (or sometimes knowingly) samples the wrong population. That is, he or she draws the sample from a set of measurements that is not the proper population of interest.

For example, suppose that we want to assess the reaction of taxpayers to a proposed park and recreation center for children. A random sample of households is selected, and interviewers are sent to those households in the sample. Unfortunately, no one is at home in 40% of the sample households, so we randomly select and substitute other households in the city to make up the deficit. The resulting sample is selected from the wrong population, and the sample is therefore said to be biased.

The specified population of interest in the household survey is the collection of opinions that would be obtained from the complete set of all households in the city. In contrast, the sample was drawn from a much smaller population or subset of this group—the set of opinions from householders who were at home when the sample was taken. It is possible that the fractions of householders favoring the park in these two populations are equal, and no damage was done by confining the sampling to those at home. But it is much more likely that those at home had small children and that this group would yield a higher fraction in favor of the park than would the city as a whole. Thus, we have a biased sample because it is loaded in favor of families with small children. Perhaps a better way to see the difficulty is to note that we unwittingly selected the sample only from a special subset of the population of interest.

Biased samples frequently result from surveys that utilize mailed questionnaires. In a sense, the investigator lets the selection and number of the sampling units depend on the interests, available time, and various other personal characteristics of the individuals who receive the questionnaires. Extremely busy and energetic people may drop the questionnaires in the nearest wastebasket; you rarely hear from those low-energy folk who are uninterested or who are engrossed with other activities. Most often, the respondents are activists—those who are highly in favor, those who are opposed, or those who have something to gain from a certain outcome of the survey.

Although numerous newscasters and analysts utilize election results as an expression of public opinion on major issues, it is a well-known fact that voting results represent a biased sample of public opinion. Those who vote represent much less than half of the eligible voters; they are individuals who desire to exercise their rights and responsibilities as citizens or are individuals who have been specially motivated to participate. The resultant subset of voters is not representative of the interests and opinions of all eligible voters in the country.

Sampling the wrong population also occurs when people attempt to extrapolate experimented results from one population to another. Numerous experimental results have been published about the effect of various products (e.g., saccharin) in inducing cancer in moles, rats, the breasts of beagles, and so forth. These results are often used to imply that humans have a high risk of developing cancer after frequent or extended exposure to the product. These inferences are not always justified, because the experimental results were not obtained on humans. It is quite possible that humans are capable of resisting much higher doses than rats, or perhaps humans may be completely resistant for some reason. Drug induction of cancer in small mammals *does* indicate a need for concern and caution by humans, but it does not prove that the drug is definitely harmful to humans. Note that we are not criticizing experimentation in various species of animals, because it is frequently the only way we can obtain any information about potential toxicity in human beings. We simply point out that the experimenter is knowingly sampling a population that is only similar (and quite likely not too similar) to the one of interest.

Engineers also test "rats" instead of "humans." Rats, in this context, are miniature models or pilot plants of a new engineering system. Experiments on the models occasionally yield results that differ substantially from the results of the larger, real systems. So again, we see a sampling from the wrong population, but it is the best the engineer can do because of the economics of the situation. Funds are not usually available to test a number of full-scale models prior to production.

Many other examples could be given of biased samples or of sampling from the wrong populations. The point is that when we communicate the results of a study or survey we should be clear about how the sample was drawn and whether it was *randomly* selected from the population of interest. If this information is not given in the published results of a survey or experiment, the reader should take the inferences with a grain of salt.

20.5 ▼ COMMUNICATION HURDLES: SAMPLE SIZE

Distortions can occur when the sample size is not discussed. For example, suppose you read that a survey indicates that approximately 75% of a sample favor a new high-rise building complex. Further investigation might reveal that the investigator sampled only four people. When three out of the four favored the project, the investigator decided to stop the survey. Of course, we exaggerate with this example; but we could also have revealed inconclusive results based on a sample of 25, even though many buyers would consider this sample size to be large enough. As you well know, very large samples are required to achieve adequate information in sampling binomial populations.

Fortunately, many publications are now providing more information about the sample size and how opinion surveys are conducted. Ten years ago, it was rare to find how many people were sampled, much less how they were sampled. Things are different now. In fact, sometimes the media have gone too far in an attempt to be completely open about how a survey was done. A case in point is the following article from the *Wall Street Journal* (September 23, 1988). How many of us understand much more than the number of persons sampled and the approximate plus or minus (confidence interval)? It would take a person well trained in statistics and survey sampling to interpret what was done. Again, the moral of the story is simple: Try to communicate in unambiguous terms.

How Poll Was Conducted

The Wall Street Journal/NBC News poll was based on nationwide telephone interviews conducted last Friday through Monday with 4,159 adults age 18 or older. There were 2,630 likely voters.

The sample was drawn from a complete list of telephone exchanges, chosen so that each region of the country was represented in proportion to its population. Households were selected by a method that gave all telephone numbers, listed and unlisted, a proportionate chance of being included. The results of the survey were weighted to adjust for variations in the sample relating to education, age, race, gender, and religion.

Chances are 19 of 20 that if all adults in the United States had been surveyed using the same questionnaire, the findings would differ from these poll results by no more than two percentage points in either direction. The margin of error for subgroups may be larger.

20.6 ▼ THE STATISTICAL REPORT

Now that you have seen the various ways to communicate the results of a statistical analysis and some of the more common hurdles to effectively communicating these results, let us address the content of a statistical report that would appear as part of an internal project report in a book or scientific journal article. Obviously, companies and journals do not all abide by the same outline for a statistical report; however, based on what we present, you should have sufficient material to rearrange some sections and delete others to satisfy specific requirements.

A statistical report of the results of a study or experiment should clearly reflect all stages of making sense of data, much as we have emphasized these stages throughout the textbook. A general outline for a statistical report is shown in Table 20.1 along with a brief description of the content for each major section of the report.

T A B L E   20.1
**General Outline for a
Statistical Report**

Outline	Stage of Making Sense of Data	Description
Title		
Summary		Sometimes this section comprises the abstract of a journal article or scientific paper. It is a short summary of the study results, preferably one page or less.
Introduction	Data-gathering stage	The introduction gives a brief summary of the background and rationale for the study (survey) run.
Study design and procedures	Data-gathering stage	The study design and procedures section gives the study (survey) objective, the study (survey) design, and a summary of procedures for conducting the study (survey), including details about the data-gathering stage.
Descriptive statistics	Data-summarization stage	This section includes the main descriptive techniques (such as histograms, scatter plots, means, standard errors, etc.) used to summarize the study (survey) data.
Statistical methodology	Data-analysis stage	This section gives a description of the methods used to analyze the study (survey) data. This would include a brief description of the statistical tests and estimation procedures used to address the objectives of the study (survey), including those on which the major conclusions are drawn. References for the tests and estimation procedures may also be included, especially if the intended audience may not be aware of the techniques.
Results and conclusions	Data-analysis stage	In this section, the authors address the main results of the statistical analyses and the conclusions that can be drawn from these analyses in light of the study (survey) objectives.

(continues)

Outline	Stage of Making Sense of Data	Description
Discussion		The discussion section of a statistical report provides an opportunity to interpret the results of the statistical analyses and to put the conclusions in the context of previous studies (surveys). Do the results confirm or contradict previous study (survey) results? If they contradict previous results, an attempt should be made to offer a viable explanation for the difference in results. The discussion section allows one to make recommendations for further studies or surveys, when appropriate.
Data listings (optional)		Sometimes it is appropriate to provide listings of the data upon which the summaries and analyses are based. This section could also include computer output for some (all) of the statistical analyses done in the data-analysis stage of the study.

20.7 ▼ DOCUMENTATION AND STORAGE OF RESULTS

The final part of this cycle of data processing, analysis, and summarization concerns the documentation and storage of results. For formal statistical analyses that are subject to careful scrutiny by others, it is important to provide detailed documentation for all data processing and the statistical analyses so that the data trail is clear and the data base or work files are readily accessible. Then the reviewer can follow what has been done, redo it, or extend the analyses. Before we list the elements of a documentation and storage file, we'll discuss several different categories of analyses.

primary analyses
backup analyses

Primary analyses are those used to address the objectives of the study and the analyses on which conclusions are drawn. **Backup analyses** include alternative methods for examining the data that confirm the results of the primary analyses; they may also include new statistical methods that are not as readily accepted as the more standard methods. Several guidelines for analyses are presented here.

Preliminary, Primary, and Backup Analyses

1. Analyses should be performed with software that has been extensively tested.
2. Computer output should be labeled to reflect which study is analyzed, what subjects (animals, patients, etc.) are used in the analysis, and a brief description of the analysis preferred. For example, TITLE statements in SAS are very helpful.
3. Variable labels and value labels (e.g., 0 = none, 1 = mild) should appear on the output.
4. A list of the data used in each analysis should be provided.
5. The output for all analyses should be checked *carefully*. Did the job run successfully? Are the sample sizes, means, and degrees of freedom correct? Other checks may be necessary as well.
6. All preliminary, primary, and backup analyses that provide the informational base from which study conclusions are drawn should be saved.

The elements of a documentation and storage file depend on the particular setting in which you work. The contents for a general documentation storage file are shown next.

Study Documentation and Storage File

1. Statistical report
2. Study description
3. Random code (used to assign subjects to treatment groups)
4. Important correspondence
5. File creation information
6. Preliminary, primary, and backup analyses
7. Raw data source
8. A data management sheet, which includes the log as well as information on the storage of the data files

The major thrust behind the documentation and storage file is that we want to provide a clear data and analysis "trail" for our own use or for someone else's use, should there be a need to revisit the data. For any given situation, you must ask yourself whether such documentation is necessary and, if so, how detailed it must be. A good test of the completeness and understandability of your documentation is to ask a colleague, unfamiliar with your project but knowledgeable in your field, to try to reconstruct and even redo the primary analyses you did. If he or she can navigate through your documentation trail, you have done the job.

20.8 ▼ SUMMARY

In this chapter, we have discussed how to present the results of a statistical analysis of data and some of the problems with effectively communicating these results to the intended audience. The task is not easy. Some of the obstacles or hurdles standing in the way of effectively communicating the results of statistical analyses include graphical distortions, biased sampling, and omitting a discussion of the sample size and sampling technique. With some understanding of these obstacles, we can better critique and understand communications aimed at us and also do a better job communicating the results of our analyses to others.

The final topic in this chapter dealt with the documentation and storage of results. Having completed your analyses, drawn conclusions, and communicated these results to the intended audience, the temptation is to postpone or eliminate the documentation and storage of results. However, it is worth your time to assess the potential for revisiting your analyses in the future and determine what steps should be taken to facilitate this process (for you or for others).

Finally, we should put our statistical analyses in the context of the practical problem(s) being addressed. The report of statistical analyses will not necessarily be the answer to an important question; it is only *part* of the answer. For example, we may demonstrate that a tablet delivers a drug more quickly than a capsule, but this is not the only consideration in the decision to market the capsule or tablet form. Such factors as cost, palatability, and stability will also be considered. Some of the relevant analyses addressing these considerations are not statistical.

▼ SUPPLEMENTARY EXERCISE

Class Project

Have each person in the class choose a commercial he or she has heard or seen lately and critique it for possible distortions. Also make suggestions for improvement with regard to clarity of message, etc. (Note: We didn't ask you to improve it from a commercial standpoint.) Present these findings.

APPENDIX A:
STATISTICAL TABLES

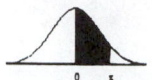

TABLE 1 Normal Curve Areas

z	.00	.01	.02	.03	.04	.05	.06	.07	.08	.09
0.00	.0000	.0040	.0080	.0120	.0160	.0199	.0239	.0279	.0319	.0359
0.10	.0398	.0438	.0478	.0517	.0557	.0596	.0636	.0675	.0714	.0753
0.20	.0793	.0832	.0871	.0910	.0948	.0987	.1026	.1064	.1103	.1141
0.30	.1179	.1217	.1255	.1293	.1331	.1368	.1406	.1443	.1480	.1517
0.40	.1554	.1591	.1628	.1664	.1700	.1736	.1772	.1808	.1844	.1879
0.50	.1915	.1950	.1985	.2019	.2054	.2088	.2123	.2157	.2190	.2224
0.60	.2257	.2291	.2324	.2357	.2389	.2422	.2454	.2486	.2517	.2549
0.70	.2580	.2611	.2642	.2673	.2704	.2734	.2764	.2794	.2823	.2852
0.80	.2881	.2910	.2939	.2967	.2995	.3023	.3051	.3078	.3106	.3133
0.90	.3159	.3186	.3212	.3238	.3264	.3289	.3315	.3340	.3365	.3389
1.00	.3413	.3438	.3461	.3485	.3508	.3531	.3554	.3577	.3599	.3621
1.10	.3643	.3665	.3686	.3708	.3729	.3749	.3770	.3790	.3810	.3830
1.20	.3849	.3869	.3888	.3907	.3925	.3944	.3962	.3980	.3997	.4015
1.30	.4032	.4049	.4066	.4082	.4099	.4115	.4131	.4147	.4162	.4177
1.40	.4192	.4207	.4222	.4236	.4251	.4265	.4279	.4292	.4306	.4319
1.50	.4332	.4345	.4357	.4370	.4382	.4394	.4406	.4418	.4429	.4441
1.60	.4452	.4463	.4474	.4484	.4495	.4505	.4515	.4525	.4535	.4545
1.70	.4554	.4564	.4573	.4582	.4591	.4599	.4608	.4616	.4625	.4633
1.80	.4641	.4649	.4656	.4664	.4671	.4678	.4686	.4693	.4699	.4706
1.90	.4713	.4719	.4726	.4732	.4738	.4744	.4750	.4756	.4761	.4767
2.00	.4772	.4778	.4783	.4788	.4793	.4798	.4803	.4808	.4812	.4817
2.10	.4821	.4826	.4830	.4834	.4838	.4842	.4846	.4850	.4854	.4857
2.20	.4861	.4864	.4868	.4871	.4875	.4878	.4881	.4884	.4887	.4890
2.30	.4893	.4896	.4898	.4901	.4904	.4906	.4909	.4911	.4913	.4916
2.40	.4918	.4920	.4922	.4925	.4927	.4929	.4931	.4932	.4934	.4936
2.50	.4938	.4940	.4941	.4943	.4945	.4946	.4948	.4949	.4951	.4952
2.60	.4953	.4955	.4956	.4957	.4959	.4960	.4961	.4962	.4963	.4964
2.70	.4965	.4966	.4967	.4968	.4969	.4970	.4971	.4972	.4973	.4974
2.80	.4974	.4975	.4976	.4977	.4977	.4978	.4979	.4979	.4980	.4981
2.90	.4981	.4982	.4982	.4983	.4984	.4984	.4985	.4985	.4986	.4986
3.00	.4987	.4987	.4987	.4988	.4988	.4989	.4989	.4989	.4990	.4990

z	Area
3.50	.49976737
4.00	.49996833
4.50	.49999660
5.00	.49999971

Source: Computed by P. J. Hildebrand.

T A B L E 2   Upper-tail Areas for the Normal Curve

z	.00	.01	.02	.03	.04	.05	.06	.07	.08	.09
0.00	.5000	.4960	.4920	.4880	.4840	.4801	.4761	.4721	.4681	.4641
0.10	.4602	.4562	.4522	.4483	.4443	.4404	.4364	.4325	.4286	.4247
0.20	.4207	.4168	.4129	.4090	.4052	.4013	.3974	.3936	.3897	.3859
0.30	.3821	.3783	.3745	.3707	.3669	.3632	.3594	.3557	.3520	.3483
0.40	.3446	.3409	.3372	.3336	.3300	.3264	.3228	.3192	.3156	.3121
0.50	.3085	.3050	.3015	.2981	.2946	.2912	.2877	.2843	.2810	.2776
0.60	.2743	.2709	.2676	.2643	.2611	.2578	.2546	.2514	.2483	.2451
0.70	.2420	.2389	.2358	.2327	.2296	.2266	.2236	.2206	.2177	.2148
0.80	.2119	.2090	.2061	.2033	.2005	.1977	.1949	.1922	.1894	.1867
0.90	.1841	.1814	.1788	.1762	.1736	.1711	.1685	.1660	.1635	.1611
1.00	.1587	.1562	.1539	.1515	.1492	.1469	.1446	.1423	.1401	.1379
1.10	.1357	.1335	.1314	.1292	.1271	.1251	.1230	.1210	.1190	.1170
1.20	.1151	.1131	.1112	.1093	.1075	.1056	.1038	.1020	.1003	.0985
1.30	.0968	.0951	.0934	.0918	.0901	.0885	.0869	.0853	.0838	.0823
1.40	.0808	.0793	.0778	.0764	.0749	.0735	.0721	.0708	.0694	.0681
1.50	.0668	.0655	.0643	.0630	.0618	.0606	.0594	.0582	.0571	.0559
1.60	.0548	.0537	.0526	.0516	.0505	.0495	.0485	.0475	.0465	.0455
1.70	.0446	.0436	.0427	.0418	.0409	.0401	.0392	.0384	.0375	.0367
1.80	.0359	.0351	.0344	.0336	.0329	.0322	.0314	.0307	.0301	.0294
1.90	.0287	.0281	.0274	.0268	.0262	.0256	.0250	.0244	.0239	.0233
2.00	.0228	.0222	.0217	.0212	.0207	.0202	.0197	.0192	.0188	.0183
2.10	.0179	.0174	.0170	.0166	.0162	.0158	.0154	.0150	.0146	.0143
2.20	.0139	.0136	.0132	.0129	.0125	.0122	.0119	.0116	.0113	.0110
2.30	.0107	.0104	.0102	.0099	.0096	.0094	.0091	.0089	.0087	.0084
2.40	.0082	.0080	.0078	.0075	.0073	.0071	.0069	.0068	.0066	.0064
2.50	.0062	.0060	.0059	.0057	.0055	.0054	.0052	.0051	.0049	.0048
2.60	.0047	.0045	.0044	.0043	.0041	.0040	.0039	.0038	.0037	.0036
2.70	.0035	.0034	.0033	.0032	.0031	.0030	.0029	.0028	.0027	.0026
2.80	.0026	.0025	.0024	.0023	.0023	.0022	.0021	.0021	.0020	.0019
2.90	.0019	.0018	.0018	.0017	.0016	.0016	.0015	.0015	.0014	.0014
3.00	.0013	.0013	.0013	.0012	.0012	.0011	.0011	.0011	.0010	.0010

z	Area
3.500	.00023263
4.000	.00003167
4.500	.00000340
5.000	.00000029

Source: Computed by J. W. Stegeman using SAS.

TABLE 3 Critical Values of T_L and T_U for the Wilcoxon Rank Sum Test: Independent Samples
(test statistic is rank sum associated with smaller sample (if equal sample sizes, either rank sum can be used))

a. $\alpha = .025$ one-tailed; $\alpha = .05$ two-tailed

n_2 \ n_1	3 T_L	3 T_U	4 T_L	4 T_U	5 T_L	5 T_U	6 T_L	6 T_U	7 T_L	7 T_U	8 T_L	8 T_U	9 T_L	9 T_U	10 T_L	10 T_U
3	5	16	6	18	6	21	7	23	7	26	8	28	8	31	9	33
4	6	18	11	25	12	28	12	32	13	35	14	38	15	41	16	44
5	6	21	12	28	18	37	19	41	20	45	21	49	22	53	24	56
6	7	23	12	32	19	41	26	52	28	56	29	61	31	65	32	70
7	7	26	13	35	20	45	28	56	37	68	39	73	41	78	43	83
8	8	28	14	38	21	49	29	61	39	73	49	87	51	93	54	98
9	8	31	15	41	22	53	31	65	41	78	51	93	63	108	66	114
10	9	33	16	44	24	56	32	70	43	83	54	98	66	114	79	131

b. $\alpha = .05$ one-tailed; $\alpha = .10$ two-tailed

n_2 \ n_1	3 T_L	3 T_U	4 T_L	4 T_U	5 T_L	5 T_U	6 T_L	6 T_U	7 T_L	7 T_U	8 T_L	8 T_U	9 T_L	9 T_U	10 T_L	10 T_U
3	6	15	7	17	7	20	8	22	9	24	9	27	10	29	11	31
4	7	17	12	24	13	27	14	30	15	33	16	36	17	39	18	42
5	7	20	13	27	19	36	20	40	22	43	24	46	25	50	26	54
6	8	22	14	30	20	40	28	50	30	54	32	58	33	63	35	67
7	9	24	15	33	22	43	30	54	39	66	41	71	43	76	46	80
8	9	27	16	36	24	46	32	58	41	71	52	84	54	90	57	95
9	10	29	17	39	25	50	33	63	43	76	54	90	66	105	69	111
10	11	31	18	42	26	54	35	67	46	80	57	95	69	111	83	127

Source: From F. Wilcoxon and R. A. Wilcox, *Some Rapid Approximate Statistical Procedures* (Pearl River, N.Y. Lederle Laboratories, 1964), pp. 20–23. Reproduced with the permission of American Cyanamid Company.

TABLE 4 Percentage Points of the *t* Distribution

df	a = .1	a = .05	a = .025	a = .01	a = .005	a = .001
1	3.078	6.314	12.706	31.821	63.657	318.309
2	1.886	2.920	4.303	6.965	9.925	22.327
3	1.638	2.353	3.182	4.541	5.841	10.215
4	1.533	2.132	2.776	3.747	4.604	7.173
5	1.476	2.015	2.571	3.365	4.032	5.893
6	1.440	1.943	2.447	3.143	3.707	5.208
7	1.415	1.895	2.365	2.998	3.499	4.785
8	1.397	1.860	2.306	2.896	3.355	4.501
9	1.383	1.833	2.262	2.821	3.250	4.297
10	1.372	1.812	2.228	2.764	3.169	4.144
11	1.363	1.796	2.201	2.718	3.106	4.025
12	1.356	1.782	2.179	2.681	3.055	3.930
13	1.350	1.771	2.160	2.650	3.012	3.852
14	1.345	1.761	2.145	2.624	2.977	3.787
15	1.341	1.753	2.131	2.602	2.947	3.733
16	1.337	1.746	2.120	2.583	2.921	3.686
17	1.333	1.740	2.110	2.567	2.898	3.646
18	1.330	1.734	2.101	2.552	2.878	3.610
19	1.328	1.729	2.093	2.539	2.861	3.579
20	1.325	1.725	2.086	2.528	2.845	3.552
21	1.323	1.721	2.080	2.518	2.831	3.527
22	1.321	1.717	2.074	2.508	2.819	3.505
23	1.319	1.714	2.069	2.500	2.807	3.485
24	1.318	1.711	2.064	2.492	2.797	3.467
25	1.316	1.708	2.060	2.485	2.787	3.450
26	1.315	1.706	2.056	2.479	2.779	3.435
27	1.314	1.703	2.052	2.473	2.771	3.421
28	1.313	1.701	2.048	2.467	2.763	3.408
29	1.311	1.699	2.045	2.462	2.756	3.396
30	1.310	1.697	2.042	2.457	2.750	3.385
40	1.303	1.684	2.021	2.423	2.704	3.307
60	1.296	1.671	2.000	2.390	2.660	3.232
120	1.289	1.658	1.980	2.358	2.617	3.160
240	1.285	1.651	1.970	2.342	2.596	3.125
inf.	1.282	1.645	1.960	2.326	2.576	3.090

Source: Computed by P. J. Hildebrand.

χ^2_a

TABLE 5   Percentage Points of the Chi-square Distribution

df	a = .999	a = .995	a = .99	a = .975	a = .95	a = .9
1	.000002	.000039	.000157	.000982	.003932	.01579
2	.002001	.01003	.02010	.05064	.1026	.2107
3	.02430	.07172	.1148	.2158	.3518	.5844
4	.09080	.2070	.2971	.4844	.7107	1.064
5	.2102	.4117	.5543	.8312	1.145	1.610
6	.3811	.6757	.8721	1.237	1.635	2.204
7	.5985	.9893	1.239	1.690	2.167	2.833
8	.8571	1.344	1.646	2.180	2.733	3.490
9	1.152	1.735	2.088	2.700	3.325	4.168
10	1.479	2.156	2.558	3.247	3.940	4.865
11	1.834	2.603	3.053	3.816	4.575	5.578
12	2.214	3.074	3.571	4.404	5.226	6.304
13	2.617	3.565	4.107	5.009	5.892	7.042
14	3.041	4.075	4.660	5.629	6.571	7.790
15	3.483	4.601	5.229	6.262	7.261	8.547
16	3.942	5.142	5.812	6.908	7.962	9.312
17	4.416	5.697	6.408	7.564	8.672	10.09
18	4.905	6.265	7.015	8.231	9.390	10.86
19	5.407	6.844	7.633	8.907	10.12	11.65
20	5.921	7.434	8.260	9.591	10.85	12.44
21	6.447	8.034	8.897	10.28	11.59	13.24
22	6.983	8.643	9.542	10.98	12.34	14.04
23	7.529	9.260	10.20	11.69	13.09	14.85
24	8.085	9.886	10.86	12.40	13.85	15.66
25	8.649	10.52	11.52	13.12	14.61	16.47
26	9.222	11.16	12.20	13.84	15.38	17.29
27	9.803	11.81	12.88	14.57	16.15	18.11
28	10.39	12.46	13.56	15.31	16.93	18.94
29	10.99	13.12	14.26	16.06	17.71	19.77
30	11.59	13.79	14.95	16.79	18.49	20.60
40	17.92	20.71	22.16	24.43	26.51	29.05
50	24.67	27.99	29.71	32.36	34.76	37.69
60	31.74	35.53	37.48	40.48	43.19	46.46
70	39.04	43.28	45.44	48.76	51.74	55.33
80	46.52	51.17	53.54	57.15	60.39	64.28
90	54.16	59.20	61.75	65.65	69.13	73.29
100	61.92	67.33	70.06	74.22	77.93	82.36
120	77.76	83.85	86.92	91.57	95.70	100.62
240	177.95	187.32	191.99	198.98	205.14	212.39

TABLE 5   *(continued)*

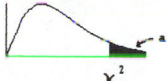

x_a^2

a = .1	a = .05	a = .025	a = .01	a = .005	a = .001	df
2.706	3.841	5.024	6.635	7.879	10.83	1
4.605	5.991	7.378	9.210	10.60	13.82	2
6.251	7.815	9.348	11.34	12.84	16.27	3
7.779	9.488	11.14	13.28	14.86	18.47	4
9.236	11.07	12.83	15.09	16.75	20.52	5
10.64	12.59	14.45	16.81	18.55	22.46	6
12.02	14.07	16.01	18.48	20.28	24.32	7
13.36	15.51	17.53	20.09	21.95	26.12	8
14.68	16.92	19.02	21.67	23.59	27.88	9
15.99	18.31	20.48	23.21	25.19	29.59	10
17.28	19.68	21.92	24.72	26.76	31.27	11
18.55	21.03	23.34	26.22	28.30	32.91	12
19.81	22.36	24.74	27.69	29.82	34.53	13
21.06	23.68	26.12	29.14	31.32	36.12	14
22.31	25.00	27.49	30.58	32.80	37.70	15
23.54	26.30	28.85	32.00	34.27	39.25	16
24.77	27.59	30.19	33.41	35.72	40.79	17
25.99	28.87	31.53	34.81	37.16	42.31	18
27.20	30.14	32.85	36.19	38.58	43.82	19
28.41	31.41	34.17	37.57	40.00	45.31	20
29.62	32.67	35.48	38.93	41.40	46.80	21
30.81	33.92	36.78	40.29	42.80	48.27	22
32.01	35.17	38.08	41.64	44.18	49.73	23
33.20	36.42	39.36	42.98	45.56	51.18	24
34.38	37.65	40.65	44.31	46.93	52.62	25
35.56	38.89	41.92	45.64	48.29	54.05	26
36.74	40.11	43.19	46.96	49.65	55.48	27
37.92	41.34	44.46	48.28	50.99	56.89	28
39.09	42.56	45.72	49.59	52.34	58.30	29
40.26	43.77	46.98	50.89	53.67	59.70	30
51.81	55.76	59.34	63.69	66.77	73.40	40
63.17	67.50	71.42	76.15	79.49	86.66	50
74.40	79.08	83.30	88.38	91.95	99.61	60
85.53	90.53	95.02	100.43	104.21	112.32	70
96.58	101.88	106.63	112.33	116.32	124.84	80
107.57	113.15	118.14	124.12	128.30	137.21	90
118.50	124.34	129.56	135.81	140.17	149.45	100
140.23	146.57	152.21	158.95	163.65	173.62	120
268.47	277.14	284.80	293.89	300.18	313.44	240

Source: Computed by P. J. Hildebrand.

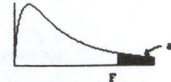

T A B L E   6   Percentage Points of the F Distribution (df_2 Between 1 and 6)

df_2	a	df_1 1	2	3	4	5	6	7	8	9	10
1	.25	5.83	7.50	8.20	8.58	8.82	8.98	9.10	9.19	9.26	9.32
	.10	39.86	49.50	53.59	55.83	57.24	58.20	58.91	59.44	59.86	60.19
	.05	161.4	199.5	215.7	224.6	230.2	234.0	236.8	238.9	240.5	241.9
	.025	647.8	799.5	864.2	899.6	921.8	937.1	948.2	956.7	963.3	968.6
	.01	4052	5000	5403	5625	5764	5859	5928	5981	6022	6056
2	.25	2.57	3.00	3.15	3.23	3.28	3.31	3.34	3.35	3.37	3.38
	.10	8.53	9.00	9.16	9.24	9.29	9.33	9.35	9.37	9.38	9.39
	.05	18.51	19.00	19.16	19.25	19.30	19.33	19.35	19.37	19.38	19.40
	.025	38.51	39.00	39.17	39.25	39.30	39.33	39.36	39.37	39.39	39.40
	.01	98.50	99.00	99.17	99.25	99.30	99.33	99.36	99.37	99.39	99.40
	.005	198.5	199.0	199.2	199.2	199.3	199.3	199.4	199.4	199.4	199.4
	.001	998.5	999.0	999.2	999.2	999.3	999.3	999.4	999.4	999.4	999.4
3	.25	2.02	2.28	2.36	2.39	2.41	2.42	2.43	2.44	2.44	2.44
	.10	5.54	5.46	5.39	5.34	5.31	5.28	5.27	5.25	5.24	5.23
	.05	10.13	9.55	9.28	9.12	9.01	8.94	8.89	8.85	8.81	8.79
	.025	17.44	16.04	15.44	15.10	14.88	14.73	14.62	14.54	14.47	14.42
	.01	34.12	30.82	29.46	28.71	28.24	27.91	27.67	27.49	27.35	27.23
	.005	55.55	49.80	47.47	46.19	45.39	44.84	44.43	44.13	43.88	43.69
	.001	167.0	148.5	141.1	137.1	134.6	132.8	131.6	130.6	129.9	129.2
4	.25	1.81	2.00	2.05	2.06	2.07	2.08	2.08	2.08	2.08	2.08
	.10	4.54	4.32	4.19	4.11	4.05	4.01	3.98	3.95	3.94	3.92
	.05	7.71	6.94	6.59	6.39	6.26	6.16	6.09	6.04	6.00	5.96
	.025	12.22	10.65	9.98	9.60	9.36	9.20	9.07	8.98	8.90	8.84
	.01	21.20	18.00	16.69	15.98	15.52	15.21	14.98	14.80	14.66	14.55
	.005	31.33	26.28	24.26	23.15	22.46	21.97	21.62	21.35	21.14	20.97
	.001	74.14	61.25	56.18	53.44	51.71	50.53	49.66	49.00	48.47	48.05
5	.25	1.69	1.85	1.88	1.89	1.89	1.89	1.89	1.89	1.89	1.89
	.10	4.06	3.78	3.62	3.52	3.45	3.40	3.37	3.34	3.32	3.30
	.05	6.61	5.79	5.41	5.19	5.05	4.95	4.88	4.82	4.77	4.74
	.025	10.01	8.43	7.76	7.39	7.15	6.98	6.85	6.76	6.68	6.62
	.01	16.26	13.27	12.06	11.39	10.97	10.67	10.46	10.29	10.16	10.05
	.005	22.78	18.31	16.53	15.56	14.94	14.51	14.20	13.96	13.77	13.62
	.001	47.18	37.12	33.20	31.09	29.75	28.83	28.16	27.65	27.24	26.92
6	.25	1.62	1.76	1.78	1.79	1.79	1.78	1.78	1.78	1.77	1.77
	.10	3.78	3.46	3.29	3.18	3.11	3.05	3.01	2.98	2.96	2.94
	.05	5.99	5.14	4.76	4.53	4.39	4.28	4.21	4.15	4.10	4.06
	.025	8.81	7.26	6.60	6.23	5.99	5.82	5.70	5.60	5.52	5.46
	.01	13.75	10.92	9.78	9.15	8.75	8.47	8.26	8.10	7.98	7.87
	.005	18.63	14.54	12.92	12.03	11.46	11.07	10.79	10.57	10.39	10.25
	.001	35.51	27.00	23.70	21.92	20.80	20.03	19.46	19.03	18.69	18.41

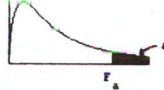

TABLE 6 (continued)

					df$_1$							
12	15	20	24	30	40	60	120	240	inf.	a	df$_2$
9.41	9.49	9.58	9.63	9.67	9.71	9.76	9.80	9.83	9.85	.25	1
60.71	61.22	61.74	62.00	62.26	62.53	62.79	63.06	63.19	63.33	.10	
243.9	245.9	248.0	249.1	250.1	251.1	252.2	253.3	253.8	254.3	.05	
976.7	984.9	993.1	997.2	1001	1006	1010	1014	1016	1018	.025	
6106	6157	6209	6235	6261	6287	6313	6339	6353	6366	.01	
3.39	3.41	3.43	3.43	3.44	3.45	3.46	3.47	3.47	3.48	.25	2
9.41	9.42	9.44	9.45	9.46	9.47	9.47	9.48	9.49	9.49	.10	
19.41	19.43	19.45	19.45	19.46	19.47	19.48	19.49	19.49	19.50	.05	
39.41	39.43	39.45	39.46	39.46	39.47	39.48	39.49	39.49	39.50	.025	
99.42	99.43	99.45	99.46	99.47	99.47	99.48	99.49	99.50	99.50	.01	
199.4	199.4	199.4	199.5	199.5	199.5	199.5	199.5	199.5	199.5	.005	
999.4	999.4	999.4	999.5	999.5	999.5	999.5	999.5	999.5	999.5	.001	
2.45	2.46	2.46	2.46	2.47	2.47	2.47	2.47	2.47	2.47	.25	3
5.22	5.20	5.18	5.18	5.17	5.16	5.15	5.14	5.14	5.13	.10	
8.74	8.70	8.66	8.64	8.62	8.59	8.57	8.55	8.54	8.53	.05	
14.34	14.25	14.17	14.12	14.08	14.04	13.99	13.95	13.92	13.90	.025	
27.05	26.87	26.69	26.60	26.50	26.41	26.32	26.22	26.17	26.13	.01	
43.39	43.08	42.78	42.62	42.47	42.31	42.15	41.99	41.91	41.83	.005	
128.3	127.4	126.4	125.9	125.4	125.0	124.5	124.0	123.7	123.5	.001	
2.08	2.08	2.08	2.08	2.08	2.08	2.08	2.08	2.08	2.08	.25	4
3.90	3.87	3.84	3.83	3.82	3.80	3.79	3.78	3.77	3.76	.10	
5.91	5.86	5.80	5.77	5.75	5.72	5.69	5.66	5.64	5.63	.05	
8.75	8.66	8.56	8.51	8.46	8.41	8.36	8.31	8.28	8.26	.025	
14.37	14.20	14.02	13.93	13.84	13.75	13.65	13.56	13.51	13.46	.01	
20.70	20.44	20.17	20.03	19.89	19.75	19.61	19.47	19.40	19.32	.005	
47.41	46.76	46.10	45.77	45.43	45.09	44.75	44.40	44.23	44.05	.001	
1.89	1.89	1.88	1.88	1.88	1.88	1.87	1.87	1.87	1.87	.25	5
3.27	3.24	3.21	3.19	3.17	3.16	3.14	3.12	3.11	3.10	.10	
4.68	4.62	4.56	4.53	4.50	4.46	4.43	4.40	4.38	4.36	.05	
6.52	6.43	6.33	6.28	6.23	6.18	6.12	6.07	6.04	6.02	.025	
9.89	9.72	9.55	9.47	9.38	9.29	9.20	9.11	9.07	9.02	.01	
13.38	13.15	12.90	12.78	12.66	12.53	12.40	12.27	12.21	12.14	.005	
26.42	25.91	25.39	25.13	24.87	24.60	24.33	24.06	23.92	23.79	.001	
1.77	1.76	1.76	1.75	1.75	1.75	1.74	1.74	1.74	1.74	.25	6
2.90	2.87	2.84	2.82	2.80	2.78	2.76	2.74	2.73	2.72	.10	
4.00	3.94	3.87	3.84	3.81	3.77	3.74	3.70	3.69	3.67	.05	
5.37	5.27	5.17	5.12	5.07	5.01	4.96	4.90	4.88	4.85	.025	
7.72	7.56	7.40	7.31	7.23	7.14	7.06	6.97	6.92	6.88	.01	
10.03	9.81	9.59	9.47	9.36	9.24	9.12	9.00	8.94	8.88	.005	
17.99	17.56	17.12	16.90	16.67	16.44	16.21	15.98	15.86	15.75	.001	

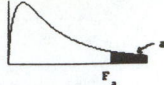

TABLE 6 Percentage Points of the *F* Distribution (df$_2$ Between 7 and 12) *(continued)*

df$_2$	a	df$_1$ 1	2	3	4	5	6	7	8	9	10
7	.25	1.57	1.70	1.72	1.72	1.71	1.71	1.70	1.70	1.69	1.69
	.10	3.59	3.26	3.07	2.96	2.88	2.83	2.78	2.75	2.72	2.70
	.05	5.59	4.74	4.35	4.12	3.97	3.87	3.79	3.73	3.68	3.64
	.025	8.07	6.54	5.89	5.52	5.29	5.12	4.99	4.90	4.82	4.76
	.01	12.25	9.55	8.45	7.85	7.46	7.19	6.99	6.84	6.72	6.62
	.005	16.24	12.40	10.88	10.05	9.52	9.16	8.89	8.68	8.51	8.38
	.001	29.25	21.69	18.77	17.20	16.21	15.52	15.02	14.63	14.33	14.08
8	.25	1.54	1.66	1.67	1.66	1.66	1.65	1.64	1.64	1.63	1.63
	.10	3.46	3.11	2.92	2.81	2.73	2.67	2.62	2.59	2.56	2.54
	.05	5.32	4.46	4.07	3.84	3.69	3.58	3.50	3.44	3.39	3.35
	.025	7.57	6.06	5.42	5.05	4.82	4.65	4.53	4.43	4.36	4.30
	.01	11.26	8.65	7.59	7.01	6.63	6.37	6.18	6.03	5.91	5.81
	.005	14.69	11.04	9.60	8.81	8.30	7.95	7.69	7.50	7.34	7.21
	.001	25.41	18.49	15.83	14.39	13.48	12.86	12.40	12.05	11.77	11.54
9	.25	1.51	1.62	1.63	1.63	1.62	1.61	1.60	1.60	1.59	1.59
	.10	3.36	3.01	2.81	2.69	2.61	2.55	2.51	2.47	2.44	2.42
	.05	5.12	4.26	3.86	3.63	3.48	3.37	3.29	3.23	3.18	3.14
	.025	7.21	5.71	5.08	4.72	4.48	4.32	4.20	4.10	4.03	3.96
	.01	10.56	8.02	6.99	6.42	6.06	5.80	5.61	5.47	5.35	5.26
	.005	13.61	10.11	8.72	7.96	7.47	7.13	6.88	6.69	6.54	6.42
	.001	22.86	16.39	13.90	12.56	11.71	11.13	10.70	10.37	10.11	9.89
10	.25	1.49	1.60	1.60	1.59	1.59	1.58	1.57	1.56	1.56	1.55
	.10	3.29	2.92	2.73	2.61	2.52	2.46	2.41	2.38	2.35	2.32
	.05	4.96	4.10	3.71	3.48	3.33	3.22	3.14	3.07	3.02	2.98
	.025	6.94	5.46	4.83	4.47	4.24	4.07	3.95	3.85	3.78	3.72
	.01	10.04	7.56	6.55	5.99	5.64	5.39	5.20	5.06	4.94	4.85
	.005	12.83	9.43	8.08	7.34	6.87	6.54	6.30	6.12	5.97	5.85
	.001	21.04	14.91	12.55	11.28	10.48	9.93	9.52	9.20	8.96	8.75
11	.25	1.47	1.58	1.58	1.57	1.56	1.55	1.54	1.53	1.53	1.52
	.10	3.23	2.86	2.66	2.54	2.45	2.39	2.34	2.30	2.27	2.25
	.05	4.84	3.98	3.59	3.36	3.20	3.09	3.01	2.95	2.90	2.85
	.025	6.72	5.26	4.63	4.28	4.04	3.88	3.76	3.66	3.59	3.53
	.01	9.65	7.21	6.22	5.67	5.32	5.07	4.89	4.74	4.63	4.54
	.005	12.23	8.91	7.60	6.88	6.42	6.10	5.86	5.68	5.54	5.42
	.001	19.69	13.81	11.56	10.35	9.58	9.05	8.66	8.35	8.12	7.92
12	.25	1.46	1.56	1.56	1.55	1.54	1.53	1.52	1.51	1.51	1.50
	.10	3.18	2.81	2.61	2.48	2.39	2.33	2.28	2.24	2.21	2.19
	.05	4.75	3.89	3.49	3.26	3.11	3.00	2.91	2.85	2.80	2.75
	.025	6.55	5.10	4.47	4.12	3.89	3.73	3.61	3.51	3.44	3.37
	.01	9.33	6.93	5.95	5.41	5.06	4.82	4.64	4.50	4.39	4.30
	.005	11.75	8.51	7.23	6.52	6.07	5.76	5.52	5.35	5.20	5.09
	.001	18.64	12.97	10.80	9.63	8.89	8.38	8.00	7.71	7.48	7.29

TABLE 6 *(continued)*

					df_1							
12	15	20	24	30	40	60	120	240	inf.	a	df_2	
1.68	1.68	1.67	1.67	1.66	1.66	1.65	1.65	1.65	1.65	.25	7
2.67	2.63	2.59	2.58	2.56	2.54	2.51	2.49	2.48	2.47	.10	
3.57	3.51	3.44	3.41	3.38	3.34	3.30	3.27	3.25	3.23	.05	
4.67	4.57	4.47	4.41	4.36	4.31	4.25	4.20	4.17	4.14	.025	
6.47	6.31	6.16	6.07	5.99	5.91	5.82	5.74	5.69	5.65	.01	
8.18	7.97	7.75	7.64	7.53	7.42	7.31	7.19	7.13	7.08	.005	
13.71	13.32	12.93	12.73	12.53	12.33	12.12	11.91	11.80	11.70	.001	
1.62	1.62	1.61	1.60	1.60	1.59	1.59	1.58	1.58	1.58	.25	8
2.50	2.46	2.42	2.40	2.38	2.36	2.34	2.32	2.30	2.29	.10	
3.28	3.22	3.15	3.12	3.08	3.04	3.01	2.97	2.95	2.93	.05	
4.20	4.10	4.00	3.95	3.89	3.84	3.78	3.73	3.70	3.67	.025	
5.67	5.52	5.36	5.28	5.20	5.12	5.03	4.95	4.90	4.86	.01	
7.01	6.81	6.61	6.50	6.40	6.29	6.18	6.06	6.01	5.95	.005	
11.19	10.84	10.48	10.30	10.11	9.92	9.73	9.53	9.43	9.33	.001	
1.58	1.57	1.56	1.56	1.55	1.54	1.64	1.53	1.53	1.53	.25	9
2.38	2.34	2.30	2.28	2.25	2.23	2.21	2.18	2.17	2.16	.10	
3.07	3.01	2.94	2.90	2.86	2.83	2.79	2.75	2.73	2.71	.05	
3.87	3.77	3.67	3.61	3.56	3.51	3.45	3.39	3.36	3.33	.025	
5.11	4.96	4.81	4.73	4.65	4.57	4.48	4.40	4.35	4.31	.01	
6.23	6.03	5.83	5.73	5.62	5.52	5.41	5.30	5.24	5.19	.005	
9.57	9.24	8.90	8.72	8.55	8.37	8.19	8.00	7.91	7.81	.001	
1.54	1.53	1.52	1.52	1.51	1.51	1.50	1.49	1.49	1.48	.25	10
2.28	2.24	2.20	2.18	2.16	2.13	2.11	2.08	2.07	2.06	.10	
2.91	2.85	2.77	2.74	2.70	2.66	2.62	2.58	2.56	2.54	.05	
3.62	3.52	3.42	3.37	3.31	3.26	3.20	3.14	3.11	3.08	.025	
4.71	4.56	4.41	4.33	4.25	4.17	4.08	4.00	3.95	3.91	.01	
5.66	5.47	5.27	5.17	5.07	4.97	4.86	4.75	4.69	4.64	.005	
8.45	8.13	7.80	7.64	7.47	7.30	7.12	6.94	6.85	6.76	.001	
1.51	1.50	1.49	1.49	1.48	1.47	1.47	1.46	1.45	1.45	.25	11
2.21	2.17	2.12	2.10	2.08	2.05	2.03	2.00	1.99	1.97	.10	
2.79	2.72	2.65	2.61	2.57	2.53	2.49	2.45	2.43	2.40	.05	
3.43	3.33	3.23	3.17	3.12	3.06	3.00	2.94	2.91	2.88	.025	
4.40	4.25	4.10	4.02	3.94	3.86	3.78	3.69	3.65	3.60	.01	
5.24	5.05	4.86	4.76	4.65	4.55	4.45	4.34	4.28	4.23	.005	
7.63	7.32	7.01	6.85	6.68	6.52	6.35	6.18	6.09	6.00	.001	
1.49	1.48	1.47	1.46	1.45	1.45	1.44	1.43	1.43	1.42	.25	12
2.15	2.10	2.06	2.04	2.01	1.99	1.96	1.93	1.92	1.90	.10	
2.69	2.62	2.54	2.51	2.47	2.43	2.38	2.34	2.32	2.30	.05	
3.28	3.18	3.07	3.02	2.96	2.91	2.85	2.79	2.76	2.72	.025	
4.16	4.01	3.86	3.78	3.70	3.62	3.54	3.45	3.41	3.36	.01	
4.91	4.72	4.53	4.43	4.33	4.23	4.12	4.01	3.96	3.90	.005	
7.00	6.71	6.40	6.25	6.09	5.93	5.76	5.59	5.51	5.42	.001	

TABLE 6 **Percentage Points of the F Distribution (df₂ Between 13 and 18) (continued)**

| df₂ | a | \multicolumn{10}{c}{df₁} |
		1	2	3	4	5	6	7	8	9	10
13	.25	1.45	1.55	1.55	1.53	1.52	1.51	1.50	1.49	1.49	1.48
	.10	3.14	2.76	2.56	2.43	2.35	2.28	2.23	2.20	2.16	2.14
	.05	4.67	3.81	3.41	3.18	3.03	2.92	2.83	2.77	2.71	2.67
	.025	6.41	4.97	4.35	4.00	3.77	3.60	3.48	3.39	3.31	3.25
	.01	9.07	6.70	5.74	5.21	4.86	4.62	4.44	4.30	4.19	4.10
	.005	11.37	8.19	6.93	6.23	5.79	5.48	5.25	5.08	4.94	4.82
	.001	17.82	12.31	10.21	9.07	8.35	7.86	7.49	7.21	6.98	6.80
14	.25	1.44	1.53	1.53	1.52	1.51	1.50	1.49	1.48	1.47	1.46
	.10	3.10	2.73	2.52	2.39	2.31	2.24	2.19	2.15	2.12	2.10
	.05	4.60	3.74	3.34	3.11	2.96	2.85	2.76	2.70	2.65	2.60
	.025	6.30	4.86	4.24	3.89	3.66	3.50	3.38	3.29	3.21	3.15
	.01	8.86	6.51	5.56	5.04	4.69	4.46	4.28	4.14	4.03	3.94
	.005	11.06	7.92	6.68	6.00	5.56	5.26	5.03	4.86	4.72	4.60
	.001	17.14	11.78	9.73	8.62	7.92	7.44	7.08	6.80	6.58	6.40
15	.25	1.43	1.52	1.52	1.51	1.49	1.48	1.47	1.46	1.46	1.45
	.10	3.07	2.70	2.49	2.36	2.27	2.21	2.16	2.12	2.09	2.06
	.05	4.54	3.68	3.29	3.06	2.90	2.79	2.71	2.64	2.59	2.54
	.025	6.20	4.77	4.15	3.80	3.58	3.41	3.29	3.20	3.12	3.06
	.01	8.68	6.36	5.42	4.89	4.56	4.32	4.14	4.00	3.89	3.80
	.005	10.80	7.70	6.48	5.80	5.37	5.07	4.85	4.67	4.54	4.42
	.001	16.59	11.34	9.34	8.25	7.57	7.09	6.74	6.47	6.26	6.08
16	.25	1.42	1.51	1.51	1.50	1.48	1.47	1.46	1.45	1.44	1.44
	.10	3.05	2.67	2.46	2.33	2.24	2.18	2.13	2.09	2.06	2.03
	.05	4.49	3.63	3.24	3.01	2.85	2.74	2.66	2.59	2.54	2.49
	.025	6.12	4.69	4.08	3.73	3.50	3.34	3.22	3.12	3.05	2.99
	.01	8.53	6.23	5.29	4.77	4.44	4.20	4.03	3.89	3.78	3.69
	.005	10.58	7.51	6.30	5.64	5.21	4.91	4.69	4.52	4.38	4.27
	.001	16.12	10.97	9.01	7.94	7.27	6.80	6.46	6.19	5.98	5.81
17	.25	1.42	1.51	1.50	1.49	1.47	1.46	1.45	1.44	1.43	1.43
	.10	3.03	2.64	2.44	2.31	2.22	2.15	2.10	2.06	2.03	2.00
	.05	4.45	3.59	3.20	2.96	2.81	2.70	2.61	2.55	2.49	2.45
	.025	6.04	4.62	4.01	3.66	3.44	3.28	3.16	3.06	2.98	2.92
	.01	8.40	6.11	5.18	4.67	4.34	4.10	3.93	3.79	3.68	3.59
	.005	10.38	7.35	6.16	5.50	5.07	4.78	4.56	4.39	4.25	4.14
	.001	15.72	10.66	8.73	7.68	7.02	6.56	6.22	5.96	5.75	5.58
18	.25	1.41	1.50	1.49	1.48	1.46	1.45	1.44	1.43	1.42	1.42
	.10	3.01	2.62	2.42	2.29	2.20	2.13	2.08	2.04	2.00	1.98
	.05	4.41	3.55	3.16	2.93	2.77	2.66	2.58	2.51	2.46	2.41
	.025	5.98	4.56	3.95	3.61	3.38	3.22	3.10	3.01	2.93	2.87
	.01	8.29	6.01	5.09	4.58	4.25	4.01	3.84	3.71	3.60	3.51
	.005	10.22	7.21	6.03	5.37	4.96	4.66	4.44	4.28	4.14	4.03
	.001	15.38	10.39	8.49	7.46	6.81	6.35	6.02	5.76	5.56	5.39

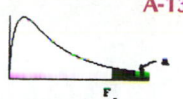

TABLE 6 *(continued)*

12	15	20	24	30	40	60	120	240	inf.	a	df$_2$
				df$_1$							
1.47	1.46	1.45	1.44	1.43	1.42	1.42	1.41	1.40	1.40	.25	13
2.10	2.05	2.01	1.98	1.96	1.93	1.90	1.88	1.86	1.85	.10	
2.60	2.53	2.46	2.42	2.38	2.34	2.30	2.25	2.23	2.21	.05	
3.15	3.05	2.95	2.89	2.84	2.78	2.72	2.66	2.63	2.60	.025	
3.96	3.82	3.66	3.59	3.51	3.43	3.34	3.25	3.21	3.17	.01	
4.64	4.46	4.27	4.17	4.07	3.97	3.87	3.76	3.70	3.65	.005	
6.52	6.23	5.93	5.78	5.63	5.47	5.30	5.14	5.05	4.97	.001	
1.45	1.44	1.43	1.42	1.41	1.41	1.40	1.39	1.38	1.38	.25	14
2.05	2.01	1.96	1.94	1.91	1.89	1.86	1.83	1.81	1.80	.10	
2.53	2.46	2.39	2.35	2.31	2.27	2.22	2.18	2.15	2.13	.05	
3.05	2.95	2.84	2.79	2.73	2.67	2.61	2.55	2.52	2.49	.025	
3.80	3.66	3.51	3.43	3.35	3.27	3.18	3.09	3.05	3.00	.01	
4.43	4.25	4.06	3.96	3.86	3.76	3.66	3.55	3.49	3.44	.005	
6.13	5.85	5.56	5.41	5.25	5.10	4.94	4.77	4.69	4.60	.001	
1.44	1.43	1.41	1.41	1.40	1.39	1.38	1.37	1.36	1.36	.25	15
2.02	1.97	1.92	1.90	1.87	1.85	1.82	1.79	1.77	1.76	.10	
2.48	2.40	2.33	2.29	2.25	2.20	2.16	2.11	2.09	2.07	.05	
2.96	2.86	2.76	2.70	2.64	2.59	2.52	2.46	2.43	2.40	.025	
3.67	3.52	3.37	3.29	3.21	3.13	3.05	2.96	2.91	2.87	.01	
4.25	4.07	3.88	3.79	3.69	3.58	3.48	3.37	3.32	3.26	.005	
5.81	5.54	5.25	5.10	4.95	4.80	4.64	4.47	4.39	4.31	.001	
1.43	1.41	1.40	1.39	1.38	1.37	1.36	1.35	1.35	1.34	.25	16
1.99	1.94	1.89	1.87	1.84	1.81	1.78	1.75	1.73	1.72	.10	
2.42	2.35	2.28	2.24	2.19	2.15	2.11	2.06	2.03	2.01	.05	
2.89	2.79	2.68	2.63	2.57	2.51	2.45	2.38	2.35	2.32	.025	
3.55	3.41	3.26	3.18	3.10	3.02	2.93	2.84	2.80	2.75	.01	
4.10	3.92	3.73	3.64	3.54	3.44	3.33	3.22	3.17	3.11	.005	
5.55	5.27	4.99	4.85	4.70	4.54	4.39	4.23	4.14	4.06	.001	
1.41	1.40	1.39	1.38	1.37	1.36	1.35	1.34	1.33	1.33	.25	17
1.96	1.91	1.86	1.84	1.81	1.78	1.75	1.72	1.70	1.69	.10	
2.38	2.31	2.23	2.19	2.15	2.10	2.06	2.01	1.99	1.96	.05	
2.82	2.72	2.62	2.56	2.50	2.44	2.38	2.32	2.28	2.25	.025	
3.46	3.31	3.16	3.08	3.00	2.92	2.83	2.75	2.70	2.65	.01	
3.97	3.79	3.61	3.51	3.41	3.31	3.21	3.10	3.04	2.98	.005	
5.32	5.05	4.78	4.63	4.48	4.33	4.18	4.02	3.93	3.85	.001	
1.40	1.39	1.38	1.37	1.36	1.35	1.34	1.33	1.32	1.32	.25	18
1.93	1.89	1.84	1.81	1.78	1.75	1.72	1.69	1.67	1.66	.10	
2.34	2.27	2.19	2.15	2.11	2.06	2.02	1.97	1.94	1.92	.05	
2.77	2.67	2.56	2.50	2.44	2.38	2.32	2.26	2.22	2.19	.025	
3.37	3.23	3.08	3.00	2.92	2.84	2.75	2.66	2.61	2.57	.01	
3.86	3.68	3.50	3.40	3.30	3.20	3.10	2.99	2.93	2.87	.005	
5.13	4.87	4.59	4.45	4.30	4.15	4.00	3.84	3.75	3.67	.001	

TABLE 6 Percentage Points of the *F* Distribution (df$_2$ Between 19 and 24) *(continued)*

df$_2$	a	1	2	3	4	5	6	7	8	9	10
19	.25	1.41	1.49	1.49	1.47	1.46	1.44	1.43	1.42	1.41	1.41
	.10	2.99	2.61	2.40	2.27	2.18	2.11	2.06	2.02	1.98	1.96
	.05	4.38	3.52	3.13	2.90	2.74	2.63	2.54	2.48	2.42	2.38
	.025	5.92	4.51	3.90	3.56	3.33	3.17	3.05	2.96	2.88	2.82
	.01	8.18	5.93	5.01	4.50	4.17	3.94	3.77	3.63	3.52	3.43
	.005	10.07	7.09	5.92	5.27	4.85	4.56	4.34	4.18	4.04	3.93
	.001	15.08	10.16	8.28	7.27	6.62	6.18	5.85	5.59	5.39	5.22
20	.25	1.40	1.49	1.48	1.47	1.45	1.44	1.43	1.42	1.41	1.40
	.10	2.97	2.59	2.38	2.25	2.16	2.09	2.04	2.00	1.96	1.94
	.05	4.35	3.49	3.10	2.87	2.71	2.60	2.51	2.45	2.39	2.35
	.025	5.87	4.46	3.86	3.51	3.29	3.13	3.01	2.91	2.84	2.77
	.01	8.10	5.85	4.94	4.43	4.10	3.87	3.70	3.56	3.46	3.37
	.005	9.94	6.99	5.82	5.17	4.76	4.47	4.26	4.09	3.96	3.85
	.001	14.82	9.95	8.10	7.10	6.46	6.02	5.69	5.44	5.24	5.08
21	.25	1.40	1.48	1.48	1.46	1.44	1.43	1.42	1.41	1.40	1.39
	.10	2.96	2.57	2.36	2.23	2.14	2.08	2.02	1.98	1.95	1.92
	.05	4.32	3.47	3.07	2.84	2.68	2.57	2.49	2.42	2.37	2.32
	.025	5.83	4.42	3.82	3.48	3.25	3.09	2.97	2.87	2.80	2.73
	.01	8.02	5.78	4.87	4.37	4.04	3.81	3.64	3.51	3.40	3.31
	.005	9.83	6.89	5.73	5.09	4.68	4.39	4.18	4.01	3.88	3.77
	.001	14.59	9.77	7.94	6.95	6.32	5.88	5.56	5.31	5.11	4.95
22	.25	1.40	1.48	1.47	1.45	1.44	1.42	1.41	1.40	1.39	1.39
	.10	2.95	2.56	2.35	2.22	2.13	2.06	2.01	1.97	1.93	1.90
	.05	4.30	3.44	3.05	2.82	2.66	2.55	2.46	2.40	2.34	2.30
	.025	5.79	4.38	3.78	3.44	3.22	3.05	2.93	2.84	2.76	2.70
	.01	7.95	5.72	4.82	4.31	3.99	3.76	3.59	3.45	3.35	3.26
	.005	9.73	6.81	5.65	5.02	4.61	4.32	4.11	3.94	3.81	3.70
	.001	14.38	9.61	7.80	6.81	6.19	5.76	5.44	5.19	4.99	4.83
23	.25	1.39	1.47	1.47	1.45	1.43	1.42	1.41	1.40	1.39	1.38
	.10	2.94	2.55	2.34	2.21	2.11	2.05	1.99	1.95	1.92	1.89
	.05	4.28	3.42	3.03	2.80	2.64	2.53	2.44	2.37	2.32	2.27
	.025	5.75	4.35	3.75	3.41	3.18	3.02	2.90	2.81	2.73	2.67
	.01	7.88	5.66	4.76	4.26	3.94	3.71	3.54	3.41	3.30	3.21
	.005	9.63	6.73	5.58	4.95	4.54	4.26	4.05	3.88	3.75	3.64
	.001	14.20	9.47	7.67	6.70	6.08	5.65	5.33	5.09	4.89	4.73
24	.25	1.39	1.47	1.46	1.44	1.43	1.41	1.40	1.39	1.38	1.38
	.10	2.93	2.54	2.33	2.19	2.10	2.04	1.98	1.94	1.91	1.88
	.05	4.26	3.40	3.01	2.78	2.62	2.51	2.42	2.36	2.30	2.25
	.025	5.72	4.32	3.72	3.38	3.15	2.99	2.87	2.78	2.70	2.64
	.01	7.82	5.61	4.72	4.22	3.90	3.67	3.50	3.36	3.26	3.17
	.005	9.55	6.66	5.52	4.89	4.49	4.20	3.99	3.83	3.69	3.59
	.001	14.03	9.34	7.55	6.59	5.98	5.55	5.23	4.99	4.80	4.64

TABLE 6 (continued)

12	15	20	24	30	40	60	120	240	inf.	a	df_1 / df_2
1.40	1.38	1.37	1.36	1.35	1.34	1.33	1.32	1.31	1.30	.25	19
1.91	1.86	1.81	1.79	1.76	1.73	1.70	1.67	1.65	1.63	.10	
2.31	2.23	2.16	2.11	2.07	2.03	1.98	1.93	1.90	1.88	.05	
2.72	2.62	2.51	2.45	2.39	2.33	2.27	2.20	2.17	2.13	.025	
3.30	3.15	3.00	2.92	2.84	2.76	2.67	2.58	2.54	2.49	.01	
3.76	3.59	3.40	3.31	3.21	3.11	3.00	2.89	2.83	2.78	.005	
4.97	4.70	4.43	4.29	4.14	3.99	3.84	3.68	3.60	3.51	.001	
1.39	1.37	1.36	1.35	1.34	1.33	1.32	1.31	1.30	1.29	.25	20
1.89	1.84	1.79	1.77	1.74	1.71	1.68	1.64	1.63	1.61	.10	
2.28	2.20	2.12	2.08	2.04	1.99	1.95	1.90	1.87	1.84	.05	
2.68	2.57	2.46	2.41	2.35	2.29	2.22	2.16	2.12	2.09	.025	
3.23	3.09	2.94	2.86	2.78	2.69	2.61	2.52	2.47	2.42	.01	
3.68	3.50	3.32	3.22	3.12	3.02	2.92	2.81	2.75	2.69	.005	
4.82	4.56	4.29	4.15	4.00	3.86	3.70	3.54	3.46	3.38	.001	
1.38	1.37	1.35	1.34	1.33	1.32	1.31	1.30	1.29	1.28	.25	21
1.87	1.83	1.78	1.75	1.72	1.69	1.66	1.62	1.60	1.59	.10	
2.25	2.18	2.10	2.05	2.01	1.96	1.92	1.87	1.84	1.81	.05	
2.64	2.53	2.42	2.37	2.31	2.25	2.18	2.11	2.08	2.04	.025	
3.17	3.03	2.88	2.80	2.72	2.64	2.55	2.46	2.41	2.36	.01	
3.60	3.43	3.24	3.15	3.05	2.95	2.84	2.73	2.67	2.61	.005	
4.70	4.44	4.17	4.03	3.88	3.74	3.58	3.42	3.34	3.26	.001	
1.37	1.36	1.34	1.33	1.32	1.31	1.30	1.29	1.28	1.28	.25	22
1.86	1.81	1.76	1.73	1.70	1.67	1.64	1.60	1.59	1.57	.10	
2.23	2.15	2.07	2.03	1.98	1.94	1.89	1.84	1.81	1.78	.05	
2.60	2.50	2.39	2.33	2.27	2.21	2.14	2.08	2.04	2.00	.025	
3.12	2.98	2.83	2.75	2.67	2.58	2.50	2.40	2.35	2.31	.01	
3.54	3.36	3.18	3.08	2.98	2.88	2.77	2.66	2.60	2.55	.005	
4.58	4.33	4.06	3.92	3.78	3.63	3.48	3.32	3.23	3.15	.001	
1.37	1.35	1.34	1.33	1.32	1.31	1.30	1.28	1.28	1.27	.25	23
1.84	1.80	1.74	1.72	1.69	1.66	1.62	1.59	1.57	1.55	.10	
2.20	2.13	2.05	2.01	1.96	1.91	1.86	1.81	1.79	1.76	.05	
2.57	2.47	2.36	2.30	2.24	2.18	2.11	2.04	2.01	1.97	.025	
3.07	2.93	2.78	2.70	2.62	2.54	2.45	2.35	2.31	2.26	.01	
3.47	3.30	3.12	3.02	2.92	2.82	2.71	2.60	2.54	2.48	.005	
4.48	4.23	3.96	3.82	3.68	3.53	3.38	3.22	3.14	3.05	.001	
1.36	1.35	1.33	1.32	1.31	1.30	1.29	1.28	1.27	1.26	.25	24
1.83	1.78	1.73	1.70	1.67	1.64	1.61	1.57	1.55	1.53	.10	
2.18	2.11	2.03	1.98	1.94	1.89	1.84	1.79	1.76	1.73	.05	
2.54	2.44	2.33	2.27	2.21	2.15	2.08	2.01	1.97	1.94	.025	
3.03	2.89	2.74	2.66	2.58	2.49	2.40	2.31	2.26	2.21	.01	
3.42	3.25	3.06	2.97	2.87	2.77	2.66	2.55	2.49	2.43	.005	
4.39	4.14	3.87	3.74	3.59	3.45	3.29	3.14	3.05	2.97	.001	

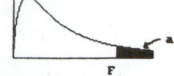

TABLE 6 Percentage Points of the *F* Distribution (df$_2$ Between 25 and 30) *(continued)*

df$_2$	a	1	2	3	4	5	6	7	8	9	10
						df$_1$					
25	.25	1.39	1.47	1.46	1.44	1.42	1.41	1.40	1.39	1.38	1.37
	.10	2.92	2.53	2.32	2.18	2.09	2.02	1.97	1.93	1.89	1.87
	.05	4.24	3.39	2.99	2.76	2.60	2.49	2.40	2.34	2.28	2.24
	.025	5.69	4.29	3.69	3.35	3.13	2.97	2.85	2.75	2.68	2.61
	.01	7.77	5.57	4.68	4.18	3.85	3.63	3.46	3.32	3.22	3.13
	.005	9.48	6.60	5.46	4.84	4.43	4.15	3.94	3.78	3.64	3.54
	.001	13.88	9.22	7.45	6.49	5.89	5.46	5.15	4.91	4.71	4.56
26	.25	1.38	1.46	1.45	1.44	1.42	1.41	1.39	1.38	1.37	1.37
	.10	2.91	2.52	2.31	2.17	2.08	2.01	1.96	1.92	1.88	1.86
	.05	4.23	3.37	2.98	2.74	2.59	2.47	2.39	2.32	2.27	2.22
	.025	5.66	4.27	3.67	3.33	3.10	2.94	2.82	2.73	2.65	2.59
	.01	7.72	5.53	4.64	4.14	3.82	3.59	3.42	3.29	3.18	3.09
	.005	9.41	6.54	5.41	4.79	4.38	4.10	3.89	3.73	3.60	3.49
	.001	13.74	9.12	7.36	6.41	5.80	5.38	5.07	4.83	4.64	4.48
27	.25	1.38	1.46	1.45	1.43	1.42	1.40	1.39	1.38	1.37	1.36
	.10	2.90	2.51	2.30	2.17	2.07	2.00	1.95	1.91	1.87	1.85
	.05	4.21	3.35	2.96	2.73	2.57	2.46	2.37	2.31	2.25	2.20
	.025	5.63	4.24	3.65	3.31	3.08	2.92	2.80	2.71	2.63	2.57
	.01	7.68	5.49	4.60	4.11	3.78	3.56	3.39	3.26	3.15	3.06
	.005	9.34	6.49	5.36	4.74	4.34	4.06	3.85	3.69	3.56	3.45
	.001	13.61	9.02	7.27	6.33	5.73	5.31	5.00	4.76	4.57	4.41
28	.25	1.38	1.46	1.45	1.43	1.41	1.40	1.39	1.38	1.37	1.36
	.10	2.89	2.50	2.29	2.16	2.06	2.00	1.94	1.90	1.87	1.84
	.05	4.20	3.34	2.95	2.71	2.56	2.45	2.36	2.29	2.24	2.19
	.025	5.61	4.22	3.63	3.29	3.06	2.90	2.78	2.69	2.61	2.55
	.01	7.64	5.45	4.57	4.07	3.75	3.53	3.36	3.23	3.12	3.03
	.005	9.28	6.44	5.32	4.70	4.30	4.02	3.81	3.65	3.52	3.41
	.001	13.50	8.93	7.19	6.25	5.66	5.24	4.93	4.69	4.50	4.35
29	.25	1.38	1.45	1.45	1.43	1.41	1.40	1.38	1.37	1.36	1.35
	.10	2.89	2.50	2.28	2.15	2.06	1.99	1.93	1.89	1.86	1.83
	.05	4.18	3.33	2.93	2.70	2.55	2.43	2.35	2.28	2.22	2.18
	.025	5.59	4.20	3.61	3.27	3.04	2.88	2.76	2.67	2.59	2.53
	.01	7.60	5.42	4.54	4.04	3.73	3.50	3.33	3.20	3.09	3.00
	.005	9.23	6.40	5.28	4.66	4.26	3.98	3.77	3.61	3.48	3.38
	.001	13.39	8.85	7.12	6.19	5.59	5.18	4.87	4.64	4.45	4.29
30	.25	1.38	1.45	1.44	1.42	1.41	1.39	1.38	1.37	1.36	1.35
	.10	2.88	2.49	2.28	2.14	2.05	1.98	1.93	1.88	1.85	1.82
	.05	4.17	3.32	2.92	2.69	2.53	2.42	2.33	2.27	2.21	2.16
	.025	5.57	4.18	3.59	3.25	3.03	2.87	2.75	2.65	2.57	2.51
	.01	7.56	5.39	4.51	4.02	3.70	3.47	3.30	3.17	3.07	2.98
	.005	9.18	6.35	5.24	4.62	4.23	3.95	3.74	3.58	3.45	3.34
	.001	13.29	8.77	7.05	6.12	5.53	5.12	4.82	4.58	4.39	4.24

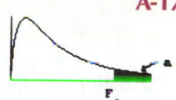

TABLE 6 *(continued)*

12	15	20	24	30	40	60	120	240	inf.	a	df_2
1.36	1.34	1.33	1.32	1.31	1.29	1.28	1.27	1.26	1.25	.25	25
1.82	1.77	1.72	1.69	1.66	1.63	1.59	1.56	1.54	1.52	.10	
2.16	2.09	2.01	1.96	1.92	1.87	1.82	1.77	1.74	1.71	.05	
2.51	2.41	2.30	2.24	2.18	2.12	2.05	1.98	1.94	1.91	.025	
2.99	2.85	2.70	2.62	2.54	2.45	2.36	2.27	2.22	2.17	.01	
3.37	3.20	3.01	2.92	2.82	2.72	2.61	2.50	2.44	2.38	.005	
4.31	4.06	3.79	3.66	3.52	3.37	3.22	3.06	2.98	2.89	.001	
1.35	1.34	1.32	1.31	1.30	1.29	1.28	1.26	1.26	1.25	.25	26
1.81	1.76	1.71	1.68	1.65	1.61	1.58	1.54	1.52	1.50	.10	
2.15	2.07	1.99	1.95	1.90	1.85	1.80	1.75	1.72	1.69	.05	
2.49	2.39	2.28	2.22	2.16	2.09	2.03	1.95	1.92	1.88	.025	
2.96	2.81	2.66	2.58	2.50	2.42	2.33	2.23	2.18	2.13	.01	
3.33	3.15	2.97	2.87	2.77	2.67	2.56	2.45	2.39	2.33	.005	
4.24	3.99	3.72	3.59	3.44	3.30	3.15	2.99	2.90	2.82	.001	
1.35	1.33	1.32	1.31	1.30	1.28	1.27	1.26	1.25	1.24	.25	27
1.80	1.75	1.70	1.67	1.64	1.60	1.57	1.53	1.51	1.49	.10	
2.13	2.06	1.97	1.93	1.88	1.84	1.79	1.73	1.70	1.67	.05	
2.47	2.36	2.25	2.19	2.13	2.07	2.00	1.93	1.89	1.85	.025	
2.93	2.78	2.63	2.55	2.47	2.38	2.29	2.20	2.15	2.10	.01	
3.28	3.11	2.93	2.83	2.73	2.63	2.52	2.41	2.35	2.29	.005	
4.17	3.92	3.66	3.52	3.38	3.23	3.08	2.92	2.84	2.75	.001	
1.34	1.33	1.31	1.30	1.29	1.28	1.27	1.25	1.24	1.24	.25	28
1.79	1.74	1.69	1.66	1.63	1.59	1.56	1.52	1.50	1.48	.10	
2.12	2.04	1.96	1.91	1.87	1.82	1.77	1.71	1.68	1.65	.05	
2.45	2.34	2.23	2.17	2.11	2.05	1.98	1.91	1.87	1.83	.025	
2.90	2.75	2.60	2.52	2.44	2.35	2.26	2.17	2.12	2.06	.01	
3.25	3.07	2.89	2.79	2.69	2.59	2.48	2.37	2.31	2.25	.005	
4.11	3.86	3.60	3.46	3.32	3.18	3.02	2.86	2.78	2.69	.001	
1.34	1.32	1.31	1.30	1.29	1.27	1.26	1.25	1.24	1.23	.25	29
1.78	1.73	1.68	1.65	1.62	1.58	1.55	1.51	1.49	1.47	.10	
2.10	2.03	1.94	1.90	1.85	1.81	1.75	1.70	1.67	1.64	.05	
2.43	2.32	2.21	2.15	2.09	2.03	1.96	1.89	1.85	1.81	.025	
2.87	2.73	2.57	2.49	2.41	2.33	2.23	2.14	2.09	2.03	.01	
3.21	3.04	2.86	2.76	2.66	2.56	2.45	2.33	2.27	2.21	.005	
4.05	3.80	3.54	3.41	3.27	3.12	2.97	2.81	2.73	2.64	.001	
1.34	1.32	1.30	1.29	1.28	1.27	1.26	1.24	1.23	1.23	.25	30
1.77	1.72	1.67	1.64	1.61	1.57	1.54	1.50	1.48	1.46	.10	
2.09	2.01	1.93	1.89	1.84	1.79	1.74	1.68	1.65	1.62	.05	
2.41	2.31	2.20	2.14	2.07	2.01	1.94	1.87	1.83	1.79	.025	
2.84	2.70	2.55	2.47	2.39	2.30	2.21	2.11	2.06	2.01	.01	
3.18	3.01	2.82	2.73	2.63	2.52	2.42	2.30	2.24	2.18	.005	
4.00	3.75	3.49	3.36	3.22	3.07	2.92	2.76	2.68	2.59	.001	

TABLE 6   **Percentage Points of the F Distribution (df$_2$ at Least 40) (continued)**

df$_2$	a	1	2	3	4	5	6	7	8	9	10
40	.25	1.36	1.44	1.42	1.40	1.39	1.37	1.36	1.35	1.34	1.33
	.10	2.84	2.44	2.23	2.09	2.00	1.93	1.87	1.83	1.79	1.76
	.05	4.08	3.23	2.84	2.61	2.45	2.34	2.25	2.18	2.12	2.08
	.025	5.42	4.05	3.46	3.13	2.90	2.74	2.62	2.53	2.45	2.39
	.01	7.31	5.18	4.31	3.83	3.51	3.29	3.12	2.99	2.89	2.80
	.005	8.83	6.07	4.98	4.37	3.99	3.71	3.51	3.35	3.22	3.12
	.001	12.61	8.25	6.59	5.70	5.13	4.73	4.44	4.21	4.02	3.87
60	.25	1.35	1.42	1.41	1.38	1.37	1.35	1.33	1.32	1.31	1.30
	.10	2.79	2.39	2.18	2.04	1.95	1.87	1.82	1.77	1.74	1.71
	.05	4.00	3.15	2.76	2.53	2.37	2.25	2.17	2.10	2.04	1.99
	.025	5.29	3.93	3.34	3.01	2.79	2.63	2.51	2.41	2.33	2.27
	.01	7.08	4.98	4.13	3.65	3.34	3.12	2.95	2.82	2.72	2.63
	.005	8.49	5.79	4.73	4.14	3.76	3.49	3.29	3.13	3.01	2.90
	.001	11.97	7.77	6.17	5.31	4.76	4.37	4.09	3.86	3.69	3.54
90	.25	1.34	1.41	1.39	1.37	1.35	1.33	1.32	1.31	1.30	1.29
	.10	2.76	2.36	2.15	2.01	1.91	1.84	1.78	1.74	1.70	1.67
	.05	3.95	3.10	2.71	2.47	2.32	2.20	2.11	2.04	1.99	1.94
	.025	5.20	3.84	3.26	2.93	2.71	2.55	2.43	2.34	2.26	2.19
	.01	6.93	4.85	4.01	3.53	3.23	3.01	2.84	2.72	2.61	2.52
	.005	8.28	5.62	4.57	3.99	3.62	3.35	3.15	3.00	2.87	2.77
	.001	11.57	7.47	5.91	5.06	4.53	4.15	3.87	3.65	3.48	3.34
120	.25	1.34	1.40	1.39	1.37	1.35	1.33	1.31	1.30	1.29	1.28
	.10	2.75	2.35	2.13	1.99	1.90	1.82	1.77	1.72	1.68	1.65
	.05	3.92	3.07	2.68	2.45	2.29	2.18	2.09	2.02	1.96	1.91
	.025	5.15	3.80	3.23	2.89	2.67	2.52	2.39	2.30	2.22	2.16
	.01	6.85	4.79	3.95	3.48	3.17	2.96	2.79	2.66	2.56	2.47
	.005	8.18	5.54	4.50	3.92	3.55	3.28	3.09	2.93	2.81	2.71
	.001	11.38	7.32	5.78	4.95	4.42	4.04	3.77	3.55	3.38	3.24
240	.25	1.33	1.39	1.38	1.36	1.34	1.32	1.30	1.29	1.27	1.27
	.10	2.73	2.32	2.10	1.97	1.87	1.80	1.74	1.70	1.65	1.63
	.05	3.88	3.03	2.64	2.41	2.25	2.14	2.04	1.98	1.92	1.87
	.025	5.09	3.75	3.17	2.84	2.62	2.46	2.34	2.25	2.17	2.10
	.01	6.74	4.69	3.86	3.40	3.09	2.88	2.71	2.59	2.48	2.40
	.005	8.03	5.42	4.38	3.82	3.45	3.19	2.99	2.84	2.71	2.61
	.001	11.10	7.11	5.60	4.78	4.25	3.89	3.62	3.41	3.24	3.09
inf.	.25	1.32	1.39	1.37	1.35	1.33	1.31	1.29	1.28	1.27	1.25
	.10	2.71	2.30	2.08	1.94	1.85	1.77	1.72	1.67	1.63	1.60
	.05	3.84	3.00	2.60	2.37	2.21	2.10	2.01	1.94	1.88	1.83
	.025	5.02	3.69	3.12	2.79	2.57	2.41	2.29	2.19	2.11	2.05
	.01	6.63	4.61	3.78	3.32	3.02	2.80	2.64	2.51	2.41	2.32
	.005	7.88	5.30	4.28	3.72	3.35	3.09	2.90	2.74	2.62	2.52
	.001	10.83	6.91	5.42	4.62	4.10	3.74	3.47	3.27	3.10	2.96

TABLE 6   *(continued)*

12	15	20	24	30	40	60	120	240	inf.	a	df$_2$
1.31	1.30	1.28	1.26	1.25	1.24	1.22	1.21	1.20	1.19	.25	40
1.71	1.66	1.61	1.57	1.54	1.51	1.47	1.42	1.40	1.38	.10	
2.00	1.92	1.84	1.79	1.74	1.69	1.64	1.58	1.54	1.51	.05	
2.29	2.18	2.07	2.01	1.94	1.88	1.80	1.72	1.68	1.64	.025	
2.66	2.52	2.37	2.29	2.20	2.11	2.02	1.92	1.86	1.80	.01	
2.95	2.78	2.60	2.50	2.40	2.30	2.18	2.06	2.00	1.93	.005	
3.64	3.40	3.14	3.01	2.87	2.73	2.57	2.41	2.32	2.23	.001	
1.29	1.27	1.25	1.24	1.22	1.21	1.19	1.17	1.16	1.15	.25	60
1.66	1.60	1.54	1.51	1.48	1.44	1.40	1.35	1.32	1.29	.10	
1.92	1.84	1.75	1.70	1.65	1.59	1.53	1.47	1.43	1.39	.05	
2.17	2.06	1.94	1.88	1.82	1.74	1.67	1.58	1.53	1.48	.025	
2.50	2.35	2.20	2.12	2.03	1.94	1.84	1.73	1.67	1.60	.01	
2.74	2.57	2.39	2.29	2.19	2.08	1.96	1.83	1.76	1.69	.005	
3.32	3.08	2.83	2.69	2.55	2.41	2.25	2.08	1.99	1.89	.001	
1.27	1.25	1.23	1.22	1.20	1.19	1.17	1.15	1.13	1.12	.25	90
1.62	1.56	1.50	1.47	1.43	1.39	1.35	1.29	1.26	1.23	.10	
1.86	1.78	1.69	1.64	1.59	1.53	1.46	1.39	1.35	1.30	.05	
2.09	1.98	1.86	1.80	1.73	1.66	1.58	1.48	1.43	1.37	.025	
2.39	2.24	2.09	2.00	1.92	1.82	1.72	1.60	1.53	1.46	.01	
2.61	2.44	2.25	2.15	2.05	1.94	1.82	1.68	1.61	1.52	.005	
3.11	2.88	2.63	2.50	2.36	2.21	2.05	1.87	1.77	1.66	.001	
1.26	1.24	1.22	1.21	1.19	1.18	1.16	1.13	1.12	1.10	.25	120
1.60	1.55	1.48	1.45	1.41	1.37	1.32	1.26	1.23	1.19	.10	
1.83	1.75	1.66	1.61	1.55	1.50	1.43	1.35	1.31	1.25	.05	
2.05	1.94	1.82	1.76	1.69	1.61	1.53	1.43	1.38	1.31	.025	
2.34	2.19	2.03	1.95	1.86	1.76	1.66	1.53	1.46	1.38	.01	
2.54	2.37	2.19	2.09	1.98	1.87	1.75	1.61	1.52	1.43	.005	
3.02	2.78	2.53	2.40	2.26	2.11	1.95	1.77	1.66	1.54	.001	
1.25	1.23	1.21	1.19	1.18	1.16	1.14	1.11	1.09	1.07	.25	240
1.57	1.52	1.45	1.42	1.38	1.33	1.28	1.22	1.18	1.13	.10	
1.79	1.71	1.61	1.56	1.51	1.44	1.37	1.29	1.24	1.17	.05	
2.00	1.89	1.77	1.70	1.63	1.55	1.46	1.35	1.29	1.21	.025	
2.26	2.11	1.96	1.87	1.78	1.68	1.57	1.43	1.35	1.25	.01	
2.45	2.28	2.09	1.99	1.89	1.77	1.64	1.49	1.40	1.28	.005	
2.88	2.65	2.40	2.26	2.12	1.97	1.80	1.61	1.49	1.35	.001	
1.24	1.22	1.19	1.18	1.16	1.14	1.12	1.08	1.06	1.00	.25	inf.
1.55	1.49	1.42	1.38	1.34	1.30	1.24	1.17	1.12	1.00	.10	
1.75	1.67	1.57	1.52	1.46	1.39	1.32	1.22	1.15	1.00	.05	
1.94	1.83	1.71	1.64	1.57	1.48	1.39	1.27	1.19	1.00	.025	
2.18	2.04	1.88	1.79	1.70	1.59	1.47	1.32	1.22	1.00	.01	
2.36	2.19	2.00	1.90	1.79	1.67	1.53	1.36	1.25	1.00	.005	
2.74	2.51	2.27	2.13	1.99	1.84	1.66	1.45	1.31	1.00	.001	

df$_1$

Source: Computed by P. J. Hildebrand.

TABLE 7   Poisson Probabilities (μ Between .1 and 4.0)

y	.1	.2	.3	.4	μ .5	.6	.7	.8	.9	1.0
0	.9048	.8187	.7408	.6703	.6065	.5488	.4966	.4493	.4066	.3679
1	.0905	.1637	.2222	.2681	.3033	.3293	.3476	.3595	.3659	.3679
2	.0045	.0164	.0333	.0536	.0758	.0988	.1217	.1438	.1647	.1839
3	.0002	.0011	.0033	.0072	.0126	.0198	.0284	.0383	.0494	.0613
4	.0000	.0001	.0003	.0007	.0016	.0030	.0050	.0077	.0111	.0153
5	.0000	.0000	.0000	.0001	.0002	.0004	.0007	.0012	.0020	.0031
6	.0000	.0000	.0000	.0000	.0000	.0000	.0001	.0002	.0003	.0005

y	1.1	1.2	1.3	1.4	μ 1.5	1.6	1.7	1.8	1.9	2.0
0	.3329	.3012	.2725	.2466	.2231	.2019	.1827	.1653	.1496	.1353
1	.3662	.3614	.3543	.3452	.3347	.3230	.3106	.2975	.2842	.2707
2	.2014	.2169	.2303	.2417	.2510	.2584	.2640	.2678	.2700	.2707
3	.0738	.0867	.0998	.1128	.1255	.1378	.1496	.1607	.1710	.1804
4	.0203	.0260	.0324	.0395	.0471	.0551	.0636	.0723	.0812	.0902
5	.0045	.0062	.0084	.0111	.0141	.0176	.0216	.0260	.0309	.0361
6	.0008	.0012	.0018	.0026	.0035	.0047	.0061	.0078	.0098	.0120
7	.0001	.0002	.0003	.0005	.0008	.0011	.0015	.0020	.0027	.0034
8	.0000	.0000	.0001	.0001	.0001	.0002	.0003	.0005	.0006	.0009

y	2.1	2.2	2.3	2.4	μ 2.5	2.6	2.7	2.8	2.9	3.0
0	.1225	.1108	.1003	.0907	.0821	.0743	.0672	.0608	.0550	.0498
1	.2572	.2438	.2306	.2177	.2052	.1931	.1815	.1703	.1596	.1494
2	.2700	.2681	.2652	.2613	.2565	.2510	.2450	.2384	.2314	.2240
3	.1890	.1966	.2033	.2090	.2138	.2176	.2205	.2225	.2237	.2240
4	.0992	.1082	.1169	.1254	.1336	.1414	.1488	.1557	.1622	.1680
5	.0417	.0476	.0538	.0602	.0668	.0735	.0804	.0872	.0940	.1008
6	.0146	.0174	.0206	.0241	.0278	.0319	.0362	.0407	.0455	.0504
7	.0044	.0055	.0068	.0083	.0099	.0118	.0139	.0163	.0188	.0216
8	.0011	.0015	.0019	.0025	.0031	.0038	.0047	.0057	.0068	.0081
9	.0003	.0004	.0005	.0007	.0009	.0011	.0014	.0018	.0022	.0027
10	.0001	.0001	.0001	.0002	.0002	.0003	.0004	.0005	.0006	.0008
11	.0000	.0000	.0000	.0000	.0000	.0001	.0001	.0001	.0002	.0002

y	3.1	3.2	3.3	3.4	μ 3.5	3.6	3.7	3.8	3.9	4.0
0	.0450	.0408	.0369	.0334	.0302	.0273	.0247	.0224	.0202	.0183
1	.1397	.1304	.1217	.1135	.1057	.0984	.0915	.0850	.0789	.0733
2	.2165	.2087	.2008	.1929	.1850	.1771	.1692	.1615	.1539	.1465
3	.2237	.2226	.2209	.2186	.2158	.2125	.2087	.2046	.2001	.1954
4	.1733	.1781	.1823	.1858	.1888	.1912	.1931	.1944	.1951	.1954
5	.1075	.1140	.1203	.1264	.1322	.1377	.1429	.1477	.1522	.1563
6	.0555	.0608	.0662	.0716	.0771	.0826	.0881	.0936	.0989	.1042

T A B L E 7 Poisson Probabilities (μ Between 3.1 and 10.0) (continued)

y	3.1	3.2	3.3	3.4	3.5	3.6	3.7	3.8	3.9	4.0
7	.0246	.0278	.0312	.0348	.0385	.0425	.0466	.0508	.0551	.0595
8	.0095	.0111	.0129	.0148	.0169	.0191	.0215	.0241	.0269	.0298
9	.0033	.0040	.0047	.0056	.0066	.0076	.0089	.0102	.0116	.0132
10	.0010	.0013	.0016	.0019	.0023	.0028	.0033	.0039	.0045	.0053
11	.0003	.0004	.0005	.0006	.0007	.0009	.0011	.0013	.0016	.0019
12	.0001	.0001	.0001	.0002	.0002	.0003	.0003	.0004	.0005	.0006
13	.0000	.0000	.0000	.0000	.0001	.0001	.0001	.0001	.0002	.0002

y	4.1	4.2	4.3	4.4	4.5	4.6	4.7	4.8	4.9	5.0
0	.0166	.0150	.0136	.0123	.0111	.0101	.0091	.0082	.0074	.0067
1	.0679	.0630	.0583	.0540	.0500	.0462	.0427	.0395	.0365	.0337
2	.1393	.1323	.1254	.1188	.1125	.1063	.1005	.0948	.0894	.0842
3	.1904	.1852	.1798	.1743	.1687	.1631	.1574	.1517	.1460	.1404
4	.1951	.1944	.1933	.1917	.1898	.1875	.1849	.1820	.1789	.1755
5	.1600	.1633	.1662	.1687	.1708	.1725	.1738	.1747	.1753	.1755
6	.1093	.1143	.1191	.1237	.1281	.1323	.1362	.1398	.1432	.1462
7	.0640	.0686	.0732	.0778	.0824	.0869	.0914	.0959	.1002	.1044
8	.0328	.0360	.0393	.0428	.0463	.0500	.0537	.0575	.0614	.0653
9	.0150	.0168	.0188	.0209	.0232	.0255	.0281	.0307	.0334	.0363
10	.0061	.0071	.0081	.0092	.0104	.0118	.0132	.0147	.0164	.0181
11	.0023	.0027	.0032	.0037	.0043	.0049	.0056	.0064	.0073	.0082
12	.0008	.0009	.0011	.0013	.0016	.0019	.0022	.0026	.0030	.0034
13	.0002	.0003	.0004	.0005	.0006	.0007	.0008	.0009	.0011	.0013
14	.0001	.0001	.0001	.0001	.0002	.0002	.0003	.0003	.0004	.0005
15	.0000	.0000	.0000	.0000	.0001	.0001	.0001	.0001	.0001	.0002

y	5.5	6.0	6.5	7.0	7.5	8.0	8.5	9.0	9.5	10.0
0	.0041	.0025	.0015	.0009	.0006	.0003	.0002	.0001	.0001	.0000
1	.0225	.0149	.0098	.0064	.0041	.0027	.0017	.0011	.0007	.0005
2	.0618	.0446	.0318	.0223	.0156	.0107	.0074	.0050	.0034	.0023
3	.1133	.0892	.0688	.0521	.0389	.0286	.0208	.0150	.0107	.0076
4	.1558	.1339	.1118	.0912	.0729	.0573	.0443	.0337	.0254	.0189
5	.1714	.1606	.1454	.1277	.1094	.0916	.0752	.0607	.0483	.0378
6	.1571	.1606	.1575	.1490	.1367	.1221	.1066	.0911	.0764	.0631
7	.1234	.1377	.1462	.1490	.1465	.1396	.1294	.1171	.1037	.0901
8	.0849	.1033	.1188	.1304	.1373	.1396	.1375	.1318	.1232	.1126
9	.0519	.0688	.0858	.1014	.1144	.1241	.1299	.1318	.1300	.1251
10	.0285	.0413	.0558	.0710	.0858	.0993	.1104	.1186	.1235	.1251
11	.0143	.0225	.0330	.0452	.0585	.0722	.0853	.0970	.1067	.1137
12	.0065	.0113	.0179	.0263	.0366	.0481	.0604	.0728	.0844	.0948
13	.0028	.0052	.0089	.0142	.0211	.0296	.0395	.0504	.0617	.0729
14	.0011	.0022	.0041	.0071	.0113	.0169	.0240	.0324	.0419	.0521
15	.0004	.0009	.0018	.0033	.0057	.0090	.0136	.0194	.0265	.0347

TABLE 7 Poisson Probabilities (μ Between 5.5 and 20.0) *(continued)*

y	μ 5.5	6.0	6.5	7.0	7.5	8.0	8.5	9.0	9.5	10.0
16	.0001	.0003	.0007	.0014	.0026	.0045	.0072	.0109	.0157	.0217
17	.0000	.0001	.0003	.0006	.0012	.0021	.0036	.0058	.0088	.0128
18	.0000	.0000	.0001	.0002	.0005	.0009	.0017	.0029	.0046	.0071
19	.0000	.0000	.0000	.0001	.0002	.0004	.0008	.0014	.0023	.0037
20	.0000	.0000	.0000	.0000	.0001	.0002	.0003	.0006	.0011	.0019
21	.0000	.0000	.0000	.0000	.0000	.0001	.0001	.0003	.0005	.0009
22	.0000	.0000	.0000	.0000	.0000	.0000	.0001	.0001	.0002	.0004
23	.0000	.0000	.0000	.0000	.0000	.0000	.0000	.0000	.0001	.0002

y	μ 11.0	12.0	13.0	14.0	15.0	16.0	17.0	18.0	19.0	20.0
0	.0000	.0000	.0000	.0000	.0000	.0000	.0000	.0000	.0000	.0000
1	.0002	.0001	.0000	.0000	.0000	.0000	.0000	.0000	.0000	.0000
2	.0010	.0004	.0002	.0001	.0000	.0000	.0000	.0000	.0000	.0000
3	.0037	.0018	.0008	.0004	.0002	.0001	.0000	.0000	.0000	.0000
4	.0102	.0053	.0027	.0013	.0006	.0003	.0001	.0001	.0000	.0000
5	.0224	.0127	.0070	.0037	.0019	.0010	.0005	.0002	.0001	.0001
6	.0411	.0255	.0152	.0087	.0048	.0026	.0014	.0007	.0004	.0002
7	.0646	.0437	.0281	.0174	.0104	.0060	.0034	.0019	.0010	.0005
8	.0888	.0655	.0457	.0304	.0194	.0120	.0072	.0042	.0024	.0013
9	.1085	.0874	.0661	.0473	.0324	.0213	.0135	.0083	.0050	.0029
10	.1194	.1048	.0859	.0663	.0486	.0341	.0230	.0150	.0095	.0058
11	.1194	.1144	.1015	.0844	.0663	.0496	.0355	.0245	.0164	.0106
12	.1094	.1144	.1099	.0984	.0829	.0661	.0504	.0368	.0259	.0176
13	.0926	.1056	.1099	.1060	.0956	.0814	.0658	.0509	.0378	.0271
14	.0728	.0905	.1021	.1060	.1024	.0930	.0800	.0655	.0514	.0387
15	.0534	.0724	.0885	.0989	.1024	.0992	.0906	.0786	.0650	.0516
16	.0367	.0543	.0719	.0866	.0960	.0992	.0963	.0884	.0772	.0646
17	.0237	.0383	.0550	.0713	.0847	.0934	.0963	.0936	.0863	.0760
18	.0145	.0255	.0397	.0554	.0706	.0830	.0909	.0936	.0911	.0844
19	.0084	.0161	.0272	.0409	.0557	.0699	.0814	.0887	.0911	.0888
20	.0046	.0097	.0177	.0286	.0418	.0559	.0692	.0798	.0866	.0888
21	.0024	.0055	.0109	.0191	.0299	.0426	.0560	.0684	.0783	.0846
22	.0012	.0030	.0065	.0121	.0204	.0310	.0433	.0560	.0676	.0769
23	.0006	.0016	.0037	.0074	.0133	.0216	.0320	.0438	.0559	.0669
24	.0003	.0008	.0020	.0043	.0083	.0144	.0226	.0328	.0442	.0557
25	.0001	.0004	.0010	.0024	.0050	.0092	.0154	.0237	.0336	.0446
26	.0000	.0002	.0005	.0013	.0029	.0057	.0101	.0164	.0246	.0343
27	.0000	.0001	.0002	.0007	.0016	.0034	.0063	.0109	.0173	.0254
28	.0000	.0000	.0001	.0003	.0009	.0019	.0038	.0070	.0117	.0181
29	.0000	.0000	.0001	.0002	.0004	.0011	.0023	.0044	.0077	.0125
30	.0000	.0000	.0000	.0001	.0002	.0006	.0013	.0026	.0049	.0083
31	.0000	.0000	.0000	.0000	.0001	.0003	.0007	.0015	.0030	.0054
32	.0000	.0000	.0000	.0000	.0001	.0001	.0004	.0009	.0018	.0034
33	.0000	.0000	.0000	.0000	.0000	.0001	.0002	.0005	.0010	.0020

Source: Computed by D. K. Hildebrand.

TABLE 8   Random Numbers

Line/Col.	(1)	(2)	(3)	(4)	(5)	(6)	(7)	(8)	(9)	(10)	(11)	(12)	(13)	(14)
1	10480	15011	01536	02011	81647	91646	69179	14194	62590	36207	20969	99570	91291	90700
2	22368	46573	25595	85393	30995	89198	27982	53402	93965	34095	52666	19174	39615	99505
3	24130	48360	22527	97265	76393	64809	15179	24830	49340	32081	30680	19655	63348	58629
4	42167	93093	06243	61680	07856	16376	39440	53537	71341	57004	00849	74917	97758	16379
5	37570	39975	81837	16656	06121	91782	60468	81305	49684	60672	14110	06927	01263	54613
6	77921	06907	11008	42751	27756	53498	18602	70659	90655	15053	21916	81825	44394	42880
7	99562	72905	56420	69994	98872	31016	71194	18738	44013	48840	63213	21069	10634	12952
8	96301	91977	05463	07972	18876	20922	94595	56869	69014	60045	18425	84903	42508	32307
9	89579	14342	63661	10281	17453	18103	57740	84378	25331	12566	58678	44947	05585	56941
10	85475	36857	53342	53988	53060	59533	38867	62300	08158	17983	16439	11458	18593	64952
11	28918	69578	88231	33276	70997	79936	56865	05859	90106	31595	01547	85590	91610	78188
12	63553	40961	48235	03427	49626	69445	18663	72695	52180	20847	12234	90511	33703	90322
13	09429	93969	52636	92737	88974	33488	36320	17617	30015	08272	84115	27156	30613	74952
14	10365	61129	87529	85689	48237	52267	67689	93394	01511	26358	85104	20285	29975	89868
15	07119	97336	71048	08178	77233	13916	47564	81056	97735	85977	29372	74461	28551	90707
16	51085	12765	51821	51259	77452	16308	60756	92144	49442	53900	70960	63990	75601	40719
17	02368	21382	52404	60268	89368	19885	55322	44819	01188	65255	64835	44919	05944	55157
18	01011	54092	33362	94904	31273	04146	18594	29852	71585	85030	51132	01915	92747	64951
19	52162	53916	46369	58586	23216	14513	83149	98736	23495	64350	94738	17752	35156	35749
20	07056	97628	33787	09998	42698	06691	76988	13602	51851	46104	88916	19509	25625	58104
21	48663	91245	85828	14346	09172	30168	90229	04734	59193	22178	30421	61666	99904	32812
22	54164	58492	22421	74103	47070	25306	76468	26384	58151	06646	21524	15227	96909	44592
23	32639	32363	05597	24200	13363	38005	94342	28728	35806	06912	17012	64161	18296	22851
24	29334	27001	87637	87308	58731	00256	45834	15398	46557	41135	10367	07684	36188	18510
25	02488	33062	28834	07351	19731	92420	60952	61280	50001	67658	32586	86679	50720	94953

Abridged from William H. Beyer, ed., *Handbook of Tables for Probability and Statistics,* 2nd ed. © The Chemical Rubber Co., 1968. Used by permission of CRC Press, Inc.

T A B L E 9 Critical Values for the Wilcoxon Signed-rank Test [$n = 5(1)54$]

One-Sided	Two-Sided	n = 5	n = 6	n = 7	n = 8	n = 9
p = .1	p = .2	2	3	5	8	10
p = .05	p = .1	0	2	3	5	8
p = .025	p = .05		0	2	3	5
p = .01	p = .02			0	1	3
p = .005	p = .01				0	1
p = .0025	p = .005					0
p = .001	p = .002					

One-Sided	Two-Sided	n = 15	n = 16	n = 17	n = 18	n = 19
p = .1	p = .2	36	42	48	55	62
p = .05	p = .1	30	35	41	47	53
p = .025	p = .05	25	29	34	40	46
p = .01	p = .02	19	23	27	32	37
p = .005	p = .01	15	19	23	27	32
p = .0025	p = .005	12	15	19	23	27
p = .001	p = .002	8	11	14	18	21

One-Sided	Two-Sided	n = 25	n = 26	n = 27	n = 28	n = 29
p = .1	p = .2	113	124	134	145	157
p = .05	p = .1	100	110	119	130	140
p = .025	p = .05	89	98	107	116	126
p = .01	p = .02	76	84	92	101	110
p = .005	p = .01	68	75	83	91	100
p = .0025	p = .005	60	67	74	82	90
p = .001	p = .002	51	58	64	71	79

One-Sided	Two-Sided	n = 35	n = 36	n = 37	n = 38	n = 39
p = .1	p = .2	235	250	265	281	297
p = .05	p = .1	213	227	241	256	271
p = .025	p = .05	195	208	221	235	249
p = .01	p = .02	173	185	198	211	224
p = .005	p = .01	159	171	182	194	207
p = .0025	p = .005	146	157	168	180	192
p = .001	p = .002	131	141	151	162	173

One-Sided	Two-Sided	n = 45	n = 46	n = 47	n = 48	n = 49
p = .1	p = .2	402	422	441	462	482
p = .05	p = .1	371	389	407	426	446
p = .025	p = .05	343	361	378	396	415
p = .01	p = .02	312	328	345	362	379
p = .005	p = .01	291	307	322	339	355
p = .0025	p = .005	272	287	302	318	334
p = .001	p = .002	249	263	277	292	307

Source: Computed by P. J. Hildebrand.

TABLE 9 *(continued)*

One-Sided	Two-Sided	$n = 10$	$n = 11$	$n = 12$	$n = 13$	$n = 14$
$p = .1$	$p = .2$	14	17	21	26	31
$p = .05$	$p = .1$	10	13	17	21	25
$p = .025$	$p = .05$	8	10	13	17	21
$p = .01$	$p = .02$	5	7	9	12	15
$p = .005$	$p = .01$	3	5	7	9	12
$p = .0025$	$p = .005$	1	3	5	7	9
$p = .001$	$p = .002$	0	1	2	4	6

One-Sided	Two-Sided	$n = 20$	$n = 21$	$n = 22$	$n = 23$	$n = 24$
$p = .1$	$p = .2$	69	77	86	94	104
$p = .05$	$p = .1$	60	67	75	83	91
$p = .025$	$p = .05$	52	58	65	73	81
$p = .01$	$p = .02$	43	49	55	62	69
$p = .005$	$p = .01$	37	42	48	54	61
$p = .0025$	$p = .005$	32	37	42	48	54
$p = .001$	$p = .002$	26	30	35	40	45

One-Sided	Two-Sided	$n = 30$	$n = 31$	$n = 32$	$n = 33$	$n = 34$
$p = .1$	$p = .2$	169	181	194	207	221
$p = .05$	$p = .1$	151	163	175	187	200
$p = .025$	$p = .05$	137	147	159	170	182
$p = .01$	$p = .02$	120	130	140	151	162
$p = .005$	$p = .01$	109	118	128	138	148
$p = .0025$	$p = .005$	98	107	116	126	136
$p = .001$	$p = .002$	86	94	103	112	121

One-Sided	Two-Sided	$n = 40$	$n = 41$	$n = 42$	$n = 43$	$n = 44$
$p = .1$	$p = .2$	313	330	348	365	384
$p = .05$	$p = .1$	286	302	319	336	353
$p = .025$	$p = .05$	264	279	294	310	327
$p = .01$	$p = .02$	238	252	266	281	296
$p = .005$	$p = .01$	220	233	247	261	276
$p = .0025$	$p = .005$	204	217	230	244	258
$p = .001$	$p = .002$	185	197	209	222	235

One-Sided	Two-Sided	$n = 50$	$n = 51$	$n = 52$	$n = 53$	$n = 54$
$p = .1$	$p = .2$	503	525	547	569	592
$p = .05$	$p = .1$	466	486	507	529	550
$p = .025$	$p = .05$	434	453	473	494	514
$p = .01$	$p = .02$	397	416	434	454	473
$p = .005$	$p = .01$	373	390	408	427	445
$p = .0025$	$p = .005$	350	367	384	402	420
$p = .001$	$p = .002$	323	339	355	372	389

TABLE 10   Percentage Points of the Studentized Range

Error df	α	\multicolumn{10}{c}{t = number of treatment means}
		2	3	4	5	6	7	8	9	10	11
5	.05	3.64	4.60	5.22	5.67	6.03	6.33	6.58	6.80	6.99	7.17
	.01	5.70	6.98	7.80	8.42	8.91	9.32	9.67	9.97	10.24	10.48
6	.05	3.46	4.34	4.90	5.30	5.63	5.90	6.12	6.32	6.49	6.65
	.01	5.24	6.33	7.03	7.56	7.97	8.32	8.61	8.87	9.10	9.30
7	.05	3.34	4.16	4.68	5.06	5.36	5.61	5.82	6.00	6.16	6.30
	.01	4.95	5.92	6.54	7.01	7.37	7.68	7.94	8.17	8.37	8.55
8	.05	3.26	4.04	4.53	4.89	5.17	5.40	5.60	5.77	5.92	6.05
	.01	4.75	5.64	6.20	6.62	6.96	7.24	7.47	7.68	7.86	8.03
9	.05	3.20	3.95	4.41	4.76	5.02	5.24	5.43	5.59	5.74	5.87
	.01	4.60	5.43	5.96	6.35	6.66	6.91	7.13	7.33	7.49	7.65
10	.05	3.15	3.88	4.33	4.65	4.91	5.12	5.30	5.46	5.60	5.72
	.01	4.48	5.27	5.77	6.14	6.43	6.67	6.87	7.05	7.21	7.36
11	.05	3.11	3.82	4.26	4.57	4.82	5.03	5.30	5.35	5.49	5.61
	.01	4.39	5.15	5.62	5.97	6.25	6.48	6.67	6.84	6.99	7.13
12	.05	3.08	3.77	4.20	4.52	4.75	4.95	5.12	5.27	5.39	5.51
	.01	4.32	5.05	5.50	5.84	6.10	6.32	6.51	6.67	6.81	6.94
13	.05	3.06	3.73	4.15	4.45	4.69	4.88	5.05	5.19	5.32	5.43
	.01	4.26	4.96	5.40	5.73	5.98	6.19	6.37	6.53	6.67	6.79
14	.05	3.03	3.70	4.11	4.41	4.64	4.83	4.99	5.13	5.25	5.36
	.01	4.21	4.89	5.32	5.63	5.88	6.08	6.26	6.41	6.54	6.66
15	.05	3.01	3.67	4.08	4.37	4.59	4.78	4.94	5.08	5.20	5.31
	.01	4.17	4.84	5.25	5.56	5.80	5.99	6.16	6.31	6.44	6.55
16	.05	3.00	3.65	4.05	4.33	4.56	4.74	4.90	5.03	5.15	5.26
	.01	4.13	4.79	5.19	5.49	5.72	5.92	6.08	6.22	6.35	6.46
17	.05	2.98	3.63	4.02	4.30	4.52	4.70	4.86	4.99	5.11	5.21
	.01	4.10	4.74	5.14	5.43	5.66	5.85	6.01	6.15	6.27	6.38
18	.05	2.97	3.61	4.00	4.28	4.49	4.67	4.82	4.96	5.07	5.17
	.01	4.07	4.70	5.09	5.38	5.60	5.79	5.94	6.08	6.20	6.31
19	.05	2.96	3.59	3.98	4.25	4.47	4.65	4.79	4.92	5.04	5.14
	.01	4.05	4.67	5.05	5.33	5.55	5.73	5.89	6.02	6.14	6.25
20	.05	2.95	3.58	3.96	4.23	4.45	4.62	4.77	4.90	5.01	5.11
	.01	4.02	4.64	5.02	5.29	5.51	5.69	5.84	5.97	6.09	6.19
24	.05	2.92	3.53	3.90	4.17	4.37	4.54	4.68	4.81	3.92	5.01
	.01	3.96	4.55	4.91	5.17	5.37	5.54	5.69	5.81	5.92	6.02
30	.05	2.89	3.49	3.85	4.10	4.30	4.46	4.60	4.72	4.82	4.92
	.01	3.89	4.45	4.80	5.05	5.24	5.40	5.54	5.65	5.76	5.85
40	.05	2.86	3.44	3.79	4.04	4.23	4.39	4.52	4.63	4.73	4.82
	.01	3.82	4.37	4.70	4.93	5.11	5.26	5.39	5.50	5.60	5.69
60	.05	2.83	3.40	3.74	3.98	4.16	4.31	4.44	4.55	4.65	4.73
	.01	3.76	4.28	4.59	4.82	4.99	5.13	5.25	5.36	5.45	5.53
120	.05	2.80	3.36	3.68	3.92	4.10	4.24	4.36	4.47	4.56	4.64
	.01	3.70	4.20	4.50	4.71	4.87	5.01	5.12	5.21	5.30	5.37
∞	.05	2.77	3.31	3.63	3.86	4.03	4.17	4.29	4.39	4.47	4.55
	.01	3.64	4.12	4.40	4.60	4.76	4.88	4.99	5.08	5.16	5.23

TABLE 10 *(continued)*

12	13	14	15	16	17	18	19	20	α	Error df
				t = number of treatment means						
7.32	7.47	7.60	7.72	7.83	7.93	8.03	8.12	8.21	.05	5
10.70	10.89	11.08	11.24	11.40	11.55	11.68	11.81	11.93	.01	
6.79	6.92	7.03	7.14	7.24	7.34	7.43	7.51	7.59	.05	6
9.48	9.65	9.81	9.95	10.08	10.21	10.32	10.43	10.54	.01	
6.43	6.55	6.66	6.76	6.85	6.94	7.02	7.10	7.17	.05	7
8.71	8.86	9.00	9.12	9.24	9.35	9.46	9.55	9.65	.01	
6.18	6.29	6.39	6.48	6.57	6.65	6.73	6.80	6.87	.05	8
8.18	8.31	8.44	8.55	8.66	8.76	8.85	8.94	9.03	.01	
5.98	6.09	6.19	6.28	6.36	6.44	6.51	6.58	6.64	.05	9
7.78	7.91	8.03	8.13	8.23	8.33	8.41	8.49	8.57	.01	
5.83	5.93	6.03	6.11	6.19	6.27	6.34	6.40	6.47	.05	10
7.49	7.60	7.71	7.81	7.91	7.99	8.08	8.15	8.23	.01	
5.71	5.81	5.90	5.98	6.06	6.13	6.20	6.27	6.33	.05	11
7.25	7.36	7.46	7.56	7.65	7.73	7.81	7.88	7.95	.01	
5.61	5.71	5.80	5.88	5.95	6.02	6.09	6.15	6.21	.05	12
7.06	7.17	7.26	7.36	7.44	7.52	7.59	7.66	7.73	.01	
5.53	5.63	5.71	5.79	5.86	5.93	5.99	6.05	6.11	.05	13
6.90	7.01	7.10	7.19	7.27	7.35	7.42	7.48	7.55	.01	
5.46	5.55	5.64	5.71	5.79	5.85	5.91	5.97	6.03	.05	14
6.77	6.87	6.96	7.05	7.13	7.20	7.27	7.33	7.39	.01	
5.40	5.49	5.57	5.65	5.72	5.78	5.85	5.90	5.96	.05	15
6.66	6.76	6.84	6.93	7.00	7.07	7.14	7.20	7.26	.01	
5.35	5.44	5.52	5.59	5.66	5.73	5.79	5.84	5.90	.05	16
6.56	6.66	6.74	6.82	6.90	6.97	7.03	7.09	7.15	.01	
5.31	5.39	5.47	5.54	5.61	5.67	5.73	5.79	5.84	.05	17
6.48	6.57	6.66	6.73	6.81	6.87	6.94	7.00	7.05	.01	
5.27	5.35	5.43	5.50	5.57	5.63	5.69	5.74	5.79	.05	18
6.41	6.50	6.58	6.65	6.73	6.79	6.85	6.91	6.97	.01	
5.23	5.31	5.39	5.46	5.53	5.59	5.65	5.70	5.75	.05	19
6.34	6.43	6.51	6.58	6.65	6.72	6.78	6.84	6.89	.01	
5.20	5.28	5.36	5.43	5.49	5.55	5.61	5.66	5.71	.05	20
6.28	6.37	6.45	6.52	6.59	6.65	6.71	6.77	6.82	.01	
5.10	5.18	5.25	5.32	5.38	5.44	5.49	5.55	5.59	.05	24
6.11	6.19	6.26	6.33	6.39	6.45	6.51	6.56	6.61	.01	
5.00	5.08	5.15	5.21	5.27	5.33	5.38	5.43	5.47	.05	30
5.93	6.01	6.08	6.14	6.20	6.26	6.31	6.36	6.41	.01	
4.90	4.98	5.04	5.11	5.16	5.22	5.27	5.31	5.36	.05	40
5.76	5.83	5.90	5.96	6.02	6.07	6.12	6.16	6.21	.01	
4.81	4.88	4.94	5.00	5.06	5.11	5.15	5.20	5.24	.05	60
5.60	5.67	5.73	5.78	5.84	5.89	5.93	5.97	6.01	.01	
4.71	4.78	4.84	4.90	4.95	5.00	5.04	5.09	5.13	.05	120
5.44	5.50	5.56	5.61	5.66	5.71	5.75	5.79	5.83	.01	
4.62	4.68	4.74	4.80	4.85	4.89	4.93	4.97	5.01	.05	∞
5.29	5.35	5.40	5.45	5.49	5.54	5.57	5.61	5.65	.01	

TABLE 11 **Percentage Points of the Duncan New Multiple Range Test**

Error df	α	r = number of ordered steps between means													
		2	3	4	5	6	7	8	9	10	12	14	16	18	20
1	.05	18.0	18.0	18.0	18.0	18.0	18.0	18.0	18.0	18.0	18.0	18.0	18.0	18.0	18.0
	.01	90.0	90.0	90.0	90.0	90.0	90.0	90.0	90.0	90.0	90.0	90.0	90.0	90.0	90.0
2	.05	6.09	6.09	6.09	6.09	6.09	6.09	6.09	6.09	6.09	6.09	6.09	6.09	6.09	6.09
	.01	14.0	14.0	14.0	14.0	14.0	14.0	14.0	14.0	14.0	14.0	14.0	14.0	14.0	14.0
3	.05	4.50	4.50	4.50	4.50	4.50	4.50	4.50	4.50	4.50	4.50	4.50	4.50	4.50	4.50
	.01	8.26	8.5	8.6	8.7	8.8	8.9	8.9	9.0	9.0	9.0	9.1	9.2	9.3	9.3
4	.05	3.93	4.01	4.02	4.02	4.02	4.02	4.02	4.02	4.02	4.02	4.02	4.02	4.02	4.02
	.01	6.51	6.8	6.9	7.0	7.1	7.1	7.2	7.2	7.3	7.3	7.4	7.4	7.5	7.5
5	.05	3.64	3.74	3.79	3.83	3.83	3.83	3.83	3.83	3.83	3.83	3.83	3.83	3.83	3.83
	.01	5.70	5.96	6.11	6.18	6.26	6.33	6.40	6.44	6.5	6.6	6.6	6.7	6.7	6.8
6	.05	3.46	3.58	3.64	3.68	3.68	3.68	3.68	3.68	3.68	3.68	3.68	3.68	3.68	3.68
	.01	5.24	5.51	5.65	5.73	5.83	5.81	5.95	6.00	6.0	6.1	6.2	6.2	6.3	6.3
7	.05	3.35	3.47	3.54	3.58	3.60	3.61	3.61	3.61	3.61	3.61	3.61	3.61	3.61	3.61
	.01	4.95	5.22	5.37	5.45	5.53	5.61	5.69	5.73	5.8	5.8	5.9	5.9	6.0	6.0
8	.05	3.26	3.39	3.47	3.52	3.55	3.56	3.56	3.56	3.56	3.56	3.56	3.56	3.56	3.56
	.01	4.74	5.00	5.14	5.23	5.32	5.40	5.47	5.51	5.5	5.6	5.7	5.7	5.8	5.8
9	.05	3.20	3.34	3.41	3.47	3.50	3.52	3.52	3.52	3.52	3.52	3.52	3.52	3.52	3.52
	.01	4.60	4.86	4.99	5.08	5.17	5.25	5.32	5.36	5.4	5.5	5.5	5.6	5.7	5.7
10	.05	3.15	3.30	3.37	3.43	3.46	3.47	3.47	3.47	3.47	3.47	3.47	3.47	3.47	3.48
	.01	4.48	4.73	4.88	4.96	5.06	5.13	5.20	5.24	5.28	5.36	5.42	5.48	5.54	5.55
11	.05	3.11	3.27	3.35	3.39	3.43	3.44	3.45	3.46	3.46	3.46	3.46	3.46	3.47	3.48
	.01	4.39	4.63	4.77	4.86	4.94	5.01	5.06	5.12	5.15	5.24	5.28	5.34	5.38	5.39
12	.05	3.08	3.23	3.33	3.36	3.40	3.42	3.44	3.44	3.46	3.46	3.46	3.46	3.47	3.48
	.01	4.32	4.55	4.68	4.76	4.84	4.92	4.96	5.02	5.07	5.13	5.17	5.22	5.23	5.26
13	.05	3.06	3.21	3.30	3.35	3.38	3.41	3.42	3.44	3.45	3.45	3.46	3.46	3.47	3.47
	.01	4.26	4.48	4.62	4.69	4.74	4.84	4.88	4.94	4.98	5.04	5.08	5.13	5.14	5.15
14	.05	3.03	3.18	3.27	3.33	3.37	3.39	3.41	3.42	3.44	3.45	3.46	3.46	3.47	3.47
	.01	4.21	4.42	4.55	4.63	4.70	4.78	4.83	4.87	4.91	4.96	5.00	5.04	5.06	5.07
15	.05	3.01	3.16	3.25	3.31	3.36	3.38	3.40	3.42	3.43	3.44	3.45	3.46	3.47	3.47
	.01	4.17	4.37	4.50	4.58	4.64	4.72	4.77	4.81	4.84	4.90	4.94	4.97	4.99	5.00

TABLE 11 (continued)

| Error | | r = number of ordered steps between means | | | | | | | | | | | | | |
df	α	2	3	4	5	6	7	8	9	10	12	14	16	18	20
16	.05	3.00	3.15	3.23	3.30	3.34	3.37	3.39	3.41	3.43	3.44	3.45	3.46	3.47	3.47
	.01	4.13	4.34	4.45	4.54	4.60	4.67	4.72	4.76	4.79	4.84	4.88	4.91	4.93	4.94
17	.05	2.98	3.13	3.22	3.28	3.33	3.36	3.38	3.40	3.42	3.44	3.45	3.46	3.47	3.47
	.01	4.10	4.30	4.41	4.50	4.56	4.63	4.68	4.72	4.75	4.80	4.83	4.86	4.88	4.89
18	.05	2.97	3.12	3.21	3.27	3.32	3.35	3.37	3.39	3.41	3.43	3.45	3.46	3.47	3.47
	.01	4.07	4.27	4.38	4.46	4.53	4.59	4.64	4.68	4.71	4.76	4.79	4.82	4.84	4.85
19	.05	2.96	3.11	3.19	3.26	3.31	3.35	3.37	3.39	3.41	3.43	3.44	3.46	3.47	3.47
	.01	4.05	4.24	4.35	4.43	4.50	4.56	4.61	4.64	4.67	4.72	4.76	4.79	4.81	4.82
20	.05	2.95	3.10	3.18	3.25	3.30	3.34	3.36	3.38	3.40	3.43	3.44	3.46	3.46	3.47
	.01	4.02	4.22	4.33	4.40	4.47	4.53	4.58	4.61	4.65	4.69	4.73	4.76	4.78	4.79
22	.05	2.93	3.08	3.17	3.24	3.29	3.32	3.35	3.37	3.39	3.42	3.44	3.45	3.46	3.47
	.01	3.99	4.17	4.28	4.36	4.42	4.48	4.53	4.57	4.60	4.65	4.68	4.71	4.74	4.75
24	.05	2.92	3.07	3.15	3.22	3.28	3.31	3.34	3.37	3.38	3.41	3.44	3.45	3.46	3.47
	.01	3.96	4.14	4.24	4.33	4.39	4.44	4.49	4.53	4.57	4.62	4.64	4.67	4.70	4.72
26	.05	2.91	3.06	3.14	3.21	3.27	3.30	3.34	3.36	3.38	3.41	3.43	3.45	3.46	3.47
	.01	3.93	4.11	4.21	4.30	4.36	4.41	4.46	4.50	4.53	4.58	4.62	4.65	4.67	4.69
28	.05	2.90	3.04	3.13	3.20	3.26	3.30	3.33	3.35	3.37	3.40	3.43	3.45	3.46	3.47
	.01	3.91	3.08	4.18	4.28	4.34	4.39	4.43	4.47	4.51	4.56	4.60	4.62	4.65	4.67
30	.05	2.89	3.04	3.12	3.20	3.25	3.29	3.32	3.35	3.37	3.40	3.43	3.44	3.46	3.47
	.01	3.89	4.06	4.16	4.22	4.32	4.36	4.41	4.45	4.48	4.54	4.58	4.61	4.63	4.65
40	.05	2.86	3.01	3.10	3.17	3.22	3.27	3.30	3.33	3.35	3.39	3.42	3.44	3.46	3.47
	.01	3.82	3.99	4.10	4.17	4.24	4.30	4.34	4.37	4.41	4.46	4.51	4.54	4.57	4.59
60	.05	2.83	2.98	3.08	3.14	3.20	3.24	3.28	3.31	3.33	3.37	3.40	3.43	3.45	3.47
	.01	3.76	3.92	4.03	4.12	4.17	4.23	4.27	4.31	4.34	4.39	4.44	4.47	4.50	4.53
100	.05	2.80	2.95	3.05	3.12	3.18	3.22	3.26	3.29	3.32	3.36	3.40	3.42	3.45	3.47
	.01	3.71	3.86	3.93	4.06	4.11	4.17	4.21	4.25	4.29	4.35	4.38	4.42	4.45	4.48
∞	.05	2.77	2.92	3.02	3.09	3.15	3.19	3.23	3.26	3.29	3.34	3.38	3.41	3.44	3.47
	.01	3.64	3.80	3.90	3.98	4.04	4.09	4.14	4.17	4.20	4.26	4.31	4.34	4.38	4.41

Reproduced from D. B. Duncan, "Multiple Range and Multiple F Tests," *Biometrics*, 11: 1 42, 1955. With permission from the Biometric Society and the author.

TABLE 12 Waller–Duncan k Ratio Test Values of t_c for $k = 100$

df$_1$	df$_2$												
	6	8	10	12	14	16	18	20	24	30	40	60	120
$F = 1.2$ (a = .913, b = 2.449)													
2–6	*	*	*	*	*	*	*	*	*	*	*	*	*
8	2.91	2.94	2.96	2.97	2.98	2.99	2.99	2.99	3.00	3.00	3.00	3.00	3.00
10	2.93	2.98	3.01	3.04	3.05	3.06	3.07	3.08	3.09	3.10	3.10	3.11	3.12
12	2.95	3.01	3.05	3.08	3.10	3.12	3.13	3.14	3.16	3.17	3.19	3.20	3.21
14	2.96	3.03	3.08	3.12	3.14	3.16	3.18	3.19	3.21	3.23	3.25	3.27	3.29
16	2.97	3.05	3.11	3.15	3.18	3.20	3.22	3.24	3.26	3.28	3.31	3.33	3.36
20	2.99	3.08	3.14	3.19	3.23	3.26	3.28	3.30	3.33	3.37	3.40	3.44	3.47
40	3.02	3.13	3.22	3.29	3.35	3.39	3.43	3.47	3.52	3.58	3.64	3.72	3.79
100	3.04	3.17	3.28	3.36	3.44	3.50	3.55	3.59	3.67	3.76	3.86	3.98	4.11
∞	3.05	3.20	3.32	3.42	3.50	3.58	3.64	3.70	3.80	3.91	4.06	4.24	4.45
$F = 1.4$ (a = .845, b = 1.871)													
2–4	*	*	*	*	*	*	*	*	*	*	*	*	*
6	2.85	2.84	2.83	2.82	2.82	2.81	2.80	2.80	2.79	2.78	2.77	2.75	2.74
8	2.88	2.89	2.90	2.90	2.90	2.89	2.89	2.89	2.88	2.88	2.87	2.86	2.85
10	2.90	2.93	2.94	2.95	2.95	2.96	2.96	2.96	2.95	2.95	2.95	2.94	2.93
12	2.92	2.95	2.98	2.99	3.00	3.00	3.01	3.01	3.01	3.01	3.01	3.00	2.99
14	2.93	2.97	3.00	3.02	3.03	3.04	3.04	3.05	3.05	3.06	3.06	3.05	3.05
16	2.94	2.99	3.02	3.04	3.06	3.07	3.08	3.08	3.09	3.09	3.10	3.10	3.09
20	2.95	3.01	3.05	3.08	3.10	3.11	3.12	3.13	3.14	3.15	3.16	3.16	3.16
40	2.98	3.06	3.12	3.16	3.19	3.22	3.24	3.25	3.28	3.30	3.31	3.32	3.32
100	2.99	3.09	3.16	3.22	3.26	3.29	3.32	3.34	3.38	3.41	3.43	3.45	3.42
∞	3.01	3.12	3.20	3.26	3.31	3.35	3.39	3.42	3.46	3.50	3.53	3.54	3.46
$F = 1.7$ (a = .767, b = 1.558)													
2	*	*	*	*	*	*	*	*	*	*	*	*	*
4	*	*	*	*	*	2.61	2.59	2.58	2.56	2.54	2.52	2.50	2.48
6	2.82	2.79	2.76	2.74	2.72	2.71	2.70	2.69	2.67	2.65	2.63	2.61	2.58
8	2.84	2.83	2.81	2.80	2.78	2.77	2.76	2.75	2.74	2.72	2.70	2.68	2.65
10	2.86	2.86	2.85	2.84	2.83	2.82	2.81	2.80	2.79	2.77	2.75	2.73	2.70
12	2.87	2.88	2.88	2.87	2.86	2.85	2.84	2.84	2.82	2.81	2.79	2.76	2.73
14	2.88	2.90	2.90	2.89	2.89	2.88	2.87	2.86	2.85	2.83	2.81	2.79	2.75
16	2.89	2.91	2.91	2.91	2.90	2.90	2.89	2.89	2.87	2.86	2.84	2.81	2.77
20	2.90	2.93	2.93	2.94	2.93	2.93	2.92	2.92	2.91	2.89	2.87	2.84	2.80
40	2.93	2.97	2.99	3.00	3.00	3.00	3.00	2.99	2.98	2.97	2.94	2.89	2.83
100	2.94	2.99	3.02	3.04	3.05	3.05	3.05	3.05	3.04	3.02	2.98	2.92	2.83
∞	2.95	3.01	3.05	3.07	3.08	3.09	3.09	3.08	3.07	3.05	3.01	2.93	2.81

TABLE 12 (continued)

df$_1$	\multicolumn{13}{c}{df$_2$}
	6	8	10	12	14	16	18	20	24	30	40	60	120
\multicolumn{14}{l}{$F = 2.0$ (a = .707, b = 1.414)}
2	*	*	*	*	*	*	*	*	*	*	*	*	*
4	2.74	2.67	2.63	2.59	2.56	2.54	2.52	2.51	2.49	2.46	2.44	2.41	2.39
6	2.79	2.74	2.70	2.67	2.64	2.62	2.60	2.59	2.57	2.54	2.52	2.49	2.46
8	2.81	2.77	2.74	2.71	2.69	2.67	2.65	2.64	2.62	2.59	2.56	2.53	2.49
10	2.83	2.80	2.77	2.74	2.72	2.70	2.69	2.67	2.65	2.62	2.59	2.56	2.52
12	2.84	2.82	2.79	2.77	2.75	2.73	2.71	2.70	2.67	2.64	2.61	2.57	2.53
14	2.85	2.83	2.81	2.79	2.77	2.75	2.73	2.72	2.69	2.66	2.63	2.59	2.54
16	2.85	2.84	2.82	2.80	2.78	2.76	2.74	2.73	2.70	2.67	2.64	2.59	2.54
20	2.86	2.85	2.84	2.82	2.80	2.78	2.77	2.75	2.72	2.69	2.65	2.61	2.55
40	2.88	2.89	2.88	2.86	2.85	2.83	2.81	2.80	2.77	2.73	2.68	2.62	2.55
100	2.89	2.91	2.90	2.89	2.88	2.86	2.84	2.82	2.79	2.75	2.69	2.62	2.53
∞	2.90	2.92	2.92	2.91	2.90	2.88	2.86	2.85	2.81	2.76	2.69	2.61	2.52
\multicolumn{14}{l}{$F = 2.4$ (a = .645, b = 1.309)}
2	*	*	*	*	*	*	*	*	*	*	*	*	2.18
4	2.71	2.63	2.57	2.53	2.49	2.47	2.44	2.43	2.40	2.37	2.34	2.31	2.28
6	2.75	2.68	2.63	2.58	2.55	2.52	2.50	2.48	2.46	2.42	2.39	2.36	2.32
8	2.77	2.71	2.66	2.62	2.59	2.56	2.54	2.52	2.49	2.45	2.42	2.38	2.34
10	2.79	2.73	2.68	2.64	2.61	2.58	2.56	2.54	2.50	2.47	2.43	2.39	2.34
12	2.79	2.74	2.70	2.66	2.62	260	2.57	2.55	2.52	2.48	2.44	2.39	2.35
14	2.80	2.75	2.71	2.67	2.64	2.61	2.58	2.56	2.53	2.49	2.44	2.40	2.35
16	2.81	2.76	2.72	2.68	2.65	2.62	2.59	2.57	2.53	2.49	2.45	2.40	2.34
20	2.82	2.77	2.73	2.69	2.66	2.63	2.60	2.58	2.54	2.50	2.45	2.40	2.34
40	2.83	2.80	2.76	2.72	2.69	2.66	2.63	2.60	2.56	2.51	2.46	2.39	2.33
100	2.84	2.81	2.78	2.74	2.71	2.67	2.64	2.62	2.57	2.51	2.45	2.39	2.32
∞	2.85	2.83	2.79	2.76	2.72	2.68	2.65	2.62	2.57	2.51	2.45	2.38	2.31
\multicolumn{14}{l}{$F = 3.0$ (a = .577, b = 1.225)}
2	*	*	2.41	2.36	2.32	2.29	2.27	2.25	2.22	2.20	2.17	2.14	2.11
4	2.68	2.57	2.50	2.45	2.41	2.38	2.35	2.33	2.30	2.27	2.24	2.20	2.17
6	2.71	2.61	2.54	2.49	2.44	2.41	2.39	2.36	2.33	2.29	2.26	2.22	2.18
8	2.72	2.63	2.56	2.51	2.47	2.43	2.40	2.38	2.34	2.31	2.27	2.22	2.18
10	2.74	2.65	2.58	2.52	2.48	2.44	2.41	2.39	2.35	2.31	2.27	2.22	2.18
12	2.74	2.66	2.59	2.53	2.49	2.45	2.42	2.40	2.36	2.31	2.27	2.22	2.18
14	2.75	2.66	2.60	2.54	2.49	2.46	2.43	2.40	2.36	2.32	2.27	2.22	2.17
16	2.75	2.67	2.60	2.55	2.50	2.46	2.43	2.40	2.36	2.32	2.27	2.22	2.17
20	2.76	2.68	2.61	2.55	2.51	2.47	2.43	2.41	2.36	2.32	2.27	2.22	2.17
40	2.77	2.70	2.63	2.57	2.52	2.48	2.44	2.41	2.37	2.32	2.26	2.21	2.16
100	2.78	2.71	2.64	2.58	2.53	2.49	2.45	2.42	2.37	2.31	2.26	2.21	2.16
∞	2.79	2.71	2.65	2.59	2.53	2.49	2.45	2.42	2.37	2.31	2.26	2.20	2.15

*All differences not significant. a $= 1/F^{1/2}$; b $= [F/(F - 1)]^{1/2}$.

From "Corrigenda" by R. Waller and D. Duncan, *Journal of the American Statistical Association*, 67, 1972, 253–55, Table A2. Reproduced by permission of the American Statistical Association.

TABLE 12 *(continued)*

| df$_1$ | \multicolumn{13}{c}{df$_2$} |
	6	8	10	12	14	16	18	20	24	30	40	60	120
\multicolumn{14}{l}{$F = 4.0$ (a = .500, b = 1.155)}
2	2.58	2.44	2.35	2.29	2.25	2.22	2.20	2.18	2.15	2.12	2.09	2.06	2.03
4	2.63	2.50	2.41	2.35	2.30	2.27	2.24	2.22	2.18	2.15	2.12	2.08	2.05
6	2.65	2.52	2.43	2.37	2.32	2.28	2.25	2.23	2.19	2.16	2.12	2.08	2.04
10	2.67	2.55	2.46	2.39	2.34	2.30	2.26	2.24	2.20	2.16	2.12	2.08	2.04
20	2.69	2.57	2.47	2.40	2.35	2.30	2.27	2.24	2.20	2.15	2.11	2.07	2.03
∞	2.71	2.59	2.49	2.42	2.36	2.31	2.27	2.24	2.19	2.15	2.11	2.06	2.02
\multicolumn{14}{l}{$F = 6.0$ (a = .408, b = 1.095)}
2	2.53	2.37	2.27	2.21	2.16	2.13	2.10	2.08	2.05	2.02	1.99	1.96	1.93
4	2.56	2.40	2.30	2.23	2.18	2.14	2.12	2.09	2.06	2.02	1.99	1.96	1.93
6	2.58	2.42	2.31	2.24	2.19	2.15	2.12	2.09	2.06	2.02	1.99	1.95	1.92
10	2.59	2.43	2.32	2.24	2.19	2.15	2.12	2.09	2.06	2.02	1.99	1.95	1.92
20	2.60	2.44	2.32	2.25	2.19	2.15	2.12	2.09	2.05	2.02	1.98	1.95	1.92
∞	2.61	2.44	2.33	2.25	2.19	2.15	2.12	2.09	2.05	2.02	1.98	1.95	1.92
\multicolumn{14}{l}{$F = 10.0$ (a = .316, b = 1.054)}
2	2.48	2.30	2.19	2.12	2.07	2.04	2.01	1.99	1.96	1.93	1.90	1.87	1.85
4	2.49	2.31	2.20	2.13	2.08	2.04	2.01	1.99	1.96	1.93	1.90	1.87	1.84
6	2.50	2.31	2.20	2.13	2.08	2.04	2.01	1.99	1.96	1.93	1.90	1.87	1.84
≥10	2.51	2.32	2.20	2.13	2.08	2.04	2.01	1.99	1.96	1.93	1.90	1.87	1.84
\multicolumn{14}{l}{$F = 25.0$ (a = .200, b = 1.021)}
2–4	2.40	2.20	2.10	2.03	1.99	1.95	1.93	1.91	1.88	1.86	1.83	1.80	1.78
≥6	2.41	2.21	2.10	2.03	1.99	1.95	1.93	1.91	1.88	1.86	1.83	1.80	1.78
\multicolumn{14}{l}{$F = \infty$ (a = 0, b = 1)}
≥2	2.33	2.13	2.03	1.97	1.93	1.90	1.88	1.86	1.84	1.81	1.79	1.76	1.74

TABLE 13   Waller–Duncan k Ratio Test Values of t_c for $k = 500$

df₁	6	8	10	12	14	16	18	20	24	30	40	60	120
$F = 1.2$ (a = .913, b = 2.449)													
2–16	*	*	*	*	*	*	*	*	*	*	*	*	*
20	4.70	4.82	4.89	*	*	*	*	*	*	*	*	*	*
40	4.75	4.91	5.03	5.12	5.20	5.25	5.30	5.34	5.41	5.48	5.55	5.61	5.67
100	4.79	4.98	5.13	5.25	5.34	5.43	5.50	5.56	5.65	5.76	5.89	6.02	6.13
∞	4.81	5.03	5.20	5.34	5.46	5.56	5.65	5.73	5.86	6.02	6.20	6.41	6.56
$F = 1.4$ (a = .845, b = 1.871)													
2–14	*	*	*	*	*	*	*	*	*	*	*	*	*
16	4.61	4.66	4.68	4.69	4.69	4.69	4.69	4.68	4.67	4.65	4.62	4.58	4.53
20	4.64	4.70	4.73	4.75	4.76	4.77	4.77	4.76	4.76	4.74	4.72	4.68	4.62
40	4.68	4.78	4.85	4.89	4.92	4.94	4.96	4.96	4.97	4.97	4.95	4.90	4.81
∞	4.74	4.88	4.99	5.06	5.12	5.17	5.20	5.23	5.26	5.28	5.26	5.16	4.82
$F = 1.7$ (a = .767, b = 1.558)													
2–8	*	*	*	*	*	*	*	*	*	*	*	*	*
10	*	*	*	*	*	*	*	*	*	4.08	4.02	3.95	3.87
12	4.50	4.46	4.42	4.38	4.34	4.30	4.27	4.24	4.19	4.14	4.07	3.99	3.90
20	4.55	4.54	4.52	4.49	4.46	4.43	4.40	4.37	4.32	4.26	4.18	4.08	3.95
40	4.59	4.61	4.61	4.60	4.57	4.55	4.52	4.49	4.44	4.36	4.26	4.12	3.93
∞	4.64	4.69	4.71	4.72	4.71	4.69	4.66	4.63	4.57	4.46	4.31	4.07	3.76
$F = 2.0$ (a = .707, b = 1.414)													
2–6	*	*	*	*	*	*	*	*	*	*	*	*	*
8	*	*	*	*	*	3.98	3.93	3.89	3.83	3.76	3.69	3.60	3.51
10	4.41	4.31	4.22	4.15	4.08	4.03	3.98	3.94	3.88	3.80	3.72	3.63	3.53
20	4.48	4.41	4.34	4.27	4.21	4.16	4.10	4.06	3.98	3.89	3.78	3.65	3.51
40	4.51	4.47	4.41	4.35	4.29	4.23	4.17	4.12	4.03	3.92	3.78	3.62	3.44
∞	4.55	4.53	4.49	4.43	4.37	4.31	4.25	4.19	4.07	3.93	3.75	3.54	3.33
$F = 2.4$ (a = .645, b = 1.309)													
2–8	*	*	*	*	*	*	*	*	*	*	*	*	*
6	*	*	*	*	3.77	3.71	3.65	3.61	3.54	3.47	3.39	3.30	3.22
8	4.31	4.14	4.01	3.91	3.83	3.76	3.70	3.66	3.58	3.50	3.41	3.32	3.22
10	4.33	4.18	4.05	3.95	3.87	3.79	3.73	3.68	3.60	3.51	3.42	3.31	3.21
20	4.39	4.26	4.14	4.04	3.95	3.87	3.80	3.74	3.64	3.53	3.41	3.28	3.15
∞	4.45	4.35	4.25	4.14	4.03	3.94	3.85	3.78	3.64	3.50	3.34	3.18	3.04

*All differences not significant. $a = 1/F^{1/2}$; $b = [F/(F - 1)]^{1/2}$.

From "Corrigenda" by R. Waller and D. Duncan, *Journal of the American Statistical Association*, 67 (1972): 253–255, Table A2. Reproduced by permission of the American Statistical Association.

T A B L E   1 3   *(continued)*

df$_1$	6	8	10	12	14	16	18	20	24	30	40	60	120
						df$_2$							
$F = 3.0$ (a = .577, b = 1.225)													
2	*	*	*	*	*	*	*	*	*	*	*	*	*
4	*	*	*	*	*	3.43	3.38	3.33	3.26	3.19	3.12	3.04	2.97
6	4.19	3.95	3.79	3.66	3.56	3.49	3.43	3.37	3.30	3.21	3.13	3.04	2.95
10	4.24	4.02	3.85	3.72	3.62	3.53	3.46	3.40	3.31	3.21	3.12	3.02	2.92
20	4.28	4.08	3.91	3.77	3.65	3.56	3.48	3.41	3.31	3.20	3.09	2.98	2.87
∞	4.33	4.15	3.97	3.82	3.69	3.57	3.48	3.40	3.28	3.15	3.03	2.92	2.82
$F = 4.0$ (a = .500, b = 1.155)													
2	*	*	*	*	*	*	*	*	*	*	*	2.81	2.75
4	*	3.74	3.54	3.40	3.30	3.22	3.16	3.11	3.04	2.96	2.89	2.81	2.74
6	4.08	3.78	3.58	3.43	3.32	3.24	3.17	3.12	3.04	2.95	2.87	2.79	2.71
10	4.12	3.83	3.62	3.46	3.34	3.25	3.17	3.11	3.03	2.94	2.85	2.77	2.69
20	4.15	3.86	3.64	3.48	3.35	3.25	3.17	3.10	3.01	2.92	2.83	2.74	2.66
∞	4.19	3.90	3.67	3.49	3.35	3.24	3.15	3.09	2.99	2.89	2.80	2.72	2.65
$F = 6.0$ (a = .408, b = 1.095)													
2	*	*	3.28	3.14	3.04	2.97	2.91	2.87	2.81	2.74	2.68	2.62	2.56
4	3.90	3.54	3.32	3.17	3.06	2.98	2.92	2.87	2.80	2.73	2.66	2.60	2.53
6	3.93	3.57	3.33	3.18	3.06	2.98	2.91	2.86	2.79	2.72	2.65	2.58	2.52
10	3.95	3.59	3.34	3.18	3.06	2.97	2.91	2.85	2.78	2.71	2.64	2.57	2.51
20	3.97	3.60	3.35	3.18	3.06	2.97	2.90	2.84	2.77	2.70	2.63	2.56	2.51
∞	3.99	3.62	3.36	3.18	3.05	2.96	2.89	2.83	2.76	2.69	2.62	2.56	2.50
$F = 10.0$ (a = .316, b = 1.054)													
2	3.72	3.33	3.10	2.96	2.86	2.79	2.74	2.70	2.64	2.58	2.52	2.47	2.42
4	3.75	3.35	3.11	2.96	2.86	2.79	2.73	2.69	2.63	2.57	2.51	2.46	2.41
10	3.78	3.36	3.11	2.96	2.85	2.78	2.72	2.68	2.62	2.56	2.50	2.45	2.40
20	3.79	3.36	3.11	2.96	2.85	2.78	2.72	2.68	2.62	2.56	2.50	2.45	2.40
∞	3.80	3.37	3.11	2.95	2.85	2.77	2.72	2.67	2.61	2.56	2.50	2.45	2.40
$F = 25.0$ (a = .200, b = 1.021)													
2	3.55	3.14	2.92	2.79	2.70	2.64	2.59	2.56	2.51	2.46	2.41	2.36	2.32
10	3.57	3.14	2.92	2.79	2.70	2.64	2.59	2.55	2.50	2.45	2.41	2.36	2.32
∞	3.57	3.14	2.92	2.78	2.70	2.63	2.59	2.55	2.50	2.45	2.41	2.36	2.32
$F = ∞$ (a = 0, b = 1)													
≥2	3.39	3.00	2.80	2.69	2.61	2.55	2.51	2.48	2.44	2.39	2.35	2.31	2.27

TABLE 14   Percentage Points of $F_{max} = s_{max}^2/s_{min}^2$

Upper 5% points

df_2 \ t	2	3	4	5	6	7	8	9	10	11	12
2	39.0	87.5	142	202	266	333	403	475	550	626	704
3	15.4	27.8	39.2	50.7	62.0	72.9	83.5	93.9	104	114	124
4	9.60	15.5	20.6	25.2	29.5	33.6	37.5	41.1	44.6	48.0	51.4
5	7.15	10.8	13.7	16.3	18.7	20.8	22.9	24.7	26.5	28.2	29.9
6	5.82	8.38	10.4	12.1	13.7	15.0	16.3	17.5	18.6	19.7	20.7
7	4.99	6.94	8.44	9.70	10.8	11.8	12.7	13.5	14.3	15.1	15.8
8	4.43	6.00	7.18	8.12	9.03	9.78	10.5	11.1	11.7	12.2	12.7
9	4.03	5.34	6.31	7.11	7.80	8.41	8.95	9.45	9.91	10.3	10.7
10	3.72	4.85	5.67	6.34	6.92	7.42	7.87	8.28	8.66	9.01	9.34
12	3.28	4.16	4.79	5.30	5.72	6.09	6.42	6.72	7.00	7.25	7.48
15	2.86	3.54	4.01	4.37	4.68	4.95	5.19	5.40	5.59	5.77	5.93
20	2.46	2.95	3.29	3.54	3.76	3.94	4.10	4.24	4.37	4.49	4.59
30	2.07	2.40	2.61	2.78	2.91	3.02	3.12	3.21	3.29	3.36	3.39
60	1.67	1.85	1.96	2.04	2.11	2.17	2.22	2.26	2.30	2.33	2.36
∞	1.00	1.00	1.00	1.00	1.00	1.00	1.00	1.00	1.00	1.00	1.00

Upper 1% points

df_2 \ t	2	3	4	5	6	7	8	9	10	11	12
2	199	448	729	1036	1362	1705	2063	2432	2813	3204	3605
3	47.5	85	120	151	184	21(6)	24(9)	28(1)	31(0)	33(7)	36(1)
4	23.2	37	49	59	69	79	89	97	106	113	120
5	14.9	22	28	33	38	42	46	50	54	57	60
6	11.1	15.5	19.1	22	25	27	30	32	34	36	37
7	8.89	12.1	14.5	16.5	18.4	20	22	23	24	26	27
8	7.50	9.9	11.7	13.2	14.5	15.8	16.6	17.9	18.9	19.8	21
9	6.54	8.5	9.9	11.1	12.1	13.1	13.9	14.7	15.3	16.0	16.6
10	5.85	7.4	8.6	9.6	10.4	11.1	11.8	12.4	12.9	13.4	13.9
12	4.91	6.1	6.9	7.6	8.2	8.7	9.1	9.5	9.9	10.2	10.6
15	4.07	4.9	5.5	6.0	6.4	6.7	7.1	7.3	7.5	7.8	8.0
20	3.32	3.8	4.3	4.6	4.9	5.1	5.3	5.5	5.6	5.8	5.9
30	2.63	3.0	3.3	3.4	3.6	3.7	3.8	3.9	4.0	4.1	4.2
60	1.96	2.2	2.3	2.4	2.4	2.5	2.5	2.6	2.6	2.7	2.7
∞	1.00	1.0	1.0	1.0	1.0	1.0	1.0	1.0	1.0	1.0	1.0

s_{max}^2 is the largest and s_{min}^2 the smallest in a set of t independent mean squares, each based on $df_2 = n - 1$ degrees of freedom. Values in the column $t = 2$ and in the rows $df_2 = 2$ and ∞ are exact. Elsewhere the third digit may be in error by a few units for the 5% points and several units for the 1% points. The third digit figures in parentheses for $df_2 = 3$ are the most uncertain.
From *Biometrika Tables for Statisticians*, 3rd ed., Vol. 1, edited by E. S. Pearson and H. O. Hartley (New York: Cambridge University Press, 1966), Table, p. 202. Reproduced by permission of the *Biometrika Trustees*.

TABLE 15 Values of 2 arcsin $\sqrt{\hat{\pi}}$

$\hat{\pi}$		$\hat{\pi}$		$\hat{\pi}$		$\hat{\pi}$		$\hat{\pi}$	
.001	.0633	.041	.4078	.36	1.2870	.76	2.1177	.971	2.7993
.002	.0895	.042	.4128	.37	1.3078	.77	2.1412	.972	2.8053
.003	.1096	.043	.4178	.38	1.3284	.78	2.1652	.973	2.8115
.004	.1266	.044	.4227	.39	1.3490	.79	2.1895	.974	2.8177
.005	.1415	.045	.4275	.40	1.3694	.80	2.2143	.975	2.8240
.006	.1551	.046	.4323	.41	1.3898	.81	2.2395	.976	2.8305
.007	.1675	.047	.4371	.42	1.4101	.82	2.2653	.977	2.8371
.008	.1791	.048	.4418	.43	1.4303	.83	2.2916	.978	2.8438
.009	.1900	.049	.4464	.44	1.4505	.84	2.3186	.979	2.8507
.010	.2003	.050	.4510	.45	1.4706	.85	2.3462	.980	2.8578
.011	.2101	.06	.4949	.46	1.4907	.86	2.3746	.981	2.8650
.012	.2195	.07	.5355	.47	1.5108	.87	2.4039	.982	2.8725
.013	.2285	.08	.5735	.48	1.5308	.88	2.4341	.983	2.8801
.014	.2372	.09	.6094	.49	1.5508	.89	2.4655	.984	2.8879
.015	.2456	.10	.6435	.50	1.5708	.90	2.4981	.985	2.8960
.016	.2537	.11	.6761	.51	1.5908	.91	2.5322	.986	2.9044
.017	.2615	.12	.7075	.52	1.6108	.92	2.5681	.987	2.9131
.018	.2691	.13	.7377	.53	1.6308	.93	2.6062	.988	2.9221
.019	.2766	.14	.7670	.54	1.6509	.94	2.6467	.989	2.9315
.020	.2838	.15	.7954	.55	1.6710	.95	2.6906	.990	2.9413
.021	.2909	.16	.8230	.56	1.6911	.951	2.6952	.991	2.9516
.022	.2978	.17	.8500	.57	1.7113	.952	2.6998	.992	2.9625
.023	.3045	.18	.8763	.58	1.7315	.953	2.7045	.993	2.9741
.024	.3111	.19	.9021	.59	1.7518	.954	2.7093	.994	2.9865
.025	.3176	.20	.9273	.60	1.7722	.955	2.7141	.995	3.0001
.026	.3239	.21	.9521	.61	1.7926	.956	2.7189	.996	3.0150
.027	.3301	.22	.9764	.62	1.8132	.957	2.7238	.997	3.0320
.028	.3363	.23	1.0004	.63	1.8338	.958	2.7288	.998	3.0521
.029	.3423	.24	1.0239	.64	1.8338	.959	2.7338	.999	3.0783
.030	.3482	.25	1.0472	.65	1.8546	.960	2.7389		
.031	.3540	.26	1.0701	.66	1.8965	.961	2.7440		
.032	.3597	.27	1.0928	.67	1.9177	.962	2.7492		
.033	.3654	.28	1.1152	.68	1.9391	.963	2.7545		
.034	.3709	.29	1.1374	.69	1.9606	.964	2.7598		
.035	.3764	.30	1.1593	.70	1.9823	.965	2.7652		
.036	.3818	.31	1.1810	.71	2.0042	.966	2.7707		
.037	.3871	.32	1.2025	.72	2.0264	.967	2.7762		
.038	.3924	.33	1.2239	.73	2.0488	.968	2.7819		
.039	.3976	.34	1.2451	.74	2.0715	.969	2.7876		
.040	.4027	.35	1.2661	.75	2.0944	.970	2.7934		

From *Experimental Design: Procedures for the Behavioral Sciences*, by Roger E. Kirk. Copyright © 1968 by Wadsworth Publishing Company, Inc. Reprinted by permission of the publisher, Brooks/Cole Publishing Company, Belmont, Calif.

TABLE 16 Values of d_n Used in Control Limits for μ

n	d_n
2	1.128
3	1.693
4	2.059
5	2.326
6	2.534
7	2.704
8	2.847
9	2.970
10	3.078
11	3.173
12	3.258

TABLE 17 Values of D_n and D_n' Used in Control Limits for Product Variability

Number of Observations in Sample, n	D_n	D_n'
2	0	3.267
3	0	2.575
4	0	2.282
5	0	2.115
6	0	2.004
7	0.076	1.924
8	0.136	1.864
9	0.184	1.816
10	0.223	1.777
11	0.256	1.744
12	0.284	1.716
13	0.308	1.692
14	0.329	1.671
15	0.348	1.652
16	0.364	1.636
17	0.379	1.621
18	0.392	1.608
19	0.404	1.596
20	0.414	1.586
21	0.425	1.575
22	0.434	1.566
23	0.443	1.557
24	0.452	1.548
25	0.459	1.541
Over 25	$3/\sqrt{n}$	$3/\sqrt{n}$

APPENDIX B: CLINICAL TRIAL DATABASE

The data presented here are from a clinical trial that was conducted to compare the safety and efficacy of three different compounds (A, B, and C) and a placebo (D) in the treatment of patients who exhibited characteristic signs and symptoms of depression. Certain predrug (baseline) determinations were made on each of the 100 patients to determine their suitability for the study. Each patient who qualified for the study was assigned at random to one of the four treatment groups and was dispensed medication for the duration of the study. Neither the investigator nor the patient knew which medication had been assigned.

At the end of the study, scores on numerous anxiety and depression scales were made. Descriptions of the variables measured and their codes are as follows.

Variable Descriptions

___

PATIENT:	patient number
AGE:	age (years)
MAR_STAT:	marital status
	1 = single
	2 = married
	3 = separated or divorced
	4 = widowed
COFF_TEA:	coffee/tea consumption (cups/day)
TOBACCO:	tobacco consumption
	0 = none
	1 = <1 pack daily
	2 = 1 pack daily
	3 = >1 pack daily
ALCOHOL:	alcohol consumption
	0 = none
	1 = social drinker (<1 drink weekly)
	2 = social drinker (1 to 2 drinks weekly)
	3 = 1 to 2 drinks most days
	4 = 3 or more drinks most days
TRT_EMOT:	previous treatment for emotional problems
	1 = psychiatrist
	2 = nonpsychiatrist physician
	3 = both
	4 = other
HOSPITAL:	hospitalization for emotional problems
	0 = no
	1 = yes
PSY_DIAG:	psychiatrist diagnosis
	1 = major depressive disorder, single episode
	2 = major depressive disorder, recurrent episode
	3 = bipolar affective disorder
	4 = chronic depressive disorder
	5 = atypical depressive disorder
	6 = adjustment disorder with depressed mood

ANXIETY: HAM-D anxiety score

RETARDTN: HAM-D retardation score

SLEEP: HAM-D sleep disturbance score

TOTAL: HAM-D total score

OBRIST: HOPKINS OBRIST cluster total

APPETITE: appetite disturbance score

CHANGED: how much the patient has changed

               1 = very much improved

               2 = much improved

               3 = minimally improved

               4 = no change

               5 = minimally worse

               6 = much worse

               7 = very much worse

THER_EFF: therapeutic effect

               1 = marked

               2 = moderate

               3 = minimal

               4 = unchanged

               5 = worse

ADV_EFF: adverse effects

               1 = none

               2 = does not significantly interfere with patient's functioning

               3 = significantly interferes with patient's functioning

               4 = nullifies therapeutic effect

TREATMNT: drug treatment group

               A

               B

               C

               D—placebo (control) group

PATIENT	AGE	MAR-STAT	COFF-TEA	TOBACCO	ALCOHOL	TRT-EMOT	HOSPITAL	PSY-DIAG	ANXIETY	RETARDTN	SLEEP	TOTAL	OBRIST	APPETITE	CHANGED	THER-EFF	ADV-EFF	TREATMNT
1	23	2	3	1	0	0	0	1	0.33	0.75	0.00	16	56	91.0	4	4	1	D
2	18	1	0	2	0	1	0	2	0.33	1.25	0.33	12	57	42.5	3	3	1	A
3	36	2	2	2	2	1	0	2	0.50	0.25	0.33	6	40	91.0	1	1	2	B
4	51	4	5	3	0	0	0	1	0.17	0.75	0.67	6	39	61.0	2	1	1	A
5	24	1	6	2	1	1	0	2	1.00	1.00	1.00	13	49	1.5	3	3	1	B
6	59	4	3	1	0	1	0	4	0.33	1.50	0.00	11	40	72.0	2	2	1	A
7	56	1	2	0	0	1	1	4	1.67	1.75	0.00	21	44	7.5	2	2	1	B
8	70	4	1	0	0	1	0	2	0.50	1.75	0.00	12	39	92.5	2	2	2	A
9	30	3	4	3	2	1	0	2	0.83	1.00	0.67	15	49	2.0	2	2	4	D
10	55	4	2	0	2	0	0	1	0.33	1.00	1.00	11	44	31.0	2	2	1	D
11	40	2	4	2	0	1	0	2	0.83	1.50	1.33	23	79	92.0	4	4	1	C
12	61	2	2	0	1	1	1	2	0.50	0.75	0.00	8	30	1.0	2	2	2	C
13	64	2	3	2	0	1	0	2	0.33	1.50	0.00	9	48	11.5	2	2	3	A
14	19	1	10	2	2	1	0	1	0.50	0.75	0.00	7	42	91.5	2	2	1	B
15	46	3	2	0	1	1	0	1	0.17	0.25	0.00	4	35	92.0	1	1	1	A
16	36	2	10	3	0	1	0	2	0.50	1.50	0.00	9	42	72.0	2	2	1	C,
17	30	2	2	0	2	0	0	1	0.00	0.50	0.67	4	35	41.0	2	1	1	B
18	34	2	8	3	1	1	0	1	0.50	1.00	0.33	14	56	91.5	3	3	1	D
19	28	1	2	0	1	0	0	1	0.67	1.00	0.33	12	43	72.5	1	2	2	B
20	33	3	3	2	1	1	0	2	0.67	0.75	0.33	8	39	93.0	2	2	2	C
21	51	3	7	3	2	1	1	5	0.83	2.00	0.00	21	99	42.5	4	4	2	A
22	51	2	0	0	0	1	0	4	0.83	1.00	0.00	12	68	61.0	3	3	3	C
23	54	2	3	0	0	0	1	2	0.83	1.00	1.00	16	49	11.0	3	3	2	B
24	35	3	5	0	2	1	0	4	0.67	1.50	0.00	11	42	61.0	2	2	3	A
25	46	2	3	0	1	0	1	2	0.67	2.00	1.33	17	63	61.5	4	4	1	D
26	34	2	7	0	0	1	1	2	0.33	0.25	0.33	6	51	12.5	2	2	2	C
27	27	3	2	1	2	1	0	5	0.83	0.75	0.00	11	58	11.0	2	2	2	A
28	23	2	0	1	0	0	0	1	0.33	0.75	0.00	6	47	91.5	2	2	2	C
29	35	3	6	0	3	1	0	5	0.33	0.50	0.00	9	68	1.0	3	2	2	B
30	19	1	2	0	0	0	0	4	0.67	0.50	0.00	11	47	2.5	3	3	2	B
31	40	3	3	0	3	0	0	1	1.00	1.75	0.00	17	71	62.0	4	4	1	D
32	52	2	5	0	0	1	1	2	0.83	1.75	0.67	19	41	91.5	3	3	2	B
33	51	3	6	0	0	1	0	2	0.83	1.00	1.67	20	63	11.0	3	4	2	D
34	34	3	0	0	2	1	0	4	0.17	0.50	0.33	5	37	13.0	2	2	1	C
35	59	2	4	2	0	1	0	1	0.67	2.50	0.00	17	54	62.5	3	3	2	C

PATIENT	AGE	MAR-STAT	C OFF TEA	TOBACCO	ALCOHOL	TRT-EMOT	HOSPITAL	PSY-DIAG	ANXIETY	RETARDTN	SLEEP	TOTAL	OBRIST	APPETITE	CHANGED	THER-EFF	ADV-EFF	TREATMNT
36	31	3	2	0	1	1	1	5	0.50	1.75	0.33	12	51	11.5	3	2	2	C
37	54	2	10	0	2	1	0	2	1.17	1.25	1.33	22	96	92.5	5	5	1	A
38	63	4	2	0	2	1	0	1	0.33	0.25	0.33	7	58	32.5	2	2	2	B
39	34	2	1	2	2	0	1	1	0.50	0.25	0.33	27	90	92.0	5	5	1	D
40	30	1	1	0	1	1	0	1	1.00	0.50	0.67	13	59	33.0	2	2	1	A
41	32	3	2	3	1	1	0	2	0.67	1.25	0.67	14	58	61.0	2	2	2	C
42	21	1	2	0	3	1	0	2	0.83	1.00	1.00	20	60	41.5	3	3	2	B
43	42	2	1	2	1	1	1	2	1.00	1.50	1.00	24	85	62.0	3	2	1	B
44	60	2	0	3	0	1	0	4	0.17	0.75	0.33	5	39	1.5	1	1	1	A
45	53	2	2	0	1	1	0	2	0.67	1.00	0.67	10	38	31.0	2	2	2	B
46	54	4	4	0	1	1	0	4	1.50	1.75	0.00	14	42	93.5	2	2	1	C
47	38	2	2	1	0	1	0	2	1.50	1.75	1.00	24	85	15.0	3	4	2	D
48	41	2	4	0	0	1	1	2	0.33	0.75	0.00	11	47	2.5	2	2	1	A
49	32	3	0	0	1	1	0	4	1.00	1.00	1.00	20	35	42.5	3	2	2	B
50	43	2	4	0	0	1	0	4	0.83	1.25	1.33	21	44	31.5	4	4	1	B
51	51	2	1	0	1	1	0	2	0.83	2.25	0.00	20	80	61.0	5	5	1	A
52	23	2	0	1	3	1	0	2	1.33	1.25	0.67	20	39	31.5	3	3	1	A
53	55	2	2	0	0	1	0	1	0.83	1.25	0.00	16	52	12.5	3	3	1	C
54	45	1	3	3	0	1	0	2	0.33	1.25	0.00	14	46	41.5	2	2	1	C
55	30	2	1	0	0	0	0	1	1.17	1.75	0.00	17	64	2.0	3	3	2	C
56	53	4	1	3	4	0	0	1	0.83	0.50	0.00	19	82	62.0	3	4	2	D
57	45	1	3	1	3	0	0	4	0.83	0.50	0.00	8	40	2.0	1	2	1	A
58	48	2	10	1	2	1	0	4	0.33	1.00	0.00	8	32	41.0	2	2	2	B
59	49	1	4	0	3	1	0	5	0.67	1.75	0.33	16	68	12.0	2	2	2	A
60	55	2	6	0	2	0	1	4	0.50	1.00	0.00	9	42	91.0	2	2	2	A
61	33	2	1	0	1	1	0	2	1.17	2.25	2.00	32	112	42.0	4	4	1	D
62	27	1	1	2	3	1	0	5	0.17	0.00	0.00	3	34	72.5	1	1	2	C
63	30	1	2	1	3	1	0	4	0.67	0.00	0.00	6	37	73.5	1	1	2	A
64	35	2	4	3	0	1	0	2	0.50	1.50	0.33	16	43	11.0	3	3	1	A
65	55	2	4	0	0	1	0	2	1.00	2.00	0.33	24	37	92.0	3	3	1	B
66	22	3	0	1	1	1	0	2	1.00	1.50	0.33	20	46	11.5	3	3	1	D
67	37	3	1	2	0	1	0	2	0.50	0.75	0.00	11	32	41.5	2	2	1	C
68	49	2	6	3	0	1	0	2	0.50	0.75	0.00	13	54	2.5	2	1	1	C
69	21	1	0	1	0	1	0	2	1.17	2.50	2.00	34	74	11.5	5	5	1	A
70	33	3	1	3	2	1	0	1	0.17	0.50	0.00	8	34	42.5	1	1	1	C

PATIENT	AGE	MAR—STAT	COFF—TEA	TOBACCO	ALCOHOL	TRT—EMOT	HOSPITAL	PSY—DIAG	ANXIETY	RETARDTN	SLEEP	TOTAL	OBRIST	APPETITE	CHANGED	THER——EFF	ADV——EFF	TREATMNT
71	35	3	10	2	0	1	0	2	1.17	2.50	0.00	24	39	92.5	4	4	2	B
72	39	3	0	0	0	1	0	2	0.50	1.25	0.00	22	66	41.0	2	3	1	D
73	34	1	0	0	0	1	0	2	0.33	0.75	0.00	14	48	71.0	2	2	1	C
74	53	3	3	1	0	1	0	2	0.50	1.00	0.00	12	39	43.5	2	2	1	A
75	34	2	0	0	0	1	0	2	0.33	0.50	0.00	10	36	3.0	2	2	1	A
76	35	2	2	0	1	1	0	2	0.83	1.00	0.33	14	35	92.0	2	2	1	B
77	32	2	0	0	0	1	0	2	0.50	1.00	0.00	12	39	91.0	2	2	1	B
78	43	2	2	2	0	1	0	2	0.17	0.50	0.67	5	57	41.5	2	2	1	C
79	64	2	3	0	0	0	1	1	1.33	1.50	1.67	26	42	72.5	5	5	1	D
80	31	3	2	2	0	1	0	2	0.17	0.00	0.00	2	35	43.0	2	1	1	B
81	41	2	2	2	1	0	0	1	0.83	1.25	0.00	16	66	41.5	3	3	1	C
82	53	2	2	0	0	1	0	2	1.67	1.50	1.67	23	47	32.0	6	5	4	D
83	61	2	3	2	1	1	0	2	0.83	1.25	0.00	15	43	31.0	3	2	1	C
84	36	2	2	2	0	1	0	4	1.00	1.75	0.00	16	63	42.0	3	3	1	C
85	29	1	0	0	0	0	1	1	0.33	0.25	0.00	5	53	12.0	1	1	1	A
86	25	2	8	3	2	1	0	1	1.17	1.25	0.00	14	80	91.0	4	4	4	A
87	55	2	1	0	0	1	0	2	0.67	0.50	2.00	16	68	31.0	2	2	2	D
88	34	2	3	0	0	0	0	1	0.67	0.00	0.00	5	31	1.0	1	1	2	C
89	20	1	0	0	0	1	0	2	0.50	0.75	1.67	15	36	43.0	3	2	1	B
90	33	2	1	2	2	1	1	2	0.83	1.25	0.67	16	63	61.5	2	2	1	B
91	44	2	0	0	0	1	0	2	0.33	1.75	1.00	16	55	42.5	4	4	1	B
92	58	2	3	0	0	1	1	2	1.17	1.75	1.00	21	75	11.5	3	4	1	D
93	46	4	3	0	2	1	0	1	0.83	1.25	1.00	16	83	13.5	3	3	2	D
94	31	2	2	0	1	1	0	4	1.17	1.50	0.00	21	65	61.5	4	4	2	D
95	29	3	2	2	2	0	0	1	0.50	1.00	0.67	20	92	32.5	4	4	2	D
96	50	3	3	0	2	1	0	1	0.83	0.25	0.67	18	50	91.0	2	1	2	D
97	27	2	0	0	0	1	0	2	0.67	1.00	2.00	17	64	62.5	3	3	2	D
98	31	3	5	0	1	1	0	2	0.50	0.75	0.67	19	74	61.5	3	3	1	D
99	69	4	4	1	2	1	0	1	1.67	2.25	1.33	26	87	63.5	5	5	3	D
100	41	2	0	0	0	1	0	2	0.00	0.50	0.00	3	37	92.5	2	2	1	D

REFERENCES

BOOKS AND ARTICLES

Anderson, M.J., R.C. Lamb, C.H. Mekelsen, and R.L. Wiscombe (1974). Feeding liquid whey to dairy cattle. *Journal of Dairy Science* 57.

Andrews, D.F., and A.M., Herzberg (1985). *Data: A collection of problems from many fields for the student and research worker.* New York: Springer-Verlag.

Bancroft, T.A. (1968). *Topics in intermediate statistical methods.* Vol. 1. Ames, Iowa: Iowa State University Press.

Barr, A.J., J.H. Goodnight, J.P. Sall, and J.T. Helwig (1976). *A user's guide to SAS 76.* Raleigh, N.C.: SAS Institute, Inc.

Bishop, Y.M.M., S.E. Fineberg, and P.W. Holland (1975). *Discrete multivariate analysis: theory and practice.* Cambridge, Mass.: MIT Press.

Boardman, T.J., and D.R. Moffitt (1971). Graphical Monte Carlo type 1 error rates for multiple comparison procedures. *Biometrics* 27: 738–744.

Bock, R.D. (1975). *Multivariate statistical methods in behavioral research.* Chap. 8. New York: McGraw-Hill.

Brown, S.S., M.J.R. Healy, and M. Kearns (1981). Report on the interlaboratory trial of the reference method for the determination of total calcium in serum. *Journal of Clinical Chemistry and Clinical Biochemistry* 19: 395–426.

Brownlee, K.A. (1965). *Statistical theory and methodology in science and engineering.* 2nd ed. New York: Wiley.

Buncher, C.R., and J-Y Tsay, editors (1981). *Statistics in the Pharmaceutical Industry.* New York: Marcel Dekker.

Bureau of Labor Statistics, *Handbook of Methods.* Vols. I and II (1982). Washington, D.C.: U.S. Department of Labor.

Carmer, S.G., and M.R. Swanson (1973). An evaluation of ten pairwise multiple comparison procedures by Monte Carlo methods. *Journal of the American Statistical Association* 68: 66–74.

Carter, R.L. (1981). Restricted maximum likelihood estimation of bias and reliability in the comparison of several measuring methods. *Biometrics* 37: 733–741.

Chew, V. (1976). *Comparing treatment means—A compendium.* Technical report no. 105. Agricultural Research Service of the U.S. Department of Agriculture and the Department of Statistics, University of Florida.

Cochran, W.G. (1954). Some methods for strengthening the common χ^2 test. *Biometrics* 10: 417–451.

———— (1964). Approximate significance levels of the Behrens–Fisher test. *Biometrics* 20: 191–195.

Cochran, W.G., and G.M. Cox (1957). *Experimental design.* 2nd ed. New York: Wiley.

Conover, W.J. (1980). *Practical nonparametric statistics.* 2nd ed. New York: Wiley.

Conover, W.J., and R.L. Iman (1981). Rank transformations as a bridge between parametric and nonparametric statistics. *The American Statistician* 35: 124–133.

Cornell, J.A. (1971). A review of multiple comparison procedures for comparing a set of k population means. *Proceedings of the Soil and Crop Science Society of Florida* 31: 92–97.

Cox, D.R. (1970). *The analysis of binary data.* New York: Halsted Press.

CRC standard mathematical tables (1961). 12th ed. Cleveland, Ohio: The Chemical Rubber Co.

Davies, O.L. (1956). *The design and analysis of experiments.* New York: Hafner.

Deming, W.E. (1982). *Quality, Productivity and Competitive Position.* MIT-CAES. Cambridge, Mass.

Deming, W.E. (1986). *Out of the Crisis.* MIT-CAES. Cambridge, Mass.

Dixon, W.J., and F.J. Massey, Jr. (1969). *Introduction to statistical analysis.* 3rd ed. New York: McGraw-Hill.

Draper, N.R., and H. Smith (1981). *Applied regression analysis*. 2nd ed. New York: Wiley.

Duncan, A.J. (1959). *Quality control and industrial statistics*. 2nd ed. Chap. 21. Homewood, Ill.: Irwin.

Duncan, D.B. (1955). Multiple range and multiple F tests. *Biometrics* 11: 1–42.

———— (1957). Multiple range tests for correlated and heteroscedastic means. *Biometrics* 13: 164–176.

———— (1975). T tests and intervals for comparisons suggested by the data. *Biometrics* 31: 339–359.

Dunnett, C.W. (1955). A multiple comparison procedure for comparing several treatments with a control. *Journal of the American Statistical Association* 50: 1096–1121.

———— (1980a). Pairwise multiple comparisons in the homogeneous variance, unequal sample size case. *Journal of the American Statistical Association* 75: 789–795.

———— (1980b). Pairwise multiple comparisons in the unequal variance case. *Journal of the American Statistical Association* 75: 796–800.

Durbin, J., and G.S. Watson (1951). Testing for Serial Correlation in Least Squares, II. *Biometrika* 38: 159–178.

Fisher, R.A. (1949). *The design of experiments*. Edinburgh: Oliver and Boyd.

Forthofer, R.N., and G.G. Koch (1973). An analysis of compounded functions of categorical data. *Biometrics* 29: 143–157.

Forthofer, R.N., and R.G. Lehnen (1981). *Public program analysis: A new categorical data approach*. Belmont, Calif.: Wadsworth.

Forthofer, R.N., C.F. Starmer, and J.E. Grizzle (1971). A program for the analysis of categorical data by linear models. *Journal of Biomedical Systems* 2: 3–48.

Gaines, P.A., H.J. Kaselman, and J.C. Rogan (1981). Simultaneous pairwise multiple comparison procedures for means where sample sizes are unequal. *Psychological Bulletin* 90: 594–598.

Geisser, S., and S.W. Greenhouse (1958). An extension of Box's results on the use of the F distribution in multivariate analysis. *Annals of Mathematical Statistics* 29: 885–891.

Goodman, L.A. (1971). The analysis of multidimensional contingency tables: stepwise procedures and direct estimation methods for building models for multiple classifications. *Technometrics* 13: 33–61.

Graybill, F.A. (1976). *Theory and application of the linear model.* Boston, Mass.: Duxbury Press.

Greenhouse, S.W., and S. Geisser (1959). On methods in the analysis of profile data. *Psychometrika* 24: 95–112.

Grizzle, J.E., C.F. Starmer, and G.G. Koch (1969). Analysis of categorical data by linear models. *Biometrics* 25: 489–504.

Grizzle, J.E., and O.D. Williams (1972). Log linear models and tests of independence for contingency tables. *Biometrics* 28: 137–156.

Haberman, S.J. (1973). *C-tab analysis of multidimensional contingency tables by log-linear models.* Ann Arbor, Mich.: National Educational Resources, Inc.

Hammer, M. (1990). Re-engineering Work: Don't Automate, Obliterate. *Harvard Business Review:* July–August.

Harnett, D.L. (1970). *Introduction to statistical methods.* Reading, Mass.: Addison-Wesley.

Harter, H.L. (1957). Error rates and sample sizes for range tests in multiple comparisons. *Biometrics* 13: 511–598.

Hicks, C.R. (1973). *Fundamental concepts in experimental design.* 2nd ed. New York: Holt, Rinehart and Winston.

Hildebrand, D.K., and L. Ott (1991). *Statistical thinking for managers.* 3rd ed. Boston, Mass.: Duxbury Press.

Hogg, R.V., and A.J. Craig (1978). *Introduction to mathematical statistics.* 4th ed. New York: Macmillan.

Hollander, M., and D.A. Wolfe (1973). *Nonparametric statistical methods.* New York: Wiley.

Hurlburt, R.J., and D.K. Spiegel (1976). Dependence of F ratios sharing a common denominator mean square. *American Statistician* 30: 74–78.

Huynh, H., and L.S. Feldt (1970). Conditions under which mean square ratios in repeated measurement designs have fixed F-distributions. *Journal of the American Statistical Association* 65: 1582–1589.

Kirk, R.E. (1968). *Experimental design: Procedures for the behavioral sciences.* Belmont, Calif.: Brooks/Cole.

Kramer, C.Y. (1956). Extension of multiple range tests to group means with unequal numbers of replications. *Biometrics* 12: 307–310.

———— (1957). Extension of multiple range tests to group correlated adjusted means. *Biometrics* 13: 13–18.

Kritzer, H. (1979). Approaches to the analysis of complex contingency tables: A guide for the perplexed. *Sociological Methods and Research* 7: 305–329.

Ku, H.H., R.N. Varner, and S. Kullback (1971). Analysis of multidimensional contingency tables. *Journal of the American Statistical Association* 66: 55–64.

Landis, J.R., M.M. Cooper, G.G. Koch, and J. Kennedy (1978). Computer program for testing average partial association in three-way contingency tables (Parcat). *Computer Programs in Biomedicine* 9: 223–246.

Landis, J.R., E.R. Heyman, and G.G. Koch (1978). Average partial association in three-way contingency tables: A review and discussion of alternative tests. *International Statistical Review* 46: 237–544.

Landis, J.R., W.M. Stanish, J.L. Freeman, and G.G. Koch (1976). A computer program for the generalized chi-square analysis of categorical data using weighted least squares (Gencat). *Computer Programs in Biomedicine* 6: 196–231.

Landis, J.R., W.M. Stanish, and G.G. Koch (1976). A computer program for the generalized chi-square analysis of categorical data using weighted least squares to compute Wald statistics (Gencat). *Biostatistics Technical Report No. 8.* Ann Arbor: Department of Biostatistics, University of Michigan.

Mallows, C.L. (1973). Some comments on Cp. *Technometrics* 15: 661–675.

Mantel, N., and W. Haenszel (1959). Statistical aspects of the analysis of data from retrospective studies of disease. *Journal of the National Cancer Institute* 22: 719–748.

Mendenhall, W. (1968). *Introduction to linear models and the design and analysis of experiments.* Belmont, Calif.: Wadsworth.

———— (1987). Introduction to probability and statistics. 7th ed. Boston, Mass.: Duxbury Press.

Miller, R.G., Jr. (1964). A trustworthy jackknife. *Annals of Mathematical Statistics* 35: 1594–1605.

Neave, H.R. (1966). A development of Tukey's quick test for location. *Journal of the American Statistical Association* 61: 897–1262.

———— (1975). A quick and simple technique for general slippage problems. *Journal of the American Statistical Association* 70: 721–726.

Omstead, P.S., and J.W. Tukey (1947). A corner test for association. *Annals of Mathematical Statistics* 18: 495–513.

Ostle, B. (1963). *Statistics in research.* 2nd ed. Ames, Iowa: Iowa State University Press.

Ott, L., R. Larson, C. Rexroat, and W. Mendenhall (1992). *Statistics: A tool for the social sciences.* 5th ed. Boston, Mass.: Duxbury Press.

Ott, L., and W. Mendenhall (1985). *Understanding statistics.* 4th ed. Boston, Mass.: Duxbury Press.

Pearson, E.S., and H.O. Hartley (1942). The probability integral of the range in samples of *n* observations from a normal population. *Biometrics* 32: 301–310.

———— (1943). Tables of the probability integral of the Studentized range. *Biometrics* 33: 89–99.

———— (1954). *Biometrika tables for statisticians.* Vol. I. Cambridge: Cambridge University Press.

———— (1966). *Biometrika tables for statisticians.* 3rd ed. Vol. 1. London: Cambridge University Press.

Rowe, N.H., R.H. Anderson, and L.A. Wanninger (1974). Effects of read-to-eat breakfast cereals on dental caries experience in adolescent children: A three-year study. *Journal of Dental Research* 53: 33.

Satterthwaite, F.E. (1946). An approximate distribution of estimates of variance components. *Biometric Bulletin* 2: 110–114.

Scheaffer, R.L., W. Mendenhall, and L. Ott (1990). *Elementary Survey Sampling.* 4th ed. Boston: PWS-KENT.

Scheffé, H. (1953). A method for judging all contrasts in an analysis of variance. *Biometrika* 40: 87–104.

———— (1959). *The analysis of variance.* New York: Wiley.

Searle, S.R. (1971). *Linear models.* New York: Wiley.

Siegel, S. (1956). *Nonparametric statistics for the behavioral sciences.* New York: McGraw-Hill.

Snedecor, G.W., and W.G. Cochran (1980). *Statistical methods.* 7th ed. Ames, Iowa: Iowa State University Press.

Snowdon, C.T., and B.A. Sanderson (1974). Lead pica produced in rats. *Science* 183: 92–94.

Sokal, R.R., and F.J. Rohlf (1969). *Biometry.* San Francisco: Freeman.

Stedl, J.L. (1981). The effect of sample size on the true α level with Funcat—A simulation study. *Proceedings of the Sixth Annual SAS User's Group.* Cary, N.C.: SAS Institute, Inc.

Steel, R.G.D., and J.H. Torrie (1980). *Principles and procedures of statistics,* 2nd ed. New York: McGraw-Hill.

"Student" (1908). The probable error of a mean. *Biometrika* 6: 1–25.

Sugiura, N., and M. Otake (1974). An extension of the Mantel-Haenszel procedure to k $2 \times c$ contingency tables and the relation to the logit model. *Communications in Statistics* 3(9): 829–842.

Tanur, J.M., F. Mosteller, W.H. Kruskal, R.S. Pieters, and G.R. Rising (1972). *Statistics: A guide to the unknown.* San Francisco: Holden-Day.

Tukey, J.W. (1953). *The problem of multiple comparisons.* Mimeographed. Princeton, N.J.: Princeton University.

——— (1959). A quick compact two-sample test to Duckworth's specifications. *Technometrics* 1: 31–48.

——— (1977). *Exploratory Data Analysis.* Reading, Mass: Addison-Wesley.

Waller, R.A., and D.B. Duncan (1969). A Bayes rule for the symmetric multiple comparison problem. *Journal of the American Statistical Association* 64: 1484–1503.

——— (1972). Corrigenda. *Journal of the American Statistical Association* 67: 253–255.

Welch, B.L. (1938). The Significance of the Differences Between Two Means When the Population Variances are Unequal. *Biometrika* 29: 350–362.

——— (1956). On linear combinations of several variances. *Journal of the American Statistical Association* 51: 132–148.

Wilcoxon, F., and R.A. Wilcox (1964). *Some rapid approximate statistical procedures.* Pearl River, N.Y.: Lederle Laboratories.

Winer, B.J. (1971). *Statistical principles in experimental design.* New York: McGraw-Hill.

Zahn, D.A. (1974). *Documentation for Contab—A computer program to aid in the analysis of multidimensional contingency tables using log-linear models.* FSU statistical report M292. Tallahassee: Department of Statistics, Florida State University.

▼ STATISTICAL SOFTWARE PACKAGES

(BMDP)
Dixon, W.J., chief editor, M.B. Brown, L. Engelman, and R.I. Jennrich, assistant editors (1990). *BMDP Statistical Software Manual,* 1990 Revision. Berkeley, CA: University of California Press.

(Execustat)
Strategy Plus, Inc. (1990). *EXECUSTAT.* Boston: PWS-KENT, 1990.

(Minitab)
Schaefer, R.L., and R.B. Anderson (1989). *The Student Edition of Minitab.* Reading, Mass.: Addison-Wesley.

(SAS)
SAS Institute (1985). *SAS User's Guide: Basics, Version 5 Edition.* Cary, NC: SAS Institute, Inc.

(SPSS-X)
SPSS, Inc. (1988). *SPSS-X User's Guide, 3rd ed.* Chicago: SPSS, Inc.

(Statgraphics)
STSC, Inc. (1986). *Statgraphics User's Guide.* Rockville, MD: STSC, Inc.

SELECTED ANSWERS

▼ CHAPTER 3

3.1 **a.**

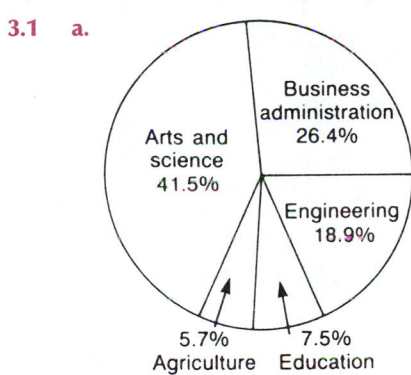

b.

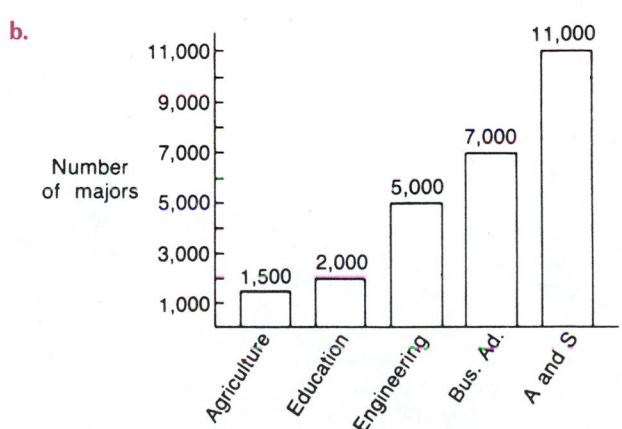

3.2 **a.** No

b.

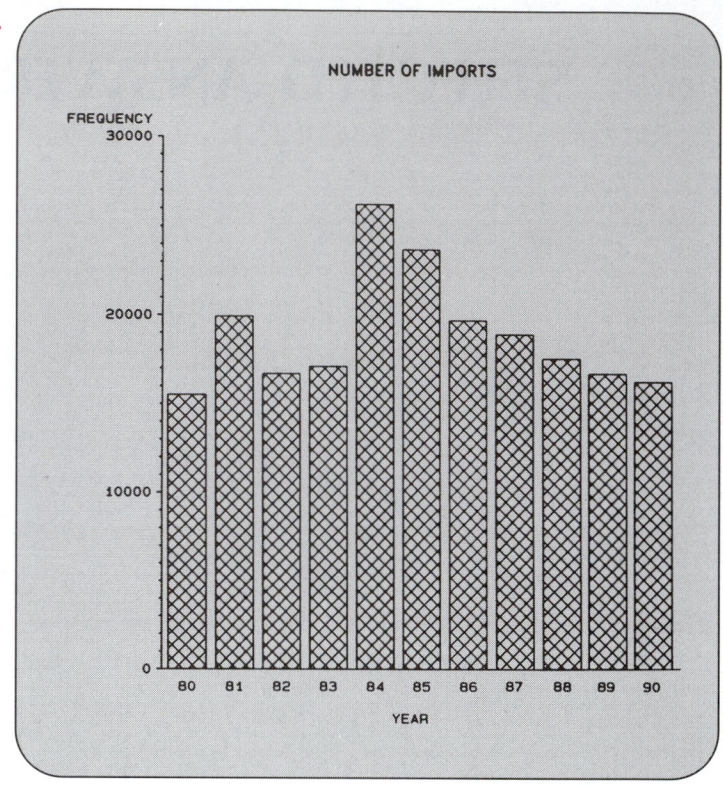

3.3   Pie chart is better.

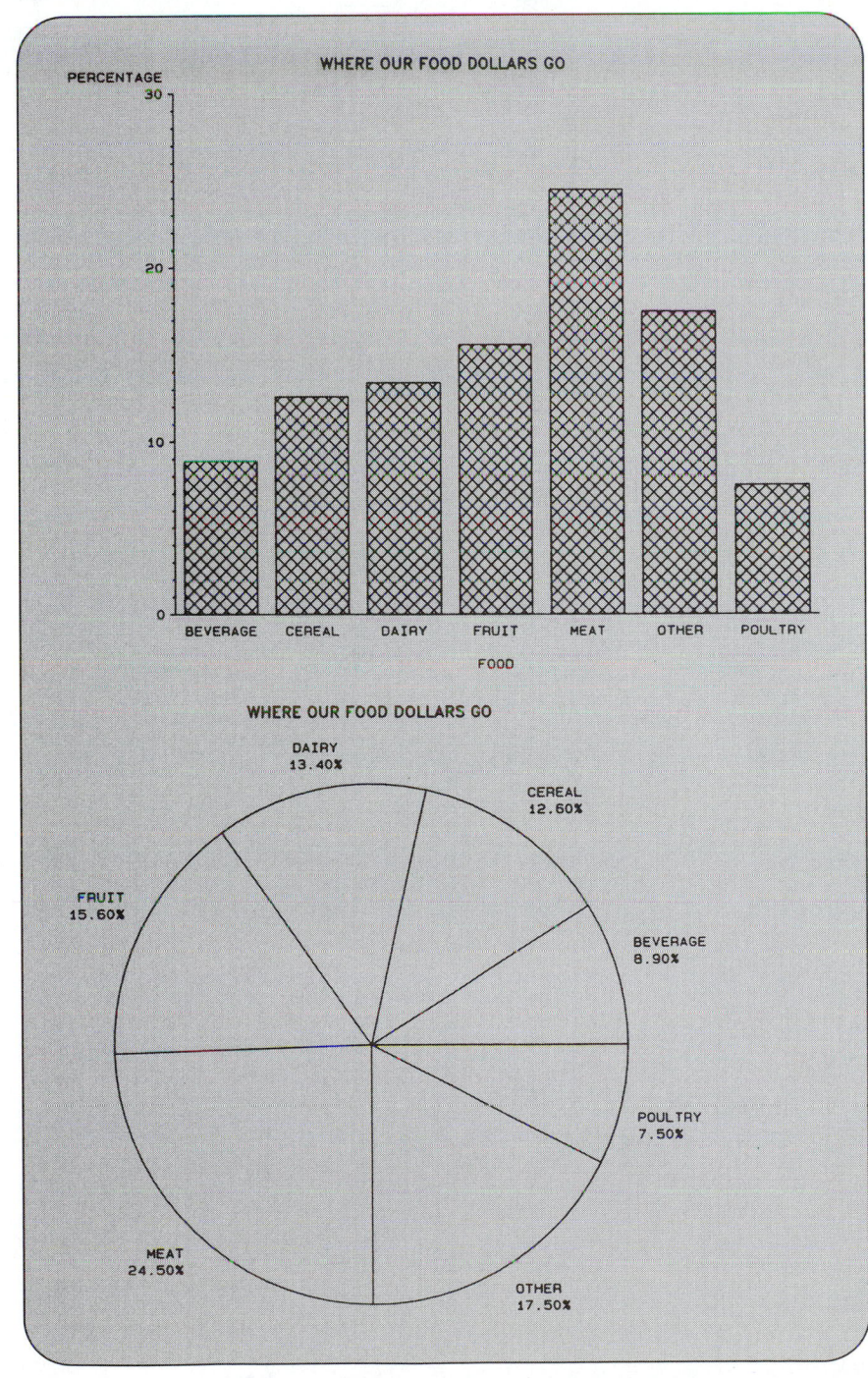

3.9

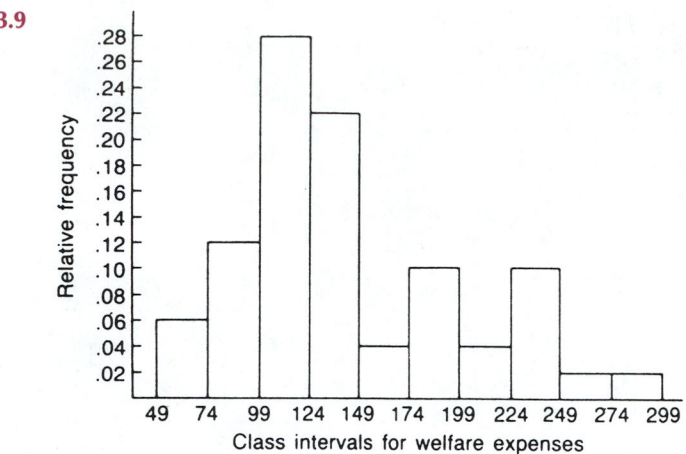

Class intervals for welfare expenses

Dollars	Relative Frequency
50– 74	.06
75– 99	.12
100–124	.28
125–149	.22
150–174	.04
175–199	.1
200–224	.04
225–249	.1
250–274	.02
275–299	.02

3.10

CLASS	FREQUENCY	PERCENT	CUMULATIVE FREQUENCY	CUMULATIVE PERCENT
22.5 – 26.5	1	4.8	1	4.8
26.5 – 28.5	1	4.8	2	9.5
28.5 – 30.5	10	47.6	12	57.1
30.5 – 32.5	9	42.9	21	100.0

3.11

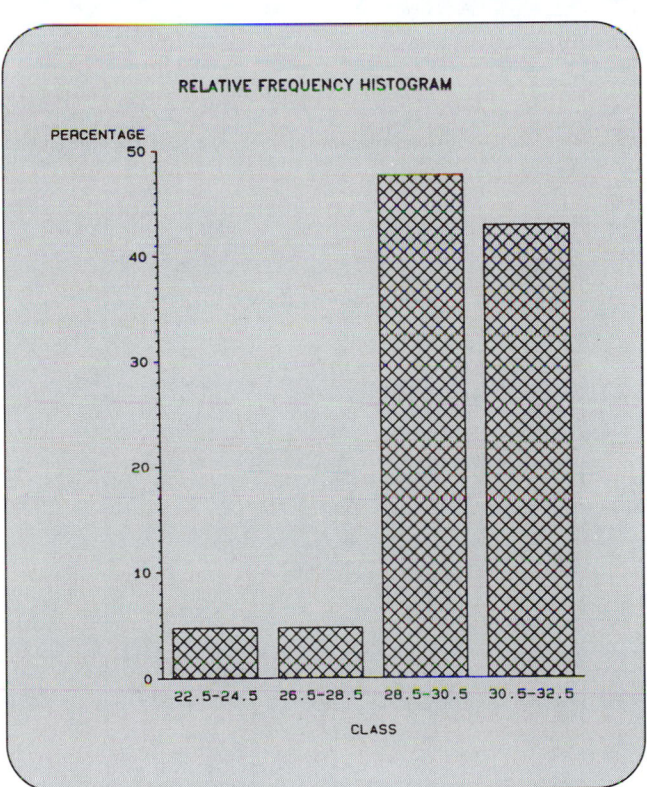

3.13

```
STEM LEAF                          #
   4  5                            1
   4  144                          3
   3  55689                        5
   3  0                            1
   2  556789                       6
   2  23                           2
   1  569                          3
   1  2                            1
   0  89                           2
      - - - - + - - - - + - - - - + - - - - +
```

3.14

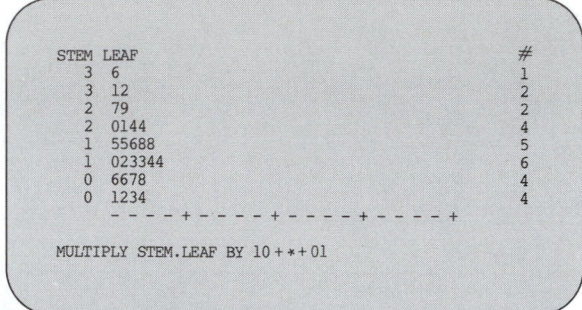

```
STEM LEAF                                        #
  3  6                                           1
  3  12                                          2
  2  79                                          2
  2  0144                                        4
  1  55688                                       5
  1  023344                                      6
  0  6678                                        4
  0  1234                                        4
     - - - - + - - - - + - - - - + - - - - +

MULTIPLY STEM.LEAF BY 10 + * + 01
```

3.15   The stem-and-leaf plot because it displays the magnitudes of the values as well as the frequencies.

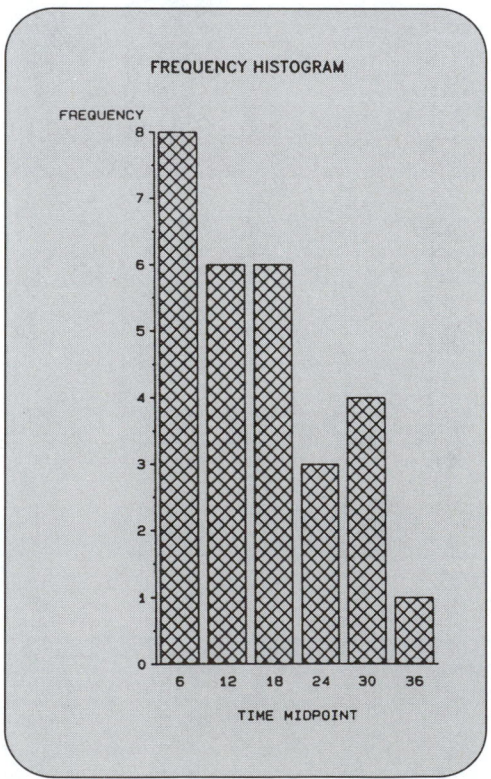

3.16   For the years 1967–1990, the difference between males and females has remained relatively constant, whereas there appears to have been a dramatic drop in verbal scores for both sexes from 1970 on, with more of a drop in females' scores.

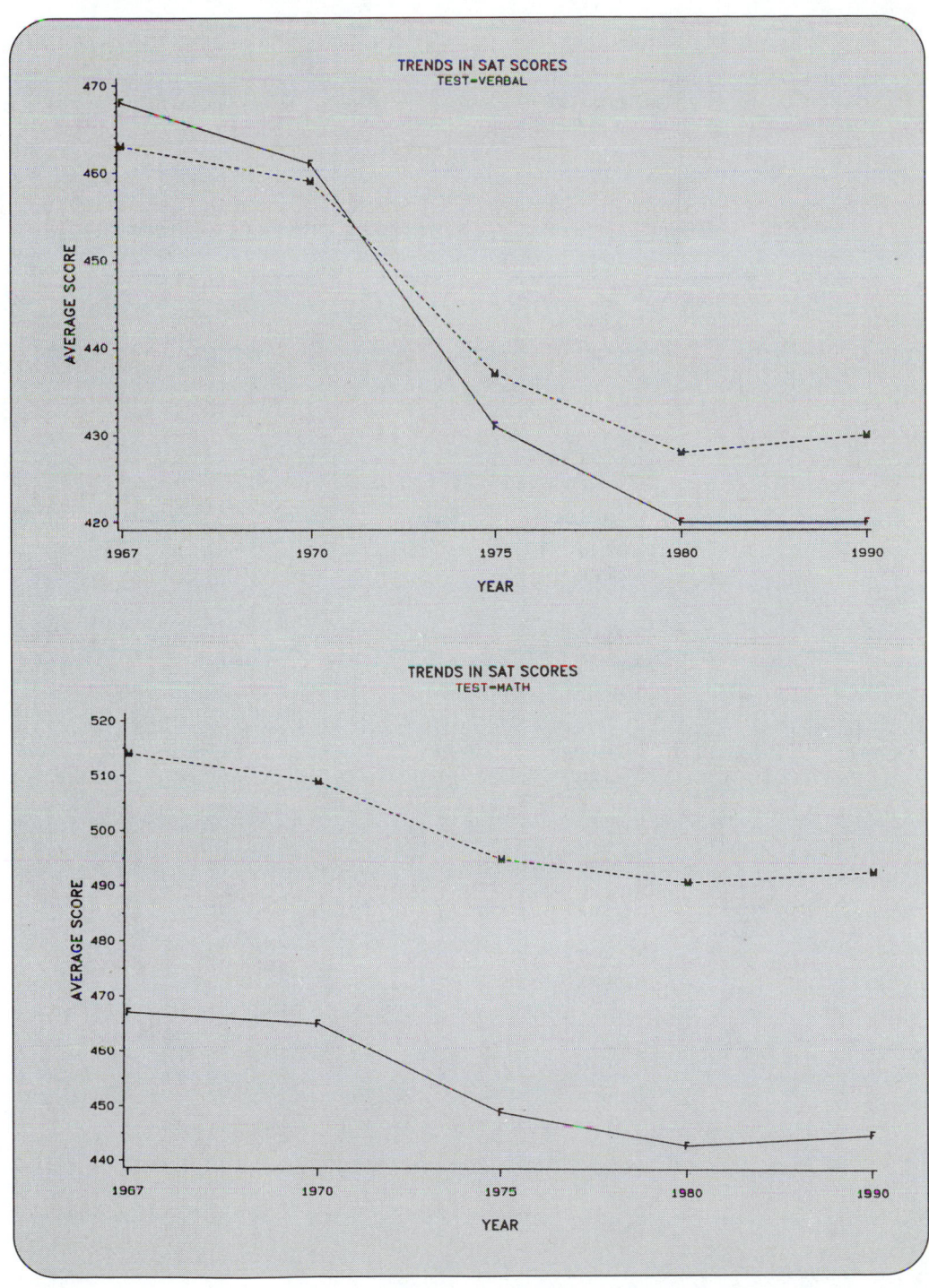

3.21

```
3 | 50
4 | 50 70 70 80 80 80
5 | 00 10 20 20 30 30 30 40 40 40 40 40 50 50 50 60 60 60 60 60 70 70 70 70 70 70 70 80 80 80 80 90 90
6 | 00 10 10 10 20 20 30 50 50
7 | 20
```

The frequency histogram and the stem-and-leaf plot for these data have the same shape.

3.25 Mean = 15.2
Median = 14.5
Mode = 18

3.26 Mean = 17.7
Median = 14.5
Mode = 18

3.28 Mode = 5
Median = 5.5
Mean = 6

3.29 Mean = 7.6
Median = 7.4
Mode = 7

3.30 Mode = 37.4
Median = 36.23
Mean = 36.03

3.31 Median = 906
Mean = 970.1

3.32 Mean = 8.465
Median = 8.6

3.34   **a.** The rounded measurements are

2.10	1.98	1.99	2.77
2.47	2.70	2.10	2.36
1.75	2.43	2.67	1.80
2.94	2.17	2.65	3.09
1.69	2.80	2.06	2.20
2.75	2.82	2.55	2.93
2.82	2.38	2.22	1.85
2.52	2.68	2.92	2.28
2.77	2.39	3.05	1.96

     b. Mode = 2.10, 2.77, 2.82
     c. Median = 2.45
     d. Mean = 2.43

3.36   Since 10 of the 15 measurements are less than the sample mean, the distribution appears to be skewed to the right. The sample median is a more appropriate measure of central tendency for these data; median = 61.61.

3.37   **a.** Mean group 1 = 2.923
        Mean group 2 = 1.592
        Mean group 3 = .797
        Group 1 median = 2.805
        Group 2 median = 1.565
        Group 3 median = 0.755
        Group 1 mode = 0
        Group 2 modes = 1.57, 1.55
        Group 3 mode = .7
     b. Mean = 1.7707
        Median = 1.565
        Modes = 1.57, 1.55, 0.7

3.38   Average mean for the three groups = 1.7707
Average median for the three groups = 1.708
Average mode for the three groups = 1.27

The average of the three net group means and the mean of the complete set of measurements are the same. This will be true whenever the groups have the same number of measurements. But, due to the way the mode and median are computed, the average of group modes and medians should differ from the overall median and mode. This is true for these data.

3.40 **b.** 5.5
c. 1.049
d. $(-.598, 3.598)$ This interval contains all the measurements in this sample.

3.42 $s^2 = 70.0996$
$s = 8.3763$

3.44 **a.**

Class Intervals	Freq.	Relative Freq.
1.5–2.5	1	.0143
2.5–3.5	1	.0143
3.5–4.5	3	.0429
4.5–5.5	5	.0714
5.5–6.5	5	.0714
6.5–7.5	12	.1714
7.5–8.5	18	.2571
8.5–9.5	15	.2143
9.5–10.5	6	.0857
10.5–11.5	3	.0429
11.5–12.5	0	.0000
12.5–13.5	1	.0143

b. 7.7286
c. (5.744, 9.713), 71.43%
(3.759, 11.698), 95.71%
(1.774, 13.683), 100%

3.46

Data set 1:	Data set 2:	Data set 3:
$\bar{y} = 2$	$\bar{y} = .02$	$\bar{y} = 1002$
$s^2 = 1$	$s^2 = .0001$	$s^2 = 1$
$s = 1$	$s = .01$	$s = 1$

3.49 Median = 20
Lower quartile = 17
Upper quartile = 24

3.52 Urban   30%
Suburban   50%
Rural   20%

3.53 Resigned   24%
Transferred   26.4%
Retired   49.6%

3.55 **a.** A:22
B:17.75
C:26
D:30
c. Hospital B

3.56 **a.** Yes

b.

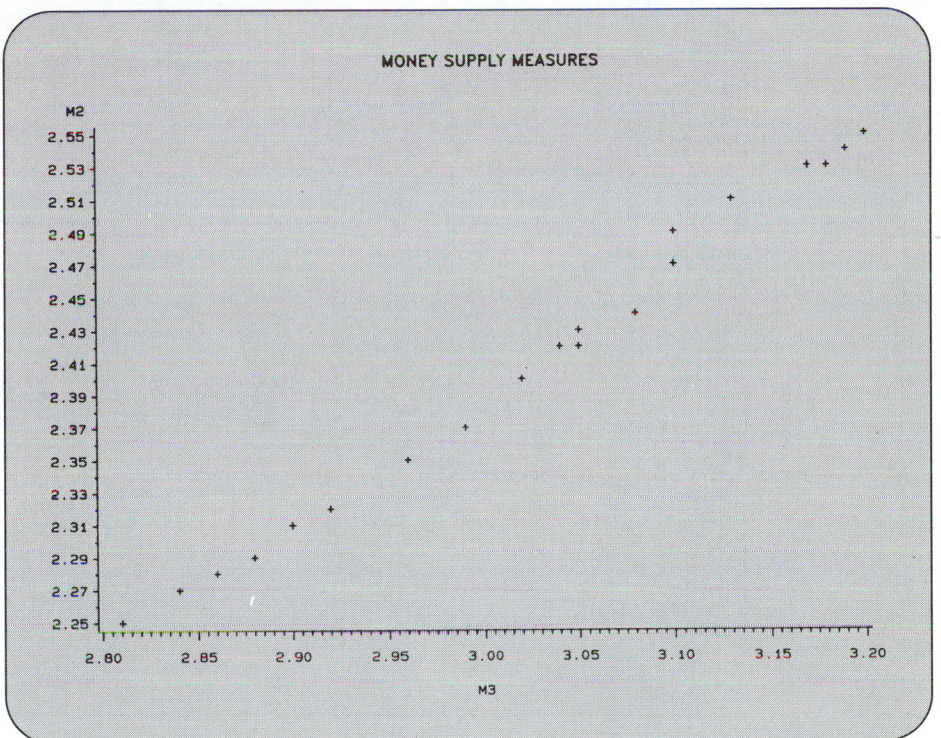

3.57 M2 and M3 move simultaneously and seem to reflect the same changes in the money supply.

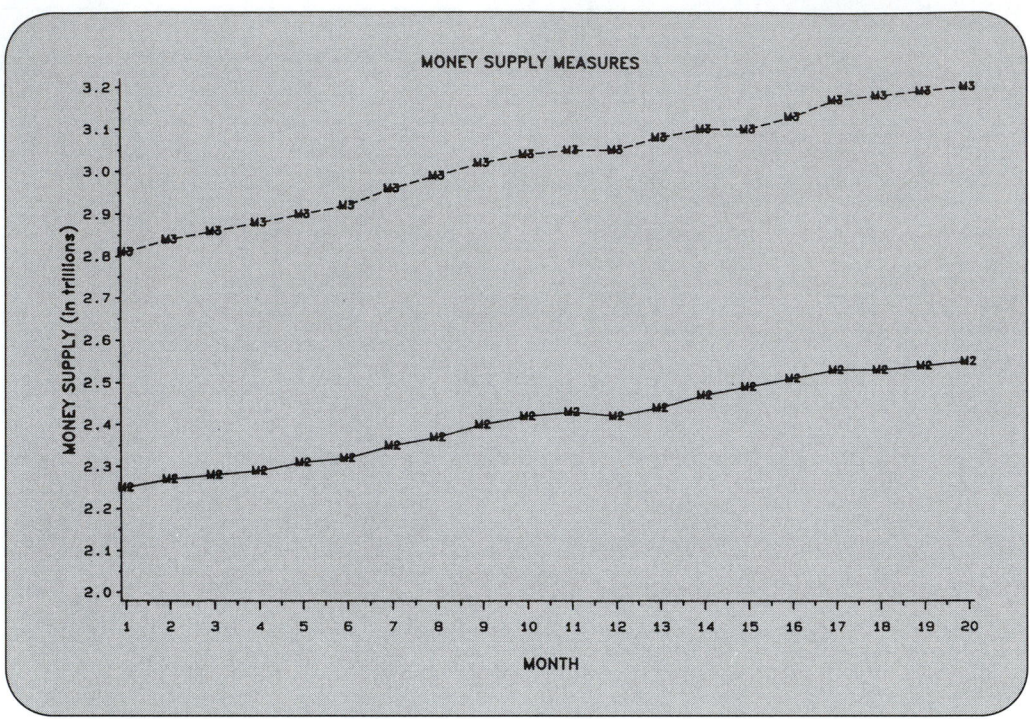

3.59 **a.** Mode = 5
Median = 15
Mean = 15.96
b. $s \approx 30/4 = 7.5$
c. $s = 8.39$
d. No, they are not mound-shaped.

3.61 In coded units, $\bar{y} = -1.08$.
$\bar{y} = 4.292$
In coded units, $s_c^2 = 9.59$.
$s_c = 3.10$
$s = 0.31$
These answers agree with those in Exercise 3.60.

3.63 **a.** Mean = .65
Median = .55
Mode = .5
b. Mean = 1.21
Median = .55
Mode = .5
Median and mode are the same as in part a.

3.65 $\bar{y}_c = 6.5$; $\bar{y} = 6.5/10 = 0.65$
$s_c = 2.76$; $s = 2.76/10 = 0.276$

3.67 $Q_1 \approx 525$
$Q_3 \approx 574$
$IQR = 49$
Median $= 560$

3.69 **a.** Mode $= 2.5$
Median $= 7.04$
b. Mean $= 8.3$
c. Since the distribution is skewed to the right, the median provides a better measure of the center of the distribution.

3.71 $\bar{y}_c = -.078$
$\bar{y} = 8.34$
$s_c^2 = 7.4658$
$s_c = 2.73$
$s = 5.46$

3.73 **a.** Median $= .93$
Mode $= 1.0$
b. Mean $= 1.33$
c. The distribution is skewed to the right.

3.76 **a.** For the south: $\bar{y} = 597.7$
For the north: $\bar{y} = 489.1$
For the west: $\bar{y} = 679.4$
b. 588.83
c. 588.83

3.79

71	1
70	0 1 2 3 3 3 8
69	1 4 7 9
68	8 8 8 8 8
67	7 9
66	4 7 7
65	6
64	
63	0
62	5
61	
60	
59	
58	
57	
56	
55	
54	7

3.81 **a.** Member 1: 93.75
Member 2: 98.65
Member 3: 113.3125
Member 4: 124.86
Member 5: 131.9

b. Member 1: 78.5
Member 2: 95
Member 3: 100
Member 4: 112.5
Member 5: 128.5

3.83 **a.** Old policy:
$\bar{y} = 4.6$
$s^2 = 6.8$
$s = 2.6$
New policy:
$\bar{y} = 2.3$
$s^2 = 10.6$
$s = 3.3$

3.85 **b.**

	Plants	Arrests
Mean	677,794.6	95.0
10% trimmed mean	142,328	59.7
20% trimmed mean	133,039	41.3

3.87

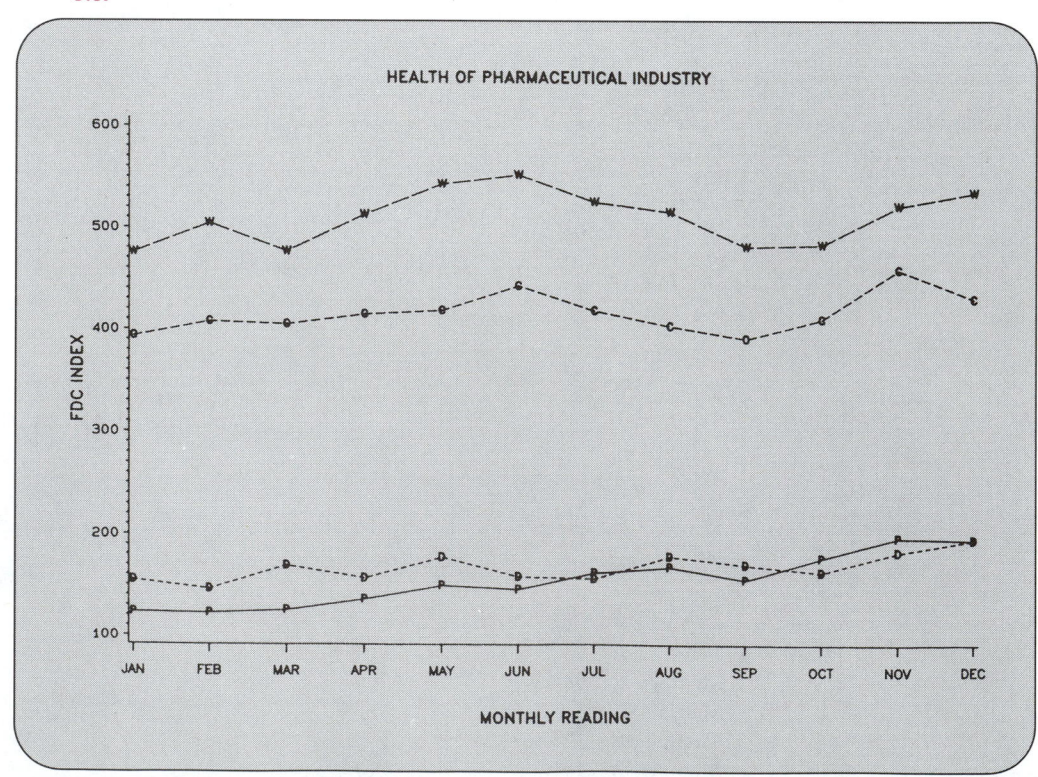

3.88

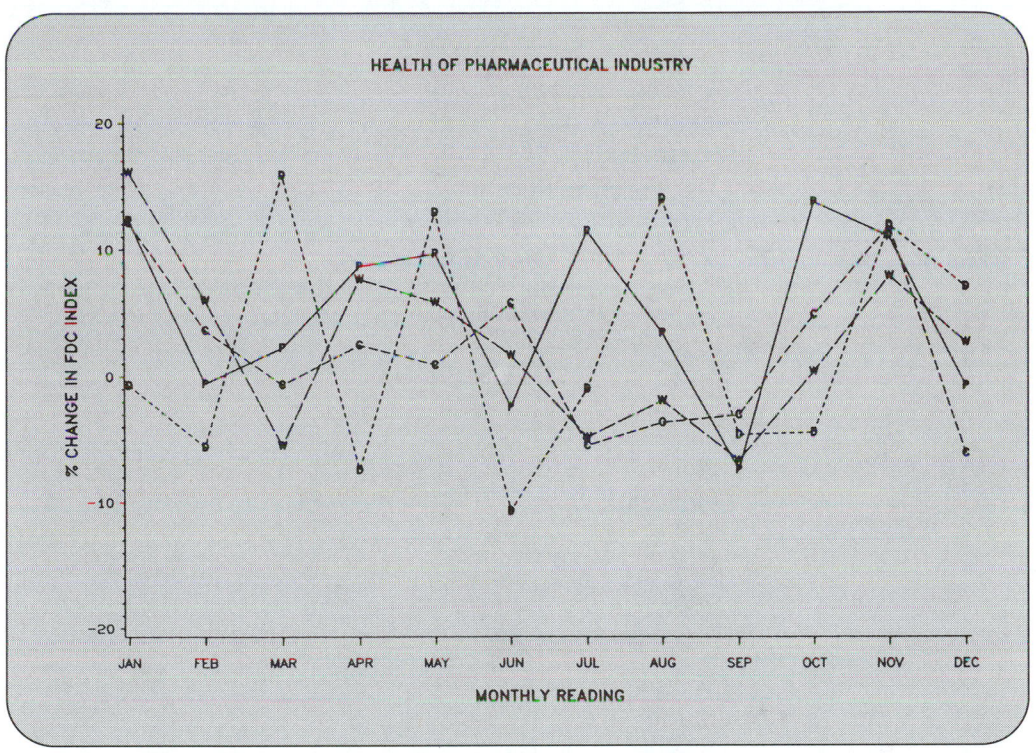

3.89   **a.** 141.125
       b. 2.429
       c. To reflect the relative importance of the component companies.

▼ CHAPTER 4

4.1   Personal or subjective probability
       b. Relative frequency concept of probability
       c. Classical interpretation of probability

4.3   HHH, HHT, HTH, HTT, THH, THT, TTH, TTT

4.4   **a.** $\frac{3}{8}$
       b. $\frac{7}{8}$
       c. $\frac{1}{8}$

4.5   **a.** $P(\bar{A}) = \frac{5}{8}$
          $P(\bar{B}) = \frac{1}{8}$
          $P(\bar{C}) = \frac{7}{8}$

4.8   **a.** $\frac{1}{6}$
       b. $\frac{1}{2}$
       c. $\frac{2}{3}$
       d. $\frac{1}{3}$

4.10   **a.** $P(A) = .65,\ P(B) = .40$
       b. .30
       c. .75

4.11   No, since $P(A \cap B) = .30 \neq 0$

4.12   There are 12 possible schedules.

4.13   **a.** Generator 1 does notwork properly.
       b. Generator 2 works properly given that generator 1 works properly.
       c. Generator 1 works properly or generator 2 works properly.

4.15   **b.** .30
       c. .60
       d. .10

4.16 $P(A) = .436$
       $P(B) = .291$
       $P(A \cap B) = .109$

4.17   **a.** No
       b. $P(B|A) = .25$
          $P(B|\bar{A}) = .323$
          No, they are not equal.

4.19   **a.** .49
       b. .91

4.20   **a.** .665
       b. .27
       c. .065

4.23   .05

4.25   .40

4.28 $P(0) = .001;\ P(1) = .01;\ P(2) = .045;\ P(3) = .12$

4.29   **a.** .201
       b. .3456
       c. .204

4.30   **a.** .8263
       b. .1737
       c. .9993
       d. .0086

4.31   **b.** .50
       c. .13
       d. .71

4.32   **a.** .64
       b. .60
       c. .07

4.37   **a.** .006
       b. .2508

       c. .6330

       d. .0001

4.38   **a.** 0.0

       b. 0.0368

       c. 0.0473

       d. 0.0282

4.39 $P(Y = 0) = .0016$

       $P(Y = 1) = .0256$

       $P(Y \leq 1) = .0272$

4.46   No; no identical trials.

4.50   **a.** .4032

       b. .4713

4.52   **a.** .4015

       b. .2794

4.54   **a.** 0.0542

       b. .1857

4.56   .8729

4.58   1.96

4.60   1.645

4.62   **a.** .5000

       b. .1056

       c. .9699

       d. .9270

       e. .3413

4.63   **a.** .475

       b. .025

       c. .95

       d. 84.2

4.65   **a.** .4332

       b. .4641

4.67   1.645; −1.645

4.69   **a.** .025

       b. .0136

       c. .0021

       d. .2327

4.71   **a.** .0336

       b. Since 55 is 2.67 standard deviations above $\mu = 39$ and the probability of observing a value of 55 or greater is .0038, we would likely conclude that the voucher has been lost or misplaced.

4.73   **a.** .0764

       b. .0228

       c. .1949

       d. .8472

4.75   **a.** 11.51

       b. .44

4.82

150	729	611	584	255
465	143	127	323	225
483	368	213	270	062
399	695	540	330	110
069	409	539	015	564

4.86 The sampling distribution will be approximately normal, with mean 960 and standard deviation 20. Observing a measurement more than 70 units (3.5 standard deviations) away from the mean (960) of a normal distribution is very improbable.

4.88 **a.** .7462
b. .1587
c. .0188

4.90 **a.** 125 ± 32 should contain approximately 68% of the weeks.
125 ± 64 should contain approximately 95% of the weeks.
125 ± 96 should contain approximately all the weeks.
b. .1379

4.92 **a.** .01
b. .0038

4.94 **a.** Approximately normal with mean 3.7 and standard deviation 0.06
b. Approximately normal with mean 3.7 and standard deviation 0.03
c. Essentially 0

4.96 **a.** Approximately normal with mean 1300 and standard deviation 35.78
b. 25th percentile = 1,275.85; 75th percentile = 1,324.15

4.98 **a.** 10
b. .0409; .00358

4.102 **a.** Essentially 0
b. Essentially 0

4.104 2.63

4.106 No, it has a mean breaking strength higher than the old fabric.

4.108 .03078; the probability that only one or less of the five is selected is 3 chances in 100. Thus this improbable occurrence is a rare event. Having actually observed this event, we could draw one of two conclusions. Either we have observed an unlikely event, or the board is currently admitting with a probability less than .7. We would tend to accept the latter conclusion.

4.110 **a.** .484
b. .0717
c. .4562

4.111 **a.** y = no. of individuals (out of 1,000) planning to buy a drink
b. No
c. Normal approximation

4.112 **a.** .005
b. .9044

4.113 **a.** .00
b. No

4.116 Mean = median = 3.50

4.118 Mean = 3.2; median = 3.0

▼ CHAPTER 5

5.5   **a.** 101.95–108.05
     b. 100.99–109.01

5.7   **a.** The width of the interval will be narrower by $1/\sqrt{2}$ times the previous width.
     b. The width of the interval will cut in half.

5.10 $25.56–$27.24

5.11 Point estimate of μ is 3.2; $3.2 \pm .176$

5.13 850 ± 25.3

5.15 430 ± 17.11

5.16 (3.225, 3.315)

5.17 (.164, .196)

5.18 **a.** 188; 752; 3,007
      b. Increase the sample size by four times.

5.19 246

5.20 174; 425

5.22 1,537

5.23 6,147; 385

5.24 44

5.25 UCL = 3.3761
      Center line = 2.0566
      LCL = .7511

5.26 UCL = 0.6456
      Center line = .640
      LCL = .6344

5.28 UCL = 110.5762
      Center line = 109.3125
      LCL = 108.0488

5.29 94

5.30 **a.** Reject H_0; $\mu > 38$
      b. No; Type II error is possible only when H_0 is not rejected.

5.31 Power = .8107, .9997, and 1.000

5.33 $-z_{.01} = 2.33$
      Do not reject H_0.

5.34 With $\Delta = 4$, $z_\alpha = 1.645$, $z_\beta = 1.28$, we obtain $n = 4.09 \simeq 5$.

5.36 Reject H_0; $z = 1.807$

5.37

μ_a	Power
2.1	.2296
2.2	.5636
2.3	.8577
2.5	.9979

5.40 $H_0: \mu = 0; H_a: \mu > 0$
$z = (\bar{y} - 0)/S\sqrt{35}$
Reject H_0 if $z > 1.96$.

5.45 **a.** .0314
b. .0628

5.46 $z = 1.74$; level of significance $= .0409$

5.47 $H_a: \mu \neq 16$
$z = 6.3158$
Reject H_0.

5.49 **a.** Reject H_0 if $t < -1.761$.
b. Reject H_0 if $|t| > 2.074$.
c. Reject H_0 if $t > 2.015$.

5.51 $t = 1.64$
Reject H_0 if $t > 1.74$.
Do not reject H_0.

5.52 **a.** 9.7 ± 2.21
b. We are 95% confident that this interval captures the population mean speed for all fourth-grade students.
c. 9.7 ± 2.76

5.54 $H_0: \mu = 80; H_a: \mu \neq 80; t = 1.09; p > .20t$

5.56 $t = -.31$. There is insufficient evidence to indicate that $\mu < 5.0$.

5.63 **a.** (9.534%, 10.865%)
b. No, since the mean content of 11% does not fall in the 95% confidence interval.

5.66 **a.** (205.122, 244.878)

5.67 **a.** 28.7
b. (25.498, 31.902)

5.71 (6.77, 11.63)

5.74 (3.130, 3.270)

5.76 $t = -3.2108$; reject H_0 if $|t| > 2.045$; reject H_0.

5.79 $t = -8.1$; reject H_0 if $t < -1.664$; reject H_0.

5.82 $t = -6.96$; reject H_0 if $t < -1.729$; reject H_0.

5.83 p-value $= .15 > \alpha = .01$; do not reject H_0.

5.85 $t = .5820$; reject H_0 if $t > 2.132$; do not reject H_0.

5.88 (1.898, 2.222)

5.89 (8.8895, 9.0349)

5.91 True; true; true

5.94 **a.** $H_0: \mu = 5.2; H_a: \mu < 5.2; z = -2.02$; reject H_0 if $z < -1.645$.
b. Reject H_0 and conclude that the mean dissolved oxygen count is less than 5.2 ppm.

5.96 $z = -2.36; p = .0091$

5.98 430 ± 22.53

5.100 **a.** 30.51 ± 4.09
b. 30.51 ± 5.39

5.102 22 ± 1.04

5.104 $98.4 \pm .04$

5.106   **a.** $H_a: \mu \neq 29$
       b. $p \approx .48$

5.108 $.76 \pm .04$

5.110 58 ± 4.08

5.112 $\beta = 0$ for $\mu = 40$; $\beta = .0301$ for $\mu = 38$; $\beta = .3192$ for $\mu = 36$; $\beta = .8624$ for $\mu = 34$.

5.114   **a.** $.76200$
       b. $.98$
       c. $(.7220, .8020)$; we are 98% confident that the population mean proportion of patients per hospital with group medical insurance lies in the interval .7220 to .8020.

5.116 $z = 2.39$; $p = .0168$

5.118   **a.** The t distribution is appropriate when the sample is selected from a population with a mound-shaped distribution. In this case, since $s > \bar{y}$, the distribution will be skewed, and a confidence interval for μ based on t would be inappropriate.
       b. The sample median could be used to estimate the center of the distribution.

5.120 $1,537$

5.123 $(17.48\%, 22.52\%)$

5.125 $(515.5, 604.5)$

5.126 658

5.127   **a.** $p < .001$
       b. 12,562.5 pints

▼ CHAPTER 6

6.1   **a.** $|t| > 2.064$
     b. $t > 2.624$
     c. $t < -1.86$
     d. (1) The samples are independent.
         (2) Sampling from normal population.
         (3) Equal variances.

6.2 $t = -2.6722$; reject H_0 if $t < -1.703$; reject H_0.

6.3 $.005 < p < .01$

6.5 $t = 1.6059$; reject H_0 if $|t| > 2.306$; do not reject H_0.

6.7 $(-3.6625, 7.6625)$

6.8 $t = 10.71$; reject H_0.

6.10 $t = 1.91$; $.025 < p < .05$

6.12   No, the distribution will be skewed, since $s > \bar{y}$ for the magnesium data.

6.14   **a.** No
     b. $t = 4.24$, $p < .001$
        $t' = 4.23$, $p < .001$

6.16 $(-6,487.7, -5,976.3)$

6.17   **b.** $(-22.4, 2.4)$
     c. $(-41.4, -20.6)$; does not include 0

6.18 **b.**

Age	99% C.I.
9	$(-5.2, 41.2)$
13	$(-1.7, 43.7)$
17	$(2.6, 47.4)$

6.20 **a.** Plumber 1: $\bar{y} = 88.8$, $s = 7.9$
Plumber 2: $\bar{y} = 108.9$, $s = 8.7$

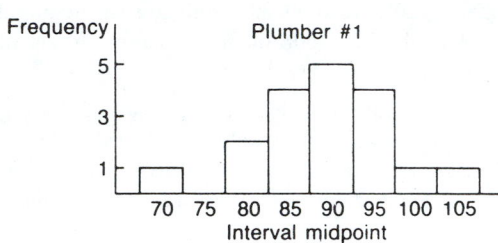

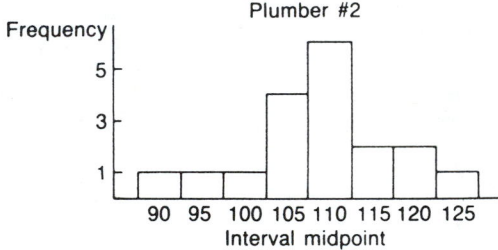

b. Since both graphs show a single, peaked, roughly symmetrical distribution, a t test appears to be appropriate.

6.22 **a.** $z = 2.92$; reject H_0.
b. Two-sample t test if the two populations have a common variance.

6.26 **a.**

	Plumber 1	Plumber 2
Largest	102.6	126.8
Smallest	71.4	90.2

$C = 26$; at $\alpha = .05$, reject H_0.

b. The t test, rank sum test, and Tukey–Duckworth test all have the same conclusion, to reject H_0.

6.30 $t = 4.91$

6.32 865

6.34 **a.** $p = .0001$. We reject H_0—the hypothesis that the population mean scores for those not exposed and those exposed to a minority environment are identical.
b. The conclusion here is the same. In this case, it does not make a difference if you use a t test or a Wilcoxon's signal rank test.

6.36 $T = 16$; reject H_0.

6.37   (3.723, 6.877)

6.38   **a.** $t = 2.68$; reject H_0 if $t > 1.782$; reject H_0.
        b. $.01 < p < .025$

6.40 $t = 3.656$; $.02 < p < .05$

6.43 $(-233.5, -191.84)$

6.47 $t = -.17025$; reject if $|t| > 2.228$; do not reject H_0.

6.50 $t = 2.4568$; reject H_0 if $|t| > 2.776$; do not reject H_0.

6.53 $(-.3378, 5.5378)$

6.56   **a.** $s_p^2 = 1.16$; $s_p = 1.07708$
        b. $(-1.262, .562)$
        c. No, since 0 is contained in the confidence interval as a believable value for $\mu_1 - \mu_2$.

6.60   95% C.I.: (5.36, 10.64)
        90% C.I.: (5.789, 10.211)

6.62   **b.** (i) Reject H_0 if $|t| > 2.306$.
           (ii) Do not reject H_0.
        c. (i) Reject H_0 if $t > 1.86$.
           (ii) Reject H_0.

6.64   **a.** One-tailed test because women are suspected of spending less.
        b. $H_0: \mu_1 - \mu_2 = 0$
            $H_a: \mu_1 - \mu_2 < 0$

6.66   **a.** One-tailed test because we think the night-shift time exceeds that of the day shift.
        b. 3.39853
        c. $H_0: \mu_1 - \mu_2 = 0$; $H_a: \mu_1 - \mu_2 > 0$; $t = 1.9733$; reject H_0 if $t > 1.734$; reject H_0.

6.68 $(-238.3, -187.1)$

6.73 $t = -1.1083$; reject H_0 if $|t| > 2.12$; do not reject H_0.

6.77   44

6.79 -23.40 ± 12.02

6.81   **a.** There are not enough observations to draw any conclusions concerning normality. There is no suggestion that the population variances are different.

6.83   **a.** $p < .001$; we are rejecting the hypothesis that there is no difference between the average symptom scores of patients on drug A and those on the placebo.
        b. Differences existing at the end of the study may be due to baseline differences. To guard against baseline differences, both treatment groups should have the same type of patients. For example, one treatment group should not include all the severe patients.

6.91   **a.** HMOs in general have a shorter length of stay.
        b. $t = 14.93$; $p < .001$

▼ CHAPTER 7

7.1   **a.** .01
        b. .90
        c. .01

7.3 $\chi^2_{.025} = 284.80$ (using Table); more precisely — 324.99
       $\chi^2_{.975} = 198.98$ (using Table); more precisely — 232.79

7.5 $\chi^2 = 46.464$; reject if $\chi^2 < 43.19$; do not reject H_0.

7.7   **a.** $\bar{y} = 3.99$; $s = 0.016$
       b. $\chi^2 = 52.5$; reject H_0 if $\chi^2 > 37.65$; reject H_0.

7.8   (.01276, .02085)

7.12   (2.0434, 15.7067)

7.14   UCL $= 26.052$
       Center line $= 13$
       LCL $= 0$

7.16   **a.** 2.91
       b. 3.71
       c. 2.35
       d. 4.00
       e. 2.99

7.18 $F = .5827$; reject H_0 if $F > 3.29$; do not reject H_0. Assume the populations are normal and independent.

7.22 $F = .37085$; reject H_0 if $F > 4.03$; do not reject H_0.

7.24   No; a t test would be inappropriate because the two samples did not have a common variance.

7.26   **a.** (10.4203, 58.7436)
       b. The critical assumption is that the random sample is drawn from a normal population.

7.28   **a.** (22.5, 27.5)
       b. A graph of the data shows a single-peak mound-shaped distribution that supports the normality assumption.
       c. $\chi^2 = 40.058$; reject if $\chi^2 > 42.56$; do not reject H_0.

7.30   (3.586, 13.226); the 95% confidence interval supports the test findings of the consumer group, since $\sigma^2 = 4$ is in the interval.

7.32 $t = -7.29$; $p < .01$; conclude that differences in the average returns for the two portfolios do exist. The method used was the test for $\mu_1 - \mu_2$, which assumes independent samples, normality, and equal variances. From previous information and work, we know we are dealing with independent samples, and we did not reject the hypothesis for equal variances; however, there is not enough information to make a decision concerning normality.

7.34   (73.35, 315.47); (8.56, 17.76)

7.36 $F = 3.5652$; reject H_0 if $F > 5.35$; do not reject H_0.

7.39 $F = 3.125$; reject H_0 if $F > 2.69$; reject H_0.

7.40 H_0: $\sigma_1^2 = \sigma_2^2$; H_a: $\sigma_1^2 > \sigma_2^2$

7.44 H_0: $\sigma_1^2 = \sigma_2^2$; $F = 2.22$; reject H_0 if $F > 2.51$; do not reject H_0. H_0: $\mu_1 - \mu_2 = 0$; $t = 15.636$; reject H_0 if $|t| > 2.021$; reject H_0.

▼ CHAPTER 8

8.3   **a.** $\chi^2 > 7.815$
       b. $\chi^2 > 21.67$
       c. $\chi^2 > 24.32$

8.5 $\chi^2 = 5.28$; do not reject H_0.

8.6 $\chi^2 = 9.08$; do not reject H_0. The goodness-of-fit test is insensitive to changes in the cell probabilities under H_0.

8.7 $\chi^2 = 49.067$; $p < .001$

8.8 $\chi^2 = 6.95$; do not reject H_0.

8.10 $\chi^2 = 6.00$; $p = .05$; reject H_0.

8.12 $\chi^2 = 120.71$; $p < .001$.

8.14 $\chi^2 = 2.40$; do not reject H_0.

8.15 $\chi^2 = 7.608$; do not reject H_0.

8.17 **a.** Yes

b. (.15, .25)

8.19 **a.** (.78, .82)

b. (.783, .817)

8.21 **a.** Yes

b. Yes, $n\pi = 16.7$ and $n(1 - \pi) = 33.3$

c. (.246, .515); increase the sample size.

8.24 (.29, .39)

8.26 **a.** A histogram using percent response as the vertical axis.

8.27 **a.** A table listed in descending order of proof of illiteracy.

8.28 **a.** Normal approximation if valid.

b. Change for breakfast is statistically significant. One would have to decide whether a 2% increase (on a national basis) represents a relatively large shift.

8.30 **a.** $\hat{\pi} = .37$; half-width of a 95% confidence interval is .024.

b. 8,955

8.34

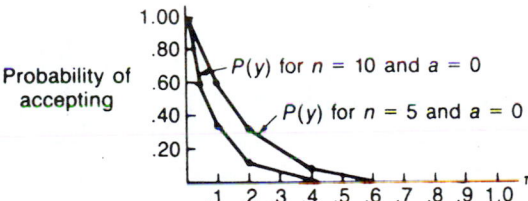

8.36 Yes, none of the percentages was outside the range 0.01 to 0.328.

8.38 $.025 \pm .169$

8.40 $z = 2.3$; $p = .011$

8.42 **a.** $z = 3.403$; $p < .001$

b. What was the amount of hair growth? What side effects were observed? What characteristics distinguish responders from nonresponders?

8.43 **a.** $z = 7.91$; $p < .001$

b. If cocaine has a similar effect in humans, it is a very dangerous drug, even more so than heroin.

8.44 For $\pi = .002$, $P(y \leq 2) = .42319$

For $\pi = .003$, $P(y \leq 2) = .17358$

8.46   **a.** .0025
           b. .9826
           c. .9975

8.48   Use the Poisson approximation. $P(y \geq 1) = .3935$.

8.50 $\chi^2 = 26.80$; reject H_0 and conclude that the number of shutdowns per day does not follow a common Poisson distribution.

8.51   **b.** $\chi^2 = .67$
           c. Reject if $\chi^2 > 3.841$; do not reject H_0.

8.54   **a.** $E_{11} = 11$; $E_{12} = 11$; $E_{21} = 9$; $E_{22} = 9$
           b. H_0: The fact that a sheep is a responder or nonresponder is independent of treatment. $\chi^2 = 6.46$; reject H_0.

8.56   The results are the same.

8.57   **a.** No
           b. $\chi^2 = 1.75$; do not reject H_0.

8.61 $\chi^2 = 32.96$; $p < .001$

8.63 $C = .21$; $C_{adj} = .297$

8.66   **a.** $\chi^2 = 17.1961$; reject H_0
           b. $C = .488$

8.68 $\chi^2_{MH} = 11.55$; $p = .0007$; we reject H_0 that there is no difference, on the average, in males and females for number of alcohol-related arrests.

8.70 $< \$20,000$ 57%
              $\$20,000–40,000$ 31%
              $> \$40,000$ 12%
           $\chi^2 = 27.22$; $p < .001$; reject H_0.

8.72 $\chi^2 = 6.43$; do not reject H_0.

8.74 $\chi^2 = 17.07$; reject H_0.

8.75   (.64, .68)

8.81 $p = .0475$

8.83 $z = .85$; do not reject H_0.

8.85   **a.**

University	Low	Medium	High	Left	Right
1	22.4%	30.2%	47.4%	61.2%	38.8%
2	44%	29.6%	26.4%	50.9%	49.1%
3	35%	36%	29%	12%	88%

           b. $\chi^2_{CMMA} = 39.82$; $p < .001$

8.87   In Exercise 8.86, when we collapsed the global outcome categories, we could not reject H_0 that the treatment groups were the same. When the full data are used, we can reject H_0 and conclude that differences exist between treatment groups. The lesson that can be learned from these results is that loss of information can distort test results.

8.89   **a.** $z = -4.12$; reject H_0 that there is no difference between the sensitivities of the two fuses.
           b. $\chi^2 = 16.67$; the relationship between the z test and the chi-square test of independence is that, except for rounding errors, $z^2 = \chi^2$ for df $= 1$.

8.91 $\chi^2 = 13.57$; reject H_0.

8.93 **a.** $z = -2.86$; reject H_0 that there is no difference in the number of acceptable tablets between formulations 1 and 2.

b. $\chi^2 = 8$; reject H_0 that the number of acceptable/unacceptable tablets is independent of the formulation.

c. Except for rounding errors, $z^2 = \chi^2$ for $df = 1$.

8.95 $\chi^2 = 59.64$; reject H_0 and conclude that age and regularity of seat belt usage are not independent.

8.97 **a.** The critical assumption is the normality of the differences. Although there are only 12 observations, the data do not appear to be skewed.

b. The t test for H_0: $\mu_d = 0$ versus H_a: $\mu_d > 0$ has $t = 21.39$ and $p < .001$. We can conclude that the entry-level blood pressure and the follow-up blood pressure were not the same.

8.98 **a.** Yes

b. $\chi^2 = 373.45$; $p < .001$

8.99 **a.** >1 week: $(.431, .609)$
≤ 1 week: $(.036, .304)$

b. No

c. $\hat{\pi} = .447$; $(.367, .527)$

8.101 **a.** No

b. $\chi^2 = 21.63$; $p < .001$

8.102 **a.** $\chi^2 = 23.46$; $p < .005$

8.104 $\chi^2 = 87.94$; reject H_0.

▼ CHAPTER 9

9.2 **a.** 7.8

9.4 **a.** $\hat{y} = .05 + 1.35x$

b. 8.15

9.6 **b.** $\hat{y} = 4.70 + 1.97x$

c. Yes

d. 73.6646

9.9 **a.** Yes

b. $\hat{y} = 1.768 + .011x$

9.10 **a.**

9.12 **a.**

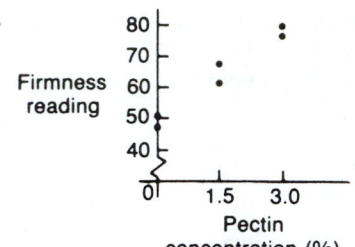

b. $\hat{y} = 48.93 + 10.33x$

9.14 **a.**

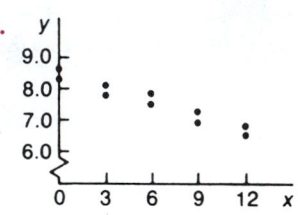

b. $\hat{y} = 8.46 - .14x$

9.16 **a.**

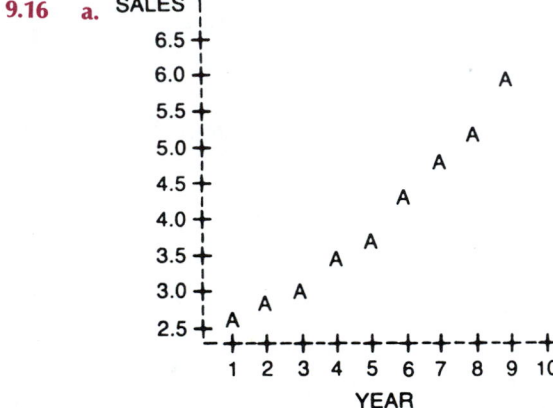

b.

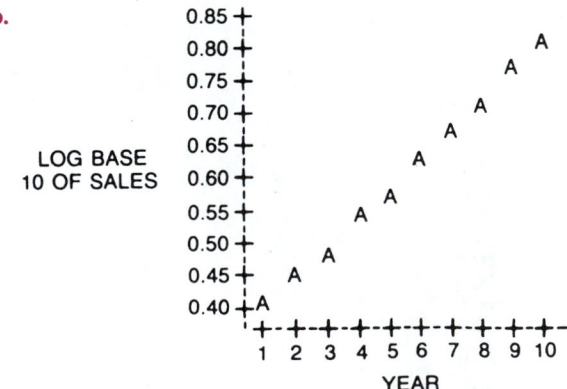

c. Log sales vs. year

9.18 Using the regression equation with sales as the dependent variable, sales in year 11 would be 6.604. Using the regression equation with log sales as the dependent variable, sales in year 11 would be 7.140. Looking at the graph of sales versus year, the forecast of 7.14 for sales in year 11 appears to be more plausible.

9.20 $r = .993$; have a strong positive linear association

9.22 **b.** $r = .809$

    c. $r^2 = .655$

9.23 **a.**

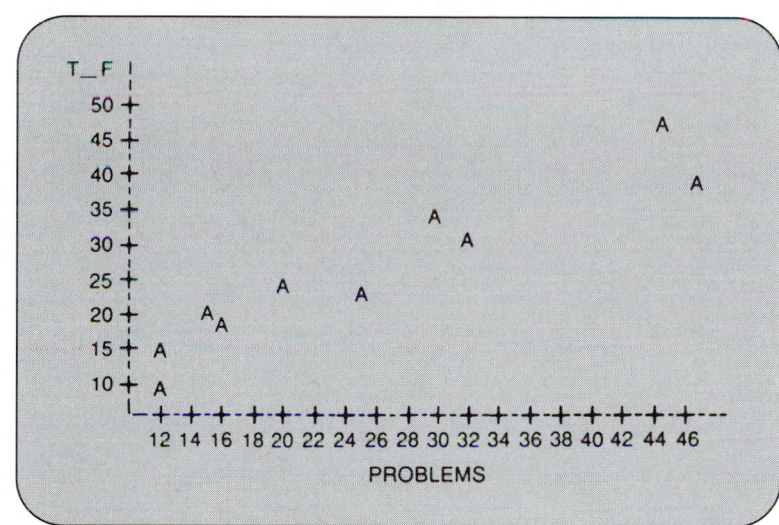

    b. $r = .948$

9.25 **a.**

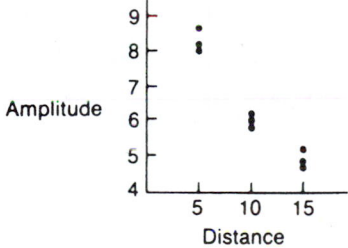

    b. $r_s = -.95$

9.27 **a.**

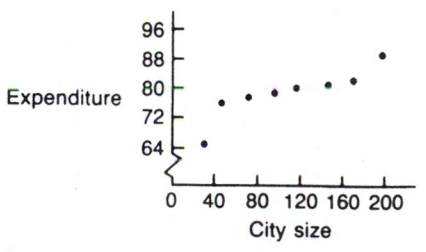

    b. $r = .8903$

9.29   **a.** $y = \beta_0 + \beta_1 x_1 + \beta_2 x_2 + \varepsilon$
   b. $y = \beta_0 + \beta_1 x_1 + \beta_2 x_1^2 + \beta_3 x_2 + \beta_4 x_2^2 + \varepsilon$
      $y = \beta_0 + \beta_1 x_1 + \beta_2 x_2 + \beta_3 x_1 x_2 + \varepsilon$

9.31

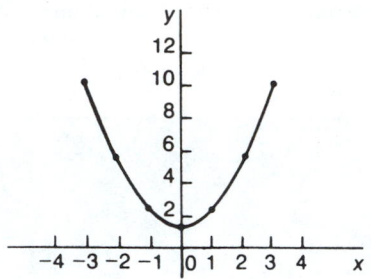

9.33

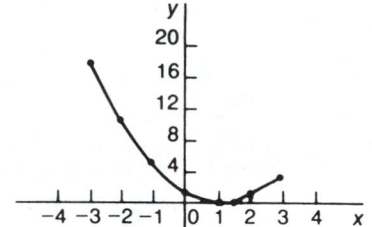

9.35

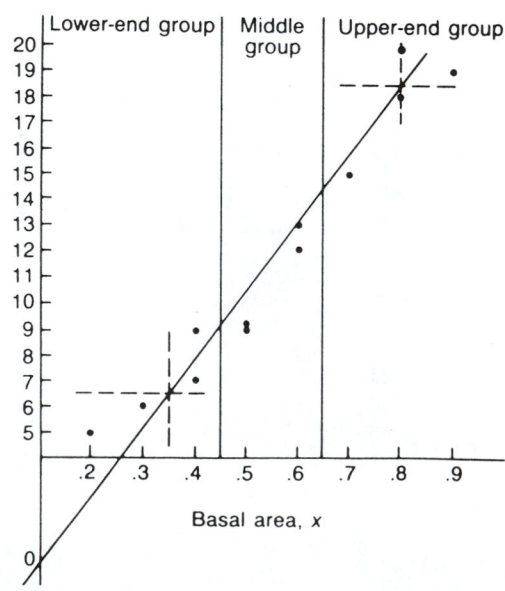

9.37 **a.** $\beta_1 x_1$, first degree   **b.** Fourth-order
$\beta_2 x_1^2$, second degree
$\beta_3 x_1^3$, third degree
$\beta_4 x_1^4$, fourth degree

9.39 **a.** $\beta_1 x_1$, first degree
$\beta_2 x_1^2$, second degree
$\beta_3 x_1^3$, third degree
$\beta_4 x_1 x_2$, second degree

b. This model is neither a first- nor second-order model. The first-order model for two independent variables is

$$y = \beta_0 + \beta_1 x_1 + \beta_2 x_2 + \varepsilon,$$

while the second-order model is

$$y = \beta_0 + \beta_1 x_1 + \beta_2 x_2 + \beta_3 x_1^2 + \beta_4 x_2^2 + \beta_5 x_1 x_2 + \varepsilon$$

9.41 **a.** Model: $y = 1.2 + x$

x	y
-1.0	.2
$-.5$	.7
0	1.2
.5	1.7
1.0	2.2

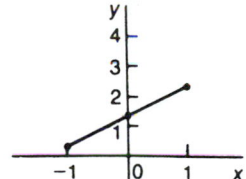

b. Model: $y = 1.2 + x + .4x^2$

x	y
-1.0	.6
$-.5$	.8
0	1.2
.5	1.8
1.0	2.6

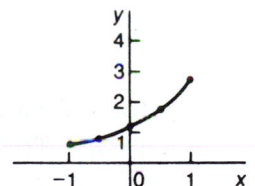

c. Model: $y = 1.2 + x + .4x^2 + .6x^3$

x	y
-1.0	0
$-.5$	.725
0	1.2
.5	1.875
1.0	3.2

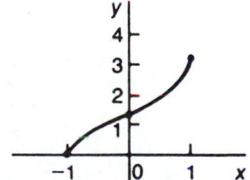

9.43 $y = \beta_0 + \beta_1 x + \varepsilon$

9.45 $y = \beta_0 + \beta_1 x + \beta_2 x^2 + \varepsilon$

9.47

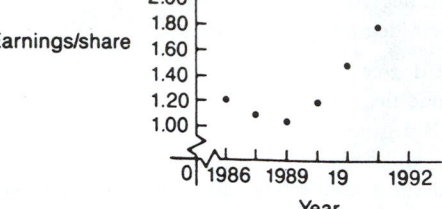

9.49 **a.**

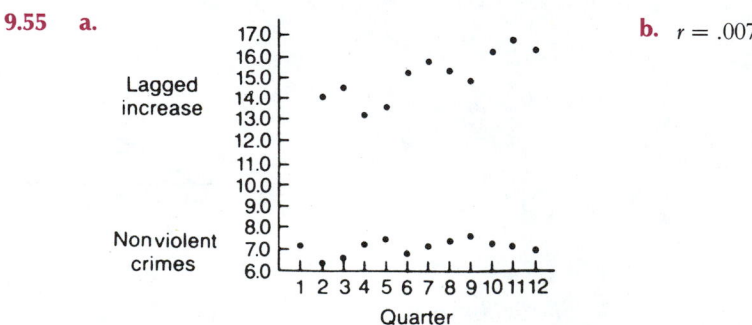

$$y = \beta_0 + \beta_1 x_1 + \beta_2 x_1^2 + \beta_3 x_2 + \beta_4 x_2^2 + \beta_5 x_1 x_2 + \varepsilon$$

b. $\hat{y} = \beta_0 + \beta_1 x + \beta_2 x^2 + \varepsilon$ **c.** $r_s = .67$

9.51 $r = .961$

9.53 Since $r > 0$, a positive linear relationship exists between the two test results. One factor that might contribute to the less than complete agreement between the two test results would be the familiarity of the students with the chosen letter. For example, suppose the first letter was a B and the second letter a Z. The students would probably have more trouble thinking of words that start with a Z than a B.

9.55 **a.**

b. $r = .007$

9.56   a.

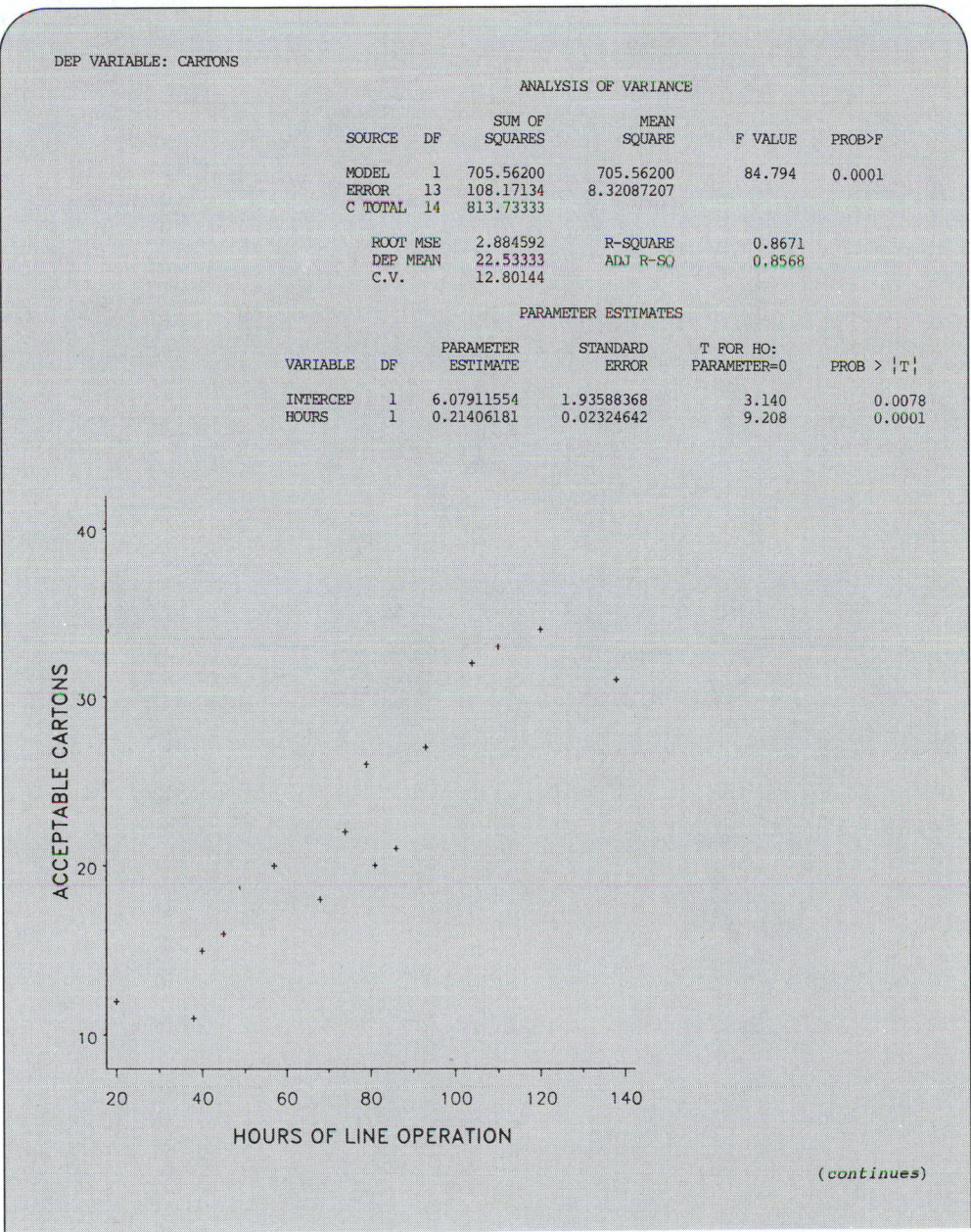

DEP VARIABLE: CARTONS

ANALYSIS OF VARIANCE

SOURCE	DF	SUM OF SQUARES	MEAN SQUARE	F VALUE	PROB>F
MODEL	1	705.56200	705.56200	84.794	0.0001
ERROR	13	108.17134	8.32087207		
C TOTAL	14	813.73333			

ROOT MSE	2.884592	R-SQUARE	0.8671
DEP MEAN	22.53333	ADJ R-SQ	0.8568
C.V.	12.80144		

PARAMETER ESTIMATES

VARIABLE	DF	PARAMETER ESTIMATE	STANDARD ERROR	T FOR H0: PARAMETER=0	PROB > |T|
INTERCEP	1	6.07911554	1.93588368	3.140	0.0078
HOURS	1	0.21406181	0.02324642	9.208	0.0001

(*continues*)

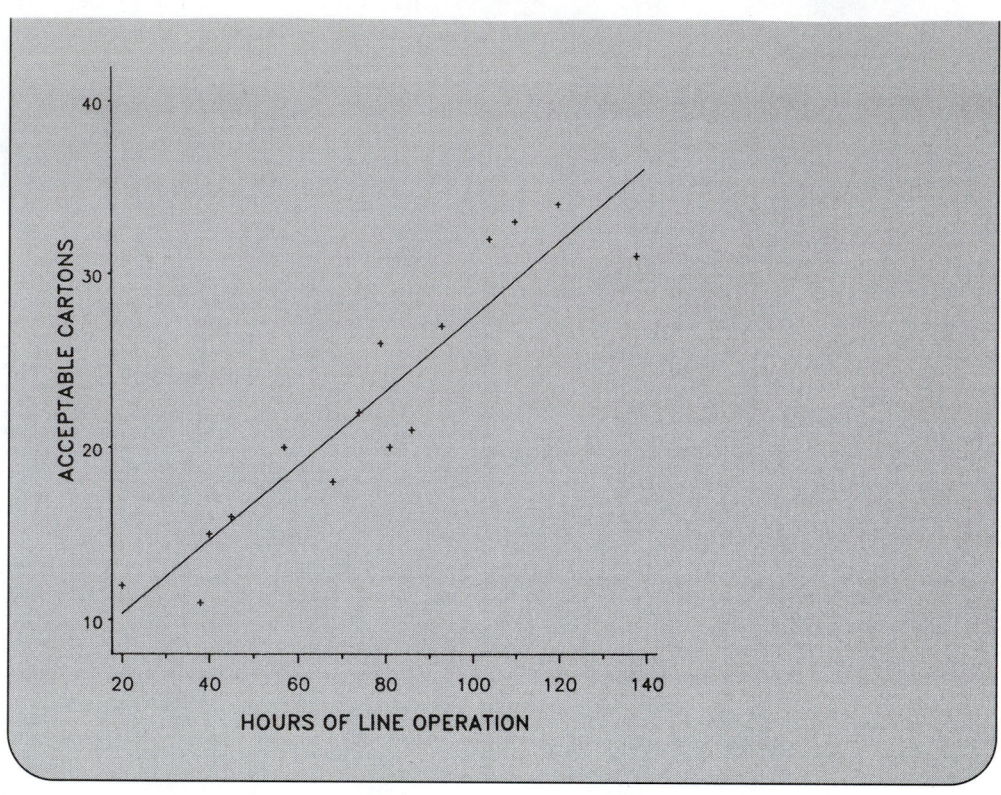

b. $\hat{y} = 6.08 + 0.21x$

9.57 **a.**

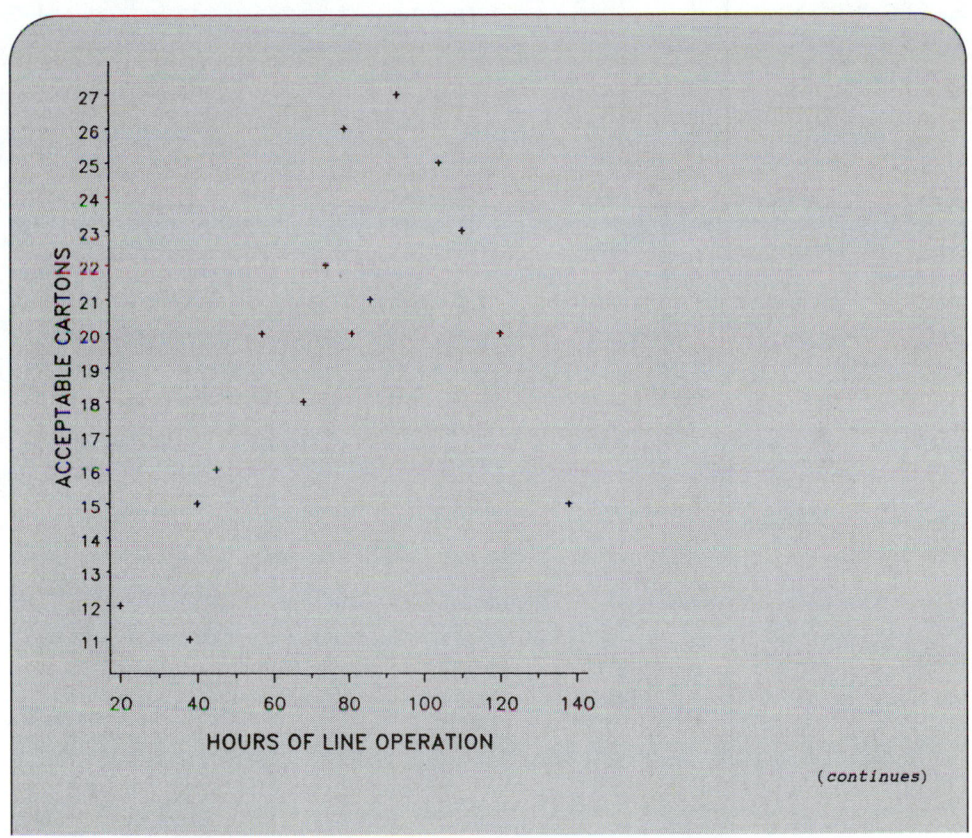

(continues)

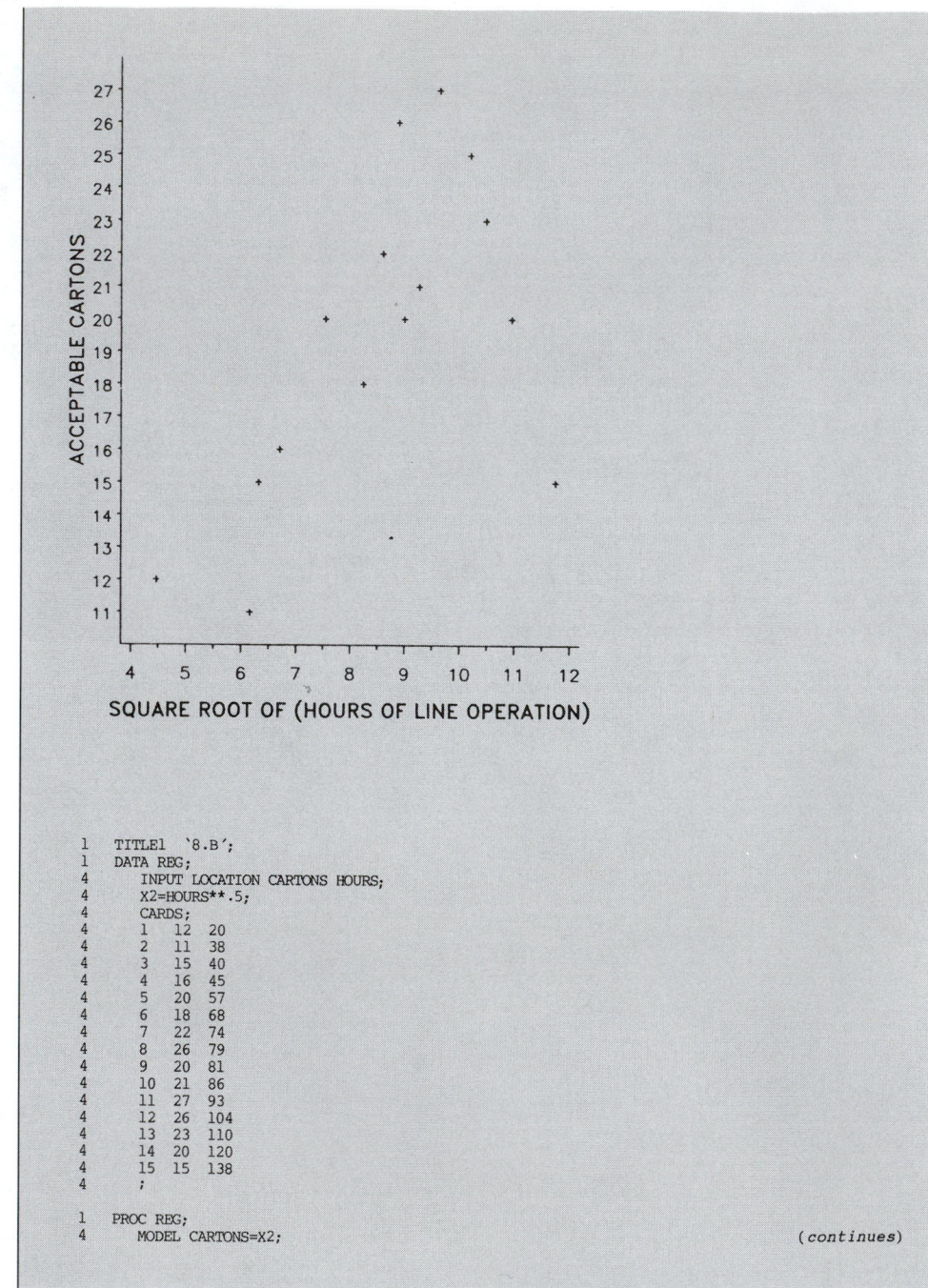

```
1    TITLE1  '8.B';
1    DATA REG;
4        INPUT LOCATION CARTONS HOURS;
4        X2=HOURS**.5;
4        CARDS;
4        1   12   20
4        2   11   38
4        3   15   40
4        4   16   45
4        5   20   57
4        6   18   68
4        7   22   74
4        8   26   79
4        9   20   81
4        10  21   86
4        11  27   93
4        12  26  104
4        13  23  110
4        14  20  120
4        15  15  138
4        ;
1    PROC REG;
4        MODEL CARTONS=X2;
```

(continues)

```
DEP VARIABLE: CARTONS                                  ANALYSIS OF VARIANCE

                                   SUM OF              MEAN
                  SOURCE    DF     SQUARES            SQUARE      F VALUE    PROB>F

                  MODEL     1     118.85538         118.85538      7.195    0.0188
                  ERROR    13     214.74462       16.51881672
                  C TOTAL  14     333.60000

                      ROOT MSE      4.064335      R-SQUARE      0.3563
                      DEP MEAN          19.4      ADJ R-SQ      0.3068
                      C.V.        20.95018

                                         PARAMETER ESTIMATES

                             PARAMETER           STANDARD        T FOR HO:
            VARIABLE    DF     ESTIMATE            ERROR        PARAMETER=0    PROB > |T|

            INTERCEP    1    6.97313075         4.75014884          1.468      0.1659
            X2          1    1.45330964         0.54179899          2.682      0.0188
```

b. No, $\sqrt{x}$

9.59

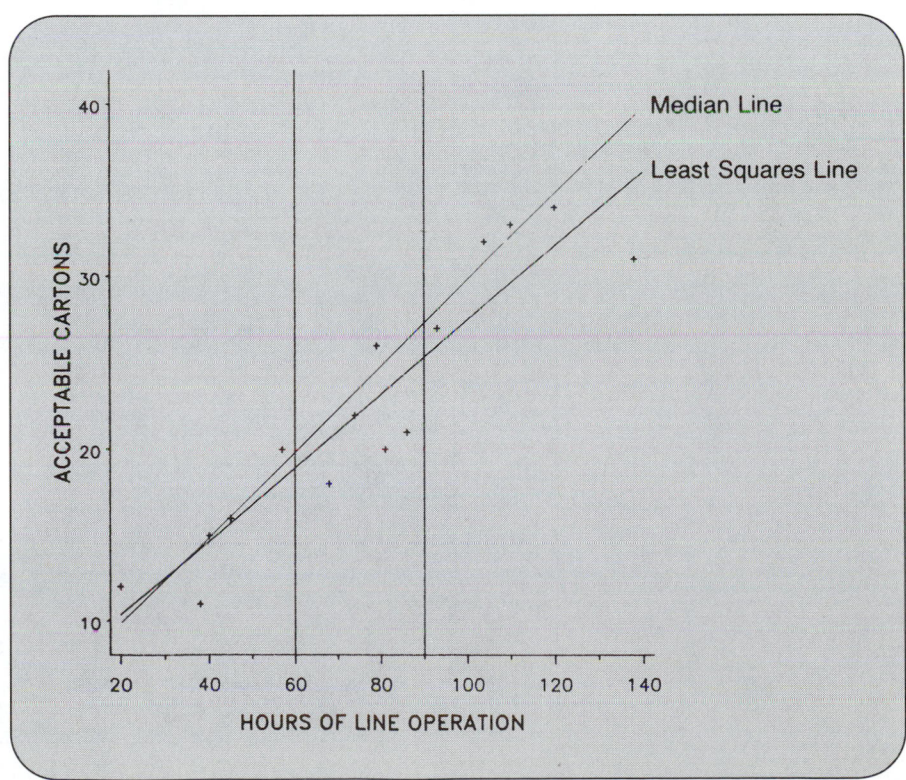

9.60 **b.** $\hat{y} = 31.33 - 7.33x$
    c. 14.471

9.62 **c.** $\hat{y} = 11.82 + 1.364x$
    d. 18.64

9.64 **b.** $\hat{y} = 1.0041 + 9.3934x$
    c. 21.6696

9.66 **b.** $\hat{y} = 11.2365 + 1.3092x$
    c. Yes
    d. .7738
    e. 50.5125

9.70 **a.** $\hat{y} = -19.8 + 2.09x_1$
       $\hat{y} = -18.8 + 1.88x_1 + .57x_2$
       $\hat{y} = 77.7 + 1.26x_1 - .363x_2 - 1.11x_3$
    b. R^2 increased from .905 to .921, but an increase is to be expected with the addition of another variable. The increase is slight and implies the simpler model.

9.71 **a.** Yes
    b. $\hat{y} = 13.90 + .56x$
    c. .962

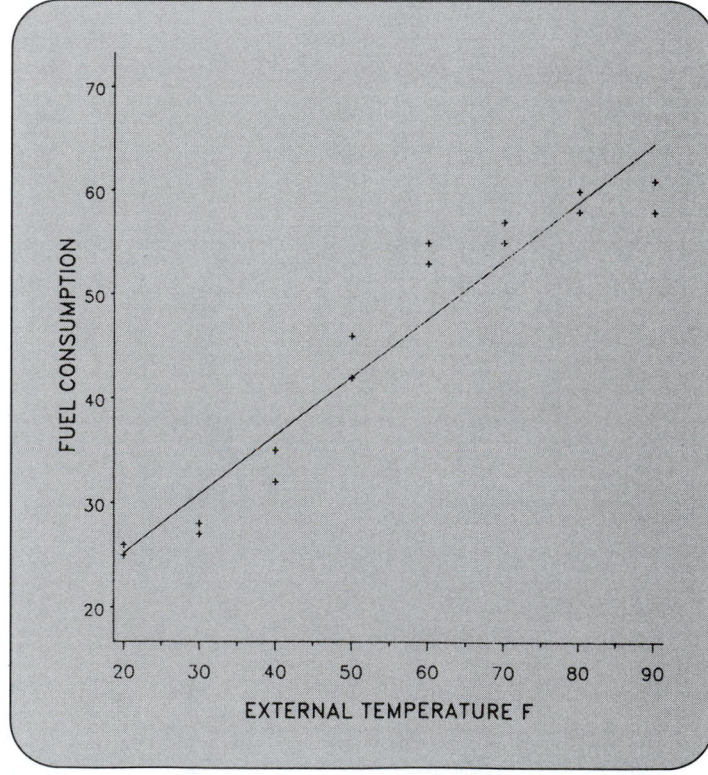

9.72 .984

9.73 **a.**

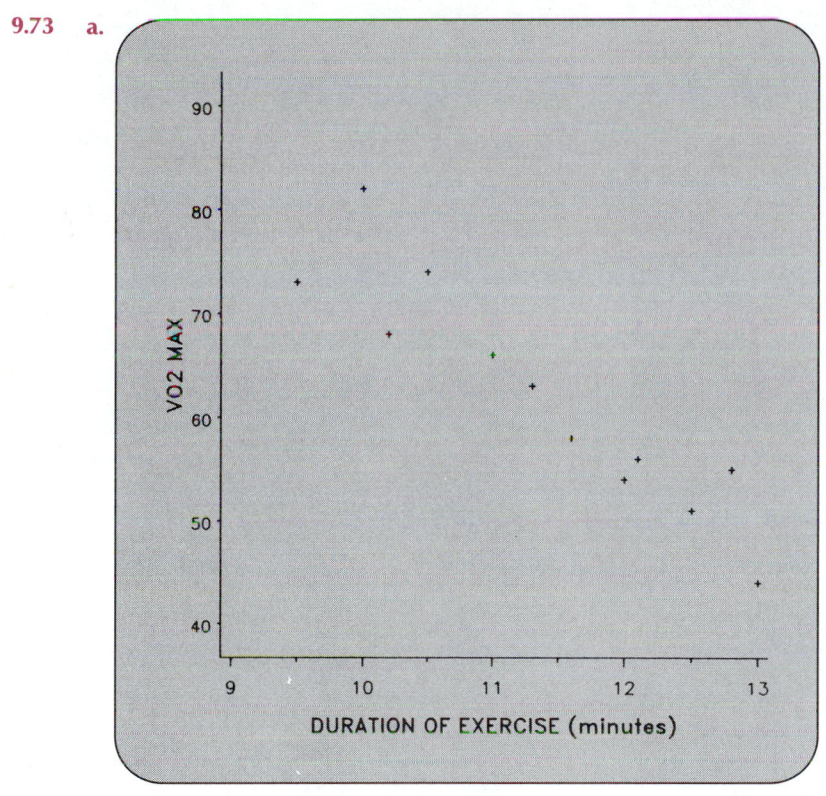

b. Yes

c. $\hat{y} = 162.57 - 8.84x$

```
DEP VARIABLE:  VO2MAX
                                                    ANALYSIS OF VARIANCE

                                    SUM OF          MEAN
                 SOURCE      DF      SQUARES         SQUARE       F VALUE      PROB>F

                 MODEL       1     1141.36689     1141.36689      61.156       0.0001
                 ERROR      10      186.63311       18.66331108
                 C TOTAL    11     1328.00000

                      ROOT MSE       4.320105      R-SQUARE       0.8595
                      DEP MEAN             62      ADJ R-SQ       0.8454
                      C.V.           6.967912

                                                    PARAMETER ESTIMATES

                                  PARAMETER        STANDARD       T FOR HO:
                 VARIABLE    DF     ESTIMATE          ERROR      PARAMETER=0    PROB >|T|

                 INTERCEP    1     162.56583      12.92006797      12.582       0.0001
                 EXERCISE    1      -8.84095189     1.13052648      -7.820       0.0001
```

9.74   **a.** $\hat{y} = 179.12 - 10.40x$

```
4    ;
1    PROC REG;
4       MODEL VO2MAX=EXERCISE;

DEP VARIABLE:  VO2MAX                                    ANALYSIS OF VARIANCE

                                    SUM OF           MEAN
                    SOURCE    DF    SQUARES          SQUARE       F VALUE    PROB>F

                    MODEL      1    1579.07362    1579.07362      36.503     0.0001
                    ERROR     10     432.59305      43.25930491
                    C TOTAL   11    2011.66667

                    ROOT MSE       6.577181       R-SQUARE       0.7850
                    DEP MEAN      60.83333        ADJ R-SQ       0.7635
                    C.V.          10.8118

                                        PARAMETER ESTIMATES

                                    PARAMETER      STANDARD      T FOR HO:
                    VARIABLE   DF    ESTIMATE       ERROR        PARAMETER=0    PROB > |T|

                    INTERCEP    1    179.12087     19.67026538      9.106       0.0001
                    EXERCISE    1    -10.39890430   1.72117949     -6.042       0.0001
```

b. $r^2 = .785$

▼ CHAPTER 10

10.1   **a.** $\sum x = 30$; $\sum y = 117$; $\sum x^2 = 110$; $\sum y^2 = 1401$; $\sum xy = 374$
    b. $\hat{\beta}_0 = 8.25$; $\hat{\beta}_1 = 1.15$; $\sigma_\varepsilon^2 = .70625$; $\sigma_\varepsilon = .84038$
    c. $\sigma_{\hat{\beta}_0} = .62$; $\sigma_{\hat{\beta}_1} = .19$

10.2   **a.** $H_a : \beta_1 \neq 0$
    b. $t = 6.12$; reject H_0 if $|t| > 2.306$; reject H_0.
    c. $p < .001$

10.3   **a.** $\hat{y} = 48.93 + 10.33x$
    b. 5.6983
    c. .7957

10.5   **a.**

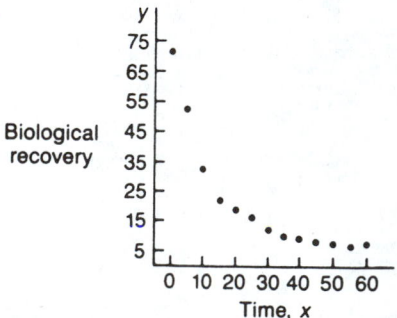

b.

Biological Recovery (%)	Log (%)
70.6	1.85
52.0	1.72
33.4	1.52
22.0	1.34
18.3	1.26
15.1	1.18
13.0	1.11
10.0	1.00
9.1	.96
8.3	.92
7.9	.90
7.7	.89
7.7	.89

10.7 $t = -9.640$; reject H_0 that $\beta_1 = 0$.

10.9 **a.**

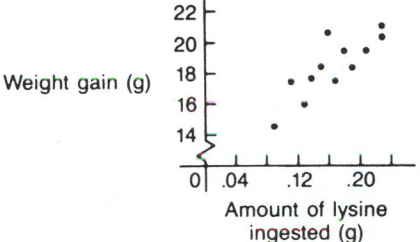

From the graph, a linear model appears to be appropriate.

b. $\hat{y} = 12.509 + 35.828x$

10.11 **a.** Since there is a nonzero, fixed percentage of lysine mixed into all feed, the mean response for no lysine eaten (β_0) should be zero. Thus, it would not make sense to give any physical interpretation to the estimate of β_0.

b. The model $y = \beta_1 x + \varepsilon$ will force the estimated regression line to pass through the origin. The model $y = \beta_0 + \beta_1 + \varepsilon$ will not necessarily pass through the origin.

10.14

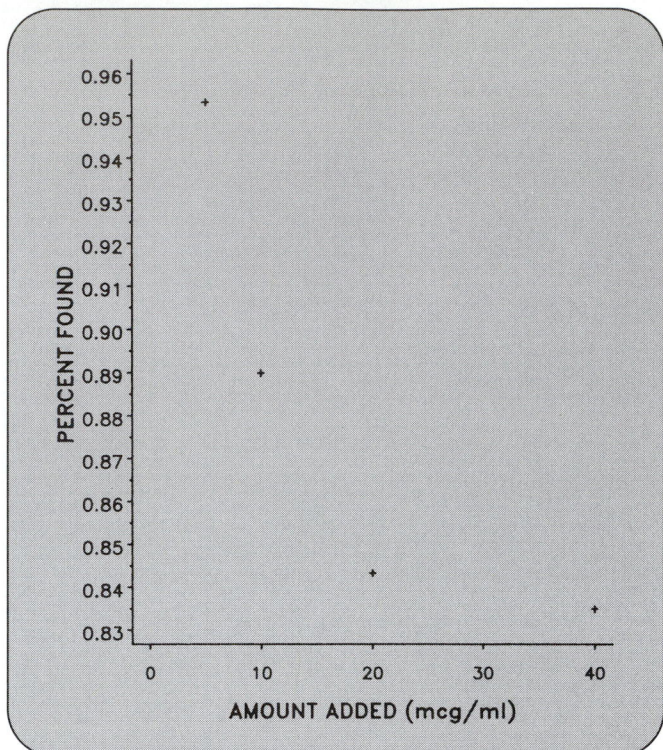

10.15 **a.**

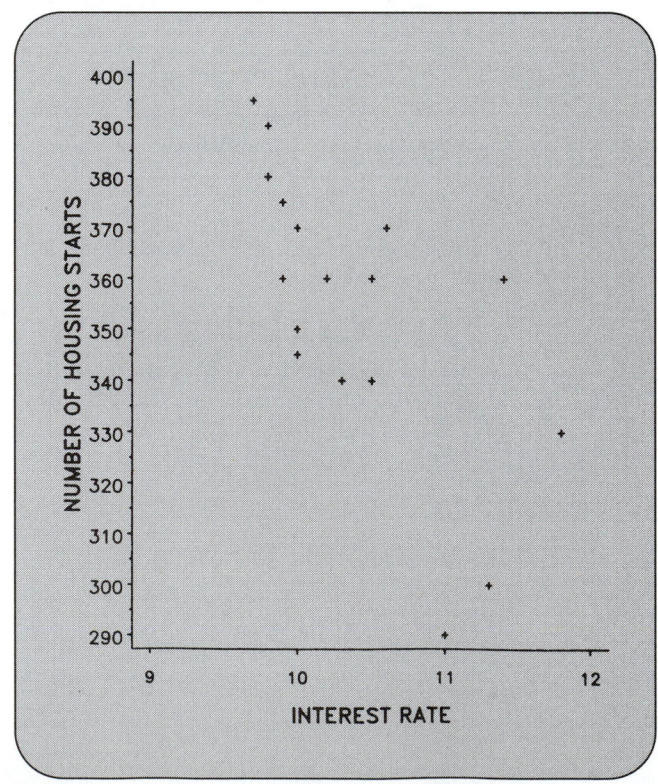

b. $\hat{y} = 672.886 - 30.654x$

c. Yes

10.16 10.2%: 360.2

9.5%: 381.7

10.17 **b.** $\hat{y} = 62.8 + 4.39x$

10.20 $c = 14$; at the $\alpha = .01$ level, we reject H_0 that the two variables are not correlated and conclude that there is a significant positive correlation between plaque weight and DNA.

10.22 **a.**

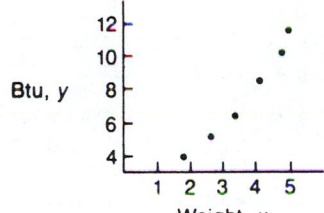

b. $c = 12$; we reject H_0 that the two variables are not correlated at the $\alpha = .05$ level.

10.24 (1.059, 1.103)

10.26 (16.889, 21.743)

10.28 **a.** $\hat{y} = -1.7333 + 1.3167x$

b. $t = 6.342$; reject H_0 and conclude that $\beta_1 > 0$.

10.30 **a.**

b. $\hat{y} = 3.370 + 4.065x$

c.

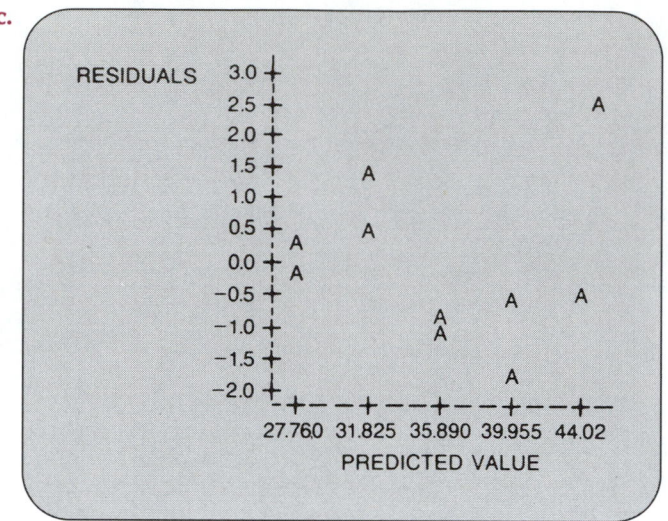

The plot of the residuals and the predicted values suggest that we may need additional terms in our model.

10.32 $F = 8.27$; reject H_0 that a linear regression model is appropriate.

10.34 **a.** $\hat{x} = 14.47$
b. (11.08, 17.63)

10.36 For 50%, $\hat{x} = 2.36703$; (0.79139, 4.44615)
For 75%, $\hat{x} = 5.86145$; (2.20339, 12.88173)

10.38 $\hat{x}_U = 5.23$, $\hat{x}_L = 4.33$

10.41 **a.** $\hat{y} = 804.85 + 2.34x$; β_0 and β_1 significantly different from 0.

10.42 $\hat{y} = 2.62x$

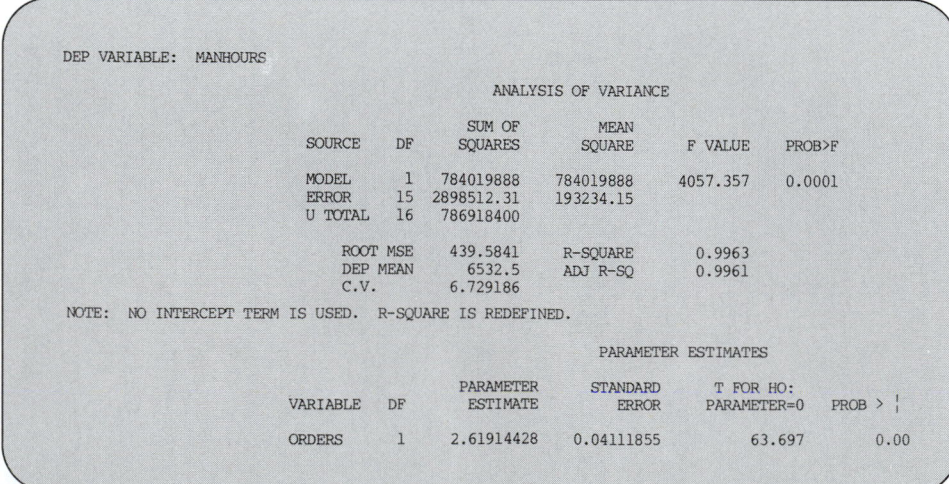

10.43 $\hat{y} = -4.505 + 3.075x$; $t = 15.59$; reject H_0: $\beta_1 = 0$.

10.45 **a.** $\hat{y} = 3.211 + .468x$

b. RESIDUALS

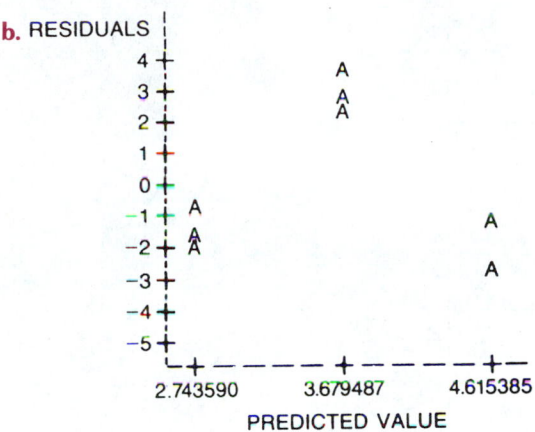

The plot of the residuals and the predicted values suggest that we may need additional terms in our model.

10.47 **a.**

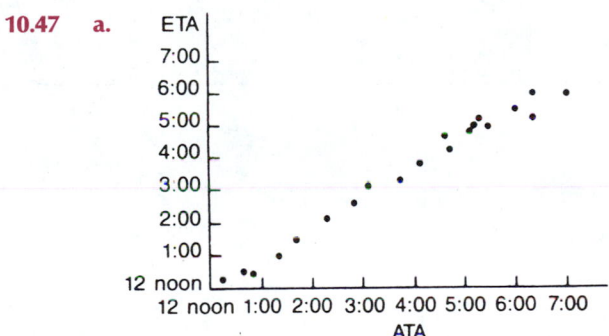

b. Yes, from the graph we can see that a linear relationship appears to be appropriate.

10.51 **a.** Yes; $\hat{y} = 10.333 + 0.267x$
b. No, the plots show no obvious pattern of lack of fit.

10.53 **a.** $y_i = \beta_0 + \beta_1 x_i + \varepsilon$; $\hat{y} = -663.9 + 144.482x$
b. $t = 4.95$; $p = .0001$; reject H_0 and conclude that β_0 is significantly different from 0.

10.54   **b.** .72

         c. Yes

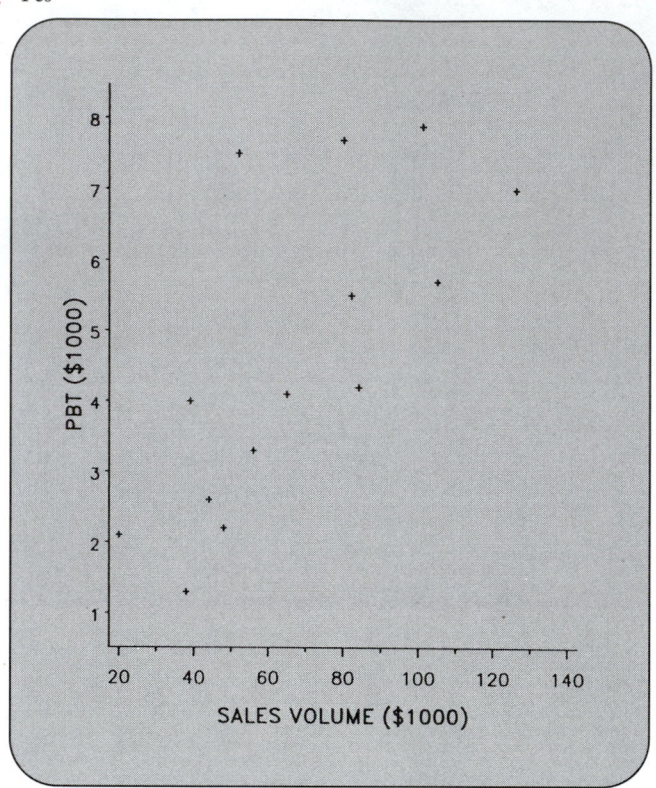

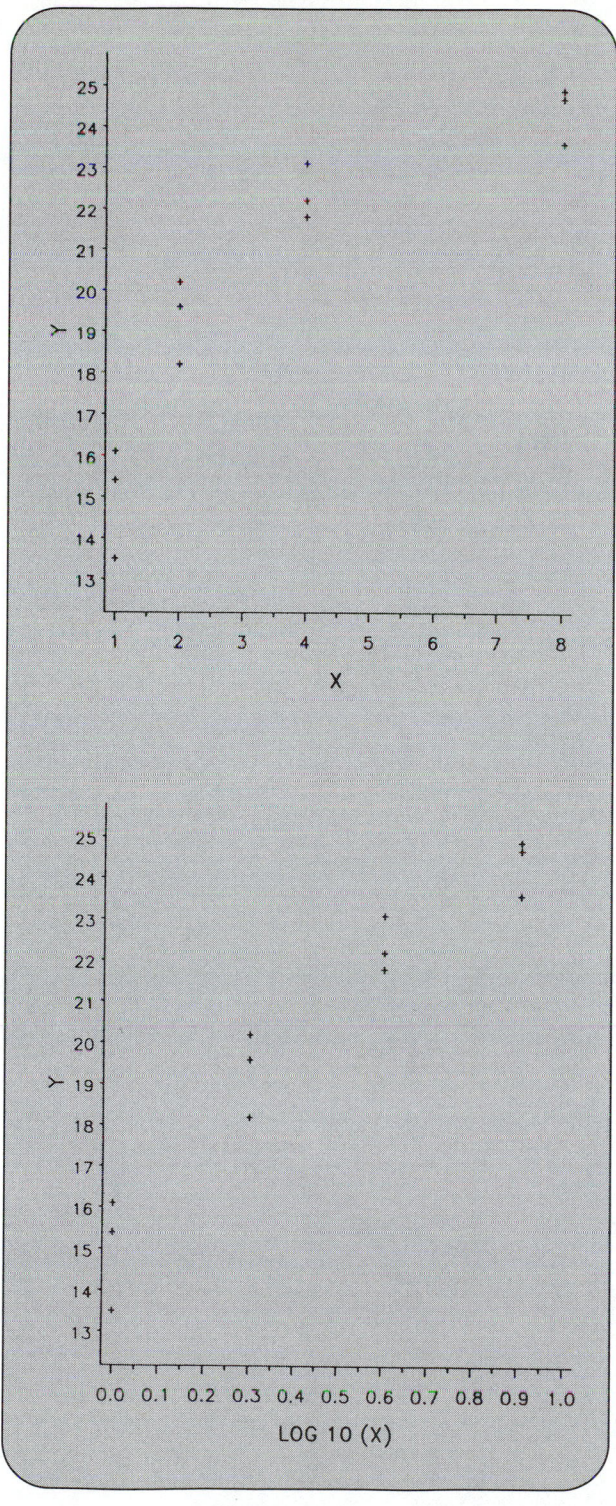

10.57 **a.** $\hat{y} = 15.81 + 1.19x$
        b. $s_\varepsilon^2 = 3.58$

10.58 **a.** $\hat{y} = 15.59 + 10.36 \log_{10} x$
        b. $s_\varepsilon^2 = 1.16$

10.59 $x = 5$; $(21.98, 23.70)$
        $x = 9$; $(24.25, 26.73)$

10.61 $\hat{y} = 41.4 + 28.2 \log_{10} x$
        $p < .001$

10.62 .05

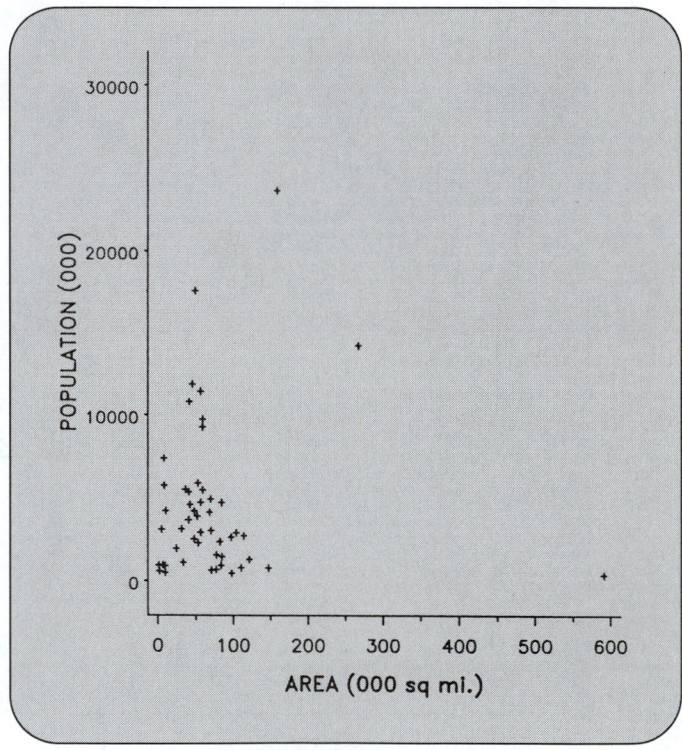

10.63 **b.** $\hat{y} = 2.3765 + .0005x$
        c. $p = .0126$

10.64 $.92763$; $p = .0077$

▼ CHAPTER 11

11.1 **a.** $y = \beta_0 + \beta_1 x_1 + \beta_2 x_2 + \beta_3 x_3 + \varepsilon$
        b. The general linear model for relating a response y to three independent quantitative variables would be the same as the model in part a.

11.3 $y = \beta_0 + \beta_1 x_1 + \beta_2 x_2 + \beta_3 x_3 + \beta_4 x_4 + \beta_5 x_5 + \beta_6 x_6 + \beta_7 x_7 + \beta_8 x_8 + \beta_9 x_9 + \varepsilon$,

where $\quad x_1 = x_1 \qquad x_4 = x_2^2 \qquad x_7 = x_1 x_2$

$\qquad\qquad x_2 = x_1^2 \qquad x_5 = x_3 \qquad x_8 = x_1 x_3$

$\qquad\qquad x_3 = x_2 \qquad x_6 = x_3^2 \qquad x_9 = x_2 x_3$

11.5 **a.**

Location		1	
Treatment	1	2	3
	$E(y) = \beta_0$	$E(y) = \beta_0 + \beta_1$	$E(y) = \beta_0 + \beta_2$
Location		2	
Treatment	1	2	3
	$E(y) = \beta_0 + \beta_3$	$E(y) = \beta_0 + \beta_1 + \beta_3$	$E(y) = \beta_0 + \beta_2 + \beta_3$

β_0—mean of treatment 1, location 1

β_1—difference of the mean of treatment 2 and the mean of treatment 1 for a given location

β_2—difference of the mean of treatment 3 and the mean of treatment 1 for a given location

β_3—difference of the mean of treatment 1, location 1, and the mean of treatment 1, location 2, for a given treatment

b. $\beta_0 + \beta_1 + \beta_3 - (\beta_0 + \beta_2 + \beta_3) = \beta_1 - \beta_2$. It is the same for location 1.

11.7 **a.**

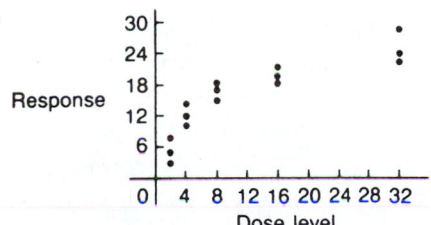

b. $\hat{y} = 8.67 + 0.58x$ **c.** A quadratic model

d. The quadratic model. The quadratic model has a much lower SSE than the linear model. The quadratic term is significant, $p = .0062$. The plot of the predicted values versus the residuals for the liner model has more of an upward trend than the same plot for the quadratic model.

11.9 **a.**

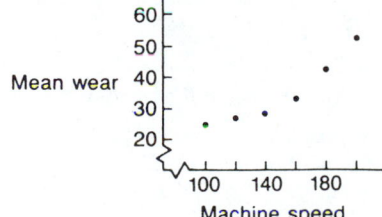

b. Quadratic model

c. The quadratic model. For the quadratic model the plot of the predicted values versus the residuals does not show any problems. The same plot for the linear model does show a trend. The R-squares for the quadratic and the cubic models are almost equivalent. However, the cubic term of the cubic model is not significant, $p = .1794$. All terms of the quadratic model are significant.

11.11 $\hat{y} = -19.41 + 29.10x_1 + 3.29x_2$
$\hat{y} = -23.81 + 31.30x_1 + 3.84x_2 - 0.28x_1x_2$
The model without the interaction term is more appropriate since the interaction term is not significant, $p = .3697$

11.13 **a.** $\bar{y} = 1.85 - .035x + .0003x^2$
b. $s_\varepsilon = .023$; $s_{\beta 1} = .0013$; $s_{\beta 2} = .00002$

11.15 **a.** β_0 would be the biological recovery at time zero.
b. No, the biological recovery is not zero at time zero. If we used this model we would be assuming that regression through the origin would be appropriate.

11.17 **a.** $r_{yx_1} = .856$; $r_{yx_2} = .870$; $r_{x_1x_2} = .894$
b. $r_{yx_1}^2 + r_{yx_2}^2 = 1.49 \neq .788 = R_{y \cdot x_1x_2}^2$. The independent variables x_1 and x_2 are correlated since $R_{y \cdot x_1x_2}^2 \neq r_{yx_1}^2 + r_{yx_2}^2$.

11.19 **a.** $\hat{y} = -2.781 + 1.048x$
b. $p = .0001$; we reject H_0 that $\beta_1 = 0$.
c. $R^2 = .8989$; $R = .948$; $r = .948$

11.21 The prediction band would be wider for a higher degree of certainty. The bands would be smaller for a degree of less certainty.

11.23 (16.889, 21.743); the 95% prediction interval computed in this exercise is wider than the confidence interval computed in Exercise 10.25 because we are predicting the value of a random variable instead of estimating a parameter.

11.25 $F_{2,8} = 30.795$; reject H_0 that $\beta_2 = \beta_3 = 0$ at $\alpha = .05$.

11.27 **a.** $\hat{y} = 48.93 + 10.33x$ **b.** .79599 **c.** 2.3881

11.29 **a.** $\hat{y} = 1.854 - .036x + .0003x^2$ **b.** .0006
c. .0170, .0013, .00002

11.31 $1.854 \pm .038$

11.36 $R^2 = .81$

11.38 $p = .0176$; reject H_0: $\beta_2 = 0$.

11.40 **a.** $\hat{y} = -44.013 + 5.087x_1 + 3.019x_2 + .205x_3 - 5.409x_4$
$\qquad\qquad\qquad (1.970)\quad (2.204)\quad (0.373)\quad (3.850)$
b. $R^2 = .9629$; 1.778

11.42 **a.** $y = \beta_0 + \beta_1x_1 + \beta_2x_2 + \varepsilon$
b. $y = \beta_0 + \beta_1x_1 + \beta_2x_1^2 + \beta_3x_2 + \beta_4x_2^2 + \beta_5x_1x_2 + \varepsilon$

11.44 **a.** $y = \beta_0 + \beta_1x_1 + \beta_2x_2 + \varepsilon$
b. $\hat{y} = 1.04 + 1.59x_1 + .22x_2$ **c.** 2.720

11.55   a. $\mathbf{A}' = \begin{bmatrix} 2 & 3 \\ 1 & 2 \end{bmatrix}$ $\mathbf{B}' = \begin{bmatrix} 2 & 1 \\ 1 & 1 \end{bmatrix}$

     b. $\mathbf{A}^{-1} = \begin{bmatrix} 2 & -1 \\ -3 & 2 \end{bmatrix}$

11.57   a. $\begin{bmatrix} 3 & 0 & 2 \\ 0 & 2 & 0 \\ 2 & 0 & 2 \end{bmatrix}$ b. 4

11.59   a. $\begin{bmatrix} 1 & 1 & 1 \\ 1 & 2 & 3 \end{bmatrix}$

     b. $\begin{bmatrix} 3 & 6 \\ 6 & 14 \end{bmatrix}$

11.61 $\begin{bmatrix} \frac{4}{3} \\ 1 \end{bmatrix}$

11.63   a. 4   b. $\begin{bmatrix} -\frac{1}{4} & 1 & -\frac{1}{4} \\ -\frac{3}{4} & 0 & \frac{1}{4} \\ \frac{2}{4} & -1 & \frac{2}{4} \end{bmatrix}$

▼ CHAPTER 12

12.3   If we run a best subset regression procedure with the C_p criterion, the following variables would be included in the model: Age, Income, and Sex ($C_p = 3.00$).

12.7 $\hat{y} = 288.06 + 402.53x_1 - 0.34x_2 - 1.99x_1x_2$, where $x_1 = \log(\text{dose})$ and $x_2 = \text{weight}$. From examination of the residual plots, it can be seen that there is some lack of fit for this model; the residuals appear to be larger for smaller values of $\hat{y}$ or equivalently for smaller values of weights (x_2) or log dose (x_1).

12.9   Based on the observed versus predicted plots and the residual plots, the second model seems to provide a better fit to the data. The first model does a poor job predicting the larger concentrations and misses the peak concentration. The second model does a better job at fitting the peak and fitting the larger concentrations while still doing an adequate job at the lower concentrations observed at the initial and late post-dose times.

12.11   There appears to be no evidence of serial correlation for the difference model. A plot of the residuals versus the month bears this out.

12.13   a.   The prediction equation is as follows:

$$\hat{y} = -29.88 + 2.33x_1 + 3.82x_2 - 0.20x_1x_2$$

     By examination of the residual plots, the model appears to fit the data.

     b.   1. Zero expectation: The model appears to fit the data thus this assumption should hold.

       2. Constant variance: As can be seen from the plot of the residuals versus the predicted values, there appears to be no problem with the assumption of constant variance.

3. Normality: From examination of the plots, no outliers appear to be present. Also a stem-and-leaf plot of the residuals does not indicate any departure from normality.
4. Independence: The Durbin–Watson statistic is 1.27. Since this value is less than 1.5, we should suspect positive serial correlation. One solution to this problem could be to run an analysis on the first differences of the variables.

12.15 As can be seen from the residual plot, the model appears to be underestimating y for small values of y and overestimating y for larger values of y. There is no evidence from this plot that any of the model assumptions have been violated. Other plots that should be considered are the residuals versus the independent variables, a histogram or stem-and-leaf plot of the residuals, and a plot of the residuals versus time if appropriate. The Durbin–Watson test for serial correlation should also be run.

12.17 **a.** $y = \beta_0 + \beta_1 x_1 + \beta_2 x_2 + \beta_3 x_3 + \beta_4 x_1 x_2 + \beta_5 x_1 x_3 + \varepsilon$

$x_1 = \log$ of drug dose $\quad\quad \beta_0 = y$-intercept for product A regression line
$x_2 = 1$ if product B $\quad\quad\quad \beta_1 = $ slope of product A regression line
$x_2 = 0$ otherwise $\quad\quad\quad\quad \beta_2 = $ difference in y-intercepts for products B and A
$x_3 = 1$ if product C $\quad\quad\quad \beta_3 = $ difference in y-intercepts for products C and A
$x_3 = 0$ otherwise $\quad\quad\quad\quad \beta_4 = $ difference in slopes for products B and A
$\quad\quad\quad\quad\quad\quad\quad\quad\quad\quad \beta_5 = $ difference in slopes for products C and A

b. $y = \beta_0 + \beta_1 x_1 + \beta_4 x_1 x_2 + \beta_5 x_1 x_3 + \varepsilon$

12.19

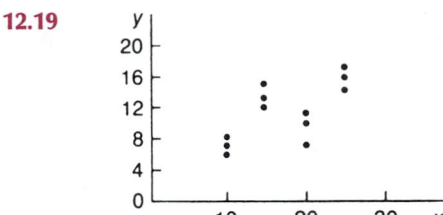

An appropriate polynomial model would be

$$y = \beta_0 = \beta_1 x + \beta_2 x^2 + \beta_3 x^3 + \varepsilon$$

12.21 **a.** $\hat{y} = -91.00 + 19.217x - 1.163x^2 + .023x^3$

b. The model of Exercise 12.20 appears to fit the data well. The model of Exercise 12.21 fits the data well except for one point that appears to be underestimated.

12.23 **a.**

Dose Level	Log$_{10}$ Dose
2	.30103
4	.60206
8	.90309
16	1.20412
32	1.50515

b. $\hat{y} = 1.20 + 16.17x_1$

c. The results in part b agree with the answers in the computer printout.

d. The linear and quadratic models of Exercise 12.22 have values of the Durbin–Watson statistic less than 1.5, suggesting possible serial correlation. The residual plots for these two models also suggest a serial correlation. In contrast, the model with the logarithmic transformation provides a much better fit to the sample data. This model has the highest value for R^2, and the Durbin–Watson statistic is larger.

12.25 **a.** No. There are only six data points and the total degree of freedom is only five; thus, we could not fit a model with all these terms.
b. $F_{1,1} = 1.127$; at the $\alpha = .05$ level, we have insufficient evidence to indicate a lack of fit for the model.

12.27

Comparison	χ_1^2	p
A vs. B	1.7884	.1811
A vs. C	.1118	.7381
A vs. D	1.0444	.3068
B vs. D	5.5914	.0180
C vs. D	1.8671	.1718

12.29 $\hat{y} = 60.48 - .71x_1 + .003x_1^2 + 8.88x_2$
$\hat{y} = 42.28 - .42x_1 + .002x_1^2 + 69.54x_2 - .95x_1x_2 + .003x_1^2x_2$

12.31 The model given would not be more appropriate. The linear model of Exercise 12.30 gives the best fit of the data. The residual plots bear this out as well as the high R^2-value.

12.33 **a.** y = number of sales per month
x_1 = price per gallon
x_2 = interest rate
$$x_3 = \begin{cases} 1 & \text{if standard} \\ 0 & \text{if luxury} \end{cases}$$
b. The prediction equation for the complete model is as follows:
For $x_3 = 0$ (luxury),
$$\hat{y} = 40.567 + 12.201x_1 - 4.595x_1^2 - 5.328x_2 + 0.154x_2^2$$
For $x_3 = 1$ (standard),
$$\hat{y} = -59.903 + 297.394x_1 - 120.608x_1^2 - 12.976x_2 + 0.372x_2^2$$
c. The regression coefficients for the two models appear to be quite different. However, care has to be taken in interpreting these differences since the interaction terms have relatively large standard errors. Perhaps we should consider a model having fewer interaction terms involving the dummy variable, x_3, but more interaction terms involving the two qualitative independent variables, x_1 and x_2.

12.35 One problem may be the presence of nonconstant variance. The plot of the residuals versus the predicted values shows that as $\hat{y}$ increases the residual also increases. A transformation of the data could help to stabilize the variance.

12.37 **a.** $\hat{y} = 44.1824 - 0.4940x + 0.0014x^2$
b. $F_{3,9} = 1.2$; since our value does not exceed the critical F-value, we have insufficient evidence to indicate a lack of fit for our model.
c. Examination of the residual plots does not reveal any problems with the model assumptions.

12.39 $F_{2,10} = 5.45$; we reject H_0 at the $\alpha = .05$ level and conclude that there are additional polynomial terms that should be included in the model.

12.41 Test for $\beta_3 = 0$; $t = 4.69$; $p = .0001$

12.43 **a.**

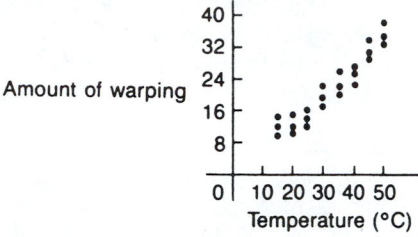

b. $\hat{y} = -1.5397 + 0.7063x$

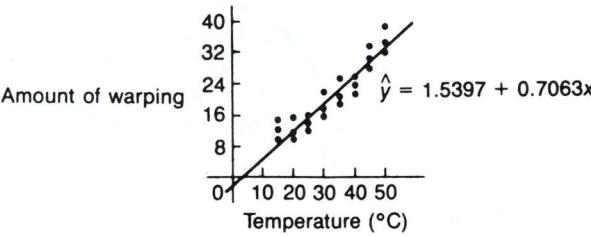

c. $\hat{y} = 9.1786 - .0468x + .0116x^2$

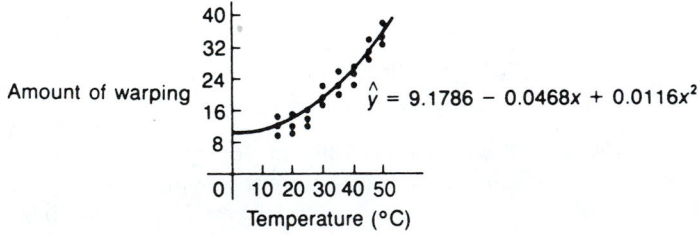

Both equations give a good fit to the data, but the quadratic one fits the lower values of the independent variable better than the linear model.

d. Using the linear equation, $\hat{y} = 17.5304$. Using the equation for the quadratic model, $\hat{y} = 16.3714$.

▼ CHAPTER 13

13.1 **b.** $F_{4,45} = 13.96$; we reject H_0 at the $\alpha = .05$ level and conclude that a difference exists between the mean nicotine tar content of the five brands.

13.3

Source	SS	df	MS	F	p
Location	.1902	3	.0634	1.01	0.409
Error	1.2559	20	.0628		
Totals	1.4679	23			

13.5 $H' = 10.01$; since the observed value of H' does not exceed 13.2767, we cannot reject H_0 at $\alpha = .01$.

13.7 A nonparametric alternative is the Kruskal–Wallis test. $H' = 8.525$; we reject H_0 at the $\alpha = .05$ level. This is the same conclusion as that of Exercise 13.6. We can conclude that differences do exist between diet I and diet II.

13.9 $H' = 4.326$; we accept H_0 at the $\alpha = .05$ level. This test confirms the results of Exercise 13.8. If the results differed, the nonparametric approach would be the analysis to use.

13.11 **a.** $\chi_3^2 = 26.62$; $p < .001$.
 b. $F_{3,28} = 55.67$; $p = .0001$. Both tests indicate rejection of the null hypothesis that all means are equal.

13.12 The analysis of variance is less sensitive to deviations from normality and equality of variances among the groups when the sample sizes are equal. Since the sample sizes are not equal in this exercise, we choose to do a Kruskal–Wallis test. $\chi_3^2 = 21.15$; $p < .001$. We reject the null hypothesis.

13.14 **a.** $y_{ij} = \mu + \alpha_i + \varepsilon_{ij}$; $i = 1, 2, 3$; $j = 1, 2, \ldots, n_i$; $n_1 = 6$, $n_2 = 5$, $n_3 = 4$
  $y_{ij} = j$th sample measurement of ith nutrient
  $\mu = $ overall mean
  $\alpha_i = $ effect of ith nutrient
  $\varepsilon_{ij} = $ random error associated with jth observation from ith nutrient
 b. $F_{2,12} = 24.31$; $p < .001$

13.16 **a.** $y_{ij} = \mu + \alpha_i + \varepsilon_{ij}$; $i = 1, 2, 3, 4$; $j = 1, 2, \ldots, 8$
 b. $F_{3,28} = 11.05$; $p < .001$; reject H_0

13.18 $F_{2,23} = 6.65$; $p < .01$

13.20 $\chi_2^2 = 9.996$; $p < .01$. Both tests rejected the null hypothesis that all means are equal at $\alpha = .01$.

13.22 **a.** $F_{3,20} = 10.99$; $p < .001$
 b. At least one of the pillow types differs from the others.

13.26 If you run an AOV on the combined ranks of the data, the conclusions will be the same as a Kruskal–Wallis test. (Refer to an article by Conover and Iman, *The American Statistician*, August 1981.) If we do this for the data of 13.25, we obtain $F_{4,25} = 46.08$, $p < .001$. These results are the same as the results of 13.25. We can conclude that differences do exist among the 5 diets.

13.28 **a.** $F_{4,20} = 6.03$; $p < .005$
 b. We reject H_0, that the days of the week are the same.

13.30 If we replace 9.8 with 15.8 we obtain the following by running an AOV on the combined ranks: $F_{4,25} = 46.08$; $p < .001$. This conclusion is the same as a Kruskal–Wallis test. In some cases you may wish to run both a Kruskal–Wallis test and an analysis of variance.

If an extreme value exists in the data it may affect the outcome of the analysis of variance. A Kruskal-Wallis test is not sensitive to extreme values because we are looking at the ranks of the data.

▼ CHAPTER 14

14.1   **a.** Yes   **b.** No

14.3 $\hat{I}_1 = -2\bar{y}_1 - \bar{y}_2 + \bar{y}_4 + 2\bar{y}_5$
$\hat{I}_2 = 2\bar{y}_1 - \bar{y}_2 - 2\bar{y}_3 - \bar{y}_4 + 2\bar{y}_5$
$\hat{I}_3 = -\bar{y}_1 + 2\bar{y}_2 - 2\bar{y}_4 + \bar{y}_5$
$\hat{I}_4 = \bar{y}_1 - 4\bar{y}_2 + 6\bar{y}_3 - 4\bar{y}_4 + \bar{y}_5$

14.5 $F_{4,45} = 15.069$; $p < .001$; reject H_0 that all means are equal.

14.7   The analysis of variance of Exercise 14.5 is identical to the analysis of this exercise; $F_{4,45} = 15.069$

14.9   **a.** $\hat{I} = -2.11$; reject H_0 that $I = 0$ **b.** $\hat{I} = -.12$; cannot reject H_0: $I = 0$ at $\alpha = .05$
  c. $\hat{I} = .46$; cannot reject H_0: $I = 0$ at $\alpha = .05$ **d.** $\hat{I} = .90$; cannot reject H_0: $I = 0$ at $\alpha = .05$

14.11   **a.** LSD $= 4.708$; A B C
  b. $W = 5.734$; A B C. Conclusions are the same.
  c. A B C. Conclusions are the same.

14.13 $W_2 = 3.538$; $W_3 = 3.714$; 3 2 1

14.15   LSD $= .83$

$$4 \quad \underline{3 \quad 2} \quad \underline{1 \quad 5}$$

14.17   Fisher's LSD procedure. Summary: 4 3 2 1. Per-comparison error rate is controlled.

14.19   LSD $= 12.72$. Summary: $\underline{1 \quad 2}$ 3

14.21 $F_{2,147} = 85.88$; we reject H_0 at the $\alpha = .05$ level. We conclude that differences exist among the methods of grass seed preparation.

14.23 $F_{3,16} = 4.35$; we reject H_0 at the $\alpha = .05$ level. We conclude that differences exist among the gasoline blends.

14.25 $\hat{I}_3 = 6.7$, $s = 15.84$; we cannot reject H_0, that $I_3 = 0$ at the $\alpha = .05$ level.

14.27 $W = .9523$

14.29   **a.**

  b. Additive A   Additive B
    $\hat{y} = 4.76 + .74x$ $\hat{y} = 5.81 + 1.19x$

▼ CHAPTER 15

15.1   **a.** Treatments $F = 5.62$
        Blocks $F = 351.56$
      b. 160.3

15.3   **a.** Blocks are investigators and treatments are mixtures.
      b. Mixtures are applied at random to experimental units within an investigator, with the stipulation that each mixture appears once in an investigator.

15.5

Source	SS	df	MS	F	p
Year	0.0655	1	0.0655	1.10	0.3188
Curricula	198.6782	10	19.8678	334.17	0.001
Error	0.5945	10	0.0595		
Totals	199.3382	21			

15.6   **a.** $F = 5.505$, $p \approx .01$
      b. $F = 6.914$, $p < .001$; it seems appropriate to block on people.

15.7   **a.** Formulation 1 appears different from formulations 2 and 3.

15.8   **a.** $y_{ijk} = \mu + \alpha_i + \beta_j + \gamma_k + \varepsilon_{ijk}$;
        $i = 1, 2, 3, 4$; $j = 1, 2, 3, 4$; $k = 1, 2, 3, 4$

      b.

Source	SS	df	MS	F	p
Rows	.00085	3	.000283	2.26	.18
Columns	.01235	3	.004117	32.94	<.01
Treatments	.48015	3	.16005	1280.4	<.01
Error	.00075	6	.000125		
Totals	.49410	15			

      We cannot reject the null hypothesis that rows are equal but we can reject the hypotheses that columns and treatments are equal.

15.12   **a.** Randomized block design

15.17   A 3×4 factorial experiment

15.19   **a.** Yes   **b.** $F = 28.16$; $p < .01$

15.21   **a.**

Source	SS	df	MS	F
A	2227.4583	3	742.4861	17.56
B	3996.0833	2	1998.0416	47.24
AB	456.9166	6	76.1528	1.80
Error	507.5	12	42.29167	
Total	7187.9583	23		

      b. The p-value for the F test on the variety-by-treatment interaction is 0.1815. Thus, we have insufficient evidence to indicate an interaction between the two main effects. We may now proceed with the tests for main effects.
      c. The p-values listed for both varieties and pesticides are very small. Hence, we would conclude that there are significant differences among the varieties and among pesticides.

d. The results from the output differ slightly from those in the SAS output and Example 15.9 because of rounding-off errors. However, all three solutions result in the same conclusions.

15.23 a.

Source	SS	df	MS
Copper	171,856.19	3	57,285.396
Manganese	3,745,496.19	3	1,248,498.7
Interaction	9,994.56	9	1,110.5069
Totals	3,927,346.94	15	

We cannot test interaction.

b. $y = \beta_0 + \beta_1 x_1 + \beta_2 x_1^2 + \beta_3 x_1^3 + \beta_4 x_2 + \beta_5 x_2^2 + \beta_6 x_2^3 + \varepsilon$

15.25 a.

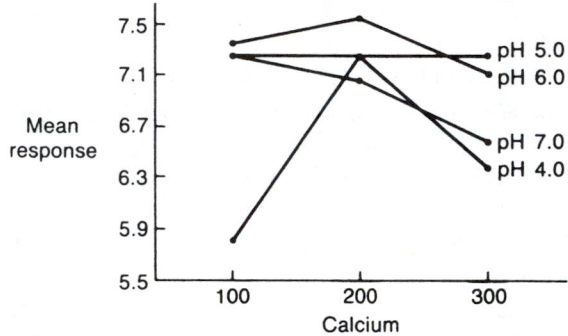

Sample Means Calcium

pH	100	200	300
4.0	5.8	7.3	6.4
5.0	7.3	7.3	7.3
6.0	7.4	7.6	7.2
7.0	7.3	7.1	6.6

b. $y_{ijk} = \mu + \alpha_i + \beta_j + \alpha\beta_{ij} + \varepsilon_{ijk};\ i = 1, 2, 3, 4;\ j = 1, 2, 3;\ k = 1, 2, 3$

c.

Source	SS	df	MS	F	p
pH	4.4608	3	1.4869	21.93	<.01
Calcium	1.4672	2	.7336	10.82	<.01
Interaction	3.2550	6	.5425	8.00	<.01
Error	1.6267	24	.0678		
Totals	10.8097	35			

15.27 Summary: M_n

20	50	80	110
1,506.5	1,948.25	2,470	2,767.5

15.29 **a.** $y_{ijk} = \mu + \alpha_i + \beta_j + \alpha\beta_{ij} + \varepsilon_{ijk}$; μ = overall mean; α_i = effect of investigator i, $i = 1$, 2, 3; β_j = effect of treatment j, $j = 1, 2, \ldots, 9$; $\alpha\beta_{ij}$ = interaction between investigator i and treatment j; ε_{ijk} = error, $k = 1, 2, \ldots, 5$

b.

Source	SS	df	MS	F	p
Investigators	.3628	2	.1814	7.37	<.01
Treatments	3.5117	8	.4390	17.85	<.01
Interaction	.5012	16	.0313	1.27	.23
Error	2.6520	108	.0246		
Totals	7.0277	134			

15.31 **a.** Completely randomized design; $y_{ij} = \mu + \alpha_i + \varepsilon_{ij}$; $i = 1, 2, 3, 4$; $j = 1, 2, \ldots, 7$

b.

Source	SS	df	MS	F	p
Class	892.7143	3	297.5714	11.055	<.01
Error	646.0000	24	26.9167		
Totals	1,538.7143	27			

15.33 **a.** Randomized block design; $y_{ij} = \mu + \alpha_i + \beta_j + \varepsilon_{ij}$; $i = 1, 2, \ldots, 5$; $j = 1, 2, \ldots, 5$

b.

Source	SS	df	MS	F	p
Temperature	16.4216	4	4.1054	42.92	<.01
Pane	7.1176	4	1.7794	18.60	<.01
Error	1.5304	16	.09565		
Totals	25.0696	24			

15.35 **a.** $4 \times 3 \times 3$ factorial
b. $y_{ijk} = \mu + \alpha_{i\cdot} + \beta_j + \gamma_k + \alpha\beta_{ij} + \beta\gamma_{ik} + \varepsilon_{jk}$; $i = 1, 2, 3, 4$; $j = 1, 2, 3$; $k = 1, 2, 3$

c.

Source	SS	df	MS	F	p
Main effects pH	4.4608	3	1.4869	28.48	<.01
calcium	1.4672	2	.7336	14.05	<.01
grove	.0089	2	.0044	.08	.92
Interaction pH-calcium	3.2550	6	.5425	10.39	<.01
pH-grove	.4000	6	.0667	1.28	.34
calcium-grove	.5911	4	.1478	2.83	.07
Error	.6267	12	.0522		
Totals	10.8097	35			

15.37 **a.** $3 \times 2 \times 2$ factorial laid off in 6 blocks

▼ CHAPTER 16

16.1 **a.** The complete and reduced models for testing treatments are as follows:
complete model: $y_{ij} = \mu + \alpha_i + \beta_j + \varepsilon_{ij}$
reduced model: $y_{ij} = \mu + \beta_j + \varepsilon_{ij}$

b. The analysis of variance table for testing the effect of treatments is shown next.

Source	SS	df	MS	F	p
Blocks	2.61464912	4	———	———	———
Treatments (adj.)	140.80083333	3	46.93361111	904.41411	0.0001
Error	.5708333	11	.05189394		
Total	143.98631579	18			

Since our computed value of F greatly exceeds the critical value and our value of p is so small, we would reject H_0 and conclude that one of the adjusted treatment means differs significantly from the rest.

c. The complete model for testing blocks would be the same complete model as in section a. The reduced model is

$$y_{ij} = \mu + \alpha_i + \varepsilon_{ij}$$

d. The AOV table for testing blocks would be

Source	SS	df	MS	F	p
Blocks (adj.)	2.11266667	4	.52816666	10.17781	.0014
Treatments	141.30281579	3	———	———	———
Error	.57083333	11	.05189394		
Total	143.98631579	18			

Our computed value of F also exceeds its critical value; hence, we will conclude that at least one of the dairies (blocks) differs significantly from the rest.

16.3

Source	SS	df	MS	F	p
Plots	386.89	3	128.96	27.44	<.01
Insecticides	1,932.14	2	966.07	205.55	<.01
Error	23.49	5	4.70		
Totals	2,342.52	10			

16.5 From a one-way analysis of variance on treatments:

TSS = 143.9863
SST = 141.3028
SSE = TSS − SST = 2.6835 = SSE_2

From Exercise 16.1, we know that $SSE_1 = .5708$. Thus,

$$SS_{drop} = SSE_2 - SSE_1 = 2.6835 - .5708 = 2.1127$$

This figure agrees with that shown in the Type III sum of squares column of the computer output in Exercise 16.1.

16.7 **a.** AOV for testing mixtures:

Source	SS	df	MS	F	p
Investigators	8,910.03	4			
Mixtures (adj.)	210,236.35	3	70,078.78	1,014.16	<.01
Error	760.15	11	69.10		
Totals	219,906.53	18			

AOV for testing investigators:

Source	SS	df	MS	F	p
Investigators (adj.)	344.60	4	86.15	1.25	>.05
Mixtures	218,801.78	3			
Error	760.15	11	69.10		
Totals	219,906.53	18			

b. Use complete and reduced models excluding this observation also.

16.9 The AOV table for testing the effect of treatment adjusted for rows and columns would be:

Source	SS	df	MS	F
Rows	22.3285	4	———	———
Columns (adjusted for rows)	2.1340	4	———	———
Treatments (adjusted for rows and columns)	165.4943	4	41.3736	315.35
Error	1.4432	11	.1312	
Totals	191.4000	23		

The F test for treatments adjusted by rows and columns is significant; i.e., the adjusted treatment means differ significantly.

Source	SS	F
Rows (adjusted for columns and treatments)	14.3668	27.38
Columns (adjusted for rows and treatments)	0.9428	1.80

The F test for rows adjusted by columns and treatments is significant; however, the other test for columns is not significant. Hence, investigators are significantly different, but days are not. Our conclusions in this exercise agree exactly with those in Exercise 16.8.

16.11

Source	SS	df	MS	F	p
Operator (adjusted for model and blend)	7.492	3	2.497	.70	> .05
Model (adjusted for operator and blend)	695.087	3	231.696	64.68	< .01
Blend (adjusted for operator and model)	82.297	3	27.432	7.66	> .05
Error	17.908	5	3.582		

16.13

Source	SS	df	MS	F	p
Persons	1,034.8	9			
Treatments (adj.)	1,747.056	5	349.411	11.55	< .01
Error	453.611	15	30.241		
Totals	3,235.467	29			

16.15 Same answer as Exercise 16.13.

16.17 Use the Type III (adjusted) sum of squares for blocks.

16.19 $W = 17.72$

16.21 **a.** An appropriate dependent variable would be the number of defects.

b. $y_{ijk} = \mu + \alpha_i + \beta_j + \gamma_{k(i)} + \alpha\beta_{ij} + \beta\gamma_{jk(i)} + \varepsilon_{ijk}$

μ: An overall mean that is an unknown constant
α_i: An effect due to training i: $i = 1, 2$
β_j: An effect due to inspector j: $j = 1, 2$
$\gamma_{k(i)}$: An effect due to assembly line k in training i: $k = 1, 2, 3$
ε_{ijk}: A random error associated with training i, inspector j and assembly line k

c.

Source	df
Training	1
Line (training)	4
Inspector	1
Inspector × Training	1
Inspector × Line (training)	4
Error	12
Total	23

16.23 **a.** No, the model does not change. We can analyze the data by fitting complete and reduced models.

b. Using the output for the complete and reduced models, we can test for the interaction Inspector × Line (train) and also for the interaction Inspector × Train.
For Inspector × Line (train): $F_{4,10} = 0.06$; we cannot reject H_0 at the $\alpha = .05$ level, and we conclude that the interaction is not significant.
For Inspector × Train: $F_{1,14} = 0.14$; we cannot reject H_0 at the $\alpha = .05$ level, and we conclude that the interaction is not significant.

16.25 A dependent variable that would account for both major and minor defects would be the ratio of major defects to minor defects. You may want to transform the data before calculating the ratios since there are several zero values for major defects.

▼ CHAPTER 17

17.1 **a.** $y_{ij} = \mu + \alpha_i + \varepsilon_{ij}$; $i = 1, 2, \ldots, 5$; $j = 1, 2, 3, 4$

$y_{ij} = j$th observation from vat i

$\mu = $ overall unknown mean

$\alpha_i = $ random effect due to the ith vat; α_i is normally distributed with mean 0 and variance σ_α^2

$\varepsilon_{ij} = $ random error associated with response j on ith vat

b.

Source	SS	df	MS	EMS	F
Vats	11.948	4	2.987	$\sigma^2 + 4\sigma_\alpha^2$	32.53
Error	1.3775	15	.09183	σ^2	
Totals	13.3255	19			

17.3 **a.** $y_{ij} = \mu + \alpha_i + \beta_j + \varepsilon_{ij}$; $i = 1, 2, 3$; $j = 1, 2, 3$

$y_{ij} = $ observation from jth subject in analyst i

$\mu = $ overall unknown concentration mean

$\alpha_i = $ random effect due to the ith analyst

$\beta_j = $ random effect due to the jth subject

$\varepsilon_{ij} = $ random error associated with response in analyst i, subject j

b.

Source	EMS
Analysts	$\sigma^2 + 3\sigma_\alpha^2$
Subjects	$\sigma^2 + 3\sigma_\beta^2$
Error	σ^2

17.5 **a.** $y_{ijk} = \mu + \alpha_i + \beta_j + \alpha\beta_{ij} + \varepsilon_{ijk}$; $i = 1, 2, 3, 4$; $j = 1, 2, 3$

$y_{ijk} = $ response (sales volume) for the kth observation at the ith level of SMSA and jth level of chain store

$\mu = $ overall unknown sales volume mean

$\alpha_i = $ random effect due to the ith SMSA

$\beta_j = $ random effect due to the jth store

$\alpha\beta_{ij} = $ random effect due to the ith level of SMSA and the jth level of store

$\varepsilon_{ijk} = $ random error associated with kth response at SMSA i, store j

b.

Source	SS	df	MS	EMS	F	p
SMSA	136,644.125	3	45,548.042	$\sigma^2 + 2\sigma_{\alpha\beta}^2 + 6\sigma_\alpha^2$	5.049	<.05
Store	62,973.000	2	31,486.5	$\sigma^2 + 2\sigma_{\alpha\beta}^2 + 8\sigma_\beta^2$	3.490	>.05
Interaction	54,127.000	6	9,021.1667	$\sigma^2 + 2\sigma_{\alpha\beta}^2$	22.03	<.05
Error	4,913.500	12	409.4583	σ^2		
Totals	258,657.625	23				

17.7

Source	SS	df	MS	EMS	F	p
Physicians	3.8115	4	.9529	$\sigma^2 + 2\sigma_{\alpha\beta}^2 + 8\theta_A$	.71	> .05
Patients	180.13275	3	60.0442	$\sigma^2 + 10\sigma_B^2$	173.41	< .05
Interaction	16.1585	12	1.3465	$\sigma^2 + 2\sigma_{\alpha\beta}^2$	3.89	< .05
Error	6.925	20	.34625	σ^2		
Totals	207.02775	39				

17.9 Yes

$H_0: \theta_A = 0;\ F = MSA/MSAC$
$H_0: \theta_B = 0;\ F = MSB/MSBC$
$H_0: \sigma_\gamma^2 = 0;\ F = MSC/MSE$
$H_0: \theta_{AB} = 0;\ F = MSAB/MSABC$
$H_0: \sigma_{\alpha\gamma}^2 = 0;\ F = MSAC/MSE$
$H_0: \sigma_{\beta\gamma}^2 = 0;\ F = MSBC/MSE$
$H_0: \sigma_{\alpha\beta\gamma}^2 = 0;\ F = MSABC/MSE$

17.11 **a.**

Source	EMS
A	$\sigma^2 + n\sigma_{ABC}^2 + cn\sigma_{AB}^2 + bn\sigma_{AC}^2 + bcn\sigma_A^2$
B	$\sigma^2 + n\sigma_{ABC}^2 + cn\sigma_{AB}^2 + an\sigma_{BC}^2 + acn\sigma_B^2$
C	$\sigma^2 + n\sigma_{ABC}^2 + bn\sigma_{AC}^2 + an\sigma_{BC}^2 + abn\sigma_C^2$
AB	$\sigma^2 + n\sigma_{ABC}^2 + cn\sigma_{AB}^2$
AC	$\sigma^2 + n\sigma_{ABC}^2 + bn\sigma_{AC}^2$
BC	$\sigma^2 + n\sigma_{ABC}^2 + an\sigma_{BC}^2$
ABC	$\sigma^2 + n\sigma_{ABC}^2$
Error	σ^2

b.

Source	F Test	df_1	df_2
ABC	MSABC/MSE	$(a-1)(b-1)(c-1)$	$abc(n-1)$
BC	MSBC/MSABC	$(b-1)(c-1)$	$(a-1)(b-1)(c-1)$
AC	MSAC/MSABC	$(a-1)(c-1)$	$(a-1)(b-1)(c-1)$
AB	MSAB/MSABC	$(a-1)(b-1)$	$(a-1)(b-1)(c-1)$

There are no exact F tests for σ_A^2, σ_B^2, and σ_C^2.

17.13 **a.** $y_{ij} = \mu + \alpha_i + \varepsilon_{ij};\ i = 1, 2, \ldots, 6;\ j = 1, 2$
$y_{ij} = j$th observation from week i
$\mu = $ overall unknown mean
$\alpha_i = $ random effect due to the ith week
$\varepsilon_{ij} = $ random error associated with response j at week i

b.

Source	SS	df	MS	EMS	F	p
Weeks	255,089.42	5	51,017.88	$\sigma^2 + 2\sigma_\alpha^2$	26.76	< .05
Error	11,439.5	6	1,906.58	σ^2		
Totals	266,528.92	11				

17.15 **a.** $y_{ijk} = \mu + \alpha_i + \beta_j + \alpha\beta_{ij} + \varepsilon_{ijk}$; $i = 1, 2, \ldots, 5$; $j = 1, 2, 3, 4$; $k = 1, 2$; in this model, α_i is a random effect due to the ith physician, while in Exercise 17.6 it was a fixed effect. In this model, α_i, β_j, and $\alpha\beta_{ij}$ are all independent.

b.

Source	SS	df	MS	EMS	F	p
Physicians	3.8115	4	.9529	$\sigma^2 + 2\sigma_{\alpha\beta}^2 + 8\sigma_\alpha^2$	.71	> .05
Patients	180.1328	3	60.0442	$\sigma^2 + 2\sigma_{\alpha\beta}^2 + 10\sigma_\beta^2$	44.59	< .05
Interaction	16.1585	12	1.3465	$\sigma^2 + 2\sigma_{\alpha\beta}^2$	3.89	
Error	6.9250	20	.34625	σ^2		
Totals	207.0278	39				

17.17 **a.** $y_{ij} = \mu + \alpha_i + \beta_j + \varepsilon_{ij}$; $i = 1, 2, 3, 4$; $j = 1, 2, \ldots, 5$

y_{ij} = observation from jth investigator in mixture i

μ = overall unknown mean

α_i = fixed effect due to ith mixture

β_j = random effect due to jth investigator; β_j is normally distributed with mean 0 and variance σ_β^2; the βs are independent

ε_{ij} = random error associated with response in mixture i, investigator j; the ε_{ij}s are independent

b.

Source	SS	df	MS	EMS	F	p
Mixtures	261,260.95	3	87,086.983	$\sigma^2 + 5\theta_A$	1,264.73	< .05
Investigators	452.50	4	113.125	$\sigma^2 + 4\sigma_\beta^2$	1.64	> .05
Error	826.30	12	68.858	σ^2		
Totals	262,539.75	19				

17.19

Source	EMS
Rows	$\sigma^2 + 4\sigma_\alpha^2$
Columns	$\sigma^2 + 4\sigma_\beta^2$
Treatments	$\sigma^2 + 4\theta_c$
Error	σ^2

17.21 **a.**

Source	EMS
Investigators	$\sigma^2 + 45\sigma_2$
Treatments	$\sigma^2 + 5\sigma_{\alpha\beta}^2 + 15\theta_B$
Interaction	$\sigma^2 + 5\sigma_{\alpha\beta}^2$
Error	σ^2

b. In the F test for differences among treatments in this exercise, the T.S. is MST/MS$_{\text{interaction}}$ with df$_1$ = 8 and df$_2$ = 16. In Exercise 15.30 the T.S. was MST/MSE with df$_1$ = 8 and df$_2$ = 108. If the test for investigators was significant in this exercise, we would conclude that σ_α^2 is greater than 0, whereas in Exercise 15.30 the conclusion would be that the means are unequal.

17.23 $\hat{\sigma}_\varepsilon^2 = 409.46$; $\hat{\sigma}_\beta^2 = 2,808.17$; $\hat{\sigma}_{\alpha\beta}^2 = 4,305.85$; $\hat{\sigma}_\alpha^2 = 6,087.81$

▼ CHAPTER 18

18.2

Source	SS	df
Between patients		
Depression group	962	3
Patients in group	582	20
Within patients		
Time	165	13
Time × group	120	39
Error	65	260

18.3

Treatment	Time	Mean	Std. Dev.
1	1	20.70	23.98
	2	28.57	12.00
	3	31.24	14.30
	4	29.44	12.65
	8	25.63	14.26
2	1	−0.76	12.26
	2	12.55	10.43
	3	18.23	10.83
	4	24.79	6.91
	8	17.57	7.83

18.6 **a.** 2-period crossover design

      c. No interaction between period and drug

18.8 **a.** Binomial or paired t test ignoring order.

▼ CHAPTER 19

19.1 $y = \beta_0 + \beta_1 x_1 + \beta_2 x_2 + \beta_3 x_3 + \beta_4 x_4 + \beta_5 x_5 + \beta_6 x_1 x_2 + \beta_7 x_1 x_3 + \beta_8 x_1 x_4 + \beta_9 x_1 x_5 + \varepsilon$

$x_1 =$ covariate

$x_2 = 1$ if treatment 2 is applied

$x_2 = 0$ otherwise

$x_3 = 1$ if treatment 3 is applied

$x_3 = 0$ otherwise

$x_4 = 1$ if treatment 4 is applied

$x_4 = 0$ otherwise

$x_5 = 1$ if treatment 5 is applied

$x_5 = 0$ otherwise

19.3 The assumption of constant slope for the 5 treatment groups can be tested using the following null hypothesis:

$$H_0: \quad \beta_6 = \beta_7 = \beta_8 = \beta_9 = 0$$

The complete model is given in Exercise 19.1 and the reduced model is $y = \beta_0 + \beta_1 x_1 + \beta_2 x_2 + \beta_3 x_3 + \beta_4 x_4 + \beta_5 x_5 + \varepsilon$. The test statistic is $F = MS_{drop}/MSE_1$, with $df_1 = 4$ and $df_2 = 20$.

19.5 **b.** Test for parallelism: $F_{2,12} = 4.133$. The test for adjusted treatment means is not valid since the slopes are not constant for the three groups.

c. Yes, the plot suggests the lines are not parallel.

19.7 **a.** Obtain SSE_1 from the complete model described in Exercise 19.6. Obtain SSE_2 from the reduced model:

$$y = \beta_0 + \beta_1 x_1 + \beta_2 x_2 + \beta_3 x_3 + \beta_4 x_4 + \beta_5 x_5 + \beta_6 x_6$$
$$+ \beta_7 x_7 + \beta_8 x_8 + \beta_9 x_9 + \beta_{10} x_{10} + \varepsilon$$

$SSE_{drop} = SSE_2 - SSE_1$; $MS_{drop} = SS_{drop}/9$; $F = MS_{drop}/MSE_1$, with $df_1 = 9$ and $df_2 =$ degrees of freedom for error in the complete model

b. Use the reduced model described in part a as the complete model and set $\beta_8 = \beta_9 = \beta_{10}$ equal to 0 for the reduced model.

19.9 **a.** Randomized block design with antidepressants as treatments and age–sex combinations as blocks.

b. $y = \beta_0 + \beta_1 x_1 + \beta_2 x_2 + \beta_3 x_3 + \beta_4 x_4 + \beta_5 x_5 + \beta_6 x_6 + \beta_7 x_7 + \beta_8 x_8 + \beta_9 x_1 x_2$
$$+ \beta_{10} x_1 x_3 + \beta_{11} x_1 x_4 + \beta_{12} x_1 x_5 + \beta_{13} x_1 x_6 + \beta_{14} x_1 x_7 + \beta_{15} x_1 x_8 + \varepsilon$$

19.11 **a.** $F_{5,9} = .174$; cannot reject H_0 that the blocks are equal.

b. Use five orthogonal contrasts:

(1) Males versus females
(2) <20 versus 20–40
(3) <20 and 20–40 versus 2(>40)
(4) Interaction between sex and first 2 age groups
(5) Interaction between sex and all age groups

c. $y = \beta_0 + \beta_1 x_1 + \beta_2 x_2 + \beta_3 x_3 + \beta_4 x_4 + \beta_5 x_5 + \beta_6 x_6 + \beta_7 x_4 x_5 + \beta_8 x_4 x_6 + \varepsilon$

$x_1 =$ covariate
$x_2 = 1$ if treatment B
$x_2 = 0$ otherwise
$x_3 = 1$ if treatment C
$x_3 = 0$ otherwise
$x_4 = 1$ if female
$x_4 = -1$ if male
$x_5 = 1$ if 20–40
$x_5 = 0$ otherwise
$x_6 = 0$ otherwise
$x_6 = 1$ if greater than 40

Source		SS	df
x_1 (covariate)		25.799	1
x_2 (treatments)		59.260	1
x_3		8.224	1
x_4 (sex)		.183	1
x_5 (age)		1.077	1
x_6		4.875	1
x_7 (age × sex)		0.366	1
x_8		0.134	1
Error		68.581	9
Totals		168.500	17

Although the single-degree-of-freedom sums of squares are given in the AOV table, no additional tests need be performed since the tests for blocks (sex, age, age × sex) of part a failed to show any significant effects

19.13 **a.**

Source	SS	df	MS	F	p
x_1	.294	1	.294	.004	>.05
x_2	726.744	1	726.744	10.865	<.01
$x_1 x_2$	825.577	1	825.577	12.343	<.01
Error	1,070.185	16	66.887		
Totals	2,622.8				

b. $F_{1,16} = 12.343$; reject H_0 that the lines are parallel.

c. The test for a difference between the 2 populations is not valid because the lines are not parallel. The assumptions for a covariance analysis are not met.

INDEX